PROPERTIES OF PROBABILITY (*continued*)

2. $p_1 + p_2 + p_3 + \cdots + p_n = 1$
(the sum of all probabilities for a sample space is 1);

3. $P(S) = 1$;

4. $P(\emptyset) = 0$.

6.2 UNION RULE

For any two events E and F from a sample space S,

$$P(E \cup F) = P(E) + P(F) - P(E \cap F).$$

6.3 PRODUCT RULE

For any events E and F,

$$P(E \cap F) = P(F) \cdot P(E|F).$$

6.4 BAYES' FORMULA

$$P(F_i|E) = \frac{P(F_i) \cdot P(E|F_i)}{P(F_1) \cdot P(E|F_1) + P(F_2) \cdot P(E|F_2) + \cdots + P(F_n) \cdot P(E|F_n)}$$

6.6 PROBABILITY IN A BERNOULLI EXPERIMENT

If p is the probability of success in a single trial of a Bernoulli experiment, the probability of x successes and $n - x$ failures in n independent repeated trials of the experiment is

$$\binom{n}{x} \cdot p^x \cdot (1 - p)^{n-x}.$$

7.2 EXPECTED VALUE

Suppose the random variable x can take on the n values $x_1, x_2, x_3, \ldots, x_n$. Also, suppose the probabilities that these values occur are respectively $p_1, p_2, p_3, \ldots, p_n$. Then the expected value of the random variable is

$$E(x) = x_1 p_1 + x_2 p_2 + x_3 p_3 + \cdots + x_n p_n.$$

7.3 VARIANCE AND STANDARD DEVIATION

The variance of a set of n numbers $x_1, x_2, x_3, \ldots, x_n$, with mean \bar{x}, is

$$\text{Var}(x) = \frac{\Sigma(x - \bar{x})^2}{n}.$$

The standard deviation of the set is

$$\sigma = \sqrt{\frac{\Sigma(x - \bar{x})^2}{n}}.$$

8.2 PROPERTIES OF REGULAR MARKOV CHAINS

Suppose a regular Markov chain has a transition matrix P.

1. For any initial probability vector v, the product $v \cdot P^n$ approaches a unique vector V as n gets larger and larger. Vector V is called the equilibrium vector or fixed vector.

2. Vector V has the property that $VP = V$.

3. To find V, solve a system of equations obtained from the matrix equation $VP = V$, and from the fact that the sum of the entries of V is 1.

4. The powers P^n come closer and closer to a matrix whose rows are made up of the entries of the equilibrium vector V.

TO THE STUDENT

If you want further help with this course, you may wish to buy a copy of the *Student's Solution Manual* (ISBN 0-673-46764-3) that accompanies this textbook. This manual provides detailed, step-by-step solutions to the odd-numbered exercises in the textbook and can help you study and understand the course material. Also included for each chapter is a practice test, with answers so you can check your work. Your college bookstore either has this manual or can order it for you.

If you would like to find out how spreadsheets can help you learn and explore some of the topics discussed in this text, you might want to obtain a copy of Sam Spero's *Topics in Finite Mathematics: An Introduction to the Electronic Spreadsheet* (ISBN 0-065-00300-4) and *The Electronic Spreadsheet and Elementary Calculus* (ISBN 0-673-46595-0) from your college bookstore. These booklets will help you get started using spreadsheets to enhance your problem-solving skills. Knowledge of spreadsheets is not assumed, and the material is adaptable to any spreadsheet program that may be available to you.

Ask your instructor for the following programs that will help you approach the exercises marked with ▨. *GraphExplorer* contains a comprehensive graphing utility and is available in IBM and Macintosh formats. David Schneider's *Explorations in Finite Mathematics* includes a wide selection of routines, including Gaussian elimination, matrix operations, graphical and simplex methods for linear programming problems, probability, binomial distribution, simple and compound interest, loan and annuity analysis, finance tables, difference equations, and more. Currently, *Explorations* is available in IBM format only.

TO THE INSTRUCTOR

We will make available the following brief overview: *Introduction to the Graphing Calculator*. Please place your order with your HarperCollins sales representative. This item is free to you and your students.

Finite Mathematics and Calculus with Applications

Fourth Edition

Margaret L. Lial
American River College

Charles D. Miller

Raymond N. Greenwell
Hofstra University

■ HarperCollins*CollegePublishers*

Sponsoring Editor: George Duda
Developmental Editor: Adam Bryer
Project Editor: Janet Tilden
Art Director: Julie Anderson
Text and Cover Designer: Jeanne Calabrese
Photo Researcher: Karen Koblik
Photo Acknowledgments: Unless otherwise acknowledged, all photographs are the property of Scott, Foresman and Company. *Cover:* Peter Aaron/Esto Photographics; *page xxi:* Peter Aaron/Esto Photographics; *page 1:* T. Zimmerman/F.P.G.; *page 68:* Wendy Rosin Malecki/PhotoEdit; *page 153:* Michael Hayman/Stock Boston; *page 265:* John Coletti/Stock Boston; *page 313:* Spencer Grant/F.P.G.; *page 352:* Scott Markewitz/F.P.G.; *page 405:* Frank DiGiacomo/The Star Ledger; *page 442:* Bob Daemmrich/Stock Boston; *page 575:* David Young—Wolff/PhotoEdit; *page 622:* Bill Pierce/Rainbow; *page 679:* Cecile Brunswick/Tony Stone Worldwide; *page 737:* NASA; *page 863:* James T. Jones/Frazier Photolibrary.
Production Administrator: Brian Branstetter
Manufacturing Manager: Paula Keller
Compositor: Waldman Graphics, Inc.
Printer and Binder: R. R. Donnelley & Sons Company
Cover Printer: The Lehigh Press, Inc.

Finite Mathematics and Calculus with Applications, Fourth Edition

Library of Congress Cataloging-in-Publication Data
Lial, Margaret L.
 Finite mathematics and calculus with applications / Margaret L. Lial, Charles D. Miller, Raymond N. Greenwell. — 4th ed.
 p. cm.
 Includes index.
 ISBN 0-673-46757-0
 1. Mathematics I. Miller, Charles David. II. Greenwell, Raymond N. III. Title
 QA37.2.L48 1992
510—dc20 92-27439
 CIP

96 97 98 99 9 8 7 6 5

PREFACE

Finite Mathematics and Calculus with Applications is a solid, application-oriented text for students majoring in business, management, economics, or the life or social sciences. A prerequisite of two to three semesters of high school algebra is assumed. Many new features, including new exercises, new applications, and motivational section-opening questions, make using this edition of the text easier and more enjoyable. Application exercises are designed to be as believable and realistic as possible.

This edition continues to offer the many popular features of the third edition: applications grouped by subject matter with subheadings indicating the specific application; extended applications to motivate student interest; careful exposition; fully developed examples with side comments; carefully graded exercises; and an algebra reference, designed to be used either in class or individually as necessary. The index of appications shows the abundant variety of applications included in the text and allows direct reference to a particular topic.

▶ NEW AND ENHANCED FEATURES

Conceptual and Writing Exercises To complement the drill and application exercises, several exercises that strengthen conceptual understanding are included in almost every exercise set. Also included are exercises that require the student to respond by writing a few sentences. (Some writing exercises are also conceptual in nature.)

Connections Some exercise sets and many review sections include one or more exercises (labeled "Connections") that integrate concepts and skills introduced earlier with those just introduced in the section or chapter, or that integrate different concepts presented in the chapter.

Summaries A few chapters include a summary of rules or formulas designed to help the student sort out the different ideas introduced in the chapter. These are included in chapters where students traditionally have trouble and become confused about when to use one of several techniques.

Section Openers Most sections open with a thought-provoking question that is answered in an application within the section or in the exercises.

Margin Reviews This feature is designed to help students better understand the ideas being presented. These notes give short explanations or comments reminding students of skills or techniques learned earlier that are needed at this point. Some include a few practice exercises; some include reference to material presented earlier in the text.

Cautions and Notes Common student difficulties and errors are now highlighted graphically and identified with the heading "Caution." Important comments and asides are treated similarly and given the heading "Notes."

Actuarial and CPA Exercises Exercises taken from actuarial or CPA exams are included in appropriate exercise sets throughout the book. These exercises enable students to see how the mathematics introduced in the book is used by professionals in these fields.

KEY CONTENT CHANGES

The use of a scientific calculator is assumed in this edition, and references to calculator use occur throughout as appropriate. Although a graphing calculator is not required, its usefulness is mentioned, and some clearly labeled computer or graphing calculator exercises are included in a few sections. These exercises showcase the ability of many graphing calculators to perform operations with matrices and to execute programs.

The fourth edition preserves the basic format and order of topics of the third edition. A few changes were made, however.

▶ A new section on graphing techniques of translations and reflections has been added to Chapter 1 on Functions and Graphs.

▶ The graphical approach to linear programming now presents the technique in one section and applications in the next.

▶ The row operations used in solving linear systems and in the simplex method are now presented so as to minimize the need for fractions. Our experience is that students find this method easier because it reduces the errors that occur so often when fractions are used. In addition, the simplex method for mixed constraints has been revised to be more straightforward.

▶ Combinations and permutations now appear in Chapter 6, where they are first used.

▶ Probability distributions and expected value are also included in Chapter 6 so that they may be covered easily even if the statistics chapter is omitted.

▶ The necessary information about geometric sequences has been incorporated into the discussion on annuities, and coverage of arithmetic sequences has been omitted.

- ▶ A new section covers the use of linear programming in game theory, connecting these seemingly unrelated topics.

- ▶ In Chapter 10, the material on continuity has been combined with the introduction of limits in Section 10.1.

- ▶ Limits at infinity are now included in Chapter 11 in a new section discussing curve sketching.

- ▶ Applications of the derivative, presented in one long chapter in earlier editions, are now covered in Chapters 11 and 12.

- ▶ Some discussion on math of finance has been included in the section on exponential functions, and Individual Retirement Accounts are presented as an extended application. Growth and decay applications of exponential and logarithmic functions are presented in Section 13.3, before the derivatives are introduced. The first three sections require no calculus and can be presented earlier with the functions discussed in Chapter 1.

- ▶ In Chapter 15, the last section now includes elementary and separable differential equations. This is a combination of the last two sections from the techniques and applications chapter in the previous edition.

The flexibility of the text is indicated in the following chart of chapter prerequisites. As shown, the course could begin with either Chapter 1 or Chapter 5, and Chapter 9 on the mathematics of finance could be covered at any time.

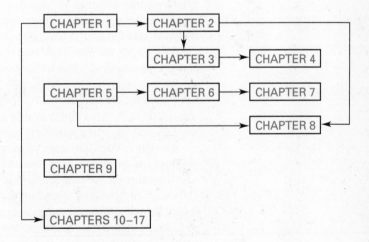

Please note: Although this text is a combination of *Finite Mathematics*, Fifth Edition, and *Calculus with Applications*, Fifth Edition, these two books have additional topics not included here. *Finite Mathematics* also includes Constructing Linear Models and Absorbing Markov Chains. *Calculus with Applications* also includes Linear First-Order Differential Equations, Euler's Method, Applications of Differential Equations, and the Trigonometric Functions.

For the Instructor The *Instructor's Guide and Solution Manual* contains solutions to even-numbered exercises; a multiple-choice version and a short-answer version of a final exam, with answers; at least 100 extra test questions per chapter, with answers; and a set of concise teaching tips.

The *Instructor's Answer Manual* provides answers to every exercise in the text.

The *HarperCollins Test Generator* is the foremost product of its kind, enabling the instructor to create multiple versions of tests, to randomly regenerate "variables," and to insert questions of his or her own devising. The Test Generator comes free to adopters.

Accompanying two-color *Overhead Transparencies*, given free to adopters, can help enhance lectures.

A brief overview entitled *Introduction to the Graphing Calculator* will also be of interest. Should you desire to cover this topic, our overview is available to you and your students at no charge.

For the Student The *Student's Solution Manual* (ISBN 0-673-46764-3) provides solutions to odd-numbered exercises and sample chapter tests, with answers.

Topics in Finite Mathematics: An Introduction to the Electronic Spreadsheet, by Sam Spero, Cuyahoga Community College, (ISBN 0-065-00300-4) is a user-friendly guide designed to introduce students to the various ways one can approach problem solving with spreadsheets. Knowledge of spreadsheets is not assumed, and the approach is adaptable to all spreadsheet programs.

The Electronic Spreadsheet and Elementary Calculus, by Sam Spero, Cuyahoga Community College, (ISBN 0-673-46595-0) is a companion to *Topics in Finite Mathematics: An Introduction to the Electronic Spreadsheet.* This guide helps students get started controlling graphing and problem solving by means of the spreadsheet. As with the companion volume, knowledge of spreadsheets is not assumed, and the approach is adaptable to all spreadsheet programs.

GraphExplorer provides students and instructors with a comprehensive graphing utility, and is available in IBM and Macintosh formats.

Explorations in Finite Mathematics (IBM format only), by David Schneider, University of Maryland, contains on one disk a wider selection of routines than in any similar software supplement. Included are utilities for Gaussian elimination, matrix operations, graphical and simplex methods for linear programming problems, probability, binomial distribution, simple and compound interest, loan and annuity analysis, finance table, difference equations, and more. Refined monitor display for fractions, color capabilities, choice of exact or approximate calculations with matrices, and refined printing capabilities further set this apart from other programs.

Acknowledgments We wish to thank the following professors for their contributions in reviewing portions of this text.

Rick Armstrong, *St. Louis Community College at Florissant Valley*
Judy Barclay, *Cuesta College*
Louise Bernauer, *County College of Morris*
Steven Blasberg, *West Valley College*
Jerry Bloomberg, *Essex Community College*
Dana Dwight Clahane, *El Camino College*
G. B. Costa, *Seton Hall University*
Gregory J. Davis, *University of Wisconsin—Green Bay*
William L. Drezdzon, *Oakton Community College*
Jerry R. Ehman, *Franklin University*
H. Joseph Faber, *East Tennessee State University*
R. E. Goad, *Sam Houston State University*
Madelyn T. Gould, *Dekalb College*
Joseph W. Guest, *Cleveland State Community College*
Frances F. Gulick, *University of Maryland—College Park*
Philip Gwanyama, *Northeastern Illinois University*
Berthe Habib, *Bellevue Community College*
Vivian Heigl, *University of Wisconsin—Parkside*
Denise Hennicke, *Collin County Community College*
Tracey J. Hoy, *College of Lake County*
Robert Keever, *State University of New York—Plattsburgh*
Jeff Knisley, *East Tennessee State University*
Martin Kotler, *Pace University Pleasantville—Briarcliff Campus*
W. David Laverell, *Calvin College*
Jaclyn LeFebvre, *Illinois Central College*
Susan S. Lenker, *Central Michigan University*
Giles Wilson Maloof, *Boise State University*
Joseph L. Mazanec, *Delta College*
Billy R. Nail, *Clayton State College*
James H. Parker, Jr., *Husson College*
K. S. Ramadurai, *Saint Francis Xavier University*
Derald D. Rothman, *Moorhead State University*
Arnold L. Schroeder, *Long Beach City College*
Gordon Shilling, *University of Texas at Arlington*
Joseph D. Sloan, *Lander College*
C. Donald Smith, *Louisiana State University in Shreveport*
Lowell Stultz, *Kalamazoo Valley Community College*
Walter W. Turner, *Western Michigan University*
Lyndon C. Weberg, *University of Wisconsin—River Falls*
Barbara W. Worley, *Des Moines Area Community College*

Paul Eldersveld, College of DuPage, deserves our gratitude for doing an excellent job coordinating all of the print ancillaries for us—an enormous and time-consuming task. Paul Van Erden, of American River College, has created an accurate and complete index for us, and Jim Walker, also of American River College, has compiled the application index and teaching tips. Barbara Bohannon, Peter Grassi, and David Knee, at Hofstra University, have provided many helpful

comments. For their invaluable help in maintaining high standards of accuracy in the answer section, we thank Michael Karelius of American River College, Jaclyn LeFebvre of Illinois Central College, C. Donald Smith of Louisiana State University in Shreveport, and Stephen Ryan and Margo McCullagh of Hofstra University. We also want to thank Karla Harby for her editorial assistance and the staff at HarperCollins for their contributions in bringing this project to a successful conclusion: George Duda, Adam Bryer, Linda Youngman, Janet Tilden, and Julie Anderson.

Finally, we want to express our deep appreciation to W. Weston Meyer, Richard E. Klabunde, Allen Blaurock, Alan Opsahl, Jane Gillum, and Henry Robinson, who provided the materials for the "Calculus at Work" essays that give additional perspective on the use of calculus in the workplace, and to Laurie Golson, who edited the essays.

Margaret L. Lial
Raymond N. Greenwell

CONTENTS

6 ▶ COUNTING PRINCIPLES; FURTHER PROBABILITY TOPICS *265*

7 ▶ STATISTICS *313*

8 ▶ MARKOV CHAINS AND GAME THEORY *352*

 APPLICATIONS OF THE DERIVATIVE 575

 EXPONENTIAL AND LOGARITHMIC FUNCTIONS *622*

 INTEGRATION *679*

Tables

Answers to Selected Exercises *898*

Index *950*

A brief overview entitled *Introduction to the Graphing Calculator* will be provided free to adopters who request it.

INDEX OF APPLICATIONS

General Interest

ALGEBRA REFERENCE

The study of calculus is impossible without a thorough knowledge of elementary algebra. The word *algebra* is derived from the Arabic word *al-jabr*, which appeared in the title of a book by Arab mathematician al-Khowarizmi in the early ninth century A.D. The study of algebra in modern times has grown to include many abstract ideas, but in this text we will be concerned primarily with the topics of elementary algebra found in al-Khowarizmi's book.

This algebra reference is designed for self-study; you can study it all at once or refer to it when needed throughout the course. Since this is a review, answers to all exercises are given in the answer section at the back of the book.

► R.1 POLYNOMIALS

A **polynomial** is an expression of the form

$$a_n x^n + a_{n-1} x^{n-1} + \cdots + a_1 x + a_0,$$

where $a_0, a_1, a_2, \ldots, a_n$ are real numbers, n is a natural number, and $a_n \neq 0$. Examples of polynomials include

$$5x^4 + 2x^3 + 6x, \qquad 8m^3 + 9m^2 - 6m + 3, \qquad 10p, \qquad \text{and} \qquad -9.$$

Adding and Subtracting Polynomials An expression such as $9p^4$ is a **term;** the number 9 is the **coefficient,** p is the **variable,** and 4 is the **exponent.** The expression p^4 means $p \cdot p \cdot p \cdot p$, while p^2 means $p \cdot p$, and so on. Terms having the same variable and the same exponent, such as $9x^4$ and $-3x^4$, are **like terms.** Terms that do not have both the same variable and the same exponent, such as m^2 and m^4, are **unlike terms.**

Polynomials can be added or subtracted by using the **distributive property,** shown below.

If a, b, and c are real numbers, then

$$a(b + c) = ab + ac \qquad \text{and} \qquad (b + c)a = ba + ca.$$

Only like terms may be added or subtracted. For example,

$$12y^4 + 6y^4 = (12 + 6)y^4 = 18y^4,$$

and

$$-2m^2 + 8m^2 = (-2 + 8)m^2 = 6m^2,$$

but the polynomial $8y^4 + 2y^5$ cannot be further simplified. To subtract polynomials, use the fact that $-(a + b) = -a - b$. In the next example, we show how to add and subtract polynomials.

EXAMPLE 1 Add or subtract as indicated.

(a) $(8x^3 - 4x^2 + 6x) + (3x^3 + 5x^2 - 9x + 8)$

Combine like terms.

$$(8x^3 - 4x^2 + 6x) + (3x^3 + 5x^2 - 9x + 8)$$
$$= (8x^3 + 3x^3) + (-4x^2 + 5x^2) + (6x - 9x) + 8$$
$$= 11x^3 + x^2 - 3x + 8$$

(b) $(-4x^4 + 6x^3 - 9x^2 - 12) + (-3x^3 + 8x^2 - 11x + 7)$
$$= -4x^4 + 3x^3 - x^2 - 11x - 5$$

(c) $(2x^2 - 11x + 8) - (7x^2 - 6x + 2)$
$$= (2x^2 - 11x + 8) + (-7x^2 + 6x - 2)$$
$$= -5x^2 - 5x + 6 \quad \blacktriangleleft$$

Multiplying Polynomials The distributive property is used also when multiplying polynomials, as shown in the next example.

EXAMPLE 2 Multiply.

(a) $8x(6x - 4)$

$$8x(6x - 4) = 8x(6x) - 8x(4)$$
$$= 48x^2 - 32x$$

(b) $(3p - 2)(p^2 + 5p - 1)$

$$(3p - 2)(p^2 + 5p - 1)$$
$$= 3p(p^2) + 3p(5p) + 3p(-1) - 2(p^2) - 2(5p) - 2(-1)$$
$$= 3p^3 + 15p^2 - 3p - 2p^2 - 10p + 2$$
$$= 3p^3 + 13p^2 - 13p + 2 \quad \blacktriangleleft$$

When two binomials are multiplied, the FOIL method (First, Outer, Inner, Last) is used as a shortcut. This method is shown below.

EXAMPLE 3 Find $(2m - 5)(m + 4)$ using the FOIL method.

$$\begin{array}{cccc} \text{F} & \text{O} & \text{I} & \text{L} \end{array}$$
$$(2m - 5)(m + 4) = (2m)(m) + (2m)(4) + (-5)(m) + (-5)(4)$$
$$= 2m^2 + 8m - 5m - 20$$
$$= 2m^2 + 3m - 20 \quad \blacktriangleleft$$

EXAMPLE 4 Find $(2k - 5)^2$.
Use FOIL.

$$(2k - 5)^2 = (2k - 5)(2k - 5)$$
$$= 4k^2 - 10k - 10k + 25$$
$$= 4k^2 - 20k + 25$$

Notice that the product of the square of a binomial is the square of the first term, $(2k)^2$, plus twice the product of the two terms, $(2)(2k)(-5)$, plus the square of the last term, $(-5)^2$. $\quad \blacktriangleleft$

Caution Avoid the common error of writing $(x + y)^2 = x^2 + y^2$. As Example 4 shows, the square of a binomial has three terms, so

$$(x + y)^2 = x^2 + 2xy + y^2.$$

Furthermore, higher powers of a binomial also result in more than two terms. For example, verify by multiplication that

$$(x + y)^3 = x^3 + 3x^2y + 3xy^2 + y^3.$$

R.1 Exercises

Perform the indicated operations.

1. $(2x^2 - 6x + 11) + (-3x^2 + 7x - 2)$

2. $(-4y^2 - 3y + 8) - (2y^2 - 6y - 2)$

3. $-3(4q^2 - 3q + 2) + 2(-q^2 + q - 4)$

4. $2(3r^2 + 4r + 2) - 3(-r^2 + 4r - 5)$

5. $(.613x^2 - 4.215x + .892) - .47(2x^2 - 3x + 5)$

6. $.83(5r^2 - 2r + 7) - (7.12r^2 + 6.423r - 2)$

7. $-9m(2m^2 + 3m - 1)$

8. $(6k - 1)(2k - 3)$

9. $(5r - 3s)(5r + 4s)$

10. $(9k + q)(2k - q)$

11. $\left(\dfrac{2}{5}y + \dfrac{1}{8}z\right)\left(\dfrac{3}{5}y + \dfrac{1}{2}z\right)$

12. $\left(\dfrac{3}{4}r - \dfrac{2}{3}s\right)\left(\dfrac{5}{4}r + \dfrac{1}{3}s\right)$

13. $(12x - 1)(12x + 1)$

14. $(6m + 5)(6m - 5)$

15. $(3p - 1)(9p^2 + 3p + 1)$

16. $(2p - 1)(3p^2 - 4p + 5)$

17. $(2m + 1)(4m^2 - 2m + 1)$

18. $(k + 2)(12k^3 - 3k^2 + k + 1)$

19. $(m - n + k)(m + 2n - 3k)$

20. $(r - 3s + t)(2r - s + t)$

► R.2 FACTORING

Multiplication of polynomials relies on the distributive property. The reverse process, where a polynomial is written as a product of other polynomials, is called **factoring.** For example, one way to factor the number 18 is to write it as the product $9 \cdot 2$. When 18 is written as $9 \cdot 2$, both 9 and 2 are called **factors** of 18. It is true that $18 = 36 \cdot 1/2$, but 36 and 1/2 are not considered factors of 18; only integers are used as factors. The number 18 also can be written with three integer factors as $2 \cdot 3 \cdot 3$. The integer factors of 18 are ± 1, ± 2, ± 3, ± 6, ± 9, ± 18.

The Greatest Common Factor To factor the algebraic expression $15m + 45$, first note that both $15m$ and 45 can be divided by 15. In fact, $15m = 15 \cdot m$ and $45 = 15 \cdot 3$. Thus, the distributive property can be used to write

$$15m + 45 = 15 \cdot m + 15 \cdot 3 = 15(m + 3).$$

Both 15 and $m + 3$ are factors of $15m + 45$. Since 15 divides into all terms of $15m + 45$ (and is the largest number that will do so), 15 is the **greatest common factor** for the polynomial $15m + 45$. The process of writing $15m + 45$ as $15(m + 3)$ is often called **factoring out** the greatest common factor.

EXAMPLE 1

Factor out the greatest common factor.

(a) $12p - 18q$

Both $12p$ and $18q$ are divisible by 6. Therefore,

$$12p - 18q = 6 \cdot 2p - 6 \cdot 3q = 6(2p - 3q).$$

(b) $8x^3 - 9x^2 + 15x$

Each of these terms is divisible by x.

$$8x^3 - 9x^2 + 15x = (8x^2) \cdot x - (9x) \cdot x + 15 \cdot x$$
$$= x(8x^2 - 9x + 15) \quad \text{or} \quad (8x^2 - 9x + 15)x \quad \blacktriangleleft$$

Caution When factoring out the greatest common factor in an expression like $2x^2 + x$, be careful to remember the 1 in the second term.

$$2x^2 + x = 2x^2 + 1x = x(2x + 1), \text{ not } x(2x)$$

Factoring Trinomials A polynomial that has no greatest common factor (other than 1) may still be factorable. For example, the polynomial $x^2 + 5x + 6$ can be factored as $(x + 2)(x + 3)$. To see that this is correct, find the product $(x + 2)(x + 3)$; you should get $x^2 + 5x + 6$. To factor a polynomial of three terms such as $x^2 + 5x + 6$, where the coefficient of x^2 is 1, proceed as shown in the following example.

EXAMPLE 2

Factor $y^2 + 8y + 15$.

Since the coefficient of y^2 is 1, factor by finding two numbers whose *product* is 15 and whose *sum* is 8. Since the constant and the middle term are positive, the numbers must both be positive. Begin by listing all pairs of positive integers having a product of 15. As you do this, also form the sum of each pair of numbers.

Products	Sums
$15 \cdot 1 = 15$	$15 + 1 = 16$
$5 \cdot 3 = 15$	$5 + 3 = 8$

The numbers 5 and 3 have a product of 15 and a sum of 8. Thus, $y^2 + 8y + 15$ factors as

$$y^2 + 8y + 15 = (y + 5)(y + 3).$$

The answer also can be written as $(y + 3)(y + 5)$. \blacktriangleleft

If the coefficient of the squared term is *not* 1, work as shown below.

EXAMPLE 3

Factor $2x^2 + 9xy - 5y^2$.

The factors of $2x^2$ are $2x$ and x; the possible factors of $-5y^2$ are $-5y$ and y, or $5y$ and $-y$. Try various combinations of these factors until one works (if, indeed, any work). For example, try the product $(2x + 5y)(x - y)$.

$$(2x + 5y)(x - y) = 2x^2 - 2xy + 5xy - 5y^2$$
$$= 2x^2 + 3xy - 5y^2$$

This product is not correct, so try another combination.

$$(2x - y)(x + 5y) = 2x^2 + 10xy - xy - 5y^2$$
$$= 2x^2 + 9xy - 5y^2$$

Since this combination gives the correct polynomial,

$$2x^2 + 9xy - 5y^2 = (2x - y)(x + 5y). \quad \blacktriangleleft$$

Special Factorizations Four special factorizations occur so often that they are listed here for future reference.

SPECIAL FACTORIZATIONS		
	$x^2 - y^2 = (x + y)(x - y)$	**Difference of two squares**
	$x^2 + 2xy + y^2 = (x + y)^2$	**Perfect square**
	$x^3 - y^3 = (x - y)(x^2 + xy + y^2)$	**Difference of two cubes**
	$x^3 + y^3 = (x + y)(x^2 - xy + y^2)$	**Sum of two cubes**

A polynomial that cannot be factored is called a **prime polynomial.**

EXAMPLE 4

Factor each of the following.

(a) $64p^2 - 49q^2 = (8p)^2 - (7q)^2 = (8p + 7q)(8p - 7q)$

(b) $x^2 + 36$ is a prime polynomial.

(c) $x^2 + 12x + 36 = (x + 6)^2$

(d) $9y^2 - 24yz + 16z^2 = (3y - 4z)^2$

(e) $y^3 - 8 = y^3 - 2^3 = (y - 2)(y^2 + 2y + 4)$

(f) $m^3 + 125 = m^3 + 5^3 = (m + 5)(m^2 - 5m + 25)$

(g) $8k^3 - 27z^3 = (2k)^3 - (3z)^3 = (2k - 3z)(4k^2 + 6kz + 9z^2) \quad \blacktriangleleft$

Caution In factoring, always look for a common factor first. Since $36x^2 - 4y^2$ has a common factor of 4,

$$36x^2 - 4y^2 = 4(9x^2 - y^2) = 4(3x + y)(3x - y).$$

It would be incomplete to factor it as

$$36x^2 - 4y^2 = (6x + 2y)(6x - 2y)$$

since each factor can be factored still further. To *factor* means to factor completely, so that each polynomial factor is prime.

R.2 Exercises

Factor each of the following. If a polynomial cannot be factored, write prime. *Factor out the greatest common factor as necessary.*

1. $8a^3 - 16a^2 + 24a$

2. $3y^3 + 24y^2 + 9y$

3. $25p^4 - 20p^3q + 100p^2q^2$

4. $60m^4 - 120m^3n + 50m^2n^2$

5. $m^2 + 9m + 14$

6. $x^2 + 4x - 5$

7. $z^2 + 9z + 20$

8. $b^2 - 8b + 7$

9. $a^2 - 6ab + 5b^2$

10. $s^2 + 2st - 35t^2$

11. $y^2 - 4yz - 21z^2$

12. $6a^2 - 48a - 120$

13. $3m^3 + 12m^2 + 9m$

14. $2x^2 - 5x - 3$

15. $3a^2 + 10a + 7$

16. $2a^2 - 17a + 30$

17. $15y^2 + y - 2$

18. $21m^2 + 13mn + 2n^2$

19. $24a^4 + 10a^3b - 4a^2b^2$

20. $32z^5 - 20z^4a - 12z^3a^2$

21. $x^2 - 64$

22. $9m^2 - 25$

23. $121a^2 - 100$

24. $9x^2 + 64$

25. $z^2 + 14zy + 49y^2$

26. $m^2 - 6mn + 9n^2$

27. $9p^2 - 24p + 16$

28. $a^3 - 216$

29. $8r^3 - 27s^3$

30. $64m^3 + 125$

R.3 RATIONAL EXPRESSIONS

Many algebraic fractions are **rational expressions,** which are quotients of polynomials with nonzero denominators. Examples include

$$\frac{8}{x - 1}, \qquad \frac{3x^2 + 4x}{5x - 6}, \qquad \text{and} \qquad \frac{2y + 1}{y^2}.$$

Methods for working with rational expressions are summarized below.

PROPERTIES OF RATIONAL EXPRESSIONS

For all mathematical expressions P, Q, R, and S, with Q and $S \neq 0$:

$$\frac{P}{Q} = \frac{PS}{QS} \qquad \text{Fundamental property}$$

$$\frac{P}{Q} \cdot \frac{R}{S} = \frac{PR}{QS} \qquad \text{Multiplication}$$

$$\frac{P}{Q} + \frac{R}{Q} = \frac{P + R}{Q} \qquad \text{Addition}$$

$$\frac{P}{Q} - \frac{R}{Q} = \frac{P - R}{Q} \qquad \text{Subtraction}$$

$$\frac{P}{Q} \div \frac{R}{S} = \frac{P}{Q} \cdot \frac{S}{R} \quad (R \neq 0) \qquad \text{Division}$$

EXAMPLE 1

Write each rational expression in lowest terms.

(a) $\dfrac{8x + 16}{4} = \dfrac{8(x + 2)}{4} = \dfrac{4 \cdot 2(x + 2)}{4} = 2(x + 2)$

The first step is to factor both numerator and denominator in order to identify any common factors. The answer also could be written as $2x + 4$, if desired.

(b) $\dfrac{k^2 + 7k + 12}{k^2 + 2k - 3} = \dfrac{(k + 4)(k + 3)}{(k - 1)(k + 3)} = \dfrac{k + 4}{k - 1}$

The answer cannot be further reduced. ◄

Caution One of the most common errors in algebra involves incorrect use of the fundamental property of rational expressions. Only common *factors* may be divided or "canceled." It is essential to factor rational expressions before writing them in lowest terms. In Example 1(b), for instance, it is not correct to "cancel" k^2 (or cancel k, or divide 12 by -3) because the additions and subtraction must be performed first. Here they cannot be performed, so it is not possible to divide. Factoring allows the expression to be reduced, however.

EXAMPLE 2 Perform each operation.

(a) $\dfrac{3y + 9}{6} \cdot \dfrac{18}{5y + 15}$

Factor where possible, then multiply numerators and denominators and reduce to lowest terms.

$$\frac{3y + 9}{6} \cdot \frac{18}{5y + 15} = \frac{3(y + 3)}{6} \cdot \frac{18}{5(y + 3)}$$

$$= \frac{3 \cdot 18(y + 3)}{6 \cdot 5(y + 3)}$$

$$= \frac{3 \cdot 6 \cdot 3(y + 3)}{6 \cdot 5(y + 3)} = \frac{3 \cdot 3}{5} = \frac{9}{5}$$

(b) $\dfrac{m^2 + 5m + 6}{m + 3} \cdot \dfrac{m}{m^2 + 3m + 2} = \dfrac{(m + 2)(m + 3)}{m + 3} \cdot \dfrac{m}{(m + 2)(m + 1)}$

$$= \frac{m(m + 2)(m + 3)}{(m + 3)(m + 2)(m + 1)} = \frac{m}{m + 1}$$

(c) $\dfrac{9p - 36}{12} \div \dfrac{5(p - 4)}{18}$

$$= \frac{9p - 36}{12} \cdot \frac{18}{5(p - 4)} \qquad \text{Invert and multiply.}$$

$$= \frac{9(p - 4)}{12} \cdot \frac{18}{5(p - 4)} = \frac{27}{10}$$

(d) $\dfrac{4}{5k} - \dfrac{11}{5k}$

As shown in the list of properties above, when two rational expressions have the same denominators, we subtract by subtracting the numerators.

$$\frac{4}{5k} - \frac{11}{5k} = \frac{4 - 11}{5k} = -\frac{7}{5k}$$

(e) $\dfrac{7}{p} + \dfrac{9}{2p} + \dfrac{1}{3p}$

These three fractions cannot be added until their denominators are the same. A common denominator into which p, $2p$, and $3p$ all divide is $6p$. Use the fundamental property to rewrite each rational expression with a denominator of $6p$.

$$\frac{7}{p} + \frac{9}{2p} + \frac{1}{3p} = \frac{6 \cdot 7}{6 \cdot p} + \frac{3 \cdot 9}{3 \cdot 2p} + \frac{2 \cdot 1}{2 \cdot 3p}$$

$$= \frac{42}{6p} + \frac{27}{6p} + \frac{2}{6p}$$

$$= \frac{42 + 27 + 2}{6p}$$

$$= \frac{71}{6p} \quad \blacktriangleleft$$

R.3 Exercises

Write each rational expression in lowest terms.

1. $\dfrac{7z^2}{14z}$

2. $\dfrac{25p^3}{10p^2}$

3. $\dfrac{8k + 16}{9k + 18}$

4. $\dfrac{3(t + 5)}{(t + 5)(t - 3)}$

5. $\dfrac{8x^2 + 16x}{4x^2}$

6. $\dfrac{36y^2 + 72y}{9y}$

7. $\dfrac{m^2 - 4m + 4}{m^2 + m - 6}$

8. $\dfrac{r^2 - r - 6}{r^2 + r - 12}$

9. $\dfrac{x^2 + 3x - 4}{x^2 - 1}$

10. $\dfrac{z^2 - 5z + 6}{z^2 - 4}$

11. $\dfrac{8m^2 + 6m - 9}{16m^2 - 9}$

12. $\dfrac{6y^2 + 11y + 4}{3y^2 + 7y + 4}$

Perform the indicated operations.

13. $\dfrac{9k^2}{25} \cdot \dfrac{5}{3k}$

14. $\dfrac{15p^3}{9p^2} \div \dfrac{6p}{10p^2}$

15. $\dfrac{a + b}{2p} \cdot \dfrac{12}{5(a + b)}$

16. $\dfrac{a - 3}{16} \div \dfrac{a - 3}{32}$

17. $\dfrac{2k + 8}{6} \div \dfrac{3k + 12}{2}$

18. $\dfrac{9y - 18}{6y + 12} \cdot \dfrac{3y + 6}{15y - 30}$

19. $\dfrac{4a + 12}{2a - 10} \div \dfrac{a^2 - 9}{a^2 - a - 20}$

20. $\dfrac{6r - 18}{9r^2 + 6r - 24} \cdot \dfrac{12r - 16}{4r - 12}$

21. $\dfrac{k^2 - k - 6}{k^2 + k - 12} \cdot \dfrac{k^2 + 3k - 4}{k^2 + 2k - 3}$

22. $\dfrac{m^2 + 3m + 2}{m^2 + 5m + 4} \div \dfrac{m^2 + 5m + 6}{m^2 + 10m + 24}$

23. $\dfrac{2m^2 - 5m - 12}{m^2 - 10m + 24} \div \dfrac{4m^2 - 9}{m^2 - 9m + 18}$

24. $\dfrac{6n^2 - 5n - 6}{6n^2 + 5n - 6} \cdot \dfrac{12n^2 - 17n + 6}{12n^2 - n - 6}$

25. $\dfrac{a + 1}{2} - \dfrac{a - 1}{2}$

26. $\dfrac{3}{p} + \dfrac{1}{2}$

27. $\dfrac{2}{y} - \dfrac{1}{4}$

28. $\dfrac{1}{6m} + \dfrac{2}{5m} + \dfrac{4}{m}$

29. $\dfrac{1}{m - 1} + \dfrac{2}{m}$

30. $\dfrac{6}{r} - \dfrac{5}{r - 2}$

31. $\dfrac{8}{3(a - 1)} + \dfrac{2}{a - 1}$

32. $\dfrac{2}{5(k - 2)} + \dfrac{3}{4(k - 2)}$

33. $\dfrac{2}{x^2 - 2x - 3} + \dfrac{5}{x^2 - x - 6}$

34. $\dfrac{2y}{y^2 + 7y + 12} - \dfrac{y}{y^2 + 5y + 6}$

35. $\dfrac{3k}{2k^2 + 3k - 2} - \dfrac{2k}{2k^2 - 7k + 3}$

36. $\dfrac{4m}{3m^2 + 7m - 6} - \dfrac{m}{3m^2 - 14m + 8}$

R.4 EQUATIONS

Linear Equations Equations that can be written in the form $ax + b = 0$, where a and b are real numbers, with $a \neq 0$, are **linear equations.** Examples of linear equations include $5y + 9 = 16, 8x = 4$, and $-3p + 5 = -8$. Equations that are *not* linear include absolute value equations such as $|x| = 4$. The following properties are used to solve linear equations.

PROPERTIES OF REAL NUMBERS

For all real numbers a, b, and c:

$a(b + c) = ab + ac$ **Distributive property**

If $a = b$, then $a + c = b + c$. **Addition property of equality**
(The same number may be added to both sides of an equation.)

If $a = b$, then $ac = bc$. **Multiplication property of equality**
(The same number may be multiplied on both sides of an equation.)

The following example shows how these properties are used to solve linear equations. Of course, the solutions should always be checked by substitution in the original equation.

EXAMPLE 1

Solve $2x - 5 + 8 = 3x + 2(2 - 3x)$.

$$2x - 5 + 8 = 3x + 4 - 6x$$ Distributive property

$$2x + 3 = -3x + 4$$ Combine like terms.

$$5x + 3 = 4$$ Add $3x$ to both sides.

$$5x = 1$$ Add -3 to both sides.

$$x = \frac{1}{5}$$ Multiply on both sides by $\frac{1}{5}$.

Check by substituting in the original equation. ◀

Absolute Value Equations Recall the definition of absolute value:

$$|x| = \begin{cases} x \text{ if } x \text{ is greater than or equal to } 0 \\ -x \text{ if } x \text{ is less than } 0. \end{cases}$$

By this definition, $|5| = 5$ and $|-5| = 5$. Think of the absolute value of a number as the distance between 0 and the number. The distance is always positive.

Although equations involving absolute value are not linear equations, sometimes they may be reduced to linear equations by using the following property.

EQUATIONS WITH ABSOLUTE VALUE

If k is greater than or equal to 0, then $|ax + b| = k$ is equivalent to

$$ax + b = k \quad \text{or} \quad -(ax + b) = k.$$

Solve each equation.

(a) $|2x - 5| = 7$

Rewrite $|2x - 5| = 7$ as follows.

$$
\begin{array}{lll}
2x - 5 = 7 & \text{or} & -(2x - 5) = 7 \\
2x = 12 & & -2x + 5 = 7 \\
x = 6 & & -2x = 2 \\
& & x = -1
\end{array}
$$

Check both answers by substituting them into the original equation.

(b) $|3x + 1| = |x - 1|$

Use a variation of the property given above to rewrite the equation.

$$
\begin{array}{lll}
3x + 1 = x - 1 & \text{or} & -(3x + 1) = x - 1 \\
2x = -2 & & -3x - 1 = x - 1 \\
x = -1 & & -4x = 0 \\
& & x = 0
\end{array}
$$

Be sure to check both answers. ◄

Quadratic Equations An equation with 2 as the highest exponent is a *quadratic equation*. A **quadratic equation** has the form $ax^2 + bx + c = 0$, where a, b, and c are real numbers and $a \neq 0$. A quadratic equation written in the form $ax^2 + bx + c = 0$ is said to be in **standard form.**

The simplest way to solve a quadratic equation, but one that is not always applicable, is by factoring. This method depends on the **zero-factor property.**

ZERO-FACTOR PROPERTY

If a and b are real numbers, with $ab = 0$, then

$$a = 0, \ b = 0, \text{ or both.}$$

Solve $6r^2 + 7r = 3$.

First write the equation in standard form.

$$6r^2 + 7r - 3 = 0$$

Now factor $6r^2 + 7r - 3$ to get

$$(3r - 1)(2r + 3) = 0.$$

By the zero-factor property, the product $(3r - 1)(2r + 3)$ can equal 0 only if

$$3r - 1 = 0 \quad \text{or} \quad 2r + 3 = 0.$$

Solve each of these equations separately to find that the solutions are 1/3 and $-3/2$. Check these solutions by substituting them in the original equation. ◀

Caution Remember, the zero-factor property requires that the product of two (or more) factors be equal to *zero*, not some other quantity. It would be incorrect to use the zero-factor property with an equation in the form $(x + 3)(x - 1) = 4$, for example.

If a quadratic equation cannot be solved easily by factoring, use the *quadratic formula*. (The derivation of the quadratic formula is given in most algebra books.)

QUADRATIC FORMULA

The solutions of the quadratic equation $ax^2 + bx + c = 0$, where $a \neq 0$, are given by

$$x = \frac{-b \pm \sqrt{b^2 - 4ac}}{2a}.$$

EXAMPLE 4 ▶ Solve $x^2 - 4x - 5 = 0$ by the quadratic formula.

The equation is already in standard form (it has 0 alone on one side of the equals sign), so the values of a, b, and c from the quadratic formula are easily identified. The coefficient of the squared term gives the value of a; here, $a = 1$. Also, $b = -4$ and $c = -5$. (Be careful to use the correct signs.) Substitute these values into the quadratic formula.

$$x = \frac{-(-4) \pm \sqrt{(-4)^2 - 4(1)(-5)}}{2(1)} \qquad \text{Let } a = 1, b = -4, c = -5.$$

$$x = \frac{4 \pm \sqrt{16 + 20}}{2} \qquad (-4)^2 = (-4)(-4) = 16$$

$$x = \frac{4 \pm 6}{2} \qquad \sqrt{16 + 20} = \sqrt{36} = 6$$

The \pm sign represents the two solutions of the equation. To find all of the solutions, first use $+$ and then use $-$.

$$x = \frac{4 + 6}{2} = \frac{10}{2} = 5 \quad \text{or} \quad x = \frac{4 - 6}{2} = \frac{-2}{2} = -1$$

The two solutions are 5 and -1. ◀

Caution Notice in the quadratic formula that the square root is added to or subtracted from the value of $-b$ *before* dividing by $2a$.

EXAMPLE 5 ▸ Solve $x^2 + 1 = 4x$.

First, add $-4x$ on both sides of the equals sign in order to get the equation in standard form.

$$x^2 - 4x + 1 = 0$$

Now identify the letters a, b, and c. Here $a = 1$, $b = -4$, and $c = 1$. Substitute these numbers into the quadratic formula.

$$x = \frac{-(-4) \pm \sqrt{(-4)^2 - 4(1)(1)}}{2(1)}$$

$$= \frac{4 \pm \sqrt{16 - 4}}{2}$$

$$= \frac{4 \pm \sqrt{12}}{2}$$

Simplify the solutions by writing $\sqrt{12}$ as $\sqrt{4 \cdot 3} = \sqrt{4} \cdot \sqrt{3} = 2\sqrt{3}$. Substituting $2\sqrt{3}$ for $\sqrt{12}$ gives

$$x = \frac{4 \pm 2\sqrt{3}}{2}$$

$$= \frac{2(2 \pm \sqrt{3})}{2} \qquad \text{Factor } 4 \pm 2\sqrt{3}.$$

$$= 2 \pm \sqrt{3}.$$

The two solutions are $2 + \sqrt{3}$ and $2 - \sqrt{3}$.

The exact values of the solutions are $2 + \sqrt{3}$ and $2 - \sqrt{3}$. The $\sqrt{\ }$ key on a calculator gives decimal approximations of these solutions (to the nearest thousandth).

$$2 + \sqrt{3} \approx 2 + 1.732 = 3.732*$$

$$2 - \sqrt{3} \approx 2 - 1.732 = .268 \quad ◀$$

Note Sometimes the quadratic formula will give a result with a negative number under the radical sign, such as $2 \pm \sqrt{-3}$. A solution of this type is not a real number. Since this text deals only with real numbers, such solutions cannot be used.

Equations with Fractions When an equation includes fractions, first eliminate all denominators by multiplying both sides of the equation by a **common denominator,** a number that can be divided (with no remainder) by each denominator in the equation. When an equation involves fractions with variable denominators, it is *necessary* to check all solutions in the original equation to be sure that no solution will lead to a zero denominator.

*The symbol \approx means "is approximately equal to."

EXAMPLE 6

Solve each equation.

(a) $\dfrac{r}{10} - \dfrac{2}{15} = \dfrac{3r}{20} - \dfrac{1}{5}$

The denominators are 10, 15, 20, and 5. Each of these numbers can be divided into 60, so 60 is a common denominator. Multiply both sides of the equation by 60 and use the distributive property. (If a common denominator cannot be found easily, all the denominators in the problem can be multiplied together to produce one.)

$$\frac{r}{10} - \frac{2}{15} = \frac{3r}{20} - \frac{1}{5}$$

$$60\left(\frac{r}{10} - \frac{2}{15}\right) = 60\left(\frac{3r}{20} - \frac{1}{5}\right)$$

$$60\left(\frac{r}{10}\right) - 60\left(\frac{2}{15}\right) = 60\left(\frac{3r}{20}\right) - 60\left(\frac{1}{5}\right)$$

$$6r - 8 = 9r - 12$$

Add $-6r$ and 12 to both sides.

$$6r - 8 + (-6r) + 12 = 9r - 12 + (-6r) + 12$$

$$4 = 3r$$

Multiply both sides by 1/3 to get

$$r = \frac{4}{3}.$$

(b) $\dfrac{3}{x^2} - 12 = 0$

Begin by multiplying both sides of the equation by x^2 to get $3 - 12x^2 = 0$. This equation could be solved by using the quadratic formula with $a = -12$, $b = 0$, and $c = 3$. Another method, which works well for the type of quadratic equation in which $b = 0$, is shown below.

$$3 - 12x^2 = 0$$

$$3 = 12x^2 \qquad \text{Add } 12x^2.$$

$$\frac{1}{4} = x^2 \qquad \text{Multiply by } \frac{1}{12}.$$

$$\pm\frac{1}{2} = x \qquad \text{Take square roots.}$$

There are two solutions, $-1/2$ and $1/2$.

(c) $\dfrac{2}{k} - \dfrac{3k}{k + 2} = \dfrac{k}{k^2 + 2k}$

Factor $k^2 + 2k$ as $k(k + 2)$. The common denominator for all the fractions is $k(k + 2)$. Multiplying both sides by $k(k + 2)$ gives the following.

$$2(k + 2) - 3k(k) = k$$
$$2k + 4 - 3k^2 = k$$
$$-3k^2 + k + 4 = 0 \qquad \text{Add } -k; \text{ rearrange terms.}$$
$$3k^2 - k - 4 = 0 \qquad \text{Multiply by } -1.$$
$$(3k - 4)(k + 1) = 0 \qquad \text{Factor.}$$

$$3k - 4 = 0 \quad \text{or} \quad k + 1 = 0$$
$$k = \frac{4}{3} \qquad\qquad k = -1$$

Verify that the solutions are 4/3 and −1. ◀

Caution It is possible to get, as a solution of a rational equation, a number that makes one or more of the denominators in the original equation equal to zero. That number is not a solution, so it is *necessary* to check all potential solutions of rational equations.

7 Solve $\dfrac{2}{x - 3} + \dfrac{1}{x} = \dfrac{6}{x(x - 3)}$.

The common denominator is $x(x - 3)$. Multiply both sides by $x(x - 3)$ and solve the resulting equation.

$$\frac{2}{x - 3} + \frac{1}{x} = \frac{6}{x(x - 3)}$$
$$2x + x - 3 = 6$$
$$3x = 9$$
$$x = 3$$

Checking this potential solution by substitution in the original equation shows that 3 makes two denominators 0. Thus 3 cannot be a solution, so there is no solution for this equation. ◀

R.4 Exercises

Solve each equation.

1. $.2m - .5 = .1m + .7$

2. $\dfrac{5}{6}k - 2k + \dfrac{1}{3} = \dfrac{2}{3}$

3. $3r + 2 - 5(r + 1) = 6r + 4$

4. $2[m - (4 + 2m) + 3] = 2m + 2$

5. $|3x + 2| = 9$

6. $|4 - 7x| = 15$

7. $|2x + 8| = |x - 4|$

8. $|5x + 2| = |8 - 3x|$

Solve each of the following equations by factoring or by using the quadratic formula. If the solutions involve square roots, give both the exact solutions and the approximate solutions to three decimal places.

9. $x^2 + 5x + 6 = 0$

10. $x^2 = 3 + 2x$

11. $m^2 + 16 = 8m$

12. $2k^2 - k = 10$

13. $6x^2 - 5x = 4$

14. $m(m - 7) = -10$

15. $9x^2 - 16 = 0$

16. $z(2z + 7) = 4$

17. $12y^2 - 48y = 0$

18. $3x^2 - 5x + 1 = 0$

19. $2m^2 = m + 4$

20. $p^2 + p - 1 = 0$

21. $k^2 - 10k = -20$

22. $2x^2 + 12x + 5 = 0$

23. $2r^2 - 7r + 5 = 0$

24. $2x^2 - 7x + 30 = 0$

25. $3k^2 + k = 6$

26. $5m^2 + 5m = 0$

Solve each of the following equations.

27. $\dfrac{3x - 2}{7} = \dfrac{x + 2}{5}$

28. $\dfrac{x}{3} - 7 = 6 - \dfrac{3x}{4}$

29. $\dfrac{4}{x - 3} - \dfrac{8}{2x + 5} + \dfrac{3}{x - 3} = 0$

30. $\dfrac{5}{2p + 3} - \dfrac{3}{p - 2} = \dfrac{4}{2p + 3}$

31. $\dfrac{2}{m} + \dfrac{m}{m + 3} = \dfrac{3m}{m^2 + 3m}$

32. $\dfrac{2y}{y - 1} = \dfrac{5}{y} + \dfrac{10 - 8y}{y^2 - y}$

33. $\dfrac{1}{x - 2} - \dfrac{3x}{x - 1} = \dfrac{2x + 1}{x^2 - 3x + 2}$

34. $\dfrac{5}{a} + \dfrac{-7}{a + 1} = \dfrac{a^2 - 2a + 4}{a^2 + a}$

35. $\dfrac{2b^2 + 5b - 8}{b^2 + 2b} + \dfrac{5}{b + 2} = -\dfrac{3}{b}$

R.5 INEQUALITIES

To write that one number is greater than or less than another number, we use the following symbols.

INEQUALITY SYMBOLS

$<$ means *is less than*

$>$ means *is greater than*

\leq means *is less than or equal to*

\geq means *is greater than or equal to*

Linear Inequalities An equation states that two expressions are equal; an **inequality** states that they are unequal. A **linear inequality** is an inequality that can be simplified to the form $ax < b$. (Properties introduced in this section are given only for $<$, but they are equally valid for $>$, \leq, or \geq.) Linear equalities are solved with the following properties.

PROPERTIES OF INEQUALITY

For all real numbers a, b, and c:

1. If $a < b$, then $a + c < b + c$.

2. If $a < b$ and if $c > 0$, then $ac < bc$.

3. If $a < b$ and if $c < 0$, then $ac > bc$.

Pay careful attention to property 3; it says that if both sides of an inequality are multiplied by a negative number, the direction of the inequality symbol must be reversed.

EXAMPLE **1** Solve $4 - 3y \leq 7 + 2y$.
Use the properties of inequality.

$$4 - 3y + (-4) \leq 7 + 2y + (-4) \quad \text{Add } -4 \text{ to both sides.}$$

$$-3y \leq 3 + 2y$$

Remember that *adding* the same number to both sides never changes the direction of the inequality symbol.

$$-3y + (-2y) \leq 3 + 2y + (-2y) \quad \text{Add } -2y \text{ to both sides.}$$
$$-5y \leq 3$$

Multiply both sides by $-1/5$. Since $-1/5$ is negative, change the direction of the inequality symbol.

$$-\frac{1}{5}(-5y) \geq -\frac{1}{5}(3)$$
$$y \geq -\frac{3}{5} \quad \blacktriangleleft$$

Caution It is a common error to forget to reverse the direction of the inequality sign when multiplying or dividing by a negative number. For example, to solve $-4x \leq 12$, we must multiply by $-1/4$ on both sides *and* reverse the inequality symbol to get $x \geq -3$.

The solution $y \geq -3/5$ in Example 1 represents an interval on the number line. **Interval notation** often is used for writing intervals. With interval notation, $y \geq -3/5$ is written as $[-3/5, \infty)$. This is an example of a **half-open interval,** since one endpoint, $-3/5$, is included. The **open interval** $(2, 5)$ corresponds to $2 < x < 5$, with neither endpoint included. The **closed interval** $[2, 5]$ includes both endpoints and corresponds to $2 \leq x \leq 5$.

The **graph** of an interval shows all points on a number line that correspond to the numbers in the interval. To graph the interval $[-3/5, \infty)$, for example, use a bracket at $-3/5$, since $-3/5$ is part of the solution. To show that the solution includes all real numbers greater than or equal to $-3/5$, draw a heavy arrow pointing to the right (the positive direction). See Figure 1.

FIGURE 1

EXAMPLE 2 Solve $-2 < 5 + 3m < 20$. Graph the solution.

The inequality $-2 < 5 + 3m < 20$ says that $5 + 3m$ is *between* -2 and 20. Solve this inequality with an extension of the properties given above. Work as follows, first adding -5 to each part.

$$-2 + (-5) < 5 + 3m + (-5) < 20 + (-5)$$
$$-7 < 3m < 15$$

Now multiply each part by $1/3$.

$$-\frac{7}{3} < m < 5$$

A graph of the solution is given in Figure 2; here parentheses are used to show that $-7/3$ and 5 are *not* part of the graph. \blacktriangleleft

FIGURE 2

Quadratic Inequalities A **quadratic inequality** has the form $ax^2 + bx + c > 0$ (or $<$, or \leq, or \geq). The highest exponent is 2. The next few examples show how to solve quadratic inequalities.

EXAMPLE **3**

Solve the quadratic inequality $x^2 - x - 12 < 0$.

This inequality is solved with values of x that make $x^2 - x - 12$ negative (< 0). The quantity $x^2 - x - 12$ changes from positive to negative or from negative to positive at the point where it equals 0. For this reason, first solve the *equation* $x^2 - x - 12 = 0$.

$$x^2 - x - 12 = 0$$
$$(x - 4)(x + 3) = 0$$
$$x = 4 \quad \text{or} \quad x = -3$$

FIGURE 3

Locating -3 and 4 on a number line, as shown in Figure 3, determines three intervals A, B, and C. Decide which intervals include numbers that make $x^2 - x - 12$ negative by substituting a number from each interval in the polynomial. For example,

choose -4 from interval A: $(-4)^2 - (-4) - 12 = 8 > 0$;

choose 0 from interval B: $0^2 - 0 - 12 = -12 < 0$;

choose 5 from interval C: $5^2 - 5 - 12 = 8 > 0$.

FIGURE 4

Only numbers in interval B satisfy the given inequality, so the solution is $(-3, 4)$. A graph of this solution is shown in Figure 4. ◀

EXAMPLE **4**

Solve the quadratic inequality $r^2 + 3r \geq 4$.

First solve the equation $r^2 + 3r = 4$.

$$r^2 + 3r = 4$$
$$r^2 + 3r - 4 = 0$$
$$(r - 1)(r + 4) = 0$$
$$r = 1 \quad \text{'or} \quad r = -4$$

These two solutions determine three intervals on the number line: $(-\infty, -4)$, $(-4, 1)$, and $(1, \infty)$. Substitute a number from each interval into the original inequality to determine the solution. The solution includes the numbers less than or equal to -4, together with the numbers greater than or equal to 1. This solution is written in interval notation as

$$(-\infty, -4] \cup [1, \infty).*$$

FIGURE 5

A graph of the solution is given in Figure 5. ◀

Inequalities with Fractions Inequalities with fractions can be solved in a manner similar to the way quadratic inequalities are solved.

EXAMPLE

Solve $\dfrac{2x - 3}{x} \geq 1$.

*The symbol \cup indicates the *union* of two sets, which includes all elements in either set.

As with quadratic inequalities, first solve the corresponding equation.

$$\frac{2x - 3}{x} = 1$$

$$2x - 3 = x$$

$$x = 3$$

The solution, $x = 3$, determines the intervals on the number line where the fraction may change from greater than 1 to less than 1. This change also may occur on either side of a number that makes the denominator equal 0. Here, the x-value that makes the denominator 0 is $x = 0$. Test each of the three intervals determined by the numbers 0 and 3.

For $(-\infty, 0)$, choose -1: $\dfrac{2(-1) - 3}{-1} = 5 \geq 1.$

For $(0, 3)$, choose 1: $\dfrac{2(1) - 3}{1} = -1 \not\geq 1.$

For $(3, \infty)$, choose 4: $\dfrac{2(4) - 3}{4} = \dfrac{5}{4} \geq 1.$

The symbol $\not\geq$ means "is *not* greater than or equal to." Testing the endpoints 0 and 3 shows that the solution is $(-\infty, 0) \cup [3, \infty)$. ◀

EXAMPLE 6 Solve $\dfrac{x^2 - 3x}{x^2 - 9} < 4.$

Begin by solving the corresponding equation.

$$\frac{x^2 - 3x}{x^2 - 9} = 4$$

$$x^2 - 3x = 4x^2 - 36 \qquad \text{Multiply by } x^2 - 9.$$

$$0 = 3x^2 + 3x - 36 \qquad \text{Get 0 on one side.}$$

$$0 = x^2 + x - 12 \qquad \text{Multiply by } \frac{1}{3}.$$

$$0 = (x + 4)(x - 3) \qquad \text{Factor.}$$

$$x = -4 \quad \text{or} \quad x = 3$$

Now set the denominator equal to 0 and solve that equation.

$$x^2 - 9 = 0$$

$$(x - 3)(x + 3) = 0$$

$$x = 3 \quad \text{or} \quad x = -3$$

The intervals determined by the three (different) solutions are $(-\infty, -4)$, $(-4, -3)$, $(-3, 3)$, and $(3, \infty)$. Testing a number from each interval and the endpoints of the intervals in the given inequality shows that the solution is

$$(-\infty, -4) \cup (-3, 3) \cup (3, \infty). \quad ◀$$

Caution Don't forget to solve the equation formed by setting the denominator equal to zero. Any number that makes the denominator zero always creates two intervals on the number line. For instance, in Example 5, 0 makes the denominator of the rational inequality equal to 0, so we know that there may be a sign change from one side of 0 to the other (as was indeed the case).

R.5 Exercises

Solve each inequality and graph the solution.

1. $-3p - 2 \geq 1$

2. $6k - 4 < 3k - 1$

3. $m - (4 + 2m) + 3 < 2m + 2$

4. $-2(3y - 8) \geq 5(4y - 2)$

5. $3p - 1 < 6p + 2(p - 1)$

6. $x + 5(x + 1) > 4(2 - x) + x$

7. $-7 < y - 2 < 4$

8. $8 \leq 3r + 1 \leq 13$

9. $-4 \leq \dfrac{2k - 1}{3} \leq 2$

10. $-1 \leq \dfrac{5y + 2}{3} \leq 4$

11. $\dfrac{3}{5}(2p + 3) \geq \dfrac{1}{10}(5p + 1)$

12. $\dfrac{8}{3}(z - 4) \leq \dfrac{2}{9}(3z + 2)$

Solve each of the following quadratic inequalities. Graph each solution.

13. $(m + 2)(m - 4) < 0$

14. $(t + 6)(t - 1) \geq 0$

15. $y^2 - 3y + 2 < 0$

16. $2k^2 + 7k - 4 > 0$

17. $q^2 - 7q + 6 \leq 0$

18. $2k^2 - 7k - 15 \leq 0$

19. $6m^2 + m > 1$

20. $10r^2 + r \leq 2$

21. $2y^2 + 5y \leq 3$

22. $3a^2 + a > 10$

23. $x^2 \leq 25$

24. $p^2 - 16p > 0$

Solve the following inequalities.

25. $\dfrac{m - 3}{m + 5} \leq 0$

26. $\dfrac{r + 1}{r - 1} > 0$

27. $\dfrac{k - 1}{k + 2} > 1$

28. $\dfrac{a - 5}{a + 2} < -1$

29. $\dfrac{2y + 3}{y - 5} \leq 1$

30. $\dfrac{a + 2}{3 + 2a} \leq 5$

31. $\dfrac{7}{k + 2} \geq \dfrac{1}{k + 2}$

32. $\dfrac{5}{p + 1} > \dfrac{12}{p + 1}$

33. $\dfrac{3x}{x^2 - 1} < 2$

34. $\dfrac{8}{p^2 + 2p} > 1$

35. $\dfrac{z^2 + z}{z^2 - 1} \geq 3$

36. $\dfrac{a^2 + 2a}{a^2 - 4} \leq 2$

R.6 EXPONENTS

Integer Exponents Recall that $a^2 = a \cdot a$, while $a^3 = a \cdot a \cdot a$, and so on. In this section a more general meaning is given to the symbol a^n.

DEFINITION OF EXPONENT If n is a natural number, then

$$a^n = a \cdot a \cdot a \cdot \,\cdots\, \cdot a,$$

where a appears as a factor n times.

In the expression a^n, n is the **exponent** and a is the **base.** This definition can be extended by defining a^n for zero and negative integer values of n.

ZERO AND NEGATIVE EXPONENTS

If a is any nonzero real number, and if n is a positive integer, then

$$a^0 = 1 \quad \text{and} \quad a^{-n} = \frac{1}{a^n}.$$

(The symbol 0^0 is meaningless.)

EXAMPLE **1**

(a) $6^0 = 1$

(b) $(-9)^0 = 1$

(c) $3^{-2} = \frac{1}{3^2} = \frac{1}{9}$

(d) $9^{-1} = \frac{1}{9^1} = \frac{1}{9}$

(e) $\left(\frac{3}{4}\right)^{-1} = \frac{1}{(3/4)^1} = \frac{1}{3/4} = \frac{4}{3}$ ◀

The following properties follow from the definitions of exponents given above.

PROPERTIES OF EXPONENTS

For any integers m and n, and any real numbers a and b for which the following exist:

1. $a^m \cdot a^n = a^{m+n}$

2. $\dfrac{a^m}{a^n} = a^{m-n}$

3. $(a^m)^n = a^{mn}$

4. $(ab)^m = a^m \cdot b^m$

5. $\left(\dfrac{a}{b}\right)^m = \dfrac{a^m}{b^m}$

EXAMPLE **2**

Use the properties of exponents to simplify each of the following. Leave answers with positive exponents. Assume that all variables represent positive real numbers.

(a) $7^4 \cdot 7^6 = 7^{4+6} = 7^{10}$ Property 1

(b) $\dfrac{9^{14}}{9^6} = 9^{14-6} = 9^8$ Property 2

(c) $\dfrac{r^9}{r^{17}} = r^{9-17} = r^{-8} = \dfrac{1}{r^8}$ Property 2

(d) $(2m^3)^4 = 2^4 \cdot (m^3)^4 = 16m^{12}$ Properties 3 and 4

(e) $(3x)^4 = 3^4 \cdot x^4$ Property 4

(f) $\left(\dfrac{9}{7}\right)^6 = \dfrac{9^6}{7^6}$ Property 5

(g) $\dfrac{2^{-3} \cdot 2^5}{2^4 \cdot 2^{-7}} = \dfrac{2^2}{2^{-3}} = 2^{2-(-3)} = 2^5$

(h) $2^{-1} + 3^{-1} = \dfrac{1}{2} + \dfrac{1}{3} = \dfrac{5}{6}$ ◀

Caution If Example 2(e) were written $3x^4$, the properties of exponents would not apply. When no parentheses are used, the exponent refers only to the factor closest to it. Also notice in Examples 2(g) and 2(h) that a negative exponent does *not* indicate a negative number.

Roots For *even* values of n, the expression $a^{1/n}$ is defined to be the **positive nth root** of a. For example, $a^{1/2}$ denotes the positive second root, or **square root,** of a, while $a^{1/4}$ is the positive fourth root of a. When n is *odd*, there is only one nth root, which has the same sign as a. For example, $a^{1/3}$, the **cube root** of a, has the same sign as a. By definition, if $b = a^{1/n}$, then $b^n = a$.

EXAMPLE **3** (A calculator will be helpful here.)

(a) $121^{1/2} = 11$, since 11 is positive and $11^2 = 121$.

(b) $625^{1/4} = 5$, since $5^4 = 625$.

(c) $256^{1/4} = 4$

(d) $64^{1/6} = 2$

(e) $27^{1/3} = 3$

(f) $(-32)^{1/5} = -2$

(g) $128^{1/7} = 2$

(h) $(-49)^{1/2}$ is not a real number. ◀

Rational Exponents In the following definition, the domain of an exponent is extended to include all rational numbers.

DEFINITION OF $a^{m/n}$

For all real numbers a for which the indicated roots exist, and for any rational number m/n,

$$a^{m/n} = (a^{1/n})^m.$$

EXAMPLE **4** **(a)** $27^{2/3} = (27^{1/3})^2 = 3^2 = 9$

(b) $32^{2/5} = (32^{1/5})^2 = 2^2 = 4$

(c) $64^{4/3} = (64^{1/3})^4 = 4^4 = 256$

(d) $25^{3/2} = (25^{1/2})^3 = 5^3 = 125$ ◀

All the properties for integer exponents given in this section also apply to any rational exponent on a nonnegative real-number base.

EXAMPLE 5

(a) $\dfrac{27^{1/3} \cdot 27^{5/3}}{27^3} = \dfrac{27^{1/3+5/3}}{27^3} = \dfrac{27^2}{27^3} = 27^{2-3} = 27^{-1} = \dfrac{1}{27}$

(b) $m^{2/3}(m^{7/3} + 2m^{1/3}) = m^{2/3+7/3} + 2m^{2/3+1/3} = m^3 + 2m$ ◀

In calculus, it is often necessary to factor expressions involving fractional exponents, as shown in the next example.

EXAMPLE 6

Factor $(x^2 + 5)(3x - 1)^{-1/2}(2) + (3x - 1)^{1/2}(2x)$.

There is a common factor of 2. Also, $(3x - 1)^{-1/2}$ and $(3x - 1)^{1/2}$ have a common factor. Always factor out the quantity to the *smallest* exponent. Here $-1/2 < 1/2$, so the common factor is $2(3x - 1)^{-1/2}$ and the factored form is

$$2(3x - 1)^{-1/2}[(x^2 + 5) + (3x - 1)x] = 2(3x - 1)^{-1/2}(4x^2 - x + 5). \quad ◀$$

R.6 Exercises

Evaluate each expression. Write all answers without exponents.

1. 8^{-2}
2. 3^{-4}
3. 6^{-3}
4. 5^0

5. $(-12)^0$
6. $2^{-1} + 4^{-1}$
7. -2^{-4}
8. $(-2)^{-4}$

9. $-(-3)^{-2}$
10. $-(-3^{-2})$
11. $\left(\dfrac{5}{8}\right)^2$
12. $\left(\dfrac{6}{7}\right)^3$

13. $\left(\dfrac{1}{2}\right)^{-3}$
14. $\left(\dfrac{1}{5}\right)^{-3}$
15. $\left(\dfrac{2}{7}\right)^{-2}$
16. $\left(\dfrac{4}{3}\right)^{-3}$

Simplify each expression. Assume that all variables represent positive real numbers. Write answers with only positive exponents.

17. $\dfrac{7^5}{7^9}$
18. $\dfrac{3^{-4}}{3^2}$
19. $\dfrac{2^{-5}}{2^{-2}}$
20. $\dfrac{6^{-1}}{6}$

21. $4^{-3} \cdot 4^6$
22. $\dfrac{8^9 \cdot 8^{-7}}{8^{-3}}$
23. $\dfrac{10^8 \cdot 10^{-10}}{10^4 \cdot 10^2}$
24. $\left(\dfrac{5^{-6} \cdot 5^3}{5^{-2}}\right)^{-1}$

25. $\dfrac{x^4 \cdot x^3}{x^5}$
26. $\dfrac{y^9 \cdot y^7}{y^{13}}$
27. $\dfrac{(4k^{-1})^2}{2k^{-5}}$
28. $\dfrac{(3z^2)^{-1}}{z^5}$

29. $\dfrac{2^{-1}x^3y^{-3}}{xy^{-2}}$
30. $\dfrac{5^{-2}m^2y^{-2}}{5^2m^{-1}y^{-2}}$
31. $\left(\dfrac{a^{-1}}{b^2}\right)^{-3}$
32. $\left(\dfrac{2c^2}{d^3}\right)^{-2}$

Evaluate each expression, assuming that $a = 2$ and $b = -3$.

33. $a^{-1} + b^{-1}$
34. $b^{-2} - a$
35. $\dfrac{2b^{-1} - 3a^{-1}}{a + b^2}$

36. $\dfrac{3a^2 - b^2}{b^{-3} + 2a^{-1}}$
37. $\left(\dfrac{a}{3}\right)^{-1} + \left(\dfrac{b}{2}\right)^{-2}$
38. $\left(\dfrac{2b}{5}\right)^2 - 3\left(\dfrac{a^{-1}}{4}\right)$

Write each number without exponents.

39. $81^{1/2}$

40. $27^{1/3}$

41. $8^{2/3}$

42. $1000^{2/3}$

43. $32^{2/5}$

44. $-125^{2/3}$

45. $\left(\dfrac{4}{9}\right)^{1/2}$

46. $\left(\dfrac{64}{27}\right)^{1/3}$

47. $16^{-5/4}$

48. $625^{-1/4}$

49. $\left(\dfrac{27}{64}\right)^{-1/3}$

50. $\left(\dfrac{121}{100}\right)^{-3/2}$

Simplify each expression. Write all answers with only positive exponents. Assume that all variables represent positive real numbers.

51. $2^{1/2} \cdot 2^{3/2}$

52. $27^{2/3} \cdot 27^{-1/3}$

53. $\dfrac{4^{2/3} \cdot 4^{5/3}}{4^{1/3}}$

54. $\dfrac{3^{-5/2} \cdot 3^{3/2}}{3^{7/2} \cdot 3^{-9/2}}$

55. $\dfrac{7^{-1/3} \cdot 7r^{-3}}{7^{2/3} \cdot (r^{-2})^2}$

56. $\dfrac{12^{3/4} \cdot 12^{5/4} \cdot y^{-2}}{12^{-1} \cdot (y^{-3})^{-2}}$

57. $\dfrac{6k^{-4} \cdot (3k^{-1})^{-2}}{2^3 \cdot k^{1/2}}$

58. $\dfrac{8p^{-3} \cdot (4p^2)^{-2}}{p^{-5}}$

59. $\dfrac{a^{4/3} \cdot b^{1/2}}{a^{2/3} \cdot b^{-3/2}}$

60. $\dfrac{x^{1/3} \cdot y^{2/3} \cdot z^{1/4}}{x^{5/3} \cdot y^{-1/3} \cdot z^{3/4}}$

61. $\dfrac{k^{-3/5} \cdot h^{-1/3} \cdot t^{2/5}}{k^{-1/5} \cdot h^{-2/3} \cdot t^{1/5}}$

62. $\dfrac{m^{7/3} \cdot n^{-2/5} \cdot p^{3/8}}{m^{-2/3} \cdot n^{3/5} \cdot p^{-5/8}}$

Factor each expression.

63. $(x^2 + 2)(x^2 - 1)^{-1/2}(x) + (x^2 - 1)^{1/2}(2x)$

64. $(3x - 1)(5x + 2)^{1/2}(15) + (5x + 2)^{-1/2}(5)$

65. $(2x + 5)^2 \left(\dfrac{1}{2}\right)(x^2 - 4)^{-1/2}(2x) + (x^2 - 4)^{1/2}(2)(2x + 5)$

66. $(4x^2 + 1)^2(2x - 1)^{-1/2} + (2x - 1)^{1/2}(2)(4x^2 + 1)$

▶ R.7 RADICALS

We have defined $a^{1/n}$ as the positive *n*th root of *a* for appropriate values of *a* and *n*. An alternative notation for $a^{1/n}$ uses radicals.

RADICALS

If *n* is an even natural number and $a > 0$, or *n* is an odd natural number, then

$$a^{1/n} = \sqrt[n]{a}.$$

The symbol $\sqrt[n]{}$ is a **radical sign,** the number *a* is the **radicand,** and *n* is the **index** of the radical. The familiar symbol \sqrt{a} is used instead of $\sqrt[2]{a}$.

(a) $\sqrt[4]{16} = 16^{1/4} = 2$

(b) $\sqrt[5]{-32} = -2$

(c) $\sqrt[3]{1000} = 10$

(d) $\sqrt[6]{\dfrac{64}{729}} = \dfrac{2}{3}$ ◀

With $a^{1/n}$ written as $\sqrt[n]{a}$, $a^{m/n}$ also can be written using radicals.

$$a^{m/n} = (\sqrt[n]{a})^m \quad \text{or} \quad a^{m/n} = \sqrt[n]{a^m}$$

The following properties of radicals depend on the definitions and properties of exponents.

PROPERTIES OF RADICALS For all real numbers a and b and natural numbers m and n such that $\sqrt[n]{a}$ and $\sqrt[n]{b}$ are real numbers:

1. $(\sqrt[n]{a})^n = a$

2. $\sqrt[n]{a^n} = \begin{cases} |a| & \text{if } n \text{ is even} \\ a & \text{if } n \text{ is odd} \end{cases}$

3. $\sqrt[n]{a} \cdot \sqrt[n]{b} = \sqrt[n]{ab}$

4. $\dfrac{\sqrt[n]{a}}{\sqrt[n]{b}} = \sqrt[n]{\dfrac{a}{b}} \quad (b \neq 0)$

5. $\sqrt[m]{\sqrt[n]{a}} = \sqrt[mn]{a}$

Property 3 can be used to simplify certain radicals. For example, since $48 = 16 \cdot 3$,

$$\sqrt{48} = \sqrt{16 \cdot 3} = \sqrt{16} \cdot \sqrt{3} = 4\sqrt{3}.$$

EXAMPLE **2**

(a) $\sqrt{1000} = \sqrt{100 \cdot 10} = \sqrt{100} \cdot \sqrt{10} = 10\sqrt{10}$

(b) $\sqrt{128} = \sqrt{64 \cdot 2} = 8\sqrt{2}$

(c) $\sqrt{108} = \sqrt{36 \cdot 3} = 6\sqrt{3}$

(d) $\sqrt[3]{54} = \sqrt[3]{27 \cdot 2} = \sqrt[3]{27} \cdot \sqrt[3]{2} = 3\sqrt[3]{2}$

(e) $\sqrt{288m^5} = \sqrt{144 \cdot m^4 \cdot 2m} = 12m^2\sqrt{2m}$

(f) $2\sqrt{18} - 5\sqrt{32} = 2\sqrt{9 \cdot 2} - 5\sqrt{16 \cdot 2}$

$$= 2\sqrt{9} \cdot \sqrt{2} - 5\sqrt{16} \cdot \sqrt{2}$$

$$= 2(3)\sqrt{2} - 5(4)\sqrt{2} = -14\sqrt{2} \quad \blacktriangleleft$$

Rationalizing Denominators The next example shows how to *rationalize* (remove all radicals from) the denominator in an expression containing radicals.

EXAMPLE **3** Simplify each of the following expressions by rationalizing the denominator.

(a) $\dfrac{4}{\sqrt{3}}$

To rationalize the denominator, multiply by $\sqrt{3}/\sqrt{3}$ (or 1) so that the denominator of the product is a rational number.

$$\frac{4}{\sqrt{3}} \cdot \frac{\sqrt{3}}{\sqrt{3}} = \frac{4\sqrt{3}}{3}$$

(b) $\dfrac{2}{\sqrt[3]{x}}$

Here, we need a perfect cube under the radical sign to rationalize the denominator. Multiplying by $\sqrt[3]{x^2}/\sqrt[3]{x^2}$ gives

$$\frac{2}{\sqrt[3]{x}} \cdot \frac{\sqrt[3]{x^2}}{\sqrt[3]{x^2}} = \frac{2\sqrt[3]{x^2}}{\sqrt[3]{x^3}} = \frac{2\sqrt[3]{x^2}}{x}.$$

(c) $\dfrac{1}{1 - \sqrt{2}}$

The best approach here is to multiply both numerator and denominator by the number $1 + \sqrt{2}$. The expressions $1 + \sqrt{2}$ and $1 - \sqrt{2}$ are conjugates.*
Doing so gives

$$\frac{1}{1 - \sqrt{2}} = \frac{1(1 + \sqrt{2})}{(1 - \sqrt{2})(1 + \sqrt{2})} = \frac{1 + \sqrt{2}}{1 - 2} = -1 - \sqrt{2}. \blacktriangleleft$$

Sometimes it is advantageous to rationalize the *numerator* of a rational expression.

EXAMPLE 4

Rationalize the numerator of $\dfrac{\sqrt{2} - \sqrt{3}}{5}$.

Multiply numerator and denominator by the conjugate of the numerator, $\sqrt{2} + \sqrt{3}$.

$$\frac{\sqrt{2} - \sqrt{3}}{5} \cdot \frac{\sqrt{2} + \sqrt{3}}{\sqrt{2} + \sqrt{3}} = \frac{2 - 3}{5(\sqrt{2} + \sqrt{3})} = \frac{-1}{5(\sqrt{2} + \sqrt{3})} \blacktriangleleft$$

EXAMPLE 5

Simplify $\sqrt{m^2 - 4m + 4}$.
Factor the polynomial as $m^2 - 4m + 4 = (m - 2)^2$. Then

$$\sqrt{(m - 2)^2} = m - 2 \quad \text{if} \quad m - 2 \geq 0.$$

If $m - 2 \leq 0$, $\sqrt{(m - 2)^2} = -(m - 2) = 2 - m. \blacktriangleleft$

Caution Avoid the common error of writing $\sqrt{a + b}$ as $\sqrt{a} + \sqrt{b}$. We must add a^2 and b^2 *before* taking the square root. For example $\sqrt{16 + 9} = \sqrt{25} = 5$, *not* $\sqrt{16} + \sqrt{9} = 4 + 3 = 7$. This idea applies as well to higher roots:

$$\sqrt[3]{a^3 + b^3} \neq \sqrt[3]{a^3} + \sqrt[3]{b^3},$$
$$\sqrt[4]{a^4 + b^4} \neq \sqrt[4]{a^4} + \sqrt[4]{b^4},$$

and so on.

*If a and b are real numbers, the *conjugate* of $a + b$ is $a - b$.

R.7 Exercises

Simplify each expression. Assume that all variables represent positive real numbers.

1. $\sqrt[3]{125}$

2. $\sqrt[4]{1296}$

3. $\sqrt[5]{-3125}$

4. $\sqrt{50}$

5. $\sqrt{2000}$

6. $\sqrt{32y^5}$

7. $7\sqrt{2} - 8\sqrt{18} + 4\sqrt{72}$

8. $4\sqrt{3} - 5\sqrt{12} + 3\sqrt{75}$

9. $2\sqrt{5} - 3\sqrt{20} + 2\sqrt{45}$

10. $3\sqrt{28} - 4\sqrt{63} + \sqrt{112}$

11. $\sqrt[3]{2} - \sqrt[3]{16} + 2\sqrt[3]{54}$

12. $2\sqrt[3]{3} + 4\sqrt[3]{24} - \sqrt[3]{81}$

13. $\sqrt[3]{32} - 5\sqrt[3]{4} + 2\sqrt[3]{108}$

14. $\sqrt{2x^3y^2z^4}$

15. $\sqrt{98r^3s^4t^{10}}$

16. $\sqrt[3]{16z^5x^8y^4}$

17. $\sqrt[4]{x^8y^7z^{11}}$

18. $\sqrt{a^3b^5} - 2\sqrt{a^7b^3} + \sqrt{a^3b^9}$

19. $\sqrt{p^7q^3} - \sqrt{p^5q^9} + \sqrt{p^9q}$

Rationalize each denominator. Assume that all radicands represent positive real numbers.

20. $\dfrac{5}{\sqrt{7}}$

21. $\dfrac{-2}{\sqrt{3}}$

22. $\dfrac{-3}{\sqrt{12}}$

23. $\dfrac{4}{\sqrt{8}}$

24. $\dfrac{3}{1 - \sqrt{5}}$

25. $\dfrac{5}{2 - \sqrt{6}}$

26. $\dfrac{-2}{\sqrt{3} - \sqrt{2}}$

27. $\dfrac{1}{\sqrt{10} + \sqrt{3}}$

28. $\dfrac{1}{\sqrt{r} - \sqrt{3}}$

29. $\dfrac{5}{\sqrt{m} - \sqrt{5}}$

30. $\dfrac{y - 5}{\sqrt{y} - \sqrt{5}}$

31. $\dfrac{z - 11}{\sqrt{z} - \sqrt{11}}$

32. $\dfrac{\sqrt{x} + \sqrt{x + 1}}{\sqrt{x} - \sqrt{x + 1}}$

33. $\dfrac{\sqrt{p} + \sqrt{p^2 - 1}}{\sqrt{p} - \sqrt{p^2 - 1}}$

Rationalize each numerator. Assume that all radicands represent positive real numbers.

34. $\dfrac{1 + \sqrt{2}}{2}$

35. $\dfrac{1 - \sqrt{3}}{3}$

36. $\dfrac{\sqrt{x} + \sqrt{x + 1}}{\sqrt{x} - \sqrt{x + 1}}$

37. $\dfrac{\sqrt{p} + \sqrt{p^2 - 1}}{\sqrt{p} - \sqrt{p^2 - 1}}$

Simplify each root, if possible.

38. $\sqrt{16 - 8x + x^2}$

39. $\sqrt{4y^2 + 4y + 1}$

40. $\sqrt{4 - 25z^2}$

41. $\sqrt{9k^2 + h^2}$

Applications at a glance . . .

*Torque on a Porsche . . . average global temperature . . .
estimating the passage of time . . . postage rates . . .*

CHAPTER 1

FUNCTIONS AND GRAPHS

(See page 65.)

Figure 1(a) shows the speed in miles per hour of a Porsche 928 over elapsed time in seconds as the car accelerates from rest.* For example, the figure shows that 15 seconds after starting, the car is going 90 miles per hour. Figure 1(b) shows the variation in blood pressure for a typical person.† (Systolic and diastolic pressures are the upper and lower limits in the periodic changes in pressure that produce the pulse. The length of time between peaks is called the period of the pulse.) After .8 seconds, the blood pressure is the same as its starting value, 80 millimeters. Figures 1(a) and 1(b) illustrate *functions*, which are rules or procedures that yield just one value of one variable from any given value of another variable.

(a) (b)

FIGURE 1

Functions are useful in describing many situations involving two variables because each value of one variable corresponds to only one value of the other variable. We can find the exact speed of the Porsche after 10 seconds because the car cannot have two different speeds at the same moment. Similarly, a person can have only one blood pressure at a given instant. A function used in this way, as a mathematical description of a real-world situation, is called a **mathematical model.** Constructing a mathematical model requires a solid understanding of the situation to be modeled, along with a good knowledge of the possible mathematical ideas that can be used to construct the model. In this chapter we discuss a variety of functions useful in modeling.

*From *Road & Track*, April and May 1978. Reprinted with permission.
†From *Calculus for the Life Sciences* by Rodolfo De Sapio. Copyright © 1976, 1978 by W. H. Freeman and Company. Reprinted by permission.

1.1 FUNCTIONS

 How has the amount of energy used in the United States varied with time, and how much money has been saved as a result of decreased use of energy?

After developing the concept of function, we will answer this question in one of the exercises.

The cost of leasing an office in an office building depends on the number of square feet of space in the office. Typical associations between the area of an office in square feet and the monthly rent in dollars are shown in Figure 2.

FIGURE 2

Another association might be set up between investments in a mutual fund and the corresponding earnings, assuming an annual return of 6%. Typical associations are shown in Figure 3.

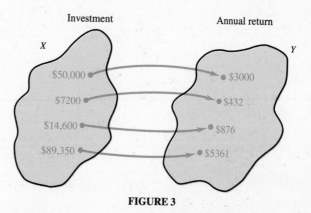

FIGURE 3

In this second example, we could use a formula to show how the numbers in set X are used to obtain the numbers in set Y. If x is a dollar amount from set X, then the corresponding annual return y in dollars from set Y can be found with the formula

$$y = .06 \times x \qquad \text{or} \qquad y = .06x.$$

In this example, x, the investment in the mutual fund, is called the **independent variable.** Because the annual return *depends* on the amount of the investment, y is called the **dependent variable.** When a specific number, say 2000, is substituted for x, then y takes on one specific value—in this case, $.06 \times 2000 = 120$. The variable y is said to be a function of x.

This pair of numbers, one for x and one for y, can be written as an **ordered pair** (x, y), where the order of the numbers is important. Using ordered pairs, the information shown in Figure 3 can be expressed as the set

$$\{(50{,}000, 3000), (7200, 432), (14{,}600, 876), (89{,}350, 5361)\}.$$

Alternatively, the ordered pairs can be given in a table, as shown in Figure 4.

x	y
7,200	432
14,600	876
50,000	3000
89,350	5361

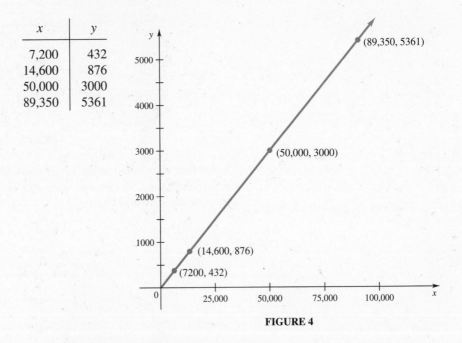

FIGURE 4

The set of points in a plane that correspond to the ordered pairs of a function is called the **graph** of the function. The function defined by $y = .06x$ that contains the ordered pairs shown in the table also can be illustrated with a graph, as shown in Figure 4.

The following definition of a function summarizes the discussion above.

DEFINITION OF FUNCTION	A **function** is a rule that assigns to each element of a set X exactly one element from a set Y.

You may wonder precisely what we mean by a rule. An alternative definition of a function is the following:

ALTERNATIVE DEFINITION OF FUNCTION	A **function** is a set of ordered pairs in which no two ordered pairs have the same first number.

For instance, in the ordered pairs in the investment example, notice that each one has a different first number: 50,000, 7200, 14,600, or 89,350. A function can have two ordered pairs with the same *second* number, however, as will be seen in Example 1(c).

Which of the following are functions?

(a)

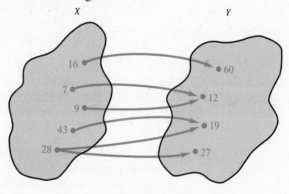

FIGURE 5

Figure 5 shows that an x-value of 28 corresponds to *two* y-values, 19 and 27. In a function, each x must correspond to exactly one y, so this correspondence is not a function.

(b) The optical reader at the checkout counter in many stores that converts codes to prices

For each code, the reader produces exactly one price, so this is a function.

(c) The x^2 key on a calculator

This correspondence between input and output is a function because the calculator produces just one x^2 (one y-value) for each x-value entered. Notice also that two x-values, such as 3 and -3, produce the same y-value of 9, but this does not violate the definition of a function.

(d)

x	1	1	2	2	3	3
y	3	-3	5	-5	8	-8

Since at least one x-value corresponds to more than one y-value, this table does not define a function.

(e) The set of ordered pairs with first elements mothers and second elements their children

Here the mother is the independent variable and the child is the dependent variable. For a given mother, there may be several children, so this correspondence is not a function.

(f) The set of ordered pairs with first elements children and second elements their birth mothers

In this case the child is the independent variable and the mother is the dependent variable. Since each child has only one birth mother, this is a function. ◄

As shown in Example 1, there are many ways to define functions. Almost every function we will use in this book will be defined by an equation, such as the equation $y = .06x$ discussed earlier in this section.

As we have seen, a function is a correspondence between the elements of two sets. These sets are given special names.

DOMAIN AND RANGE

The set of all possible values for the independent variable in a function is called the **domain** of the function; the set of all possible values for the dependent variable is the **range**.

The domain and range may or may not be the same set. For example, the domain of the function in Figure 2 is a set of positive whole numbers representing area, and the range is a set of positive whole numbers representing dollars. For the function in Figure 3, both the domain and range are a set of positive decimal numbers representing dollars and cents.

EXAMPLE 2 Decide whether each of the following equations represents a function. (Assume that x represents the independent variable here, an assumption we shall make throughout this book.) Give the domain and range of any functions.

(a) $y = -4x + 11$

For a given value of x, calculating $-4x + 11$ produces exactly one value of y. (For example, if $x = -7$, then $y = -4(-7) + 11 = 39$.) Since one value of the independent variable leads to exactly one value of the dependent variable, $y = -4x + 11$ is a function. Both x and y may take on any real-number values, so both the domain and range are the set of all real numbers.

(b) $y^2 = x$

Suppose $x = 36$. Then $y^2 = x$ becomes $y^2 = 36$, from which $y = 6$ or $y = -6$. Since one value of the independent variable can lead to two values of the dependent variable, $y^2 = x$ does not represent a function. ◄

The following agreement on domains is customary.

AGREEMENT ON DOMAINS Unless otherwise stated, assume that the domain of all functions defined by an equation is the largest set of real numbers that are meaningful replacements for the independent variable.

For example, suppose

$$y = \frac{-4x}{2x - 3}.$$

Any real number can be used for x except $x = 3/2$, which makes the denominator equal 0. By the agreement on domains, the domain of this function is the set of all real numbers except $3/2$.

Caution When finding the domain of a function, there are two operations to avoid: (1) dividing by zero; and (2) taking the square root (or any even root) of a negative number. Later chapters will present other functions, such as logarithms, which require further restrictions on the domain. For now, just remember these two restrictions on the domain.

From now on, we will write domains and ranges in **interval notation.** With this notation, the set of real numbers less than 4 is written $(-\infty, 4)$. The symbol ∞ (the symbol for infinity) is not a real number; it shows that the interval includes *all* real numbers less than 4.

A square bracket is used to indicate that a number is included in the interval. For example, $2 \le x < 5$ is written $[2, 5)$, indicating that 2 is included but 5 is not. Similarly, $x \ge 10$ is written $[10, \infty)$. Using interval notation, the set of all real numbers is the interval $(-\infty, \infty)$. Parentheses and brackets also are used to graph intervals on the number line.

The Inequalities section of the Algebra Reference discusses the graphs of intervals. For example, the interval $[2, 5)$ is graphed as shown.

EXAMPLE 3 Find the domain and range for each of the functions defined as follows.

(a) $y = x^2$

Any number may be squared, so the domain is the set of all real numbers, written $(-\infty, \infty)$. Since $x^2 \ge 0$ for every value of x, the range is $[0, \infty)$.

(b) $y = \sqrt{6 - x}$

For y to be a real number, $6 - x$ must be nonnegative. This happens only when $6 - x \ge 0$, or $6 \ge x$, making the domain $(-\infty, 6]$. The range is $[0, \infty)$ because $\sqrt{6 - x}$ is always nonnegative.

The Inequalities section of the Algebra Reference demonstrates the method for solving a quadratic inequality. To solve $2x^2 + 5x - 12 \ge 0$, factor the quadratic to get $(2x - 3)(x + 4) \ge 0$. Setting each factor equal to 0 gives $x = 3/2$ or $x = -4$, leading to the intervals $(-\infty, -4]$, $[-4, 3/2]$, and $[3/2, \infty)$. Testing a number from each interval shows that the solution is $(-\infty, -4] \cup [3/2, \infty)$.

(c) $y = \sqrt{2x^2 + 5x - 12}$

The domain includes only those values of x satisfying $2x^2 + 5x - 12 \ge 0$. Using the methods for solving a quadratic inequality produces the domain

$$(-\infty, -4] \cup [3/2, \infty).*$$

As in part (b), the range is $[0, \infty)$.

*The *union* of sets A and B, written $A \cup B$, is defined as the set of all elements in A or B or both.

(d) $y = \dfrac{1}{x + 3}$

Since the denominator cannot be zero, $x \neq -3$ and the domain is

$$(-\infty, -3) \cup (-3, \infty).$$

Because the numerator can never be zero, $y \neq 0$. There are no other restrictions on y, so the range is $(-\infty, 0) \cup (0, \infty)$. ◀

Function Notation The letters f, g, and h frequently are used to represent functions. For example, f might be used to name the function defined by $y = 5 - 3x$. For a given value of x in the domain of a function f, there is exactly one corresponding value of y in the range. To emphasize that y is obtained by applying function f to the element x, replace y with the symbol $f(x)$, read "f of x" or "f at x." Here x is the independent variable; either y or $f(x)$ represents the dependent variable.

Using $f(x)$ to replace y in the equation $y = 5 - 3x$ gives

$$f(x) = 5 - 3x.$$

If 2 is chosen as a value of x, $f(x)$ becomes

$$f(2) = 5 - 3 \cdot 2$$
$$f(2) = -1.$$

In a similar manner,

$$f(-4) = 5 - 3(-4) = 17, \qquad f(0) = 5, \qquad f(-6) = 23,$$

and so on.

To understand how a function works, think of a function f as a machine—for example, a calculator or computer—that takes an input x from the domain and uses it to produce an output $f(x)$ (which represents the y value), as shown in Figure 6.

FIGURE 6

EXAMPLE **4** Let $g(x) = -x^2 + 4x - 5$. Find each of the following.

(a) $g(3)$

Replace x with 3.

$$g(3) = -3^2 + 4 \cdot 3 - 5 = -9 + 12 - 5 = -2$$

(b) $g(a)$

Replace x with a to get $g(a) = -a^2 + 4a - 5$.

This replacement of one variable with another is important in later chapters.

(c) $g(x + h) = -(x + h)^2 + 4(x + h) - 5$

$$= -(x^2 + 2xh + h^2) + 4(x + h) - 5$$

$$= -x^2 - 2xh - h^2 + 4x + 4h - 5$$

(d) $g\left(\dfrac{2}{r}\right) = -\left(\dfrac{2}{r}\right)^2 + 4\left(\dfrac{2}{r}\right) - 5 = -\dfrac{4}{r^2} + \dfrac{8}{r} - 5$ ◀

Caution Notice from part (c) that $g(x + h)$ is *not* the same as $g(x) + h$, which equals $-x^2 + 4x - 5 + h$. There is a significant difference between applying a function to the quantity $x + h$ and applying a function to x and adding h afterwards.

If you tend to get confused when replacing x with $x + h$, as in Example 4(c), you might try replacing the x in the original function with a box, like this:

$$g\left(\boxed{}\right) = -\left(\boxed{}\right)^2 + 4\left(\boxed{}\right) - 5.$$

Then, to compute $g(x + h)$, just enter $x + h$ into the box:

$$g\left(\boxed{x + h}\right) = -\left(\boxed{x + h}\right)^2 + 4\left(\boxed{x + h}\right) - 5$$

and proceed as in Example 4(c).

Graphs As mentioned earlier, it often is useful to draw a graph of the ordered pairs produced by a function. To graph a function, we use a **Cartesian coordinate system,** as shown in Figure 7. The horizontal number line, or **x-axis,** represents the elements from the domain of the function, and the vertical or **y-axis** represents the elements from the range. The point where the number lines cross is the zero point on both of these number lines; this point is called the **origin.**

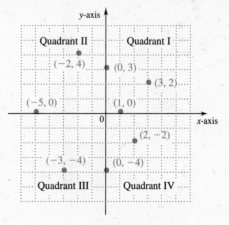

FIGURE 7

The name "Cartesian" honors René Descartes (1596–1650), a brilliant but sickly man. According to legend, Descartes was lying in bed when he noticed an insect crawling on the ceiling and realized that if he could determine the distance from the bug to each of two perpendicular walls, he could describe its position at any given moment. The same idea can be used to locate a point in a plane.

Each point in a Cartesian coordinate system corresponds to an ordered pair of real numbers. Several points and their corresponding ordered pairs are shown in Figure 7. In the point $(-2, 4)$, for example, -2 is the **x-coordinate** and 4 is the **y-coordinate.** From now on, instead of referring to "the point corresponding to the ordered pair $(-2, 4)$," we will say "the point $(-2, 4)$."

The x-axis and y-axis divide the plane into four parts or **quadrants.** For example, quadrant I includes points whose x- and y-coordinates are both positive. The quadrants are numbered as shown in Figure 7. The points of the axes themselves belong to no quadrant.

EXAMPLE **5** Let $f(x) = 3 - 2x$, with domain $\{-2, -1, 0, 1, 2, 3, 4\}$. Graph the ordered pairs produced by this function.

If $x = -2$, then $f(-2) = 3 - 2(-2) = 7$, giving the corresponding ordered pair $(-2, 7)$. Using additional values of x in the equation gives the ordered pairs listed below.

x	-2	-1	0	1	2	3	4
y	7	5	3	1	-1	-3	-5
Ordered Pair	$(-2, 7)$	$(-1, 5)$	$(0, 3)$	$(1, 1)$	$(2, -1)$	$(3, -3)$	$(4, -5)$

These ordered pairs are graphed in Figure 8. ◀

FIGURE 8

FIGURE 9

If a graph is to represent a function, each value of x from the domain must lead to exactly one value of y. In the graph in Figure 9, the domain value x_1 leads to *two* y-values, y_1 and y_2. Since the given x-value corresponds to two different y-values, this is not the graph of a function. This example suggests the **vertical line test** for the graph of a function.

VERTICAL LINE TEST If a vertical line intersects a graph in more than one point, the graph is not the graph of a function.

EXAMPLE 6 Use the vertical line test to decide which of the graphs in Figure 10 are graphs of functions.

(a) (b) (c)

FIGURE 10

(a) Every vertical line intersects this graph in at most one point, so this is the graph of a function.

(b) Again, each vertical line intersects the graph in at most one point, showing that this is the graph of a function.

(c) It is possible for a vertical line to intersect the graph in part (c) twice. This is not the graph of a function. ◀

EXAMPLE 7 An overnight delivery service charges $25 for a package weighing up to 2 pounds. For each additional pound there is an additional charge of $3. Let $D(x)$ represent the cost to send a package weighing x pounds. Graph $D(x)$ for x in the interval $(0, 6]$.

For x in the interval $(0, 2]$, $y = 25$. For x in $(2, 3]$, $y = 25 + 3 = 28$. For x in $(3, 4]$, $y = 28 + 3 = 31$, and so on. The graph is shown in Figure 11. ◀

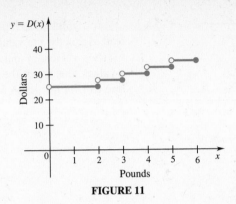

FIGURE 11

The function discussed in Example 7 is called a **step function.** Many real-life situations are best modeled by step functions. Additional examples are given in the exercises.

1.1 Exercises

Which of the following rules define y as a function of x?

1.

2.

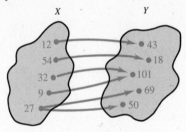

3.

x	y
3	9
2	4
1	1
0	0
−1	1
−2	4
−3	9

4.

x	y
9	3
4	2
1	1
0	0
1	−1
4	−2
9	−3

5. $y = x^3$ **6.** $y = \sqrt{x}$ **7.** $x = |y|$ **8.** $x = y^4 - 1$

List the ordered pairs obtained from each equation, given $\{-2, -1, 0, 1, 2, 3\}$ as domain. Graph each set of ordered pairs. Give the range.

9. $y = x - 1$ **10.** $y = 2x + 3$ **11.** $y = -4x + 9$ **12.** $y = -6x + 12$

13. $2x + y = 9$ **14.** $3x + y = 16$ **15.** $2y - x = 5$ **16.** $6x - y = -3$

17. $y = x(x + 1)$ **18.** $y = (x - 2)(x - 3)$ **19.** $y = x^2$ **20.** $y = -2x^2$

21. $y = \dfrac{1}{x + 3}$ **22.** $y = \dfrac{-2}{x + 4}$ **23.** $y = \dfrac{3x - 3}{x + 5}$ **24.** $y = \dfrac{2x + 1}{x + 3}$

Write each expression in interval notation. Graph each interval.

25. $x < 0$ **26.** $x \geq -3$ **27.** $1 \leq x < 2$

28. $-5 < x \leq -4$ **29.** $-9 > x$ **30.** $6 \leq x$

Using the variable x, write each interval as an inequality.

31. $(-4, 3)$ **32.** $[2, 7)$ **33.** $(-\infty, -1]$ **34.** $(3, \infty)$

35.

36.

37.

38.

Give the domain of each function defined as follows.

39. $f(x) = 2x$

40. $f(x) = x + 2$

41. $f(x) = x^4$

42. $f(x) = (x - 2)^2$

43. $f(x) = \sqrt{16 - x^2}$

44. $f(x) = |x - 1|$

45. $f(x) = (x - 3)^{1/2}$

46. $f(x) = (3x + 5)^{1/2}$

47. $f(x) = \dfrac{2}{x^2 - 4}$

48. $f(x) = \dfrac{-8}{x^2 - 36}$

49. $f(x) = -\sqrt{\dfrac{2}{x^2 + 9}}$

50. $f(x) = -\sqrt{\dfrac{5}{x^2 + 36}}$

51. $f(x) = \sqrt{x^2 - 4x - 5}$

52. $f(x) = \sqrt{15x^2 + x - 2}$

53. $f(x) = \dfrac{1}{\sqrt{x^2 - 6x + 8}}$

54. $f(x) = \sqrt{\dfrac{x + 1}{x - 1}}$

Give the domain and the range of each function. Where arrows are drawn, assume the function continues in the indicated direction.

55.

56.

57.

58.

For each function, find **(a)** $f(4)$, **(b)** $f(-3)$, **(c)** $f(-1/2)$, **(d)** $f(a)$, *and* **(e)** $f(2/m)$.

59. $f(x) = 3x + 2$

60. $f(x) = 5x - 6$

61. $f(x) = -x^2 + 5x + 1$

62. $f(x) = (x + 3)(x - 4)$

63. $f(x) = \dfrac{2x + 1}{x - 2}$

64. $f(x) = \dfrac{3x - 5}{2x + 3}$

Use each graph to find **(a)** $f(-2)$, **(b)** $f(0)$, **(c)** $f(1/2)$, *and* **(d)** $f(4)$.

65.

66.

67.

68.

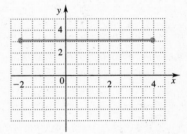

Let $f(x) = 6x - 2$ *and* $g(x) = x^2 - 2x + 5$ *to find the following values.*

69. $f(m - 3)$

70. $f(2r - 1)$

71. $g(r + h)$

72. $g(z - p)$

73. $g\left(\dfrac{3}{q}\right)$

74. $g\left(-\dfrac{5}{z}\right)$

Decide whether each graph represents a function.

75.

76.

77.

78.

79.

80.

For each function defined as follows, find **(a)** $f(x + h)$, **(b)** $f(x + h) - f(x)$, *and* **(c)** $\dfrac{f(x + h) - f(x)}{h}$.

81. $f(x) = x^2 - 4$

82. $f(x) = 8 - 3x^2$

83. $f(x) = 6x + 2$

84. $f(x) = 4x - 11$

85. $f(x) = \dfrac{1}{x}$

86. $f(x) = -\dfrac{1}{x^2}$

Applications

General Interest

87. Energy Use The graph* shows two functions of time. The first, which we will designate $I(t)$, shows the energy intensity of the United States, which is defined as the thousands of BTUs of energy consumed per dollar of Gross National Product (GNP) in year t (in 1982 dollars). The second, which we will designate $S(t)$, represents the national savings per year (also in 1982 dollars) due to a decreased use of energy.

(a) Find $I(1976)$. **(b)** Find $S(1984)$.

(c) Find $I(t)$ and $S(t)$ at the time when the two graphs cross. In what year does this occur?

(d) What significance, if any, is there to the point at which the two graphs cross?

Business and Economics

88. Saw Rental A chain-saw rental firm charges $7 per day or fraction of a day to rent a saw, plus a fixed fee of $4 for resharpening the blade. Let $S(x)$ represent the cost of renting a saw for x days.
Find each of the following.

(a) $S\left(\dfrac{1}{2}\right)$ **(b)** $S(1)$ **(c)** $S\left(1\dfrac{1}{4}\right)$

(d) $S\left(3\dfrac{1}{2}\right)$ **(e)** $S(4)$ **(f)** $S\left(4\dfrac{1}{10}\right)$

(g) What does it cost to rent a saw for $4\%_{10}$ days?

(h) A portion of the graph of $y = S(x)$ is shown here. Explain how the graph could be continued.

(i) What is the independent variable?

(j) What is the dependent variable?

(k) Write a sentence or two explaining what part (f) and its answer represent.

(l) We have left $x = 0$ out of the graph. Discuss why it should or shouldn't be included. If it were included, how would you define $S(0)$?

*Graph entitled "Energy Use Down, Savings Up" from "The Greenhouse Effect" by the Union of Concerned Scientists. Reprinted by permission of the Union of Concerned Scientists.

89. Postage In 1991, it cost 29¢ in the United States to mail a first-class letter weighing up to 1 oz, and 23¢ for each additional ounce or fraction of an ounce. Let $C(x)$ represent the postage for a first-class letter weighing x oz. Find each of the following.

(a) $C\left(\dfrac{2}{3}\right)$ **(b)** $C\left(1\dfrac{1}{3}\right)$ **(c)** $C(2)$ **(d)** $C\left(3\dfrac{1}{8}\right)$

(e) What does it cost to send a letter that weighs $2\frac{5}{8}$ oz?

(f) Write a sentence or two explaining what part (e) and its answer represent.

(g) Graph the function defined by $y = C(x)$ for $0 < x \le 4$.

(h) What is the independent variable?

(i) What is the dependent variable?

(j) We have left $x = 0$ out of the domain. Discuss why it should or shouldn't be included. If it were included, how would you define $S(0)$?

1.2 LINEAR FUNCTIONS

How much vinyl siding should a supplier produce each month, and how much should the supplier charge for it?

In this section, we will learn one method for solving such problems.

Many practical situations can be described (or at least approximated) with a *linear function*. Some examples are the relationships between Fahrenheit and Celsius temperatures, between price and supply of some consumer goods, and between demand and supply of some commodities.

LINEAR FUNCTION

A function f is **linear** if its equation can be written as

$$f(x) = ax + b$$

for real numbers a and b.

FIGURE 12

As the name implies, every linear function has a graph that is a straight line. An important characteristic of a straight line is its *slope*, a numerical measure of the steepness of the line. To find this measure, start with the line through the two distinct points (x_1, y_1) and (x_2, y_2), as shown in Figure 12. (Assume $x_1 \ne x_2$.) The difference

$$x_2 - x_1$$

is called the *change in x* and written with the symbol Δx (read "delta x"), where Δ is the Greek letter delta. In the same way, the *change in y* is

$$\Delta y = y_2 - y_1.$$

The **slope** of a nonvertical line is defined as the quotient of the change in y and the change in x.

SLOPE OF A LINE

The slope m of the nonvertical line through the distinct points (x_1, y_1) and (x_2, y_2) is

$$m = \frac{\Delta y}{\Delta x} = \frac{y_2 - y_1}{x_2 - x_1}.$$

Similar triangles can be used to show that the slope does not depend on which pair of points on the line is used to calculate slope. That is, the same slope will be obtained for any two points on the line.

The slope of a line can be found only if the line is nonvertical, because $x_2 \neq x_1$ for a nonvertical line, so that the denominator $x_2 - x_1 \neq 0$. The slope of a vertical line is not defined.

EXAMPLE 1 Find the slope of the line through each of the following pairs of points.

(a) $(-4, 8)$ and $(2, -3)$

Choosing $x_1 = -4$, $y_1 = 8$, $x_2 = 2$, and $y_2 = -3$ gives $\Delta y = -3 - 8 = -11$ and $\Delta x = 2 - (-4) = 6$. By definition, the slope is

$$m = \frac{\Delta y}{\Delta x} = -\frac{11}{6}.$$

(b) $(2, 7)$ and $(2, -4)$

A sketch shows that the line through $(2, 7)$ and $(2, -4)$ is vertical. As mentioned above, the slope of a vertical line is undefined. (An attempt to use the definition of slope here would produce a zero denominator.)

(c) $(5, -3)$ and $(-2, -3)$

By the definition of slope,

$$m = \frac{-3 - (-3)}{-2 - 5} = 0. \quad \blacktriangleleft$$

Drawing a graph of the line in Example 1(c) shows that it is horizontal, which suggests that the slope of a horizontal line is 0.

The slope of a horizontal line is 0.

The slope of a vertical line is not defined.

Caution The phrase "no slope" should be avoided; specify instead whether the slope is zero or undefined.

Figure 13 (on the next page) shows lines of various slopes. As suggested by the figure, a line with a positive slope goes up from left to right, and a line with a negative slope goes down.

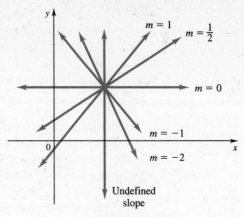

FIGURE 13

EXAMPLE **2** Graph the line that passes through $(-1, 5)$ and has slope $-5/3$.

First, locate the point $(-1, 5)$, as shown in Figure 14. Since the slope of this line is $-5/3$, a change of 3 units horizontally produces a change of -5 units vertically, giving a second point, $(2, 0)$, which is used to complete the graph. ◀

FIGURE 14

Equations of a Line The slope of a line, together with a point on the line, can be used to find an equation of the line. The procedure for finding the equation depends on whether or not the line is vertical. The vertical line through the point $(a, 0)$ passes through all points of the form (a, y), making the equation $x = a$.

VERTICAL LINE An equation of the vertical line through the point $(a, 0)$ is $x = a$.

Let m be the slope of a nonvertical line. Assume that the fixed point (x_1, y_1) is on the line. Let (x, y) represent any other point on the line. The point (x, y) can be on the line if and only if the slope of the line through (x_1, y_1) and (x, y) is m; that is, if

$$\frac{y - y_1}{x - x_1} = m.$$

Multiplying both sides by $x - x_1$ gives

$$y - y_1 = m(x - x_1).$$

This result is summarized below.

POINT-SLOPE FORM

The line with slope m passing through the point (x_1, y_1) has an equation

$$y - y_1 = m(x - x_1).$$

This equation is called the **point-slope form** of the equation of a line.

EXAMPLE **3** Write an equation of each line.

(a) Through $(-4, 1)$, with slope -3

Use the point-slope form of the equation of a line, with $x_1 = -4$, $y_1 = 1$, and $m = -3$.

$$y - 1 = -3[x - (-4)]$$
$$y - 1 = -3(x + 4)$$
$$y - 1 = -3x - 12$$
$$y = -3x - 11 \quad \text{or} \quad 3x + y = -11$$

(b) Through $(-3, 2)$ and $(2, -4)$

First, find the slope with the definition of slope:

$$m = \frac{-4 - 2}{2 - (-3)} = \frac{-6}{5}.$$

Either $(-3, 2)$ or $(2, -4)$ can be used for (x_1, y_1). Choosing $x_1 = -3$ and $y_1 = 2$ gives

$$y - 2 = \frac{-6}{5}[x - (-3)]$$
$$5(y - 2) = -6(x + 3)$$
$$5y - 10 = -6x - 18$$
$$5y = -6x - 8 \quad \text{or} \quad 6x + 5y = -8.$$

Verify that using $(2, -4)$ instead of $(-3, 2)$ leads to the same result. ◀

Any value of x where a graph crosses the x-axis is called an **x-intercept** for the graph. Any value of y where the graph crosses the y-axis is called a **y-intercept** for the graph. The graph in Figure 15 has x-intercepts x_1, x_2, and x_3 and y-intercept y_1. As suggested by the graph, x-intercepts can be found by letting $y = 0$ and y-intercepts by letting $x = 0$.

FIGURE 15

Figure 16 shows a line with y-intercept b; the line goes through $(0, b)$. If the slope of the line is m, then by the point-slope form, an equation of the line is

$$y - y_1 = m(x - x_1)$$
$$y - b = m(x - 0)$$
$$y = mx + b.$$

This result, which shows both the slope and the y-intercept, is the **slope-intercept form** of the equation of a line. Reversing these steps shows that any equation of the form $y = mx + b$ has a graph that is a line with slope m that passes through the point $(0, b)$.

FIGURE 16

SLOPE-INTERCEPT FORM

The line with slope m passing through the point $(0, b)$ has an equation

$$y = mx + b.$$

This equation is called the slope-intercept form of the equation of a line.

This result, together with the fact that vertical lines have equations of the form $x = a$, shows that every line has an equation of the form $ax + by + c = 0$, where a and b are not both 0. Conversely, assuming $b \neq 0$ and solving $ax + by + c = 0$ for y gives $y = (-a/b)x - c/b$. By the result above, this equation is a line with slope $-a/b$ and y-intercept $-c/b$. If $b = 0$, solve for x to get $x = -c/a$, a vertical line. In any case, the equation $ax + by + c = 0$ has a straight line for its graph.

If a and b are not both 0, then the equation $ax + by + c = 0$ has a line for its graph. Also, any line has an equation of the form $ax + by + c = 0$.

Graph $3x + 2y = 6$.

This equation has a line for its graph because it can be written as $3x + 2y - 6 = 0$. Two distinct points on the line are enough to locate the graph. The intercepts often provide the necessary points. To find the x-intercept, let $y = 0$.

For review on solving a linear equation, see Section R.4 of the Algebra Reference.

$$3x + 2(0) = 6$$
$$3x = 6$$
$$x = 2$$

The x-intercept is 2. Let $x = 0$ to find that the y-intercept is 3. These two intercepts were used to get the graph shown in Figure 17.

Alternatively, solve $3x + 2y = 6$ for y to get

$$3x + 2y = 6$$
$$2y = -3x + 6$$
$$y = -\frac{3}{2}x + 3.$$

By the slope-intercept form, the graph of this equation is the line with y-intercept 3 and slope $-3/2$. (This means that the line goes down 3 units for each 2 units it goes to the right.) ◀

FIGURE 17

Graph $y = -3$.

For a line to have an x-intercept, there must be a value of x that makes $y = 0$. Here, however, $y = -3 \neq 0$. This means the line has no x-intercept, a situation that can happen only if the line is parallel to the x-axis (see Figure 18). ◀

FIGURE 18

Example 5 suggests the following generalization.

HORIZONTAL LINE An equation of the horizontal line through $(0, a)$ is $y = a$.

Parallel and Perpendicular Lines One application of slope involves deciding whether two lines are parallel. Since two parallel lines are equally "steep," they should have the same slope. Also, two lines with the same "steepness" are parallel.

PARALLEL LINES Two nonvertical lines are parallel if and only if they have the same slope.

EXAMPLE **6** Find the equation of the line that passes through the point (3, 5) and is parallel to the line $2x + 5y = 4$.

The slope of $2x + 5y = 4$ can be found by writing the equation in slope-intercept form.

$$2x + 5y = 4$$

$$y = -\frac{2}{5}x + \frac{4}{5}$$

This result shows that the slope is $-2/5$. Since the lines are parallel, $-2/5$ is also the slope of the line whose equation is needed. This line passes through (3, 5). Substituting $m = -2/5$, $x_1 = 3$, and $y_1 = 5$ into the point-slope form gives

$$y - y_1 = m(x - x_1)$$

$$y - 5 = -\frac{2}{5}(x - 3)$$

$$5(y - 5) = -2(x - 3)$$

$$5y - 25 = -2x + 6$$

$$2x + 5y = 31. \quad \blacktriangleleft$$

As mentioned above, two nonvertical lines are parallel if and only if they have the same slope. Two lines having slopes with a product of -1 are perpendicular. A proof of this fact, which depends on similar triangles from geometry, is given as Exercise 61 in this section.

PERPENDICULAR LINES Two lines, neither of which is vertical, are perpendicular if and only if their slopes have a product of -1.

EXAMPLE **7** Find the slope of the line L perpendicular to the line having the equation $5x - y = 4$.

To find the slope, write $5x - y = 4$ in slope-intercept form:

$$y = 5x - 4.$$

The slope is 5. Since the lines are perpendicular, if line L has slope m, then

$$5m = -1$$

$$m = -\frac{1}{5}. \quad \blacktriangleleft$$

Supply and Demand Linear functions are often good choices for **supply and demand curves.** Typically, as the price of an item increases, the demand for the item decreases, while the supply increases. On the other hand, when demand for an item increases, so does its price, causing the supply of the item to decrease.

For example, several years ago the price of gasoline increased rapidly. As the price continued to escalate, most buyers became more and more prudent in their use of gasoline in order to restrict their demand to an affordable amount. Consequently, the overall demand for gasoline decreased and the supply increased, to a point where there was an oversupply of gasoline. This caused prices to fall until supply and demand were approximately balanced. Many other factors were involved in the situation, but the relationship between price, supply, and demand was nonetheless typical. Some commodities, however, such as medical care, college tuition, and certain luxury items, may be exceptions to these typical relationships.

Although economists consider price to be the independent variable, they have the unfortunate habit of plotting price, usually denoted by p, on the vertical axis, while the usual custom is to graph the independent variable on the horizontal axis. This practice originated with the English economist Alfred Marshall (1842–1924). To conform to this custom, we will write p, the price, as a function of q, the quantity produced, and plot p on the vertical axis. But remember that it is really price that determines how much consumers demand and producers supply, not the other way around.

Supply and demand functions are not necessarily linear (the simplest kind of function). Most functions are *approximately* linear if a small enough piece of the graph is taken, however, allowing applied mathematicians to often use linear functions for simplicity. We will take that approach in this section and the next.

EXAMPLE 8 Suppose that Greg Okjakjian, an economist, has studied the supply and demand for vinyl siding and has determined that the price in dollars per square yard, p, and the quantity demanded monthly in thousands of square yards, q, are related by the linear function

$$p = D(q) = 60 - \frac{3}{4}q, \quad \text{Demand}$$

while the price p and the supply q are related by

$$p = S(q) = \frac{3}{4}q. \quad \text{Supply}$$

(a) Find the demand at a price of $45 and at a price of $18.

Start with the demand function

$$p = 60 - \frac{3}{4}q,$$

and replace p with 45.

$$45 = 60 - \frac{3}{4}q$$

Solve this equation to find that

$$q = 20.$$

Thus, at a price of $45 per square yard, 20,000 square yards are demanded per month.

Similarly, replace 18 for p to find the demand when the price is $18. Verify that this leads to $q = 56$. When the price is lowered from $45 to $18, the demand increases from 20,000 square yards to 56,000 square yards.

(b) Find the supply at a price of $60 and at a price of $12.

Substitute 60 for p in the supply equation

$$p = \frac{3}{4}q$$

to find that $q = 80$, so the supply is 80,000 square yards. Similarly, replacing p with 12 in the supply equation gives a supply of 16,000 square yards. If the price decreases from $60 to $12 per square yard, the supply also decreases, from 80,000 square yards to 16,000 square yards.

(c) Graph both functions on the same axes.

The results of part (a) are written as the ordered pairs (20, 45) and (56, 18). The line through the corresponding points is the graph of $p = 60 - (3/4)q$, shown in black in Figure 19.

Use the ordered pairs (80, 60) and (16, 12) from the work in part (b) to get the supply graph shown in color in Figure 19. ◄

FIGURE 19

In Example 8 supply and demand are determined by

$$p = \frac{3}{4}q \quad \text{and} \quad p = 60 - \frac{3}{4}q,$$

respectively. To find a price where supply and demand are equal, solve the equation

$$\frac{3}{4}q = 60 - \frac{3}{4}q,$$

getting

$$q = 40.$$

Supply and demand will be equal when $q = 40$. This happens at a price of

$$p = \frac{3}{4}q$$

$$p = \frac{3}{4}(40) = 30,$$

or $30. (Find the same result by using the demand function.) If the price of the item is more than $30, the supply will exceed the demand. At a price less than $30, the demand will exceed the supply. Only at a price of $30 will demand and supply be equal. For this reason, $30 is called the *equilibrium price*. When the price is $30, demand and supply both equal 40,000 square yards, the *equilibrium quantity*.

Generalizing, the **equilibrium price** of a commodity is the price at the point where the supply and demand graphs for that commodity cross. The **equilibrium quantity** is the demand and the supply at the same point.

Another important issue is how, in practice, the equations of the supply and demand functions can be found. This issue is important for many problems involving linear functions in this section and the next. Data must be collected, and if the plotted points lie perfectly along a line, the equation can easily be found using any two points. What usually happens, however, is that the data are scattered, and there is no line that goes through all the points. In this case, a line must be found that approximates the linear trend in the data as much as possible (assuming the points lie approximately along a line). This is usually done by the *method of least squares,* also referred to as *linear regression.* A discussion of this method can be found in the chapter on functions of several variables, or in any elementary statistics text.

1.2 Exercises

Find the slope of each line that has a slope.

1. Through $(4, 5)$ and $(-1, 2)$
2. Through $(5, -4)$ and $(1, 3)$
3. Through $(8, 4)$ and $(8, -7)$
4. Through $(1, 5)$ and $(-2, 5)$
5. $y = 2x$
6. $y = 3x - 2$
7. $5x - 9y = 11$
8. $4x + 7y = 1$
9. $x = -6$
10. The x-axis
11. A line parallel to $2y - 4x = 7$
12. A line perpendicular to $6x = y - 3$
13. Through $(-1.978, 4.806)$ and $(3.759, 8.125)$
14. Through $(11.72, 9.811)$ and $(-12.67, -5.009)$

Find the slope of each of the lines shown.

15.

16.

17.

18.

Write an equation in the form ax + by = c for each line.

19. Through $(1, 3)$, $m = -2$

20. Through $(2, 4)$, $m = -1$

21. Through $(6, 1)$, $m = 0$

22. Through $(-8, 1)$, with undefined slope

23. Through $(4, 2)$ and $(1, 3)$

24. Through $(8, -1)$ and $(4, 3)$

25. Through $(1/2, 5/3)$ and $(3, 1/6)$

26. Through $(-2, 3/4)$ and $(2/3, 5/2)$

27. x-intercept 3, y-intercept -2

28. x-intercept -2, y-intercept 4

29. Vertical, through $(-6, 5)$

30. Horizontal, through $(8, 7)$

31. Through $(-1.76, 4.25)$, with slope -5.081

32. Through $(5.469, 11.08)$, with slope 4.723

Graph each line.

33. Through $(-1, 3)$, $m = 3/2$

34. Through $(-2, 8)$, $m = -1$

35. Through $(3, -4)$, $m = -1/3$

36. Through $(-2, -3)$, $m = -3/4$

37. $3x + 5y = 15$　　**38.** $2x - 3y = 12$

39. $4x - y = 8$　　**40.** $x + 3y = 9$

41. $x + 2y = 0$　　**42.** $3x - y = 0$

43. $x = -1$　　**44.** $y + 2 = 0$

45. $y = -3$　　**46.** $x = 5$

Write an equation for each line in Exercises 47–54.

47. Through $(-1, 4)$, parallel to $x + 3y = 5$

48. Through $(2, -5)$, parallel to $y - 4 = 2x$

49. Through $(3, -4)$, perpendicular to $x + y = 4$

50. Through $(-2, 6)$, perpendicular to $2x - 3y = 5$

51. x-intercept -2, parallel to $y = 2x$

52. y-intercept 3, parallel to $x + y = 4$

53. The line with y-intercept 2 and perpendicular to $3x + 2y = 6$

54. The line with x-intercept $-2/3$ and perpendicular to $2x - y = 4$

55. Do the points $(4, 3)$, $(2, 0)$, and $(-18, -12)$ lie on the same line? (*Hint:* Find the equation of the line through two of the points.)

56. Find k so that the line through $(4, -1)$ and $(k, 2)$ is
 (a) parallel to $3y + 2x = 6$,
 (b) perpendicular to $2y - 5x = 1$.

57. Use slopes to show that the quadrilateral with vertices at $(1, 3)$, $(-5/2, 2)$, $(-7/2, 4)$, and $(2, 1)$ is a parallelogram.

58. Use slopes to show that the square with vertices at $(-2, 3)$, $(4, 3)$, $(4, -1)$, and $(-2, -1)$ has diagonals that are perpendicular.

59. Use similar triangles from geometry to show that the slope of a line is the same, no matter which two distinct points on the line are chosen to compute it.

60. Suppose that $(0, b)$ and (x_1, y_1) are distinct points on the line $y = mx + b$. Show that $(y_1 - b)/x_1$ is the slope of the line, and that $m = (y_1 - b)/x_1$.

61. To show that two perpendicular lines, neither of which is vertical, have slopes with a product of -1, go through the following steps. Let line L_1 have equation $y = m_1 x + b_1$, and let L_2 have equation $y = m_2 x + b_2$. Assume that L_1

and L_2 are perpendicular, and use right triangle *MPN* shown in the figure. Prove each of the following statements.

 (a) *MQ* has length m_1.
 (b) *QN* has length $-m_2$.
 (c) Triangles *MPQ* and *PNQ* are similar.
 (d) $m_1/1 = 1/-m_2$ and $m_1 m_2 = -1$.

Applications

Business and Economics

62. Supply and Demand Let the supply and demand functions for sugar be given in dollars by the following equations.
$$S(q) = 1.4q - .6 \qquad D(q) = -2q + 3.2$$
 (a) Graph these equations on the same axes.
 (b) Find the equilibrium quantity.
 (c) Find the equilibrium price.

63. Supply and Demand Let the supply and demand functions for wool sweaters be given in dollars by the following equations.
$$S(q) = \frac{2}{5}q \qquad D(q) = 100 - \frac{2}{5}q$$
 (a) Graph these equations on the same axes.
 (b) Find the equilibrium quantity.
 (c) Find the equilibrium price.

64. Cost Analysis A company finds that it can make a total of 20 small trailers for $13,900 and 10 for $7500. Let y be the total cost to produce x trailers. Assume that the rela-

tionship between the number of trailers produced and the cost is linear.
 (a) Find the slope and an equation for the line in the form $y = mx + b$.
 (b) Use your answer from part (a) to find out how many small trailers the company can make for $20,000.
 (c) Use your answer from part (a) to find out how many small trailers the company can make for $1000.

65. Rental Car Cost The cost to rent a midsized car is $27 per day or fraction of a day. If the car is picked up in Lansing and dropped off in West Lafayette, there is a fixed $25 dropoff charge. Let $C(x)$ represent the cost of renting the car for x days, taking it from Lansing to West Lafayette. Find each of the following.

 (a) $C(3/4)$ (b) $C(9/10)$ (c) $C(1)$ (d) $C\left(1\frac{5}{8}\right)$
 (e) Find the cost of renting the car for 2.4 days.
 (f) Graph $y = C(x)$.
 (g) Is C a function?
 (h) Is C a linear function?

Life Sciences

66. Pollution When a certain industrial pollutant is introduced into a river, the reproduction of catfish declines. In a given period of time, depositing 3 tons of the pollutant results in a fish population of 37,000. Also, 12 tons of pollutant produce a fish population of 28,000. Let y be the fish population when x tons of pollutant are introduced into the river.

(a) Assuming a linear relationship, find the slope and an equation of the line in the form $y = mx + b$.

(b) Use your answer from part (a) to find out how many tons of the pollutant would result in the elimination of the fish population.

(c) Use your answer from part (a) to find out how much pollution should be allowed to maintain a fish population of 45,000.

67. Human Growth A person's tibia bone goes from ankle to knee. A male with a tibia 40 cm in length has a height of 177 cm, while a tibia 43 cm in length corresponds to a height of 185 cm.

(a) Write a linear equation showing how the height of a male, h, relates to the length of his tibia, t.

(b) Suppose an archaeologist finds a collection of tibias having lengths of 38–45 cm. Assuming these belonged to males, estimate what their heights would have been.

(c) Estimate the length of the tibia for a male whose height is 190 cm.

68. Human Growth The radius bone goes from the wrist to the elbow. A female whose radius bone is 24 cm long is 167 cm tall, while a radius of 26 cm corresponds to a height of 174 cm.

(a) Write a linear equation showing how the height of a female, h, corresponds to the length of her radius bone, r.

(b) Suppose an archaeologist finds a collection of radius bones having lengths of 23–27 cm. Assuming these belonged to females, estimate what their heights would have been.

(c) Estimate the length of a radius bone for a height of 170 cm.

Social Sciences

69. Voting Trends According to research done by the political scientist James March, if the Democrats win 45% of the two-party vote for the House of Representatives, they win 42.5% of the seats. If the Democrats win 55% of the vote, they win 67.5% of the seats.

(a) Let y be the percent of seats won and x the percent of the two-party vote. Assuming a linear relationship, find the slope and an equation of the line in the form $y = mx + b$.

(b) Use your answer from part (a) to find the percent of the vote the Democrats would need to capture 60% of the seats in the House of Representatives.

(c) Could the equation found in part (a) be valid for every percent of the vote between 0% and 100%? Explain.

70. Voting Trends If the Republicans win 45% of the two-party vote, they win 32.5% of the seats (see Exercise 69). If they win 60% of the vote, they get 70% of the seats. Let y represent the percent of the seats and x the percent of the vote.

(a) Assume the relationship is linear and find the slope and an equation of the line in the form $y = mx + b$.

(b) Use your answer from part (a) to find the percent of the vote the Republicans would need to capture 60% of the seats in the House of Representatives.

(c) Could the equation found in part (a) be valid for every possible percent of the vote between 0% and 100%? Explain.

Physical Sciences

71. Global Warming In 1990, the Intergovernmental Panel on Climate Change predicted that the average temperature on the earth would rise .3° C per decade in the absence of international controls on greenhouse emissions.* Let t measure the time in years since 1970, when the average global temperature was 15° C.

(a) Find an equation giving the average global temperature in degrees Celsius as a function of t, the number of years since 1970.

(b) Scientists have estimated that the sea level will rise by 65 cm if the average global temperature rises to 19° C. According to your answer to part (a), when would this occur?

*See *Science News*, June 23, 1990, p. 391.

1.3 LINEAR MATHEMATICAL MODELS

 What diameter drill bit should a homeowner purchase, if the required length is known?

Later in this section, this question will be answered using a linear function.

Throughout this book, we construct mathematical models that describe real-world situations. In this section we describe some situations that lead to linear functions as mathematical models.

Temperature One of the most common linear relationships found in everyday situations deals with temperature. Recall that water freezes at 32° Fahrenheit and 0° Celsius, while it boils at 212° Fahrenheit and 100° Celsius. The ordered pairs (0, 32) and (100, 212) are graphed in Figure 20 on axes showing Fahrenheit (*F*) as a function of Celsius (*C*). The line joining them is the graph of the function.

FIGURE 20

EXAMPLE 1 Derive an equation relating *F* and *C*.

To derive the required linear equation, first find the slope using the given ordered pairs, (0, 32) and (100, 212).

$$m = \frac{212 - 32}{100 - 0} = \frac{9}{5}$$

The *F*-intercept of the graph is 32, so by the slope-intercept form, the equation of the line is

$$F = \frac{9}{5}C + 32.$$

With simple algebra this equation can be rewritten to give *C* in terms of *F*:

$$C = \frac{5}{9}(F - 32). \quad \blacktriangleleft$$

Sales Analysis It is common to compare the change in sales of two companies by comparing the rates at which these sales change. If the sales of the two companies can be approximated by linear functions, we can use the work of the last section to find the rate of change of the dependent variable compared to the change in the independent variable.

EXAMPLE 2

The chart below shows sales in two different years for two different companies.

Company	Sales in 1988	Sales in 1991
A	$10,000	$16,000
B	5,000	14,000

A study of company records suggests that the sales of both companies have increased linearly (that is, the sales can be closely approximated by a linear function).

(a) Find a linear equation describing the sales for Company A.

To find a linear equation describing the sales, let $x = 0$ represent 1988, so that 1991 corresponds to $x = 3$. Then, by the chart above, the line representing the sales for Company A passes through the points (0, 10,000) and (3, 16,000). The slope of the line through these points is

$$\frac{16,000 - 10,000}{3 - 0} = 2000.$$

Using the point-slope form of the equation of a line gives

$$y - 10,000 = 2000(x - 0)$$
$$y = 2000x + 10,000$$

as an equation describing the sales of Company A.

(b) Find a linear equation describing the sales for Company B.

Since the sales for Company B also have increased linearly, they can be described by a line through (0, 5000) and (3, 14,000). Using the same procedure as in part (a) gives

$$y = 3000x + 5000$$

as the equation describing the sales of Company B. ◀

Average Rate of Change Notice that the sales for Company A in Example 2 increased from $10,000 to $16,000 over the period from 1988 to 1991, representing a total increase of $6000 in three years.

$$\text{Average rate of change in sales} = \frac{\$6000}{3} = \$2000 \text{ per year}$$

This is the same as the slope found in part (a) of the example. Verify that the average annual rate of change in sales for Company B over the three-year period also agrees with the slope found in part (b). Management needs to watch the rate of change in sales closely in order to be aware of any unfavorable trends. If the rate of change is decreasing, then sales growth is slowing down, and this trend may require some response.

AVERAGE RATE OF CHANGE

For $y = f(x)$, as x changes from x to $x + \Delta x$, the **average rate of change in y with respect to x** is given by

$$\frac{\text{Change in } y}{\text{Change in } x} = \frac{f(x + \Delta x) - f(x)}{(x + \Delta x) - x}$$

$$= \frac{f(x + \Delta x) - f(x)}{\Delta x}$$

$$= \frac{\Delta y}{\Delta x}.$$

For data that can be modeled with a linear function, the average rate of change, which is the same as the slope of the line, is constant.

Cost Analysis The cost of manufacturing an item commonly consists of two parts. The first is a **fixed cost** for designing the product, setting up a factory, training workers, and so on. Within broad limits, the fixed cost is constant for a particular product and does not change as more items are made. The second part is a *cost per item* for labor, materials, packing, shipping, and so on. The total value of this second cost *does* depend on the number of items made.

 EXAMPLE 3 Suppose that the cost of producing clock-radios can be approximated by

$$C(x) = 12x + 100,$$

where $C(x)$ is the cost in dollars to produce x radios. The cost to produce 0 radios is

$$C(0) = 12(0) + 100 = 100,$$

or $100. This sum, $100, is the fixed cost.

Once the company has invested the fixed cost into the clock-radio project, what then will be the additional cost per radio? As an example, we first find the cost of a total of 5 radios:

$$C(5) = 12(5) + 100 = 160,$$

or $160. The cost of 6 radios is

$$C(6) = 12(6) + 100 = 172,$$

or $172.

The sixth radio itself costs $172 - $160 = $12 to produce. In the same way, the 81st radio costs $C(81) - C(80) = \$1072 - \$1060 = \$12$ to produce. In fact, the $(n + 1)$st radio costs

$$C(n + 1) - C(n) = [12(n + 1) + 100] - (12n + 100)$$
$$= 12,$$

or $12, to produce. The number 12 is also the slope of the graph of the cost function $C(x) = 12x + 100$. ◀

In economics, **marginal cost** is the rate of change of cost $C(x)$ at a level of production x. The marginal cost is equal to the slope of the cost function. It approximates the cost of producing one additional item. In fact, some books define the marginal cost to be the cost of producing one additional item. With *linear functions*, these two definitions are equivalent, and the marginal cost, which is equal to the slope of the cost function, is constant. For instance, in the clock-radio example, the marginal cost of each radio is $12. For other types of functions, these two definitions are only approximately equal. Marginal cost is important to management in making decisions in areas such as cost control, pricing, and production planning.

The work in Example 3 can be generalized. Suppose the total cost to make x items is given by the cost function $C(x) = mx + b$. The fixed cost is found by letting $x = 0$:

$$C(0) = m \cdot 0 + b = b;$$

thus, the fixed cost is b dollars. The additional cost of the $(n + 1)$st item is

$$C(n + 1) - C(n) = [m(n + 1) + b] - (mn + b)$$
$$= mn + m + b - mn - b$$
$$= m.$$

This is exactly equal to the slope of the line $C(x) = mx + b$, which represents the marginal cost.

COST FUNCTION

In a cost function of the form $C(x) = mx + b$, m represents the marginal cost per item and b the fixed cost. Conversely, if the fixed cost of producing an item is b and the marginal cost is m, then the **cost function** $C(x)$ for producing x items is $C(x) = mx + b$.

Another important concept is the average cost function $\overline{C}(x)$, pronounced "C bar."

AVERAGE COST FUNCTION

If $C(x)$ is the total cost to manufacture x items, then the **average cost** per item is given by

$$\overline{C}(x) = \frac{C(x)}{x}.$$

In Example 3 the average cost per clock-radio is

$$\overline{C}(x) = \frac{C(x)}{x} = \frac{12x + 100}{x} = 12 + \frac{100}{x}.$$

The second term, $100/x$, shows that as more and more items are produced and the fixed cost is spread over a larger number of items, the average cost per item decreases.

EXAMPLE 4 ▶ The marginal cost for raising a certain type of frog for laboratory study is $12 per unit of frogs, while the cost to produce 100 units is $1500.

(a) Find the cost function $C(x)$, given that it is linear.

 Since the cost function is linear, it can be expressed in the form $C(x) = mx + b$. The marginal cost is $12 per unit, which gives the value for m, leading to $C(x) = 12x + b$. To find b, use the fact that the cost of producing 100 units of frogs is $1500, or $C(100) = 1500$. Substituting $C(x) = 1500$ and $x = 100$ into $C(x) = 12x + b$ gives

$$C(x) = 12x + b$$
$$1500 = 12 \cdot 100 + b$$
$$1500 = 1200 + b$$
$$300 = b.$$

The model is given by $C(x) = 12x + 300$, where the fixed cost is $300.

(b) Find the average cost per item to produce 50 units and 300 units.

 The average cost per item is

$$\overline{C}(x) = \frac{C(x)}{x} = \frac{12x + 300}{x} = 12 + \frac{300}{x}.$$

If 50 units of frogs are produced, the average cost is

$$\overline{C}(50) = 12 + \frac{300}{50} = 18,$$

or $18 per unit. Producing 300 units of frogs will lead to an average cost of

$$\overline{C}(300) = 12 + \frac{300}{300} = 13,$$

or $13 per unit. ◀

Break-Even Analysis The **revenue** $R(x)$ from selling x units of a product is the product of the price per unit p and the number of units sold (demand) x, so that

$$R(x) = px.$$

The corresponding **profit** $P(x)$ is the difference between revenue $R(x)$ and cost $C(x)$. That is,

$$P(x) = R(x) - C(x).$$

A company can make a profit only if the revenue received from its customers exceeds the cost of producing and selling its goods and services. The number of units at which revenue just equals cost is the **break-even point.**

EXAMPLE 5 A firm producing poultry feed finds that the total cost $C(x)$ of producing and selling x units is given by

$$C(x) = 20x + 100.$$

Management plans to charge $24 per unit for the feed.

(a) How many units must be sold for the firm to break even?

The firm will break even (no profit and no loss) as long as revenue just equals cost, or $R(x) = C(x)$. From the given information, since $R(x) = px$ and $p = \$24$,

$$R(x) = 24x.$$

Substituting for $R(x)$ and $C(x)$ in the equation $R(x) = C(x)$ gives

$$24x = 20x + 100,$$

from which $x = 25$. The firm breaks even by selling 25 units. The graphs of $C(x) = 20x + 100$ and $R(x) = 24x$ are shown in Figure 21. The break-even point (where $x = 25$) is shown on the graph. If the company sells more than 25 units (if $x > 25$), it makes a profit. If it sells less than 25 units, it loses money.

FIGURE 21

(b) What is the profit if 100 units of feed are sold?

Use the formula for profit $P(x)$.

$$P(x) = R(x) - C(x)$$
$$= 24x - (20x + 100)$$
$$= 4x - 100$$

Then $P(100) = 4(100) - 100 = 300$. The firm will make a profit of $300 from the sale of 100 units of feed. ◄

EXAMPLE 6

A do-it-yourself homeowner purchased a set of drill bits. Each drill bit is 1/64 inch larger in diameter than the previous bit, and as the bits increase in diameter, their lengths also increase in a linear way. This means the length of a bit is a linear function of its diameter. The smallest bit has a diameter of 1/16 inch and a length of $1\frac{5}{8}$ inches. The largest bit has a diameter of 1/4 inch and a length of 4 inches. Suppose the homeowner wants a bit with a length of at least $5\frac{1}{4}$ inches. What diameter bit is needed, assuming that bit diameters are available in multiples of 1/64 inch, and that the relationship between a bit's diameter and its length remains constant? (For those who want more drill problems, we hope this example will satisfy them, at least a bit.)

Let x be the diameter of the bit and y its length. The smallest and largest bits in the set then correspond to the ordered pairs $(1/16, 1\frac{5}{8})$ and $(1/4, 4)$. The slope of the line through these two points is

$$m = \frac{4 - 1\frac{5}{8}}{1/4 - 1/16} = \frac{19/8}{3/16} = \frac{38}{3}.$$

Using the point-slope form of a linear equation and the point $(1/4, 4)$, we have

$$y - 4 = \frac{38}{3}\left(x - \frac{1}{4}\right).$$

Verify that this simplifies to

$$y = \frac{38}{3}x + \frac{5}{6}.$$

Now let $y = 5\frac{1}{4}$ in this equation. Go through the details to verify that $x = 53/152$. Unfortunately, the bits are sold only in multiples of 1/64 inch. To find how many 64ths equal 53/152, solve the equation

$$\frac{n}{64} = \frac{53}{152}$$

by multiplying both sides by 64, giving $n = 424/19$, which is approximately 22.3. Since n must be an integer, and a bit *at least* $5\frac{1}{4}$ inches long is needed, let $n = 23$. In other words, the homeowner should request a bit with a diameter of 23/64 inch. ◀

1.3 Exercises

Write a cost function for each situation. Identify all variables used.

1. A chain-saw rental firm charges $12 plus $1 per hour.
2. A trailer-hauling service charges $45 plus $2 per mile.
3. A parking garage charges 50¢ plus 35¢ per half-hour.
4. For a one-day rental, a car rental firm charges $44 plus 28¢ per mile.

Assume that each situation can be expressed as a linear cost function, and find the appropriate cost function.

5. Fixed cost, $100; 50 items cost $1600 to produce

6. Fixed cost, $400; 10 items cost $650 to produce

7. Fixed cost, $1000; 40 items cost $2000 to produce

8. Fixed cost, $8500; 75 items cost $11,875 to produce

9. Marginal cost, $50; 80 items cost $4500 to produce

10. Marginal cost, $120; 100 items cost $15,800 to produce

11. Marginal cost, $90; 150 items cost $16,000 to produce

12. Marginal cost, $120; 700 items cost $96,500 to produce

Applications

Business and Economics

13. **Sales Analysis** Suppose the sales of a particular brand of electric guitar satisfy the relationship

$$S(x) = 300x + 2000,$$

where $S(x)$ represents the number of guitars sold in year x, with $x = 0$ corresponding to 1987.

Find the sales in each of the following years.

(a) 1987 (b) 1990 (c) 1991

(d) The manufacturer needs sales to reach 4000 guitars by 1996 to pay off a loan. Will sales reach that goal?

(e) Find the annual rate of change of sales.

14. **Sales Analysis** Assume that the sales of a certain appliance dealer are approximated by a linear function. Suppose that sales were $850,000 in 1982 and $1,262,500 in 1987. Let $x = 0$ represent 1982.

(a) Find an equation giving the dealer's yearly sales.

(b) Estimate the sales in 1994.

(c) The dealer estimates that a new store will be necessary once sales exceed $2,170,000. When is this expected to occur?

15. **Sales Analysis** Assume that the sales of a certain automobile parts company are approximated by a linear function. Suppose that sales were $200,000 in 1981 and $1,000,000 in 1988. Let $x = 0$ represent 1981 and $x = 7$ represent 1988.

(a) Find the equation giving the company's yearly sales.

(b) Estimate the sales in 1992.

(c) The company wants to negotiate a new contract with the automobile company once sales reach $2,000,000. When is this expected to occur?

16. **Marginal Cost** In deciding whether to set up a new manufacturing plant, company analysts have established that a reasonable function for the total cost to produce x items is

$$C(x) = 500,000 + 4.75x.$$

(a) Find the total cost to produce 100,000 items.

(b) Find the marginal cost of the items to be produced in this plant.

17. **Marginal Cost** The manager of a local restaurant has found that his cost function for producing coffee is $C(x) = .097x$, where $C(x)$ is the total cost in dollars of producing x cups. (He is ignoring the cost of the coffee pot and the cost of labor.)

Find the total cost of producing the following numbers of cups of coffee.

(a) 1000 cups (b) 1001 cups

(c) Find the marginal cost of the 1001st cup.

(d) What is the marginal cost for *any* cup?

Average Cost *Let $C(x)$ be the total cost in dollars to manufacture x items. Find the average cost per item in Exercises 18 and 19.*

18. $C(x) = 500,000 + 4.75x$

(a) $x = 1000$ (b) $x = 5000$ (c) $x = 10,000$

19. $C(x) = 800 + 20x$

(a) $x = 10$ (b) $x = 50$ (c) $x = 200$

20. **Break-Even Analysis** The cost to produce x units of squash is $C(x) = 100x + 6000$, while the revenue is $R(x) = 500x$. Find the break-even point.

21. **Break-Even Analysis** The cost to produce x units of wire is $C(x) = 50x + 5000$, while the revenue is $R(x) = 60x$. Find the break-even point and the revenue at the break-even point.

Break-Even Analysis *Suppose that you are the manager of a firm. You are considering the manufacture of a new product, so you ask the accounting department to produce cost estimates and the sales department to produce estimates for revenue and sales. After you receive the data, you must decide whether to go ahead with production of the new product. Analyze the data in Exercises 22–25 (find a break-even point), and then decide what you would do.*

22. $C(x) = 105x + 6000$; $R(x) = 250x$; not more than 400 units can be sold

23. $C(x) = 85x + 900$; $R(x) = 105x$; not more than 38 units can be sold

24. $C(x) = 1000x + 5000$; $R(x) = 900x$ (*Hint:* What does a negative break-even point mean?)

25. $C(x) = 70x + 500$; $R(x) = 60x$

26. **Break-Even Analysis** The graph shows the productivity of U.S. and Japanese workers in appropriate units over a 35-year period. Estimate the break-even point (the point at which workers in the two countries produced the same amounts).*

27. **Break-Even Analysis** The graph gives U.S. imports and exports in billions of dollars over a five-year period. Estimate the break-even point.*

Life Sciences

28. **Medical Expenses** Over a recent three-year period, medical expenses in the U.S. rose in a linear pattern from 10.3% of the gross national product in year 0 to 11.2% in year 3.

 (a) Assuming that this change in medical expenses continues to be linear, write a linear function describing the percent of the gross national product devoted to medical expenses, y, in terms of the year, x.

 (b) Give the average rate of change of medical expenses from year 0 to 3. Compare it with the slope of the line in part (a). What do you find?

29. **Life Span** Some scientists believe there is a limit to how long humans can live.[†] One supporting argument is that during the last century, life expectancy from age 65 has increased more slowly than life expectancy from birth, so eventually these two will be equal, at which point, according to these scientists, life expectancy should increase no further. In 1900, life expectancy at birth was 46 years, and life expectancy at age 65 was 76. In 1975, these figures had risen to 75 and 80, respectively. In both cases, the increase in life expectancy has been linear. Using these assumptions and the data given, find the maximum life expectancy for humans.

30. **Population** The number of children in the U.S. from 5 to 13 years old decreased from 31.2 million in 1980 to 30.3 million in 1986.

 (a) Write a linear function describing this population y in terms of year x for the given period.

 (b) What was the average rate of change in this population over the period from 1980 to 1986?

*Figures, "Manufacturing Productivity Index" and "Manufacturing Trade Deficit" as appeared in *The Sacramento Bee*, December 21, 1987. Reprinted by permission of The Associated Press.
[†]See *Science*, November 15, 1991, pp. 936–38.

MANUFACTURING PRODUCTIVITY INDEX

Japan

U.S.

1977 equals 100

Source: U.S. Bureau of Labor Statistics and Salomon Brothers Inc.

EXERCISE 26

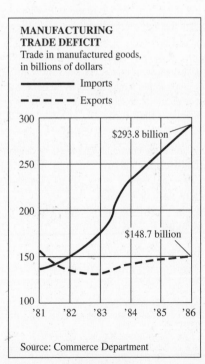

MANUFACTURING TRADE DEFICIT
Trade in manufactured goods, in billions of dollars

—— Imports

– – – Exports

$293.8 billion

$148.7 billion

Source: Commerce Department

EXERCISE 27

31. Ponies Trotting A 1991 study found that the peak vertical force on a trotting Shetland pony increased linearly with the pony's speed, and that when the force reached a critical level, the pony switched from a trot to a gallop.* For one pony, the critical force was 1.16 times its body weight. It experienced a force of .75 times its body weight at a speed of 2 meters per second, and a force of .93 times its body weight at 3 meters per second. At what speed did the pony switch from a trot to a gallop?

Social Sciences

32. Stimulus Effect In psychology, the just-noticeable-difference (JND) for some stimulus is defined as the amount by which the stimulus must be increased so that a person will perceive it as having barely been increased. For example, suppose a research study indicates that a line 40 centimeters in length must be increased to 42 cm before a subject thinks that it is longer. In this case, the JND would be $42 - 40 = 2$ cm. In a particular experiment, the JND is given by

$$y = .03x,$$

where x represents the original length of the line and y the JND.

Find the JND for lines having the following lengths.

(a) 10 cm (b) 20 cm (c) 50 cm (d) 100 cm

(e) Find the rate of change in the JND with respect to the original length of the line.

33. Passage of Time Most people are not very good at estimating the passage of time. Some people's estimations are too fast, and those of others are too slow. One psychologist has constructed a mathematical model for actual time as a function of estimated time: if y represents actual time and x estimated time, then

$$y = mx + b,$$

where m and b are constants that must be determined experimentally for each person. Suppose that for a particular person, $m = 1.25$ and $b = -5$. Find y for each of the following values of x.

(a) 30 min (b) 60 min (c) 120 min
(d) 180 min

34. Passage of Time In Exercise 33, suppose that for another person, $m = .85$ and $b = 1.2$. Find y for each of the following values of x.

(a) 15 min (b) 30 min (c) 60 min
(d) 120 min

For this same person, find x for each of the following values of y.

(e) 60 min (f) 90 min

Physical Sciences

35. Temperature Use the formulas for conversion between Fahrenheit and Celsius derived in Example 1 to convert each temperature.

(a) 58° F to Celsius

(b) 50° C to Fahrenheit

36. Temperature Use the formulas for conversion between Fahrenheit and Celsius derived in Example 1 to convert each temperature.

(a) 98.6° F to Celsius

(b) 20° C to Fahrenheit

37. Derive a formula giving the Celsius temperature (C) as a linear function of the Fahrenheit temperature (F).

38. Find the temperature at which the Celsius and Fahrenheit temperatures are numerically equal.

General Interest

39. Class Size Let $R(x) = -8x + 240$ represent the number of students present in a large business calculus class, where x represents the number of hours of study required weekly.

Find the number of students present at each of the following levels of required study.

(a) $x = 0$ (b) $x = 5$ (c) $x = 10$

(d) What is the rate of change of the number of students in the class with respect to the number of hours of study? Interpret the negative sign in the answer.

(e) The professor in charge of the class likes to have exactly 16 students. How many hours of study must he require in order to have exactly 16 students?

40. Drill Bits For the drill bit example in the text, find the diameter of the bit that should be ordered for each of the following minimum lengths. Remember that the diameter must be a multiple of 1/64 inches.

(a) 4¼ inches (b) 5½ inches

*See *Science*, July 19, 1991, pp. 306–8.

1.4 QUADRATIC FUNCTIONS

 How much should a company charge for its seminars? When Power and Money, Inc., charges $600 for a seminar on management techniques, it attracts 1000 people. For each $20 decrease in the fee, an additional 100 people will attend the seminar. The managers are wondering how much to charge for the seminar to maximize their revenue.

In this section we will see how knowledge of *quadratic functions* will help provide an answer to the question above.

In this section you will need to know how to solve a quadratic equation by factoring and by the quadratic formula, which are covered in Sections R.2 and R.4 of the Algebra Reference. Factoring is usually easiest; when a polynomial is set equal to zero and factored, then a solution is found by setting any one factor equal to zero. But factoring is not always possible. The quadratic formula will provide the solution to *any* quadratic equation.

A linear function was defined by

$$f(x) = ax + b,$$

for real numbers a and b. In a *quadratic function* the independent variable is squared. A quadratic function is an especially good model for many situations with a maximum or a minimum function value. Quadratic functions also may be used to describe supply and demand curves; cost, revenue, and profit; as well as other quantities. Next to linear functions, they are the simplest type of function, and so well worth studying thoroughly.

QUADRATIC FUNCTION

A **quadratic function** is defined by

$$f(x) = ax^2 + bx + c,$$

where a, b, and c are real numbers, with $a \neq 0$.

The simplest quadratic function has $f(x) = x^2$, with $a = 1$, $b = 0$, and $c = 0$. This function can be graphed by choosing several values of x and then finding the corresponding values of $f(x)$. We then plot the resulting ordered pairs $(x, f(x))$, and draw a smooth curve through them, as in Figure 22. This graph is called a **parabola.** Every quadratic function has a parabola as its graph. The lowest (or highest) point on a parabola is the **vertex** of the parabola. The vertex of the parabola in Figure 22 is (0, 0).

x	$f(x)$
-2	4
-1	1
0	0
1	1
2	4

FIGURE 22

If the graph in Figure 22 were folded in half along the y-axis, the two halves of the parabola would match exactly. This means that the graph of a quadratic function is *symmetric* with respect to a vertical line through the vertex; this line is the **axis** of the parabola.

There are many real-world instances of parabolas. For example, cross sections of spotlight reflectors or radar dishes form parabolas. Also, a projectile thrown in the air follows a parabolic path.

EXAMPLE 1

Graph $y = x^2 - 4$.

Each value of y will be 4 less than the corresponding value of y in $y = x^2$. The graph of $y = x^2 - 4$ has the same shape as that of $y = x^2$ but is 4 units lower. See Figure 23. The vertex of the parabola (on this parabola, the *lowest* point) is at $(0, -4)$. The x-intercepts can be found by letting $y = 0$ to get

$$0 = x^2 - 4,$$

from which $x = 2$ and $x = -2$ are the x-intercepts. The axis of the parabola is the vertical line $x = 0$. ◀

FIGURE 23

Example 1 suggests that the effect of c in $ax^2 + bx + c$ is to lower the graph if c is negative and to raise the graph if c is positive.

EXAMPLE 2

Graph $y = -2x^2$.

By plotting points, we see that since a is negative, the graph opens downward. Since a is larger than 1, the graph is steeper than the original graph. See Figure 24. ◀

FIGURE 24

Example 2 shows that the sign of a in $ax^2 + bx + c$ determines whether the parabola opens upward or downward, and the magnitude of a determines how steeply the graph increases. The effect of b in $ax^2 + bx + c$ is more complicated; it causes the vertex to shift to the left or right, as well as to be raised or lowered, depending upon the values of a and c. For a quadratic function with $b \neq 0$, the best way to understand what the graph looks like is by first using the quadratic formula to find the x-intercepts, when they exist. Recall from algebra that if $ax^2 + bx + c = 0$, where $a \neq 0$, then

$$x = \frac{-b \pm \sqrt{b^2 - 4ac}}{2a}.$$

Notice that this is the same as

$$x = \frac{-b}{2a} \pm \frac{\sqrt{b^2 - 4ac}}{2a} = \frac{-b}{2a} \pm Q,$$

where $Q = \sqrt{b^2 - 4ac}/2a$. Since a parabola is symmetric with respect to its axis, the vertex is halfway between its two x-intercepts. Halfway between $x = -b/(2a) + Q$ and $x = -b/(2a) - Q$ is $x = -b/(2a)$. Once we have the x-coordinate of the vertex, we can easily get the y-coordinate by substituting the x-coordinate into the original equation. The following example illustrates the complete process.

EXAMPLE 3

Graph $y = -3x^2 - 2x + 1$.

Because the x^2 term has a coefficient of -3, the graph is a parabola opening downward. The x-intercepts are found by taking

$$x = \frac{-b \pm \sqrt{b^2 - 4ac}}{2a}$$

$$= \frac{-(-2) \pm \sqrt{(-2)^2 - 4(-3)(1)}}{2(-3)}$$

$$= \frac{2 \pm \sqrt{16}}{-6}$$

$$x = -1 \quad \text{or} \quad \frac{1}{3}.$$

(These solutions could also have been found by factoring.) The x-coordinate of the vertex is

$$x = \frac{-b}{2a} = \frac{-(-2)}{2(-3)} = -\frac{1}{3}.$$

We get the y-coordinate by substituting this x-value into the equation.

$$y = -3\left(-\frac{1}{3}\right)^2 - 2\left(-\frac{1}{3}\right) + 1$$

$$= \frac{4}{3}.$$

Thus, the vertex is $(-1/3, 4/3)$. The axis goes through the vertex; it is the line $x = -1/3$. Finally, the y-intercept, which is c, or 1 in this case, should always be plotted. Plotting the vertex, the two x-intercepts, and the y-intercept yields the graph in Figure 25. ◀

In many examples the quadratic will not factor, and the solutions given by the quadratic formula will be irrational numbers. If this happens, you can use a calculator to approximate the solutions. Another situation that may arise is the absence of any x-intercepts, as in the next example.

$y = -3x^2 - 2x + 1$

FIGURE 25

EXAMPLE **4** Graph $y = x^2 + 4x + 6$.

This does not appear to factor, so we'll try the quadratic formula.

$$x = \frac{-b \pm \sqrt{b^2 - 4ac}}{2a}$$

$$= \frac{-4 \pm \sqrt{4^2 - 4(1)(6)}}{2(1)} = \frac{-4 \pm \sqrt{-8}}{2}$$

As soon as we see the negative under the square root sign, we know there are no x-intercepts. Nevertheless, the vertex is still at

$$x = \frac{-b}{2a} = \frac{-4}{2} = -2.$$

Substituting this into the equation gives

$$y = (-2)^2 + 4(-2) + 6 = 2.$$

The y-intercept is at $(0, 6)$, 2 units to the right of the parabola's axis $x = -2$. We can also plot the mirror image of this point on the opposite side of the parabola's axis: at $x = -4$, 2 units to the left of the axis, y is also equal to 6. Plotting the vertex, the y-intercept, and the point $(-4, 6)$ gives the graph in Figure 26. ◀

$(-4, 6)$ $(0, 6)$

$y = x^2 + 4x + 6$

$(-2, 2)$

FIGURE 26

We will now return to the example with which we started this section.

EXAMPLE **5** When Power and Money, Inc., charges $600 for a seminar on management techniques, it attracts 1000 people. For each $20 decrease in the fee, an additional 100 people will attend the seminar. The managers are wondering how much to charge for the seminar to maximize their revenue.

Let x be the number of $20 decreases in the price. Then the price charged per person will be

$$\text{Price per person} = 600 - 20x,$$

and the number of people in the seminar will be

$$\text{Number of people} = 1000 + 100x.$$

The total revenue, $R(x)$, is given by the product of the price and the number of people attending, or

$$R(x) = (600 - 20x)(1000 + 100x)$$
$$= 600,000 + 40,000x - 2000x^2.$$

We see by the negative in the x^2 term that this defines a parabola opening downward, so the maximum revenue is at the vertex. The x-coordinate of the vertex is

$$x = \frac{-b}{2a} = \frac{-40,000}{2(-2000)} = 10.$$

The y-coordinate is then

$$y = 600,000 + 40,000(10) - 2000(10^2)$$
$$= 800,000.$$

Therefore, the maximum revenue is $800,000, which is achieved by charging $600 - 20x = 600 - 20(10) = 400 per person. ◀

Notice in this last example that the maximum revenue was achieved by charging less than the current price of $600, which was more than made up for by the increase in sales. This is typical of many applications. Mathematics is a powerful tool for solving such problems, since the answer is not what one might have guessed intuitively.

The main difficulty in learning to solve such problems is deriving expressions for the price per person and the number of people. You will find this process easier if you notice that both expressions are linear functions of x. Notice also that you know the constant term in these functions because the original problem told what happens if $x = 0$ (i.e., if there is no $20 decrease). Finally, notice that the slope of these linear functions is just the amount of change each time x increases by 1. (In this example, an increase in x by 1 is equivalent to a $20 decrease in the price.) Once you understand these ideas (after getting some practice), you will be able to solve any such maximization problem.

EXAMPLE 6 Ms. Tilden owns and operates Aunt Emma's Pie Shop. She has hired a consultant to analyze her business operations. The consultant tells her that her profit $P(x)$ from the sale of x units of pies is given by

$$P(x) = 120x - x^2.$$

How many units of pies must be sold in order to maximize the profit? What is the maximum possible profit?

Because of the $-x^2$ term, the profit function defines a parabola opening downward with the maximum at the vertex. It can be factored as

$$P(x) = x(120 - x).$$

The two x-intercepts are therefore 0 and 120, so the x-coordinate of the vertex is halfway between those values, at $x = 60$. Since $P(60) = 3600$, the maximum profit of $3600 is reached when 60 units of pies are sold. In this case, profit increases as more pies are sold up to 60 units and then decreases as more pies are sold past this point. See Figure 27. ◀

$P(x) = 120x - x^2$

(60, 3600)

FIGURE 27

Although many examples in this book involve maximization of profit, we do not mean to imply that this should always be the only goal of a company. Far-sighted companies may seek to minimize their impact on the environment or maximize their market share. U.S. companies have been criticized for seeking only to maximize short-term profits, which in the long run may cause them to lose out to foreign competitors who take a longer view.

1.4 Exercises

1. Graph the functions in parts (a)–(d) on the same coordinate system.

 (a) $f(x) = 2x^2$ (b) $f(x) = 3x^2$ (c) $f(x) = \frac{1}{2}x^2$ (d) $f(x) = \frac{1}{3}x^2$

 (e) How does the coefficient affect the shape of the graph?

2. Graph the functions in parts (a)–(d) on the same coordinate system.

 (a) $y = \frac{1}{2}x^2$ (b) $y = -\frac{1}{2}x^2$ (c) $y = 4x^2$ (d) $y = -4x^2$

 (e) What effect does the negative sign have on the graph?

3. Graph the functions in parts (a)–(d) on the same coordinate system.

 (a) $f(x) = x^2 + 2$ (b) $f(x) = x^2 - 1$ (c) $f(x) = x^2 + 1$ (d) $f(x) = x^2 - 2$

 (e) How do these graphs differ from the graph of $f(x) = x^2$?

4. Graph the functions in parts (a)–(d) on the same coordinate system.

 (a) $f(x) = (x - 2)^2$ (b) $f(x) = (x + 1)^2$ (c) $f(x) = (x + 3)^2$ (d) $f(x) = (x - 4)^2$

 (e) How do these graphs differ from the graph of $f(x) = x^2$?

Graph each parabola and give its vertex, axis, x-intercepts, and y-intercept.

5. $y = x^2 + 6x + 5$

6. $y = x^2 - 10x + 21$

7. $y = x^2 - 4x + 4$

8. $y = x^2 + 8x + 16$

9. $y = -2x^2 - 12x - 16$

10. $y = -3x^2 + 12x - 11$

11. $y = 2x^2 + 12x - 16$

12. $y = x^2 - 2x + 3$

13. $y = 3x^2 + 6x + 2$

14. $y = -x^2 - 4x + 2$

15. $y = -x^2 + 6x - 6$

16. $y = 2x^2 - 4x + 5$

17. $y = -3x^2 + 24x - 36$

18. $y = -\frac{1}{3}x^2 + 2x + 4$

19. $y = \frac{5}{2}x^2 + 10x + 8$

20. $y = -\frac{1}{2}x^2 - x - \frac{7}{2}$

21. $y = \frac{2}{3}x^2 - \frac{8}{3}x + \frac{5}{3}$

22. $y = \frac{1}{2}x^2 + 2x + \frac{7}{2}$

23. Let x be in the interval $[0, 1]$. Use a graph to suggest that the product $x(1 - x)$ is always less than or equal to $1/4$. For what values of x does the product equal $1/4$?

Applications

Business and Economics

24. **Minimizing Cost** George Duda runs a sandwich shop. By studying data concerning his past costs, he has found that the cost of operating his shop is given by

 $$C(x) = 2x^2 - 20x + 360,$$

 where $C(x)$ is the daily cost in dollars to make x batches of sandwiches.

 (a) Find the number of batches George must sell to minimize the cost.

 (b) What is the minimum cost?

 (c) Find a formula for the average daily cost per batch of sandwiches as a function of x.

 (d) Find the average daily cost per batch of sandwiches when 4 batches are produced.

 (e) Find the average daily cost per batch of sandwiches when 7 batches are produced.

25. **Maximizing Revenue** The revenue of a charter bus company depends on the number of unsold seats. If the revenue $R(x)$ is given by

 $$R(x) = 5000 + 50x - x^2,$$

 where x is the number of unsold seats, find the maximum revenue and the number of unsold seats that corresponds to maximum revenue.

26. **Maximizing Revenue** A charter flight charges a fare of $200 per person plus $4 per person for each unsold seat on the plane. The plane holds 100 passengers. Let x represent the number of unsold seats.

 (a) Find an expression for the total revenue received for the flight $R(x)$. (*Hint:* Multiply the number of people flying, $100 - x$, by the price per ticket.)

 (b) Graph the expression from part (a).

 (c) Find the number of unsold seats that will produce the maximum revenue.

 (d) What is the maximum revenue?

 (e) Some managers might be concerned about the empty seats, arguing that it doesn't make economic sense to leave any seats empty. Write a few sentences explaining why this is not necessarily so.

27. **Maximizing Revenue** The demand for a certain type of cosmetic is given by

$$p = 500 - x,$$

 where p is the price in dollars when x units are demanded.

 (a) Find the revenue $R(x)$ that would be obtained at a price of p. (*Hint:* Revenue = Demand × Price)

 (b) Graph the revenue function $R(x)$.

 (c) From the graph of the revenue function, estimate the price that will produce maximum revenue.

 (d) What is the maximum revenue?

28. **Revenue** The manager of a peach orchard is trying to decide when to arrange for picking the peaches. If they are picked now, the average yield per tree will be 100 lb, which can be sold for 40¢ per pound. Past experience shows that the yield per tree will increase about 5 lb per week, while the price will decrease about 2¢ per pound per week.

 (a) Let x represent the number of weeks that the manager should wait. Find the income per pound.

 (b) Find the number of pounds per tree.

 (c) Find the total revenue from a tree.

 (d) When should the peaches be picked in order to produce maximum revenue?

 (e) What is the maximum revenue?

29. **Income** The manager of an 80-unit apartment complex is trying to decide what rent to charge. Experience has shown that at a rent of $200, all the units will be full. On the average, one additional unit will remain vacant for each $20 increase in rent.

 (a) Let x represent the number of $20 increases. Find an expression for the rent for each apartment.

 (b) Find an expression for the number of apartments rented.

 (c) Find an expression for the total revenue from all rented apartments.

 (d) What value of x leads to maximum revenue?

 (e) What is the maximum revenue?

Life Sciences

30. **Maximizing Population** The number of mosquitoes $M(x)$, in millions, in a certain area of Kentucky depends on the June rainfall x, in inches, approximately as follows.

$$M(x) = 10x - x^2$$

 Find the rainfall that will produce the maximum number of mosquitoes.

31. **Maximizing Chlorophyll Production** For the months of June through October, the percent of maximum possible chlorophyll production in a leaf is approximated by $C(x)$, where

$$C(x) = 10x + 50.$$

 Here x is time in months with $x = 1$ representing June. From October through December, $C(x)$ is approximated by

$$C(x) = -20(x - 5)^2 + 100,$$

 with x as above. Find the percent of maximum possible chlorophyll production in each of the following months.

 (a) June (b) July (c) September

 (d) October (e) November (f) December

32. **Maximizing Chlorophyll Production** Use your results from Exercise 31 to sketch a graph of $y = C(x)$, from June through December. In what month is chlorophyll production at a maximum?

Physical Sciences

33. **Maximizing the Height of an Object** If an object is thrown upward with an initial velocity of 32 ft/sec, then its height after t sec is given by

$$h = 32t - 16t^2.$$

 Find the maximum height attained by the object. Find the number of seconds it takes the object to hit the ground.

34. **Maximizing the Height of an Object** If an object is thrown upward with an initial velocity of 64 ft/sec, then its height after t sec is given by

$$h = -16t^2 + 64t$$

 Find the maximum height attained by the object. Find the number of seconds it takes the object to hit the ground.

General Interest

35. Maximizing Area Glenview Community College wants to construct a rectangular parking lot on land bordered on one side by a highway. It has 320 ft of fencing to use along the other three sides. What should be the dimensions of the lot if the enclosed area is to be a maximum? (*Hint:* Let x represent the width of the lot, and let $320 - 2x$ represent the length.)

36. Maximizing Area What would be the maximum area that could be enclosed by the college's 320 ft of fencing if it decided to close the entrance by enclosing all four sides of the lot? (See Exercise 35.)

37. Number Analysis Find two numbers whose sum is 20 and whose product is a maximum. (*Hint:* Let x and $20 - x$ be the two numbers, and write an equation for the product.)

38. Number Analysis Find two numbers whose sum is 45 and whose product is a maximum.

39. Parabolic Arch An arch is shaped like a parabola. It is 30 m wide at the base and 15 m high. How wide is the arch 10 m from the ground?

40. Parabolic Culvert A culvert is shaped like a parabola, 18 ft across the top and 12 ft deep. How wide is the culvert 8 ft from the top?

1.5 POLYNOMIAL AND RATIONAL FUNCTIONS

In the previous sections we discussed linear and quadratic functions and their graphs. Both of these functions are special types of *polynomial functions*.

POLYNOMIAL FUNCTION

A **polynomial function** of degree n, where n is a nonnegative integer, is defined by

$$f(x) = a_n x^n + a_{n-1} x^{n-1} + \cdots + a_1 x + a_0,$$

where $a_n, a_{n-1}, \ldots, a_1$, and a_0 are real numbers, with $a_n \neq 0$.

For $n = 1$, a polynomial function takes the form

$$f(x) = a_1 x + a_0,$$

a linear function. A linear function, therefore, is a polynomial function of degree 1. (Note, however, that a linear function of the form $f(x) = a_0$ for a real number a_0 is a polynomial function of degree 0.) A polynomial function of degree 2 is a quadratic function.

Accurate graphs of polynomial functions of degree 3 or higher require methods of calculus to be discussed later. Meanwhile, point plotting can be used to get reasonable sketches of such graphs. A graphing calculator or computer program is helpful for plotting many points rapidly, but such technology requires thought and care, as shall be seen.

The simplest polynomial functions of higher degree are those of the form $f(x) = x^n$. To graph $f(x) = x^3$, for example, find several ordered pairs that satisfy $y = x^3$, then plot the points and connect them with a smooth curve. The graph of $f(x) = x^3$ is shown as a black curve in Figure 28. This same figure also shows the graph of $f(x) = x^5$ in color.

$f(x) = x^3$

x	$f(x)$
-2	-8
-1	-1
0	0
1	1
2	8

$f(x) = x^5$

x	$f(x)$
-1.5	-7.6
-1	-1
0	0
1	1
1.5	7.6

FIGURE 28

The graphs of $f(x) = x^4$ and $f(x) = x^6$ can be sketched in a similar manner. Figure 29 shows $f(x) = x^4$ as a black curve and $f(x) = x^6$ in color. These graphs have symmetry about the y-axis, as does the graph of $f(x) = ax^2$ for a nonzero real number a. As with the graph of $f(x) = ax^2$, the value of a in $f(x) = ax^n$ affects the direction of the graph. When $a > 0$, the graph has the same general appearance as the graph of $f(x) = x^n$. However, if $a < 0$, the graph is rotated $180°$ about the x-axis.

$f(x) = x^4$

x	$f(x)$
-2	16
-1	1
0	0
1	1
2	16

$f(x) = x^6$

x	$f(x)$
-1.5	11.4
-1	1
0	0
1	1
1.5	11.4

FIGURE 29

EXAMPLE 1

Graph $f(x) = 8x^3 - 12x^2 + 2x + 1$.

Letting x take values from -3 through 3 leads to the values of $f(x)$ given in the following table.

x	-3	-2	-1	0	1	2	3
$f(x)$	-329	-115	-21	1	-1	21	115

$$f(x) = 8x^3 - 12x^2 + 2x + 1$$

FIGURE 30

Plot the points with reasonable y-values. You will probably need to calculate additional ordered pairs between these points to get a good idea of how the graph should look. See Figure 30. ◄

We will check our work by graphing this function on a graphing calculator. We press the Graph key, and type in the function, but no graph appears on the screen. Let's check the domain and range over which the calculator is plotting the graph. Here's the problem: the domain is set as $-360 \le x \le 360$ and the range as $-1.6 \le y \le 1.6$, settings left over from a previous graph. They are unsuitable for this function, however. For such large x values, the corresponding y values will be far beyond what the screen can show. If we change the range to $-1{,}000{,}000 \le y \le 1{,}000{,}000$, we find that this also is not a wise choice. The graph fits on the screen, but it looks similar in shape to the graph of $y = x^3$ in Figure 28. The two "bumps" around $x = 0$ and $x = 1$ in Figure 30 don't appear. (These "bumps" will be referred to as **turning points.** Later another term will be used: *relative extrema.*) And no wonder they don't appear: their heights are minuscule compared with 1,000,000.

Getting a good graph with a computer or calculator requires care in choosing the domain and range. Later we will see how calculus helps us locate the interesting parts of the graph. It will also tell us what is happening between the points we plot by hand, by calculator, or by computer.

EXAMPLE 2 Graph $f(x) = 3x^4 - 14x^3 + 54x - 3$.
Complete a table of ordered pairs, and plot the points.

x	-3	-2	-1	0	1	2	3
$f(x)$	456	49	-40	-3	40	41	24

The graph is shown in Figure 31. ◄

As suggested by the graphs above, the domain of a polynomial function is the set of all real numbers. The range of a polynomial function of odd degree is also the set of all real numbers. Some typical graphs of polynomial functions of odd degree are shown in Figure 32. These graphs suggest that for every polynomial function f of odd degree there is at least one real value of x for which $f(x) = 0$. Such a value of x is called a **real zero** of f; these values are also the x-intercepts of the graph.

$$f(x) = 3x^4 - 14x^3 + 54x - 3$$

FIGURE 31

Degree 3; three real zeros

Degree 3; one real zero

Degree 5; three real zeros

FIGURE 32

Polynomial functions of even degree have a range that takes either the form $(-\infty, k]$ or the form $[k, \infty)$ for some real number k. Figure 33 shows two typical graphs of polynomial functions of even degree.

Degree 4;
two real zeros

Degree 6;
four real zeros

FIGURE 33

A fifth-degree polynomial can have four turning points, as in the last graph of Figure 32, or no turning points, as in Figure 28. By examining the figures in this section, you may notice that the graph of a polynomial of degree n has at most $n - 1$ turning points. In a later chapter we will use calculus to see why this is true. Meanwhile, you can learn much about a polynomial by examining its graph. For example, if you are presented with the second graph in Figure 33 and told it is a polynomial, you know immediately that it is of even degree, because the range is of the form $(-\infty, k]$. It has five turning points, so it must be of degree 6 or higher. It could be of degree 8 or 10 or 12, etc., but you can't be sure from the graph alone.

The shape of a third-degree polynomial is similar to the first graph in Figure 32 if the x^3 term is positive, and to the second graph in Figure 32 if the x^3 term is negative. The turning points may also be flattened out, as they are in Figure 28.

Rational Functions Many situations require mathematical models that are quotients. A common model for such situations is a *rational function*.

RATIONAL FUNCTION A function defined by

$$f(x) = \frac{p(x)}{q(x)},$$

where $p(x)$ and $q(x)$ are polynomial functions and $q(x) \neq 0$, is called a **rational function.**

Since any values of x such that $q(x) = 0$ are excluded from the domain, a rational function usually has a graph with one or more breaks.

EXAMPLE 3

Graph $y = \dfrac{1}{x}$.

This function is undefined for $x = 0$, since 0 is not allowed in the denominator of a fraction. For this reason, the graph of this function will not intersect the vertical line $x = 0$, which is the y-axis. Since x can take on any value except 0, the values of x can approach 0 as closely as desired from either side of 0.

x approaches 0 ↓

x	$-.5$	$-.2$	$-.1$	$-.01$	$.01$	$.1$	$.2$	$.5$
$y = \dfrac{1}{x}$	-2	-5	-10	-100	100	10	5	2

↑ |y| gets larger and larger

The table above suggests that as x gets closer and closer to 0, $|y|$ gets larger and larger. This is true in general: as the denominator gets smaller, the fraction gets larger. Thus, the graph of the function approaches the vertical line $x = 0$ (the y-axis) without ever touching it.

As $|x|$ gets larger and larger, $y = 1/x$ gets closer and closer to 0, as shown in the table below. This is also true in general: as the denominator gets larger, the fraction gets smaller.

x	-100	-10	-4	-1	1	4	10	100
$y = \dfrac{1}{x}$	$-.001$	$-.01$	$-.25$	-1	1	$.25$	$.01$	$.001$

The graph of the function approaches the horizontal line $y = 0$ (the x-axis). The information from the table supports the graph in Figure 34. ◄

FIGURE 34

In Example 3, the vertical line $x = 0$ and the horizontal line $y = 0$ are *asymptotes*, defined as follows.

ASYMPTOTES

If a number k makes the denominator 0 in a rational function, then the line $x = k$ is a **vertical asymptote.***

If the values of y approach a number k as $|x|$ gets larger and larger, the line $y = k$ is a **horizontal asymptote.**

EXAMPLE 4

Graph $y = \dfrac{3x + 2}{2x + 4}$.

The value $x = -2$ makes the denominator 0, so the line $x = -2$ is a vertical asymptote. To find a horizontal asymptote, find y as x gets larger and larger, as in the following chart.

x	$y = \dfrac{3x + 2}{2x + 4}$	Ordered Pair
10	$\dfrac{32}{24} = 1.33$	(10, 1.33)
20	$\dfrac{62}{44} = 1.41$	(20, 1.41)
100	$\dfrac{302}{204} = 1.48$	(100, 1.48)
100,000	$\dfrac{300,002}{200,004} = 1.49998$	(100,000, 1.49998)

The chart suggests that as x gets larger and larger, $(3x + 2)/(2x + 4)$ gets closer and closer to 1.5, or 3/2, making the line $y = 3/2$ a horizontal asymptote. Use a calculator to show that as x gets more negative and takes on the values -10, -100, -1000, $-1,000,000$, and so on, the graph again approaches the line $y = 3/2$. The intercepts should also be noted. When $x = 0$, $y = 2/4 = 1/2$ (the y-intercept). To make a fraction 0, the numerator must be 0; so to make $y = 0$, it is necessary that $3x + 2 = 0$. Solve this for x to get $x = -2/3$ (the x-intercept). These two points, the asymptotes, and several other plotted points (at least one to the left of the vertical asymptote to pin down that side of the graph) produce the graph in Figure 35. ◀

*Actually, we should make sure that $x = k$ does not also make the numerator 0. If both the numerator and denominator are 0, then there may be no vertical asymptote at k.

x	y
3	1.1
2	1
1	5/6
0	1/2
−1	−1/2
−3/2	−5/2
−5/2	11/2
−3	7/2
−4	5/2
−6	2

FIGURE 35

In Example 4, $y = 3/2$ was the horizontal asymptote for the rational function $y = (3x + 2)/(2x + 4)$. An equation for the horizontal asymptote can also be found by asking what happens when x gets large. If x is very large, then $3x + 2 \approx 3x$,[†] because the 2 is very small by comparison. In other words, just keep the larger term ($3x$ in this case) and discard the smaller term. Similarly, the denominator is approximately equal to $2x$, so $y \approx 3x/2x = 3/2$. This means that the line $y = 3/2$ is a horizontal asymptote. A more precise way of approaching this idea will be seen later, when limits at infinity are discussed.

Rational functions occur often in practical applications. In many situations involving environmental pollution, much of the pollutant can be removed from the air or water at a fairly reasonable cost, but the last small part of the pollutant can be very expensive to remove. Cost as a function of the percentage of pollutant removed from the environment can be calculated for various percentages of removal, with a curve fitted through the resulting data points. This curve then leads to a mathematical model of the situation. Rational functions are often a good choice for these **cost-benefit models** because they rise rapidly as they approach a vertical asymptote.

Suppose a cost-benefit model is given by

$$y = \frac{18x}{106 - x},$$

where y is the cost (in thousands of dollars) of removing x percent of a certain pollutant. The domain of x is the set of all numbers from 0 to 100 inclusive; any

[†]The symbol \approx means *approximately equal to*.

amount of pollutant from 0% to 100% can be removed. Find the cost to remove the following amounts of the pollutant: 100%, 95%, 90%, and 80%. Graph the function.

Removal of 100% of the pollutant would cost

$$y = \frac{18(100)}{106 - 100} = 300,$$

or $300,000. Check that 95% of the pollutant can be removed for $155,000, 90% for $101,000, and 80% for $55,000. Using these points, as well as others that could be obtained from the function above, gives the graph shown in Figure 36. ◄

FIGURE 36

1.5 Exercises

Each of the following is the graph of a polynomial function. Give the possible values for the degree of the polynomial, and give the sign (+ or −) for the x^n term.

1.

2.

3.

4.

5.

6.

7.

8.

Find the horizontal and vertical asymptotes for each of the following rational functions. Draw the graph of each function, including any x- or y-intercepts.

9. $y = \dfrac{-4}{x - 3}$

10. $y = \dfrac{-1}{x + 3}$

11. $y = \dfrac{2}{3 + 2x}$

12. $y = \dfrac{4}{5 + 3x}$

13. $y = \dfrac{3x}{x - 1}$

14. $y = \dfrac{4x}{3 - 2x}$

15. $y = \dfrac{x + 1}{x - 4}$

16. $y = \dfrac{x - 3}{x + 5}$

17. $y = \dfrac{1 - 2x}{5x + 20}$

18. $y = \dfrac{6 - 3x}{4x + 12}$

19. $y = \dfrac{-x - 4}{3x + 6}$

20. $y = \dfrac{-x + 8}{2x + 5}$

Applications

Business and Economics

21. Average Cost Suppose the average cost per unit $C(x)$, in dollars, to produce x units of margarine is given by

$$C(x) = \frac{500}{x + 30}.$$

(a) Find $C(10)$, $C(20)$, $C(50)$, $C(75)$, and $C(100)$.

(b) Which of the intervals $(0, \infty)$ and $[0, \infty)$ would be a more reasonable domain for C? Why?

(c) Graph $y = C(x)$.

22. Cost Analysis In a recent year, the cost per ton, y, to build an oil tanker of x thousand deadweight tons was approximated by

$$y = \frac{110,000}{x + 225}.$$

(a) Find y for $x = 25$, $x = 50$, $x = 100$, $x = 200$, $x = 300$, and $x = 400$.

(b) Graph the function.

Tax Rates *Exercises 23 to 26 refer to the* Laffer curve, *originated by the economist Arthur Laffer. It has been a center of controversy. An idealized version of this curve is shown here. According to this curve, decreasing a tax rate, say from x_2 percent to x_1 percent on the graph, can actually lead to an increase in government revenue. The theory is that people will work harder and earn more money if they are taxed at a lower rate, so the government ends up with more revenue than it would at a higher tax rate. All economists agree on the endpoints—0 revenue at tax rates of both 0% and 100%—but there is much disagreement on the location of the tax rate x_1 that produces the maximum revenue.*

23. A function that might describe the entire Laffer curve is

$$y = x(100 - x)(x^2 + 500),$$

where y is government revenue in hundreds of thousands of dollars from a tax rate of x percent, with the function valid for $0 \le x \le 100$. Find the revenue from the following tax rates.

 (a) 10% (b) 40% (c) 50% (d) 80%
 (e) Graph the function.

24. Find the equations of two quadratic functions that could describe the Laffer curve by having roots at $x = 0$ and $x = 100$. Give the first a maximum of 100 and the second a maximum of 250, then multiply them together to get a new Laffer curve with a maximum of 25,000. Plot the resulting function.

25. Suppose an economist studying the Laffer curve produced the rational function

$$y = \frac{60x - 6000}{x - 120},$$

where y is government revenue in millions from a tax rate of x percent, with the function valid for $50 \le x \le 100$. Find the revenue from the following tax rates.

 (a) 50% (b) 60% (c) 80% (d) 100%
 (e) Graph the function.

26. Suppose the economist in Exercise 25 studies a different tax, this time producing

$$y = \frac{80x - 8000}{x - 110},$$

where y is the government revenue in tens of millions of dollars for a tax rate of x percent, with the function valid for $55 \le x \le 100$. Find the revenue from the following tax rates.

 (a) 55% (b) 60% (c) 70% (d) 90%
 (e) 100% (f) Graph the function.

27. **Cost-Benefit Model** Suppose a cost-benefit model is given by

$$y = \frac{6.7x}{100 - x}.$$

where y is the cost in thousands of dollars of removing x percent of a given pollutant.

Find the cost of removing each of the following percents of pollutants.

 (a) 50% (b) 70% (c) 80% (d) 90%
 (e) 95% (f) 98% (g) 99%
 (h) Is it possible, according to this function, to remove *all* the pollutant?
 (i) Graph the function.

28. **Cost-Benefit Model** Suppose a cost-benefit model is given by

$$y = \frac{6.5x}{102 - x},$$

where y is the cost in thousands of dollars of removing x percent of a certain pollutant.

Find the cost of removing each of the following percents of pollutants.

 (a) 0% (b) 50% (c) 80% (d) 90%
 (e) 95% (f) 99% (g) 100%
 (h) Graph the function.

Life Sciences

29. **Cardiac Output** A technique for measuring cardiac output depends on the concentration of a dye after a known amount is injected into a vein near the heart. In a normal heart, the concentration of the dye at time x (in seconds) is given by the function

$$g(x) = -.006x^4 + .140x^3 - .053x^2 + 1.79x.$$

 (a) 🖩 Graph $g(x)$ by point plotting or by using a graphing calculator or computer. ✦

(b) In your graph from part (a), notice that the function initially increases. Considering the form of $g(x)$, do you think it can keep increasing forever? Explain.

(c) Write a short paragraph describing the extent to which the concentration of dye might be described by the function $g(x)$.

30. Population Variation During the early part of the twentieth century, the deer population of the Kaibab Plateau in Arizona experienced a rapid increase, because hunters had reduced the number of natural predators. The increase in population depleted the food resources and eventually caused the population to decline. For the period from 1905 to 1930, the deer population was approximated by

$$D(x) = -.125x^5 + 3.125x^4 + 4000,$$

where x is time in years from 1905.

(a) ▦ Use a calculator to find enough points to graph $D(x)$, or use a graphing calculator or computer to graph the function.

(b) From the graph, over what period of time (from 1905 to 1930) was the population increasing? relatively stable? decreasing?

31. Alcohol Concentration The polynomial function

$$A(x) = -.015x^3 + 1.058x$$

gives the approximate alcohol concentration (in tenths of a percent) in an average person's bloodstream x hours after drinking about eight ounces of 100-proof whiskey. The function is approximately valid for x in the interval $[0, 8]$.

(a) Graph $A(x)$.

(b) Using the graph you drew for part (a), estimate the time of maximum alcohol concentration.

(c) In one state, a person is legally drunk if the blood alcohol concentration exceeds .08%. Use the graph from part (a) to estimate the period in which this average person is legally drunk.

32. Drug Dosage To calculate the drug dosage for a child, pharmacists may use the formula

$$d(x) = \frac{Dx}{x + 12},$$

where x is the child's age in years and D is the adult dosage. Let $D = 70$, the adult dosage of the drug Naldecon.

(a) What is the vertical asymptote for this function?

(b) What is the horizontal asymptote for this function?

(c) Graph $d(x)$.

33. Population Biology The function

$$f(x) = \frac{\lambda x}{1 + (ax)^b}$$

is used in population models to give the size of the next generation $(f(x))$ in terms of the current generation (x).*

(a) What is a reasonable domain for this function, considering what x represents?

(b) Graph this function for $\lambda = a = b = 1$.

(c) Graph this function for $\lambda = a = 1$ and $b = 2$.

(d) What is the effect of making b larger?

34. Growth Models The function

$$f(x) = \frac{Kx}{A + x}$$

is used in biology to give the growth rate of a population in the presence of a quantity x of food. This is called Michaelis-Menten kinetics.†

(a) What is a reasonable domain for this function, considering what x represents?

(b) Graph this function for $K = 5$ and $A = 2$.

(c) Show that $y = K$ is a horizontal asymptote.

(d) What do you think K represents?

(e) Show that A represents the quantity of food for which the growth rate is half of its maximum.

Physical Sciences

35. Electronics In electronics, the circuit gain is given by

$$G(R) = \frac{R}{r + R},$$

where R is the resistance of a temperature sensor in the circuit and r is constant. Let $r = 1000$ ohms.

(a) Find any vertical asymptotes of the function.

(b) Find any horizontal asymptotes of the function.

(c) Graph $G(R)$.

36. Oil Pressure The pressure of the oil in a reservoir tends to drop with time. By taking sample pressure readings, petroleum engineers have found that the *change* in pressure in a particular oil reservoir is given by

$$P(t) = t^3 - 25t^2 + 200t,$$

where t is time in years from the date of the first reading.

(a) Graph $P(t)$.

(b) For what time period is the *change* in pressure (the drop in pressure) increasing? decreasing?

*See J. Maynard Smith, *Models in Ecology*, Cambridge University Press, 1974.

†See Leah Edelstein-Keshet, *Mathematical Models in Biology*, Random House, 1988.

37. Catastrophe Theory An area of mathematics known as catastrophe theory studies how functions suddenly change their behavior. In catastrophe theory, the family of functions

$$f_t(u) = u^3 + tu$$

arises, where t can take on various values to produce various functions.*

Graph $f_t(u)$ for

(a) $t = 1$ (b) $t = -1$

38. Catastrophe theory has been applied to thermodynamics, in which the following family of functions arises:[†]

$$f(x) = \frac{x^4}{4} + \frac{a}{2}x^2 + bx$$

Graph $f(x)$ for

(a) $a = -5$ and $b = 1$; (b) $a = 1$ and $b = -6$.

General Interest

39. Antique Cars Antique car fans often enter their cars in a *concours d'élègance* in which a maximum of 100 points can be awarded to a particular car. Points are awarded for the general attractiveness of the car. Based on a recent article in *Business Week*, we constructed the following mathematical model for the cost, in thousands of dollars, of restoring a car so that it will win x points.

$$C(x) = \frac{10x}{49(101 - x)}$$

Find the cost of restoring a car so that it will win the following numbers of points.

(a) 99 (b) 100

Computer/Graphing Calculator

Make an intelligent guess about the shape of each polynomial function. Then sketch the graph of the function by using a graphing calculator or computer program, and compare the graph with what you expected it to look like.

40. $f(x) = x^3 - 7x - 9$ **41.** $f(x) = -x^3 + 4x^2 + 3x - 8$ **42.** $f(x) = -6x^3 - 11x^2 + x + 6$

43. $f(x) = 2x^3 + 4x - 1$ **44.** $f(x) = x^4 - 5x^2 + 7$ **45.** $f(x) = x^4 + x^3 - 2$

46. $f(x) = -8x^4 + 2x^3 + 47x^2 + 52x + 15$ **47.** $f(x) = -x^4 - 2x^3 + 3x^2 + 3x + 5$

48. $f(x) = x^5 - 2x^4 - x^3 + 3x^2 + x + 2$ **49.** $f(x) = -x^5 + 6x^4 - 11x^3 + 6x^2 + 5$

50. Write a paragraph or two describing what you learned about the graphs of polynomial functions from Exercises 40–49.

Find the horizontal and vertical asymptotes for each of the following rational functions. Use this information to make an intelligent guess about the shape of each function. Then sketch the graph of the function by using a graphing calculator or computer program, and compare the graph with what you expected it to look like.

51. $f(x) = \dfrac{-2x^2 + x - 1}{2x + 3}$ **52.** $f(x) = \dfrac{3x + 2}{x^2 - 4}$ **53.** $f(x) = \dfrac{2x^2 - 5}{x^2 - 1}$

54. $f(x) = \dfrac{4x^2 - 1}{x^2 + 1}$ **55.** $f(x) = \dfrac{-2x^2}{x^2 - 10}$ **56.** $f(x) = \dfrac{5x + 4}{2x^2 - 1}$

57. Write a paragraph or two describing what you learned about the graphs of rational functions from Exercises 51–56.

*See T. Poston and I. Stewart, *Catastrophe Theory and Its Applications*, Pitman Publishing, 1978, p. 105.

[†]Ibid, p. 337.

1.6 TRANSLATIONS AND REFLECTIONS OF FUNCTIONS

We saw in the section on quadratics that the effect of adding a constant to a quadratic function is to raise the function if the constant is positive, and to lower it if the constant is negative. This is true for any function; the movement up or down is referred to as a **vertical translation** of the function.

EXAMPLE 1

Graph $f(x) = x^3 - 1$.

The graph of $y = x^3$ is familiar from the previous section. Just lower the graph by 1 to get the graph in Figure 37. ◄

$y = x^3 - 1$

FIGURE 37

The next example illustrates a **horizontal translation** of a function, which is a shift to the left or right.

EXAMPLE 2

Graph $f(x) = (x - 4)^2$.

Notice here that 4 is subtracted from the x *before* the squaring occurs. This is not the same as taking the graph of $y = x^2$ and translating it downward by 4. Consider that the vertex of $y = x^2$ is the lowest value of the function, $y = 0$, which occurs when $x = 0$. In the current function, this value of y occurs when $x = 4$. Therefore, the new graph is just the graph of $y = x^2$ shifted 4 units to the right, as shown in Figure 38. ◄

$x = 4$

$y = (x - 4)^2$

$(4, 0)$

FIGURE 38

It was shown in the section on quadratic functions that multiplying a function by a negative number flips it upside down. This is called a **vertical reflection** of the graph. The following table summarizes what we have seen about translations and reflections, and adds a new feature: **horizontal reflections.**

TRANSLATIONS AND REFLECTIONS OF FUNCTIONS

Let $f(x)$ be any function, and let a, h, and k be positive constants (Figure 39). The graph of $y = f(x) + k$ is the graph of $y = f(x)$ translated upward by an amount k (Figure 40).

FIGURE 39

k

FIGURE 40

FIGURE 41

The graph of $y = f(x) - k$ is the graph of $y = f(x)$ translated downward by an amount k (Figure 41).

The graph of $y = f(x - h)$ is the graph of $y = f(x)$ translated to the right by an amount h (Figure 42).

The graph of $y = f(x + h)$ is the graph of $y = f(x)$ translated to the left by an amount h (Figure 43).

FIGURE 42 **FIGURE 43**

The graph of $y = -f(x)$ is the graph of $y = f(x)$ reflected vertically, that is, turned upside down (Figure 44).

The graph of $y = f(-x)$ is the graph of $y = f(x)$ reflected horizontally, that is, its mirror image (Figure 45).

FIGURE 44

FIGURE 45

It is also possible to consider multiplying x or $f(x)$ by a constant a, as in $f(ax)$ or $af(x)$. When a is positive, this does not change the basic appearance of the graph, except to compress it or stretch it. When a is negative, it also results in a reflection. More details are included in the examples and exercises.

FIGURE 46

EXAMPLE 3

Graph $f(x) = -\sqrt{4 - x} + 3$.

Begin with the simplest possible function, then add each piece in turn. Start with the graph of $f(x) = \sqrt{x}$. As Figure 46 reveals, this is just one-half of the graph of $f(x) = x^2$ lying on its side.

$f(x) = \sqrt{-x}$

FIGURE 47

Now add another component of the original function, the negative in front of the x, giving $f(x) = \sqrt{-x}$. This is a horizontal reflection of the $f(x) = \sqrt{x}$ graph, as shown in Figure 47.

Next, include the 4 under the square root sign. To get $4 - x$ into the form $f(x - h)$ or $f(x + h)$, we need to factor out the negative: $\sqrt{4 - x} = \sqrt{-(x - 4)}$. Now the 4 is subtracted, so this function is a translation to the right of the function $f(x) = \sqrt{-x}$ by 4 units, as Figure 48 reveals.

$f(x) = \sqrt{4 - x}$

FIGURE 48

$f(x) = -\sqrt{4 - x}$

FIGURE 49

The effect of the negative in the front is a vertical reflection, as in Figure 49, which shows the graph of $f(x) = -\sqrt{4 - x}$.

Finally, adding the constant 3 raises the entire graph by 3 units, giving the graph of $f(x) = -\sqrt{4 - x} + 3$ in Figure 50. ◀

$f(x) = -\sqrt{4 - x} + 3$

FIGURE 50

A quadratic function can be analyzed using the process described in Example 3 if it is in the form

$$y = a(x - h)^2 + k.$$

A quadratic equation not given in this form can be converted by a process called **completing the square**, which is explained next.

EXAMPLE 4

Graph $f(x) = -3x^2 - 2x + 1$.
To rewrite $-3x^2 - 2x + 1$ in the form $a(x - h)^2 + k$, first factor -3 from $-3x^2 - 2x$ to get

$$y = -3\left(x^2 + \frac{2}{3}x\right) + 1.$$

One-half the coefficient of x is $1/3$, and $(1/3)^2 = 1/9$. Add and subtract $1/9$ inside the parentheses as follows:

$$y = -3\left(x^2 + \frac{2}{3}x + \frac{1}{9} - \frac{1}{9}\right) + 1.$$

Using the distributive property and simplifying gives

$$y = -3\left(x^2 + \frac{2}{3}x + \frac{1}{9}\right) - 3\left(-\frac{1}{9}\right) + 1 = -3\left(x^2 + \frac{2}{3}x + \frac{1}{9}\right) + \frac{4}{3}.$$

$$\left(-\frac{1}{3}, \frac{4}{3}\right)$$

$$y = -3x^2 - 2x + 1$$

FIGURE 51

Factor to get

$$y = -3\left(x + \frac{1}{3}\right)^2 + \frac{4}{3}.$$

This result shows that the graph is the parabola $y = x^2$ translated $1/3$ unit to the left, flipped upside down, and translated $4/3$ units upward. This puts the vertex at $(-1/3, 4/3)$. The 3 in front of the squared term causes the parabola to be stretched vertically by a factor of 3. These results are shown in Figure 51. ◀

1.6 Exercises

Use the ideas in this section to graph each of the following functions.

1. $f(x) = (x + 2)^3$

2. $f(x) = (x - 3)^3$

3. $f(x) = (4 - x)^3 + 10$

4. $f(x) = (1 - x)^3 - 3$

5. $f(x) = -2(x + 1)^3 + 5$

6. $f(x) = -3(x - 2)^3 - 2$

7. $f(x) = -2x^5 + 5$

8. $f(x) = (x - 1)^6 - 4$

9. $f(x) = \dfrac{2}{x + 3} - 2$

10. $f(x) = \dfrac{3}{x - 2} + 3$

11. $f(x) = \dfrac{-1}{x - 1} + 2$

12. $f(x) = \dfrac{-1}{x + 4} - 1$

13. $f(x) = \sqrt{x - 1} + 3$

14. $f(x) = \sqrt{x + 1} - 4$

15. $f(x) = -\sqrt{-4 - x} - 2$

16. $f(x) = -\sqrt{2 - x} + 2$

Use the method of completing the square to graph each of the following quadratic functions.

17. $f(x) = x^2 - 2x - 2$

18. $f(x) = x^2 + 4x + 6$

19. $f(x) = x^2 - 4x - 2$

20. $f(x) = x^2 + 6x + 8$

21. $f(x) = 2x^2 + 4x - 1$

22. $f(x) = 3x^2 - 12x + 10$

23. $f(x) = -3x^2 + 24x - 55$

24. $f(x) = -2x^2 - 8x - 13$

Drawing a figure similar to Figures 39–45, show the graph of $f(ax)$ where a satisfies the given condition.

25. $0 < a < 1$

26. $1 < a$

27. $-1 < a < 0$

28. $a < -1$

Drawing a figure similar to Figures 39–45, show the graph of $af(x)$ where a satisfies the given condition.

29. $0 < a < 1$

30. $1 < a$

31. $-1 < a < 0$

32. $a < -1$

33. One observation arising from these methods is that all rational functions in which both the numerator and denominator are linear functions are just translations and reflections of the function $f(x) = c/x$.

(a) Show that if $f(x) = c/x$, then $af(x - h) + k$ is also a rational function.

(b) Going the other way is a bit trickier. Here is an example:

$$f(x) = \frac{2x + 3}{4x + 1} = \frac{2x + 1/2}{4x + 1} + \frac{5/2}{4x + 1} = \frac{1}{2} + \frac{5}{2} \cdot \frac{1}{4x + 1}.$$

How would you know to break up the 3 as $1/2 + 5/2$?

(c) Use the result of part (b) to graph $f(x) = \dfrac{2x + 3}{4x + 1}$.

(d) Use this method to graph $f(x) = \dfrac{2x + 5}{x + 2}$.

Applications

Business and Economics

34. Cost Analysis Ali finds that the cost in dollars to make x chocolate chip cookies in his bakery is approximately equal to

$$C(x) = 10\sqrt{x} + 50.$$

(a) Find the cost to make 100 cookies.

(b) What is the fixed cost (i.e., the cost of making no cookies)?

(c) Sketch the graph of $C(x)$.

(d) Find a formula for $A(x)$, the average cost per cookie.

(e) What is the average cost per cookie when 100 cookies are made?

(f) What is the average cost per cookie when 1600 cookies are made?

35. Advertising A study done by an advertising agency reveals that when x thousands of dollars are spent on advertising, it results in a sales increase in thousands of dollars given by the function

$$S(x) = \frac{1}{25}(x - 10)^3 + 40.$$

(a) Find the increase in sales when no money is spent on advertising.

(b) Find the increase in sales when $10,000 is spent on advertising.

(c) Sketch the graph of $S(x)$.

(d) Find a formula for $A(x)$, the average increase in sales per amount spent on advertising.

(e) What is the average increase in sales per amount spent on advertising when $10,000 is spent on advertising?

(f) What is the average increase in sales per amount spent on advertising when $50,000 is spent on advertising?

Chapter Summary Key Terms

To understand the concepts presented in this chapter, you should know the meaning and use of the following words. For easy reference, the section in the chapter where a word (or expression) was first used is given with each item.

1.1 function
 mathematical model
 independent variable
 dependent variable
 ordered pair
 domain
 range
 interval notation
 Cartesian coordinate system
 x-axis
 y-axis
 origin
 x-coordinate
 y-coordinate
 quadrant
 vertical line test
 step function

1.2 linear function
 slope
 point-slope form
 slope-intercept form
 supply and demand curves
 equilibrium price
 equilibrium quantity
1.3 average rate of change
 fixed cost
 marginal cost
 cost function
 break-even point
1.4 quadratic function
 parabola
 vertex
 axis

1.5 polynomial function
 turning point
 real zero
 rational function
 vertical asymptote
 horizontal asymptote
 cost-benefit model
1.6 vertical translation
 horizontal translation
 vertical reflection
 horizontal reflection
 completing the square

Chapter 1	**Review Exercises**

1. What is a function? A linear function? A quadratic function? A rational function?
2. How do you find a vertical asymptote? A horizontal asymptote?
3. What is the marginal cost? The fixed cost?
4. What can you tell about the graph of a polynomial function of degree n before you plot any points?

List the ordered pairs obtained from each of the following if the domain of x for each exercise is $\{-3, -2, -1, 0, 1, 2, 3\}$. *Graph each set of ordered pairs. Give the range.*

5. $2x - 5y = 10$
6. $3x + 7y = 21$
7. $y = (2x + 1)(x - 1)$
8. $y = (x + 4)(x + 3)$
9. $y = -2 + x^2$
10. $y = 3x^2 - 7$
11. $y = \dfrac{2}{x^2 + 1}$
12. $y = \dfrac{-3 + x}{x + 10}$
13. $y + 1 = 0$
14. $y = 3$

In Exercises 15–18, find **(a)** $f(6)$, **(b)** $f(-2)$, **(c)** $f(-4)$, *and* **(d)** $f(r + 1)$.

15. $f(x) = 4x - 1$
16. $f(x) = 3 - 4x$
17. $f(x) = -x^2 + 2x - 4$
18. $f(x) = 8 - x - x^2$
19. Let $f(x) = 5x - 3$ and $g(x) = -x^2 + 4x$. Find each of the following.
 (a) $f(-2)$ **(b)** $g(3)$ **(c)** $g(-4)$ **(d)** $f(5)$ **(e)** $g(-k)$ **(f)** $g(3m)$ **(g)** $g(k - 5)$ **(h)** $f(3 - p)$

Graph each of the following.

20. $y = 6 - 2x$
21. $y = 4x + 3$
22. $2x + 7y = 14$
23. $3x - 5y = 15$
24. $y = 1$
25. $x + 2 = 0$
26. $x + 3y = 0$
27. $y = 2x$

Find the slope for each line that has a slope.

28. Through $(4, -1)$ and $(3, -3)$
29. Through $(-2, 5)$ and $(4, 7)$
30. Through the origin and $(0, 7)$
31. Through the origin and $(11, -2)$
32. $4x - y = 7$
33. $2x + 3y = 15$
34. $3y - 1 = 14$
35. $x + 4 = 9$

Find an equation in the form ax + by = c for each line.

36. Through $(8, 0)$, with slope $-1/4$
37. Through $(5, -1)$, with slope $2/3$
38. Through $(2, -3)$ and $(-3, 4)$
39. Through $(5, -2)$ and $(1, 3)$
40. Slope 0, through $(-2, 5)$
41. Undefined slope, through $(-1, 4)$
42. Through $(2, -1)$, parallel to $3x - y = 1$
43. Through $(0, 5)$, perpendicular to $8x + 5y = 3$
44. Through $(2, -10)$, perpendicular to a line with undefined slope
45. Through $(3, -5)$, parallel to $y = 4$
46. Through $(-7, 4)$, perpendicular to $y = 8$

Graph each line.

47. Through $(2, -4)$, $m = 3/4$
48. Through $(0, 5)$, $m = -2/3$
49. Through $(-4, 1)$, $m = 3$
50. Through $(-3, -2)$, $m = -1$

Graph each of the following.

51. $y = x^2 - 4$

52. $y = 6 - x^2$

53. $y = 3x^2 + 6x - 2$

54. $y = -\frac{1}{4}x^2 + x + 2$

55. $y = x^2 - 4x + 2$

56. $y = -3x^2 - 12x - 1$

57. $f(x) = x^3 + 5$

58. $f(x) = 1 - x^4$

59. $y = -(x - 1)^3 + 4$

60. $y = -(x + 2)^4 - 2$

61. $y = 2\sqrt{x + 3} + 1$

62. $y = -\sqrt{x - 4} + 2$

63. $f(x) = \frac{8}{x}$

64. $f(x) = \frac{2}{3x - 1}$

65. $f(x) = \frac{4x - 2}{3x + 1}$

66. $f(x) = \frac{6x}{x + 2}$

In Exercises 67–70, complete the square for the indicated exercise, then use the methods of Section 1.6 to graph the function.

67. Exercise 53

68. Exercise 54

69. Exercise 55

70. Exercise 56

Applications

Business and Economics

71. Supply and Demand The supply and demand for crab meat in a local fish store are related by the equations

Supply: $p = S(q) = 6q + 3$

and Demand: $p = D(q) = 19 - 2q$,

where p represents the price in dollars per pound and q represents the quantity of crab meat in pounds per day. Find the supply and the demand at each of the following prices.

(a) $10 (b) $15 (c) $18

(d) Graph both the supply and the demand functions on the same axes.

(e) Find the equilibrium price.

(f) Find the equilibrium quantity.

72. Supply For boxes of computer disks, 72 boxes will be supplied at a price of $6, while 104 boxes will be supplied at a price of $10. Write a supply function for this product. Assume it is a linear function.

Cost *Find the linear cost functions and average cost functions in Exercises 73–75.*

73. Eight units cost $300; fixed cost is $60

74. Twelve units cost $445; 50 units cost $1585

75. Thirty units cost $1500; 120 units cost $5640

76. Cost The cost of producing x bird cages is $C(x)$, where

$$C(x) = 20x + 100.$$

The product sells for $40 per cage.

(a) Find the break-even point.

(b) What revenue will the company receive if it sells just that number of cages?

77. Car Rental To rent a midsized car from one agency costs $40 per day or fraction of a day. If you pick up the car in Boston and drop it off in Utica, there is a fixed $40 charge. Let $C(x)$ represent the cost of renting the car for x days and taking it from Boston to Utica.

Find each of the following.

(a) $C\left(\frac{3}{4}\right)$ (b) $C\left(\frac{9}{10}\right)$ (c) $C(1)$

(d) $C\left(1\frac{5}{8}\right)$ (e) $C\left(2\frac{1}{9}\right)$

(f) Graph the function defined by $y = C(x)$ for $0 < x \leq 5$.

(g) What is the independent variable?

(h) What is the dependent variable?

Life Sciences

78. Fever A certain viral infection causes a fever that typically lasts 6 days. A model of the fever (in °F) on day x, $1 \leq x \leq 6$, is

$$F(x) = -\frac{2}{3}x^2 + \frac{14}{3}x + 96.$$

According to the model, on what day should the maximum fever occur? What is the maximum fever?

79. AIDS HIV is the virus responsible for AIDS. The following graph shows an estimation of the HIV infection rate in the United States.[*]

[*]"Reconstruction and Future Trends of the AIDS Epidemic in the United States" by Dr. Ronald Brookmeyer from *Science*, Volume 253, July 5, 1991, p. 37. Copyright © 1991 by the American Association for the Advancement of Science. Reprinted by permission of the AAAS and Dr. Brookmeyer.

(a) When was the rate the highest?

(b) What was the maximum rate?

(c) What is the range of the function?

80. **Pollution** The cost to remove x percent of a pollutant is

$$y = \frac{7x}{100 - x},$$

in thousands of dollars.

Find the cost of removing each of the following percents of pollution.

(a) 80% **(b)** 50% **(c)** 90%

(d) Graph the function.

(e) Can all of the pollutant be removed?

81. **Medicine Dosage** Different rules exist for determining medicine dosages for children and adults. Two methods for finding dosages for children are given here. If the child's age in years is A, the adult dosage is d, and the child's dosage is c, then

$$c = \frac{A}{A + 12}d \quad \text{or} \quad c = \frac{A + 1}{24}d.$$

(a) At what age are the two dosages equal?

(b) Carefully graph $c = f(A) = \dfrac{A}{A + 12}$ and $c =$

$g(A) = \dfrac{A + 1}{24}$ for the interval $2 \le A \le 13$ on the same axes by plotting points. Give the intervals where $f(A) > g(A)$ and where $g(A) > f(A)$.

(c) From the graphs in part (b), at what ages do the functions f and g appear to differ the most?

General Interest

82. **Postage** Assume that it costs 30¢ to mail a letter weighing one ounce or less, with each additional ounce, or portion of an ounce, costing 27¢. Let $C(x)$ represent the cost to mail a letter weighing x oz.

Find the costs of mailing letters of the following weights.

(a) 3.4 oz **(b)** 1.02 oz **(c)** 5.9 oz **(d)** 10 oz

(e) Graph C.

(f) Give the domain and range of C.

83. **Automobiles** The following graph shows the horsepower and torque for a 1991 Porsche 911 Turbo as a function of the engine speed.[†]

(a) Find the engine speed giving the maximum horsepower.

(b) Find the maximum horsepower.

(c) Find the torque when the horsepower is at its maximum.

Connections

84. **Cost** Suppose the cost in dollars to produce x hundreds of nails is given by

$$C(x) = x^2 + 4x + 7.$$

(a) Sketch a graph of $C(x)$.

(b) Find a formula for $C(x + 1) - C(x)$, the cost to produce an additional hundred nails when x hundred are already produced. (This quantity is approximately equal to the marginal cost.)

(c) Find a formula for $A(x)$, the average cost per hundred nails.

(d) Find a formula for $A(x + 1) - A(x)$, the change in the average cost per nail when one additional batch of

[†]Porsche product information kit for the 1991 Turbo Performance and Torque 1991 model year. Reprinted by permission of Porsche Cars North America, Inc.

100 nails is produced. (This quantity is approximately equal to the marginal average cost, discussed in the chapter on the derivative.)

85. Cost Suppose the cost in dollars to produce x posters is given by

$$C(x) = \frac{5x + 3}{x + 1}.$$

(a) Sketch a graph of $C(x)$.

(b) Find a formula for $C(x + 1) - C(x)$, the cost to produce an additional poster when x posters are already produced.

(c) Find a formula for $A(x)$, the average cost per poster.

(d) Find a formula for $A(x + 1) - A(x)$, the change in the average cost per poster when one additional poster is produced. (This quantity is approximately equal to the marginal average cost, discussed in the chapter on the derivative.)

86. Salaries According to Hofstra University's 1991 faculty contract, the increase in faculty salaries in the fourth year of the contract would be the percent increase in the Consumer Price Index (CPI), increased or decreased by 1/4% for each 1% increase or decrease in student enrollment. The minimum increase would be 6%, while the maximum would be 8%.

Suppose the percent change in the CPI that year is 6.5%. Denote the percent increase in faculty salaries by $I(x)$, where x is the percent change in student enrollment.

(a) Find $I(-3)$.

(b) Find $I(-2)$.

(c) Find the percent increase in salaries if enrollment increases by 4%.

(d) Find the percent increase in salaries if enrollment increases by 7%.

(e) Graph $I(x)$ for $-3 \leq x \leq 7$.

(f) What is the range of I?

Extended Application / Marginal Cost—Booz, Allen & Hamilton

Booz, Allen & Hamilton Inc. is a large management consulting firm.* One service it provides to client companies is profitability studies showing ways in which the client can increase profit levels. The client company requesting the analysis presented in this case is a large producer of a staple food. The company buys from farmers and then processes the food in its mills, resulting in a finished product. The company sells some food at retail under its own brands and some in bulk to other companies who use the product in the manufacture of convenience foods.

The client company has been reasonably profitable in recent years, but the management retained Booz, Allen & Hamilton to see whether its consultants could suggest ways of increasing company profits. The management of the company had long operated with the philosophy of trying to process and sell as much of its product as possible, since, they felt, this would lower the average processing cost per unit sold. The consultants found, however, that the client's fixed mill costs were quite low, and that, in fact, processing extra units made the cost per unit start to increase. (There are several reasons for this: the company must run three shifts, machines break down more often, and so on.)

In this application, we shall discuss the marginal cost of two of the company's products. The marginal cost (approximate cost of producing an extra unit) of production for product A was found by the consultants to be approximated by the linear function

$$y = .133x + 10.09,$$

where x is the number of units produced (in millions) and y is the marginal cost. (Here the marginal cost is not a constant, as it was in the examples in the text.)

For example, at a level of production of 3.1 million units, an additional unit of product A would cost about

$$y = .133(3.1) + 10.09 \approx \$10.50.$$

At a level of production of 5.7 million units, an extra unit costs $10.85. Figure 52 shows a graph of the marginal cost function from $x = 3.1$ to $x = 5.7$, the domain over which the function above was found to apply.

The selling price for product A is $10.73 per unit, so that, as shown in Figure 52, the company was losing money on many units of the product that it sold. Since the selling price could

*This case was supplied by John R. Dowdle of the Chicago office of Booz, Allen & Hamilton Inc. Reprinted by permission.

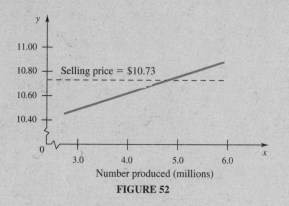

FIGURE 52

The consultants ran similar cost analyses of other products made by the company and then issued their recommendation: the company should reduce total production by 2.1 million units. The analysts predicted that this would raise profits for the products under discussion from $8.3 million annually to $9.6 million—which is very close to what actually happened when the client took the advice.

Exercises

1. At what level of production, x, was the marginal cost of a unit of product A equal to the selling price?

2. Graph the marginal cost function for product B from $x = 3.1$ million units to $x = 5.7$ million units.

3. Find the number of units for which marginal costs equals the selling price for product B.

4. For product C, the marginal cost of production is

$$y = .133x + 9.46.$$

(a) Find the marginal costs at production levels of 3.1 million units and 5.7 million units.
(b) Graph the marginal cost function.
(c) For a selling price of $9.57, find the production level for which the cost equals the selling price.

not be raised if the company were to remain competitive, the consultants recommended that production of product A be cut.

For product B, the Booz, Allen & Hamilton consultants found a marginal cost function given by

$$y = .0667x + 10.29,$$

with x and y as defined above. Verify that at a production level of 3.1 million units, the marginal cost is about $10.50, while at a production level of 5.7 million units, the marginal cost is about $10.67. Since the selling price of this product is $9.65, the consultants again recommended a cutback in production.

Applications at a glance . . .
Male Versus Female Educational Attainment . . . choosing investments . . . models of small economies . . .

CHAPTER 2

SYSTEMS OF LINEAR EQUATIONS AND MATRICES

(See page 97.)

Many mathematical models require finding the solutions of two or more equations. The solutions must satisfy *all* of the equations in the model. A set of equations related in this way is called a **system of equations.** In this chapter we will discuss systems of equations, introduce the idea of a *matrix*, and then show how matrices are used to solve systems of equations.

2.1 SOLUTION OF LINEAR SYSTEMS BY THE ECHELON METHOD

How much of each ingredient should be used in an animal feed to meet dietary requirements?

Suppose that an animal feed is made from three ingredients: corn, soybeans, and cottonseed. One unit of each ingredient provides the number of units of protein, fat, and fiber shown in the table. For example, the entries in the first column, .25, .4, and .3, indicate that one unit of corn provides twenty-five hundredths (one-fourth) of a unit of protein, four-tenths of a unit of fat, and three-tenths of a unit of fiber.

	Corn	Soybeans	Cottonseed
Protein	.25	.4	.2
Fat	.4	.2	.3
Fiber	.3	.2	.1

Now suppose we need to know the number of units of each ingredient that should be used to make a feed that contains 22 units of protein, 28 units of fat, and 18 units of fiber. To find out, we let x represent the required number of units of corn, y the number of units of soybeans, and z the number of units of cottonseed. Since the total amount of protein is to be 22 units,

$$.25x + .4y + .2z = 22.$$

The feed must supply 28 units of fat, so

$$.4x + .2y + .3z = 28,$$

and 18 units of fiber, so

$$.3x + .2y + .1z = 18.$$

To solve this problem, we must find values of x, y, and z that satisfy this system of equations. Verify that $x = 40$, $y = 15$, and $z = 30$ is a solution of the system, since these numbers satisfy all three equations. In fact, this is the only solution of this system. Many practical problems lead to such systems of *first-degree equations.*

A **first-degree equation** in n unknowns is any equation of the form

$$a_1x_1 + a_2x_2 + \cdots + a_nx_n = k,$$

where a_1, a_2, \ldots, a_n and k are all real numbers.* Each of the three equations from the animal feed problem is a first-degree equation in three unknowns. A **solution** of the first-degree equation

$$a_1x_1 + a_2x_2 + \cdots + a_nx_n = k$$

is a sequence of numbers s_1, s_2, \ldots, s_n such that

$$a_1s_1 + a_2s_2 + \cdots + a_ns_n = k.$$

A solution of an equation is usually written between parentheses as (s_1, s_2, \ldots, s_n). For example, $(1, 6, 2)$ is a solution of the equation $3x_1 + 2x_2 - 4x_3 = 7$, since $3(1) + 2(6) - 4(2) = 7$. This is an extension of the idea of an ordered pair, which was introduced in Chapter 1. A solution of a first-degree equation in two unknowns is an ordered pair, and the graph of the equation is a straight line. For this reason all first-degree equations are also called linear equations. In this section we develop a method for solving a system of first-degree equations. Although the discussion will be confined to equations with only a few variables, the method of solution can be extended to systems with many variables.

Because the graph of a linear equation in two unknowns is a straight line, there are three possibilities for the solutions of a system of two linear equations in two unknowns.

1. The two graphs are lines intersecting at a single point. The system has one solution, and it is given by the coordinates of this point. See Figure 1(a).

2. The graphs are distinct parallel lines. When this is the case, the system is **inconsistent;** that is, there is no solution common to both equations. See Figure 1(b).

3. The graphs are the same line. In this case, the equations are said to be **dependent,** since any solution of one equation is also a solution of the other. There are infinitely many solutions. See Figure 1(c).

(a) (b) (c)

FIGURE 1

*a_1 is read "a-sub-one." The notation a_1, a_2, \ldots, a_n represents n real-number coefficients (some of which may be equal), and the notation x_1, x_2, \ldots, x_n represents n different variables, or unknowns.

In larger systems, with more equations and more variables, there also may be exactly one solution, no solutions, or infinitely many solutions. If no solution satisfies every equation in the system, the system is inconsistent, and if there are infinitely many solutions that satisfy all the equations in the system, the equations are dependent.

Transformations To solve a linear system of equations, we use properties of algebra to change, or transform, the system into a simpler *equivalent* system. An **equivalent system** is one that has the same solutions as the given system. Algebraic properties are the basis of the transformations listed below.

TRANSFORMATIONS OF A SYSTEM

The following transformations can be applied to a system of equations to get an equivalent system:

1. exchanging any two equations;
2. multiplying both sides of an equation by any nonzero real number;
3. adding to a multiple of any equation a multiple of some other equation.

Use of these transformations leads to an equivalent system because each transformation can be reversed or "undone," allowing a return to the original system.

The Echelon Method A systematic approach for solving systems of equations using the three transformations is called the **echelon method.** The goal of the echelon method is to use the transformations to rewrite the equations of the system until the system has a triangular form. For a system of three equations in three variables, for example, the system should have the form

$$x + ay + bz = c$$
$$y + dz = e$$
$$z = f,$$

where a, b, c, d, e, and f are constants. Then the value of z from the third equation can be substituted into the second equation to find y, and the values of y and z can be substituted into the first equation to find x. This is called **back-substitution.**

EXAMPLE 1 Solve the system

$$2x + 3y = 12 \tag{1}$$
$$3x - 4y = 1. \tag{2}$$

We will first use transformation 3 to eliminate the x term from equation (2). Multiply equation (1) by 3 and add the results to -2 times equation (2).

$$6x + 9y + (-6x + 8y) = 36 + (-2)$$
$$17y = 34$$

We will indicate this process by the notation $3R_1 + (-2)R_2 \rightarrow R_2$. (R stands for row.) The new system is

$$2x + 3y = 12 \tag{1}$$
$$3R_1 + (-2)R_2 \rightarrow R_2 \qquad 17y = 34. \tag{3}$$

Now use transformation 2 to make the coefficient of the first term in each row equal to 1. Here, we must multiply equation (1) by 1/2 and equation (3) by 1/17 to accomplish this.

We get the system

$$\tfrac{1}{2}R_1 \rightarrow R_1 \qquad x + \tfrac{3}{2}y = 6$$
$$\tfrac{1}{17}R_2 \rightarrow R_2 \qquad y = 2.$$

Back-substitution gives

$$x + \tfrac{3}{2}(2) = 6$$
$$x + 3 = 6$$
$$x = 3.$$

The solution of the system is (3, 2). The graphs of the two equations in Figure 2 suggest that (3, 2) indeed satisfies both equations in the system. ◀

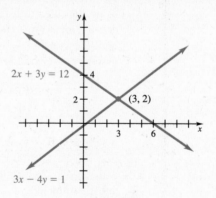

FIGURE 2

EXAMPLE 2 Solve the system

$$3x - 2y = 4 \tag{1}$$
$$-6x + 4y = 7. \tag{2}$$

Eliminate x in equation (2) to get the system

$$3x - 2y = 4 \tag{1}$$
$$2R_1 + 1R_2 \rightarrow R_2 \qquad 0 = 15. \tag{3}$$

In equation (3), both variables have been eliminated, leaving a false statement. This is a signal that these two equations have no common solution. This system is inconsistent and has no solution. As Figure 3 shows, the graph of the system is made up of two distinct parallel lines. ◀

There is no point where the two lines intersect.

$-6x + 4y = 7$

$3x - 2y = 4$

FIGURE 3

EXAMPLE 3 ▶ Solve the system

$$-4x + y = 2 \qquad \text{(1)}$$
$$8x - 2y = -4. \qquad \text{(2)}$$

Use transformation 3 to eliminate x in equation (2), getting the system

$$-4x + y = 2 \qquad \text{(1)}$$
$$2R_1 + R_2 \rightarrow R_2 \qquad 0 = 0. \qquad \text{(3)}$$

The system becomes

$$x - \frac{1}{4}y = -\frac{1}{2} \qquad \text{Multiply by } -\tfrac{1}{4}. \qquad \text{(4)}$$
$$0 = 0. \qquad \text{(5)}$$

The true statement in equation (5) indicates that the two equations have the same graph, which means that there is an infinite number of solutions for the system. For each value of y, $x = (1/4)y - (1/2)$ from equation (4), and all the ordered pairs of the form

$$\left(\frac{1}{4}y - \frac{1}{2}, y \right)$$

are solutions. For example, some solutions are $(\tfrac{1}{4}(6) - \tfrac{1}{2}, 6) = (1, 6)$, $(0, 2)$, and $(-3, -10)$. See Figure 4. ◀

There are infinitely many points where the two lines intersect.

$-4x + y = 2$
and
$8x - 2y = -4$

FIGURE 4

Additional steps are needed to solve a system with three equations in three unknowns by the echelon method.

EXAMPLE 4

Solve the system

$$2x + y - z = 2 \tag{1}$$
$$x + 3y + 2z = 1 \tag{2}$$
$$x + y + z = 2. \tag{3}$$

As in the previous examples, we begin by eliminating the term with x, this time from equations (2) and (3), as follows.

$$2x + y - z = 2 \tag{1}$$
$$1R_1 + (-2)R_2 \rightarrow R_2 \qquad -5y - 5z = 0 \tag{4}$$
$$1R_1 + (-2)R_3 \rightarrow R_3 \qquad -y - 3z = -2 \tag{5}$$

In the same way, use equation (4) to eliminate y in equation (5). The new system is

$$2x + y - z = 2$$
$$-5y - 5z = 0$$
$$1R_2 + (-5)R_3 \rightarrow R_3 \qquad 10z = 10.$$

Make the coefficient of the first term in each equation equal 1.

$$\frac{1}{2}R_1 \rightarrow R_1 \qquad x + \frac{1}{2}y - \frac{1}{2}z = 1 \tag{6}$$
$$\left(-\frac{1}{5}\right)R_2 \rightarrow R_2 \qquad y + z = 0 \tag{7}$$
$$\frac{1}{10}R_3 \rightarrow R_3 \qquad z = 1$$

Substitute 1 for z in equation (7) to get $y = -1$. Finally, substitute 1 for z and -1 for y in equation (6) to get $x = 2$. The solution of the system is $(2, -1, 1)$. Note the triangular form of the last system. This is the typical echelon form. ◀

In summary, to solve a linear system in n variables by the echelon method, perform the following steps using the three transformations given earlier.

THE ECHELON METHOD OF SOLVING A LINEAR SYSTEM

1. If possible, arrange the equations so that there is an x_1 term in the first equation, an x_2 term in the second equation, and so on.
2. Eliminate the x_1 term in all equations after the first equation.
3. Eliminate the x_2 term in all equations after the second equation.
4. Eliminate the x_3 term in all equations after the third equation.
5. Continue in this way until the last equation has the form $ax_n = k$, for constants a and k.
6. Multiply each equation by the reciprocal of the coefficient of its first term.
7. Use back-substitution to find the value of each variable.

Systems of Equations with No Unique Solution The systems of equations discussed so far have had the same number of equations as variables. Such systems have either one solution, no solution, or infinitely many solutions. Some mathematical models lead to systems of equations with fewer equations than variables, however. Such systems have either infinitely many solutions or no solution.

EXAMPLE 5 Solve the system

$$2x + 3y + 4z = 6 \tag{1}$$
$$x - 2y + z = 9 \tag{2}$$
$$3x - 6y + 3z = 27. \tag{3}$$

Since the third equation here is a multiple of the second equation, this system really has only the two equations

$$2x + 3y + 4z = 6 \tag{1}$$
$$x - 2y + z = 9. \tag{2}$$

Use the steps of the echelon method as far as possible. Eliminate the x term in equation (2).

$$2x + 3y + 4z = 6 \tag{1}$$
$1R_1 + (-2)R_2 \to R_2$
$$7y + 2z = -12 \tag{3}$$

Since there are only two equations, it is not possible to continue with the echelon method as before. To complete the solution, make the coefficient of the first term in each equation equal 1.

$\dfrac{1}{2}R_1 \to R_1$
$$x + \frac{3}{2}y + 2z = 3 \tag{4}$$

$\dfrac{1}{7}R_2 \to R_2$
$$y + \frac{2}{7}z = -\frac{12}{7} \tag{5}$$

Solve equation (5) for y in terms of z.

$$y = -\frac{2}{7}z - \frac{12}{7}$$

Substitute the expression for y in equation (4) and solve the equation for x.

$$x + \frac{3}{2}\left(-\frac{2}{7}z - \frac{12}{7}\right) + 2z = 3$$

$$x - \frac{3}{7}z - \frac{18}{7} + 2z = 3$$

$$x + \frac{11}{7}z = \frac{39}{7}$$

$$x = -\frac{11}{7}z + \frac{39}{7}$$

The solution is

$$\left(-\frac{11}{7}z + \frac{39}{7}, \; -\frac{2}{7}z - \frac{12}{7}, \; z\right).$$

Each choice of a value for z leads to values for x and y.

If $z = 1$, then $x = -\frac{11}{7} \cdot 1 + \frac{39}{7} = 4$ and $y = -\frac{2}{7} \cdot 1 - \frac{12}{7} = -2,$

so one solution is $(4, -2, 1)$.

If $z = -6$, then $x = 15$ and $y = 0$,

so another solution is $(15, 0, -6)$.

If $z = 0$, then $x = \frac{39}{7}$ and $y = -\frac{12}{7},$

so a third solution is $(39/7, -12/7, 0)$. ◀

Since both x and y in Example 5 were expressed in terms of z, the variable z is called a **parameter.** If we solved the system in a different way, x or y could be the parameter.

Some models produce a system of equations with *more* equations than variables. One example of a model of this type of system is discussed in Section 8.2, and a method of solution is shown there.

Applications The mathematical techniques in this text will be useful to you only if you are able to apply them to practical problems. To do this, always begin by reading the problem carefully. Next, identify what must be found. Let each unknown quantity be represented by a variable. (It is a good idea to *write down* exactly what each variable represents.) Now reread the problem, looking for all necessary data. Write those down, too. Finally, look for one or more sentences that lead to equations or inequalities. The next example illustrates these steps.

EXAMPLE 6

A company produces three color television sets: models X, Y, and Z. Each model X set requires 2 hours of electronics work, 2 hours of assembly time, and 1 hour of finishing time. Each model Y requires 1, 3, and 1 hours of electronics, assembly, and finishing time, respectively. Each model Z requires 3, 2, and 2 hours of the same work, respectively. There are 100 hours available for electronics, 100 hours available for assembly, and 65 hours available for finishing per week. How many of each model should be produced each week if all available time must be used?

Let $x =$ the number of model X produced per week;

$y =$ the number of model Y produced per week;

$z =$ the number of model Z produced per week.

A chart is useful for organizing the information in a problem of this type.

	Model X	Model Y	Model Z	Totals
Hours of Electronics Work	2	1	3	100
Hours of Assembly Time	2	3	2	100
Hours of Finishing Time	1	1	2	65

The x model X sets require $2x$ hours of electronics, the y model Y sets require $1y$ (or y) hours of electronics, and the z model Z sets require $3z$ hours of electronics. Since 100 hours are available for electronics,

$$2x + y + 3z = 100.$$

Similarly, from the fact that 100 hours are available for assembly,

$$2x + 3y + 2z = 100,$$

and the fact that 65 hours are available for finishing leads to the equation

$$x + y + 2z = 65.$$

Solve the system

$$2x + y + 3z = 100$$
$$2x + 3y + 2z = 100$$
$$x + y + 2z = 65$$

to find $x = 15$, $y = 10$, and $z = 20$. The company should produce 15 model X, 10 model Y, and 20 model Z sets per week. ◀

Notice the advantage of setting up the chart in Example 6. By reading across, we can easily determine the coefficients and the constants in the system.

2.1 Exercises

Use the echelon method to solve each system of two equations in two unknowns. Check your answers.

1. $x + y = 9$
$\quad 2x - y = 0$

2. $4x + y = 9$
$\quad 3x - y = 5$

3. $5x + 3y = 7$
$\quad 7x - 3y = -19$

4. $2x + 7y = -8$
$\quad -2x + 3y = -12$

5. $3x + 2y = -6$
$\quad 5x - 2y = -10$

6. $-6x + 2y = 8$
$\quad 5x - 2y = -8$

7. $2x - 3y = -7$
$\quad 5x + 4y = 17$

8. $4m + 3n = -1$
$\quad 2m + 5n = 3$

9. $5p + 7q = 6$
$\quad 10p - 3q = 46$

10. $12s - 5t = 9$
$\quad 3s - 8t = -18$

11. $6x + 7y = -2$
$\quad 7x - 6y = 26$

12. $2a + 9b = 3$
$\quad 5a + 7b = -8$

13. $3x + 2y = 5$
$6x + 4y = 8$

14. $9x - 5y = 1$
$-18x + 10y = 1$

15. $4x - y = \quad 9$
$-8x + 2y = -18$

16. $3x + 5y + 2 = 0$
$9x + 15y + 6 = 0$

17. An inconsistent system has _____ solutions.

18. The solution of a 2×2 system with two dependent equations is _____ .

Use the echelon method to solve each system. Check your answers.

19. $\dfrac{x}{2} + \dfrac{y}{3} = 8$

$\dfrac{2x}{3} + \dfrac{3y}{2} = 17$

20. $\dfrac{x}{5} + 3y = 31$

$2x - \dfrac{y}{5} = 8$

21. $\dfrac{x}{2} + y = \dfrac{3}{2}$

$\dfrac{x}{3} + y = \dfrac{1}{3}$

22. $x + \dfrac{y}{3} = -6$

$\dfrac{x}{5} + \dfrac{y}{4} = -\dfrac{7}{4}$

Use the echelon method to solve each system of three equations in three unknowns. Check your answers.

23. $x + y + z = \quad 2$
$2x + y - z = \quad 5$
$x - y + z = -2$

24. $2x + y + z = 9$
$-x - y + z = 1$
$3x - y + z = 9$

25. $x + 3y + 4z = 14$
$2x - 3y + 2z = 10$
$3x - y + z = \quad 9$

26. $4x - y + 3z = -2$
$3x + 5y - z = \quad 15$
$-2x + y + 4z = \quad 14$

27. $x + 2y + 3z = 8$
$3x - y + 2z = 5$
$-2x - 4y - 6z = 5$

28. $3x - 2y - 8z = \quad 1$
$9x - 6y - 24z = -2$
$x - y + z = \quad 1$

29. $2x - 4y + z = -4$
$x + 2y - z = \quad 0$
$-x + y + z = \quad 6$

30. $4x - 3y + z = 9$
$3x + 2y - 2z = 4$
$x - y + 3z = 5$

31. $x + 4y - z = \quad 6$
$2x - y + z = \quad 3$
$3x + 2y + 3z = 16$

32. $3x + y - z = \quad 7$
$2x - 3y + z = -7$
$x - 4y + 3z = -6$

33. $5m + n - 3p = -6$
$2m + 3n + p = \quad 5$
$-3m - 2n + 4p = \quad 3$

34. $2r - 5s + 4t = -35$
$5r + 3s - t = \quad 1$
$r + s + t = \quad 1$

35. $a - 3b - 2c = -3$
$3a + 2b - c = \quad 12$
$-a - b + 4c = \quad 3$

36. $2x + 2y + 2z = \quad 6$
$3x - 3y - 4z = -1$
$x + y + 3z = \quad 11$

37. In your own words, describe the echelon method as used to solve a system of three equations in three variables.

Solve each of the following systems of equations. Let z be the parameter in Exercises 38–41 and w be the parameter in Exercises 42 and 43.

38. $5x + 3y + 4z = \quad 19$
$3x - y + z = -4$

39. $3x + y - z = \quad 0$
$2x - y + 3z = -7$

40. $x + 2y + 3z = 11$
$2x - y + z = \quad 2$

41. $-x + y - z = -7$
$2x + 3y + z = \quad 7$

42. $x + y - z + 2w = -20$
$2x - y + z + w = \quad 11$
$3x - 2y + z - 2w = \quad 27$

43. $4x - 3y + z + w = \quad 21$
$-2x - y + 2z + 7w = \quad 2$
$x + 4y + 2z - w = -12$

44. The examples in this section did not use the first transformation. How might this transformation be used in the echelon method?

Applications

Business and Economics

45. Production The Gonzalez Company makes ROM chips and RAM chips for computers. Both require time on two assembly lines. Each unit of ROM chips requires 1 hr on line A and 2 hr on line B. Each unit of RAM chips re-
quires 3 hr on line A and 1 hr on line B. Both assembly lines operate 15 hr per day. How many units of each product can be produced in a day under these conditions?

46. Production A company produces two models of bicycles, model 201 and model 301. Model 201 requires 2 hr of assembly time, and model 301 requires 3 hr of assembly time. The parts for model 201 cost $25 per bike, and the parts for model 301 cost $30 per bike. If the company has a total of 34 hr of assembly time and $365 available per day for these two models, how many of each can be made in a day?

47. Banking A bank teller has a total of 70 bills in five-, ten- and twenty-dollar denominations. The number of fives is three times the number of tens, while the total value of the money is $960. Find the number of each type of bill.

48. Investments Nui invests $10,000, received from her grandmother, in three ways. With one part, she buys mutual funds that offer a return of 6.5% per year. She uses the second part, which amounts to twice the first, to buy government bonds at 6% per year. She puts the rest in the bank at 5% annual interest. The first year her investments bring a return of $605. How much did she invest in each way?

49. (a) Loans To get the necessary funds for a planned expansion, a small company took out three loans totaling $25,000. Company owners were able to borrow some of the money at 13%. They borrowed $2000 more than one-half the amount of the 13% loan at 14%, and the rest at 12%. The total annual interest on the loans was $3240. How much did they borrow at each rate?

(b) Suppose we drop the condition that the amount borrowed at 14% is $2000 more than one-half the amount borrowed at 13%. How is the solution changed?

(c) Suppose the company can borrow only $6000 at 12%. Is a solution possible that still meets the conditions given in part (a)?

Life Sciences

50. (a) Dietetics A hospital dietician is planning a special diet for a certain patient. The total amount per meal of food groups A, B, and C must equal 400 grams. The diet should include one-third as much of group A as of group B, and the sum of the amounts of group A and group C should equal twice the amount of group B. How many grams of each food group should be included?

(b) Suppose we drop the requirement that the diet include one-third as much of group A as of group B. Describe the set of all possible solutions.

(c) Suppose that, in addition to the conditions given in part (a), foods A and B cost 2 cents per gram and food C costs 3 cents per gram, and that a meal must cost $8. Is a solution possible?

2.2 SOLUTION OF LINEAR SYSTEMS BY THE GAUSS-JORDAN METHOD

How can an auto manufacturer with more than one factory and several dealers decide how many cars to send to each dealer from each factory?

Questions like this are called *transportation problems*; they frequently lead to a system of equations that must be satisfied. In this section we use a further refinement of the echelon method to answer this question. When we use the echelon method, since the variables are always the same, we really need to keep track of just the coefficients and the constants. For example, look at the system solved in Example 4 of the previous section.

$$2x + y - z = 2$$
$$x + 3y + 2z = 1$$
$$x + y + z = 2$$

This system can be written in an abbreviated form as

$$\begin{bmatrix} 2 & 1 & -1 & 2 \\ 1 & 3 & 2 & 1 \\ 1 & 1 & 1 & 2 \end{bmatrix}.$$

Such a rectangular array of numbers enclosed by brackets is called a **matrix** (plural: **matrices**). Each number in the array is an **element** or **entry.** To separate the constants in the last column of the matrix from the coefficients of the variables, we use a vertical line, producing the following **augmented matrix.**

$$\left[\begin{array}{rrr|r} 2 & 1 & -1 & 2 \\ 1 & 3 & 2 & 1 \\ 1 & 1 & 1 & 2 \end{array}\right]$$

The rows of the augmented matrix can be transformed in the same way as the equations of the system, since the matrix is just a shortened form of the system. The following **row operations** on the augmented matrix correspond to the transformations of systems of equations given earlier.

ROW OPERATIONS

For any augmented matrix of a system of equations, the following operations produce the augmented matrix of an equivalent system:

1. interchanging any two rows;

2. multiplying the elements of a row by any nonzero real number;

3. adding a multiple of the elements of one row to the corresponding elements of a multiple of some other row.

Row operations, like the transformations of systems of equations, are reversible. If they are used to change matrix A to matrix B, then it is possible to use row operations to transform B back into A. In addition to their use in solving equations, row operations are very important in the simplex method to be described in Chapter 4.

In the examples in this section, we will use the same notation as in Section 1 to show the row operation used. The notation R_1 indicates row 1 of the previous matrix, for example, and $-3R_1 + R_2$ means that row 1 is multiplied by -3 and added to row 2.

By the first row operation, the matrix

$$\left[\begin{array}{rrr|r} 1 & 3 & 5 & 6 \\ 0 & 1 & 2 & 3 \\ 2 & 1 & -2 & -5 \end{array}\right] \quad \text{becomes} \quad \left[\begin{array}{rrr|r} 0 & 1 & 2 & 3 \\ 1 & 3 & 5 & 6 \\ 2 & 1 & -2 & -5 \end{array}\right] \quad \begin{array}{l} \text{Interchange} \\ R_1 \text{ and } R_2. \end{array}$$

by interchanging the first two rows. Row 3 is left unchanged.

The second row operation allows us to change

$$\left[\begin{array}{rrr|r} 1 & 3 & 5 & 6 \\ 0 & 1 & 2 & 3 \\ 2 & 1 & -2 & -5 \end{array}\right] \quad \text{to} \quad \left[\begin{array}{rrr|r} -2 & -6 & -10 & -12 \\ 0 & 1 & 2 & 3 \\ 2 & 1 & -2 & -5 \end{array}\right] \quad -2R_1$$

by multiplying the elements of row 1 of the original matrix by -2. Note that rows 2 and 3 are left unchanged.

Using the third row operation, we change

$$\begin{bmatrix} 1 & 3 & 5 & | & 6 \\ 0 & 1 & 2 & | & 3 \\ 2 & 1 & -2 & | & -5 \end{bmatrix} \quad \text{to} \quad \begin{bmatrix} -1 & 2 & 7 & | & 11 \\ 0 & 1 & 2 & | & 3 \\ 2 & 1 & -2 & | & -5 \end{bmatrix} \quad -1R_3 + R_1$$

by first multiplying each element in row 3 of the original matrix by -1 and then adding the results to the corresponding elements in the first row of that matrix. Work as follows.

$$\begin{bmatrix} 1 + 2(-1) & 3 + 1(-1) & 5 + (-2)(-1) & | & 6 + (-5)(-1) \\ 0 & 1 & 2 & | & 3 \\ 2 & 1 & -2 & | & -5 \end{bmatrix} = \begin{bmatrix} -1 & 2 & 7 & | & 11 \\ 0 & 1 & 2 & | & 3 \\ 2 & 1 & -2 & | & -5 \end{bmatrix}$$

Again rows 2 and 3 are left unchanged, *even though the elements of row 3 were used to transform row 1.*

The Gauss-Jordan Method The *Gauss-Jordan method* is an extension of the echelon method of solving systems. Before the Gauss-Jordan method can be used, the system must be in proper form: the terms with variables should be on the left and the constants on the right in each equation, with the variables in the same order in each equation. The following example illustrates the use of the Gauss-Jordan method to solve a system of equations.

EXAMPLE **1** Solve the system

$$3x - 4y = 1 \tag{1}$$
$$5x + 2y = 19. \tag{2}$$

The system is already in the proper form for use of the Gauss-Jordan method. The steps used are essentially the same as in the echelon method, but we go further by eliminating the term with y in equation (1) before using transformation 2. By doing this, the final solution can be read directly from the last matrix. To begin, change the 5 in row 2 to 0. (Notice that the same notation is used to indicate each transformation, as in the previous section.)

$$\begin{bmatrix} 3 & -4 & | & 1 \\ 5 & 2 & | & 19 \end{bmatrix} \quad \text{Given matrix}$$

$$5R_1 + (-3)R_2 \to R_2 \quad \begin{bmatrix} 3 & -4 & | & 1 \\ 0 & -26 & | & -52 \end{bmatrix}$$

Change the -4 in row 1 to 0.

$$-4R_2 + 26R_1 \to R_1 \quad \begin{bmatrix} 78 & 0 & | & 234 \\ 0 & -26 & | & -52 \end{bmatrix}$$

Change the first nonzero number in each row to 1.

$$\begin{aligned} \frac{1}{78} R_1 &\rightarrow R_1 \\ -\frac{1}{26} R_2 &\rightarrow R_2 \end{aligned} \qquad \begin{bmatrix} 1 & 0 & 3 \\ 0 & 1 & 2 \end{bmatrix}$$

The last matrix corresponds to the system

$$x = 3$$
$$y = 2,$$

so the solution can be read directly from the last column of the final matrix. Check the solution by substitution in the equations of the original system. ◀

Note If your solution does not check, the most efficient way to find the error is to substitute back through the equations that correspond to each matrix, starting with the last matrix. When you find a system that is not satisfied by your (incorrect) answers, you have probably reached the matrix just before the error occurred. Look for the error in the transformation to the next matrix.

When the Gauss-Jordan method is used to solve a system, the final matrix always will have zeros above and below the diagonal of 1's on the left of the vertical bar. To transform the matrix, it is best to work column by column from left to right. For each column, first perform the steps that give the zeros. When all columns have zeros in place, multiply each row by the reciprocal of the coefficient of the remaining nonzero number in that row to get the required 1's. With dependent equations or inconsistent systems, it will not be possible to get the complete diagonal of 1's.

EXAMPLE 2 Use the Gauss-Jordan method to solve the system

$$x + 5z = -6 + y$$
$$3x + 3y = 10 + z$$
$$x + 3y + 2z = 5.$$

First, rewrite the system in proper form as follows.

$$\begin{aligned} x - y + 5z &= -6 \\ 3x + 3y - z &= 10 \\ x + 3y + 2z &= 5 \end{aligned}$$

Begin the solution by writing the augmented matrix of the linear system.

$$\begin{bmatrix} 1 & -1 & 5 & -6 \\ 3 & 3 & -1 & 10 \\ 1 & 3 & 2 & 5 \end{bmatrix}$$

Row transformations will be used to rewrite this matrix in the form

$$\begin{bmatrix} 1 & 0 & 0 & | & m \\ 0 & 1 & 0 & | & n \\ 0 & 0 & 1 & | & p \end{bmatrix},$$

where m, n, and p are real numbers. From this final form of the matrix, the solution can be read: $x = m$, $y = n$, $z = p$, or (m, n, p).

In the first column, we need zeros in the second and third rows. Multiply the first row by -3 and add to the second row to get a zero there. Then multiply the first row by -1 and add to the third row to get that zero.

$$\begin{array}{c} -3R_1 + 1R_2 \rightarrow R_2 \\ -1R_1 + 1R_3 \rightarrow R_3 \end{array} \qquad \begin{bmatrix} 1 & -1 & 5 & | & -6 \\ 0 & 6 & -16 & | & 28 \\ 0 & 4 & -3 & | & 11 \end{bmatrix}$$

Now get zeros in the second column in a similar way. We want zeros in the first and third rows. Row two will not change.

$$\begin{array}{c} 1R_2 + 6R_1 \rightarrow R_1 \\ \\ 2R_2 + (-3)R_3 \rightarrow R_3 \end{array} \qquad \begin{bmatrix} 6 & 0 & 14 & | & -8 \\ 0 & 6 & -16 & | & 28 \\ 0 & 0 & -23 & | & 23 \end{bmatrix}$$

Next, we want zeros in the first and second rows of the third column. Row three will not change.

$$\begin{array}{c} 14R_3 + 23R_1 \rightarrow R_1 \\ 16R_3 + (-23)R_2 \rightarrow R_2 \end{array} \qquad \begin{bmatrix} 138 & 0 & 0 & | & 138 \\ 0 & -138 & 0 & | & -276 \\ 0 & 0 & -23 & | & 23 \end{bmatrix}$$

Finally, get 1's in each row by multiplying the row by the reciprocal of (or dividing the row by) the number in the diagonal position.

$$\begin{array}{c} (1/138)R_1 \rightarrow R_1 \\ (-1/138)R_2 \rightarrow R_2 \\ (-1/23)R_3 \rightarrow R_3 \end{array} \qquad \begin{bmatrix} 1 & 0 & 0 & | & 1 \\ 0 & 1 & 0 & | & 2 \\ 0 & 0 & 1 & | & -1 \end{bmatrix}$$

The linear system associated with the final augmented matrix is

$$\begin{aligned} x &= 1 \\ y &= 2 \\ z &= -1, \end{aligned}$$

and the solution is $(1, 2, -1)$. ◄

In summary, the Gauss-Jordan method of solving a linear system requires the following steps.

THE GAUSS-JORDAN METHOD OF SOLVING A LINEAR SYSTEM

1. Write each equation so that variable terms are in the same order on the left side of the equals sign and constants are on the right.

2. Write the augmented matrix that corresponds to the system.

3. Use row operations to transform the first column so that all elements except the element in the first row are zero.

4. Use row operations to transform the second column so that all elements except the element in the second row are zero.

5. Use row operations to transform the third column so that all elements except the element in the third row are zero.

6. Continue in this way until the last row is written in the form

$$[0 \quad 0 \quad 0 \quad \cdots \quad 0 \quad j \,|\, k],$$

where j and k are constants.

7. Multiply each row by the reciprocal of the nonzero element in that row.

Systems with Non-Unique Solutions If some of the given steps cannot be performed, the solution cannot be completed with the Gauss-Jordan method. The next examples illustrate how such cases may be handled.

EXAMPLE 3 Use the Gauss-Jordan method to solve the system

$$x + y = 2$$
$$2x + 2y = 5.$$

Begin by writing the augmented matrix.

$$\begin{bmatrix} 1 & 1 & | & 2 \\ 2 & 2 & | & 5 \end{bmatrix}$$

To get a zero for the second element in column 1, multiply the numbers in row 1 by -2 and add the results to the corresponding elements in row 2.

$$-2R_1 + R_2 \rightarrow R_2 \qquad \begin{bmatrix} 1 & 1 & | & 2 \\ 0 & 0 & | & 1 \end{bmatrix}$$

This matrix corresponds to the system

$$x + y = 2$$
$$0x + 0y = 1.$$

Since the second equation is $0 = 1$, the system is inconsistent and therefore has no solution. The row $[0 \quad 0 \,|\, 1]$ is a signal that the given system is inconsistent. ◀

EXAMPLE 4

Use the Gauss-Jordan method to solve the system

$$x + 2y - z = 0$$
$$3x - y + z = 6$$
$$-2x - 4y + 2z = 0.$$

The augmented matrix is

$$\begin{bmatrix} 1 & 2 & -1 & \bigm| & 0 \\ 3 & -1 & 1 & \bigm| & 6 \\ -2 & -4 & 2 & \bigm| & 0 \end{bmatrix}.$$

Get zeros in the second and third rows of column 1.

$$\begin{array}{c} -3R_1 + R_2 \rightarrow R_2 \\ 2R_1 + R_3 \rightarrow R_3 \end{array} \quad \begin{bmatrix} 1 & 2 & -1 & \bigm| & 0 \\ 0 & -7 & 4 & \bigm| & 6 \\ 0 & 0 & 0 & \bigm| & 0 \end{bmatrix}$$

To continue, get a zero in the first row of column 2 using the second row, as usual.

$$2R_2 + 7R_1 \rightarrow R_1 \quad \begin{bmatrix} 7 & 0 & 1 & \bigm| & 12 \\ 0 & -7 & 4 & \bigm| & 6 \\ 0 & 0 & 0 & \bigm| & 0 \end{bmatrix}$$

We cannot get a zero for the first row, third column element without changing the form of the first two columns. We must multiply each of the first two rows by the reciprocal of the first nonzero number.

$$\begin{array}{c} \frac{1}{7}R_1 \rightarrow R_1 \\[2mm] -\frac{1}{7}R_2 \rightarrow R_2 \end{array} \quad \begin{bmatrix} 1 & 0 & \frac{1}{7} & \bigm| & \frac{12}{7} \\[2mm] 0 & 1 & -\frac{4}{7} & \bigm| & -\frac{6}{7} \\[2mm] 0 & 0 & 0 & \bigm| & 0 \end{bmatrix}$$

To complete the solution, write the equations that correspond to the first two lines of the matrix.

$$x + \frac{1}{7}z = \frac{12}{7}$$

$$y - \frac{4}{7}z = -\frac{6}{7}$$

Since both equations involve z, solve the first equation for x and the second equation for y to get

$$x = -\frac{1}{7}z + \frac{12}{7} \quad \text{and} \quad y = \frac{4}{7}z - \frac{6}{7}.$$

As shown in the previous section, the general solution is

$$\left(-\frac{1}{7}z + \frac{12}{7}, \frac{4}{7}z - \frac{6}{7}, z\right). \quad \blacktriangleleft$$

Although the examples have used only systems with two equations in two unknowns or three equations in three unknowns, the Gauss-Jordan method can be used for any system with n equations and n unknowns. In fact, it can be used with n equations and m unknowns, as in Example 4. The system in Example 4 actually had just two equations with three unknowns. The method does become tedious even with three equations in three unknowns; however, it is very suitable for use by computers. A computer can produce the solution to a fairly large system very quickly. Most computer programs get the matrix in the echelon form working column by column from the first column on. The 1 in each column is produced first, then the zeros in that column. We produced the 1's in the last step to avoid fractions or decimals, which make the work tedious to do by hand, but which cause a computer no difficulty.

2.2 Exercises

Write the augmented matrix for each of the following systems. **Do not solve.**

1. $2x + 3y = 11$
 $\quad x + 2y = 8$

2. $3x + 5y = -13$
 $\quad 2x + 3y = -9$

3. $2x + y + z = 3$
 $\quad 3x - 4y + 2z = -7$
 $\quad x + y + z = 2$

4. $4x - 2y + 3z = 4$
 $\quad 3x + 5y + z = 7$
 $\quad 5x - y + 4z = 7$

Write the system of equations associated with each of the following augmented matrices. **Do not solve.**

5. $\begin{bmatrix} 1 & 0 & 2 \\ 0 & 1 & 3 \end{bmatrix}$

6. $\begin{bmatrix} 1 & 0 & 5 \\ 0 & 1 & -3 \end{bmatrix}$

7. $\begin{bmatrix} 1 & 0 & 0 & 2 \\ 0 & 1 & 0 & 3 \\ 0 & 0 & 1 & -2 \end{bmatrix}$

8. $\begin{bmatrix} 1 & 0 & 0 & 4 \\ 0 & 1 & 0 & 2 \\ 0 & 0 & 1 & 3 \end{bmatrix}$

9. _____ on a matrix correspond to transformations of a system of equations.

10. Describe in your own words what $2R_1 + R_3 \rightarrow R_3$ means.

Use the indicated row operations to change each matrix.

11. Replace R_2 by $R_1 + (-2)R_2$.

$\begin{bmatrix} 2 & 3 & 8 & 20 \\ 1 & 4 & 6 & 12 \\ 0 & 3 & 5 & 10 \end{bmatrix}$

12. Replace R_3 by $-1R_1 + 3R_3$.

$\begin{bmatrix} 3 & 2 & 6 & 18 \\ 2 & -2 & 5 & 7 \\ 1 & 0 & 5 & 20 \end{bmatrix}$

13. Replace R_1 by $-4R_2 + R_1$.

$\begin{bmatrix} 1 & 4 & 2 & 9 \\ 0 & 1 & 5 & 14 \\ 0 & 3 & 8 & 16 \end{bmatrix}$

14. Replace R_1 by $R_3 + (-3)R_1$.

$\begin{bmatrix} 1 & 0 & 4 & 21 \\ 0 & 6 & 5 & 30 \\ 0 & 0 & 12 & 15 \end{bmatrix}$

15. Replace R_1 by $\frac{1}{3}R_1$.

$\begin{bmatrix} 3 & 0 & 0 & 18 \\ 0 & 5 & 0 & 9 \\ 0 & 0 & 4 & 8 \end{bmatrix}$

16. Replace R_3 by $\frac{1}{4}R_3$.

$\begin{bmatrix} 1 & 0 & 0 & 6 \\ 0 & 1 & 0 & 5 \\ 0 & 0 & 4 & 12 \end{bmatrix}$

Use the Gauss-Jordan method to solve the following systems of equations.

17. $x + y = 5$
$x - y = -1$

18. $x + 2y = 5$
$2x + y = -2$

19. $3x + 3y = -9$
$2x - 5y = -6$

20. $3x - 2y = 4$
$3x + y = -2$

21. $2x = 10 + 3y$
$2y = 5 - 2x$

22. $y = 5 - 4x$
$2x = 3 - y$

23. $2x - 5y = 10$
$4x - 5y = 15$

24. $4x - 2y = 3$
$-2x + 3y = 1$

25. $2x - 3y = 2$
$4x - 6y = 1$

26. $x + 2y = 1$
$2x + 4y = 3$

27. $6x - 3y = 1$
$-12x + 6y = -2$

28. $x - y = 1$
$-x + y = -1$

29. $2x - 2y = -2$
$y + z = 4$
$x + z = 1$

30. $x - z = -3$
$y + z = 9$
$-x + z = 3$

31. $4x + 4y - 4z = 24$
$2x - y + z = -9$
$x - 2y + 3z = 1$

32. $x + 3y - 6z = 7$
$2x - y + 2z = 0$
$x + y + 2z = -1$

33. $y = x - 1$
$y = 6 + z$
$z = -1 - x$

34. $x = 1 - y$
$2x = z$
$2z = -2 - y$

35. $3x + 5y - z = 0$
$4x - y + 2z = 1$
$-6x - 10y + 2z = 0$

36. $3x - 6y + 3z = 15$
$2x + y - z = 2$
$-2x + 4y - 2z = 2$

37. $2x + 3y + z = 9$
$4x - y + 3z = -1$
$6x + 2y - 4z = -8$

38. $3x + 2y - z = -16$
$6x - 4y + 3z = 12$
$3x + 3y + z = -11$

39. $5x - 4y + 2z = 4$
$5x + 3y - z = 17$
$15x - 5y + 3z = 25$

40. $4x - 2y - 3z = -23$
$-4x + 3y + z = 11$
$8x - 5y + 4z = 6$

41. $x + 2y - w = 3$
$2x + 4z + 2w = -6$
$x + 2y - z = 6$
$2x - y + z + w = -3$

42. $x + 3y - 2z - w = 9$
$2x + 4y + 2w = 10$
$-3x - 5y + 2z - w = -15$
$x - y - 3z + 2w = 6$

Applications

Business and Economics

43. Transportation An auto manufacturer sends cars from two plants, I and II, to dealerships A and B located in a midwestern city. Plant I has a total of 28 cars to send, and plant II has 8. Dealer A needs 20 cars, and dealer B needs 16. Transportation costs per car, based on the distance of each dealership from each plant, are $220 from I to A, $300 from I to B, $400 from II to A, and $180 from II to B. The manufacturer wants to limit transportation costs to $10,640. How many cars should be sent from each plant to each of the two dealerships? Use the Gauss-Jordan method to find the solution.

44. Transportation A knitting shop orders yarn from three suppliers, I, II, and III. One month the shop ordered a total of 100 units of yarn from these suppliers. The delivery costs were $80, $50, and $65 per unit for the orders from suppliers I, II, and III, respectively, with total delivery costs of $5990. The shop ordered the same amount from suppliers I and III. How many units were ordered from each supplier? Use the Gauss-Jordan method to find the solution.

45. Transportation A manufacturer purchases a part for use at both of its plants—one at Roseville, California, the other at Akron, Ohio. The part is available in limited quantities from two suppliers. Each supplier has 75 units available. The Roseville plant needs 40 units, and the Akron plant requires 75 units. The first supplier charges $70 per unit delivered to Roseville and $90 per unit delivered to Akron. Corresponding costs from the second supplier are $80 and $120. The manufacturer wants to order a total of 75 units from the first, less expensive, supplier, with the remaining 40 units to come from the second supplier. If the company spends $10,750 to purchase the required number of units

for the two plants, find the number of units that should be purchased from each supplier for each plant as follows:

(a) Assign variables to the four unknowns.

(b) Write a system of five equations with the four variables. (Not all equations will involve all four variables.)

(c) Use the Gauss-Jordan method to solve the system of equations.

Social Sciences

46. Traffic Control At rush hours, substantial traffic congestion is encountered at the traffic intersections shown in the figure. (The streets are one-way, as shown by the arrows.)

The city wishes to improve the signals at these corners so as to speed the flow of traffic. The traffic engineers first gather data. As the figure shows, 700 cars per hour come down M Street to intersection A, and 300 cars per hour come down 10th Street to intersection A. A total of x_1 of these cars leave A on M Street, and x_4 cars leave A on 10th Street. The number of cars entering A must equal the number leaving, so that

$$x_1 + x_4 = 700 + 300$$

or

$$x_1 + x_4 = 1000.$$

For intersection B, x_1 cars enter on M street and x_2 on 11th Street. The figure shows that 900 cars leave B on 11th and 200 on M. We have

$$x_1 + x_2 = 900 + 200$$
$$x_1 + x_2 = 1100.$$

At intersection C, 400 cars enter on N Street and 300 on 11th Street, while x_2 leave on 11th Street and x_3 on N Street. This gives

$$x_2 + x_3 = 400 + 300$$
$$x_2 + x_3 = 700.$$

Finally, intersection D has x_3 cars entering on N and x_4 on 10th. There are 400 leaving D on 10th and 200 on N, so that

$$x_3 + x_4 = 400 + 200$$
$$x_3 + x_4 = 600.$$

(a) Use the four equations to set up an augmented matrix, and then use the Gauss-Jordan method to solve it. (*Hint:* Keep going until you get a row of all zeros.)

(b) Since you got a row of all zeros, the system of equations does not have a unique solution. Write three equations, corresponding to the three nonzero rows of the matrix.

(c) Solve each of the equations for x_4.

(d) One of your equations should have been $x_4 = 1000 - x_1$. What is the largest possible value of x_1 so that x_4 is not negative? What is the largest value of x_4 so that x_1 is not negative?

(e) Your second equation should have been $x_4 = x_2 - 100$. Find the smallest possible value of x_2 so that x_4 is not negative.

(f) For the third equation, $x_4 = 600 - x_3$, find the largest possible values of x_3 and x_4 so that neither variable is negative.

(g) Look at your answers for parts (d)–(f). What is the maximum value of x_4 so that all the equations are satisfied and all variables are nonnegative? Of x_3? Of x_2? Of x_1?

Computer/Graphing Calculator

If you have access to a computer and software or a graphing calculator that uses the Gauss-Jordan method, try solving the following linear systems. In Exercises 49–52, you must first set up a system of equations to solve the problem.

47. $10.47x + 3.52y + 2.58z - 6.42w = 218.65$
$8.62x - 4.93y - 1.75z + 2.83w = 157.03$
$4.92x + 6.83y - 2.97z + 2.65w = 462.3$
$2.86x + 19.10y - 6.24z - 8.73w = 398.4$

48. $28.6x + 94.5y + 16.0z - 2.94w = 198.3$
$16.7x + 44.3y - 27.3z + 8.9w = 254.7$
$12.5x - 38.7y + 92.5z + 22.4w = 562.7$
$40.1x - 28.3y + 17.5z - 10.2w = 375.4$

49. Mixing Plant Foods Natural Brand plant food is made from three chemicals. The mix must include 10.8% of the first chemical, and the other two chemicals must be in a ratio of 4 to 3 as measured by weight. How much of each chemical is required to make 750 kg of the plant food?

50. Bacterial Food Requirements Three species of bacteria are fed three foods, I, II, and III. A bacterium of the first species consumes 1.3 units each of foods I and II and 2.3 units of food III each day. A bacterium of the second species consumes 1.1 units of food I, 2.4 units of food II, and 3.7 units of food III each day. A bacterium of the third species consumes 8.1 units of I, 2.9 units of II, and 5.1 units of III each day. If 16,000 units of I, 28,000 units of II, and 44,000 units of III are supplied each day, how many of each species can be maintained in this environment?

51. Fish Food Requirements A lake is stocked each spring with three species of fish, A, B, and C. Three foods, I, II,

and III, are available in the lake. Each fish of species A requires an average of 1.32 units of food I, 2.9 units of food II, and 1.75 units of food III each day. Species B fish each require 2.1 units of food I, .95 unit of food II, and .6 unit of food III daily. Species C fish require .86, 1.52, and 2.01 units of I, II, and III per day, respectively. If 490 units of food I, 897 units of food II, and 653 units of food III are available daily, how many of each species should be stocked?

52. Packaging A company produces three combinations of mixed vegetables that sell in 1-kg packages. Italian style combines .3 kg of zucchini, .3 of broccoli, and .4 of carrots. French style combines .6 kg of broccoli and .4 of carrots. Oriental style combines .2 kg of zucchini, .5 of broccoli, and .3 of carrots. The company has a stock of 16,200 kg of zucchini, 41,400 kg of broccoli, and 29,400 kg of carrots. How many packages of each style should it prepare to use up existing supplies?

2.3 ADDITION AND SUBTRACTION OF MATRICES

? A company sends monthly shipments to its warehouses in several cities. How might the company keep track of the shipments to each warehouse most efficiently?

In the previous section, matrices were used to store information about systems of linear equations. In this section, we begin a study of matrices and show additional uses of matrix notation that will answer the question posed above. The use of matrices has gained increasing importance in the fields of management, natural science, and social science because matrices provide a convenient way to organize data, as Example 1 demonstrates.

EXAMPLE **1** The EZ Life Company manufactures sofas and armchairs in three models, A, B, and C. The company has regional warehouses in New York, Chicago, and San Francisco. In its August shipment, the company sends 10 model-A sofas, 12 model-B sofas, 5 model-C sofas, 15 model-A chairs, 20 model-B chairs, and 8 model-C chairs to each warehouse.

To organize this data, we might first list it as follows.

Sofas	10 model-A	12 model-B	5 model-C
Chairs	15 model-A	20 model-B	8 model-C

Alternatively, we might tabulate the data in a chart.

		Model		
		A	B	C
Furniture Type	Sofas	10	12	5
	Chairs	15	20	8

With the understanding that the numbers in each row refer to the furniture type (sofa, chair) and the numbers in each column refer to the model (A, B, C), the same information can be given by a matrix, as follows.

$$M = \begin{bmatrix} 10 & 12 & 5 \\ 15 & 20 & 8 \end{bmatrix} \blacktriangleleft$$

Matrices often are named with capital letters, as in Example 1. Matrices are classified by size; that is, by the number of rows and columns they contain. For example, matrix M above has two rows and three columns. This matrix is a 2×3 (read "2 by 3") matrix. By definition, a matrix with m rows and n columns is an $m \times n$ matrix. The number of rows is always given first.

EXAMPLE 2

(a) The matrix $\begin{bmatrix} 6 & 5 \\ 3 & 4 \\ 5 & -1 \end{bmatrix}$ is a 3×2 matrix.

(b) $\begin{bmatrix} 5 & 8 & 9 \\ 0 & 5 & -3 \\ -4 & 0 & 5 \end{bmatrix}$ is a 3×3 matrix.

(c) $[1 \quad 6 \quad 5 \quad -2 \quad 5]$ is a 1×5 matrix.

(d) $\begin{bmatrix} 3 \\ -5 \\ 0 \\ 2 \end{bmatrix}$ is a 4×1 matrix. \blacktriangleleft

A matrix with the same number of rows as columns is called a **square matrix.** The matrix in Example 2(b) is a square matrix.

A matrix containing only one row is called a **row matrix** or a **row vector.** The matrix in Example 2(c) is a row matrix, as are

$$[5 \quad 8], \qquad [6 \quad -9 \quad 2], \qquad \text{and} \qquad [-4 \quad 0 \quad 0 \quad 0].$$

A matrix of only one column, as in Example 2(d), is a **column matrix** or a **column vector.**

Equality for matrices is defined as follows.

MATRIX EQUALITY Two matrices are equal if they are the same size and if each pair of corresponding elements is equal.

By this definition,

$$\begin{bmatrix} 2 & 1 \\ 3 & -5 \end{bmatrix} \quad \text{and} \quad \begin{bmatrix} 1 & 2 \\ -5 & 3 \end{bmatrix}$$

are not equal (even though they contain the same elements and are the same size) since the corresponding elements differ.

EXAMPLE **3** (a) From the definition of matrix equality given above, the only way that the statement

$$\begin{bmatrix} 2 & 1 \\ p & q \end{bmatrix} = \begin{bmatrix} x & y \\ -1 & 0 \end{bmatrix}$$

can be true is if $2 = x$, $1 = y$, $p = -1$, and $q = 0$.

(b) The statement

$$\begin{bmatrix} x \\ y \end{bmatrix} = \begin{bmatrix} 1 \\ 4 \\ 0 \end{bmatrix}$$

can never be true, since the two matrices are different sizes. (One is 2×1 and the other is 3×1.) ◀

Addition The matrix given in Example 1,

$$M = \begin{bmatrix} 10 & 12 & 5 \\ 15 & 20 & 8 \end{bmatrix},$$

shows the August shipment from the EZ Life plant to each of its warehouses. If matrix N below gives the September shipment to the New York warehouse, what is the total shipment of each item of furniture to the New York warehouse for these two months?

$$N = \begin{bmatrix} 45 & 35 & 20 \\ 65 & 40 & 35 \end{bmatrix}$$

If 10 model-A sofas were shipped in August and 45 in September, then altogether $10 + 45 = 55$ model-A sofas were shipped in the two months. The other corresponding entries can be added in a similar way to get a new matrix, Q, that represents the total shipment for the two months.

$$Q = \begin{bmatrix} 55 & 47 & 25 \\ 80 & 60 & 43 \end{bmatrix}$$

It is convenient to refer to Q as the sum of M and N.

The way these two matrices were added illustrates the following definition of addition of matrices.

ADDITION OF MATRICES The **sum** of two $m \times n$ matrices X and Y is the $m \times n$ matrix $X + Y$ in which each element is the sum of the corresponding elements of X and Y.

Caution It is important to remember that only matrices that are the same size can be added.

EXAMPLE 4

Find each sum if possible.

(a) $\begin{bmatrix} 5 & -6 \\ 8 & 9 \end{bmatrix} + \begin{bmatrix} -4 & 6 \\ 8 & -3 \end{bmatrix} = \begin{bmatrix} 5 + (-4) & -6 + 6 \\ 8 + 8 & 9 + (-3) \end{bmatrix} = \begin{bmatrix} 1 & 0 \\ 16 & 6 \end{bmatrix}$

(b) The matrices

$$A = \begin{bmatrix} 5 & 8 \\ 6 & 2 \end{bmatrix} \quad \text{and} \quad B = \begin{bmatrix} 3 & 9 & 1 \\ 4 & 2 & 5 \end{bmatrix}$$

are different sizes. Therefore, the sum $A + B$ does not exist. ◄

EXAMPLE 5

The September shipments from the EZ Life Company to the New York, San Francisco, and Chicago warehouses are given in matrices N, S, and C below.

$$N = \begin{bmatrix} 45 & 35 & 20 \\ 65 & 40 & 35 \end{bmatrix} \quad S = \begin{bmatrix} 30 & 32 & 28 \\ 43 & 47 & 30 \end{bmatrix} \quad C = \begin{bmatrix} 22 & 25 & 38 \\ 31 & 34 & 35 \end{bmatrix}$$

What was the total amount shipped to the three warehouses in September?

The total of the September shipments is represented by the sum of the three matrices N, S, and C.

$$N + S + C = \begin{bmatrix} 45 & 35 & 20 \\ 65 & 40 & 35 \end{bmatrix} + \begin{bmatrix} 30 & 32 & 28 \\ 43 & 47 & 30 \end{bmatrix} + \begin{bmatrix} 22 & 25 & 38 \\ 31 & 34 & 35 \end{bmatrix}$$

$$= \begin{bmatrix} 97 & 92 & 86 \\ 139 & 121 & 100 \end{bmatrix}$$

For example, this sum shows that the total number of model-C sofas shipped to the three warehouses in September was 86. ◄

The additive inverse of the real number a is $-a$; a similar definition applies to matrices.

ADDITIVE INVERSE

The **additive inverse** (or **negative**) of a matrix X is the matrix $-X$ in which each element is the additive inverse of the corresponding element of X.

If

$$A = \begin{bmatrix} 1 & 2 & 3 \\ 0 & -1 & 5 \end{bmatrix} \quad \text{and} \quad B = \begin{bmatrix} -2 & 3 & 0 \\ 1 & -7 & 2 \end{bmatrix},$$

then by the definition of the additive inverse of a matrix,

$$-A = \begin{bmatrix} -1 & -2 & -3 \\ 0 & 1 & -5 \end{bmatrix} \quad \text{and} \quad -B = \begin{bmatrix} 2 & -3 & 0 \\ -1 & 7 & -2 \end{bmatrix}.$$

Compare this with the identity property for real numbers: for any real number a, $a + 0 = 0 + a = a$. Exercises 34–37 give other properties of matrices that are parallel to the properties of real numbers.

By the definition of matrix addition, for each matrix X the sum $X + (-X)$ is a **zero matrix,** O, whose elements are all zeros. There is an $m \times n$ zero matrix for each pair of values of m and n. Such a matrix serves as an $m \times n$ **additive identity,** similar to the additive identity 0 for any real number. Zero matrices have the following identity property.

ZERO MATRIX

If O is an $m \times n$ zero matrix, and A is any $m \times n$ matrix, then

$$A + O = O + A = A.$$

Subtraction The **subtraction** of matrices is defined in a manner comparable to subtraction of real numbers.

SUBTRACTION OF MATRICES

For two $m \times n$ matrices X and Y, the **difference $X - Y$** is the $m \times n$ matrix defined by

$$X - Y = X + (-Y).$$

This definition means that matrix subtraction can be performed by subtracting corresponding elements. For example, with A, B, and $-B$ as defined above,

$$A - B = A + (-B) = \begin{bmatrix} 1 & 2 & 3 \\ 0 & -1 & 5 \end{bmatrix} + \begin{bmatrix} 2 & -3 & 0 \\ -1 & 7 & -2 \end{bmatrix}$$

$$= \begin{bmatrix} 3 & -1 & 3 \\ -1 & 6 & 3 \end{bmatrix}.$$

EXAMPLE 6

(a) $[8 \quad 6 \quad -4] - [3 \quad 5 \quad -8] = [5 \quad 1 \quad 4]$

(b) The matrices

$$\begin{bmatrix} -2 & 5 \\ 0 & 1 \end{bmatrix} \quad \text{and} \quad \begin{bmatrix} 3 \\ 5 \end{bmatrix}$$

are different sizes and cannot be subtracted. ◄

EXAMPLE 7

During September the Chicago warehouse of the EZ Life Company shipped out the following numbers of each model.

$$K = \begin{bmatrix} 5 & 10 & 8 \\ 11 & 14 & 15 \end{bmatrix}$$

What was the Chicago warehouse inventory on October 1, taking into account only the number of items received and sent out during the month?

The number of each kind of item received during September is given by matrix C from Example 5; the number of each model sent out during September is given by matrix K. The October 1 inventory will be represented by the matrix $C - K$:

$$\begin{bmatrix} 22 & 25 & 38 \\ 31 & 34 & 35 \end{bmatrix} - \begin{bmatrix} 5 & 10 & 8 \\ 11 & 14 & 15 \end{bmatrix} = \begin{bmatrix} 17 & 15 & 30 \\ 20 & 20 & 20 \end{bmatrix}. \quad \blacktriangleleft$$

2.3 Exercises

Mark each statement as true *or* false. *If false, tell why.*

1. $\begin{bmatrix} 1 & 3 \\ 5 & 7 \end{bmatrix} = \begin{bmatrix} 1 & 5 \\ 3 & 7 \end{bmatrix}$

2. $\begin{bmatrix} 1 \\ 2 \\ 3 \end{bmatrix} = \begin{bmatrix} 1 & 2 & 3 \end{bmatrix}$

3. $\begin{bmatrix} x \\ y \end{bmatrix} = \begin{bmatrix} 3 \\ 5 \end{bmatrix}$ if $x = 3$ and $y = 5$.

4. $\begin{bmatrix} 3 & 5 & 2 & 8 \\ 1 & -1 & 4 & 0 \end{bmatrix}$ is a 4 × 2 matrix.

5. $\begin{bmatrix} 1 & 9 & -4 \\ 3 & 7 & 2 \\ -1 & 1 & 0 \end{bmatrix}$ is a square matrix.

6. $\begin{bmatrix} 2 & 4 & -1 \\ 3 & 7 & 5 \\ 0 & 0 & 0 \end{bmatrix} = \begin{bmatrix} 2 & 4 & -1 \\ 3 & 7 & 5 \end{bmatrix}$

Find the size of each matrix. Identify any square, column, or row matrices. Give the additive inverse of each matrix.

7. $\begin{bmatrix} -4 & 8 \\ 2 & 3 \end{bmatrix}$

8. $\begin{bmatrix} -9 & 6 & 2 \\ 4 & 1 & 8 \end{bmatrix}$

9. $\begin{bmatrix} -6 & 8 & 0 & 0 \\ 4 & 1 & 9 & 2 \\ 3 & -5 & 7 & 1 \end{bmatrix}$

10. $\begin{bmatrix} 8 & -2 & 4 & 6 & 3 \end{bmatrix}$

11. $\begin{bmatrix} 2 \\ 4 \end{bmatrix}$

12. $\begin{bmatrix} -9 \end{bmatrix}$

13. The sum of an $n \times m$ matrix and its additive inverse is _____ .

14. If A is a 5 × 2 matrix and $A + K = A$, what do you know about K?

Find the values of the variables in each equation.

15. $\begin{bmatrix} 2 & 1 \\ 4 & 8 \end{bmatrix} = \begin{bmatrix} x & 1 \\ y & z \end{bmatrix}$

16. $\begin{bmatrix} -5 \\ y \end{bmatrix} = \begin{bmatrix} -5 \\ 8 \end{bmatrix}$

17. $\begin{bmatrix} x + 6 & y + 2 \\ 8 & 3 \end{bmatrix} = \begin{bmatrix} -9 & 7 \\ 8 & k \end{bmatrix}$

18. $\begin{bmatrix} 9 & 7 \\ r & 0 \end{bmatrix} = \begin{bmatrix} m - 3 & n + 5 \\ 8 & 0 \end{bmatrix}$

19. $\begin{bmatrix} -7 + z & 4r & 8s \\ 6p & 2 & 5 \end{bmatrix} + \begin{bmatrix} -9 & 8r & 3 \\ 2 & 5 & 4 \end{bmatrix} = \begin{bmatrix} 2 & 36 & 27 \\ 20 & 7 & 12a \end{bmatrix}$

20. $\begin{bmatrix} a + 2 & 3z + 1 & 5m \\ 4k & 0 & 3 \end{bmatrix} + \begin{bmatrix} 3a & 2z & 5m \\ 2k & 5 & 6 \end{bmatrix} = \begin{bmatrix} 10 & -14 & 80 \\ 10 & 5 & 9 \end{bmatrix}$

Perform the indicated operations where possible.

21. $\begin{bmatrix} 1 & 2 & 5 & -1 \\ 3 & 0 & 2 & -4 \end{bmatrix} + \begin{bmatrix} 8 & 10 & -5 & 3 \\ -2 & -1 & 0 & 0 \end{bmatrix}$

22. $\begin{bmatrix} 1 & 5 \\ 2 & -3 \\ 3 & 7 \end{bmatrix} + \begin{bmatrix} 2 & 3 \\ 8 & 5 \\ -1 & 9 \end{bmatrix}$

23. $\begin{bmatrix} 1 & 3 & -2 \\ 4 & 7 & 1 \end{bmatrix} + \begin{bmatrix} 3 & 0 \\ 6 & 4 \\ -5 & 2 \end{bmatrix}$

24. $\begin{bmatrix} 1 & 3 & -2 \\ 4 & 7 & 1 \end{bmatrix} - \begin{bmatrix} 3 & 6 & -5 \\ 0 & 4 & 2 \end{bmatrix}$

25. $\begin{bmatrix} 2 & 8 & 12 & 0 \\ 7 & 4 & -1 & 5 \\ 1 & 2 & 0 & 10 \end{bmatrix} - \begin{bmatrix} 1 & 3 & 6 & 9 \\ 2 & -3 & -3 & 4 \\ 8 & 0 & -2 & 17 \end{bmatrix}$

26. $\begin{bmatrix} 2 & 1 \\ 5 & -3 \\ -7 & 2 \\ 9 & 0 \end{bmatrix} + \begin{bmatrix} 1 & -8 & 0 \\ 5 & 3 & 2 \\ -6 & 7 & -5 \\ 2 & -1 & 0 \end{bmatrix}$

27. $\begin{bmatrix} 2 & 3 \\ -2 & 4 \end{bmatrix} + \begin{bmatrix} 4 & 3 \\ 7 & 8 \end{bmatrix} - \begin{bmatrix} 3 & 2 \\ 1 & 4 \end{bmatrix}$

28. $\begin{bmatrix} 4 & 3 \\ 1 & 2 \end{bmatrix} - \begin{bmatrix} 1 & 1 \\ 1 & 0 \end{bmatrix} + \begin{bmatrix} 1 & 1 \\ 1 & 4 \end{bmatrix}$

29. $\begin{bmatrix} 1 & 5 \\ -3 & 7 \end{bmatrix} - \begin{bmatrix} 6 & 3 \\ 2 & 4 \end{bmatrix} + \begin{bmatrix} 8 & 10 \\ -1 & 0 \end{bmatrix}$

30. $\begin{bmatrix} -1 & -1 \\ -1 & 0 \end{bmatrix} + \begin{bmatrix} 4 & 3 \\ 1 & 2 \end{bmatrix} + \begin{bmatrix} 1 & 1 \\ 1 & 4 \end{bmatrix}$

31. $\begin{bmatrix} -4x + 2y & -3x + y \\ 6x - 3y & 2x - 5y \end{bmatrix} + \begin{bmatrix} -8x + 6y & 2x \\ 3y - 5x & 6x + 4y \end{bmatrix}$

32. $\begin{bmatrix} 4k - 8y \\ 6z - 3x \\ 2k + 5a \\ -4m + 2n \end{bmatrix} - \begin{bmatrix} 5k + 6y \\ 2z + 5x \\ 4k + 6a \\ 4m - 2n \end{bmatrix}$

33. For matrix X given below, find the matrix $-X$.

Using matrices $O = \begin{bmatrix} 0 & 0 \\ 0 & 0 \end{bmatrix}$, $P = \begin{bmatrix} m & n \\ p & q \end{bmatrix}$, $T = \begin{bmatrix} r & s \\ t & u \end{bmatrix}$, and $X = \begin{bmatrix} x & y \\ z & w \end{bmatrix}$, *verify the statements in Exercises 34–37.*

34. $X + T = T + X$ (commutative property of addition of matrices)

35. $X + (T + P) = (X + T) + P$ (associative property of addition of matrices)

36. $X + (-X) = O$ (inverse property of addition of matrices)

37. $P + O = P$ (identity property of addition of matrices)

38. Which of the above properties are valid for matrices that are not square?

Applications

Business and Economics

39. Management A toy company has plants in Boston, Chicago, and Seattle that manufacture toy phones and calculators. The matrix below gives the production costs (in dollars) for each item at the Boston plant:

	Phones	Calculators
Material	4.27	6.94
Labor	3.45	3.65

(a) In Chicago, a phone costs $4.05 for materials and $3.27 for labor; a calculator costs $7.01 for material and $3.51 for labor. In Seattle, material costs are $4.40 for a phone and $6.90 for a calculator; labor costs are $3.54 for a phone and $3.76 for a calculator. Write the production cost matrices for Chicago and Seattle.

(b) Assume each plant makes the same number of each item. Write a matrix that expresses the average production costs for all three plants.

(c) Suppose labor costs increase by \$.11 per item in Chicago and material costs there increase by \$.37 for a phone and \$.42 for a calculator. What is the new production cost matrix for Chicago?

(d) After the Chicago cost increases, the Boston plant is closed and production divided evenly between the other two plants. What is the matrix that now expresses the average production costs for the entire country?

40. Management There are three convenience stores in Folsom. This week, Store I sold 88 loaves of bread, 48 quarts of milk, 16 jars of peanut butter, and 112 pounds of cold cuts. Store II sold 105 loaves of bread, 72 quarts of milk, 21 jars of peanut butter, and 147 pounds of cold cuts. Store III sold 60 loaves of bread, 40 quarts of milk, no peanut butter, and 50 pounds of cold cuts.

(a) Use a 3×4 matrix to express the sales information for the three stores.

(b) During the following week, sales on these products at Store I increased by 25%; sales at Store II increased by 1/3; and sales at Store III increased by 10%. Write the sales matrix for that week.

(c) Write a matrix that represents total sales over the two-week period.

Life Sciences

41. Dietetics A dietician prepares a diet specifying the amounts a patient should eat of four basic food groups: group I, meats; group II, fruits and vegetables; group III, breads and starches; group IV, milk products. Amounts are given in "exchanges" that represent 1 oz (meat), 1/2 cup (fruits and vegetables), 1 slice (bread), 8 oz (milk), or other suitable measurements.

(a) The number of "exchanges" for breakfast for each of the four food groups, respectively, are 2, 1, 2, and 1; for lunch, 3, 2, 2, and 1; and for dinner, 4, 3, 2, and 1. Write a 3×4 matrix using this information.

(b) The amounts of fat, carbohydrates, and protein (in appropriate units) in each food group respectively are as follows.

> Fat: 5, 0, 0, 10
> Carbohydrates: 0, 10, 15, 12
> Protein: 7, 1, 2, 8

Use this information to write a 4×3 matrix.

(c) There are 8 calories per exchange of fat, 4 calories per exchange of carbohydrates, and 5 calories per exchange of protein; summarize this data in a 3×1 matrix.

42. Animal Growth At the beginning of a laboratory experiment, five baby rats measured 5.6, 6.4, 6.9, 7.6, and 6.1 centimeters in length, and weighed 144, 138, 149, 152, and 146 grams, respectively.

(a) Write a 2×5 matrix using this information.

(b) At the end of two weeks, their lengths in centimeters were 10.2, 11.4, 11.4, 12.7, and 10.8, and their weights (in grams) were 196, 196, 225, 250, and 230. Write a 2×5 matrix with this information.

(c) Use matrix subtraction and the matrices found in parts (a) and (b) to write a matrix that gives the amount of change in length and weight for each rat.

(d) During the third week the rats grew by the amounts shown in the matrix below.

$$\begin{array}{c} \text{Length} \\ \text{Weight} \end{array} \begin{bmatrix} 1.8 & 1.5 & 2.3 & 1.8 & 2.0 \\ 25 & 22 & 29 & 33 & 20 \end{bmatrix}$$

What were their lengths and weights at the end of this week?

43. Testing Medication A drug company is testing 200 patients to see if Painfree (a new headache medicine) is effective. Half the patients receive Painfree and half receive a placebo. The data on the first 50 patients is summarized in this matrix:

Pain Relief Obtained

$$\begin{array}{c} \\ \text{Painfree} \\ \text{Placebo} \end{array} \begin{array}{cc} \text{Yes} & \text{No} \\ \begin{bmatrix} 22 & 3 \\ 8 & 17 \end{bmatrix} \end{array}$$

(a) Of those who took the placebo, how many got relief?

(b) Of those who took the new medication, how many got no relief?

(c) The test was repeated on three more groups of 50 patients each, with the results summarized by these matrices.

$$\begin{bmatrix} 21 & 4 \\ 6 & 19 \end{bmatrix} \quad \begin{bmatrix} 19 & 6 \\ 10 & 15 \end{bmatrix} \quad \begin{bmatrix} 23 & 2 \\ 3 & 22 \end{bmatrix}$$

Find the total results for all 200 patients.

(d) On the basis of these results, does it appear that Painfree is effective?

44. Educational Attainment The following table gives the educational attainment of the U.S. population 25 years and older.*

	Male		Female	
	Four Years of High School or More	Four Years of College or More	Four Years of High School or More	Four Years of College or More
1940	22.7%	5.5%	26.3%	3.8%
1950	32.6	7.3	36.0	5.2
1959	42.2	10.3	45.2	6.0
1970	55.0	14.1	55.4	8.2
1980	69.1	20.8	68.1	13.5
1987	76.0	23.6	75.3	16.5

(a) Write a matrix for the educational attainment of males.

(b) Write a matrix for the educational attainment of females.

(c) Use the matrices from (a) and (b) to write a matrix showing how much more (or less) education males have attained than females.

*"Educational Attainment by Percentage of Population 25 Years and Older and by Sex, 1940–87" from U.S. Bureau of the Census, *Educational Attainment in the U.S.: March 1987 and 1986.*

2.4 MULTIPLICATION OF MATRICES

What is a contractor's total cost for materials required for various types of model homes?

Matrix multiplication will be used to answer this question in Example 5. We begin by defining the product of a real number and a matrix. In work with matrices, a real number is called a **scalar.**

PRODUCT OF A MATRIX AND A SCALAR

The **product** of a scalar k and a matrix X is the matrix kX, each of whose elements is k times the corresponding element of X.

For example,

$$(-3)\begin{bmatrix} 2 & -5 \\ 1 & 7 \end{bmatrix} = \begin{bmatrix} -6 & 15 \\ -3 & -21 \end{bmatrix}.$$

Finding the product of two matrices is more involved, but such multiplication is important in solving practical problems. To understand the reasoning behind matrix multiplication, it may be helpful to consider another example concerning

the EZ Life Company discussed in the previous section. Suppose sofas and chairs of the same model are often sold as sets. Matrix W shows the number of sets of each model in each warehouse.

$$\begin{array}{c} \\ \text{New York} \\ \text{Chicago} \\ \text{San Francisco} \end{array} \begin{array}{ccc} \text{A} & \text{B} & \text{C} \\ \begin{bmatrix} 10 & 7 & 3 \\ 5 & 9 & 6 \\ 4 & 8 & 2 \end{bmatrix} \end{array} = W$$

If the selling price of a model-A set is \$800, of a model-B set \$1000, and of a model-C set \$1200, the total value of the sets in the New York warehouse is found as follows.

Type	Number of Sets		Price of Set		Total
A	10	×	\$800	=	\$8000
B	7	×	\$1000	=	\$7000
C	3	×	\$1200	=	\$3600
		(Total for New York)			\$18,600

The total value of the three kinds of sets in New York is \$18,600.

The work done in the table above is summarized as follows:

$$10(\$800) + 7(\$1000) + 3(\$1200) = \$18,600.$$

In the same way, we find that the Chicago sets have a total value of

$$5(\$800) + 9(\$1000) + 6(\$1200) = \$20,200,$$

and in San Francisco, the total value of the sets is

$$4(\$800) + 8(\$1000) + 2(\$1200) = \$13,600.$$

The selling prices can be written as a column matrix, P, and the total value in each location as another column matrix, V.

$$\begin{bmatrix} 800 \\ 1000 \\ 1200 \end{bmatrix} = P \qquad \begin{bmatrix} 18,600 \\ 20,200 \\ 13,600 \end{bmatrix} = V$$

Look at the elements of W and P below; multiplying the first, second, and third elements of the first row of W by the first, second, and third elements, respectively, of the column matrix P and then adding these products gives the first element in V. Doing the same thing with the second row of W gives the second element of V; the third row of W leads to the third element of V, suggesting that it is reasonable to write the product of matrices

$$W = \begin{bmatrix} 10 & 7 & 3 \\ 5 & 9 & 6 \\ 4 & 8 & 2 \end{bmatrix} \quad \text{and} \quad P = \begin{bmatrix} 800 \\ 1000 \\ 1200 \end{bmatrix}$$

as
$$WP = \begin{bmatrix} 10 & 7 & 3 \\ 5 & 9 & 6 \\ 4 & 8 & 2 \end{bmatrix} \begin{bmatrix} 800 \\ 1000 \\ 1200 \end{bmatrix} = \begin{bmatrix} 18,600 \\ 20,200 \\ 13,600 \end{bmatrix} = V.$$

The product was found by multiplying the elements of *rows* of the matrix on the left and the corresponding elements of the *column* of the matrix on the right, and then finding the sum of these separate products. Notice that the product of a 3×3 matrix and a 3×1 matrix is a 3×1 matrix.

The **product** AB of an $m \times n$ matrix A and an $n \times k$ matrix B is found as follows. Multiply each element of the *first row* of A by the corresponding element of the *first column* of B. The sum of these n products is the *first row, first column* element of AB. Similarly, the sum of the products found by multiplying the elements of the *first row* of A by the corresponding elements of the *second column* of B gives the *first-row, second-column* element of AB, and so on.

PRODUCT OF TWO MATRICES

Let A be an $m \times n$ matrix and let B be an $n \times k$ matrix. To find the element in the ith row and jth column of the **product matrix** AB, multiply each element in the ith row of A by the corresponding element in the jth column of B, then add these products. The product matrix AB is an $m \times k$ matrix.

EXAMPLE 1

Find the product AB of matrices

$$A = \begin{bmatrix} 2 & 3 & -1 \\ 4 & 2 & 2 \end{bmatrix} \quad \text{and} \quad B = \begin{bmatrix} 1 \\ 8 \\ 6 \end{bmatrix}.$$

Step 1 Multiply the elements of the first row of A and the corresponding elements of the column of B.

$$\begin{bmatrix} 2 & 3 & -1 \\ 4 & 2 & 2 \end{bmatrix} \begin{bmatrix} 1 \\ 8 \\ 6 \end{bmatrix} \qquad 2 \cdot 1 + 3 \cdot 8 + (-1) \cdot 6 = 20$$

Thus, 20 is the first-row entry of the product matrix AB.

Step 2 Multiply the elements of the second row of A and the corresponding elements of B.

$$\begin{bmatrix} 2 & 3 & -1 \\ 4 & 2 & 2 \end{bmatrix} \begin{bmatrix} 1 \\ 8 \\ 6 \end{bmatrix} \qquad 4 \cdot 1 + 2 \cdot 8 + 2 \cdot 6 = 32$$

The second-row entry of the product matrix AB is 32.

Step 3 Write the product as a column matrix using the two entries found above.

$$AB = \begin{bmatrix} 2 & 3 & -1 \\ 4 & 2 & 2 \end{bmatrix} \begin{bmatrix} 1 \\ 8 \\ 6 \end{bmatrix} = \begin{bmatrix} 20 \\ 32 \end{bmatrix} \blacktriangleleft$$

EXAMPLE 2

Find the product CD of matrices

$$C = \begin{bmatrix} -3 & 4 & 2 \\ 5 & 0 & 4 \end{bmatrix} \quad \text{and} \quad D = \begin{bmatrix} -6 & 4 \\ 2 & 3 \\ 3 & -2 \end{bmatrix}.$$

Step 1

$$\begin{bmatrix} -3 & 4 & 2 \\ 5 & 0 & 4 \end{bmatrix}\begin{bmatrix} -6 & 4 \\ 2 & 3 \\ 3 & -2 \end{bmatrix} \quad (-3) \cdot (-6) + 4 \cdot 2 + 2 \cdot 3 = 32$$

Step 2

$$\begin{bmatrix} -3 & 4 & 2 \\ 5 & 0 & 4 \end{bmatrix}\begin{bmatrix} -6 & 4 \\ 2 & 3 \\ 3 & -2 \end{bmatrix} \quad (-3) \cdot 4 + 4 \cdot 3 + 2 \cdot (-2) = -4$$

Step 3

$$\begin{bmatrix} -3 & 4 & 2 \\ 5 & 0 & 4 \end{bmatrix}\begin{bmatrix} -6 & 4 \\ 2 & 3 \\ 3 & -2 \end{bmatrix} \quad 5 \cdot (-6) + 0 \cdot 2 + 4 \cdot 3 = -18$$

Step 4

$$\begin{bmatrix} -3 & 4 & 2 \\ 5 & 0 & 4 \end{bmatrix}\begin{bmatrix} -6 & 4 \\ 2 & 3 \\ 3 & -2 \end{bmatrix} \quad 5 \cdot 4 + 0 \cdot 3 + 4 \cdot (-2) = 12$$

Step 5 The product is

$$CD = \begin{bmatrix} -3 & 4 & 2 \\ 5 & 0 & 4 \end{bmatrix}\begin{bmatrix} -6 & 4 \\ 2 & 3 \\ 3 & -2 \end{bmatrix} = \begin{bmatrix} 32 & -4 \\ -18 & 12 \end{bmatrix}.$$

Here the product of a 2×3 matrix and a 3×2 matrix is a 2×2 matrix. ◄

As the definition of matrix multiplication shows,

the product AB of two matrices A and B can be found only if the number of columns of A is the same as the number of rows of B.

The final product will have as many rows as A and as many columns as B.

EXAMPLE 3

Suppose matrix A is 2×2 and matrix B is 2×4. Can the product AB be calculated? What is the size of the product?

The following diagram helps decide the answers to these questions.

Matrix A
size

2 × 2

Matrix B
size

2 × 4

must match

Size of AB
is 2 × 4.

The product of A and B can be found because A has two columns and B has two rows. The size of the product is 2 × 4. ◄

EXAMPLE 4

Find BA, given

$$A = \begin{bmatrix} 1 & -3 \\ 7 & 2 \end{bmatrix} \quad \text{and} \quad B = \begin{bmatrix} 1 & 0 & -1 \\ 3 & 1 & 4 \end{bmatrix}.$$

Since B is a 2 × 3 matrix and A is a 2 × 2 matrix, the product BA cannot be found. ◄

In Example 4, the product BA could not be found. The product AB, however, is given by

$$AB = \begin{bmatrix} 1 & -3 \\ 7 & 2 \end{bmatrix} \cdot \begin{bmatrix} 1 & 0 & -1 \\ 3 & 1 & 4 \end{bmatrix}$$

$$= \begin{bmatrix} -8 & -3 & -13 \\ 13 & 2 & 1 \end{bmatrix}.$$

This means that matrix multiplication is *not* commutative. Even if both A and B are square matrices, in general, matrices AB and BA are not equal. (See Exercise 31.)

Caution Since matrix multiplication is not commutative, always be careful to multiply matrices in the correct order.

Matrix multiplication *is* associative, however. For example, if

$$C = \begin{bmatrix} 2 & 1 \\ 3 & 4 \\ 1 & 5 \end{bmatrix},$$

$(AB)C = A(BC)$, where A and B are the matrices given in Example 4. (Verify this.) Also, there is a distributive property of matrices such that, for appropriate matrices A, B, and C,

$$A(B + C) = AB + AC.$$

(See Exercises 32 and 33.) Other properties of matrix multiplication involving scalars are included in the exercises. Multiplicative inverses and multiplicative identities are defined in the next section.

EXAMPLE 5

A contractor builds three kinds of houses, models A, B, and C, with a choice of two styles, Spanish or contemporary. Matrix P shows the number of each kind of house planned for a new 100-home subdivision. The amounts for each of the exterior materials depend primarily on the style of the house. These amounts are shown in matrix Q. (Concrete is in cubic yards, lumber in units of 1000 board feet, brick in 1000's, and shingles in units of 100 square feet.) Matrix R gives the cost in dollars for each kind of material.

$$\begin{array}{c}\text{Spanish} \quad \text{Contemporary}\end{array}$$
$$\begin{array}{c}\text{Model A}\\ \text{Model B}\\ \text{Model C}\end{array}\begin{bmatrix}0 & 30\\ 10 & 20\\ 20 & 20\end{bmatrix}=P$$

$$\begin{array}{c}\text{Concrete} \quad \text{Lumber} \quad \text{Brick} \quad \text{Shingles}\end{array}$$
$$\begin{array}{c}\text{Spanish}\\ \text{Contemporary}\end{array}\begin{bmatrix}10 & 2 & 0 & 2\\ 50 & 1 & 20 & 2\end{bmatrix}=Q$$

$$\begin{array}{c}\text{Cost per Unit}\end{array}$$
$$\begin{array}{c}\text{Concrete}\\ \text{Lumber}\\ \text{Brick}\\ \text{Shingles}\end{array}\begin{bmatrix}20\\ 180\\ 60\\ 25\end{bmatrix}=R$$

(a) What is the total cost of these materials for each model?

To find the cost for each model, first find PQ, which shows the amount of each material needed for each model.

$$PQ=\begin{bmatrix}0 & 30\\ 10 & 20\\ 20 & 20\end{bmatrix}\begin{bmatrix}10 & 2 & 0 & 2\\ 50 & 1 & 20 & 2\end{bmatrix}$$

$$=\begin{bmatrix}1500 & 30 & 600 & 60\\ 1100 & 40 & 400 & 60\\ 1200 & 60 & 400 & 80\end{bmatrix}\begin{array}{l}\text{Model A}\\ \text{Model B}\\ \text{Model C}\end{array}$$

with headers Concrete, Lumber, Brick, Shingles.

Now multiply PQ and R, the cost matrix, to get the total cost of the exterior materials for each model.

$$\begin{bmatrix}1500 & 30 & 600 & 60\\ 1100 & 40 & 400 & 60\\ 1200 & 60 & 400 & 80\end{bmatrix}\begin{bmatrix}20\\ 180\\ 60\\ 25\end{bmatrix}=\begin{bmatrix}72,900\\ 54,700\\ 60,800\end{bmatrix}\begin{array}{l}\text{Model A}\\ \text{Model B}\\ \text{Model C}\end{array}$$

The total cost of materials is $72,900 for Model A, $54,700 for Model B, and $60,800 for Model C.

(b) How much of each of the four kinds of material must be ordered?

The totals of the columns of matrix PQ will give a matrix whose elements represent the total amounts of each material needed for the subdivision. Call this matrix T, and write it as a row matrix.

$$T=[3800 \quad 130 \quad 1400 \quad 200]$$

Thus, 3800 cubic yards of concrete, 130,000 board feet of lumber, 1,400,000 bricks, and 20,000 square feet of shingles are needed.

(c) What is the total cost for exterior materials?

For the total cost of all the exterior materials, find the product of matrix T, the matrix showing the total amount of each material, and matrix R, the cost matrix. (To multiply these and get a 1×1 matrix representing total cost, we need a 1×4 matrix multiplied by a 4×1 matrix. This is why T was written as a row matrix in (b) above.)

$$TR = [3800 \quad 130 \quad 1400 \quad 200] \begin{bmatrix} 20 \\ 180 \\ 60 \\ 25 \end{bmatrix} = [188,400]$$

The total cost for exterior materials is $188,400.

(d) Suppose the contractor builds the same number of homes in five subdivisions. Calculate the total amount of each exterior material for each model for all five subdivisions.

Multiply PQ by the scalar 5, as follows.

$$5 \begin{bmatrix} 1500 & 30 & 600 & 60 \\ 1100 & 40 & 400 & 60 \\ 1200 & 60 & 400 & 80 \end{bmatrix} = \begin{bmatrix} 7500 & 150 & 3000 & 300 \\ 5500 & 200 & 2000 & 300 \\ 6000 & 300 & 2000 & 400 \end{bmatrix}$$

The total amount of concrete needed for Model A homes, for example, is 7500 cubic yards. ◀

Choosing Matrix Notation It is helpful to use a notation that keeps track of the quantities a matrix represents. For example, let matrix P, from Example 5, represent models/styles, matrix Q represent styles/materials, and matrix R represent materials/cost. In each case, write the meaning of the rows first and the columns second. In the product PQ from Example 5, the rows of the matrix represent models and the columns represent materials. Therefore, the matrix product PQ represents models/materials. Note that the common quantity, styles, in both P and Q was eliminated in the product PQ. By this method, the product $(PQ)R$ represents models/cost.

In practical problems this notation helps us decide in which order to multiply matrices so that the results are meaningful. In Example 5(c) either RT or TR could have been found. Since T represents subdivisions/materials and R represents materials/cost, the product TR gives subdivisions/cost.

2.4 Exercises

In Exercises 1–6, the sizes of two matrices A and B are given. Find the sizes of the product AB and the product BA, whenever these products exist.

1. A is 2×2, and B is 2×2.

2. A is 3×3, and B is 3×3.

3. A is 3×5, and B is 5×2.

4. A is 4×3, and B is 3×6.

5. A is 4×2, and B is 3×4.

6. A is 7×3, and B is 2×7.

Let $A = \begin{bmatrix} -2 & 4 \\ 0 & 3 \end{bmatrix}$ and $B = \begin{bmatrix} -6 & 2 \\ 4 & 0 \end{bmatrix}$.

Find each of the following.

7. $2A$

8. $-3B$

9. $-4B$

10. $5A$

11. $-4A + 5B$

12. $3A - 10B$

13. To find the product matrix AB, the number of _____ of A must be the same as the number of _____ of B.

14. The product matrix AB has the same number of _____ as A and the same number of _____ as B.

Find each of the matrix products in Exercises 15–30 if possible.

15. $\begin{bmatrix} 1 & 2 \\ 3 & 4 \end{bmatrix}\begin{bmatrix} -1 \\ 7 \end{bmatrix}$

16. $\begin{bmatrix} -1 & 5 \\ 7 & 0 \end{bmatrix}\begin{bmatrix} 6 \\ 2 \end{bmatrix}$

17. $\begin{bmatrix} 1 & 5 & 3 \\ -1 & 2 & 7 \end{bmatrix}\begin{bmatrix} 4 \\ 2 \\ -3 \end{bmatrix}$

18. $\begin{bmatrix} 5 & 2 \\ 7 & 6 \\ 1 & 0 \end{bmatrix}\begin{bmatrix} 1 & 4 & 0 \\ 2 & -1 & 2 \end{bmatrix}$

19. $\begin{bmatrix} 5 & 1 \\ 2 & 3 \end{bmatrix}\begin{bmatrix} 3 & -1 & 0 \\ 1 & 0 & 2 \end{bmatrix}$

20. $\begin{bmatrix} 6 & 0 & -4 \\ 1 & 2 & 5 \\ 10 & -1 & 3 \end{bmatrix}\begin{bmatrix} 1 \\ 2 \\ 0 \end{bmatrix}$

21. $\begin{bmatrix} 2 & 2 & -1 \\ 3 & 0 & 1 \end{bmatrix}\begin{bmatrix} 0 & 2 \\ -1 & 4 \\ 0 & 2 \end{bmatrix}$

22. $\begin{bmatrix} -9 & 2 & 1 \\ 3 & 0 & 0 \end{bmatrix}\begin{bmatrix} 2 \\ -1 \\ 4 \end{bmatrix}$

23. $\begin{bmatrix} 1 & 2 \\ 3 & 4 \end{bmatrix}\begin{bmatrix} -1 & 5 \\ 7 & 0 \end{bmatrix}$

24. $\begin{bmatrix} -1 & 5 \\ 7 & 0 \end{bmatrix}\begin{bmatrix} 1 & 2 \\ 3 & 4 \end{bmatrix}$

25. $\begin{bmatrix} -2 & -3 & 7 \\ 1 & 5 & 6 \end{bmatrix}\begin{bmatrix} 1 \\ 2 \\ 3 \end{bmatrix}$

26. $\begin{bmatrix} 6 \\ 5 \\ 4 \end{bmatrix}[-1 \quad 1 \quad 1]$

27. $\left(\begin{bmatrix} 4 & 3 \\ 1 & 2 \\ 0 & -5 \end{bmatrix}\begin{bmatrix} 2 & -2 \\ 1 & -1 \end{bmatrix}\right)\begin{bmatrix} 10 \\ 0 \end{bmatrix}$

28. $\begin{bmatrix} 4 & 3 \\ 1 & 2 \\ 0 & -5 \end{bmatrix}\left(\begin{bmatrix} 2 & -2 \\ 1 & -1 \end{bmatrix}\begin{bmatrix} 10 \\ 0 \end{bmatrix}\right)$

29. $\begin{bmatrix} 2 & -2 \\ 1 & -1 \end{bmatrix}\left(\begin{bmatrix} 4 & 3 \\ 1 & 2 \end{bmatrix} + \begin{bmatrix} 7 & 0 \\ -1 & 5 \end{bmatrix}\right)$

30. $\begin{bmatrix} 2 & -2 \\ 1 & -1 \end{bmatrix}\begin{bmatrix} 4 & 3 \\ 1 & 2 \end{bmatrix} + \begin{bmatrix} 2 & -2 \\ 1 & -1 \end{bmatrix}\begin{bmatrix} 7 & 0 \\ -1 & 5 \end{bmatrix}$

31. Let $A = \begin{bmatrix} -2 & 4 \\ 1 & 3 \end{bmatrix}$ and $B = \begin{bmatrix} -2 & 1 \\ 3 & 6 \end{bmatrix}$.

 (a) Find AB.

 (b) Find BA.

 (c) Did you get the same answer in parts (a) and (b)?

 (d) In general, for matrices A and B such that AB and BA both exist, does AB always equal BA?

Given matrices $P = \begin{bmatrix} m & n \\ p & q \end{bmatrix}$, $X = \begin{bmatrix} x & y \\ z & w \end{bmatrix}$, and $T = \begin{bmatrix} r & s \\ t & u \end{bmatrix}$, verify that the statements in Exercises 32–35

are true. The statements are valid for any matrices whenever matrix multiplication and addition can be carried out. This, of course, depends on the size of the matrices.

32. $(PX)T = P(XT)$ (associative property: see Exercises 27 and 28.)

33. $P(X + T) = PX + PT$ (distributive property: see Exercises 29 and 30.)

34. $k(X + T) = kX + kT$ for any real number k.

35. $(k + h)P = kP + hP$ for any real numbers k and h.

36. Let I be the matrix $I = \begin{bmatrix} 1 & 0 \\ 0 & 1 \end{bmatrix}$, and let matrices P, X, and T be defined as for Exercises 32–35.

(a) Find IP, PI, and IX.

(b) Without calculating, guess what the matrix IT might be.

(c) Suggest a reason for naming a matrix such as I an *identity matrix*.

37. Show that the system of linear equations

$$2x_1 + 3x_2 + x_3 = 5$$
$$x_1 - 4x_2 + 5x_3 = 8$$

can be written as the matrix equation

$$\begin{bmatrix} 2 & 3 & 1 \\ 1 & -4 & 5 \end{bmatrix} \begin{bmatrix} x_1 \\ x_2 \\ x_3 \end{bmatrix} = \begin{bmatrix} 5 \\ 8 \end{bmatrix}.$$

38. Let $A = \begin{bmatrix} 1 & 2 \\ -3 & 5 \end{bmatrix}$, $X = \begin{bmatrix} x_1 \\ x_2 \end{bmatrix}$, and $B = \begin{bmatrix} -4 \\ 12 \end{bmatrix}$.

Show that the equation $AX = B$ represents a linear system of two equations in two unknowns. Solve the system and substitute into the matrix equation to check your results.

Applications

Business and Economics

39. Cost Analysis The four departments of Spangler Enterprises need to order the following amounts of the same products.

	Paper	Tape	Printer Ribbon	Memo Pads	Pens
Department 1	10	4	3	5	6
Department 2	7	2	2	3	8
Department 3	4	5	1	0	10
Department 4	0	3	4	5	5

The unit price (in dollars) of each product is given below for two suppliers.

	Supplier A	Supplier B
Paper	2	3
Tape	1	1
Printer Ribbon	4	3
Memo Pads	3	3
Pens	1	2

(a) Use matrix multiplication to get a matrix showing the comparative costs for each department for the products from the two suppliers.

(b) Find the total cost to buy products from each supplier. From which supplier should the company make the purchase?

40. Cost Analysis The Mundo Candy Company makes three types of chocolate candy: Cheery Cherry, Mucho Mocha, and Almond Delight. The company produces its products

in San Diego, Mexico City, and Managua using two main ingredients: chocolate and sugar.

(a) Each kilogram of Cheery Cherry requires .5 kg of sugar and .2 kg of chocolate; each kilogram of Mucho Mocha requires .4 kg of sugar and .3 kg of chocolate; and each kilogram of Almond Delight requires .3 kg of sugar and .3 kg of chocolate. Put this information into a 2×3 matrix, labeling the rows and columns.

(b) The cost of 1 kg of sugar is $3 in San Diego, $2 in Mexico City, and $1 in Managua. The cost of 1 kg of chocolate is $3 in San Diego, $3 in Mexico City, and $4 in Managua. Put this information into a matrix in such a way that when you multiply it with your matrix from part (a), you get a matrix representing the ingredient cost of producing each type of candy in each city.

(c) Multiply the matrices in parts (a) and (b), labeling the product matrix.

(d) From part (c) what is the combined sugar-and-chocolate cost to produce 1 kg of Mucho Mocha in Managua?

(e) Mundo Candy needs to quickly produce a special shipment of 100 kg of Cheery Cherry, 200 kg of Mucho Mocha, and 500 kg of Almond Delight, and it decides to select one factory to fill the entire order. Use matrix multiplication to determine in which city the total sugar-and-chocolate cost to produce the order is the smallest.

Social Sciences

41. **World Population** The 1987 birth and death rates per million for several regions and the world population (in millions) by region are given below.*

	Births	Deaths
Asia	.027	.009
Latin America	.030	.007
North America	.015	.009
Europe	.013	.011
Soviet Union	.019	.011

	Asia	Latin America	North America	Europe	USSR
1960	1596	218	199	425	214
1970	1996	286	226	460	243
1980	2440	365	252	484	266
1990	2906	455	277	499	291

(a) Write the information in each table as a matrix.

(b) Use the matrices from part (a) to find the total number (in millions) of births and deaths in each year.

(c) Using the results of part (b), compare the number of births in 1960 and in 1990. Also compare the birth rates from part (a). Which gives better information?

(d) Using the results of part (b), compare the number of deaths in 1980 and in 1990. Discuss how this comparison differs from comparing death rates from part (a).

Life Sciences

42. **Dietetics** In Exercise 41 from Section 2.3, label the matrices found in parts (a), (b), and (c), respectively, X, Y, and Z.

(a) Find the product matrix XY. What do the entries of this matrix represent?

(b) Find the product matrix YZ. What do the entries represent?

Computer/Graphing Calculator

Use a computer or graphing calculator and the following matrices to find the matrix products and sums in Exercises 43–45.

$$A = \begin{bmatrix} 2 & 3 & -1 & 5 & 10 \\ 2 & 8 & 7 & 4 & 3 \\ -1 & -4 & -12 & 6 & 8 \\ 2 & 5 & 7 & 1 & 4 \end{bmatrix} \qquad B = \begin{bmatrix} 9 & 3 & 7 & -6 \\ -1 & 0 & 4 & 2 \\ -10 & -7 & 6 & 9 \\ 8 & 4 & 2 & -1 \\ 2 & -5 & 3 & 7 \end{bmatrix}$$

$$C = \begin{bmatrix} -6 & 8 & 2 & 4 & -3 \\ 1 & 9 & 7 & -12 & 5 \\ 15 & 2 & -8 & 10 & 11 \\ 4 & 7 & 9 & 6 & -2 \\ 1 & 3 & 8 & 23 & 4 \end{bmatrix} \qquad D = \begin{bmatrix} 5 & -3 & 7 & 9 & 2 \\ 6 & 8 & -5 & 2 & 1 \\ 3 & 7 & -4 & 2 & 11 \\ 5 & -3 & 9 & 4 & -1 \\ 0 & 3 & 2 & 5 & 1 \end{bmatrix}$$

43. (a) Find AC. (b) Find CA. (c) Does $AC = CA$?

44. (a) Find CD. (b) Find DC. (c) Does $CD = DC$?

45. (a) Find $C + D$. (b) Find $(C + D)B$. (c) Find CB. (d) Find DB. (e) Find $CB + DB$.

(f) Does $(C + D)B = CB + DB$?

46. Which properties of matrices do Exercises 43–45 illustrate?

*"Vital Events and Rates by Region and Development Category, 1987" and "World Population by Region and Development Category, 1950–2025" from U.S. Bureau of the Census, *World Population Profile: 1987.*

► 2.5 MATRIX INVERSES

 An investor wants to invest a specific amount of money in three types of bonds with restrictions on each investment. How much money should be invested in each type?

This question is answered in Exercise 53 for this section by solving a matrix equation. In this section, we introduce the idea of a matrix inverse, which is comparable to the reciprocal of a real number. This will allow us to solve a matrix equation.

Earlier, we defined a zero matrix as an additive identity matrix with properties similar to those of the real number 0, the additive identity for real numbers. The real number 1 is the *multiplicative* identity for real numbers: for any real number $a, a \cdot 1 = 1 \cdot a = a$. In this section, we define a *multiplicative identity matrix I* that has properties similar to those of the number 1. We then use the definition of matrix I to find the *multiplicative inverse* of any square matrix that has an inverse.

If I is to be the identity matrix, both of the products AI and IA must equal A. This means that an identity matrix exists only for square matrices. Otherwise, IA and AI could not both be found. The **2 × 2 identity matrix** that satisfies these conditions is

$$I = \begin{bmatrix} 1 & 0 \\ 0 & 1 \end{bmatrix}.$$

To check that I, as defined above, is really the 2 × 2 identity matrix, let

$$A = \begin{bmatrix} a & b \\ c & d \end{bmatrix}.$$

Then AI and IA should both equal A.

$$AI = \begin{bmatrix} a & b \\ c & d \end{bmatrix}\begin{bmatrix} 1 & 0 \\ 0 & 1 \end{bmatrix} = \begin{bmatrix} a(1) + b(0) & a(0) + b(1) \\ c(1) + d(0) & c(0) + d(1) \end{bmatrix} = \begin{bmatrix} a & b \\ c & d \end{bmatrix} = A$$

$$IA = \begin{bmatrix} 1 & 0 \\ 0 & 1 \end{bmatrix}\begin{bmatrix} a & b \\ c & d \end{bmatrix} = \begin{bmatrix} 1(a) + 0(c) & 1(b) + 0(d) \\ 0(a) + 1(c) & 0(b) + 1(d) \end{bmatrix} = \begin{bmatrix} a & b \\ c & d \end{bmatrix} = A$$

This verifies that I has been defined correctly. (It can be shown that I is the only 2 × 2 identity matrix.)

The identity matrices for 3 × 3 matrices and 4 × 4 matrices, respectively, are

$$I = \begin{bmatrix} 1 & 0 & 0 \\ 0 & 1 & 0 \\ 0 & 0 & 1 \end{bmatrix} \quad \text{and} \quad I = \begin{bmatrix} 1 & 0 & 0 & 0 \\ 0 & 1 & 0 & 0 \\ 0 & 0 & 1 & 0 \\ 0 & 0 & 0 & 1 \end{bmatrix}.$$

By generalizing, we can find an $n \times n$ identity matrix for any value of n.

Recall that the multiplicative inverse of the nonzero real number a is $1/a$. The product of a and its multiplicative inverse $1/a$ is 1. Given a matrix A, can a **multiplicative inverse matrix A^{-1}** (read "A-inverse") be found satisfying both

$$AA^{-1} = I \quad \text{and} \quad A^{-1}A = I?$$

For a given matrix, we often can find an inverse matrix by using the row operations of Section 2.2.

Note A^{-1} does not mean $1/A$; here, A^{-1} is just the notation for the multiplicative inverse of matrix A. Also, only square matrices can have inverses because both $A^{-1}A$ and AA^{-1} must exist and be equal to I.

If an inverse exists, it is unique. That is, any given square matrix has no more than one inverse. The proof of this is left to Exercise 50 in this section.

As an example, let us find the inverse of

$$A = \begin{bmatrix} 2 & 4 \\ 1 & -1 \end{bmatrix}.$$

Let the unknown inverse matrix be

$$A^{-1} = \begin{bmatrix} x & y \\ z & w \end{bmatrix}.$$

By the definition of matrix inverse, $AA^{-1} = I$, or

$$AA^{-1} = \begin{bmatrix} 2 & 4 \\ 1 & -1 \end{bmatrix} \begin{bmatrix} x & y \\ z & w \end{bmatrix} = \begin{bmatrix} 1 & 0 \\ 0 & 1 \end{bmatrix}.$$

By matrix multiplication,

$$\begin{bmatrix} 2x + 4z & 2y + 4w \\ x - z & y - w \end{bmatrix} = \begin{bmatrix} 1 & 0 \\ 0 & 1 \end{bmatrix}.$$

Setting corresponding elements equal gives the system of equations

$$2x + 4z = 1 \tag{1}$$
$$2y + 4w = 0 \tag{2}$$
$$x - z = 0 \tag{3}$$
$$y - w = 1. \tag{4}$$

Since equations (1) and (3) involve only x and z, while equations (2) and (4) involve only y and w, these four equations lead to two systems of equations,

$$\begin{array}{ccc} 2x + 4z = 1 & & 2y + 4w = 0 \\ & \text{and} & \\ x - z = 0 & & y - w = 1. \end{array}$$

Writing the two systems as augmented matrices gives

$$\begin{bmatrix} 2 & 4 & | & 1 \\ 1 & -1 & | & 0 \end{bmatrix} \quad \text{and} \quad \begin{bmatrix} 2 & 4 & | & 0 \\ 1 & -1 & | & 1 \end{bmatrix}.$$

Each of these systems can be solved by the Gauss-Jordan method. Notice, however, that the elements to the left of the vertical bar are identical. The two systems can be combined into the single matrix

$$\begin{bmatrix} 2 & 4 & 1 & 0 \\ 1 & -1 & 0 & 1 \end{bmatrix}$$

and solved simultaneously as follows.

$$R_1 + (-2)R_2 \rightarrow R_2 \quad \begin{bmatrix} 2 & 4 & 1 & 0 \\ 0 & 6 & 1 & -2 \end{bmatrix}$$ Get 0 in the second row, first column position.

$$-2R_2 + 3R_1 \rightarrow R_1 \quad \begin{bmatrix} 6 & 0 & 1 & 4 \\ 0 & 6 & 1 & -2 \end{bmatrix}$$ Get 0 in the first row, second column position.

$$\frac{1}{6}R_1 \rightarrow R_1 \quad \begin{bmatrix} 1 & 0 & \frac{1}{6} & \frac{2}{3} \\ 0 & 1 & \frac{1}{6} & -\frac{1}{3} \end{bmatrix}$$ Get 1's down the diagonal.
$$\frac{1}{6}R_2 \rightarrow R_2$$

The numbers in the first column to the right of the vertical bar give the values of x and z. The second column gives the values of y and w. That is,

$$\begin{bmatrix} 1 & 0 & x & y \\ 0 & 1 & z & w \end{bmatrix} = \begin{bmatrix} 1 & 0 & \frac{1}{6} & \frac{2}{3} \\ 0 & 1 & \frac{1}{6} & -\frac{1}{3} \end{bmatrix}$$

so that

$$A^{-1} = \begin{bmatrix} x & y \\ z & w \end{bmatrix} = \begin{bmatrix} \frac{1}{6} & \frac{2}{3} \\ \frac{1}{6} & -\frac{1}{3} \end{bmatrix}.$$

To check, multiply A by A^{-1}. The result should be I.

$$AA^{-1} = \begin{bmatrix} 2 & 4 \\ 1 & -1 \end{bmatrix}\begin{bmatrix} \frac{1}{6} & \frac{2}{3} \\ \frac{1}{6} & -\frac{1}{3} \end{bmatrix} = \begin{bmatrix} \frac{1}{3} + \frac{2}{3} & \frac{4}{3} - \frac{4}{3} \\ \frac{1}{6} - \frac{1}{6} & \frac{2}{3} + \frac{1}{3} \end{bmatrix} = \begin{bmatrix} 1 & 0 \\ 0 & 1 \end{bmatrix} = I$$

Verify that $A^{-1}A = I$, also. Finally,

$$A^{-1} = \begin{bmatrix} \frac{1}{6} & \frac{2}{3} \\ \frac{1}{6} & -\frac{1}{3} \end{bmatrix}.$$

FINDING A MULTIPLICATIVE INVERSE MATRIX

To obtain A^{-1} for any $n \times n$ matrix A for which A^{-1} exists, follow these steps.

1. Form the augmented matrix $[A|I]$, where I is the $n \times n$ identity matrix.
2. Perform row operations on $[A|I]$ to get a matrix of the form $[I|B]$.
3. Matrix B is A^{-1}.

EXAMPLE 1

Find A^{-1} if $A = \begin{bmatrix} 1 & 0 & 1 \\ 2 & -2 & -1 \\ 3 & 0 & 0 \end{bmatrix}$.

Write the augmented matrix $[A|I]$.

$$[A|I] = \begin{bmatrix} 1 & 0 & 1 & | & 1 & 0 & 0 \\ 2 & -2 & -1 & | & 0 & 1 & 0 \\ 3 & 0 & 0 & | & 0 & 0 & 1 \end{bmatrix}$$

Begin by selecting the row operation that produces a zero for the first element in row 2.

$$\begin{array}{c} -2R_1 + R_2 \rightarrow R_2 \\ -3R_1 + R_3 \rightarrow R_3 \end{array} \begin{bmatrix} 1 & 0 & 1 & | & 1 & 0 & 0 \\ 0 & -2 & -3 & | & -2 & 1 & 0 \\ 0 & 0 & -3 & | & -3 & 0 & 1 \end{bmatrix}$$ Get zeros in the first column.

Column 2 already has zeros in the required positions, so we work on column 3.

$$\begin{array}{c} R_3 + 3R_1 \rightarrow R_1 \\ R_3 + (-1)R_2 \rightarrow R_2 \end{array} \begin{bmatrix} 3 & 0 & 0 & | & 0 & 0 & 1 \\ 0 & 2 & 0 & | & -1 & -1 & 1 \\ 0 & 0 & -3 & | & -3 & 0 & 1 \end{bmatrix}$$ Get zeros in the third column.

Now get 1's down the main diagonal.

$$\begin{array}{c} \frac{1}{3}R_1 \rightarrow R_1 \\ \frac{1}{2}R_2 \rightarrow R_2 \\ -\frac{1}{3}R_3 \rightarrow R_3 \end{array} \begin{bmatrix} 1 & 0 & 0 & | & 0 & 0 & \frac{1}{3} \\ 0 & 1 & 0 & | & -\frac{1}{2} & -\frac{1}{2} & \frac{1}{2} \\ 0 & 0 & 1 & | & 1 & 0 & -\frac{1}{3} \end{bmatrix}$$ Get 1's down the diagonal.

From the last transformation, the desired inverse is

$$A^{-1} = \begin{bmatrix} 0 & 0 & \frac{1}{3} \\ -\frac{1}{2} & -\frac{1}{2} & \frac{1}{2} \\ 1 & 0 & -\frac{1}{3} \end{bmatrix}.$$

Confirm this by forming the products $A^{-1}A$ and AA^{-1}, both of which should equal I. ◄

EXAMPLE 2

Find A^{-1} if $A = \begin{bmatrix} 2 & -4 \\ 1 & -2 \end{bmatrix}$.

Using row operations to transform the first column of the augmented matrix

$$\begin{bmatrix} 2 & -4 & | & 1 & 0 \\ 1 & -2 & | & 0 & 1 \end{bmatrix}$$

gives the following results.

$$R_1 + (-2)R_2 \rightarrow R_2 \quad \begin{bmatrix} 2 & -4 & | & 1 & 0 \\ 0 & 0 & | & 1 & -2 \end{bmatrix}$$

Because the last row has all zeros to the left of the vertical bar, there is no way to complete the process of finding the inverse matrix. What is wrong? Just as the real number 0 has no multiplicative inverse, some matrices do not have inverses. Matrix A is an example of a matrix that has no inverse: there is no matrix A^{-1} such that $AA^{-1} = A^{-1}A = I$. ◄

Solving Systems of Equations with Inverses We used matrices to solve systems of linear equations by the Gauss-Jordan method in Section 2.2. Another way to use matrices to solve linear systems is to write the system as a matrix equation $AX = B$, where A is the matrix of the coefficients of the variables of the system, X is the matrix of the variables, and B is the matrix of the constants. Matrix A is called the **coefficient matrix.**

To solve the matrix equation $AX = B$, first see if A^{-1} exists. Assuming A^{-1} exists and using the facts that $A^{-1}A = I$ and $IX = X$ gives

$$AX = B$$
$$A^{-1}(AX) = A^{-1}B \qquad \text{Multiply both sides by } A^{-1}.$$
$$(A^{-1}A)X = A^{-1}B \qquad \text{Associative property}$$
$$IX = A^{-1}B \qquad \text{Multiplicative inverse property}$$
$$X = A^{-1}B. \qquad \text{Identity property}$$

Caution When multiplying by matrices on both sides of a matrix equation, be careful to multiply in the same order on both sides of the equation, since multiplication of matrices is not commutative (unlike multiplication of real numbers).

The work above leads to the following method of solving a system of equations written as a matrix equation.

SOLVING A SYSTEM $AX = B$ USING MATRIX INVERSES

To solve a system of equations $AX = B$, where A is the matrix of coefficients, X is the matrix of variables, and B is the matrix of constants, first find A^{-1}. Then $X = A^{-1}B$.

This method is most practical in solving several systems that have the same coefficient matrix but different constants, as in Example 4 in this section. Then just one inverse matrix must be found.

EXAMPLE
 3

Use the inverse of the coefficient matrix to solve the linear system

$$2x - 3y = 4$$
$$x + 5y = 2.$$

To represent the system as a matrix equation, use the coefficient matrix of the system together with the matrix of variables and the matrix of constants:

$$A = \begin{bmatrix} 2 & -3 \\ 1 & 5 \end{bmatrix}, \qquad X = \begin{bmatrix} x \\ y \end{bmatrix}, \qquad \text{and} \qquad B = \begin{bmatrix} 4 \\ 2 \end{bmatrix}.$$

The system can now be written in matrix form as the equation $AX = B$ since

$$AX = \begin{bmatrix} 2 & -3 \\ 1 & 5 \end{bmatrix} \begin{bmatrix} x \\ y \end{bmatrix} = \begin{bmatrix} 2x - 3y \\ x + 5y \end{bmatrix} = \begin{bmatrix} 4 \\ 2 \end{bmatrix} = B.$$

To solve the system, first find A^{-1}. Do this by using row operations on matrix $[A|I]$ to get

$$\begin{bmatrix} 1 & 0 & \frac{5}{13} & \frac{3}{13} \\ 0 & 1 & -\frac{1}{13} & \frac{2}{13} \end{bmatrix}.$$

From this result,

$$A^{-1} = \begin{bmatrix} \frac{5}{13} & \frac{3}{13} \\ -\frac{1}{13} & \frac{2}{13} \end{bmatrix}.$$

Next, find the product $A^{-1}B$.

$$A^{-1}B = \begin{bmatrix} \frac{5}{13} & \frac{3}{13} \\ -\frac{1}{13} & \frac{2}{13} \end{bmatrix}\begin{bmatrix} 4 \\ 2 \end{bmatrix} = \begin{bmatrix} 2 \\ 0 \end{bmatrix}$$

Since $X = A^{-1}B$,

$$X = \begin{bmatrix} x \\ y \end{bmatrix} = \begin{bmatrix} 2 \\ 0 \end{bmatrix}.$$

The solution of the system is $(2, 0)$. ◀

EXAMPLE 4 Use the inverse of the coefficient matrix to solve the following systems.

(a) $-x - 2y + 2z = 9$
$2x + y - z = -3$
$3x - 2y + z = -6$

(b) $-x - 2y + 2z = 3$
$2x + y - z = 3$
$3x - 2y + z = 7$

(c) $-x - 2y + 2z = 12$
$2x + y - z = 0$
$3x - 2y + z = 18$

(d) $-x - 2y + 2z = 1$
$2x + y - z = 7$
$3x - 2y + z = 17$

Notice that the four systems all have the same matrix of coefficients and the same matrix of variables:

$$A = \begin{bmatrix} -1 & -2 & 2 \\ 2 & 1 & -1 \\ 3 & -2 & 1 \end{bmatrix} \quad \text{and} \quad X = \begin{bmatrix} x \\ y \\ z \end{bmatrix}.$$

Find A^{-1} first, then use it to solve all four systems.

To find A^{-1}, we start with matrix

$$[A|I] = \begin{bmatrix} -1 & -2 & 2 & 1 & 0 & 0 \\ 2 & 1 & -1 & 0 & 1 & 0 \\ 3 & -2 & 1 & 0 & 0 & 1 \end{bmatrix}$$

and use row operations to get $[I|A^{-1}]$, from which

$$A^{-1} = \begin{bmatrix} \frac{1}{3} & \frac{2}{3} & 0 \\ \frac{5}{3} & \frac{7}{3} & -1 \\ \frac{7}{3} & \frac{8}{3} & -1 \end{bmatrix}.$$

Now we can solve each of the four systems by using $X = A^{-1}B$.

(a) Here the matrix of constants is

$$B = \begin{bmatrix} 9 \\ -3 \\ -6 \end{bmatrix}.$$

Since $X = A^{-1}B$,

$$X = \begin{bmatrix} \frac{1}{3} & \frac{2}{3} & 0 \\ \frac{5}{3} & \frac{7}{3} & -1 \\ \frac{7}{3} & \frac{8}{3} & -1 \end{bmatrix} \begin{bmatrix} 9 \\ -3 \\ -6 \end{bmatrix} = \begin{bmatrix} 1 \\ 14 \\ 19 \end{bmatrix}.$$

From this result, $x = 1$, $y = 14$, and $z = 19$, and the solution is $(1, 14, 19)$.

(b) The matrix of constants is $B = \begin{bmatrix} 3 \\ 3 \\ 7 \end{bmatrix}$.

$$X = A^{-1}B = \begin{bmatrix} \frac{1}{3} & \frac{2}{3} & 0 \\ \frac{5}{3} & \frac{7}{3} & -1 \\ \frac{7}{3} & \frac{8}{3} & -1 \end{bmatrix} \begin{bmatrix} 3 \\ 3 \\ 7 \end{bmatrix} = \begin{bmatrix} 3 \\ 5 \\ 8 \end{bmatrix}.$$

The solution is $(3, 5, 8)$.

By following the same procedure, verify that the answers to parts (c) and (d) are $(4, 2, 10)$ and $(5, 1, 4)$, respectively. ◀

In Example 4, using the matrix inverse method of solving the systems involved considerably less work than using row operations for each of the four systems.

EXAMPLE **5**

Use the inverse of the coefficient matrix to solve the system

$$2x + 5y = 20$$
$$6x + 15y = 30.$$

The matrices needed are

$$A = \begin{bmatrix} 2 & 5 \\ 6 & 15 \end{bmatrix}, \qquad X = \begin{bmatrix} x \\ y \end{bmatrix}, \qquad \text{and} \qquad B = \begin{bmatrix} 20 \\ 30 \end{bmatrix}.$$

Find A^{-1}.

$$\left[\begin{array}{cc|cc} 2 & 5 & 1 & 0 \\ 6 & 15 & 0 & 1 \end{array} \right]$$

$$-3R_1 + R_2 \rightarrow R_2 \quad \left[\begin{array}{cc|cc} 2 & 5 & 1 & 0 \\ 0 & 0 & -3 & 1 \end{array} \right] \quad \begin{array}{l} \text{Get 0 in the second row} \\ \text{of the first column.} \end{array}$$

As mentioned above, the zeros in the second row indicate that matrix A does not have an inverse. This means that the given system either has no solution or has dependent equations. Another method must be used to determine which is the case. Verify that this system is inconsistent and has no solution. ◀

2.5 Exercises

In Exercises 1–8, decide whether the given matrices are inverses of each other. (Check to see if their product is the identity matrix I.)

1. $\begin{bmatrix} 2 & 3 \\ 1 & 1 \end{bmatrix}$ and $\begin{bmatrix} -1 & 3 \\ 1 & -2 \end{bmatrix}$

2. $\begin{bmatrix} 5 & 7 \\ 2 & 3 \end{bmatrix}$ and $\begin{bmatrix} 3 & -7 \\ -2 & 5 \end{bmatrix}$

3. $\begin{bmatrix} 2 & 1 \\ 3 & 2 \end{bmatrix}$ and $\begin{bmatrix} 2 & 1 \\ -3 & 2 \end{bmatrix}$

4. $\begin{bmatrix} -1 & 2 \\ 3 & -5 \end{bmatrix}$ and $\begin{bmatrix} -5 & -2 \\ -3 & -1 \end{bmatrix}$

5. $\begin{bmatrix} 1 & 2 & 0 \\ 0 & 1 & 0 \\ 0 & 1 & 0 \end{bmatrix}$ and $\begin{bmatrix} 1 & -2 & 0 \\ 0 & 1 & 0 \\ 0 & -1 & 1 \end{bmatrix}$

6. $\begin{bmatrix} 0 & 1 & 0 \\ 0 & 0 & -2 \\ 1 & -1 & 0 \end{bmatrix}$ and $\begin{bmatrix} 1 & 0 & 1 \\ 1 & 0 & 0 \\ 0 & -1 & 0 \end{bmatrix}$

7. $\begin{bmatrix} 1 & 3 & 3 \\ 1 & 4 & 3 \\ 1 & 3 & 4 \end{bmatrix}$ and $\begin{bmatrix} 7 & -3 & -3 \\ -1 & 1 & 0 \\ -1 & 0 & 1 \end{bmatrix}$

8. $\begin{bmatrix} -1 & 0 & 2 \\ 3 & 1 & 0 \\ 0 & 2 & -3 \end{bmatrix}$ and $\begin{bmatrix} -\frac{1}{5} & \frac{4}{15} & -\frac{2}{15} \\ \frac{3}{5} & \frac{1}{5} & \frac{2}{5} \\ \frac{2}{5} & \frac{2}{15} & -\frac{1}{15} \end{bmatrix}$

9. Does a matrix with a row of all zeros have an inverse? Why?

10. Matrix A has A^{-1} as its inverse. What does $(A^{-1})^{-1}$ equal? (*Hint:* Experiment with a few matrices to see what you get.)

Find the inverse, if it exists, for each matrix.

11. $\begin{bmatrix} 1 & -1 \\ 2 & 0 \end{bmatrix}$

12. $\begin{bmatrix} -1 & 2 \\ -2 & -1 \end{bmatrix}$

13. $\begin{bmatrix} 3 & -1 \\ -5 & 2 \end{bmatrix}$

14. $\begin{bmatrix} -1 & -2 \\ 3 & 4 \end{bmatrix}$

15. $\begin{bmatrix} -6 & 4 \\ -3 & 2 \end{bmatrix}$

16. $\begin{bmatrix} 5 & 10 \\ -3 & -6 \end{bmatrix}$

17. $\begin{bmatrix} 1 & 0 & 0 \\ 0 & -1 & 0 \\ 1 & 0 & 1 \end{bmatrix}$

18. $\begin{bmatrix} 1 & 0 & 1 \\ 0 & -1 & 0 \\ 2 & 1 & 1 \end{bmatrix}$

19. $\begin{bmatrix} -1 & -1 & -1 \\ 4 & 5 & 0 \\ 0 & 1 & -3 \end{bmatrix}$

20. $\begin{bmatrix} 2 & 0 & 4 \\ 3 & 1 & 5 \\ -1 & 1 & -2 \end{bmatrix}$

21. $\begin{bmatrix} 1 & 2 & 3 \\ -3 & -2 & -1 \\ -1 & 0 & 1 \end{bmatrix}$

22. $\begin{bmatrix} 2 & 0 & 4 \\ 1 & 0 & -1 \\ 3 & 0 & -2 \end{bmatrix}$

23. $\begin{bmatrix} 2 & 4 & 6 \\ -1 & -4 & -3 \\ 0 & 1 & -1 \end{bmatrix}$

24. $\begin{bmatrix} 2 & 2 & -4 \\ 2 & 6 & 0 \\ -3 & -3 & 5 \end{bmatrix}$

25. $\begin{bmatrix} 1 & -2 & 3 & 0 \\ 0 & 1 & -1 & 1 \\ -2 & 2 & -2 & 4 \\ 0 & 2 & -3 & 1 \end{bmatrix}$

26. $\begin{bmatrix} 1 & 1 & 0 & 2 \\ 2 & -1 & 1 & -1 \\ 3 & 3 & 2 & -2 \\ 1 & 2 & 1 & 0 \end{bmatrix}$

Solve each system of equations by using the inverse of the coefficient matrix.

27. $\begin{aligned} 2x + 3y &= 10 \\ x - y &= -5 \end{aligned}$

28. $\begin{aligned} -x + 2y &= 15 \\ -2x - y &= 20 \end{aligned}$

29. $\begin{aligned} 2x + y &= 5 \\ 5x + 3y &= 13 \end{aligned}$

30. $\begin{aligned} -x - 2y &= 8 \\ 3x + 4y &= 24 \end{aligned}$

31. $\begin{aligned} -x + y &= 1 \\ 2x - y &= 1 \end{aligned}$

32. $\begin{aligned} 3x - 6y &= 1 \\ -5x + 9y &= -1 \end{aligned}$

33. $\begin{aligned} -x - 8y &= 12 \\ 3x + 24y &= -36 \end{aligned}$

34. $\begin{aligned} x + 3y &= -14 \\ 2x - y &= 7 \end{aligned}$

Solve each system of equations by using the inverse of the coefficient matrix. (The inverses for the first four problems were found in Exercises 19, 20, 23, and 24 above.)

35. $\begin{aligned} -x - y - z &= 1 \\ 4x + 5y &= -2 \\ y - 3z &= 3 \end{aligned}$

36. $\begin{aligned} 2x + 4z &= -8 \\ 3x + y + 5z &= 2 \\ -x + y - 2z &= 4 \end{aligned}$

37. $\begin{aligned} 2x + 4y + 6z &= 4 \\ -x - 4y - 3z &= 8 \\ y - z &= -4 \end{aligned}$

38. $\begin{aligned} 2x + 2y - 4z &= 12 \\ 2x + 6y &= 16 \\ -3x - 3y + 5z &= -20 \end{aligned}$

39. $\begin{aligned} 2x - 2y &= 5 \\ 4y + 8z &= 7 \\ x + 2z &= 1 \end{aligned}$

40. $\begin{aligned} x + z &= 3 \\ y + 2z &= 8 \\ -x + y &= 4 \end{aligned}$

Solve the systems of equations in Exercises 41–42 by using the inverse of the coefficient matrix. (The inverses were found in Exercises 25 and 26.)

41.
$$\begin{aligned} x - 2y + 3z &= 4 \\ y - z + w &= -8 \\ -2x + 2y - 2z + 4w &= 12 \\ 2y - 3z + w &= -4 \end{aligned}$$

42.
$$\begin{aligned} x + y + 2w &= 3 \\ 2x - y + z - w &= 3 \\ 3x + 3y + 2z - 2w &= 5 \\ x + 2y + z &= 3 \end{aligned}$$

Let $A = \begin{bmatrix} a & b \\ c & d \end{bmatrix}$ in Exercises 43–48.

43. Show that $IA = A$.

44. Show that $AI = A$.

45. Show that $A \cdot 0 = 0$.

46. Find A^{-1}.
(Assume $ad - bc \neq 0$.)

47. Show that $A^{-1}A = I$.

48. Show that $AA^{-1} = I$.

49. Using the definitions and properties listed in this section, show that for square matrices A and B of the same size, if $AB = O$ and if A^{-1} exists, then $B = O$.

50. Prove that, if it exists, the inverse of a matrix is unique. (*Hint:* Assume there are two inverses B and C for some matrix A, so that $AB = BA = I$ and $AC = CA = I$. Multiply the first equation by C and the second by B.)

Applications

Business and Economics

Solve the following problems by using the inverse of the coefficient matrix to solve a system of equations.

51. Analysis of Orders The Bread Box Bakery sells three types of cake, each requiring the amounts of the basic ingredients shown in the following matrix.

		Type of Cake		
		I	II	III
Ingredient	Flour (in cups)	2	4	2
	Sugar (in cups)	2	1	2
	Eggs	2	1	3

To fill its daily orders for these three kinds of cake, the bakery uses 72 cups of flour, 48 cups of sugar, and 60 eggs.

(a) Write a 3×1 matrix for the amounts used daily.

(b) Let the number of daily orders for cakes be a 3×1 matrix X with entries x_1, x_2, and x_3. Write a matrix equation that can be solved for X, using the given matrix and the matrix from part (a).

(c) Solve the equation from part (b) to find the number of daily orders for each type of cake.

52. Production Requirements An electronics company produces transistors, resistors, and computer chips. Each transistor requires 3 units of copper, 1 unit of zinc, and 2 units of glass. Each resistor requires 3, 2, and 1 units of the three materials, and each computer chip requires 2, 1, and 2 units of these materials, respectively. How many of each product can be made with the following amounts of materials?

(a) 810 units of copper, 410 units of zinc, and 490 units of glass

(b) 765 units of copper, 385 units of zinc, and 470 units of glass

(c) 1010 units of copper, 500 units of zinc, and 610 units of glass

53. Investments An investment firm recommends that a client invest in AAA, A, and B rated bonds. The average yield on AAA bonds is 6%, on A bonds 7%; and on B bonds 10%. The client wants to invest twice as much in AAA bonds as in B bonds. How much should be invested in each type of bond under the following conditions?

(a) The total investment is $25,000, and the investor wants an annual return of $1810 on the three investments.

(b) The values in part (a) are changed to $30,000 and $2150, respectively.

(c) The values in part (a) are changed to $40,000 and $2900, respectively.

Computer/Graphing Calculator

Use matrices C and D in Exercises 54–58.

$$C = \begin{bmatrix} -6 & 8 & 2 & 4 & -3 \\ 1 & 9 & 7 & -12 & 5 \\ 15 & 2 & -8 & 10 & 11 \\ 4 & 7 & 9 & 6 & -2 \\ 1 & 3 & 8 & 23 & 4 \end{bmatrix}, \quad D = \begin{bmatrix} 5 & -3 & 7 & 9 & 2 \\ 6 & 8 & -5 & 2 & 1 \\ 3 & 7 & -4 & 2 & 11 \\ 5 & -3 & 9 & 4 & -1 \\ 0 & 3 & 2 & 5 & 1 \end{bmatrix}$$

54. Find C^{-1}. **55.** Find $(CD)^{-1}$. **56.** Find D^{-1}. **57.** Is $C^{-1}D^{-1} = (CD)^{-1}$? **58.** Is $D^{-1}C^{-1} = (CD)^{-1}$?

Solve the matrix equation $AX = B$ for X by finding A^{-1}, given A and B as follows.

59. $A = \begin{bmatrix} 2 & 3 & 5 \\ 1 & 7 & 9 \\ -3 & 2 & 10 \end{bmatrix}$, $B = \begin{bmatrix} 3 \\ 4 \\ 1 \end{bmatrix}$

60. $A = \begin{bmatrix} 2 & 5 & 7 & 9 \\ 1 & 3 & -4 & 6 \\ -1 & 0 & 5 & 8 \\ 2 & -2 & 4 & 10 \end{bmatrix}$, $B = \begin{bmatrix} 3 \\ 7 \\ -1 \\ 5 \end{bmatrix}$

61. $A = \begin{bmatrix} 3 & 2 & -1 & -2 & 6 \\ -5 & 17 & 4 & 3 & 15 \\ 7 & 9 & -3 & -7 & 12 \\ 9 & -2 & 1 & 4 & 8 \\ 1 & 21 & 9 & -7 & 25 \end{bmatrix}$, $B = \begin{bmatrix} -2 \\ 5 \\ 3 \\ -8 \\ 25 \end{bmatrix}$

2.6 INPUT-OUTPUT MODELS

? **What production levels are needed to keep an economy going and to supply demands from outside the economy?**

A method for solving such questions is developed in this section.

Nobel prize winner Wassily Leontief developed an interesting application of matrix theory to economics. His matrix models for studying the interdependencies in an economy are called *input-output* models. In practice these models are very complicated, with many variables. Only simple examples with a few variables are discussed here.

Input-output models are concerned with the production and flow of goods (and perhaps services). In an economy with *n* basic commodities, or sectors, the production of each commodity uses some (perhaps all) of the commodities in the economy as inputs. The amounts of each commodity used in the production of one unit of each commodity can be written as an $n \times n$ matrix A, called the **technological matrix** or **input-output matrix** of the economy.

 EXAMPLE 1 Suppose a simplified economy involves just three commodity categories: agriculture, manufacturing, and transportation, all in appropriate units. Production of 1 unit of agriculture requires 1/2 unit of manufacturing and 1/4 unit of trans-

portation; production of 1 unit of manufacturing requires 1/4 unit of agriculture and 1/4 unit of transportation; and production of 1 unit of transportation requires 1/3 unit of agriculture and 1/4 unit of manufacturing. Give the input-output matrix for this economy.

The matrix is shown below.

$$
\begin{array}{c}
\\
\text{Agriculture} \\
\text{Manufacturing} \\
\text{Transportation}
\end{array}
\begin{array}{ccc}
\text{Agriculture} & \text{Manufacturing} & \text{Transportation}
\end{array}
\begin{bmatrix}
0 & \frac{1}{4} & \frac{1}{3} \\
\frac{1}{2} & 0 & \frac{1}{4} \\
\frac{1}{4} & \frac{1}{4} & 0
\end{bmatrix} = A
$$

The first column of the input-output matrix represents the amount of each of the three commodities consumed in the production of 1 unit of agriculture. The second column gives the amounts required to produce 1 unit of manufacturing, and the last column gives the amounts required to produce 1 unit of transportation. (Although it is perhaps unrealistic that production of a unit of each commodity requires none of that commodity, the simpler matrix involved is useful for our purposes.) ◄

Another matrix used with the input-output matrix is the matrix giving the amount of each commodity produced, called the **production matrix,** or the matrix of gross output. In an economy producing n commodities, the production matrix can be represented by a column matrix X with entries $x_1, x_2, x_3, \ldots, x_n$.

EXAMPLE 2 In Example 1, suppose the production matrix is

$$
X = \begin{bmatrix} 60 \\ 52 \\ 48 \end{bmatrix}.
$$

Then 60 units of agriculture, 52 units of manufacturing, and 48 units of transportation are produced. Because 1/4 unit of agriculture is used for each unit of manufacturing produced, $1/4 \times 52 = 13$ units of agriculture must be used in the "production" of manufacturing. Similarly, $1/3 \times 48 = 16$ units of agriculture will be used in the "production" of transportation. Thus, $13 + 16 = 29$ units of agriculture are used for production in the economy. Look again at the matrices A and X. Since X gives the number of units of each commodity produced and A gives the amount (in units) of each commodity used to produce 1 unit of each of the various commodities, the matrix product AX gives the amount of each commodity used in the production process.

$$
AX = \begin{bmatrix} 0 & \frac{1}{4} & \frac{1}{3} \\ \frac{1}{2} & 0 & \frac{1}{4} \\ \frac{1}{4} & \frac{1}{4} & 0 \end{bmatrix} \begin{bmatrix} 60 \\ 52 \\ 48 \end{bmatrix} = \begin{bmatrix} 29 \\ 42 \\ 28 \end{bmatrix}
$$

From this result, 29 units of agriculture, 42 units of manufacturing, and 28 units of transportation are used to produce 60 units of agriculture, 52 units of manufacturing, and 48 units of transportation. ◄

The matrix product AX represents the amount of each commodity used in the production process. The remainder (if any) must be enough to satisfy the demand for the various commodities from outside the production system. In an n-commodity economy, this demand can be represented by a **demand matrix** D with entries d_1, d_2, \ldots, d_n. The difference between the production matrix X and the amount AX used in the production process must equal the demand D, or

$$D = X - AX.$$

In Example 2,

$$D = \begin{bmatrix} 60 \\ 52 \\ 48 \end{bmatrix} - \begin{bmatrix} 29 \\ 42 \\ 28 \end{bmatrix} = \begin{bmatrix} 31 \\ 10 \\ 20 \end{bmatrix},$$

so production of 60 units of agriculture, 52 units of manufacturing, and 48 units of transportation would satisfy a demand of 31, 10, and 20 units of each commodity, respectively.

Another way to state the relationship between production X and demand D is to express X as $X = D + AX$. In practice, A and D usually are known and X must be found. That is, we need to decide what amounts of production are needed to satisfy the required demands. Matrix algebra can be used to solve the equation $D = X - AX$ for X.

$$D = X - AX$$
$$D = IX - AX \qquad \text{Identity property}$$
$$D = (I - A)X \qquad \text{Distributive property}$$

If the matrix $I - A$ has an inverse, then

$$X = (I - A)^{-1}D.$$

EXAMPLE 3 Suppose, in the three-commodity economy from Examples 1 and 2, there is a demand for 516 units of agriculture, 258 units of manufacturing, and 129 units of transportation. What should production of each commodity be?

The demand matrix is

$$D = \begin{bmatrix} 516 \\ 258 \\ 129 \end{bmatrix}.$$

To find the production matrix X, first calculate $I - A$.

$$I - A = \begin{bmatrix} 1 & 0 & 0 \\ 0 & 1 & 0 \\ 0 & 0 & 1 \end{bmatrix} - \begin{bmatrix} 0 & \frac{1}{4} & \frac{1}{3} \\ \frac{1}{2} & 0 & \frac{1}{4} \\ \frac{1}{4} & \frac{1}{4} & 0 \end{bmatrix} = \begin{bmatrix} 1 & -\frac{1}{4} & -\frac{1}{3} \\ -\frac{1}{2} & 1 & -\frac{1}{4} \\ -\frac{1}{4} & -\frac{1}{4} & 1 \end{bmatrix}$$

Use row operations to find the inverse of $I - A$ (the entries are rounded to two decimal places).

$$(I - A)^{-1} = \begin{bmatrix} 1.40 & .50 & .59 \\ .84 & 1.36 & .62 \\ .56 & .47 & 1.30 \end{bmatrix}$$

Since $X = (I - A)^{-1}D$,

$$X = \begin{bmatrix} 1.40 & .50 & .59 \\ .84 & 1.36 & .62 \\ .56 & .47 & 1.30 \end{bmatrix} \begin{bmatrix} 516 \\ 258 \\ 129 \end{bmatrix} = \begin{bmatrix} 928 \\ 864 \\ 578 \end{bmatrix}.$$

(Each entry in X has been rounded to the nearest whole number).

The last result shows that production of 928 units of agriculture, 864 units of manufacturing, and 578 units of transportation is required to satisfy demands of 516, 258, and 129 units respectively. ◀

EXAMPLE 4 ▶ An economy depends on two basic products, wheat and oil. To produce 1 metric ton of wheat requires .25 metric tons of wheat and .33 metric tons of oil. Production of 1 metric ton of oil consumes .08 metric tons of wheat and .11 metric tons of oil. Find the production that will satisfy a demand for 500 metric tons of wheat and 1000 metric tons of oil.

The input-output matrix is

$$A = \begin{bmatrix} .25 & .08 \\ .33 & .11 \end{bmatrix}.$$

Also,

$$I - A = \begin{bmatrix} .75 & -.08 \\ -.33 & .89 \end{bmatrix}.$$

Next, calculate $(I - A)^{-1}$.

$$(I - A)^{-1} = \begin{bmatrix} 1.39 & .13 \\ .51 & 1.17 \end{bmatrix} \quad \text{(rounded)}$$

To find the production matrix X, use the equation $X = (I - A)^{-1}D$, with

$$D = \begin{bmatrix} 500 \\ 1000 \end{bmatrix}.$$

The production matrix is

$$X = \begin{bmatrix} 1.39 & .13 \\ .51 & 1.17 \end{bmatrix} \begin{bmatrix} 500 \\ 1000 \end{bmatrix} = \begin{bmatrix} 815 \\ 1425 \end{bmatrix}.$$

Production of 815 metric tons of wheat and 1425 metric tons of oil is required to satisfy the indicated demand. ◀

The input-output model discussed above is referred to as an **open model,** since it allows for a surplus from the production equal to D. In the **closed model,** all the production is consumed internally in the production process, so that $X = AX$. There is nothing left over to satisfy any outside demands from other parts of the economy or from other economies. In this case, the sum of each column in the input-output matrix equals 1.

To solve the equation $X = AX$ for X, first let O represent a column matrix with n rows, in which each element is equal to 0. Write $X = AX$ or $X - AX = O$; then

$$IX - AX = O \qquad \text{Identity property}$$

$$(I - A)X = O. \qquad \text{Distributive property}$$

Parameters were discussed in the first section of this chapter. As mentioned there, parameters are required when a system has infinitely many solutions.

The system of equations that corresponds to $(I - A)X = O$ does not have a single unique solution, but it can be solved in terms of a parameter.

EXAMPLE 5 Use matrix A below to find the production of each commodity in a closed model.

$$A = \begin{bmatrix} \frac{1}{2} & \frac{1}{4} & \frac{1}{3} \\ 0 & \frac{1}{4} & \frac{1}{3} \\ \frac{1}{2} & \frac{1}{2} & \frac{1}{3} \end{bmatrix}$$

Find the value of $I - A$, then set $(I - A)X = O$ to find X.

$$I - A = \begin{bmatrix} \frac{1}{2} & -\frac{1}{4} & -\frac{1}{3} \\ 0 & \frac{3}{4} & -\frac{1}{3} \\ -\frac{1}{2} & -\frac{1}{2} & \frac{2}{3} \end{bmatrix}$$

$$(I - A)X = \begin{bmatrix} \frac{1}{2} & -\frac{1}{4} & -\frac{1}{3} \\ 0 & \frac{3}{4} & -\frac{1}{3} \\ -\frac{1}{2} & -\frac{1}{2} & \frac{2}{3} \end{bmatrix} \begin{bmatrix} x_1 \\ x_2 \\ x_3 \end{bmatrix} = \begin{bmatrix} 0 \\ 0 \\ 0 \end{bmatrix}$$

Multiply to get

$$\begin{bmatrix} \frac{1}{2}x_1 - \frac{1}{4}x_2 - \frac{1}{3}x_3 \\ 0\,x_1 + \frac{3}{4}x_2 - \frac{1}{3}x_3 \\ -\frac{1}{2}x_1 - \frac{1}{2}x_2 + \frac{2}{3}x_3 \end{bmatrix} = \begin{bmatrix} 0 \\ 0 \\ 0 \end{bmatrix}.$$

The last matrix equation corresponds to the following system.

$$\tfrac{1}{2}x_1 - \tfrac{1}{4}x_2 - \tfrac{1}{3}x_3 = 0$$

$$\tfrac{3}{4}x_2 - \tfrac{1}{3}x_3 = 0$$

$$-\tfrac{1}{2}x_1 - \tfrac{1}{2}x_2 + \tfrac{2}{3}x_3 = 0$$

Solving the sytem with x_3 as the parameter gives the solution of the system

$$(\tfrac{8}{9}x_3, \tfrac{4}{9}x_3, x_3).$$

If $x_3 = 9$, then $x_1 = 8$ and $x_2 = 4$, so the production of the three commodities should be in the ratio 8:4:9. ◀

2.6 Exercises

Find the production matrix for the following input-output and demand matrices using the open model.

1. $A = \begin{bmatrix} .5 & .4 \\ .25 & .2 \end{bmatrix}, \quad D = \begin{bmatrix} 2 \\ 4 \end{bmatrix}$

2. $A = \begin{bmatrix} .2 & .04 \\ .6 & .05 \end{bmatrix}, \quad D = \begin{bmatrix} 3 \\ 10 \end{bmatrix}$

3. $A = \begin{bmatrix} .1 & .03 \\ .07 & .6 \end{bmatrix}, \quad D = \begin{bmatrix} 5 \\ 10 \end{bmatrix}$

4. $A = \begin{bmatrix} .01 & .03 \\ .05 & .05 \end{bmatrix}, \quad D = \begin{bmatrix} 100 \\ 200 \end{bmatrix}$

5. $A = \begin{bmatrix} .4 & 0 & .3 \\ 0 & .8 & .1 \\ 0 & .2 & .4 \end{bmatrix}, \quad D = \begin{bmatrix} 1 \\ 3 \\ 2 \end{bmatrix}$

6. $A = \begin{bmatrix} .1 & .5 & 0 \\ 0 & .3 & .4 \\ .1 & .2 & .1 \end{bmatrix}, \quad D = \begin{bmatrix} 10 \\ 4 \\ 2 \end{bmatrix}$

Find the ratios of products A, B, *and* C *using a closed model.*

7.
$$\begin{array}{c} \\ A \\ B \\ C \end{array} \begin{array}{c} \begin{array}{ccc} A & B & C \end{array} \\ \begin{bmatrix} .3 & .1 & .8 \\ .5 & .6 & .1 \\ .2 & .3 & .1 \end{bmatrix} \end{array}$$

8.
$$\begin{array}{c} \\ A \\ B \\ C \end{array} \begin{array}{c} \begin{array}{ccc} A & B & C \end{array} \\ \begin{bmatrix} .2 & .1 & .5 \\ .4 & .3 & .4 \\ .4 & .6 & .1 \end{bmatrix} \end{array}$$

Applications

Business and Economics

Input-Output Open Model *In Exercises 9 and 10, refer to Example 4.*

9. If the demand is changed to 690 metric tons of wheat and 920 metric tons of oil, how many units of each commodity should be produced?

10. Change the technological matrix so that production of 1 ton of wheat requires 1/5 metric ton of oil (and no wheat), and production of 1 metric ton of oil requires 1/3 metric ton of wheat (and no oil). To satisfy the same demand matrix, how many units of each commodity should be produced?

Input-Output Open Model *In Exercises 11–14, refer to Example 3.*

11. If the demand is changed to 516 units of each commodity, how many units of each commodity should be produced?

12. Suppose 1/3 unit of manufacturing (no agriculture or transportation) is required to produce 1 unit of agriculture, 1/4 unit of transportation is required to produce 1 unit of manufacturing, and 1/2 unit of agriculture is required to produce 1 unit of transportation. How many units of each commodity should be produced to satisfy a demand of 1000 units for each commodity?

13. Suppose 1/4 unit of manufacturing and 1/2 unit of transportation are required to produce 1 unit of agriculture, 1/2 unit of agriculture and 1/4 unit of transportation to produce 1 unit of manufacturing, and 1/4 unit of agriculture and 1/4 unit of manufacturing to produce 1 unit of transportation. How many units of each commodity should be produced to satisfy a demand of 1000 units for each commodity?

14. If the input-output matrix is changed so that 1/4 unit of manufacturing and 1/2 unit of transportation are required to produce 1 unit of agriculture, 1/2 unit of agriculture and 1/4 unit of transportation are required to produce 1 unit of manufacturing, and 1/4 unit each of agriculture and manufacturing are required to produce 1 unit of transportation, find the number of units of each commodity that should be produced to satisfy a demand for 500 units of each commodity.

15. A primitive economy depends on two basic goods, yams and pork. Production of 1 bushel of yams requires 1/4 bushel of yams and 1/2 of a pig. To produce 1 pig requires 1/6 bushel of yams. Find the amount of each commodity that should be produced to get the following.

(a) 1 bushel of yams and 1 pig

(b) 100 bushels of yams and 70 pigs

16. **Input-Output Closed Model** Use the input-output matrix

$$\begin{array}{c} \\ \text{Yams} \\ \text{Pigs} \end{array} \begin{array}{c} \begin{array}{cc} \text{Yams} & \text{Pigs} \end{array} \\ \begin{bmatrix} \frac{1}{4} & \frac{1}{2} \\ \frac{3}{4} & \frac{1}{2} \end{bmatrix} \end{array}$$

and the closed model to find the ratios of yams and pigs produced.

Computer/Graphing Calculator

Use a computer to find the production matrix X, given the following input-output and demand matrices.

17. $A = \begin{bmatrix} .25 & .25 & .25 & .05 \\ .01 & .02 & .01 & .1 \\ .3 & .3 & .01 & .1 \\ .2 & .01 & .3 & .01 \end{bmatrix}$ $D = \begin{bmatrix} 2930 \\ 3570 \\ 2300 \\ 580 \end{bmatrix}$ 18. $A = \begin{bmatrix} .2 & .1 & .2 & .2 \\ .3 & .05 & .07 & .02 \\ .1 & .03 & .02 & .01 \\ .05 & .3 & .05 & .03 \end{bmatrix}$ $D = \begin{bmatrix} 5000 \\ 1000 \\ 8000 \\ 500 \end{bmatrix}$

19. A simple economy depends on three commodities: oil, corn, and coffee. Production of 1 unit of oil re-quires .1 unit of oil, .2 unit of corn, and no units of coffee. To produce 1 unit of corn requires .2 unit of oil, .1 unit of corn, and .05 unit of coffee. To produce 1 unit of coffee requires .1 unit of oil, .05 unit of corn, and .1 unit of coffee. Find the gross production required to give a net production of 1000 units each of oil, corn, coffee.

Chapter Summary Key Terms

To understand the concepts presented in this chapter, you should know the meaning and use of the following terms. For easy reference, the section in the chapter where a word (or expression) was first used is given with each item.

system of equations
2.1 first-degree equation in n unknowns
inconsistent system
dependent equations
equivalent system
echelon method
back-substitution
parameter
2.2 matrix (matrices)

element (entry)
augmented matrix
row operations
Gauss-Jordan method
2.3 square matrix
row matrix (row vector)
column matrix (column vector)
additive inverse (negative) of a matrix
zero matrix

additive identity
2.4 scalar
2.5 identity matrix
multiplicative inverse matrix
2.6 input-output (technological) matrix
production matrix
demand matrix
open model
closed model

Chapter 2 Review Exercises

1. What is true about the number of solutions to a system of m linear equations in n unknowns if $m = n$? If $m < n$? If $m > n$?

2. Suppose someone says that a more reasonable way to multiply two matrices than the method presented in the text is to multiply corresponding elements. For example, the result of

$$\begin{bmatrix} 1 & 2 \\ 3 & 4 \end{bmatrix} \cdot \begin{bmatrix} 3 & 5 \\ 7 & 11 \end{bmatrix} \quad \text{should be} \quad \begin{bmatrix} 3 & 10 \\ 21 & 44 \end{bmatrix},$$

according to this person. How would you respond?

Solve each system by the echelon method.

3. $\begin{aligned} 2x + 3y &= 10 \\ -3x + y &= 18 \end{aligned}$

4. $\begin{aligned} \frac{x}{2} + \frac{y}{4} &= 3 \\ \frac{x}{4} - \frac{y}{2} &= 4 \end{aligned}$

5. $\begin{aligned} 2x - 3y + z &= -5 \\ x + 4y + 2z &= 13 \\ 5x + 5y + 3z &= 14 \end{aligned}$

6. $\begin{aligned} x - y &= 3 \\ 2x + 3y + z &= 13 \\ 3x - 2z &= 21 \end{aligned}$

Solve each system by the Gauss-Jordan method.

7. $\begin{aligned} 2x + 4y &= -6 \\ -3x - 5y &= 12 \end{aligned}$

8. $\begin{aligned} x + 2y &= -9 \\ 4x + 9y &= 41 \end{aligned}$

9. $\begin{aligned} x - y + 3z &= 13 \\ 4x + y + 2z &= 17 \\ 3x + 2y + 2z &= 1 \end{aligned}$

10.
$$x - 2z = 5$$
$$3x + 2y = 8$$
$$-x + 2z = 10$$

11.
$$3x - 6y + 9z = 12$$
$$-x + 2y - 3z = -4$$
$$x + y + 2z = 7$$

Find the size of each matrix, find the values of any variables, and identify any square, row, or column matrices.

12. $\begin{bmatrix} 2 & 3 \\ 5 & q \end{bmatrix} = \begin{bmatrix} a & b \\ c & 9 \end{bmatrix}$

13. $\begin{bmatrix} 2 & x \\ y & 6 \\ 5 & z \end{bmatrix} = \begin{bmatrix} a & -1 \\ 4 & 6 \\ p & 7 \end{bmatrix}$

14. $[m \quad 4 \quad z \quad -1] = [12 \quad k \quad -8 \quad r]$

15. $\begin{bmatrix} a + 5 & 3b & 6 \\ 4c & 2 + d & -3 \\ -1 & 4p & q - 1 \end{bmatrix} = \begin{bmatrix} -7 & b + 2 & 2k - 3 \\ 3 & 2d - 1 & 4l \\ m & 12 & 8 \end{bmatrix}$

Given the matrices

$$A = \begin{bmatrix} 4 & 10 \\ -2 & -3 \\ 6 & 9 \end{bmatrix}, \quad B = \begin{bmatrix} 2 & 3 & -2 \\ 2 & 4 & 0 \\ 0 & 1 & 2 \end{bmatrix}, \quad C = \begin{bmatrix} 5 & 0 \\ -1 & 3 \\ 4 & 7 \end{bmatrix},$$

$$D = \begin{bmatrix} 6 \\ 1 \\ 0 \end{bmatrix}, \quad E = [1 \quad 3 \quad -4], \quad F = \begin{bmatrix} -1 & 4 \\ 3 & 7 \end{bmatrix}, \quad G = \begin{bmatrix} 2 & 5 \\ 1 & 6 \end{bmatrix},$$

find each of the following, if it exists.

16. $A + C$ **17.** $2G - 4F$ **18.** $3C + 2A$ **19.** $B - A$ **20.** $2A - 5C$

21. AF **22.** AC **23.** DE **24.** ED **25.** BD

26. EA **27.** F^{-1} **28.** B^{-1} **29.** $(A + C)^{-1}$

Find the inverse of each of the following matrices that has an inverse.

30. $\begin{bmatrix} 2 & 1 \\ 5 & 3 \end{bmatrix}$ **31.** $\begin{bmatrix} -4 & 2 \\ 0 & 3 \end{bmatrix}$ **32.** $\begin{bmatrix} 2 & 0 \\ -1 & 5 \end{bmatrix}$ **33.** $\begin{bmatrix} 6 & 4 \\ 3 & 2 \end{bmatrix}$

34. $\begin{bmatrix} 2 & -1 & 0 \\ 1 & 0 & 1 \\ 1 & -2 & 0 \end{bmatrix}$ **35.** $\begin{bmatrix} 2 & 0 & 4 \\ 1 & -1 & 0 \\ 0 & 1 & -2 \end{bmatrix}$ **36.** $\begin{bmatrix} 1 & 3 & 6 \\ 4 & 0 & 9 \\ 5 & 15 & 30 \end{bmatrix}$ **37.** $\begin{bmatrix} 2 & 3 & 5 \\ -2 & -3 & -5 \\ 1 & 4 & 2 \end{bmatrix}$

Solve the matrix equation $AX = B$ for X using the given matrices.

38. $A = \begin{bmatrix} 2 & 4 \\ -1 & -3 \end{bmatrix}$, $B = \begin{bmatrix} 8 \\ 3 \end{bmatrix}$

39. $A = \begin{bmatrix} 1 & 2 \\ 2 & 4 \end{bmatrix}$, $B = \begin{bmatrix} 5 \\ 10 \end{bmatrix}$

40. $A = \begin{bmatrix} 1 & 0 & 2 \\ -1 & 1 & 0 \\ 3 & 0 & 4 \end{bmatrix}$, $B = \begin{bmatrix} 8 \\ 4 \\ -6 \end{bmatrix}$

41. $A = \begin{bmatrix} 2 & 4 & 0 \\ 1 & -2 & 0 \\ 0 & 0 & 3 \end{bmatrix}$, $B = \begin{bmatrix} 72 \\ -24 \\ 48 \end{bmatrix}$

Solve each of the following systems of equations by inverses.

42.
$$2x + y = 5$$
$$3x - 2y = 4$$

43.
$$5x + 10y = 80$$
$$3x - 2y = 120$$

44.
$$x + y + z = 1$$
$$2x + y = -2$$
$$3y + z = 2$$

Find each production matrix, given the following input-output and demand matrices.

45. $A = \begin{bmatrix} .01 & .05 \\ .04 & .03 \end{bmatrix}$, $D = \begin{bmatrix} 200 \\ 300 \end{bmatrix}$

46. $A = \begin{bmatrix} .2 & .1 & .3 \\ .1 & 0 & .2 \\ 0 & 0 & .4 \end{bmatrix}$, $D = \begin{bmatrix} 500 \\ 200 \\ 100 \end{bmatrix}$

Applications

Business and Economics

Write each of Exercises 47–50 as a system of equations and solve.

47. Scheduling Production An office supply manufacturer makes two kinds of paper clips, standard and extra large. To make 1000 standard paper clips requires 1/4 hr on a cutting machine and 1/2 hr on a machine that shapes the clips. One thousand extra large paper clips require 1/3 hr on each machine. The manager of paper clip production has 4 hr per day available on the cutting machine and 6 hr per day on the shaping machine. How many of each kind of clip can he make?

48. Investment Gretchen Schmidt plans to buy shares of two stocks. One costs $32 per share and pays dividends of $1.20 per share. The other costs $23 per share and pays dividends of $1.40 per share. She has $10,100 to spend and wants to earn dividends of $540. How many shares of each stock should she buy?

49. Production Requirements The Waputi Indians make woven blankets, rugs, and skirts. Each blanket requires 24 hr for spinning the yarn, 4 hr for dying the yarn, and 15 hr for weaving. Rugs require 30, 5, and 18 hr and skirts 12, 3, and 9 hr, respectively. If there are 306, 59, and 201 hr available for spinning, dying, and weaving, respectively, how many of each item can be made? (*Hint:* Simplify the equations you write, if possible, before solving the system.)

50. Distribution An oil refinery in Tulsa sells 50% of its production to a Chicago distributor, 20% to a Dallas distributor, and 30% to an Atlanta distributor. Another refinery in New Orleans sells 40% of its production to the Chicago distributor, 40% to the Dallas distributor, and 20% to the Atlanta distributor. A third refinery in Ardmore sells the same distributors 30%, 40%, and 30% of its production. The three distributors received 219,000, 192,000, and 144,000 gallons of oil, respectively. How many gallons of oil were produced at each of the three plants?

51. Stock Reports The New York Stock Exchange reports in daily newspapers give the dividend, price-to-earnings ratio, sales (in hundreds of shares), last price, and change in price for each company. Write the following stock reports as a 4×5 matrix: American Telephone & Telegraph: 5, 7, 2532, $52\frac{3}{8}$, $-\frac{1}{4}$; General Electric: 3, 9, 1464, 56, $+\frac{1}{8}$; Mobil Oil: 2.50, 5, 4974, 41, $-1\frac{1}{2}$; Sears: 1.36, 10, 1754, $18\frac{7}{8}$, $+\frac{1}{2}$.

52. Filling Orders A printer has three orders for pamphlets that require three kinds of paper, as shown in the following matrix.

$$
\begin{array}{c}
\\
\\
\text{Paper}
\end{array}
\begin{array}{c}
\\
\text{High-grade} \\
\text{Medium-grade} \\
\text{Coated}
\end{array}
\overset{\text{Order}}{
\begin{array}{ccc}
\text{I} & \text{II} & \text{III} \\
\end{array}}
\begin{bmatrix}
10 & 5 & 8 \\
12 & 0 & 4 \\
0 & 10 & 5
\end{bmatrix}
$$

The printer has on hand 3170 sheets of high-grade paper, 2360 sheets of medium-grade paper, and 1800 sheets of coated paper. All the paper must be used in preparing the order.

(a) Write a 3×1 matrix for the amounts of paper on hand.

(b) Write a matrix of variables to represent the number of pamphlets that must be printed in each of the three orders.

(c) Write a matrix equation using the given matrix and your matrices from parts (a) and (b).

(d) Solve the equation from part (c).

53. Input-Output An economy depends on two commodities, goats and cheese. It takes 2/3 of a unit of goats to produce 1 unit of cheese and 1/2 unit of cheese to produce 1 unit of goats.

(a) Write the input-output matrix for this economy.

(b) Find the production required to satisfy a demand of 400 units of cheese and 800 units of goats.

Life Sciences

54. Animal Activity The activities of a grazing animal can be classified roughly into three categories: grazing, moving, and resting. Suppose horses spend 8 hr grazing, 8 moving, and 8 resting, cattle spend 10 grazing, 5 moving and 9 resting; sheep spend 7 grazing, 10 moving, and 7 resting; and goats spend 8 grazing, 9 moving, and 7 resting. Write this information as a 4×3 matrix.

Connections

55. The following system of equations is given.

$$x + 2y + z = 7$$
$$2x - y - z = 2$$
$$3x - 3y + 2z = -5$$

(a) Solve by the echelon method.

(b) The *Gaussian method* for solving a linear system uses row operations, and it parallels the echelon method exactly. With the Gaussian method, we change only the entries below the main diagonal to zeros and the

diagonal numbers to 1's, then use back-substitution. Use the Gaussian method to solve the system above and compare the steps with the echelon method.

(c) Solve by the Gauss-Jordan method. Compare with the Gaussian method.

(d) Write the system as a matrix equation, $AX = B$.

(e) Find the inverse of matrix A from part (d).

(f) Solve the system using A^{-1} from part (e).

Extended Application / Leontief's Model of the American Economy

In the April 1965 issue of *Scientific American*, Wassily Leontief explained his input-output system, using the 1958 American economy as an example.* He divided the economy into 81 sectors, grouped into six families of related sectors. In order to keep the discussion reasonably simple, we will treat each family of sectors as a single sector and so, in effect, work with a six-sector model. The sectors are listed in Table 1.

Table 1

Sector	Examples
Final nonmetal (FN)	Furniture, processed food
Final metal (FM)	Household appliances, motor vehicles
Basic metal (BM)	Machine-shop products, mining
Basic nonmetal (BN)	Agriculture, printing
Energy (E)	Petroleum, coal
Services (S)	Amusements, real estate

Table 2

	FN	FM	BM	BN	E	S
FN	.170	.004	0	.029	0	.008
FM	.003	.295	.018	.002	.004	.016
BM	.025	.173	.460	.007	.011	.007
BN	.348	.037	.021	.403	.011	.048
E	.007	.001	.039	.025	.358	.025
S	.120	.074	.104	.123	.173	.234

The workings of the American economy in 1958 are described in the input-output table (Table 2) based on Leontief's figures. We will demonstrate the meaning of Table 2 by considering the first left-hand column of numbers. The numbers in this column mean that 1 unit of final nonmetal production requires the consumption of .170 unit of (other) final nonmetal production, .003 unit of final metal output, .025 unit of basic metal products, and so on down the column. Since the unit of measurement that Leontief used for this table is millions of dollars, we conclude that $1 million worth of final nonmetal production consumes $.170 million, or $170,000, worth of other final nonmetal products, $3000 of final metal production, $25,000 of basic metal products, and so on. Similarly, the entry in the column headed FM and opposite S of .074 means that $74,000 worth of output from the service industries goes into the production of $1 million worth of final metal products, and the number .358 in the column headed E and opposite E means that $358,000 worth of energy must be consumed to produce $1 million worth of energy.

By the underlying assumption of Leontief's model, the production of n units (n = any number) of final nonmetal production consumes $.170n$ units of final nonmetal output, $.003n$ units of final metal output, $.025n$ units of basic metal production, and so on. Thus, production of $50 million worth of products from the final nonmetal section of the 1958 American economy required $(.170)(50) = 8.5$ units ($8.5 million) worth of final nonmetal output, $(.003)(50) = .15$ unit of final metal output, $(.025)(50) = 1.25$ units of basic metal production, and so on.

*From *Applied Finite Mathematics* by Robert F. Brown and Brenda W. Brown. (Belmont, California: Wadsworth Publishing Company, Inc., 1977). Reprinted by permission of the authors.

Example 1

According to the simplified input-output table for the 1958 American economy, how many dollars' worth of final metal products, basic nonmetal products, and services are required to produce $120 million worth of basic metal products?

Each unit ($1 million worth) of basic metal products requires .018 unit of final metal products because the number in the BM column opposite FM is .018. Thus, $120 million, or 120 units, requires $(.018)(120) = 2.16$ units, or $2.16 million of final metal products. Similarly, 120 units of basic metal production uses $(.021)(120) = 2.52$ units of basic nonmetal production and $(.104)(120) = 12.48$ units of services, or $2.52 million and $12.48 million of basic nonmetal output and services, respectively.

The Leontief model also involves a *bill* of demands, that is, a list of requirements for units of output beyond that required for its inner workings as described in the input-output table. These demands represent exports, surpluses, government and individual consumption, and the like. The bill of demands for the simplified version of the 1958 American economy we have been using was (in millions)

FN	$99,640
FM	$75,548
BM	$14,444
BN	$33,501
E	$23,527
S	$263,985.

We can now use the methods developed earlier to answer the following question: How many units of output from each sector are needed in order to run the economy and fill the bill of demands? The units of output from each sector required to run the economy and fill the bill of demands is unknown, so we denote them by variables. In our example, there are six quantities that are, at the moment, unknown. The number of units of final nonmetal production required to solve the problem will be our first unknown, because this sector is represented by the first row of the input-output matrix. The unknown quantity of final nonmetal units will be represented by the symbol x_1. Following the same pattern, we represent the unknown quantities in the following manner:

$x_1 = $ units of final nonmetal production required;

$x_2 = $ units of final metal production required;

$x_3 = $ units of basic metal production required;

$x_4 = $ units of basic nonmetal production required;

$x_5 = $ units of energy required;

$x_6 = $ units of services required.

These six numbers are the quantities we are attempting to calculate.

To find these numbers, first let A be the 6×6 matrix corresponding to the input-output table.

$$A = \begin{bmatrix} .170 & .004 & 0 & .029 & 0 & .008 \\ .003 & .295 & .018 & .002 & .004 & .016 \\ .025 & .173 & .460 & .007 & .011 & .007 \\ .348 & .037 & .021 & .403 & .011 & .048 \\ .007 & .001 & .039 & .025 & .358 & .025 \\ .120 & .074 & .104 & .123 & .173 & .234 \end{bmatrix}$$

A is the input-output matrix. The bill of demands leads to a 6×1 demand matrix D, and X is the matrix of unknowns.

$$D = \begin{bmatrix} 99,640 \\ 75,548 \\ 14,444 \\ 33,501 \\ 23,527 \\ 263,985 \end{bmatrix} \quad \text{and} \quad X = \begin{bmatrix} x_1 \\ x_2 \\ x_3 \\ x_4 \\ x_5 \\ x_6 \end{bmatrix}$$

Now we need to find $I - A$.

$$I - A = \begin{bmatrix} 1 & 0 & 0 & 0 & 0 & 0 \\ 0 & 1 & 0 & 0 & 0 & 0 \\ 0 & 0 & 1 & 0 & 0 & 0 \\ 0 & 0 & 0 & 1 & 0 & 0 \\ 0 & 0 & 0 & 0 & 1 & 0 \\ 0 & 0 & 0 & 0 & 0 & 1 \end{bmatrix}$$
$$- \begin{bmatrix} .170 & .004 & 0 & .029 & 0 & .008 \\ .003 & .295 & .018 & .002 & .004 & .016 \\ .025 & .173 & .460 & .007 & .011 & .007 \\ .348 & .037 & .021 & .403 & .011 & .048 \\ .007 & .001 & .039 & .025 & .358 & .025 \\ .120 & .074 & .104 & .123 & .173 & .234 \end{bmatrix}$$
$$= \begin{bmatrix} .830 & -.004 & 0 & -.029 & 0 & -.008 \\ -.003 & .705 & -.018 & -.002 & -.004 & -.016 \\ -.025 & -.173 & .540 & -.007 & -.011 & -.007 \\ -.348 & -.037 & -.021 & .597 & -.011 & -.048 \\ -.007 & -.001 & -.039 & -.025 & .642 & -.025 \\ -.120 & -.074 & -.104 & -.123 & -.173 & .766 \end{bmatrix}$$

Find the inverse (actually an approximation) by the methods of this chapter.

$$(I - A)^{-1} = \begin{bmatrix} 1.234 & 0.014 & .006 & .064 & .007 & .018 \\ .017 & 1.436 & .057 & .012 & .020 & .032 \\ .071 & .465 & 1.877 & .019 & .045 & .031 \\ .751 & .134 & .100 & 1.740 & .066 & .124 \\ .060 & .045 & .130 & .082 & 1.578 & .059 \\ .339 & .236 & .307 & .312 & .376 & 1.349 \end{bmatrix}$$

Therefore,

$$X = (I - A)^{-1}D = \begin{bmatrix} 1.234 & .014 & .006 & .064 & .007 & .018 \\ .017 & 1.436 & .057 & .012 & .020 & .032 \\ .071 & .465 & 1.877 & .019 & .045 & .031 \\ .751 & .134 & .100 & 1.740 & .066 & .124 \\ .060 & .045 & .130 & .082 & 1.578 & .059 \\ .339 & .236 & .307 & .312 & .376 & 1.349 \end{bmatrix} \begin{bmatrix} 99,640 \\ 75,548 \\ 14,444 \\ 33,501 \\ 23,527 \\ 263,985 \end{bmatrix} = \begin{bmatrix} 131,161 \\ 120,324 \\ 79,194 \\ 178,936 \\ 66,703 \\ 426,542 \end{bmatrix}.$$

From this result,

$$x_1 = 131,161 \quad x_2 = 120,324 \quad x_3 = 79,194 \quad x_4 = 178,936 \quad x_5 = 66,703 \quad x_6 = 426,542.$$

In other words, by this model 131,161 units ($131,161 million worth) of final nonmetal production, 120,324 units of final metal output, 79,194 units of basic metal products, and so on are required to run the 1958 American economy and completely fill the stated bill of demands.

Exercises

1. A much simplified version of Leontief's 42-sector analysis of the 1947 American economy divides the economy into just three sectors: agriculture, manufacturing, and the household (i.e., the sector of the economy that produces labor). It consists of the following input-output table:

	Agriculture	Manufacturing	Household
Agriculture	.245	.102	.051
Manufacturing	.099	.291	.279
Household	.433	.372	.011

The bill of demands (in billions of dollars) is

Agriculture	2.88
Manufacturing	31.45
Household	30.91.

(a) Write the input-output matrix A, the demand matrix D, and the matrix X.
(b) Compute $I - A$.
(c) Check that

$$(I - A)^{-1} = \begin{bmatrix} 1.454 & .291 & .157 \\ .533 & 1.763 & .525 \\ .837 & .791 & 1.278 \end{bmatrix}$$

is an approximation to the inverse of $I - A$ by calculating $(I - A)^{-1}(I - A)$.
(d) Use the matrix of part (c) to compute X.
(e) Explain the meaning of the numbers in X in dollars.

2. An analysis of the 1958 Israeli economy* is here simplified by grouping the economy into three sectors: agriculture, manufacturing, and energy. The input-output table is the following.

	Agriculture	Manufacturing	Energy
Agriculture	.293	0	0
Manufacturing	.014	.207	.017
Energy	.044	.010	.216

Exports (in thousands of Israeli pounds) were as follows.

Agriculture	138,213
Manufacturing	17,597
Energy	1,786

(a) Write the input-output matrix A and the demand (export) matrix D.
(b) Compute $I - A$.
(c) Check that

$$(I - A)^{-1} = \begin{bmatrix} 1.414 & 0 & 0 \\ .027 & 1.261 & .027 \\ .080 & .016 & 1.276 \end{bmatrix}$$

is an approximation to the inverse of $I - A$ by calculating $(I - A)^{-1}(I - A)$.
(d) Use the matrix of part (c) to determine the number of Israeli pounds' worth of agricultural products, manufactured goods, and energy required to run this model of the Israeli economy and export the stated value of products.

*From Wassily Leontief, *Input-Output Economics* (New York: Oxford University Press, 1966), pp. 54–57.

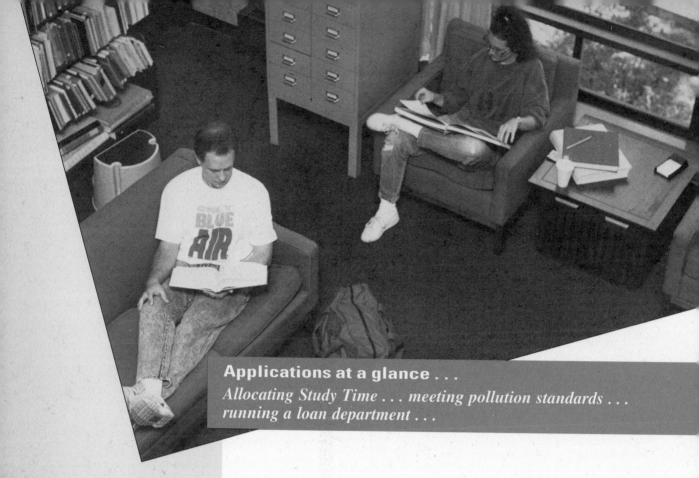

Applications at a glance...

*Allocating Study Time... meeting pollution standards...
running a loan department...*

(See page 152.)

Many realistic problems involve inequalities—a factory can manufacture *no more* than 12 items on a shift, or a medical researcher must interview *at least* a hundred patients to be sure that a new treatment for a disease is better than the old treatment. *Linear inequalities* of the form $ax + by \leq c$ (or with \geq, $<$, or $>$ instead of \leq) can be used in a process called *linear programming* to *optimize* (find the maximum or minimum value of a quantity for) a given situation.

In this chapter we introduce some *linear programming* problems that can be solved by graphical methods. Then, in Chapter 4, we discuss the simplex method, a general method for solving linear programming problems with many variables.

3.1 GRAPHING LINEAR INEQUALITIES

As mentioned above, a linear inequality is defined as follows.

LINEAR INEQUALITY

A **linear inequality** in two variables has the form

$$ax + by \leq c$$

for real numbers a, b, and c, with a and b not both 0.

Although we will generally use the symbol \leq in definitions and theorems involving inequalities, keep in mind that \leq may be replaced with \geq, $<$, or $>$.

EXAMPLE 1

Graph the linear inequality $3x - 2y \leq 6$.

Because of the "$=$" portion of \leq, the points of the line $3x - 2y = 6$ satisfy the linear inequality $3x - 2y \leq 6$ and are part of its graph. As in Chapter 1, find the intercepts by first letting $x = 0$ and then letting $y = 0$; use these points to get the graph of $3x - 2y = 6$ shown in Figure 1.

Recall from Chapter 1 that one way to sketch a line is to first let $x = 0$ to find the y-intercept, then let $y = 0$ to find the x-intercept. For example, given $3x - 2y = 6$, letting $x = 0$ yields $-2y = 6$, so $y = -3$. Letting $y = 0$ yields $3x = 6$, so $x = 2$. Plot these two points, as in Figure 1, then use a straightedge to draw a line through them.

FIGURE 1

Points satisfy $y > \frac{3}{2}x - 3$.

Points satisfy $y < \frac{3}{2}x - 3$.

FIGURE 2

The points on the line satisfy "$3x - 2y$ *equals* 6." To locate the points satisfying "$3x - 2y$ *is less than* or equal to 6," first solve $3x - 2y \le 6$ for y.

$$3x - 2y \le 6$$
$$-2y \le -3x + 6$$
$$y \ge \frac{3}{2}x - 3$$

(Recall that multiplying both sides of an inequality by a negative number reverses the direction of the inequality symbol.)

As shown in Figure 2, the points *above* the line $3x - 2y = 6$ satisfy

$$y > \frac{3}{2}x - 3,$$

while those below the line satisfy

$$y < \frac{3}{2}x - 3.$$

In summary, the inequality $3x - 2y \le 6$ is satisfied by all points *on or above* the line $3x - 2y = 6$. Indicate the points above the line by shading, as in Figure 3. The line and shaded region in Figure 3 make up the graph of the linear inequality $3x - 2y \le 6$. ◀

$3x - 2y \le 6$

FIGURE 3

Caution In this chapter, be sure to use a straightedge to draw lines, and to plot the points with care. A sloppily drawn line could give a deceptive picture of the region being considered.

In Example 1, the line $3x - 2y = 6$, which separates the points in the solution from the points that are not in the solution, is called the **boundary.**

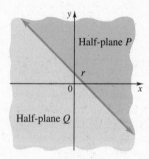

FIGURE 4

2 Graph $x + 4y < 4$.

The boundary here is the line $x + 4y = 4$. Since the points on this line do not satisfy $x + 4y < 4$, the line is drawn dashed, as in Figure 4. To decide whether to shade the region above the line or the region below the line, solve for y.

$$x + 4y < 4$$
$$4y < -x + 4$$
$$y < -\frac{1}{4}x + 1$$

Since y is less than $(-1/4)x + 1$, the solution is the region below the boundary, as shown by the shaded region in Figure 4. ◄

There is an alternative way to find the correct region to shade, or to check the method shown above. Choose as a test point any point not on the boundary line. For example, in Example 2 we could choose the point $(0, 0)$, which is not on the line $x + 4y = 4$. Substitute 0 for x and 0 for y in the given inequality.

$$x + 4y < 4$$
$$0 + 4(0) < 4 \quad \text{Let } x = 0, y = 0.$$
$$0 < 4 \quad \text{True}$$

Wait — this image is Figure 5.

FIGURE 5

Since the result $0 < 4$ is true, the test point $(0, 0)$ belongs on the side of the boundary line where all the points satisfy $x + 4y < 4$. For this reason, we shade the side containing $(0, 0)$, as in Figure 4. Choosing a point on the other side of the line, such as $(1, 5)$, would produce a false result when the values $x = 1$ and $y = 5$ were substituted into the given inequality. In such a case, we would shade the side of the line *not including* the test point.

As the examples above suggest, the graph of a linear inequality is represented by a shaded region in the plane, perhaps including the line that is the boundary of the region. Each of the shaded regions is an example of a **half-plane,** a region on one side of a line. For example, in Figure 5 line r divides the plane into half-planes P and Q. The points on r belong neither to P nor to Q. Line r is the boundary of each half-plane.

3 Graph $x \leq -1$.

Recall that the graph of $x = -1$ is the vertical line through $(-1, 0)$. To decide which half-plane belongs to the solution, choose a test point. Choosing $(0, 0)$ and replacing x with 0 gives a false statement:

$$x < -1$$
$$0 < -1. \quad \text{False}$$

The correct half-plane is the one that does *not* contain $(0, 0)$; it is shaded in Figure 6. ◄

FIGURE 6

EXAMPLE 4 Graph $x \geq 3y$.

Start by graphing the boundary, $x = 3y$. If $x = 0$, then $y = 0$, giving the point $(0, 0)$. Setting $y = 0$ would produce $(0, 0)$ again. To get a second point on the line, choose another number for x. Choosing $x = 3$ gives $y = 1$, with $(3, 1)$ a point on the line $x = 3y$. The points $(0, 0)$ and $(3, 1)$ lead to the line graphed in Figure 7. To decide which half-plane to shade, let us choose $(6, 1)$ as a test point. (Any point that is not on the line $x = 3y$ may be used.) Replacing x with 6 and y with 1 in the original inequality gives a true statement, so the half-plane containing $(6, 1)$ is shaded, as shown in Figure 7. ◀

FIGURE 7

The steps in graphing a linear inequality are summarized below.

GRAPHING A LINEAR INEQUALITY

1. Draw the graph of the boundary line. Make the line solid if the inequality involves \leq or \geq; make the line dashed if the inequality involves $<$ or $>$.
2. Decide which half-plane to shade. Use either of the following methods.
 (a) Solve the inequality for y; shade the region above the line if the inequality uses $>$ or \geq; shade the region below the line if the inequality uses $<$ or \leq.
 (b) Choose any point not on the line as a test point. Shade the half-plane that includes the test point if the test point satisfies the original inequality; otherwise, shade the half-plane on the other side of the boundary line.

Systems of Inequalities Realistic problems often involve many inequalities. For example, a manufacturing problem might produce inequalities resulting from production requirements, as well as inequalities about cost requirements. A collection of at least two inequalities is called a **system of inequalities.** The **solution** of a system of inequalities is made up of all those points that satisfy all the inequalities of the system at the same time. To graph the solution of a system of inequalities, graph all the inequalities on the same axes and identify, by heavy shading, the region common to all graphs. The next example shows how this is done.

EXAMPLE **5**

Graph the system

$$y < -3x + 12$$
$$x < 2y.$$

The heavily shaded region in Figure 8 shows all the points that satisfy both inequalities of the system. Since the points on the boundary lines are not in the solution, the boundary lines are dashed. ◄

FIGURE 8

Note When drawing regions as in Example 5 by hand, it is sometimes difficult to tell what is shaded heavily and what is shaded only lightly, particularly when more than two inequalities are involved. In such cases, an alternative technique is to shade the region *opposite* that of the inequality. In other words, the region that is *not* wanted can be shaded. Then, when the various regions are shaded, whatever is not shaded is the desired region. We will not use this technique in this text, but you may wish to try it on your own.

A region consisting of the overlapping parts of two or more graphs of inequalities in a system, such as the heavily shaded region in Figure 8, is sometimes called the **region of feasible solutions** or the **feasible region,** since it is made up of all the points that satisfy (are feasible for) each inequality of the system.

EXAMPLE **6**

Graph the feasible region for the system

$$2x - 5y \le 10$$
$$x + 2y \le 8$$
$$x \ge 0$$
$$y \ge 0.$$

FIGURE 9

On the same axes, graph each inequality by graphing the boundary and choosing the appropriate half-plane. Then find the feasible region by locating the overlap of all the half-planes. This feasible region is shaded in Figure 9. The inequalities $x \ge 0$ and $y \ge 0$ restrict the feasible region to the first quadrant. ◄

Applications As shown in the rest of this chapter, many realistic problems lead to systems of linear inequalities. The next example is typical of such problems.

EXAMPLE 7 Midtown Manufacturing Company makes plastic plates and cups, both of which require time on two machines. Manufacturing a unit of plates requires one hour on machine A and two on machine B, and producing a unit of cups requires three hours on machine A and one on machine B. Each machine is operated for at most 15 hours per day. Write a system of inequalities that expresses these restrictions, and sketch the feasible region.

Start by making a chart that summarizes the given information.

	Number Made	Time on Machine	
		A	**B**
Plates	x	1	2
Cups	y	3	1
Maximum Time Available		15	15

Here x represents the number of units of plates to be made, and y represents the number of units of cups.

On machine A, x units of plates require a total of $1 \cdot x = x$ hours, and y units of cups require $3 \cdot y = 3y$ hours. Since machine A is available no more than 15 hours per day,

$$x + 3y \le 15. \quad \text{Machine A}$$

The requirement that machine B be used no more than 15 hours per day gives

$$2x + y \le 15. \quad \text{Machine B}$$

Since it is not possible to produce a negative number of cups or plates,

$$x \ge 0 \quad \text{and} \quad y \ge 0.$$

The feasible region for this system of inequalities is shown in Figure 10. ◀

$x + 3y \le 15$
$2x + y \le 15$
$x \ge 0$
$y \ge 0$

FIGURE 10

3.1 Exercises

Graph the following linear inequalities.

1. $x + y \le 2$
2. $y \le x + 1$
3. $x \ge 3 + y$
4. $y \ge x - 3$
5. $4x - y < 6$
6. $3y + x > 4$
7. $3x + y < 6$
8. $2x - y > 2$
9. $x + 3y \ge -2$
10. $2x + 3y \le 6$
11. $4x + 3y > -3$
12. $5x + 3y > 15$
13. $2x - 4y < 3$
14. $4x - 3y < 12$
15. $x \le 5y$
16. $2x \ge y$

17. $-3x < y$ **18.** $-x > 6y$ **19.** $x + y \leq 0$ **20.** $3x + 2y \geq 0$

21. $y < x$ **22.** $y > -2x$ **23.** $x < 4$ **24.** $y > 5$

25. $y \leq -2$ **26.** $x \geq 3$

Graph the feasible region for each system of inequalities.

27. $x + y \leq 1$
$x - y \geq 2$

28. $3x - 4y < 6$
$2x + 5y > 15$

29. $2x - y < 1$
$3x + y < 6$

30. $x + 3y \leq 6$
$2x + 4y \geq 7$

31. $-x - y < 5$
$2x - y < 4$

32. $6x - 4y > 8$
$3x + 2y > 4$

33. $x + y \leq 4$
$x - y \leq 5$
$4x + y \leq -4$

34. $3x - 2y \geq 6$
$x + y \leq -5$
$y \leq 4$

35. $-2 < x < 3$
$-1 \leq y \leq 5$
$2x + y < 6$

36. $-2 < x < 2$
$y > 1$
$x - y > 0$

37. $2y + x \geq -5$
$y \leq 3 + x$
$x \geq 0$
$y \geq 0$

38. $2x + 3y \leq 12$
$2x + 3y > -6$
$3x + y < 4$
$x \geq 0$
$y \geq 0$

39. $3x + 4y > 12$
$2x - 3y < 6$
$0 \leq y \leq 2$
$x \geq 0$

40. $0 \leq x \leq 9$
$x - 2y \geq 4$
$3x + 5y \leq 30$
$y \geq 0$

Applications

Business and Economics

41. Production Scheduling A small pottery shop makes two kinds of planters, glazed and unglazed. The glazed type requires 1/2 hr to throw on the wheel and 1 hr in the kiln. The unglazed type takes 1 hr to throw on the wheel and 6 hr in the kiln. The wheel is available for at most 8 hr per day, and the kiln for at most 20 hr per day.

(a) Complete this chart.

	Number Made	Time on Wheel	Time in Kiln
Glazed	x		
Unglazed	y		
Maximum Time Available			

(b) Set up a system of inequalities and graph the feasible region.

(c) Using your graph from part (b), can 5 glazed and 2 unglazed planters be made? Can 10 glazed and 2 unglazed planters be made?

42. Time Management Carmella and Walt produce handmade shawls and afghans. They spin the yarn, dye it, and then weave it. A shawl requires 1 hr of spinning, 1 hr of dyeing, and 1 hr of weaving. An afghan needs 2 hr of spinning, 1 hr of dyeing, and 4 hr of weaving. Together, they spend at most 8 hr spinning, 6 hr dyeing, and 14 hr weaving.

(a) Complete this chart.

	Number Made	Spinning Time	Dyeing Time	Weaving Time
Shawls	x			
Afghans	y			
Maximum Time Available		8	6	14

(b) Set up a system of inequalities and graph the feasible region.

(c) Using your graph from part (b), can 3 shawls and 2 afghans be made? Can 4 shawls and 3 afghans be made?

For Exercises 43–46, perform each of the following steps.

(a) *Write a system of inequalities to express the conditions of the problem.*

(b) *Graph the feasible region of the system.*

43. Transportation Southwestern Oil supplies two distributors located in the Northwest. One distributor needs at least 3000 barrels of oil, and the other needs at least 5000 barrels. Southwestern can send out at most 10,000 barrels. Let x = the number of barrels of oil sent to distributor 1 and y = the number sent to distributor 2.

44. Finance The loan department in a bank has set aside $25 million for commercial and home loans. The bank's policy is to allocate at least four times as much money to home loans as to commercial loans. The bank's return is 12% on a home loan and 10% on a commercial loan. The manager of the loan department wants to earn a return of at least $2.8 million on these loans. Let x = the amount for home loans and y = the amount for commercial loans.

45. Transportation The California Almond Growers have 2400 boxes of almonds to be shipped from their plant in Sacramento to Des Moines and San Antonio. The Des Moines market needs at least 1000 boxes, while the San Antonio market must have at least 800 boxes. Let x = the number of boxes to be shipped to Des Moines and y = the number of boxes to be shipped to San Antonio.

46. Production Scheduling A cement manufacturer produces at least 3.2 million barrels of cement annually. He is told by the Environmental Protection Agency (EPA) that his operation emits 2.5 lb of dust for each barrel produced. The EPA has ruled that annual emissions must be reduced to 1.8 million pounds. To do this the manufacturer plans to replace the present dust collectors with two types of electronic precipitators. One type would reduce emissions to .5 lb per barrel and would cost 16¢ per barrel. The other would reduce the dust to .3 lb per barrel and would cost 20¢ per barrel. The manufacturer does not want to spend more than .8 million dollars on the precipitators. He needs to know how many barrels he could produce with each type. Let x = the number of barrels (in millions) produced with the first type and y = the number of barrels (in millions) produced with the second type.

3.2 SOLVING LINEAR PROGRAMMING PROBLEMS GRAPHICALLY

Many mathematical models designed to solve problems in business, biology, and economics involve finding the optimum value (maximum or minimum) of a function, subject to certain restrictions. In a **linear programming** problem, we must find the maximum or minimum value of a function, called the **objective function,** and also satisfy a set of restrictions, or **constraints,** given by linear inequalities. When only two variables are involved, the solution to a linear programming problem can be found by first graphing the set of constraints, then finding the feasible region as discussed in the previous section. This method is explained in the following example.

EXAMPLE 1 Find the maximum value of the objective function $z = 2x + 5y$, subject to the following constraints.

$$3x + 2y \leq 6$$
$$-x + 2y \leq 4$$
$$x \geq 0$$
$$y \geq 0$$

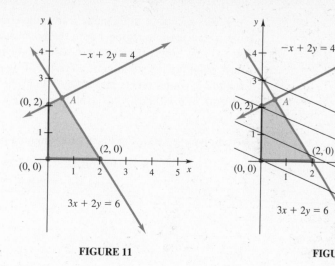

FIGURE 11 **FIGURE 12**

Recall from Chapter 2 that two equations in two unknowns can be solved by using row operations to eliminate one variable. For example, to solve the system

$$3x + 2y = 6$$
$$-x + 2y = 4,$$

we could take the first equation plus 3 times the second to eliminate x. (This is equivalent to $R_1 + 3R_2 \rightarrow R_2$ in the Gauss-Jordan method.) The result is $8y = 18$, so $y = \frac{18}{8} = \frac{9}{4}$. We can then substitute this value of y into either equation and solve for x. For example, substitution into the first equation yields

$$3x + 2\left(\frac{9}{4}\right) = 6$$

$$3x = 6 - \frac{9}{2} = \frac{3}{2}$$

$$x = \frac{1}{2}.$$

We instead could have subtracted the two original equations to eliminate y, yielding $4x = 2$, or $x = \frac{1}{2}$.

The feasible region is graphed in Figure 11. We can find the coordinates of point A, $(1/2, 9/4)$, by solving the system

$$3x + 2y = 6$$
$$-x + 2y = 4.$$

Every point in the feasible region satisfies all the constraints; however, we want to find those points that produce the maximum possible value of the objective function. To see how to find this maximum value, change the graph of Figure 11 by adding lines that represent the objective function $z = 2x + 5y$ for various sample values of z. By choosing the values 0, 5, 10, and 15 for z, the objective function becomes (in turn)

$$0 = 2x + 5y, \quad 5 = 2x + 5y, \quad 10 = 2x + 5y, \quad \text{and} \quad 15 = 2x + 5y.$$

These four lines are graphed in Figure 12. (Why are all of the lines parallel?) The figure shows that z cannot take on the value 15 because the graph for $z = 15$ is entirely outside the feasible region. The maximum possible value of z will be obtained from a line parallel to the others and between the lines representing the objective function when $z = 10$ and $z = 15$. The value of z will be as large as possible and all constraints will be satisfied if this line just touches the feasible region. This occurs at point A. We found above that A has coordinates $(1/2, 9/4)$. The value of z at this point is

$$z = 2x + 5y = 2\left(\frac{1}{2}\right) + 5\left(\frac{9}{4}\right) = 12\frac{1}{4}.$$

The maximum possible value of z is 12¼. Of all the points in the feasible region, A leads to the largest possible value of z. ◄

Points such as A in Example 1 are called corner points. A **corner point** is a point in the feasible region where the boundary lines of two constraints cross. Since corner points occur where two straight lines cross, the coordinates of a corner point are the solution of a system of two linear equations. As we saw in Example 1, corner points play a key role in the solution of linear programming problems. We will make this explicit after the following example.

Solve the following linear programming problem.

$$\text{Minimize} \qquad z = 2x + 4y$$
$$\text{subject to:} \qquad x + 2y \geq 10$$
$$3x + y \geq 10$$
$$x \geq 0$$
$$y \geq 0.$$

Figure 13 shows the feasible region and the lines that result when z in the objective function is replaced by 0, 10, 20, 40, and 50. The line representing the objective function touches the region of feasible solutions when $z = 20$. Two corner points, $(2, 4)$ and $(10, 0)$, lie on this line; both $(2, 4)$ and $(10, 0)$, as well as all the points on the boundary line between them, give the same optimum value of z. There are infinitely many equally good values of x and y that will give the same minimum value of the objective function $z = 2x + 4y$. This minimum value is 20. ◄

FIGURE 13

The feasible region in Example 1 is **bounded,** since the region is enclosed by boundary lines on all sides. Linear programming problems with bounded regions always have solutions. On the other hand, the feasible region in Example 2 is **unbounded,** and no solution will *maximize* the value of the objective function.

Some general conclusions can be drawn from the method of solution used in Examples 1 and 2. Figure 14 shows various feasible regions and the lines that

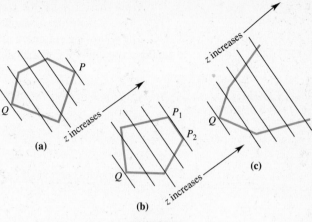

FIGURE 14

result from various values of z. (We assume the lines are in order from left to right as z increases.) In part (a) of the figure, the objective function takes on its minimum value at corner point Q and its maximum value at P. The minimum is again at Q in part (b), but the maximum occurs at P_1 or P_2, or any point on the line segment connecting them. Finally, in part (c), the minimum value occurs at Q, but the objective function has no maximum value because the feasible region is unbounded.

The preceding discussion suggests the truth of the **corner point theorem.**

CORNER POINT THEOREM If an optimum value (either a maximum or a minimum) of the objective function exists, it will occur at one or more of the corner points of the feasible region.

This theorem simplifies the job of finding an optimum value. First, we graph the feasible region and find all corner points. Then we test each corner point in the objective function. Finally, we identify the corner point producing the optimum solution. For unbounded regions, we must decide whether the required optimum can be found (see Example 2).

With the theorem, we can solve the problem in Example 1 by first identifying the four corner points in Figure 11: $(0, 0)$, $(0, 2)$, $(\frac{1}{2}, \frac{9}{4})$, and $(2, 0)$. Then we substitute each of the four points into the objective function $z = 2x + 5y$ to identify the corner point that produces the maximum value of z.

Corner Point	Value of $z = 2x + 5y$	
$(0, 0)$	$2(0) + 5(0) = 0$	
$(0, 2)$	$2(0) + 5(2) = 10$	
$(\frac{1}{2}, \frac{9}{4})$	$2(\frac{1}{2}) + 5(\frac{9}{4}) = 12\frac{1}{4}$	Maximum
$(2, 0)$	$2(2) + 5(0) = 4$	

From these results, the corner point ($\frac{1}{2}$, $\frac{9}{4}$) yields the maximum value of $12\frac{1}{4}$. This is the same as the result found earlier.

The following summary gives the steps to use in solving a linear programming problem by the graphical method.

SOLVING A LINEAR PROGRAMMING PROBLEM

1. Write the objective function and all necessary constraints.

2. Graph the feasible region.

3. Identify all corner points.

4. Find the value of the objective function at each corner point.

5. For a bounded region, the solution is given by the corner point producing the optimum value of the objective function.

6. For an unbounded region, check that a solution actually exists. If it does, it will occur at a corner point.

EXAMPLE **3** Sketch the feasible region for the following set of constraints, and then find the maximum and minimum values of the objective function $z = 5x + 2y$.

$$3y - 2x \geq 0$$
$$y + 8x \leq 52$$
$$y - 2x \leq 2$$
$$x \geq 3$$

The graph in Figure 15 shows that the feasible region is bounded. Use the corner points from the graph to find the maximum and minimum values of the objective function.

FIGURE 15

Corner Point	Value of $z = 5x + 2y$	
(3, 2)	$5(3) + 2(2) = 19$	Minimum
(6, 4)	$5(6) + 2(4) = 38$	
(5, 12)	$5(5) + 2(12) = 49$	Maximum
(3, 8)	$5(3) + 2(8) = 31$	

The minimum value of $z = 5x + 2y$ is 19 at the corner point (3, 2). The maximum value is 49 at (5, 12). ◀

3.2 Exercises

The graphs in Exercises 1–6 show regions of feasible solutions. Use these regions to find maximum and minimum values of the given objective functions.

1. $z = 3x + 5y$

2. $z = 6x - y$

3. $z = .40x + .75y$

4. $z = .35x + 1.25y$

5. **(a)** $z = 4x + 2y$
 (b) $z = 2x + 3y$
 (c) $z = 2x + 4y$
 (d) $z = x + 4y$

6. **(a)** $z = 4x + y$
 (b) $z = 5x + 6y$
 (c) $z = x + 2y$
 (d) $z = x + 6y$

Use graphical methods to solve the linear programming problems in Exercises 7–14.

7. Maximize $z = 5x + 2y$
subject to: $2x + 3y \leq 6$
$4x + y \leq 6$
$x \geq 0$
$y \geq 0.$

8. Minimize $z = x + 3y$
subject to: $x + y \leq 10$
$5x + 2y \geq 20$
$-x + 2y \geq 0$
$x \geq 0$
$y \geq 0.$

9. Maximize $z = 2x + 2y$
subject to: $3x - y \geq 12$
$x + y \leq 15$
$x \geq 2$
$y \geq 5.$

10. Maximize $z = 10x + 8y$
subject to: $2x + 3y \leq 100$
$5x + 4y \leq 200$
$x \geq 10$
$y \leq 20.$

11. Maximize $z = 4x + 2y$
subject to: $x - y \leq 10$
$5x + 3y \leq 75$
$x + y \leq 20$
$x \geq 0$
$y \geq 0.$

12. Maximize $z = 4x + 5y$
subject to: $10x - 5y \leq 100$
$20x + 10y \geq 150$
$x + y \geq 12$
$x \geq 0$
$y \geq 0.$

13. Find values of $x \geq 0$ and $y \geq 0$ that maximize $z = 10x + 12y$ subject to each of the following sets of constraints.

(a) $x + y \leq 20$
$x + 3y \leq 24$

(b) $3x + y \leq 15$
$x + 2y \leq 18$

(c) $2x + 5y \geq 22$
$4x + 3y \leq 28$
$2x + 2y \leq 17$

14. Find values of $x \geq 0$ and $y \geq 0$ that minimize $z = 3x + 2y$ subject to each of the following sets of constraints.

(a) $10x + 7y \leq 42$
$4x + 10y \geq 35$

(b) $6x + 5y \geq 25$
$2x + 6y \geq 15$

(c) $x + 2y \geq 10$
$2x + y \geq 12$
$x - y \leq 8$

***15.** You are given the following linear programming problem:

Maximize $z = c_1x_1 + c_2x_2$
subject to: $2x_1 + x_2 \leq 11$
$-x_1 + 2x_2 \leq 2$
$x_1 \geq 0, \quad x_2 \geq 0.$

If $c_2 > 0$, determine the range of c_1/c_2 for which $(x_1, x_2) = (4, 3)$ is an optimal solution.

(a) $[-2, 1/2]$ **(b)** $[-1/2, 2]$ **(c)** $[-11, -1]$ **(d)** $[1, 11]$ **(e)** $[-11, 11]$

*Problem 5 from "November 1989 Course 130 Examination Operations Research" of the *Education and Examination Committee of The Society of Actuaries*. Reprinted by permission of The Society of Actuaries.

3.3 APPLICATIONS OF LINEAR PROGRAMMING

? How many filing cabinets of different types should an office manager purchase, given a limited budget and limited office space?

We will use linear programming to answer this question in Example 1.

EXAMPLE 1 An office manager needs to purchase new filing cabinets. He knows that Ace cabinets cost \$40 each, require 6 square feet of floor space, and hold 24

cubic feet of files. On the other hand, each Excello cabinet costs $80, requires 8 square feet of floor space, and holds 36 cubic feet. His budget permits him to spend no more than $560 on files, while the office has space for no more than 72 square feet of cabinets. The manager desires the greatest storage capacity within the limitations imposed by funds and space. How many of each type of cabinet should he buy?

Let x represent the number of Ace cabinets to be bought and let y represent the number of Excello cabinets. Summarize the given information in a table.

	Number of Cabinets	Cost of Each	Space Required	Storage Capacity
Ace	x	$40	6 ft^2	24 ft^3
Excello	y	$80	8 ft^2	36 ft^3
Maximum Available		$560	72 ft^2	

The constraints imposed by cost and space are as follows.

$$40x + 80y \leq 560 \quad \text{Cost}$$
$$6x + 8y \leq 72 \quad \text{Floor space}$$

Since the number of cabinets cannot be negative, $x \geq 0$ and $y \geq 0$. The objective function to be maximized gives the amount of storage capacity provided by some combination of Ace and Excello cabinets. From the information in the table, the objective function is

$$\text{Storage space} = z = 24x + 36y.$$

In summary, the mathematical model for the given linear programming problem is as follows:

$$\text{Maximize} \quad z = 24x + 36y \tag{1}$$
$$\text{subject to:} \quad 40x + 80y \leq 560 \tag{2}$$
$$6x + 8y \leq 72 \tag{3}$$
$$x \geq 0 \tag{4}$$
$$y \geq 0. \tag{5}$$

Using the methods described in the previous section, graph the feasible region for the system of inequalities (2)–(5), as in Figure 16. Three of the corner points can be identified from the graph as (0, 0), (0, 7), and (12, 0). The fourth corner point, labeled Q in the figure, can be found by solving the system of equations

$$40x + 80y = 560$$
$$6x + 8y = 72.$$

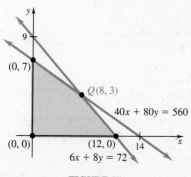

FIGURE 16

Solve this system to find that Q is the point (8, 3). Now test these four points in the objective function to determine the maximum value of z. The results are shown in the table.

Corner Point	Value of $z = 24x + 36y$
(0, 0)	0
(0, 7)	252
(12, 0)	288
(8, 3)	300 Maximum

The objective function, which represents storage space, is maximized when $x = 8$ and $y = 3$. The manager should buy 8 Ace cabinets and 3 Excello cabinets. ◀

Fortunately, the answer to the linear programming problem in Example 1 is a point with integer coordinates, as the number of each type of cabinet must be an integer. Unfortunately, there is no guarantee that this will always happen. When the solution to a linear programming problem is restricted to integers, it is an *integer programming* problem, which is more difficult than a linear programming problem. In this text, all problems in which fractional solutions are meaningless are contrived to have integer solutions.

EXAMPLE 2 ▶ A 4-H member raises only geese and pigs. She wants to raise no more than 16 animals, including no more than 10 geese. She spends $5 to raise a goose and $15 to raise a pig, and she has $180 available for this project. The 4-H member wishes to maximize her profits. Each goose produces $6 in profit and each pig $20 in profit.

First, set up a table that shows the information given in the problem.

	Number Raised	Cost to Raise	Profit (each)
Geese	x	$5	$6
Pigs	y	$15	$20
Maximum Available	16	$180	

Use the table to write the necessary constraints. Since the total number of animals cannot exceed 16, the first constraint is

$$x + y \le 16.$$

"No more than 10 geese" means

$$x \leq 10.$$

The cost to raise x geese at \$5 per goose is $5x$ dollars, while the cost for y pigs at \$15 each is $15y$ dollars. Since only \$180 is available,

$$5x + 15y \leq 180.$$

Dividing both sides by 5 gives the equivalent inequality

$$x + 3y \leq 36.$$

The number of geese and pigs cannot be negative, so

$$x \geq 0 \quad \text{and} \quad y \geq 0.$$

The 4-H member wants to know how many geese and pigs to raise in order to produce maximum profit. Each goose yields \$6 profit, and each pig \$20. If z represents total profit, then

$$z = 6x + 20y.$$

In summary, we have the following linear programming problem:

$$
\begin{aligned}
\text{Maximize} \quad & z = 6x + 20y \\
\text{subject to:} \quad & x + y \leq 16 \\
& x \leq 10 \\
& x + 3y \leq 36 \\
& x \geq 0 \\
& y \geq 0.
\end{aligned}
$$

A graph of the feasible region is shown in Figure 17. The corner points $(0, 12)$, $(0, 0)$, and $(10, 0)$ can be read directly from the graph. The coordinates of each of the other corner points can be found by solving a system of linear equations.

FIGURE 17

Test each corner point in the objective function to find the maximum profit.

Corner Point	Value of $z = 6x + 20y$
(0, 12)	$6(0) + 20(12) = 240$ Maximum
(6, 10)	$6(6) + 20(10) = 236$
(10, 6)	$6(10) + 20(6) = 180$
(10, 0)	$6(10) + 20(0) = 60$
(0, 0)	$6(0) + 20(0) = 0$

The maximum of 240 occurs at (0, 12). Thus, 12 pigs and no geese will produce a maximum profit of $240. ◀

EXAMPLE 3 Certain laboratory animals must have at least 30 grams of protein and at least 20 grams of fat per feeding period. These nutrients come from food A, which costs 18 cents per unit and supplies 2 grams of protein and 4 of fat; and food B, which costs 12 cents per unit and has 6 grams of protein and 2 of fat. Food B is bought under a long-term contract requiring that at least 2 units of B be used per serving. How much of each food must be bought to produce the minimum cost per serving?

Let x represent the required amount of food A and y the amount of food B. Use the given information to prepare the following table.

Food	Number of Units	Grams of Protein	Grams of Fat	Cost
A	x	2	4	18¢
B	y	6	2	12¢
Minimum Required		30	20	

The linear programming problem can be stated as follows.

$$\text{Minimize} \quad z = .18x + .12y$$
$$\text{subject to:} \quad 2x + 6y \geq 30 \quad \text{Protein}$$
$$4x + 2y \geq 20 \quad \text{Fat}$$
$$y \geq 2$$
$$x \geq 0.$$

(The usual constraint $y \geq 0$ is redundant because of the constraint $y \geq 2$.) A graph of the feasible region is shown in Figure 18. The corner points are (0, 10), (3, 4), and (9, 2).

FIGURE 18

Test each corner point in the objective function to find the minimum cost.

Corner Point	Value of $z = .18x + .12y$	
(0, 10)	$.18(0) + .12(10) = 1.20$	
(3, 4)	$.18(3) + .12(4) = 1.02$	Minimum
(9, 2)	$.18(9) + .12(2) = 1.86$	

The minimum of 1.02 occurs at (3, 4). Thus, 3 units of food A and 4 units of food B will produce a minimum cost of $1.02 per serving. ◀

The feasible region in Figure 18 is an *unbounded* feasible region—the region extends indefinitely to the upper right. With this region it would not be possible to *maximize* the objective function, because the total cost of the food could always be increased by encouraging the animals to eat more.

3.3 Exercises

Write Exercises 1–6 as linear inequalities. Identify all variables used. (Note: Not all of the given information is used in Exercises 5 and 6.)

1. Product A requires 2 hr on machine I, while product B needs 3 hr on the same machine. The machine is available for at most 45 hr per week.

2. A cow requires a third of an acre of pasture and a sheep needs a quarter acre. A rancher wants to use at least 120 acres of pasture.

3. Jesus Garcia needs at least 25 units of vitamin A per day. Green pills provide 4 units and red pills provide 1.

4. Pauline Wong spends 3 hr selling a small computer and 5 hr selling a larger model. She works no more than 45 hr per week.

5. Coffee costing $6 per pound is to be mixed with coffee costing $5 per pound to get at least 50 lb of a blended coffee.

6. A tank in an oil refinery holds 120 gal. The tank contains a mixture of light oil worth $1.25 per gallon and heavy oil worth $.80 per gallon.

Applications

Business and Economics

7. Transportation The Miers Company produces small engines for several manufacturers. The company receives orders from two assembly plants for their Topflight engine. Plant I needs at least 50 engines, and plant II needs at least 27 engines. The company can send at most 85 engines to these two assembly plants. It costs $20 per engine to ship to plant I and $35 per engine to ship to plant II. How many engines should be shipped to each plant to minimize shipping costs? What is the minimum cost?

8. Transportation A manufacturer of refrigerators must ship at least 100 refrigerators to its two West Coast warehouses. Each warehouse holds a maximum of 100 refrigerators. Warehouse A holds 25 refrigerators already, and warehouse B has 20 on hand. It costs $12 to ship a refrigerator to warehouse A and $10 to ship one to warehouse B. How many refrigerators should be shipped to each warehouse to minimize costs? What is the minimum cost?

9. (a) Insurance A company is considering two insurance plans with the types of coverage (in thousands of dollars) and premiums (per thousand dollars) as shown below.

		Coverage		Premium
		Fire/Theft	Liability	
Policy	A	$10	$80	$50
	B	$15	$120	$40

 The company wants at least $100,000 fire/theft insurance and at least $1,000,000 liability insurance from these plans. How many units should be purchased from each plan to minimize the cost of the premiums? What is the minimum premium?

(b) Suppose the premium for policy A is reduced to $25 per thousand dollars. Now how many units should be purchased from each plan to minimize the cost of the premiums? What is the minimum premium?

10. (a) Profit Seall Manufacturing Company makes color television sets. It produces a bargain set that sells for $100 profit and a deluxe set that sells for $150 profit. On the assembly line the bargain set requires 3 hr, and the deluxe set takes 5 hr. The cabinet shop spends 1 hr on the cabinet for the bargain set and 3 hr on the cabinet for the deluxe set. Both sets require 2 hr of time for testing and packing. On a particular production run the Seall Company has available 3900 work hours on the assembly line, 2100 work hours in the cabinet shop, and 2200 work hours in the testing and packing department. How many sets of each type should it produce to make maximum profit? What is the maximum profit?

(b) Suppose the profit on deluxe sets goes up to $170. Now how many sets of each type should be produced to make a maximum profit? What is the maximum profit?

11. Revenue A machine shop manufactures two types of bolts. Each can be made on any of three groups of machines, but the time required on each group differs, as shown in the table below.

		Machine Group		
		I	II	III
Bolts	Type 1	.1 min	.1 min	.1 min
	Type 2	.1 min	.4 min	.5 min

Production schedules are made up one day at a time. In a day there are 240, 720, and 160 min available, respectively, on these machines. Type 1 bolts sell for 10¢ and type 2 bolts for 12¢. How many of each type of bolt should be manufactured per day to maximize revenue? What is the maximum revenue?

12. Revenue The manufacturing process requires that oil refineries must manufacture at least 2 gal of gasoline for every gallon of fuel oil. To meet the winter demand for fuel oil, at least 3 million gallons a day must be produced. The demand for gasoline is no more than 6.4 million gallons per day. If the refinery sells gasoline for $1.25 per gallon and fuel oil for $1 per gallon, how much of each should be produced to maximize revenue? Find the maximum revenue.

13. (a) Revenue A candy company has 100 kg of chocolate-covered nuts and 125 kg of chocolate-covered raisins to be sold as two different mixes. One mix will contain half nuts and half raisins and will sell for $6 per kilogram. The other mix will contain 1/3 nuts and 2/3 raisins and will sell for $4.80 per kilogram. How many kilograms of each mix should the company prepare for maximum revenue? Find the maximum revenue.

(b) The company raises the price of the half-and-half mix to $8 per kilogram. Now how many kilograms of each mix should the company prepare for maximum revenue? Find the maximum revenue.

14. Profit A small country can grow only two crops for export, coffee and cocoa. The country has 500,000 hectares of land available for the crops. Long-term contracts require that at least 100,000 hectares be devoted to coffee and at least 200,000 hectares to cocoa. Cocoa must be processed locally, and production bottlenecks limit cocoa to 270,000 hectares. Coffee requires two workers per hectare, with cocoa requiring five. No more than 1,750,000 people are available for working with these crops. Coffee produces a profit of $220 per hectare and cocoa a profit of $310 per hectare. How many hectares should the country devote to each crop in order to maximize profit? Find the maximum profit.

15. Blending The Mostpure Milk Company gets milk from two dairies and then blends the milk to get the desired amount of butterfat for the company's premier product. If dairy I can supply at most 50 gal of milk averaging 3.7% butterfat, and dairy II can supply at most 80 gal of milk averaging 3.2% butterfat, how much milk from each supplier should Mostpure use to get at most 100 gal of milk with maximum butterfat? What is the maximum amount of butterfat?

16. Transportation A greeting card manufacturer has 370 boxes of a particular card in warehouse I and 290 boxes of the same card in warehouse II. A greeting card shop in San Jose orders 350 boxes of the card, and another shop in Memphis orders 300 boxes. The shipping costs per box to these shops from the two warehouses are shown in the following table.

		Destination	
		San Jose	Memphis
Warehouse	I	$.25	$.22
	II	$.20	$.21

How many boxes should be shipped to each city from each warehouse to minimize shipping costs? What is the minimum cost? (*Hint:* Use x, $350 - x$, y, and $300 - y$ as the variables.)

17. Finance A pension fund manager decides to invest at most $40 million in U.S. Treasury Bonds paying 12% annual interest and in mutual funds paying 8% annual interest. He plans to invest at least $20 million in bonds and at least $15 million in mutual funds. How much should be

invested in each to maximize annual interest? What is the maximum annual interest?

Life Science

18. Health Care Mark, who is ill, takes vitamin pills. Each day he must have at least 16 units of vitamin A, 5 units of vitamin B_1, and 20 units of vitamin C. He can choose between pill #1, which contains 8 units of A, 1 of B_1, and 2 of C; and pill #2, which contains 2 units of A, 1 of B_1, and 7 of C. Pill #1 costs 15¢, and pill #2 costs 30¢. How many of each pill should he buy in order to minimize his cost? What is the minimum cost?

19. Predator Food Requirements A certain predator requires at least 10 units of protein and 8 units of fat per day. One prey of species I provides 5 units of protein and 2 units of fat; one prey of species II provides 3 units of protein and 4 units of fat. Capturing and digesting each species II prey requires 3 units of energy, and capturing and digesting each species I prey requires 2 units of energy. How many of each prey would meet the predator's daily food requirements with the least expenditure of energy? Are the answers reasonable? How could they be interpreted?

20. Health Care Ms. Oliveras was given the following advice. She should supplement her daily diet with at least 6000 USP units of vitamin A, at least 195 milligrams of vitamin C, and at least 600 USP units of vitamin D. Ms. Oliveras finds that Mason's Pharmacy carries Brand X vitamin pills at 5¢ each and Brand Y vitamins at 4¢ each. Each Brand X pill contains 3000 USP units of A, 45 milligrams of C, and 75 USP units of D, while Brand Y pills contain 1000 USP units of A, 50 milligrams of C, and 200 USP units of D. What combination of vitamin pills should she buy to obtain the least possible cost? What is the least possible cost per day?

21. Dietetics Sing, who is dieting, requires two food supplements, I and II. He can get these supplements from two different products, A and B, as shown below.

		Grams of Supplement per Serving	
		I	II
Product	A	3	2
	B	2	4

Sing's physician has recommended that he include at least 15 g of each supplement in his daily diet. If product A costs 25¢ per serving and product B costs 40¢ per serving, how can he satisfy his dietetic requirements most economically? Find the minimum cost.

General Interest

22. Construction In a small town in South Carolina, zoning rules require that the window space (in square feet) in a house be at least one-sixth of the space used up by solid walls. The monthly cost to heat the house is 20¢ for each square foot of solid walls and 80¢ for each square foot of windows. Find the maximum total area (windows plus walls) if $160 is available to pay for heat.

23. Farming An agricultural advisor looks at the results of Example 2 and claims that it cannot possibly be correct. After all, the 4-H member is able to raise 16 animals, and she is only raising 12 animals. Surely she can earn more profit by raising all 16 animals. How would you respond?

Manufacturing *(Note: Exercises 24–26 are from qualification examinations for Certified Public Accountants.*) The Random Company manufactures two products, Zeta and Beta. Each product must pass through two processing operations. All materials are introduced at the start of Process No. 1. There are no work-in-process inventories. Random may produce either one product exclusively or various combinations of both products subject to the following constraints:*

	Process No. 1	Process No. 2	Contribution Margin (per unit)
Hours Required to Produce One Unit of:			
Zeta	1 hr	1 hr	$4.00
Beta	2 hr	3 hr	5.25
Total Capacity (in hours per day)	1,000 hr	1,275 hr	

A shortage of technical labor has limited Beta production to 400 units per day. There are no constraints on the production of Zeta other than the hour constraints in the above schedule. Assume that all relationships between capacity and production are linear.

24. Given the objective to maximize total contribution margin, what is the production constraint for Process No. 1?

 (a) Zeta + Beta ≤ 1,000. **(b)** Zeta + 2 Beta ≤ 1,000. **(c)** Zeta + Beta ≥ 1,000. **(d)** Zeta + 2 Beta ≥ 1,000.

25. Given the objective to maximize total contribution margin, what is the labor constraint for production of Beta?

 (a) Beta ≤ 400. **(b)** Beta ≥ 400. **(c)** Beta ≤ 425. **(d)** Beta ≥ 425.

26. What is the objective function of the data presented?

 (a) Zeta + 2 Beta = $9.25.

 (b) $4.00 Zeta + 3($5.25) Beta = Total Contribution Margin.

 (c) $4.00 Zeta + $5.25 Beta = Total Contribution Margin.

 (d) 2($4.00) Zeta + 3($5.25) Beta = Total Contribution Margin.

Chapter Summary	**Key Terms**	

3.1 linear inequality
boundary
half-plane
system of inequalities

region of feasible solutions
3.2 linear programming
objective function
constraints

corner point
bounded
unbounded

Chapter 3	**Review Exercises**

1. Why doesn't the graphical method work for more than two variables?

2. How many constraints are we limited to in the graphical method?

Graph each linear inequality.

3. $y \geq 2x + 3$

4. $3x - y \leq 5$

5. $3x + 4y \leq 12$

6. $2x - 6y \geq 18$

7. $y \geq x$

8. $y \leq 3$

Graph the solution of each system of inequalities. Find all corner points.

9. $x + y \leq 6$
$2x - y \geq 3$

10. $4x + y \geq 8$
$2x - 3y \leq 6$

11. $-4 \leq x \leq 2$
$-1 \leq y \leq 3$
$x + y \leq 4$

12. $2 \leq x \leq 5$
$1 \leq y \leq 7$
$x - y \leq 3$

13. $x + 3y \geq 6$
$4x - 3y \leq 12$
$x \geq 0$
$y \geq 0$

14. $x + 2y \leq 4$
$2x - 3y \leq 6$
$x \geq 0$
$y \geq 0$

Use the given regions to find the maximum and minimum values of the objective function $z = 2x + 4y$.

15.

16.

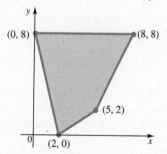

Use the graphical method to solve each linear programming problem.

17. Maximize $z = 2x + 4y$
subject to: $3x + 2y \leq 12$
$5x + y \geq 5$
$x \geq 0$
$y \geq 0.$

18. Minimize $z = 3x + 2y$
subject to: $8x + 9y \geq 72$
$6x + 8y \geq 72$
$x \geq 0$
$y \geq 0.$

19. Minimize $z = 4x + 2y$
subject to: $x + y \leq 50$
$2x + y \geq 20$
$x + 2y \geq 30$
$x \geq 0$
$y \geq 0.$

20. Maximize $z = 4x + 3y$
subject to: $2x + 7y \leq 14$
$2x + 3y \leq 10$
$x \geq 0$
$y \geq 0.$

21. Why must the solution to a linear programming problem always occur at a corner point of the feasible region?

22. Is there necessarily a unique point in the feasible region where the maximum or minimum occurs? Why or why not?

Applications

Business and Economics

23. Time Management A bakery makes both cakes and cookies. Each batch of cakes requires 2 hr in the oven and 3 hr in the decorating room. Each batch of cookies needs $1\frac{1}{2}$ hr in the oven and $\frac{2}{3}$ hr in the decorating room. The oven is available no more than 15 hr per day, and the decorating room can be used no more than 13 hr per day. Set up a system of inequalities, and then graph the solution of the system.

24. Cost Analysis A company makes two kinds of pizza, basic and plain. Basic contains cheese and beef, and plain contains onions and beef. The company sells at least 3 units a day of basic and at least 2 units of plain. The beef costs $5 per unit for basic and $4 per unit for plain. No more than $50 per day can be spent on beef. Dough for basic is $2 per unit, and dough for plain is $1 per unit. The company can spend no more than $16 per day on dough. Set up a system of inequalities, and then graph the solution of the system.

25. Profit How many batches of cakes and cookies should the bakery in Exercise 23 make in order to maximize profits if cookies produce a profit of $20 per batch and cakes produce a profit of $30 per batch?

26. Profit How many units of each kind of pizza should the company in Exercise 24 make in order to maximize profits if basic sells for $20 per unit and plain for $15 per unit?

General Interest

27. Studying Charles Cavaliere is trying to allocate his study time this weekend. He can spend time working with either his math tutor or his accounting tutor to prepare for exams in both classes the following Monday. His math tutor charges $20 per hour and his accounting tutor charges $40 per hour. He has $220 to spend on tutoring. Each hour that he spends working with his math tutor requires 2 aspirin and 1 hr of sleep to recover, while each hour he spends with his accounting tutor requires 1 aspirin and 3 hr of sleep. He only has 16 aspirin left, and he can only afford 15 hr of sleep this weekend. He expects that each hour with his math tutor will increase his score on the math exam by 3 points, while each hour with his accounting tutor will increase his score on the accounting exam by 5 points. How many hours should he spend with each tutor in order to maximize the number of points he will get on the two tests combined?

Connections

28. It is not necessary to check all corner points in a linear programming problem. This exercise illustrates an alternative procedure, which is essentially an expansion of ideas illustrated in Example 1 of Section 3.2.

Maximize $z = 2x + 5y$
subject to $3x + 2y \leq 6$
$-x + 2y \leq 4$
$x \geq 0$
$y \geq 0.$

(a) Sketch the feasible region, and add the line $z = 10$. (*Note:* 10 is chosen because the numbers work out simply, but the chosen value of z is arbitrary.)

(b) Draw a line parallel to the line $z = 10$ that is as far from the origin as possible but still touches the feasible region.

(c) The line you drew in part (b) should go through the point $(1/2, 9/4)$. Explain how you know the maximum must be located at this point.

29. Use the method described in the previous exercise to solve Exercise 11 of Section 3.2.

Applications at a glance...

Profit of a Toy Manufacturer ... illustrations in a textbook ... script for a television commercial ...

CHAPTER 4

LINEAR PROGRAMMING: THE SIMPLEX METHOD

(See page 192.)

In the previous chapter, we discussed solving linear programming problems by the graphical method. This method illustrates the basic ideas of linear programming, but it is practical only for problems with two variables. For problems with more than two variables, or problems with two variables and many constraints, the *simplex method* is used. This method grew out of a practical problem faced by George B. Dantzig in 1947. Dantzig was concerned with finding the least expensive way to allocate supplies for the United States Air Force.

The **simplex method** starts with the selection of one corner point (often the origin) from the feasible region. Then, in a systematic way, another corner point is found that improves the value of the objective function. Finally, an optimum solution is reached, or it can be seen that no such solution exists.

The simplex method requires a number of steps. In this chapter we divide the presentation of these steps into two parts. First, a problem is set up in Section 4.1 and the method started, and then, in Section 4.2, the method is completed. Special situations are discussed in the remainder of the chapter.

▶ 4.1 SLACK VARIABLES AND THE PIVOT

Because the simplex method is used for problems with many variables, it usually is not convenient to use letters such as x, y, z, or w as variable names. Instead, the symbols x_1 (read "x-sub-one"), x_2, x_3, and so on, are used. These variable names lend themselves easily to use on the computer. In the simplex method, all constraints must be expressed in the linear form

$$a_1x_1 + a_2x_2 + a_3x_3 + \cdots \leq b,$$

where x_1, x_2, x_3, ... are variables and a_1, a_2, ..., and b are constants.

In this section we will use the simplex method only for problems such as the following:

$$\begin{aligned}
\text{Maximize} \quad & z = 2x_1 - 3x_2 \\
\text{subject to:} \quad & 2x_1 + x_2 \leq 10 \\
& x_1 - 3x_2 \leq 5 \\
\text{with} \quad & x_1 \geq 0, \quad x_2 \geq 0.
\end{aligned}$$

This type of problem is said to be in *standard maximum form*.

STANDARD MAXIMUM FORM

A linear programming problem is in **standard maximum form** if the following conditions are satisfied.

1. The objective function is to be maximized.
2. All variables are nonnegative ($x_i \geq 0$).
3. All constraints involve \leq.
4. All of the constants in the constraints are nonnegative ($b \geq 0$).

(Problems that do not meet all of these conditions are discussed in Sections 4.3 and 4.4.)

To use the simplex method, we start by converting the constraints, which are linear inequalities, into linear equations. We do this by adding a nonnegative variable, called a **slack variable,** to each constraint. For example, the inequality $x_1 + x_2 \leq 10$ is converted into an equation by adding the slack variable x_3 to get

$$x_1 + x_2 + x_3 = 10, \quad \text{where } x_3 \geq 0.$$

The inequality $x_1 + x_2 \leq 10$ says that the sum $x_1 + x_2$ is less than or perhaps equal to 10. The variable x_3 "takes up any slack" and represents the amount by which $x_1 + x_2$ fails to equal 10. For example, if $x_1 + x_2$ equals 8, then x_3 is 2. If $x_1 + x_2 = 10$, the value of x_3 is 0.

| Caution | A different slack variable must be used for each constraint.

EXAMPLE **1** Restate the following linear programming problem by introducing slack variables.

$$\text{Maximize} \quad z = 3x_1 + 2x_2 + x_3$$
$$\text{subject to:} \quad 2x_1 + x_2 + x_3 \leq 150$$
$$2x_1 + 2x_2 + 8x_3 \leq 200$$
$$2x_1 + 3x_2 + x_3 \leq 320$$
$$\text{with} \quad x_1 \geq 0, \quad x_2 \geq 0, \quad x_3 \geq 0.$$

Rewrite the three constraints as equations by adding slack variables x_4, x_5, and x_6, one for each constraint. Then the problem can be restated as follows.

$$\text{Maximize} \quad z = 3x_1 + 2x_2 + x_3$$
$$\text{subject to:} \quad 2x_1 + x_2 + x_3 + x_4 \qquad\qquad = 150$$
$$2x_1 + 2x_2 + 8x_3 \qquad + x_5 \qquad = 200$$
$$2x_1 + 3x_2 + x_3 \qquad\qquad + x_6 = 320$$
$$\text{with} \quad x_1 \geq 0, \quad x_2 \geq 0, \quad x_3 \geq 0, \quad x_4 \geq 0, \quad x_5 \geq 0, \quad x_6 \geq 0. \quad \blacktriangleleft$$

Adding slack variables to the constraints converts a linear programming problem into a system of linear equations. In each of these equations, all variables should be on the left side of the equals sign and all constants on the right. All the equations in Example 1 satisfy this condition except for the objective function, $z = 3x_1 + 2x_2 + x_3$, which may be written with all variables on the left as

$$-3x_1 - 2x_2 - x_3 + z = 0.$$

Now the equations in Example 1 can be written as the following augmented matrix.

$$
\begin{array}{ccccccc}
x_1 & x_2 & x_3 & x_4 & x_5 & x_6 & z \\
\end{array}
$$

$$
\left[\begin{array}{ccccccc|c}
2 & 1 & 1 & 1 & 0 & 0 & 0 & 150 \\
2 & 2 & 8 & 0 & 1 & 0 & 0 & 200 \\
2 & 3 & 1 & 0 & 0 & 1 & 0 & 320 \\
\hline
-3 & -2 & -1 & 0 & 0 & 0 & 1 & 0
\end{array}\right]
$$

Indicators

This matrix is called the **initial simplex tableau.** The numbers in the bottom row, which are from the objective function, are called **indicators** (except for the 1 and 0 at the far right).

EXAMPLE 2

Set up the initial simplex tableau for the following problem:

A farmer has 100 acres of available land that he wishes to plant with a mixture of potatoes, corn, and cabbage. It costs him $400 to produce an acre of potatoes, $160 to produce an acre of corn, and $280 to produce an acre of cabbage. He has a maximum of $20,000 to spend. He makes a profit of $120 per acre of potatoes, $40 per acre of corn, and $60 per acre of cabbage. How many acres of each crop should he plant to maximize his profit?

Begin by summarizing the given information as follows.

Crop	Number of Acres	Cost (per acre)	Profit (per acre)
Potatoes	x_1	$400	$120
Corn	x_2	$160	$ 40
Cabbage	x_3	$280	$ 60
Maximum Available	100	$20,000	

If the number of acres allotted to each of the three crops is represented by x_1, x_2, and x_3, respectively, then the constraints of the example can be expressed as

$$
\begin{aligned}
x_1 + x_2 + x_3 &\leq 100 \qquad \text{Number of acres} \\
400x_1 + 160x_2 + 280x_3 &\leq 20{,}000 \qquad \text{Production costs}
\end{aligned}
$$

where x_1, x_2, and x_3 are all nonnegative. The first of these constraints says that $x_1 + x_2 + x_3$ is less than or perhaps equal to 100. Use x_4 as the slack variable, giving the equation

$$
x_1 + x_2 + x_3 + x_4 = 100.
$$

Here x_4 represents the amount of the farmer's 100 acres that will not be used (x_4 may be 0 or any value up to 100).

The constraint $400x_1 + 160x_2 + 280x_3 \leq 20{,}000$ can be simplified by dividing by 40, yielding

$$10x_1 + 4x_2 + 7x_3 \leq 500.$$

This inequality can also be converted into an equation by adding a slack variable, x_5.

$$10x_1 + 4x_2 + 7x_3 + x_5 = 500$$

If we had not divided by 40, the slack variable would have represented any unused portion of the farmer's \$20,000 capital. Instead, the slack variable represents 1/40 of that unused portion. (Note that x_5 may be any value from 0 to 500.)

The objective function represents the profit. The farmer wants to maximize

$$z = 120x_1 + 40x_2 + 60x_3.$$

The linear programming problem can now be stated as follows:

Maximize $z = 120x_1 + 40x_2 + 60x_3$

subject to: $x_1 + x_2 + x_3 + x_4 \qquad = 100$

$\qquad\qquad 10x_1 + 4x_2 + 7x_3 \qquad + x_5 = 500$

with $x_1 \geq 0, \quad x_2 \geq 0, \quad x_3 \geq 0, \quad x_4 \geq 0, \quad x_5 \geq 0.$

Rewrite the objective function as $-120x_1 - 40x_2 - 60x_3 + z = 0$, and complete the initial simplex tableau as follows.

$$
\begin{array}{cccccc|c}
x_1 & x_2 & x_3 & x_4 & x_5 & z & \\
\hline
1 & 1 & 1 & 1 & 0 & 0 & 100 \\
10 & 4 & 7 & 0 & 1 & 0 & 500 \\
\hline
-120 & -40 & -60 & 0 & 0 & 1 & 0
\end{array}
$$ ◀

The maximization problem in Example 2 consists of a system of two equations (describing the constraints) in five variables, together with the objective function. As with the graphical method, it is necessary to solve this system to find corner points of the region of feasible solutions. To produce a single, distinct solution, the number of variables must equal the number of equations in the system.

For example, the system of equations in Example 2 has an infinite number of solutions since there are more variables than equations. To see this, solve the system for x_4 and x_5.

$$x_4 = 100 - x_1 - x_2 - x_3$$
$$x_5 = 500 - 10x_1 - 4x_2 - 7x_3$$

Each choice of values for x_1, x_2, and x_3 gives corresponding values for x_4 and x_5 that produce a solution of the system. But only some of these solutions are feasible. In a feasible solution all variables must be nonnegative. To get a unique feasible solution, we set three of the five variables equal to 0. In general, if there

are m equations, then m variables can be nonzero. These are called **basic variables,** and the corresponding solutions are called **basic feasible solutions.** Each basic feasible solution corresponds to a corner point. In particular, if we choose the solution with $x_1 = 0$, $x_2 = 0$, and $x_3 = 0$, then $x_4 = 100$ and $x_5 = 500$. This solution, which corresponds to the corner point at the origin, is hardly optimal. It produces a profit of $0 for the farmer, since the objective function becomes

$$-120(0) - 40(0) - 60(0) + 0x_4 + 0x_5 + z = 0.$$

In the next section we will use the simplex method to start with this solution and improve it to find the maximum possible profit.

Each step of the simplex method produces a solution that corresponds to a corner point of the region of feasible solutions. These solutions can be read directly from the matrix, as shown in the next example.

 EXAMPLE 3

Read a solution from the matrix below.

$$\begin{array}{cccccc} x_1 & x_2 & x_3 & x_4 & x_5 & z \\ \end{array}$$
$$\begin{bmatrix} 2 & 0 & 8 & 5 & 2 & 0 & 17 \\ 9 & 5 & 3 & 12 & 0 & 0 & 45 \\ -2 & 0 & -4 & 0 & 0 & 3 & 90 \end{bmatrix}$$

The variables x_2 and x_5 are basic variables. They can quickly be identified because the columns for these variables have all zeros except for one nonzero entry. The nonzero entry for x_2 is a 5 in the second row. This means that if x_1, x_3, and x_4 are zero, then $5x_2 = 45$ (from the x_2 column and the last column), so $x_2 = 9$. Similarly, $2x_5 = 17$, so $x_5 = 17/2$. From the bottom row, $3z = 90$, so $z = 30$. The solution is thus $x_1 = 0$, $x_2 = 9$, $x_3 = 0$, $x_4 = 0$, and $x_5 = 17/2$, with $z = 30$. ◀

Pivots Solutions read directly from the initial simplex tableau are seldom, if ever, optimal. It is necessary to proceed to other solutions (corresponding to other corner points of the feasible region) until an optimum solution is found. To get these other solutions, use the row operations from Chapter 2 to change the tableau by using one of the nonzero entries of the tableau as a **pivot.** The row operations are performed to change to 0 all entries in the column containing the pivot (except for the pivot itself, which remains unchanged). Pivoting, explained in the next example, produces a new tableau leading to another solution of the system of equations obtained from the original problem.

The row operations discussed in Chapter 2 are:

1. interchanging any two rows;
2. multiplying the elements of a row by any nonzero real number; and
3. adding a multiple of the elements of one row to the corresponding elements of a multiple of some other row.

In this chapter we will only use operations 2 and 3; we will never interchange two rows.

Caution In this chapter, when adding a multiple of one row to a multiple of another, we will never take a negative multiple of the row being changed. For example, when changing row 2, we might use $-2R_1 + 3R_2 \rightarrow R_2$, but we will never use $2R_1 - 3R_2 \rightarrow R_2$. If you get a negative number in the rightmost column, you will know immediately that you have made an error.

23. Find $x_1 \geq 0$ and $x_2 \geq 0$ such that
$$3x_1 + x_2 \leq 12$$
$$x_1 + x_2 \leq 15$$
and $z = 2x_1 + x_2$ is maximized.

24. Find $x_1 \geq 0$ and $x_2 \geq 0$ such that
$$10x_1 + 4x_2 \leq 100$$
$$20x_1 + 10x_2 \leq 150$$
and $z = 4x_1 + 5x_2$ is maximized.

Applications

Set up Exercises 25–29 for solution by the simplex method. First express the linear constraints and objective function, then add slack variables, and then set up the initial simplex tableau. The solutions of some of these problems will be completed in the exercises for the next section.

Business and Economics

25. Profit A small boat manufacturer builds three types of fiberglass boats: prams, runabouts, and trimarans. The pram sells at a profit of $75, the runabout at a profit of $90, and the trimaran at a profit of $100. The factory is divided into two sections. Section A does the molding and construction work, while section B does the painting, finishing and equipping. The pram takes 1 hr in section A and 2 hr in section B. The runabout takes 2 hr in A and 5 hr in B. The trimaran takes 3 hr in A and 4 hr in B. This year, section A has a total of 6240 hr available, and section B has 10,800 hr available. The manufacturer has ordered a supply of fiberglass that will build at most 3000 boats, figuring the average amount used per boat. How many of each type of boat should be made to produce maximum profit? What is the maximum profit?

26. Profit Caroline's Quality Candy Confectionery is famous for fudge, chocolate cremes, and pralines. Its candy-making equipment is set up to make 100-lb batches at a time. Currently there is a chocolate shortage and the company can get only 120 lb of chocolate in the next shipment. On a week's run, the confectionery's cooking and processing equipment is available for a total of 42 machine hours. During the same period the employees have a total of 56 work hours available for packaging. A batch of fudge requires 20 lb of chocolate, and a batch of cremes uses 25 lb of chocolate. The cooking and processing take 120 min for fudge, 150 min for chocolate cremes, and 200 min for pralines. The packaging times measured in minutes per 1-lb box are 1, 2, and 3, respectively, for fudge, cremes, and pralines. Determine how many batches of each type of candy the confectionery should make, assuming that the profit per 1-lb box is 50¢ on fudge, 40¢ on chocolate cremes, and 45¢ on pralines. What is the maximum profit?

27. Costs The authors of a best-selling textbook in finite mathematics are told that, for the next edition of their

book, each simple figure would cost the project $20, each figure with additions would cost $35, and each computer-drawn sketch would cost $60. They are limited to 400 figures, for which they are allowed to spend up to $2200. The number of computer-drawn sketches must be no more than the number of the other two types combined, and there must be at least twice as many simple figures as there are figures with additions. If each simple figure increases the royalties by $95, each figure with additions increases royalties by $200, and each computer-drawn sketch increases royalties by $325, how many of each type of figure should they include in the book?

28. Profit Banal, Inc., produces art for motel rooms. Its painters can turn out mountain scenes, seascapes, and pictures of clowns. Each painting is worked on by three different artists, T, D, and H. Artist T works only 25 hr per week, while artists D and H work 45 hr and 40 hr per week, respectively. Artist T spends 1 hr on a mountain scene, 2 hr on a seascape, and 1 hr on a clown. Corresponding times for D and H are 3, 2, and 2 hr, and 2, 1, and 4 hr, respectively. Banal makes $20 on a mountain scene, $18 on a seascape, and $22 on a clown. The head painting packer can't stand clowns, so no more than 4 clown paintings may be done in a week. Find the number of each type of painting that should be made weekly in order to maximize profit. Find the maximum possible profit.

29. Profit A manufacturer of bicycles builds one-, three-, and ten-speed models. The bicycles are made of both aluminum and steel. The company has available 91,800 units of steel and 42,000 units of aluminum. The one-, three-, and ten-speed models need 17, 27, and 34 units of steel, and 12, 21, and 15 units of aluminum, respectively. How many of each type of bicycle should be made in order to maximize profit if the company makes $8 per one-speed bike, $12 per three-speed, and $22 per ten-speed? What is the maximum possible profit?

4.2 SOLVING MAXIMIZATION PROBLEMS

How many one-speed, three-speed, and ten-speed bicycles should a bicycle manufacturer make to maximize profit?

We will answer this question in Exercise 21 of this section using the simplex algorithm.

In the previous section we showed how to prepare a linear programming problem for solution. First, we converted the constraints to linear equations with slack variables; then we used the coefficients of the variables from the linear equations to write an augmented matrix. Finally, we used the pivot to go from one vertex to another vertex in the region of feasible solutions.

Now we are ready to put all this together and produce an optimum value for the objective function. To see how this is done, let us complete Example 2 from Section 4.1, the example about the farmer.

In the previous section we set up the following simplex tableau.

$$
\begin{array}{c}
\begin{array}{cccccc} x_1 & x_2 & x_3 & x_4 & x_5 & z \end{array} \\
\left[
\begin{array}{cccccc|c}
1 & 1 & 1 & 1 & 0 & 0 & 100 \\
10 & 4 & 7 & 0 & 1 & 0 & 500 \\
\hline
-120 & -40 & -60 & 0 & 0 & 1 & 0
\end{array}
\right]
\end{array}
$$

This tableau leads to the solution $x_1 = 0$, $x_2 = 0$, $x_3 = 0$, $x_4 = 100$, and $x_5 = 500$, with x_4 and x_5 as the basic variables. These values produce a value of 0 for z. Since a value of 0 for the farmer's profit is not an optimum, we will try to improve this value.

The coefficients of x_1, x_2, and x_3 in the objective function are nonzero, so the profit could be improved by making any one of these variables take on a nonzero value in a solution. To decide which variable to use, look at the indicators in the initial simplex tableau above. (Recall that the indicators are the numbers in the bottom row, except for the two numbers at the far right.) The coefficient of x_1, -120, is the most negative of the indicators. This means that x_1 has the largest coefficient in the objective function, so that profit is increased the most by increasing x_1.

As we saw earlier, because there are two equations in the system (for the two constraints), only two of the five variables can be basic variables (and be nonzero). If x_1 is nonzero in the solution, then x_1 will be a basic variable. This means that either x_4 or x_5 no longer will be a basic variable. To decide which variable will no longer be basic, start with the equations of the system,

$$
\begin{aligned}
x_1 + x_2 + x_3 + x_4 &= 100 \\
10x_1 + 4x_2 + 7x_3 \quad + x_5 &= 500,
\end{aligned}
$$

and solve for x_4 and x_5 respectively.

$$
\begin{aligned}
x_4 &= 100 - x_1 - x_2 - x_3 \\
x_5 &= 500 - 10x_1 - 4x_2 - 7x_3
\end{aligned}
$$

Only x_1 is being changed to a nonzero value; both x_2 and x_3 keep the value 0. Replacing x_2 and x_3 with 0 gives

$$x_4 = 100 - x_1$$
$$x_5 = 500 - 10x_1.$$

Since both x_4 and x_5 must remain nonnegative, there is a limit to how much the value of x_1 can be increased. The equation $x_4 = 100 - x_1$ (or $x_4 = 100 - 1x_1$) shows that x_1 cannot exceed $100/1$, or 100. The second equation, $x_5 = 500 - 10x_1$, shows that x_1 cannot exceed $500/10$, or 50. To satisfy both these conditions, x_1 cannot exceed 50, the smaller of 50 and 100. If we let x_1 take the value 50, then $x_1 = 50$, $x_2 = 0$, $x_3 = 0$, and $x_5 = 0$. Since $x_4 = 100 - x_1$, then

$$x_4 = 100 - 50 = 50.$$

Therefore, x_4 is still a basic variable, while x_5 is no longer a basic variable, having been replaced in the set of basic variables by x_1. We say that x_5 is a **departing variable,** while x_1 is an **entering variable.** This solution gives a profit of

$$z = 120x_1 + 40x_2 + 60x_3 + 0x_4 + 0x_5$$
$$= 120(50) + 40(0) + 60(0) + 0(50) + 0(0) = 6000,$$

or $6000.

The same result could have been found from the initial simplex tableau given above. To use the tableau, select the most negative indicator. (If no indicator is negative, then the value of the objective function cannot be improved.)

Basic variables

x_1	x_2	x_3	x_4	x_5	z	
1	1	1	1	0	0	100
10	4	7	0	1	0	500
-120	-40	-60	0	0	1	0

Most negative indicator

The most negative indicator identifies the variable whose value is to be made nonzero. To find the variable that is now basic and will become nonbasic, calculate the quotients that were found above. Do this by dividing each number from the right side of the tableau by the corresponding number from the column with the most negative indicator.

Basic variables

Quotients		x_1	x_2	x_3	x_4	x_5	z	
$100/1 = 100$		1	1	1	1	0	0	100
Smaller $\rightarrow 500/10 = 50$		**10**	4	7	0	1	0	500
		-120	-40	-60	0	0	1	0

Entering variable Departing variable

Notice that we do not form a quotient for the bottom row. Of the two quotients found, the smallest is 50 (from the second row), so 10 is the pivot. Using 10 as the pivot, perform the appropriate row operations to get zeros in the rest of the column. The new tableau is as follows.

Basic variables

$$
\begin{array}{c}
-R_2 + 10R_1 \rightarrow R_1 \\
\\
12R_2 + R_3 \rightarrow R_3
\end{array}
\begin{array}{ccccccc}
x_1 & x_2 & x_3 & x_4 & x_5 & z \\
\left[\begin{array}{cccccc|c}
0 & 6 & 3 & 10 & -1 & 0 & 500 \\
10 & 4 & 7 & 0 & 1 & 0 & 500 \\
\hline
0 & 8 & 24 & 0 & 12 & 1 & 6000
\end{array}\right]
\end{array}
$$

The solution read from this tableau is

$$x_1 = 50, \quad x_2 = 0, \quad x_3 = 0, \quad x_4 = 50, \quad x_5 = 0,$$

with $z = 6000$, the same as the result found above.

None of the indicators in the final simplex tableau is negative, which means that the value of z cannot be improved beyond \$6000. To see why, recall that the last row gives the coefficients of the objective function so that

$$0x_1 + 8x_2 + 24x_3 + 0x_4 + 12x_5 + z = 6000,$$

or
$$z = 6000 - 0x_1 - 8x_2 - 24x_3 - 0x_4 - 12x_5.$$

Since x_2, x_3, and x_5 are zero, $z = 6000$, but if any of these three variables were to increase, z would decrease.

This result suggests that the optimal solution has been found as soon as no indicators are negative. As long as an indicator is negative, the value of the objective function can be improved. If any indicators are negative, we just find a new pivot and use row operations, repeating the process until no negative indicators remain.

We can finally state the solution to the problem about the farmer: the optimum value of z is 6000, where $x_1 = 50$, $x_2 = 0$, $x_3 = 0$, $x_4 = 50$, and $x_5 = 0$. That is, the farmer will make a maximum profit of \$6000 by planting 50 acres of potatoes. Another 50 acres should be left unplanted. It may seem strange that leaving assets unused can produce a maximum profit, but such results actually occur often.

Note The reading of the final tableau can be simplified by creating a 1 in the columns corresponding to the basic variables and z. In the previous example, this is accomplished by dividing rows 1 and 2 by 10.

$$
\begin{array}{c}
R_1/10 \rightarrow R_1 \\
R_2/10 \rightarrow R_2
\end{array}
\begin{array}{ccccccc}
x_1 & x_2 & x_3 & x_4 & x_5 & z \\
\left[\begin{array}{cccccc|c}
0 & 6/10 & 3/10 & 1 & -1/10 & 0 & 50 \\
1 & 4/10 & 7/10 & 0 & 1/10 & 0 & 50 \\
\hline
0 & 8 & 24 & 0 & 12 & 1 & 6000
\end{array}\right]
\end{array}
$$

It is now easy to read the solution from this tableau:

$$x_1 = 50, \quad x_2 = 0, \quad x_3 = 0, \quad x_4 = 50, \quad x_5 = 0,$$

with $z = 6000$, the same as before.

In summary, the following steps are involved in solving a standard maximum linear programming problem by the simplex method.

SIMPLEX METHOD

1. Determine the objective function.
2. Write all the necessary constraints.
3. Convert each constraint into an equation by adding slack variables.
4. Set up the initial simplex tableau.
5. Locate the most negative indicator. If there are two such indicators, choose the one farthest to the left.
6. Form the necessary quotients to find the pivot. Disregard any negative quotients or quotients with a 0 denominator. The smallest nonnegative quotient gives the location of the pivot. If all quotients must be disregarded, no maximum solution exists. If two quotients are both equal and smallest, choose the pivot in the row nearest the top of the matrix.
7. Use row operations to change all other numbers in the pivot column to zero. All operations are of the form $c_1R_i + c_2R_j \rightarrow R_j$, where row i is the pivot row and $c_2 > 0$.
8. If the indicators are all positive or 0, this is the final tableau. If not, go back to Step 5 above and repeat the process until a tableau with no negative indicators is obtained.
9. Read the solution from the final tableau.

In Steps 5 and 6, the choice of the column farthest to the left or the row closest to the top is arbitrary. You may choose another row or column in case of a tie, and you will get the same final answer, but your intermediate results will be different.

EXAMPLE **1**

To compare the simplex method with the graphical method, use the simplex method to solve the problem in Example 1, Section 3.3. The graph is shown again in Figure 1. The objective function to be maximized was

$$z = 24x_1 + 36x_2.$$ Storage space

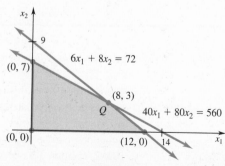

FIGURE 1

(Since we are using the simplex method, we use x_1 and x_2 as variables instead of x and y.) The constraints were as follows:

$$40x_1 + 80x_2 \leq 560 \qquad \text{Cost}$$
$$6x_1 + 8x_2 \leq 72 \qquad \text{Floor space}$$

with

$$x_1 \geq 0, \, x_2 \geq 0.$$

To simplify the arithmetic, divide the first inequality on both sides by 40 and the second by 2, which gives

$$x_1 + 2x_2 \leq 14$$
$$3x_1 + 4x_2 \leq 36$$

with

$$x_1 \geq 0, \, x_2 \geq 0.$$

Add a slack variable to each constraint:

$$x_1 + 2x_2 + x_3 \qquad = 14$$
$$3x_1 + 4x_2 \qquad + x_4 = 36$$

with

$$x_1 \geq 0, \, x_2 \geq 0, \, x_3 \geq 0, \, x_4 \geq 0.$$

Write the initial tableau.

$$
\begin{array}{ccccc}
x_1 & x_2 & x_3 & x_4 & z
\end{array}
$$
$$
\left[
\begin{array}{ccccc|c}
1 & 2 & 1 & 0 & 0 & 14 \\
3 & 4 & 0 & 1 & 0 & 36 \\
\hline
-24 & -36 & 0 & 0 & 1 & 0
\end{array}
\right]
$$

This tableau leads to the solution $x_1 = 0$, $x_2 = 0$, $x_3 = 14$, and $x_4 = 36$, with $z = 0$, which corresponds to the origin in Figure 1. The most negative indicator is -36, which is in column 2 of row 3. The quotients of the numbers in the right-hand column and in column 2 are

$$\frac{14}{2} = 7 \qquad \text{and} \qquad \frac{36}{4} = 9.$$

The smaller quotient is 7, giving 2 as the pivot. Use row operations to get the new tableau. For clarity, we will continue to label the columns with x_1, x_2, etc., although this is not necessary in practice.

$$
\begin{array}{ccccc}
x_1 & x_2 & x_3 & x_4 & z
\end{array}
$$
$$
\begin{array}{l}
\\
-2R_1 + R_2 \rightarrow R_2 \\
18R_1 + R_3 \rightarrow R_3
\end{array}
\left[
\begin{array}{ccccc|c}
1 & 2 & 1 & 0 & 0 & 14 \\
1 & 0 & -2 & 1 & 0 & 8 \\
\hline
-6 & 0 & 18 & 0 & 1 & 252
\end{array}
\right]
$$

The solution from this tableau is $x_1 = 0$ and $x_2 = 7$, with $z = 252$. (From now on, we will only list the original variables when giving the solution.) This corresponds to the corner point $(0, 7)$ in Figure 1. Because of the indicator -6, the value of z can be improved. We compare quotients as before and choose the 1 in row 2, column 1 as pivot to get the final tableau.

$$-R_2 + R_1 \to R_1 \qquad \begin{bmatrix} 0 & 2 & 3 & -1 & 0 & 6 \\ 1 & 0 & -2 & 1 & 0 & 8 \\ 0 & 0 & 6 & 6 & 1 & 300 \end{bmatrix}$$

$$6R_2 + R_3 \to R_3$$

There are no more negative indicators, so the optimum solution has been achieved. Create a 1 in column 2 by dividing row 1 by 2.

$$\begin{array}{ccccc} x_1 & x_2 & x_3 & x_4 & z \end{array}$$

$$R_1/2 \to R_1 \qquad \begin{bmatrix} 0 & 1 & 3/2 & -1/2 & 0 & 3 \\ 1 & 0 & -2 & 1 & 0 & 8 \\ 0 & 0 & 6 & 6 & 1 & 300 \end{bmatrix}$$

Here the solution is $x_1 = 8$ and $x_2 = 3$, with $z = 300$. This solution, which corresponds to the corner point $(8, 3)$ in Figure 1, is the same as the solution found earlier.

Each simplex tableau above gave a solution corresponding to one of the corner points of the feasible region. As shown in Figure 2, the first solution corresponded to the origin, with $z = 0$. By choosing the appropriate pivot, we moved systematically to a new corner point, $(0, 7)$, which improved the value of z to 252. The next tableau took us to $(8, 3)$, producing the optimum value of $z = 300$. There was no reason to test the last corner point, $(12, 0)$, since the optimum value of z was found before that point was reached. ◀

FIGURE 2

Caution Never choose a zero or a negative number as the pivot. The reason for this is explained in the next example.

EXAMPLE 2

Find the pivot for the following initial simplex tableau.

$$\begin{array}{cccccc} x_1 & x_2 & x_3 & x_4 & x_5 & z \end{array}$$

$$\begin{bmatrix} 1 & -2 & 1 & 0 & 0 & 0 & 100 \\ 3 & 4 & 0 & 1 & 0 & 0 & 200 \\ 5 & 0 & 0 & 0 & 1 & 0 & 150 \\ -10 & -25 & 0 & 0 & 0 & 1 & 0 \end{bmatrix}$$

The most negative indicator is -25. To find the pivot, find the quotients formed by the entries in the rightmost column and in the x_2 column: $100/(-2)$, $200/4$, and $150/0$. Since division by 0 is meaningless, disregard the quotient $150/0$, which comes from the equation

$$5x_1 + 0x_2 + 0x_3 + 0x_4 + x_5 = 150.$$

If $x_1 = 0$, the equation reduces to

$$0x_2 + x_5 = 150 \qquad \text{or} \qquad x_5 = 150 - 0x_2.$$

Since the pivot is in the x_2 column, x_5 cannot be 0. In general, disregard quotients with 0 denominators.

The quotients predict the value of a variable in the solution, and thus cannot be negative. In this example, the quotient $100/(-2) = -50$ comes from the equation

$$x_1 - 2x_2 + x_3 + 0x_4 + 0x_5 = 100.$$

For $x_1 = 0$, this becomes

$$x_3 = 100 + 2x_2.$$

If x_3 is to be nonnegative, then

$$100 + 2x_2 \geq 0$$
$$x_2 \geq -50.$$

Since x_2 must be nonnegative anyway, this equation tells us nothing new. For this reason, disregard negative quotients also.

The only usable quotient is $200/4 = 50$, making 4 the pivot. If all the quotients either are negative or have zero denominators, no unique optimum solution will be found. Such a situation indicates an unbounded feasible region. The quotients, then, determine whether or not an optimum solution exists. ◀

Caution If there is a 0 in the right-hand column, do not disregard that row. In fact, such a row gives a quotient of 0, so it will automatically have the smallest ratio. It will not cause an increase in z, but it may lead to another tableau in which z can be further increased.

In many real-life problems, the number of variables and constraints may be in the hundreds, if not the thousands, in which case a computer is used to implement the simplex algorithm. Computer programs for the simplex algorithm differ in some ways from the algorithm we have shown. For example, it is not necessary for a computer to divide common factors out of inequalities to simplify the arithmetic. In fact, computer versions of the algorithm do not keep all the numbers as integers. In Step 7 of the algorithm, most computer programs divide the pivot row by the value of the pivot to change the pivot to a 1. This may introduce decimals, which makes the arithmetic more difficult to do by hand, but creates no problem for a computer other than round-off error.

4.2 Exercises

In Exercises 1–6, the initial tableau of a linear programming problem is given. Use the simplex method to solve each problem.

1.

x_1	x_2	x_3	x_4	x_5	z	
1	2	4	1	0	0	8
2	2	1	0	1	0	10
−2	−5	−1	0	0	1	0

2.

x_1	x_2	x_3	x_4	x_5	z	
2	2	1	1	0	0	10
1	2	3	0	1	0	15
−3	−2	−1	0	0	1	0

3.

x_1	x_2	x_3	x_4	x_5	z	
1	3	1	0	0	0	12
2	1	0	1	0	0	10
1	1	0	0	1	0	4
−2	−1	0	0	0	1	0

4.

x_1	x_2	x_3	x_4	x_5	x_6	z	
2	2	1	1	0	0	0	50
1	1	3	0	1	0	0	40
4	2	5	0	0	1	0	80
−2	−3	−5	0	0	0	1	0

5.

$$\begin{array}{ccccccc}
x_1 & x_2 & x_3 & x_4 & x_5 & x_6 & z \\
\end{array}$$

$$\begin{bmatrix}
2 & 2 & 8 & 1 & 0 & 0 & 0 & 40 \\
4 & -5 & 6 & 0 & 1 & 0 & 0 & 60 \\
2 & -2 & 6 & 0 & 0 & 1 & 0 & 24 \\
-14 & -10 & -12 & 0 & 0 & 0 & 1 & 0
\end{bmatrix}$$

6.

$$\begin{array}{cccccc}
x_1 & x_2 & x_3 & x_4 & x_5 & z \\
\end{array}$$

$$\begin{bmatrix}
3 & 2 & 4 & 1 & 0 & 0 & 18 \\
2 & 1 & 5 & 0 & 1 & 0 & 8 \\
-1 & -4 & -2 & 0 & 0 & 1 & 0
\end{bmatrix}$$

Use the simplex method to solve each linear programming problem.

7. Maximize $z = 4x_1 + 3x_2$

subject to: $2x_1 + 3x_2 \leq 11$
$x_1 + 2x_2 \leq 6,$

with $x_1 \geq 0, x_2 \geq 0.$

8. Maximize $z = 2x_1 + 3x_2$

subject to: $3x_1 + 5x_2 \leq 29$
$2x_1 + x_2 \leq 10,$

with $x_1 \geq 0, x_2 \geq 0.$

9. Maximize $z = 10x_1 + 12x_2$

subject to: $4x_1 + 2x_2 \leq 20$
$5x_1 + x_2 \leq 50$
$2x_1 + 2x_2 \leq 24$

with $x_1 \geq 0, x_2 \geq 0.$

10. Maximize $z = 1.2x_1 + 3.5x_2$

subject to: $2.4x_1 + 1.5x_2 \leq 10$
$1.7x_1 + 1.9x_2 \leq 15$

with $x_1 \geq 0, x_2 \geq 0.$

11. Maximize $z = 8x_1 + 3x_2 + x_3$

subject to: $x_1 + 6x_2 + 8x_3 \leq 118$
$x_1 + 5x_2 + 10x_3 \leq 220$

with $x_1 \geq 0, x_2 \geq 0, x_3 \geq 0.$

12. Maximize $z = 12x_1 + 15x_2 + 5x_3$

subject to: $2x_1 + 2x_2 + x_3 \leq 8$
$x_1 + 4x_2 + 3x_3 \leq 12$

with $x_1 \geq 0, x_2 \geq 0, x_3 \geq 0.$

13. Maximize $z = x_1 + 2x_2 + x_3 + 5x_4$

subject to: $x_1 + 2x_2 + x_3 + x_4 \leq 50$
$3x_1 + x_2 + 2x_3 + x_4 \leq 100$

with $x_1 \geq 0, x_2 \geq 0, x_3 \geq 0, x_4 \geq 0.$

14. Maximize $z = x_1 + x_2 + 4x_3 + 5x_4$

subject to: $x_1 + 2x_2 + 3x_3 + x_4 \leq 115$
$2x_1 + x_2 + 8x_3 + 5x_4 \leq 200$
$x_1 + x_3 \leq 50$

with $x_1 \geq 0, x_2 \geq 0, x_3 \geq 0, x_4 \geq 0.$

15. The simplex algorithm still works if an indicator other than the most negative one is chosen. (Try it!) List the disadvantages that might occur if this is done.

16. What goes wrong if a quotient other than the smallest nonnegative quotient is chosen in the simplex algorithm? (Try it!)

Applications

Set up and solve Exercises 17–23 by the simplex method.

Business and Economics

17. Charitable Contributions Jayanta is working to raise money for the homeless by sending information letters and making follow-up calls to local labor organizations and church groups. She discovered that each church group requires 2 hr of letter writing and 1 hr of follow-up, while for each labor union she needs 2 hr of letter writing and 3 hr of follow-up. Jayanta can raise $100 from each church group and $200 from each union local, and she has a maximum of 16 hr of letter-writing time and a maximum of 12 hr of follow-up time available per month. Determine the most profitable mixture of groups she should contact and the most money she can raise in a month.

18. Profit Seall Manufacturing Company makes color television sets. It produces a bargain set that sells for $100 profit and a deluxe set that sells for $150 profit. On the assembly line the bargain set requires 3 hr, and the deluxe set takes 5 hr. The cabinet shop spends 1 hr on the cabinet for the bargain set and 3 hr on the cabinet for the deluxe set. Both sets require 2 hr of time for testing and packing. On a particular production run the Seall Company has available 3900 work hours on the assembly line, 2100 work hours in the cabinet shop, and 2200 work hours in the testing and packing department. How many sets of each type should it produce to make maximum profit? What is the maximum profit? (See Exercise 10 in Section 3.3.)

19. **Profit** The Soul Sounds Recording Company produces three main types of musical albums: jazz, blues, and reggae. Each jazz album requires 4 hr of recording, 2 hr of mixing, and 6 hr of editing; each blues album requires 4 hr of recording, 8 hr of mixing, and 2 hr of editing; and each reggae album requires 10 hr of recording, 4 hr of mixing, and 6 hr of editing. The recording studio is available 80 hr per week, staff to operate the mixing board are available 52 hr a week, and the editing crew has at most 54 hr per week for work. Soul Sounds makes 80¢ per jazz album, 60¢ per blues album, and $1.20 for each reggae album. Use the simplex method to determine how many of each type of album the company should produce to maximize weekly profit. Find the maximum profit.

20. **Income** A baker has 150 units of flour, 90 of sugar, and 150 of raisins. A loaf of raisin bread requires 1 unit of flour, 1 of sugar, and 2 of raisins, while a raisin cake needs 5, 2, and 1 units, respectively. If raisin bread sells for $1.75 a loaf and raisin cake for $4.00 each, how many of each should be baked so that gross income is maximized? What is the maximum gross income?

21. **Profit** A manufacturer of bicycles builds one-, three-, and ten-speed models. The bicycles are made of both aluminum and steel. The company has available 91,800 units of steel and 42,000 units of aluminum. The one-, three-, and ten-speed models need 17, 27, and 34 units of steel, and 12, 21, and 15 units of aluminum, respectively. How many of each type of bicycle should be made in order to maximize profit if the company makes $8 per one-speed bike, $12 per three-speed, and $22 per ten-speed? What is the maximum possible profit? (See Exercise 29 in Section 4.1.)

22. **Profit** Caroline's Quality Candy Confectionery is famous for fudge, chocolate cremes, and pralines. Its candy-making equipment is set up to make 100-lb batches at a time. Currently there is a chocolate shortage and the company can get only 120 lb of chocolate in the next shipment. On a week's run, the confectionery's cooking and processing equipment is available for a total of 42 machine hours. During the same period the employees have a total of 56 work hours available for packaging. A batch of fudge requires 20 lb of chocolate, and a batch of cremes uses 25 lb chocolate. The cooking and processing take 120 min for fudge, 150 min for chocolate cremes, and 200 min for pralines. The packaging times measured in minutes per one-pound box are 1, 2, and 3, respectively, for fudge, cremes and pralines. Determine how many batches of each type of candy the confectionery should make, assuming that the profit per pound box is 50¢ on fudge, 40¢ on chocolate cremes, and 45¢ on pralines. Also, find the maximum profit for the week. (See Exercise 26 in Section 4.1.)

23. **Revenue** A candy company has 100 kg of chocolate-covered nuts and 125 kg of chocolate-covered raisins to be sold as two different mixtures. One mix will contain half nuts and half raisins and will sell for $6 per kilogram. The other mix will contain 1/3 nuts and 2/3 raisins and will sell for $4.80 per kilogram. How many kilograms of each mix should the company prepare for maximum revenue? Find the maximum revenue. (See Exercise 13 in Section 3.3.)

*The next two problems come from past CPA examinations.**
Select the appropriate answer for each question.

24. **Profit** The Ball Company manufactures three types of lamps, labeled A, B, and C. Each lamp is processed in two departments, I and II. Total available man-hours per day for departments I and II are 400 and 600, respectively. No additional labor is available. Time requirements and profit per unit for each lamp type is as follows:

	A	B	C
Man-hours in I	2	3	1
Man-hours in II	4	2	3
Profit per Unit	$5	$4	$3

The company has assigned you as the accounting member of its profit planning committee to determine the numbers of types of A, B, and C lamps that it should produce in order to maximize its total profit from the sale of lamps. The following questions relate to a linear programming model that your group has developed.

(a) The coefficients of the objective function would be
 (1) 4, 2, 3. (2) 2, 3, 1.
 (3) 5, 4, 3. (4) 400, 600.

(b) The constraints in the model would be
 (1) 2, 3, 1. (2) 5, 4, 3.
 (3) 4, 2, 3. (4) 400, 600.

(c) The constraint imposed by the available man-hours in department I could be expressed as
 (1) $4X_1 + 2X_2 + 3X_3 \leq 400$.
 (2) $4X_1 + 2X_2 + 3X_3 \geq 400$.
 (3) $2X_1 + 3X_2 + 1X_3 \leq 400$.
 (4) $2X_1 + 3X_2 + 1X_3 \geq 400$.

25. **Profit** The Golden Hawk Manufacturing Company wants to maximize the profits on products A, B, and C. The contribution margin for each product follows:

Product	Contribution Margin
A	$2
B	$5
C	$4

The production requirements and departmental capacities, by departments, are as follows:

Department	Production Requirements by Product (hours)			Departmental Capacity (total hours)
	A	B	C	
Assembling	2	3	2	30,000
Painting	1	2	2	38,000
Finishing	2	3	1	28,000

(a) What is the profit-maximization formula for the Golden Hawk Company?
 (1) $2A + $5B + $4C = X$ (where X = profit)
 (2) $5A + 8B + 5C ≤ 96,000$
 (3) $2A + $5B + $4C ≤ X$
 (4) $2A + $5B + $4C = 96,000$

(b) What is the constraint for the painting department of the Golden Hawk Company?
 (1) $1A + 2B + 2C ≥ 38,000$
 (2) $2A + $5B + $4C ≥ 38,000$
 (3) $1A + 2B + 2C ≤ 38,000$
 (4) $2A + 3B + 2C ≤ 30,000$

Set up and solve Exercises 26–27 by the simplex method.

Life Sciences

26. **Blending Nutrients** A biologist has 500 kg of nutrient A, 600 kg of nutrient B, and 300 kg of nutrient C. These nutrients will be used to make 4 types of food, whose contents (in percent of nutrient per kilogram of food) and whose "growth values" are as shown below.

Food	Nutrient (%)			Growth Value
	A	B	C	
P	0	0	100	90
Q	0	75	25	70
R	37.5	50	12.5	60
S	62.5	37.5	0	50

How many kilograms of each food should be produced in order to maximize total growth value? Find the maximum growth value.

Social Sciences

27. **Politics** A political party is planning a half-hour television show. The show will have 3 min of direct requests for money from viewers. Three of the party's politicians will be on the show—a senator, a congresswoman, and a governor. The senator, a party "elder statesman," demands that he be on screen at least twice as long as the governor. The total time taken by the senator and the governor must be at least twice the time taken by the congresswoman. Based on a pre-show survey, it is believed that 40, 60, and 50 (in thousands) viewers will watch the program for each minute the senator, congresswoman, and governor, respectively, are on the air. Find the time that should be allotted to each politician in order to get the maximum number of viewers. Find the maximum number of viewers.

Computer/Graphing Calculator

Determine the constraints and the objective functions for Exercises 28 and 29, and then solve each problem.

28. (a) **Profit** A manufacturer makes two products, toy trucks and toy fire engines. Both are processed in four different departments, each of which has a limited capacity. The sheet metal department can handle at least $1\frac{1}{2}$ times as many trucks as fire engines; the truck assembly department can handle at most 6700 trucks per week; and the fire engine assembly department assembles at most 5500 fire engines weekly. The painting department, which finishes both toys, has a maximum capacity of 12,000 per week. If the profit is $8.50 for a toy truck and $12.10 for a toy fire engine, how many of each item should the company produce to maximize profit?

(b) Find a value for the profit for a toy truck and a value for the profit for a toy fire engine that would result in no fire engines being manufactured to maximize profit, given the constraints in part (a).

(c) Find a value for the profit for a toy truck and a value for the profit for a toy fire engine that would result in no toy trucks being manufactured to maximize the profit, given the constraints in part (a).

29. (a) **Resource Management** The average weights of the three species stocked in the lake referred to in Section 2.2, Exercise 51, are 1.62, 2.14, and 3.01 kg for spe-

cies A, B, and C, respectively. If the largest amounts of food that can be supplied each day are given as in Exercise 51, how should the lake be stocked to maximize the weight of the fish supported by the lake?

(b) Find a value for each of the average weights of the three species that would result in none of species B or C being stocked to maximize the weight of the fish

supported by the lake, given the constraints in part (a).

(c) Find a value for each of the average weights of the three species that would result in none of species A or B being stocked to maximize the weight of the fish supported by the lake, given the constraints in part (a).

4.3 NONSTANDARD PROBLEMS; MINIMIZATION

 How many cars should an auto manufacturer send from each of its two plants to each of two dealerships in order to minimize the cost while meeting each dealership's needs?

We will learn techniques in this section for answering questions like the one above.

So far we have used the simplex method to solve linear programming problems in standard maximum form only. In this section, this work is extended to include linear programming problems with mixed \leq and \geq constraints. The solution of these problems then gives a method of solving minimization problems. (An alternative approach for solving minimization problems is given in the next section.)

Problems with \leq and \geq Constraints Suppose a new constraint is added to the farmer problem from Example 2 of Section 4.1: to satisfy orders from regular buyers, the farmer must plant a total of at least 60 acres of the three crops. This constraint introduces the new inequality

$$x_1 + x_2 + x_3 \geq 60.$$

As before, this inequality must be rewritten as an equation in which the variables all represent nonnegative numbers. The inequality $x_1 + x_2 + x_3 \geq 60$ means that

$$x_1 + x_2 + x_3 - x_6 = 60$$

for some nonnegative variable x_6. (Remember that x_4 and x_5 are the slack variables in the problem.)

The new variable, x_6, is called a **surplus variable.** The value of this variable represents the excess number of acres (over 60) that may be planted. Since the total number of acres planted is to be no more than 100 but at least 60, the value of x_6 can vary from 0 to 40.

We must now solve the system of equations

$$
\begin{aligned}
x_1 + x_2 + x_3 + x_4 & = 100 \\
10x_1 + 4x_2 + 7x_3 + x_5 & = 500 \\
x_1 + x_2 + x_3 - x_6 & = 60 \\
-120x_1 - 40x_2 - 60x_3 + z & = 0,
\end{aligned}
$$

with $x_1, x_2, x_3, x_4, x_5,$ and x_6 all nonnegative.

Set up the initial simplex tableau.

$$
\begin{array}{ccccccc}
x_1 & x_2 & x_3 & x_4 & x_5 & x_6 & z \\
\end{array}
$$

$$
\left[
\begin{array}{ccccccc|c}
1 & 1 & 1 & 1 & 0 & 0 & 0 & 100 \\
10 & 4 & 7 & 0 & 1 & 0 & 0 & 500 \\
1 & 1 & 1 & 0 & 0 & -1 & 0 & 60 \\
\hline
-120 & -40 & -60 & 0 & 0 & 0 & 1 & 0 \\
\end{array}
\right]
$$

This tableau gives the solution

$$
x_1 = 0, \quad x_2 = 0, \quad x_3 = 0, \quad x_4 = 100, \quad x_5 = 20{,}000, \quad x_6 = -60.
$$

But this is not a feasible solution, since x_6 is negative. All the variables in any feasible solution must be nonnegative if the solution is to correspond to a corner point of the region of feasible solutions.

When a negative value of a variable appears in the solution, row operations are used to transform the matrix until a solution is found in which all variables are nonnegative. Here the problem is the -1 in a column corresponding to a basic variable. If the number in that row of the right-hand column were 0, we could simply multiply this row by -1 to remove the negative from the column. (This corresponds to changing an expression such as $x_1 - x_6 \geq 0$ into $-x_1 + x_6 \leq 0$, in which the basic variable has the value 0 in either case.) But we cannot do this with a 60 in the right-hand column. Instead, we find the positive entry that is farthest to the left in the third row (the row containing the -1); namely, the 1 in row 3, column 1. We will pivot using this column. (Actually, any column with a positive entry in row 3 will do; we chose the column farthest to the left arbitrarily.) Use quotients as before to find the pivot, which is the 10 in row 2, column 1. Then use row operations to get the following tableau.

$$
\begin{array}{ccccccc}
& x_1 & x_2 & x_3 & x_4 & x_5 & x_6 & z \\
\end{array}
$$

$$
\begin{array}{l}
-R_2 + 10R_1 \rightarrow R_1 \\
\\
-R_2 + 10R_3 \rightarrow R_3 \\
12R_2 + R_4 \rightarrow R_4
\end{array}
\left[
\begin{array}{ccccccc|c}
0 & 6 & 3 & 10 & -1 & 0 & 0 & 500 \\
10 & 4 & 7 & 0 & 1 & 0 & 0 & 500 \\
0 & 6 & 3 & 0 & -1 & -10 & 0 & 100 \\
0 & 8 & 24 & 0 & 12 & 0 & 1 & 6000 \\
\end{array}
\right]
$$

Notice from the x_6 column that $-10x_6 = 100$, so x_6 is still negative. We therefore apply the procedure again. The 6 in row 3, column 2, is the positive entry farthest to the left in row 3, and by investigating quotients, we see that it is also the pivot. This leads to the following tableau.

$$
\begin{array}{ccccccc}
& x_1 & x_2 & x_3 & x_4 & x_5 & x_6 & z \\
\end{array}
$$

$$
\begin{array}{l}
-R_3 + R_1 \rightarrow R_1 \\
-2R_3 + 3R_2 \rightarrow R_2 \\
\\
-4R_3 + 3R_4 \rightarrow R_4
\end{array}
\left[
\begin{array}{ccccccc|c}
0 & 0 & 0 & 10 & 0 & 10 & 0 & 400 \\
30 & 0 & 15 & 0 & 5 & 20 & 0 & 1{,}300 \\
0 & 6 & 3 & 0 & -1 & -10 & 0 & 100 \\
0 & 0 & 60 & 0 & 40 & 40 & 3 & 17{,}600 \\
\end{array}
\right]
$$

The value of x_6 is now 0. If there were any negative numbers in the bottom row, we would continue as before. Since there are none, we have merely to create a 1 in each column corresponding to a basic variable or z.

$$
\begin{array}{c}
R_1/10 \rightarrow R_1 \\
R_2/30 \rightarrow R_2 \\
R_3/6 \rightarrow R_3 \\
R_4/3 \rightarrow R_4
\end{array}
\begin{array}{cccccccc}
x_1 & x_2 & x_3 & x_4 & x_5 & x_6 & z & \\
\left[\begin{array}{ccccccc|c}
0 & 0 & 0 & 1 & 0 & 1 & 0 & 40 \\
1 & 0 & 1/2 & 0 & 1/6 & 2/3 & 0 & 130/3 \\
0 & 1 & 1/2 & 0 & -1/6 & -5/3 & 0 & 50/3 \\
0 & 0 & 20 & 0 & 40/3 & 40/3 & 1 & 17,600/3
\end{array}\right]
\end{array}
$$

The solution is

$$x_1 = 130/3 = 43\frac{1}{3}, \quad x_2 = 50/3 = 16\frac{2}{3}, \quad x_3 = 0, \quad z = 17,600/3 = 5866.67.$$

For maximum profit with this new constraint, the farmer should plant 43⅓ acres of potatoes, 16⅔ acres of corn, and no cabbage. The profit will be $5866.67, less than the $6000 profit if the farmer were to plant only 50 acres of potatoes. Because of the additional constraint that at least 60 acres must be planted, the profit is reduced.

Note If we ever reach a point where a surplus variable still has a negative solution, but there are no positive elements left in the row, then the problem has no feasible solution.

The procedure we have followed is a simplified version of the **two-phase method,** which is widely used for solving problems with mixed constraints. To see the complete method, including how to handle some complications that may arise, see *Elementary Linear Programming with Applications* by Bernard Kolman and Robert E. Beck, Academic Press, 1980.

The steps for finding a feasible solution are now summarized.

FINDING A FEASIBLE SOLUTION

1. If any basic variable has a negative value, locate the nonzero number in that variable's column, and note what row it is in.

2. In the row located in Step 1, find the positive entry that is farthest to the left, and note what column it is in.

3. In the column found in Step 2, choose a pivot by investigating quotients.

4. Use row operations to change the other numbers in the pivot column to 0.

5. Continue Steps 1 through 4 until all basic variables are nonnegative. If it ever becomes impossible to continue, then the problem has no feasible solution.

6. Once a feasible solution has been found, continue to use the simplex method until the optimal solution is found.

Minimization Problems The definition of a problem in standard maximum form was given earlier in this chapter. Now we can define a linear programming problem in *standard minimum form*, as follows.

STANDARD MINIMUM FORM

A linear programming problem is in **standard minimum form** if the following conditions are satisfied.

1. The objective function is to be minimized.

2. All variables are nonnegative.

3. All constraints involve \geq.

4. All of the constants in the constraints are nonnegative.

The difference between maximization and minimization problems is in conditions 1 and 3: in problems stated in standard minimum form the objective function is to be *minimized*, rather than maximized, and all constraints must have \geq instead of \leq.

Problems in standard minimum form can be solved with the method of surplus variables presented above. To solve a problem in standard minimum form, first observe that the minimum of an objective function is the same number as the *maximum* of the *negative* of the function. For example, if $w = 3y_1 + 2y_2$, then $z = -w = -3y_1 - 2y_2$. For corner points (0, 3), (6, 0), and (2, 1), the same ordered pair that minimizes w maximizes z, as shown below.

	$w = 3y_1 + 2y_2$		$z = -3y_1 - 2y_2$	
(0, 3)	6	Minimum	-6	Maximum
(6, 0)	18		-18	
(2, 1)	8		-8	

See Figure 3.

FIGURE 3

The next example illustrates the procedure described above. We use y_1 and y_2 as variables and w for the objective function as a reminder that this is a minimizing problem.

EXAMPLE 1

Minimize $w = 3y_1 + 2y_2$
subject to: $y_1 + 3y_2 \geq 6$
 $2y_1 + y_2 \geq 3$
with $y_1 \geq 0, \quad y_2 \geq 0.$

Change this to a maximization problem by letting z equal the *negative* of the objective function: $z = -w$. Then find the *maximum* value of z.

$$z = -w = -3y_1 - 2y_2$$

The problem can now be stated as follows.

Maximize $z = -3y_1 - 2y_2$
subject to: $y_1 + 3y_2 \geq 6$
 $2y_1 + y_2 \geq 3$
with $y_1 \geq 0, \quad y_2 \geq 0.$

For phase I of the solution, add surplus variables and set up the first tableau.

$$\begin{array}{ccccc} y_1 & y_2 & y_3 & y_4 & z \end{array}$$
$$\left[\begin{array}{ccccc|c} 1 & 3 & -1 & 0 & 0 & 6 \\ 2 & 1 & 0 & -1 & 0 & 3 \\ \hline 3 & 2 & 0 & 0 & 1 & 0 \end{array}\right]$$

The solution, $y_1 = 0$, $y_2 = 0$, $y_3 = -6$, and $y_4 = -3$, contains negative numbers. Row operations must be used to get a tableau with a feasible solution. We start with y_3, which has a -1 in row 1. The positive entry farthest to the left in row 1 is the 1 in column 1. The element in column 1 which gives the smallest quotient is 2, so it becomes the pivot. The following matrix results.

$$\begin{array}{ccccc} y_1 & y_2 & y_3 & y_4 & z \end{array}$$
$$\begin{array}{l} -R_2 + 2R_1 \rightarrow R_1 \\ \\ -3R_2 + 2R_3 \rightarrow R_3 \end{array} \left[\begin{array}{ccccc|c} 0 & 5 & -2 & 1 & 0 & 9 \\ 2 & 1 & 0 & -1 & 0 & 3 \\ \hline 0 & 1 & 0 & 3 & 2 & -9 \end{array}\right]$$

At this point $-2y_3 = 9$, so y_3 is still negative. This time the 5 in column 2 is the positive entry farthest to the left in row 1, and it also gives the smallest quotient in column 2. Pivoting on that element gives the following matrix.

$$\begin{array}{ccccc} y_1 & y_2 & y_3 & y_4 & z \end{array}$$
$$\begin{array}{l} -R_1 + 5R_2 \rightarrow R_2 \\ -R_1 + 5R_3 \rightarrow R_3 \end{array} \left[\begin{array}{ccccc|c} 0 & 5 & -2 & 1 & 0 & 9 \\ 10 & 0 & 2 & -6 & 0 & 6 \\ \hline 0 & 0 & 2 & 14 & 10 & -54 \end{array}\right]$$

Now y_3 is 0, as is y_4. Furthermore, there are no negative numbers in the bottom row. Divide row 1 by 5, row 2 by 10, and row 3 by 10 to find that the final solution is $y_1 = 3/5$ and $y_2 = 9/5$. Since $z = -w = -27/5$, the minimum value is $w = 27/5$. ◀

In summary, the following steps are involved in solving the nonstandard problems in this section.

SOLVING NONSTANDARD PROBLEMS

1. If necessary, convert the problem to a maximization problem.
2. Add slack variables and subtract surplus variables as needed.
3. Write the initial simplex tableau.
4. If the solution from this tableau is not feasible, use row operations to get a feasible solution.
5. After a feasible solution is reached, continue to use the simplex method as long as there are negative numbers in the last row.

An important application of linear programming is the problem of minimizing the cost of transporting goods. This type of problem is often referred to as a *transportation problem* or *warehouse problem*. Some problems of this type were included in the exercise sets in previous chapters. The next example illustrates the use of the simplex method in solving a transportation problem.

EXAMPLE 2 In the exercises for Section 2.2, we saw the following problem:

An auto manufacturer sends cars from two plants, I and II, to dealerships A and B located in a midwestern city. Plant I has a total of 28 cars to send, and plant II has 8. Dealer A needs 20 cars, and dealer B needs 16. Transportation costs per car based on the distance of each dealership from each plant are $220 from I to A, $300 from I to B, $400 from II to A, and $180 from II to B. The manufacturer wants to limit transportation costs to $10,640. How many cars should be sent from each plant to each of the two dealerships? Use the Gauss-Jordan method to find the solution.

This problem is a bit contrived. In practice, the manufacturer would not want to make the cost exactly equal to $10,640; he would want to make the cost as small as possible, while meeting the constraints.

To begin, let

$$y_1 = \text{the number of cars shipped from I to A;}$$
$$y_2 = \text{the number of cars shipped from I to B;}$$
$$y_3 = \text{the number of cars shipped from II to A;}$$

and

$$y_4 = \text{the number of cars shipped from II to B.}$$

Plant I has only 28 cars to ship, so

$$y_1 + y_2 \leq 28.$$

Similarly,

$$y_3 + y_4 \leq 8.$$

Since dealership A needs 20 cars and dealership B needs 16 cars,

$$y_1 + y_3 \geq 20 \quad \text{and} \quad y_2 + y_4 \geq 16.$$

The manufacturer wants to minimize transportation costs, so the objective function is

$$w = 220y_1 + 300y_2 + 400y_3 + 180y_4.$$

Now write the problem as a system of linear equations, adding slack or surplus variables as needed, and let $z = -w$.

$$
\begin{aligned}
y_1 + y_2 && + y_5 && &= 28 \\
y_3 + y_4 && + y_6 && &= 8 \\
y_1 \quad\quad + y_3 && - y_7 && &= 20 \\
y_2 \quad\quad + y_4 && - y_8 && &= 16 \\
220y_1 + 300y_2 + 400y_3 + 180y_4 && + z && &= 0
\end{aligned}
$$

Set up the first simplex tableau.

$$
\begin{array}{ccccccccc}
y_1 & y_2 & y_3 & y_4 & y_5 & y_6 & y_7 & y_8 & z \\
\end{array}
$$

$$
\left[
\begin{array}{ccccccccc|c}
1 & 1 & 0 & 0 & 1 & 0 & 0 & 0 & 0 & 28 \\
0 & 0 & 1 & 1 & 0 & 1 & 0 & 0 & 0 & 8 \\
1 & 0 & 1 & 0 & 0 & 0 & -1 & 0 & 0 & 20 \\
0 & 1 & 0 & 1 & 0 & 0 & 0 & -1 & 0 & 16 \\
\hline
220 & 300 & 400 & 180 & 0 & 0 & 0 & 0 & 1 & 0
\end{array}
\right]
$$

Because $y_7 = -20$, we choose the positive entry farthest to the left in row 3, which is the 1 in column 1. After forming the necessary quotients, we find that the 1 is also the pivot, leading to the following tableau.

$$
\begin{array}{cccccccccc}
& y_1 & y_2 & y_3 & y_4 & y_5 & y_6 & y_7 & y_8 & z \\
\end{array}
$$

$$
\begin{array}{l}
-R_3 + R_1 \rightarrow R_1 \\
\\
\\
-220R_3 + R_5 \rightarrow R_5
\end{array}
\left[
\begin{array}{ccccccccc|c}
0 & 1 & -1 & 0 & 1 & 0 & 1 & 0 & 0 & 8 \\
0 & 0 & 1 & 1 & 0 & 1 & 0 & 0 & 0 & 8 \\
1 & 0 & 1 & 0 & 0 & 0 & -1 & 0 & 0 & 20 \\
0 & 1 & 0 & 1 & 0 & 0 & 0 & -1 & 0 & 16 \\
0 & 300 & 180 & 180 & 0 & 0 & 220 & 0 & 1 & -4400
\end{array}
\right]
$$

We still have $y_8 = -16$. Verify that the 1 in row 1, column 2, is the next pivot, leading to the following tableau.

$$
\begin{array}{cccccccccc}
& y_1 & y_2 & y_3 & y_4 & y_5 & y_6 & y_7 & y_8 & z \\
\end{array}
$$

$$
\begin{array}{l}
\\
\\
\\
-R_1 + R_4 \rightarrow R_4 \\
-300R_1 + R_5 \rightarrow R_5
\end{array}
\left[
\begin{array}{ccccccccc|c}
0 & 1 & -1 & 0 & 1 & 0 & 1 & 0 & 0 & 8 \\
0 & 0 & 1 & 1 & 0 & 1 & 0 & 0 & 0 & 8 \\
1 & 0 & 1 & 0 & 0 & 0 & -1 & 0 & 0 & 20 \\
0 & 0 & 1 & 1 & -1 & 0 & -1 & -1 & 0 & 8 \\
0 & 0 & 480 & 180 & -300 & 0 & -80 & 0 & 1 & -6800
\end{array}
\right]
$$

We still have $y_8 = -8$. Choosing column 3 to pivot, there is a tie between rows 2 and 4. Observe that if we choose row 4, we will remove y_4 from the set of basic variables. This would be a smart next move. But our algorithm says to choose the row nearest the top, so we will do this and see where it leads.

$$
\begin{array}{c}
\\
R_2 + R_1 \to R_1 \\
\\
-R_2 + R_3 \to R_3 \\
-R_2 + R_4 \to R_4 \\
-480R_2 + R_5 \to R_5
\end{array}
\begin{array}{cccccccccc}
y_1 & y_2 & y_3 & y_4 & y_5 & y_6 & y_7 & y_8 & z & \\
\left[\begin{array}{ccccccccc|c}
0 & 1 & 0 & 1 & 1 & 1 & 1 & 0 & 0 & 16 \\
0 & 0 & 1 & 1 & 0 & 1 & 0 & 0 & 0 & 8 \\
1 & 0 & 0 & -1 & 0 & -1 & -1 & 0 & 0 & 12 \\
0 & 0 & 0 & 0 & -1 & -1 & -1 & -1 & 0 & 0 \\
0 & 0 & 0 & -300 & -300 & -480 & -80 & 0 & 1 & -10{,}640
\end{array}\right]
\end{array}
$$

Now $y_8 = 0$. There is still a -1 in column 8, but this can be removed by multiplying row 4 by -1. We then have the feasible solution

$$y_1 = 12, \quad y_2 = 16, \quad y_3 = 8, \quad y_4 = 0, \quad y_5 = 0, \quad y_6 = 0, \quad y_7 = 0, \quad y_8 = 0,$$

with $w = 10{,}640$. This is the solution to the problem in Sec 2.2. But there are still negatives in the bottom row, so we can keep going. After three more tableaus, we find that

$$y_1 = 20, \quad y_2 = 8, \quad y_3 = 0, \quad y_4 = 8,$$

with $w = 8240$. Therefore, send 20 cars from plant I to dealership A and 8 cars to dealership B. From plant II, send 8 cars to dealership B and none to dealership A. The transportation cost will be $8240, a savings of $2400 over the original stated cost. ◀

When one or more of the constraints in a linear programming problem is an equation, rather than an inequality, there is no need for a slack or surplus variable. The simplex method requires an additional variable, however, for *each* constraint. To meet this condition, an **artificial variable** is added to each equation. These variables are called artificial variables because they have no meaning in the context of the original problem. The first goal of the simplex method is to eliminate any artificial variables as basic variables, since they must have a value of 0 in the solution.

EXAMPLE 3 In the transportation problem discussed in Example 2, it would be more realistic for the dealerships to order exactly 20 and 16 cars, respectively. Solve the problem with these two equality constraints.

Using the same variables, we can state the problem as follows.

$$
\begin{array}{ll}
\text{Minimize} & w = 220y + 300y + 400y + 180y \\
\text{subject to:} & y_1 + y_2 \le 28 \\
& y_3 + y_4 \le 8 \\
& y_1 + y_3 = 20 \\
& y_2 + y_4 = 16
\end{array}
$$

with all variables nonnegative.

The corresponding system of equations requires slack variables y_5 and y_6 and two artificial variables that we shall call a_1 and a_2, to remind us that they require special handling. The system

$$
\begin{aligned}
y_1 + \quad y_2 \qquad\qquad\qquad\qquad + y_5 \qquad\qquad\qquad &= 28 \\
y_3 + \quad y_4 \qquad + y_6 \qquad\qquad\qquad &= 8 \\
y_1 \qquad\qquad + \quad y_3 \qquad\qquad\qquad + a_1 \qquad\quad &= 20 \\
y_2 \qquad\qquad + \quad y_4 \qquad\qquad\qquad\qquad + a_2 &= 16 \\
220y_1 + 300y_2 + 400y_3 + 180y_4 \qquad\qquad\qquad\qquad + z &= \ 0
\end{aligned}
$$

produces a tableau exactly the same as in Example 2, except that the columns labeled y_7 and y_8 in that example are now labeled a_1 and a_2. We proceed as we did in Example 2, except for one difference: as soon as we get to the point where $a_1 = 0$, we drop the a_1 column. Similarly, as soon as $a_2 = 0$, we drop the a_2 column. The solution will proceed as in Example 2 and will give the same result. In other problems, however, equality constraints can result in a higher cost. ◀

Two linear programming models in actual use, one on making ice cream, the other on merit pay, are presented at the end of this chapter. These models illustrate the usefulness of linear programming. In most real applications, the number of variables is so large that these problems could not be solved without using methods (like the simplex method) that can be adapted to computers.

4.3 Exercises

Rewrite each system of inequalities, adding slack variables or subtracting surplus variables as necessary.

1. $2x_1 + 3x_2 \le 8$
$\quad x_1 + 4x_2 \ge 7$

2. $5x_1 + 8x_2 \le 10$
$\quad 6x_1 + 2x_2 \ge 7$

3. $x_1 + x_2 + x_3 \le 100$
$\quad x_1 + x_2 + x_3 \ge 75$
$\quad x_1 + x_2 \quad\ \ge 27$

4. $2x_1 \quad\quad + x_3 \le 40$
$\quad x_1 + x_2 \quad\quad \ge 18$
$\quad x_1 \quad\quad + x_3 \ge 20$

Convert the following problems into maximization problems.

5. Minimize $\quad w = 4y_1 + 3y_2 + 2y_3$

subject to: $\quad y_1 + y_2 + \ y_3 \ge 5$
$\qquad\qquad\ \ y_1 + y_2 \qquad\quad \ge 4$
$\qquad\qquad\ \ 2y_1 + y_2 + 3y_3 \ge 15$

with $\qquad\qquad y_1 \ge 0, \quad y_2 \ge 0, \quad y_3 \ge 0.$

6. Minimize $\quad w = 8y_1 + 3y_2 + y_3$

subject to: $\quad 7y_1 + 6y_2 + \ 8y_3 \ge 18$
$\qquad\qquad\ \ 4y_1 + 5y_2 + 10y_3 \ge 20$

with $\qquad\qquad y_1 \ge 0, \quad y_2 \ge 0, \quad y_3 \ge 0.$

7. Minimize $\quad w = y_1 + 2y_2 + y_3 + 5y_4$

subject to: $\quad y_1 + y_2 + \ y_3 + y_4 \ge 50$
$\qquad\qquad\ \ 3y_1 + y_2 + 2y_3 + y_4 \ge 100$

with $\qquad\qquad y_1 \ge 0, \quad y_2 \ge 0, \quad y_3 \ge 0, \quad y_4 \ge 0.$

8. Minimize $\quad w = y_1 + y_2 + 4y_3$

subject to: $\quad y_1 + 2y_2 + 3y_3 \ge 115$
$\qquad\qquad\ \ 2y_1 + \ y_2 + \ y_3 \le 200$
$\qquad\qquad\ \ y_1 \qquad\quad + \ y_3 \ge 50$

with $\qquad\qquad y_1 \ge 0, \quad y_2 \ge 0, \quad y_3 \ge 0.$

Use the simplex method to solve each of the following.

9. Find $x_1 \ge 0$ and $x_2 \ge 0$ such that

$$x_1 + 2x_2 \ge 24$$
$$x_1 + \ x_2 \le 40$$

and $z = 12x_1 + 10x_2$ is maximized.

10. Find $x_1 \ge 0$ and $x_2 \ge 0$ such that

$$3x_1 + 4x_2 \ge 48$$
$$2x_1 + 4x_2 \le 60$$

and $z = 6x_1 + 8x_2$ is maximized.

11. Find $x_1 \geq 0$, $x_2 \geq 0$, and $x_3 \geq 0$ such that

$$x_1 + x_2 + x_3 \leq 150$$
$$x_1 + x_2 + x_3 \geq 100$$

and $z = 2x_1 + 5x_2 + 3x_3$ is maximized.

12. Find $x_1 \geq 0$, $x_2 \geq 0$, and $x_3 \geq 0$ such that

$$x_1 + x_2 + 2x_3 \leq 38$$
$$2x_1 + x_2 + x_3 \geq 24$$

and $z = 3x_1 + 2x_2 + 2x_3$ is maximized.

13. Find $x_1 \geq 0$ and $x_2 \geq 0$ such that

$$x_1 + x_2 \leq 100$$
$$x_1 + x_2 \geq 50$$
$$2x_1 + x_2 \leq 110$$

and $z = -2x_1 + 3x_2$ is maximized.

14. Find $x_1 \geq 0$ and $x_2 \geq 0$ such that

$$x_1 + 2x_2 \leq 18$$
$$x_1 + 3x_2 \geq 12$$
$$2x_1 + 2x_2 \leq 24$$

and $z = 5x_1 - 10x_2$ is maximized.

15. Find $y_1 \geq 0$, $y_2 \geq 0$ such that

$$10y_1 + 5y_2 \geq 100$$
$$20y_1 + 10y_2 \geq 150$$

and $w = 4y_1 + 5y_2$ is minimized.

16. Minimize $w = 3y_1 + 2y_2$

subject to:
$$2y_1 + 3y_2 \geq 60$$
$$y_1 + 4y_2 \geq 40$$

with $y_1 \geq 0$, $y_2 \geq 0$.

17. Minimize $w = 2y_1 + y_2 + 3y_3$

subject to:
$$y_1 + y_2 + y_3 \geq 100$$
$$2y_1 + y_2 \geq 50$$

with $y_1 \geq 0$, $y_2 \geq 0$, $y_3 \geq 0$.

18. Minimize $w = 3y_1 + 2y_2$

subject to:
$$y_1 + 2y_2 \geq 10$$
$$y_1 + y_2 \geq 8$$
$$2y_1 + y_2 \geq 12$$

with $y_1 \geq 0$, $y_2 \geq 0$.

Solve each of the following using artificial variables.

19. Maximize $z = 3x_1 + 2x_2$

subject to:
$$x_1 + x_2 = 50$$
$$4x_1 + 2x_2 \geq 120$$
$$5x_1 + 2x_2 \leq 200$$

with $x_1 \geq 0$, $x_2 \geq 0$.

20. Maximize $z = 10x_1 + 9x_2$

subject to:
$$x_1 + x_2 = 30$$
$$x_1 + x_2 \geq 25$$
$$2x_1 + x_2 \leq 40$$

with $x_1 \geq 0$, $x_2 \geq 0$.

21. Minimize $w = 15y_1 + 12y_2$

subject to:
$$y_1 + 2y_2 \leq 12$$
$$3y_1 + y_2 \geq 18$$
$$y_1 + y_2 = 10$$

with $y_1 \geq 0$, $y_2 \geq 0$.

22. Minimize $w = 32y_1 + 40y_2$

subject to:
$$20y_1 + 10y_2 = 200$$
$$25y_1 + 40y_2 \leq 500$$
$$18y_1 + 24y_2 \geq 300$$

with $y_1 \geq 0$, $y_2 \geq 0$.

23. Explain how, in any linear programming problem, the value of the objective function can be found without using the number in the lower right-hand corner of the final tableau.

24. Explain why, for a maximization problem, you write the negative of the coefficients of the objective function on the bottom row, while, for a minimization problem, you write the coefficients themselves.

Applications

Use the simplex method to solve Exercises 25–35.

Business and Economics

25. **Transportation** Southwestern Oil supplies two distributors in the Northwest from two outlets, S_1 and S_2. Distributor D_1 needs at least 3000 barrels of oil, and distributor D_2 needs at least 5000 barrels. The two outlets can each furnish 5000 barrels of oil. The costs per barrel to ship the oil are given at the right.

		Distributors	
		D$_1$	D$_2$
Outlets	S$_1$	$30	$20
	S$_2$	$25	$22

How should the oil be supplied to minimize shipping costs?

26. **Finance** A bank has set aside a maximum of $25 million for commercial and home loans. The bank's policy is to loan at least four times as much for home loans as for commercial loans. Because of prior commitments, at least $10 million will be used for these two types of loans. The bank earns 12% on home loans and 10% on commercial loans. What amount of money should be loaned out for each type of loan to maximize the interest income?

27. **Production Costs** Brand X Canners produces canned whole tomatoes and tomato sauce. This season, the company has available 3,000,000 kg of tomatoes for these two products. To meet the demands of regular customers, it must produce at least 80,000 kg of sauce and 800,000 kg of whole tomatoes. The cost per kilogram is $4 to produce canned whole tomatoes and $3.25 to produce tomato sauce. How many kilograms of tomatoes should Brand X use for each product to minimize cost?

28. **Production Costs** A brewery produces regular beer and a lower-carbohydrate "light" beer. Steady customers of the brewery buy 12 units of regular beer and 10 units of light beer monthly. While setting up the brewery to produce the beers, the management decides to produce extra beer, beyond that needed to satisfy the steady customers. The cost per unit of regular beer is $36,000 and the cost per unit of light beer is $48,000. The number of units of light beer should not exceed twice the number of units of regular beer. At least 20 additional units of beer can be sold. How much of each type of beer should be made so as to minimize total production costs?

29. **Supply Costs** The chemistry department at a local college decides to stock at least 800 small test tubes and 500 large test tubes. It wants to buy at least 1500 test tubes to take advantage of a special price. Since the small tubes are broken twice as often as the larger, the department will order at least twice as many small tubes as large. If the small test tubes cost 15¢ each and large ones, made of a cheaper glass, cost 12¢ each, how many of each size should be ordered to minimize cost?

30. **Blending Seed** Topgrade Turf lawn seed mixture contains three types of seeds: bluegrass, rye, and bermuda. The costs per pound of the three types of seed are 20¢, 15¢, and 5¢, respectively. In each batch there must be at least 20% bluegrass seed, and the amount of bermuda must be no more than the amount of rye. To fill current orders, the company must make at least 5000 pounds of the mixture. How much of each kind of seed should be used to minimize cost?

31. **Investments** Virginia Keleske has decided to invest a $100,000 inheritance in government securities that earn 7% per year, municipal bonds that earn 6% per year, and mutual funds that earn an average of 10% per year. She will spend at least $40,000 on government securities, and she wants at least half the inheritance to go to bonds and mutual funds. How much should be invested in each way to maximize the interest yet meet the constraints? What is the maximum interest she can earn?

32. **Transportation** The manufacturer of a popular personal computer has orders from two dealers. Dealer D_1 wants at least 32 computers, and dealer D_2 wants at least 20 computers. The manufacturer can fill the orders from either of two warehouses, W_1 or W_2. W_1 has 25 of the computers on hand, and W_2 has 30. The costs (in dollars) to ship one computer to each dealer from each warehouse are given below.

		Dealer	
		D_1	D_2
Warehouse	W_1	$14	$22
	W_2	$12	$10

How should the orders be filled to minimize shipping costs?

Life Sciences

33. **Health Care** Mark, who is ill, takes vitamin pills. Each day he must have at least 16 units of vitamin A, 5 units of vitamin B_1, and 20 units of vitamin C. He can choose between pill #1, which costs 10 cents and contains 8 units of A, 1 of B_1, and 2 of C; and pill #2, which costs 20 cents and contains 2 units of A, 1 of B_1, and 7 of C. How many of each pill should he buy in order to minimize his cost? (See Exercise 18 in Section 3.3.)

34. **Dietetics** Sing, who is dieting, requires two food supplements, I and II. He can get these supplements from two different products, A and B, as shown below.

Grams of
Supplement
(per serving)

$$\text{Product} \begin{array}{c} A \\ B \end{array} \begin{bmatrix} \text{I} & \text{II} \\ 3 & 2 \\ 2 & 4 \end{bmatrix}$$

Sing's physician has recommended that he include at least 15 g of each supplement in his daily diet. If product A costs 25¢ per serving and product B costs 40¢ per serving, how can he satisfy his requirements most economically? Find the minimum cost. (See Exercise 21 in Section 3.3.)

35. Blending Nutrients A biologist must make a nutrient of her algae. The nutrient must contain the three basic elements D, E, and F, and must contain at least 10 kg of D, 12 kg of E, and 20 kg of F. The nutrient is made from three ingredients, I, II, and III. The quantity of D, E, and F in one unit of each of the ingredients is as given in the following chart.

Ingredient	Kilograms of Elements (per unit of ingredients)			Cost per Unit
	D	E	F	
I	4	3	0	4
II	1	2	4	7
III	10	1	5	5

How many units of each ingredient are required to meet the biologist's needs at minimum cost?

 Computer/Graphing Calculator

Determine the constraints and the objective function for Exercises 36–38, and then solve each problem.

36. Blending Chemicals Natural Brand plant food is made from three chemicals. (See Section 2.2, Exercise 49.) In a batch of the plant food there must be at least 81 kg of the first chemical, and the other two chemicals must be in the ratio of 4 to 3. If the three chemicals cost $1.09, $.87 and $.65 per kilogram, respectively, how much of each should be used to minimize the cost of producing at least 750 kg of the plant food?

37. Blending Gasoline A company is developing a new additive for gasoline. The additive is a mixture of three liquid ingredients, I, II, and III. For proper performance, the total amount of additive must be at least 10 oz per gallon of gasoline. However, for safety reasons, the amount of additive should not exceed 15 oz per gallon of gasoline. At least 1/4 oz of ingredient I must be used for every ounce of ingredient II, and at least 1 oz of ingredient III must be used for every ounce of ingredient I. If the costs of I, II, and III are $.30, $.09, and $.27 per ounce, respectively, find the mixture of the three ingredients that produces the minimum cost of the additive. How much of the additive should be used per gallon of gasoline?

38. Blending a Soft Drink A popular soft drink called Sugarlo, which is advertised as having a sugar content of no more than 10%, is blended from five ingredients, each of which has some sugar content. Water may also be added to dilute the mixture. The sugar content of the ingredients and their costs per gallon are given below.

	Ingredient					
	1	2	3	4	5	Water
Sugar Content (%)	.28	.19	.43	.57	.22	0
Cost ($/gal)	.48	.32	.53	.28	.43	.04

At least .01 of the content of Sugarlo must come from ingredients 3 or 4, .01 must come from ingredients 2 or 5, and .01 from ingredients 1 or 4. How much of each ingredient should be used in preparing 15,000 gal of Sugarlo to minimize the cost?

4.4 DUALITY

? **How many units of different types of feed should a dog breeder purchase to meet the nutrient requirements of her beagles at a minimum cost?**

Using the method of duals, we will learn to answer this and other questions.

In this section we discuss an alternative method of solving minimization problems in standard form. An interesting connection exists between standard maximization and standard minimization problems: any solution of a standard maximization problem produces the solution of an associated standard minimization problem, and vice versa. Each of these associated problems is called the **dual** of the other. One advantage of duals is that standard minimization problems

can be solved by the simplex method discussed in the first two sections of this chapter. Let us explain the idea of a dual with an example. (This is similar to Example 1 in the previous section; compare this method of solution with the one given there.)

EXAMPLE 1

Minimize $w = 8y_1 + 16y_2$

subject to: $y_1 + 5y_2 \geq 9$

$2y_1 + 2y_2 \geq 10$

with $y_1 \geq 0, \, y_2 \geq 0.$

(As mentioned earlier, y_1 and y_2 are used as variables and w as the objective function as a reminder that this is a minimization problem.)

Without considering slack variables just yet, write the augmented matrix of the system of inequalities, and include the coefficients of the objective function (not their negatives) as the last row in the matrix.

$$\begin{bmatrix} 1 & 5 & 9 \\ 2 & 2 & 10 \\ 8 & 16 & 0 \end{bmatrix}$$

Now look at the following matrix, which we obtain from the one above by interchanging rows and columns.

$$\begin{bmatrix} 1 & 2 & 8 \\ 5 & 2 & 16 \\ 9 & 10 & 0 \end{bmatrix}$$

The *rows* of the first matrix (for the minimization problem) are the *columns* of the second matrix.

The entries in this second matrix could be used to write the following maximization problem in standard form (again ignoring the fact that the numbers in the last row are not negative):

Maximize $z = 9x_1 + 10x_2$

subject to: $x_1 + 2x_2 \leq 8$

$5x_1 + 2x_2 \leq 16.$

with all variables nonnegative.

Figure 4(a) shows the region of feasible solutions for the minimization problem given above, while Figure 4(b) shows the region of feasible solutions for the maximization problem produced by exchanging rows and columns. The solutions of the two problems are given below.

Corner Point	$w = 8y_1 + 16y_2$		Corner Point	$z = 9x_1 + 10x_2$	
(0, 5)	80		(0, 0)	0	
(4, 1)	48	Minimum	(0, 4)	40	
(9, 0)	72		(2, 3)	48	Maximum
			($^{16}/_5$, 0)	28.8	

The minimum is 48 when $y_1 = 4$ and $y_2 = 1.$ The maximum is 48 when $x_1 = 2$ and $x_2 = 3.$

FIGURE 4

The two feasible regions in Figure 4 are different and the corner points are different, but the values of the objective functions are equal—both are 48. An even closer connection between the two problems is shown by using the simplex method to solve the maximization problem given above.

Maximization Problem

$$
\begin{array}{ccccc}
x_1 & x_2 & x_3 & x_4 & z \\
\end{array}
$$

$$
\left[
\begin{array}{ccccc|c}
1 & 2 & 1 & 0 & 0 & 8 \\
5 & 2 & 0 & 1 & 0 & 16 \\
\hline
-9 & -10 & 0 & 0 & 1 & 0 \\
\end{array}
\right]
$$

$$
\begin{array}{ccccc}
x_1 & x_2 & x_3 & x_4 & z \\
\end{array}
$$

$-R_1 + R_2 \rightarrow R_2$
$5R_1 + R_3 \rightarrow R_3$

$$
\left[
\begin{array}{ccccc|c}
1 & 2 & 1 & 0 & 0 & 8 \\
4 & 0 & -1 & 1 & 0 & 8 \\
\hline
-4 & 0 & 5 & 0 & 1 & 40 \\
\end{array}
\right]
$$

$$
\begin{array}{ccccc}
x_1 & x_2 & x_3 & x_4 & z \\
\end{array}
$$

$-R_2 + 4R_1 \rightarrow R_1$

$R_2 + R_3 \rightarrow R_3$

$$
\left[
\begin{array}{ccccc|c}
0 & 8 & 5 & -1 & 0 & 24 \\
4 & 0 & -1 & 1 & 0 & 8 \\
\hline
0 & 0 & 4 & 1 & 1 & 48 \\
\end{array}
\right]
$$

$$
\begin{array}{ccccc}
x_1 & x_2 & x_3 & x_4 & z \\
\end{array}
$$

$R_1/8 \rightarrow R_1$
$R_2/4 \rightarrow R_2$

$$
\left[
\begin{array}{ccccc|c}
0 & 1 & 5/8 & -1/8 & 0 & 3 \\
1 & 0 & -1/4 & 1/4 & 0 & 2 \\
\hline
0 & 0 & 4 & 1 & 1 & 48 \\
\end{array}
\right]
$$

The maximum is 48 when
$x_1 = 2$ and $x_2 = 3$.

Notice that the solution to the *minimization problem* is found in the bottom row and slack variable columns of the final simplex tableau for the maximization problem. This result suggests that standard minimization problems can be solved by forming the dual standard maximization problem, solving it by the simplex method, and then reading the solution for the minimization problem from the bottom row of the final simplex tableau. ◀

Before using this method to actually solve a minimization problem, let us find the duals of some typical linear programming problems. The process of exchanging the rows and columns of a matrix, which is used to find the dual, is called **transposing** the matrix, and each of the two matrices is the **transpose** of the other.

EXAMPLE 2

Find the transpose of each matrix.

(a) $A = \begin{bmatrix} 2 & -1 & 5 \\ 6 & 8 & 0 \\ -3 & 7 & -1 \end{bmatrix}$

Write the rows of matrix A as the columns of the transpose.

$$\text{Transpose of } A = \begin{bmatrix} 2 & 6 & -3 \\ -1 & 8 & 7 \\ 5 & 0 & -1 \end{bmatrix}$$

(b) $B = \begin{bmatrix} 1 & 2 & 4 & 0 \\ 2 & 1 & 7 & 6 \end{bmatrix}$

$$\text{Transpose of } B = \begin{bmatrix} 1 & 2 \\ 2 & 1 \\ 4 & 7 \\ 0 & 6 \end{bmatrix} \blacktriangleleft$$

EXAMPLE 3

Write the dual of each of the following standard maximization linear programming problems.

(a) Maximize $z = 2x_1 + 5x_2$

subject to: $x_1 + x_2 \leq 10$

$2x_1 + x_2 \leq 8$

with $x_1 \geq 0, \quad x_2 \geq 0.$

Begin by writing the augmented matrix for the given problem.

$$\left[\begin{array}{cc|c} 1 & 1 & 10 \\ 2 & 1 & 8 \\ \hline 2 & 5 & 0 \end{array} \right]$$

Form the transpose of this matrix:

$$\left[\begin{array}{cc|c} 1 & 2 & 2 \\ 1 & 1 & 5 \\ \hline 10 & 8 & 0 \end{array} \right].$$

The dual problem is stated from this second matrix as follows (using y instead of x):

$$\text{Minimize} \qquad w = 10y_1 + 8y_2$$
$$\text{subject to:} \qquad y_1 + 2y_2 \geq 2$$
$$y_1 + y_2 \geq 5$$
$$\text{with} \qquad y_1 \geq 0, \quad y_2 \geq 0.$$

(b) Minimize $\qquad w = 7y_1 + 5y_2 + 8y_3$

subject to: $\qquad 3y_1 + 2y_2 + y_3 \geq 10$

$$y_1 + y_2 + y_3 \geq 8$$
$$4y_1 + 5y_2 \qquad \geq 25$$

with $\qquad y_1 \geq 0, \quad y_2 \geq 0, \quad y_3 \geq 0.$

The dual problem is stated as follows.

$$\text{Maximize} \qquad z = 10x_1 + 8x_2 + 25x_3$$
$$\text{subject to:} \qquad 3x_1 + x_2 + 4x_3 \leq 7$$
$$2x_1 + x_2 + 5x_3 \leq 5$$
$$x_1 + x_2 \qquad \leq 8$$
$$\text{with} \qquad x_1 \geq 0, \quad x_2 \geq 0, \quad x_3 \geq 0. \quad \blacktriangleleft$$

In Example 3, all the constraints of the given standard maximization problems were \leq inequalities, while all those in the dual minimization problems were \geq inequalities. This is generally the case; inequalities are reversed when the dual problem is stated.

The following table shows the close connection between a problem and its dual.

Given Problem	Dual Problem
m variables	n variables
n constraints	m constraints
Coefficients from objective function	Constants
Constants	Coefficients from objective function

The next theorem, whose proof requires advanced methods, guarantees that a standard minimization problem can be solved by forming a dual standard maximization problem.

THEOREM OF DUALITY The objective function w of a minimization linear programming problem takes on a minimum value if and only if the objective function z of the corresponding dual maximization problem takes on a maximum value. The maximum value of z equals the minimum value of w.

This method is illustrated in the following example. (This is Example 1 of the previous section; compare this solution with the one given there.)

Minimize $w = 3y_1 + 2y_2$

subject to: $y_1 + 3y_2 \geq 6$

$2y_1 + y_2 \geq 3$

with $y_1 \geq 0, \quad y_2 \geq 0.$

Use the given information to write the matrix.

$$\begin{bmatrix} 1 & 3 & | & 6 \\ 2 & 1 & | & 3 \\ \hline 3 & 2 & | & 0 \end{bmatrix}$$

Transpose to get the following matrix for the dual problem.

$$\begin{bmatrix} 1 & 2 & | & 3 \\ 3 & 1 & | & 2 \\ \hline 6 & 3 & | & 0 \end{bmatrix}$$

Write the dual problem from this matrix, as follows:

Maximize $z = 6x_1 + 3x_2$

subject to: $x_1 + 2x_2 \leq 3$

$3x_1 + x_2 \leq 2$

with $x_1 \geq 0, \quad x_2 \geq 0.$

Solve this standard maximization problem using the simplex method. Start by introducing slack variables to give the system

$$x_1 + 2x_2 + x_3 \qquad\qquad = 3$$
$$3x_1 + x_2 \qquad + x_4 \qquad = 2$$
$$-6x_1 - 3x_2 - 0x_3 - 0x_4 + z = 0$$

with $x_1 \geq 0, \quad x_2 \geq 0, \quad x_3 \geq 0, \quad x_4 \geq 0.$

The first tableau for this system is given below, with the pivot as indicated.

Quotients		x_1	x_2	x_3	x_4	z	
3/1 = 3		1	2	1	0	0	3
2/3		**3**	1	0	1	0	2
		−6	−3	0	0	1	0

The simplex method gives the following as the final tableau.

x_1	x_2	x_3	x_4	z	
0	1	3/5	−1/5	0	7/5
1	0	−1/5	2/5	0	1/5
0	0	3/5	9/5	1	27/5

So long as a 1 has been created in the z column, the last row of this final tableau gives the solution to the minimization problem. The minimum value of $w = 3y_1 + 2y_2$, subject to the given constraints, is 27/5 and occurs when $y_1 = 3/5$ and $y_2 = 9/5$. The minimum value of w, 27/5, is the same as the maximum value of z. ◄

Let us summarize the steps in solving a standard minimization linear programming problem by the method of duals.

SOLVING MINIMUM PROBLEMS WITH DUALS

1. Find the dual standard maximization problem.
2. Solve the maximization problem using the simplex method.
3. The minimum value of the objective function w is the maximum value of the objective function z.
4. The optimum solution is given by the entries in the bottom row of the columns corresponding to the slack variables, so long as the entry in the z column is equal to 1.

Caution If the final entry in the z column is a value other than 1, divide the bottom row through by that value so that it will become 1. Only then can the solution of the minimization problem be found in the bottom row of the columns corresponding to the slack variables.

Note If the objective function is written below the constraints (except those stating that each variable must be greater than or equal to 0), and the variables are lined up vertically, it is easy to go from the original problem to the dual: the coefficients in any column of the original problem become the coefficients of the corresponding row in the dual. In Example 4, for instance, if the objective function is written below the constraints, the coefficients in the first column are 1, 2, 3, leading to the first row of the dual: $x_1 + 2x_2 \leq 3$. The last column is 6, 3, so the last row of the dual is: Maximize $z = 6x_1 + 3x_2$.

Further Uses of the Dual The dual is useful not only in solving minimization problems, but also in seeing how small changes in one variable will affect the value of the objective function. For example, suppose a dog breeder needs at least 6 units per day of nutrient A and at least 3 units of nutrient B for her beagles, and that the breeder can choose between two different feeds, feed 1 and feed 2. Find the minimum cost for the breeder if each bag of feed 1 costs \$3 and provides 1 unit of nutrient A and 2 units of B, while each bag of feed 2 costs \$2 and provides 3 units of nutrient A and 1 of B.

If y_1 represents the number of bags of feed 1 and y_2 represents the number of bags of feed 2, the given information leads to the following problem.

$$\text{Minimize} \qquad w = 3y_1 + 2y_2$$
$$\text{subject to:} \qquad y_1 + 3y_2 \geq 6$$
$$2y_1 + y_2 \geq 3$$
$$\text{with} \qquad y_1 \geq 0, \quad y_2 \geq 0.$$

This standard minimization linear programming problem is the one solved in Example 4 of this section. In that example, the dual was formed and the following tableau was found.

$$\begin{array}{ccccc} x_1 & x_2 & x_3 & x_4 & z \end{array}$$
$$\left[\begin{array}{ccccc|c} 0 & 1 & 3/5 & -1/5 & 0 & 7/5 \\ 1 & 0 & -1/5 & 2/5 & 0 & 1/5 \\ \hline 0 & 0 & 3/5 & 9/5 & 1 & 27/5 \end{array}\right]$$

This final tableau shows that the breeder will obtain minimum feed costs by using 3/5 bag of feed 1 and 9/5 bags of feed 2 per day, for a daily cost of $27/5 = 5.40$ dollars.

Now look at the data from the problem shown in the table below.

	Unit of Nutrient (per bag):		Cost (per Bag)
	A	B	
Feed 1	1	2	$3
Feed 2	3	1	$2
	6	3	

If x_1 and x_2 are the costs per unit of nutrients A and B, the constraints of the dual problem can be stated as follows.

$$\text{Cost of feed 1:} \quad x_1 + 2x_2 \leq 3$$
$$\text{Cost of feed 2:} \quad 3x_1 + x_2 \leq 2$$

The solution of the dual problem, which maximizes nutrients, can be read from the final tableau:

$$x_1 = \frac{1}{5} = .20 \quad \text{and} \quad x_2 = \frac{7}{5} = 1.40,$$

which means that a unit of nutrient A costs $0.20, while a unit of nutrient B costs $1.40. The minimum daily cost, $5.40, is found by the following procedure.

$$($.20 \text{ per unit of A}) \times (6 \text{ units of A}) = $1.20$$
$$\underline{+ ($1.40 \text{ per unit of B}) \times (3 \text{ units of B}) = $4.20}$$
$$\text{Minimum daily cost} = $5.40$$

The numbers .20 and 1.40 are called the **shadow costs** of the nutrients. These two numbers from the dual, $.20 and $1.40, also allow the breeder to estimate feed costs for small changes in nutrient requirements. For example, an increase of one unit in the requirement for each nutrient would produce a total cost of

$5.40	6 units of A, 3 of B
.20	1 extra unit of A
1.40	1 extra unit of B
$7.00.	Total cost per day

4.4 Exercises

Find the transpose of each matrix.

1. $\begin{bmatrix} 1 & 2 & 3 \\ 3 & 2 & 1 \\ 1 & 10 & 0 \end{bmatrix}$

2. $\begin{bmatrix} 2 & 5 & 8 & 6 & 0 \\ 1 & -1 & 0 & 12 & 14 \end{bmatrix}$

3. $\begin{bmatrix} -1 & 4 & 6 & 12 \\ 13 & 25 & 0 & 4 \\ -2 & -1 & 11 & 3 \end{bmatrix}$

4. $\begin{bmatrix} 1 & 11 & 15 \\ 0 & 10 & -6 \\ 4 & 12 & -2 \\ 1 & -1 & 13 \\ 2 & 25 & -1 \end{bmatrix}$

State the dual problem for each of the following.

5. Maximize $z = 4x_1 + 3x_2 + 2x_3$

subject to: $x_1 + x_2 + x_3 \le 5$
$x_1 + x_2 \le 4$
$2x_1 + x_2 + 3x_3 \le 15$

with $x_1 \ge 0, \quad x_2 \ge 0, \quad x_3 \ge 0.$

6. Maximize $z = 8x_1 + 3x_2 + x_3$

subject to: $7x_1 + 6x_2 + 8x_3 \le 18$
$4x_1 + 5x_2 + 10x_3 \le 20$

with $x_1 \ge 0, \quad x_2 \ge 0, \quad x_3 \ge 0.$

7. Minimize $w = y_1 + 2y_2 + y_3 + 5y_4$

subject to: $y_1 + y_2 + y_3 + y_4 \ge 50$
$3y_1 + y_2 + 2y_3 + y_4 \ge 100$

with $y_1 \ge 0, \quad y_2 \ge 0, \quad y_3 \ge 0, \quad y_4 \ge 0.$

8. Minimize $w = y_1 + y_2 + 4y_3$

subject to: $y_1 + 2y_2 + 3y_3 \ge 115$
$2y_1 + y_2 + 8y_3 \ge 200$
$y_1 + y_3 \ge 50$

with $y_1 \ge 0, \quad y_2 \ge 0, \quad y_3 \ge 0.$

Use the simplex method to solve Exercises 9–14.

9. Find $y_1 \ge 0$ and $y_2 \ge 0$ such that

$2y_1 + 3y_2 \ge 6$
$2y_1 + y_2 \ge 7$

and $w = 5y_1 + 2y_2$ is minimized.

11. Find $y_1 \ge 0$ and $y_2 \ge 0$ such that

$10y_1 + 5y_2 \ge 100$
$20y_1 + 10y_2 \ge 150$

and $w = 4y_1 + 5y_2$ is minimized.

13. Minimize $w = 2y_1 + y_2 + 3y_3$

subject to: $y_1 + y_2 + y_3 \ge 100$
$2y_1 + y_2 \ge 50$

with $y_1 \ge 0, y_2 \ge 0, y_3 \ge 0.$

10. Find $y_1 \ge 0$ and $y_2 \ge 0$ such that

$3y_1 + y_2 \ge 12$
$y_1 + 4y_2 \ge 16$

and $w = 2y_1 + y_2$ is minimized.

12. Minimize $w = 3y_1 + 2y_2$

subject to: $2y_1 + 3y_2 \ge 60$
$y_1 + 4y_2 \ge 40$

with $y_1 \ge 0, y_2 \ge 0.$

14. Minimize $w = 3y_1 + 2y_2$

subject to: $y_1 + 2y_2 \ge 10$
$y_1 + y_2 \ge 8$
$2y_1 + y_2 \ge 12$

with $y_1 \ge 0, y_2 \ge 0.$

***15.** You are given the following linear programming problem (P):

Minimize $z = x_1 + 2x_2$

subject to $-2x_1 + x_2 \ge 1$
$x_1 - 2x_2 \ge 1$

$x_1 \ge 0, \quad x_2 \ge 0.$

The dual of (P) is (D).
Which of the statements in the next column is true?

(a) (P) has no feasible solution and the objective function of (D) is unbounded.

(b) (D) has no feasible solution and the objective function of (P) is unbounded.

(c) The objective functions of both (P) and (D) are unbounded.

(d) Both (P) and (D) have optimal solutions.

(e) Neither (P) nor (D) has feasible solutions.

*Problem 2 from "November 1989 Course 130 Examination Operations Research" of the *Education and Examination Committee of The Society of Actuaries.* Reprinted by permission of The Society of Actuaries.

Applications

Solve each of the following by using the method of duals.

Business and Economics

16. Profit Refer to the problem about the farmer solved at the beginning of Section 4.2.

 (a) Give the dual problem.

 (b) Use the shadow values to estimate the farmer's profit if land is cut to 90 acres but capital increases to $21,000.

 (c) Suppose the farmer has 110 acres but only $19,000. Find the optimum profit and the planting strategy that will produce this profit.

17. Profit A small toy manufacturing firm has 200 squares of felt, 600 oz of stuffing, and 90 ft of trim available to make two types of toys, a small bear and a monkey. The bear requires 1 square of felt and 4 oz of stuffing. The monkey requires 2 squares of felt, 3 oz of stuffing, and 1 ft of trim. The firm makes $1 profit on each bear and $1.50 profit on each monkey. The linear program to maximize profit is

$$\text{Maximize} \qquad x_1 + 1.5x_2 = z$$
$$\text{subject to:} \qquad x_1 + 2x_2 \leq 200$$
$$4x_1 + 3x_2 \leq 600$$
$$x_2 \leq 90.$$

The final simplex tableau is shown below.

$$\begin{bmatrix} 0 & 1 & .8 & -.2 & 0 & 0 & 40 \\ 1 & 0 & -.6 & .4 & 0 & 0 & 120 \\ 0 & 0 & -.8 & .2 & 1 & 0 & 50 \\ \hline 0 & 0 & .6 & .1 & 0 & 1 & 180 \end{bmatrix}$$

 (a) What is the corresponding dual problem?

 (b) What is the optimal solution to the dual problem?

 (c) Use the shadow values to estimate the profit the firm will make if its supply of felt increases to 210 squares.

 (d) How much profit will the firm make if its supply of stuffing is cut to 590 oz and its supply of trim is cut to 80 ft?

18. Production Costs Brand X Canners produces canned whole tomatoes and tomato sauce. This season, the company has available 3,000,000 kg of tomatoes for these two products. To meet the demands of regular customers, it

must produce at least 80,000 kg of sauce and 800,000 kg of whole tomatoes. The cost per kilogram is $4 to produce canned whole tomatores and $3.25 to produce tomato sauce. How many kilograms of tomatoes should Brand X use for each product to minimize cost? (See Exercise 27 in Section 4.3.)

19. Food Cost A store sells two brands of snacks. A package of Sun Hill costs $3 and contains 10 oz of peanuts, 4 oz of raisins, and 2 oz of rolled oats. A package of Bear Valley costs $2 and contains 2 oz of peanuts, 4 oz of raisins, and 8 oz of rolled oats. If you wish to make a mixture that contains at least 20 oz of peanuts, 24 oz of raisins, and 24 oz of rolled oats, how many packages of each should you buy to minimize the cost?

20. Interview Time Joan McKee has a part-time job conducting public opinion interviews. She has found that a political interview takes 45 min and a market interview takes 55 min. She needs to minimize the time she spends doing interviews to allow more time for her full-time job. Unfortunately, to keep her part-time job, she must complete at least 8 interviews each week. Also, she must earn at least $60 per week at this job; she earns $8 for each political interview and $10 for each market interview. Finally, to stay in good standing with her supervisor, she must earn at least 40 bonus points per week; she receives 6 bonus points for each political interview and 5 points for each market interview. How many of each interview should she do each week to minimize the time spent?

Life Sciences

21. Feed Costs Refer to the example at the end of this section on minimizing the daily cost of feeds.

 (a) Find a combination of feeds that will cost $7.00 and give 7 units of A and 4 units of B.

 (b) Use the dual variables to predict the daily cost of feed if the requirements change to 5 units of A and 4 units of B. Find a combination of feeds to meet these requirements at the predicted price.

Computer/Graphing Calculator

Determine the constraints and the objective functions for Exercises 22 and 23, and then use the method of duals to solve each problem by the simplex method.

22. **Blending Chemicals** Natural Brand plant food is made from three chemicals. In a batch of the plant food there must be at least 81 kg of the first chemical, and the other two chemicals must be in the ratio of 4 to 3. If the three chemicals cost \$1.09, \$.87, and \$.65 per kilogram, respectively, how much of each should be used to minimize the cost of producing at least 750 kg of the plant food? (See Exercise 36 in Section 4.3.)

23. **Blending Gasoline** A company is developing a new additive for gasoline. The additive is a mixture of three liquid ingredients, I, II, III. For proper performance, the total amount of additive must be at least 10 oz per gallon of gasoline. For safety reasons, the amount of additive should not exceed 15 oz per gallon of gasoline. At least 1/4 oz of ingredient I must be used for every ounce of ingredient II, and at least 1 oz of ingredient III must be used for every ounce of ingredient I. If the costs of I, II, and III are \$.30, \$.09, and \$.27 per ounce, respectively, find the mixture of the three ingredients that produces the minimum cost of the additive. How much of the additive should be used per gallon of gasoline? (See Exercise 37 in Section 4.3.)

Chapter Summary Key Terms

simplex method
4.1 standard maximum form
slack variable
simplex tableau
indicators
basic variables

feasible solution
pivot
4.2 departing variable
entering variable
4.3 surplus variable
two-phase method

standard minimum form
artificial variable
4.4 dual
transpose
shadow costs

Chapter 4 Review Exercises

1. When is it necessary to use the simplex method rather than the graphical method?

2. What can you conclude if a surplus variable cannot be made nonnegative?

For Exercises 3–6, (a) add slack variables or subtract surplus variables, and (b) set up the initial simplex tableau.

3. Maximize $z = 5x_1 + 3x_2$
 subject to: $2x_1 + 5x_2 \le 50$
 $x_1 + 3x_2 \le 25$
 $4x_1 + x_2 \le 18$
 $x_1 + x_2 \le 12$
 with $x_1 \ge 0, \quad x_2 \ge 0.$

4. Maximize $z = 25x_1 + 30x_2$
 subject to: $3x_1 + 5x_2 \le 47$
 $x_1 + x_2 \le 25$
 $5x_1 + 2x_2 \le 35$
 $2x_1 + x_2 \le 30$
 with $x_1 \ge 0, \quad x_2 \ge 0.$

5. Maximize $z = 5x_1 + 8x_2 + 6x_3$
 subject to: $x_1 + x_2 + x_3 \le 90$
 $2x_1 + 5x_2 + x_3 \le 120$
 $x_1 + 3x_2 \ge 80$
 with $x_1 \ge 0, \quad x_2 \ge 0, \quad x_3 \ge 0.$

6. Maximize $z = 2x_1 + 3x_2 + 4x_3$
 subject to: $x_1 + x_2 + x_3 \ge 100$
 $2x_1 + 3x_2 \le 500$
 $x_1 + 2x_3 \le 350$
 with $x_1 \ge 0, \quad x_2 \ge 0, \quad x_3 \ge 0.$

Use the simplex method to solve the maximization linear programming problems with initial tableaus as given in Exercises 7–10.

7.

x_1	x_2	x_3	x_4	x_5	z	
1	2	3	1	0	0	28
2	4	1	0	1	0	32
-5	-2	-3	0	0	1	0

8.

x_1	x_2	x_3	x_4	z	
2	1	1	0	0	10
1	3	0	1	0	16
-2	-3	0	0	1	0

9.

x_1	x_2	x_3	x_4	x_5	x_6	z	
1	2	2	1	0	0	0	50
3	1	0	0	1	0	0	20
1	0	2	0	0	-1	0	15
-5	-3	-2	0	0	0	1	0

10.

x_1	x_2	x_3	x_4	x_5	z	
3	6	-1	0	0	0	28
1	1	0	1	0	0	12
2	1	0	0	1	0	16
-1	-2	0	0	0	1	0

Convert the problems of Exercises 11–13 into maximization problems.

11. Minimize $w = 10y_1 + 15y_2$

subject to: $y_1 + y_2 \geq 17$

$5y_1 + 8y_2 \geq 42$

with $y_1 \geq 0, \quad y_2 \geq 0.$

12. Minimize $w = 20y_1 + 15y_2 + 18y_3$

subject to: $2y_1 + y_2 + y_3 \geq 112$

$y_1 + y_2 + y_3 \geq 80$

$y_1 + y_2 \qquad \geq 45$

with $y_1 \geq 0, \quad y_2 \geq 0, \quad y_3 \geq 0.$

13. Minimize $w = 7y_1 + 2y_2 + 3y_3$

subject to: $y_1 + y_2 + 2y_3 \geq 48$

$y_1 + y_2 \qquad \geq 12$

$y_3 \geq 10$

$3y_1 \qquad + y_3 \geq 30$

with $y_1 \geq 0, \quad y_2 \geq 0, y_3 \geq 0.$

The tableaus in Exercises 14–16 are the final tableaus of minimization problems. State the solution and the minimum value of the objective function for each problem.

14.

y_1	y_2	y_3	y_4	y_5	y_6	z	
1	0	0	3	1	2	0	12
0	0	1	4	5	3	0	5
0	1	0	-2	7	-6	0	8
0	0	0	5	7	3	1	-172

15.

y_1	y_2	y_3	y_4	y_5	y_6	z	
0	0	3	0	1	1	0	2
1	0	-2	0	2	0	0	8
0	1	7	0	0	0	0	12
0	0	1	1	-4	0	0	1
0	0	5	0	8	0	1	-62

16.

y_1	y_2	y_3	y_4	y_5	z	
5	1	0	7	-1	0	100
-2	0	1	1	3	0	27
12	0	0	7	2	1	-640

17. What types of problems can be solved using slack, surplus, and artificial variables?

18. What kind of problem can be solved using the method of duals?

19. In solving a linear programming problem, you are given the following initial tableau.

4	2	3	1	0	0	9
5	4	1	0	1	0	10
-6	-7	-5	0	0	1	0

(a) What is the problem being solved?

(b) If the 1 in row 1, column 4 were a -1 rather than a 1, how would it change your answer to part (a)?

(c) After several steps of the simplex algorithm, the following tableau results.

$$\begin{bmatrix} 3 & 0 & 5 & 2 & -1 & 0 & 8 \\ 11 & 10 & 0 & -1 & 3 & 0 & 21 \\ 47 & 0 & 0 & 13 & 11 & 10 & 227 \end{bmatrix}$$

What is the solution? (List only the values of the original variables and the objective function. Do not include slack or surplus variables.)

(d) What is the dual of the problem you found in part (a)?

(e) What is the solution of the dual you found in part (d)? [Do not perform any steps of the simplex algorithm; just examine the tableau given in part (c).]

Applications

For Exercises 20–23, (a) select appropriate variables; (b) write the objective functions; (c) write the constraints as inequalities.

Business and Economics

20. Profit Roberta Hernandez sells three items, A, B, and C, in her gift shop. Each unit of A costs her $5 to buy, $1 to sell, and $2 to deliver. For each unit of B, the costs are $3, $2, and $1, respectively; and for each unit of C the costs are $6, $2, and $5, respectively. The profit on A is $4; on B, $3; and on C, $3. How many of each should she order to maximize her profit if she can spend $1200 on buying costs, $800 on selling costs, and $500 on delivery costs?

21. Investments An investor is considering three types of investments: a high-risk venture into oil leases with a potential return of 15%, a medium-risk investment in bonds with a 9% return, and a relatively safe stock investment with a 5% return. He has $50,000 to invest. Because of the risk, he will limit his investment in oil leases and bonds to 30% and his investment in oil leases and stock to 50%. How much should he invest in each to maximize his return, assuming investment returns are as expected?

22. Profit The Aged Wood Winery makes two white wines, Fruity and Crystal, from two kinds of grapes and sugar. The wines require the following amounts of each ingredient per gallon and produce a profit per gallon as shown below.

	Grape A (bushels)	Grape B (bushels)	Sugar (pounds)	Profit (dollars)
Fruity	2	2	2	12
Crystal	1	3	1	15

The winery has available 110 bushels of grape A, 125 bushels of grape B, and 90 lb of sugar. How much of each wine should be made to maximize profit?

23. Profit A company makes three sizes of plastic bags: 5-gallon, 10-gallon and 20-gallon. The production time (in hours) for cutting, sealing, and packaging a unit of each size is shown below.

Size	Cutting Time	Sealing Time	Packaging Time
5 gallon	1	1	2
10 gallon	1.1	1.2	3
20 gallon	1.5	1.3	4

There are at most 8 hr available each day for each of the three operations. If the profit on a unit of 5-gal bags is $1; 10-gal bags, $.90; and 20-gal bags, $.95, how many of each size should be made per day to maximize profit?

24. Solve Exercise 20. **25.** Solve Exercise 21.

26. Solve Exercise 22. **27.** Solve Exercise 23.

Connections

28. In Chapter 2, we wrote a system of linear equations using matrix notation. We can do the same thing for the system of linear inequalities in this chapter.

(a) Find matrices A, B, C, and X such that the maximization problem in Example 1 of Section 4.1 can be written as

$$\begin{aligned} \text{Maximize} \quad & C^T X \\ \text{subject to:} \quad & AX \le B \\ \text{with} \quad & X \ge O, \end{aligned}$$

where C^T represents the transpose of C. (*Hint:* Let B, C and X be column matrices. Then C^T is a row matrix.)

(b) Show that the dual of the problem in part (a) can be written as

$$\begin{aligned} \text{Minimize} \quad & B^T Y \\ \text{subject to:} \quad & A^T Y \ge C \\ \text{with} \quad & Y \ge O. \end{aligned}$$

Extended Application / Making Ice Cream*

The first step in the commercial manufacture of ice cream is to blend several ingredients (such as dairy products, eggs, and sugar) to obtain a mix that meets the necessary minimum quality restrictions regarding butterfat content, serum solids, and so on.

Usually many different combinations of ingredients may be blended to obtain a mix of the necessary quality. Within this range of possible substitutes, the firm desires the combination that produces minimum total cost. This problem is quite suitable for solution by linear programming methods.

A "mid-quality" line of ice cream requires the following minimum percentages of constituents by weight.

Fat	16
Serum solids	8
Sugar solids	16
Egg solids	.35
Stabilizer	.25
Emulsifier	.15
Total	40.75%

The balance of the mix is water. A batch of mix is made by blending a number of ingredients, each of which contains one or more of the necessary constituents, and nothing else. The following chart shows the possible ingredients and their costs.

Ingredients	1	2	3	4	5	6	7	8	9
	40% Cream	23% Cream	Butter	Plastic Cream	Butter Oil	4% Milk	Skim Condensed Milk	Skim Milk Powder	Liquid Sugar
Cost ($/lb)	.298	.174	.580	.576	.718	.045	.052	.165	.061
Constituents									
(1) Fat	.400	.230	.805	.800	.998	.040			
(2) Serum solids	.054	.069		.025		.078	.280	.970	
(3) Sugar solids									.707
(4) Egg solids									
(5) Stabilizer									
(6) Emulsifier									

Ingredients	10	11	12	13	14	Requirements
	Sugared Egg Yolk	Powdered Egg Yolk	Stabilizer	Emulsifier	Water	
Cost ($/lb)	.425	1.090	.600	.420	0	
Constituents						
(1) Fat	.500	.625				16
(2) Serum solids						8
(3) Sugar solids	.100					16
(4) Egg solids	.350	.315				.35
(5) Stabilizer			1			.25
(6) Emulsifier				1		.15

*From *Linear Programming in Industry: Theory and Application, An Introduction* by Sven Danø, Fourth revised and enlarged edition. Copyright © 1974 by Springer-Verlag/Wien. Reprinted by permission of Springer-Verlag New York, Inc.

In setting up the mathematical model, use c_1, c_2, \ldots, c_{14} as the cost per unit of the ingredients, and x_1, x_2, \ldots, x_{14} for the quantities of ingredients. The table shows the composition of each ingredient. For example, one pound of ingredient 1 (the 40% cream) contains .400 pounds of fat and .054 pounds of serum solids, with the balance being water.

The ice cream mix is made up in batches of 100 pounds at a time. For a batch to contain 16% fat, at least $100(.16) = 16$ pounds of fat is necessary. Fat is contained in ingredients 1, 2, 3, 4, 5, 6, 10, and 11. The requirement of at least 16 pounds of fat produces the constraint

$$.400x_1 + .230x_2 + .805x_3 + .800x_4 + .998x_5$$
$$+ .040x_6 + .500x_{10} + .625x_{11} \geq 16.$$

Similar constraints can be obtained for the other constituents. Because of the minimum requirements, the total of the ingredients will be at least 40.75 pounds, or

$$x_1 + x_2 + \cdots + x_{13} \geq 40.75.$$

The balance of the 100 pounds is water, making $x_{14} \leq 59.25$. Also, for a 100-pound batch,

$$x_1 + x_2 + \cdots x_{13} \leq 100.$$

The table shows that ingredients 12 and 13 must be used—there is no alternate way of getting these constituents into the final mix. For a 100-pound batch of ice cream, $x_{12} = .25$ pound and $x_{13} = .15$ pound. Removing x_{12} and x_{13} as variables permits a substantial simplification of the model. The problem is now reduced to the following system of five constraints:

$$.400x_1 + .230x_2 + .805x_3 + .800x_4 + .998x_5 + .040x_6 + .500x_{10} + .625x_{11} \geq 16$$
$$.054x_1 + .069x_2 + .025x_4 + .078x_6 + .280x_7 + .970x_8 \geq 8$$
$$.707x_9 + .100x_{10} \geq 16$$
$$.350x_{10} + .315x_{11} \geq .35$$
$$x_1 + x_2 + x_3 + x_4 + x_5 + x_6 + x_7 + x_8 + x_9 + x_{10} + x_{11} \leq 99.6.$$

The objective function, which is to be minimized, is

$$c = .298x_1 + .174x_2 + \cdots + 1.090x_{11}.$$

As usual, $x_1 \geq 0, x_2 \geq 0, \ldots, x_{14} \geq 0$.

Solving this linear programming problem with the simplex method gives

$$x_2 = 67.394, \quad x_8 = 3.459, \quad x_9 = 22.483, \quad x_{10} = 1.000.$$

The total cost of these ingredients is $14.094. To this, we must add the cost of ingredients 12 and 13, producing total cost of

$$14.094 + (.600)(.250) + (.420)(.150) = 14.307,$$

or $14.31.

Exercises

1. Find the mix of ingredients needed for a 100-pound batch of the following grades of ice cream.
 (a) "Generic," sold in a white box, with at least 14% fat and at least 6% serum solids (all other ingredients the same);
 (b) "Premium," sold with a funny foreign name, with at least 17% fat and at least 16.5% sugar solids (all other ingredients the same).

2. Solve the problem in the text by using a computer.

Extended Application / Merit Pay—The UpJohn Company

Individuals doing the same job within the management of a company often receive different salaries. These salaries may differ because of length of service, productivity of an individual worker, and so on. However, for each job there is usually an established minimum and maximum salary.

Many companies make annual reviews of the salary of each of their management employees. At these reviews, an employee may receive a general cost of living increase, an increase based on merit, both, or neither.

In this case, we look at a mathematical model for distributing merit increases in an optimum way.* The salary situation for an individual who is due for salary review may be described as shown in Figure 5, where i represents the number of the employee whose salary is being reviewed. Here the salary ceiling is the maximum salary for the job classification, x_i is the merit increase to be awarded to the individual ($x_i \geq 0$), d_i is the present distance of the current salary from the salary ceiling, and the difference, $d_i - x_i$, is the remaining gap.

FIGURE 5

We let w_i be a measure of the relative worth of the individual to the company. This is the most difficult variable of the model to actually calculate. One way to evaluate w_i is to give a rating sheet to a number of co-workers and supervisors of employee i. An average rating can be obtained and then divided by the highest rating received by an employee with that same job. This number then gives the worth of employee i relative to all other employees with that job.

The best way to allocate money available for merit pay increases is to minimize $w_1(d_1 - x_1) + w_2(d_2 - x_2) + \cdots +$

$w_n(d_n - x_n)$. This sum can be written in a shorthand notation called *summation notation* as

$$\sum_{i=1}^{n} w_i(d_i - x_i).$$

The sum is found by multiplying the relative worth of employee i and the distance of employee i from the maximum salary, after any merit increase. Here n represents the total number of employees who have this job. One constraint here is that the total of all merit increases cannot exceed P, the total amount available for merit increases. That is,

$$\sum_{i=1}^{n} x_i \leq P.$$

Also, the increases for an employee must not put that employee over the maximum salary. That is, for employee i,

$$x_i \leq d_i.$$

We can simplify the objective function by using rules from algebra.

$$\sum_{i=1}^{n} w_i(d_i - x_i) = \sum_{i=1}^{n} (w_i d_i - w_i x_i)$$

$$= \sum_{i=1}^{n} w_i d_i - \sum_{i=1}^{n} w_i x_i$$

For a given individual, w_i and d_i are constant. Therefore, $\sum_{i=1}^{n} w_i d_i$ is some constant, say Z, and

$$\sum_{i=1}^{n} w_i(d_i - x_i) = Z - \sum_{i=1}^{n} w_i x_i.$$

We want to minimize the sum on the left; we can do so by *maximizing* the sum on the right. (Why?) Thus, the original model simplifies to maximizing

$$\sum_{i=1}^{n} w_i x_i,$$

subject to: $\sum_{i=1}^{n} x_i \leq P$ and $x_i \leq d_i$ for each i.

*Based on part of a paper by Jack Northam, Head, Mathematical Services Department, The Upjohn Company, Kalamazoo, Michigan.

Exercises

Here are the current salary information and job evaluation averages for six employees who have the same job. The salary ceiling is $2700 per month.

Employee Number	Evaluation Average	Current Salary
1	570	$2600
2	500	$2550
3	450	$2500
4	600	$2610
5	520	$2530
6	565	$2420

1. Find w_i for each employee by dividing that employee's evaluation average by the highest evaluation average.

2. Use the simplex method with a computer to find the merit increase for each employee. Assume that P is 400.

Applications at a glance...

Failure of a Back-up Computer ... unmarried but eligible population in the U.S.... effectiveness of seat belts ...

CHAPTER 5

SETS AND PROBABILITY

(See page 249.)

In this chapter and the next, we introduce the basic ideas of probability theory, a branch of mathematics that has become increasingly important in management and in the biological and social sciences. Probability theory is valuable because it provides a way to deal with uncertainty. Since the language of sets and set operations are used in the study of probability, we begin there.

5.1 SETS

Think of a **set** as a collection of objects. A set of coins might include one of each type of coin now put out by the United States government. Another set might be made up of all the students in your English class. In mathematics, sets are often made up of numbers. The set consisting of the numbers 3, 4, and 5 is written

$$\{3, 4, 5\},$$

with set braces, { }, enclosing the numbers belonging to the set. The numbers 3, 4, and 5 are called the **elements** or **members** of this set. To show that 4 is an element of the set $\{3, 4, 5\}$, we use the symbol \in and write

$$4 \in \{3, 4, 5\},$$

read "4 is an element of the set containing 3, 4, and 5," Also, $5 \in \{3, 4, 5\}$. To show that 8 is *not* an element of this set, place a slash through the symbol:

$$8 \notin \{3, 4, 5\}.$$

Sets often are named with capital letters, so that if

$$B = \{5, 6, 7\},$$

then, for example $6 \in B$ and $10 \notin B$.

It is possible to have a set with no elements. Some examples are the set of counting numbers less than one, the set of female presidents of the United States, and the set of men more than 10 feet tall. A set with no elements is called the **empty set** and is written \emptyset.

Caution Be careful to distinguish between the symbols 0, \emptyset, $\{0\}$, and $\{\emptyset\}$. The symbol 0 represents a *number;* \emptyset represents a *set* with 0 elements; $\{0\}$ represents a set with one element, 0; and $\{\emptyset\}$ represents a set with one element, \emptyset.

Two sets are **equal** if they contain the same elements. The sets $\{5, 6, 7\}$, $\{7, 6, 5\}$, and $\{6, 5, 7\}$ all contain exactly the same elements and are equal. In symbols,

$$\{5, 6, 7\} = \{7, 6, 5\} = \{6, 5, 7\}.$$

Sets that do not contain exactly the same elements are *not equal*. For example, the sets $\{5, 6, 7\}$ and $\{7, 8, 9\}$ do not contain exactly the same elements and thus are not equal. To show that these sets are not equal, we write

$$\{5, 6, 7\} \neq \{7, 8, 9\}.$$

Sometimes we are interested in a common property of the elements in a set, rather than a list of the elements. This common property can be expressed by using **set-builder notation;** for example,

$$\{x|x \text{ has property } P\}$$

(read "the set of all elements x such that x has property P") represents the set of all elements x having some stated property P.

 EXAMPLE 1 Write the elements belonging to each of the following sets.

(a) $\{x|x \text{ is a natural number less than } 5\}$

The natural numbers less than 5 make up the set $\{1, 2, 3, 4\}$.

(b) $\{x|x \text{ is a state that borders Florida}\} = \{\text{Alabama, Georgia}\}$ ◀

The **universal set** for a particular discussion is a set that includes all the objects being discussed. In elementary school arithmetic, for instance, the set of whole numbers might be the universal set, while in a college algebra class the universal set might be the set of real numbers. The universal set will be specified when necessary, or it will be clearly understandable from the context of the problem.

Subsets Sometimes every element of one set also belongs to another set. For example, if

$$A = \{3, 4, 5, 6\}$$

and

$$B = \{2, 3, 4, 5, 6, 7, 8\},$$

then every element of A is also an element of B. This is an example of the following definition.

SUBSET A set A is a **subset** of a set B (written $A \subseteq B$) if every element of A is also an element of B.

To indicate that A is *not* a subset of B, we write $A \nsubseteq B$.

 EXAMPLE 2 Decide whether the following statements are true or false.

(a) $\{3, 4, 5, 6\} = \{4, 6, 3, 5\}$

Both sets contain exactly the same elements, so the sets are equal and the given statement is true. (The fact that the elements are listed in a different order does not matter.)

(b) $\{5, 6, 9, 12\} \subseteq \{5, 6, 7, 8, 9, 10, 11\}$

The first set is not a subset of the second because it contains an element, 12, that does not belong to the second set. Therefore, the statement is false. ◀

By the definition of subset, the empty set (which contains no elements) is a subset of every set. That is, if A is any set, and the symbol \emptyset represents the empty

set, then $\emptyset \subseteq A$. Also, the definition of subset can be used to show that every set is a subset of itself; that is, if A is any set, then $A \subseteq A$.

For any set A,

$$\emptyset \subseteq A \quad \textbf{and} \quad A \subseteq A.$$

EXAMPLE 3 List all possible subsets for each of the following sets.

(a) $\{7, 8\}$

There are 4 subsets of $\{7, 8\}$:

$$\emptyset, \quad \{7\}, \quad \{8\}, \quad \text{and} \quad \{7, 8\}.$$

(b) $\{a, b, c\}$

There are 8 subsets of $\{a, b, c\}$:

$$\emptyset, \quad \{a\}, \quad \{b\}, \quad \{c\}, \quad \{a, b\}, \quad \{a, c\}, \quad \{b, c\}, \quad \text{and} \quad \{a, b, c\}. \quad \blacktriangleleft$$

In Example 3, all the subsets of $\{7, 8\}$ and all the subsets of $\{a, b, c\}$ were found by trial and error. An alternate method uses a **tree diagram,** a systematic way of listing all the subsets of a given set. Figure 1 shows tree diagrams for finding the subsets of $\{7, 8\}$ and $\{a, b, c\}$.

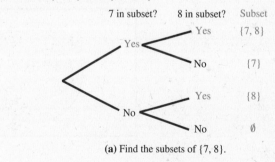

(a) Find the subsets of $\{7, 8\}$.

(b) Find the subsets of $\{a, b, c\}$.

FIGURE 1

As Figure 1 shows, there are two possibilities for each element (either it's in the subset or it's not), so a set with 2 elements has $2 \cdot 2 = 2^2 = 4$ subsets, and a set with 3 elements has $2^3 = 8$ subsets. This idea can be extended to a set with any finite number of elements, which leads to the following conclusion.

> A set of n distinct elements has 2^n subsets.

EXAMPLE 4 Find the number of subsets for each of the following sets.

(a) $\{3, 4, 5, 6, 7\}$

This set has 5 elements; thus, it has 2^5 or 32 subsets.

(b) $\{-1, 2, 3, 4, 5, 6, 12, 14\}$

This set has 8 elements and therefore has 2^8 or 256 subsets.

(c) \emptyset

Since the empty set has 0 elements, it has $2^0 = 1$ subset—itself. ◀

Set Operations Figure 2 shows a set A that is a subset of a set B. The rectangle represents the universal set, U. Such diagrams, called **Venn diagrams**, are used to help illustrate relationships among sets.

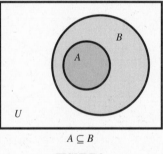

$A \subseteq B$

FIGURE 2

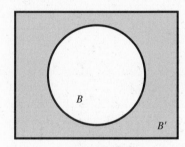

FIGURE 3

Given a set A and a universal set U, the set of all elements of U that do *not* belong to A is called the **complement** of set A. For example, if set A is the set of all the female students in a class, and U is the set of all students in the class, then the complement of A would be the set of all male students in the class. The complement of set A is written A', read "A-prime." The Venn diagram in Figure 3 shows a set B. Its complement, B', is shown in color.

EXAMPLE 5 Let $U = \{1, 2, 3, 4, 5, 6, 7\}$, $A = \{1, 3, 5, 7\}$, and $B = \{3, 4, 6\}$. Find each of the following sets.

(a) A'

Set A' contains the elements of U that are not in A.

$$A' = \{2, 4, 6\}$$

(b) $B' = \{1, 2, 5, 7\}$

(c) $\emptyset' = U$ and $U' = \emptyset$

(d) $(A')' = A$ ◀

Given two sets A and B, the set of all elements belonging to *both* set A and set B is called the **intersection** of the two sets, written $A \cap B$. For example, the elements that belong to both set $A = \{1, 2, 4, 5, 7\}$ and set $B = \{2, 4, 5, 7, 9, 11\}$ are 2, 4, 5, and 7, so that

$$A \cap B = \{1, 2, 4, 5, 7\} \cap \{2, 4, 5, 7, 9, 11\} = \{2, 4, 5, 7\}.$$

The Venn diagram in Figure 4 shows two sets A and B, with their intersection, $A \cap B$, shown in color.

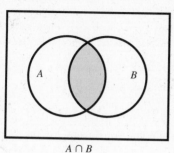

$A \cap B$

FIGURE 4

 EXAMPLE **6**

(a) $\{9, 15, 25, 36\} \cap \{15, 20, 25, 30, 35\} = \{15, 25\}$

The elements 15 and 25 are the only ones belonging to both sets.

(b) $\{-2, -3, -4, -5, -6\} \cap \{-4, -3, -2, -1, 0, 1, 2\} = \{-4, -3, -2\}$ ◀

Two sets that have no elements in common are called **disjoint sets.** For example, there are no elements common to both $\{50, 51, 54\}$ and $\{52, 53, 55, 56\}$, so these two sets are disjoint, and

$$\{50, 51, 54\} \cap \{52, 53, 55, 56\} = \emptyset.$$

This result can be generalized.

DISJOINT SETS

For any sets A and B, if A and B are disjoint sets, then $A \cap B = \emptyset$.

Figure 5 shows a pair of disjoint sets.

The set of all elements belonging to set A, to set B, or to both sets is called the **union** of the two sets, written $A \cup B$. For example,

$$\{1, 3, 5\} \cup \{3, 5, 7, 9\} = \{1, 3, 5, 7, 9\}.$$

The Venn diagram in Figure 6 shows two sets A and B; their union, $A \cup B$, is shown in color.

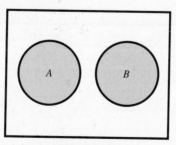

A and B are disjoint sets.

FIGURE 5

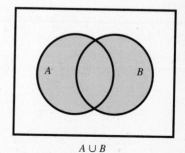

$A \cup B$

FIGURE 6

EXAMPLE 7

(a) Find the union of $\{1, 2, 5, 9, 14\}$ and $\{1, 3, 4, 8\}$.

Begin by listing the elements of the first set, $\{1, 2, 5, 9, 14\}$. Then include any elements from the second set *that are not already listed.* Doing this gives

$$\{1, 2, 5, 9, 14\} \cup \{1, 3, 4, 8\} = \{1, 2, 5, 9, 14, 3, 4, 8\}$$
$$= \{1, 2, 3, 4, 5, 8, 9, 14\}.$$

(b) $\{1, 3, 5, 7\} \cup \{2, 4, 6\} = \{1, 3, 5, 7, 2, 4, 6\} = \{1, 2, 3, 4, 5, 6, 7\}$ ◀

Note As Example 7 shows, when forming unions, we do not list the same element more than once in a set.

Finding the complement of a set, the intersection of two sets, and the union of two sets are examples of *set operations*. These are similar to operations on numbers, such as addition, subtraction, multiplication, and division. The set operations are summarized below.

SET OPERATIONS

Let A and B be any sets, with U representing the universal set. Then the complement of A, written A', is

$$A' = \{x | x \notin A \text{ and } x \in U\};$$

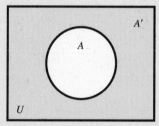

the intersection of A and B is

$$A \cap B = \{x | x \in A \text{ and } x \in B\};$$

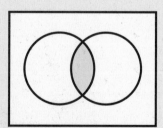

and the union of A and B is

$$A \cup B = \{x | x \in A \text{ or } x \in B\}.$$

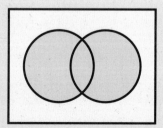

Caution As shown in the definitions given on the previous page, an element is in the *intersection* of sets A and B if it is in A *and* B. On the other hand, an element is in the *union* of sets A and B if it is in A *or* B (or both).

EXAMPLE 8

The table below gives the current dividend, price-to-earnings ratio (the quotient of the price per share and the annual earnings per share), and price change at the end of a day for six companies, as listed on the New York Stock Exchange.

Stock	Dividend	Price-to-Earnings Ratio	Price Change
ATT	1.32	15	$+\frac{1}{4}$
GE	2.28	14	-1
Hershey	.98	20	$-\frac{1}{2}$
IBM	4.84	13	$+1\frac{1}{8}$
Mobil	3.20	13	$-\frac{1}{4}$
Nike	.56	13	$-\frac{1}{8}$

Let set U include all stocks listed, A include all stocks with a dividend greater than \$3, B all stocks with a price-to-earnings ratio of at least 13, and C all stocks with a positive price change. Find the following.

(a) A'

Set A' contains all the listed stocks outside set A, those with a dividend less than or equal to \$3, so $A' = \{$ATT, GE, Hershey, Nike$\}$.

(b) $A \cap B$

The intersection of A and B will contain those stocks that offer a dividend greater than \$3 *and* have a price-to-earnings ratio of at least 13.

$$A \cap B = \{\text{IBM, Mobil}\}$$

(c) $A \cup C$

The union of A and C is the set of all stocks with a dividend greater than \$3 *or* a positive price change (or both).

$$A \cup C = \{\text{ATT, IBM, Mobil}\} \quad \blacktriangleleft$$

EXAMPLE 9

A department store classifies credit applicants by sex, marital status, and employment status. Let the universal set be the set of all applicants, M be the set of male applicants, S be the set of single applicants, and E be the set of employed applicants. Describe the following sets in words.

(a) $M \cap E$

The set $M \cap E$ includes all applicants who are both male *and* employed; that is, employed male applicants.

(b) $M' \cup S$

This set includes all applicants who are female (not male) *or* single. All female applicants and all single applicants are in this set.

(c) $M' \cap S'$

These applicants are female *and* married (not single); thus, $M' \cap S'$ is the set of all married female applicants.

(d) $M \cup E'$

$M \cup E'$ is the set of applicants that are male *or* unemployed. The set includes *all* male applicants and *all* unemployed applicants. ◄

5.1 Exercises

Write true *or* false *for each statement.*

1. $3 \in \{2, 5, 7, 9, 10\}$

2. $6 \in \{-2, 6, 9, 5\}$

3. $9 \notin \{2, 1, 5, 8\}$

4. $3 \notin \{7, 6, 5, 4\}$

5. $\{2, 5, 8, 9\} = \{2, 5, 9, 8\}$

6. $\{3, 7, 12, 14\} = \{3, 7, 12, 14, 0\}$

7. {all whole numbers greater than 7 and less than 10} = $\{8, 9\}$

8. {all counting numbers not greater than 3} = $\{0, 1, 2\}$

9. $\{x | x$ is an odd integer; $6 \leq x \leq 18\} = \{7, 9, 11, 15, 17\}$

10. $\{x | x$ is an even counting number; $x \leq 9\} = \{0, 2, 4, 6, 8\}$

11. Which of the following notations represents the empty set?

(a) 0 **(b)** \emptyset **(c)** { } **(d)** $\{\emptyset\}$

12. What is set-builder notation? Give an example.

Let $A = \{2, 4, 6, 8, 10, 12\}$; $B = \{2, 4, 8, 10\}$; $C = \{4, 8, 12\}$; $D = \{2, 10\}$; $E = \{6\}$; *and* $U = \{2, 4, 6, 8, 10, 12, 14\}$. *Insert* \subseteq *or* \nsubseteq *to make the statement true.*

13. $A _ U$

14. $E _ A$

15. $A _ E$

16. $B _ C$

17. $\emptyset _ A$

18. $\{0, 2\} _ D$

19. $D _ B$

20. $A _ C$

Insert a number in each blank to make the statement true.

21. There are exactly __ subsets of A.

22. There are exactly __ subsets of B.

23. There are exactly __ subsets of C.

24. There are exactly __ subsets of D.

Find the number of subsets for each set.

25. $\{4, 5, 6\}$

26. $\{3, 7, 9, 10\}$

27. $\{x | x$ is a counting number between 6 and 12$\}$

28. $\{x | x$ is a whole number between 8 and 12$\}$

29. Describe the intersection and union of sets. How do they differ?

Insert \cap *or* \cup *to make each statement true.*

30. $\{5, 7, 9, 19\} _ \{7, 9, 11, 15\} = \{7, 9\}$

31. $\{8, 11, 15\} _ \{8, 11, 19, 20\} = \{8, 11\}$

32. $\{2, 1, 7\} _ \{1, 5, 9\} = \{1\}$

33. $\{6, 12, 14, 16\} _ \{6, 14, 19\} = \{6, 14\}$

34. $\{3, 5, 9, 10\} _ \emptyset = \emptyset$

35. $\{3, 5, 9, 10\} _ \emptyset = \{3, 5, 9, 10\}$

36. $\{1, 2, 4\} _ \{1, 2, 4\} = \{1, 2, 4\}$

37. $\{1, 2, 4\} _ \{1, 2\} = \{1, 2, 4\}$

38. Is it possible for two nonempty sets to have the same intersection and union? If so, give an example.

Let $U = \{2, 3, 4, 5, 7, 9\}$; $X = \{2, 3, 4, 5\}$; $Y = \{3, 5, 7, 9\}$; *and* $Z = \{2, 4, 5, 7, 9\}$.

List the members of each of the following sets, using set braces.

39. $X \cap Y$

40. $X \cup Y$

41. X'

42. Y'

43. $X' \cap Y'$

44. $X' \cap Z$

45. $X \cup (Y \cap Z)$

46. $Y \cap (X \cup Z)$

Let U = {all students in this school}; M = {all students taking this course}; N = {all students taking accounting}; P = {all students taking zoology}.

Describe each of the following sets in words.

47. M' **48.** $M \cup N$ **49.** $N \cap P$ **50.** $N' \cap P'$

51. Refer to the sets listed for Exercises 13–20. Which pairs of sets are disjoint?

52. Refer to the sets listed for Exercises 39–46. Which pairs are disjoint?

Refer to Example 8 in the text. Describe each of the sets in Exercises 53–56 in words; then list the elements of each set.

53. B' **54.** $A \cup B$ **55.** $(A \cap B)'$ **56.** $(A \cup C)'$

Applications

General Interest

Cable Television *The top five pay-cable services in 1988 are listed below.* Use this information for Exercises 57–62.*

Network	Subscribers (in thousands)	Content
Home Box Office (HBO)	16,500	Movies, variety, sports, documentaries, etc.
Showtime	6,100	Movies, variety, comedy specials
Cinemax	5,100	Movies, comedy, music specials
The Disney Channel	4,000	Original and classic movies, cartoons, etc.
The Movie Channel	2,500	Movies, film festivals, etc.

List the elements of the following sets.

57. F, the set of networks with more than 5000 subscribers **58.** G, the set of networks that show comedy (not cartoons)

59. H, the set of networks that show movies

60. $F \cup G$ **61.** $H \cap G$ **62.** G'

Life Sciences

Health *The table below shows some symptoms of an overactive thyroid and an underactive thyroid.*

Underactive Thyroid	Overactive Thyroid
Sleepiness, s	Insomnia, i
Dry hands, d	Moist hands, m
Intolerance of cold, c	Intolerance of heat, h
Goiter, g	Goiter, g

Let U be the smallest possible set that includes all the symptoms listed, N be the set of symptoms for an underactive thyroid, and O be the set of symptoms for an overactive thyroid. Find each of the following sets.

63. O' **64.** N' **65.** $N \cap O$ **66.** $N \cup O$ **67.** $N \cap O'$

*"Top Five Pay-Cable Services, 1988" from *The Universal Almanac, 1990*, edited by John W. Wright (Kansas City and New York: Andrews and McMeel, 1990).

5.2 APPLICATIONS OF VENN DIAGRAMS

The responses to a survey of 100 households show that 21 have a home computer, 56 have a videocassette recorder, and 12 have both. How many have neither a home computer nor a videocassette recorder?

It is difficult to answer this question from the given information. In this section we show how a Venn diagram can be used to sort out such information, and later in this section, we are able to answer this question. Venn diagrams were used in the previous section to illustrate set union and intersection. The rectangular region of a Venn diagram represents the universal set U. Including only a single set A inside the universal set, as in Figure 7, divides U into two regions. Region 1 represents those elements of U outside set A, and region 2 represents those elements belonging to set A. (Our numbering of these regions is arbitrary.)

The Venn diagram in Figure 8 shows two sets inside U. These two sets divide the universal set into four regions. As labeled in Figure 8, region 1 represents the set whose elements are outside both set A and set B. Region 2 shows the set whose elements belong to A and not to B. Region 3 represents the set whose elements belong to both A and B. Which set is represented by region 4? (Again, the labeling is arbitrary.)

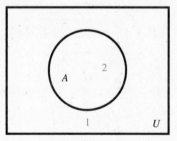

One set leads to 2 regions
(numbering is arbitrary).

FIGURE 7

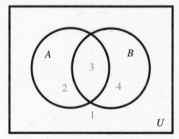

Two sets lead to 4 regions
(numbering is arbitrary).

FIGURE 8

EXAMPLE **1** Draw Venn diagrams similar to Figure 8 and shade the regions representing the following sets.

(a) $A' \cap B$

Set A' contains all the elements outside set A. As labeled in Figure 8, A' is represented by regions 1 and 4. Set B is represented by regions 3 and 4. The intersection of sets A' and B, the set $A' \cap B$, is given by the region common to regions 1 and 4 and regions 3 and 4. The result is the set represented by region 4, which is shaded in Figure 9.

FIGURE 9

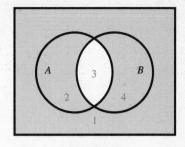

FIGURE 10

(b) $A' \cup B'$

Again, set A' is represented by regions 1 and 4. and set B' by regions 1 and 2. To find $A' \cup B'$, identify the region that represents the set of all elements in A', B', or both. The result, which is shaded in Figure 10, includes regions 1, 2, and 4. ◀

Venn diagrams also can be drawn with three sets inside U. These three sets divide the universal set into eight regions, which can be numbered (arbitrarily) as in Figure 11.

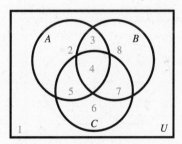

Three sets lead to 8 regions

FIGURE 11

FIGURE 12

2 On a Venn diagram, shade the region that represents $A' \cup (B \cap C')$.

First find $B \cap C'$. Set B is represented by regions 3, 4, 7, and 8, and set C' by regions 1, 2, 3, and 8. The overlap of these regions (regions 3 and 8) represents the set $B \cap C'$. Set A' is represented by regions 1, 6, 7, and 8. The union of the set represented by regions 3 and 8 and the set represented by regions 1, 6, 7, and 8 is the set represented by regions 1, 3, 6, 7, and 8, which are shaded in Figure 12. ◀

Applications We can now use a Venn diagram to answer the question posed at the beginning of this section. A researcher collecting data on 100 households finds that

21 have a home computer;
56 have a videocassette recorder (VCR); and
12 have both.

The researcher wants to answer the following questions.

(a) How many do not have a VCR?
(b) How many have neither a computer nor a VCR?
(c) How many have a computer but not a VCR?

A Venn diagram like the one in Figure 13 will help sort out the information. In Figure 13(a), we put the number 12 in the region common to both a VCR and a computer, because 12 households have both. Of the 21 with a home computer, $21 - 12 = 9$ have no VCR, so in Figure 13(b) we put 9 in the region for a computer but no VCR. Similarly, $56 - 12 = 44$ households have a VCR but not a computer, so we put 44 in that region. Finally, the diagram shows that $100 - 44 - 12 - 9 = 35$ households have neither a VCR nor a computer. Now we can answer the questions:

(a) $35 + 9 = 44$ do not have a VCR;
(b) 35 have neither,
(c) 9 have a computer but not a VCR.

(a)

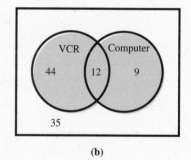

(b)

FIGURE 13

EXAMPLE **3** A survey of 60 freshman business students at a large university produced the following results.

19 of the students read *Business Week*;
18 read *The Wall Street Journal*;
50 read *Fortune*;
13 read *Business Week* and *The Wall Street Journal*;
11 read *The Wall Street Journal* and *Fortune*;
13 read *Business Week* and *Fortune*;
 9 read all three.

Use this information to answer the following questions.

(a) How many students read none of the publications?

(b) How many read only *Fortune?*

(c) How many read *Business Week* and *The Wall Street Journal*, but not *Fortune?*

Since 9 students read all three publications, begin by placing 9 in the area that belongs to all three regions, as shown in Figure 14. Of the 13 students who read *Business Week* and *Fortune*, 9 also read *The Wall Street Journal*. Therefore, only 13 − 9 = 4 read just *Business Week* and *Fortune*. Place the number 4 in the area of Figure 14 common only to *Business Week* and *Fortune* readers. In the same way, place 4 in the region common only to *Business Week* and *The Wall Street Journal*, and 2 in the region common only to *Fortune* and *The Wall Street Journal*.

The data show that 19 students read *Business Week*. However, 4 + 9 + 4 = 17 readers already have been placed in the region representing *Business Week*. The balance of this region will contain only 19 − 17 = 2 students. These 2 students read *Business Week* only—not *Fortune* and not *The Wall Street Journal*. In the same way, 3 students read only *The Wall Street Journal* and 35 read only *Fortune*.

A total of 2 + 4 + 3 + 4 + 9 + 2 + 35 = 59 students are placed in the three circles in Figure 14. Since 60 students were surveyed, 60 − 59 = 1 student reads none of the three publications, and 1 is placed outside all the three regions.

Now Figure 14 can be used to answer the questions asked above.

(a) Only 1 student reads none of the three publications.

(b) From Figure 14, there are 35 students who read only *Fortune*.

(c) The overlap of the regions representing readers of *Business Week* and *The Wall Street Journal* shows that 4 students read *Business Week* and *The Wall Street Journal* but not *Fortune*. ◀

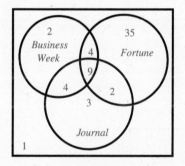

FIGURE 14

EXAMPLE 4 Jeff Friedman is a section chief for an electric utility company. The employees in his section cut down trees, climb poles, and splice wire. Friedman reported the following information to the management of the utility.

"Of the 100 employees in my section,
45 can cut trees;
50 can climb poles;
57 can splice wire;
28 can cut trees and climb poles;
20 can climb poles and splice wire;
25 can cut trees and splice wire;
11 can do all three;
 9 can't do any of the three (management trainees)."

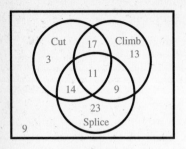

FIGURE 15

The data supplied by Friedman lead to the numbers shown in Figure 15. Add the numbers from all of the regions to get the total number of employees:

$$9 + 3 + 14 + 23 + 11 + 9 + 17 + 13 = 99.$$

Friedman claimed to have 100 employees, but his data indicate only 99. The management decided that Friedman didn't qualify as a section chief, and he was reassigned as a night-shift meter reader in Guam. (Moral: he should have taken this course.) ◀

Caution Note that in all the examples above, we started with the innermost region, the intersection of all categories. This is usually the best way to begin solving problems of this type.

We use the symbol $n(A)$ to indicate the *number* of elements in a set A. For example, if $A = \{a, b, c, d, e\}$, then $n(A) = 5$. The following statement about the number of elements in the union of two sets will be used later in our study of probability.

UNION RULE FOR SETS	$n(A \cup B) = n(A) + n(B) - n(A \cap B)$

To prove this statement, let $x + y$ represent $n(A)$, y represent $n(A \cap B)$, and $y + z$ represent $n(B)$, as shown in Figure 16. Then

$$n(A \cup B) = x + y + z,$$
$$n(A) + n(B) - n(A \cap B) = (x + y) + (y + z) - y = x + y + z,$$

so
$$n(A \cup B) = n(A) + n(B) - n(A \cap B).$$

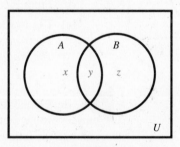

FIGURE 16

EXAMPLE 5 A group of 10 students meet to plan a school function. All are majors in accounting or economics or both. Five of the students are economics majors and 7 are majors in accounting. How many major in both subjects?

Use the union rule, with $n(A) = 5$, $n(B) = 7$, and $n(A \cup B) = 10$. We must find $n(A \cap B)$.

$$n(A \cup B) = n(A) + n(B) - n(A \cap B)$$
$$10 = 5 + 7 - n(A \cap B),$$

so
$$n(A \cap B) = 5 + 7 - 10 = 2. \quad ◀$$

5.2 Exercises

Sketch a Venn diagram like the one in the figure, and use shading to show each of the following sets.

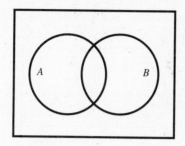

1. $B \cap A'$

2. $A \cup B'$

3. $A' \cup B$

4. $A' \cap B'$

5. $B' \cup (A' \cap B')$

6. $(A \cap B) \cup B'$

7. U'

8. \emptyset'

9. Three sets divide the universal set into _____ regions.

Sketch a Venn diagram like the one shown, and use shading to show each of the following sets.

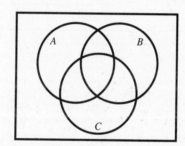

10. $(A \cap B) \cap C$

11. $(A \cap C') \cup B$

12. $A \cap (B \cup C')$

13. $A' \cap (B \cap C)$

14. $(A' \cap B') \cap C$

15. $(A \cap B') \cup C$

16. $(A \cap B') \cap C$

17. $A' \cap (B' \cup C)$

18. What does the notation $n(A)$ represent?

Use the union rule to answer the following questions.

19. If $n(A) = 5$, $n(B) = 8$, and $n(A \cap B) = 4$, what is $n(A \cup B)$?

20. If $n(A) = 12$, $n(B) = 27$, and $n(A \cup B) = 30$, what is $n(A \cap B)$?

21. Suppose $n(B) = 7$, $n(A \cap B) = 3$, and $n(A \cup B) = 20$. What is $n(A)$?

22. Suppose $n(A \cap B) = 5$, $n(A \cup B) = 35$, and $n(A) = 13$. What is $n(B)$?

Draw a Venn diagram and use the given information to fill in the number of elements for each region.

23. $n(U) = 38$, $n(A) = 16$, $n(A \cap B) = 12$, $n(B') = 20$

24. $n(A) = 26$, $n(B) = 10$, $n(A \cup B) = 30$, $n(A') = 17$

25. $n(A \cup B) = 17$, $n(A \cap B) = 3$, $n(A) = 8$, $n(A' \cup B') = 21$

26. $n(A') = 28$, $n(B) = 25$, $n(A' \cup B') = 45$, $n(A \cap B) = 12$

27. $n(A) = 28$, $n(B) = 34$, $n(C) = 25$, $n(A \cap B) = 14$, $n(B \cap C) = 15$, $n(A \cap C) = 11$, $n(A \cap B \cap C) = 9$, $n(U) = 59$

28. $n(A) = 54$, $n(A \cap B) = 22$, $n(A \cup B) = 85$, $n(A \cap B \cap C) = 4$, $n(A \cap C) = 15$, $n(B \cap C) = 16$, $n(C) = 44$, $n(B') = 63$

29. $n(A \cap B) = 6$, $n(A \cap B \cap C) = 4$, $n(A \cap C) = 7$, $n(B \cap C) = 4$, $n(A \cap C') = 11$, $n(B \cap C') = 8$, $n(C) = 15$, $n(A' \cap B' \cap C') = 5$

30. $n(A) = 13$, $n(A \cap B \cap C) = 4$, $n(A \cap C) = 6$, $n(A \cap B') = 6$, $n(B \cap C) = 6$, $n(B \cap C') = 11$, $n(B \cup C) = 22$, $n(A' \cap B' \cap C') = 5$

*In Exercises 31–34, show that the statement is true by drawing Venn diagrams and shading the regions representing the sets on each side of the equals sign.**

31. $(A \cup B)' = A' \cap B'$

32. $(A \cap B)' = A' \cup B'$

33. $A \cap (B \cup C) = (A \cap B) \cup (A \cap C)$

34. $A \cup (B \cap C) = (A \cup B) \cap (A \cup C)$

Applications

Business and Economics

Use Venn diagrams to answer the following questions.

35. **Cooking Preferences** Jeff Friedman, of Example 4 in the text, was again reassigned, this time to the home economics department of the electric utility. He interviewed 140 people in a suburban shopping center to discover some of their cooking habits. He obtained the following results:

 58 use microwave ovens;

 63 use electric ranges;

 58 use gas ranges;

 19 use microwave ovens and electric ranges;

 17 use microwave ovens and gas ranges;

 4 use both gas and electric ranges;

 1 uses all three;

 2 use none of the three.

Should he be reassigned one more time? Why?

36. **Harvesting Fruit** Toward the middle of the harvesting season, peaches for canning come in three types, early, late, and extra late, depending on the expected date of ripening. During a certain week, the following data were recorded at a fruit delivery station:

 34 trucks went out carrying early peaches;

 61 carried late peaches;

 50 carried extra late;

 25 carried early and late;

 30 carried late and extra late;

 8 carried early and extra late;

 6 carried all three;

 9 carried only figs (no peaches at all).

 (a) How many trucks carried only late variety peaches?

 (b) How many carried only extra late?

 (c) How many carried only one type of peach?

 (d) How many trucks (in all) went out during the week?

37. **Cola Consumption** Market research showed that the adult residents of a certain small town in Georgia fit the following categories of cola consumption.

Age	Drink Regular Cola (R)	Drink Diet Cola (D)	Drink No Cola (N)	Totals
21–25 (Y)	40	15	15	70
26–35 (M)	30	30	20	80
Over 35 (O)	10	50	10	70
Totals	80	95	45	220

Using the letters given in the table, find the number of people in each of the following sets.

 (a) $Y \cap R$ **(b)** $M \cap D$ **(c)** $M \cup (D \cap Y)$

 (d) $Y' \cap (D \cup N)$ **(e)** $O' \cup N$ **(f)** $M' \cap (R' \cap N')$

 (g) Describe the set $M \cup (D \cap Y)$ in words.

38. **Investment Habits** The following table shows the results of a survey taken by a bank in a medium-sized town in Tennessee. The survey asked questions about the investment habits of bank customers.

Age	Stocks (S)	Bonds (B)	Savings Accounts (A)	Totals
18–29 (Y)	6	2	15	23
30–49 (M)	14	5	14	33
50 or over (O)	32	20	12	64
Totals	52	27	41	120

*The statements in Exercises 31 and 32 are known as De Morgan's laws. They are named for the English mathematician Augustus De Morgan (1806–1871).

Using the letters given in the table, find the number of people in each of the following sets.

(a) $Y \cap B$ **(b)** $M \cup A$ **(c)** $Y \cap (S \cup B)$

(d) $O' \cup (S \cup A)$ **(e)** $(M' \cup O') \cap B$

(f) Describe the set $Y \cap (S \cup B)$ in words.

Life Sciences

39. Genetics After a genetics experiment, the number of pea plants having certain characteristics was tallied, with the following results.

> 22 were tall;
> 25 had green peas;
> 39 had smooth peas;
> 9 were tall and had green peas;
> 17 were tall and had smooth peas;
> 20 had green peas and smooth peas;
> 6 had all three characteristics;
> 4 had none of the characteristics.

(a) Find the total number of plants counted.

(b) How many plants were tall and had peas that were neither smooth nor green?

(c) How many plants were not tall but had peas that were smooth and green?

40. Blood Antigens Human blood can contain either no antigens, the A antigen, the B antigen, or both the A and B antigens. A third antigen, called the Rh antigen, is important in human reproduction, and again may or may not be present in an individual. Blood is called type A-positive if the individual has the A and Rh, but not the B antigen. A person having only the A and B antigens is said to have type AB-negative blood. A person having only the Rh antigen has type O-positive blood. Other blood types are defined in a similar manner. Identify the blood types of the individuals in regions (a)–(g) below. (One region has no letter assigned to it.)

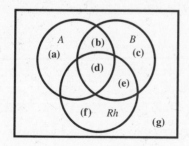

41. Blood Antigens (Use the diagram from Exercise 40.) In a certain hospital, the following data were recorded.

> 25 patients had the A antigen;
> 17 had the A and B antigens;
> 27 had the B antigen;
> 22 had the B and Rh antigens;
> 30 had the Rh antigen;
> 12 had none of the antigens;
> 16 had the A and Rh antigens;
> 15 had all three antigens.

How many patients

(a) were represented? **(b)** had exactly one antigen?

(c) had exactly two antigens? **(d)** had O-positive blood?

(e) had AB-positive blood? **(f)** had B-negative blood?

(g) had O-negative blood? **(h)** had A-positive blood?

General Interest

42. Native American Ceremonies At a pow-wow in Arizona, Native Americans from all over the Southwest came to participate in the ceremonies. A coordinator of the pow-wow took a survey and found that

> 15 families brought food, costumes, and crafts;
> 25 families brought food and crafts;
> 42 families brought food;
> 20 families brought costumes and food;
> 6 families brought costumes and crafts, but not food;
> 4 families brought crafts, but neither food nor costumes;
> 10 families brought none of the three items;
> 18 families brought costumes but not crafts.

(a) How many families were surveyed?

(b) How many families brought costumes?

(c) How many families brought crafts, but not costumes?

(d) How many families did not bring crafts?

(e) How many families brought food or costumes?

43. Chinese New Year A survey of people attending a Lunar New Year celebration in Chinatown yielded the following results:

> 120 were women;
> 150 spoke Cantonese;
> 170 lit firecrackers;
> 108 of the men spoke Cantonese;
> 100 of the men did not light firecrackers;
> 18 of the non-Cantonese-speaking women lit firecrackers;
> 78 non-Cantonese-speaking men did not light firecrackers;
> 30 of the women who spoke Cantonese lit firecrackers.

(a) How many attended?

(b) How many of those who attended did not speak Cantonese?

(c) How many women did not light firecrackers?

(d) How many of those who lit firecrackers were Cantonese-speaking men?

44. Poultry Analysis A chicken farmer surveyed his flock with the following results. The farmer had

 9 fat red roosters;
 2 fat red hens;
 37 fat chickens;
 26 fat roosters;

 7 thin brown hens;
 18 thin brown roosters;
 6 thin red roosters;
 5 thin red hens.

Answer the following questions about the flock. [*Hint:* You need a Venn diagram with regions for fat, for male (a rooster is a male; a hen is a female), and for red (assume that brown and red are opposites in the chicken world).] How many chickens were

(a) fat? (b) red?

(c) male? (d) fat, but not male?

(e) brown, but not fat? (f) red and fat?

5.3 INTRODUCTION TO PROBABILITY

If you go to a supermarket and buy 5 pounds of peaches at 54 cents per pound, you can easily find the *exact* price of your purchase: $2.70. On the other hand, the produce manager of the market is faced with the problem of ordering peaches. The manager may have a good estimate of the number of pounds of peaches that will be sold during the day, but it is impossible to predict the *exact* amount. The number of pounds that customers will purchase during a day is *random:* the quantity cannot be predicted exactly. A great many problems that come up in applications of mathematics involve random phenomena—those for which exact prediction is impossible. The best that we can do is determine the *probability* of the possible outcomes.

Sample Spaces In probability, an **experiment** is an activity or occurrence with an observable result. Each repetition of an experiment is called a **trial.** The possible results of each trial are called **outcomes.** The set of all possible outcomes for an experiment is the **sample space** for that experiment. A sample space for the experiment of tossing a coin is made up of the outcomes heads (*h*) and tails (*t*). If *S* represents this sample space, then

$$S = \{h, t\}.$$

EXAMPLE 1 Give the sample space for each experiment.

(a) A spinner like the one in Figure 17 is spun.

The three outcomes are 1, 2, or 3, so the sample space is

$$\{1, 2, 3\}.$$

(b) For the purposes of a public opinion poll, respondents are classified as young, middle-aged, or older, and as male or female.

A sample space for this poll could be written as a set of ordered pairs:

$$\{(\text{young, male}), (\text{young, female}), (\text{middle-aged, male}),$$
$$(\text{middle-aged, female}), (\text{older, male}), (\text{older, female})\}.$$

FIGURE 17

(c) An experiment consists of studying the numbers of boys and girls in families with exactly 3 children. Let *b* represent *boy* and *g* represent *girl*.

A three-child family can have 3 boys, written *bbb*, 3 girls, *ggg*, or various combinations, such as *bgg*. A sample space with four outcomes (not equally likely) is

$$S_1 = \{3 \text{ boys}, 2 \text{ boys and } 1 \text{ girl}, 1 \text{ boy and } 2 \text{ girls}, 3 \text{ girls}\}.$$

Notice that a family with 3 boys or 3 girls can occur in just one way, but a family of 2 boys and 1 girl or 1 boy and 2 girls can occur in more than one way. If the *order* of the births is considered, so that *bgg* is different from *gbg* or *ggb*, for example, another sample space is

$$S_2 = \{bbb, bbg, bgb, gbb, bgg, gbg, ggb, ggg\}.$$

The second sample space, S_2, has equally likely outcomes; S_1 does not, since there is more than one way to get a family with 2 boys and 1 girl or a family with 2 girls and 1 boy, but only one way to get 3 boys or 3 girls. ◀

Caution An experiment may have more than one sample space, as shown in Example 1(c). The most useful sample spaces have equally likely outcomes, but it is not always possible to choose such a sample space.

Events An **event** is a subset of a sample space. If the sample space for tossing a coin is $S = \{h, t\}$, then one event is $E = \{h\}$, which represents the outcome "heads."

An ordinary die is a cube whose six different faces show the following numbers of dots: 1, 2, 3, 4, 5, and 6. If the die is fair (not "loaded" to favor certain faces over others), then any one of the faces is equally likely to come up when the die is rolled. The sample space for the experiment of rolling a single fair die is $S = \{1, 2, 3, 4, 5, 6\}$. Some possible events are listed below.

The die shows an even number: $E_1 = \{2, 4, 6\}$.

The die shows a 1: $E_2 = \{1\}$.

The die shows a number less than 5: $E_3 = \{1, 2, 3, 4\}$.

The die shows a multiple of 3: $E_4 = \{3, 6\}$.

EXAMPLE 2 For the sample space S_2 in Example 1(c), write the following events.

(a) Event *H*: the family has exactly 2 girls

Families with three children can have exactly 2 girls with either *bgg*, *gbg*, or *ggb*, so event *H* is

$$H = \{bgg, gbg, ggb\}.$$

(b) Event *K*: the three children are the same sex

Two outcomes satisfy this condition: all boys or all girls.

$$K = \{bbb, ggg\}$$

(c) Event *J*: the family has three girls

Only *ggg* satisfies this condition, so

$$J = \{ggg\}. \quad ◀$$

In Example 2(c), event *J* had only one possible outcome, *ggg*. Such an event, with only one possible outcome, is a **simple event.** If event *E* equals the sample space *S*, then *E* is called a **certain event.** If event $E = \emptyset$, then *E* is called an **impossible event.**

3 Suppose a fair die is rolled. As shown above, the sample space is {1, 2, 3, 4, 5, 6}.

(a) The event "the die shows a 4," {4}, has only one possible outcome. It is a simple event.

(b) The event "the number showing is less than 10" equals the sample space. $S = \{1, 2, 3, 4, 5, 6\}$. This event is a certain event; if a die is rolled, the number showing (either 1, 2, 3, 4, 5, or 6), must be less than 10.

(c) The event "the die shows a 7" is the empty set, \emptyset; this is an impossible event. ◀

Probability For sample spaces with *equally likely* outcomes, the probability of an event is defined as follows.

BASIC PROBABILITY PRINCIPLE

Let *S* be a sample space with *n* equally likely outcomes. Let event *E* contain *m* of these outcomes. Then the **probability that event *E* occurs,** written *P(E)*, is

$$P(E) = \frac{m}{n}.$$

By this definition, the probability of an event is a number that indicates the relative likelihood of the event.

4 Suppose a single fair die is rolled. Use the sample space $S = \{1, 2, 3, 4, 5, 6\}$ and give the probability of each of the following events.

(a) *E*: the die shows an even number

Here, $E = \{2, 4, 6\}$, a set with three elements. Since *S* contains six elements,

$$P(E) = \frac{3}{6} = \frac{1}{2}.$$

(b) *F*: the die shows a number greater than 4

Since *F* contains two elements, 5 and 6,

$$P(F) = \frac{2}{6} = \frac{1}{3}.$$

(c) *G*: the die shows a number less than 10

Event *G* is a certain event, with

$$G = \{1, 2, 3, 4, 5, 6\},$$

so that

$$P(G) = \frac{6}{6} = 1.$$

(d) *H*: the die shows an 8
This event is impossible, so

$$P(H) = 0. \blacktriangleleft$$

A standard deck of 52 cards has four suits, hearts, clubs, diamonds, and spades, with 13 cards in each suit. Each suit has an ace, a king, a queen, a jack, and cards numbered from 2 to 10. The hearts and diamonds are red, and the spades and clubs are black. The king, queen, and jack are called *face cards*. See Figure 18. We will refer to this standard deck of cards often in our discussion of probability.

FIGURE 18

EXAMPLE 5 If a single playing card is drawn at random from a standard 52-card deck, find the probability of each of the following events.

(a) Drawing an ace
There are 4 aces in the deck. The event "drawing an ace" is

{heart ace, diamond ace, club ace, spade ace}.

Therefore,

$$P(\text{ace}) = \frac{4}{52} = \frac{1}{13}.$$

(b) Drawing a face card
Since there are 12 face cards (three in each of the four suits),

$$P(\text{face card}) = \frac{12}{52} = \frac{3}{13}.$$

(c) Drawing a spade
The deck contains 13 spades, so

$$P(\text{spade}) = \frac{13}{52} = \frac{1}{4}.$$

(d) Drawing a spade or a heart

Besides the 13 spades, the deck contains 13 hearts, so

$$P(\text{spade or heart}) = \frac{26}{52} = \frac{1}{2}. \quad \blacktriangleleft$$

In the preceding examples, the probability of each event was a number between 0 and 1. The same thing is true in general. If an event E contains m of the n possible outcomes, then $0 \le m \le n$, and $0 \le m/n \le 1$. Since $P(E) = m/n$, $P(E)$ is between 0 and 1.

For any event E, $0 \le P(E) \le 1$.

5.3 Exercises

1. What is meant by a "fair" coin or die?

2. What is the sample space for an experiment?

Write sample spaces for the experiments in Exercises 3–10.

3. A month of the year is chosen for a wedding.

4. A day in April is selected for a bicycle race.

5. A student is asked how many points she earned on a recent 80-point test.

6. A person is asked the number of hours (to the nearest hour) he watched television yesterday.

7. The management of an oil company must decide whether to go ahead with a new oil shale plant or to cancel it.

8. A record is kept for three days about whether a particular stock goes up or down.

9. A coin is tossed, and a die is rolled.

10. A box contains five balls, numbered 1, 2, 3, 4, and 5. A ball is drawn at random, the number on it recorded, and the ball replaced. The box is shaken, a second ball is drawn, and its number is recorded.

11. Define an event.

12. What is a simple event?

For the experiments in Exercises 13–18, write out the sample space, and then write the indicated events in set notation.

13. A committee of 2 people is selected from 5 executives: Alam, Bartolini, Chinn, Dickson, and Ellsberg.

 (a) Chinn is on the committee.

 (b) Dickson and Ellsberg are not both on the committee.

 (c) Both Alam and Chinn are on the committee.

14. A marble is drawn at random from a bowl containing 3 yellow, 4 white, and 8 blue marbles.

 (a) A yellow marble is drawn. **(b)** A blue marble is drawn.

 (c) A white marble is drawn. **(d)** A black marble is drawn.

15. Slips of paper marked with the numbers 1, 2, 3, 4, and 5 are placed in a box. After being mixed, two slips are drawn without replacement.

 (a) Both slips are marked with even numbers.

 (b) One slip is marked with an odd number and the other is marked with an even number.

 (c) Both slips are marked with the same number.

16. An unprepared student takes a three-question, true/false quiz in which he guesses the answers to all three questions.

 (a) The student gets three answers wrong. **(b)** The student gets exactly two answers correct.

 (c) The student gets only the first answer correct.

17. A die is tossed twice, with the tosses recorded as ordered pairs.

 (a) The first die shows a 3. **(b)** The sum of the numbers showing is 8.

 (c) The sum of the numbers showing is 13.

18. One jar contains four balls, labeled 1, 2, 3, and 4. A second jar contains five balls, labeled 1, 2, 3, 4, and 5. An experiment consists of taking one ball from the first jar, and then taking a ball from the second jar.

 (a) The number on the first ball is even. **(b)** The number on the second ball is even.

 (c) The sum of the numbers on the two balls is 5. **(d)** The sum of the numbers on the two balls is 1.

A single fair die is rolled. Find the probabilities of the following events.

19. Getting a 2

20. Getting an odd number

21. Getting a number less than 5

22. Getting a number greater than 2

23. Getting a 3 or a 4

24. Getting any number except 3

A card is drawn from a well-shuffled deck of 52 cards. Find the probability of drawing each of the following.

25. A 9

26. A black card

27. A black 9

28. A heart

29. The 9 of hearts

30. A face card

31. A 2 or a queen

32. A black 7 or a red 8

33. A red face card

34. A heart or a spade

A jar contains 2 white, 3 orange, 5 yellow, and 8 black marbles. If a marble is drawn at random, find the probability that it is the following.

35. White

36. Orange

37. Yellow

38. Black

39. Not black

40. Orange or yellow

Applications

Business and Economics

41. Survey of Workers The management of a firm wishes to check on the opinions of its assembly line workers. Before the workers are interviewed, they are divided into various categories. Define events E, F, and G as follows.

 E: worker is female

 F: worker has worked less than 5 years

 G: worker contributes to a voluntary retirement plan

Describe each of the following events in words.

 (a) E' **(b)** $E \cap F$ **(c)** $E \cup G'$

42. Survey of Workers Refer to Exercise 41. Describe each of the following events in words.

 (a) F' **(b)** $F \cup G$ **(c)** $F' \cap G'$

43. Research Funding According to an article in *Business Week*, in 1989 funding for university research in the United States totaled $15 billion. Support came from various sources, as shown in the next column.*

1989 Funding for University Research

Source	Amount (in billions of dollars)
Federal government	9.0
State and local government	1.2
Institutional	2.7
Industry	1.0
Other	1.1

Find the probability that funds for a particular project came from each of the following sources.

 (a) Federal government **(b)** Industry

 (c) The institution

*"University Research: The Squeeze Is On," reprinted from May 20, 1991 issue of *Business Week* by special permission. Copyright © 1991 by McGraw-Hill, Inc.

Life Sciences

44. Medical Survey For a medical experiment, people are classified as to whether they smoke, have a family history of heart disease, or are overweight. Define events E, F, and G as follows.

E: person smokes
F: person has a family history of heart disease
G: person is overweight

Describe each of the following events in words.

(a) G' **(b)** $F \cap G$ **(c)** $E \cup G'$

45. Medical Survey Refer to Exercise 44. Describe each of the following events in words.

(a) $E \cup F$ **(b)** $E' \cap F$ **(c)** $F' \cup G'$

Social Sciences

46. U.S. Population The population of the United States by race in 1990, and the projected population by race for the year 2000 are given below.*

U.S. Population by Race (in thousands)		
Race	1990	2000
White	191,594	197,634
Black	30,915	35,440
Hispanic	21,854	30,295
Other	7,930	11,110

Find the probability of a randomly selected person being each of the following.

(a) Hispanic in 1990 **(b)** White in 2000

(c) Black in 2000

*"Projections of the Population, by Race and Hispanic Origin, 1990–2000" from U.S. Bureau of the Census, *Projections of the Population of the U.S. by Age, Sex, and Race, 1983–2080.*

5.4 BASIC CONCEPTS OF PROBABILITY

What is the probability that a customer entering a store will spend less than $10?

We discuss the probability of this and other more complex events in this section. Since events are sets, we can use set operations to find unions, intersections, and complements of events.

EXAMPLE 1

Suppose a die is tossed. Let E be the event "the die shows a number greater than 3," and let F be the event "the die shows an even number." Then

$$E = \{4, 5, 6\} \quad \text{and} \quad F = \{2, 4, 6\}.$$

Find each of the following.

(a) $E \cap F$

The event $E \cap F$ is "the die shows a number that is greater than 3 *and* is even." Event $E \cap F$ includes the outcomes common to *both* E and F.

$$E \cap F = \{4, 6\}$$

(b) $E \cup F$

The event $E \cup F$ is "the die shows a number that is greater than 3 *or* is even." Event $E \cup F$ includes the outcomes of E *or* F, or both.

$$E \cup F = \{2, 4, 5, 6\}$$

(c) E'

Event E' is "the die shows a number that is *not* greater than 3." E' includes the elements of the sample space that are not in E. (Event E' is the complement of event E.)

$$E' = \{1, 2, 3\}$$

The Venn diagrams in Figure 19 show the events $E \cap F$, $E \cup F$, and E'. ◄

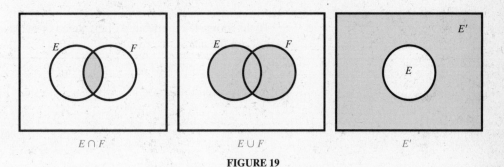

FIGURE 19

Two events that cannot both occur at the same time, such as getting both a head and a tail on the same toss of a coin, are called **mutually exclusive events**.

MUTUALLY EXCLUSIVE EVENTS

Events E and F are mutually exclusive events if $E \cap F = \emptyset$.

For any event E, E and E' are mutually exclusive. See Figure 19. By definition, mutually exclusive events are disjoint sets.

EXAMPLE 2

Let $S = \{1, 2, 3, 4, 5, 6\}$, the sample space for tossing a single die. Let $E = \{4, 5, 6\}$, and let $G = \{1, 2\}$. Then E and G are mutually exclusive events since they have no outcomes in common: $E \cap G = \emptyset$. See Figure 20. ◄

$E \cap G = \emptyset$

FIGURE 20

A summary of the set operations for events is given below.

SET OPERATIONS FOR EVENTS

Let E and F be events for a sample space S.
$E \cap F$ occurs when both E and F occur;
$E \cup F$ occurs when E or F or both occur;
E' occurs when E does not occur;
E and F are mutually exclusive if $E \cap F = \emptyset$.

To determine the probability of the union of two events E and F in a sample space S, use the union rule for sets,

$$n(E \cup F) = n(E) + n(F) - n(E \cap F),$$

which was proved in Section 5.2. Dividing both sides by $n(S)$ shows that

$$\frac{n(E \cup F)}{n(S)} = \frac{n(E)}{n(S)} + \frac{n(F)}{n(S)} - \frac{n(E \cap F)}{n(S)}$$

$$P(E \cup F) = P(E) + P(F) - P(E \cap F).$$

This result is called the **union rule for probability.**

UNION RULE FOR PROBABILITY

For any events E and F from a sample space S,

$$P(E \cup F) = P(E) + P(F) - P(E \cap F).$$

(Although the union rule applies to any events E and F from any sample space, the derivation we have given is valid only for sample spaces with equally likely simple events.)

EXAMPLE 3 If a single card is drawn from an ordinary deck of cards, find the probability that it will be a red or a face card.

Let R represent the event "red card" and F the event "face card." There are 26 red cards in the deck, so $P(R) = 26/52$. There are 12 face cards in the deck, so $P(F) = 12/52$. Since there are 6 red face cards in the deck, $P(R \cap F) = 6/52$. By the union rule, the probability of the card being red or a face card is

$$P(R \cup F) = P(R) + P(F) - P(R \cap F)$$

$$= \frac{26}{52} + \frac{12}{52} - \frac{6}{52} = \frac{32}{52} = \frac{8}{13}. \quad \blacktriangleleft$$

EXAMPLE 4 Suppose two fair dice are rolled. Find each of the following probabilities.

(a) The first die shows a 2, or the sum of the results is 6 or 7.

The sample space for the throw of two dice is shown in Figure 21, where 1-1 represents the event "the first die shows a 1 and the second die shows a 1," 1-2 represents "the first die shows a 1 and the second die shows a 2," and so on. Let A represent the event "the first die shows a 2," and B represent the event

FIGURE 21

"the sum of the results is 6 or 7." These events are indicated in Figure 21. From the diagram, event A has 6 elements, B has 11 elements, and the sample space has 36 elements. Thus,

$$P(A) = \frac{6}{36}, \quad P(B) = \frac{11}{36}, \quad \text{and} \quad P(A \cap B) = \frac{2}{36}.$$

By the union rule,

$$P(A \cup B) = P(A) + P(B) - P(A \cap B)$$

$$P(A \cup B) = \frac{6}{36} + \frac{11}{36} - \frac{2}{36} = \frac{15}{36} = \frac{5}{12}.$$

(b) The sum of the results is 11, or the second die shows a 5.

$P(\text{sum is } 11) = 2/36$, $P(\text{second die shows a } 5) = 6/36$, and $P(\text{sum is } 11$ and second die shows a 5$) = 1/36$, so

$$P(\text{sum is } 11 \text{ or second die shows a } 5) = \frac{2}{36} + \frac{6}{36} - \frac{1}{36} = \frac{7}{36}. \quad \blacktriangleleft$$

If events E and F are mutually exclusive, then $E \cap F = \emptyset$ by definition; hence, $P(E \cap F) = 0$. Applying the union rule yields this useful fact.

For mutually exclusive events E and F,

$$P(E \cup F) = P(E) + P(F).$$

Caution You may wonder why we did not use $S = \{2, 3, 4, 5, \ldots, 12\}$ as the sample space in Example 4. Remember, we prefer to use a sample space with equally likely outcomes. The outcomes in set S above are not equally likely—a sum of 2 can occur in just one way, a sum of 3 in two ways, a sum of 4 in three ways, and so on, as shown in Figure 21.

By definition of E', for any event E from a sample space S,

$$E \cup E' = S \quad \text{and} \quad E \cap E' = \emptyset.$$

Since $E \cap E' = \emptyset$, events E and E' are mutually exclusive, so that

$$P(E \cup E') = P(E) + P(E').$$

However, $E \cup E' = S$, the sample space, and $P(S) = 1$. Thus

$$P(E \cup E') = P(E) + P(E') = 1.$$

Rearranging these terms gives the following useful rule.

COMPLEMENT RULE $P(E) = 1 - P(E')$ and $P(E') = 1 - P(E).$

EXAMPLE 5 If a fair die is rolled, what is the probability that any number but 5 will come up?

If E is the event that 5 comes up, then E' is the event that any number but 5 comes up. Since $P(E) = \frac{1}{6}$, we have $P(E') = 1 - \frac{1}{6} = \frac{5}{6}$. ◄

EXAMPLE 6 In Example 4, find the probability that the sum of the numbers rolled is greater than 3.

To calculate this probability directly, we must find the probabilities that the sum is 4, 5, 6, 7, 8, 9, 10, 11, or 12 and then add them. It is much simpler to first find the probability of the complement, the event that the sum is less than or equal to 3.

$$P(\text{sum} \le 3) = P(\text{sum is 2}) + P(\text{sum is 3})$$

$$= \frac{1}{36} + \frac{2}{36}$$

$$= \frac{3}{36} = \frac{1}{12}$$

Now use the fact that $P(E) = 1 - P(E')$ to get

$$P(\text{sum} > 3) = 1 - P(\text{sum} \le 3)$$

$$= 1 - \frac{1}{12} = \frac{11}{12}.$$ ◄

Odds Sometimes probability statements are given in terms of **odds,** a comparison of $P(E)$ with $P(E')$. For example, suppose $P(E) = 4/5$. Then $P(E') = 1 - 4/5 = 1/5$. These probabilities predict that E will occur 4 out of 5 times and E' will occur 1 out of 5 times. Then we say the **odds in favor** of E are 4 to 1.

ODDS The **odds in favor** of an event E are defined as the ratio of $P(E)$ to $P(E')$, or

$$\frac{P(E)}{P(E')}, \quad P(E') \neq 0.$$

EXAMPLE 7 Suppose the weather forecaster says that the probability of rain tomorrow is 1/3. Find the odds in favor of rain tomorrow.

Let E be the event "rain tomorrow." Then E' is the event "no rain tomorrow." Since $P(E) = 1/3$, $P(E') = 2/3$. By the definition of odds, the odds in favor of rain are

$$\frac{1/3}{2/3} = \frac{1}{2}, \qquad \text{written} \qquad 1 \text{ to } 2, \text{ or } 1{:}2.$$

On the other hand, the odds that it will *not* rain, or the *odds against* rain, are

$$\frac{2/3}{1/3} = \frac{2}{1}, \qquad \text{written} \qquad 2 \text{ to } 1, \text{ or } 2{:}1. \quad \blacktriangleleft$$

If the odds in favor of an event are, say, 3 to 5, then the probability of the event is $3/8$, while the probability of the complement of the event is $5/8$. (Odds of 3 to 5 indicate 3 outcomes in favor of the event out of a total of 8 possible outcomes.) This example suggests the following generalization.

If the odds favoring event E are m to n, then

$$P(E) = \frac{m}{m + n} \qquad \text{and} \qquad P(E') = \frac{n}{m + n}.$$

EXAMPLE 8

The odds that a particular bid will be the low bid are 4 to 5.

(a) Find the probability that the bid will be the low bid.

Odds of 4 to 5 show 4 favorable chances out of $4 + 5 = 9$ chances altogether:

$$P(\text{bid will be low bid}) = \frac{4}{4 + 5} = \frac{4}{9}.$$

(b) Find the odds against that bid being the low bid.

There is a $5/9$ chance that the bid will not be the low bid, so the odds against a low bid are

$$\frac{P(\text{bid will not be low})}{P(\text{bid will be low})} = \frac{5/9}{4/9} = \frac{5}{4},$$

or $5{:}4$. $\quad \blacktriangleleft$

EXAMPLE 9

If the odds in favor of a particular horse's winning a race are 5 to 7, what is the probability that the horse will win the race?

The odds indicate chances of 5 out of 12 ($5 + 7 = 12$) that the horse will win, so

$$P(\text{winning}) = \frac{5}{12}.$$

Race tracks generally give odds *against* a horse winning. In this case, the track would give the odds as 7 to 5. $\quad \blacktriangleleft$

Empirical Probability In many realistic problems, it is not possible to establish exact probabilities for events. Useful approximations, however, often can be found by using past experience as a guide to the future. The next example shows one approach to such **empirical probabilities.**

EXAMPLE 10

The manager of a store has decided to make a study of the amounts of money spent by people coming into the store. To begin, he chooses a day that seems fairly typical and gathers the following data. (Purchases have been rounded to the nearest dollar, with sales tax ignored.)

Amount Spent	Number of Customers
$0	158
$1–$5	94
$6–$9	203
$10–$19	126
$20–$49	47
$50–$99	38
$100 and over	53

First, the manager might add the numbers of customers to find that 719 people came into the store that day. Of these 719 people, $126/719 \approx .175$ made a purchase of at least $10 but no more than $19. Also, $53/719 \approx .074$ of the customers spent $100 or more. For this day, the probability that a customer entering the store will spend from $10 to $19 is .175, and the probability that the customer will spend $100 or more is .074. Probabilities for the other purchase amounts can be assigned in the same way, giving the results shown in the following table.

Amount Spent	Probability
$0	.220
$1–$5	.131
$6–$9	.282
$10–$19	.175
$20–$49	.065
$50–$99	.053
$100 and over	.074
	Total: 1.000

The categories in the table are mutually exclusive simple events. Thus, by the union rule,

$$P(\text{spending} < \$10) = .220 + .131 + .282 = .633.$$

From the table, .282 of the customers spend from \$6 to \$9, inclusive. Since this price range attracts more than a quarter of the customers, perhaps the store's advertising should emphasize items in, or near, this price range.

The manager should use this table of probabilities to help in predicting the results on other days only if the manager is reasonably sure that the day when the measurements were made is fairly typical of the other days the store is open. For example, on the last few days before Christmas, the probabilities might be quite different. ◀

A table of probabilities, as in Example 10, sets up a **probability distribution;** that is, for each possible outcome of an experiment, a number, called the probability of that outcome, is assigned. This assignment may be done in any reasonable way (on an empirical basis, as in Example 11, or by theoretical reasoning, as in Section 5.3), provided that it satisfies the following conditions.

PROPERTIES OF PROBABILITY

Let S be a sample space consisting of n distinct outcomes, s_1, s_2, \ldots, s_n. An acceptable probability assignment consists of assigning to each outcome s_i a number p_i (the probability of s_i) according to these rules.

1. The probability of each outcome is a number between 0 and 1.

$$0 \le p_1 \le 1, \quad 0 \le p_2 \le 1, \ldots, \quad 0 \le p_n \le 1$$

2. The sum of the probabilities of all possible outcomes is 1.

$$p_1 + p_2 + p_3 + \cdots + p_n = 1$$

Probability distributions are discussed further in the next two chapters.

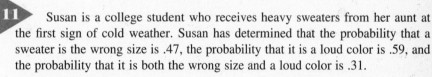

EXAMPLE 11 Susan is a college student who receives heavy sweaters from her aunt at the first sign of cold weather. Susan has determined that the probability that a sweater is the wrong size is .47, the probability that it is a loud color is .59, and the probability that it is both the wrong size and a loud color is .31.

(a) Find the probability that the sweater is the correct size and not a loud color.

Let W represent the event "wrong size," and L represent "loud color." Place the given information on a Venn diagram, starting with .31 in the intersection of the regions for W and L (see Figure 22). As stated earlier, event W has probability .47. Since .31 has already been placed inside the intersection of W and L,

$$.47 - .31 = .16$$

goes inside region W, but outside the intersection of W and L. In the same way,

$$.59 - .31 = .28$$

goes inside the region for L, and outside the overlap.

Using regions W and L, the event we want is $W' \cap L'$. From the Venn diagram in Figure 22, the labeled regions have a total probability of

$$.16 + .31 + .28 = .75.$$

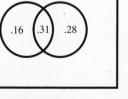

FIGURE 22

Since the entire region of the Venn diagram must have probability 1, the region outside W and L, or $W' \cap L'$, has probability

$$1 - .75 = .25.$$

The probability is .25 that the sweater is the correct size and not a loud color.

(b) Find the probability that the sweater is the correct size or is not loud.

The corresponding region, $W' \cup L'$, has probability

$$.25 + .16 + .28 = .69. \quad \blacktriangleleft$$

Another example of empirical probability occurs in weather forecasting. In one method, the forecaster compares atmospheric conditions with similar conditions in the past, and predicts the weather according to what happened previously under such conditions. For example, if certain kinds of atmospheric conditions have been followed by rain 80 times out of 100, the forecaster will announce an 80% chance of rain under those conditions.

One difficulty with empirical probability is that different people may assign different probabilities to the same event. Nevertheless, empirical probabilities can be used in many situations where the objective approach to probability cannot be used. In the next chapter the use of empirical probability is discussed in more detail.

5.4 Exercises

1. Define mutually exclusive events in your own words.

Decide whether the events in Exercises 2–7 are mutually exclusive.

2. Owning a car and owning a truck

3. Wearing glasses and wearing sandals

4. Being married and being over 30 years old

5. Being a teenager and being over 30 years old

6. Rolling a die once and getting a 4 and an odd number

7. Being a male and being a postal worker

Two dice are rolled. Find the probabilities of rolling the given sums.

8. (a) 2 **(b)** 4 **(c)** 5 **(d)** 6

9. (a) 8 **(b)** 9 **(c)** 10 **(d)** 13

10. (a) 9 or more **(b)** Less than 7 **(c)** Between 5 and 8 (exclusive)

11. (a) Not more than 5 **(b)** Not less than 8 **(c)** Between 3 and 7 (exclusive)

One card is drawn from an ordinary deck of 52 cards. Find the probabilities of drawing the following cards.

12. (a) A 9 or 10

(b) A red card or a 3

(c) A 9 or a black 10

(d) A heart or a black card

13. (a) Less than a 4 (count aces as ones)

(b) A diamond or a 7

(c) A black card or an ace

(d) A heart or a jack

Ms. Elliott invites 10 relatives to a party: her mother, 2 aunts, 3 uncles, 2 brothers, 1 male cousin, and 1 female cousin. If the chances of any 1 guest arriving first are equally likely, find the probabilities that the first guest to arrive is as follows.

14. (a) A brother or an uncle

 (b) A brother or a cousin

 (c) A brother or her mother

15. (a) An uncle or a cousin

 (b) A male or a cousin

 (c) A female or a cousin

The numbers 1, 2, 3, 4, and 5 are written on slips of paper, and 2 slips are drawn at random without replacement. Find each of the probabilities in Exercises 16 and 17.

16. (a) The sum of the numbers is 9.

 (b) The sum of the numbers is 5 or less.

 (c) The first number is 2 or the sum is 6.

17. (a) Both numbers are even.

 (b) One of the numbers is even or greater than 3.

 (c) The sum is 5 or the second number is 2.

Use Venn diagrams to work Exercises 18 and 19.

18. Suppose $P(E) = .26$, $P(F) = .41$, and $P(E \cap F) = .17$. Find each of the following.

 (a) $P(E \cup F)$ (b) $P(E' \cap F)$ (c) $P(E \cap F')$ (d) $P(E' \cup F')$

19. Let $P(Z) = .42$, $P(Y) = .38$, and $P(Z \cup Y) = .61$. Find each of the following probabilities.

 (a) $P(Z' \cap Y')$ (b) $P(Z' \cup Y')$ (c) $P(Z' \cup Y)$ (d) $P(Z \cap Y')$

20. Define what is meant by odds.

A single fair die is rolled. Find the odds in favor of getting the following results.

21. 5 **22.** 3, 4, or 5 **23.** 1, 2, 3, or 4 **24.** Some number less than 2

25. A marble is drawn from a box containing 3 yellow, 4 white, and 8 blue marbles. Find the odds in favor of drawing the following.

 (a) A yellow marble (b) A blue marble (c) A white marble

26. Find the odds of *not* drawing a white marble in Exercise 25.

27. The probability that a company will make a profit this year is .74. Find the odds against the company making a profit.

28. If the odds that it will rain are 4 to 7, what is the probability of rain?

29. If the odds that a given candidate will win an election are 3 to 2, what is the probability that the candidate will lose?

30. On page 134 of Roger Staubach's autobiography, *First Down, Lifetime to Go,* Staubach makes the following statement regarding his experience in Vietnam:

 ''Odds against a direct hit are very low but when your life is in danger, you don't worry too much about the odds.''

Is this wording consistent with our definition of odds, for and against? How could it have been said so as to be technically correct?

Which of the following are examples of empirical probability?

31. The probability of heads on 5 consecutive tosses of a coin

32. The probability that a freshman entering college will graduate with a degree

33. The probability that a person is allergic to penicillin

34. The probability of drawing an ace from a standard deck of 52 cards

35. The probability that a person will get lung cancer from smoking cigarettes

36. A weather forecast that predicts a 70% chance of rain tomorrow

37. A gambler's claim that a roll of a fair die, $P(\text{even}) = 1/2$

38. A surgeon's prediction that a patient has a 90% chance of a full recovery

39. What is a probability distribution?

The table below gives a certain golfer's probabilities of scoring in various ranges on a par-70 course.

Range	Probability
Below 60	.01
60–64	.08
65–69	.15
70–74	.28
75–79	.22
80–84	.08
85–89	.06
90–94	.04
95–99	.02
100 or more	.06

In a given round, find the probability that the golfer's score will be as follows.

40. (a) 90 or higher

 (b) In the 90s

 (c) Below par of 70

41. (a) In the 70s

 (b) Not in the 60s

 (c) Not in the 60s or 70s

The table below shows the probability that a customer of a department store will make a purchase in the indicated range.

Amount Spent	Probability
Below $10	.07
$10–$24.99	.18
$25–$49.99	.21
$50–$74.99	.16
$75–$99.99	.11
$100–$199.99	.09
$200–$349.99	.07
$350–$499.99	.08
$500 or more	.03

Find the probabilities that a customer makes a purchase in the following ranges.

42. (a) Less than $25

 (b) More than $24.99

 (c) $50 to $199.99

43. (a) Less than $350

 (b) $75 or more

 (c) $200 or more

An experiment is conducted for which the sample space is $S = \{s_1, s_2, s_3, s_4, s_5\}$. Which of the probability assignments in Exercises 44–49 is possible for this experiment? If an assignment is not possible, tell why.

44.

Outcomes	s_1	s_2	s_3	s_4	s_5
Probabilities	.09	.32	.21	.25	.13

45.

Outcomes	s_1	s_2	s_3	s_4	s_5
Probabilities	.92	.03	0	.02	.03

46.

Outcomes	s_1	s_2	s_3	s_4	s_5
Probabilities	$\frac{1}{3}$	$\frac{1}{4}$	$\frac{1}{6}$	$\frac{1}{8}$	$\frac{1}{10}$

47.

Outcomes	s_1	s_2	s_3	s_4	s_5
Probabilities	$\frac{1}{5}$	$\frac{1}{3}$	$\frac{1}{4}$	$\frac{1}{5}$	$\frac{1}{10}$

48.

Outcomes	s_1	s_2	s_3	s_4	s_5
Probabilities	.64	$-.08$.30	.12	.02

49.

Outcomes	s_1	s_2	s_3	s_4	s_5
Probabilities	.05	.35	.5	.2	$-.3$

Applications

Business and Economics

50. Defective Merchandise Suppose that 8% of a certain batch of calculators have a defective case, and that 11% have defective batteries. Also, 3% have both a defective case and defective batteries. A calculator is selected from the batch at random. Find the probability that the calculator has a good case and good batteries.

51. Credit Charges The table shows the probabilities of a person accumulating specific amounts of credit card charges over a 12-month period.

Charges	Probability
Under $100	.31
$100–$499	.18
$500–$999	.18
$1000–$1999	.13
$2000–$2999	.08
$3000–$4999	.05
$5000–$9999	.06
$10,000 or more	.01

Find the probabilities that a person's total charges during the period are the following.

(a) $500 or more **(b)** Less than $1000

(c) $500 to $2999 **(d)** $3000 or more

Life Sciences

52. Body Types A study on body types gave the following results: 45% were short; 25% were short and overweight; and 24% were tall and not overweight. Find the probabilities that a person is the following.

(a) Overweight

(b) Short, but not overweight

(c) Tall and overweight

53. Color Blindness Color blindness is an inherited characteristic that is more common in males than in females. If M represents male and C represents red-green color blindness, we use the relative frequencies of the incidences of males and red-green color blindness as probabilities to get $P(C) = .049$, $P(M \cap C) = .042$, $P(M \cup C) = .534$. Find the following probabilities.

(a) $P(C')$ **(b)** $P(M)$ **(c)** $P(M')$

(d) $P(M' \cap C')$ **(e)** $P(C \cap M')$ **(f)** $P(C \cup M')$

54. Genetics Gregor Mendel, an Austrian monk, was the first to use probability in the study of genetics. In an effort to understand the mechanism of character transmittal from one generation to the next in plants, he counted the number of occurrences of various characteristics. Mendel found that the flower color in certain pea plants obeyed this scheme:

Pure red crossed with pure white produces red.

From its parents, the red offspring received genes for both red (R) and white (W), but in this case red is *dominant* and white *recessive*, so the offspring exhibits the color red. However, the offspring still carries both genes, and when two such offspring are crossed, several things can happen in the third generation. The table below, which is called a *Punnet square*, shows the equally likely outcomes.

		Second Parent	
		R	W
First Parent	R	RR	RW
	W	WR	WW

Use the fact that red is dominant over white to find each of the following.

(a) $P(\text{red})$ **(b)** $P(\text{white})$

55. Genetics Mendel found no dominance in snapdragons, with one red gene and one white gene producing pink-flowered offspring. These second-generation pinks, however, still carry one red and one white gene, and when they are crossed, the next generation still yields the Punnet square shown above. Find each of the following probabilities.

(a) $P(\text{red})$ **(b)** $P(\text{pink})$ **(c)** $P(\text{white})$

(Mendel verified these probability ratios experimentally and did the same for many characteristics other than flower color. His work, published in 1866, was not recognized until 1890.)

56. Genetics In most animals and plants, it is very unusual for the number of main parts of the organism (such as arms, legs, toes, or flower petals) to vary from generation to generation. Some species, however, have *meristic variability*, in which the number of certain body parts varies from generation to generation. One researcher studied the front feet of certain guinea pigs and produced the following probabilities.*

$$P(\text{only four toes, all perfect}) = .77$$

$$P(\text{one imperfect toe and four good ones}) = .13$$

$$P(\text{exactly five good toes}) = .10$$

Find the probability of each of the following events.

(a) No more than four good toes

(b) Five toes, whether perfect or not

Social Sciences

57. Earnings The following data were gathered for 130 adult U.S. workers: 55 were women; 3 women earned more than $40,000; and 62 men earned less than $40,000. Find the probabilities that an individual is

(a) a woman earning less than $40,000;

(b) a man earning more than $40,000;

(c) a man or is earning more than $40,000;

(d) a woman or is earning less than $40,000.

58. Expenditures for Music A survey of 100 people about their music expenditures gave the following information: 38 bought rock music; 20 were teenagers who bought rock music; and 26 were teenagers. Find the probabilities that a person is

(a) a teenager who buys nonrock music;

(b) someone who buys rock music or is a teenager;

(c) not a teenager;

(d) not a teenager, but a buyer of rock music.

59. Refugees In a refugee camp in southern Mexico, it was found that 90% of the refugees came to escape political oppression, 80% came to escape abject poverty, and 70% came to escape both. What is the probability that a refugee in the camp was not poor nor seeking political asylum?

60. Community Activities At the first meeting of a committee to plan a local Lunar New Year celebration, the persons attending are 3 Chinese men, 4 Chinese women, 3 Vietnamese women, 2 Vietnamese men, 4 Korean women, and 2 Korean men. A chairperson is selected at random. Find the probabilities that the chairperson is the following.

(a) Chinese **(b)** Korean or a woman

(c) a man or Vietnamese **(d)** Chinese or Vietnamese

(e) Korean and a woman

📠 Computer/Graphing Calculator

One way to solve a probability problem is to repeat the experiment (or a simulation of the experiment) many times, keeping track of the results. Then the probability can be approximated using the basic definition of the probability of an event E: $P(E) = m/n$, where m favorable outcomes occur in n trials of an experiment. This is called the Monte Carlo method of finding probabilities.

61. Suppose a coin is tossed 5 times. Use the Monte Carlo method to approximate the following probabilities. Then calculate the theoretical probabilities using the methods of the text and compare the results.

(a) $P(4 \text{ heads})$

(b) $P(2 \text{ heads}, 1 \text{ tail}, 2 \text{ heads})$ (in the order given)

62. Use the Monte Carlo method to approximate the following probabilities if 4 cards are drawn from 52.

(a) $P(\text{any 2 cards and then 2 kings})$ **(b)** $P(2 \text{ kings})$

Use the Monte Carlo method to approximate the following probabilities.

63. A jeweler received 8 identical watches, boxed individually, with the serial number of the watch on each box. An assistant, who does not know that the boxes are marked, is told to polish the watches and then put them back in the boxes. She puts them in the boxes at random. What is the probability that she gets at least 1 watch in the right box?

64. A checkroom attendant has 10 hats but has lost the numbers identifying them. If he gives them back randomly, what is the probability that at least 1 hat is given back correctly?

*From "An Analysis of Variability in Guinea Pigs" by J. R. Wright, in *Genetics* 19, pp. 506–536. Reprinted by permission.

5.5 CONDITIONAL PROBABILITY; INDEPENDENT EVENTS

What is the probability that a broker who uses research picks stocks that go up?

The training manager for a large stockbrokerage firm has noticed that some of the firm's brokers use the firm's research advice, while other brokers tend to follow their own feelings of which stocks will go up. To see whether the research department performs better than the feelings of the brokers, the manager conducted a survey of 100 brokers, with results as shown in the following table.

	Picked Stocks That Went Up	Didn't Pick Stocks That Went Up	Totals
Used research	30	15	45
Didn't use research	30	25	55
Totals	60	40	100

Letting A represent the event "picked stocks that went up," and letting B represent the event "used research," we can find the following probabilities.

$$P(A) = \frac{60}{100} = .6 \qquad P(A') = \frac{40}{100} = .4$$

$$P(B) = \frac{45}{100} = .45 \qquad P(B') = \frac{55}{100} = .55$$

To answer the question asked at the beginning of this section, suppose we want to find the probability that a broker using research will pick stocks that go up. From the table above, of the 45 brokers who use research, 30 picked stocks that went up, with

$$P(\text{broker who uses research picks stocks that go up}) = \frac{30}{45} = .667.$$

This is a different number than the probability that a broker picks stocks that go up, .6, since we have additional information (the broker uses research) that has *reduced the sample space*. In other words, we found the probability that a broker picks stocks that go up, A, given the additional information that the broker uses research, B. This is called the *conditional probability* of event A, given that event B has occurred, written $P(A|B)$. ($P(A|B)$ may also be read as "the probability of A given B.") In the example above,

$$P(A|B) = \frac{30}{45},$$

which can be written as

$$P(A|B) = \frac{30/100}{45/100} = \frac{P(A \cap B)}{P(B)},$$

where $P(A \cap B)$ represents, as usual, the probability that both A and B will occur.

To generalize this result, assume that E and F are two events for a particular experiment. Assume that the sample space S for this experiment has n possible equally likely outcomes. Suppose event F has m elements, and $E \cap F$ has p elements ($p \leq m$). Using the fundamental principle of probability,

$$P(F) = \frac{m}{n} \quad \text{and} \quad P(E \cap F) = \frac{p}{n}.$$

We now want $P(E|F)$, the probability that E occurs given that F has occurred. Since we assume F has occurred, reduce the sample space to F: look only at the m elements inside F. See Figure 23. Of these m elements, there are p elements where E also occurs, since $E \cap F$ has p elements. This makes

$$P(E|F) = \frac{p}{m}.$$

Divide numerator and denominator by n to get

$$P(E|F) = \frac{\dfrac{p}{n}}{\dfrac{m}{n}} = \frac{P(E \cap F)}{P(F)}.$$

This last result motivates the definition of conditional probability.

Event F has a total of m elements.

FIGURE 23

DEFINITION OF CONDITIONAL PROBABILITY

The **conditional probability** of event E given event F, written $P(E|F)$, is

$$P(E|F) = \frac{P(E \cap F)}{P(F)}, \quad \text{where } P(F) \neq 0.$$

This definition tells us that, for equally likely outcomes, conditional probability is found by *reducing the sample space to event F*, and then finding the number of outcomes in F that are also in event E. Thus,

$$P(E|F) = \frac{n(E \cap F)}{n(F)}.$$

EXAMPLE 1 Use the information given in the chart at the beginning of this section to find the following probabilities.

(a) $P(B|A)$

This represents the probability that the broker used research, given that the broker picked stocks that went up. Reduce the sample space to A. Then find $n(A \cap B)$ and $n(A)$.

$$P(B|A) = \frac{P(B \cap A)}{P(A)} = \frac{n(A \cap B)}{n(A)} = \frac{30}{60} = \frac{1}{2}$$

If a broker picked stocks that went up, then the probability is $1/2$ that the broker used research.

(b) $P(A'|B)$

In words, this is the probability that a broker picks stocks that do not go up, even though he used research.

$$P(A'|B) = \frac{n(A' \cap B)}{n(B)} = \frac{15}{45} = \frac{1}{3}$$

(c) $P(B'|A')$

Here, we want the probability that a broker who picked stocks that did not go up did not use research.

$$P(B'|A') = \frac{n(B' \cap A')}{n(A')} = \frac{25}{40} = \frac{5}{8} \;\blacktriangleleft$$

Venn diagrams are useful for illustrating problems in conditional probability. A Venn diagram for Example 1, in which the probabilities are used to indicate the number in the set defined by each region, is shown in Figure 24. In the diagram, $P(B|A)$ is found by reducing the sample space to just set A. Then $P(B|A)$ is the ratio of the number in that part of set B which is also in A to the number in set A, or $.3/.6 = .5$.

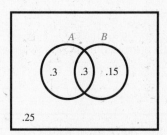

$P(A) = .3 + .3 = .6$

FIGURE 24

EXAMPLE

2 Given $P(E) = .4$, $P(F) = .5$, and $P(E \cup F) = .7$, find $P(E|F)$.
Find $P(E \cap F)$ first. By the union rule,

$$P(E \cup F) = P(E) + P(F) - P(E \cap F)$$
$$.7 = .4 + .5 - P(E \cap F)$$
$$P(E \cap F) = .2.$$

$P(E|F)$ is the ratio of the probability of that part of E which is in F to the probability of F, or

$$P(E|F) = \frac{P(E \cap F)}{P(F)} = \frac{.2}{.5} = \frac{2}{5}. \;\blacktriangleleft$$

The Venn diagram in Figure 25 illustrates Example 2.

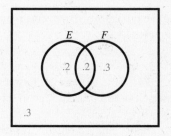

FIGURE 25

EXAMPLE
3 Two fair coins were tossed, and it is known that at least one was a head. Find the probability that both were heads.

The sample space has four equally likely outcomes, $S = \{hh, ht, th, tt\}$. Define two events:

$$E_1 = \text{at least 1 head} = \{hh, ht, th\},$$
and $$E_2 = 2 \text{ heads} = \{hh\}.$$

Since there are four equally likely outcomes, $P(E_1) = 3/4$. Also, $P(E_1 \cap E_2) = 1/4$. We want the probability that both were heads, given that at least one was a head; that is, we want to find $P(E_2|E_1)$. Because of the condition that at least one coin was a head, the reduced sample space is

$$\{hh, ht, th\}.$$

Since only one outcome in this reduced sample space is 2 heads,

$$P(E_2|E_1) = \frac{1}{3}.$$

Alternatively, use the definition given above.

$$P(E_2|E_1) = \frac{P(E_2 \cap E_1)}{P(E_1)} = \frac{1/4}{3/4} = \frac{1}{3} \blacktriangleleft$$

Product Rule If $P(E) \neq 0$ and $P(F) \neq 0$, then the definition of conditional probability shows that

$$P(E|F) = \frac{P(E \cap F)}{P(F)} \quad \text{and} \quad P(F|E) = \frac{P(F \cap E)}{P(E)}.$$

Using the fact that $P(E \cap F) = P(F \cap E)$, and solving each of these equations for $P(E \cap F)$, we obtain the following rule.

PRODUCT RULE OF PROBABILITY

If E and F are events, then $P(E \cap F)$ may be found by either of these formulas.

$$P(E \cap F) = P(F) \cdot P(E|F) \quad \text{or} \quad P(E \cap F) = P(E) \cdot P(F|E).$$

The product rule gives a method for finding the probability that events E and F both occur, as illustrated by the next few examples.

EXAMPLE 4 In a class with 2/5 women and 3/5 men, 25% of the women are business majors. Find the probability that a student chosen from the class at random is a female business major.

Let B and W represent the events "business major" and "woman," respectively. We want to find $P(B \cap W)$. By the product rule,

$$P(B \cap W) = P(W) \cdot P(B|W).$$

Using the given information, $P(W) = 2/5 = .4$ and $P(B|W) = .25$. Thus,

$$P(B \cap W) = .4(.25) = .10. \quad \blacktriangleleft$$

The next examples show how a tree diagram is used with the product rule to find the probability of a sequence of events.

EXAMPLE 5

A company needs to hire a new director of advertising. It has decided to try to hire either person A or person B, who are assistant advertising directors for its major competitor. To decide between A and B, the company does research on the campaigns managed by either A or B (no campaign is managed by both), and finds that *A* is in charge of twice as many advertising campaigns as B. Also, A's campaigns have satisfactory results 3 out of 4 times, while B's campaigns have satisfactory results only 2 out of 5 times. Suppose one of the competitor's advertising campaigns (managed by A or B) is selected randomly.

We can represent this situation schematically as follows. Let *A* denote the event "person A does the job" and *B* the event "person B does the job." Let *S* be the event "satisfactory results" and *U* the event "unsatisfactory results." Then the given information can be summarized in the tree diagram in Figure 26. Since A does twice as many jobs as B, $P(A) = 2/3$ and $P(B) = 1/3$, as noted on the first-stage branches of the tree. When A does a job, the probability of satisfactory results is 3/4, and of unsatisfactory results, 1/4, as noted on the second-stage branches. Similarly, the probabilities when *B* does the job are noted on the remaining second-stage branches. The composite branches labeled 1–4 represent the four mutually exclusive possibilities for the running and outcome of the selected campaign.

Executive	Campaign	Branch	Probability
	$\frac{3}{4}$ S	1	$\frac{2}{3} \cdot \frac{3}{4} = \frac{1}{2}$
$\frac{2}{3}$ A	$\frac{1}{4}$ U	2	$\frac{2}{3} \cdot \frac{1}{4} = \frac{1}{6}$
	$\frac{2}{5}$ S	3	$\frac{1}{3} \cdot \frac{2}{5} = \frac{2}{15}$
$\frac{1}{3}$ B	$\frac{3}{5}$ U	4	$\frac{1}{3} \cdot \frac{3}{5} = \frac{1}{5}$
			Total: 1

FIGURE 26

(a) Find the probability that A is in charge of the selected campaign and that it produces satisfactory results.

We are asked to find $P(A \cap S)$. We know that when A does the job, the probability of success is 3/4, that is, $P(S|A) = 3/4$. Hence, by the product rule,

$$P(A \cap S) = P(A) \cdot P(S|A) = \frac{2}{3} \cdot \frac{3}{4} = \frac{1}{2}.$$

The event $A \cap S$ is represented by branch 1 of the tree, and as we have just seen, its probability is the product of the probabilities of the pieces that make up that branch.

(b) Find the probability that B runs the campaign and that it produces satisfactory results.

We must find $P(B \cap S)$. The event is represented by branch 3 of the tree and, as before, its probability is the product of the probabilities of the pieces of that branch:

$$P(B \cap S) = P(B) \cdot P(S|B) = \frac{1}{3} \cdot \frac{2}{5} = \frac{2}{15}.$$

(c) What is the probability that the selected campaign is satisfactory?

The event S is the union of the mutually exclusive events $A \cap S$ and $B \cap S$, which are represented by branches 1 and 3 of the tree diagram. By the union rule,

$$P(S) = P(A \cap S) + P(B \cap S) = \frac{1}{2} + \frac{2}{15} = \frac{19}{30}.$$

Thus, the probability of an event that appears on several branches is the sum of the probabilities of each of these branches.

(d) What is the probability that the selected campaign is unsatisfactory?

$P(U)$ can be read from branches 2 and 4 of the tree.

$$P(U) = \frac{1}{6} + \frac{1}{5} = \frac{11}{30}$$

Alternatively, since U is the complement of S,

$$P(U) = 1 - P(S) = 1 - \frac{19}{30} = \frac{11}{30}.$$

(e) Find the probability that either A runs the selected campaign or the results are unsatisfactory (or possibly both).

This is $P(A \cup S)$. Event A combines branches 1 and 2, while event S combines branches 1 and 3, so use branches 1, 2, and 3.

$$P(A \cup S) = \frac{1}{2} + \frac{1}{6} + \frac{2}{15} = \frac{4}{5} \quad \blacktriangleleft$$

EXAMPLE 6 ▶ From a box containing 3 white, 2 green, and 1 red marble, 2 marbles are drawn one at a time without replacing the first before the second is drawn. Find the probability that 1 white and 1 green marble are drawn.

A tree diagram showing the various possible outcomes is given in Figure 27. In this diagram, W represents the event "drawing a white marble" and G represents "drawing a green marble." On the first draw, $P(W \text{ first}) = 3/6 = 1/2$ because 3 of the 6 marbles in the box are white. On the second draw, $P(G \text{ second}|W \text{ first}) = 2/5$. One white marble has been removed, leaving 5, of which 2 are green.

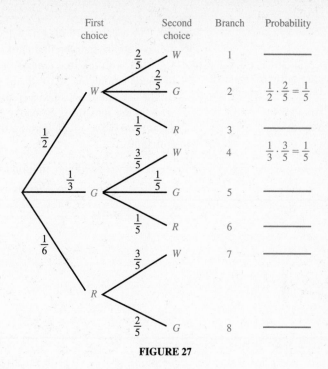

FIGURE 27

Now we want to find the probability of drawing exactly one white marble and one green marble. This event can occur in two ways: drawing a white marble first and then a green one (branch 2 of the tree diagram), or drawing a green marble first and then a white one (branch 4). For branch 2,

$$P(W \text{ first}) \cdot P(G \text{ second}|W \text{ first}) = \frac{1}{2} \cdot \frac{2}{5} = \frac{1}{5}.$$

For branch 4, where the green marble is drawn first,

$$P(G \text{ first}) \cdot P(W \text{ second}|G \text{ first}) = \frac{1}{3} \cdot \frac{3}{5} = \frac{1}{5}.$$

Since the two events are mutually exclusive, the final probability is the sum of these two probabilities, or

$$P(1\ W,\ 1\ G) = P(W \text{ first}) \cdot P(G \text{ second}|W \text{ first})$$

$$+ P(G \text{ first}) \cdot P(W \text{ second}|G \text{ first}) = \frac{2}{5}. \ \blacktriangleleft$$

The product rule is often used with *stochastic processes*, where the outcome of an experiment depends on the outcomes of previous experiments. For example, the outcome of a draw of a card from a deck depends on any cards previously drawn. (Stochastic processes are studied in more detail in a later chapter.)

You may wish to refer to the picture of a deck of cards shown in Figure 18 (Section 5.3) and the description accompanying it.

EXAMPLE **7** Two cards are drawn, without replacement, from an ordinary deck. Find the probability that the first card is a heart and the second card is red.

Start with the tree diagram in Figure 28. On the first draw, since there are 13 hearts among the 52 cards, the probability of drawing a heart is $13/52 = 1/4$. On the second draw, since a (red) heart has been drawn already, there are 25 red cards in the remaining 51 cards. Thus, the probability of drawing a red card on the second draw, given that the first is a heart, is $25/51$. By the product rule of probability,

$$P(\text{heart first and red second})$$
$$= P(\text{heart first}) \cdot P(\text{red second}|\text{heart first})$$
$$= \frac{1}{4} \cdot \frac{25}{51} = \frac{25}{204} \approx .123. \blacktriangleleft$$

FIGURE 28

EXAMPLE **8** Three cards are drawn, without replacement, from an ordinary deck. Find the probability that exactly 2 of the cards are red.

Here we need a tree diagram with 3 stages, as shown in Figure 29. The three branches indicated with arrows produce exactly 2 red cards from the three draws. Multiply the probabilities along each of these branches and then add.

$$P(\text{exactly 2 red cards}) = \frac{26}{52} \cdot \frac{25}{51} \cdot \frac{26}{50} + \frac{26}{52} \cdot \frac{26}{51} \cdot \frac{25}{50} + \frac{26}{52} \cdot \frac{26}{51} \cdot \frac{25}{50}$$

$$= \frac{50,700}{132,600} = \frac{13}{34} \approx .382 \blacktriangleleft$$

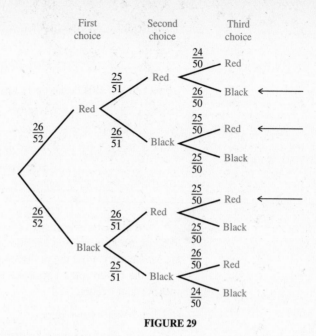

FIGURE 29

Independent Events Suppose a fair coin is tossed and shows heads. The probability of heads on the next toss is still 1/2; the fact that heads was obtained on a given toss has no effect on the outcome of the next toss. Coin tosses are *independent events*, since knowledge of the outcome of one toss does not help decide on the outcome of the next toss. Rolls of a fair die are independent events; the fact that a two came up on one roll does not increase our knowledge of the outcome of the next roll. On the other hand, the events "today is cloudy" and "today is rainy" are *dependent events;* if it is cloudy, then there is an increased chance of rain. Also, in the example at the beginning of this section, the events A (broker picked stocks that went up) and B (broker used research) are dependent events, since information about the use of research affects the probability of picking stocks that go up. That is, $P(A|B)$ is different from $P(A)$.

If events E and F are independent, then the knowledge that E has occurred gives no (probability) information about the occurrence or nonoccurrence of event F. That is, $P(F)$ is exactly the same as $P(F|E)$, or

$$P(F|E) = P(F).$$

This, in fact, is the formal definition of independent events.

INDEPENDENT EVENTS	E and F are **independent events** if		
	$$P(F	E) = P(F) \quad \text{or} \quad P(E	F) = P(E).$$

When E and F are independent events, then $P(F|E) = P(F)$ and the product rule becomes

$$P(E \cap F) = P(E) \cdot P(F|E) = P(E) \cdot P(F).$$

Conversely, if this equation holds, then it follows that $P(F) = P(F|E)$. Consequently, we have this useful fact:

PRODUCT RULE FOR INDEPENDENT EVENTS	E and F are independent events if and only if $$P(E \cap F) = P(E) \cdot P(F).$$

EXAMPLE 9

A calculator requires a keystroke assembly and a logic circuit. Assume that 99% of the keystroke assemblies are satisfactory and 97% of the logic circuits are satisfactory. Find the probability that a finished calculator will be satisfactory.

If the failure of a keystroke assembly and the failure of a logic circuit are independent events, then

P(satisfactory calculator)

$\qquad = P$(satisfactory keystroke assembly) $\cdot P$(satisfactory logic circuit)

$\qquad = (.99)(.97) \approx .96.$

(The probability of a defective calculator is $1 - .96 = .04$.) ◄

EXAMPLE 10

When black-coated mice are crossed with brown-coated mice, a pair of genes, one from each parent, determines the coat color of the offspring. Let b represent the gene for brown and B the gene for black. If a mouse carries either one B gene and one b gene (Bb or bB) or two B genes (BB), the coat will be black. If the mouse carries two b genes (bb), the coat will be brown. Find the probability that a mouse born to a brown-coated female and a black-coated male who is known to carry the Bb combination will be brown.

To be brown-coated, the offspring must receive one b gene from each parent. The brown-coated parent carries two b genes, so that the probability of getting one b gene from the mother is 1. The probability of getting one b gene from the black-coated father is $1/2$. Therefore, since these are independent events, the probability that a brown-coated offspring will be produced by these parents is $1 \cdot 1/2 = 1/2$. ◄

Caution It is common for students to confuse the ideas of *mutually exclusive* events and *independent* events. Events E and F are mutually exclusive if $E \cap F = \emptyset$. For example, if a family has exactly one child, the only possible outcomes are $B = \{\text{boy}\}$ and $G = \{\text{girl}\}$. These two events are mutually exclusive. The events are *not* independent, however, since $P(G|B) = 0$ (if a family with only one child has a boy, the probability it has a girl is then 0). Since $P(G|B) \neq P(G)$, the events are not independent.

Of all the families with exactly two children, the events $G_1 = $ {first child is a girl} and $G_2 = $ {second child is a girl} are independent, since $P(G_2|G_1)$ equals $P(G_2)$. However, G_1 and G_2 are not mutually exclusive, since $G_1 \cap G_2 = $ {both children are girls} $\neq \emptyset$.

To show that two events E and F are independent, show that $P(F|E) = P(F)$ or that $P(E|F) = P(E)$ or that $P(E \cap F) = P(E) \cdot P(F)$. Another way is to observe that knowledge of one outcome does not influence the probability of the other outcome, as we did for coin tosses.

EXAMPLE 11

On a typical January day in Manhattan the probability of snow is .10, the probability of a traffic jam is .80, and the probability of snow or a traffic jam (or both) is .82. Are the event "it snows" and the event "a traffic jam occurs" independent?

Let S represent the event "it snows" and T represent the event "a traffic jam occurs." We must determine whether

$$P(T|S) = P(T) \quad \text{or} \quad P(S|T) = P(S).$$

We know $P(S) = .10$, $P(T) = .8$, and $P(S \cup T) = .82$. We can use the union rule (or a Venn diagram) to find $P(S \cap T) = .08$, $P(T|S) = .8$, and $P(S|T) = .1$. Since

$$P(T|S) = P(T) = .8 \quad \text{and} \quad P(S|T) = P(S) = .1,$$

the events "it snows" and "a traffic jam occurs" are independent. ◀

Although we showed $P(T|S) = P(T)$ and $P(S|T) = P(S)$ in Example 11, only one of these results is needed to establish independence.

5.5 Exercises

If a single fair die is rolled, find the probabilities of the following results.

1. A 2, given that the number rolled was odd

2. A 4, given that the number rolled was even

3. An even number, given that the number rolled was 6

If two fair dice are rolled, find the probabilities of the following results.

4. A sum of 8, given that the sum is greater than 7

5. A sum of 6, given that the roll was a "double" (two identical numbers)

6. A double, given that the sum was 9

If two cards are drawn without replacement from an ordinary deck, find the probabilities of the following results.

7. The second is a heart, given that the first is a heart.

8. The second is black, given that the first is a spade.

9. The second is a face card, given that the first is a jack.

10. The second is an ace, given that the first is not an ace.

11. A jack and a 10 are drawn.

12. An ace and a 4 are drawn.

13. Two black cards are drawn.

14. Two hearts are drawn.

15. In your own words, explain how to find the conditional probability $P(E|F)$.

16. Your friend asks you to explain how the product rule for independent events differs from the product rule for dependent events. How would you respond?

17. Another friend asks you to explain how to tell whether two events are dependent or independent. How would you reply? (Use your own words.)

18. A student reasons that the probability in Example 3 of both coins being heads is just the probability that the other coin is a head, that is, 1/2. Explain why this reasoning is wrong.

Use a tree diagram in Exercises 19–22.

19. Suppose 20% of the population are 65 or over, 26% of those 65 or over have loans, and 53% of those under 65 have loans. Find the probabilities that a person fits into the following categories.

(a) 65 or over and has a loan (b) Has a loan

20. In a certain area, 15% of the population are joggers and 40% of the joggers are women. If 55% of those who do not jog are women, find the probabilities that an individual from that community fits the following descriptions.

(a) A woman jogger (b) Not a jogger

21. A survey has shown that 52% of the women in a certain community work outside the home. Of these women, 64% are married, while 86% of the women who do not work outside the home are married. Find the probabilities that a woman in that community can be categorized as follows.

(a) Married

(b) A single woman working outside the home

22. Two-thirds of the population are on a diet at least occasionally. Of this group, 4/5 drink diet soft drinks, while 1/2 of the rest of the population drink diet soft drinks. Find the probabilities that a person fits into the following categories.

(a) Drinks diet soft drinks

(b) Diets, but does not drink diet soft drinks

23. A smooth-talking young man has a 1/3 probability of talking a policeman out of giving him a speeding ticket. The probability that he is stopped for speeding during a given weekend is 1/2.
Find the probabilities of the events in parts (a) and (b).

(a) He will receive no speeding tickets on a given weekend.

(b) He will receive no speeding tickets on 3 consecutive weekends.

(c) We have assumed that what happens on the second or third weekend is the same as what happened on the first weekend. Is this realistic? Will driving habits remain the same after getting a ticket?

24. In one area, 4% of the population drive luxury cars. However, 17% of the CPAs drive luxury cars. Are the events ''person drives a luxury car'' and ''person is a CPA'' independent?

25. The probability that a key component of a space rocket will fail is .03.

(a) How many such components must be used as backups to ensure that the probability of at least one of the components' working is .999999?

(b) Is it reasonable to assume independence here?

26. A medical experiment showed that the probability that a new medicine is effective is .75, the probability that a patient will have a certain side effect is .4, and the probability that both events occur is .3. Decide whether these events are dependent or independent.

27. A teacher has found that the probability that a student studies for a test is .6, the probability that a student gets a good grade on a test is .7, and the probability that both occur is .52. Are these events independent?

Applications

Business and Economics

Banking *The Midtown Bank has found that most customers at the tellers' windows either cash a check or make a deposit. The chart below indicates the transactions for one teller for one day.*

	Cash Check	No Check	Totals
Make Deposit	50	20	70
No Deposit	30	10	40
Totals	80	30	110

Letting C represent ''cashing a check'' and D represent ''making a deposit,'' express each of the following probabilities in words and find its value.

28. $P(C|D)$ **29.** $P(D'|C)$ **30.** $P(C'|D')$ **31.** $P(C'|D)$ **32.** $P[(C \cap D)']$

33. Airline Flights According to a booklet put out by East-west Airlines, 98% of all scheduled Eastwest flights actually take place. (The other flights are canceled due to weather, equipment problems, and so on.) Assume that the event that a given flight takes place is independent of the event that another flight takes place.

(a) Elisabeta Guervara plans to visit her company's branch offices; her journey requires 3 separate flights on Eastwest Airlines. What is the probability that all of these flights will take place?

(b) Based on the reasons we gave for a flight to be canceled, how realistic is the assumption of independence that we made?

34. Backup Computers Corporations where a computer is essential to day-to-day operations, such as banks, often have a second backup computer in case the main computer fails. Suppose there is a .003 chance that the main computer will fail in a given time period, and a .005 chance that the backup computer will fail while the main computer is being repaired. Assume these failures represent independent events, and find the fraction of the time that the corporation can assume it will have computer service. How realistic is our assumption of independence?

Quality Control *A bicycle factory runs two assembly lines, A and B. If 95% of line A's products pass inspection, while only 90% of line B's products pass inspection, and 60% of the factory's bikes come off assembly line B (the rest off A), find the probabilities that one of the factory's bikes did not pass inspection and came off the following.*

35. Assembly line A

36. Assembly line B

Life Sciences

37. Genetics Both of a certain pea plant's parents had a gene for red and a gene for white flowers. (See Exercises 54 and 55 in Section 5.4.) If the offspring has red flowers, find the probability that it combined a gene for red and a gene for white (rather than 2 for red.)

Genetics *Assuming that boy and girl babies are equally likely, fill in the remaining probabilities on the tree diagram and use the following information to find the probability that a family with three children has all girls, given the following.*

38. The first is a girl.

39. The third is a girl.

40. The second is a girl.

41. At least 2 are girls.

42. At least 1 is a girl.

Color Blindness *The following table shows frequencies for red-green color blindness, where M represents "person is male" and C represents "person is color-blind." Use this table to find the following probabilities.*

	M	*M'*	Totals
C	.042	.007	.049
C'	.485	.466	.951
Totals	.527	.473	1.000

43. $P(M)$ **44.** $P(C)$

45. $P(M \cap C)$ **46.** $P(M \cup C)$

47. $P(M|C)$ **48.** $P(C|M)$

49. $P(M'|C)$

50. Are the events C and M, described above, dependent? What does this mean?

51. Color Blindness A scientist wishes to determine whether there is a relationship between color blindness (C) and deafness (D).

 (a) Given the probabilities listed in the table below, what should the findings be? (See Exercises 43–50.)

 (b) Explain what your answer tells us about color blindness and deafness.

	D	*D'*	Totals
C	.0004	.0796	.0800
C'	.0046	.9154	.9200
Totals	.0050	.9950	1.0000

52. Drug Screening In searching for a new drug with commercial possibilities, drug company researchers use the ratio

$$N_S : N_A : N_P : 1.$$

That is, if the company gives preliminary screening to N_S substances, it may find that N_A of them are worthy of further study, with N_P of these surviving into full-scale development. Finally, 1 of the substances will result in a marketable drug. Typical numbers used by Smith, Kline, and French Laboratories in planning research budgets might be 2000 : 30 : 8 : 1.* Use this ratio for (a)–(f).

 (a) Suppose a compound has been chosen for preliminary screening. Find the probability that the compound will survive and become a marketable drug.

 (b) Find the probability that the compound will not lead to a marketable drug.

 (c) Suppose the number of such compounds receiving preliminary screening is a. Set up the probability that none of them produces a marketable drug. (Assume independence throughout these exercises.)

 (d) Use your results from part (c) to find the probability that at least one of the drugs will prove marketable.

 (e) Suppose now that N scientists are employed in the preliminary screening, and that each scientist can screen c compounds per year. Find the probability that no marketable drugs will be discovered in a year.

 (f) Find the probability that at least one marketable drug will be discovered.

General Interest

53. Driver's License Test The Motor Vehicle Department has found that the probability of a person passing the test for a driver's license on the first try is .75. The probability that an individual who fails on the first test will pass on the second try is .80, and the probability that an individual who fails the first and second tests will pass the third time is .70. Find the probabilities that an individual will do the following.

 (a) Fail both the first and second test

 (b) Fail three times in a row

 (c) Require at least two tries

Pet Store Stock *A pet shop has 10 puppies, 6 of them males. There are 3 beagles (1 male), 1 cocker spaniel (male), and 6 poodles. Construct a table and find the probabilities that 1 of these puppies, chosen at random, is the following.*

54. A beagle

55. A beagle, given that it is a male

56. A male, given that it is a beagle

57. A cocker spaniel, given that it is a female

58. A poodle, given that it is a male

59. A female, given that it is a beagle

*Reprinted by permission of E. B. Pyle, III, B. Douglas, G. W. Ebright, W. J. Westlake, and A. B. Bender, "Scientific Manpower Allocation to New Drug Screening Programs," *Management Science*, Vol. 19, No. 12, August 1973. Copyright © 1973 by The Institute of Management Sciences.

5.6 BAYES' THEOREM

What is the probability that a particular defective item was produced by a new machine operator?

This question and others like it are answered using Bayes' theorem, discussed in this section. Suppose the probability that a person gets lung cancer, given that the person smokes a pack or more of cigarettes daily, is known. For a research project, it might be necessary to know the probability that a person smokes a pack or more of cigarettes daily, given that the person has lung cancer. More generally, if $P(E|F)$ is known for two events E and F, can $P(F|E)$ be found? The answer is yes, we can find $P(F|E)$ using the formula to be developed in this section. To develop this formula, we can use a tree diagram to find $P(F|E)$. Since $P(E|F)$ is known, the first outcome is either F or F'. Then for each of these outcomes, either E or E' occurs, as shown in Figure 30. The four cases have the probabilities shown on the right. By the definition of conditional probability,

$$P(F|E) = \frac{P(F \cap E)}{P(E)} = \frac{P(F) \cdot P(E|F)}{P(F) \cdot P(E|F) + P(F') \cdot P(E|F')}.$$

Notice that $P(E)$ is the sum of the first and third cases. This result is a special case of Bayes' theorem, which is generalized later in this section.

FIGURE 30

| **BAYES' THEOREM (SPECIAL CASE)** | $P(F|E) = \dfrac{P(F) \cdot P(E|F)}{P(F) \cdot P(E|F) + P(F') \cdot P(E|F')}$ |
|---|---|

EXAMPLE **1**
For a fixed length of time, the probability of a worker error on a certain production line is .1, the probability that an accident will occur when there is a worker error is .3, and the probability that an accident will occur when there is no worker error is .2. Find the probability of a worker error if there is an accident.

Let E represent the event of an accident, and let F represent the event of worker error. From the information above,

$$P(F) = .1, \qquad P(E|F) = .3, \qquad \text{and} \qquad P(E|F') = .2.$$

These probabilities are shown on the tree diagram in Figure 31.

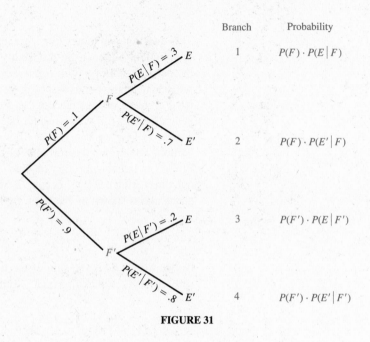

	Branch	Probability	
E	1	$P(F) \cdot P(E	F)$
E'	2	$P(F) \cdot P(E'	F)$
E	3	$P(F') \cdot P(E	F')$
E'	4	$P(F') \cdot P(E'	F')$

FIGURE 31

Find $P(F|E)$ using the tree diagram or Bayes' theorem.

$$P(F|E) = \frac{P(F) \cdot P(E|F)}{P(F) \cdot P(E|F) + P(F') \cdot P(E|F')}$$

$$= \frac{(.1)(.3)}{(.1)(.3) + (.9)(.2)} = \frac{1}{7} \approx .143 \quad \blacktriangleleft$$

The special case of Bayes' theorem can be generalized to more than two events with the tree diagram of Figure 32. This diagram shows the paths that can produce an event E. We assume that the events F_1, F_2, \ldots, F_n are mutually exclusive events (that is, disjoint events) whose union is the sample space, and that E is an event that has occurred. See Figure 33.

The probability $P(F_i|E)$, where $1 \le i \le n$, can be found by dividing the probability for the branch containing $P(E|F_i)$ by the sum of the probabilities of all the branches producing event E.

BAYES' THEOREM

$$P(F_i|E) = \frac{P(F_i) \cdot P(E|F_i)}{P(F_1) \cdot P(E|F_1) + P(F_2) \cdot P(E|F_2) + \cdots + P(F_n) \cdot P(E|F_n)}$$

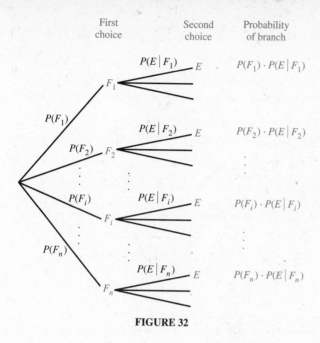

First choice Second choice Probability of branch

FIGURE 32

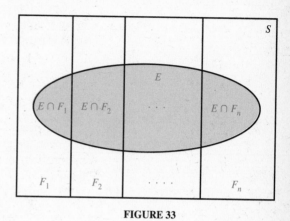

FIGURE 33

This result is known as **Bayes' theorem,** after the Reverend Thomas Bayes (1702–1761), whose paper on probability was published about 200 years ago.

The statement of Bayes' theorem can be daunting. Actually, it is easier to remember the formula by thinking of the tree diagram that produced it. Go through the following steps.

USING BAYES' THEOREM

1. Start a tree diagram with branches representing events F_1, F_2, \ldots, F_n. Label each branch with its corresponding probability.

2. From the end of each of these branches, draw a branch for event E. Label this branch with the probability of getting to it, $P(E|F_i)$.

3. You now have n different paths that result in event E. Next to each path, put its probability—the product of the probabilities that the first branch occurs, $P(F_i)$, and that the second branch occurs, $P(E|F_i)$; that is, the product $P(F_i) \cdot P(E|F_i)$.

4. $P(F_i|E)$ is found by dividing the probability of the branch for F_i by the sum of the probabilities of all the branches producing event E.

EXAMPLE 2

Based on past experience, a company knows that an experienced machine operator (one or more years of experience) will produce a defective item 1% of the time. Operators with some experience (up to one year) have a 2.5% defect rate, and new operators have a 6% defect rate. At any one time, the company has 60% experienced operators, 30% with some experience, and 10% new opera-

tors. Find the probability that a particular defective item was produced by a new operator.

Let E represent the event "item is defective," F_1 represent "item was made by an experienced operator," F_2 represent "item was made by an operator with some experience," and F_3 represent "item was made by a new operator." Then

$$P(F_1) = .60 \qquad P(E|F_1) = .01$$
$$P(F_2) = .30 \qquad P(E|F_2) = .025$$
$$P(F_3) = .10 \qquad P(E|F_3) = .06.$$

We need to find $P(F_3|E)$, the probability that an item was produced by a new operator, given that it is defective. First, draw a tree diagram using the given information, as in Figure 34. The steps leading to event E are shown in heavy type.

FIGURE 34

Find $P(F_3|E)$ with the bottom branch of the tree in Figure 34: divide the probability for this branch by the sum of the probabilities of all the branches leading to E, or

$$P(F_3|E) = \frac{.10(.06)}{.60(.01) + .30(.025) + .10(.06)} = \frac{.006}{.0195} = \frac{4}{13} \approx .308.$$

In a similar way,

$$P(F_2|E) = \frac{.30(.025)}{.60(.01) + .30(.025) + .10(.06)} = \frac{.0075}{.0195} = \frac{5}{13} \approx .385.$$

Finally, $P(F_1|E) = 4/13 \approx .308$. Check that $P(F_1|E) + P(F_2|E) + P(F_3|E) = 1$ (that is, the defective item was made by *someone*). ◀

EXAMPLE 3

A manufacturer buys items from 6 different suppliers. The fraction of the total number of items obtained from each supplier, along with the probability that an item purchased from that supplier is defective, are shown in the following chart.

Supplier	Fraction of Total Supplied	Probability of Defect
1	.05	.04
2	.12	.02
3	.16	.07
4	.23	.01
5	.35	.03
6	.09	.05

Find the probability that a defective item came from supplier 5.

Let F_1 be the event that an item came from supplier 1, with F_2, F_3, F_4, F_5, and F_6 defined in a similar manner. Let E be the event that an item is defective. We want to find $P(F_5|E)$. Use the probabilities in the table above to prepare a tree diagram, or work with the rows of the table to get

$$P(F_5|E) = \frac{(.35)(.03)}{(.05)(.04) + (.12)(.02) + (.16)(.07) + (.23)(.01) + (.35)(.03) + (.09)(.05)}$$

$$= \frac{.0105}{.0329} \approx .319.$$

There is about a 32% chance that a defective item came from supplier 5. ◀

Even though supplier 5 has only 3% defectives, his probability of being "guilty" is relatively high, about 32%, because of the large fraction supplied by 5.

5.6 Exercises

For two events M and N, $P(M) = .4$, $P(N|M) = .3$, and $P(N|M') = .4$. Find each of the following.

1. $P(M|N)$ **2.** $P(M'|N)$

For mutually exclusive events R_1, R_2, and R_3, we have $P(R_1) = .05$, $P(R_2) = .6$, and $P(R_3) = .35$. Also, $P(Q|R_1) = .40$, $P(Q|R_2) = .30$, and $P(Q|R_3) = .60$. Find each of the following.

3. $P(R_1|Q)$ **4.** $P(R_2|Q)$ **5.** $P(R_3|Q)$ **6.** $P(R_1'|Q)$

Suppose you have three jars with the following contents: 2 black balls and 1 white ball in the first, 1 black ball and 2 white balls in the second, and 1 black ball and 1 white ball in the third. One jar is to be selected, and then 1 ball is to be drawn from the selected jar. If the probabilities of selecting the first, second, or third jar are 1/2, 1/3, and 1/6, respectively, find the probabilities that if a white ball is drawn, it came from the following jars.

7. The second jar **8.** The third jar

Applications

Business and Economics

9. **Employment Test** A manufacturing firm finds that 70% of its new hires turn out to be good workers and 30% become poor workers. All current workers are given a reasoning test. Of the good workers, 80% pass it; 40% of the poor workers pass it. Assume that these figures will hold true in the future. If the company makes the test part of its hiring procedure and only hires people who meet the previous requirements and also pass the test, what percent of the new hires will turn out to be good workers?

Job Qualifications *Of all the people applying for a certain job, 70% are qualified and 30% are not. The personnel manager claims that she approves qualified people 85% of the time; she approves an unqualified person 20% of the time. Find each of the following probabilities.*

10. A person is qualified if he or she was approved by the manager.

11. A person is unqualified if he or she was approved by the manager.

Quality Control *A building contractor buys 70% of his cement from supplier A, and 30% from supplier B. A total of 90% of the bags from A arrive undamaged, while 95% of the bags from B arrive undamaged. Give the probabilities that a damaged bag is from each of the following sources.*

12. Supplier A 13. Supplier B

Credit *The probability that a customer of a local department store will be a "slow pay" is .02. The probability that a "slow pay" will make a large down payment when buying a refrigerator is .14. The probability that a person who is not a "slow pay" will make a large down payment when buying a refrigerator is .50. Suppose a customer makes a large down payment on a refrigerator, Find the probabilities that the customer is in the following categories.*

14. A "slow pay" 15. Not a "slow pay"

Appliance Reliability *Companies A, B, and C produce 15%, 40%, and 45%, respectively, of the major appliances sold in a certain area. In that area, 1% of the company A appliances, 1½% of the company B appliances, and 2% of the company C appliances need service within the first year. Suppose a defective appliance is chosen at random; find the probabilities that it was manufactured by the following companies.*

16. Company A 17. Company B

Television Advertising *On a given weekend in the fall, a tire company can buy television advertising time for a college football game, a baseball game, or a professional football game. If the company sponsors the college football game, there is a 70% chance of a high rating, a 50% chance if they sponsor a baseball game, and a 60% chance if they sponsor a professional football game. The probabilities of the company sponsoring these various games are .5, .2, and .3, respectively. Suppose the company does get a high rating; find the probabilities that it sponsored the following.*

18. A college football game 19. A professional football game

Economy Versus Investments *According to readings in business publications, there is a 50% chance of a booming economy next summer, a 20% chance of a mediocre economy, and a 30% chance of a recession. The probabilities that a particular investment strategy will produce a huge profit under each of these possibilities are .1, .6, and .3, respectively. Suppose the strategy does produce huge profits; find the probabilities that the economy was in the following conditions.*

20. Booming 21. In recession

22. **Shipping Errors** The following information pertains to three shipping terminals operated by Krag Corp.:*

Terminal	Percentage of cargo handled	Percentage of error
Land	50	2
Air	40	4
Sea	10	14

Krag's internal auditor randomly selects one set of shipping documents, ascertaining that the set selected contains an error. Which of the following gives the probability that the error occurred in the Land Terminal?

(a) .02 (b) .10 (c) .25 (d) .50

23. **Mortgage Defaults** A bank finds that the relationship between mortgage defaults and the size of the down payment is given by this table.

| | **Down Payment** | | | |
	5%	10%	20%	25%
Number of mortgages with this down payment	1260	700	560	280
Probability of default	.05	.03	.02	.01

(a) If a default occurs, what is the probability that it is on a mortgage with a 5% down payment?

(b) What is the probability that a mortgage that is paid to maturity has a 10% down payment?

Life Sciences

24. **Toxemia Test** A magazine article described a new test for toxemia, a disease that affects pregnant women. To perform the test, the woman lies on her left side and then rolls onto her back. The test is considered positive if there is a 20 millimeter rise in her blood pressure within one minute. The article gives the following probabilities, where T represents having toxemia at some time during the pregnancy, and N represents a negative test.

$$P(T'|N) = .90 \quad \text{and} \quad P(T|N') = .75$$

Assume that $P(N') = .11$, and find each of the following.

(a) $P(N|T)$ (b) $P(N'|T)$

25. **AIDS Testing** Suppose the probability that an individual has AIDS is .01, the probability of a person testing posi-

tive if he or she has AIDS is .95, and the probability of a person testing positive if he or she does not have AIDS is .05.

(a) Find the probability that a person who tests positive has AIDS.

(b) It has been argued that everyone should be tested for AIDS. Based on the results of part (a), how useful would the results of such testing be?

26. **Hepatitis Blood Test** The probability that a person with certain symptoms has hepatitis is .8. The blood test used to confirm this diagnosis gives positive results for 90% of people with the disease and 5% of those without the disease. What is the probability that an individual who has the symptoms and who reacts positively to the test actually has hepatitis?

Social Sciences

Never-Married Adults by Age Group *The following table gives the proportions of adult men and women in the U.S. population, and the proportions of adult men and women who have never married, in a recent year.*†

| | **Men** | |
Age	**Proportion of Men**	**Proportion Never Married**
18–24	.151	.875
25–29	.126	.433
30–34	.126	.250
35–39	.110	.140
40 or over	.487	.054

| | **Women** | |
Age	**Proportion of Women**	**Proportion Never Married**
18–24	.142	.752
25–29	.117	.295
30–34	.116	.161
35–39	.103	.090
40 or over	.522	.033

27. Find the probability that a randomly selected man who has never married is between 30 and 34 years old (inclusive).

28. Find the probability that a randomly selected woman who has been married is between 18 and 24 (inclusive).

29. Find the probability that a randomly selected woman who has never been married is between 35 and 39 (inclusive).

*Uniform CPA Examination, November 1989

†From the *Sacramento Bee*, September 10, 1987.

Seat Belt Effectiveness *A federal study showed that in 1990, 49% of all those involved in a fatal car crash wore seat belts. Of those in a fatal crash who wore seat belts, 44% were injured and 27% were killed. For those not wearing seat belts, the comparable figures were 41% and 50%, respectively.* *

30. Find the probability that a randomly selected person who was killed in a car crash was wearing a seat belt.

31. Find the probability that a randomly selected person who was unharmed in a fatal crash was not wearing a seat belt.

Voting *The following table shows the proportion of people over 18 who are in various age categories, along with the probabilities that a person in a given age category will vote in a general election.*

Age	Proportion of Voting Age Population	Probability of a Person of This Age Voting
18–21	11.0%	.48
22–24	7.6%	.53
25–44	37.6%	.68
45–64	28.3%	.64
65 or over	15.5%	.74

Suppose a voter is picked at random. Find the probabilities that the voter is in the following age categories.

32. 18–21 **33.** 65 or over

34. Find the probability that a person who did not vote was in the age category 45–64.

◆ **PROBABILITY SUMMARY**

BASIC PROBABILITY PRINCIPLE

Let S be a sample space with n equally likely outcomes. Let event E contain m of these outcomes. Then the probability that event E occurs, written $P(E)$, is

$$P(E) = \frac{m}{n}.$$

UNION RULE

For any events E and F from a sample space S.

$$P(E \cup F) = P(E) + P(F) - P(E \cap F).$$

For mutually exclusive events E and F.

$$P(E \cup F) = P(E) + P(F).$$

COMPLEMENT RULE $P(E) = 1 - P(E')$ and $P(E') = 1 - P(E)$

*National Highway Traffic Safety Administration, Office of Driver and Pedestrian Research: "Occupant Protection Trends in 19 Cities" (November 1989) and "Use of Automatic Safety Belt Systems in 19 Cities" (February 1991).

ODDS

The odds in favor of event E are $\dfrac{P(E)}{P(E')}$, $P(E') \neq 0$.

PROPERTIES OF PROBABILITY

1. For any event E in sample space S, $0 \leq P(E) \leq 1$.

2. The sum of the probabilities of all possible distinct outcomes is 1.

CONDITIONAL PROBABILITY

The conditional probability of event E, given that event F has occurred, is

$$P(E|F) = \frac{P(E \cap F)}{P(F)}, \quad \text{where } P(F) \neq 0.$$

For equally likely outcomes, conditional probability is found by reducing the sample space to event F; then

$$P(E|F) = \frac{n(E \cap F)}{n(F)}.$$

PRODUCT RULE OF PROBABILITY

If E and F are events, then $P(E \cap F)$ may be found by either of these formulas.

$$P(E \cap F) = P(F) \cdot P(E|F) \quad \text{or} \quad P(E \cap F) = P(E) \cdot P(F|E).$$

If E and F are independent events, then $P(E \cap F) = P(E) \cdot P(F)$.

Chapter Summary Key Terms

5.1 **set**
 element (member)
 empty set
 set-builder notation
 universal set
 subset
 tree diagram
 Venn diagram
 complement
 intersection
 disjoint sets
 union
 set operations

5.2 **union rule for sets**
5.3 **experiment**
 trial
 outcome
 sample space
 event
 simple event
 certain event
 impossible event

5.4 **mutually exclusive events**
 union rule for probability
 odds
 empirical probability
 probability distribution
5.5 **conditional probability**
 product rule
 independent events
5.6 **Bayes' theorem**

Chapter 5 Review Exercises

Write true *or* false *for each statement.*

1. $9 \in \{8, 4, -3, -9, 6\}$
2. $4 \notin \{3, 9, 7\}$
3. $2 \notin \{0, 1, 2, 3, 4\}$
4. $0 \in \{0, 1, 2, 3, 4\}$
5. $\{3, 4, 5\} \subseteq \{2, 3, 4, 5, 6\}$
6. $\{1, 2, 5, 8\} \subseteq \{1, 2, 5, 10, 11\}$
7. $\{3, 6, 9, 10\} \subseteq \{3, 9, 11, 13\}$
8. $\emptyset \subseteq \{1\}$
9. $\{2, 8\} \not\subseteq \{2, 4, 6, 8\}$
10. $0 \subseteq \emptyset$

In Exercises 11–20, let $U = \{a, b, c, d, e, f, g\}$, $K = \{c, d, f, g\}$, *and* $R = \{a, c, d, e, g\}$. *Find the following.*

11. The number of subsets of K
12. The number of subsets of R
13. K'
14. R'
15. $K \cap R$
16. $K \cup R$
17. $(K \cap R)'$
18. $(K \cup R)'$
19. \emptyset'
20. U'

In Exercises 21–26, let $U = \{$all employees of the K. O. Brown Company$\}$;
$A = \{$employees in the accounting department$\}$;
$B = \{$employees in the sales department$\}$;
$C = \{$employees with at least 10 years in the company$\}$;
$D = \{$employees with an MBA degree$\}$.

Describe the following sets in words.

21. $A \cap C$
22. $B \cap D$
23. $A \cup D$
24. $A' \cap D$
25. $B' \cap C'$
26. $(B \cup C)'$

Draw a Venn diagram and shade each set in Exercises 27–30.

27. $A \cup B'$
28. $A' \cap B$
29. $(A \cap B) \cup C$
30. $(A \cup B)' \cap C$

Write the sample space for each of the following experiments.

31. Rolling a die
32. Drawing a card from a deck containing only the 13 spades
33. Measuring the weight of a person to the nearest half pound (the scale will not measure more than 300 lb)
34. Tossing a coin 4 times

A jar contains 5 balls labeled 3, 5, 7, 9, and 11, respectively, while a second jar contains 4 red and 2 green balls. An experiment consists of pulling 1 ball from each jar, in turn. In Exercises 35–37, write each set using set notation.

35. The sample space
36. Event E: the number on the first ball is greater than 5
37. Event F: the second ball is green
38. Are the outcomes in the sample space in Exercise 35 equally likely?

When a single card is drawn from an ordinary deck, find the probabilities that it will be the following.

39. A heart
40. A red queen
41. A face card
42. Black or a face card
43. Red, given that it is a queen
44. A jack, given that it is a face card
45. A face card, given that it is a king
46. Describe what is meant by disjoint sets.
47. Describe what is meant by mutually exclusive events.
48. How are disjoint sets and mutually exclusive events related?

49. Define independent events.

50. Are independent events always mutually exclusive? Are they ever mutually exclusive?

51. An uproar has raged since September 1990 over the answer to a puzzle* published in *Parade* magazine, a supplement of the Sunday newspaper. In the "Ask Marilyn" column, Marilyn vos Savant answered the following question: "Suppose you're on a game show, and you're given the choice of three doors. Behind one door is a car; behind the others, goats. You pick a door, say number 1, and the host, who knows what's behind the other doors, opens another door, say number 3, which has a goat. He then says to you, 'Do you want to pick door number 2?' Is it to your advantage to take the switch?"

 Ms. vos Savant estimates that she has since received some 10,000 letters; most of them, including many from mathematicians and statisticians, disagreed with her answer. Her answer has been debated by both professionals and amateurs, and tested in classes at all levels, from grade school to graduate school. But by performing the experiment repeatedly, it can be shown that vos Savant's answer was correct. Find the probabilities of getting the car if you switch or do not switch, and then answer the question yourself. (*Hint:* Consider the sample space.)

Find the odds in favor of a card drawn from an ordinary deck being each of the following.

52. A club **53.** A black jack **54.** A red face card or a queen

Find the probabilities of getting the following sums when 2 fair dice are rolled.

55. 8 **56.** 0 **57.** At least 10 **58.** No more than 5

59. An odd number greater than 8 **60.** 12, given that the sum is greater than 10

61. 7, given that at least one die shows a 4 **62.** At least 9, given that at least one die shows a 5

63. Suppose $P(E) = .51$, $P(F) = .37$, and $P(E \cap F) = .22$. Find each of the following.
 (a) $P(E \cup F)$ **(b)** $P(E \cap F')$ **(c)** $P(E' \cup F)$ **(d)** $P(E' \cap F')$

64. Box A contains 5 red balls and 1 black ball; box B contains 2 red balls and 3 black balls. A box is chosen, and a ball is selected from it. The probability of choosing box A is $3/8$. If the selected ball is black, what is the probability that it came from box A?

65. Find the probability that the ball in Exercise 64 came from box B, given that it is red.

Applications

Business and Economics

Appliance Repairs *Of the appliance repair shops listed in the phone book, 80% are competent and 20% are not. A competent shop can repair an appliance correctly 95% of the time; an incompetent shop can repair an appliance correctly 60% of the time. Suppose an appliance was repaired correctly. Find the probabilities that it was repaired by the following.*

66. A competent shop **67.** An incompetent shop

Suppose an appliance was repaired incorrectly. Find the probabilities that it was repaired by the following.

68. A competent shop **69.** An incompetent shop

70. Sales A company sells typewriters and copiers. Let E be the event "a customer buys a typewriter," and let F be the event "a customer buys a copier." Write each of the following using \cap, \cup, or ' as necessary.

(a) A customer buys neither machine.

(b) A customer buys at least one of the machines.

71. Defective Items A sample shipment of five hair dryers is chosen at random. The probability of exactly 0, 1, 2, 3, 4, or 5 hair dryers being defective is given in the following table.

Number Defective	0	1	2	3	4	5
Probability	.31	.25	.18	.12	.08	.06

Find the probabilities that the following numbers of hair dryers are defective.

(a) No more than 3 (b) At least 3

72. Defective Items A manufacturer buys items from four different suppliers. The fraction of the total number of items that is obtained from each supplier, along with the probability that an item purchased from that supplier is defective, is shown in the following table.

Supplier	Fraction of Total Supplied	Probability of Defective
1	.17	.04
2	.39	.02
3	.35	.07
4	.09	.03

(a) Find the probability that a defective item came from supplier 4.

(b) Find the probability that a defective item came from supplier 2.

Life Sciences

73. Sickle Cell Anemia The square shows the four possible (equally likely) combinations when both parents are carriers of the sickle cell anemia trait. Each carrier parent has normal cells (N) and trait cells (T).

		Second Parent	
		N_2	T_2
First Parent	N_1 T_1		$N_1 T_2$

(a) Complete the table.

(b) If the disease occurs only when two trait cells combine, find the probability that a child born to these parents will have sickle cell anemia.

(c) The child will carry the trait but not have the disease if a normal cell combines with a trait cell. Find this probability.

(d) Find the probability that the child is neither a carrier nor has the disease.

Social Sciences

74. Television Viewing Habits A telephone survey of television viewers revealed the following information.

 20 watch situation comedies;
 19 watch game shows;
 27 watch movies;
 5 watch both situation comedies and game shows;
 8 watch both game shows and movies;
 10 watch both situation comedies and movies;
 3 watch all three;
 6 watch none of these.

(a) How many viewers were interviewed?

(b) How many viewers watch comedies and movies but not game shows?

(c) How many viewers watch only movies?

(d) How many viewers watch comedies and game shows but not movies?

75. Randomized Response Method for Getting Honest Answers to Sensitive Questions.* Basically, this is a method to guarantee that an individual who answers sensitive questions will remain anonymous, thus encouraging a truthful response. This method is, in effect, an application of the formula for finding the probability of an intersection, and operates as follows. Questions A and B are posed, one of which is sensitive and the other not. The probability of receiving a ''yes'' to the nonsensitive question must be known. For example, one could ask

 A: Does your Social Security number end in an odd digit? (Nonsensitive)
 B: Have you ever intentionally cheated on your income taxes? (Sensitive)

We know that $P(\text{answer yes}|\text{answer } A) = 1/2$. We wish to approximate $P(\text{answer yes}|\text{answer } B)$. The subject is asked to flip a coin and answer A if the coin comes up heads and otherwise to answer B. In this way, the interviewer does not know which question the subject is answering. Thus, a ''yes'' answer is not incriminating. There is no way for the interviewer to know whether the subject is saying ''Yes, my Social Security number ends in an odd digit'' or ''Yes, I have intentionally cheated on my income taxes.'' The percentage of subjects in the group answering ''yes'' is used to approximate $P(\text{answer yes})$.

*From *Applied Statistics With Probability* by J. S. Milton and J. J. Corbet. Copyright © 1979 by Litton Educational Publishing, Inc. Reprinted by permission of Brooks/Cole Publishing Company, Monterey, California.

(a) Use the fact that the event "answer yes" is the union of the event "answer yes and answer A" with the event "answer yes and answer B" to prove that

$P(\text{answer yes}|\text{answer } B)$

$$= \frac{P(\text{answer yes}) - P(\text{answer yes}|\text{answer } A) \cdot P(\text{answer } A)}{P(\text{answer } B)}$$

(b) If this technique is tried on 100 subjects and 60 answered "yes," what is the approximate probability that a person randomly selected from the group has intentionally cheated on income taxes?

General Interest

76. **Making a First Down** A first down is desirable in football—it guarantees four more plays by the team making it, assuming no score or turnover occurs in the plays. After getting a first down, a team can get another by advancing the ball at least 10 yards. During the four plays given by a first down, a team's position will be indicated by a phrase such as "third and 4," which means that the team has already had two of its four plays, and that 4 more yards are needed to get the 10 yards necessary for another first down. An article in a management journal[†] offers the following results for 189 games of a recent National Football League season. "Trials" represents the number of times a team tried to make a first down, given that it was currently playing either a third or a fourth down. Here, n represents the number of yards still needed for a first down.

n	Trials	Successes	Probability of Making First Down with n Yards to Go
1	543	388	
2	327	186	
3	356	146	
4	302	97	
5	336	91	

(a) Complete the table.

(b) Why is the sum of the answers in the table not equal to 1?

Connections

77. The table below shows the results of a survey of buyers of a certain model of car.

Car Type	Satisfied	Not Satisfied	Totals
New	300	100	
Used	450		600
Totals		250	

(a) Complete the table.

(b) How many buyers were surveyed?

(c) How many bought a new car and were satisfied?

(d) How many were not satisfied?

(e) How many bought used cars?

(f) How many of those who were not satisfied had bought a used car?

(g) Rewrite the event stated in part (f) using the expression "given that."

(h) Find the probability of the outcome in parts (f) and (g).

(i) Find the probability that a used-car buyer is not satisfied.

(j) You should have different answers in parts (h) and (i). Explain why.

[†]Reprinted by permission of Virgil Carter and Robert Machols, "Optimal Strategies on Fourth Down," *Management Science*, vol. 24, no. 16, December 1978. Copyright © 1978 by The Institute of Management Sciences.

Extended Application / Medical Diagnosis

When a patient is examined, information (typically incomplete) is obtained about his state of health. Probability theory provides a mathematical model appropriate for this situation, as well as a procedure for quantitatively interpreting such partial information to arrive at a reasonable diagnosis.*

To develop a model, we list the states of health that can be distinguished in such a way that the patient can be in one and only one state at the time of the examination. For each state of health H, we associate a number, $P(H)$, between 0 and 1 such that the sum of all these numbers is 1. This number $P(H)$ represents the probability, before examination, that a patient is in the state of health H, and $P(H)$ may be chosen subjectively from medical experience, using any information available prior to the examination. The probability may be most conveniently established from clinical records; that is, a mean probability is established for patients in general, although the number would vary from patient to patient. Of course, the more information that is brought to bear in establishing $P(H)$, the better the diagnosis.

For example, limiting the discussion to the condition of a patient's heart, suppose there are exactly 3 states of health, with probabilities as follows.

State of Health, H		$P(H)$
H_1	Patient has a normal heart	.8
H_2	Patient has minor heart irregularities	.15
H_3	Patient has a severe heart condition	.05

Having selected $P(H)$, the information from the examination is processed. First, the results of the examination must be classified. The examination itself consists of observing the state of a number of characteristics of the patient. Let us assume that the examination for a heart condition consists of a stethoscope examination and a cardiogram. The outcome of such an examination, C, might be one of the following:

$C_1 =$ stethoscope shows normal heart and cardiogram shows normal heart;

$C_2 =$ stethoscope shows normal heart and cardiogram shows minor irregularities;

and so on.

It remains to assess for each state of health H the conditional probability $P(C|H)$ of each examination outcome C using only the knowledge that a patient is in a given state of health. (This may be based on the medical knowledge and clinical experience of the doctor.) The conditional probabilities $P(C|H)$ will not vary from patient to patient (although they should be reviewed periodically), so that they may be built into a diagnostic system.

Suppose the result of the examination is C_1. Let us assume the following probabilities:

$$P(C_1|H_1) = .9$$
$$P(C_1|H_2) = .4$$
$$P(C_1|H_3) = .1.$$

Now, for a given patient, the appropriate probability associated with each state of health H, after examination, is $P(H|C)$, where C is the outcome of the examination. This can be calculated by using Bayes' theorem. For example, to find $P(H_1|C_1)$—that is, the probability that the patient has a normal heart given that the examination showed a normal stethoscope examination and a normal cardiogram—we use Bayes' theorem as follows:

$$P(H_1|C_1) = \frac{P(C_1|H_1)P(H_1)}{P(C_1|H_1)P(H_1) + P(C_1|H_2)P(H_2) + P(C_1|H_3)P(H_3)}$$

$$= \frac{(.9)(.8)}{(.9)(.8) + (.4)(.15) + (.1)(.05)} \approx .92.$$

Hence, the probability is about .92 that the patient has a normal heart on the basis of the examination results. This means that in 8 out of 100 patients, some abnormality will be present and not be detected by the stethoscope or the cardiogram.

Exercises

1. Find $P(H_2|C_1)$.
2. Assuming the following probabilities, find $P(H_1|C_2)$.
 $P(C_2|H_1) = .2$ $P(C_2|H_2) = .8$ $P(C_2|H_3) = .3$
3. Assuming the probabilities of Exercise 2, find $P(H_3|C_2)$.

*From Roger Wright, "Probabilistic Medical Diagnosis" in Robert M. Thrall, ed., *Some Mathematical Models in Biology*, rev. ed., University of Michigan, 1967. Used by permission of Robert M. Thrall.

Applications at a glance . . .

Damage from Seeding a Hurricane . . . winning a lottery . . . births of twins . . .

CHAPTER 6

COUNTING PRINCIPLES; FURTHER PROBABILITY TOPICS

(See page 308.)

In this chapter, we continue our discussion of probability theory. To use the basic definition of probability, $P(E) = n(E)/n(S)$ (where S is the sample space with equally likely outcomes) up to now we have simply listed the outcomes in S and in E. However, when S has many outcomes, listing them all becomes very tedious. In the first two sections of this chapter, we introduce methods for counting the number of outcomes in a set without actually listing them, and then we use this approach in the third section to find probabilities. In the section on *Bernoulli trials* (repeated independent trials of an experiment), we introduce a formula for finding the probability of a certain number of successes in a large number of trials. The final section continues the discussion of probability distributions that we began in Chapter 5.

6.1 THE MULTIPLICATION PRINCIPLE; PERMUTATIONS

In how many ways can seven panelists be seated in a row of seven chairs?

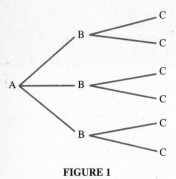

FIGURE 1

This question will be answered later in this section using *permutations*. Let us begin with a simple example. If there are 3 roads from town A to town B and 2 roads from town B to town C, in how many ways can a person travel from A to C by way of B? For each of the 3 roads from A there are 2 different routes leading from B to C, or a total of $3 \cdot 2 = 6$ different ways for the trip, as shown in Figure 1. This example illustrates a general principle of counting, called the **multiplication principle**.

MULTIPLICATION PRINCIPLE	Suppose n choices must be made, with

MULTIPLICATION PRINCIPLE

Suppose n choices must be made, with

$$m_1 \text{ ways to make choice 1,}$$

and for each of these ways,

$$m_2 \text{ ways to make choice 2,}$$

and so on, with

$$m_n \text{ ways to make choice } n.$$

Then there are

$$m_1 \cdot m_2 \cdot \cdots \cdot m_n$$

different ways to make the entire sequence of choices.

EXAMPLE **1** A certain combination lock can be set to open to any one 3-letter sequence. How many such sequences are possible?

Since there are 26 letters in the alphabet, there are 26 choices for each of the 3 letters. By the multiplication principle, there are $26 \cdot 26 \cdot 26 = 17{,}576$ different sequences. ◀

EXAMPLE **2** Morse code uses a sequence of dots and dashes to represent letters and words. How many sequences are possible with at most 3 symbols?

"At most 3" means "1 or 2 or 3" here. Each symbol may be either a dot or a dash. Thus the following number of sequences are possible in each case.

Number of Symbols	Number of Sequences
1	**2**
2	$2 \cdot 2 = 4$
3	$2 \cdot 2 \cdot 2 = 8$

Altogether, $2 + 4 + 8 = 14$ different sequences are possible. ◀

EXAMPLE **3** An ancient Chinese philosophical work known as the *I Ching* (*Book of Changes*) is often used as an oracle from which people can seek and obtain advice. The philosophy describes the duality of the universe in terms of two primary forces: *yin* (passive, dark, receptive) and *yang* (active, light, creative). See Figure 2. The yin energy is represented by a broken line (– –) and the yang by a solid line (—) These lines are written on top of one another in groups of three, known as *trigrams*. For example, the trigram is called *Tui*, the Joyous, and has the image of a lake.

(a) How many trigrams are there altogether?

Think of choosing between the 2 types of lines for each of the 3 positions in the trigram. There will be 2 choices for each position, so there are $2 \cdot 2 \cdot 2 = 8$ different trigrams.

(b) The trigrams are grouped together, one on top of the other, in pairs known as *hexagrams*. Each hexagram represents one aspect of the *I Ching* philosophy. How many hexagrams are there?

For each position in the hexagram there are 8 possible trigrams, giving $8 \cdot 8 = 64$ hexagrams. ◀

yin yang

FIGURE 2

EXAMPLE **4** A teacher has 5 different books that he wishes to arrange side by side. How many different arrangements are possible?

Five choices will be made, one for each space that will hold a book. Any of the 5 books could be chosen for the first space. There are 4 choices for the second

space, since 1 book has already been placed in the first space; there are 3 choices for the third space, and so on. By the multiplication principle, the number of different possible arrangements is $5 \cdot 4 \cdot 3 \cdot 2 \cdot 1 = 120.$ ◄

The use of the multiplication principle often leads to products such as $5 \cdot 4 \cdot 3 \cdot 2 \cdot 1$, the product of all the natural numbers from 5 down to 1. If n is a natural number, the symbol $n!$ (read "n factorial") denotes the product of all the natural numbers from n down to 1. If $n = 1$, this formula is understood to give $1! = 1$.

FACTORIAL NOTATION	For any natural number n, $$n! = n(n-1)(n-2) \cdots (3)(2)(1).$$ Also, $$0! = 1.$$

With this symbol, the product $5 \cdot 4 \cdot 3 \cdot 2 \cdot 1$ can be written as $5!$. Also, $3! = 3 \cdot 2 \cdot 1 = 6$. The definition of $n!$ could be used to show that $n[(n - 1)!] = n!$ for all natural numbers $n \geq 2$. It is helpful if this result also holds for $n = 1$. This can happen only if $0!$ equals 1, as defined above.

Some calculators have an $n!$ key. A calculator with a 10-digit display and scientific notation capability will usually give the exact value of $n!$ for $n \leq 13$, and approximate values of $n!$ for $14 \leq n \leq 69$.

 EXAMPLE 5 Suppose the teacher in Example 4 wishes to place only 3 of the 5 books on his desk. How many arrangements of 3 books are possible?

The teacher again has 5 ways to fill the first space, 4 ways to fill the second space, and 3 ways to fill the third. Since he wants to use only 3 books, only 3 spaces can be filled (3 events) instead of 5, for $5 \cdot 4 \cdot 3 = 60$ arrangements. ◄

Permutations The answer 60 in Example 5 is called the number of *permutations* of 5 things taken 3 at a time. A **permutation** of r (where $r \geq 1$) elements from a set of n elements is any specific ordering or arrangement, *without repetition*, of the r elements. Each rearrangement of the r elements is a different permutation. The number of permutations of n things taken r at a time (with $r \leq n$) is written $P(n, r)$. Based on the work in Example 5,

$$P(5, 3) = 5 \cdot 4 \cdot 3 = 60.$$

Factorial notation can be used to express this product as follows.

$$5 \cdot 4 \cdot 3 = 5 \cdot 4 \cdot 3 \cdot \frac{2 \cdot 1}{2 \cdot 1} = \frac{5 \cdot 4 \cdot 3 \cdot 2 \cdot 1}{2 \cdot 1} = \frac{5!}{2!} = \frac{5!}{(5 - 3)!}$$

This example illustrates the general rule of permutations, which can be stated as follows.

PERMUTATIONS If $P(n, r)$ (where $r \leq n$) is the number of permutations of n elements taken r at a time, then

$$P(n, r) = \frac{n!}{(n - r)!}.$$

Many calculators have a key for finding permutations.

Caution The letter P here represents *permutations*, not *probability*. In probability notation, the quantity in parentheses describes an *event*. In permutations notation, the quantity in parentheses always comprises *two numbers*.

The proof of the permutations rule follows the discussion in Example 5. There are n ways to choose the first of the r elements, $n - 1$ ways to choose the second, and $n - r + 1$ ways to choose the rth element, so that

$$P(n, r) = n(n - 1)(n - 2) \cdots (n - r + 1).$$

Now multiply on the right by $(n - r)!/(n - r)!$.

$$P(n, r) = n(n - 1)(n - 2) \cdots (n - r + 1) \cdot \frac{(n - r)!}{(n - r)!}$$

$$= \frac{n(n - 1)(n - 2) \cdots (n - r + 1)(n - r)!}{(n - r)!}$$

$$= \frac{n!}{(n - r)!}$$

To find $P(n, r)$, we can use either the permutations formula or direct application of the multiplication principle, as the following example shows.

EXAMPLE 6 Find the number of permutations of 8 elements taken 3 at a time.

Since there are 3 choices to be made, the multiplication principle gives $P(8, 3) = 8 \cdot 7 \cdot 6 = 336$. Alternatively, use the permutations formula to get

$$P(8, 3) = \frac{8!}{(8 - 3)!} = \frac{8!}{5!} = \frac{8 \cdot 7 \cdot 6 \cdot 5!}{5!} = 8 \cdot 7 \cdot 6 = 336. \quad \blacktriangleleft$$

Caution When calculating the number of permutations with the formula, do not try to cancel unlike factorials. For example,

$$\frac{8!}{4!} \neq 2! = 2 \cdot 1 = 2.$$

$$\frac{8!}{4!} = \frac{8 \cdot 7 \cdot 6 \cdot 5 \cdot 4 \cdot 3 \cdot 2 \cdot 1}{4 \cdot 3 \cdot 2 \cdot 1} = 8 \cdot 7 \cdot 6 \cdot 5 = 1680.$$

Always write out the factors first, then cancel where appropriate.

EXAMPLE 7

Find each of the following.

(a) The number of permutations of the letters A, B, and C
By the formula for $P(n, r)$ with both n and r equal to 3,

$$P(3, 3) = \frac{3!}{(3 - 3)!} = \frac{3!}{0!} = \frac{3!}{1} = 3 \cdot 2 \cdot 1 = 6.$$

The 6 permutations (or arrangements) are

$$ABC, \quad ACB, \quad BAC, \quad BCA, \quad CAB, \quad CBA.$$

(b) The number of permutations if just 2 of the letters A, B, and C are to be used
Find $P(3, 2)$.

$$P(3, 2) = \frac{3!}{(3 - 2)!} = \frac{3!}{1!} = 3! = 6$$

This result is exactly the same answer as in part (a). This is because, in the case of $P(3, 3)$, after the first 2 choices are made, the third is already determined, as shown in the table below.

First Two Letters	AB	AC	BA	BC	CA	CB
Third Letter	C	B	C	A	B	A

EXAMPLE 8

A televised talk show will include 4 women and 3 men as panelists.

(a) In how many ways can the panelists be seated in a row of 7 chairs?
Find $P(7, 7)$, the total number of ways to seat 7 panelists in 7 chairs.

$$P(7, 7) = \frac{7!}{(7 - 7)!} = \frac{7!}{0!} = \frac{7!}{1} = 7 \cdot 6 \cdot 5 \cdot 4 \cdot 3 \cdot 2 \cdot 1 = 5040$$

There are 5040 ways to seat the 7 panelists.

(b) In how many ways can the panelists be seated if the men and women are to be alternated?
Use the multiplication principle. In order to alternate men and women, a woman must be seated in the first chair (since there are 4 women and only 3 men), any of the men next, and so on. Thus there are 4 ways to fill the first seat, 3 ways to fill the second seat, 3 ways to fill the third seat (with any of the 3 remaining women), and so on. This gives

$$4 \cdot 3 \cdot 3 \cdot 2 \cdot 2 \cdot 1 \cdot 1 = 144$$

ways to seat the panelists.

If the n objects in a permutation are not all distinguishable—that is, if there are n_1 of type 1, n_2 of type 2, and so on for r different types, then the number of **distinguishable permutations** is

$$\frac{n!}{n_1!n_2! \cdots n_r!}.$$

For example, suppose we want to find the number of permutations of the numbers 1, 1, 4, 4, 4. We cannot distinguish between the two 1's or among the three 4's, so using 5! would give too many distinguishable arrangements. Since the two 1's are indistinguishable and account for 2! of the permutations, we divide 5! by 2!. Similarly, we also divide by 3! to account for the three indistinguishable 4's. This gives

$$\frac{5!}{2!3!} = 10$$

permutations.

EXAMPLE 9

In how many ways can the letters in the word *Mississippi* be arranged? This word contains 1 m, 4 i's, 4 s's, and 2 p's. To use the formula, let $n = 11$, $n_1 = 1$, $n_2 = 4$, $n_3 = 4$, and $n_4 = 2$ to get

$$\frac{11!}{1!4!4!2!} = 34,650$$

arrangements. ◀

Note If Example 9 had asked for the number of ways that the letters in a word with 11 *different* letters could be arranged, the answer would be 11! = 39,916,800.

6.1 Exercises

Evaluate the following factorials and permutations.

1. $P(4, 2)$ **2.** 3! **3.** 6! **4.** $P(8, 0)$ **5.** 0!

6. $P(7, 1)$ **7.** 4! **8.** $P(4, 4)$ **9.** $P(13, 2)$ **10.** $P(12, 3)$

11. $P(25, 12)$ **12.** $P(38, 17)$ **13.** $P(14, 5)$ **14.** $P(17, 12)$

Use the multiplication principle to solve the following problems.

15. How many different types of homes are available if a builder offers a choice of 5 basic plans, 3 roof styles, and 2 exterior finishes?

16. An automobile manufacturer produces 7 models, each available in 6 different exterior colors, with 4 different upholstery fabrics and 5 interior colors. How many varieties of automobile are available?

17. A menu offers a choice of 3 salads, 8 main dishes, and 5 desserts. How many different meals consisting of one salad, one main dish, and one dessert are possible?

18. A couple has narrowed down the choice of a name for their new baby to 3 first names and 5 middle names. How many different first- and middle-name arrangements are possible?

19. A concert to raise money for an economics prize is to consist of 5 works: 2 overtures, 2 sonatas, and a piano concerto.

 (a) In how many ways can the program be arranged?

 (b) In how many ways can the program be arranged if an overture must come first?

Use permutations to solve each of the following problems.

20. In an experiment on social interaction, 6 people will sit in 6 seats in a row. In how many ways can this be done?

21. In how many ways can 7 of 10 monkeys be arranged in a row for a genetics experiment?

22. A business school gives courses in typing, shorthand, transcription, business English, technical writing, and accounting. In how many ways can a student arrange a schedule if 3 courses are taken?

23. If your college offers 400 courses, 20 of which are in mathematics, and your counselor arranges your schedule of 4 courses by random selection, how many schedules are possible that do not include a math course?

24. In a club with 15 members, how many ways can a slate of 3 officers consisting of president, vice-president, and secretary/treasurer be chosen?

25. A baseball team has 20 players. How many 9-player batting orders are possible?

26. A chapter of union Local 715 has 35 members. In how many different ways can the chapter select a president, a vice-president, a treasurer, and a secretary?

27. A zydeco band from Louisiana will play 5 traditional and 3 original Cajun compositions at a concert. In how many ways can they arrange the program if

 (a) they begin with a traditional piece?

 (b) an original piece will be played last?

28. In an election with 3 candidates for one office and 5 candidates for another office, how many different ballots may be printed?

29. The television schedule for a certain evening shows 8 choices from 8 to 9 P.M., 5 choices from 9 to 10 P.M., and 6 choices from 10 to 11 P.M. In how many different ways could a person schedule that evening of television viewing from 8 to 11 P.M.? (Assume each program that is selected is watched for an entire hour.)

30. Define *permutation* in your own words.

31. In Example 7, there are six 3-letter permutations of the letters A, B, and C. How many 3-letter subsets (unordered groups of letters) are there?

32. In Example 7, how many unordered 2-letter subsets of the letters A, B, and C are there?

33. Find the number of distinguishable permutations of the letters in each of the following words.

 (a) initial (b) little (c) decreed

34. A printer has 5 A's, 4 B's, 2 C's, and 2 D's. How many different ''words'' are possible that use all these letters? (A ''word'' does not have to have any meaning here.)

35. Wing has different books to arrange on a shelf: 4 blue, 3 green, and 2 red.

 (a) In how many ways can the books be arranged on a shelf?

 (b) If books of the same color are to be grouped together, how many arrangements are possible?

 (c) In how many distinguishable ways can the books be arranged if books of the same color are identical but need not be grouped together?

36. A child has a set of differently shaped plastic objects. There are 3 pyramids, 4 cubes, and 7 spheres.

 (a) In how many ways can she arrange the objects in a row if each is a different color?

 (b) How many arrangements are possible if objects of the same shape must be grouped together?

 (c) In how many distinguishable ways can the objects be arranged in a row if objects of the same shape are also the same color, but need not be grouped together?

Applications

Life Sciences

37. **Drug Sequencing** Eleven drugs have been found to be effective in the treatment of a disease. It is believed that the sequence in which the drugs are administered is important in the effectiveness of the treatment. In how many different sequences can 5 of the 11 drugs be administered?

38. **Insect Classification** A biologist is attempting to classify 52,000 species of insects by assigning 3 initials to each species. Is it possible to classify all the species in this way? If not, how many initials should be used?

General Interest

39. Radio Station Call Letters How many different 4-letter radio station call letters can be made if

(a) the first letter must be K or W and no letter may be repeated?

(b) repeats are allowed, but the first letter is K or W?

(c) the first letter is K or W, there are no repeats, and the last letter is R?

40. Telephone Numbers How many 7-digit telephone numbers are possible if the first digit cannot be zero and

(a) only odd digits may be used?

(b) the telephone number must be a multiple of 10 (that is, it must end in zero)?

(c) the telephone number must be a multiple of 100?

(d) the first 3 digits are 481?

(e) no repetitions are allowed?

Telephone Area Codes *The United States is rapidly running out of telephone numbers. In large cities, telephone companies introduced new area codes as numbers were used up; as a result, only three unallocated area codes are left: 210, 810, and 910.*

41. (a) At present, all area codes have a 0 or a 1 as the middle digit and the first digit cannot be 0 or 1. How many area codes are there with this arrangement? How many telephone numbers does the current 7-digit sequence permit per area code? (The 3-digit sequence that follows the area code cannot start with 0 or 1. Assume there are no other restrictions.)

(b) The actual number of area codes available is 152. Explain the discrepancy between this number and your answer to part (a).

42. How many area codes would be possible if all restrictions on the second digit were removed?

43. A problem with the plan in Exercise 42 is that the second digit in the area code now tells the phone company equipment that a long-distance call is being made. To avoid changing all equipment, an alternative plan proposes a 4-digit area code and restricting the first and second digits as before. How many area codes would this plan provide?

44. Still another alternative solution is to increase the local dialing sequence to 8 digits instead of 7. How many additional numbers would this plan create? (Assume the same restrictions.)

45. License Plates For many years, the state of California used 3 letters followed by 3 digits on its automobile license plates.

(a) How many different license plates are possible with this arrangement?

(b) When the state ran out of new numbers, the order was reversed to 3 digits followed by 3 letters. How many new license plate numbers were then possible?

(c) Several years ago, the numbers described in (b) were also used up. The state then issued plates with 1 letter followed by 3 digits and then 3 letters. How many new license plate numbers will this provide?

46. Social Security Numbers A social security number has 9 digits. How many social security numbers are there? The United States population in 1989 was about 248 million. Is it possible for every U.S. resident to have a unique social security number? (Assume no restrictions.)

47. Postal Zip Codes The United States Postal Service currently uses 5-digit zip codes in most areas. How many zip codes are possible if there are no restrictions on the digits used? How many would be possible if the first number could not be 0?

48. The Postal Service is encouraging the use of 9-digit zip codes in some areas, adding 4 digits after the usual 5-digit code. How many such zip codes are possible with no restrictions?

◄ 6.2 COMBINATIONS

In how many ways can a manager select 4 employees for promotion from 12 eligible employees?

As we shall see, permutations cannot be used to answer this question, but combinations will provide the answer.

In the previous section, we saw that there are 60 ways that a teacher can arrange 3 of 5 different books on his desk. That is, there are 60 permutations of 5 books taken 3 at a time. Suppose now that the teacher does not wish to arrange the books on his desk, but rather wishes to choose, without regard to order, any 3 of the 5 books for a book sale to raise money for his school. In how many ways can this be done?

At first glance, we might say 60 again, but this is incorrect. The number 60 counts all possible *arrangements* of 3 books chosen from 5. The following 6 arrangements, however, would all lead to the same set of 3 books being given to the book sale.

mystery-biography-textbook	biography-textbook-mystery
mystery-textbook-biography	textbook-biography-mystery
biography-mystery-textbook	textbook-mystery-biography

The list shows 6 different *arrangements* of 3 books, but only one *subset* of 3 books. A subset of items selected *without regard to order* is called a **combination.** The number of combinations of 5 things taken 3 at a time is written $\binom{5}{3}$, and read "5 over 3" or "5 choose 3."* Since they are subsets, combinations are *not ordered.*

To evaluate $\binom{5}{3}$, start with the $5 \cdot 4 \cdot 3$ *permutations* of 5 things taken 3 at a time. Since combinations are not ordered, find the number of combinations by dividing the number of permutations by the number of ways each group of 3 can be ordered; that is, divide by 3!.

$$\binom{5}{3} = \frac{5 \cdot 4 \cdot 3}{3!} = \frac{5 \cdot 4 \cdot 3}{3 \cdot 2 \cdot 1} = 10.$$

There are 10 ways that the teacher can choose 3 books for the book sale.

Generalizing this discussion gives the following formula for the number of combinations of n elements taken r at a time:

$$\binom{n}{r} = \frac{P(n, r)}{r!}.$$

Another version of this formula is found as follows.

$$\binom{n}{r} = \frac{P(n, r)}{r!}$$

$$= \frac{n!}{(n - r)!} \cdot \frac{1}{r!}$$

$$= \frac{n!}{(n - r)!r!}$$

The steps above lead to the following result.

*Other common notations for $\binom{n}{r}$ are $_nC_r$, C_r^n, and $C(n, r)$.

COMBINATIONS

If $\binom{n}{r}$ denotes the number of combinations of n elements taken r at a time, where $r \leq n$, then

$$\binom{n}{r} = \frac{n!}{(n-r)!r!}.$$

 The table of combinations in the Appendix gives the values of $\binom{n}{r}$ for $n \leq 20$. Also, many calculators have a key for finding combinations.

EXAMPLE 1

How many committees of 3 people can be formed from a group of 8 people? A committee is an unordered group, so use the combinations formula for $\binom{8}{3}$.

$$\binom{8}{3} = \frac{8!}{5!3!} = \frac{8 \cdot 7 \cdot 6 \cdot 5 \cdot 4 \cdot 3 \cdot 2 \cdot 1}{5 \cdot 4 \cdot 3 \cdot 2 \cdot 1 \cdot 3 \cdot 2 \cdot 1} = \frac{8 \cdot 7 \cdot 6}{3 \cdot 2 \cdot 1} = 56 \quad \blacktriangleleft$$

EXAMPLE 2

Three lawyers are to be selected from a group of 30 to work on a special project.

(a) In how many different ways can the lawyers be selected?

Here we wish to know the number of 3-element combinations that can be formed from a set of 30 elements. (We want combinations, not permutations, since order within the group of 3 doesn't matter.)

$$\binom{30}{3} = \frac{30!}{27!3!} = \frac{30 \cdot 29 \cdot 28 \cdot 27!}{27! \cdot 3 \cdot 2 \cdot 1}$$

$$= \frac{30 \cdot 29 \cdot 28}{3 \cdot 2 \cdot 1}$$

$$= 4060$$

There are 4060 ways to select the project group.

(b) In how many ways can the group of 3 be selected if a certain lawyer must work on the project?

Since 1 lawyer already has been selected for the project, the problem is reduced to selecting 2 more from the remaining 29 lawyers.

$$\binom{29}{2} = \frac{29!}{27!2!} = \frac{29 \cdot 28 \cdot 27!}{27! \cdot 2 \cdot 1} = \frac{29 \cdot 28}{2 \cdot 1} = 29 \cdot 14 = 406$$

In this case, the project group can be selected in 406 ways.

(c) In how many ways can a nonempty group of at most 3 lawyers be selected from these 30 lawyers?

Here, by "at most 3" we mean "1 or 2 or 3." Find the number of ways for each case.

Case	Number of Ways
1	$\binom{30}{1} = \dfrac{30!}{29!1!} = \dfrac{30 \cdot 29!}{29!(1)} = 30$
2	$\binom{30}{2} = \dfrac{30!}{28!2!} = \dfrac{30 \cdot 29 \cdot 28!}{28! \cdot 2 \cdot 1} = 435$
3	$\binom{30}{3} = \dfrac{30!}{27!3!} = \dfrac{30 \cdot 29 \cdot 28 \cdot 27!}{27! \cdot 3 \cdot 2 \cdot 1} = 4060$

The total number of ways to select at most 3 lawyers will be the sum

$$30 + 435 + 4060 = 4525.$$ ◀

EXAMPLE 3 A salesman has 10 accounts in a certain city.

(a) In how many ways can he select 3 accounts to call on?

Within a selection of 3 accounts, the arrangement of the calls is not important, so there are

$$\binom{10}{3} = \frac{10!}{7!3!} = \frac{10 \cdot 9 \cdot 8}{3 \cdot 2 \cdot 1} = 120$$

ways he can make a selection of 3 accounts.

(b) In how many ways can he select at least 8 of the 10 accounts to use in preparing a report?

"At least 8" means "8 or more," which is "8 or 9 or 10." First find the number of ways to choose in each case.

Case	Number of Ways
8	$\binom{10}{8} = \dfrac{10!}{2!8!} = \dfrac{10 \cdot 9}{2 \cdot 1} = 45$
9	$\binom{10}{9} = \dfrac{10!}{1!9!} = \dfrac{10}{1} = 10$
10	$\binom{10}{10} = \dfrac{10!}{0!10!} = 1$

He can select at least 8 of the 10 accounts in $45 + 10 + 1 = 56$ ways. ◀

The formulas for permutations and combinations given in this section and in the previous section will be very useful in solving probability problems in the next section. Any difficulty in using these formulas usually comes from being unable to differentiate between them. Both permutations and combinations give the number of ways to choose r objects from a set of n objects. The differences between permutations and combinations are outlined in the following table.

Permutations	Combinations
Different orderings or arrangements of the r objects are different permutations. $$P(n, r) = \frac{n!}{(n - r)!}$$ Clue words: arrangement, schedule, order	Each choice or subset of r objects gives one combination. Order within the group of r objects does not matter. $$\binom{n}{r} = \frac{n!}{(n - r)!r!}$$ Clue words: group, committee, sample

In the next examples, concentrate on recognizing which formula should be applied.

EXAMPLE 4 For each of the following problems, tell whether permutations or combinations should be used to solve the problem.

(a) How many 4-digit code numbers are possible if no digits are repeated?

Since changing the order of the 4 digits results in a different code, use permutations.

(b) A sample of 3 light bulbs is randomly selected from a batch of 15. How many different samples are possible?

The order in which the 3 light bulbs are selected is not important. The sample is unchanged if the items are rearranged, so combinations should be used.

(c) In a baseball conference with 8 teams, how many games must be played so that each team plays every other team exactly once?

Selection of two teams for a game is an *unordered* subset of 2 from the set of 8 teams. Use combinations again.

(d) In how many ways can 4 patients be assigned to 6 hospital rooms so that each patient has a private room?

The room assignments are an *ordered* selection of 4 rooms from the 6 rooms. Exchanging the rooms of any two patients within a selection of 4 rooms gives a different assignment, so permutations should be used.

(e) Solve the problems in parts (a)–(d) above. The answers are given in the footnote.* ◄

EXAMPLE 5 A manager must select 4 employees for promotion; 12 employees are eligible.

(a) In how many ways can 4 be chosen?

Since there is no reason to differentiate among the 4 who are selected, use combinations.

$$\binom{12}{4} = \frac{12!}{8!4!} = 495$$

*(a) 5040 (b) 455 (c) 28 (d) 360

(b) In how many ways can 4 employees be chosen (from 12) to be placed in 4 different jobs?

In this case, once a group of 4 is selected, they can be assigned in many different ways (or arrangements) to the 4 jobs. Therefore, this problem requires permutations.

$$P(12, 4) = \frac{12!}{8!} = 11{,}880 \quad \blacktriangleleft$$

EXAMPLE 6 In how many ways can a full house of aces and eights (3 aces and 2 eights) occur in 5-card poker?

Examples 6 and 7 involve a standard deck of 52 playing cards, as shown in Figure 18 in the preceding chapter. Recall the discussion that accompanies the photograph.

The arrangement of the 3 aces or the 2 eights does not matter, so we use combinations and the multiplication principle. There are $\binom{4}{3}$ ways to get 3 aces from the 4 aces in the deck, and $\binom{4}{2}$ ways to get 2 eights. By the multiplication principle, the number of ways to get 3 aces and 2 eights is

$$\binom{4}{3} \cdot \binom{4}{2} = 4 \cdot 6 = 24. \quad \blacktriangleleft$$

EXAMPLE 7 Five cards are dealt from a standard 52-card deck.

(a) How many such hands have only face cards?

The face cards are the king, queen, and jack of each suit. Since there are 4 suits, there are 12 face cards. The arrangement of the 5 cards is not important, so use combinations to get

$$\binom{12}{5} = \frac{12!}{7!5!} = 792.$$

(b) How many such hands have exactly 2 hearts?

There are 13 hearts in the deck, so the 2 hearts will be selected from those 13 cards. The other 3 cards must come from the remaining 39 cards that are not hearts. Use combinations and the multiplication principle to get

$$\binom{13}{2}\binom{39}{3} = 78 \cdot 9139 = 712{,}842.$$

Notice that the two top numbers in the combinations add up to 52, the total number of cards, and the two bottom numbers add up to 5, the number of cards in a hand.

(c) How many such hands have cards of a single suit?

The total number of ways that 5 cards of a particular suit of 13 cards can occur is $\binom{13}{5}$. Since the arrangement of the 5 cards is not important, use combinations. There are four different suits, so the multiplication principle gives

$$4 \cdot \binom{13}{5} = 4 \cdot 1287 = 5148$$

ways to deal 5 cards of the same suit. \blacktriangleleft

As Example 7 shows, often both combinations and the multiplication principle must be used in the same problem.

EXAMPLE 8 To illustrate the differences between permutations and combinations in another way, suppose 2 cans of soup are to be selected from 4 cans on a shelf: noodle (N), bean (B), mushroom (M), and tomato (T). As shown in Figure 3(a), there are 12 ways to select 2 cans from the 4 cans if the order matters (if noodle first and bean second is considered different from bean, then noodle, for example). On the other hand, if order is unimportant, then there are 6 ways to choose 2 cans of soup from the 4, as illustrated in Figure 3(b). ◄

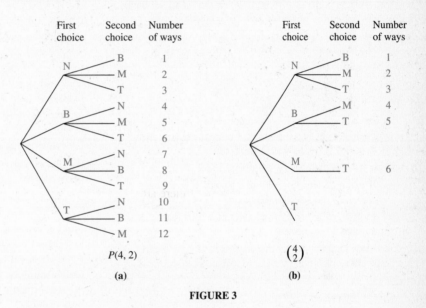

FIGURE 3

Caution It should be stressed that not all counting problems lend themselves to either permutations or combinations. Whenever a tree diagram or the multiplication principle can be used directly, it's best to use it.

6.2 Exercises

Evaluate the following combinations.

1. $\binom{8}{3}$ **2.** $\binom{8}{1}$ **3.** $\binom{12}{5}$ **4.** $\binom{10}{8}$ **5.** $\binom{6}{0}$ **6.** $\binom{5}{3}$

7. $\binom{21}{10}$ **8.** $\binom{34}{25}$ **9.** $\binom{25}{16}$ **10.** $\binom{30}{15}$ **11.** $\binom{27}{10}$ **12.** $\binom{28}{5}$

Use combinations to solve Exercises 13–18.

13. How many different 13-card bridge hands can be selected from an ordinary deck?

14. In how many ways can a hand of 6 clubs be chosen from an ordinary deck?

15. Five cards are marked with the numbers 1, 2, 3, 4, and 5, then shuffled, and 2 cards are drawn.

 (a) How many different 2-card combinations are possible?

 (b) How many 2-card hands contain a number less than 3?

16. Five cards are chosen from an ordinary deck. In how many ways is it possible to get the following results?

 (a) All queens

 (b) No face card

 (c) Exactly 2 face cards

 (d) At least 2 face cards

 (e) 1 heart, 2 diamonds, and 2 clubs

17. If a baseball coach has 5 good hitters and 4 poor hitters on the bench and chooses 3 players at random, in how many ways can he choose at least 2 good hitters?

18. An economics club has 30 members.

 (a) If a committee of 4 is to be selected, in how many ways can the selection be made?

 (b) In how many ways can a committee of at least 1 and at most 3 be selected?

19. Use a tree diagram to find the number of ways 2 letters can be chosen from the set {L, M, N} if order is important and

 (a) if repetition is allowed;

 (b) if no repeats are allowed.

 (c) Find the number of combinations of 3 elements taken 2 at a time. Does this answer differ from (a) or (b)?

20. Repeat Exercise 19 using the set {L, M, N, P}.

21. Explain the difference between a permutation and a combination.

22. Padlocks with digit dials are often referred to as "combination locks." According to the mathematical definition of combination, is this an accurate description? Explain.

Decide whether each of the following problems involves permutations or combinations, and then solve the problem.

23. In a club with 8 male and 11 female members, how many 5-member committees can be chosen that have

 (a) all men? **(b)** all women?

 (c) 3 men and 2 women?

24. In Exercise 23, how many committees can be selected that have

 (a) at least 4 women? **(b)** no more than 2 men?

25. A group of 3 students is to be selected from a group of 12 students to take part in a class in cell biology.

 (a) In how many ways can this be done?

 (b) In how many ways can the group who will not take part be chosen?

26. The coach of the Morton Valley Softball Team has 6 good hitters and 8 poor hitters. He chooses 3 hitters at random.

 (a) In how many ways can he choose 2 good hitters and 1 poor hitter?

 (b) In how many ways can he choose 3 good hitters?

 (c) In how many ways can he choose at least 2 good hitters?

27. In a game of musical chairs, 12 children will sit in 11 chairs arranged in a row (one will be left out). In how many ways can the 11 children find seats?

28. Marbles are drawn without replacement from a bag containing 15 marbles.

 (a) How many samples of 2 marbles can be drawn?

 (b) How many samples of 4 marbles can be drawn?

 (c) If the bag contains 3 yellow, 4 white, and 8 blue marbles, how many samples of 2 marbles can be drawn in which both marbles are blue?

29. From a pool of 7 secretaries, 3 are selected to be assigned to 3 managers. In how many ways can they be selected?

30. From 10 names on a ballot, 4 will be elected to a political party committee. In how many ways can the committee of 4 be formed if each person will have a different responsibility?

31. There are 5 rotten apples in a crate of 25 apples.

 (a) How many samples of 3 apples can be drawn from the crate?

 (b) How many samples of 3 could be drawn in which all 3 are rotten?

 (c) How many samples of 3 could be drawn in which there are two good apples and one rotten one?

32. A bag contains 5 black, 1 red, and 3 yellow jelly beans; you take 3 at random. How many samples are possible in which the jelly beans are

 (a) all black? **(b)** all red?

 (c) all yellow? **(d)** 2 black and 1 red?

 (e) 2 black and 1 yellow? **(f)** 2 yellow and 1 black?

 (g) 2 red and 1 yellow?

33. In how many ways can 5 out of 9 plants be arranged in a row on a windowsill?

34. A salesperson has the names of 6 prospects.
 (a) In how many ways can she arrange her schedule if she calls on all 6?
 (b) In how many ways can she arrange her schedule if she can call on only 4 of the 6?

35. Five orchids from a collection of 20 are to be selected for a flower show.

(a) In how many ways can this be done?

(b) In how many ways can the 5 be selected if 2 special plants must be included?

36. Show that $\begin{pmatrix} n \\ r \end{pmatrix} = \begin{pmatrix} n \\ n-r \end{pmatrix}$.

37. Derive a formula for $\begin{pmatrix} n \\ n-1 \end{pmatrix}$.

Applications

Business and Economics

38. **Worker Grievances** A group of 7 workers decides to send a delegation of 2 to their supervisor to discuss their grievances.
 (a) How many delegations are possible?
 (b) If it is decided that a particular worker must be in the delegation, how many different delegations are possible?
 (c) If there are 2 women and 5 men in the group, how many delegations would include at least 1 woman?

39. **Hamburger Variety** McDonald's sells hamburgers with regular meat or a McLean burger with a special low-fat meat. Either type can include cheese, relish, lettuce, tomato, mustard, or catsup.
 (a) How many different hamburgers can be ordered with exactly three extras?
 (b) How many different hamburgers with regular meat can be ordered with exactly three extras?
 (c) How many different hamburgers with regular meat can be ordered with at least five extras?

40. **Assembly Line Sampling** Five items are to be randomly selected from the first 50 items on an assembly line to determine the defect rate. How many different samples of 5 items can be chosen?

Life Sciences

41. **Research Participants** From a group of 16 smokers and 20 nonsmokers, a researcher wants to randomly select 8 smokers and 8 nonsmokers for a study. In how many ways can the study group be selected?

42. **Plant Hardiness** In an experiment on plant hardiness, a researcher gathers 6 wheat plants, 3 barley plants, and 2 rye plants. She wishes to select 4 plants at random.

(a) In how many ways can this be done?

(b) In how many ways can this be done if 2 wheat plants must be included?

Social Sciences

43. **Legislative Committee** A legislative committee consists of 5 Democrats and 4 Republicans. A delegation of 3 is to be selected to visit a small Pacific Island republic.
 (a) How many different delegations are possible?
 (b) How many delegations would have all Democrats?
 (c) How many delegations would have 2 Democrats and 1 Republican?
 (d) How many delegations would include at least 1 Republican?

General Interest

44. **Ice Cream Flavors** Baskin-Robbins advertises that it has 31 flavors of ice cream.
 (a) How many different double-scoop cones can be made?
 (b) How many different triple-scoop cones can be made?

45. **Lottery** A state lottery game requires that you pick 6 different numbers from 1 to 99. If you pick all 6 winning numbers, you win $1 million.
 (a) How many ways are there to choose 6 numbers if order is not important?
 (b) How many ways are there to choose 6 numbers if order matters?

46. **Lottery** In Exercise 45, if you pick 5 of the 6 numbers correctly, you win $250,000. In how many ways can you pick exactly 5 of the 6 winning numbers without regard to order?

6.3 PROBABILITY APPLICATIONS OF COUNTING PRINCIPLES

If three engines are tested from a shipping container packed with 12 diesel engines, 2 of which are defective, what is the probability that at least one of the defective engines will be found (in which case the container will not be shipped)?

This problem theoretically could be solved with a tree diagram, but it would require a tree with a large number of branches. Many of the probability problems involving *dependent* events that were solved earlier by using tree diagrams can also be solved by using combinations. Combinations are especially helpful when the numbers involved are large.

To compare the method of using combinations with the method of tree diagrams used in Section 5.5, the first example repeats Example 6 from that section.

EXAMPLE 1 From a box containing 3 white, 2 green, and 1 red marble, 2 marbles are drawn one at a time without replacement. Find the probability that 1 white and 1 green marble are drawn.

The use of combinations to solve probability problems depends on the basic probability principle introduced earlier and repeated here:

Let S be a sample space with n equally likely outcomes. Let event E contain m of these outcomes. Then the probability that event E occurs, written $P(E)$, is

$$P(E) = \frac{m}{n}.$$

In Example 6 of Section 5.5, it was necessary to consider the order in which the marbles were drawn. With combinations, this is not necessary. Simply count the number of ways in which 1 white and 1 green marble can be drawn from the given selection. The white marble can be drawn from the 3 white marbles in $\binom{3}{1}$ ways, and the green marble can be drawn from the 2 green marbles in $\binom{2}{1}$ ways. By the multiplication principle, both results can occur in

$$\binom{3}{1} \cdot \binom{2}{1} \text{ ways,}$$

giving the numerator of the probability fraction, $P(E) = m/n$. For the denominator, 2 marbles are to be drawn from a total of 6 marbles. This can occur in $\binom{6}{2}$ ways. The required probability is

$$P(1 \text{ white and } 1 \text{ green}) = \frac{\binom{3}{1}\binom{2}{1}}{\binom{6}{2}} = \frac{\frac{3!}{2!1!} \cdot \frac{2!}{1!1!}}{\frac{6!}{4!2!}} = \frac{6}{15} = \frac{2}{5}.$$

This agrees with the answer found earlier. ◀

EXAMPLE 2 From a group of 22 nurses, 4 are to be selected to present a list of grievances to management.

(a) In how many ways can this be done?

Four nurses from a group of 22 can be selected in $\binom{22}{4}$ ways. (Use combinations, since the group of 4 is an unordered set.)

$$\binom{22}{4} = \frac{22!}{18!4!} = \frac{(22)(21)(20)(19)}{(4)(3)(2)(1)} = 7315$$

There are 7315 ways to choose 4 people from 22.

(b) One of the nurses is Michael Branson. Find the probability that Branson will be among the 4 selected.

The probability that Branson will be selected is given by m/n, where m is the number of ways the chosen group includes him, and n is the total number of ways the group of 4 can be chosen. If Branson must be one of the 4 selected, the problem reduces to finding the number of ways that the 3 additional nurses can be chosen. The 3 are chosen from 21 nurses; this can be done in

$$\binom{21}{3} = \frac{21!}{18!3!} = 1330$$

ways, so $m = 1330$. Since n is the number of ways 4 nurses can be selected from 22,

$$n = \binom{22}{4} = 7315.$$

The probability that Branson will be one of the 4 chosen is

$$P(\text{Branson is chosen}) = \frac{1330}{7315} \approx .182.$$

(c) Find the probability that Branson will not be selected.

The probability that he will not be chosen is $1 - .182 = .818.$ ◀

EXAMPLE

3 When shipping diesel engines abroad, it is common to pack 12 engines in one container that is then loaded on a rail car and sent to a port. Suppose that a company has received complaints from its customers that many of the engines arrive in nonworking condition. To help solve this problem, the company decides to make a spot check of containers after loading. The company will test 3 engines from a container at random; if any of the 3 are nonworking, the container will not be shipped until each engine in it is checked. Suppose a given container has 2 nonworking engines. Find the probability that the container will not be shipped.

The container will not be shipped if the sample of 3 engines contains 1 or 2 defective engines. If $P(1 \text{ defective})$ represents the probability of exactly 1 defective engine in the sample, then

$$P(\text{not shipping}) = P(1 \text{ defective}) + P(2 \text{ defective}).$$

There are $\binom{12}{3}$ ways to choose the 3 engines for testing:

$$\binom{12}{3} = \frac{12!}{9!3!} = 220.$$

There are $\binom{2}{1}$ ways of choosing 1 defective engine from the 2 in the container, and for each of these ways, there are $\binom{10}{2}$ ways of choosing 2 good engines from among the 10 in the container. By the multiplication principle, there are

$$\binom{2}{1}\binom{10}{2} = \frac{2!}{1!1!} \cdot \frac{10!}{8!2!} = 90$$

ways of choosing a sample of 3 engines containing 1 defective engine with

$$P(1 \text{ defective}) = \frac{90}{220}.$$

There are $\binom{2}{2}$ ways of choosing 2 defective engines from the 2 defective engines in the container, and $\binom{10}{1}$ ways of choosing 1 good engine from among the 10 good engines, for

$$\binom{2}{2}\binom{10}{1} = \frac{2!}{0!2!} \cdot \frac{10!}{9!1!} = 10$$

ways of choosing a sample of 3 engines containing 2 defective engines. Finally,

$$P(2 \text{ defective}) = \frac{10}{220}$$

and

$$P(\text{not shipping}) = P(1 \text{ defective}) + P(2 \text{ defective})$$
$$= \frac{90}{220} + \frac{10}{220} = \frac{100}{220} \approx .455.$$

Notice that the probability is $1 - .455 = .545$ that the container will be shipped, even though it has 2 defective engines. The management must decide whether this probability is acceptable; if not, it may be necessary to test more than 3 engines from a container. ◀

Recall that if E and E' are complements, then $P(E') = 1 - P(E)$. In Example 3, the event "0 defective in the sample" is the complement of the event "1 or 2 defective in the sample," since there are only 0 or 1 or 2 defective engines possible in the sample of 3 engines.

Instead of finding the sum $P(1 \text{ defective}) + P(2 \text{ defective})$, the result in Example 3 could be found as $1 - P(0 \text{ defective})$.

$$P(\text{not shipping}) = 1 - P(0 \text{ defective in sample})$$
$$= 1 - \frac{\binom{2}{0}\binom{10}{3}}{\binom{12}{3}}$$
$$= 1 - \frac{1(120)}{220}$$
$$= 1 - \frac{120}{220} = \frac{100}{220} \approx .455$$

EXAMPLE 4

In a common form of the card game *poker*, a hand of 5 cards is dealt to each player from a deck of 52 cards. There are a total of

$$\binom{52}{5} = \frac{52!}{47!5!} = 2{,}598{,}960$$

such hands possible. Find the probability of getting each of the following hands.

(a) A hand containing only hearts, called a *heart flush*

There are 13 hearts in a deck, with

$$\binom{13}{5} = \frac{13!}{8!5!} = \frac{(13)(12)(11)(10)(9)}{(5)(4)(3)(2)(1)} = 1287$$

different hands containing only hearts. The probability of a heart flush is

$$P(\text{heart flush}) = \frac{1287}{2{,}598{,}960} \approx .000495.$$

(b) A flush of any suit (5 cards of the same suit)

There are 4 suits in a deck, so

$$P(\text{flush}) = 4 \cdot P(\text{heart flush}) = 4 \cdot .000495 \approx .00198.$$

(c) A full house of aces and eights (3 aces and 2 eights)

There are $\binom{4}{3}$ ways to choose 3 aces from among the 4 in the deck, and $\binom{4}{2}$ ways to choose 2 eights.

$$P(\text{3 aces, 2 eights}) = \frac{\binom{4}{3} \cdot \binom{4}{2}}{2{,}598{,}960} = \frac{4 \cdot 6}{2{,}598{,}960} \approx .00000923$$

(d) Any full house (3 cards of one value, 2 of another)

The 13 values in a deck give 13 choices for the first value, leaving 12 choices for the second value (order *is* important here, since a full house of 3 aces and 2 eights is not the same as a full house of 3 eights and 2 aces). From part (c), the probability for any *particular* full house is .00000923; the probability of *any* full house is

$$P(\text{full house}) = 13 \cdot 12 \cdot .00000923 \approx .00144. \quad \blacktriangleleft$$

EXAMPLE 5

A music teacher has 3 violin pupils, Fred, Carl, and Helen. For a recital, the teacher selects a first violinist and a second violinist. The third pupil will play with the others, but not solo. If the teacher selects randomly, what is the probability that Helen is first violist, Carl is second violinist, and Fred does not solo?

Use *permutations* to find the number of arrangements in the sample space.

$$P(3, 3) = 3! = 6$$

(Think of this as filling the positions of first violin, second violin, and no solo.) The 6 arrangements are equally likely, since the teacher will select randomly. Thus, the required probability is 1/6. \blacktriangleleft

EXAMPLE 6

Suppose a group of n people is in a room. Find the probability that at least 2 of the people have the same birthday.

"Same birthday" refers to the month and the day, not necessarily the same year. Also, ignore leap years, and assume that each day in the year is equally likely as a birthday. To see how to proceed, we first find the probability that *no 2 people* from among 5 people have the same birthday. There are 365 different birthdays possible for the first of the 5 people, 364 for the second (so that the people have different birthdays), 363 for the third, and so on. The number of ways the 5 people can have different birthdays is thus the number of permutations of 365 days taken 5 at a time or

$$P(365, 5) = 365 \cdot 364 \cdot 363 \cdot 362 \cdot 361.$$

The number of ways that 5 people can have the same or different birthdays is

$$365 \cdot 365 \cdot 365 \cdot 365 \cdot 365 = (365)^5.$$

Finally, the *probability* that none of the 5 people have the same birthday is

$$\frac{P(365, 5)}{(365)^5} = \frac{365 \cdot 364 \cdot 363 \cdot 362 \cdot 361}{365 \cdot 365 \cdot 365 \cdot 365 \cdot 365} \approx .973.$$

The probability that at least 2 of the 5 people *do* have the same birthday is $1 - .973 = .027$.

Now this result can be extended to more than 5 people. Generalizing, the probability that no 2 people among n people have the same birthday is

$$\frac{P(365, n)}{(365)^n}.$$

The probability that at least 2 of the n people *do* have the same birthday is

$$1 - \frac{P(365, n)}{(365)^n}.$$

The following table shows this probability for various values of n.

Number of People, n	Probability that Two Have the Same Birthday
5	.027
10	.117
15	.253
20	.411
22	.476
23	.507
25	.569
30	.706
35	.814
40	.891
50	.970
365	1

The probability that 2 people among 23 have the same birthday is .507, a little more than half. Many people are surprised at this result; it seems that a larger number of people should be required. ◀

6.3 Exercises

A basket contains 6 red apples and 4 yellow apples. A sample of 3 apples is drawn. Find the probabilities that the sample contains the following.

1. All red apples

2. All yellow apples

3. 2 yellow and 1 red apple

4. More red than yellow apples

Two cards are drawn at random from an ordinary deck of 52 cards.

5. How many 2-card hands are possible?

Find the probability that the 2-card hand described above contains the following.

6. 2 aces

7. At least 1 ace

8. All spades

9. 2 cards of the same suit

10. Only face cards

11. No face cards

12. No card higher than 8 (count ace as 1)

Twenty-six slips of paper are each marked with a different letter of the alphabet and placed in a basket. A slip is pulled out, its letter recorded (in the order in which the slip was drawn), and the slip is replaced. This is done 5 times. Find the probabilities that the following "words" are formed.

13. Chuck

14. A word that starts with p

15. A word with no repetition of letters

16. A word that contains no x, y, or z

Find the probabilities of the following hands at poker. Assume aces are either high or low.

17. Royal flush (5 highest cards of a single suit)

18. Straight flush (5 in a row in a single suit, but not a royal flush)

19. Four of a kind (4 cards of the same value)

20. Straight (5 cards in a row, not all of the same suit), with ace either high or low

A bridge hand is made up of 13 cards from a deck of 52. Set up the probabilities that a hand chosen at random contains the following.

21. Only hearts

22. 4 aces

23. Exactly 3 aces and exactly 3 kings

24. 6 of one suit, 5 of another, and 2 of another

25. At a conference of black writers in Detroit, special-edition books were selected to be given away in contests. There were 9 books written by Langston Hughes, 5 books by James Baldwin, and 7 books by Toni Morrison. The judge of one contest selected 6 books at random for prizes. Find the probabilities that the selection consisted of

 (a) 3 Hughes and 3 Morrison books;

 (b) exactly 4 Baldwin books;

 (c) 2 Hughes, 3 Baldwin, and 1 Morrison book;

 (d) at least 4 Hughes books;

 (e) exactly 4 books written by males (Morrison is female).

 (f) no more than 2 books written by Baldwin.

26. At the first meeting of a committee to plan a Northern California pow-wow, there were 3 women and 3 men from the Miwok tribe, 2 men and 3 women from the Hoopa tribe, and 4 women and 5 men from the Pomo tribe. If the ceremony subcouncil consists of 5 people, and is randomly selected, find the probabilities that the subcouncil contains

 (a) 3 men and 2 women;

 (b) exactly 3 Miwoks and 2 Pomos;

 (c) 2 Miwoks, 2 Hoopas, and a Pomo;

 (d) 2 Miwoks, 2 Hoopas, and 2 Pomos;

 (e) more women than men;

 (f) exactly 3 Hoopas;

 (g) at least 2 Pomos.

27. Discuss the relative merits of using tree diagrams versus combinations to solve probability problems. When would each approach be most appropriate?

28. Several examples in this section used the rule $P(E') = 1 - P(E)$. Explain the advantage (especially in Example 6) of using this rule.

For Exercises 29–32, refer to Example 6 in this section.

29. Set up the probability that at least 2 of the 41 men who have served as president of the United States before have had the same birthday.*

30. What is the probability that at least 2 of the 100 U.S. Senators have the same birthday.

31. What is the probability that 2 of the 435 members of the House of Representatives have the same birthday?

32. Show that the probability that in a group of n people *exactly one pair* have the same birthday is
$$\binom{n}{2} \cdot \frac{P(365, n - 1)}{(365)^n}.$$

33. An elevator has 4 passengers and stops at 7 floors. It is equally likely that a person will get off at any one of the 7 floors. Find the probability that no 2 passengers leave at the same floor.

34. On National Public Radio, the "Weekend Edition" program on Sunday, September 7, 1991, posed the following probability problem: Given a certain number of balls, of which some are blue, pick 5 at random. The probability that all 5 are blue is 1/2. Determine the original number of balls and decide how many were blue.

Applications

Business and Economics

Quality Control *A shipment of 9 typewriters contains 2 that are defective. Find the probabilities that a sample of each of the following sizes, drawn from the 9, will not contain a defective typewriter.*

35. 1 **36.** 2 **37.** 3 **38.** 4

Refer to Example 3. The managers feel that the probability of .545 that a container will be shipped even though it contains 2 defective engines is too high. They decide to increase the sample size chosen. Find the probabilities that a container will be shipped even though it contains 2 defective engines, if the sample size is increased to the following.

39. 4 **40.** 5

General Interest

41. **Lottery** In the previous section, we found the number of ways to pick 6 different numbers from 1 to 99 in a state lottery. Assuming order is unimportant, what is the probability of picking all 6 numbers correctly to win the big prize?

42. **Lottery** In Exercise 41, what is the probability of picking exactly 5 of the 6 numbers correctly?

43. **Football** The Big Eight Conference during the 1988 college football season ended the season in a "perfect progression", as shown in the table below.†

Won	Lost	Team
7	0	Nebraska (NU)
6	1	Oklahoma (OU)
5	2	Oklahoma State (OSU)
4	3	Colorado (CU)
3	4	Iowa State (ISU)
2	5	Missouri (MU)
1	6	Kansas (KU)
0	7	Kansas State (KSU)

Someone wondered what the probability of such an outcome might be.

(a) Assuming no ties and that each team had an equally likely probability of winning each game, find the probability of the perfect progression shown above.

(b) Find a general expression for the probability of a perfect progression in an n-team league with the same assumptions.

*In fact, James Polk and Warren Harding were both born on November 2.
†From Richard Madsen, "On the Probability of a Perfect Progression," *The American Statistician,*
August 1991, vol. 45, no. 3, p. 214.

📊 **Computer/Graphing Calculator**

44. Use a computer or a graphing calculator and the Monte Carlo method with $n = 50$ to estimate the probabilities of the following hands at poker. Assume aces are either high or low. Since each hand has 5 cards, you will need $50 \cdot 5 = 250$ random numbers to "look at" 50 hands. Compare these experimental results with the theoretical results.

(a) A pair of aces

(b) Any two cards of the same value

(c) Three of a kind

45. Use a computer or a graphing calculator and the Monte Carlo method with $n = 20$ to estimate the probabilities of the following 13-card bridge hands. Since each hand has 13 cards, you will need $20 \cdot 13 = 260$ random numbers to "look at" 20 hands.

(a) No aces

(b) 2 kings and 2 aces

(c) No cards of any one suit—that is, only 3 suits represented

▶ 6.4 BERNOULLI TRIALS

 What is the probability that 3 of 6 people prefer Diet Supercola over its competitors?

This question involves an experiment that is repeated 6 times. Many probability problems are concerned with experiments in which an event is repeated many times. Other examples include finding the probability of getting 7 heads in 8 tosses of a coin, of hitting a target 6 times out of 6, and of finding 1 defective item in a sample of 15 items. Probability problems of this kind are called **Bernoulli trials** problems, or **Bernoulli processes,** named after the Swiss mathematician Jakob Bernoulli (1654–1705), who is well known for his work in probability theory. In each case, some outcome is designated a success, and any other outcome is considered a failure. Thus, if the probability of a success in a single trial is p, the probability of failure will be $1 - p$. A Bernoulli trials problem, or **binomial experiment**, must satisfy the following conditions.

BERNOULLI TRIALS

1. The same experiment is repeated several times.

2. There are only two possible outcomes, success and failure.

3. The repeated trials are independent, so that the probability of success remains the same for each trial.

Consider a typical problem of this type: find the probability of getting 5 ones on 5 rolls of a die. Here, the experiment, rolling a die, is repeated five times. If getting a one is designated as a success, then getting any other outcome is a failure. The five trials are independent; the probability of success (getting a one) is $1/6$ on each trial, while the probability of a failure (any other result) is $5/6$.

Thus, the required probability is

$$P(5 \text{ ones on } 5 \text{ rolls}) = P(1) \cdot P(1) \cdot P(1) \cdot P(1) \cdot P(1) = \left(\frac{1}{6}\right)^5$$

$$\approx .00013.$$

Now suppose the problem is changed to that of finding the probability of getting a one exactly 4 times in 5 rolls of the die. Again, a success on any roll of the die is defined as getting a one, and the trials are independent, with the same probability of success each time. The desired outcome for this experiment can occur in more than one way, as shown below, where *s* represents a success (getting a one), and *f* represents a failure (getting any other result), and each row represents a different outcome.

$$
\begin{array}{ccccc}
s & s & s & s & f \\
s & s & s & f & s \\
s & s & f & s & s \\
s & f & s & s & s \\
f & s & s & s & s
\end{array}
$$

The probability of each of these five outcomes is

$$\left(\frac{1}{6}\right)^4\left(\frac{5}{6}\right).$$

Recall the union rule for mutually exclusive events:

$$P(A \text{ or } B) = P(A \cup B)$$
$$= P(A) + P(B).$$

In the example, the probabilities of the five events are the same, so multiplying by 5 gives the same result as adding them.

Since the five outcomes represent mutually exclusive events, add the probabilities to get

$$P(4 \text{ ones in } 5 \text{ rolls}) = 5\left(\frac{1}{6}\right)^4\left(\frac{5}{6}\right) = \frac{5^2}{6^5} \approx .0032.$$

In the same way, we can compute the probability of rolling a one exactly 3 times in 5 rolls of a die. The probability of any one way of achieving 3 successes and 2 failures will be

$$\left(\frac{1}{6}\right)^3\left(\frac{5}{6}\right)^2.$$

Again, the desired outcome can occur in more than one way. To see this, let the set {1, 2, 3, 4, 5} represent the first, second, third, fourth, and fifth tosses. The number of 3-element subsets of this set will correspond to the number of ways in which 3 successes and 2 failures can occur. Using combinations, there are $\binom{5}{3}$ such subsets. Since $\binom{5}{3} = 5!/(2!3!) = 10$,

$$P(3 \text{ ones in } 5 \text{ rolls}) = 10\left(\frac{1}{6}\right)^3\left(\frac{5}{6}\right)^2$$

$$= \frac{250}{6^5} \approx .032.$$

A similar argument works in the general case.

BINOMIAL PROBABILITY

If p is the probability of success in a single trial of a binomial experiment, the probability of x successes and $n - x$ failures in n independent repeated trials of the experiment is

$$\binom{n}{x} \cdot p^x \cdot (1 - p)^{n-x}.$$

This formula plays an important role in the study of *binomial distributions*, which are discussed in the next chapter.

EXAMPLE 1 The advertising agency that handles the Diet Supercola account believes that 40% of all consumers prefer this product over its competitors. Suppose a sample of 6 people is chosen. Assume that all responses are independent of each other. Find the probability of each of the following.

(a) Exactly 3 of the 6 people prefer Diet Supercola.

Think of the 6 responses as 6 independent trials. A success occurs if a person prefers Diet Supercola. Then this is a binomial experiment with $p = P(\text{success}) = P(\text{prefer Diet Supercola}) = .4$. The sample is made up of 6 people, so $n = 6$. To find the probability that exactly 3 people prefer this drink, let $x = 3$ and use the formula in the box.

$$P(\text{exactly } 3) = \binom{6}{3}(.4)^3(1 - .4)^{6-3}$$
$$= 20(.4)^3(.6)^3$$
$$= 20(.064)(.216)$$
$$= .27648$$

(b) None of the 6 people prefer Diet Supercola.

Let $x = 0$.

$$P(\text{exactly } 0) = \binom{6}{0}(.4)^0(1 - .4)^6 = 1(1)(.6)^6 \approx .0467 \quad \blacktriangleleft$$

EXAMPLE 2 Find the probability of getting exactly 7 heads in 8 tosses of a fair coin.

The probability of success (getting a head in a single toss) is $1/2$. The probability of a failure (getting a tail) is $1 - 1/2 = 1/2$. Thus,

$$P(7 \text{ heads in } 8 \text{ tosses}) = \binom{8}{7}\left(\frac{1}{2}\right)^7\left(\frac{1}{2}\right)^1 = 8\left(\frac{1}{2}\right)^8 = .03125. \quad \blacktriangleleft$$

3 Assuming that selection of items for a sample can be treated as independent trials, and that the probability that any 1 item is defective is .01, find the following.

(a) The probability of 1 defective item in a random sample of 15 items from a production line

Here, a "success" is a defective item. Since selecting each item for the sample is assumed to be an independent trial, the binomial probability formula applies. The probability of success (a defective item) is .01, while the probability of failure (an acceptable item) is .99. This makes

$$P(1 \text{ defective in 15 items}) = \binom{15}{1}(.01)^1(.99)^{14}$$
$$= 15(.01)(.99)^{14}$$
$$\approx .130.$$

(b) The probability of at most 1 defective item in a random sample of 15 items from a production line

"At most 1" means 0 defective items or 1 defective item. Since 0 defective items is equivalent to 15 acceptable items,

$$P(0 \text{ defective}) = (.99)^{15} \approx .860.$$

Use the union rule, noting that 0 defective and 1 defective are mutually exclusive events, to get

$$P(\text{at most 1 defective}) = P(0 \text{ defective}) + P(1 \text{ defective})$$
$$\approx .860 + .130$$
$$= .990. \blacktriangleleft$$

4 A new style of shoe is sweeping the country. In one area, 30% of all the shoes sold are of this type. Assume that these sales are independent events, and find the following probabilities using Bernoulli trials.

(a) Of 10 people who buy shoes, at least 8 buy the new shoe style.

Let success be "buy the new style," so that $P(\text{success}) = .3$. For at least 8 people out of 10 to buy the shoe, it must be sold to 8, 9, or 10 people. Thus,

$$P(\text{at least 8}) = P(8) + P(9) + P(10)$$
$$= \binom{10}{8}(.3)^8(.7)^2 + \binom{10}{9}(.3)^9(.7)^1 + \binom{10}{10}(.3)^{10}(.7)^0$$
$$\approx .0014467 + .0001378 + .0000059$$
$$= .0015904.$$

(b) Of 10 people who buy shoes, no more than 7 buy the new shoe style.

"No more than 7" means 0, 1, 2, 3, 4, 5, 6, or 7 people buy the shoe. The required probability could be found by adding $P(0)$, $P(1)$, and so on, but it is

easier to use the formula $P(E) = 1 - P(E')$. The complement of "no more than 7" is "8 or more," so

$$P(\text{no more than } 7) = 1 - P(8 \text{ or more})$$
$$= 1 - .0015904 \qquad \text{Answer from part (a)}$$
$$= .9984096. \blacktriangleleft$$

The triangular array of numbers shown below is called **Pascal's triangle** in honor of the seventeenth-century mathematician Blaise Pascal (1623–1662), who was one of the first to use it extensively.

PASCAL'S TRIANGLE

```
               1
            1     1
         1     2     1
      1     3     3     1
   1     4     6     4     1
1  .  5  . 10  . 10  .  5  . 1
   :     :     :     :     :
```

This array provides a quick way to find binomial probabilities. The nth row of the triangle, where $n = 0, 1, 2, 3, \ldots$, gives the coefficients $\binom{n}{r}$ for $r = 0, 1, 2, 3, \ldots, n$. For example, for $n = 4$, $1 = \binom{4}{0}$, $4 = \binom{4}{1}$, $6 = \binom{4}{2}$, and so on. Each number in the triangle is the sum of the two numbers directly above it. For example, in the row for $n = 4$, 1 is the sum of 1, the only number above it, 4 is the sum of 1 and 3, 6 is the sum of 3 and 3, and so on. Adding in this way gives the sixth row:

$$1 \qquad 6 \qquad 15 \qquad 20 \qquad 15 \qquad 6 \qquad 1.$$

EXAMPLE 5 Use Pascal's triangle to find the probability in Example 4 that of 6 people who buy shoes, at least 3 buy the new style.

The probability of success is .3. Since at least 3 means 3, 4, 5, or 6,

$$P(\text{at least } 3) = P(3) + P(4) + P(5) + P(6)$$

$$= \binom{6}{3}(.3)^3(.7)^3 + \binom{6}{4}(.3)^4(.7)^2$$

$$+ \binom{6}{5}(.3)^5(.7)^1 + \binom{6}{6}(.3)^6(.7)^0.$$

Use the sixth row of Pascal's triangle for the combinations to get

$$P(\text{at least } 3) = 20(.3)^3(.7)^3 + 15(.3)^4(.7)^2 + 6(.3)^5(.7)^1 + 1(.3)^6(.7)^0$$

$$= .1852 + .0595 + .0044 + .0007$$

$$= .2498. \blacktriangleleft$$

6.4 Exercises

Suppose that a family has 5 children. Also, suppose that the probability of having a girl is 1/2. Find the probabilities that the family has the following children.

1. Exactly 2 girls and 3 boys

2. Exactly 3 girls and 2 boys

3. No girls

4. No boys

5. At least 4 girls

6. At least 3 boys

7. No more than 3 boys

8. No more than 4 girls

A die is rolled 12 times. Find the probabilities of rolling the following.

9. Exactly 12 ones

10. Exactly 6 ones

11. Exactly 1 one

12. Exactly 2 ones

13. No more than 3 ones

14. No more than 1 one

A coin is tossed 5 times. Find the probabilities of getting the following.

15. All heads

16. Exactly 3 heads

17. No more than 3 heads

18. At least 3 heads

19. How do you identify a probability problem that involves a binomial experiment?

20. How is Pascal's triangle used to find probabilities?

Applications

Business and Economics

Fast Food *According to a recent article in a business publication, only 20% of the population of the United States has never had a Big Burg hamburger at a major fast-food chain. Assume independence and find the probabilities that a random sample of 10 people produces the following data.*

21. Exactly 2 people have never had a Big Burg.

22. Exactly 5 people have never had a Big Burg.

23. Three people or fewer have never had a Big Burg.

24. Four people or more *have* had a Big Burg.

Insurance Rating *The insurance industry has found that the probability is .9 that a life insurance applicant will qualify at the regular rates. Find the probabilities that of the next 10 applicants for life insurance, the following numbers will qualify at the regular rates.*

25. Exactly 10

26. Exactly 9

27. At least 9

28. Less than 8

Personnel Screening *A company gives prospective workers a 6-question, multiple-choice test. Each question has 5 possible answers, so that there is a 1/5 or 20% chance of answering a question correctly just by guessing. Find the probabilities of getting the following results by chance.*

29. Exactly 2 correct answers

30. No correct answers

31. At least 4 correct answers

32. No more than 3 correct answers

33. **Customer Satisfaction** Over the last decade, 10% of all clients of J. K. Loss & Company have lost their life savings. Suppose a sample of 3 of the current clients of the firm is chosen. Assuming independence, find the probability that exactly 1 of the 3 clients will lose everything.

Quality Control *A factory tests a random sample of 20 transistors for defective transistors. The probability that a particular transistor will be defective has been established by past experience as .05.*

34. What is the probability that there are no defective transistors in the sample?

35. What is the probability that the number of defective transistors in the sample is at most 2?

Life Sciences

Drug Effectiveness *A new drug cures 70% of the people taking it. Suppose 20 people take the drug; find the probabilities of the following.*

36. Exactly 18 people are cured.

37. Exactly 17 people are cured.

38. At least 17 people are cured.

39. At least 18 people are cured.

Births of Twins *The probability that a birth will result in twins is .012. Assuming independence (perhaps not a valid assumption), what are the probabilities that out of 100 births in a hospital, there will be the following numbers of sets of twins?*

40. Exactly 2 sets of twins

41. At most 2 sets of twins

Vitamin A Deficiency *Six mice from the same litter, all suffering from a vitamin A deficiency, are fed a certain dose of carrots. If the probability of recovery under such treatment is .70, find the probabilities of the following results.*

42. None of the mice recover.

43. Exactly 3 of the 6 mice recover.

44. All of the mice recover.

45. No more than 3 mice recover.

46. Effects of Radiation In an experiment on the effects of a radiation dose on cells, a beam of radioactive particles is aimed at a group of 10 cells. Find the probability that 8 of the cells will be hit by the beam, if the probability that any single cell will be hit is .6. (Assume independence.)

47. Effects of Radiation The probability of a mutation of a given gene under a dose of 1 roentgen of radiation is approximately 2.5×10^{-7}. What is the probability that in 10,000 genes, at least 1 mutation occurs?

48. Drug Side Effects A new drug being tested causes a serious side effect in 5 out of 100 patients. What is the probability that no side effects occur in a sample of 10 patients taking the drug?

Social Sciences

49. Minorities in Schools A study recently released by the College Board and the Western Interstate Commission for Higher Education predicts that by 1995, one-third of U.S. public elementary and secondary school students will belong to ethnic minorities. Assuming independence, find the probability that of 5 students selected at random in 1995 from schools throughout the nation, no more than 3 belong to ethnic minorities.

Testing *In a 10-question, multiple-choice biology test with 5 choices for each question, an unprepared student guesses the answer to each item. Find the probabilities of the following results.*

50. Exactly 6 correct answers

51. Exactly 7 correct answers

52. At least 8 correct answers

53. Less than 8 correct answers

54. Community College Population According to the state of California, 33% of all state community college students belong to ethnic minorities. Find the probabilities of the following results in a random sample of 10 California community college students.

(a) Exactly 2 belong to an ethnic minority.

(b) Three or fewer belong to an ethnic minority.

(c) Exactly 5 do not belong to an ethnic minority.

(d) Six or more do not belong to an ethnic minority.

Computer/Graphing Calculator

Use a computer or a calculator to compute each of the probabilities in Exercises 55–58.

55. Flu Inoculations A flu vaccine has a probability of 80% of preventing a person who is inoculated from getting the flu. A county health office inoculates 134 people. Find the probabilities of the following.

(a) Exactly 10 of the people inoculated get the flu.

(b) No more than 10 of the people inoculated get the flu.

(c) None of the people inoculated get the flu.

56. Color Blindness The probability that a male will be color-blind is .042. Find the probabilities that in a group of 53 men, the following will be true.

(a) Exactly 5 are color-blind.

(b) No more than 5 are color-blind.

(c) At least 1 is color-blind.

57. Quality Control The probability that a certain machine turns out a defective item is .05. Find the probabilities that in a run of 75 items, the following results are obtained.

(a) Exactly 5 defective items

(b) No defective items

(c) At least 1 defective item

58. Survey Results A company is taking a survey to find out whether people like its product. Its last survey indicated that 70% of the population like the product. Based on that, in a sample of 58 people, find the probabilities of the following.

(a) All 58 like the product.

(b) From 28 to 30 (inclusive) like the product.

6.5 PROBABILITY DISTRIBUTIONS; EXPECTED VALUE

What is the expected value of winning a prize for someone who buys one ticket in a raffle?

In this section we shall see that the *expected value* of a probability distribution is a type of average. Probability distributions were introduced briefly in the chapter on sets and probability. Now we take a more complete look at probability distributions. A probability distribution depends on the idea of a *random variable*, so we begin with that.

Random Variables Suppose a bank is interested in improving its services to the public. The manager decides to begin by finding the amount of time tellers spend on each transaction, rounded to the nearest minute. To each transaction, then, will be assigned one of the numbers 0, 1, 2, 3, 4, That is, if *t* represents an outcome of the experiment of timing a transaction, then *t* may take on any of the values from the list 0, 1, 2, 3, 4, Since the value that *t* takes on for a particular transaction is random, *t* is called a random variable.

RANDOM VARIABLE	A **random variable** is a function that assigns a real number to each outcome of an experiment.

Table 1

Time	Frequency
1	3
2	5
3	9
4	12
5	15
6	11
7	10
8	6
9	3
10	1
	Total: 75

Probability Distribution Suppose that the bank manager in our example records the times for 75 different transactions, with results as shown in Table 1. As the table shows, the shortest transaction time was 1 minute, with 3 transactions of 1-minute duration. The longest time was 10 minutes; only one transaction took that long.

In Table 1, the first column gives the ten values assumed by the random variable *t*, and the second column shows the number of occurrences corresponding to each of these values, or the **frequency** of that value. Table 1 is an example of a **frequency distribution,** a table listing the frequencies for each value a random variable may assume.

Now suppose that several weeks after starting new procedures to speed up transactions, the manager takes another survey. This time she observes 57 transactions, and she records their times as shown in the frequency distribution in Table 2. (Table 1 is repeated to allow comparison of the two tables.)

Table 1

Time	Frequency
1	3
2	5
3	9
4	12
5	15
6	11
7	10
8	6
9	3
10	1
	Total: 75

Table 2

Time	Frequency
1	4
2	5
3	8
4	10
5	12
6	17
7	0
8	1
9	0
10	0
	Total: 57

Do the results in Table 2 indicate an improvement? It is hard to compare the two tables, since one is based on 75 transactions and the other on 57. To make them comparable, we can add a column to each table that gives the *relative frequency* of each transaction time. These results are shown in Tables 3 and 4. (Where necessary, decimals are rounded to the nearest hundredth.) To find a **relative frequency,** divide each frequency by the total of the frequencies. Here the individual frequencies are divided by 75 or 57.

Table 3

Time	Frequency	Relative Frequency
1	3	$3/75 = .04$
2	5	$5/75 \approx .07$
3	9	$9/75 = .12$
4	12	$12/75 = .16$
5	15	$15/75 = .20$
6	11	$11/75 \approx .15$
7	10	$10/75 \approx .13$
8	6	$6/75 = .08$
9	3	$3/75 = .04$
10	1	$1/75 \approx .01$

Table 4

Time	Frequency	Relative Frequency
1	4	$4/57 \approx .07$
2	5	$5/57 \approx .09$
3	8	$8/57 \approx .14$
4	10	$10/57 \approx .18$
5	12	$12/57 \approx .21$
6	17	$17/57 \approx .30$
7	0	$0/57 = 0$
8	1	$1/57 \approx .02$
9	0	$0/57 = 0$
10	0	$0/57 = 0$

Whether the differences in relative frequency between the distributions in Tables 3 and 4 are interpreted as desirable or undesirable depends on management goals. If the manager wanted to eliminate the most time-consuming transactions, the results appear to be desirable. However, before the new procedures were followed, the largest relative frequency of transactions, .20, was for transactions of 5 minutes. With the new procedures, the largest relative frequency, .30, corresponds to a transaction time of 6 minutes. At any rate, the results shown in the two tables are easier to compare using relative frequencies.

The relative frequencies in Tables 3 and 4 can be considered as probabilities. Such a table that lists all the possible values of a random variable, together with the corresponding probabilities, is called a **probability distribution.** The sum of the probabilities in a probability distribution must always equal 1. (The sum in an actual distribution may vary slightly from 1 because of rounding.)

EXAMPLE 1 Many plants have seed pods with variable numbers of seeds. One variety of green beans has no more than 6 seeds per pod. Suppose that examination of 30 such bean pods gives the results shown in Table 5. Here the random variable x tells the number of seeds per pod. Give a probability distribution for these results.

The probabilities are found by computing the relative frequencies. A total of 30 bean pods were examined, so each frequency should be divided by 30 to get the probability distribution shown as Table 6. (Some of the results have been rounded to the nearest hundredth.)

Table 5

x	Frequency
0	3
1	4
2	6
3	8
4	5
5	3
6	1
	Total: 30

Table 6

x	Frequency	Probability
0	3	$3/30 = .10$
1	4	$4/30 \approx .13$
2	6	$6/30 = .20$
3	8	$8/30 \approx .27$
4	5	$5/30 \approx .17$
5	3	$3/30 = .10$
6	1	$1/30 \approx .03$
	Total: 30	

As shown in Table 6, the probability that the random variable, x, takes on the value 2 is 6/30, or .20. This is often written as

$$P(x = 2) = .20.$$

Also, $P(x = 5) = .10$, and $P(x = 6) \approx .03$. ◄

Instead of writing the probability distribution in Example 1 as a table, we could write the same information as a set of ordered pairs:

$$\{(0, .10), (1, .13), (2, .20), (3, .27), (4, .17), (5, .10), (6, .03)\}.$$

There is just one probability for each value of the random variable. Thus, a probability distribution defines a function, called a **probability distribution function,** or simply a **probability function.** We shall use the terms ''probability distribution'' and ''probability function'' interchangeably.

The information in a probability distribution is often displayed graphically as a special kind of bar graph called a **histogram.** The bars of a histogram all have the same width, usually 1. The heights of the bars are determined by the

frequencies. A histogram for the data in Table 3 is given in Figure 4. A histogram shows important characteristics of a distribution that may not be readily apparent in tabular form, such as the relative sizes of the probabilities and any symmetry in the distribution.

The area of the bar above $t = 1$ in Figure 4 is the product of 1 and .04, or $.04 \cdot 1 = .04$. Since each bar has a width of 1, its area is equal to the probability that corresponds to that value of t. The probability that a particular value will occur is thus given by the area of the appropriate bar of the graph. For example, the probability that a transaction will take less than four minutes is the sum of the areas for $t = 1$, $t = 2$, and $t = 3$. This area, shown in color in Figure 5, corresponds to 23% of the total area, since

$$P(t < 4) = P(t = 1) + P(t = 2) + P(t = 3)$$
$$= .04 + .07 + .12 = .23.$$

FIGURE 4

FIGURE 5

EXAMPLE 2 Construct a histogram for the probability distribution in Example 1. Find the area that gives the probability that the number of seeds will be more than 4.

A histogram for this distribution is shown in Figure 6. The portion of the histogram in color represents

$$P(x > 4) = P(x = 5) + P(x = 6)$$
$$= .10 + .03 = .13,$$

or 13% of the total area. ◀

FIGURE 6

EXAMPLE 3 **(a)** Give the probability distribution for the number of heads showing when two coins are tossed.

Let x represent the random variable "number of heads." Then x can take on the values 0, 1, or 2. Now find the probability of each outcome. The results are shown in Table 7.

(b) Draw a histogram for the distribution in Table 7. Find the probability that at least one coin comes up heads.

The histogram is shown in Figure 7. The portion in color represents

$$P(x \geq 1) = P(x = 1) + P(x = 2)$$
$$= \frac{3}{4}.$$

◀

Table 7

x	$P(x)$
0	1/4
1	1/2
2	1/4

FIGURE 7

Expected Value In working with experimental data, it is often useful to have a typical or "average" number that represents the entire set of data. For example, we compare our heights and weights to those of the typical or "average" person on weight charts. Students are familiar with the "class average" and their own "average" in a given course.

In a recent year, a citizen of the United States could expect to complete about 13 years of school, to be a member of a household earning $30,800 per year, and to live in a household of 3.17 people. What do we mean here by "expect"? Many people have completed less than 13 years of school; many others have completed more. Many households have less income than $30,800 per year; many others have more. The idea of a household of 3.17 people is a little hard to swallow. The numbers all refer to *averages*. When the term "expect" is used in this way, it refers to *mathematical expectation*, which is a kind of average.

As an example, let us find the average number of offspring for a certain species of pheasant, given the probability distribution in Table 8.

Table 8

Number of Offspring	Frequency	Probability
0	8	.08
1	14	.14
2	29	.29
3	32	.32
4	17	.17
	Total: 100	

We might be tempted to find the typical number of offspring by averaging the numbers 0, 1, 2, 3, and 4, which represent the numbers of offspring possible. This won't work, however, since the various numbers of offspring do not occur with equal probability: for example, a brood with 3 offspring is much more common than one with 0 or 1 offspring. The differing probabilities of occurrence can be taken into account with a **weighted average,** found by multiplying each of the possible numbers of offspring by its corresponding probability, as follows:

$$\text{Average number of offspring} = \frac{\text{Total offspring}}{\text{Number of broods}}$$

$$= \frac{0 \cdot 8 + 1 \cdot 14 + 2 \cdot 29 + 3 \cdot 32 + 4 \cdot 17}{100}$$

$$= 0 \cdot \frac{8}{100} + 1 \cdot \frac{14}{100} + 2 \cdot \frac{29}{100} + 3 \cdot \frac{32}{100} + 4 \cdot \frac{17}{100}$$

$$= 0(.08) + 1(.14) + 2(.29) + 3(.32) + 4(.17)$$

$$= 0 + .14 + .58 + .96 + .68$$

$$= 2.36.$$

Based on the data given above, a typical brood of pheasants includes 2.36 offspring.

It is certainly not possible for a pair of pheasants to produce 2.36 offspring. If the numbers of offspring produced by many different pairs of pheasants are found, however, the average of these numbers will be about 2.36.

We can use the results of this example to define the *expected value* of a probability distribution as follows.

EXPECTED VALUE Suppose the random variable x can take on the n values $x_1, x_2, x_3, \ldots, x_n$. Also, suppose the probabilities that these values occur are, respectively, $p_1, p_2, p_3, \ldots, p_n$. Then the **expected value** of the random variable is

$$E(x) = x_1 p_1 + x_2 p_2 + x_3 p_3 + \cdots + x_n p_n.$$

As in the example above, the expected value of a random variable may be a number that can never occur in any one trial of the experiment.

$E(x) = 2.36$

Offspring

FIGURE 8

Physically, the expected value of a probability distribution represents a balance point. Figure 8 shows a histogram for the distribution of the pheasant offspring. If the histogram is thought of as a series of weights with magnitudes represented by the heights of the bars, then the system would balance if supported at the point corresponding to the expected value.

EXAMPLE 4

Suppose a local church decides to raise money by raffling a microwave oven worth $400. A total of 2000 tickets are sold at $1 each. Find the expected value of winning for a person who buys one ticket in the raffle.

Here the random variable represents the possible amounts of net winnings, where net winnings = amount won − cost of ticket. The net winnings of the person winning the oven are $400 (amount won) − $1 (cost of ticket) = $399. The net winnings for each losing ticket are $0 − $1 = −$1.

The probability of winning is 1 in 2000, or 1/2000, while the probability of losing is 1999/2000. See Table 9. The expected winnings for a person buying one ticket are

$$399\left(\frac{1}{2000}\right) + (-1)\left(\frac{1999}{2000}\right) = \frac{399}{2000} - \frac{1999}{2000}$$

$$= -\frac{1600}{2000}$$

$$= -.80.$$

On the average, a person buying one ticket in the raffle will lose $.80, or 80 cents.

It is not possible to lose 80¢ in this raffle: either you lose $1, or you win a $400 prize. If you bought tickets in many such raffles over a long period of time, however, you would lose 80¢ per ticket, on the average. ◄

Table 9

Outcome (Net Winnings)	Probability
$399	$\dfrac{1}{2000}$
−$1	$\dfrac{1999}{2000}$

EXAMPLE 5

What is the expected number of girls in a family with exactly 3 children? Some families with 3 children have 0 girls, others have 1 girl, and so on. We need to find the probabilities associated with 0, 1, 2, or 3 girls in a family of 3 children. To find these probabilities, first write the sample space, S, of all possible 3-child families: $S = \{ggg, ggb, bgg, gbg, gbb, bgb, bbg, bbb\}$. This sample

Table 10

Outcome (Number of Girls)	Probability
0	1/8
1	3/8
2	3/8
3	1/8

space gives the probabilities shown in Table 10, assuming that the probability of a girl at each birth is 1/2. The expected number of girls can now be found by multiplying each outcome (number of girls) by its corresponding probability and finding the sum of these values.

$$\text{Expected number of girls} = 0 \cdot \frac{1}{8} + 1 \cdot \frac{3}{8} + 2 \cdot \frac{3}{8} + 3 \cdot \frac{1}{8}$$

$$= \frac{3}{8} + \frac{6}{8} + \frac{3}{8}$$

$$= \frac{12}{8}$$

$$= \frac{3}{2} = 1.5$$

On the average, a 3-child family will have 1.5 girls. This result agrees with our intuition that, on the average, half the children born will be girls. ◀

EXAMPLE 6

Each day Donna and Mary toss a coin to see who buys the coffee (40 cents a cup). One tosses and the other calls the outcome. If the person who calls the outcome is correct, the other buys the coffee; otherwise the caller pays. Find Donna's expected winnings.

Assume that an honest coin is used, that Mary tosses the coin, and that Donna calls the outcome. The possible results and corresponding probabilities are shown below.

	Possible Results			
Result of toss	Heads	Heads	Tails	Tails
Call	Heads	Tails	Heads	Tails
Caller wins?	Yes	No	No	Yes
Probability	1/4	1/4	1/4	1/4

Donna wins a 40¢ cup of coffee whenever the results and calls match, and she loses a 40¢ cup when there is no match. Her expected winnings are

$$(.40)\left(\frac{1}{4}\right) + (-.40)\left(\frac{1}{4}\right) + (-.40)\left(\frac{1}{4}\right) + (.40)\left(\frac{1}{4}\right) = 0.$$

On the average, over the long run, Donna neither wins nor loses. ◀

A game with an expected value of 0 (such as the one in Example 6) is called a **fair game.** Casinos do not offer fair games. If they did, they would win (on the average) $0, and have a hard time paying the help! Casino games have expected winnings for the house that vary from 1.5 cents per dollar to 60 cents per dollar.

Exercises 36–41 at the end of the section ask you to find the expected winnings for certain games of chance.

The idea of expected value can be very useful in decision making, as shown by the next example.

EXAMPLE 7

At age 50, you receive a letter from the Mutual of Mauritania Insurance Company. According to the letter, you must tell the company immediately which of the following two options you will choose: take $20,000 at age 60 (if you are alive, $0 otherwise) or $30,000 at age 70 (again, if you are alive, $0 otherwise). Based *only* on the idea of expected value, which should you choose?

Life insurance companies have constructed elaborate tables showing the probability of a person living a given number of years into the future. From a recent such table, the probability of living from age 50 to age 60 is .88, while the probability of living from age 50 to 70 is .64. The expected values of the two options are given below.

$$\text{First option: } (20,000)(.88) + (0)(.12) = 17,600$$
$$\text{Second option: } (30,000)(.64) + (0)(.36) = 19,200$$

Based strictly on expected values, choose the second option. ◀

6.5 Exercises

In Exercises 1–6, **(a)** *give the probability distribution, and* **(b)** *sketch its histogram.*

1. In a seed viability test 50 seeds were placed in 10 rows of 5 seeds each. After a period of time, the number that germinated in each row were counted, giving the results in the table.

Number Germinated	Frequency
0	0
1	0
2	1
3	3
4	4
5	2
	Total: 10

2. At a large supermarket during the 5 P.M. rush, the manager counted the number of customers waiting in each of 10 checkout lines. The results are shown in the table.

Number Waiting	Frequency
2	1
3	2
4	4
5	2
6	0
7	1
	Total: 10

3. At a training program for police officers, each member of a class of 25 took 6 shots at a target. The numbers of bull's-eyes shot are shown in the table.

Number of Bull's-eyes	Frequency
0	0
1	1
2	0
3	4
4	10
5	8
6	2
	Total: 25

4. A class of 42 students took a 10-point quiz. The frequency of scores is given in the table.

Number of Points	Frequency
5	2
6	5
7	10
8	15
9	7
10	3
	Total: 42

5. Five mice are inoculated against a disease. After an incubation period, the number that contract the disease is noted. The experiment is repeated 20 times, with the results shown in the table.

Number With the Disease	Frequency
0	3
1	5
2	6
3	3
4	2
5	1
	Total: 20

6. The telephone company kept track of the calls for the correct time during a 24-hour period for two weeks. The results are shown in the table.

Number of Calls	Frequency
28	1
29	1
30	2
31	3
32	2
33	2
34	2
35	1
	Total: 14

For each of the experiments described below, let x determine a random variable, and use your knowledge of probability to prepare a probability distribution.

7. Four coins are tossed, and the number of heads is noted.

8. Two dice are rolled, and the total number of points is recorded.

9. Three cards are drawn from a deck. The number of aces is counted.

10. Two balls are drawn from a bag in which there are 4 white balls and 2 black balls. The number of black balls is counted.

Draw a histogram for each of the following, and shade the region that gives the indicated probability.

11. Exercise 7; $P(x \leq 2)$

12. Exercise 8; $P(x \geq 11)$

13. Exercise 9; P(at least one ace)

14. Exercise 10; P(at least one black ball)

15. Exercise 7; $P(x = 2 \text{ or } x = 3)$

16. Exercise 8; $P(4 \leq x \leq 9)$

Find the expected value for each random variable.

17.

x	2	3	4	5
$P(x)$.1	.4	.3	.2

18.

y	4	6	8	10
$P(y)$.4	.4	.05	.15

19.

z	9	12	15	18	21
$P(z)$.14	.22	.36	.18	.10

20.

x	30	32	36	38	44
$P(x)$.31	.30	.29	.06	.04

Find the expected values for the random variables x having the probability functions graphed below.

21.

22.

23.

24.

25. For the game in Example 6, find Mary's expected winnings. Is it a fair game?

26. Suppose one day Mary brings a 2-headed coin and uses it to toss for the coffee. Since Mary tosses, Donna calls.

 (a) Is this still a fair game?

 (b) What is Donna's expected gain if she calls heads?

 (c) What is Donna's expected gain if she calls tails?

27. A raffle offers a first prize of $100 and 2 second prizes of $40 each. One ticket costs $1, and 500 tickets are sold. Find the expected winnings for a person who buys 1 ticket. Is this a fair game?

28. A raffle offers a first prize of $1000, 2 second prizes of $300 each, and 20 third prizes of $10 each. If 10,000 tickets are sold at 50¢ each, find the expected winnings for a person buying 1 ticket. Is this a fair game?

Many of the following exercises require the use of combinations.

29. If 3 marbles are drawn from a bag containing 3 yellow and 4 white marbles, what is the expected number of yellow marbles in the sample?

30. If 5 apples in a barrel of 25 apples are known to be rotten, what is the expected number of rotten apples in a sample of 2 apples?

31. A delegation of 3 is selected from a city council made up of 5 liberals and 4 conservatives.

 (a) What is the expected number of liberals in the delegation?

 (b) What is the expected number of conservatives in the delegation?

32. From a group of 2 women and 5 men, a delegation of 2 is selected. Find the expected number of women in the delegation.

33. In a club with 20 senior and 10 junior members, what is the expected number of junior members on a 3-member committee?

34. If 2 cards are drawn at one time from a deck of 52 cards, what is the expected number of diamonds?

35. Suppose someone offers to pay you $5 if you draw 2 diamonds in the game in Exercise 34. He says that you should pay 50 cents for the chance to play. Is this a fair game?

Find the expected winnings for the games of chance described in Exercises 36–41.

36. A state lottery requires you to choose 1 heart, 1 club, 1 diamond, and 1 spade in that order from the 13 cards in

each suit. If all four choices are correct, you win $5000. It costs $1 to play.

37. If exactly 3 of the 4 selections in Exercise 36 are correct, the player wins $200. (It still costs $1 to play.)

38. In one form of roulette, you bet $1 on "even." If one of the 18 even numbers comes up, you get your dollar back, plus another one. If one of the 20 noneven (18 odd, 0, and 00) numbers comes up, you lose your dollar.

39. In another form of roulette, there are only 19 noneven numbers (no 00).

40. *Numbers* is a game in which you bet $1 on any three-digit number from 000 to 999. If your number comes up, you get $500.

41. In one form of the game *Keno*, the house has a pot containing 80 balls, each marked with a different number from 1 to 80. You buy a ticket for $1 and mark one of the 80 numbers on it. The house then selects 20 numbers at random. If your number is among the 20, you get $3.20 (for a net winning of $2.20).

42. Your friend missed class the day probability distributions were discussed. How would you explain to him what a probability distribution is?

43. How is the relative frequency of a random variable found?

44. The expected value of a random variable is

(a) the most frequent value of the random variable;

(b) the average of all possible values of the random variable;

(c) the sum of the products formed by multiplying each value of the random variable by its probability;

(d) the product of the sum of all values of the random variable and the sum of the probabilities.

Applications

Business and Economics

45. Complaints A local used-car dealer gets complaints about his cars as shown in the following table.

Number of Complaints per Day	0	1	2	3	4	5	6
Probability	.01	.05	.15	.26	.33	.14	.06

Find the expected number of complaints per day.

46. Payout on Insurance Policies An insurance company has written 100 policies of $10,000, 500 of $5000, and 1000 of $1000 for people of age 20. If experience shows that the probability that a person will die at age 20 is .001, how much can the company expect to pay out during the year the policies were written?

47. Rating Sales Accounts Levi Strauss and Company* uses expected value to help its salespeople rate their accounts. For each account, a salesperson estimates potential additional volume and the probability of getting it. The product of these figures gives the expected value of the potential, which is added to the existing volume. The totals are then classified as A, B, or C, as follows: $40,000 or below, class C; from $40,000 up to and including $55,000, class B; above $55,000, class A. Complete the following chart for one salesperson.

Account Number	Existing Volume	Potential Additional Volume	Probability of Getting It	Expected Value of Potential	Existing Volume + Expected Value of Potential	Class
1	$15,000	$10,000	.25	$2,500	$17,500	C
2	$40,000	$0	—	—	$40,000	C
3	$20,000	$10,000	.20			
4	$50,000	$10,000	.10			
5	$5,000	$50,000	.50			
6	$0	$100,000	.60			
7	$30,000	$20,000	.80			

*This example was supplied by James McDonald, Levi Strauss and Company, San Francisco.

Life Sciences

48. Animal Offspring In a certain animal species, the probability that a healthy adult female will have no offspring in a given year is .31, while the probabilities of 1, 2, 3, or 4 offspring are, respectively, .21, .19, .17, and .12. Find the expected number of offspring.

Physical Sciences

49. Seeding Storms One of the few methods that can be used in an attempt to cut the severity of a hurricane is to *seed* the storm. In this process, silver iodide crystals are dropped into the storm. Unfortunately, silver iodide crystals sometimes cause the storm to *increase* its speed. Wind speeds may also increase or decrease even with no seeding. Use the tree diagram to answer the following.*

(a) Find the expected amount of damage under each option, "seed" and "do not seed."

(b) To minimize total expected damage, what option should be chosen?

	Change in wind speed	Property damage (millions of dollars)
Seed .038	+32%	335.8
.143	+16%	191.1
.392	0	100.0
.255	−16%	46.7
.172	−34%	16.3
Do not seed .054	+32%	335.8
.206	+16%	191.1
.480	0	100.0
.206	−16%	46.7
.054	−34%	16.3

General Interest

50. Golf Tournament At the end of play in a major golf tournament, two players, an "old pro" and a "new kid," are tied. Suppose first prize is $80,000 and second prize is $20,000. Find the expected winnings for the old pro if

(a) both players are of equal ability;

(b) the new kid will freeze up, giving the old pro a 3/4 chance of winning.

51. Contests A magazine distributor offers a first prize of $100,000, two second prizes of $40,000 each, and two third prizes of $10,000 each. A total of 2,000,000 entries are received in the contest. Find the expected winnings if you submit one entry to the contest. If it would cost you 50¢ in time, paper, and stamps to enter, would it be worth it?

52. Contests A contest at a fast-food restaurant offered the following cash prizes and probabilities of winning on one visit.

Prize	Probability
$100,000	$\dfrac{1}{176,402,500}$
$25,000	$\dfrac{1}{39,200,556}$
$5,000	$\dfrac{1}{17,640,250}$
$1,000	$\dfrac{1}{1,568,022}$
$100	$\dfrac{1}{282,244}$
$5	$\dfrac{1}{7,056}$
$1	$\dfrac{1}{588}$

Suppose you spend $1 to buy a bus pass that lets you go to 25 different restaurants in the chain and pick up entry forms. Find your expected value.

*The probabilities and amounts of property damage in the tree diagram for Exercise 49 are from R. A. Howard, J. E. Matheson, and D. W. North, "The Decision to Seed Hurricanes," *Science*, vol. 176, no. 16, June 1972, pp. 1191–1202. Copyright © 1972 by the American Association for the Advancement of Science. Reprinted by permission.

Chapter Summary

Key Terms

6.1 multiplication principle
factorial notation
permutations
6.2 combinations
6.4 Bernoulli trials
binomial experiment

binomial probability
Pascal's triangle
6.5 random variable
frequency distribution
relative frequency
probability distribution

probability function
histogram
weighted average
expected value
fair game

Chapter 6

Review Exercises

1. In how many ways can 6 shuttle vans line up at the airport?

2. How many variations in first-, second-, and third-place finishes are possible in a 100-yard dash with 6 runners?

3. In how many ways can a sample of 3 oranges be taken from a bag of a dozen oranges?

4. If 2 of the oranges in Exercise 3 are rotten, in how many ways can the sample of 3 include

 (a) 1 rotten orange?

 (b) 2 rotten oranges?

 (c) no rotten oranges?

 (d) at most 2 rotten oranges?

5. In how many ways can 2 pictures, selected from a group of 5 different pictures, be arranged in a row on a wall?

6. In how many ways can the 5 pictures in Exercise 5 be arranged in a row if a certain one must be first?

7. In how many ways can the 5 pictures in Exercise 5 be arranged if 2 are landscapes and 3 are puppies and if

 (a) like types must be kept together?

 (b) landscapes and puppies are alternated?

8. In a Chinese restaurant the menu lists 8 items in column A and 6 items in column B.

 (a) To order a dinner, the diner is told to select 3 items from column A and 2 from column B. How many dinners are possible?

 (b) How many dinners are possible if the diner can select up to 3 from column A and up to 2 from column B? Assume at least one item must be included from either A or B.

9. A representative is to be selected from each of 3 departments in a small college. There are 7 people in the first department, 5 in the second department, and 4 in the third department.

 (a) How many different groups of 3 representatives are possible?

 (b) How many groups are possible if any number (at least 1) up to 3 representatives can form a group?

10. Explain under what circumstances a permutation should be used in a probability problem, and under what circumstances a combination should be used.

11. Discuss under what circumstances the binomial probability formula should be used in a probability problem.

A basket contains 4 black, 2 blue, and 5 green balls. A sample of 3 balls is drawn. Find the probabilities that the sample contains the following.

12. All black balls

13. All blue balls

14. 2 black balls and 1 green ball

15. Exactly 2 black balls

16. Exactly 1 blue ball

17. 2 green balls and 1 blue ball

Suppose a family plans 6 children, and the probability that a particular child is a girl is 1/2. Find the probabilities that the 6-child family has the following children.

18. Exactly 3 girls

19. All girls

20. At least 4 girls

21. No more than 2 boys

Suppose 2 cards are drawn without replacement from an ordinary deck of 52. Find the probabilities of the following results.

22. Both cards are red.

23. Both cards are spades.

24. At least 1 card is a spade.

25. One is a face card and the other is not.

26. At least one is a face card.

27. At most one is a queen.

In Exercises 28–32, **(a)** *give a probability distribution,* **(b)** *sketch its histogram, and* **(c)** *find the expected value.*

28.

x	1	2	3	4	5
Frequency	3	7	9	3	2

29.

x	8	9	10	11	12	13	14
Frequency	1	0	2	5	8	4	3

30. A coin is tossed 3 times and the number of heads is recorded.

31. A pair of dice is rolled and the sum of the results for each roll is recorded.

32. Patients in groups of 5 are given a new treatment for a fatal disease. The experiment is repeated 10 times with the following results.

Number Who Survived	Frequency
0	1
1	1
2	2
3	3
4	3
5	0
	Total: 10

In Exercises 33 and 34, give the probability that corresponds to the shaded region of each histogram.

33.

34.

35. You pay $6 to play in a game where you will roll a die, with payoffs as follows: $8 for a 6, $7 for a 5, and $4 for any other results. What are your expected winnings? Is the game fair?

36. A lottery has a first prize of $5000, two second prizes of $1000 each, and two $100 third prizes. A total of 10,000 tickets is sold, at $1 each. Find the expected winnings of a person buying 1 ticket.

37. Find the expected number of girls in a family of 5 children.

38. Three cards are drawn from a standard deck of 52 cards.

 (a) What is the expected number of aces?

 (b) What is the expected number of clubs?

39. Suppose someone offers to pay you $100 if you draw 3 cards from a standard deck of 52 cards and all the cards are clubs. What should you pay for the chance to win if it is a fair game?

Applications

Business and Economics

Quality Control *A certain machine that is used to manufacture screws produces a defect rate of .01. A random sample of 20 screws is selected. Find the probabilities that the sample contains the following.*

40. Exactly 4 defective screws

41. Exactly 3 defective screws

42. No more than 4 defective screws

43. *Set up* the probability that the sample has 12 or more defective screws. (Do not evaluate.)

44. **Land Development** A developer can buy a piece of property that will produce a profit of $16,000 with probability .7, or a loss of $9000 with probability .3. What is the expected profit?

General Interest

45. **Contests** At one time, game boards for a United Airlines contest could be obtained by sending a self-addressed, stamped envelope to a certain address. The prize was a ticket for any city to which United flies. Assume that the value of the ticket was $1000 (we might as well go first-class), and that the probability that a particular game board would win was 1/4000. If the stamps to enter the contest cost 30¢ and envelopes cost 1¢ each, find the expected winnings for a person ordering 1 game board.

*Exercises 46 and 47 are taken from actuarial examinations given by the Society of Actuaries.**

46. A company is considering the introduction of a new product that is believed to have probability .5 of being successful and probability .5 of being unsuccessful. Successful products pass quality control 80% of the time. Unsuccessful products pass quality control 25% of the time. If the product is successful, the net profit to the company will be $40 million; if unsuccessful, the net loss will be $15 million. Determine the expected net profit if the product passes quality control.

 (a) $23 million (b) $24 million (c) $25 million

 (d) $26 million (e) $27 million

47. A merchant buys boxes of fruit from a grower and sells them. Each box of fruit is either Good or Bad. A Good box contains 80% excellent fruit and will earn $200 profit on the retail market. A Bad box contains 30% excellent fruit and will produce a loss of $1000. The a priori probability of receiving a Good box of fruit is .9. Before the merchant decides to put the box on the market, he can sample one piece of fruit to test whether it is excellent. Based on that sample, he has the option of rejecting the box without paying for it. Determine the expected value of the right to sample. (*Hint:* If the merchant samples the fruit, what are the probabilities of accepting a Good box, accepting a Bad box, and not accepting the box? What are these probabilities if he does not sample the fruit?)

 (a) 0 (b) $16 (c) $34 (d) $72 (e) $80

Connections

48. The March 1982 issue of *Mathematics Teacher* included "Overbooking Airline Flights," an article by Joe Dan Austin. In this article, Austin developed a model for the expected income for an airline flight. With appropriate assumptions, the probability that exactly x of n people with reservations show up at the airport to buy a ticket is given by the binomial probability formula. Assume the following: 6 reservations have been accepted for 3 seats, $p = .6$ is the probability that a person with a reservation will show up, a ticket costs $100, and the airline must pay $100 to anyone with a reservation who does not get a ticket. Complete the following table.

Number Who Show Up (x)	0	1	2	3	4	5	6
Airline's Income							
$P(x)$							

 (a) Use the table to find $E(I)$, the expected airline income from the 3 seats.

 (b) Find $E(I)$ for $n = 3$, $n = 4$, and $n = 5$. Compare these answers with $E(I)$ for $n = 6$. For these values of n, how many reservations should the airline book for the 3 seats in order to maximize the expected revenue?

*From Course 130 Examination, Operations Research, November 1989.

Extended Application / Optimal Inventory For a Service Truck

For many different items it is difficult or impossible to take the item to a central repair facility when service is required. Washing machines, large television sets, office copiers, and computers are only a few examples of such items. Service for items of this type is commonly performed by sending a repair person to the item, with the person driving to the location in a truck containing various parts that might be required in repairing the item. Ideally, the truck should contain all the parts that might be required. However, most parts would be needed only infrequently, so that inventory costs for the parts would be high.

An optimum policy for deciding on which parts to stock on a truck would require that the probability of not being able to repair an item without a trip back to the warehouse for needed parts be as low as possible, consistent with minimum inventory costs. An analysis similar to the one below was developed at the Xerox Corporation.*

To set up a mathematical model for deciding on the optimum truck-stocking policy, let us assume that a broken machine might require one of 5 different parts (we could assume any number of different parts—we use 5 to simplify the notation). Suppose also that the probability that a particular machine requires part 1 is p_1; that it requires part 2 is p_2; and so on. Assume also that failures of different part types are independent, and that at most one part of each type is used on a given job.

Suppose that, on the average, a repair person makes N service calls per time period. If the repair person is unable to make a repair because at least one of the parts is unavailable, there is a penalty cost, L, corresponding to wasted time for the repair person, an extra trip to the parts depot, customer unhappiness, and so on. For each of the parts carried on the truck, an average inventory cost is incurred. Let H_i be the average inventory cost for part i, where $1 \le i \le 5$.

Let M_1 represent a policy of carrying only part 1 on the repair truck, M_{24} represent a policy of carrying only parts 2 and 4, with M_{12345} and M_0 representing policies of carrying all parts and no parts, respectively.

For policy M_{35}, carrying parts 3 and 5 only, the expected cost per time period per repair person, written $C(M_{35})$, is

$$C(M_{35}) = (H_3 + H_5) + NL[1 - (1 - p_1)(1 - p_2)(1 - p_4)].$$

(The expression in brackets represents the probability of needing at least one of the parts not carried, 1, 2, or 4 here.) As further examples,

$$C(M_{125}) = (H_1 + H_2 + H_5) + NL[1 - (1 - p_3)(1 - p_4)],$$

while

$$C(M_{12345}) = (H_1 + H_2 + H_3 + H_4 + H_5) + NL[1 - 1]$$
$$= H_1 + H_2 + H_3 + H_4 + H_5,$$

and

$$C(M_0) = NL[1 - (1 - p_1)(1 - p_2)(1 - p_3)(1 - p_4)(1 - p_5)].$$

To find the best policy, evaluate $C(M_0)$, $C(M_1)$, ..., $C(M_{12345})$, and choose the smallest result. (A general solution method is in the *Management Science* paper.)

Example

Suppose that for a particular item, only 3 possible parts might need to be replaced. By studying past records of failures of the item, and finding necessary inventory costs, suppose that the following values have been found.

p_1	p_2	p_3	H_1	H_2	H_3
.09	.24	.17	$15	$40	$9

Suppose $N = 3$ and L is $54. Then, as an example,

$$C(M_1) = H_1 + NL[1 - (1 - p_2)(1 - p_3)]$$
$$= 15 + 3(54)[1 - (1 - .24)(1 - .17)]$$
$$= 15 + 3(54)[1 - (.76)(.83)]$$
$$\approx 15 + 59.81 = 74.81.$$

Thus, if policy M_1 is followed (carrying only part 1 on the truck), the expected cost per repair person per time period is $74.81. Also,

$$C(M_{23}) = H_2 + H_3 + NL[1 - (1 - p_1)]$$
$$= 40 + 9 + 3(54)(.09) = 63.58,$$

so that M_{23} is a better policy than M_1. By finding the expected values for all other possible policies (see the exercises), the optimum policy may be chosen.

Exercises

1. Refer to the example and find each of the following.
 (a) $C(M_0)$ (b) $C(M_2)$ (c) $C(M_3)$ (d) $C(M_{12})$
 (e) $C(M_{13})$ (f) $C(M_{123})$

2. Which policy leads to the lowest expected cost?

3. In the example, $p_1 + p_2 + p_3 = .09 + .24 + .17 = .50$. Why is it not necessary that the probabilities add up to 1?

4. Suppose an item to be repaired might need one of n different parts. How many different policies would then need to be evaluated?

*Reprinted by permission of Stephen Smith, John Chambers, and Eli Shlifer, "Optimal Inventories Based on Job Completion Rate for Repairs Requiring Multiple Items," *Management Science*, Vol. 26, No. 8, August 1980, copyright © 1980 by The Institute of Management Sciences.

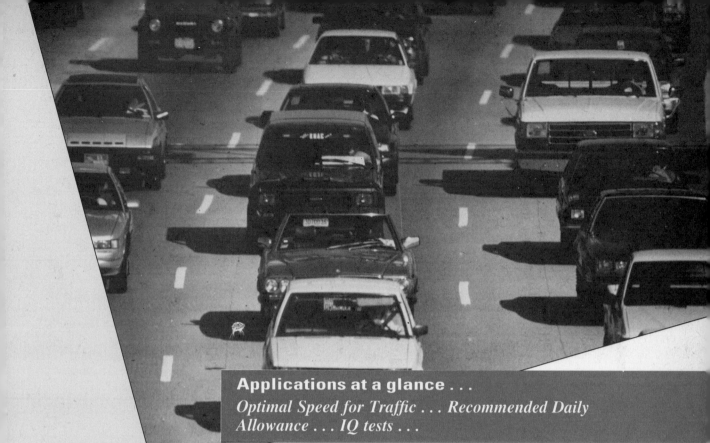

Applications at a glance . . .

Optimal Speed for Traffic . . . Recommended Daily Allowance . . . IQ tests . . .

CHAPTER 7

STATISTICS

(See page 341.)

Statistics is a branch of mathematics that deals with the collection and summarization of data. Methods of statistical analysis make it possible for us to draw conclusions about a population based on data from a sample of the population. Statistical models have become increasingly useful in manufacturing, government, agriculture, medicine, and the social sciences, and in all types of research. In this chapter we give a brief introduction to some of the key topics from statistical theory.

In the previous chapter, we saw that a frequency distribution can be transformed into a probability distribution by using the relative frequency of each value of the random variable as its probability. Sometimes it is convenient to work directly with a frequency distribution.

7.1 FREQUENCY DISTRIBUTIONS; MEASURES OF CENTRAL TENDENCY

? How can the results of a survey of business executives on the number of college courses in management needed by a business major best be organized to provide useful information?

Frequency distributions can provide an answer to this question.

Often, a researcher wishes to learn something about a characteristic of a population, but because the population is very large or mobile, it is not possible to examine all of its elements. Instead, a limited sample drawn from the population is studied to determine the characteristics of the population. For these inferences to be correct, the sample chosen must be a **random sample.** Random samples are representative of the population because they are chosen so that every element of the population is equally likely to be selected. A hand dealt from a well-shuffled deck of cards is a random sample.

Once a sample has been chosen and all data of interest are collected, the data must be organized so that conclusions may be more easily drawn. One method of organization is to group the data into intervals; equal intervals are usually chosen.

EXAMPLE 1 A survey asked 30 business executives for their recommendations as to the number of college units in management that a business major should have. The results are shown below. Group the data into intervals and find the frequency of each interval.

3	25	22	16	0	9	14	8	34	21
15	12	9	3	8	15	20	12	28	19
17	16	23	19	12	14	29	13	24	18

The highest number in the list is 34 and the lowest is 0; one convenient way to group the data is in intervals of size 5, starting with 0–4 and ending with 30–34. This gives an interval for each number in the list and results in seven

equal intervals of a convenient size. Too many intervals of smaller size would not simplify the data enough, while too few intervals of larger size would conceal information that the data might provide. A rule of thumb is to use from six to fifteen intervals.

First tally the number of college units falling into each interval. Then total the tallies in each interval, as in the table below. This table is an example of a **grouped frequency distribution.**

College Units	Tally	Frequency							
0–4					3				
5–9						4			
10–14							6		
15–19									8
20–24						5			
25–29					3				
30–34			1						
	Total:	30							

The frequency distribution in Example 1 shows information about the data that might not have been noticed before. For example, the interval with the largest number of recommended units is 15–19, and 19 executives (more than half) recommended between 9 and 25 units. Also, the frequency in each interval increases rather evenly (up to 8) and then decreases at about the same pace. However, some information has been lost; for example, we no longer know how many executives recommended 12 units.

The information in a grouped frequency distribution can be displayed in a histogram similar to the histograms for probability distributions in the previous chapter. The intervals determine the widths of the bars; if equal intervals are used, all the bars have the same width. The heights of the bars are determined by the frequencies.

Remember that in a histogram, the widths of the bars must be the same.

A **frequency polygon** is another form of graph that illustrates a grouped frequency distribution. The polygon is formed by joining consecutive midpoints of the tops of the histogram bars with straight line segments. The midpoints of the first and last bars are joined to endpoints on the horizontal axis where the next midpoint would appear.

EXAMPLE 2 A grouped frequency distribution of college units was found in Example 1. Draw a histogram and a frequency polygon for this distribution.

First draw a histogram, shown in black in Figure 1 on the next page. To get a frequency polygon, connect consecutive midpoints of the tops of the bars. The frequency polygon is shown in color.

FIGURE 1

Mean The average value of a probability distribution is the expected value of the distribution. Three measures of central tendency, or ''averages,'' are used with frequency distributions: the mean, the median, and the mode. The most important of these is the mean, which is similar to the expected value of a probability distribution. The **arithmetic mean** (the **mean**) of a set of numbers is the sum of the numbers, divided by the total number of numbers. We write the sum of n numbers $x_1, x_2, x_3, \ldots, x_n$ in a compact way using **summation notation**, also called **sigma notation**. With the Greek letter Σ (sigma), the sum

$$x_1 + x_2 + x_3 + \cdots + x_n$$

is written

$$x_1 + x_2 + x_3 + \cdots + x_n = \sum_{i=1}^{n} x_i.$$

In statistics, $\sum_{i=1}^{n} x_i$ is often abbreviated as just $\Sigma\, x$. The symbol \bar{x} (read x-bar) is used to represent the mean of a sample.

MEAN

The mean of the n numbers $x_1, x_2, x_3, \ldots, x_n$ is

$$\bar{x} = \frac{\sum x}{n}.$$

EXAMPLE **3**

The number of bankruptcy petitions (in thousands) filed in the United States in the fiscal years 1985–1990 are given in the table below.* Find the mean number of bankruptcy petitions filed annually during this period.

Year	Petitions Filed
1985	365
1986	478
1987	561
1988	594
1989	643
1990	725

Let $x_1 = 365$, $x_2 = 478$, and so on. Here, $n = 6$, since there are 6 numbers.

$$\bar{x} = \frac{365 + 478 + 561 + 594 + 643 + 725}{6}$$

$$\bar{x} = \frac{3366}{6} = 561$$

The mean number of bankruptcy petitions filed during the given years is 561,000. ◀

The mean of data that have been arranged in a frequency distribution is found in a similar way. For example, suppose the following data are collected.

Value	Frequency
84	2
87	4
88	7
93	4
99	3
Total:	20

The value 84 appears twice, 87 four times, and so on. To find the mean, first add 84 two times, 87 four times, and so on; or get the same result faster by multiplying 84 by 2, 87 by 4, and so on, and then adding the results. Dividing the sum by 20, the total of the frequencies, gives the mean.

$$\bar{x} = \frac{(84 \cdot 2) + (87 \cdot 4) + (88 \cdot 7) + (93 \cdot 4) + (99 \cdot 3)}{20}$$

$$= \frac{168 + 348 + 616 + 372 + 297}{20} = \frac{1801}{20}$$

$$\bar{x} = 90.05$$

*From *The World Almanac and Book of Facts, 1992.*

Find the mean for the data shown in the following frequency distribution.

Value	Frequency	Value × Frequency
30	6	$30 \cdot 6 = 180$
32	9	$32 \cdot 9 = 288$
33	7	$33 \cdot 7 = 231$
37	12	$37 \cdot 12 = 444$
42	6	$42 \cdot 6 = 252$
Total:	40	Total: 1395

A new column, "Value × Frequency," has been added to the frequency distribution. Adding the products from this column gives a total of 1395. The total from the frequency column is 40. The mean is

$$\bar{x} = \frac{1395}{40} = 34.875. \quad \blacktriangleleft$$

Note The mean in Example 4 is found in the same way as was the expected value of a probability distribution in the previous chapter. In fact, the words *mean* and *expected value* are often used interchangeably.

The mean of grouped data is found in a similar way. For grouped data, intervals are used, rather than single values. To calculate the mean, it is assumed that all these values are located at the midpoint of the interval. The letter x is used to represent the midpoints and f represents the frequencies, as shown in the next example.

Find the mean for the following grouped frequency distribution.

Interval	Midpoint, x	Frequency, f	Product, xf
40–49	44.5	2	89
50–59	54.5	4	218
60–69	64.5	7	451.5
70–79	74.5	9	670.5
80–89	84.5	5	422.5
90–99	94.5	3	283.5
		Total: 30	Total: 2135

A column for the midpoint of each interval has been added. The numbers in this column are found by adding the endpoints of each interval and dividing by 2. For the interval 40–49, the midpoint is $(40 + 49)/2 = 44.5$. The numbers in the product column on the right are found by multiplying frequencies and cor-

responding midpoints. Finally, we divide the total of the product column by the total of the frequency column to get

$$\bar{x} = \frac{2135}{30} = 71.2 \text{ (to the nearest tenth).} \blacktriangleleft$$

The formula for the mean of a grouped frequency distribution is given below.

MEAN OF A GROUPED DISTRIBUTION

The mean of a distribution, where x represents the midpoints, f the frequencies, and $n = \Sigma f$, is

$$\bar{x} = \frac{\sum xf}{n}.$$

Median Asked by a reporter to give the average height of the players on his team, a Little League coach lined up his 15 players by increasing height. He picked the player in the middle and pronounced that player to be of average height. This kind of average, called the **median,** is defined as the middle entry in a set of data arranged in either increasing or decreasing order. If there is an even number of entries, the median is defined to be the mean of the two center entries.

Odd Number of Entries	Even Number of Entries
8	2
7	3
Median = 4	4
3	7 } Median = $\frac{4 + 7}{2}$ = 5.5
1	9
	12

EXAMPLE 6 Find the median for each of the following list of numbers.

(a) 11, 12, 17, 20, 23, 28, 29

The median is the middle number; in this case, 20. (Note that the numbers are already arranged in numerical order.) In this list, three numbers are smaller than 20 and three are larger.

(b) 15, 13, 7, 11, 19, 30, 39, 5, 10

First arrange the numbers in numerical order, from smallest to largest.

5, 7, 10, 11, **13,** 15, 19, 30, 39

The middle number, or median, can now be determined; it is 13.

(c) 47, 59, 32, 81, 74, 153

Write the numbers in numerical order.

32, 47, **59, 74,** 81, 153

There are 6 numbers here; the median is the mean of the 2 middle numbers.

$$\text{Median} = \frac{59 + 74}{2} = \frac{133}{2} = 66\frac{1}{2} \quad \blacktriangleleft$$

In some situations, the median gives a truer representative or typical element of the data than the mean. For example, suppose in an office there are 10 sales-persons, 4 secretaries, the sales manager, and Ms. Daly, who owns the business. Their annual salaries are as follows: secretaries, $15,000 each; salespersons, $25,000 each; manager, $35,000; and owner, $200,000. The mean salary is

$$\overline{x} = \frac{(15,000)4 + (25,000)10 + 35,000 + 200,000}{16} = \$34,062.50.$$

However, since 14 people earn less than $34,062.50 and only 2 earn more, this does not seem very representative. The median salary is found by ranking the salaries by size: $15,000, $15,000, $15,000, $15,000, $25,000, $25,000, . . . , $200,000. Since there are 16 salaries (an even number) in the list, the mean of the eighth and ninth entries will give the value of the median. The eighth and ninth entries are both $25,000, so the median is $25,000. In this example, the median gives a truer average than the mean.

Mode Sue's scores on ten class quizzes include one 7, two 8's, six 9's, and one 10. She claims that her average grade on quizzes is 9, because most of her scores are 9's. This kind of "average," found by selecting the most frequent entry, is called the **mode.**

EXAMPLE **7** Find the mode for each list of numbers.

(a) 57, 38, 55, 55, 80, 87, 98
The number 55 occurs more often than any other, so it is the mode. It is not necessary to place the numbers in numerical order when looking for the mode.

(b) 182, 185, 183, 185, 187, 187, 189
Both 185 and 187 occur twice. This list has *two* modes.

(c) 10,708, 11,519, 10,972, 17,546, 13,905, 12,182
No number occurs more than once. This list has no mode. ◀

The mode has the advantages of being easily found and not being influenced by data that are very large or very small compared to the rest of the data. It is often used in samples where the data to be "averaged" are not numerical. A major disadvantage of the mode is that we cannot always locate exactly one mode for a set of values. There can be more than one mode, in the case of ties, or there can be no mode if all entries occur with the same frequency.

The mean is the most commonly used measure of central tendency. Its advantages are that it is easy to compute, it takes all the data into consideration, and it is reliable—that is, repeated samples are likely to give very similar means. A disadvantage of the mean is that it is influenced by extreme values, as illustrated in the salary example above.

The median can be easy to compute and is influenced very little by extremes. Like the mode, the median can be found in situations where the data are not numerical. For example, in a taste test, people are asked to rank five soft drinks from the one they like best to the one they like least. The combined rankings then produce an ordered sample, from which the median can be identified. A disadvantage of the median is the need to rank the data in order; this can be difficut when the number of items is large.

7.1 Exercises

For Exercises 1–4,

(a) *group the data as indicated;*
(b) *prepare a frequency distribution with a column for intervals and frequencies;*
(c) *construct a histogram;*
(d) *construct a frequency polygon;*

1. Use six intervals, starting with 0–24.

74	133	4	127	20	30
103	27	139	118	138	121
149	132	64	141	130	76
42	50	95	56	65	104
4	140	12	88	119	64

2. Use seven intervals, starting with 30–39.

79	71	78	87	69	50	63	51
60	46	65	65	56	88	94	56
74	63	87	62	84	76	82	67
59	66	57	81	93	93	54	88
55	69	78	63	63	48	89	81
98	42	91	66	60	70	64	70
61	75	82	65	68	39	77	81
67	62	73	49	51	76	94	54
83	71	94	45	73	95	72	66
71	77	48	51	54	57	69	87

3. Use 70–74 as the first interval.

79	84	88	96	102	104	110	108	106	106	
104	99	97	92	94	90	82	74	72	83	
85	92	100	99	101	107	111	102	97	94	92

4. Use 140–149 as the first interval.

174	190	172	182	179	186	171	152	174	185
180	170	160	173	163	177	165	157	149	167
169	182	178	158	182	169	181	173	183	176
170	162	159	147	150	192	179	165	148	188

5. How does a frequency polygon differ from a histogram?

6. Discuss the advantages and disadvantages of the mean as a measure of central tendency.

Find the mean for each list of numbers. Round to the nearest tenth.

7. 8, 10, 16, 21, 25

8. 44, 41, 25, 36, 67, 51

9. 21,900, 22,850, 24,930, 29,710, 28,340, 40,000

10. 38,500, 39,720, 42,183, 21,982, 43,250

11. 9.4, 11.3, 10.5, 7.4, 9.1, 8.4, 9.7, 5.2, 1.1, 4.7

12. 30.1, 42.8, 91.6, 51.2, 88.3, 21.9, 43.7, 51.2

Find the mean for each of the following. Round to the nearest tenth.

13. Value	Frequency
3	4
5	2
9	1
12	3

14. Value	Frequency
9	3
12	5
15	1
18	1

15. Value	Frequency
12	4
13	2
15	5
19	3
22	1
23	5

16. Value	Frequency
25	1
26	2
29	5
30	4
32	3
33	5

Find the median for each of the following lists of numbers.

17. 12, 18, 32, 51, 58, 92, 106

18. 596, 604, 612, 683, 719

19. 100, 114, 125, 135, 150, 172

20. 1072, 1068, 1093, 1042, 1056, 1005, 1009

21. 28.4, 9.1, 3.4, 27.6, 59.8, 32.1, 47.6, 29.8

22. .6, .4, .9, 1.2, .3, 4.1, 2.2, .4, .7, .1

Find the mode or modes for each of the following lists of numbers.

23. 4, 9, 8, 6, 9, 2, 1, 3

24. 21, 32, 46, 32, 49, 32, 49

25. 74, 68, 68, 68, 75, 75, 74, 74, 70

26. 158, 162, 165, 162, 165, 157, 163

27. 6.8, 6.3, 6.3, 6.9, 6.7, 6.4, 6.1, 6.0

28. 12.75, 18.32, 19.41, 12.75, 18.30, 19.45, 18.33

29. When is the median the most appropriate measure of central tendency?

30. Under what circumstances would the mode be an appropriate measure of central tendency?

For grouped data, the modal class *is the interval containing the most data values. Give the mean and modal class for each of the following collections of grouped data.*

31. Use the distribution in Exercise 3.

32. Use the distribution in Exercise 4.

Applications

Business and Economics

33. Absenteeism A firm took a random sample of the number of days absent in a year for 40 employees, with results as shown below.

Days Absent	Frequency
0–2	23
3–5	11
6–8	5
9–11	0
12–14	1

Sketch a histogram and a frequency polygon for the data.

Wheat Production *U.S. wheat prices and production figures for a recent decade are given below.*

Year	Price ($ per bushel)	Production (millions of bushels)
1982	2.70	2200
1983	2.30	2000
1984	2.95	1750
1985	3.80	2200
1986	3.90	2400
1987	3.60	2800
1988	3.55	2800
1989	3.50	2450
1990	3.35	2600
1991	3.20	2750

Find the mean for each of the following.

34. Price per bushel of wheat

35. Wheat production

Life Science

36. Home Range The size of the home ranges (in square kilometers) of several pandas were surveyed over a year's time, with the following results.

Home Range	Frequency
.1–.5	11
.6–1.0	12
1.1–1.5	7
1.6–2.0	6
2.1–2.5	2
2.6–3.0	1
3.1–3.5	1

Sketch a histogram and frequency polygon for the data.

Social Sciences

37. Population The histogram below shows the percent of the U.S. population in each age group in 1980.* What percent of the population was in each of the following age groups?

(a) 10–19 **(b)** 60–69

(c) What age group had the largest percent of the population?

38. Population The histogram below shows the estimated percent of the U.S. population in each age group in the year 2000.* What percent of the population is estimated to be in each of the following age groups then?

(a) 20–29 **(b)** 70+

(c) What age group will have the largest percent of the population?

(d) Compare the histogram in Exercise 37 with the histogram below. What seems to be true of the U.S. population?

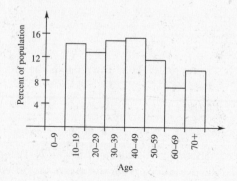

General Interest

39. Temperature The table below gives the average monthly temperatures in degrees Fahrenheit for a certain area.

Month	Maximum	Minimum
January	39	16
February	39	18
March	44	21
April	50	26
May	60	32
June	69	37
July	79	43
August	78	42
September	70	37
October	51	31
November	47	24
December	40	20

Find the mean for each of the following.

(a) The maximum temperature

(b) The minimum temperature

40. Olympics The number of nations participating in the winter Olympic games, from 1968 to 1992, is given below.

Year	Nations Participating
1968	37
1972	37
1976	37
1980	37
1984	49
1988	57
1992	65

Find the following measures for the data.

(a) Mean **(b)** Median **(c)** Mode

*Data from Census Bureau statistics.

7.2 MEASURES OF VARIATION

How can we tell when a manufacturing process is out of control?

To answer this question, we need to understand measures of variation, which tell us how much the numbers in a sample vary from the mean.

The mean gives a measure of central tendency of a list of numbers, but tells nothing about the *spread* of the numbers in the list. For example, look at the following three samples.

I	3	5	6	3	3
II	4	4	4	4	4
III	10	1	0	0	9

Each of these three samples has a mean of 4, and yet they are quite different; the amount of dispersion or variation within the samples is different. Therefore, in addition to a measure of central tendency, another kind of measure is needed that describes how much the numbers vary.

The largest number in sample I is 6, while the smallest is 3, a difference of 3. In sample II this difference is 0; in sample III, it is 10. The difference between the largest and smallest number in a sample is called the **range,** one example of a measure of variation. The range of sample I is 3, of sample II, 0, and of sample III, 10. The range has the advantage of being very easy to compute, and gives a rough estimate of the variation among the data in the sample. It depends only on the two extremes, however, and tells nothing about how the other data are distributed between the extremes.

 EXAMPLE 1 Find the range for each list of numbers.

(a) 12, 27, 6, 19, 38, 9, 42, 15

The highest number here is 42; the lowest is 6. The range is the difference between these numbers, or $42 - 6 = 36$.

(b) 74, 112, 59, 88, 200, 73, 92, 175

$$\text{Range} = 200 - 59 = 141 \blacktriangleleft$$

The most useful measure of variation is the *standard deviation.* Before defining it, however, we must find the **deviations from the mean,** the differences found by subtracting the mean from each number in a distribution.

 EXAMPLE 2 Find the deviations from the mean for the numbers

32, 41, 47, 53, 57.

Adding these numbers and dividing by 5 gives a mean of 46. To find the deviations from the mean, subtract 46 from each number in the list. For example, the first deviation from the mean is $32 - 46 = -14$; the last is $57 - 46 = 11$.

Number	Deviation from Mean
32	-14
41	-5
47	1
53	7
57	11

To check your work, find the sum of these deviations. It should always equal 0. (The answer is always 0 because the positive and negative numbers cancel each other.) ◄

To find a measure of variation, we might be tempted to use the mean of the deviations. As mentioned above, however, this number is always 0, no matter how widely the data are dispersed. One way to solve this problem is to use absolute value and find the mean of the absolute values of the deviations from the mean.

$$\frac{|-14| + |-5| + |1| + |7| + |11|}{5} = \frac{38}{5} = 7.6$$

Absolute value is awkward to work with algebraically, however.

Another way to get a list of positive numbers is to square each deviation, as shown in the table below, and then find the mean.

Number	Deviation from Mean	Square of Deviation
32	-14	196
41	-5	25
47	1	1
53	7	49
57	11	121

To find the mean of the squared deviations, statisticians prefer to divide by $n - 1$, rather than n, to create what is known as an *unbiased estimator,* a concept that is beyond the scope of this text. For the distribution above, this gives

$$\frac{196 + 25 + 1 + 49 + 121}{5 - 1} = \frac{392}{4} = 98.$$

This number, 98, is called the **variance** of the distribution. Since it is found by averaging a list of squares, the variance of a sample is represented by s^2.

For a sample of n numbers $x_1, x_2, x_3, \ldots, x_n$, with mean \bar{x}, the variance is

$$s^2 = \frac{\Sigma (x - \bar{x})^2}{n - 1}.$$

The following shortcut formula for the variance can be derived algebraically from the formula given above.

VARIANCE

The variance of a sample of n numbers, $x_1, x_2, x_3, \ldots, x_n$, with mean \bar{x}, is

$$s^2 = \frac{\Sigma x^2 - n\bar{x}^2}{n - 1}.$$

To find the variance, we squared the deviations from the mean, so the variance is in squared units. To return to the same units as the data, we use the *square root* of the variance, called the **standard deviation.**

STANDARD DEVIATION

The standard deviation of the n numbers $x_1, x_2, x_3, \ldots, x_n$ with mean \bar{x}, is

$$s = \sqrt{\frac{\Sigma x^2 - n\bar{x}^2}{n - 1}}.$$

As its name indicates, the standard deviation is the most commonly used measure of variation. The standard deviation is a measure of the variation from the mean. The size of the standard deviation tells us something about how spread out the data are from the mean.

EXAMPLE **3**

Find the standard deviation of the numbers

$$7, \ 9, \ 18, \ 22, \ 27, \ 29, \ 32, \ 40.$$

The mean of the numbers is

$$\frac{7 + 9 + 18 + 22 + 27 + 29 + 32 + 40}{8} = 23.$$

Arrange the work in columns, as shown in the table. The total of the second column gives $\Sigma x^2 = 5132$. Now find the quantity under the radical. The variance is

$$s^2 = \frac{\Sigma x^2 - n\bar{x}^2}{n - 1}$$

$$= \frac{5132 - 8(23)^2}{8 - 1}$$

$$= 128.6,$$

Number	Square of the Number
7	49
9	81
18	324
22	484
27	729
29	841
32	1024
40	1600
Total:	5132

rounded, and the standard deviation is

$$\sqrt{128.6} = 11.3 \quad \blacktriangleleft$$

Caution Be careful to divide by $n - 1$, not n, when calculating the standard deviation of a sample.

One way to interpret the standard deviation uses the fact that, for many populations, most of the data are within three standard deviations of the mean. (See Section 7.3.) This implies that, in Example 3, most of the population from which this sample is taken are between

$$\bar{x} - 3s = 23 - 3(11.3) = -10.9$$

and

$$\bar{x} + 3s = 23 + 3(11.3) = 56.9.$$

This has important implications for quality control. If the sample in Example 3 represents measurements of a product that the manufacturer wants to be between 5 and 45, the standard deviation is too large, even though all the numbers are within these bounds.

For data in a grouped frequency distribution, a slightly different formula for the standard deviation is used.

STANDARD DEVIATION FOR A GROUPED DISTRIBUTION

The standard deviation for a distribution with mean \bar{x}, where x is an interval midpoint with frequency f, and $n = \Sigma f$, is

$$s = \sqrt{\frac{\Sigma \, fx^2 - n\bar{x}^2}{n - 1}}.$$

The formula indicates that the product fx^2 is to be found for each interval. Then these products are summed, n times the sum of the mean is subtracted, and the difference is divided by one less than the total frequency; that is, by $n - 1$. The square root of this result is s, the standard deviation.

Caution In calculating the standard deviation for either a grouped or ungrouped distribution, using a rounded value for the mean may produce an inaccurate value.

EXAMPLE 4

Find s for the grouped data of Example 5, Section 7.1.

Begin by including columns for x^2 and fx^2. Recall from Section 7.1 that $\bar{x} = 71.2$.

Interval	f	x	x^2	fx^2
40–49	2	44.5	1980.25	3,960.50
50–59	4	54.5	2970.25	11,881.00
60–69	7	64.5	4160.25	29,121.75
70–79	9	74.5	5550.25	49,952.25
80–89	5	84.5	7140.25	35,701.25
90–99	3	94.5	8930.25	26,790.75
Total:	30		Total:	157,407.50

Use the formula above with $n = 30$ to find s.

$$s = \sqrt{\frac{\Sigma\, fx^2 - n\bar{x}^2}{n - 1}}$$

$$= \sqrt{\frac{157,407.50 - 30(71.2)^2}{30 - 1}}$$

$$\approx 13.5 \quad \blacktriangleleft$$

Note A calculator is almost a necessity for finding a standard deviation. A good procedure to follow is to first calculate \bar{x}. Then for each x, square that number, then multiply the result by the appropriate frequency. If your calculator has a key that accumulates a sum, use it to accumulate the total in the last column of the table.*

Chebyshev's Theorem So far, in this section and the last, we have developed formulas for the mean and standard deviation of a sample. The mean and standard deviation of the population from which the sample is drawn are designated μ and σ, respectively. These numbers often are unknown, and the sample statistics are used to estimate them. If μ and σ are known, then there are methods for estimating the range of values in the distribution. In the next section, we discuss *normal distributions,* in which more than 99% of the values are within three standard deviations of the mean. For other distributions, a rough estimate of the fraction of values that lie within a specified distance of the mean is given by **Chebyshev's theorem,** named after the Russian mathematician P. L. Chebyshev (1821–1894).

*Some calculators are equipped with statistical keys that compute the variance and standard deviation of a list of numbers directly, without your having to do any arithmetic or take any square roots. Some of these calculators use $n - 1$ and others use n for these computations; some may have keys for both. Check the instruction book before using a statistical calculator for the exercises.

CHEBYSHEV'S THEOREM

For any distribution of numbers with mean μ and standard deviation σ, the probability that a number will lie within k standard deviations of the mean is at least

$$1 - \frac{1}{k^2}.$$

That is,

$$P(\mu - k\sigma \leq x \leq \mu + k\sigma) \geq 1 - \frac{1}{k^2}.$$

By Chebyshev's theorem, at least

$$1 - \frac{1}{3^2} = 1 - \frac{1}{9} = \frac{8}{9},$$

or about 89%, of the numbers in any distribution lie within 3 standard deviations of the mean. Figure 2 shows a geometric interpretation of this result. ◀

At least 89% of the distribution falls
in this interval.

FIGURE 2

Suppose a distribution has mean 52 and standard deviation 3.5. Then "3 standard deviations" is $3 \times 3.5 = 10.5$, and "3 standard deviations from the mean" is

$$52 - 10.5 \quad \text{to} \quad 52 + 10.5,$$

or 41.5 to 62.5.

By Example 5, at least 89% of the values in this distribution will lie between 41.5 and 62.5.

Chebyshev's theorem gets much of its importance from the fact that it applies to *any* distribution; only the mean and the standard deviation must be known. Other results given later produce more accurate estimates, because more is known about the nature of the distribution.

EXAMPLE 6

The Forever Power Company claims that its batteries have a mean life of 26.2 hours, with a standard deviation of 4.1 hours. In a shipment of 100 batteries, what minimum number will have a life within 2 standard deviations of the mean—that is, between $26.2 - (4.1 \times 2) = 18$, and $26.2 + (4.1 \times 2) = 34.4$ hours?

Use Chebyshev's theorem with $k = 2$. At least

$$1 - \frac{1}{2^2} = 1 - \frac{1}{4} = \frac{3}{4},$$

or 75%, of the batteries should have a life within 2 standard deviations of the mean. At least $75\% \times 100 = 75$ of the batteries can be expected to last between 18 and 34.4 hours. ◀

EXAMPLE 7

Statistical process control is a method of determining when a manufacturing process is out of control, producing defective items. The procedure involves taking samples of a measurement on a product over a production run and calculating the mean and standard deviation of each sample. These results are used to determine when the manufacturing process is out of control. For example, three sample measurements from a manufacturing process on each of four days are given in the table below. The mean \bar{x} and standard deviation s are calculated for each sample.

Day	1			2			3			4		
Sample Number	1	2	3	1	2	3	1	2	3	1	2	3
Measurements	-3	0	4	5	-2	4	3	-1	0	4	-2	1
	0	5	3	4	0	3	-2	0	0	3	0	3
	2	2	2	3	1	4	0	1	-2	3	-1	0
\bar{x}	$-1/3$	$7/3$	3	4	$-1/3$	$11/3$	$1/3$	0	$-2/3$	$10/3$	-1	$4/3$
s	2.5	2.5	1	1	1.5	.6	2.5	1	1.2	.6	1	1.5

Next, the mean of the 12 sample means, \overline{X}, and the mean of the 12 sample standard deviations, \bar{s}, are found (using the formula for \bar{x}). Here, these measures are

$$\overline{X} = 1.3 \quad \text{and} \quad \bar{s} = 1.41.$$

The control limits for the sample means are given by

$$\overline{X} \pm k_1 \bar{s},$$

where k_1 is a constant found from a manual. For samples of size 3, $k_1 = 1.954$, so the control limits for the sample means are

$$1.3 \pm (1.954)(1.41).$$

The upper control limit is 4.06, and the lower control limit is -1.46.

Similarly, the control limits for the sample standard deviations are given by $k_2 \cdot \bar{s}$ and $k_3 \cdot \bar{s}$, where k_2 and k_3 also are values given in the same manual. Here, $k_2 = 2.568$ and $k_3 = 0$, with the upper and lower control limits for the sample standard deviations equal to $2.568(1.41)$ and $0(1.41)$, or 3.62 and 0. As long as the sample means are between -1.46 and 4.06 and the sample standard deviations are between 0 and 3.62, the process is in control. ◀

7.2 Exercises

1. How are the variance and the standard deviation related?
2. Why can't we use the sum of the deviations from the mean as a measure of dispersion of a distribution?

Find the range and standard deviation for each set of numbers.

3. 42, 38, 29, 74, 82, 71, 35
4. 122, 132, 141, 158, 162, 169, 180
5. 241, 248, 251, 257, 252, 287
6. 51, 58, 62, 64, 67, 71, 74, 78, 82, 93
7. 3, 7, 4, 12, 15, 18, 19, 27, 24, 11
8. 15, 42, 53, 7, 9, 12, 28, 47, 63, 14

Find the standard deviation for the following grouped data.

9. (From Exercise 1, Section 7.1)

Interval	Frequency
0–24	4
25–49	3
50–74	6
75–99	3
100–124	5
125–149	9

10. (From Exercise 2, Section 7.1)

Interval	Frequency
30–39	1
40–49	6
50–59	13
60–69	22
70–79	17
80–89	13
90–99	8

11. The grouped data in the answer for Exercise 3, Section 7.1
12. The grouped data in the answer for Exercise 4, Section 7.1

Find the fraction of all the numbers of a data set lying within the following numbers of standard deviations from the mean, using Chebyshev's theorem.

13. 2 14. 4 15. 5

In a certain distribution of numbers, the mean is 50 with a standard deviation of 6. Use Chebyshev's theorem to tell what percent of the numbers are in the following intervals.

16. Between 38 and 62 17. Between 32 and 68 18. Between 26 and 74
19. Less than 38 or more than 62 20. Less than 32 or more than 68

21. The weekly wages of the six employees at Harold's Hardware Store are $300, $320, $380, $420, $500, and $2000.
 (a) Find the mean and standard deviation of this distribution.
 (b) How many of the workers earn wages within 1 standard deviation of the mean?
 (c) How many earn wages within 2 standard deviations of the mean?
 (d) What does Chebyshev's theorem give as the minimum fraction earning wages within 2 standard deviations of the mean?

22. Discuss what the standard deviation tells us about a distribution.

Applications

Business and Economics

23. Battery Life Forever Power Company analysts conducted tests on the life of its batteries and those of a competitor (Brand X). They found that their batteries had a mean life (in hours) of 26.2, with a standard deviation of 4.1 (see Example 6). Their results for a sample of 10 Brand X batteries were as follows: 15, 18, 19, 23, 25, 25, 28, 30, 34, 38.

(a) Find the mean and standard deviation for the sample of Brand X batteries.

(b) Which batteries have a more uniform life in hours?

(c) Which batteries have the highest average life in hours?

24. Sales Promotion The Quaker Oats Company conducted a survey to determine whether a proposed premium, to be included in boxes of cereal, was appealing enough to generate new sales.* Four cities were used as test markets, where the cereal was distributed with the premium, and four cities as control markets, where the cereal was distributed without the premium. The eight cities were chosen on the basis of their similarity in terms of population, per-capita income, and total cereal purchase volume. The results were as follows.

	City	Percent Change in Average Market Share per Month
Test Cities	1	+18
	2	+15
	3	+7
	4	+10
Control Cities	1	+1
	2	−8
	3	−5
	4	0

(a) Find the mean of the change in market share for the four test cities.

(b) Find the mean of the change in market share for the four control cities.

(c) Find the standard deviation of the change in market share for the test cities.

(d) Find the standard deviation of the change in market share for the control cities.

(e) Find the difference between the means of (a) and (b). This difference represents the estimate of the percent change in sales due to premium.

(f) The two standard deviations from (c) and (d) were used to calculate an "error" of ± 7.95 for the estimate in (e). With this amount of error, what are the smallest and largest estimates of the increase in sales?

On the basis of the interval estimate of part (f), the company decided to mass-produce the premium and distribute it nationally.

25. Process Control The following table gives 10 samples, of three measurements, made during a production run.

				Sample Number					
1	2	3	4	5	6	7	8	9	10
2	3	−2	−3	−1	3	0	−1	2	0
−2	−1	0	1	2	2	1	2	3	0
1	4	1	2	4	2	2	3	2	2

Use the information in Example 7 to find the following.

(a) Find the mean \bar{x} for each sample of three measurements.

(b) Find the standard deviation s for each sample of three measurements.

(c) Find the mean \bar{X} of the sample means.

(d) Find the mean \bar{s} of the sample standard deviations.

(e) Using $k_1 = 1.954$, find the upper and lower control limits for the sample means.

(f) Using $k_2 = 2.568$ and $k_3 = 0$, find the upper and lower control limits for the sample standard deviations.

26. Process Control Given the following measurements from later samples on the process in Exercise 25, decide whether the process is out of control. (*Hint:* Use the results of Exercise 25(e) and (f).)

		Sample Number			
1	2	3	4	5	6
3	−4	2	5	4	0
−5	2	0	1	−1	1
2	1	1	−4	−2	−6

*This example was supplied by Jeffery S. Berman, Senior Analyst, Marketing Information, Quaker Oats Company.

General Interest

27. Unemployment The number of unemployed workers in the United States in recent years (in millions) is given below.*

Year	Number Unemployed
1984	8.54
1985	8.31
1986	8.24
1987	7.43
1988	6.70
1989	6.53
1990	6.87

(a) Find the mean number unemployed (in millions) in this period. Which year has unemployment closest to the mean?

(b) Find the standard deviation for the data.

(c) In how many of these years is unemployment within 1 standard deviation of the mean?

28. Immigration The numbers of immigrants to the United States (in thousands) from selected parts of the world in 1990 are shown in the table at the top of the next column.*

Region	Immigrants
Europe	124.0
Asia	321.9
Canada	24.6
Mexico	680.2
Other America	346.2

(a) Find the mean number of immigrants from these regions. Which region produces the number of immigrants closest to the mean?

(b) Find the standard deviation. Are any of the data more than 2 standard deviations from the mean? If so, which ones? Are any of the data more than 1 standard deviation from the mean? If so, which ones?

Computer/Graphing Calculator

Use a computer or a calculator with a standard deviation key to solve the problems in Exercises 29–30.

29. An assembly-line machine turns out washers with the following thicknesses (in millimeters).

1.20	1.01	1.25	2.20	2.58	2.19	1.29	1.15
2.05	1.46	1.90	2.03	2.13	1.86	1.65	2.27
1.64	2.19	2.25	2.08	1.96	1.83	1.17	2.24

Find the mean and standard deviation of these thicknesses.

30. A medical laboratory tested 21 samples of human blood for acidity on the pH scale, with the following results.

7.1	7.5	7.3	7.4	7.6	7.2	7.3
7.4	7.5	7.3	7.2	7.4	7.3	7.5
7.5	7.4	7.4	7.1	7.3	7.4	7.4

Find the mean and standard deviation.

*Excerpts from "Employment and Unemployment in the U.S." and "Immigration by Country of Last Residence 1820–1990" from *The World Almanac and Book of Facts 1992*. Copyright © 1991 by Pharos Books. Reprinted by permission of World Almanac, an imprint of Pharos Books, a Scripps Howard Company.

7.3 THE NORMAL DISTRIBUTION

What is the probability that a salesperson drives between 1200 miles and 1600 miles per month?

Recall that the general idea of a probability distribution was discussed in Section 6.5. In this section, we consider a specific type of probability distribution.

This question can be answered by using the normal probability distribution introduced in this section.

Figure 3(a) shows a histogram and frequency polygon for the bank-transaction example in Section 6.5. The heights of the bars are the probabilities. The transaction times in the example were given to the nearest minute. Theoretically at least, they could have been timed to the nearest tenth of a minute, or hundredth of a minute, or even more precisely. In each case, a histogram and frequency polygon could be drawn. If the times are measured with smaller and smaller units, there are more bars in the histogram and the frequency polygon begins to look more and more like the curve in Figure 3(b) instead of a polygon. Actually, it is possible for the transaction times to take on any real number value greater than 0. A distribution in which the outcomes can take any real number value within some interval is a **continuous distribution.** The graph of a continuous distribution is a curve.

(a) **(b)**

FIGURE 3

The distribution of heights (in inches) of college women is another example of a continuous distribution, since these heights include infinitely many possible measurements, such as 53, 58.5, 66.3, 72.666, ..., and so on. Figure 4 shows the continuous distribution of heights of college women. Here the most frequent heights occur near the center of the interval shown.

Another continuous curve, which approximates the distribution of yearly incomes in the United States, is given in Figure 5. The graph shows that the most frequent incomes are grouped near the low end of the interval. This kind of distribution, where the peak is not at the center, is called **skewed.**

Heights of female college freshman (in inches)

FIGURE 4

Income in the United States (in dollars)

FIGURE 5

Many natural and social phenomena produce continuous probability distributions whose graphs are bell-shaped curves, such as those shown in Figure 6. Such distributions are called **normal distributions** and their graphs are called **normal curves.** Examples of normal distributions are the heights of college women and the errors made in filling 1-pound cereal boxes. We use the Greek letters μ (mu) to denote the mean, and σ (sigma) to denote the standard deviation, of a normal distribution.

Three normal distributions

FIGURE 6

There are many normal distributions. Some of the corresponding normal curves are tall and thin and others are short and wide, as shown in Figure 6. But every normal curve has the following properties.

1. Its peak occurs directly above the mean μ.
2. The curve is symmetric about the vertical line through the mean (that is, if you fold the page along this line, the left half of the graph will fit exactly on the right half).
3. The curve never touches the x-axis—it extends indefinitely in both directions.
4. The area under the curve (and above the horizontal axis) is always 1. (This agrees with the fact that the sum of the probabilities in any distribution is 1.)

It can be shown that a normal distribution is completely determined by its mean μ and standard deviation σ.* A small standard deviation leads to a tall, narrow curve like the one in the center of Figure 6. A large standard deviation produces a flat, wide curve like the one on the right in Figure 6.

Since the area under a normal curve is 1, parts of this area can be used to determine certain probabilities. For instance, Figure 7(a) is the probability distribution of the annual rainfall in a certain region. Calculus can be used to show that the probability that the annual rainfall will be between 25 inches and 35 inches is the area under the curve from 25 to 35. The general case, shown in Figure 7(b), can be stated as shown on the next page.

(a)

(b)

FIGURE 7

*As is shown in more advanced courses, its graph is the graph of the function

$$f(x) = \frac{1}{\sigma\sqrt{2\pi}}e^{-(x-\mu)^2/(2\sigma^2)},$$

where $e \approx 2.71828$ is a real number.

The area of the shaded region under the normal curve from a to b is the probability that an observed data value will be between a and b.

To use normal curves effectively, we must be able to calculate areas under portions of these curves. These calculations have already been done for the normal curve with mean $\mu = 0$ and standard deviation $\sigma = 1$ (which is called the **standard normal curve**) and are available in a table in the Appendix. The following examples demonstrate how to use the table to find such areas. Later we shall see how the standard normal curve may be used to find areas under any normal curve.

EXAMPLE 1 The horizontal axis of the standard normal curve is usually labeled z. Find the following areas under the standard normal curve.

(a) The area to the left of $z = 1.25$

Look up 1.25 in the normal curve table. (Find 1.2 in the left-hand column and .05 at the top, then locate the intersection of the corresponding row and column.) The specified area is .8944, so the shaded area shown in Figure 8 is .8944. This area represents 89.44% of the total area under the normal curve, and so the probability that $z \leq 1.25$ is .8944.

(b) The area to the right of $z = 1.25$

From part (a), the area to the left of $z = 1.25$ is .8944. The total area under the normal curve is 1, so the area to the right of $z = 1.25$ is

$$1 - .8944 = .1056.$$

See Figure 9, where the shaded area represents 10.56% of the total area under the normal curve, and the probability that $z \geq 1.25$ is .1056.

(c) Between $z = -1.02$ and $z = .92$

To find this area, which is shaded in Figure 10, start with the area to the left of $z = .92$ and subtract the area to the left of $z = -1.02$. See the two shaded regions in Figure 11. The result is $.8212 - .1539 = .6673$, which represents $P(-1.02 \leq z \leq .92)$. ◀

Shaded area
= .8944

$z = 1.25$

FIGURE 8

Shaded area
$= 1 - .8944$
$= .1056$

$z = 1.25$

FIGURE 9

$z = -1.02$ $z = .92$

FIGURE 10

$z = .92$ $z = -1.02$

FIGURE 11

2 Find a value of z satisfying the following conditions.

(a) 12.1% of the area is to the left of z.

Use the table backwards. Look in the body of the table for an area of .1210, and find the corresponding z using the left column and the top column of the table. You should find that $z = -1.17$ corresponds to an area of .1210.

(b) 20% of the area is to the right of z.

If 20% of the area is to the right, 80% is to the left. Find the z corresponding to an area of .8000. The closest value is $z = .84$. ◀

The key to finding areas under *any* normal curve is to express each number x on the horizontal axis in terms of standard deviations above or below the mean. The **z-score** for x is the number of standard deviations that x lies from the mean (positive if x is above the mean, negative if x is below the mean).

3 If a normal distribution has mean 50 and standard deviation 4, find the following z-scores.

(a) The z-score for $x = 46$

Since 46 is 4 units below 50 and the standard deviation is 4, 46 is 1 standard deviation below the mean. So, its z-score is -1.

(b) The z-score for $x = 60$

The z-score is 2.5 because 60 is 10 units above the mean (since $60 - 50 = 10$), and 10 units is 2.5 standard deviations (since $10/4 = 2.5$). ◀

In Example 3(b), we found the z-score by taking the difference between 60 and the mean and dividing this difference by the standard deviation. The same procedure works in the general case.

If a normal distribution has mean μ and standard deviation σ, then the z-score for the number x is

$$z = \frac{x - \mu}{\sigma}.$$

The importance of z-scores lies in the following fact.

AREA UNDER A NORMAL CURVE

The area under a normal curve between $x = a$ and $x = b$ is the same as the area under the standard normal curve between the z-score for a and the z-score for b.

Therefore, by converting to z-scores and using the table for the standard normal curve, we can find areas under any normal curve. Since these areas are probabilities, we can now handle a variety of applications.

EXAMPLE 4 Dixie Office Supplies finds that its sales force drives an average of 1200 miles per month per person, with a standard deviation of 150 miles. Assume that the number of miles driven by a salesperson is closely approximated by a normal distribution.

(a) Find the probability that a salesperson drives between 1200 miles and 1600 miles per month.

Here $\mu = 1200$ and $\sigma = 150$, and we must find the area under the normal distribution curve between $x = 1200$ and $x = 1600$. We begin by finding the z-score for $x = 1200$.

$$z = \frac{x - \mu}{\sigma} = \frac{1200 - 1200}{150} = \frac{0}{150} = 0$$

The z-score for $x = 1600$ is

$$z = \frac{x - \mu}{\sigma} = \frac{1600 - 1200}{150} = \frac{400}{150} = 2.67.$$

From the table, the area to the left of $z = 2.67$ is .9962, the area to the left of $z = 0$ is .5000, and

$$.9962 - .5000 = .4962.$$

Therefore, the probability that a salesperson drives between 1200 miles and 1600 miles per month is .4962. See Figure 12.

(b) Find the probability that a salesperson drives between 1000 miles and 1500 miles per month.

As shown in Figure 13, z-scores for both $x = 1000$ and $x = 1500$ are needed.

$\mu = 1200$ $x = 1600$

FIGURE 12

$x = 1000$ $x = 1500$

$\mu = 1200$

FIGURE 13

For $x = 1000$,
$$z = \frac{1000 - 1200}{150}$$
$$= \frac{-200}{150}$$
$$z \approx -1.33.$$

For $x = 1500$,
$$z = \frac{1500 - 1200}{150}$$
$$= \frac{300}{150}$$
$$z = 2.00.$$

From the table, $z = -1.33$ leads to an area of .0918, while $z = 2.00$ corresponds to .9772. A total of .9772 − .0918 = .8854, or 88.54%, of the drivers travel between 1000 and 1500 miles per month. The probability that a driver travels between 1000 miles and 1500 miles per month is .8854. ◄

As mentioned above, z-scores are standard deviations, so $z = 1$ corresponds to 1 standard deviation above the mean, and so on. Looking up $z = 1.00$ and $z = -1.00$ in the table shows that

$$.8413 - .1587 = .6826,$$

or 68.3% of the area under a normal curve lies within 1 standard deviation of the mean. Also,

$$.9772 - .0228 = .9544,$$

or 95.4% of the area lies within 2 standard deviations of the mean. These results, summarized in Figure 14, can be used to get a quick estimate of results when working with normal curves.

FIGURE 14

Suppose a normal distribution has $\mu = 1000$ and $\sigma = 150$. Then about 95% of all values lie within 2 standard deviations of the mean; that is, between

$$1000 - (2 \times 150) = 700 \quad \text{and} \quad 1000 + (2 \times 150) = 1300.$$

Chebyshev's theorem (Section 7.2) says that *at least*

$$1 - \frac{1}{2^2} = 1 - \frac{1}{4} = \frac{3}{4},$$

or 75% of the values lie between 700 and 1300. The difference between 95.44% and "at least 75%," from Chebyshev's theorem, comes from the fact that Chebyshev's theorem applies to *any* distribution, while the methods of this section apply only to *normal* distributions. Thus it should not be surprising that having more information (a normal distribution) should produce more accurate results (95.44% instead of "at least 75%"). ◀

7.3 Exercises

1. The peak in a normal curve occurs directly above _____ .
2. The total area under a normal curve (above the horizontal axis) is _____ .
3. How are z-scores found for normal distributions where $\mu \neq 0$ or $\sigma \neq 1$?
4. How is the standard normal curve used to find probabilities for normal distributions?

Find the percent of the area under a normal curve between the mean and the given number of standard deviations from the mean.

5. 2.50 6. .81 7. -1.71 8. -2.04

Find the percent of the total area under the standard normal curve between each pair of z-scores.

9. $z = 1.41$ and $z = 2.83$ 10. $z = .64$ and $z = 2.11$
11. $z = -2.48$ and $z = -.05$ 12. $z = -1.74$ and $z = -1.02$
13. $z = -3.11$ and $z = 1.44$ 14. $z = -2.94$ and $z = -.43$
15. $z = -.42$ and $z = .42$ 16. $z = -1.98$ and $z = 1.98$

Find a z-score satisfying the following conditions.

17. 5% of the total area is to the left of z. 18. 1% of the total area is to the left of z.
19. 15% of the total area is to the right of z. 20. 25% of the total area is to the right of z.

Assume the distributions in Exercises 21–26 are normal, and use the areas under the normal curve given in the Appendix to answer the questions.

21. A machine produces bolts with an average diameter of .25 inch and a standard deviation of .02 inch. What is the probability that a bolt will be produced with a diameter greater than .3 inch?

22. A machine that fills quart milk cartons is set up to average 32.2 oz per carton, with a standard deviation of 1.2 oz. What is the probability that a filled carton will contain less than 32 oz of milk?

23. At the Discount Market, the average weekly grocery bill is $52.25, with a standard deviation of $19.50. What are the largest and smallest amounts spent by the middle 50% of this market's customers?

24. The mean clotting time of blood is 7.45 sec, with a standard deviation of 3.6 sec. What is the probability that an individual's blood clotting time will be less than 7 sec or greater than 8 sec?

25. The average size of the fish in Lake Amotan is 12.3 inches, with a standard deviation of 4.1 inches. Find the probability of catching a fish longer than 18 inches in Lake Amotan.

26. To be graded extra large, an egg must weigh at least 2.2 oz. If the average weight for an egg is 1.5 oz, with a standard deviation of .4 oz, how many eggs in a sample of five dozen would you expect to grade extra large?

Applications

Business and Economics

Life of Light Bulbs *A certain type of light bulb has an average life of 500 hr, with a standard deviation of 100 hr. The length of life of the bulb can be closely approximated by a normal curve. An amusement park buys and installs 10,000 such bulbs. Find the total number that can be expected to last for each of the following periods of time.*

27. At least 500 hr 28. Less than 500 hr 29. Between 650 and 780 hr
30. Between 290 and 540 hr 31. Less than 740 hr 32. More than 300 hr

Quality Control *A box of oatmeal must contain 16 oz. The machine that fills the oatmeal boxes is set so that, on the average, a box contains 16.5 oz. The boxes filled by the machine have weights that can be closely approximated by a normal curve. What fraction of the boxes filled by the machine are underweight if the standard deviation is as follows?*

33. .5 oz **34.** .3 oz **35.** .2 oz **36.** .1 oz

Quality Control *The chickens at Colonel Thompson's Ranch have a mean weight of 1850 g, with a standard deviation of 150 g. The weights of the chickens are closely approximated by a normal curve. Find the percent of all chickens having weights in the following ranges.*

37. More than 1700 g

38. Less than 1800 g

39. Between 1750 and 1900 g

40. Between 1600 and 2000 g

41. Less than 1550 g

42. More than 2100 g

Life Sciences

Vitamin Requirements *In nutrition, the Recommended Daily Allowance of vitamins is a number set by the government as a guide to an individual's daily vitamin intake. Actually, vitamin needs vary drastically from person to person, but the needs are very closely approximated by a normal curve. To calculate the Recommended Daily Allowance, the government first finds the average need for vitamins among people in the population, and the standard deviation. The Recommended Daily Allowance is then defined as the mean plus 2.5 times the standard deviation.*

43. What percent of the population will receive adequate amounts of vitamins under this plan?

Find the Recommended Daily Allowance for each vitamin in Exercises 44–46.

44. Mean = 1800 units; standard deviation = 140 units

45. Mean = 159 units; standard deviation = 12 units

46. Mean = 1200 units; standard deviation = 92 units

Social Sciences

Speed Limits *New studies by Federal Highway Administration traffic engineers suggest that speed limits on many thoroughfares are set arbitrarily and often are artificially low. According to traffic engineers, the ideal limit should be the "85th percentile speed." This means the speed at or below which 85 percent of the traffic moves. Assuming speeds are normally distributed, find the 85th percentile speed for roads with the following conditions.*

47. The mean speed is 50 mph with a standard deviation of 10 mph.

48. The mean speed is 30 mph with a standard deviation of 5 mph.

Education *One professor uses the following system for assigning letter grades in a course.*

Grade	Total Points
A	Greater than $\mu + \frac{3}{2}\sigma$
B	$\mu + \frac{1}{2}\sigma$ to $\mu + \frac{3}{2}\sigma$
C	$\mu - \frac{1}{2}\sigma$ to $\mu + \frac{1}{2}\sigma$
D	$\mu - \frac{3}{2}\sigma$ to $\mu - \frac{1}{2}\sigma$
F	Below $\mu - \frac{3}{2}\sigma$

What percent of the students receive the following grades?

49. A **50.** B **51.** C

52. Do you think this system would be more likely to be fair in a large freshman class in psychology or in a graduate seminar of five students? Why?

Education *A teacher gives a test to a large group of students. The results are closely approximated by a normal curve. The mean is 74, with a standard deviation of 6. The teacher wishes to give A's to the top 8% of the students and F's to the bottom 8%. A grade of B is given to the next 15%, with D's given similarly. All other students get C's. Find the bottom cutoff (rounded to the nearest whole number) for the following grades.*

53. A **54.** B **55.** C **56.** D

Computer/Graphing Calculator

Use a computer to find the following probabilities by finding the comparable area under a standard normal curve.

57. $P(1.372 \leq z \leq 2.548)$ **58.** $P(-2.751 \leq z \leq 1.693)$

59. $P(z > -2.476)$ **60.** $P(z < 1.692)$

Use a computer to find the following probabilities for a distribution with a mean of 35.693 and a standard deviation of 7.104.

61. $P(12.275 < x < 28.432)$ **62.** $P(x > 38.913)$

63. $P(x < 17.462)$ **64.** $P(17.462 \leq x \leq 53.106)$

7.4 THE BINOMIAL DISTRIBUTION

What is the probability of finding three or fewer defective bolts in a sample of 100 produced by a certain machine?

This is a Bernoulli trials problem with a large number (100) of trials. The normal curve is used to approximate the binomial distribution to solve this problem.

Many practical experiments have only two possible outcomes, *success* or *failure*. Such experiments are called *binomial experiments* or *Bernoulli trials* and were first studied in Section 6.4. Examples of Bernoulli trials include flipping a coin (with heads being a success, for instance, and tails a failure) or testing televisions coming off the assembly line to see whether or not they are defective.

A **binomial distribution** is a probability distribution that satisfies the following conditions.

1. The experiment is a series of repeated independent trials.

2. There are only two possible outcomes for each trial, success or failure.

3. The probability of success is constant from trial to trial.

For example, suppose a fair die is tossed 5 times, with 1 or 2 considered a success and 3, 4, 5, or 6 considered a failure. Each trial is independent of the others and the probability of success in each trial is $p = 2/6 = 1/3$. In 5 tosses there can be any number of successes from 0 through 5. But these 6 possible results are not equally likely, as shown in the chart on the next page, which was obtained by using the following formula from Section 6.4.

x	$P(x)$	
0	$\binom{5}{0}\left(\frac{1}{3}\right)^0\left(\frac{2}{3}\right)^5$	$= \dfrac{32}{243}$
1	$\binom{5}{1}\left(\frac{1}{3}\right)^1\left(\frac{2}{3}\right)^4$	$= \dfrac{80}{243}$
2	$\binom{5}{2}\left(\frac{1}{3}\right)^2\left(\frac{2}{3}\right)^3$	$= \dfrac{80}{243}$
3	$\binom{5}{3}\left(\frac{1}{3}\right)^3\left(\frac{2}{3}\right)^2$	$= \dfrac{40}{243}$
4	$\binom{5}{4}\left(\frac{1}{3}\right)^4\left(\frac{2}{3}\right)^1$	$= \dfrac{10}{243}$
5	$\binom{5}{5}\left(\frac{1}{3}\right)^5\left(\frac{2}{3}\right)^0$	$= \dfrac{1}{243}$

$$P(x) = \binom{n}{x}p^x(1-p)^{n-x},$$

where n is the number of trials (here, $n = 5$), x is the number of successes, p is the probability of success in a single trial (here, $p = 1/3$), and $P(x)$ is the probability that exactly x of the n trials result in success.

As we saw in Section 6.4, the expected value of this experiment (that is, the expected number of successes in 5 trials) is the sum

$$0 \cdot P(0) + 1 \cdot P(1) + 2 \cdot P(2) + 3 \cdot P(3) + 4 \cdot P(4) + 5 \cdot P(5)$$

$$= 0\left(\frac{32}{243}\right) + 1\left(\frac{80}{243}\right) + 2\left(\frac{80}{243}\right) + 3\left(\frac{40}{243}\right) + 4\left(\frac{10}{243}\right) + 5\left(\frac{1}{243}\right)$$

$$= \frac{405}{243} = 1\frac{2}{3} \quad \text{or} \quad \frac{5}{3}.$$

This result agrees with our intuition since there is, on average, 1 success in 3 trials; so we would expect 2 successes in 6 trials, and a bit less than 2 in 5 trials. From another point of view, since the probability of success is 1/3 in each trial and 5 trials were performed, the expected number of successes should be $5 \cdot (1/3) = 5/3$. More generally, we can show the following.

The expected number of successes in n binomial trials is np, where p is the probability of success in a single trial.

The expected value of any probability distribution is actually its *mean*, as we now show. Suppose an experiment has the possible numerical outcomes $x_1, x_2 \ldots, x_k$, which occur with frequencies f_1, f_2, \ldots, f_k, respectively. If n is the total number of items, then (as shown in Section 7.1) the mean of this grouped distribution is

$$\mu = \frac{x_1f_1 + x_2f_2 + \cdots + x_kf_k}{n} = \frac{x_1f_1}{n} + \frac{x_2f_2}{n} + \cdots + \frac{x_kf_k}{n}$$

$$= x_1\left(\frac{f_1}{n}\right) + x_2\left(\frac{f_2}{n}\right) + \cdots + x_k\left(\frac{f_k}{n}\right).$$

But in each case, the fraction f_i/n (frequency over total number) is the probability p_i of outcome x_i. Hence,

$$\mu = x_1p_1 + x_2p_2 + \cdots + x_kp_k,$$

which is the expected value.

We saw above that the expected value (mean) of a binomial distribution is np. It can be shown that the variance and standard deviation of a binomial distribution are also given by convenient formulas:

$$\sigma^2 = np(1-p) \quad \text{and} \quad \sigma = \sqrt{np(1-p)}.$$

Substituting the appropriate values for n and p from the die-toss example into this new formula gives

$$\sigma^2 = 5\left(\frac{1}{3}\right)\left(\frac{2}{3}\right) = \frac{10}{9} \quad \text{and} \quad \sigma = \sqrt{\frac{10}{9}} = \frac{\sqrt{10}}{3}.$$

A summary of these results follows.

BINOMIAL DISTRIBUTION

Suppose an experiment is a series of n independent, repeated trials, where the probability of a success in a single trial is always p. Let x be the number of successes in the n trials. Then the probability that exactly x successes will occur in n trials is given by

$$\binom{n}{x}p^x(1 - p)^{n-x}.$$

The mean μ and variance σ^2 of a binomial distribution are, respectively,

$$\mu = np \quad \text{and} \quad \sigma^2 = np(1 - p).$$

The standard deviation is

$$\sigma = \sqrt{np(1 - p)}.$$

EXAMPLE 1

The probability of picking a defective plate at random from a china factory's assembly line is .01. A sample of 3 plates is to be selected. Write the distribution for the number of defective plates in the sample, and give its mean and standard deviation.

Since 3 plates will be selected, the possible number of defective plates ranges from 0 to 3. Here, n (the number of trials) is 3, and p (the probability of selecting a defective plate on a single trial) is .01. The distribution and the probability of each outcome are shown below.

x	$P(x)$
0	$\binom{3}{0}(.01)^0(.99)^3 = .970299$
1	$\binom{3}{1}(.01)(.99)^2 = .029403$
2	$\binom{3}{2}(.01)^2(.99) = .000297$
3	$\binom{3}{3}(.01)^3(.99)^0 = .000001$

The mean of the distribution is

$$\mu = np = 3(.01) = .03.$$

The standard deviation is

$$\sigma = \sqrt{np(1 - p)} = \sqrt{3(.01)(.99)} = \sqrt{.0297} \approx .17. \quad \blacktriangleleft$$

Binomial distributions are very useful, but the probability calculations become difficult if n is large. However, the standard normal distribution of the previous section (which shall be referred to as the normal distribution from now on) can be used to get a good approximation of the binomial distributions. This approximation was first discovered in 1718 by Abraham DeMoivre (1667–1754) for the case $p = 1/2$. The result was generalized by the French mathematician Pierre-Simon Laplace (1749–1827) in a book published in 1812.

As an example, to show how the normal distribution is used, consider the distribution for the expected number of heads if one coin is tossed 15 times. This is a binomial distribution with $p = 1/2$ and $n = 15$. The distribution is shown in Figure 15. The mean of the distribution is

$$\mu = np = 15\left(\frac{1}{2}\right) = 7.5.$$

The standard deviation is

$$\sigma = \sqrt{np(1 - p)} = \sqrt{15\left(\frac{1}{2}\right)\left(1 - \frac{1}{2}\right)}$$

$$= \sqrt{15\left(\frac{1}{2}\right)\left(\frac{1}{2}\right)} = \sqrt{3.75} \approx 1.94.$$

x	$P(x)$
0	.00003
1	.00046
2	.00320
3	.01389
4	.04166
5	.09164
6	.15274
7	.19638
8	.19638
9	.15274
10	.09164
11	.04166
12	.01389
13	.00320
14	.00046
15	.00003

FIGURE 15

In Figure 15 we have superimposed the normal curve with $\mu = 7.5$ and $\sigma = 1.94$ over the histogram of the distribution.

As shown in the distribution, the probability of getting exactly 9 heads in the 15 tosses is approximately .153. This answer is about the same fraction that would be found by dividing the area of the bar in color in Figure 15 by the total area of all 16 bars in the graph. (Some of the bars at the extreme left and right ends of the graph are too short to show up.)

As the graph suggests, the area in color is approximately equal to the area under the normal curve from $x = 8.5$ to $x = 9.5$. The normal curve is higher than the top of the bar in the left half but lower in the right half.

To find the area under the normal curve from $x = 8.5$ to $x = 9.5$, first find z-scores, as in the previous section. Use the mean and the standard deviation for the distribution, which we have already calculated, to get z-scores for $x = 8.5$ and $x = 9.5$.

For $x = 8.5$,

$$z = \frac{8.5 - 7.5}{1.94}$$

$$= \frac{1.00}{1.94}$$

$$z \approx .52.$$

For $x = 9.5$,

$$z = \frac{9.5 - 7.5}{1.94}$$

$$= \frac{2.00}{1.94}$$

$$z \approx 1.03.$$

From the table in the Appendix, $z = .52$ gives an area of .6985, and $z = 1.03$ gives .8485. The difference between these two numbers is the desired result.

$$.8485 - .6985 = .1500$$

This answer (.1500) is not far from the exact answer, .153, found above.

EXAMPLE 2

About 6% of the bolts produced by a certain machine are defective.

(a) Find the probability that in a sample of 100 bolts, 3 or fewer are defective.

This problem satisfies the conditions of the definition of a binomial distribution, so the normal curve approximation can be used. First find the mean and the standard deviation using $n = 100$ and $p = 6\% = .06$.

$$\mu = 100(.06)$$
$$= 6$$

$$\sigma = \sqrt{100(.06)(1 - .06)}$$
$$= \sqrt{100(.06)(.94)}$$
$$= \sqrt{5.64} \approx 2.37$$

As the graph in Figure 16 shows, we need to find the area to the left of $x = 3.5$ (since we want 3 or fewer defective bolts). The z-score corresponding to $x = 3.5$ is

$$z = \frac{3.5 - 6}{2.37} = \frac{-2.5}{2.37} \approx -1.05.$$

From the table, $z = -1.05$ leads to an area of .1469, so the probability of getting 3 or fewer defective bolts in a sample of 100 bolts is .1469, or 14.69%.

$x = 3.5$ $\mu = 6$

FIGURE 16

FIGURE 17

(b) Find the probability of getting exactly 11 defective bolts in a sample of 100 bolts.

As Figure 17 shows, we need to find the area between $x = 10.5$ and $x = 11.5$.

$$\text{If } x = 10.5, \text{ then } z = \frac{10.5 - 6}{2.37} \approx 1.90.$$

$$\text{If } x = 11.5, \text{ then } z = \frac{11.5 - 6}{2.37} \approx 2.32.$$

Use the table to find that $z = 1.90$ gives an area of .9713, and $z = 2.32$ yields .9898. The final answer is the difference of these numbers, or

$$.9898 - .9713 = .0185.$$

The probability of having exactly 11 defective bolts is about .0185. ◄

Caution The normal curve approximation to a binomial distribution is quite accurate *provided that n is large and p is not close to 0 or 1*. As a rule of thumb, the normal curve approximation can be used as long as both np and $n(1 - p)$ are at least 5.

7.4 Exercises

1. What three conditions must be satisfied for a distribution to be binomial?

2. What must be known to find the mean and standard deviation of a binomial distribution?

In Exercises 3–8, several binomial experiments are described. For each one, give **(a)** *the distribution;* **(b)** *the mean; and* **(c)** *the standard deviation.*

3. A die is rolled six times, and the number of 1's that come up is tallied. Write the distribution of 1's that can be expected to occur.

4. A 6-item multiple-choice test has 4 possible answers for each item. A student selects all his answers randomly. Give the distribution of correct answers.

5. To maintain quality control on its production line, the Bright Lite Company randomly selects 3 light bulbs each day for testing. Experience has shown a defective rate of .02. Write the distribution for the number of defective bulbs in the daily samples.

6. In a taste test, each member of a panel of 4 is given 2 glasses of Supercola, one made using the old formula and one with the new formula, and asked to identify the new formula. Assuming the panelists operate independently, write the distribution of the number of successful identifications, if each judge actually guesses.

7. The probability that a radish seed will germinate is .7. Joe's mother gives him 4 seeds to plant. Write the distribution for the number of seeds that germinate.

8. Five patients in Ward 8 of Memorial Hospital have a disease with a known mortality rate of .1. Write the distribution of the number who survive.

Work the following problems involving binomial experiments.

9. The probability that an infant will die in the first year of life is about .025. In a group of 500 babies, what are the mean and standard deviation of the number of babies who can be expected to die in their first year of life?

10. The probability that a particular kind of mouse will have a brown coat is 1/4. In a litter of 8, assuming independence, how many could be expected to have a brown coat? With what standard deviation?

11. A certain drug is effective 80% of the time. Give the mean and standard deviation of the number of patients using the drug who recover, out of a group of 64 patients.

12. The probability that a newborn infant will be a girl is .49. If 50 infants are born on Susan B. Anthony's birthday, how many can be expected to be girls? With what standard deviation?

13. What is the rule of thumb for using the normal distribution to approximate a binomial distribution?

For the remaining exercises, use the normal curve approximation to the binomial distribution.

Suppose 16 coins are tossed. Find the probability of getting each of the following results.

14. Exactly 8 heads

15. Exactly 7 heads

16. More than 10 tails

17. Less than 12 tails

Suppose 1000 coins are tossed. Find the probability of getting each of the following results.

18. Exactly 500 heads

19. Exactly 510 heads

20. 480 heads or more

21. Less than 470 tails

A die is tossed 120 times. Find the probability of getting each of the following results.

22. Exactly twenty 5's

23. Exactly twenty-four 6's

24. More than eighteen 3's

25. Fewer than twenty-two 6's

Applications

Business and Economics

Quality Control *Two percent of the quartz heaters produced in a certain plant are defective. Suppose the plant produced 10,000 such heaters last month. Find the probability that among these heaters, the following numbers were defective.*

26. Fewer than 170

27. More than 222

Life Sciences

28. Nest Predation For certain bird species, with appropriate assumptions, the number of nests escaping predation has a binomial distribution.* Suppose the probability of success (that is, a nest escaping predation) is .3. Find the probability that at least half of 26 nests escape predation.

29. Food Consumption Under certain appropriate assumptions, the probability of a competing young animal eating x units of food is binomially distributed, with n equal to the maximum number of food units the animal can acquire, and p equal to the probability per time unit that an animal eats a unit of food.† Suppose $n = 120$ and $p = .6$.

(a) Find the probability that an animal consumes 80 units of food.

(b) Suppose the animal must consume at least 70 units of food to survive. What is the probability that this happens?

30. Drug Side Effects An experimental drug causes a rash in 15% of all people taking it. If the drug is given to 12,000 people, find the probability that more than 1700 people will get the rash.

Drug Effectiveness *A new drug cures 80% of the patients to whom it is administered. It is given to 25 patients. Find the probability that among these patients, the following results occur.*

31. Exactly 20 are cured.

32. All are cured.

33. No one is cured.

34. Twelve or fewer are cured.

Social Sciences

35. Straw Votes In one state, 55% of the voters expect to vote for Edison Diest. Suppose 1400 people are asked the name of the person for whom they expect to vote. Find the probability that at least 750 people will say that they expect to vote for Diest.

Computer/Graphing Calculator

Use the area under the normal curve to approximate each of the following probabilities.

36. Flu Inoculations A flu vaccine has a probability of 80% of preventing a person who is inoculated from getting the flu. A county health office inoculates 134 people. Find the probabilities of the following.

(a) Exactly 10 of the people inoculated get the flu.

(b) No more than 10 of the people inoculated get the flu.

(c) None of the people inoculated get the flu.

37. Color Blindness The probability that a male will be color-blind is .042. Find the probabilities that in a group of 53 men, the following will be true.

(a) Exactly 5 are color-blind.

(b) No more than 5 are color-blind.

(c) At least 1 is color-blind.

*From H. M. Wilbur, *American Naturalist*, vol. 111.
†From G. deJong, *American Naturalist*, vol. 110.

38. **Quality Control** The probability that a certain machine turns out a defective item is .05. Find the probabilities that in a run of 75 items, the following results are obtained.

 (a) Exactly 5 defectives (b) No defectives (c) At least 1 defective

39. **Survey Results** A company is taking a survey to find out whether people like its product. Their last survey indicated that 70% of the population like the product. Based on that, of a sample of 58 people, find the probabilities of the following.

 (a) All 58 like the product. (b) From 28 to 30 (inclusive) like the product.

Chapter Summary Key Terms

7.1 random sample
grouped frequency distribution
frequency polygon
(arithmetic) mean
summation (sigma) notation
median
mode

7.2 range
variance
standard deviation
Chebyshev's theorem
7.3 continuous distribution
skewed distribution

normal distribution
normal curve
standard normal curve
z-score
7.4 binomial distribution

Chapter 7 Review Exercises

1. Discuss some reasons for organizing data into a grouped frequency distribution.

2. What is the rule of thumb for an appropriate interval in a grouped frequency distribution?

In Exercises 3 and 4, **(a)** *write a frequency distribution;* **(b)** *draw a histogram;* **(c)** *draw a frequency polygon.*

3. The following numbers give the sales (in dollars) for the lunch hour at a local hamburger stand for the last 20 Fridays. Use intervals 450–474, 475–499, and so on.

 480 451 501 478 512 473 509 515 458 566
 516 535 492 558 488 547 461 475 492 471

4. The number of units carried in one semester by students in a business mathematics class was as follows. Use intervals 9–10, 11–12, 13–14, 15–16.

 10 9 16 12 13 15 13 16 15 11 13
 12 12 15 12 14 10 12 14 15 15 13

Find the mean for each of the following.

5. 41, 60, 67, 68, 72,
 74, 78, 83, 90, 97

6. 105, 108, 110, 115, 106,
 110, 104, 113, 117

7.
Interval	Frequency
10–19	6
20–29	12
30–39	14
40–49	10
50–59	8

8.
Interval	Frequency
40–44	2
45–49	5
50–54	7
55–59	10
60–64	4
65–69	1

9. What do the mean, median, and mode of a distribution have in common? How do they differ? Describe each in a sentence or two.

Find the median and the mode (or modes) for each of the following.

10. 32, 35, 36, 44, 46, 46, 59

11. 38, 36, 42, 44, 38, 36, 48, 35

Find the modal class for the indicated distributions.

12. Exercise 7

13. Exercise 8

14. What is meant by the range of a distribution?

15. How are the variance and the standard deviation of a distribution related? What is measured by the standard deviation?

Find the range and standard deviation for each of the following distributions.

16. 14, 17, 18, 19, 32

17. 26, 43, 51, 29, 37, 56, 29, 82, 74, 93

Find the standard deviation for the following.

18. Exercise 7

19. Exercise 8

20. Describe the characteristics of a normal distribution.

21. What is meant by a skewed distribution?

22. A probability distribution has an expected value of 28 and a standard deviation of 4. Use Chebyshev's theorem to decide what percent of the distribution is as follows.

 (a) Between 20 and 36 (b) Less than 23.2 or greater than 32.8

23. (a) Find the percent of the area under a normal curve within 2.5 standard deviations of the mean.

 (b) Compare your answer for part (a) with the result found by using Chebyshev's theorem.

Find the following areas under the standard normal curve.

24. Between $z = 0$, and $z = 1.27$

25. To the left of $z = .41$

26. Between $z = -1.88$ and $z = 2.10$

27. Between $z = 1.53$ and $z = 2.82$

28. Find a z-score such that 8% of the area under the curve is to the right of z.

29. Why is the normal distribution not a good approximation of a binomial distribution that has a value of p close to 0 or 1?

Applications

Business and Economics

30. **Stock Returns** The annual returns of two stocks for three years are given below.

Stock	1989	1990	1991
Stock I	11%	−1%	14%
Stock II	9%	5%	10%

 (a) Find the mean and standard deviation for each stock over the three-year period.

 (b) If you are looking for security (hence, less variability) with an average 8% return, which of these stocks should you choose?

31. **Quality Control** A machine that fills quart orange juice cartons is set to fill them with 32.1 oz. If the actual contents of the cartons vary normally, with a standard deviation of .1 oz, what percent of the cartons contain less than a quart (32 oz)?

32. **Quality Control** The probability that a can of peaches from a certain cannery is defective is .005. A sample of 4 cans is selected at random. Write a distribution for the number of defective cans in the sample, and give its mean and standard deviation.

33. **Quality Control** About 6% of the frankfurters produced by a certain machine are overstuffed and thus defective. Find the following probabilities for a sample of 500 frankfurters.

 (a) Twenty-five or fewer are overstuffed.

 (b) Exactly 30 are overstuffed.

 (c) More than 40 are overstuffed.

34. **Bankruptcy** The probability that a small business will go bankrupt in its first year is .21. For 50 such small businesses, find the following probabilities.

 (a) Exactly 8 go bankrupt.

 (b) No more than 2 go bankrupt.

Life Sciences

35. **Rat Diets** The weight gains of 2 groups of 10 rats fed different experimental diets were as follows.

Diet	Weight Gains									
A	1	0	3	7	1	1	5	4	1	4
B	2	1	1	2	3	2	1	0	1	0

Compute the mean and standard deviation for each group.

(a) Which diet produced the greatest mean gain?

(b) Which diet produced the most consistent gain?

Chemical Effectiveness *White flies are devastating California crops. An area infested with white flies is to be sprayed with a chemical which is known to be 98% effective for each application. Assume a sample of 100 flies is checked. Use the normal distribution approximation.*

36. Find the probability that 95% of the flies are killed in one application.

37. Find the probability that at least 95% of the flies are killed in one application.

38. Find the probability that at least 90% of the flies are killed in one application.

39. Find the probability that all the flies are killed in one application.

Social Sciences

Commuting Times *The average resident of a certain East-Coast suburb spends 42 minutes per day commuting, with a standard deviation of 12 minutes. Assume a normal distribution. Find the percent of all residents of this suburb who have the following commuting times.*

40. At least 50 minutes per day

41. No more than 35 minutes per day

42. Between 32 and 40 minutes per day

43. Between 38 and 60 minutes per day

44. I.Q. Scores On standard IQ tests, the mean is 100, with a standard deviation of 15. The results are very close to fitting a normal curve. Suppose an IQ test is given to a very large group of people. Find the percent of those people whose IQ scores are as follows.

(a) More than 130 (b) Less than 85

(c) Between 85 and 115

Connections

45. Much of our work in Chapters 6 and 7 is interrelated. Note the similarities in the following parallel treatments of a frequency distribution and a probability distribution.

Frequency Distribution

Complete the table below for the following data. (Recall that x is the midpoint of the interval.)

14, 7, 1, 11, 2, 3, 11, 6, 10, 13, 11, 11, 16, 12, 9,
11, 9, 10, 7, 12, 9, 6, 4, 5, 9, 16, 12, 12, 11, 10,
14, 9, 13, 10, 15, 11, 11, 1, 12, 12, 6, 7, 8, 2, 9,
12, 10, 15, 9, 3

Interval	x	Tally	f	$x \cdot f$
1–3				
4–6				
7–9				
10–12				
13–15				
16–18				

Probability Distribution

A binomial distribution has n = 10 and p = .5. Complete the table below.

x	$P(x)$	$x \cdot P(x)$
0	.001	
1	.010	
2	.044	
3	.117	
4		
5		
6		
7		
8		
9		
10		

(a) Find the mean (or expected value) for each distribution.

(b) Find the standard deviation for each distribution.

(c) For both distributions, use Chebyshev's theorem to find the interval that contains 75% of the distribution.

(d) Use the normal approximation of the Bernoulli probability distribution to find the interval that contains 95.44% of that distribution.

(e) Why can't we use the normal distribution to answer probability questions about the frequency distribution?

Applications at a glance . . .
English-Australian Cricket Matches . . . patterns in language . . . random walk . . .

CHAPTER 8

MARKOV CHAINS AND GAME THEORY

(See page 362.)

In Chapter 5 we touched on *stochastic processes,* mathematical models in which the outcome of an experiment depends on the outcomes of previous experiments. In this chapter we study a special kind of stochastic process called a *Markov chain,* where the outcome of an experiment depends only on the outcome of the previous experiment. In other words, the next **state** of the system depends only on the present state, not on preceding states. Such experiments are common enough in applications to make their study worthwhile. Markov chains are named after the Russian mathematician A. A. Markov, 1856–1922, who started the theory of stochastic processes. Game theory, discussed later, uses matrices and expected value to clarify the options when a difficult decision must be made.

8.1 BASIC PROPERTIES OF MARKOV CHAINS

If we know the probability that the child of a lower-class parent becomes middle-class or upper-class, and we know similar information for the child of a middle-class or upper-class parent, what is the probability that the grandchild or great-grandchild of a lower-class parent is middle- or upper-class?

Using Markov chains, we will learn the answers to such questions.

Transition Matrix In sociology, it is convenient to classify people by income as *lower-class, middle-class,* and *upper-class.* Sociologists have found that the strongest determinant of the income class of an individual is the income class of the individual's parents. For example, if an individual in the lower-income class is said to be in *state 1*, an individual in the middle-income class is in *state 2*, and an individual in the upper-income class is in *state 3*, then the following probabilities of change in income class from one generation to the next might apply.*

	Next Generation		
State	1	2	3
Current 1	.65	.28	.07
Generation 2	.15	.67	.18
3	.12	.36	.52

Recall from Chapter 5 that the probability of an event is a real number between 0 and 1. The closer the probability is to 1, the more likely the event will occur. Also, the sum of the probabilities of all possible mutually exclusive events equals 1.

 This table shows that if an individual is in state 1 (lower-income class) then there is a probability of .65 that any offspring will be in the lower-income class, a probability of .28 that offspring will be in the middle-income class, and a probability of .07 that offspring will be in the upper-income class.

*For an example with actual data, see D. V. Glass and J. R. Hall, ''Social Mobility in Great Britain: A Study of Intergenerational Changes in Status,'' in *Social Mobility in Great Britain*, D. V. Glass, ed., Routledge & Kegan Paul, 1954. This data is analyzed using Markov chains in *Finite Markov Chains* by John G. Kemeny and J. Laurie Snell, Springer-Verlag, 1976.

The symbol p_{ij} will be used for the probability of transition from state i to state j in one generation. For example, p_{23} represents the probability that a person in state 2 will have offspring in state 3; from the table above.

$$p_{23} = .18.$$

Also from the table, $p_{31} = .12$, $p_{22} = .67$, and so on.

The information from the table can be written in other forms. Figure 1 is a **transition diagram** that shows the three states and the probabilities of going from one state to another.

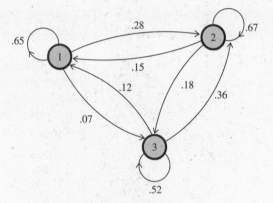

FIGURE 1

In a **transition matrix,** the states are indicated at the side and the top. If P represents the transition matrix for the table above, then

$$
\begin{array}{c}
\quad\;\; 1 \quad\; 2 \quad\; 3 \\
\begin{array}{c} 1 \\ 2 \\ 3 \end{array}
\left[\begin{array}{ccc}
.65 & .28 & .07 \\
.15 & .67 & .18 \\
.12 & .36 & .52
\end{array}\right] = P.
\end{array}
$$

A transition matrix has several features:

1. It is square, since all possible states must be used both as rows and as columns.

2. All entries are between 0 and 1, inclusive; this is because all entries represent probabilities.

3. The sum of the entries in any row must be 1, since the numbers in the row give the probability of changing from the state at the left to one of the states indicated across the top.

Markov Chains A transition matrix, such as matrix P above, also shows two key features of a Markov chain.

MARKOV CHAIN	A sequence of trials of an experiment is a **Markov chain** if

1. the outcome of each experiment is one of a set of discrete states;

2. the outcome of an experiment depends only on the present state, and not on any past states.

For example, in transition matrix P, a person is assumed to be in one of three discrete states (lower, middle, or upper income), with each offspring in one of these same three discrete states.

EXAMPLE 1 A small town has only two dry cleaners, Johnson and NorthClean. Johnson's manager hopes to increase the firm's market share by conducting an extensive advertising campaign. After the campaign, a market research firm finds that there is a probability of .8 that a customer of Johnson's will bring his next batch of dirty clothes to Johnson, and a .35 chance that a NorthClean customer will switch to Johnson for his next batch. Write a transition matrix showing this information.

We must assume that the probability that a customer comes to a given dry cleaner depends only on where the last batch of clothes was taken. If there is a probability of .8 that a Johnson customer will return to Johnson, then there must be a $1 - .8 = .2$ chance that the customer will switch to NorthClean. In the same way, there is a $1 - .35 = .65$ chance that a NorthClean customer will return to NorthClean. These probabilities give the following transition matrix.

$$
\begin{array}{c}
\textit{Second Batch} \\
\begin{array}{cc}
\text{Johnson} & \text{NorthClean}
\end{array}
\end{array}
$$

$$
\textit{First Batch}\quad
\begin{array}{c}
\text{Johnson} \\
\text{NorthClean}
\end{array}
\begin{bmatrix}
.8 & .2 \\
.35 & .65
\end{bmatrix}
$$

We shall come back to this transition matrix later in this section (Example 4).

Figure 2 shows a transition diagram with the probabilities of using each dry cleaner for the second batch of dirty clothes. ◀

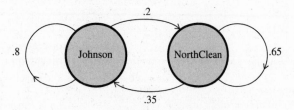

FIGURE 2

Look again at transition matrix P for income-class changes.

$$
\begin{array}{c c}
 & \begin{array}{ccc} 1 & 2 & 3 \end{array} \\
\begin{array}{c} 1 \\ 2 \\ 3 \end{array} &
\left[\begin{array}{ccc}
.65 & .28 & .07 \\
.15 & .67 & .18 \\
.12 & .36 & .52
\end{array}\right] = P
\end{array}
$$

This matrix shows the probability of change in income class from one generation to the next. Now let us investigate the probabilities for changes in income class over *two* generations. For example, if a parent is in state 3 (the upper-income class), what is the probability that a grandchild will be in state 2?

To find out, start with a tree diagram, as shown in Figure 3; the various probabilities come from transition matrix P.

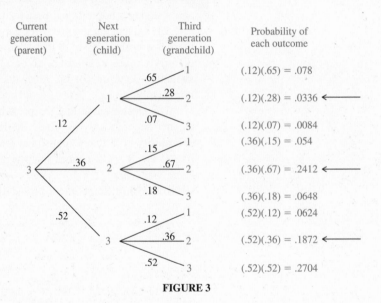

FIGURE 3

Multiplication of matrices was covered in Chapter 2. To get the entry in row i, column j of a product, multiply row i of the first matrix times column j of the second matrix and add up the products. For example, to get the element in row 1, column 1 of P^2, where

$$
P = \begin{bmatrix}
.65 & .28 & .07 \\
.15 & .67 & .18 \\
.12 & .36 & .52
\end{bmatrix},
$$

we calculate $(.65)(.65) + (.28)(.15) + (.07)(.12) = .4729 \approx .47$. To get row 3, column 2, the computation is $(.12)(.28) + (.36)(.67) + (.52)(.36) = .462 \approx .46$. You should review matrix multiplication by working out the rest of P^2 and verifying that it agrees with the result given in Example 2.

The arrows point to the outcomes "grandchild in state 2"; the grandchild can get to state 2 after having had parents in either state 1, state 2, or state 3. The probability that a parent in state 3 will have a grandchild in state 2 is given by the sum of the probabilities indicated with arrows, or

$$.0336 + .2412 + .1872 = .4620.$$

We used p_{ij} to represent the probability of changing from state i to state j in one generation. This notation can be used to write the probability that a parent in state 3 will have a grandchild in state 2:

$$p_{31} \cdot p_{12} + p_{32} \cdot p_{22} + p_{33} \cdot p_{32}.$$

This sum of products of probabilities should remind you of matrix multiplication—it is nothing more than one step in the process of multiplying matrix P by itself. In particular, it is row 3 of P times column 2 of P. If P^2 represents the

matrix product $P \cdot P$, then P^2 gives the probabilities of a transition from one state to another in *two* repetitions of an experiment. Generalizing,

P^k gives the probabilities of a transition from one state to another in k repetitions of an experiment.

EXAMPLE 2

For transition matrix P (income-class changes),

$$P^2 = \begin{bmatrix} .65 & .28 & .07 \\ .15 & .67 & .18 \\ .12 & .36 & .52 \end{bmatrix}\begin{bmatrix} .65 & .28 & .07 \\ .15 & .67 & .18 \\ .12 & .36 & .52 \end{bmatrix} \approx \begin{bmatrix} .47 & .39 & .13 \\ .22 & .56 & .22 \\ .19 & .46 & .34 \end{bmatrix}.$$

(The numbers in the product have been rounded to the same number of decimal places as in matrix P.) The entry in row 3, column 2 of P^2 gives the probability that a person in state 3 will have a grandchild in state 2; that is, that an upper-class person will have a middle-class grandchild. This number, .46, is the result (rounded to two decimal places) found through using the tree diagram.

Row 1, column 3 of P^2 gives the number .13, the probability that a person in state 1 will have a grandchild in state 3; that is, that a lower-class person will have an upper-class grandchild. How would the entry .47 be interpreted? ◀

EXAMPLE 3

In the same way that matrix P^2 gives the probabilities of income-class changes after *two* generations, the matrix $P^3 = P \cdot P^2$ gives the probabilities of change after *three* generations.

For matrix P,

$$P^3 = P \cdot P^2 = \begin{bmatrix} .65 & .28 & .07 \\ .15 & .67 & .18 \\ .12 & .36 & .52 \end{bmatrix}\begin{bmatrix} .47 & .39 & .13 \\ .22 & .56 & .22 \\ .19 & .46 & .34 \end{bmatrix} \approx \begin{bmatrix} .38 & .44 & .17 \\ .25 & .52 & .23 \\ .23 & .49 & .27 \end{bmatrix}.$$

(The rows of P^3 don't necessarily total 1 exactly because of rounding errors.) Matrix P^3 gives a probability of .25 that a person in state 2 will have a great-grandchild in state 1. The probability is .52 that a person in state 2 will have a great-grandchild in state 2. ◀

EXAMPLE 4

Let us return to the transition matrix for the dry cleaners.

Second Batch

Johnson NorthClean

First Batch Johnson $\begin{bmatrix} .8 & .2 \\ .35 & .65 \end{bmatrix}$
NorthClean

As this matrix shows, there is a .8 chance that a person bringing his first batch to Johnson will also bring his second batch to Johnson, and so on. To find the probabilities for the third batch, the second stage of this Markov chain, find the square of the transition matrix. If C represents the transition matrix, then

$$C^2 = C \cdot C = \begin{bmatrix} .8 & .2 \\ .35 & .65 \end{bmatrix}\begin{bmatrix} .8 & .2 \\ .35 & .65 \end{bmatrix} = \begin{bmatrix} .71 & .29 \\ .51 & .49 \end{bmatrix}.$$

From C^2, the probability that a person bringing his first batch of clothes to Johnson will also bring his third batch to Johnson is .71; the probability that a person bringing his first batch to NorthClean will bring his third batch to NorthClean is .49.

The cube of matrix C gives the probabilities for the fourth batch, the third step in our experiment.

$$C^3 = C \cdot C^2 = \begin{bmatrix} .67 & .33 \\ .58 & .42 \end{bmatrix}$$

The probability is .58, for example, that a person bringing his first batch to NorthClean will bring his fourth batch to Johnson. ◀

Distribution of States Look again at the transition matrix for income-class changes:

$$P = \begin{bmatrix} .65 & .28 & .07 \\ .15 & .67 & .18 \\ .12 & .36 & .52 \end{bmatrix}.$$

Suppose the following table gives the initial distribution of people in the three income classes.

Class	State	Proportion
Lower	1	21%
Middle	2	68%
Upper	3	11%

To see how these proportions would change after one generation, use the tree diagram in Figure 4. For example, to find the proportion of people in state 2 after one generation, add the numbers indicated with arrows.

$$.0588 + .4556 + .0396 = .5540$$

In a similar way, the proportion of people in state 1 after one generation is

$$.1365 + .1020 + .0132 = .2517,$$

and the proportion of people in state 3 after one generation is

$$.0147 + .1224 + .0572 = .1943.$$

The initial distribution of states, 21%, 68%, and 11%, becomes, after one generation, 25.17% in state 1, 55.4% in state 2, and 19.43% in state 3. These distributions can be written as *probability vectors* (where the percents have been changed to decimals rounded to the nearest hundredth)

$$[.21 \quad .68 \quad .11] \qquad \text{and} \qquad [.25 \quad .55 \quad .19],$$

respectively. A **probability vector** is a matrix of only one row, having non-negative entries, with the sum of the entries equal to 1.

FIGURE 4

The work with the tree diagram to find the distribution of states after one generation is exactly the work required to multiply the initial probability vector, [.21 .68 .11], and the transition matrix P:

$$[.21 \quad .68 \quad .11] \begin{bmatrix} .65 & .28 & .07 \\ .15 & .67 & .18 \\ .12 & .36 & .52 \end{bmatrix} \approx [.25 \quad .55 \quad .19].$$

In a similar way, the distribution of income classes after two generations can be found by multiplying the initial probability vector and the square of P, the matrix P^2. Using P^2 from above,

$$[.21 \quad .68 \quad .11] \begin{bmatrix} .47 & .39 & .13 \\ .22 & .56 & .22 \\ .19 & .46 & .34 \end{bmatrix} \approx [.27 \quad .51 \quad .21].$$

In the next section we will develop a long-range prediction for the proportions of the population in each income class. The work in this section is summarized below.

Suppose a Markov chain has initial probability vector

$$I = [i_1 \quad i_2 \quad i_3 \cdots i_n]$$

and transition matrix P. The probability vector after n repetitions of the experiment is

$$I \cdot P^n.$$

8.1 Exercises

Decide whether each of the matrices in Exercises 1–9 could be a probability vector.

1. $[\frac{2}{3} \quad \frac{1}{2}]$

2. $[\frac{1}{2} \quad 1]$

3. $[0 \quad 1]$

4. $[.1 \quad .1]$

5. $[.4 \quad .2 \quad 0]$

6. $[\frac{1}{4} \quad \frac{1}{8} \quad \frac{5}{8}]$

7. $[.07 \quad .04 \quad .37 \quad .52]$

8. $[.3 \quad -.1 \quad .8]$

9. $[0 \quad -.2 \quad .6 \quad .6]$

Decide whether each of the matrices in Exercises 10–15 could be a transition matrix, by definition. Sketch a transition diagram for any transition matrices.

10. $\begin{bmatrix} .5 & 0 \\ 0 & .5 \end{bmatrix}$

11. $\begin{bmatrix} \frac{2}{3} & \frac{1}{3} \\ 1 & 0 \end{bmatrix}$

12. $\begin{bmatrix} \frac{1}{4} & \frac{3}{4} \\ \frac{1}{2} & \frac{1}{2} \end{bmatrix}$

13. $\begin{bmatrix} \frac{1}{4} & \frac{3}{4} & 0 \\ 2 & 0 & 1 \\ 1 & \frac{2}{3} & 3 \end{bmatrix}$

14. $\begin{bmatrix} \frac{1}{3} & \frac{1}{3} & \frac{1}{3} \\ 0 & 1 & 0 \\ \frac{1}{2} & 0 & \frac{1}{2} \end{bmatrix}$

15. $\begin{bmatrix} \frac{1}{3} & \frac{1}{2} & 1 \\ 0 & 1 & 0 \\ \frac{1}{2} & \frac{1}{2} & 1 \end{bmatrix}$

In Exercises 16–18, write any transition diagrams as transition matrices.

16.

17.

18.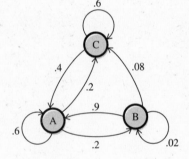

Find the first three powers of each of the transition matrices in Exercises 19–24 (for example, A, A^2, and A^3 in Exercise 19). For each transition matrix, find the probability that state 1 changes to state 2 after three repetitions of the experiment.

19. $A = \begin{bmatrix} 1 & 0 \\ .8 & .2 \end{bmatrix}$

20. $B = \begin{bmatrix} .7 & .3 \\ 0 & 1 \end{bmatrix}$

21. $C = \begin{bmatrix} .5 & .5 \\ .72 & .28 \end{bmatrix}$

22. $D = \begin{bmatrix} .3 & .2 & .5 \\ 0 & 0 & 1 \\ .6 & .1 & .3 \end{bmatrix}$

23. $E = \begin{bmatrix} .8 & .1 & .1 \\ .3 & .6 & .1 \\ 0 & 1 & 0 \end{bmatrix}$

24. $F = \begin{bmatrix} .01 & .9 & .09 \\ .72 & .1 & .18 \\ .34 & 0 & .66 \end{bmatrix}$

25. **(a)** Verify that $I \cdot P^n$ can be computed in two ways: (1) by first multiplying P by itself n times, then multiplying I times this result; and (2) by multiplying IP, multiplying this result by P, and continuing to multiply by P a total of n times. (*Hint:* Use the fact that matrix multiplication is associative.)

(b) Which of the two methods in part (a) is simpler? Explain your answer.

Applications

Business and Economics

26. Dry Cleaning The dry cleaning example in the text used the following transition matrix:

$$
\begin{array}{cc}
& \begin{array}{cc} \text{Johnson} & \text{NorthClean} \end{array} \\
\begin{array}{c} \text{Johnson} \\ \text{NorthClean} \end{array} &
\begin{bmatrix} .8 & .2 \\ .35 & .65 \end{bmatrix}
\end{array}
$$

Suppose now that each customer brings in one batch of clothes per week. Use various powers of the transition matrix to find the probability that a customer initially bringing a batch of clothes to Johnson also brings a batch to Johnson after the following time periods.

(a) 1 wk (b) 2 wk (c) 3 wk (d) 4 wk

(e) What is the probability that a customer initially bringing a batch of clothes to NorthClean brings a batch to Johnson after 2 wk?

27. Dry Cleaning Suppose Johnson has a 40% market share initially, with NorthClean having a 60% share. Use this information to write a probability vector; use this vector, along with the transition matrix above, to find the share of the market for each firm after each of the following time periods. (As in Exercise 26, assume that customers bring in one batch of dry cleaning per week.)

(a) 1 wk (b) 2 wk (c) 3 wk (d) 4 wk

28. Insurance An insurance company classifies its drivers into three groups: G_0 (no accidents), G_1 (one accident), and G_2 (more than one accident). The probability that a G_0 driver will remain a G_0 after one year is .85, that the driver will become a G_1 is .10, and that the driver will become a G_2 is .05. A G_1 driver cannot become a G_0 (this company has a long memory). There is a .8 probability that a G_1 driver will remain a G_1. A G_2 driver must remain a G_2. Write a transition matrix using this information.

29. Insurance Suppose that the company in Exercise 28 accepts 50,000 new policyholders, all of whom are G_0 drivers. Find the number in each group after the following time periods.

(a) 1 yr (b) 2 yr (c) 3 yr (d) 4 yr

30. Insurance The difficulty with the mathematical model in Exercises 28 and 29 is that no "grace period" is provided; there should be a certain positive probability of moving from G_1 or G_2 back to G_0 (say, after four years with no

accidents). A new system with this feature might produce the following transition matrix.

$$
\begin{bmatrix} .85 & .10 & .05 \\ .15 & .75 & .10 \\ .10 & .30 & .60 \end{bmatrix}
$$

Suppose that when this new policy is adopted, the company has 50,000 policyholders, all in G_0. Find the number in each group after the following time periods.

(a) 1 yr (b) 2 yr (c) 3 yr

31. Market Share Research done by Gulf Oil Corporation* produced the following transition matrix for the probabilities that during a given year, a person with one system of home heating will keep the same system or switch to another.

$$
\begin{array}{cc}
& \begin{array}{c} \textit{Next Year's} \\ \textit{Heating System} \end{array} \\
\begin{array}{c} \textit{This Year's} \\ \textit{Heating System} \end{array} &
\begin{array}{c}
\begin{array}{ccc} \text{Oil} & \text{Gas} & \text{Electric} \end{array} \\
\begin{array}{c} \text{Oil} \\ \text{Gas} \\ \text{Electric} \end{array}
\begin{bmatrix} .825 & .175 & 0 \\ .060 & .919 & .021 \\ .049 & 0 & .951 \end{bmatrix}
\end{array}
\end{array}
$$

The current share of the market held by these three heating systems is given by the vector [.26 .60 .14]. Assuming that these trends continue, find the share of the market held by each heating system after the following time periods.

(a) 1 yr (b) 2 yr (c) 3 yr

32. Land Use In one state, a Board of Realtors land use survey showed that 35% of all land was used for agricultural purposes, while 10% was urban. Ten years later, of the agricultural land, 15% had become urbanized and 80% had remained agricultural. (The remainder lay idle.) Of the idle land, 20% had become urbanized and 10% had been converted for agricultural use. Of the urban land, 90% remained urban and 10% was idle. Assume that these trends continue.

(a) Write a transition matrix using this information.

(b) Write a probability vector for the initial distribution of land.

Find the land use pattern after the following time periods.

(c) 10 yr (d) 20 yr

*From Ali Ezzati, "Forecasting Market Shares of Alternative Home Heating Units," *Management Science,* vol. 21, no. 4, December 1974. Copyright © 1974 by The Institute of Management Sciences. Reprinted by permission of the author.

33. Queuing Chain In the queuing chain, we assume that people are queuing up to be served by, say, a bank teller. For simplicity, let us assume that once two people are in line, no one else can enter the line. Let us further assume that one person is served every minute, as long as someone is in line. Assume further that in any minute, there is a probability of $\frac{1}{2}$ that no one enters the line, a probability of $\frac{1}{3}$ that exactly one person enters the line, and a probability of $\frac{1}{6}$ that exactly two people enter the line, assuming there is room. If there is not enough room for two people, then the probability that one person enters the line is $\frac{1}{2}$. Let the state be given by the number of people in line.

(a) Verify that the transition matrix is

$$
\begin{array}{c}
 \\
0 \\
1 \\
2
\end{array}
\begin{array}{ccc}
0 & 1 & 2 \\
\end{array}
\left[
\begin{array}{ccc}
\frac{1}{2} & \frac{1}{3} & \frac{1}{6} \\
\frac{1}{2} & \frac{1}{3} & \frac{1}{6} \\
0 & \frac{1}{2} & \frac{1}{2}
\end{array}
\right].
$$

(b) Find the transition matrix for a two-minute period.

(c) Use your result from part (b) to find the probability that a queue with no one in line has two people in line two minutes later.

Social Sciences

34. Housing Patterns In a survey investigating changes in housing patterns in one urban area, it was found that 75% of the population lived in single-family dwellings and 25% in multiple housing of some kind. Five years later, in a follow-up survey, of those who had been living in single-family dwellings, 90% still did so, but 10% had moved to multiple-family dwellings. Of those in multiple-family housing, 95% were still living in that type of housing, while 5% had moved to single-family dwellings. Assume that these trends continue.

(a) Write a transition matrix for this information.

(b) Write a probability vector for the initial distribution of housing.

What percent of the population can be expected in each category after the following time periods?

(c) 5 yr (d) 10 yr

35. Voting Trends At the end of June in a presidential election year, 40% of the voters were registered as liberal, 45% as conservative, and 15% as independent. Over a one-month period, the liberals retained 80% of their constituency, while 15% switched to conservative and 5% to independent. The conservatives retained 70%, and lost

20% to the liberals. The independents retained 60% and lost 20% each to the conservatives and liberals. Assume that these trends continue.

(a) Write a transition matrix using this information.

(b) Write a probability vector for the initial distribution.

Find the percent of each type of voter at the end of each of the following months.

(c) July (d) August (e) September

(f) October

General Interest

36. Cricket The results of cricket matches between England and Australia have been found to be modeled by a Markov chain.* The probability that England wins, loses, or draws is based on the result of the previous game, with the following transition matrix:

	Wins	Loses	Draws
Wins	.443	.364	.193
Loses	.277	.436	.287
Draws	.266	.304	.430

(a) Compute the transition matrix for the game after the next one, based on the result of the last game.

(b) Use your answer from part (a) to find the probability that, if England won the last game, England will win the game after the next one.

(c) Use your answer from part (a) to find the probability that, if Australia won the last game, England will win the game after the next one.

Computer/Graphing Calculator

For each of the following transition matrices, find the first five powers of the matrix. Then find the probability that state 2 changes to state 4 after 5 repetitions of the experiment.

37.
$$
\begin{bmatrix}
.1 & .2 & .2 & .3 & .2 \\
.2 & .1 & .1 & .2 & .4 \\
.2 & .1 & .4 & .2 & .1 \\
.3 & .1 & .1 & .2 & .3 \\
.1 & .3 & .1 & .1 & .4
\end{bmatrix}
$$

38.
$$
\begin{bmatrix}
.3 & .2 & .3 & .1 & .1 \\
.4 & .2 & .1 & .2 & .1 \\
.1 & .3 & .2 & .2 & .2 \\
.2 & .1 & .3 & .2 & .2 \\
.1 & .1 & .4 & .2 & .2
\end{bmatrix}
$$

*From Derek Colwell, Brian Jones, and Jack Gillett, ''A Markov Chain in Cricket,'' *The Mathematical Gazette*, June 1991.

39. Company Training Program A company with a new training program classified each worker into one of the following four states: s_1, never in the program; s_2, currently in the program; s_3, discharged; s_4, completed the program. The transition matrix for this company is given below.

$$
\begin{array}{c c}
 & \begin{array}{cccc} s_1 & s_2 & s_3 & s_4 \end{array} \\
\begin{array}{c} s_1 \\ s_2 \\ s_3 \\ s_4 \end{array} &
\left[\begin{array}{cccc}
.40 & .20 & .05 & .35 \\
0 & .45 & .05 & .50 \\
0 & 0 & 1.00 & 0 \\
0 & 0 & 0 & 1.00
\end{array} \right]
\end{array}
$$

(a) What proportion of workers who had never been in the program (state s_1) completed the program (state s_4) after the program had been offered 5 times?

(b) If the initial proportions of workers in each state were [.5 .5 0 0], find the corresponding proportions after the program had been offered 4 times.

8.2 REGULAR MARKOV CHAINS

Given the transition probabilities for two dry cleaners, how can we calculate the long-range proportion of customers using each dry cleaner if we do not know the current proportions?

In Example 2 of this section we will see how to answer this question.

If we start with a transition matrix P and an initial probability vector, we can use the nth power of P to find the probability vector for n repetitions of an experiment. In this section, we try to decide what happens to an initial probability vector in the long run; that is, as n gets larger and larger.

For example, let us use the transition matrix associated with the dry cleaning example in the previous section.

$$
\begin{bmatrix}
.8 & .2 \\
.35 & .65
\end{bmatrix}
$$

The initial probability vector, which gives the market share for each firm at the beginning of the experiment, is [.4 .6]. The market shares shown in the following table were found by using powers of the transition matrix. (See Exercise 27 in the previous section.)

Weeks After Start	Johnson	NorthClean
0	.4	.6
1	.53	.47
2	.59	.41
3	.62	.38
4	.63	.37
5	.63	.37
12	.64	.36

The results seem to approach the numbers in the probability vector [.64 .36].

What happens if the initial probability vector is different from [.4 .6]? Suppose [.75 .25] is used; the same powers of the transition matrix as above give the following results.

Weeks After Start	Johnson	NorthClean
0	.75	.25
1	.69	.31
2	.66	.34
3	.65	.35
4	.64	.36
5	.64	.36
6	.64	.36

The results again seem to be approaching the numbers in the probability vector [.64 .36], the same numbers approached with the initial probability vector [.4 .6]. In either case, the long-range trend is for a market share of about 64% for Johnson and 36% for NorthClean. The example above suggests that this long range trend does not depend on the initial distribution of market shares.

Regular Transition Matrices One of the many applications of Markov chains is in finding long-range predictions. It is not possible to make long-range predictions with all transition matrices, but for a large set of transition matrices, long-range predictions *are* possible. Such predictions are always possible with **regular transition matrices**. A transition matrix is **regular** if some power of the matrix contains all positive entries. A Markov chain is a **regular Markov chain** if its transition matrix is regular.

 EXAMPLE 1 Decide whether the following transition matrices are regular.

(a) $A = \begin{bmatrix} .75 & .25 & 0 \\ 0 & .5 & .5 \\ .6 & .4 & 0 \end{bmatrix}$

Square A.

$$A^2 = \begin{bmatrix} .5625 & .3125 & .125 \\ .3 & .45 & .25 \\ .45 & .35 & .2 \end{bmatrix}$$

Since all entries in A^2 are positive, matrix A is regular.

(b) $B = \begin{bmatrix} .5 & 0 & .5 \\ 0 & 1 & 0 \\ 0 & 0 & 1 \end{bmatrix}$

Find various powers of B.

$$B^2 = \begin{bmatrix} .25 & 0 & .75 \\ 0 & 1 & 0 \\ 0 & 0 & 1 \end{bmatrix}; \quad B^3 = \begin{bmatrix} .125 & 0 & .875 \\ 0 & 1 & 0 \\ 0 & 0 & 1 \end{bmatrix}; \quad B^4 = \begin{bmatrix} .0625 & 0 & .9375 \\ 0 & 1 & 0 \\ 0 & 0 & 1 \end{bmatrix}$$

Further powers of B will still give the same zero entries, so no power of matrix B contains all positive entries. For this reason, B is not regular. ◀

Note If a transition matrix P has some zero entries, and P^2 does as well, you

regular. The answer is that if zeros occur in the identical places in both P^k and P^{k+1} for any k, they will appear in those places for all higher powers of P, so P is not regular.

Suppose that v is a probability vector. It can be shown that for a regular Markov chain with a transition matrix P, there exists a single vector V such that $v \cdot P^n$ gets closer and closer to V as n gets larger and larger.

EQUILIBRIUM VECTOR OF A MARKOV CHAIN	If a Markov chain with transition matrix P is regular, then there is a unique vector V such that, for any probability vector v and for large values of n, $$v \cdot P^n \approx V.$$ Vector V is called the **equilibrium vector** or the **fixed vector** of the Markov chain.

In the example with Johnson Cleaners, the equilibrium vector V is approximately $[.64 \quad .36]$. Vector V can be determined by finding P^n for larger and larger values of n, and then looking for a vector that the product $v \cdot P^n$ approaches. Such an approach can be very tedious, however, and is prone to error. To find a better way, start with the fact that for a large value of n,

$$v \cdot P^n \approx V,$$

as mentioned above. From this result, $v \cdot P^n \cdot P \approx V \cdot P$, so that

$$v \cdot P^n \cdot P = v \cdot P^{n+1} \approx VP.$$

Since $v \cdot P^n \approx V$ for large values of n, it is also true that $v \cdot P^{n+1} \approx V$ for large values of n (the product $v \cdot P^n$ approaches V, so that $v \cdot P^{n+1}$ must also approach V). Thus, $v \cdot P^{n+1} \approx V$ and $v \cdot P^{n+1} \approx VP$, which suggests that

$$VP = V.$$

If a Markov chain with transition matrix P is regular, then there exists a probability vector V such that

$$VP = V.$$

This vector V gives the long-range trend of the Markov chain. Vector V is found by solving a system of linear equations, as shown in the next examples.

2 Find the long-range trend for the Markov chain in the dry cleaning example with transition matrix

$$\begin{bmatrix} .8 & .2 \\ .35 & .65 \end{bmatrix}.$$

This matrix is regular since all entries are positive. Let P represent this transition matrix, and let V be the probability vector $[v_1 \quad v_2]$. We want to find V such that

$$VP = V,$$

or
$$[v_1 \quad v_2]\begin{bmatrix} .8 & .2 \\ .35 & .65 \end{bmatrix} = [v_1 \quad v_2].$$

Use matrix multiplication on the left.

$$[.8v_1 + .35v_2 \quad .2v_1 + .65v_2] = [v_1 \quad v_2]$$

Set corresponding entries from the two matrices equal to get

$$.8v_1 + .35v_2 = v_1 \quad \text{and} \quad .2v_1 + .65v_2 = v_2.$$

Simplify each of these equations.

$$-.2v_1 + .35v_2 = 0 \quad \text{and} \quad .2v_1 - .35v_2 = 0$$

These last two equations are really the same. (The equations in the system obtained from $VP = V$ are always dependent.) To find the values of v_1 and v_2, recall that $V = [v_1 \quad v_2]$ is a probability vector, so that

$$v_1 + v_2 = 1.$$

To find v_1 and v_2, solve the system

$$-.2v_1 + .35v_2 = 0$$
$$v_1 + v_2 = 1.$$

From the second equation, $v_1 = 1 - v_2$. Substitute $1 - v_2$ for v_1 in the first equation.

$$-.2(1 - v_2) + .35v_2 = 0$$
$$-.2 + .2v_2 + .35v_2 = 0$$
$$.55v_2 = .2$$
$$v_2 = \frac{4}{11} \approx .364$$

Since $v_1 = 1 - v_2$, $v_1 = 7/11 \approx .636$, and the equilibrium vector is $V = [7/11 \quad 4/11] \approx [.636 \quad .364]$. ◀

Some powers of the transition matrix P in Example 1 (the dry cleaning example) in the previous section (with entries rounded to two decimal places) are shown here.

$$P^2 = \begin{bmatrix} .71 & .29 \\ .51 & .49 \end{bmatrix} \qquad P^3 = \begin{bmatrix} .67 & .33 \\ .58 & .42 \end{bmatrix} \qquad P^4 = \begin{bmatrix} .65 & .35 \\ .62 & .38 \end{bmatrix}$$

$$P^5 = \begin{bmatrix} .65 & .35 \\ .63 & .37 \end{bmatrix} \qquad P^6 = \begin{bmatrix} .64 & .36 \\ .63 & .37 \end{bmatrix} \qquad P^{10} = \begin{bmatrix} .64 & .36 \\ .64 & .36 \end{bmatrix}$$

As these results suggest, higher and higher powers of the transition matrix P approach a matrix having all rows identical; these identical rows have as entries the entries of the equilibrium vector V. This agrees with the statement above: the initial state does not matter. Regardless of the initial probability vector, the system will approach a fixed vector V. This unexpected and remarkable fact is the basic property of regular Markov chains—*the limiting distribution is independent of the initial distribution*. This happens because some power of the transition matrix has all positive entries, so that all the initial probabilities are thoroughly mixed.

Let us summarize the results of this section.

PROPERTIES OF REGULAR MARKOV CHAINS

Suppose a regular Markov chain has a transition matrix P.

1. As n gets larger and larger, the product $v \cdot P^n$ approaches a unique vector V for any initial probability vector v. Vector V is called the *equilibrium vector* or *fixed vector*.

2. Vector V has the property that $VP = V$.

3. To find V, solve a system of equations obtained from the matrix equation $VP = V$, and from the fact that the sum of the entries of V is 1.

4. The powers P^n come closer and closer to a matrix whose rows are made up of the entries of the equilibrium vector V.

EXAMPLE 3

Find the equilibrium vector for the transition matrix

$$K = \begin{bmatrix} .2 & .6 & .2 \\ .1 & .1 & .8 \\ .3 & .3 & .4 \end{bmatrix}.$$

Matrix K has all positive entries and thus is regular. For this reason, an equilibrium vector V must exist such that $VK = V$. Let $V = [v_1 \quad v_2 \quad v_3]$. Then

$$[v_1 \quad v_2 \quad v_3] \begin{bmatrix} .2 & .6 & .2 \\ .1 & .1 & .8 \\ .3 & .3 & .4 \end{bmatrix} = [v_1 \quad v_2 \quad v_3].$$

Use matrix multiplication on the left.

$$[.2v_1 + .1v_2 + .3v_3 \quad .6v_1 + .1v_2 + .3v_3 \quad .2v_1 + .8v_2 + .4v_3]$$
$$= [v_1 \quad v_2 \quad v_3]$$

The Gauss-Jordan process was discussed in Chapter 2. We will review it here by solving the system given in the text. Begin by ridding the matrix of decimals by multiplying by 10, and then dividing out any common factors:

$$\begin{matrix} 10R_2 \to R_2 \\ 10R_3/3 \to R_3 \\ 10R_4/2 \to R_4 \end{matrix} \begin{bmatrix} 1 & 1 & 1 & 1 \\ -8 & 1 & 3 & 0 \\ 2 & -3 & 1 & 0 \\ 1 & 4 & -3 & 0 \end{bmatrix}$$

Next clear column 1 by combining multiples of row 1 with multiples of the other rows.

$$\begin{matrix} 8R_1 + R_2 \to R_2 \\ -2R_1 + R_3 \to R_3 \\ -R_1 + R_4 \to R_4 \end{matrix} \begin{bmatrix} 1 & 1 & 1 & 1 \\ 0 & 9 & 11 & 8 \\ 0 & -5 & -1 & -2 \\ 0 & 3 & -4 & -1 \end{bmatrix}$$

Similarly clear columns 2 and 3 using multiplies of rows 2 and 3 as shown below.

$$\begin{matrix} -R_2 + 9R_1 \to R_1 \\ \\ 5R_2 + 9R_3 \to R_3 \\ -R_2 + 3R_4 \to R_4 \end{matrix} \begin{bmatrix} 9 & 0 & -2 & 1 \\ 0 & 9 & 11 & 8 \\ 0 & 0 & 46 & 22 \\ 0 & 0 & -23 & -11 \end{bmatrix}$$

$$\begin{matrix} R_3 + 23R_1 \to R_1 \\ -11R_3 + 46R_2 \to R_2 \\ R_3/2 \to R_3 \\ R_3 + 2R_4 \to R_4 \end{matrix} \begin{bmatrix} 207 & 0 & 0 & 45 \\ 0 & 414 & 0 & 126 \\ 0 & 0 & 23 & 11 \\ 0 & 0 & 0 & 0 \end{bmatrix}$$

Thus $v_1 = 45/207 = 5/23$; $v_2 = 126/414 = 7/23$; and $v_3 = 11/23$.

Setting corresponding entries equal gives the following equations:

$$.2v_1 + .1v_2 + .3v_3 = v_1$$
$$.6v_1 + .1v_2 + .3v_3 = v_2$$
$$.2v_1 + .8v_2 + .4v_3 = v_3.$$

Simplifying these equations gives

$$-.8v_1 + .1v_2 + .3v_3 = 0$$
$$.6v_1 - .9v_2 + .3v_3 = 0$$
$$.2v_1 + .8v_2 - .6v_3 = 0.$$

Since V is a probability vector,

$$v_1 + v_2 + v_3 = 1.$$

This gives a system of four equations in three unknowns:

$$v_1 + v_2 + v_3 = 1$$
$$-.8v_1 + .1v_2 + .3v_3 = 0$$
$$.6v_1 - .9v_2 + .3v_3 = 0$$
$$.2v_1 + .8v_2 - .6v_3 = 0.$$

This system can be solved with the Gauss-Jordan method presented earlier. Start with the augmented matrix

$$\begin{bmatrix} 1 & 1 & 1 & 1 \\ -.8 & .1 & .3 & 0 \\ .6 & -.9 & .3 & 0 \\ .2 & .8 & -.6 & 0 \end{bmatrix}.$$

The solution of this system is $v_1 = 5/23$, $v_2 = 7/23$, $v_3 = 11/23$, and

$$V = \begin{bmatrix} \frac{5}{23} & \frac{7}{23} & \frac{11}{23} \end{bmatrix} \approx [.22 \quad .30 \quad .48]. \blacktriangleleft$$

8.2 Exercises

Which of the transition matrices in Exercises 1–6 are regular?

1. $\begin{bmatrix} .2 & .8 \\ .9 & .1 \end{bmatrix}$

2. $\begin{bmatrix} .22 & .78 \\ .43 & .57 \end{bmatrix}$

3. $\begin{bmatrix} 1 & 0 \\ .6 & .4 \end{bmatrix}$

4. $\begin{bmatrix} .55 & .45 \\ 0 & 1 \end{bmatrix}$

5. $\begin{bmatrix} 0 & 1 & 0 \\ .4 & .2 & .4 \\ 1 & 0 & 0 \end{bmatrix}$

6. $\begin{bmatrix} .3 & .5 & .2 \\ 1 & 0 & 0 \\ .5 & .1 & .4 \end{bmatrix}$

Find the equilibrium vector for each transition matrix in Exercises 7–14.

7. $\begin{bmatrix} \frac{1}{4} & \frac{3}{4} \\ \frac{1}{2} & \frac{1}{2} \end{bmatrix}$

8. $\begin{bmatrix} \frac{2}{3} & \frac{1}{3} \\ \frac{1}{8} & \frac{7}{8} \end{bmatrix}$

9. $\begin{bmatrix} .3 & .7 \\ .4 & .6 \end{bmatrix}$

10. $\begin{bmatrix} .8 & .2 \\ .1 & .9 \end{bmatrix}$

11. $\begin{bmatrix} .1 & .1 & .8 \\ .4 & .4 & .2 \\ .1 & .2 & .7 \end{bmatrix}$ **12.** $\begin{bmatrix} .5 & .2 & .3 \\ .1 & .4 & .5 \\ .2 & .2 & .6 \end{bmatrix}$ **13.** $\begin{bmatrix} .25 & .35 & .4 \\ .1 & .3 & .6 \\ .55 & .4 & .05 \end{bmatrix}$ **14.** $\begin{bmatrix} .16 & .28 & .56 \\ .43 & .12 & .45 \\ .86 & .05 & .09 \end{bmatrix}$

Find the equilibrium vector for each transition matrix in Exercises 15–20. These matrices were first used in the exercises for Section 8.1. (Note: Not all of these transition matrices are regular, but equilibrium vectors still exist. Why doesn't this contradict the work of this section?)

15. Insurance categories (8.1 Exercise 28)

$\begin{bmatrix} .85 & .10 & .05 \\ 0 & .80 & .20 \\ 0 & 0 & 1 \end{bmatrix}$

16. Modified insurance categories (8.1 Exercise 30)

$\begin{bmatrix} .85 & .10 & .05 \\ .15 & .75 & .10 \\ .10 & .30 & .60 \end{bmatrix}$

17. Home heating systems (8.1 Exercise 31)

$\begin{bmatrix} .825 & .175 & 0 \\ .060 & .919 & .021 \\ .049 & 0 & .951 \end{bmatrix}$

18. Land use (8.1 Exercise 32)

$\begin{bmatrix} .80 & .15 & .05 \\ 0 & .90 & .10 \\ .10 & .20 & .70 \end{bmatrix}$

19. Voter registration (8.1 Exercise 35)

$\begin{bmatrix} .80 & .15 & .05 \\ .20 & .70 & .10 \\ .20 & .20 & .60 \end{bmatrix}$

20. Housing patterns (8.1 Exercise 34)

$\begin{bmatrix} .90 & .10 \\ .05 & .95 \end{bmatrix}$

21. Find the equilibrium vector for the transition matrix

$$\begin{bmatrix} p & 1-p \\ 1-q & q \end{bmatrix},$$

where $0 < p < 1$ and $0 < q < 1$. Under what conditions is this matrix regular?

22. Show that the transition matrix

$$K = \begin{bmatrix} \frac{1}{4} & 0 & \frac{3}{4} \\ 0 & 1 & 0 \\ 0 & 0 & 1 \end{bmatrix}$$

has more than one vector V such that $VK = V$. Why does this not violate the statements of this section?

23. Let

$$P = \begin{bmatrix} a_{11} & a_{12} \\ a_{21} & a_{22} \end{bmatrix}$$

be a regular transition matrix having *column* sums of 1. Show that the equilibrium vector for P is $[1/2 \quad 1/2]$.

24. Notice in Example 3 that the system of equations $VK = V$, with the extra equation that the sum of the elements of V must equal 1, had exactly one solution. What can you say about the number of solutions to the system $VK = V$?

Applications

Business and Economics

25. Quality Control The probability that a complex assembly line works correctly depends on whether the line worked correctly the last time it was used. There is a .9 chance that the line will work correctly if it worked correctly the time before, and a .7 chance that it will work correctly if it did *not* work correctly the time before. Set up a transition matrix with this information and find the long-run probability that the line will work correctly.

26. Quality Control Suppose improvements are made in the assembly line of Exercise 25, so that the transition matrix becomes

	Works	Doesn't Work
Works	$\begin{bmatrix} .95 \\ .80 \end{bmatrix}$	$\begin{matrix} .05 \\ .20 \end{matrix} \Big]$
Doesn't Work		

Find the new long-range probability that the line will work properly.

27. **Sales Management** Each month, a sales manager classifies her salespeople as low, medium, or high producers. There is a .4 chance that a low producer one month will become a medium producer the following month, and a .1 chance that a low producer will become a high producer. A medium producer will become a low or high producer, respectively, with probabilities .25 and .3. A high producer will become a low or medium producer, respectively, with probabilities .05 and .4. Find the long-range trend for the proportion of low, medium, and high producers.

Life Sciences

28. **Research with Mice** A large group of mice is kept in a cage having connected compartments A, B, and C. Mice in compartment A move to B with probability .3 and to C with probability .4. Mice in B move to A or C with probabilities of .15 and .55, respectively. Mice in C move to A or B with probabilities of .3 and .6, respectively. Find the long-range prediction for the fraction of mice in each of the compartments.

Physical Sciences

29. **Weather** The weather in a certain spot is classified as fair, cloudy without rain, or rainy. A fair day is followed by a fair day 60% of the time, and by a cloudy day 25% of the time. A cloudy day is followed by a cloudy day 35% of the time, and by a rainy day 25% of the time. A rainy day is followed by a cloudy day 40% of the time, and by another rainy day 25% of the time. What proportion of days are expected to be fair, cloudy, and rainy over the long term?

Social Sciences

30. **Education** At one liberal arts college, students are classified as humanities majors, science majors, or undecided. There is a 20% chance that a humanities major will change to a science major from one year to the next, and a 45% chance that a humanities major will change to undecided. A science major will change to humanities with probability .15, and to undecided with probability .35. An undecided will switch to humanities or science with probabilities of .5 and .3, respectively.

 (a) Find the long-range prediction for the fraction of students in each of these three majors.

 (b) Compare the result of part (a) with the result of Exercise 23. Make a conjecture, and describe how this conjecture, if true, would allow you to predict the answer to part (a) with very little computation.

31. **Rumors** The manager of the slot machines at a major casino makes a decision about whether or not to "loosen up" the slots so that the customers get a larger payback. The manager tells only one other person, a person whose word cannot be trusted. In fact, there is only a probability p, where $0 < p < 1$, that this person will tell the truth. Suppose this person tells several other people, each of whom tells several people, what the manager's decision is. Suppose there is always a probability p that the decision is passed on as heard. Find the long-range prediction for the fraction of the people who will hear the decision correctly. (*Hint:* Use a transition matrix; let the first row be $[p \quad 1 - p]$ and the second row be $[1 - p \quad p]$.)

General Interest

32. **Ehrenfest Chain** The model for the Ehrenfest chain consists of 2 boxes containing a total of n balls, where n is any integer greater than or equal to 2. In each turn, a ball is picked at random and moved from whatever box it is in to the other box. Let the state of the Markov process be the number of balls in the first box.

 (a) Verify that the probability of going from state i to state j is given by the following.

$$
p_{ij} = \begin{cases} \dfrac{i}{n} & \text{if } i \geq 1 \text{ and } j = i - 1 \\[2mm] 1 - \dfrac{i}{n} & \text{if } i \leq n - 1 \text{ and } j = i + 1 \\[2mm] 1 & \text{if } i = 0 \text{ and } j = 1 \text{ or } i = n \text{ and } j = n - 1 \\[2mm] 0 & \text{otherwise.} \end{cases}
$$

 (b) Verify that the transition matrix is given by

$$
\begin{array}{c@{}c}
 & \begin{array}{ccccc} 0 & 1 & 2 & \cdots & n \end{array} \\
\begin{array}{c} 0 \\ 1 \\ 2 \\ \vdots \\ n \end{array} &
\left[\begin{array}{ccccc}
0 & 1 & 0 & \cdots & 0 \\
\dfrac{1}{n} & 0 & 1 - \dfrac{1}{n} & \cdots & 0 \\
0 & \dfrac{2}{n} & 0 & \cdots & 0 \\
\vdots & \vdots & \vdots & \ddots & \vdots \\
0 & 0 & 0 & \cdots & 0
\end{array}\right].
\end{array}
$$

 (c) Write the transition matrix for the case $n = 2$.

 (d) Determine whether the transition matrix in part (c) is a regular transition matrix.

 (e) Determine an equilibrium vector for the matrix in part (c). Explain what the result means.

33. **Language** One of Markov's own applications was a 1913 study of how often a vowel is followed by another vowel or a consonant by another consonant in Russian text. A

similar study of a passage of English text revealed the following transition matrix.

$$
\begin{array}{c}
 \\
\text{Vowel} \\
\text{Consonant}
\end{array}
\begin{array}{cc}
\text{Vowel} & \text{Consonant} \\
\begin{bmatrix} .12 & .88 \\ .54 & .46 \end{bmatrix}
\end{array}
$$

Find the percent of letters in English text that are expected to be vowels.

34. Random Walk Many phenomena can be viewed as examples of a random walk. Consider the following simple example. A security guard can stand in front of any one of three doors 20 ft apart in front of a building, and every minute he decides whether to move to another door chosen at random. If he is at the middle door, he is equally likely to stay where he is, move to the door to the left, or move to the door to the right. If he is at the door on either end, he is equally likely to stay where he is or move to the middle door.

(a) Verify that the transition matrix is given by

(b) Find the long-range trend for the fraction of time the guard spends in front of each door.

Find the equilibrium vector for transition matrices 35 and 36 by taking powers of the matrix.

$$
35. \begin{bmatrix}
.1 & .2 & .2 & .3 & .2 \\
.2 & .1 & .1 & .2 & .4 \\
.2 & .1 & .4 & .2 & .1 \\
.3 & .1 & .1 & .2 & .3 \\
.1 & .3 & .1 & .1 & .4
\end{bmatrix}
\qquad
36. \begin{bmatrix}
.3 & .2 & .3 & .1 & .1 \\
.4 & .2 & .1 & .2 & .1 \\
.1 & .3 & .2 & .2 & .2 \\
.2 & .1 & .3 & .2 & .2 \\
.1 & .1 & .4 & .2 & .2
\end{bmatrix}
$$

37. (a) Using the initial probability vector [.4 .6], find the probability vector for the next 9 weeks in Example 2 (the dry cleaning example). Compute your answers to at least 6 decimal places.

(b) Using your results from part (a), find the difference between the equilibrium proportion of customers going to Johnson and the proportion going there for each of the first 10 weeks. Be sure to compute the equilibrium proportions to at least 6 decimal places.

(c) Find the ratio between each difference calculated in part (b) and the difference for the previous week.

(d) Using your results from part (c), explain how the probability vector approaches the equilibrium vector.

(e) Repeat parts (a)–(d) of this exercise using the initial probability vector [.75 .25].

8.3 DECISION MAKING

? How should a farmer decide whether or not to use smudge pots in his citrus orchard, given that he doesn't know whether it will freeze or not?

? Freezing temperatures are endangering the orange crop in central California. A farmer can protect his crop by burning smudge pots—the heat from the pots keeps the oranges from freezing. Burning the pots is expensive, however; the cost is $4000. The farmer knows that if he burns smudge pots, he will be able to sell his crop for a net profit (after smudge pot costs are deducted) of $10,000, provided that the freeze does develop and wipes out many of the other orange growers in California. If he does nothing, he will either lose $2000 in planting costs if there is a freeze, or make a profit of $9600 if there is no freeze. (If there is no freeze, there will be a large supply of oranges, and thus his profit will be lower than if there was a small supply.)

What should the farmer do? He should begin by carefully defining the problem. First he must decide on the **states of nature,** the possible alternatives over

which he has no control. Here there are two: freezing temperatures, or no freezing temperatures. Next, the farmer should list the things he can control—his actions or **strategies.** The farmer has two possible strategies: to use smudge pots, or not to use smudge pots. The consequences of each action under each state of nature, called **payoffs,** can be summarized in a **payoff matrix,** as shown below. The payoffs in this case represent the profit for each possible combination of events.

$$\begin{array}{c} & \text{States of Nature} \\ & \begin{array}{cc} \text{Freeze} & \text{No Freeze} \end{array} \\ \textit{Farmer's Strategies} \quad \begin{array}{c} \text{Use Smudge Pots} \\ \text{Do Not Use Pots} \end{array} & \begin{bmatrix} \$10,000 & \$5,600 \\ -\$2,000 & \$9,600 \end{bmatrix} \end{array}$$

To get the $5600 entry in the payoff matrix, we took the profit if there is no freeze, $9600, and subtracted the $4000 cost of using the smudge pots.

Once the farmer makes the payoff matrix, what then? The farmer might be an optimist (some might call him a gambler); in this case he might assume that the best will happen and go for the biggest number on the matrix ($10,000). To get this profit, he must adopt the strategy ''use smudge pots.''

On the other hand, if the farmer were a pessimist, he would want to minimize the worst thing that could happen. If he uses smudge pots, the worst that can happen to him would be a profit of $5600, which will result if there is no freeze. If he does not use smudge pots, he might lose $2000. To minimize the worst, he once again should adopt the strategy ''use smudge pots.''

Suppose the farmer decides that he is neither an optimist nor a pessimist, but would like further information before choosing a strategy. For example, he might call the weather forecaster and ask for the probability of a freeze. Further, suppose the forecaster says that this probability is only .1. What should the farmer do? He should recall our earlier discussion of expected value and calculate the expected profit for each of his two possible strategies. If the probability of a freeze is .1, then the probability that there will be no freeze is .9. This information gives the following expected values.

<div style="margin-left: 2em;">

Recall from Sec. 6.5 that if a random variable x can take on the n values x_1, x_2, x_3, \ldots, x_n with probabilities $p_1, p_2, p_3, \ldots, p_n$, then the expected value of the random variable is

$E(x) = x_1p_1 + x_2p_2 + x_3p_3 + \cdots + x_np_n.$

In the smudge-pot example, $n = 2$, $x_1 = 10,000$, $x_2 = 5600$, $p_1 = .1$, and $p_2 = .9$.

</div>

$$\text{If smudge pots are used: } 10,000(.1) + 5600(.9) = \$6040$$
$$\text{If no smudge pots are used: } -2000(.1) + 9600(.9) = \$8440$$

Here the maximum expected profit, $8440, is obtained if smudge pots are *not* used. If the probability of a freeze is .6, the expected profit from the strategy ''use pots'' would be $8240, and from ''use no pots,'' $2640. As the example shows, the farmer's beliefs about the probabilities of a freeze affect his choice of strategy.

EXAMPLE 1

A small manufacturer of Christmas cards must decide in February what type of cards to emphasize in her fall line. She has three possible strategies: emphasize modern cards, emphasize old-fashioned cards, or emphasize a mixture of the two. Her success is dependent on the state of the economy in December. If the economy is strong, she will do well with her modern cards, while in a weak economy people long for the old days and buy old-fashioned cards. In an in-between economy, her mixture of lines would do the best. She first prepares a

payoff matrix for all three possibilities. The numbers in the matrix represent her profits in thousands of dollars.

States of Nature

		Weak Economy	In-Between	Strong Economy
	Modern	40	85	120
Strategies	Old-Fashioned	90	45	85
	Mixture	75	105	65

(a) If the manufacturer is an optimist, she should aim for the biggest number on the matrix, 120 (representing $120,000 in profit). Her strategy in this case would be to produce modern cards.

(b) A pessimistic manufacturer wants to avoid the worst of all bad things that can happen. If she produces modern cards, the worst that can happen is a profit of $40,000. For old-fashioned cards, the worst is a profit of $45,000, while the worst that can happen from a mixture is a profit of $65,000. Her strategy here is to use a mixture.

(c) Suppose the manufacturer reads in a business magazine that leading experts feel there is a 50% chance of a weak economy at Christmas, a 20% chance of an in-between economy, and a 30% chance of a strong economy. The manufacturer can now find her expected profit for each possible strategy.

Modern: $40(.50) + 85(.20) + 120(.30) = 73$

Old-Fashioned: $90(.50) + 45(.20) + 85(.30) = 79.5$

Mixture: $75(.50) + 105(.20) + 65(.30) = 78$

Here the best strategy is old-fashioned cards; the expected profit is 79.5, or $79,500. ◄

Sometimes the numbers (or payoffs) in a payoff matrix do not represent money (profits or costs, for example), but *utility*. A **utility** is a number that measures the satisfaction (or lack of it) that results from a certain action. The numbers must be assigned by each individual, depending on how he or she feels about a situation. For example, one person might assign a utility of $+20$ to a week's vacation in San Francisco, but only -6 if the vacation were moved to Sacramento. Matrices with utility payoffs are treated the same as those with dollar payoffs. (See Exercises 9 and 10 in this section.)

When a decision is made again and again, the meaning of the term *strategy* is more general than in this section: a strategy is then a rule for determining which choice is made each time the game is played. One possible strategy for the Christmas card manufacturer in Example 1 is to always choose modern cards; another is to rotate, using modern cards one year, old-fashioned the next, a mixture the third year, and so on; a third strategy is to make a random choice each year. If a decision is only made once, however, this meaning of the term *strategy* reduces to that of making a choice. In the section on mixed strategies, we will see how more complex strategies can be selected.

8.3 Exercises

Applications

Business and Economics

1. **Investments** An investor has $20,000 to invest in stocks. She has two possible strategies: buy conservative blue-chip stocks or buy highly speculative stocks. There are two states of nature: the market goes up or the market goes down. The following payoff matrix shows the net amounts she will have under the various circumstances.

	Market Up	Market Down
Buy Blue-Chip	$25,000	$18,000
Buy Speculative	$30,000	$11,000

What should the investor do if she is

(a) an optimist? (b) a pessimist?

(c) Suppose there is a .7 probability of the market going up. What is the best strategy? What is the expected profit?

(d) What is the best strategy if the probability of a market rise is .2?

2. **Land Development** A developer has $100,000 to invest in land. He has a choice of two parcels (at the same price): one on the highway and one on the coast. With both parcels, his ultimate profit depends on whether he faces light opposition from environmental groups or heavy opposition. He estimates that the payoff matrix is as follows (the numbers represent his profit).

	Opposition	
	Light	Heavy
Parcels Highway	$70,000	$30,000
Coast	$150,000	−$40,000

What should the developer do if he is

(a) an optimist? (b) a pessimist?

(c) Suppose the probability of heavy opposition is .8. What is his best strategy? What is the expected profit?

(d) What is the best strategy if the probability of heavy opposition is only .4?

3. **Concert Preparations** Hillsdale College has sold out all tickets for a jazz concert to be held in the stadium. If it rains, the show will have to be moved to the gym, which has a much smaller capacity. The dean must decide in advance whether to set up the seats and the stage in the gym or in the stadium, or both, just in case. The payoff matrix below shows the net profit in each case.

	States of Nature	
	Rain	No Rain
Strategies Set Up in Stadium	−$1550	$1500
Set Up in Gym	$1000	$1000
Set Up in Both	$750	$1400

What strategy should the dean choose if she is

(a) an optimist? (b) a pessimist?

(c) If the weather forecaster predicts rain with a probability of .6, what strategy should the dean choose to maximize expected profit? What is the maximum expected profit?

4. **Machine Repairs** An analyst must decide what fraction of the items produced by a certain machine are defective. He has already decided that there are three possibilities for the fraction of defective items: .01, .10, and .20. He may recommend two courses of action: repair the machine or make no repairs. The payoff matrix below represents the *costs* to the company in each case.

	States of Nature		
	.01	.10	.20
Strategies Repair	$130	$130	$130
No Repair	$25	$200	$500

What strategy should the analyst recommend if he is

(a) an optimist? (b) a pessimist?

(c) Suppose the analyst is able to estimate probabilities for the three states of nature, as follows.

Fraction of Defective Items	Probability
.01	.70
.10	.20
.20	.10

Which strategy should he recommend? Find the expected cost to the company if this strategy is chosen.

5. **Marketing** The research department of the Allied Manufacturing Company has developed a new process that it believes will result in an improved product. Management must decide whether to go ahead and market the new product. The new product may or may not be better than the old one. If the new product is better and the company decides to market it, sales should increase by $50,000. If it is not better and they replace the old product with the new product on the market, they will lose $25,000 to competitors. If they decide not to market the new product, they will lose a total of $40,000 if it is better, and just research costs of $10,000 if it is not.

 (a) Prepare a payoff matrix.

 (b) If management believes there is a probability of .4 that the new product is better, find the expected profits under each strategy and determine the best action.

6. **Machinery Overhaul** A businessman is planning to ship a used machine to his plant in Nigeria. He would like to use it there for the next four years. He must decide whether to overhaul the machine before sending it. The cost of overhaul is $2600. If the machine fails when in operation in Nigeria, it will cost him $6000 in lost production and repairs. He estimates the probability that it will fail at .3 if he does not overhaul it, and .1 if he does overhaul it. Neglect the possibility that the machine might fail more than once in the next four years.

 (a) Prepare a payoff matrix.

 (b) What should the businessman do to minimize his expected costs?

7. **Contractor Bidding** A contractor prepares to bid on a job. If all goes well, his bid should be $30,000, which will cover his costs plus his usual profit margin of $4500. If a threatened labor strike actually occurs, however, his bid should be $40,000 to give him the same profit. If there is a strike and he bids $30,000, he will lose $5500. If he bids $40,000 and there is no strike, his bid will be too high, and he will lose the job entirely.

 (a) Prepare a payoff matrix.

 (b) If the contractor believes that the probability of a strike is .6, how much should he bid?

Life Sciences

8. **Anti-Smoking Campaign** A community is considering an anti-smoking campaign.* The city council will choose one of three possible strategies: a campaign for everyone over

age 10 in the community, a campaign for youths only, or no campaign at all. The two states of nature are a true cause-effect relationship between smoking and cancer, and no cause-effect relationship. The first year costs to the community (including loss of life and productivity) in each case are as shown below.

	States of Nature	
	Cause-Effect Relationship	No Cause-Effect Relationship
Campaign for All	$100,000	$800,000
Strategies Campaign for Youth	$2,820,000	$20,000
No Campaign	$3,100,100	$0

What action should the city council choose if it is

 (a) optimistic? (b) pessimistic?

 (c) If the director of public health estimates that the probability of a true cause-effect relationship is .8, which strategy should the city council choose?

Social Sciences

9. **Reelection Strategy** A politician must plan her reelection strategy. She can emphasize jobs or she can emphasize the environment. The voters may be concerned mainly about jobs or about the environment. A payoff matrix showing the utility of each possible outcome is shown below.

	Voters	
	Jobs	Environment
Candidate Jobs	+25	−10
Environment	−15	+30

The political analysts feel that there is a .35 chance that the voters will emphasize jobs. What strategy should the candidate adopt? What is its expected utility?

10. **Education** In an accounting class, the instructor permits the students to bring a calculator or a reference book (but not both) to an examination. The examination itself can emphasize either numerical problems or definitions. In trying to decide which aid to take to an examination, a student first decides on the utilities shown in the following payoff matrix.

	Exam's Emphasis	
	Numbers	Definitions
Student's Choice Calculator	+50	0
Book	+10	+40

 (a) What strategy should the student choose if the probability that the examination will emphasize numbers is .6? What is the expected utility in this case?

 (b) Suppose the probability that the examination emphasizes numbers is .4. What strategy should be chosen by the student?

*This problem is based on an article by B. G. Greenberg in the September 1969 issue of the *Journal of the American Statistical Association*.

8.4 STRICTLY DETERMINED GAMES

 How can a football team decide which play to run when the play might have varying degrees of success, depending upon what the opposing team does?

In this section, we will learn strategies for making such decisions.

The word *game* in the title of this chapter may have led you to think of chess, or perhaps some card game. While *game theory* does have some application to these recreational games, it was developed in the 1940s to analyze competitive situations in business, warfare, and social situations. Game theory was invented by John von Neumann (1903–1957), who, with Oskar Morgenstern, wrote the book entitled *Theory of Games and Economic Behavior* in 1944. **Game theory** is the study of how to make decisions when competing with an aggressive opponent.

A game can be set up with a payoff matrix, such as the one shown below. This game involves the two players A and B, and is called a **two-person game.** Player A can choose either row 1 or row 2, while player B can choose either column 1 or column 2. A player's choice is called a **strategy,** just as before. The payoff is at the intersection of the row and column selected. As a general agreement, a positive number represents a payoff from B to A; a negative number represents a payoff from A to B. For example, if A chooses row 2 and B chooses column 2, then B pays $4 to A.

$$
\begin{array}{cc}
 & \begin{array}{cc} & B \\ 1 & 2 \end{array} \\
A\ \begin{array}{c} 1 \\ 2 \end{array} & \begin{bmatrix} 2 & -1 \\ -3 & 4 \end{bmatrix}
\end{array}
$$

EXAMPLE 1 In the payoff matrix just shown, suppose A chooses row 1 and B chooses column 2. Who gets what?

Row 1 and column 2 lead to the number −1. This number represents a payoff of $1 from A to B. ◀

While the numbers in this payoff matrix represent money, they could just as easily represent goods or other property.

In the game above, no money enters the game from the outside; whenever one player wins, the other loses. Such a game is called a **zero-sum game.** The stock market is not a zero-sum game. Stocks can go up or down as a result of outside forces. Therefore, it is possible that all investors can make or lose money.

Only two-person zero-sum games are discussed in the rest of this chapter. Each player can have many different options. In particular, an $m \times n$ matrix game is one in which player A has m strategies (rows) and player B has n strategies (columns). We will always use rows for player A and columns for player B, so labels are not necessary on the matrix.

Dominated Strategies In the rest of this section, the best possible strategy for each player is determined. Let us begin with the 3×3 game defined by the following matrix.

$$\begin{array}{c} & \begin{array}{ccc} 1 & 2 & 3 \end{array} \\ \begin{array}{c} 1 \\ 2 \\ 3 \end{array} & \left[\begin{array}{ccc} -3 & -6 & 10 \\ 3 & 0 & -9 \\ 5 & -4 & -8 \end{array}\right] \end{array}$$

From B's viewpoint, strategy 2 is better than strategy 1 no matter which strategy A selects. This can be seen by comparing the two columns. If A chooses row 1, receiving $6 from A is better than receiving $3; in row 2, breaking even is better than paying $3; and in row 3, getting $4 from A is better than paying $5. Therefore, B should never select strategy 1. Strategy 2 is said to *dominate* strategy 1, and strategy 1 (the dominated strategy) can be removed from consideration, producing the following reduced matrix.

$$\begin{array}{c} & \begin{array}{cc} 2 & 3 \end{array} \\ \begin{array}{c} 1 \\ 2 \\ 3 \end{array} & \left[\begin{array}{cc} -6 & 10 \\ 0 & -9 \\ -4 & -8 \end{array}\right] \end{array}$$

Either player may have dominated strategies. In fact, after a dominated strategy for one player is removed, the other player may then have a dominated strategy where there was none before.

DOMINATED STRATEGIES

A row for A **dominates** another row if every entry in the first row is *larger* than the corresponding entry in the second row. For a column for B to dominate another, each entry must be *smaller*.

In the 3×2 matrix above, neither player now has a dominated strategy. From A's viewpoint, strategy 1 is best if B chooses strategy 3, while strategy 2 is best if B chooses strategy 1. Verify that there are no dominated strategies for either player.

To find any dominated strategy, start with the first two rows. If the first row has an entry larger than the corresponding entry in the second row and another entry smaller than the corresponding entry in the second row, then neither row dominates. If this is not the case, then one row is dominated and may be removed. Continue this process for every pair of rows and every pair of columns.

 EXAMPLE 2 Find any dominated strategies in the games with the given payoff matrices.

(a)
$$\begin{array}{c} & \begin{array}{cccc} 1 & 2 & 3 & 4 \end{array} \\ \begin{array}{c} 1 \\ 2 \end{array} & \left[\begin{array}{cccc} -8 & -4 & -6 & -9 \\ -3 & 0 & -9 & 12 \end{array}\right] \end{array}$$

Here every entry in column 3 is smaller than the corresponding entry in column 2. Thus, column 3 dominates column 2. (Notice that column 1 also dom-

inates column 2.) By removing the dominated column 2, the final game is as follows.

$$\begin{array}{c} \\ 1 \\ 2 \end{array} \begin{array}{ccc} 1 & 3 & 4 \\ \begin{bmatrix} -8 & -6 & -9 \\ -3 & -9 & 12 \end{bmatrix} \end{array}$$

(b) $\begin{array}{c} \\ 1 \\ 2 \\ 3 \end{array} \begin{array}{cc} 1 & 2 \\ \begin{bmatrix} 3 & -2 \\ 0 & 8 \\ 6 & 4 \end{bmatrix} \end{array}$

Each entry in row 3 is greater than the corresponding entry in row 1, so that row 3 dominates row 1. Removing row 1 gives the following game.

$$\begin{array}{c} \\ 2 \\ 3 \end{array} \begin{array}{cc} 1 & 2 \\ \begin{bmatrix} 0 & 8 \\ 6 & 4 \end{bmatrix} \end{array} \blacktriangleleft$$

Strictly Determined Games Which strategies should the players choose in the following game?

$$\begin{array}{c} \\ \\ A \begin{array}{c} 1 \\ 2 \\ 3 \end{array} \end{array} \begin{array}{ccc} & B & \\ 1 & 2 & 3 \\ \begin{bmatrix} -9 & -11 & 7 \\ 2 & 3 & 5 \\ -1 & 6 & -3 \end{bmatrix} \end{array}$$

The goal of game theory is to find **optimum strategies:** those that are the most profitable to the respective players. The payoff that results from each player's choosing the optimum strategy is called the **value** of the game.

The simplest strategy for a player is to consistently choose a certain row (or column). Such a strategy is called a **pure strategy,** in contrast to strategies requiring the random choice of a row (or column); these alternative strategies are discussed in the next section.*

To choose a pure strategy in the game above, player A could choose row 1, in hopes of getting the payoff of $7. Player B would quickly discover this, however, and start playing column 2. By playing column 2, B would receive $11 from A. If A were to choose row 2 consistently, then B would again minimize outgo by choosing column 1 (a payoff of $2 by B to A is better than paying $3 or $5, respectively, to A). By choosing row 3 consistently, A would cause B to choose column 3. The table shows what B will do when A chooses a given row consistently.

If A Chooses Pure Strategy:	Then B Will Choose:	With Payoff:
Row 1	Column 2	$11 to B
Row 2	Column 1	$2 to A
Row 3	Column 3	$3 to B

*In this section we solve (find the optimum strategies for) only games that have optimum *pure* strategies.

Based on these results, A's optimum strategy is to choose row 2; in this way A will get a minimum payoff of $2 per play of the game, no matter what B does.

The optimum pure strategy in this game for A (the *row* player), is found by identifying the *smallest* number in each row of the payoff matrix; the row giving the *largest* such number gives the optimum strategy.

By going through a similar analysis for player B, we find that B should choose the column that will minimize the amount A can win. In the game above, B will pay $2 to A if B consistently chooses column 1. By choosing column 2 consistently, B will pay $6 to A, and by choosing column 3, player B will pay $7 to A. The optimum strategy for B is thus to choose column 1—with each play of the game B will pay $2 to A.

The optimum pure strategy in this game for B (the column player) is to identify the *largest* number in each column of the payoff matrix, and then choose the column producing the *smallest* such number.

In the game above, the entry 2 is both the *smallest* entry in its *row* and the *largest* entry in its *column*. Such an entry is called a **saddle point.** (The seat of a saddle is the maximum from one direction and the minimum from another direction. See Figure 5.) As Example 3(b) will show, there may be more than one such entry, but then the entries will have the same value.

FIGURE 5

OPTIMUM PURE STRATEGY	In a game with a saddle point, the optimum pure strategy for player A is to choose the row containing the saddle point, while the optimum pure strategy for B is to choose the column containing the saddle point.

A game with a saddle point is called a **strictly determined game.** By using these optimum strategies, A and B will ensure that the same amount always changes hands with each play of the game; this amount, given by the saddle point, is the value of the game. The value of the game above is $2. A game having a value of 0 is a **fair game;** the game above is not fair. It is also not much fun for player B, who will lose $2 to player A every time. Finally, it is not very interesting to play the game, because players A and B will choose strategies 2 and 1, respectively, every time. Games in which there is no optimum pure strategy are more interesting to play again and again; we will study these in the next section.

To find the saddle point easily, underline the smallest number in each row, and circle the largest number in each column. If a number is both underlined and circled, it is the saddle point. In case of a tie, underline or circle all numbers that equal the smallest number in the row or the largest number in the column.

EXAMPLE 3 Find the saddle points in the following games.

(a) $\begin{array}{c} \\ 1 \\ 2 \\ 3 \\ 4 \end{array} \begin{array}{cc} 1 & 2 \\ \left[\begin{array}{cc} 2 & 2 \\ 0 & 4 \\ 1 & 6 \\ 3 & 7 \end{array}\right] \end{array}$

Underlining the smallest number in each row and circling the largest value in each column yields the following:

$$
\begin{array}{c c c}
 & 1 & 2 \\
1 & \underline{2} & 2 \\
2 & \underline{0} & 4 \\
3 & \underline{1} & 6 \\
4 & \boxed{\circled{\underline{3}}} & \circled{7}
\end{array}
$$

The number that is both circled and underlined is 3. Thus, 3 is the saddle point, and the game has a value of 3. The strategies producing the saddle point can be written (4, 1). (Player A's strategy is written first.)

(b)
$$
\begin{array}{c}
1 \\
2
\end{array}
\begin{bmatrix}
\;\;4 & \;\;6 & 4 & 12 \\
-8 & -9 & 3 & \;\;2
\end{bmatrix}
\quad
\begin{array}{cccc}
1 & 2 & 3 & 4
\end{array}
$$

Underlining the smallest number in each row and circling the largest value in each column yields the following:

$$
\begin{array}{c}
1 \\
2
\end{array}
\begin{bmatrix}
\circled{\underline{4}} & \circled{6} & \circled{\underline{4}} & \circled{12} \\
-8 & \underline{-9} & 3 & 2
\end{bmatrix}
\quad
\begin{array}{cccc}
1 & 2 & 3 & 4
\end{array}
$$

The saddle point, 4, occurs with either of two strategies, (1, 1) or (1, 3). The value of the game is 4. (Neither of the games in parts (a) or (b) of this example is a fair game, because neither game has a value of 0.)

(c)
$$
\begin{array}{c}
1 \\
2
\end{array}
\begin{bmatrix}
3 & \;\;6 & -2 \\
8 & -3 & \;\;5
\end{bmatrix}
\quad
\begin{array}{ccc}
1 & 2 & 3
\end{array}
$$

Underlining the smallest number in each row and circling the largest value in each column yields the following.

$$
\begin{array}{c}
1 \\
2
\end{array}
\begin{bmatrix}
3 & \circled{6} & \underline{-2} \\
\circled{8} & \underline{-3} & \circled{5}
\end{bmatrix}
\quad
\begin{array}{ccc}
1 & 2 & 3
\end{array}
$$

There is no number that is both the smallest number in its row and the largest number in its column, so the game has no saddle point. Since the game has no saddle point, it is not strictly determined. In the next section, methods are given for finding optimum strategies for such games. ◀

8.4 Exercises

In the following game, decide on the payoff when the strategies of Exercises 1–6 are used.

$$
\begin{array}{cc}
 & \begin{array}{ccc} & B & \\ 1 & 2 & 3 \end{array} \\
A\;\begin{array}{c}1\\2\\3\end{array} & \begin{bmatrix} \;\;6 & -4 & \;\;0 \\ \;\;3 & -2 & \;\;6 \\ -1 & \;\;5 & 11 \end{bmatrix}
\end{array}
$$

1. (1, 1)
2. (1, 2)
3. (2, 2)
4. (2, 3)
5. (3, 1)
6. (3, 2)

7. Does the game have any dominated strategies?

8. Does it have a saddle point?

Remove any dominated strategies in the games in Exercises 9–14. (From now on, we will save space by deleting the names of the strategies.)

9. $\begin{bmatrix} 0 & -2 & 8 \\ 3 & -1 & -9 \end{bmatrix}$

10. $\begin{bmatrix} 6 & 5 \\ 3 & 8 \\ -1 & -4 \end{bmatrix}$

11. $\begin{bmatrix} 1 & 4 \\ 4 & -1 \\ 3 & 5 \\ -4 & 0 \end{bmatrix}$

12. $\begin{bmatrix} 2 & 3 & 1 & -5 \\ -1 & 5 & 4 & 1 \\ 1 & 0 & 2 & -3 \end{bmatrix}$

13. $\begin{bmatrix} 8 & 12 & -7 \\ -2 & 1 & 4 \end{bmatrix}$

14. $\begin{bmatrix} 6 & 2 \\ -1 & 10 \\ 3 & 5 \end{bmatrix}$

For each game, when the saddle point exists, find the strategies producing it and the value of the game. Identify any games that are strictly determined.

15. $\begin{bmatrix} 3 & 5 \\ 2 & -5 \end{bmatrix}$

16. $\begin{bmatrix} 7 & 8 \\ -2 & 15 \end{bmatrix}$

17. $\begin{bmatrix} 3 & -4 & 1 \\ 5 & 3 & -2 \end{bmatrix}$

18. $\begin{bmatrix} -4 & 2 & -3 & -7 \\ 4 & 3 & 5 & -9 \end{bmatrix}$

19. $\begin{bmatrix} -6 & 2 \\ -1 & -10 \\ 3 & 5 \end{bmatrix}$

20. $\begin{bmatrix} 1 & 4 & -3 & 1 & -1 \\ 2 & 5 & 0 & 4 & 10 \\ 1 & -3 & 2 & 5 & 2 \end{bmatrix}$

21. $\begin{bmatrix} 2 & 3 & 1 \\ -1 & 4 & -7 \\ 5 & 2 & 0 \\ 8 & -4 & -1 \end{bmatrix}$

22. $\begin{bmatrix} 3 & 8 & -4 & -9 \\ -1 & -2 & -3 & 0 \\ -2 & 6 & -4 & 5 \end{bmatrix}$

23. $\begin{bmatrix} -6 & 1 & 4 & 2 \\ 9 & 3 & -8 & -7 \end{bmatrix}$

24. $\begin{bmatrix} 6 & -1 \\ 0 & 3 \\ 4 & 0 \end{bmatrix}$

25. Suppose the payoff matrix for a game has at least three rows. Also, suppose that row 1 dominates row 2, and row 2 dominates row 3. Show that row 1 must dominate row 3.

Applications

Business and Economics

26. Competition Two merchants are planning to build competing stores to serve an area of three small cities. The fraction of the total population that live in each city is shown in the figure. If both merchants locate in the same city, merchant A will get 65% of the total business. If the merchants locate in different cities, each will get 80% of the business in the city it is in, and A will get 60% of the business from the city not containing B. Payoffs are measured by the number of percentage points above or below 50%. Write a payoff matrix for this game. Is this game strictly determined?

27. Competition In Exercise 26, if merchant A gets 55% of the total business when both merchants locate in the same city, and the population is otherwise unchanged, write the payoff matrix and find the strategies producing the saddle point and the value of the game.

Social Sciences

28. Border Patrol A person attempting an illegal crossing of the border into the United States has two options. He can cross into unoccupied territory on foot at night, or he can cross at a regular entry point if he is hidden in a vehicle. His chances of detection depend on whether extra border guards are sent to the unoccupied territory or to the regular entry point. His chances of getting across the unoccupied territory are 50% if there are no extra patrols and 40% if there are extra patrols. His chances of crossing at a regular entry point are 30% if there are no extra border guards and 20% if there are extra guards. Letting the payoffs be the probabilities of success, find the strategies producing the saddle point and the value of the game.

General Interest

29. War Games Two armies, A and B, are involved in a war game. Each army has available three different strategies, with payoffs as shown below. These payoffs represent square kilometers of land, with positive numbers representing gains by A.

$$\begin{bmatrix} 3 & -8 & -9 \\ 0 & 6 & -12 \\ -8 & 4 & -10 \end{bmatrix}$$

Find the strategies producing the saddle point and the value of the game.

30. Football When a football team has the ball and is planning its next play, it can choose one of several plays or strategies. The success of the chosen play depends largely on how well the other team "reads" the chosen play. Suppose a team with the ball (team A) can choose from three plays, while the opposition (team B) has four possible strategies. The numbers shown in the following payoff matrix represent yards of gain to team A.

$$\begin{bmatrix} 9 & -3 & -4 & 16 \\ 12 & 9 & 6 & 8 \\ -5 & -2 & 3 & 18 \end{bmatrix}$$

Find the strategies producing the saddle point. Find the value of the game.

31. Children's Game In the children's game *rock, paper, scissors,* two players simultaneously extend one hand in the form of a fist (*rock*), a flat palm (*paper*), or two fingers extended (*scissors*). If they both make the same choice, the game is a tie. Otherwise, *rock* beats *scissors* (because rock can crush scissors), *scissors* beats *paper* (because scissors can cut paper), and *paper* beats *rock* (because paper can cover rock). Whoever wins gains one point; in the case of a tie, no one gains any points. Find the payoff matrix for this game. Is the game strictly determined?

32. Finger Game John and Joann play a finger matching game—each shows one or two fingers, simultaneously. If the sum of the number of fingers showing is even, Joann pays John that number of dollars; for an odd sum, John pays Joann. Find the payoff matrix for this game. Is the game strictly determined?

Dang! Tied again! Ready...One, two, *three!*

6-28 © 1986 Universal Press Syndicate Larson

Before paper and fire

8.5 MIXED STRATEGIES

How can a farmer choose whether or not to spray insecticide on his crops, given that neither strategy, when used consistently, is the best strategy?

In this section, we will learn the answer to such questions.

As mentioned earlier, not every game has a saddle point. Two-person zero-sum games still have optimum strategies, however, even if the strategy is not as simple as the ones we saw earlier. In a game with a saddle point, the optimum

strategy for player A is to pick the row containing the saddle point. Such a strategy is called a *pure strategy,* since the same row is always chosen.

If there is no saddle point, then it will be necessary for both players to mix their strategies. For example, A will sometimes play row 1, sometimes row 2, and so on. If this were done in some specific pattern, the competitor would soon guess it and play accordingly.

For this reason, it is best to mix strategies according to previously determined probabilities. For example, if a player has only two strategies and has decided to play them with equal probability, the random choice could be made by tossing a fair coin, letting heads represent one strategy and tails the other. This would result in the two strategies being used about equally over the long run. On a particular play, however, it would not be possible to predetermine the strategy to be used. (Some other device, such as a spinner, is necessary if there are more than two strategies or if the probabilities are not 1/2.)

EXAMPLE 1 Suppose a game has payoff matrix

$$\begin{bmatrix} -1 & 2 \\ 1 & 0 \end{bmatrix},$$

where the entries represent dollar winnings. Suppose player A chooses row 1 with probability 1/3 and row 2 with probability 2/3, and player B chooses each column with probability 1/2. Find the expected value of the game.

Assume that rows and columns are chosen independently, so that

$$P(\text{row 1, column 1}) = P(\text{row 1}) \cdot P(\text{column 1}) = \frac{1}{3} \cdot \frac{1}{2} = \frac{1}{6}$$

$$P(\text{row 1, column 2}) = P(\text{row 1}) \cdot P(\text{column 2}) = \frac{1}{3} \cdot \frac{1}{2} = \frac{1}{6}$$

$$P(\text{row 2, column 1}) = P(\text{row 2}) \cdot P(\text{column 1}) = \frac{2}{3} \cdot \frac{1}{2} = \frac{1}{3}$$

$$P(\text{row 2, column 2}) = P(\text{row 2}) \cdot P(\text{column 2}) = \frac{2}{3} \cdot \frac{1}{2} = \frac{1}{3}.$$

The table below lists the probability of each possible outcome, along with the payoff to player A.

Outcome	Probability of Outcome	Payoff for A
Row 1, column 1	1/6	−1
Row 1, column 2	1/6	2
Row 2, column 1	1/3	1
Row 2, column 2	1/3	0

The expected value of the game is given by the sum of the products of the probabilities and the payoffs, or

$$\text{Expected value} = \frac{1}{6}(-1) + \frac{1}{6}(2) + \frac{1}{3}(1) + \frac{1}{3}(0) = \frac{1}{2}.$$

In the long run, for a great many plays of the game, the payoff to A will average $1/2$ dollar per play of the game. It is important to note that as the mixed strategies used by A and B are changed, the expected value of the game may well change. (See Example 2 below.) ◀

To generalize the work of Example 1, let the payoff matrix for a 2×2 game be

$$M = \begin{bmatrix} a_{11} & a_{12} \\ a_{21} & a_{22} \end{bmatrix}.$$

Let player A choose row 1 with probability p_1 and row 2 with probability p_2, where $p_1 + p_2 = 1$. Write these probabilities as the row matrix

$$A = [p_1 \quad p_2].$$

Let player B choose column 1 with probability q_1 and column 2 with probability q_2, where $q_1 + q_2 = 1$. Write this as the column matrix

$$B = \begin{bmatrix} q_1 \\ q_2 \end{bmatrix}.$$

The probability of choosing row 1 and column 1 is

$$P(\text{row 1, column 1}) = P(\text{row 1}) \cdot P(\text{column 1}) = p_1 \cdot q_1.$$

In the same way, the probabilities of each possible outcome are shown in the table below, along with the payoff for each outcome.

Outcome	Probability of Outcome	Payoff for A
Row 1, column 1	$p_1 \cdot q_1$	a_{11}
Row 1, column 2	$p_1 \cdot q_2$	a_{12}
Row 2, column 1	$p_2 \cdot q_1$	a_{21}
Row 2, column 2	$p_2 \cdot q_2$	a_{22}

The expected value for this game is

$$(p_1 \cdot q_1) \cdot a_{11} + (p_1 \cdot q_2) \cdot a_{12} + (p_2 \cdot q_1) \cdot a_{21} + (p_2 \cdot q_2) \cdot a_{22}.$$

This same result can be written as the matrix product

$$\text{Expected value} = [p_1 \quad p_2] \begin{bmatrix} a_{11} & a_{12} \\ a_{21} & a_{22} \end{bmatrix} \begin{bmatrix} q_1 \\ q_2 \end{bmatrix} = AMB.$$

The same method works for games larger than 2×2: let the payoff matrix for a game have dimension $m \times n$; call this matrix $M = [a_{ij}]$. Let the mixed

strategy for player A be given by the row matrix

$$A = [p_1 \quad p_2 \quad p_3 \cdots p_m]$$

and the mixed strategy for player B be given by the column matrix

$$B = \begin{bmatrix} q_1 \\ q_2 \\ \cdot \\ \cdot \\ \cdot \\ q_n \end{bmatrix}.$$

The expected value for this game is the product

$$AMB = [p_1 \quad p_2 \cdots p_m] \begin{bmatrix} a_{11} & a_{12} & \cdots & a_{1n} \\ a_{21} & a_{22} & \cdots & a_{2n} \\ \cdot & \cdot & & \cdot \\ \cdot & \cdot & & \cdot \\ a_{m1} & a_{m2} & \cdots & a_{mn} \end{bmatrix} \begin{bmatrix} q_1 \\ q_2 \\ \cdot \\ \cdot \\ q_n \end{bmatrix}.$$

EXAMPLE **2** In the game in Example 1, having payoff matrix

$$M = \begin{bmatrix} -1 & 2 \\ 1 & 0 \end{bmatrix},$$

suppose player A chooses row 1 with probability .2, and player B chooses column 1 with probability .6. Find the expected value of the game.

If A chooses row 1 with probability .2, then row 2 is chosen with probability $1 - .2 = .8$, giving

$$A = [.2 \quad .8].$$

In the same way,

$$B = \begin{bmatrix} .6 \\ .4 \end{bmatrix}.$$

The expected value of this game is given by the product AMB, or

$$AMB = [.2 \quad .8] \begin{bmatrix} -1 & 2 \\ 1 & 0 \end{bmatrix} \begin{bmatrix} .6 \\ .4 \end{bmatrix}$$

$$= [.6 \quad .4] \begin{bmatrix} .6 \\ .4 \end{bmatrix} = [.52].$$

On the average, these two strategies will produce a payoff of $.52, or 52¢, for A for each play of the game. This payoff is slightly better than the 50¢ in Example 1. ◀

In Example 2, player B could reduce the payoff to A by changing strategy. (Check this by choosing different matrices for B.) For this reason, player A needs to develop an *optimum strategy*—a strategy that will produce the best possible payoff no matter what B does. Just as in the previous section, this is done by finding the largest of the smallest possible amounts that can be won.

To find values of p_1 and p_2 so that the probability vector $[p_1 \quad p_2]$ produces an optimum strategy, start with the payoff matrix

$$M = \begin{bmatrix} -1 & 2 \\ 1 & 0 \end{bmatrix}$$

and assume that A chooses row 1 with probability p_1. If player B chooses column 1, then the player A's expectation is given by E_1, where

$$E_1 = -1 \cdot p_1 + 1 \cdot p_2 = -p_1 + p_2.$$

Since $p_1 + p_2 = 1$, we have $p_2 = 1 - p_1$, and

$$E_1 = -p_1 + 1 - p_1 = 1 - 2p_1.$$

If B chooses column 2, then A's expected value is given by E_2, where

$$E_2 = 2 \cdot p_1 + 0 \cdot p_2 = 2p_1.$$

Draw graphs of $E_1 = 1 - 2p_1$ and $E_2 = 2p_1$; see Figure 6.

FIGURE 6

As mentioned previously, A needs to maximize the smallest amounts that can be won. On the graph, the smallest amounts that can be won are represented by the points of E_2 up to the intersection point. To the right of the intersection point, the smallest amounts that can be won are represented by the points of the line E_1. Player A can maximize the smallest amounts that can be won by choosing the point of intersection itself, the peak of the heavily shaded line in Figure 6.

To find this point of intersection, find the simultaneous solution of the two equations. At the point of intersection, $E_1 = E_2$. Substitute $1 - 2p_1$ for E_1 and $2p_1$ for E_2.

$$E_1 = E_2$$
$$1 - 2p_1 = 2p_1$$
$$1 = 4p_1$$
$$\frac{1}{4} = p_1$$

By this result, player A should choose strategy 1 with probability 1/4, and strategy 2 with probability $1 - 1/4 = 3/4$. This will maximize A's expected

winnings. To find the maximum winnings (which is also the value of the game), substitute 1/4 for p_1 in either E_1 or E_2. Choosing E_2 gives

$$E_2 = 2p_1 = 2\left(\frac{1}{4}\right) = \frac{1}{2};$$

that is, 1/2 dollar, or 50¢. Going through a similar argument for player B shows that the optimum strategy for player B is to choose each column with probability 1/2; in this case the value also turns out to be 50¢. In Example 2, A's winnings were 52¢; however, that was because B was not using his optimum strategy.

In the game above, player A can maximize expected winnings by playing row 1 with probability 1/4 and row 2 with probability 3/4. Such a strategy is called a **mixed strategy.** To actually decide which row to use on a given game, player A could use a spinner, such as the one in Figure 7, or use a random number generator on a calculator or a computer. A typical random number generator gives numbers between 0 and 1 with no apparent pattern or order. Player A could choose row 1 if the number generated were less than .25, and choose row 2 if the number is at least .25.

FIGURE 7

EXAMPLE 3

Boll weevils threaten the cotton crop near Hattiesburg. Charles Dawkins owns a small farm; he can protect his crop by spraying with a potent (and expensive) insecticide. In trying to decide what to do, Dawkins first sets up a payoff matrix. The numbers in the matrix represent his profits.

		States of Nature	
		Boll Weevil Attack	No Attack
Strategies	Spray	$14,000	$7000
	Don't Spray	−$3000	$8000

Let p_1 represent the probability with which Dawkins chooses to spray, so that $1 - p_1$ is the probability with which he chooses not to spray. If nature chooses a boll weevil attack, Dawkins' expected value is

$$E_1 = 14{,}000p_1 - 3000(1 - p_1)$$
$$= 14{,}000p_1 - 3000 + 3000p_1$$
$$E_1 = 17{,}000p_1 - 3000.$$

If nature chooses not to attack, Dawkins has an expected value of

$$E_2 = 7000p_1 + 8000(1 - p_1)$$
$$= 7000p_1 + 8000 - 8000p_1$$
$$E_2 = 8000 - 1000p_1.$$

As suggested by the work above, to maximize his expected profit, Dawkins should find the value of p_1 for which $E_1 = E_2$.

$$E_1 = E_2$$
$$17{,}000p_1 - 3000 = 8000 - 1000p_1$$
$$18{,}000p_1 = 11{,}000$$
$$p_1 = 11/18$$

Thus, $p_2 = 1 - p_1 = 1 - 11/18 = 7/18$.

Dawkins will maximize his expected profit if he chooses to spray with probability 11/18 and not spray with probability 7/18. His expected profit from this mixed strategy, [11/18 7/18], can be found by substituting 11/18 for p_1 in either E_1 or E_2. If E_1 is chosen,

$$\text{Expected profit} = 17{,}000\left(\frac{11}{18}\right) - 3000 = \frac{133{,}000}{18} \approx \$74{,}000.$$

If Dawkins makes this decision only once, it is meaningless to make a choice with probability 11/18. This is because Dawkins either sprays or he doesn't spray. Because the strategy to spray has a higher probability, he should spray, and not use the spinner discussed earlier. To make a choice with a given probability only makes sense if the decision is made many times.

Notice that we do not solve this problem by the methods of the section on decision making. Based on that section, regardless of whether Dawkins is an optimist or a pessimist, he would spray. If he knew the probability of a boll weevil attack, he would choose the strategy with the highest expected value. In practice, Dawkins should be a rational player, rather than an optimist or a pessimist, and he may not know the probability of a boll weevil attack. In game theory, the other player (nature, in this case) is thought of as trying to maximize his or her gain, which is the same as minimizing the first player's gain. It may seem strange to think of nature as choosing a strategy to minimize Dawkins' profit, but that is a safe and rational strategy when no other information about the state of nature is known. ◀

To obtain a formula for the optimum strategy in a game that is not strictly determined (that is, the game has no saddle point), start with the matrix

$$M = \begin{bmatrix} a_{11} & a_{12} \\ a_{21} & a_{22} \end{bmatrix},$$

the payoff matrix of the game. Assume that A chooses row 1 with probability p_1. The expected value for A, assuming that B plays column 1, is E_1, where

$$E_1 = a_{11} \cdot p_1 + a_{21} \cdot (1 - p_1).$$

The expected value for A if B chooses column 2 is E_2, where

$$E_2 = a_{12} \cdot p_1 + a_{22} \cdot (1 - p_1).$$

As above, the optimum strategy for player A is found by letting $E_1 = E_2$.

$$a_{11} \cdot p_1 + a_{21} \cdot (1 - p_1) = a_{12} \cdot p_1 + a_{22} \cdot (1 - p_1)$$

Solve this equation for p_1.

$$a_{11} \cdot p_1 + a_{21} - a_{21} \cdot p_1 = a_{12} \cdot p_1 + a_{22} - a_{22} \cdot p_1$$

$$a_{11} \cdot p_1 - a_{21} \cdot p_1 - a_{12} \cdot p_1 + a_{22} \cdot p_1 = a_{22} - a_{21}$$

$$p_1(a_{11} - a_{21} - a_{12} + a_{22}) = a_{22} - a_{21}$$

$$p_1 = \frac{a_{22} - a_{21}}{a_{11} - a_{21} - a_{12} + a_{22}}$$

Since $p_2 = 1 - p_1$,

$$p_2 = 1 - \frac{a_{22} - a_{21}}{a_{11} - a_{21} - a_{12} + a_{22}}$$

$$= \frac{a_{11} - a_{21} - a_{12} + a_{22} - (a_{22} - a_{21})}{a_{11} - a_{21} - a_{12} + a_{22}}$$

$$= \frac{a_{11} - a_{12}}{a_{11} - a_{21} - a_{12} + a_{22}}.$$

This result is valid only if $a_{11} - a_{21} - a_{12} + a_{22} \neq 0$; this condition is satisfied if the game is not strictly determined.

There is a similar result for player B, which is included in the following summary.

OPTIMUM STRATEGIES IN A NON-STRICTLY-DETERMINED GAME

Let a non-strictly-determined game have payoff matrix

$$\begin{bmatrix} a_{11} & a_{12} \\ a_{21} & a_{22} \end{bmatrix}.$$

The optimum strategy for player A is $[p_1 \quad p_2]$, where

$$p_1 = \frac{a_{22} - a_{21}}{a_{11} - a_{21} - a_{12} + a_{22}} \quad \text{and} \quad p_2 = \frac{a_{11} - a_{12}}{a_{11} - a_{21} - a_{12} + a_{22}}.$$

The optimum strategy for player B is $\begin{bmatrix} q_1 \\ q_2 \end{bmatrix}$, where

$$q_1 = \frac{a_{22} - a_{12}}{a_{11} - a_{21} - a_{12} + a_{22}} \quad \text{and} \quad q_2 = \frac{a_{11} - a_{21}}{a_{11} - a_{21} - a_{12} + a_{22}}.$$

The value of the game is

$$\frac{a_{11}a_{22} - a_{12}a_{21}}{a_{11} - a_{21} - a_{12} + a_{22}}.$$

Note It is important to remember that the use of mixed strategies is meaningful only if a game is repeated many, many times. For one play of the game, the mixed strategy has no meaning.

EXAMPLE 4

Suppose a game has the following payoff matrix.

$$\begin{bmatrix} 5 & -2 \\ -3 & -1 \end{bmatrix}$$

Here $a_{11} = 5$, $a_{12} = -2$, $a_{21} = -3$, and $a_{22} = -1$. To find the optimum strategy for player A, first find p_1.

$$p_1 = \frac{-1 - (-3)}{5 - (-3) - (-2) + (-1)} = \frac{2}{9}$$

Player A should play row 1 with probability 2/9, and row 2 with probability $1 - 2/9 = 7/9$.

For player B,

$$q_1 = \frac{-1 - (-2)}{5 - (-3) - (-2) + (-1)} = \frac{1}{9}.$$

Player B should choose column 1 with probability 1/9, and column 2 with probability 8/9. The value of the game is

$$\frac{5(-1) - (-2)(-3)}{5 - (-3) - (-2) + (-1)} = -\frac{11}{9}.$$

On the average, B will receive 11/9 dollars from A per play of the game. ◀

8.5 Exercises

1. Suppose a game has payoff matrix $\begin{bmatrix} 3 & -4 \\ -5 & 2 \end{bmatrix}$. Suppose that player B uses the strategy $\begin{bmatrix} .3 \\ .7 \end{bmatrix}$. Find the expected value of the game if player A uses each of the following strategies.

(a) [.5 .5] **(b)** [.1 .9] **(c)** [.8 .2] **(d)** [.2 .8]

2. Suppose a game has payoff matrix $\begin{bmatrix} 0 & -4 & 1 \\ 3 & 2 & -4 \\ 1 & -1 & 0 \end{bmatrix}$. Find the expected value of the game for the following strategies for players A and B.

(a) $A = [.1 \quad .4 \quad .5]$; $B = \begin{bmatrix} .2 \\ .4 \\ .4 \end{bmatrix}$ **(b)** $A = [.3 \quad .4 \quad .3]$; $B = \begin{bmatrix} .8 \\ .1 \\ .1 \end{bmatrix}$

Find the optimum strategies for player A and player B in the games in Exercises 3–14. Find the value of each game. (Be sure to look for a saddle point first.)

3. $\begin{bmatrix} 5 & 1 \\ 3 & 4 \end{bmatrix}$ **4.** $\begin{bmatrix} -4 & 5 \\ 3 & -4 \end{bmatrix}$ **5.** $\begin{bmatrix} -2 & 0 \\ 3 & -4 \end{bmatrix}$ **6.** $\begin{bmatrix} 6 & 2 \\ -1 & 10 \end{bmatrix}$

7. $\begin{bmatrix} 4 & -3 \\ -1 & 7 \end{bmatrix}$ **8.** $\begin{bmatrix} 0 & 6 \\ 4 & 0 \end{bmatrix}$ **9.** $\begin{bmatrix} -2 & \frac{1}{2} \\ 0 & -3 \end{bmatrix}$ **10.** $\begin{bmatrix} 6 & \frac{3}{4} \\ \frac{2}{3} & -1 \end{bmatrix}$

11. $\begin{bmatrix} \frac{8}{3} & -\frac{1}{2} \\ \frac{3}{4} & -\frac{5}{12} \end{bmatrix}$
 12. $\begin{bmatrix} -\frac{1}{2} & \frac{2}{3} \\ \frac{7}{8} & -\frac{3}{4} \end{bmatrix}$
 13. $\begin{bmatrix} -1 & 2 \\ 3 & 1 \end{bmatrix}$
 14. $\begin{bmatrix} 8 & 18 \\ -4 & 2 \end{bmatrix}$

Remove any dominated strategies, and then find the optimum strategy for each player and the value of the game.

15. $\begin{bmatrix} -4 & 9 \\ 3 & -5 \\ 8 & 7 \end{bmatrix}$
 16. $\begin{bmatrix} 3 & 4 & -1 \\ -2 & 1 & 0 \end{bmatrix}$
 17. $\begin{bmatrix} 8 & 6 & 3 \\ -1 & -2 & 4 \end{bmatrix}$

18. $\begin{bmatrix} -1 & 6 \\ 8 & 3 \\ -2 & 5 \end{bmatrix}$
 19. $\begin{bmatrix} 9 & -1 & 6 \\ 13 & 11 & 8 \\ 6 & 0 & 9 \end{bmatrix}$
 20. $\begin{bmatrix} 4 & 8 & -3 \\ 2 & -1 & 1 \\ 7 & 9 & 0 \end{bmatrix}$

21. Verify that the optimum strategy for player B in a non-strictly-determined game is as given in the text.

22. Verify that the value of a non-strictly-determined game is as given in the text when players A and B play their optimum strategies.

23. Some people claim that even if a mixed strategy is used only once, the player should use a spinner to determine what choice to make. Write a few sentences giving your reaction to this claim.

24. For a game with optimum strategy $[\frac{1}{2}, \frac{1}{2}]$, a player could flip a coin to determine which choice to make. A nonrandom method would be to alternate between the two choices. Discuss any disadvantages of this nonrandom method. Would it make a difference whether or not the opposing player is intelligent? Explain.

Applications

Business and Economics

25. **Advertising** Suppose Allied Manufacturing Company decides to put its new product on the market with a big television and radio advertising campaign. At the same time, the company finds out that its major competitor, Bates Manufacturing, also has decided to launch a big advertising campaign for a similar product. The payoff matrix below shows the increased sales (in millions) for Allied, as well as the decreased sales for Bates.

Bates

$$\text{Allied} \begin{array}{c} \text{T.V.} \\ \text{Radio} \end{array} \begin{bmatrix} \overset{\text{T.V.}}{1.0} & \overset{\text{Radio}}{-.7} \\ -.5 & .5 \end{bmatrix}$$

Find the optimum strategy for Allied Manufacturing and the value of the game.

26. **Pricing** The payoffs in the next matrix represent the differences between Boeing Aircraft Company's profit and its competitor's profit for two prices (in millions) on commercial jet transports, with positive payoffs being in Boeing's favor. What should Boeing's price strategy be?*

Competitor's Price Strategy

$$\text{Boeing's Strategy} \begin{array}{c} 4.9 \\ 4.75 \end{array} \begin{bmatrix} \overset{4.75}{-4} & \overset{4.9}{2} \\ 2 & 0 \end{bmatrix}$$

27. **Purchasing** *The Huckster*[†] Merrill has a concession at Yankee Stadium for the sale of sunglasses and umbrellas. The business places quite a strain on him, the weather being what it is. He has observed that he can sell about 500 umbrellas when it rains, and about 100 when it is sunny; in the latter case he can also sell 1000 sunglasses. Umbrellas cost him 50 cents and sell for $1; sunglasses cost 20 cents and sell for 50 cents. He is willing to invest $250 in the project. Everything that is not sold is considered a total loss.

He assembles the facts regarding profit in a matrix.

*From "Pricing, Investment, and Games of Strategy," by Georges Brigham in *Management Sciences Models and Techniques*, vol. 1. Coypright © 1960 by Pergamon Press, Ltd. Reprinted with permission.

[†]From J. D. Williams, *The Compleat Strategyst*, McGraw-Hill Book Company, 1966. Reprinted by permission from The Rand Corporation. This is an excellent nontechnical book on game theory.

Selling during:

$$\begin{array}{c} \\ \textit{Buying for:} \end{array} \begin{array}{c} \text{Rain} \\ \text{Shine} \end{array} \begin{array}{cc} \text{Rain} & \text{Shine} \\ \left[\begin{array}{cc} 250 & -150 \\ -150 & 350 \end{array}\right] \end{array}$$

He immediately takes heart, for this is a mixed-strategy game, and he should be able to find a stabilizing strategy that will save him from the vagaries of the weather. Find the best mixed strategy for Merrill.

Life Sciences

28. **Choosing Medication** The number of cases of African flu has reached epidemic levels. The disease is known to have two strains with similar symptoms. Dr. De Luca has two medicines available: the first is 60% effective against the first strain and 40% effective against the second. The second medicine is completely effective against the second strain but ineffective against the first. Use the matrix below to decide which medicine she should use and the results she can expect.

$$\begin{array}{c} \\ \textit{Medicine} \end{array} \begin{array}{c} \\ 1 \\ 2 \end{array} \begin{array}{cc} \multicolumn{2}{c}{\textit{Strain}} \\ 1 \quad 2 \\ \left[\begin{array}{cc} .6 & .4 \\ 0 & 1 \end{array}\right] \end{array}$$

General Interest

29. **Cats** When two cats, Euclid and Jamie, play together, their game involves facing each other while several feet apart; each cat must then decide whether to pounce or to "freeze"—to stay motionless until the other cat pounces. Euclid weighs 2 lb more than Jamie, so if they both pounce, Jamie is squashed and Euclid gains 3 points. If they both freeze, Euclid remains in control of the area and gains 2 points. If Euclid pounces while Jamie freezes, Jamie can put up a good defense, so Euclid only gains 1

point. Euclid is poor at defense, so if he freezes while Jamie pounces, he loses 2 points. Find the optimum strategy for each cat, and find the value of the game.

30. **Coin Game** In the game of matching coins, each of two players flips a coin. If both coins match (both show heads or both show tails), player A wins $1. If there is no match, player B wins $1, as in the payoff matrix below. Find the optimum strategies for the two players and the value of the game.

$$\left[\begin{array}{cc} 1 & -1 \\ -1 & 1 \end{array}\right]$$

31. **Finger Game** Players A and B play a game in which they show either one or two fingers at the same time. If there is a match, A wins the amount of dollars equal to the total number of fingers shown. If there is no match, B wins the amount of dollars equal to the number of fingers shown.

(a) Write the payoff matrix.

(b) Find optimum strategies for A and B and the value of the game.

32. **Finger Game** Repeat Exercise 31 if each player may show either 0 or 2 fingers with the same payoffs.

▶ 8.6 GAME-THEORY AND LINEAR PROGRAMMING

How can a company decide in which cities it should advertise?

Using game theory and linear programming, we will answer this and other questions.

Until now, we have not learned how to solve games in which one player has more than two choices, but for which neither player has an optimum pure strategy.

In this section, we shall see how the techniques of linear programming can be used to solve games in which each player has an arbitrary number of choices.

If there is no saddle point, we must find the probabilities for mixing strategies to obtain an optimum solution and the value of the game. Let us see how linear programming can be applied to Example 1 in the previous section, for which the payoff matrix was

$$M = \begin{bmatrix} -1 & 2 \\ 1 & 0 \end{bmatrix}.$$

When player A chose row 1 with probability p_1 and row 2 with probability p_2, we found that his expectations when B chose columns 1 and 2 were as follows.

$$E_1 = -1 \cdot p_1 + 1 \cdot p_2 = -p_1 + p_2$$
$$E_2 = 2 \cdot p_1 + 0 \cdot p_2 = 2p_1$$

Player A wishes his expected gain to be as large as possible, so he should maximize the minimum of the expected gains.

Let g represent the minimum of the expected gains, so that

$$E_1 = -p_1 + p_2 \geq g$$
$$E_2 = 2p_1 \quad\quad \geq g.$$

The graphical method of linear programming was discussed in Chapter 3. In brief, the method involves the following steps:

1. Graph the region satisfied by all the inequalities, known as the feasible region.

2. Identify the corner points of the feasible region, and find the value of the objective function at each corner point.

3. The solution is found at the corner point producing the optimum value of the objective function.

We can simplify this system of linear inequalities by dividing both inequalities by g, yielding

$$E_1 = -\left(\frac{p_1}{g}\right) + \left(\frac{p_2}{g}\right) \geq 1$$
$$E_2 = 2\left(\frac{p_1}{g}\right) \quad\quad \geq 1.$$

Then denote p_1/g by x and p_2/g by y.

$$E_1 = -x + y \geq 1$$
$$E_2 = 2x \quad\quad \geq 1$$

To determine the objective function, note that $x + y = p_1/g + p_2/g = 1/g$. Since the goal is to maximize g, which is the same as minimizing $1/g$, we can rewrite the problem as the following linear programming problem:

$$\text{Minimize} \quad w = x + y$$
$$\text{subject to:} \quad -x + y \geq 1$$
$$2x \quad\quad \geq 1$$
$$\text{with} \quad x \geq 0, \; y \geq 0,$$

FIGURE 8

where $w = 1/g$. This linear programming problem can be solved by the graphical method described in Chapter 3. The graph is shown in Figure 8. There is only one corner point, $(1/2, 3/2)$, and the value of w there is 2. Thus the value of the game is $g = 1/w = 1/2$, and the optimum strategy for player A is $p_1 = gx = (1/2)(1/2) = 1/4$, $p_2 = gy = (1/2)(3/2) = 3/4$. Notice that this agrees with the result found in the previous section.

A similar analysis can be done for player B. When B chooses columns 1 and 2 with probabilities q_1 and q_2, respectively, then her expectations when A chooses rows 1 and 2 are

$$E_1 = -1 \cdot q_1 + 2 \cdot q_2 = -q_1 + 2q_2$$
$$E_2 = 1 \cdot q_1 + 0 \cdot q_2 = q_1.$$

Player B wishes the payoff to be as small as possible, so she should minimize the maximum of the expected gains:

$$E_1 = -q_1 + 2q_2 \leq g$$
$$E_2 = q_1 \leq g.$$

As before, we divide both inequalities by g and denote q_1/g by x and q_2/g by y, yielding:

$$E_1 = -x + 2y \leq 1$$
$$E_2 = x \leq 1.$$

As before, $x + y = q_1/g + q_2/g = 1/g$. Player B's goal is to minimize g, which is the same as maximizing $1/g$. We therefore have the following linear programming problem:

$$\text{Maximize} \quad z = x + y$$
$$\text{subject to:} \quad -x + 2y \leq 1$$
$$x \leq 1$$
$$\text{with} \quad x \geq 0, \ y \geq 0,$$

where $z = 1/g$. This problem can also be solved by the graphical method; the reader should do this, and verify that $x = y = 1$ and $z = 2$. As before, the value of the game is $g = 1/z = 1/2$, and the optimum strategy for B is $q_1 = gx = (1/2)(1) = 1/2$, $q_2 = gy = (1/2)(1) = 1/2$. The value of the game is the same for B as for A, as must be the case, and B's strategy is the same as in the previous section.

Most significantly, notice that the linear programming problem for player A is the dual of the linear programming problem for player B. It can be shown that, under circumstances to be discussed shortly, this remarkable fact is always true. Notice further that A wants to maximize the minimum value of the game, while B wants to minimize the maximum of the game. That these two values are equal is an illustration of the **minimax principle,** which follows from the theorem of duality discussed in Chapter 4.

If the value of a zero-sum game is positive, the game can always be solved by linear programming techniques, in which case the solution of the row player is the dual of the solution of the column player. To guarantee that the value of the game is positive, find the most negative number in the payoff matrix and add its absolute value to all entries in the matrix. For example, in the previous game, -1 was the most negative entry. Adding 1 to all entries yields the payoff matrix

$$N = \begin{bmatrix} 0 & 3 \\ 2 & 1 \end{bmatrix}.$$

The steps of the simplex algorithm can be stated briefly as follows.

1. Convert each constraint in the maximization problem into an equation by adding a slack variable.

2. Set up the initial tableau, with the negative of the payoffs in the bottom row.

3. Choose the column with the most negative indicator. Divide each element from that column into the corresponding element from the last column. The pivot is the element with the smallest quotient. Use row operations to change all elements in the pivot column other than the pivot to 0.

4. When there are no more negative numbers in the bottom row, the values of x_1, \ldots, x_n can be read using the appropriate columns in combination with the last column.

5. If the z column contains a number other than 1, divide the last row by that number. The value of z, which is the same as the value of w, is given in the lower right hand corner. The solution to y_1, \ldots, y_n can be read from the bottom row of the columns corresponding to the slack variables.

Solving this game as before would result in the same optimum strategies, but the value of the game would be 3/2, that is, 1 greater than it was before. We would then subtract 1 from 3/2 to get 1/2, the value of the original game.

Just as linear programming problems with more than two variables could be solved by the simplex method, we can now use the simplex method to solve games involving more than two choices for each player. Before giving such an example, let us review the steps in solving a matrix game.

SOLVING ZERO-SUM GAMES

1. Set up the payoff matrix M.

2. Remove any dominated rows or columns.

3. Check to see if there is a saddle point. If so, the optimum strategies are determined by the row and column in which the saddle point is located.

4. If there is no saddle point and negative numbers are present, add a positive number to all entries in the payoff matrix so that all entries are nonnegative. Denote the elements of the remaining $m \times n$ matrix by a_{ij}.

5. Set up the linear programming problem:

$$\text{Maximize} \quad z = x_1 + \cdots + x_n$$
$$\text{subject to} \quad a_{11}x_1 + \cdots + a_{1n}x_n \leq 1$$
$$\vdots \qquad\qquad \vdots$$
$$a_{m1}x_1 + \cdots + a_{mn}x_n \leq 1$$
$$\text{with} \qquad x_1 \geq 0, \ldots, x_n \geq 0.$$

6. After finding the solution z to the linear programming problem, the value of the modified game is given by $g = 1/z$. The optimum strategy for player B is found by multiplying the solution to the linear programming problem by g: $q_i = gx_i$, $i = 1, \ldots, n$. The optimum strategy for player A is found by multiplying the solution to the dual problem by g: $p_j = gy_j$, $j = 1, \ldots, m$.

7. Subtract from g the positive number added in step 4 to find the value of the original game.

EXAMPLE

1 We now return to the game that was introduced in the section on strictly determined games, but never solved. The payoff matrix for this game, after deleting the column that was dominated by another column, was

$$\begin{bmatrix} -6 & 10 \\ 0 & -9 \\ -4 & -8 \end{bmatrix}.$$

Verify that this game has no saddle point. Because of the negative entries, we will add 9 to all entries to make them all nonnegative. The resulting payoff matrix is

$$\begin{bmatrix} 3 & 19 \\ 9 & 0 \\ 5 & 1 \end{bmatrix}.$$

The linear programming problem to be solved is

$$\begin{array}{ll} \text{Maximize} & z = x_1 + x_2 \\ \text{subject to} & 3x_1 + 19x_2 \leq 1 \\ & 9x_1 \quad\quad\quad \leq 1 \\ & 5x_1 + \quad x_2 \leq 1 \\ \text{with} & x_1 \geq 0, \ x_2 \geq 0. \end{array}$$

The initial tableau is

$$\begin{array}{ccccccc} x_1 & x_2 & x_3 & x_4 & x_5 & z & \\ \end{array}$$
$$\begin{bmatrix} 3 & 19 & 1 & 0 & 0 & 0 & 1 \\ 9 & 0 & 0 & 1 & 0 & 0 & 1 \\ 5 & 1 & 0 & 0 & 1 & 0 & 1 \\ \hline -1 & -1 & 0 & 0 & 0 & 1 & 0 \end{bmatrix}.$$

The first and second columns both contain a -1 at the bottom; since these are equally negative, we arbitrarily choose the first column. The smallest ratio is formed by the 9 in row 2. Making this the pivot yields the following matrix.

$$\begin{array}{ccccccc} x_1 & x_2 & x_3 & x_4 & x_5 & z & \\ \end{array}$$
$$\begin{array}{l} -R_2 + 3R_1 \to R_1 \\ \\ -5R_2 + 9R_3 \to R_3 \\ R_2 + 9R_4 \to R_4 \end{array} \begin{bmatrix} 0 & 57 & 3 & -1 & 0 & 0 & 2 \\ 9 & 0 & 0 & 1 & 0 & 0 & 1 \\ 0 & 9 & 0 & -5 & 9 & 0 & 4 \\ 0 & -9 & 0 & 1 & 0 & 9 & 1 \end{bmatrix}$$

The next pivot is the 57 in row 1, column 2.

$$\begin{array}{ccccccc} x_1 & x_2 & x_3 & x_4 & x_5 & z & \\ \end{array}$$
$$\begin{array}{l} \\ \\ -3R_1 + 19R_3 \to R_3 \\ 3R_1 + 19R_4 \to R_4 \end{array} \begin{bmatrix} 0 & 57 & 3 & -1 & 0 & 0 & 2 \\ 9 & 0 & 0 & 1 & 0 & 0 & 1 \\ 0 & 0 & -9 & -92 & 171 & 0 & 70 \\ 0 & 0 & 9 & 16 & 0 & 171 & 25 \end{bmatrix}$$

Notice the 171 in the bottom of the z column. Dividing the bottom row by 171 gives a z value of $25/171$, so

$$g = \frac{1}{z} = \frac{171}{25}.$$

The values of y_1, y_2, and y_3 are read from the bottom of the columns for the three slack variables, after dividing the bottom row by 171.

$$y_1 = \frac{9}{171} = \frac{1}{19}, \qquad y_2 = \frac{16}{171}, \qquad y_3 = 0$$

We find the values of p_1, p_2, and p_3 by multiplying the values of y_j by g.

$$p_1 = \left(\frac{1}{19}\right)\left(\frac{171}{25}\right) = \frac{9}{25}, \quad p_2 = \left(\frac{16}{171}\right)\left(\frac{171}{25}\right) = \frac{16}{25}, \quad p_3 = (0)\left(\frac{171}{25}\right) = 0$$

Next find x_1 and x_2 by using the first and second columns combined with the last column.

$$x_1 = \frac{1}{9}, \quad x_2 = \frac{2}{57}$$

Find the values of q_1 and q_2 by multiplying the values of x_i by g.

$$q_1 = \left(\frac{1}{9}\right)\left(\frac{171}{25}\right) = \frac{19}{25}, \quad q_2 = \left(\frac{2}{57}\right)\left(\frac{171}{25}\right) = \frac{6}{25}$$

Finally, the value of the game is found by subtracting from g the 9 that was added at the beginning, yielding $(171/25) - 9 = -54/25$.

To summarize, the optimum strategy for player A is $(9/25, 16/25, 0)$, and the optimum strategy for player B is $(19/25, 6/25)$. When these strategies are used, the value of the game is $-54/25$. ◀

Some applications of these techniques are given in the exercises. For many realistic games, the details of using the simplex method become very tedious, and so the computer is of great help. With the aid of a program to perform the simplex algorithm, the optimum strategies for games of virtually any size can be found.

8.6 Exercises

In Exercises 1–6, use the graphical method to find the optimum strategy for players A and B and the value of the game for each payoff matrix.

1. $\begin{bmatrix} 1 & 2 \\ 3 & 1 \end{bmatrix}$

2. $\begin{bmatrix} 5 & 2 \\ 0 & 3 \end{bmatrix}$

3. $\begin{bmatrix} 4 & -2 \\ -1 & 6 \end{bmatrix}$

4. $\begin{bmatrix} -1 & 5 \\ 1 & -6 \end{bmatrix}$

5. $\begin{bmatrix} 8 & -7 \\ -2 & 4 \end{bmatrix}$

6. $\begin{bmatrix} -4 & 1 \\ 5 & 0 \end{bmatrix}$

In Exercises 7–12, use the simplex method to find the optimum strategy for players A and B and the value of the game for each payoff matrix.

7. $\begin{bmatrix} 3 & -4 & 1 \\ 5 & 3 & -2 \end{bmatrix}$

8. $\begin{bmatrix} -6 & 1 & 4 & 2 \\ 9 & 3 & -8 & -7 \end{bmatrix}$

9. $\begin{bmatrix} -1 & 2 & 4 \\ 3 & -2 & 0 \end{bmatrix}$

10. $\begin{bmatrix} 1 & 0 \\ -2 & 4 \\ -1 & -1 \end{bmatrix}$

11. $\begin{bmatrix} 1 & 0 & -1 \\ -1 & 0 & 1 \\ 2 & -1 & 2 \end{bmatrix}$

12. $\begin{bmatrix} 2 & -1 & 1 \\ 0 & 2 & 3 \\ 4 & 1 & 0 \end{bmatrix}$

Applications

In Exercises 13–19, use the graphical method when the payoff matrix is a 2 × 2 matrix or can be reduced to one after removing rows or columns that are dominated. Otherwise, use the simplex method.

Business and Economics

13. Contractor Bidding A contractor prepares to bid on a job. If all goes well, his bid should be $30,000, which will cover his costs plus his usual profit margin of $4500. If a threatened labor strike actually occurs, however, his bid should be $40,000 to give him the same profit. If there is a strike and he bids $30,000, he will lose $5500. If he bids $40,000 and there is no strike, his bid will be too high, and he will lose the job entirely. Find the optimum strategy for the contractor, and find the value of the game. (See Exercise 7 in the section on decision making.)

14. Negotiation In negotiating a labor contract, a labor union has considered four different approaches, based on different positions the union can take on various issues. Management can take three different approaches in the negotiations. The final contract will be regarded as either favorable to the union, favorable to the management, or neither, leading to the following payoff matrix.

Management Strategies

$$\text{Labor Strategies} \begin{array}{c} 1 \\ 2 \\ 3 \\ 4 \end{array} \begin{bmatrix} -1 & -1 & 1 \\ -1 & 0 & 0 \\ 1 & -1 & -1 \\ 1 & 1 & -1 \end{bmatrix}$$

Find the optimum strategies for labor and management, and find the value of the game.

15. Advertising Marketing executives for two competing companies are trying to decide whether to advertise in Atlanta, Boston, or Cleveland, given that each company can afford to target only one city at a time. General Items Company has a leading market share and has found that it will earn an additional profit of $10,000, $8,000, or $6,000 per week if they advertise in Atlanta, Boston, or Cleveland, respectively. Original Imitators, Inc., has a smaller portion of the market, but its executives can cut General Items' additional profit in half and gain that profit themselves if they run competing ads in the same city as General Items. If they run ads in a different city, their ads seem to have no impact.

(a) Set up the payoff matrix for this game.

(b) Find the optimum strategy for each company and the value of the game.

Life Sciences

16. Health Care Based on preliminary tests, a physician decides that a patient has one of three possible diseases, labeled A, B, and C. The physician can choose to begin any one of three treatments, labeled 1, 2, and 3. Each treatment is beneficial for certain diseases but is less so for others, leading to the following payoff matrix:

Disease

$$\text{Treatment} \begin{array}{c} 1 \\ 2 \\ 3 \end{array} \begin{array}{ccc} A & B & C \end{array} \begin{bmatrix} 3 & 1 & -2 \\ -1 & 2 & 1 \\ 0 & -2 & 3 \end{bmatrix}$$

Find the optimum strategy for the physician and the value of the game.

Social Sciences

17. Education In an accounting class, the instructor permits the students to bring a calculator or a reference book (but not both) to an examination. The examination will emphasize either numerical problems or definitions. In trying to decide which aid to take to an examination, a student first decides on the utilities shown in the following payoff matrix.

Exam's Emphasis

	Numbers	Definitions
Calculator	+50	0
Book	+10	40

Student's Choice

What is the student's optimum strategy, and what is the value of the game? (See Exercise 10 in the section on decision making.)

General Interest

18. **Card Games** Player A deals Player B one of three cards—
ace, king, or queen—at random, face down.* B looks at
the card. If it is an ace, B must say "ace;" if it is a king,
B can say either "king" or "ace;" if it is a queen, B can
say either "queen" or "ace." If B says "ace," A can
either believe him and give him $1, or ask him to show
his card. If it is an ace, A must pay B $2; but if it is not, B
pays A $2. If B says "king," neither side loses anything;
but if he says "queen," B must pay A $1.

 (a) Set up the payoff matrix for this game. (*Hint*: Con-
sider these two choices for A: (1) believe B when B
says "ace;" (2) ask B to show his card when B says
"ace." Consider the following four choices for B:
(1) always tell the truth; (2) lie only if the card is a
queen; (3) lie only if the card is a king; (4) lie if the
card is a queen or a king.)

 (b) Find the optimum strategy for each player and the
value of the game.

19. **Children's Game** In the children's game *rock, paper, scis-
sors,* two players simultaneously extend one hand in the
form of a fist (*rock*), a flat palm (*paper*), or two fingers
extended (*scissors*). If they both make the same choice, the
game is a tie. Otherwise, *rock*, beats *scissors* (because rock
can crush scissors), *scissors* beats *paper* (because scissors
can cut paper), and *paper* beats *rock* (because paper can
cover rock). Whoever wins gains one point; in the case of
a tie, no one gains any points. (See Exercise 31 in the sec-
tion on strictly determined games.)

 (a) Find the optimum strategy for each player and the
value of the game.

 (b) In hindsight, how could you have guessed the answer
to part (a) without doing any work?

📊 **Computer/Graphing Calculator**

20. **Marketing** Solve the Christmas card problem discussed in
the section on decision making, for which we had the fol-
lowing payoff matrix.

		States of Nature	
	Weak Economy	In-Between	Strong Economy
Modern	40	85	120
Strategies Old-Fashioned	90	45	85
Mixture	75	105	65

21. **Military Science** The Colonel Blotto game is a type of
military strategy game.[†] Two opposing armies are ap-
proaching two posts. Colonel Blotto has four regiments
under his command, while his opponent, Captain Kije, has
three regiments. Each commander must decide how many
regiments to send to each post. The army that sends more
regiments to a post not only captures that post but also
captures the losing army's regiments. If both armies send
the same number of regiments to a post, there is a stand-
off, and neither army wins. The payoff is one point for
capturing the post and one point for each regiment
captured.

 (a) Set up the payoff matrix for this game. (*Hint*: Colonel
Blotto has five choices, and Captain Kije has four.)

 (b) Find the optimum strategy for each commander and
the value of the game.

 (c) Show that if Colonel Blotto uses the strategy found in
part (b), then any strategy used by Colonel Kije re-
sults in the same payoff. (*Hint*: Show that $AM = (14/9)R$, where R is a row matrix consisting of all
1's, and then use the fact that $RB = [1]$.)

 (d) Based on the result of part (c), what can you conclude
about the uniqueness of the optimum strategy found
by linear programming?

22. **Competition** Two merchants are planning to build com-
peting stores to serve an area of three small cities. The
fraction of the total population that live in each city is
shown in the figure. If both merchants locate in the same
city, merchant A will get 65% of the total business. If the
merchants locate in different cities, each will get 80% of
the business in the city where it is located, and 20% of the
business in the city where the other merchant is located. In
the city where neither is located, A will get 60% of the
business and B will get 40%. Payoffs are measured by the
number of percentage points above or below 50%. Find an
optimum strategy for each merchant, and find the value of
this game. (See Exercise 26 in the section on strictly deter-
mined games.)

*From L. C. Thomas, *Games, Theory, and Applications*, Halsted Press, 1986.
†From Samuel Karlin, *Mathematical Methods and Theory in Games, Programming, and Economics*, Addison-Wesley, 1959.

Chapter Summary	**Key Terms**

state

8.1 transition diagram
transition matrix
Markov chain
probability vector

8.2 regular transition matrix
regular Markov chain
equilibrium (or fixed) vector

8.3 states of nature
strategies
payoff matrix
utility

8.4 game theory
two-person game
zero-sum game
dominated strategy

optimum strategy
value of the game
pure strategy
saddle point
strictly determined game
fair game

8.5 mixed strategy

8.6 minimax principle

Chapter 8	**Review Exercises**

1. How can you tell by looking at a matrix whether it represents the transition matrix for a Markov chain?

2. Under what conditions is the existence of an equilibrium vector guaranteed?

Decide whether each of the matrices in Exercise 3–6 could be a transition matrix.

3. $\begin{bmatrix} .4 & .6 \\ 1 & 0 \end{bmatrix}$

4. $\begin{bmatrix} -.2 & 1.2 \\ .8 & .2 \end{bmatrix}$

5. $\begin{bmatrix} .8 & .2 & 0 \\ 0 & 1 & 0 \\ .1 & .4 & .5 \end{bmatrix}$

6. $\begin{bmatrix} .6 & .2 & .3 \\ .1 & .5 & .4 \\ .3 & .3 & .4 \end{bmatrix}$

For each of the transition matrices in Exercise 7–10, (a) find the first three powers; and (b) find the probability that state 2 changes to state 1 after three repetitions of the experiment.

7. $C = \begin{bmatrix} .6 & .4 \\ 1 & 0 \end{bmatrix}$

8. $D = \begin{bmatrix} .3 & .7 \\ .5 & .5 \end{bmatrix}$

9. $E = \begin{bmatrix} .2 & .5 & .3 \\ .1 & .8 & .1 \\ 0 & 1 & 0 \end{bmatrix}$

10. $F = \begin{bmatrix} .14 & .12 & .74 \\ .35 & .28 & .37 \\ .71 & .24 & .05 \end{bmatrix}$

In Exercises 11–14, use the transition matrix T, along with the given initial distribution D, to find the distribution after two repetitions of the experiment. Also, predict the long-range distribution.

11. $D = [.3 \quad .7]; \ T = \begin{bmatrix} .4 & .6 \\ .5 & .5 \end{bmatrix}$

12. $D = [.8 \quad .2]; \ T = \begin{bmatrix} .7 & .3 \\ .2 & .8 \end{bmatrix}$

13. $D = [.2 \quad .4 \quad .4]; \ T = \begin{bmatrix} .6 & .2 & .2 \\ .3 & .3 & .4 \\ .5 & .4 & .1 \end{bmatrix}$

14. $D = [.1 \quad .1 \quad .8]; \ T = \begin{bmatrix} .2 & .3 & .5 \\ .1 & .1 & .8 \\ .7 & .1 & .2 \end{bmatrix}$

15. How can you tell from the transition matrix whether a Markov chain is regular or not?

Decide whether each of the following transition matrices is regular.

16. $\begin{bmatrix} 0 & 1 \\ .2 & .8 \end{bmatrix}$

17. $\begin{bmatrix} .4 & .2 & .4 \\ 0 & 1 & 0 \\ .6 & .3 & .1 \end{bmatrix}$

18. $\begin{bmatrix} 1 & 0 & 0 \\ 0 & 1 & 0 \\ .3 & .5 & .2 \end{bmatrix}$

19. How can you determine from the payoff matrix whether or not a game is strictly determined?

20. Briefly explain any advantages of removing from the payoff matrix any dominated strategies.

In Exercises 21–24, use the following payoff matrix to determine the payoff if each of the given strategies is used.

$$\begin{bmatrix} -2 & 5 & -6 & 3 \\ 0 & -1 & 7 & 5 \\ 2 & 6 & -4 & 4 \end{bmatrix}$$

21. (1, 1) **22.** (1, 4) **23.** (2, 3) **24.** (3, 4)

25. Are there any dominated strategies in this game? **26.** Is there a saddle point?

Remove any dominated strategies in the following games.

27. $\begin{bmatrix} -11 & 6 & 8 & 9 \\ -10 & -12 & 3 & 2 \end{bmatrix}$ **28.** $\begin{bmatrix} -1 & 9 & 0 \\ 4 & -10 & 6 \\ 8 & -6 & 7 \end{bmatrix}$ **29.** $\begin{bmatrix} -2 & 4 & 1 \\ 3 & 2 & 7 \\ -8 & 1 & 6 \\ 0 & 3 & 9 \end{bmatrix}$ **30.** $\begin{bmatrix} 3 & -1 & 4 \\ 0 & 4 & -1 \\ 1 & 2 & -3 \\ 0 & 0 & 2 \end{bmatrix}$

For the following games, find the strategies producing any saddle points. Give the value of the game. Identify any fair games.

31. $\begin{bmatrix} -2 & 3 \\ -4 & 5 \end{bmatrix}$ **32.** $\begin{bmatrix} -4 & 0 & 2 & -5 \\ 6 & 9 & 3 & 8 \end{bmatrix}$ **33.** $\begin{bmatrix} -4 & -1 \\ 6 & 0 \\ 8 & -3 \end{bmatrix}$

34. $\begin{bmatrix} 4 & -1 & 6 \\ -3 & -2 & 0 \\ -1 & -4 & 3 \end{bmatrix}$ **35.** $\begin{bmatrix} 8 & 1 & -7 & 2 \\ -1 & 4 & -3 & 3 \end{bmatrix}$ **36.** $\begin{bmatrix} 2 & -9 \\ 7 & 1 \\ 4 & 2 \end{bmatrix}$

Find the optimum strategies for each of the following games using the formulas from the section on mixed strategies. Find the value of the game.

37. $\begin{bmatrix} 1 & 0 \\ -2 & 3 \end{bmatrix}$ **38.** $\begin{bmatrix} 2 & -3 \\ -3 & 5 \end{bmatrix}$ **39.** $\begin{bmatrix} -3 & 5 \\ 1 & 0 \end{bmatrix}$ **40.** $\begin{bmatrix} 8 & -3 \\ -6 & 2 \end{bmatrix}$

For each of the following games, remove any dominated strategies, then solve the game using the formulas from the section on mixed strategies. Find the value of the game.

41. $\begin{bmatrix} -4 & 8 & 0 \\ -2 & 9 & -3 \end{bmatrix}$ **42.** $\begin{bmatrix} 1 & 0 & 3 & -3 \\ 4 & -2 & 4 & -1 \end{bmatrix}$ **43.** $\begin{bmatrix} 2 & -1 \\ -4 & 5 \\ -1 & -2 \end{bmatrix}$ **44.** $\begin{bmatrix} 8 & -6 \\ 4 & -8 \\ -9 & 9 \end{bmatrix}$

Applications

Business and Economics

Advertising *Currently, 35% of all hot dogs sold in one area are made by Dogkins, and 65% are made by Long Dog. Suppose that Dogkins starts a heavy advertising campaign, with the campaign producing the following transition matrix.*

		After Campaign	
		Dogkins	Long Dog
Before	Dogkins	.8	.2
Campaign	Long Dog	.4	.6

45. Find the share of the market for each company
 (a) after one campaign.
 (b) after three such campaigns.

46. Predict the long-range market share for Dogkins.

Credit Cards *A credit card company classifies its customers in three groups; nonusers in a given month, light users, and heavy users. The transition matrix for these states is*

	Nonuser	Light	Heavy
Nonuser	.8	.15	.05
Light	.25	.55	.2
Heavy	.04	.21	.75

Suppose the initial distribution for the three states is [.4 .4 .2]. *Find the distribution after each of the following periods.*

47. 1 mo **48.** 2 mo **49.** 3 mo

50. What is the long-range prediction for the distribution of users?

Labor Relations *In labor-management relations, both labor and management can adopt either a friendly or a hostile attitude. The results are shown in the following payoff matrix. The numbers give the wage gains made by an average worker.*

		Management	
		Friendly	Hostile
Labor	Friendly	$600	$800
	Hostile	$400	$950

51. Suppose the chief negotiator for labor is an optimist. What strategy should he choose?

52. What strategy should he choose if he is a pessimist?

53. The chief negotiator for labor feels that there is a 70% chance that the company will be hostile. What strategy should he adopt? What is the expected payoff?

54. Just before negotiations begin, a new management is installed in the company. There is only a 40% chance that the new management will be hostile. What strategy should be adopted by labor?

55. Find the optimum strategy for labor and management, and find the value of the game.

Life Sciences

Medical Research *A medical researcher is studying the risk of heart attack in men. She first divides men into three weight categories: thin, normal, and overweight. By studying the male ancestors, sons, and grandsons of these men, the researcher comes up with the following transition matrix.*

	Thin	Normal	Overweight
Thin	.3	.5	.2
Normal	.2	.6	.2
Overweight	.1	.5	.4

Find the probabilities of the following for a man of normal weight.

56. Thin son **57.** Thin grandson

58. Thin great-grandson

Find the probabilities of the following for an overweight man.

59. Overweight son **60.** Overweight grandson

61. Overweight great-grandson

Suppose that the distribution of men by weight is initially given by [.2 .55 .25]. *Find each of the following distributions.*

62. After 1 generation **63.** After 2 generations

64. After 3 generations

65. Find the long-range prediction for the distribution of weights.

Social Sciences

Politics *A candidate for city council can come out in favor of a new factory, be opposed to it, or waffle on the issue. The change in votes for the candidate depends on what her opponent does, with payoffs as shown.*

		Opponent		
		Favors	Waffles	Opposes
Candidate	Favors	0	−1000	−4000
	Waffles	1000	0	−500
	Opposes	5000	2000	0

66. What should the candidate do if she is an optimist?

67. What should she do if she is a pessimist?

68. Suppose the candidate's campaign manager feels there is a 40% chance that the opponent will favor the plant, and a 35% chance that he will waffle. What strategy should the candidate adopt? What is the expected change in the number of votes?

69. The opponent conducts a new poll that shows strong opposition to the new factory. This changes the probability that he will favor the factory to 0, and the probability that he will waffle to .7. What strategy should our candidate adopt? What is the expected change in the number of votes now?

70. Find the optimum strategy for the candidate and her opponent, and find the value of the game.

Connections

71. **Theology** In his book entitled *Pensées*, the French mathematician Blaise Pascal (1623–1662) described what has since been referred to as Pascal's Wager. Pascal observed that everyone must wager on whether or not God exists. If you decide that God exists and in fact God does exist, you gain "an infinity of an infinitely happy life"; if you are wrong, you lose nothing of great value, having lived a good life without receiving any reward for it. If you choose not to believe in God and God actually does exist, you will have lost something of infinite value. Pascal concludes that believing in God is by far the most rational action. Write a few sentences discussing Pascal's Wager, including how the payoff matrix might be set up and what the optimum strategy would be. What is your reaction to Pascal's conclusion?

Extended Application / Cavities and Restoration

K. H. Lu of the University of Oregon Dental School has applied Markov chains to analyze the progress of teeth from a healthy state to having cavities to being restored through a filling or becoming so decayed as to require removal.*

Dental records of 184 grade and high school students were observed at six-month intervals (a total of 1080 records). The state of the upper second premolar of each student was recorded as a five-digit number. To interpret this number, consider the tooth being roughly shaped like a box, with one side fastened to the gum. There are five sides of the tooth where cavities can occur, and each digit of the five-digit number refers to one of these sides. The first digit referred to the occlusal (or chewing) surface, the second to the mesial surface (the surface closest to the front of the mouth), the third to the distal surface (the surface farthest from the front of the mouth), the fourth to the lingual surface (next to the tongue), and the fifth to the buccal surface (next to the cheek). A 0 was used to signify a healthy surface, a 1 for a decayed surface, and a 2 for a restored surface, in which the decayed area has been filled. Thus, (00000) signifies a healthy tooth and (20100) a tooth with a filling on the occlusal surface and a cavity on the distal surface, with the other three surfaces healthy.

The total number of possible states in this Markov chain is $3^5 = 243$, but only 18 of these states occurred among the students. In the initial analysis, all 9 states with a 2 in one of the digits were grouped together under the heading "restoration." This is an absorbing state, because once a surface has been filled, the filling stayed there for the duration of the study. Another absorbing state is (11111); once the tooth reaches this point, it is lost.

Lu found the following transition matrix.

	00000	00100	01000	01100	10000	10100	11000	11100	Restoration	11111
00000	.84853	.07003	.03420	.01140	.01791	.00326	0	.00326	.01141	0
00100	0	.74725	0	.18681	0	.02198	0	0	.04396	0
01000	0	0	.70312	.26563	0	0	.03125	0	0	0
01100	0	0	0	.90510	0	0	0	.03650	.05840	0
10000	0	0	0	0	.71430	.17857	0	.03571	.03571	.03571
10100	0	0	0	0	0	.75000	0	.25000	0	0
11000	0	0	0	0	0	0	.50000	.50000	0	0
11100	0	0	0	0	0	0	0	.90000	.05000	.05000
Restoration	0	0	0	0	0	0	0	0	1	0
11111	0	0	0	0	0	0	0	0	0	1

Using a computer, he found the fundamental matrix to be as follows.

$$
\begin{bmatrix}
6.601967 & 1.829222 & .760534 & 6.522655 & .413865 & .542530 & .047533 & 4.337776 \\
0 & 3.956479 & 0 & 7.788301 & 0 & .347854 & 0 & 3.712363 \\
0 & 0 & 3.368364 & 9.428226 & 0 & 0 & .210523 & 4.493915 \\
0 & 0 & 0 & 10.537407 & 0 & 0 & 0 & 3.846153 \\
0 & 0 & 0 & 0 & 3.500175 & 2.500105 & 0 & 7.500173 \\
0 & 0 & 0 & 0 & 0 & 4.000000 & 0 & 9.999998 \\
0 & 0 & 0 & 0 & 0 & 0 & 2.000000 & 9.999998 \\
0 & 0 & 0 & 0 & 0 & 0 & 0 & 9.999998 \\
\end{bmatrix}
$$

*From K. H. Lu, "A Markov Chain Analysis of Caries Process with Consideration for the Effect of Restoration," *Archives of Oral Biology* 13, 1968, pp. 1119–32.

From the fundamental matrix and the transition matrix, it is possible to explore what happens to the students' teeth as time progresses. For example, Lu concluded from the small probabilities of entering the restoration state that the group was receiving poor dental care, and that only when cavities had done considerable damage to the teeth were restorative measures sought. Further evidence for the lack of dental care was the fact that the probability of going from state (01000), (11000), or (10100) to the restoration state is zero; students in these states did not receive professional care until more damage to the tooth had occurred. Other aspects of this study are explored in the exercises.

Exercises

1. Use a computer to verify the fundamental matrix given in the text.

2. Given a student with a healthy tooth, use the fundamental matrix to approximate the amount of time until the student has some tooth decay. [*Hint:* Find the expected number of times state (00000) is visited, and use the fact that each time represents 6 mo.]

3. Compute *FR*. (*Note:* A computer may be helpful here.)

4. Use your answer from Exercise 3 to find the probability that a healthy tooth is eventually lost.

5. For one of the five surfaces, once decay had occurred, there was a .5 probability that the tooth would be lost. Use your answer from Exercise 3 to find which surface that is.

Applications at a glance...
*Paying Off a Lottery Winner ... penalty for late taxes ...
Swiss bank accounts ...*

CHAPTER 9

MATHEMATICS OF FINANCE

(See page 434.)

Whether you are in a position to invest money or to borrow money, it is important for both business managers and consumers to understand *interest*. The formulas for interest are developed in this chapter. A calculator with a y^x key will be useful throughout the chapter.

9.1 SIMPLE INTEREST AND DISCOUNT

What amount should a person be willing to accept if a $5000 loan is paid off early?

The answer to the question above is found in this section.

Interest on loans of a year or less is usually calculated as *simple interest*, a type of interest that is charged (or paid) only on the amount borrowed (or invested), and not on past interest. The amount borrowed is called the **principal.** The **rate** of interest is given as a percent per year, expressed as a decimal. For example, $6\% = .06$ and $11\frac{1}{2}\% = .115$. The **time** the money is earning interest is calculated in years. Simple interest is the product of the principal, rate, and time.

SIMPLE INTEREST

The **simple interest** I on a principal of P dollars at a rate of interest r for a time t in years is given by

$$I = Prt.$$

EXAMPLE 1 To buy furniture for a new apartment, Theresa DePalo borrowed $5000 at 11% simple interest for 11 months. How much interest will she pay?

From the formula, $I = Prt$, with $P = 5000$, $r = .11$, and $t = 11/12$ (in years). The total interest she will pay is

$$I = 5000(.11)(11/12) = 504.17,$$

or $504.17. ◀

A deposit of P dollars today at a rate of interest r for t years produces interest of $I = Prt$. The interest, added to the original principal P, gives

$$P + Prt = P(1 + rt).$$

This amount is called the **future value** of P dollars at an interest rate r for time t in years. When loans are involved, the future value is often called the **maturity value** of the loan. This idea is summarized as follows.

FUTURE OR MATURITY VALUE FOR SIMPLE INTEREST

The **future** or **maturity value** A of P dollars at a rate of interest r for t years is

$$A = P(1 + rt).$$

EXAMPLE 2 Find the maturity value for each of the following loans at simple interest.

(a) A loan of $2500 to be repaid in 8 months with interest of 12.1%

The loan is for 8 months, or $8/12 = 2/3$ of a year. The maturity value is

$$A = P(1 + rt)$$

$$A = 2500\left[1 + .121\left(\frac{2}{3}\right)\right]$$

$$\approx 2500(1 + .08067) \approx 2701.67,$$

or $2701.67. (The answer is rounded to the nearest cent, as is customary in financial problems.) Of this maturity value,

$$\$2701.67 - \$2500 = \$201.67$$

represents interest.

(b) A loan of $11,280 for 85 days at 11% interest

It is common to assume 360 days in a year when working with simple interest. We shall usually make such an assumption in this book. The maturity value in this example is

$$A = 11,280\left[1 + .11\left(\frac{85}{360}\right)\right] \approx 11,280(1.0259722) \approx 11,572.97,$$

or $11,572.97. ◀

In part (b) of Example 1 we assumed 360 days in a year. Interest found using a 360-day year is called *ordinary interest*, and interest found using a 365-day year is called *exact interest*.

Present Value A sum of money that can be deposited today to yield some larger amount in the future is called the *present value* of that future amount. Let P be the present value of some amount A at some time t (in years) in the future. Assume a simple interest rate r. As above, the future value of this sum is

$$A = P(1 + rt).$$

Since the principal P is the present value, divide both sides of this last result by $1 + rt$ to get

$$P = \frac{A}{1 + rt}.$$

PRESENT VALUE FOR SIMPLE INTEREST

The **present value** P of a future amount of A dollars at a simple interest rate r for t years is

$$P = \frac{A}{1 + rt}.$$

EXAMPLE 3

Find the present value of $10,000 invested for 1 year at 8% simple interest. Here $A = 10,000$, $t = 1$, and $r = .08$. Use the formula for present value.

$$P = \frac{10,000}{1 + (.08)(1)} = \frac{10,000}{1.08} \approx 9259.26$$

If $9259.26 were deposited today at 8% interest, a total of $10,000 would be in the account after 1 year. These two sums—$9259.26 today, and $10,000 in a year—are equivalent (at 8%) because the first amount becomes the other after one year. ◀

EXAMPLE 4

Because of a court settlement, Charlie Dawkins owes $5000 to Arnold Parker. The money must be paid in 10 months, with no interest. Suppose Dawkins wishes to pay the money today. What amount should Parker be willing to accept? Assume simple interest of 11%.

The amount that Parker should be willing to accept today is given by the present value:

$$P = \frac{5000}{1 + (.11)\left(\frac{10}{12}\right)} = 4580.15.$$

Parker should be willing to accept $4580.15 in settlement. ◀

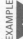

EXAMPLE 5

Suppose that in order to borrow $8000 today, you must agree to pay $8380 in 6 months to repay the loan with interest. What is the simple interest rate?

Use the formula for future value, with $A = 8380$, $P = 8000$, $t = 6/12 = 1/2$, and solve for r.

$$A = P(1 + rt)$$

$$8380 = 8000\left(1 + r \cdot \frac{1}{2}\right)$$

$$8380 = 8000 + \frac{8000r}{2} \qquad \text{Distributive property}$$

$$380 = 4000r \qquad \text{Subtract 8000 on both sides.}$$

$$r = .095$$

Thus, the interest rate is 9.5%.

Simple Discount Notes The loans discussed up to this point are called **simple interest notes,** where interest on the face value of the loan is added to the loan and paid at maturity. Another common type of note, called a **simple discount note,** has the interest deducted in advance from the amount of a loan before giving the *balance* to the borrower. The *full* value of the note must be paid at maturity. The money that is deducted is called the **bank discount** or just the **discount,** and the money actually received by the borrower is called the **proceeds.**

EXAMPLE **6**

Lillie Chalmers needs a loan and agrees to pay $8500 to her banker in 9 months. The banker subtracts a discount of 10% and gives the balance to Chalmers. Find the amounts of the discount and the proceeds.

The discount is found in the same way that simple interest is found.

$$\text{Discount} = 8500(.10)\left(\frac{9}{12}\right) = 637.50$$

The proceeds are found by subtracting the discount from the original amount.

$$\text{Proceeds} = \$8500 - \$637.50 = \$7862.50 \quad \blacktriangleleft$$

In Example 6, the borrower was charged a discount of 10%. However, 10% is *not* the interest rate paid, since 10% applies to the $8500, while the borrower actually received only $7862.50. In the next example, we find the actual rate of interest paid by the borrower.

EXAMPLE **7**

Find the actual rate of interest paid by Chalmers in Example 6.

As mentioned above, the rate of 10% stated in Example 6 is not the actual rate of interest, since it applies to the total amount of $8500 and not to the amount actually borrowed, $7862.50. To find the rate of interest paid by Chalmers on the amount borrowed, use the formula for simple interest with r as the unknown. Chalmers received $P = \$7862.50$ and paid interest of $I = \$637.50$ for 9 months, so $t = 9/12$. Substitute these values into $I = Prt$.

$$I = Prt$$
$$637.50 = 7862.50(r)\left(\frac{9}{12}\right)$$
$$\frac{637.50}{7862.50\left(\frac{9}{12}\right)} = r$$
$$.108 \approx r$$

The interest rate paid by the borrower is about 10.8%. $\quad \blacktriangleleft$

Let D represent the amount of discount on a loan. Then $D = Art$, where A is the maturity value of the loan (the amount borrowed plus interest), and r is the stated rate of interest. The amount actually received, the proceeds, can be written as $P = A - D$, or $P = A - Art$, from which $P = A(1 - rt)$.

These formulas for discount are summarized below.

DISCOUNT	If D is the discount on a loan having a maturity value A at a rate of interest r for t years, and if P represents the proceeds, then

$$P = A - D \quad \text{or} \quad P = A(1 - rt).$$

One common use of discount is in *discounting a note*, a process by which a promissory note due at some time in the future can be converted to cash now.

EXAMPLE 8

Bill Bass owes $4250 to Jenny Chinn. The loan is payable in 1 year at 10% interest. Chinn needs cash to buy a new car, so 3 months before the loan is payable, she goes to her bank to have the loan discounted. The bank will pay her the maturity value of the note less an 11% discount fee. Find the amount of cash she will receive from the bank.

First find the maturity value of the loan, the amount Bass must pay to Chinn. By the formula for maturity value,

$$A = P(1 + rt)$$
$$A = 4250[1 + (.10)(1)]$$
$$= 4250(1.10) = 4675,$$

or $4675.

The bank applies its discount rate to this total:

$$\text{Amount of discount} = 4675(.11)\left(\frac{3}{12}\right) = 128.56.$$

(Remember that the loan was discounted 3 months before it was due.) Chinn actually receives

$$\$4675 - \$128.56 = \$4546.44$$

in cash from the bank. Three months later, the bank will get $4675 from Bass. ◀

9.1 Exercises

1. What factors determine the amount of interest earned on a fixed principal?

Find the simple interest in Exercises 2–5.

2. $25,000 at 7% for 9 mo

3. $3850 at 9% for 8 mo

4. $1974 at 6.3% for 7 mo

5. $3724 at 8.4% for 11 mo

Find the simple interest. Assume a 360-day year and a 30-day month.

6. $5147.18 at 10.1% for 58 days

7. $2930.42 at 11.9% for 123 days

8. $7980 at 10%; loan made on May 7 and due September 19

9. $5408 at 12%; loan made on August 16 and due December 30

Find the simple interest. Assume 365 days in a year, and use the exact number of days in a month. (Assume 28 days in February.)

10. $7800 at 11%; made on July 7 and due October 25

11. $11,000 at 10%, made on February 19 and due May 31

12. $2579 at 9.6%; made on October 4 and due March 15

13. $37,098 at 11.2%; made on September 12 and due July 30

14. In your own words, describe the *maturity value* of a loan.

15. What is meant by the *present value* of money?

Find the present value of each of the future amounts in Exercises 16–19. Assume 360 days in a year.

16. $15,000 for 8 mo; money earns 6%

17. $48,000 for 9 mo; money earns 5%

18. $15,402 for 125 days; money earns 6.3%

19. $29,764 for 310 days; money earns 7.2%

Find the proceeds for the amounts in Exercises 20–23. Assume 360 days in a year.

20. $7150; discount rate 12%; length of loan 11 mo

21. $9450; discount rate 10%; length of loan 7 mo

22. $35,800; discount rate 9.1%; length of loan 183 days

23. $50,900; discount rate 8.2%; length of loan 238 days

24. Why might the discount rate charged on a simple discount note be different from the interest rate paid on the proceeds?

Find the interest rate to the nearest tenth on the proceeds for the following simple discount notes.

25. $6200; discount rate 10%; length of loan 8 mo

26. $5000; discount rate 8.1%; length of loan 6 mo

27. $58,000; discount rate 10.8%; length of loan 9 mo

28. $43,000; discount rate 9%; length of loan 4 mo

Applications

Business and Economics

29. **Loan Repayment** Michelle Beese borrowed $25,900 from her father to start a flower shop. She repaid him after 11 mo, with interest of 8.4%. Find the total amount she repaid.

30. **Delinquent Taxes** An accountant for a corporation forgot to pay the firm's income tax of $725,896.15 on time. The government charged a penalty of 12.7% interest for the 34 days the money was late. Find the total amount (tax and penalty) that was paid. (Use a 365-day year.)

31. **Savings** A $100,000 certificate of deposit held for 60 days is worth $101,133.33. To the nearest tenth of a percent, what interest rate was earned?

32. **Savings** Tuition of $1769 will be due when the spring term begins in 4 mo. What amount should a student deposit today, at 6.25%, to have enough to pay the tuition?

33. **Savings** A firm of accountants has ordered 7 new IBM computers at a cost of $5104 each. The machines will not be delivered for 7 mo. What amount could the firm deposit in an account paying 6.42% to have enough to pay for the machines?

34. **Loan Repayment** Sun Kang needs $5196 to pay for remodeling work on his house. He plans to repay the loan in 10 mo. His bank loans money at a discount rate of 13%. Find the amount of his loan.

35. **Loan Repayment** Mary Collins decides to go back to college. To get to school, she buys a small car for $6100. She decides to borrow the money from a bank that charges an 11.8% discount rate. If she will repay the loan in 7 months, find the amount of the loan.

36. **Loan Interest** John Matthews signs a $4200 note at the bank. The bank charges a 12.2% discount rate. Find the net proceeds if the note is for 10 mo. Find the actual interest rate (to the nearest hundredth) charged by the bank.

37. **Stock Growth** A stock that sold for $22 at the beginning of the year was selling for $24 at the end of the year. If the stock paid a dividend of $.50 per share, what is the simple interest rate on an investment in this stock? (*Hint*: Consider the interest to be the increase in value plus the dividend.)

38. **Bond Interest** A bond with a face value of $10,000 in 10 yr can be purchased now for $5988.02. What is the simple interest rate?

39. Loan Interest A building contractor gives a $13,500 note to a plumber. The note is due in 9 mo, with interest of 9%. Three months after the note is signed, the plumber discounts it at the bank. The bank charges a 10.1% discount rate. How much will the plumber receive? Will it be enough to pay a bill for $13,582?

40. Loan Interest Maria Lopez owes $7000 to the Eastside Music Shop. She has agreed to pay the amount in 7 mo at an interest rate of 10%. Two months before the loan is due, the store needs $7350 to pay a wholesaler's bill. The bank will discount the note at a rate of 10.5%. How much will the store receive? Is it enough to pay the bill?

9.2 COMPOUND INTEREST

If money can be borrowed at 11% interest compounded annually or at 10.8% compounded monthly, which loan would cost less?

We shall see how to make such comparisons in this section.

As mentioned earlier, simple interest is normally used for loans or investments of a year or less. For longer periods *compound interest* is used. With **compound interest,** interest is charged (or paid) on interest as well as on principal. For example, if $1000 is deposited at 5% interest for 1 year, at the end of the year the interest is $1000(.05)(1) = $50. The balance in the account is $1000 + $50 = $1050. If this amount is left at 5% interest for another year, the interest is calculated on $1050 instead of the original $1000, so the amount in the account at the end of the second year is $1050 + $1050(.05)(1) = $1102.50. Note that simple interest would produce a total amount of only

$$\$1000[1 + (.05)(2)] = \$1100.$$

To find a formula for compound interest, first suppose that P dollars is deposited at a rate of interest r per year. The amount on deposit at the end of the first year is found by the simple interest formula, with $t = 1$.

$$A = P(1 + r \cdot 1) = P(1 + r)$$

If the deposit earns compound interest, the interest earned during the second year is paid on the total amount on deposit at the end of the first year. Using the formula $A = P(1 + rt)$ again, with $P = P(1 + r)$ and $t = 1$, gives the total amount on deposit at the end of the second year.

$$A = [P(1 + r)](1 + r \cdot 1) = P(1 + r)^2$$

In the same way, the total amount on deposit at the end of the third year is

$$P(1 + r)^3.$$

Generalizing, in t years the total amount on deposit is

$$A = P(1 + r)^t,$$

called the **compound amount.**

Note Compare this formula for compound interest with the formula for simple interest given in the previous section:

Compound interest $A = P(1 + r)^t$

Simple interest $A = P(1 + rt).$

The important distinction between the two formulas is that in the compound interest formula, the number of years, t, is an *exponent*, so that money grows much more rapidly when interest is compounded.

Interest can be compounded more than once per year. Common compounding periods include *semiannually* (two periods per year), *quarterly* (four periods per year), *monthly* (twelve periods per year), or *daily* (usually 365 periods per year). The *interest rate per period, i,* is found by dividing the annual interest rate, r, by the number of compounding periods per year. To find the total number of compounding periods, n, we multiply the number of years, t, by the number of compounding periods per year. The following formula can be derived in the same way as the previous formula.

COMPOUND AMOUNT

If P dollars are deposited for n compounding periods at a rate of interest i per period, the compound amount A is

$$A = P(1 + i)^n.$$

EXAMPLE 1 Suppose $1000 is deposited for 6 years in an account paying 8% per year compounded annually.

(a) Find the compound amount.

In the formula above, $P = 1000$, $i = .08 \div 1$, and $n = 6(1) = 6$. The compound amount is

$$A = P(1 + i)^n$$

$$A = 1000(1.08)^6.$$

The quantity $(1.08)^6$ can be found with a calculator that has a y^x key. Enter 1.08, press the y^x key, then press 6, followed by the $=$ or EXE key. You should get $(1.08)^6 \approx 1.58687$, and so

$$A = 1000(1.58687) = 1586.87.$$

The compound amount is $1586.87.

(b) Find the actual amount of interest earned.

From the compound amount, subtract the initial deposit.

Amount of interest $= \$1586.87 - \$1000 = \$586.87$ ◄

2 Find the amount of interest earned by a deposit of $1000 for 6 years at 6% compounded quarterly.

Interest compounded quarterly is compounded 4 times a year. In 6 years, there are $6 \cdot 4 = 24$ quarters, or 24 periods. Thus, $n = 24$. Interest of 6% per year is 6%/4, or 1.5% per quarter, so $i = .015$. The compound amount is

$$1000(1 + .015)^{24} = 1000(1.015)^{24}.$$

Using a calculator, $(1.015)^{24} \approx 1.42950$, so

$$A = 1000(1.42950) = 1429.50,$$

or $1429.50. The compound amount is $1429.50, and the interest earned is $1429.50 - $1000 = $429.50. ◀

3 Suppose $24,000 is deposited at 8% interest for 9 years. Find the interest earned by (a) daily, and (b) hourly compounding.

(a) The interest rate per period will be $i = .08/365$. In 9 years there are $9 \cdot 365 = 3285$ days, so the number of compounding periods is $n = 3285$. The compound amount with daily compounding is

$$24,000\left(1 + \frac{.08}{365}\right)^{3285} = 49,302.51,$$

or $49,302.51. The interest earned is

$$\$49,302.51 - \$24,000 = \$25,302.51.$$

(b) In 1 year there are $365 \times 24 = 8760$ hours, and in 9 years there are $9 \cdot 8760 = 78,840$ hours. Thus, $n = 78,840$ and $i = .08/8760$. The compound amount is

$$24,000\left(1 + \frac{.08}{8760}\right)^{78,840} = 49,306.23,^{\dagger}$$

or $49,306.23. This amount includes interest of

$$\$49,306.23 - \$24,000 = \$25,306.23,$$

only $3.72 more than when interest is compounded daily. ◀

Effective Rate If $1 is deposited at 4% compounded quarterly, a calculator can be used to find that at the end of one year, the compound amount is $1.0406, an increase of 4.06% over the original $1. The actual increase of 4.06% in the money is somewhat higher than the stated increase of 4%. To differentiate between these two numbers, 4% is called the **nominal** or **stated** rate of interest, while 4.06% is called the **effective** rate.* To avoid confusion between stated rates

†This calculation is very sensitive to roundoff; your calculator may give a slightly different answer.
*When applied to consumer finance, the effective rate is called the annual percentage rate, or APR.

and effective rates, we shall continue to use r for the stated rate and we will use r_e for the effective rate.

4 Find the effective rate corresponding to a nominal rate of 6% compounded semiannually.

Here, $r/m = 6\%/2 = 3\%$ for 2 periods. Use a calculator to find that $(1.03)^2 \approx 1.06090$, which shows that \$1 will increase to \$1.06090, an actual increase of 6.09%. The effective rate is $r_e = 6.09\%$. ◀

Generalizing from this example, the effective rate of interest is given by the following formula.

EFFECTIVE RATE

The effective rate corresponding to a stated rate of interest r compounded m times per year is

$$r_e = \left(1 + \frac{r}{m}\right)^m - 1.$$

5 A bank pays interest of 5% compounded monthly. Find the effective rate.
Use the formula given above, with $r = .05$ and $m = 12$. The effective rate is

$$r_e = \left(1 + \frac{.05}{12}\right)^{12} - 1 \approx .0512,$$

or 5.12%. ◀

6 John Lenchek needs to borrow money. His neighborhood bank charges 11% interest compounded semiannually. A downtown bank charges 10.8% interest compounded monthly. At which bank will Lenchek pay the lesser amount of interest?

Compare the effective rates.

Neighborhood bank: $r_e = \left(1 + \frac{.11}{2}\right)^2 - 1 = .113025 = 11.3\%$

Downtown bank: $r_e = \left(1 + \frac{.108}{12}\right)^{12} - 1 = .11351 = 11.4\%$

The neighborhood bank has the lower effective rate, although it has a higher stated rate. ◀

Present Value with Compound Interest The formula for compound interest $A = P(1 + i)^n$, has four variables: A, P, i, and n. Given the values of any three of these variables, the value of the fourth can be found. In particular, if A (the future amount), i, and n are known, then P can be found. Here P is the amount that should be deposited today to produce A dollars in n periods.

EXAMPLE 7

Joan Ailanjian must pay a lump sum of $6000 in 5 years. What amount deposited today at 8% compounded annually will amount to $6000 in 5 years?

Here $A = 6000$, $i = .08$, $n = 5$, and P is unknown. Substituting these values into the formula for the compound amount gives

$$6000 = P(1.08)^5 \approx P(1.46933)$$

$$P = \frac{6000}{1.46933} \approx 4083.49,$$

or $4083.49. If Ailanjian leaves $4083.49 for 5 years in an account paying 8% compounded annually, she will have $6000 when she needs it. ◄

As Example 7 shows, $6000 in 5 years is the same as $4083.49 today (if money can be deposited at 8% compounded annually). Recall from the previous section that an amount that can be deposited today to yield a given sum in the future is called the *present value* of the future sum. Generalizing from Example 7, by solving $A = P(1 + i)^n$ for P, we get the following formula for present value.

PRESENT VALUE FOR COMPOUND INTEREST

The **present value** of A dollars compounded at an interest rate i per period for n periods is

$$P = \frac{A}{(1 + i)^n} \quad \text{or} \quad P = A(1 + i)^{-n}.$$

Compare this with the present value of an amount at simple annual interest r for t years, given in the previous section:

$$P = \frac{A}{1 + rt}.$$

EXAMPLE 8

Find the present value of $16,000 in 9 years if money can be deposited at 6% compounded semiannually.

In 9 years there are $2 \cdot 9 = 18$ semiannual periods. A rate of 6% per year is 3% in each semiannual period. Apply the formula with $A = 16,000$, $i = .03$, and $n = 18$.

$$P = \frac{A}{(1 + i)^n} = \frac{16,000}{(1.03)^{18}} \approx 9398.31$$

A deposit of $9398.31 today, at 6% compounded semiannually, will produce a total of $16,000 in 9 years. ◄

We can solve the compound amount formula for n also, as the following example shows.

EXAMPLE **9**

Suppose the general level of inflation in the economy averages 8% per year. Find the number of years it would take for the overall level of prices to double.

To find the number of years it will take for $1 worth of goods or services to cost $2, find n in the equation

$$2 = 1(1 + .08)^n,$$

where $A = 2$, $P = 1$, and $i = .08$. This equation simplifies to

$$2 = (1.08)^n.$$

By trying various values of n, we find that $n = 9$ is approximately correct, because $1.08^9 = 1.99900 \approx 2$. The exact value of n can be found quickly by using logarithms, but that is beyond the scope of this chapter. ◀

At this point, it seems helpful to summarize the notation and the most important formulas for simple and compound interest. We use the following variables.

P = principal or present value

A = future or maturity value

r = annual (stated or nominal) interest rate

t = number of years

m = number of compounding periods per year

i = interest rate per period ($i = r/m$)

r_e = effective rate

n = total number of compounding periods ($n = tm$)

Simple Interest
$$A = P(1 + rt)$$

$$P = \frac{A}{1 + rt}$$

Compound Interest
$$A = P(1 + i)^n$$

$$P = \frac{A}{(1 + i)^n} = A(1 + i)^{-n}$$

$$r_e = \left(1 + \frac{r}{m}\right)^m - 1$$

9.2 Exercises

1. Explain the difference between simple interest and compound interest.
2. As the number of compounding periods per year increases, what happens to the interest for the year?

Find the compound amount for each of the following deposits.

3. $1000 at 6% compounded annually for 8 yr
4. $1000 at 7% compounded annually for 10 yr
5. $470 at 10% compounded semiannually for 12 yr
6. $15,000 at 6% compounded semiannually for 11 yr
7. $6500 at 12% compounded quarterly for 6 yr
8. $9100 at 8% compounded quarterly for 4 yr

Find the amount of interest earned by each of the following deposits.

9. $6000 at 8% compounded annually for 8 yr

10. $21,000 at 6% compounded annually for 5 yr

11. $43,000 at 10% compounded semiannually for 9 yr

12. $7500 at 8% compounded semiannually for 5 yr

13. $2196.58 at 10.8% compounded quarterly for 4 yr

14. $4915.73 at 11.6% compounded quarterly for 3 yr

Find the present value of each of the following investments.

15. $4500 at 8% compounded annually for 9 yr

16. $11,500 at 8% compounded annually for 12 yr

17. $15,902.74 at 9.8% compounded annually for 7 yr

18. $27,159.68 at 12.3% compounded annually for 11 yr

19. $2000 at 9% compounded semiannually for 8 yr

20. $2000 at 11% compounded semiannually for 8 yr

21. $8800 at 10% compounded quarterly for 5 yr

22. $7500 at 12% compounded quarterly for 9 yr

23. How do the nominal or stated interest rate and the effective interest rate differ?

24. If interest is compounded more than once per year, which rate is higher, the stated rate or the effective rate?

Find the effective rate corresponding to each of the following nominal rates.

25. 4% compounded semiannually

26. 8% compounded quarterly

27. 8% compounded semiannually

28. 10% compounded semiannually

29. 12% compounded semiannually

30. 12% compounded quarterly

Applications

Business and Economics

31. Savings When Lindsay Branson was born, her grandfather made an initial deposit of $3000 in an account for her college education. Assuming an interest rate of 6% compounded quarterly, how much will the account be worth in 18 yr?

32. Individual Retirement Account Bill Bass has $10,000 in an Individual Retirement Account (IRA). Because of new tax laws, he decides to make no further deposits. The account earns 6% interest compounded semiannually. Find the amount on deposit in 15 yr.

33. Housing Prices Suppose the average price of a house nationally is $78,000. Housing prices are increasing at a rate of 3% per year. Assuming prices continue to increase at the same rate, what will be the average price of a house in 12 yr?

34. Natural Gas Rates Natural gas rates are rising by 5% per year. If the average monthly bill is now $52, what will the average monthly bill be in 10 yr?

35. Loan Interest A small business borrows $50,000 for expansion at 12% compounded monthly. The loan is due in 4 yr. How much interest will the business pay?

36. Loan Interest A developer needs $80,000 to buy land. He is able to borrow the money at 10% per year compounded quarterly. How much will the interest amount to if he pays off the loan in 5 yr?

37. Paying Off a Lawsuit A company has agreed to pay $2.9 million in 5 yr to settle a lawsuit. How much must they invest now in an account paying 8% compounded monthly to have that amount when it is due?

38. Buying a House George Duda wants to have $20,000 available in 5 yr for a down payment on a house. He has inherited $15,000. How much of the inheritance should he invest now to accumulate the $20,000, if he can get an interest rate of 8% compounded quarterly?

39. Comparing Investments Two partners agree to invest equal amounts in their business. One will contribute $10,000 immediately. The other plans to contribute an equivalent amount in 3 yr, when she expects to acquire a large sum of money. How much should she contribute at that time to match her partner's investment now, assuming an interest rate of 6% compounded semiannually?

40. Comparing Investments As the prize in a contest, you are offered $1000 now or $1210 in 5 yr. If money can be invested at 6% compounded annually, which is larger?

Doubling Time *Use the ideas from Example 9 to find the time it would take for the general level of prices in the economy to double at each of the following average annual inflation rates.*

41. 4% **42.** 5%

43. Doubling Time The consumption of electricity has increased historically at 6% per year. If it continues to increase at this rate indefinitely, find the number of years before the electric utilities will need to double their generating capacity.

44. Doubling Time Suppose a conservation campaign coupled with higher rates causes the demand for electricity to increase at only 2% per year, as it has recently. Find the number of years before the utilities will need to double generating capacity.

Negative Interest *Under certain conditions, Swiss banks pay negative interest: they charge you. (You didn't think all that secrecy was free?) Suppose a bank "pays" −2.4% interest compounded annually. Find the compound amount for a deposit of $150,000 after each of the following periods.*

45. 2 yr **46.** 4 yr **47.** 8 yr **48.** 12 yr

*The following problem is from an actuarial examination.**

49. Savings On January 1, 1980, Jack deposited $1000 into bank X to earn interest at the rate of *j* per annum compounded semiannually. On January 1, 1985, he transferred his account to bank Y to earn interest at the rate of *k* per annum compounded quarterly. On January 1, 1988, the balance of bank Y is $1,990.76. If Jack could have earned interest at the rate of *k* per annum compounded quarterly from January 1, 1980, through January 1, 1988, his balance would have been $2,203.76. Calculate the ratio k/j.

**Problem 5 from "Course 140 Examination, Mathematics of Compound Interest" of the Education and Examination Committee of The Society of Actuaries. Reprinted by permission of The Society of Actuaries.

9.3 ANNUITIES

If $1500 is deposited each year for 6 years in an account paying 8% interest compounded annually, how much will be in the account at the end of this period?

In this section and the next, we develop future value and present value formulas for such periodic payments. To develop these formulas, we must first discuss *sequences*.

Geometric Sequences If *a* and *r* are nonzero real numbers, the infinite list of numbers $a, ar, ar^2, ar^3, ar^4, \ldots, ar^n, \ldots$ is called a **geometric sequence**. For example, if $a = 3$ and $r = -2$, we have the sequence

$$3, 3(-2), 3(-2)^2, 3(-2)^3, 3(-2)^4, \ldots,$$

or $$3, -6, 12, -24, 48, \ldots.$$

In the sequence $a, ar, ar^2, ar^3, ar^4, \ldots$, the number *a* is called the **first term** of the sequence, *ar* is the **second term**, ar^2 is the **third term,** and so on. Thus, for any $n \geq 1$,

$$ar^{n-1} \text{ is the } n\text{th term of the sequence.}$$

Each term in the sequence is *r* times the preceding term. The number *r* is called the **common ratio** of the sequence.

EXAMPLE **1** ► Find the seventh term of the geometric sequence 5, 20, 80, 320,

Here, $a = 5$ and $r = 20/5 = 4$. We want the seventh term, so $n = 7$. Use ar^{n-1}, with $a = 5$, $r = 4$, and $n = 7$.

$$ar^{n-1} = (5)(4)^{7-1} = 5(4)^6 = 20{,}480 \quad ◄$$

EXAMPLE **2** ► Find the first five terms of the geometric sequence with $a = 10$ and $r = 2$.

The first five terms are

$$10, \ 10(2), \ 10(2)^2, \ 10(2)^3, \ 10(2)^4,$$

or

$$10, \ 20, \ 40, \ 80, \ 160. \quad ◄$$

Next, we need to find the sum S_n of the first n terms of a geometric sequence, where

$$S_n = a + ar + ar^2 + ar^3 + ar^4 + \cdots + ar^{n-1}. \tag{1}$$

If $r = 1$, then

$$S_n = \underbrace{a + a + a + a + \cdots + a}_{n \text{ terms}} = na.$$

If $r \neq 1$, multiply both sides of equation (1) by r to get

$$rS_n = ar + ar^2 + ar^3 + ar^4 + \cdots + ar^n. \tag{2}$$

Now subtract corresponding sides of equation (1) from equation (2).

$$
\begin{aligned}
rS_n &= \qquad ar + ar^2 + ar^3 + ar^4 + \cdots + ar^{n-1} + ar^n \\
-S_n &= -(a + ar + ar^2 + ar^3 + ar^4 + \cdots + ar^{n-1}) \\
\hline
rS_n - S_n &= -a + ar^n \\
S_n(r - 1) &= a(r^n - 1) \\
S_n &= \frac{a(r^n - 1)}{r - 1}
\end{aligned}
$$

This result is summarized below.

SUM OF TERMS

If a geometric sequence has first term a and common ratio r, then the sum S_n of the first n terms is given by

$$S_n = \frac{a(r^n - 1)}{r - 1}, \quad r \neq 1.$$

EXAMPLE **3**

Find the sum of the first six terms of the geometric sequence 3, 12, 48,
Here $a = 3$ and $r = 4$. Find S_6 by the formula above.

$$S_6 = \frac{3(4^6 - 1)}{4 - 1} \qquad \text{Let } n = 6, a = 3, r = 4.$$

$$= \frac{3(4096 - 1)}{3}$$

$$= 4095 \quad \blacktriangleleft$$

Ordinary Annuities A sequence of equal payments made at equal periods of time is called an **annuity.** If the payments are made at the end of the time period, and if the frequency of payments is the same as the frequency of compounding, the annuity is called an **ordinary annuity.** The time between payments is the **payment period,** and the time from the beginning of the first payment period to the end of the last period is called the **term** of the annuity. The **future value of the annuity,** the final sum on deposit, is defined as the sum of the compound amounts of all the payments, compounded to the end of the term.

For example, suppose $1500 is deposited at the end of each year for the next six years in an account paying 8% per year compounded annually. Figure 1 shows this annuity. To find the future value of the annuity, look separately at each of the $1500 payments. The first of these payments will produce a compound amount of

$$1500(1 + .08)^5 = 1500(1.08)^5.$$

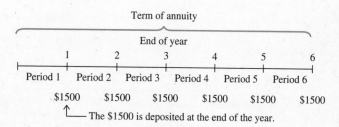

FIGURE 1

Use 5 as the exponent instead of 6 since the money is deposited at the *end* of the first year and earns interest for only five years. The second payment of $1500 will produce a compound amount of $1500(1.08)^4$. As shown in Figure 2 on the next page, the future value of the annuity is

$$1500(1.08)^5 + 1500(1.08)^4 + 1500(1.08)^3 + 1500(1.08)^2$$
$$+ 1500(1.08)^1 + 1500.$$

(The last payment earns no interest at all.)

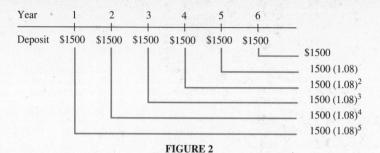

FIGURE 2

Reading this sum in reverse order, we see that it is the sum of the first six terms of a geometric sequence, with $a = 1500$, $r = 1.08$, and $n = 6$. Thus, the sum equals

$$\frac{a(r^n - 1)}{r - 1} = \frac{1500[(1.08)^6 - 1]}{1.08 - 1} = \$11{,}003.89.$$

To generalize this result, suppose that payments of R dollars each are deposited into an account at the end of each period for n periods, at a rate of interest i per period. The first payment of R dollars will produce a compound amount of $R(1 + i)^{n-1}$ dollars, the second payment will produce $R(1 + i)^{n-2}$ dollars, and so on; the final payment earns no interest and contributes just R dollars to the total. If S represents the future value (or sum) of the annuity, then (as shown in Figure 3),

$$S = R(1 + i)^{n-1} + R(1 + i)^{n-2} + R(1 + i)^{n-3} + \cdots + R(1 + i) + R,$$

or, written in reverse order,

$$S = R + R(1 + i)^1 + R(1 + i)^2 + \cdots + R(1 + i)^{n-1}.$$

FIGURE 3

This result is the sum of the first n terms of the geometric sequence having first term R and common ratio $1 + i$. Using the formula for the sum of the first n terms of a geometric sequence,

$$S = \frac{R[(1 + i)^n - 1]}{(1 + i) - 1} = \frac{R[(1 + i)^n - 1]}{i} = R\left[\frac{(1 + i)^n - 1}{i}\right].$$

The quantity in brackets is commonly written $s_{\overline{n}|i}$ (read "*s-angle-n* at *i*"), so that

$$S = R \cdot s_{\overline{n}|i}.$$

Values of $s_{\overline{n}|i}$ can be found with a calculator.

We can summarize this work as follows.

FUTURE VALUE OF AN ANNUITY

The future value, S, of an annuity of n payments of R dollars each at the end of each consecutive interest period, with interest compounded at a rate i per period, is

$$S = R\left[\frac{(1 + i)^n - 1}{i}\right] \quad \text{or} \quad S = Rs_{\overline{n}|i}.$$

EXAMPLE 4

Earl Karn is an athlete who believes that his playing career will last 7 years. To prepare for his future, he deposits $22,000 at the end of each year for 7 years in an account paying 6% compounded annually. How much will he have on deposit after 7 years?

His payments form an ordinary annuity, with $R = 22,000$, $n = 7$, and $i = .06$. The future value of this annuity (by the formula above) is

$$S = 22,000\left[\frac{(1.06)^7 - 1}{.06}\right] = 184,664.43,$$

or $184,664.43. ◀

EXAMPLE 5

Suppose $1000 is deposited at the end of each six-month period for 5 years in an account paying 5% compounded semiannually. Find the future value of the annuity.

Interest of $5\%/2 = 2.5\%$ is earned semiannually. In 5 years there are $5 \cdot 2 = 10$ semiannual periods. Since

$$s_{\overline{10}|.025} \approx 11.20338,$$

the future value is

$$S = 1000\left(\frac{(1.025)^{10} - 1}{.025}\right) \approx 1000(11.20338) = 11,203.38,$$

or $11,203.38. ◀

Annuities Due The formula developed above is for *ordinary annuities*—those with payments made at the *end* of each time period. These results can be modified slightly to apply to **annuities due**—annuities in which payments are made at the *beginning* of each time period. To find the future value of an annuity due, treat each payment as if it were made at the *end* of the *preceding* period. That is, find

$s_{\overline{n}|i}$ for *one additional period;* to compensate for this, subtract the amount of one payment.

Thus, the **future value of an annuity due** of n payments of R dollars each at the beginning of consecutive interest periods, with interest compounded at the rate of i per period, is

$$S = R\left[\frac{(1 + i)^{n+1} - 1}{i}\right] - R \quad \text{or} \quad S = Rs_{\overline{n+1}|i} - R.$$

EXAMPLE 6

Find the future value of an annuity due if payments of $500 are made at the beginning of each quarter for 7 years, in an account paying 6% compounded quarterly.

In 7 years, there are 28 quarterly periods. Add one period to get $n = 29$, and use the formula with $i = 6\%/4 = 1.5\%$.

$$S = 500\left[\frac{(1.015)^{29} - 1}{.015}\right] - 500$$

$$S = 500(35.99870) - 500 = 17,499.35$$

The account will have a total of $17,499.35 after 7 years. ◀

Sinking Funds A fund set up to receive periodic payments is called a **sinking fund.** The periodic payments, together with the interest earned by the payments, are designed to produce a given sum at some time in the future. For example, a sinking fund might be set up to receive money that will be needed to pay off the principal on a loan at some future time. If the payments are all the same amount and are made at the end of a regular time period, they form an ordinary annuity.

EXAMPLE 7

The Stockdales are close to retirement. They agree to sell an antique urn to the local museum for $17,000. Their tax adviser suggests that they defer receipt of this money until they retire, 5 years in the future. (At that time, they might well be in a lower tax bracket.) Find the amount of each payment the museum must make into a sinking fund so that it will have the necessary $17,000 in 5 years. Assume that the museum can earn 5% compounded annually on its money. Also, assume that the payments are made annually.

These payments are the periodic payments into an ordinary annuity. The annuity will amount to $17,000 in 5 years at 5% compounded annually. Using the formula,

$$17,000 = Rs_{\overline{5}|.05}$$

or

$$R = \frac{17,000}{s_{\overline{5}|.05}}$$

$$R \approx \frac{17,000}{5.52563}$$

$$R \approx 3076.57,$$

or $3076.57. If the museum deposits $3076.57 at the end of each year for 5 years in an account paying 5% compounded annually, it will have the total amount that it needs. This result is shown in the following sinking fund table. In these tables, the last payment may differ slightly from the others because of rounding errors.

Payment Number	Amount of Deposit	Interest Earned	Total in Account
1	$3076.57	$0	$3076.57
2	$3076.57	$153.83	$6306.97
3	$3076.57	$315.35	$9698.89
4	$3076.57	$484.94	$13260.40
5	$3076.58	$663.02	$17000.00

To construct the table, notice that the first payment does not earn interest until the second payment is made. Line 2 of the table shows the second payment, 5% interest of $153.83 on the first payment, and the sum of these amounts added to the total in line 1. Line 3 shows the third payment, 5% interest of $315.35 on the total from line 2, and the new total found by adding these amounts to the total in line 2. This procedure is continued to complete the table. The last payment is 1 cent larger to make the final total exactly $17,000.00. ◀

9.3 Exercises

Find the fifth term of each of the following geometric sequences. (See Example 1.)

1. $a = 3$; $r = 2$
2. $a = 5$; $r = 3$
3. $a = -8$; $r = 3$
4. $a = -6$; $r = 2$

5. $a = 1$; $r = -3$
6. $a = 12$; $r = -2$
7. $a = 1024$; $r = \frac{1}{2}$
8. $a = 729$; $r = \frac{1}{3}$

Find the sum of the first four terms for each of the following geometric sequences. (See Example 3.)

9. $a = 1$; $r = 2$
10. $a = 3$; $r = 3$
11. $a = 5$; $r = \frac{1}{5}$

12. $a = 6$; $r = \frac{1}{2}$
13. $a = 128$; $r = -\frac{3}{2}$
14. $a = 81$; $r = -\frac{2}{3}$

Find each of the following values.

15. $s\,\overline{12}|.05$
16. $s\,\overline{20}|.06$
17. $s\,\overline{16}|.04$

18. $s\,\overline{40}|.02$
19. $s\,\overline{20}|.01$
20. $s\,\overline{18}|.015$

21. Explain the difference between an ordinary annuity and an annuity due.

Find the future value of the following ordinary annuities. Interest is compounded annually. (See Example 4.)

22. $R = 100$; $i = .06$; $n = 4$
23. $R = 1000$; $i = .06$; $n = 5$
24. $R = 46,000$; $i = .06$; $n = 32$
25. $R = 29,500$; $i = .05$; $n = 15$

Find the future value of each of the following ordinary annuities, if payments are made and interest is compounded as given.

26. $R = 9200$; 10% interest compounded semiannually for 7 yr

27. $R = 3700$; 8% interest compounded semiannually for 11 yr

28. $R = 800$; 6% interest compounded semiannually for 12 yr

29. $R = 4600$; 8% interest compounded quarterly for 9 yr

30. $R = 15{,}000$; 12% interest compounded quarterly for 6 yr

31. $R = 42{,}000$; 10% interest compounded semiannually for 12 yr

Find the future value of each annuity due. Assume that interest is compounded annually.

32. $R = 600$; $i = .06$; $n = 8$ **33.** $R = 1400$; $i = .08$; $n = 10$

34. $R = 20{,}000$; $i = .08$; $n = 6$ **35.** $R = 4000$; $i = .06$; $n = 11$

Find the future value of each annuity due.

36. Payments of $1000 made at the beginning of each semiannual period for 9 yr at 8% compounded semi-annually

37. $750 deposited at the beginning of each month for 15 yr at 6% compounded monthly

38. $100 deposited at the beginning of each quarter for 9 yr at 12% compounded quarterly

39. $1500 deposited at the beginning of each semiannual period for 11 yr at 6% compounded semiannually

Find the periodic payment that will amount to each of the given sums under the given conditions.

40. $S = \$10{,}000$; interest is 5% compounded annually; payments are made at the end of each year for 12 yr

41. $S = \$100{,}000$; interest is 8% compounded semiannually; payments are made at the end of each semiannual period for 9 yr

42. What is meant by a sinking fund? Give an example of a sinking fund.

Find the amount of each payment to be made into a sinking fund so that enough will be present to accumulate the following amounts. Payments are made at the end of each period.

43. $8500; money earns 8% compounded annually; 7 annual payments

44. $2000; money earns 6% compounded annually; 5 annual payments

45. $75,000; money earns 6% compounded semiannually for $4\frac{1}{2}$ yr

46. $11,000; money earns 10% compounded semiannually for 6 yr

47. $25,000; money earns 5% compounded quarterly for $3\frac{1}{2}$ yr

48. $50,000; money earns 8% compounded quarterly for $2\frac{1}{2}$ yr

49. $9000; money earns 12% compounded monthly for $2\frac{1}{2}$ yr

50. $6000; money earns 12% compounded monthly for 3 yr

Applications

Business and Economics

51. Comparing Accounts Pat Dillon deposits $12,000 at the end of each year for 9 yr in an account paying 8% interest compounded annually.

(a) Find the final amount she will have on deposit.

(b) Pat's brother-in-law works in a bank that pays 6% compounded annually. If she deposits her money in

this bank instead of the one above, how much will she have in her account?

(c) How much would Dillon lose over 9 yr by using her brother-in-law's bank?

52. Savings Charlie Dawkins is saving for a computer. At the end of each month he puts $60 in a savings account that pays 8% interest compounded monthly. How much is in the account after 3 yr?

53. Savings Hassi is paid on the first day of the month and $80 is automatically deducted from his pay and deposited in a savings account. If the account pays 7.5% interest compounded monthly, how much will be in the account after 3 yr and 9 mo?

54. Savings A father opened a savings account for his daughter on the day she was born, depositing $1000. Each year on her birthday he deposits another $1000, making the last deposit on her twenty-first birthday. If the account pays 9.5% interest compounded annually, how much is in the account at the end of the day on the daughter's twenty-first birthday?

55. Retirement Planning A 45-year-old man puts $1000 in a retirement account at the end of each quarter until he reaches the age of 60 and makes no further deposits. If the account pays 8% interest compounded quarterly, how much will be in the account when the man retires at age 65?

56. Retirement Planning At the end of each quarter a 50-year-old woman puts $1200 in a retirement account that pays 7% interest compounded quarterly. When she reaches age 60, she withdraws the entire amount and places it in a mutual fund that pays 9% interest compounded monthly. From then on she deposits $300 in the mutual fund at the end of each month. How much is in the account when she reaches age 65?

57. Savings Jasspreet Kaur deposits $2435 at the beginning of each semiannual period for 8 yr in an account paying 6% compounded semiannually. She then leaves that money alone, with no further deposits, for an additional 5 yr. Find the final amount on deposit after the entire 13-yr period.

58. Savings Chuck Hickman deposits $10,000 at the beginning of each year for 12 years in an account paying 5% compounded annually. He then puts the total amount on deposit in another account paying 6% compounded semiannually for another 9 yr. Find the final amount on deposit after the entire 21-yr period.

59. Savings Ray Berkowitz needs $10,000 in 8 yr.

(a) What amount should he deposit at the end of each quarter at 8% compounded quarterly so that he will have his $10,000?

(b) Find Berkowitz's quarterly deposit if the money is deposited at 6% compounded quarterly.

60. Buying Equipment Harv, the owner of Harv's Meats, knows that he must buy a new deboner machine in 4 yr. The machine costs $12,000. In order to accumulate enough money to pay for the machine, Harv decides to deposit a sum of money at the end of each 6 mo in an account pay-

ing 6% compounded semiannually. How much should each payment be?

61. Buying a Car Barb Silverman wants to buy an $18,000 car in 6 yr. How much money must she deposit at the end of each quarter in an account paying 5% compounded quarterly so that she will have enough to pay for her car?

Individual Retirement Accounts *Suppose a 40-year-old person deposits* $2000 *per year in an Individual Retirement Account until age 65. Find the total in the account with the following assumptions of interest rates. (Assume semiannual compounding, with payments of* $1000 *made at the end of each semiannual period.)*

62. 6% **63.** 8% **64.** 4% **65.** 10%

Connections

66. In a recent state lottery, the jackpot was $27 million. An Australian investment firm tried to buy all possible combinations of numbers, which would have cost $7 million. (In fact, the firm ran out of time and was unable to buy all combinations.) Suppose the investment firm had accomplished its goal and held the only winning ticket in the lottery. The firm would receive the jackpot in 20 equal annual payments of $1.35 million. Assume these payments meet the conditions of an ordinary annuity.

(a) Suppose the firm can invest money at 8% interest compounded annually. How many years would it take until the investors would be further ahead than if they had simply invested the $7 million at the same rate? (*Hint:* Experiment with different values of *n*, the number of years.)

(b) How many years would it take in part (a) at an interest rate of 12%?

Computer/Graphing Calculator

67. Buying Real Estate Diane Gray sells some land in Nevada. She will be paid a lump sum of $60,000 in 7 yr. Until then, the buyer pays 8% simple interest quarterly.

(a) Find the amount of each quarterly interest payment.

(b) The buyer sets up a sinking fund so that enough money will be present to pay off the $60,000. The buyer wants to make semiannual payments into the sinking fund; the account pays 6% compounded semiannually. Find the amount of each payment into the fund.

(c) Prepare a table showing the amount in the sinking fund after each deposit.

68. Buying Rare Stamps Joe Seniw bought a rare stamp for his collection. He agreed to pay a lump sum of $4000 after 5 yr. Until then, he pays 6% simple interest semiannually.

(a) Find the amount of each semiannual interest payment.

(b) Seniw sets up a sinking fund so that enough money will be present to pay off the $4000. He wants to make annual payments into the fund. The account pays 8% compounded annually. Find the amount of each payment.

(c) Prepare a table showing the amount in the sinking fund after each deposit.

9.4 PRESENT VALUE OF AN ANNUITY; AMORTIZATION

What monthly payment will pay off a $10,000 car loan in 36 monthly payments at 12% annual interest?

The answer to this question is given later in this section. We shall see that it involves finding the present value of an annuity.

Suppose that at the end of each year, for the next 10 years, $500 is deposited in a savings account paying 7% interest compounded annually. This is an example of an ordinary annuity. The **present value** of this annuity is the amount that would have to be deposited in one lump sum today (at the same compound interest rate) in order to produce exactly the same balance at the end of 10 years. We can find a formula for the present value of an annuity as follows.

Suppose deposits of R dollars are made at the end of each period for n periods at interest rate i per period. Then the amount in the account after n periods is the future value of this annuity:

$$S = R \cdot s_{\overline{n}|i} = R\left[\frac{(1 + i)^n - 1}{i}\right].$$

On the other hand, if P dollars are deposited today at the same compound interest rate i, then at the end of n periods, the amount in the account is $P(1 + i)^n$. If P is the present value of the annuity, this amount must be the same as the amount S in the formula above; that is,

$$P(1 + i)^n = R\left[\frac{(1 + i)^n - 1}{i}\right].$$

To solve this equation for P, multiply both sides by $(1 + i)^{-n}$.

$$P = R(1 + i)^{-n}\left[\frac{(1 + i)^n - 1}{i}\right]$$

Recall that for any nonzero number a, $a^0 = 1$. Also, by the product rule for exponents, $a^x \cdot a^y = a^{x+y}$. In particular, for any nonzero number a, $a^n \cdot a^{-n} = a^{n+(-n)} = a^0 = 1$.

Use the distributive property; also recall that $(1 + i)^{-n}(1 + i)^n = 1$.

$$P = R\left[\frac{(1 + i)^{-n}(1 + i)^n - (1 + i)^{-n}}{i}\right] = R\left[\frac{1 - (1 + i)^{-n}}{i}\right]$$

The amount P is called the **present value of the annuity.** The quantity in brackets is abbreviated as $a_{\overline{n}|i}$, so

$$a_{\overline{n}|i} = \frac{1 - (1 + i)^{-n}}{i}.$$

(The symbol $a_{\overline{n}|i}$ is read "a-angle-n at i." Compare this quantity with $s_{\overline{n}|i}$ in the previous section.) The formula for the present value of an annuity is summarized below.

PRESENT VALUE OF AN ANNUITY

The present value P of an annuity of n payments of R dollars each at the end of consecutive interest periods with interest compounded at a rate of interest i per period is

$$P = R\left[\frac{1 - (1 + i)^{-n}}{i}\right] \qquad \text{or} \qquad P = Ra_{\overline{n}|i}.$$

Caution Don't confuse the formula for the present value of an annuity with the one for the future value of an annuity. Notice the difference: the numerator of the fraction in the present value formula is $1 - (1 + i)^{-n}$, but in the future value formula, it is $(1 + i)^n - 1$.

EXAMPLE 1 Mr. Bryer and Ms. Gonsalez are both graduates of the Brisbane Institute of Technology. They both agree to contribute to the endowment fund of BIT. Mr. Bryer says that he will give $500 at the end of each year for 9 years. Ms. Gonsalez prefers to give a lump sum today. What lump sum can she give that will equal the present value of Mr. Bryer's annual gifts, if the endowment fund earns 7.5% compounded annually?

Here $R = 500$, $n = 9$, and $i = .075$ and we have

$$P = R \cdot a_{\overline{9}|.075} = 500\left[\frac{1 - (1.075)^{-9}}{.075}\right]$$
$$\approx 500(6.37889) = 3189.44.$$

Therefore, Ms. Gonsalez must donate a lump sum of $3189.44 today. ◀

One of the most important uses of annuities is in determining the equal monthly payments needed to pay off a loan, as illustrated in the next example.

EXAMPLE 2 A car costs $12,000. After a down payment of $2000, the balance will be paid off in 36 monthly payments with interest of 12% per year on the unpaid balance. Find the amount of each payment.

A single lump sum payment of $10,000 today would pay off the loan. So, $10,000 is the present value of an annuity of 36 monthly payments with interest

of $12\%/12 = 1\%$ per month. Thus, $P = 10,000$, $n = 36$, $i = .01$, and we must find the monthly payment R in the formula

$$P = R\left[\frac{1 - (1 + i)^{-n}}{i}\right]$$

$$10,000 = R\left[\frac{1 - (1.01)^{-36}}{.01}\right] \approx R(30.10751)$$

$$R = \frac{10,000}{30.10751} = 332.14.$$

A monthly payment of \$332.14 will be needed. ◀

Amortization A loan is **amortized** if both the principal and interest are paid by a sequence of equal periodic payments. In Example 2 above, a loan of \$10,000 at 12% interest compounded monthly could be amortized by paying \$332.14 per month for 36 months.

The periodic payment needed to amortize a loan may be found, as in Example 2, by solving the present value equation for R.

AMORTIZATION PAYMENTS

A loan of P dollars at interest rate i per period may be amortized in n equal periodic payments of R dollars made at the end of each period, where

$$R = \frac{P}{a\,\overline{n}|i} = \frac{P}{\left[\dfrac{1 - (1 + i)^{-n}}{i}\right]} = \frac{Pi}{1 - (1 + i)^{-n}}.$$

EXAMPLE **3** The Perez family buys a house for \$94,000 with a down payment of \$16,000. They take out a 30-year mortgage for \$78,000 at an annual interest rate of 9.6%.

(a) Find the amount of the monthly payment needed to amortize this loan.

Here $P = 78,000$ and the monthly interest rate is $9.6\%/12 = .096/12 = .008$.* The number of monthly payments is $12 \cdot 30 = 360$. Therefore,

$$R = \frac{78,000}{a\,\overline{360}|.008} = \frac{78,000}{\left[\dfrac{1 - (1.008)^{-360}}{.008}\right]} \approx \frac{78,000}{117.90229} = 661.56.$$

Monthly payments of \$661.56 are required to amortize the loan.

*Mortgage rates are quoted in terms of annual interest, but it is always understood that the monthly rate is $1/12$ of the annual rate and that interest is compounded monthly.

(b) Find the total amount of interest paid when the loan is amortized over 30 years.

The Perez family makes 360 payments of $661.56 each, for a total of $238,161.60. Since the amount of the loan was $78,000, the total interest paid is

$$\$238,161.60 - \$78,000 = \$160,161.60.$$

This large amount of interest is typical of what happens with a long mortgage. A 15-year mortgage would have higher payments, but would involve significantly less interest.

(c) Find the part of the first payment that is interest and the part that is applied to reducing the debt.

As we saw in part (a), the monthly interest rate is .008. During the first month, the entire $78,000 is owed. Interest on this amount for 1 month is found by the formula for simple interest.

$$I = Prt = 78,000(.008)(1) = 624$$

At the end of the month, a payment of $661.56 is made; since $624 of this is interest, a total of

$$\$661.56 - \$624 = \$37.56$$

is applied to the reduction of the original debt. ◀

Amortization Schedules In the preceding example, 360 payments are made to amortize a $78,000 loan. The loan balance after the first payment is reduced by only $37.56, which is much less than $(1/360)(78,000) \approx \216.67. Therefore, even though equal *payments* are made to amortize a loan, the loan *balance* does not decrease in equal steps. This fact is very important if a loan is paid off early.

EXAMPLE **4**

Ellen Ryan borrows $1000 for 1 year at 12% annual interest compounded monthly. Verify that her monthly loan payment is $88.85. After making three payments, she decides to pay off the remaining balance all at once. How much must she pay?

Since nine payments remain to be paid, they can be thought of as an annuity consisting of nine payments of $88.85 at 1% interest per period. The present value of this annuity is

$$88.85 \left[\frac{1 - (1.01)^{-9}}{.01} \right] = 761.09.$$

So Ellen's remaining balance, computed by this method, is $761.09.

An alternative method of figuring the balance is to consider the payments already made as an annuity of three payments. At the beginning, the present value of this annuity was

$$88.85 \left[\frac{1 - (1.01)^{-3}}{.01} \right] = 261.31.$$

So she still owes the difference $1000 − $261.31 = $738.69. Furthermore, she owes the interest on this amount for 3 months, for a total of

$$(738.69)(1.01)^3 = \$761.07.$$

This balance due differs from the one obtained by the first method by 2 cents because the monthly payment and the other calculations were rounded to the nearest penny. ◄

Although most people would not quibble about a difference of 2 cents in the balance due in Example 4, the difference in other cases (larger amounts or longer terms) might be more than that. A bank or business must keep its books accurately to the nearest penny, so it must determine the balance due in such cases unambiguously and exactly. This is done by means of an **amortization schedule,** which lists how much of each payment is interest and how much goes to reduce the balance, as well as how much is owed after *each* payment.

EXAMPLE 5 Determine the exact amount Ellen Ryan in Example 4 owes after three monthly payments.

An amortization table for the loan is shown below. It is obtained as follows. The annual interest rate is 12% compounded monthly, so the interest rate per month is 12%/12 = 1% = .01. When the first payment is made, 1 month's interest—namely, .01(1000) = $10—is owed. Subtracting this from the $88.85 payment leaves $78.85 to be applied to repayment. Hence, the principal at the end of the first payment period is 1000 − 78.85 = $921.15, as shown in the "payment 1" line of the chart.

When payment 2 is made, 1 month's interest on $921.15 is owed, namely .01(921.15) = $9.21. Subtracting this from the $88.85 payment leaves $79.64 to reduce the principal. Hence, the principal at the end of payment 2 is 921.15 − 79.64 = $841.51. The interest portion of payment 3 is based on this amount, and the remaining lines of the table are found in a similar fashion.

Payment Number	Amount of Payment	Interest for Period	Portion to Principal	Principal at End of Period
0	—	—	—	$1000.00
1	$88.85	$10.00	$78.85	$921.15
2	$88.85	$9.21	$79.64	$841.51
3	$88.85	$8.42	$80.43	$761.08
4	$88.85	$7.61	$81.24	$679.84
5	$88.85	$6.80	$82.05	$597.79
6	$88.85	$5.98	$82.87	$514.92
7	$88.85	$5.15	$83.70	$431.22
8	$88.85	$4.31	$84.54	$346.68
9	$88.85	$3.47	$85.38	$261.30
10	$88.85	$2.61	$86.24	$175.06
11	$88.85	$1.75	$87.10	$87.96
12	$88.84	$.88	$87.96	0

The schedule shows that after three payments, she still owes $761.08, an amount that differs slightly from that obtained by either method in Example 6. ◀

The amortization schedule in Example 5 is typical. In particular, note that all payments are the same except the last one. It is often necessary to adjust the amount of the final payment to account for rounding off earlier, and to ensure that the final balance is exactly 0.

An amortization schedule also shows how the periodic payments are applied to interest and principal. The amount going to interest decreases with each payment, while the amount going to reduce the principal owed increases with each payment.

9.4 Exercises

1. Which of the following is represented by $a_{\overline{n}|i}$?

(a) $\dfrac{(1 + i)^{-n} - 1}{i}$ **(b)** $\dfrac{(1 + i)^n - 1}{i}$ **(c)** $\dfrac{1 - (1 + i)^{-n}}{i}$ **(d)** $\dfrac{1 - (1 + i)^n}{i}$

2. Which of the choices in Exercise 1 represents $s_{\overline{n}|i}$?

Find each of the following.

3. $a_{\overline{15}|.06}$

4. $a_{\overline{10}|.03}$

5. $a_{\overline{18}|.04}$

6. $a_{\overline{30}|.01}$

7. $a_{\overline{16}|.01}$

8. $a_{\overline{32}|.02}$

9. Explain the difference between the present value of an annuity and the future value of an annuity. For a given annuity, which is larger? Why?

Find the present value of each ordinary annuity.

10. Payments of $890 each year for 16 yr at 8% compounded annually

11. Payments of $1400 each year for 8 yr at 8% compounded annually

12. Payments of $10,000 semiannually for 15 yr at 10% compounded semiannually

13. Payments of $50,000 quarterly for 10 yr at 8% compounded quarterly

14. Payments of $15,806 quarterly for 3 yr at 10.8% compounded quarterly

15. Payments of $18,579 every six months for 8 yr at 9.4% compounded semiannually

Find the lump sum deposited today that will yield the same total amount as payments of $10,000 at the end of each year for 15 yr, at each of the given interest rates.

16. 4% compounded annually **17.** 5% compounded annually **18.** 6% compounded annually

19. What does it mean to amortize a loan?

Find the payment necessary to amortize each of the following loans.

20. $2500, 8% compounded quarterly, 6 quarterly payments

21. $41,000, 10% compounded semiannually, 10 semiannual payments

22. $90,000, 8% compounded annually, 12 annual payments

23. $140,000, 12% compounded quarterly, 15 quarterly payments

24. $7400, 8% compounded semiannually, 18 semiannual payments

25. $5500, 12% compounded monthly, 24 monthly payments

Use the amortization table in Example 5 to answer the questions in Exercises 26–29.

26. How much of the fourth payment is interest?

27. How much of the eleventh payment is used to reduce the debt?

28. How much interest is paid in the first 4 months of the loan?

29. How much interest is paid in the last 4 months of the loan?

30. What sum deposited today at 5% compounded annually for 8 yr will provide the same amount as $1000 deposited at the end of each year for 8 yr at 6% compounded annually?

31. What lump sum deposited today at 8% compounded quarterly for 10 yr will yield the same final amount as deposits of $4000 at the end of each six-month period for 10 yr at 6% compounded semiannually?

Find the monthly house payments necessary to amortize the following loans.

32. $149,560 at 7.75% for 25 yr

33. $170,892 at 8.11% for 30 yr

34. $153,762 at 8.45% for 30 yr

35. $96,511 at 9.57% for 25 yr

Applications

Business and Economics

36. **Installment Buying** Stereo Shack sells a stereo system for $600 down and monthly payments of $30 for the next 3 yr. If the interest rate is 1.25% per month on the unpaid balance, find

 (a) the cost of the stereo system;

 (b) the total amount of interest paid.

37. **Car Payments** Hong Le buys a car costing $6000. He agrees to make payments at the end of each monthly period for 4 years. He pays 12% interest, compounded monthly.

 (a) What is the amount of each payment?

 (b) Find the total amount of interest Le will pay.

38. **Land Purchase** A speculator agrees to pay $15,000 for a parcel of land; this amount, with interest, will be paid over 4 yr, with semiannual payments, at an interest rate of 10% compounded semiannually. Find the amount of each payment.

39. **Lottery Winnings** In the Million-Dollar Lottery, a winner is paid a million dollars at the rate of $50,000 per year for 20 yr. Assume that these payments form an ordinary annuity, and that the lottery managers can invest money at 6% compounded annually. Find the lump sum that the management must put away to pay off the million-dollar winner.

Installment Buying *In Exercises 40–43, prepare an amortization schedule showing the first 4 payments for each loan.*

40. An insurance firm pays $4000 for a new printer for its computer. It amortizes the loan for the printer in 4 annual payments at 8% compounded annually.

41. Large semitrailer trucks cost $72,000 each. Ace Trucking buys such a truck and agrees to pay for it by a loan that will be amortized with 9 semiannual payments at 10% compounded semiannually.

42. One retailer changes $1048 for a computer monitor. A firm of tax accountants buys 8 of these monitors. They make a down payment of $1200 and agree to amortize the balance with monthly payments at 12% compounded monthly for 4 yr.

43. When Teresa Flores opened her law office, she bought $14,000 worth of law books and $7200 worth of office furniture. She paid $1200 down and agreed to amortize the balance with semiannual payments for 5 yr, at 12% compounded semiannually.

44. **House Payments** Don Cole buys a house for $285,000. He pays $60,000 down and takes out a mortgage at 9.5% on the balance. Find his monthly payment and the total amount of interest he will pay if the length of the mortgage is

 (a) 15 yr; (b) 20 yr; (c) 25 yr.

45. **Inheritance** Mary Yaroslavsky has died, leaving $25,000 to her husband. He deposits the money at 6% compounded annually. He wants to make annual withdrawals from the account so that the money (principal and interest) is gone in exactly 8 yr.

 (a) Find the amount of each withdrawal.

 (b) Find the amount of each withdrawal if the money must last 12 yr.

46. Charitable Trust The trustees of a college have accepted a gift of $150,000. The donor has directed the trustees to deposit the money in an account paying 6% per year, compounded semiannually. The trustees may withdraw money at the end of each six-month period; the money must last 5 yr.

(a) Find the amount of each withdrawal.

(b) Find the amount of each withdrawal if the money must last 6 yr.

 Computer/Graphing Calculator

Amortization *Prepare an amortization schedule for each of the following loans.*

47. A loan of $37,947.50 with interest at 8.5% compounded annually, to be paid with equal annual payments over 10 years

48. A loan of $4835.80 at 9.25% interest compounded semiannually, to be repaid in 5 years in equal semiannual payments

► FINANCIAL FORMULAS SUMMARY

We have presented a lot of new formulas in this chapter. By answering the following questions, you can decide which formula to use for a particular problem.

1. Is simple or compound interest involved?

Simple interest is normally used for investments or loans of a year or less; compound interest is normally used in all other cases.

2. If simple interest is being used, what is being sought: interest amount, future value, present value, or discount?

3. If compound interest is being used, does it involve a lump sum (single payment) or an annuity (sequence of payments)?

(a) For a lump sum, what is being sought: present value, future value, number of periods at interest, or effective rate?

(b) For an annuity,

 (i) Is it an ordinary annuity (payment at the end of each period) or an annuity due (payment at the beginning of each period)?

 (ii) What is being sought: present value, future value, or payment amount?

Once you have answered these questions, choose the appropriate formula and work the problem. As a final step, consider whether the answer you get makes sense. For instance, present value should always be less than future value. The amount of interest or the payments in an annuity should be fairly small compared to the total future value.

List of Variables

r is the annual interest rate.

i is the interest rate per period.

t is the number of years.

n is the number of periods.

m is the number of periods per year.

P is the principal or present value.

A is the future value of a lump sum.

S is the future value of an annuity.

R is the periodic payment in an annuity.

	Simple Interest	*Compound Interest*
Interest	$I = Prt$	$I = A - P$
Future Value	$A = P(1 + rt)$	$A = P(1 + i)^n$
Present Value	$P = \dfrac{A}{1 + rt}$	$P = \dfrac{A}{(1 + i)^n} = A(1 + i)^{-n}$
Discount	$D = Art$	Effective $\quad r_e = \left(1 + \dfrac{r}{m}\right)^m - 1$
Proceeds	$P = A - D$	Rate
	$= A(1 - rt)$	

Ordinary Annuity

Future Value $\qquad S = R\left[\dfrac{(1 + i)^n - 1}{i}\right] = R \cdot s_{\overline{n}|i}$

Present Value $\qquad P = R\left[\dfrac{1 - (1 + i)^{-n}}{i}\right] = R \cdot a_{\overline{n}|i}$

Annuity Due

Future Value $\qquad S = R\left[\dfrac{(1 + i)^{n+1} - 1}{i}\right] - R$

Chapter Summary Key Terms

9.1 simple interest
principal
rate
time
future value
maturity value
present value
(bank) discount
proceeds

9.2 compound interest
compound amount
nominal (stated) rate
effective rate
9.3 geometric sequence
terms
common ratio
annuity
ordinary annuity
future value of an annuity
annuity due
sinking fund

9.4 present value of an annuity
amortize a loan
amortization schedule

Chapter 9 Review Exercises

Find the simple interest for the following loans.

1. $15,903 at 8% for 8 mo

2. $4902 at 9.5% for 11 mo

3. $42,368 at 5.22% for 5 mo

4. $3478 at 7.4% for 88 days
(assume a 360-day year)

5. What is meant by the present value of an amount A?

Find the present value of each of the future amounts in Exercises 6 and 7. Assume 360 *days in a year; use simple interest.*

6. $25,000 for 10 mo; money earns 5%

7. $459.57 for 7 mo; money earns 4.5%

8. Explain what happens when a borrower is charged a discount. What are the proceeds?

Find the proceeds in Exercises 9 and 10. Assume 360 *days in a year.*

9. $802.34; discount rate 8.6%; length of loan 11 mo

10. $12,000; discount rate 7.09%; length of loan 145 days

11. For a given amount of money at a given interest rate for a given time period greater than 1 year, does simple interest or compound interest produce more interest?

Find the compound amount in each of the following.

12. $2800 at 6% compounded annually for 10 yr

13. $19,456.11 at 12% compounded semiannually for 7 yr

14. $312.45 at 6% compounded semiannually for 16 yr

15. $57,809.34 at 12% compounded quarterly for 5 yr

Find the amount of interest earned by each deposit.

16. $3954 at 8% compounded annually for 12 yr

17. $12,699.36 at 10% compounded semiannually for 7 yr

18. $12,903.45 at 10.37% compounded quarterly for 29 quarters

19. $34,677.23 at 9.72% compounded monthly for 32 mo

Find the present value of the following amounts.

20. $42,000 in 7 yr, 12% compounded monthly

21. $17,650 in 4 yr, 8% compounded quarterly

22. $1347.89 in 3.5 yr, 6.77% compounded semiannually

23. $2388.90 in 44 mo, 5.93% compounded monthly

24. Write the first five terms of the geometric sequence with $a = 2$ and $r = 3$.

25. Write the first four terms of the geometric sequence with $a = 4$ and $r = 1/2$.

26. Find the sixth term of the geometric sequence with $a = -3$ and $r = 2$.

27. Find the fifth term of the geometric sequence with $a = -2$ and $r = -2$.

28. Find the sum of the first four terms of the geometric sequence with $a = -3$ and $r = 3$.

29. Find the sum of the first five terms of the geometric sequence with $a = 8000$ and $r = -1/2$.

30. Find $s_{\overline{30}|.01}$.

31. Find $s_{\overline{20}|.05}$.

32. What is meant by the future value of an annuity?

Find the future value of each annuity.

33. $500 deposited at the end of each six-month period for 8 yr; money earns 6% compounded semiannually

34. $1288 deposited at the end of each year for 14 yr; money earns 8% compounded annually

35. $4000 deposited at the end of each quarter for 7 yr; money earns 6% compounded quarterly

36. $233 deposited at the end of each month for 4 yr; money earns 12% compounded monthly

37. $672 deposited at the beginning of each quarter for 7 yr; money earns 8% compounded quarterly

38. $11,900 deposited at the beginning of each month for 13 mo; money earns 12% compounded monthly

39. What is the purpose of a sinking fund?

Find the amount of each payment that must be made into a sinking fund to accumulate the following amounts.

40. $6500; money earns 8% compounded annually; 6 annual payments

41. $57,000; money earns 6% compounded semiannually for $8\frac{1}{2}$ yr

42. $233,188; money earns 9.7% compounded quarterly for $7\frac{3}{4}$ yr

43. $1,056,788; money earns 8.12% compounded monthly for $4\frac{1}{2}$ yr

Find the present value of each ordinary annuity.

44. Payments of $850 annually for 4 yr at 8% compounded annually

45. Payments of $1500 quarterly for 7 yr at 8% compounded quarterly

46. Payments of $4210 semiannually for 8 yr at 8.6% compounded semiannually

47. Payments of $877.34 monthly for 17 mo at 9.4% compounded monthly

48. Give two examples of the types of loans that are commonly amortized.

Find the amount of the payment necessary to amortize each of the following loans.

49. $80,000 loan; 8% compounded annually; 9 annual payments

50. $3200 loan; 8% compounded quarterly; 10 quarterly payments

51. $32,000 loan; 9.4% compounded quarterly; 17 quarterly payments

52. $51,607 loan; 13.6% compounded monthly; 32 monthly payments

Find the monthly house payments for the following mortgages.

53. $56,890 at 10.74% for 25 yr **54.** $77,110 at 11.45% for 30 yr

A portion of an amortization table is given below for a $127,000 loan at 8.5% interest compounded monthly for 25 yr.

Payment Number	Amount of Payment	Interest for Period	Portion to Principal	Principal at End of Period
1	$1022.64	$899.58	$123.06	$126,876.94
2	$1022.64	$898.71	$123.93	$126,753.02
3	$1022.64	$897.83	$124.80	$126,628.21
4	$1022.64	$896.95	$125.69	$126,502.53
5	$1022.64	$896.06	$126.58	$126,375.95
6	$1022.64	$895.16	$127.48	$126,248.47
7	$1022.64	$894.26	$128.38	$126,120.09
8	$1022.64	$893.35	$129.29	$125,990.80
9	$1022.64	$892.43	$130.20	$125,860.60
10	$1022.64	$891.51	$131.13	$125,729.48
11	$1022.64	$890.58	$132.05	$125,597.42
12	$1022.64	$889.65	$132.99	$125,464.43

Use the table to answer the following questions.

55. How much of the fifth payment is interest?

56. How much of the twelfth payment is used to reduce the debt?

57. How much interest is paid in the first 3 months of the loan?

58. How much has the debt been reduced at the end of the first year?

Applications

Business and Economics

59. Personal Finance Tom Wilson owes $5800 to his mother. He has agreed to pay back the money in 10 mo, at an interest rate of 10%. Three months before the loan is due, Tom's mother discounts the loan at the bank to get cash for $6000 worth of new furniture. The bank charges a 13.45% discount rate. How much money does Tom's mother receive? Is it enough to buy the furniture?

60. Business Financing Larry DiCenso needs $9812 to buy new equipment for his business. The bank charges a discount rate of 12%. Find the amount of DiCenso's loan if he borrows the money for 7 mo.

61. Business Financing An accountant loans $28,000 at simple interest to her business. The loan is at 11.5% and earns $3255 interest. Find the time of the loan in months.

62. Business Investment A developer deposits $84,720 for 7 mo and earns $4055.46 in interest. Find the interest rate.

63. Personal Finance In 3 yr Ms. Flores must pay a pledge of $7500 to her college's building fund. What lump sum can she deposit today, at 10% compounded semiannually, so that she will have enough to pay the pledge?

64. Personal Finance According to the terms of a divorce settlement, one spouse must pay the other a lump sum of $2800 in 17 mo. What lump sum can be invested today, at 6% compounded monthly, so that enough will be available for the payment?

65. Business Investment A firm of attorneys deposits $5000 of profit-sharing money at the end of each semiannual period for $7\frac{1}{2}$ yr. Find the final amount in the account if the deposits earn 10% compounded semiannually. Find the amount of interest earned.

66. Business Financing A small resort must add a swimming pool to compete with a new resort built nearby. The pool will cost $28,000. The resort borrows the money and agrees to repay it with equal payments at the end of each quarter for $6\frac{1}{2}$ yr at an interest rate of 12% compounded quarterly. Find the amount of each payment.

67. Business Financing The owner of Eastside Hallmark borrows $48,000 to expand the business. The money will be repaid in equal payments at the end of each year for 7 yr. Interest is 10%. Find the amount of each payment.

68. Personal Finance To buy a new computer, Mark Nguyen borrows $3250 from a friend at 9% interest compounded annually for 4 yr. Find the compound amount he must pay back at the end of the 4 yr.

69. Home Financing When the Lee family bought their home, they borrowed $115,700 at 10.5% compounded monthly for 25 yr. If they make all 300 payments, repaying the loan on schedule, how much interest will they pay? (Assume the last payment is the same as the previous ones.)

Connections

70. Buying and Selling a House The Taggart family bought a house for $91,000. They paid $20,000 down and took out a 30-yr mortgage for the balance at 9%.

(a) Find their monthly payment.

(b) How much of the first payment is interest?

After 180 payments, the family sells their house for $136,000. They must pay closing costs of $3700 plus 2.5% of the sale price.

(c) Estimate the current mortgage balance at the time of the sale using one of the methods form Example 4 in Section 4.

(d) Find the total closing costs.

(e) Find the amount of money they receive from the sale after paying off the mortgage.

*The following problem is from an actuarial examination.**

71. Death Benefit The proceeds of a $10,000 death benefit are left on deposit with an insurance company for 7 yr at an annual effective interest rate of 5%. The balance at the end of 7 yr is paid to the beneficiary in 120 equal monthly payments of X, with the first payment made immediately. During the payout period, interest is credited at an annual effective interest rate of 3%. Calculate X.

(a) 117 (b) 118 (c) 129 (d) 135 (e) 158

*Problem 16 from ''Course 140 Examination, Mathematics of Compound Interest'' of the Education and Examination Committee of The Society of Actuaries. Reprinted by permission of The Society of Actuaries.

Extended Application / Present Value*

The Southern Pacific Railroad, with lines running from Oregon to Louisiana, is one of the country's most profitable railroads. The railroad has vast landholdings (granted by the government in the last half of the nineteenth century) and is diversified into trucking, pipelines, and data transmission.

The railroad was recently faced with a decision on the fate of an old bridge that crosses the Kings River in central California. The bridge is on a minor line that carries very little traffic. Just north of the Southern Pacific bridge is a bridge owned by the Santa Fe Railroad; it too carries little traffic. The Southern Pacific had two alternatives: it could replace its own bridge or it could negotiate with the Santa Fe for the rights to run trains over its bridge. In the second alternative, a yearly fee would be paid to the Santa Fe, and new connecting tracks would be built. The situation is shown in Figure 4.

To find the better of these two alternatives, the railroad used the following approach.

1. Calculate estimated expenses for each alternative.
2. Calculate annual cash flows in after-tax dollars. At a 48% corporate tax rate, $1 of expenses actually costs the railroad $.52, and $1 of revenue can bring a maximum of $.52 in profit. Cash flow for a given year is found by the following formula.

 Cash flow $= -.52$ (operating and maintenance expenses)
 $+ .52$ (savings and revenue)
 $+ .48$ (depreciation)

3. Calculate the net present values of all cash flows for future years. The formula used is

 Net present value $= \Sigma$ (cash flow in year i)$(1 + k)^{1-i}$,

 where i is a particular year in the life of the project and k is the assumed annual rate of interest. (Recall that Σ indicates a sum.) The net present value is found for interest rates from 0% to 20%.

4. The interest rate that leads to a net present value of $0 is called the **rate of return** on the project.

Let us now see how these steps worked out in practice.

Alternative 1: operate over the Santa Fe bridge. First, estimated expenses were calculated.

1976	Work done by Southern Pacific on Santa Fe track	$27,000
1976	Work by Southern Pacific on its own track	11,600
1976	Undepreciated portion of cost of current bridge	97,410
1976	Salvage value of bridge	12,690
1977	Annual maintenance of Santa Fe track	16,717
1977	Annual rental fee to Santa Fe	7,382

From these figures and others not given here, annual cash flows and net present values were calculated. The following table was then prepared.

FIGURE 4

Interest Rate, %	Net Present Value, $
0	85,731
4	67,566
8	53,332
12	42,008
16	32,875
20	25,414

Interest Rate, %	Net Present Value, $
0	399,577
4	96,784
7.8	0
8	−3,151
12	−43,688
16	−62,615
20	−72,126

Although the table does not show a net present value of $0, the interest rate that leads to that value is 44%. This is the rate of return for this alternative.

Alternative 2: build a new bridge. Again, estimated expenses were calculated.

1976	Annual maintenance	$2,870
1976	Annual inspection	120
1976	Repair trestle	17,920
1977	Install bridge	189,943
1977	Install walks and handrails	15,060
1978	Repaint steel (every 10 years)	10,000
1978	Repair trusses (every 10 years, increases by $200)	2,000
1981	Replace ties	31,000
1988	Repair concrete (every 10 years)	400
2021	Replace ties	31,000

After cash flows and net present values were calculated, the following table was prepared.

In this alternative the net present value is $0 at 7.8%, the rate of return.

Based on this analysis, the first alternative, renting from the Sante Fe, is clearly preferable.

Exercises

Find the cash flow in each of the following years. Use the formula given above.

1. Alternative 1, year 1977, operating expenses $6228, maintenance expenses $2976, savings $26,251, depreciation $10,778

2. Alternative 1, year 1984, same as Exercise 1, but depreciation is only $1347

3. Alternative 2, year 1976, maintenance $2870, operating expenses $6386, savings $26,251, no depreciation

4. Alternative 2, year 1980, operating expenses $6228, maintenance expenses $2976, savings $10,618, depreciation $6736

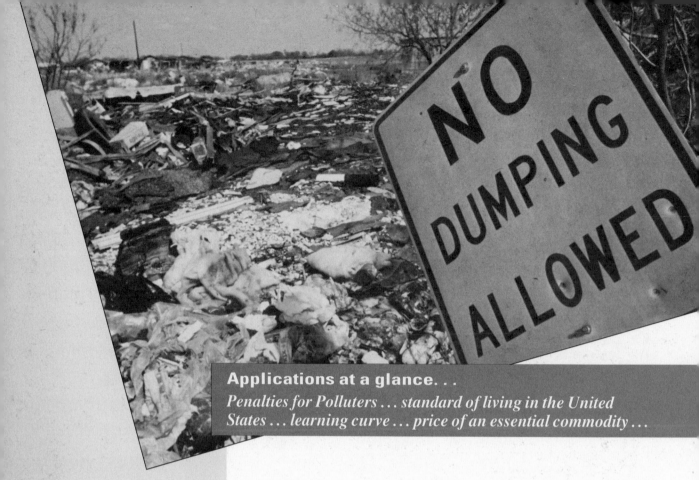

Applications at a glance...
Penalties for Polluters ... standard of living in the United States ... learning curve ... price of an essential commodity ...

C H A P T E R 1 0

THE DERIVATIVE

(See page 466.)

In Chapter 1 we were concerned with static situations. We saw how to answer questions like the following:

How much would 100 items cost?

In how many years will the population reach 10,000?

How wide is the arch 10 meters from the ground?

With calculus, we will be able to solve problems that involve changing situations.

How fast is the car moving after 20 seconds?

At what rate is the population growing?

What is the rate of change of profit when sales reach $1000?

Solving each of these problems requires a *derivative*. Since the definition of the derivative involves the idea of a *limit*, we begin with limits.

10.1 LIMITS AND CONTINUITY

Evaluating function notation, such as $f(1) = 3$, was discussed in Section 1.1. Verify that for

$$f(x) = \frac{x^2 - 4}{x - 2},$$

$f(0) = 2$, $f(3) = 5$, and $f(-1) = 1$.

We can find the value of the function defined by

$$f(x) = \frac{x^2 - 4}{x - 2}$$

when $x = 1$ by substitution:

$$f(1) = \frac{1^2 - 4}{1 - 2} = 3.$$

For values of x close to 1, the values of $f(x)$ are close to $f(1) = 3$, as the following table shows.

	x approaches 1 from the left.			↓	*x* approaches 1 from the right.						
x	.8	.9	.99	.9999	1	1.0000001	1.0001	1.001	1.01	1.05	1.1
$f(x)$	2.8	2.9	2.99	2.9999	3	3.0000001	3.0001	3.001	3.01	3.05	3.1
			f(x) approaches 3.		↑	*f(x)* approaches 3.					

At $x = 2$, however, $f(2)$ is not defined because the denominator is zero when $x = 2$. For values of x close to (but not equal to) 2, $f(x)$ *is* defined. The following table shows some values of $f(x)$ for x-values closer and closer to 2.

	x approaches 2 from the left.			↓	*x* approaches 2 from the right.					
x	1.8	1.9	1.99	1.9999	2	2.0000001	2.00001	2.001	2.05	2.1
$f(x)$	3.8	3.9	3.99	3.9999	4	4.0000001	4.00001	4.001	4.05	4.1
			f(x) approaches 4.		↑	*f(x)* approaches 4.				

The table suggests that, as x gets closer and closer to 2 from either side, $f(x)$ gets closer and closer to 4. In fact, using a calculator, we can make $f(x)$ as close to 4 as we wish by choosing a value of x "close enough" to 2. In such a case, we say "the limit of $f(x)$ as x approaches 2 is 4," which is written as

$$\lim_{x \to 2} f(x) = 4.$$

In the first example, we found that

$$\lim_{x \to 1} f(x) = 3,$$

because the values of $f(x)$ got closer and closer to 3 as x got closer and closer to 1, *from either side* of 1.

The phrase "x approaches 1 from the left" is written $x \to 1^-$. Similarly, "x approaches 1 from the right" is written $x \to 1^+$. These expressions are used to write **one-sided limits.** The **limit from the left** is written

$$\lim_{x \to 1^-} f(x) = 3,$$

and the **limit from the right** is written

$$\lim_{x \to 1^+} f(x) = 3.$$

A **two-sided limit,** such as

$$\lim_{x \to 1} f(x) = 3,$$

exists only if both one-sided limits are the same; that is, if $f(x)$ approaches the same number as x approaches a given number from *either* side.

The examples suggest the following informal definition.

LIMIT OF A FUNCTION

Let f be a function and let a and L be real numbers. If

1. as x takes values closer and closer (but not equal) to a on both sides of a, the corresponding values of $f(x)$ get closer and closer (and perhaps equal) to L; and

2. the value of $f(x)$ can be made as close to L as desired by taking values of x close enough to a;

then L is the **limit** of $f(x)$ as x approaches a, written

$$\lim_{x \to a} f(x) = L.$$

This definition is informal because the expressions "closer and closer to" and "as close as desired" have not been defined. A more formal definition would be needed to prove the rules for limits given later in this section.

Note The definition of a limit describes what happens to $f(x)$ when x is near a. It is not affected by whether or not $f(a)$ is defined. Also, the definition implies

that the function values cannot approach two different numbers, so that if a limit exists, it is unique.

Find $\lim\limits_{x \to 2} g(x)$, where $g(x) = \dfrac{x^2 + 4}{x - 2}$.

Make a table of values.

			x approaches 2 from the left.				x approaches 2 from the right.	
x	1.8	1.9	1.99	1.999	2	2.001	2.01	2.05
$g(x)$	-36.2	-76.1	-796	-7996		8004	804	164

<div align="center">Limit does
not exist.</div>

g(x) gets smaller and smaller. g(x) gets larger and larger.

$x = 2$

$g(x) = \dfrac{x^2 + 4}{x - 2}$

FIGURE 1

Both the table above and the graph in Figure 1 suggest that as $x \to 2$ from the left, $g(x)$ gets more and more negative, hence, smaller and smaller. This is indicated by writing

$$\lim_{x \to 2^-} g(x) = -\infty.$$

The symbol $-\infty$ does not represent a real number; it simply indicates that as $x \to 2^-$, $g(x)$ gets smaller without bound. In the same way, the behavior of the function as $x \to 2$ from the right is indicated by writing

$$\lim_{x \to 2^+} g(x) = \infty.$$

Since there is no real number that $g(x)$ approaches as $x \to 2$ (from either side),

$$\lim_{x \to 2} \frac{x^2 + 4}{x - 2} \text{ does not exist.} \quad \blacktriangleleft$$

Find $\lim\limits_{x \to 0} \dfrac{|x|}{x}$.

The function $f(x) = |x|/x$ is not defined when $x = 0$. When $x > 0$, the definition of absolute value shows that $f(x) = |x|/x = x/x = 1$. When $x < 0$, then $|x| = -x$ and $f(x) = -x/x = -1$. The graph of f is shown in Figure 2.

As x approaches 0 from the right, x is always positive and the corresponding value of $f(x)$ is 1, so

$$\lim_{x \to 0^+} f(x) = \textbf{1}.$$

But as x approaches 0 from the left, x is always negative and the corresponding value of $f(x)$ is -1, so

$$\lim_{x \to 0^-} f(x) = -\textbf{1}.$$

$f(x)$

FIGURE 2

Thus, as x approaches 0 from either side, the corresponding values of $f(x)$ do not get closer and closer to a *single* real number. Therefore, the limit of $|x|/x$ as x approaches 0 does not exist. ◀

The discussion up to this point can be summarized as follows.

EXISTENCE OF LIMITS

The limit of f as x approaches a may not exist.

1. If $f(x)$ becomes infinitely large or infinitely small as x approaches the number a from either side, we write $\lim\limits_{x \to a} f(x) = \infty$ or $\lim\limits_{x \to a} f(x) = -\infty$. In either case, the limit does not exist.

2. If $\lim\limits_{x \to a^-} f(x) = L$ and $\lim\limits_{x \to a^+} f(x) = M$, and $L \ne M$, then $\lim\limits_{x \to a} f(x)$ does not exist.

Figure 3 illustrates the facts listed above.

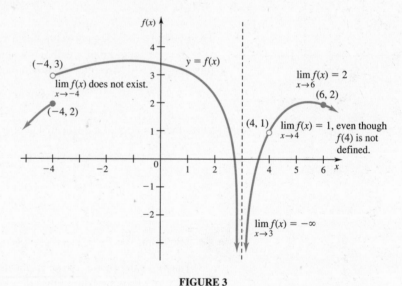

FIGURE 3

Rules for Limits As shown by the examples above, tables and graphs can be used to find limits. However, it is more efficient to use the rules for limits given on the next page. (Proofs of these rules require a formal definition of limit, which we have not given.)

RULES FOR LIMITS

Let a, k, n, A, and B be real numbers, and let f and g be functions such that

$$\lim_{x \to a} f(x) = A \qquad \text{and} \qquad \lim_{x \to a} g(x) = B.$$

1. If k is a constant, then $\lim_{x \to a} k = k$ and $\lim_{x \to a} k \cdot f(x) = k \cdot \lim_{x \to a} f(x)$.

2. $\lim_{x \to a} [f(x) \pm g(x)] = \lim_{x \to a} f(x) \pm \lim_{x \to a} g(x) = A \pm B$

 (The limit of a sum or difference is the sum or difference of the limits.)

3. If $p(x)$ is a polynomial, then $\lim_{x \to a} p(x) = p(a)$.

4. $\lim_{x \to a} [f(x) \cdot g(x)] = [\lim_{x \to a} f(x)] \cdot [\lim_{x \to a} g(x)] = A \cdot B$

 (The limit of a product is the product of the limits.)

5. $\lim_{x \to a} \dfrac{f(x)}{g(x)} = \dfrac{\lim_{x \to a} f(x)}{\lim_{x \to a} g(x)} = \dfrac{A}{B}$ if $B \neq 0$

 (The limit of a quotient is the quotient of the limits, provided the limit of the denominator is not zero.)

6. For any real number n, $\lim_{x \to a} [f(x)]^n = [\lim_{x \to a} f(x)]^n = A^n$.

7. $\lim_{x \to a} f(x) = \lim_{x \to a} g(x)$ if $f(x) = g(x)$ for all $x \neq a$.

This list may seem imposing, but most limit problems have solutions that agree with your common sense. Algebraic techniques and tables of values can also help resolve questions about limits, as some of the following examples will show.

EXAMPLE 3

Find each limit.

(a) $\lim_{x \to 3} [(x^2 + 2x) + (3x^3 - 5x + 1)]$

$$\lim_{x \to 3} [(x^2 + 2x) + (3x^3 - 5x + 1)]$$

$$= \lim_{x \to 3} (x^2 + 2x) + \lim_{x \to 3} (3x^3 - 5x + 1) \qquad \text{Rule 2}$$

$$= [3^2 + 2(3)] + [3(3^3) - 5(3) + 1] \qquad \text{Rule 3}$$

$$= 15 + 67$$

$$= 82$$

(b) $\lim\limits_{x \to -2} (x^2 - 5)(3x + 1)$

$$\lim\limits_{x \to -2} (x^2 - 5)(3x + 1)$$

$$= [\lim\limits_{x \to -2} (x^2 - 5)][\lim\limits_{x \to -2} (3x + 1)] \quad \text{Rule 4}$$

$$= [(-2)^2 - 5][3(-2) + 1] \quad \text{Rule 3}$$

$$= (-1)(-5)$$

$$= 5$$

(c) $\lim\limits_{x \to -1} 5(3x^2 + 2)$

$$\lim\limits_{x \to -1} 5(3x^2 + 2) = 5 \lim\limits_{x \to -1} (3x^2 + 2) \quad \text{Rule 1}$$

$$= 5 [3(-1)^2 + 2] \quad \text{Rule 3}$$

$$= 25$$

(d) $\lim\limits_{x \to 4} \dfrac{x}{x + 2}$

$$\lim\limits_{x \to 4} \frac{x}{x + 2} = \frac{\lim\limits_{x \to 4} x}{\lim\limits_{x \to 4} (x + 2)} \quad \text{Rule 5}$$

$$= \frac{4}{4 + 2} = \frac{2}{3}$$

(e) $\lim\limits_{x \to 9} \sqrt{4x - 11}$

As $x \to 9$, the expression $4x - 11$ approaches $4 \cdot 9 - 11 = 25$. Using Rule 6, with $n = 1/2$, gives

$$\lim\limits_{x \to 9} (4x - 11)^{1/2} = [\lim\limits_{x \to 9} (4x - 11)]^{1/2} = \sqrt{25} = 5. \quad \blacktriangleleft$$

As Example 3 suggests, the rules for limits actually mean that many limits can be found simply by evaluation. The next examples illustrate some exceptions.

 EXAMPLE 4 Find $\lim\limits_{x \to 2} \dfrac{x^2 + x - 6}{x - 2}$.

Rule 5 cannot be used here, since

$$\lim\limits_{x \to 2} (x - 2) = 0.$$

For $x \neq 2$, we can, however, simplify the function by rewriting the fraction as

$$\frac{x^2 + x - 6}{x - 2} = \frac{(x + 3)(x - 2)}{x - 2} = x + 3.$$

Now Rule 7 can be used.

$$\lim_{x \to 2} \frac{x^2 + x - 6}{x - 2} = \lim_{x \to 2} (x + 3) = 2 + 3 = 5 \quad \blacktriangleleft$$

EXAMPLE 5

Find $\lim\limits_{x \to 4} \dfrac{\sqrt{x} - 2}{x - 4}$.

As $x \to 4$, the numerator approaches 0 and the denominator also approaches 0, giving the meaningless expression 0/0. Algebra can be used to rationalize the numerator by multiplying both the numerator and the denominator by $\sqrt{x} + 2$. This gives

$$\frac{\sqrt{x} - 2}{x - 4} \cdot \frac{\sqrt{x} + 2}{\sqrt{x} + 2} = \frac{\sqrt{x} \cdot \sqrt{x} - 2\sqrt{x} + 2\sqrt{x} - 4}{(x - 4)(\sqrt{x} + 2)}$$

$$= \frac{x - 4}{(x - 4)(\sqrt{x} + 2)} = \frac{1}{\sqrt{x} + 2}$$

if $x \neq 4$. Now use rules for limits.

$$\lim_{x \to 4} \frac{\sqrt{x} - 2}{x - 4} = \lim_{x \to 4} \frac{1}{\sqrt{x} + 2} = \frac{1}{\sqrt{4} + 2} = \frac{1}{2 + 2} = \frac{1}{4} \quad \blacktriangleleft$$

EXAMPLE 6

Find $\lim\limits_{x \to 1} \dfrac{x + 1}{x^2 - 1}$.

Again, Rule 5 cannot be used since $\lim\limits_{x \to 1} x^2 - 1 = 0$. If $x \neq 1$, the function can be rewritten as

$$\frac{x + 1}{x^2 - 1} = \frac{x + 1}{(x + 1)(x - 1)} = \frac{1}{x - 1}.$$

Then

$$\lim_{x \to 1} \frac{x + 1}{x^2 - 1} = \lim_{x \to 1} \frac{1}{x - 1}$$

by Rule 7. None of the rules can be used to find

$$\lim_{x \to 1} \frac{1}{x - 1},$$

but a table of values would show that the values of $1/(x - 1)$ increase as $x \to 1^+$ and the values of $1/(x - 1)$ decrease as $x \to 1^-$. Therefore,

$$\lim_{x \to 1} \frac{1}{x - 1} \text{ does not exist.} \quad \blacktriangleleft$$

Continuity Sometimes $\lim\limits_{x \to a} f(x)$ and $f(a)$ both exist, but are not equal, as in the next example.

$$f(x) = \begin{cases} 0 \text{ if } x = 1 \\ 1 \text{ if } x \neq 1 \end{cases}$$

FIGURE 4

7 Let f be the function defined by

$$f(x) = \begin{cases} 0 & \text{if } x = 1 \\ 1 & \text{if } x \neq 1 \end{cases}$$

and graphed in Figure 4.

The graph shows that $\lim\limits_{x \to 1} f(x) = 1$, but $f(1) = 0$. For this function,

$$\lim_{x \to 1} f(x) \neq f(1). \blacktriangleleft$$

The function in Example 7 is said to be *discontinuous at $x = 1$*. A function is discontinuous at some x-value if you must lift your pencil from the paper at that point to sketch the graph. The graph of the function discussed earlier, with

$$f(x) = \frac{x^2 - 4}{x - 2},$$

is shown in Figure 5. The function is discontinuous at $x = 2$, as shown by the "hole" in the graph, because $f(2)$ is undefined. Also, the function g, whose graph is shown in Figure 1, is discontinuous at $x = 2$, because $g(2)$ is undefined. These examples suggest the next definition.

$$f(x) = \frac{x^2 - 4}{x - 2}$$

FIGURE 5

DEFINITION OF CONTINUITY AT A POINT	A function f is **continuous** at c if the following three conditions are satisfied: **1.** $f(c)$ is defined, **2.** $\lim\limits_{x \to c} f(x)$ exists, and **3.** $\lim\limits_{x \to c} f(x) = f(c)$. If f is not continuous at c, it is **discontinuous** there.

EXAMPLE 8

Tell why the functions are discontinuous at the indicated points.

(a) $f(x)$ in Figure 6 at $x = 3$

The open circle on the graph of Figure 6 at the point where $x = 3$ means that $f(3)$ is not defined. Because of this, part (1) of the definition fails.

FIGURE 6 **FIGURE 7**

(b) $h(x)$ in Figure 7 at $x = 0$

According to the graph of Figure 7, $h(0) = -1$. Also, as x approaches 0 from the left, $h(x)$ is -1. As x approaches 0 from the right, however, $h(x)$ is 1. Since there is no single number approached by the values of $h(x)$ as x approaches 0, $\lim\limits_{x\to 0} h(x)$ does not exist, and part (2) of the definition fails.

(c) $g(x)$ in Figure 8 at $x = 4$

In Figure 8, the heavy dot above 4 shows that $g(4)$ is defined. In fact, $g(4) = 1$. The graph also shows, however, that

$$\lim_{x\to 4} g(x) = -2,$$

so $\lim\limits_{x\to 4} g(x) \neq g(4)$, and part (3) of the definition fails.

FIGURE 8 **FIGURE 9**

(d) $f(x)$ in Figure 9 at $x = -2$

The function f graphed in Figure 9 is not defined at -2, and $\lim\limits_{x\to -2} f(x)$ does not exist. Either of these reasons is sufficient to show that f is not continuous at -2. (Function f *is* continuous at any value of x greater than -2, however.) ◄

The next table lists some key functions and tells where each is continuous.

Continuous Functions

Type of Function	Where It Is Continuous	Graphic Example
Polynomial Function $y = a_n x^n + a_{n-1} x^{n-1} + \cdots + a_1 x + a_0$, where a_n, $a_{n-1}, \ldots, a_1, a_0$ are real numbers, not all 0	For all x	
Rational Function $y = \dfrac{p(x)}{q(x)}$, where $p(x)$ and $q(x)$ are polynomials, with $q(x) \neq 0$	For all x where $q(x) \neq 0$	
Root Function $y = \sqrt{ax + b}$, where a and b are real numbers, with $a \neq 0$ and $ax + b \geq 0$	For all x where $ax + b > 0$	

FIGURE 10

EXAMPLE 9

A trailer rental firm charges a flat $8 to rent a hitch. The trailer itself is rented for $22 per day or fraction of a day. Let $C(x)$ represent the cost of renting a hitch and trailer for x days.

(a) Graph C.

The charge for one day is $8 for the hitch and $22 for the trailer, or $30. In fact, if $0 < x \leq 1$, then $C(x) = 30$. To rent the trailer for more than one day, but not more than two days, the charge is $8 + 2 \cdot 22 = 52$ dollars. For any value of x satisfying $1 < x \leq 2$, $C(x) = 52$. Also, if $2 < x \leq 3$, then $C(x) = 74$. These results lead to the graph in Figure 10.

(b) Find any points of discontinuity for C.

As the graph suggests, C is discontinuous at $x = 1, 2, 3, 4$, and all other positive integers. ◀

10.1 Exercises

Decide whether each limit exists. If a limit exists, find its value.

1. $\lim\limits_{x \to 3} f(x)$

2. $\lim\limits_{x \to 2} F(x)$

3. $\lim\limits_{x \to -2} f(x)$

4. $\lim\limits_{x \to 3} g(x)$

5. $\lim\limits_{x \to 1} g(x)$

6. $\lim\limits_{x \to 0} f(x)$

7. Explain why $\lim\limits_{x \to 2} F(x)$ in Exercise 2 exists, but $\lim\limits_{x \to -2} f(x)$ in Exercise 3 does not.

8. In Exercise 5, why does $\lim\limits_{x \to 1} g(x)$ exist, even though $g(1)$ is undefined?

9. Use the table of values below to estimate $\lim\limits_{x \to 1} f(x)$.

x	.9	.99	.999	.9999	1.0001	1.001	1.01	1.1
$f(x)$	1.9	1.99	1.999	1.9999	2.0001	2.001	2.01	2.1

Complete the tables and use the results to find the indicated limits. (You will need a calculator with a \sqrt{x} key or a computer for Exercises 13 and 14.)

10. If $f(x) = 2x^2 - 4x + 3$, find $\lim\limits_{x \to 1} f(x)$.

x	.9	.99	.999	1.001	1.01	1.1
$f(x)$			1.000002	1.000002		

11. If $k(x) = \dfrac{x^3 - 2x - 4}{x - 2}$, find $\lim\limits_{x \to 2} k(x)$.

x	1.9	1.99	1.999	2.001	2.01	2.1
$k(x)$						

12. If $f(x) = \dfrac{2x^3 + 3x^2 - 4x - 5}{x + 1}$, find $\lim\limits_{x \to -1} f(x)$.

x	-1.1	-1.01	-1.001	-.999	-.99	-.9
$f(x)$						

13. If $h(x) = \dfrac{\sqrt{x} - 2}{x - 1}$, find $\lim\limits_{x \to 1} h(x)$.

x	.9	.99	.999	1.001	1.01	1.1
$h(x)$						

14. If $f(x) = \dfrac{\sqrt{x} - 3}{x - 3}$, find $\lim\limits_{x \to 3} f(x)$.

x	2.9	2.99	2.999	3.001	3.01	3.1
$f(x)$						

Let $\lim\limits_{x \to 4} f(x) = 16$ *and* $\lim\limits_{x \to 4} g(x) = 8$. *Use the limit rules to find the following limits.*

15. $\lim\limits_{x \to 4} [f(x) - g(x)]$

16. $\lim\limits_{x \to 4} [g(x) \cdot f(x)]$

17. $\lim\limits_{x \to 4} \dfrac{f(x)}{g(x)}$

18. $\lim\limits_{x \to 4} [3 \cdot f(x)]$

19. $\lim\limits_{x \to 4} \sqrt{f(x)}$

20. $\lim\limits_{x \to 4} \sqrt[3]{g(x)}$

21. $\lim\limits_{x \to 4} [g(x)]^3$

22. $\lim\limits_{x \to 4} [1 + f(x)]^2$

23. $\lim\limits_{x \to 4} \dfrac{f(x) + g(x)}{2g(x)}$

24. $\lim\limits_{x \to 4} \dfrac{5g(x) + 2}{1 - f(x)}$

Use the properties of limits to help decide whether the following limits exist. If a limit exists, find its value.

25. $\lim\limits_{x \to 3} \dfrac{x^2 - 9}{x - 3}$

26. $\lim\limits_{x \to -2} \dfrac{x^2 - 4}{x + 2}$

27. $\lim\limits_{x \to -2} \dfrac{x^2 - x - 6}{x + 2}$

28. $\lim\limits_{x \to 5} \dfrac{x^2 - 3x - 10}{x - 5}$

29. $\lim\limits_{x \to 0} \dfrac{x^3 - 4x^2 + 8x}{2x}$

30. $\lim\limits_{x \to 0} \dfrac{-x^5 - 9x^3 + 8x^2}{5x}$

31. $\lim\limits_{x \to 0} \dfrac{[1/(x + 3)] - 1/3}{x}$

32. $\lim\limits_{x \to 0} \dfrac{[-1/(x + 2)] + 1/2}{x}$

33. $\lim\limits_{x \to 25} \dfrac{\sqrt{x} - 5}{x - 25}$

34. $\lim\limits_{x \to 36} \dfrac{\sqrt{x} - 6}{x - 36}$

35. $\lim\limits_{h \to 0} \dfrac{(x + h)^2 - x^2}{h}$

36. $\lim\limits_{h \to 0} \dfrac{(x + h)^3 - x^3}{h}$

37. Let $F(x) = \dfrac{3x}{(x + 2)^3}$. **(a)** Find $\lim\limits_{x \to -2} F(x)$. **(b)** Find the vertical asymptote of the graph of $F(x)$.
 (c) Compare your answers for parts (a) and (b). What can you conclude?

38. Let $G(x) = \dfrac{-6}{(x - 4)^2}$. **(a)** Find $\lim\limits_{x \to 4} G(x)$. **(b)** Find the vertical asymptote of the graph of $G(x)$.
 (c) Compare your answers for parts (a) and (b). Are they related? How?

Find all points $x = a$ *where the function is discontinuous. For each point of discontinuity, give* $f(a)$ *and* $\lim\limits_{x \to a} f(x)$.

39.

40.

41.

42.

43.

44.

Decide whether each of the following is continuous at the given values of x.

45. $g(x) = \dfrac{1}{x(x-2)};\quad x = 0, 2, 4$

46. $g(x) = \dfrac{-2x}{(2x+1)(3x+6)};\quad x = 0, -1/2, -2$

47. $k(x) = \dfrac{5+x}{2+x};\quad x = 0, -2, -5$

48. $f(x) = \dfrac{4-x}{x-9};\quad x = 0, 4, 9$

49. $g(x) = \dfrac{x^2-4}{x-2};\quad x = 0, 2, -2$

50. $h(x) = \dfrac{x^2-25}{x+5};\quad x = 0, 5, -5$

51. $p(x) = x^2 - 4x + 11;\quad x = 0, 2, -1$

52. $q(x) = -3x^3 + 2x^2 - 4x + 1;\quad x = -2, 3, 1$

53. $p(x) = \dfrac{|x+2|}{x+2};\quad x = -2, 0, 2$

54. $r(x) = \dfrac{|5-x|}{x-5};\quad x = -5, 0, 5$

Applications

Business and Economics

55. Consumer Demand When the price of an essential commodity (such as gasoline) rises rapidly, consumption drops slowly at first. If the price continues to rise, however, a "tipping" point may be reached, at which consumption takes a sudden, substantial drop. Suppose the accompanying graph shows the consumption of gasoline, $G(t)$, in millions of gallons, in a certain area. We assume that the price is rising rapidly. Here t is time in months after the price began rising. Use the graph to find the following.

(a) $\lim\limits_{t \to 12} G(t)$ **(b)** $\lim\limits_{t \to 16} G(t)$ **(c)** $G(16)$

(d) the tipping point (in months)

56. Production The graph shows the profit from the daily production of x thousand kilograms of an industrial chemical.

Use the graph to find the following limits.

(a) $\lim\limits_{x \to 6} P(x)$ **(b)** $\lim\limits_{x \to 10^-} P(x)$ **(c)** $\lim\limits_{x \to 10^+} P(x)$

(d) $\lim\limits_{x \to 10} P(x)$

(e) Where is the function discontinuous? What might account for such a discontinuity?

(f) Use the graph to estimate the number of units of the chemical that must be produced before the second shift is as profitable as the first.

57. Cost Analysis The cost to transport a mobile home depends on the distance, x, in miles that the home is moved. Let $C(x)$ represent the cost to move a mobile home x miles. One firm charges as follows.

Cost per Mile	Distance in Miles
$4.00	$0 < x \le 150$
$3.00	$150 < x \le 400$
$2.50	$400 < x$

Find the cost to move a mobile home the following distances.

(a) 130 mi **(b)** 150 mi **(c)** 210 mi

(d) 400 mi **(e)** 500 mi

(f) Where is C discontinuous?

58. **Cost Analysis** A company charges $1.20 per pound for a certain fertilizer on all orders not over 100 lb, and $1 per pound for orders over 100 lb. Let $F(x)$ represent the cost for buying x lb of the fertilizer. Find the cost of buying

(a) 80 lb; **(b)** 150 lb; **(c)** 100 lb.

(d) Where is F discontinuous?

59. **Car Rental** Recently, a car rental firm charged $30 per day or portion of a day to rent a car for a period of 1 to 5 days. Days 6 and 7 were then "free," while the charge for days 8 through 12 was again $30 per day. Let $C(t)$ represent the total cost to rent the car for t days, where $0 < t \le 12$. Find the total cost of a rental for the following number of days:

(a) 4 **(b)** 5 **(c)** 6 **(d)** 7 **(e)** 8.

(f) Find $\lim\limits_{t \to 5^-} C(t)$. **(g)** Find $\lim\limits_{t \to 5^+} C(t)$.

(h) Where is C discontinuous on the given interval?

60. **Sales Tax** Like many states, California suffered a large budget deficit in 1991. As part of the solution, officials raised the sales tax by $.0125. The graph shows the California state sales tax since it was first established in 1933. Let $T(x)$ represent the sales tax in year x. Find the following.

(a) $\lim\limits_{x \to 80} T(x)$ **(b)** $\lim\limits_{x \to 73^-} T(x)$ **(c)** $\lim\limits_{x \to 73^+} T(x)$

(d) $\lim\limits_{x \to 73} T(x)$

(e) List three years for which the graph indicates a discontinuity.

Social Sciences

61. **Learning Skills** With certain skills (such as musical skills), learning is rapid at first and then levels off, with sudden insights causing learning to take a jump. A typical graph of such learning is shown in the figure. Where is the function discontinuous?

Physical Sciences

62. **Temperature** Suppose a gram of ice is at a temperature of $-100°C$. The graph shows the temperature of the ice as increasing numbers of calories of heat are applied. It takes 80 calories to melt one gram of ice at $0°C$ into water, and 539 calories to boil one gram of water at $100°C$ into steam. Where is this graph discontinuous?

10.2 RATES OF CHANGE

How fast do the blood cells in a capillary move in the seconds after a drug is injected?

How does the manufacturing cost per item change as the number of items manufactured changes?

These questions will be answered in Examples 4 and 5 in this section as we develop a method for finding the rate of change of one variable with respect to a unit change in another variable.

Average Rate of Change Suppose we take a trip from San Francisco. Someone keeps track every half-hour of the distance traveled from San Francisco, with the following results for the first three hours.

Time in Hours	0	.5	1	1.5	2	2.5	3
Distance in Miles	0	20	48	80	104	126	150

Let f be the function giving the distance in miles from San Francisco at time t. Then $f(0) = 0$, $f(1) = 48$, $f(2.5) = 126$, and so on.

Distance equals time multiplied by rate (or speed); so the distance formula is $d = rt$. Solving for rate gives $r = d/t$, or

$$\text{Average speed} = \frac{\text{Distance}}{\text{Time}}.$$

For example, the average speed over the time interval from $t = 0$ to $t = 3$ is

$$\text{Average speed} = \frac{f(3) - f(0)}{3 - 0} = \frac{150 - 0}{3} = 50,$$

or 50 miles per hour. We can use this formula to find the average speed for any interval of time during the trip, as shown below.

Recall the formula for the slope of a line through two points (x_1, y_1) and (x_2, y_2):

$$\frac{y_2 - y_1}{x_2 - x_1}.$$

Find the slope of the line through the following points.

(.5, 20)	and	(1, 48)
(.5, 20)	and	(1.5, 80)
(1, 48)	and	(2, 104)

Compare your answers to the average speeds shown in the chart.

Time Interval	Average Speed $= \dfrac{\text{Distance}}{\text{Time}}$
$t = .5$ to $t = 1$	$\dfrac{f(1) - f(.5)}{1 - .5} = \dfrac{28}{.5} = 56$
$t = .5$ to $t = 1.5$	$\dfrac{f(1.5) - f(.5)}{1.5 - .5} = \dfrac{60}{1} = 60$
$t = 1$ to $t = 2$	$\dfrac{f(2) - f(1)}{2 - 1} = \dfrac{56}{1} = 56$
$t = 1$ to $t = 3$	$\dfrac{f(3) - f(1)}{3 - 1} = \dfrac{102}{2} = 51$
$t = a$ to $t = b$	$\dfrac{f(b) - f(a)}{b - a}$

The last line in the chart suggests a way to apply this approach to any function f to get the average rate of change of $f(x)$ with respect to x.

AVERAGE RATE OF CHANGE

The average rate of change of $f(x)$ with respect to x for a function f as x changes from a to b, where $a < b$, is

$$\frac{f(b) - f(a)}{b - a}.$$

Note The formula for the average rate of change is the same as the formula for the slope of the line through $(a, f(a))$ and $(b, f(b))$. This connection between slope and rate of change will be examined more closely in the next section.

EXAMPLE **1** A manager is testing two different advertising campaigns in different parts of the country for three months, with the intention of canceling the least effective of the two after that time. The graphs in Figure 11 show sales as a function of time (in months) since the two campaigns began. The graphs suggest that although campaign 1 consistently did better than campaign 2 for the first three months, its *rate of change in sales* is decreasing after that time. The rate of change in sales for campaign 2, however, is increasing at time $t = 3$ months.

FIGURE 11

On the interval [2, 4], for campaign 1,

$$\text{Average rate of change} = \frac{f(4) - f(2)}{4 - 2}$$

$$= \frac{30,000 - 40,000}{4 - 2} = -5000.$$

For campaign 2, on [2, 4],

$$\text{Average rate of change} = \frac{50,000 - 25,000}{4 - 2} = 12,500.$$

The negative result for campaign 1 indicates that sales were *decreasing* at the rate of 5000 per month over the two-month period, while the positive result for campaign 2 indicates *increasing* sales at the rate of 12,500 over the same two-month period. These rates show that the second campaign will be the most productive in the future, and campaign 1 should be discontinued. ◀

EXAMPLE 2

The graph in Figure 12 shows the profit $P(x)$, in hundreds of thousands of dollars, from a highly popular new computer program x months after its introduction on the market. The graph shows that profit increases until the program has been on the market for 25 months, after which profit decreases as the popularity of the program decreases. Here the average rate of change of *profit* with respect to *time* on some interval is defined as the change in profit divided by the change in time on the interval.

FIGURE 12

(a) On the interval from 15 months to 25 months the average rate of change of profit with respect to time is given by

$$\frac{P(25) - P(15)}{25 - 15} = \frac{4 - 1.4}{25 - 15} = \frac{2.6}{10} = .26.$$

On the average, each month from 15 months to 25 months shows an increase in profit of .26 hundred thousand dollars, or $26,000.

(b) From 25 months to 30 months, the average rate of change is

$$\frac{3 - 4}{30 - 25} = \frac{-1}{5} = -.20,$$

with each month in this interval showing an average *decline* in profits of .20 hundred thousand dollars, or $20,000.

(c) From 10 months to 35 months, the average rate of change of profit is

$$\frac{1 - 1}{35 - 10} = \frac{0}{25} = 0,$$

or $0. ◀

Velocity As Example 2(c) suggests, finding the average rate of change of a function over a large interval can lead to answers that are not very helpful. The results often are more useful if the average rate of change is found over a fairly narrow interval. Finding the exact rate of change at a particular *x*-value requires a continuous function. For example, suppose a car starts from a stop sign and moves along a straight road. Assume that the distance in feet traveled in *t* seconds is given by $s(t) = t^2 + 3t$. After 5 seconds, the car has traveled $s(5) = (5)^2 + 3(5) = 40$ feet. The speed of the car at exactly 10 seconds after starting can be estimated by finding the average speed over shorter and shorter time intervals.

Interval	Average Speed
$t = 10$ to $t = 10.1$	$\dfrac{s(10.1) - s(10)}{10.1 - 10} = \dfrac{132.31 - 130}{.1} = 23.1$
$t = 10$ to $t = 10.01$	$\dfrac{s(10.01) - s(10)}{10.01 - 10} = \dfrac{130.2301 - 130}{.01} = 23.01$
$t = 10$ to $t = 10.001$	$\dfrac{s(10.001) - s(10)}{10.001 - 10} = \dfrac{130.023001 - 130}{.001} = 23.001$

The results in the chart suggest that the exact speed at $t = 10$ seconds is 23 feet per second.

We can generalize this discussion by finding the average speed from $t = 10$ to $t = 10 + h$, where *h* represents some nonzero small number. The average speed for this interval is as follows.

$$\frac{s(10 + h) - s(10)}{(10 + h) - 10} = \frac{s(10 + h) - s(10)}{h}$$

$$= \frac{[(10 + h)^2 + 3(10 + h)] - 130}{h}$$

$$= \frac{100 + 20h + h^2 + 30 + 3h - 130}{h}$$

$$= \frac{23h + h^2}{h}$$

$$= \frac{h(23 + h)}{h}$$

$$= 23 + h$$

If we take the limit of $23 + h$ as *h* approaches zero, we get

$$\lim_{h \to 0} \frac{s(10 + h) - s(10)}{h} = \lim_{h \to 0} (23 + h) = 23.$$

We call this limit the *instantaneous velocity at time t.*

To generalize the results of the car example, suppose an object is moving in a straight line, with its position (distance from some fixed point) given by $s(t)$, where t represents time. The quotient

$$\frac{s(t + h) - s(t)}{h}$$

represents the average rate of change of the distance, or the average velocity of the object. The instantaneous velocity at time t (often called just the velocity at time t) is the limit of the quotient above as h approaches 0.

VELOCITY

If $v(t)$ represents the velocity at time t of an object moving in a straight line with position $s(t)$, then

$$v(t) = \lim_{h \to 0} \frac{s(t + h) - s(t)}{h},$$

provided this limit exists.

Caution Remember, $s(t + h) \neq s(t) + s(h)$. To find $s(t + h)$ replace t with $t + h$ in the expression for $s(t)$. For example, if $s(t) = t^2$,

$$s(t + h) = (t + h)^2 = t^2 + 2th + h^2,$$

but
$$s(t) + s(h) = t^2 + h^2.$$

EXAMPLE 3 The distance in feet of an object from a starting point is given by $s(t) = 2t^2 - 5t + 40$, where t is time in seconds.

(a) Find the average velocity of the object from 2 seconds to 4 seconds.

The average velocity is

$$\frac{s(4) - s(2)}{4 - 2} = \frac{52 - 38}{2} = \frac{14}{2} = 7$$

feet per second.

(b) Find the instantaneous velocity at 4 seconds.

For $t = 4$, the instantaneous velocity is

$$\lim_{h \to 0} \frac{s(4 + h) - s(4)}{h}$$

feet per second.

$$\begin{aligned}
s(4 + h) &= 2(4 + h)^2 - 5(4 + h) + 40 \\
&= 2(16 + 8h + h^2) - 20 - 5h + 40 \\
&= 32 + 16h + 2h^2 - 20 - 5h + 40 \\
&= 2h^2 + 11h + 52
\end{aligned}$$

Also,

$$s(4) = 2(4)^2 - 5(4) + 40 = 52.$$

Therefore, the instantaneous velocity at $t = 4$ is

$$\lim_{h \to 0} \frac{2h^2 + 11h}{h} = \lim_{h \to 0} (2h + 11) = 11,$$

or 11 feet per second. ◀

The next examples provide answers to the questions asked at the beginning of this section.

EXAMPLE 4

The velocity of blood cells is of interest to physicians; a slower velocity than normal might indicate a constriction, for example. Suppose the position of a red blood cell in a capillary is given by

$$s(t) = 1.2t + 5,$$

where $s(t)$ gives the position of a cell in millimeters from some initial point and t is time in seconds. Find the velocity of this cell at time t.

Evaluate the limit given above. To find $s(t + h)$, substitute $t + h$ for the variable t in $s(t) = 1.2t + 5$.

$$s(t + h) = 1.2(t + h) + 5$$

Now use the definition of velocity.

$$
\begin{aligned}
v(t) &= \lim_{h \to 0} \frac{s(t + h) - s(t)}{h} \\
&= \lim_{h \to 0} \frac{1.2(t + h) + 5 - (1.2t + 5)}{h} \\
&= \lim_{h \to 0} \frac{1.2t + 1.2h + 5 - 1.2t - 5}{h} = \lim_{h \to 0} \frac{1.2h}{h} = 1.2
\end{aligned}
$$

The velocity of the blood cell is a constant 1.2 millimeters per second. ◀

Note In Example 4, the velocity of 1.2 is the same as the slope of the linear function $s(t) = 1.2t + 5$. As mentioned earlier, we will learn more about this connection in the next section.

Instantaneous Rate of Change We can extend the ideas that led to the definition of velocity to any function f. Earlier, we saw that the average rate of change of $f(x)$ with respect to x as x changes from a to b is

$$\frac{f(b) - f(a)}{b - a}.$$

Replacing b with $x_0 + h$ and a with x_0 and letting h approach 0 gives the following definition.

INSTANTANEOUS RATE OF CHANGE

The **instantaneous rate of change** for a function f when $x = x_0$ is

$$\lim_{h \to 0} \frac{f(x_0 + h) - f(x_0)}{h},$$

provided this limit exists.

A company determines that the cost in dollars of manufacturing x units of a certain item is

$$C(x) = 100 + 5x - x^2.$$

(a) Find the average rate of change of cost per item for manufacturing between 1 and 5 items.

Use the formula for average rate of change. The cost to manufacture 1 item is

$$C(1) = 100 + 5(1) - (1)^2 = 104,$$

or $104. The cost to manufacture 5 items is

$$C(5) = 100 + 5(5) - (5)^2 = 100,$$

or $100. The average rate of change of cost is

$$\frac{C(5) - C(1)}{5 - 1} = \frac{100 - 104}{4} = -1.$$

Thus, on the average, cost is decreasing at the rate of $1 per item when production is increased from 1 to 5 items.

(b) Find the instantaneous rate of change of cost with respect to the number of items produced when just 1 item is produced.

The instantaneous rate of change for $x = 1$ is given by

$$\lim_{h \to 0} \frac{C(1 + h) - C(1)}{h}$$

$$= \lim_{h \to 0} \frac{[100 + 5(1 + h) - (1 + h)^2] - [100 + 5(1) - 1^2]}{h}$$

$$= \lim_{h \to 0} \frac{[100 + 5 + 5h - 1 - 2h - h^2 - 104]}{h}$$

$$= \lim_{h \to 0} \frac{3h - h^2}{h} \qquad \text{Combine terms.}$$

$$= \lim_{h \to 0} (3 - h) \qquad \text{Divide by } h.$$

$$= 3. \qquad \text{Calculate the limit.}$$

When 1 item is manufactured, the cost is increasing at the rate of $3 per item. Thus, the instantaneous rate of change of cost represents the approximate cost of manufacturing an additional item. As mentioned in Chapter 1, this rate of change of cost is called the marginal cost.

(c) Find the instantaneous rate of change of cost when 5 items are made. The instantaneous rate of change for $x = 5$ is given by

$$\lim_{h \to 0} \frac{C(5 + h) - C(5)}{h}$$

$$= \lim_{h \to 0} \frac{[100 + 5(5 + h) - (5 + h)^2] - [100 + 5(5) - 5^2]}{h}$$

$$= \lim_{h \to 0} \frac{[100 + 25 + 5h - 25 - 10h - h^2] - 100}{h}$$

$$= \lim_{h \to 0} \frac{-5h - h^2}{h}$$

$$= \lim_{h \to 0} (-5 - h)$$

$$= -5.$$

When 5 items are manufactured, the cost is *decreasing* at the rate of $5 per item—that is, the marginal cost when $x = 5$ is $-$5$. Notice that as the number of items produced goes up, the marginal cost goes down, as might be expected. ◀

10.2 Exercises

Find the average rate of change for each function over the given interval.

1. $y = x^2 + 2x$ between $x = 0$ and $x = 3$

2. $y = -4x^2 - 6$ between $x = 2$ and $x = 5$

3. $y = 2x^3 - 4x^2 + 6x$ between $x = -1$ and $x = 1$

4. $y = -3x^3 + 2x^2 - 4x + 1$ between $x = 0$ and $x = 1$

5. $y = \sqrt{x}$ between $x = 1$ and $x = 4$

6. $y = \sqrt{3x - 2}$ between $x = 1$ and $x = 2$

7. $y = \dfrac{1}{x - 1}$ between $x = -2$ and $x = 0$

8. $y = \dfrac{-5}{2x - 3}$ between $x = 2$ and $x = 4$

Use the properties of limits to find the following limits for $s(t) = t^2 + 5t + 2$.

9. $\lim\limits_{h \to 0} \dfrac{s(6 + h) - s(6)}{h}$

10. $\lim\limits_{h \to 0} \dfrac{s(1 + h) - s(1)}{h}$

11. $\lim\limits_{h \to 0} \dfrac{s(10 + h) - s(10)}{h}$

Use the properties of limits to find the following limits for $s(t) = t^3 + 2t + 9$.

12. $\lim\limits_{h \to 0} \dfrac{s(1 + h) - s(1)}{h}$

13. $\lim\limits_{h \to 0} \dfrac{s(4 + h) - s(4)}{h}$

Find the instantaneous rate of change for each function at the given value.

14. $f(x) = x^2 + 2x$ at $x = 0$

15. $s(t) = -4t^2 - 6$ at $t = 2$

16. $g(t) = 1 - t^2$ at $t = -1$

17. $F(x) = x^2 + 2$ at $x = 0$

18. Explain the difference between the average rate of change of y as x changes from a to b, and the instantaneous rate of change of y at $x = a$.

19. If the instantaneous rate of change of $f(x)$ with respect to x is positive when $x = 1$, is f increasing or decreasing there?

Calculus at Work

This section shows you that calculus is not just a body of theorems, concepts, and exercises without any real human dimension; there are practicing mathematicians, scientists, and engineers using this knowledge. The following essays were written by professionals who work with mathematics daily to create products and services we commonly use.

W. Weston Meyer
Mathematics Department, Research Laboratories
General Motors Corporation
"Perfecting the Skin of an Automobile"

Richard E. Klabunde, Ph.D.
Senior Cardiovascular Group Leader, Department of Pharmacology
Abbott Laboratories
Roberto Mercado, Ph.D.
Manager, Nonclinical Statistics, Pharmaceutical Products Division
Abbott Laboratories
"Drugs, Derivatives, and Heart Disease"

Jane Gillum (Summer Intern), Alan Opsahl, Allen Blaurock
Technology Center
Kraft General Foods
"Crystals, Integration, and . . . Margarine!"

Henry Robinson
Meteorologist, National Weather Service
National Oceanic and Atmospheric Administration
"Calculus in a Cornfield"

When you read these testaments, you may appreciate more the phenomenal versatility of calculus–how it is useful, in fact indispensable, in virtually countless aspects of our lives: from the margarine you spread on your breakfast toast, to the car you drive, to predicting the weather you experience.

The fact that calculus works as well as it does in as many ways as it does leads us to the following conclusion: learning about calculus is learning about an amazing pattern, one repeated in the most disparate and surprising places in our world.

Perfecting the Skin of an Automobile

W. Weston Meyer

Mathematics Department, Research Laboratories / General Motors Corporation, Warren, Michigan

As far as we know, it was Jacob Bernoulli (1654 – 1705) who first published a formula to measure curvature. Three centuries later, the same formula is put to uses Bernoulli could never have imagined. Designing the "skin" of an automobile — its outer panel surface — is one such use.

Bernoulli's formula, given below, provides an expression for the radius R of the circle whose curvature is identical to the curvature of the curve $y = f(x)$ at a fixed point (see Figure 1).

$$R(x) = \frac{\left[1 + f'(x)^2\right]^{3/2}}{|f''(x)|}$$

Since curvature of a circle is equal to the inverse of its radius, the curvature of f at a fixed point is the inverse of the expression for $R(x)$.

Traditionally, the outer panels of an automobile materialized from scores of longitudinal and transverse section lines painstakingly inscribed on parchment by drafting artisans, then transferred to a clay mass by means of wooden templates. The clay was then smoothed between section lines by hand to give an iconic model.

Today, the design process is largely computer-aided. The artisan sits before a graphic display console and chooses control parameters to generate section lines, plus the surface that interpolates them, as mathematical entities. Feedback information is immediate and visual, but a computer image does not always convey enough information for the designer to make an aesthetic judgment. A line or surface that seems impeccable on the computer screen may have very

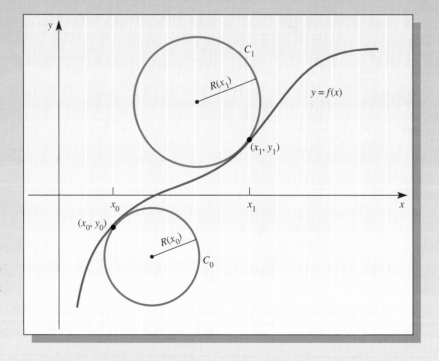

Figure 1 The curvature of circle C_0 is identical to the curvature of $y = f(x)$ at x_0; the curvature of circle C_1 is identical to the curvature of $y = f(x)$ at x_1. In each case the radius R is a function of x:

$$R(x) = \frac{\left[1 + f'(x)^2\right]^{3/2}}{|f''(x)|}.$$

minuscule undulations, flat spots, or depressions that become apparent only after full-scale manufacturing.

Curvature measurements can be especially helpful in detecting and eliminating such flaws at the design level. Figures 2(a) and 2(b) show a line "before and after" adjustment, with emanating quills to represent, by their lengths, the local curvatures according to Bernoulli. Before adjustment, the line has invisible irregularities, as evidenced by occasional spikes among the quills. After adjustment, the quill pattern is quite smooth.

Calculus at a somewhat higher level provides techniques to measure curvature of surfaces. Figure 3(a) shows a typical segment of an automobile outer panel. Superimposed on the surface in Figure 3(b) are curvature measurements, communicated by means of color coding. The presence of blue, in contrast to yellow, indicates a local flattening that needs to be eliminated. Figure 3(c) shows the result of a minor mathematical adjustment: the flattening is gone.

Figure 3 On the surface of an automobile's outer panel (a), curvature measurements coded by color indicate flattenings that need to be eliminated (b). In (c), the surface appears smooth after minor mathematical adjustments are made.

General Motors Research

(a)

General Motors Research

(b)

General Motors Research

(c)

Figure 2 The length of each quill emanating from the curve in (a) represents the local curvature according to Bernoulli's formula. In (b), the quills are quite regular, after adjustments have fixed the unseen undulations in the curve. (Figure from "The Porcupine Technique: Principle, Application, and Algorithm" by Klaus-Peter Beier, October 1987. Reprinted by permission of Klaus-Peter Beier.)

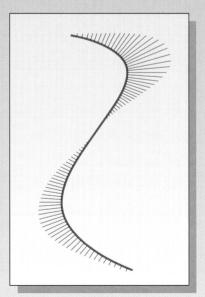

(a)

(b)

Drugs, Derivatives, and Heart Disease

Richard E. Klabunde, Ph.D.
Senior Cardiovascular Group Leader, Department of Pharmacology
Abbott Laboratories, Abbott Park, Illinois

Roberto Mercado, Ph.D.
Manager, Nonclinical Statistics, Pharmaceutical Products Division
Abbott Laboratories, Abbott Park, Illinois

(Dr. Alan Sewell served as a consultant for this article.)

A healthy heart is a flexible muscle, one that contracts vigorously to force blood out its left ventricle, then relaxes easily to repeat the process. The pressure exerted by blood leaving the heart is measured as a function of time (see Figure 1). The *rate* of change of pressure with respect to time measures the *contractility* of the heart — that is, how quickly the heart contracts and relaxes. A heart with poor contractility will demonstrate a consistently slow rate of change of pressure. Drugs are used to improve this rate.

Mathematically, we can describe the pressure of blood leaving the left ventricle as $p(t)$, where p is the pressure at a given time t. The rate of change of pressure with respect to time, then, can be described as the derivative of $p(t)$, or dp/dt. Pharmacologists are interested in both the positive rate of change, measuring the contraction of

Figure 1 Left ventricular pressure is measured in mmHg as a function of time.

Figure 2 Change in left ventricular pressure is measured in mmHg / sec as a function of time. *dp/dt* is computed using an analog differentiator.

the heart, and the negative rate of change, measuring the relaxation of the heart. The positive rate of change is denoted by $+dp/dt$, while the negative rate of change is denoted by $-dp/dt$. The critical measurements are the maximum positive rate of change and the maximum negative rate of change: the heart's "best efforts" in contracting and relaxing.

As is often the case when we use mathematics to model the real world, an explicit equation describing $p(t)$ does not exist. Pressure is clearly a function of time, and can be measured accurately. But we cannot express the relationship except in the form of data and graphs.

Since no equation explicitly describes $p(t)$, no equation explicitly describes dp/dt. In practice, dp/dt is determined by inserting a fluid-filled plastic tube into the ventricular cavity. The tube is hooked up to a gauge and an amplifier, and pressure is measured directly. A special type of computer, called an *analog differentiator,* receives the output signal of the amplifier and numerically approximates values for $+dp/dt$ and $-dp/dt$, particularly the maximum values (see Figure 2). With these results, a physician can judge the contractility of the heart and, when a drug is being used, the effectiveness of the drug in improving contractility.

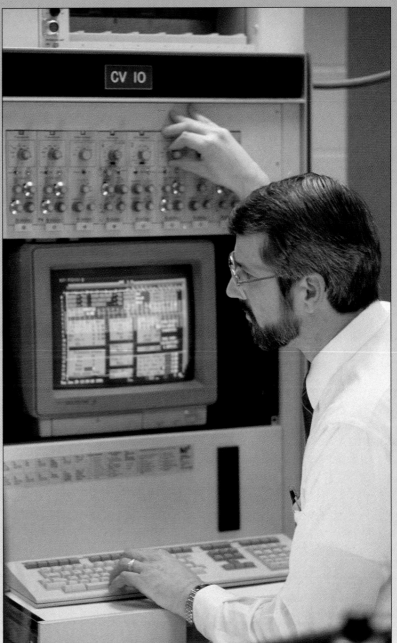

Abbott Laboratories

Pharmacologist making amplifier adjustments to the analog differentiator.

Crystals, Integration, and . . . Margarine!

Jane Gillum, Summer Intern; Alan Opsahl; Allen Blaurock

Technology Center / Kraft General Foods, Glenview, Illinois

Crystals of gems are valuable commodities, but in the manufacturing of certain foods, it is crystals of fat that are valuable. Makers of peanut butter, for example, add a small amount of a crystallizable fat — "partially hydrogenated vegetable oil" — so that a separate layer of oil doesn't form on top.

In making margarine, the rate fat crystals form is very important, because if the crystals form too rapidly, the margarine solidifies and plugs the pipes on its way to the machine where the margarine is shaped into sticks. Of course, sticks won't form at all if fat crystals form too slowly.

Crystallization is an orderly process involving two key events: (1) the creation of a crystal "seed" and (2) the growth of the crystal as oil molecules add on to the seed. The number n of crystal seeds depends on temperature, time, and the volume of oil. The size g, in meters, of the growing crystal seed depends on tem-

Figure 1 *Beta* crystals formed by very slow cooling of trilaurin, an edible oil. The feathery growth pattern started from a *single* crystal seed. Polarized-light micrograph (x 62.5).

perature and time. Given a constant temperature and a fixed volume of oil, however, both n and g are simple, constant functions of time: $n(t) = Nt$ and $g(t) = Gt$ (in meters). The rate of seed creation can thus be given as $n'(t) = N$, a constant, and the rate of size growth as $g'(t) = G$, also a constant.

The volume of the crystals is a function of the rate of seed creation, the rate of size growth, and time. It is, in fact, an integral function. Fixing the amount of oil and the temperature, the volume V, in cubic meters, of crystals can be expressed as

$$V(t) = \left(8NG^3\right)\int_0^t x^3\,dx\ .$$

Finding the antiderivative and evaluating,

$$V(t) = 8NG^3\left[x^4/4\right]_{}^{t}$$

$$= 2NG^3 t^4\ .$$

Keen-eyed readers will recognize that this expression for V cannot be valid for arbitrarily large values of t, for then the volume of the crystals would grow arbitrarily large, too. This growth does not in fact occur, however, since the amount of oil, after all, is fixed.

Thus, the expression for V does not hold true for large values of t. It is valid only for the early period of crystallization, that is, when t is small.

Since the rate N at which crystal seeds appear varies with temperature, margarine makers alter temperature to create varying textures. Lower temperatures mean more crystal seeds and a finer texture; higher temperatures mean fewer crystal seeds and a coarser texture. Spreadability and "mouthfeel" are controlled, then, by adjusting temperature and, hence, texture.

Figure 2 *Beta prime* (dim circular regions) and *alpha* crystals (brighter regions) formed by more rapid cooling. Each region ("Maltese crosses") started with a single crystal seed. Polarized-light micrograph (x 125).

Calculus in a Cornfield

Henry Robinson

Meteorologist, National Weather Service

National Oceanic and Atmospheric Administration, Silver Springs, Maryland

Predicting the weather is important for almost everybody, but for farmers it often means the difference between success and failure.

For example, agriculture specialists have found that the amount of time a crop needs to reach peak maturity is closely related to the temperature during its growing time. Predicting the daily temperature after planting can help a farmer know the optimal harvest day and schedule a manageable workload.

The relationship between temperature and growing time is expressed in terms of *growing degree days*, determined as follows: when the average temperature on a day is one degree above a certain threshold (specific to each crop); the day amounts to one growing degree day, when the average temperature on a day is two degrees above threshold, the day amounts to two growing degree days; and so on. For peas, the threshold temperature has

been found to be 39° F. Thus, if a day's average temperature is 72° F, the day amounts to 33 growing degree days for peas.

The total number of growing degree days D between the time t_p of planting and the time t_h of optimal harvest is thus a sum, expressible as an integral:

$$D = \int_{t_p}^{t_h} [T(t) - r]\,dt,$$

where $T(t)$ is the temperature at any time t and r is the threshold temperature.

Predicting t_h, the time of optimal harvest, for a crop involves knowing the threshold temperature and total number of growing degree days for that crop, and finding an explicit expression for $T(t)$, the temperature at any time t.

For virtually all cash crops, experience and experimentation has allowed agricultural specialists to determine the

threshold temperature and the total number of growing degree days. This information has not gone unnoticed by investors who deal in crop futures on the various worldwide exchanges. The missing piece of information is an explicit expression for $T(t)$.

One formulation is the Cosine Day of the Year Model for Temperature. Patterns of temperature change for individual locales tend to resemble a cosine curve, as the red line in the figure shows. Modifying the cosine function by appropriate factors results in an explicit expression of temperature as a function of time (the blue line). In general the model used is

$$T(t) = a \cdot \cos(t - b) + c,$$

where T is temperature, t is the day of the year (1 to 365), and a, b, c are constants.

Using this model allows the integral expression for the total number of growing degree days D to be evaluated, and hence the time of optimal harvest t_h to be approximated.

On the national and international scene, this information helps to predict potential shortages or oversupplies of crops. Farmers can use a simpler technique to determine the optimal harvest time, since they can reasonably consider the change in time dt to be a single day. By taking each day's temperature and subtracting the threshold, the farmer keeps a running total of growing degree days. When the total gets close to D, the farmer predicts the growing degree days on the basis of the weather forecast, thereby identifying the day for harvesting.

$$T(t) = -15 \cos\left[(t - 17)\left(\frac{2\pi}{365}\right)\right] + 10$$

Day of Year (t)

The Cosine Day of the Year Model for Temperature approximates the climatology of temperature for Des Moines, Iowa, with little error.

Applications

Business and Economics

20. Catalog Sales The graph shows the total sales in thousands of dollars from the distribution of x thousand catalogs. Find and interpret the average rate of change of sales with respect to the number of catalogs distributed for the following changes in x.

(a) 10 to 20 (b) 20 to 30 (c) 30 to 40

(d) What is happening to the average rate of change of sales as the number of catalogs distributed increases?

(e) Explain why (d) might happen.

21. Sales The graph shows annual sales (in appropriate units) of a Nintendo game. Find the average annual rate of change in sales for the following changes in years.

(a) 1 to 4 (b) 4 to 7 (c) 7 to 12

(d) What do your answers for (a) to (c) tell you about the sales of this product?

(e) Give an example of another product that might have such a sales curve.

22. Market Share IBM Europe has been steadily losing its share of the market since 1985, as shown in the figure.* Estimate the average drop in market share (in percent) over the following intervals.

(a) 1985 to 1986 (b) 1986 to 1988

(c) 1988 to 1989 (d) 1989 to 1990

(e) From your answers to parts (a)–(d), does the decline in market share seem to be increasing or tapering off?

(f) In which one-year period was the decline the greatest?

23. Profit Margin The figure* shows the profit margin (in percent) for IBM Europe. Estimate the average rate of change over the following intervals.

(a) 1985 to 1987 (b) 1987 to 1988

(c) 1988 to 1989 (d) 1989 to 1990

(e) In what time period(s) did the profit margin increase?

(f) Do the results in parts (a)–(d) indicate an improving, stable, or declining profit margin?

*Graphs entitled "IBM Europe Is Losing Market Share . . . and Feeling a Profit Pinch" reprinted from May 6, 1991 issue of *Business Week* by special permission, copyright © 1991 by McGraw-Hill, Inc.

24. Profit Suppose that the total profit in hundreds of dollars from selling x items is given by

$$P(x) = 2x^2 - 5x + 6.$$

Find the average rate of change of profit for the following changes in x.

(a) 2 to 4 (b) 2 to 3

(c) Use $P(x) = 2x^2 - 5x + 6$ to complete the following table.

h	1	.1	.01	.001	.0001
$2 + h$	3	2.1	2.01	2.001	2.0001
$P(2 + h)$	9	4.32	4.0302	4.003002	4.00030002
$P(2)$	4	4	4	4	4
$P(2 + h) - P(2)$	5	.32	.0302	.003002	.00030002
$\dfrac{P(2 + h) - P(2)}{h}$	5	3.2	___	___	___

(d) Use the bottom row of the chart to find

$$\lim_{h \to 0} \frac{P(2 + h) - P(2)}{h}.$$

(e) Use the rule for limits from Section 1 of this chapter to find

$$\lim_{h \to 0} \frac{P(2 + h) - P(2)}{h}.$$

(f) What is the instantaneous rate of change of profit with respect to the number of items produced when $x = 2$? (This number, called the *marginal profit* at $x = 2$, is the approximate profit from producing the third item.)

25. Profit Redo the chart in Exercise 24. This time, change the second line to $4 + h$, the third line to $P(4 + h)$, and so on. Then find

$$\lim_{h \to 0} \frac{P(4 + h) - P(4)}{h}.$$

(As in Exercise 24, this result is the marginal profit, the approximate profit from producing the fifth item.)

26. Revenue The revenue (in thousands of dollars) from producing x units of an item is

$$R = 10x - .002x^2.$$

(a) Find the average rate of change of revenue when production is increased from 1000 to 1001 units.

(b) Find the marginal revenue when 1000 units are produced.

(c) Find the additional revenue if production is increased from 1000 to 1001 units.

(d) Compare your answers for parts (a) and (c). What do you find?

27. Demand Suppose customers in a hardware store are willing to buy $N(p)$ boxes of nails at p dollars per box, as given by

$$N(p) = 80 - 5p^2, \quad 1 \le p \le 4.$$

(a) Find the average rate of change of demand for a change in price from $2 to $3.

(b) Find the instantaneous rate of change of demand when the price is $2.

(c) Find the instantaneous rate of change of demand when the price is $3.

(d) As the price is increased from $2 to $3, how is demand changing? Is the change to be expected?

Life Sciences

28. Bacteria Population The graph shows the population in millions of bacteria t minutes after a bactericide is introduced into a culture. Find and interpret the average rate of change of population with respect to time for the following time intervals.

(a) 1 to 2 (b) 2 to 3 (c) 3 to 4 (d) 4 to 5

(e) How long after the bactericide was introduced did the population begin to decrease?

(f) At what time did the rate of decrease of the population slow down?

29. Polluter Fines The graph* shows changes since 1981 in the amounts paid to the U.S. Treasury in fines by polluters.

(a) Find the average rate of change for civil penalties from 1987 to 1988, and from 1988 to 1989.

(b) Find the average rate of change of criminal penalties from 1987 to 1988, and from 1988 to 1989.

(c) Compare your results for parts (a) and (b). What do they tell you? What might account for the differences?

*Graph entitled "Higher EPA Fines Make Pollution Costly" by Mark Holmes from *National Geographic*, February 1991. Copyright © 1991 by National Geographic Society. Reprinted by permission of National Geographic Society.

(d) Find the average rate of change for civil penalties from 1981 to 1989. What was the general trend over this eight-year period? What might explain this trend?

Higher EPA Fines Make Pollution Costly

With stricter enforcement by the Environmental Protection Agency, polluters are paying more to the U. S. Treasury.

*Accurate data not available before 1987
Source: EPA; art by Mark Holmes, NGS staff

Social Sciences

30. Memory In an experiment, teenagers memorize a list of 15 words and then listen to rock music for certain periods of time. After t minutes of listening to rock music, the typical subject can remember approximately

$$R(t) = -.03t^2 + 15$$

of the words, where $0 \leq t \leq 20$. Find the instantaneous rate at which the typical student remembers the words after the following times have elapsed.

(a) 5 min **(b)** 15 min

31. Drug Trade The U.S. government estimates that Americans spent more than $40 billion for illicit drugs in 1990. However, as shown in the graph[†], Americans have spent

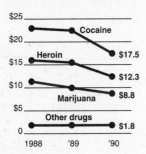

[†]Graph from ''What Americans spend on drugs'' as appeared in *The Sacramento Bee*, June 20, 1991. Reprinted by permission of Knight-Ridder/Tribune News Wire.

less on drugs since 1988. Estimate the average rate of change in the amounts spent on each of the following drugs from 1988 to 1989, and then from 1989 to 1990.

(a) Cocaine **(b)** Heroin **(c)** Marijuana

(d) Which drug had the greatest decrease in amount spent?

(e) List some possible reasons for the observed decreases and some reasons why not all drug expenditures decreased at the same rate.

Physical Sciences

32. Temperature The graph shows the temperature T in degrees Celsius as a function of the altitude h in feet when an inversion layer is over Southern California. (An inversion layer is formed when air at a higher altitude, say 3000 ft, is warmer than air at sea level, even though air normally is cooler with increasing altitude.) Estimate and interpret the average rate of change in temperature for the following changes in altitude.

(a) 1000 to 3000 ft **(b)** 1000 to 5000 ft

(c) 3000 to 9000 ft **(d)** 1000 to 9000 ft

(e) At what altitude at or below 7000 ft is the temperature highest? lowest? How would your answer change if 7000 ft is changed to 10,000 ft?

(f) At what altitude is the temperature the same as it is at 1000 ft?

33. Velocity A car is moving along a straight test track. The position in feet of the car, $s(t)$, at various times t is measured, with the following results.

t (seconds)	0	2	6	10	12	18
$s(t)$ (feet)	0	10	14	18	30	36

Find and interpret the average velocities for the following changes in t.

(a) 0 to 2 sec **(b)** 2 to 6 sec **(c)** 6 to 10 sec

(d) 10 to 12 sec **(e)** 12 to 18 sec

(f) Compare the average velocities found in parts (a)–(e), and describe when the car is slowing down, speeding up, or maintaining a constant average velocity.

34. Velocity The distance of a particle from some fixed point is given by

$$s(t) = t^2 + 5t + 2,$$

where t is time measured in seconds.
Find the average velocity of the particle over each of the following intervals.

(a) 4 to 6 sec **(b)** 4 to 5 sec

(c) Complete the following table.

h	1	.1	.01	.001	.0001
$4 + h$	5	4.1	4.01	4.001	4.0001
$s(4 + h)$	52	39.31	38.1301	38.013001	38.00130001
$s(4)$	38	38	38	38	38
$s(4 + h) - s(4)$	14	1.31	.1301	.013001	.00130001
$\dfrac{s(4 + h) - s(4)}{h}$	14	13.1			

(d) Find $\lim\limits_{h \to 0} \dfrac{s(4 + h) - s(4)}{h}$, and give the instantaneous velocity of the particle when $t = 4$.

10.3 DEFINITION OF THE DERIVATIVE

If the distances of a vehicle from a starting point are known for the corresponding elapsed times, how can the speed of the vehicle be determined?

In this section, we continue our discussion of the relationship among distance, rate, and time. We develop a formula for instantaneous velocity that will be used in Example 7 to answer the question posed above.

In the previous section, the formula

$$\lim_{h \to 0} \frac{f(x_0 + h) - f(x_0)}{h}$$

was used to calculate the instantaneous rate of change of a function f at the point where $x = x_0$. The same formula can be used to find the slope of a line *tangent* to the graph of a function f at the point $x = x_0$. (Remember that the slope of a line is really the rate of change of y with respect to x.)

In geometry, a tangent line to a circle is defined as a line that touches the circle at only one point. The semicircle in Figure 13 is the graph of a function. Comparing the various lines in Figure 14 to the tangent at point P in Figure 13 suggests that the lines in Figure 14 at P_1 and P_3 are tangent to the curve, while

FIGURE 13

FIGURE 14

the lines at P_2 and P_5 are not. The tangent lines just touch the curve, while the other lines pass through it. To decide about the line at P_4, we need to define the idea of a tangent line to the graph of a function more carefully.

To see how we might define the slope of a line tangent to the graph of a function f at a given point, let R be a fixed point with coordinates $(x_0, f(x_0))$ on the graph of a function $y = f(x)$, as in Figure 15. Choose a different point S on the graph and draw the line through R and S; this line is called a **secant line.** If S has coordinates $(x_0 + h, f(x_0 + h))$, then by the definition of slope, the slope of the secant line RS is given by

$$\text{Slope of secant} = \frac{\Delta y}{\Delta x} = \frac{f(x_0 + h) - f(x_0)}{x_0 + h - x_0} = \frac{f(x_0 + h) - f(x_0)}{h}.$$

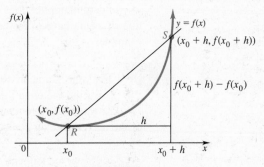

FIGURE 15

This slope corresponds to the average rate of change of y with respect to x over the interval from x_0 to $x_0 + h$. As h approaches 0, point S will slide along the curve, getting closer and closer to the fixed point R. See Figure 16, which shows successive positions S_1, S_2, S_3, and S_4 of the point S. If the slopes of the corresponding secant lines approach a limit as h approaches 0, then this limit is defined to be the slope of the tangent line at point R.

FIGURE 16

TANGENT LINE

The **tangent line** to the graph of $y = f(x)$ at the point $(x_0, f(x_0))$ is the line through this point having slope

$$\lim_{h \to 0} \frac{f(x_0 + h) - f(x_0)}{h},$$

provided this limit exists. If this limit does not exist, then there is no tangent at the point.

The slope of this line at a point is also called the **slope of the curve** at the point and corresponds to the instantaneous rate of change of y with respect to x at the point.

In certain applications of mathematics it is necessary to determine the equation of a line tangent to the graph of a function at a given point, as in the next example.

EXAMPLE 1 Find the slope of the tangent line to the graph of $f(x) = x^2 + 2$ at $x = -1$. Find the equation of the tangent line.

Use the definition given above, with $f(x) = x^2 + 2$ and $x_0 = -1$. The slope of the tangent line is given by

$$
\begin{aligned}
\text{Slope of tangent} &= \lim_{h \to 0} \frac{f(x_0 + h) - f(x_0)}{h} \\
&= \lim_{h \to 0} \frac{[(-1 + h)^2 + 2] - [(-1)^2 + 2]}{h} \\
&= \lim_{h \to 0} \frac{[1 - 2h + h^2 + 2] - [1 + 2]}{h} \\
&= \lim_{h \to 0} \frac{-2h + h^2}{h} \\
&= \lim_{h \to 0} (-2 + h) = -2.
\end{aligned}
$$

In Section 1.2, we saw that the equation of a line can be found with the point-slope form $y - y_1 = m(x - x_1)$, if the slope m and the coordinates (x_1, y_1) of a point on the line are known. Use the point-slope form to find the equation of the line with slope 3 that goes through the point $(-1, 4)$.

Let $m = 3$, $x_1 = -1$, $y_1 = 4$. Then

$$
\begin{aligned}
y - y_1 &= m(x - x_1) \\
y - 4 &= 3(x - (-1)) \\
y - 4 &= 3x + 3 \\
y &= 3x + 7.
\end{aligned}
$$

The slope of the tangent line at $(-1, f(-1)) = (-1, 3)$ is -2. The equation of the tangent line can be found with the point-slope form of the equation of a line from Chapter 1.

$$
\begin{aligned}
y - y_1 &= m(x - x_1) \\
y - 3 &= -2[x - (-1)] \\
y - 3 &= -2(x + 1) \\
y - 3 &= -2x - 2 \\
y &= -2x + 1
\end{aligned}
$$

Figure 17 shows a graph of $f(x) = x^2 + 2$, along with a graph of the tangent line at $x = -1$. ◀

FIGURE 17

The Derivative The special limit

$$\lim_{h \to 0} \frac{f(x_0 + h) - f(x_0)}{h}$$

has now been developed in two different ways—as the instantaneous rate of change of y with respect to x and also as the slope of a tangent line. This particular limit is so important that it is given a special name, the *derivative*.

DERIVATIVE	The **derivative** of the function f at x, written $f'(x)$, is defined as $$f'(x) = \lim_{h \to 0} \frac{f(x + h) - f(x)}{h},$$ provided this limit exists.

The notation $f'(x)$ is read "f-prime of x". The alternative notation y' (read "y-prime") is sometimes used for the derivative of $y = f(x)$.

Notice that the derivative is a function of x since $f'(x)$ varies as x varies. This differs from the slope of the tangent, given above, in which a specific x-value, x_0, is used. This new function given by $y' = f'(x)$ has as its domain all the points at which the specified limit exists, and its value at x is $f'(x)$. The function f' is called the derivative of f with respect to x. If x is a value in the domain of f, and if $f'(x)$ exists, then f is **differentiable** at x.

In summary, the derivative of a function f is a new function f'. The mathematical process that produces f' is called **differentiation.** This new function has several interpretations. Two of them are given here.

1. The function f' represents the *slope* of the graph of $f(x)$ at any point x. If the derivative is evaluated at the point $x = x_0$, then it represents the slope of the curve at that point.

2. The function f' represents the *instantaneous rate of change* of y with respect to x. This instantaneous rate of change could be interpreted as marginal cost, revenue, or profit (if the original function represented cost, revenue, or profit) or velocity (if the original function described displacement along a line). From now on we will use "rate of change" to mean "instantaneous rate of change."

The next few examples show how to use the definition to find the derivative of a function by means of a four-step procedure.

EXAMPLE 2

Let $f(x) = x^2$.

(a) Find the derivative.

By definition, for all values of x where the following limit exists, the derivative is given by

$$f'(x) = \lim_{h \to 0} \frac{f(x + h) - f(x)}{h}.$$

Use the following sequence of steps to evaluate this limit.

Step 1 Find $f(x + h)$.

Replace x with $x + h$ in the equation for $f(x)$. Simplify the result.

$$f(x) = x^2$$
$$f(x + h) = (x + h)^2$$
$$= x^2 + 2xh + h^2$$

(Note that $f(x + h) \neq f(x) + h$, since $f(x) + h = x^2 + h$.)

Step 2 Find $f(x + h) - f(x)$.

Since $f(x) = x^2$,

$$f(x + h) - f(x) = (x^2 + 2xh + h^2) - x^2 = 2xh + h^2.$$

Step 3 Find and simplify the quotient $\dfrac{f(x + h) - f(x)}{h}$.

$$\frac{f(x + h) - f(x)}{h} = \frac{2xh + h^2}{h}$$
$$= \frac{h(2x + h)}{h} = 2x + h$$

Step 4 Finally, find the limit as h approaches 0. In this step, h is the variable and x is fixed.

$$f'(x) = \lim_{h \to 0} \frac{f(x + h) - f(x)}{h}$$
$$= \lim_{h \to 0} (2x + h)$$
$$= 2x + 0 = 2x$$

(b) Calculate and interpret $f'(3)$. Use the function defined by $f'(x) = 2x$.

$$f'(3) = 2 \cdot 3 = 6$$

The number 6 is the slope of the tangent line to the graph of $f(x) = x^2$ at the point where $x = 3$, that is, at $(3, f(3)) = (3, 9)$. See Figure 18. ◀

FIGURE 18

> **Caution** In Example 2(b), do not confuse $f(3)$ and $f'(3)$. The value $f(3)$ is the y-value that corresponds to $x = 3$. It is found by substituting 3 for x in $f(x)$; $f(3) = 3^2 = 9$. On the other hand, $f'(3)$ is the slope of the tangent line to the curve at $x = 3$; as Example 2 shows, $f'(3) = 2 \cdot 3 = 6$.

FINDING $f'(x)$ FROM THE DEFINITION OF DERIVATIVE

The four steps used when finding the derivative $f'(x)$ for a function $y = f(x)$ are summarized here.

1. Find $f(x + h)$.

2. Find and simplify $f(x + h) - f(x)$.

3. Divide by h to get $\dfrac{f(x + h) - f(x)}{h}$.

4. Let $h \to 0$; $f'(x) = \lim\limits_{h \to 0} \dfrac{f(x + h) - f(x)}{h}$ if this limit exists.

EXAMPLE **3** Let $f(x) = 2x^3 + 4x$. Find $f'(x)$, $f'(2)$, and $f'(-3)$.
Go through the four steps used above to find $f'(x)$.

Step 1 Find $f(x + h)$ by replacing x with $x + h$.

$$\begin{aligned}
f(x + h) &= 2(x + h)^3 + 4(x + h) \\
&= 2(x^3 + 3x^2h + 3xh^2 + h^3) + 4(x + h) \\
&= 2x^3 + 6x^2h + 6xh^2 + 2h^3 + 4x + 4h
\end{aligned}$$

Step 2 $f(x + h) - f(x) = 2x^3 + 6x^2h + 6xh^2 + 2h^3 + 4x + 4h - 2x^3 - 4x$

$$= 6x^2h + 6xh^2 + 2h^3 + 4h$$

Step 3 $\dfrac{f(x + h) - f(x)}{h} = \dfrac{6x^2h + 6xh^2 + 2h^3 + 4h}{h}$

$$= \dfrac{h(6x^2 + 6xh + 2h^2 + 4)}{h}$$

$$= 6x^2 + 6xh + 2h^2 + 4$$

Step 4 Now use the rules for limits to get

$$f'(x) = \lim_{h \to 0} \frac{f(x + h) - f(x)}{h}$$

$$= \lim_{h \to 0} (6x^2 + 6xh + 2h^2 + 4)$$

$$= 6x^2 + 6x(0) + 2(0)^2 + 4$$

$$f'(x) = 6x^2 + 4.$$

Use this result to find $f'(2)$ and $f'(-3)$.

$$f'(2) = 6 \cdot 2^2 + 4$$

$$= 28$$

$$f'(-3) = 6 \cdot (-3)^2 + 4 = 58 \quad \blacktriangleleft$$

EXAMPLE **4** Let $f(x) = \dfrac{4}{x}$. Find $f'(x)$.

Step 1 $f(x + h) = \dfrac{4}{x + h}$

Step 2 $f(x + h) - f(x) = \dfrac{4}{x + h} - \dfrac{4}{x}$

$$= \dfrac{4x - 4(x + h)}{x(x + h)} \qquad \text{Find a common denominator.}$$

$$= \dfrac{4x - 4x - 4h}{x(x + h)} \qquad \text{Simplify the numerator.}$$

$$= \dfrac{-4h}{x(x + h)}$$

Step 3 $\dfrac{f(x + h) - f(x)}{h} = \dfrac{\dfrac{-4h}{x(x + h)}}{h}$

$$= \dfrac{-4h}{x(x + h)} \cdot \dfrac{1}{h} \qquad \text{Invert and multiply.}$$

$$= \dfrac{-4}{x(x + h)}$$

Step 4 $f'(x) = \lim\limits_{h\to 0} \dfrac{f(x + h) - f(x)}{h}$

$\qquad\qquad = \lim\limits_{h\to 0} \dfrac{-4}{x(x + h)}$

$\qquad\qquad = \dfrac{-4}{x(x + 0)}$

$\qquad f'(x) = \dfrac{-4}{x(x)} = \dfrac{-4}{x^2}$ ◀

Notice that in Example 4 neither $f(x)$ nor $f'(x)$ is defined when $x = 0$.

EXAMPLE 5 Let $f(x) = \sqrt{x}$. Find $f'(x)$.

Step 1 $f(x + h) = \sqrt{x + h}$

Step 2 $f(x + h) - f(x) = \sqrt{x + h} - \sqrt{x}$

Step 3 $\dfrac{f(x + h) - f(x)}{h} = \dfrac{\sqrt{x + h} - \sqrt{x}}{h}$

At this point, in order to be able to divide by h, multiply both numerator and denominator by $\sqrt{x + h} + \sqrt{x}$; that is, rationalize the *numerator*.

$\dfrac{f(x + h) - f(x)}{h} = \dfrac{\sqrt{x + h} - \sqrt{x}}{h} \cdot \dfrac{\sqrt{x + h} + \sqrt{x}}{\sqrt{x + h} + \sqrt{x}}$

$\qquad\qquad = \dfrac{(\sqrt{x + h})^2 - (\sqrt{x})^2}{h(\sqrt{x + h} + \sqrt{x})}$

$\qquad\qquad = \dfrac{x + h - x}{h(\sqrt{x + h} + \sqrt{x})}$

$\qquad\qquad = \dfrac{1}{\sqrt{x + h} + \sqrt{x}}$

Step 4 $f'(x) = \lim\limits_{h\to 0} \dfrac{1}{\sqrt{x + h} + \sqrt{x}} = \dfrac{1}{\sqrt{x} + \sqrt{x}} = \dfrac{1}{2\sqrt{x}}$ ◀

EXAMPLE 6 The cost in dollars to manufacture x graphing calculators is given by $C(x) = -.005x^2 + 20x + 150$. Find the rate of change of cost with respect to the number manufactured when 100 calculators are made and when 1000 calculators are made.

The rate of change of cost is given by the derivative of the cost function,

$$C'(x) = \lim\limits_{h\to 0} \dfrac{C(x + h) - C(x)}{h}.$$

Going through the steps for finding $C'(x)$ gives

$$C'(x) = -.01x + 20.$$

When $x = 100$,

$$C'(100) = -.01(100) + 20 = 19.$$

This rate of change of cost per calculator gives the marginal cost at $x = 100$, which means the approximate cost of producing the 101st calculator is $19. When 1000 calculators are made, the marginal cost is

$$C'(1000) = -.01(1000) + 20 = 10,$$

or $10. ◄

EXAMPLE 7 A sales representative for a textbook publishing company frequently makes a four-hour drive from her home in a large city to a university in another city. During several trips, she periodically jotted down the distance (in miles) she had traveled from home. This data is shown by the points in Figure 19. She was a mathematics major in college, so she decided to find a function that described her distance from home at time t (in hours). The position of the points suggested the curve shown in the figure, with equation $s(t) = -5t^3 + 30t^2$.

FIGURE 19

Notice that this function is suitable only over the domain $[0, 4]$. Use the function to answer the following questions.

(a) How far from home will she be after 1 hour? After 1 1/2 hours?

Her distance from home after 1 hour is

$$s(1) = -5(1)^3 + 30(1)^2 = 25,$$

or 25 miles. After 1 1/2 (or 3/2) hours, her distance from home is

$$s\left(\frac{3}{2}\right) = -5\left(\frac{3}{2}\right)^3 + 30\left(\frac{3}{2}\right)^2 = \frac{405}{8} = 50.625,$$

or 50.625 miles.

(b) How far apart are the two cities?

Since the trip takes 4 hours and the distance is given by $s(t)$, the university city is $s(4) = 160$ miles from her home.

(c) How fast is she driving 1 hour into the trip? $1\frac{1}{2}$ hours into the trip?

Velocity (or speed) is the instantaneous rate of change in position with respect to time. We need to find the value of the derivative $s'(t)$ at $t = 1$ and $t = 1\frac{1}{2}$. Going through the four steps for finding the derivative gives $s'(t) = -15t^2 + 60t$ as the velocity of the car at any time t. At $t = 1$, the velocity is

$$s'(1) = -15(1)^2 + 60(1) = 45,$$

or 45 miles per hour. At $t = 1\frac{1}{2}$, the velocity is

$$s'\left(\frac{3}{2}\right) = -15\left(\frac{3}{2}\right)^2 + 60\left(\frac{3}{2}\right) \approx 56,$$

or about 56 miles per hour.

(d) Does she ever exceed the speed limit of 65 miles per hour on the trip?

To find the maximum velocity, notice that the graph of the velocity function $s'(t) = -15t^2 + 60t$ is a parabola opening downward. The maximum velocity will occur at the vertex. Verify that the vertex of this parabola is (2, 60). Thus, her maximum velocity during the trip is 60 miles per hour, so she never exceeds the speed limit. ◄

Existence of the Derivative The definition of the derivative included the phrase "provided this limit exists." If the limit used to define the derivative does not exist, then of course the derivative does not exist. For example, a derivative cannot exist at a point where the function itself is not defined. If there is no function value for a particular value of x, there can be no tangent line for that value. This was the case in Example 4—there was no tangent line (and no derivative) when $x = 0$.

Derivatives also do not exist at "corners" or "sharp points" on a graph. For example, the function graphed in Figure 20 is the **absolute value function,** defined as

$$f(x) = \begin{cases} x & \text{if } x \geq 0 \\ -x & \text{if } x < 0, \end{cases}$$

FIGURE 20

and written $f(x) = |x|$. By the definition of derivative, the derivative at any value of x is given by

$$\lim_{h \to 0} \frac{f(x + h) - f(x)}{h},$$

provided this limit exists. To find the derivative at 0 for $f(x) = |x|$, replace x with 0 and $f(x)$ with $|0|$ to get

$$\lim_{h \to 0} \frac{|0 + h| - |0|}{h} = \lim_{h \to 0} \frac{|h|}{h}.$$

FIGURE 21

In Example 2 in the first section of this chapter, we showed that

$$\lim_{h \to 0} \frac{|h|}{h} \text{ does not exist;}$$

therefore, the derivative does not exist at 0. However, the derivative does exist for all values of x other than 0.

A graph of the function $f(x) = x^{1/3}$ is shown in Figure 21. As the graph suggests, the tangent line is vertical when $x = 0$. Since a vertical line has an undefined slope, the derivative of $f(x) = x^{1/3}$ cannot exist when $x = 0$.

Figure 22 summarizes the various ways that a derivative can fail to exist.

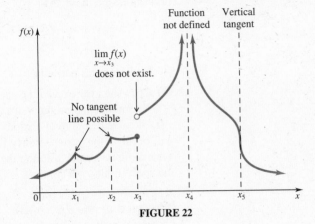

FIGURE 22

EXAMPLE 8 ▶ A nova is a star whose brightness suddenly increases and then gradually fades. The cause of the sudden increase in brightness is thought to be an explosion of some kind. The intensity of light emitted by a nova as a function of time is shown in Figure 23.* Notice that although the graph is a continuous curve, it is not differentiable at the point of the explosion. ◀

FIGURE 23

*"The Light Curve of a Nova" from *Astronomy: The Structure of the Universe* by William J. Kaufmann, III. Reprinted by permission of William J. Kaufmann, III.

10.3 Exercises

Find the slope of the tangent line to each curve when x has the given value. (Hint for Exercise 5: In Step 3, multiply numerator and denominator by $\sqrt{16 + h} + \sqrt{16}$.)

1. $f(x) = -4x^2 + 11x;\quad x = -2$ **2.** $f(x) = 6x^2 - 4x;\quad x = -1$ **3.** $f(x) = -2/x;\quad x = 4$

4. $f(x) = 6/x;\quad x = -1$ **5.** $f(x) = \sqrt{x};\quad x = 16$ **6.** $f(x) = -3\sqrt{x};\quad x = 1$

Find the equation of the tangent line to each curve when x has the given value.

7. $f(x) = x^2 + 2x;\quad x = 3$ **8.** $f(x) = 6 - x^2;\quad x = -1$ **9.** $f(x) = 5/x;\quad x = 2$

10. $f(x) = -3/(x + 1);\quad x = 1$ **11.** $f(x) = 4\sqrt{x};\quad x = 9$ **12.** $f(x) = \sqrt{x};\quad x = 25$

Estimate the slope of the tangent line to each curve at the given point (x, y).

13.

14.

15.

16.

17.

18.

19. By considering, but not calculating, the slope of the tangent line, give the derivative of each of the following.

 (a) $f(x) = 5$ **(b)** $f(x) = x$ **(c)** $f(x) = -x$ **(d)** the line $x = 3$ **(e)** the line $y = mx + b$

20. **(a)** Suppose $g(x) = \sqrt[3]{x}$. Find $g'(0)$

 (b) Explain why the derivative of a function does not exist at a point where the tangent line is vertical.

21. If $f(x) = \dfrac{x^2 - 1}{x + 2}$, where is f not differentiable?

Find $f'(x)$, $f'(2)$, $f'(0)$, and $f'(-3)$ for each of the following.

22. $f(x) = -4x^2 + 11x$ **23.** $f(x) = 6x^2 - 4x$ **24.** $f(x) = 8x + 6$ **25.** $f(x) = -9x - 5$

26. $f(x) = -\dfrac{2}{x}$ **27.** $f(x) = \dfrac{6}{x}$ **28.** $f(x) = \sqrt{x}$ **29.** $f(x) = -3\sqrt{x}$

Find the x-values where the following do not have derivatives.

30.

31.

32.

33.

34.

35.

36. If the rate of change of $f(x)$ is zero when $x = a$, what can be said about the tangent line to the graph of $f(x)$ at $x = a$?

37. For the function shown in the sketch, give the intervals or points on the x-axis where the rate of change of $f(x)$ with respect to x is **(a)** positive; **(b)** negative; and **(c)** zero.

In Exercises 38–39, tell which graph, (a) or (b), represents velocity and which represents distance from a starting point. (Hint: Consider where the derivative is zero, positive, or negative.)

38. (a)

(b)

39. (a)

(b)

Applications

Business and Economics

40. Demand Suppose the demand for a certain item is given by $D(p) = -2p^2 + 4p + 6$, where p represents the price of the item in dollars.

(a) Given $D'(p) = -4p + 4$, find the rate of change of demand with respect to price.

(b) Find and interpret the rate of change of demand when the price is $10.

41. Debt The figure gives the percent of outstanding credit card loans that were at least 30 days past due from June 1989 to March 1991.* Assume the function changes smoothly.

Source: Moody's Investors Service

(a) Estimate and interpret the rate of change in the debt level at the indicated points on the curve.

(b) During which months in 1990 did the rate of change indicate a decreasing debt level?

(c) The overall direction of the curve is upward. What does this say about the average rate of change over the time period shown in the figure?

42. Profit The profit (in dollars) from the expenditure of x thousand dollars on advertising is given by $P(x) = 1000 + 32x - 2x^2$, with $P'(x) = 32 - 4x$. Find the marginal profit at the following expenditures. In each case, decide whether the firm should increase the expenditure.

(a) $8000 (b) $6000 (c) $12,000 (d) $20,000

43. Revenue The revenue in dollars generated from the sale of x picnic tables is given by $R(x) = 20x - \dfrac{x^2}{500}$; therefore, $R'(x) = 20 - \dfrac{1}{250}x$.

(a) Find the marginal revenue when 1000 tables are sold.

(b) Estimate the revenue from the sale of the 1001st table by finding $R'(1000)$.

(c) Determine the actual revenue from the sale of the 1001st table.

(d) Compare your answers for (b) and (c). What do you find?

44. Cost The cost of producing x tacos is $C(x) = 1000 + .24x^2$, $0 \le x \le 30{,}000$; therefore, $C'(x) = .48x$.

(a) Find the marginal cost.

(b) Find and interpret the marginal cost at a production level of 100 tacos.

(c) Find the exact cost to produce the 101st taco.

(d) Compare the answers to (b) and (c). How are they related?

*Credit Card Defaults Pose Worry for Banks'' as appeared in *The Sacramento Bee*, June 1991. Reprinted by permission of The Associated Press.

Life Sciences

45. Bacteria Population A biologist has estimated that if a bactericide is introduced into a culture of bacteria, the number of bacteria, $B(t)$, present at time t (in hours) is given by $B(t) = 1000 + 50t - 5t^2$ million. If $B'(t) = 50 - 10t$, find the rate of change of the number of bacteria with respect to time after each of the following numbers of hours: **(a)** 3 **(b)** 5 **(c)** 6.

(d) When does the population of bacteria start to decline?

46. Shellfish Population In one research study, the population of a certain shellfish in an area at time t was closely approximated by the graph below. Estimate and interpret the derivative at each of the marked points.

47. Flight Speed The graph shows the relationship between the speed of the arctic tern in flight and the required power expended by its flight muscles.* Several significant flight speeds are indicated on the curve.

(a) The speed V_{mp} minimizes energy costs per unit of time. What is the slope of the line tangent to the curve at the point corresponding to V_{mp}? What is the physical significance of the slope at that point?

(b) The speed V_{mr} minimizes the energy costs per unit of distance covered. Estimate the slope of the curve at the point corresponding to V_{mr}. Give the significance of the slope at that point.

(c) The speed V_{opt} minimizes the total duration of the migratory journey. Estimate the slope of the curve at the point corresponding to V_{opt}. Relate the significance of this slope to the slopes found in (a) and (b).

(d) By looking at the shape of the curve, describe how the power level decreases and increases for various speeds.

Physical Sciences

48. Temperature The graph shows the temperature in degrees Celsius as a function of the altitude h in feet when an inversion layer is over Southern California. (See Exercise 32 in the previous section.) Estimate and interpret the derivatives of $T(h)$ at the marked points.

*"Bird Flight and Optimal Migration" by Thomas Alerstam from *Trends in Ecology and Evolution* July 1991. Copyright © 1991 by Elsevier Trends Journals. Reprinted by permission of Elsevier Trends Journals and Thomas Alerstam.

Connections

49. In Exercise 46 a graph was given of a function f that approximated the population of a certain shellfish at time t. In that exercise, the derivative was estimated at each of the marked points.

(a) Fill in the chart below with the estimate of the derivative at each of the points indicated.

t	2	10	13
$f'(t)$			

(b) Plot the points found in (a) and connect them with a smooth curve.

(c) The graph in part (b) is a graph of the derivative of the original function. Discuss what this graph tells of the behavior of the original function.

50. Use the graph given in Exercise 48 to repeat the steps in Exercise 49. Use h-values of 500, 1500, 3500, and 5000. You will need to draw the tangent line at $h = 3500$ yourself.

10.4 TECHNIQUES FOR FINDING DERIVATIVES

How can a manager determine the best production level if the relationship between profit and production is known?

From the shape of a tumor, how can a physician determine how fast the tumor is growing?

These questions can be answered by finding the derivative of an appropriate function. We shall return to them at the end of this section in Examples 7 and 8.

Using the definition to calculate the derivative of a function is a very involved process even for simple functions. In this section we develop rules that make the calculation of derivatives much easier. Keep in mind that even though the process of finding a derivative will be greatly simplified with these rules, *the interpretation of the derivative will not change*. But first, a few words about notation are in order.

Several alternative notations for the derivative are used. In the previous section the symbols y' and $f'(x)$ were used to represent the derivative of $y = f(x)$. Sometimes it is important to show that the derivative is taken with respect to a particular variable; for example, if y is a function of x, the notation

$$\frac{dy}{dx} \quad \text{or} \quad D_x y$$

(both read "the derivative of y with respect to x") can be used for the derivative of y with respect to x. The dy/dx notation for the derivative is sometimes referred to as *Leibniz notation*, named after one of the co-inventors of the calculus, Gottfried Wilhelm von Leibniz (1646–1716). (The other was Sir Isaac Newton, 1642–1727.)

Another notation is used to write a derivative without functional symbols such as f or f'. With this notation, the derivative of $y = 2x^3 + 4x$, for example, which was found in the last section to be $y' = 6x^2 + 4$, would be written

$$\frac{d}{dx}[2x^3 + 4x] = 6x^2 + 4, \quad \text{or} \quad D_x[2x^3 + 4x] = 6x^2 + 4.$$

Either $\dfrac{d}{dx}[f(x)]$ or $D_x[f(x)]$ represents the derivative of the function f with respect to x.

NOTATIONS FOR THE DERIVATIVE

The derivative of $y = f(x)$ may be written in any of the following ways:

$$f'(x), \qquad y', \qquad \frac{dy}{dx}, \qquad \frac{d}{dx}[f(x)], \qquad \text{or} \qquad D_x[f(x)].$$

A variable other than x often may be used as the independent variable. For example, if $y = f(t)$ gives population growth as a function of time, then the derivative of y with respect to t could be written

$$f'(t), \qquad \frac{dy}{dt}, \qquad \frac{d}{dt}[f(t)], \qquad \text{or} \qquad D_t[f(t)].$$

Other variables also may be used to name the function, as in $g(x)$ or $h(t)$.

Now we will use the definition

$$f'(x) = \lim_{h \to 0} \frac{f(x + h) - f(x)}{h}$$

to develop some rules for finding derivatives more easily than by the four-step process given in the previous section.

The first rule tells how to find the derivative of a constant function defined by $f(x) = k$, where k is a constant real number. Since $f(x + h)$ is also k, by definition $f'(x)$ is

$$f'(x) = \lim_{h \to 0} \frac{f(x + h) - f(x)}{h}$$

$$= \lim_{h \to 0} \frac{k - k}{h} = \lim_{h \to 0} \frac{0}{h} = \lim_{h \to 0} 0 = 0,$$

establishing the following rule.

CONSTANT RULE

If $f(x) = k$, where k is any real number, then

$$f'(x) = 0.$$

(The derivative of a constant is 0.)

FIGURE 24

Figure 24 illustrates this constant rule geometrically; it shows a graph of the horizontal line $y = k$. At any point P on this line, the tangent line at P is the line itself. Since a horizontal line has a slope of 0, the slope of the tangent line is 0. This agrees with the result above: the derivative of a constant is 0.

EXAMPLE 1

(a) If $f(x) = 9$, then $f'(x) = 0$.

(b) If $y = \pi$, then $y' = 0$.

(c) If $y = 2^3$, then $dy/dx = 0$. ◀

Functions of the form $y = x^n$, where n is a real number, are very common in applications. To get a rule for finding the derivative of such a function, we can use the definition to work out the derivatives for various special values of n. This was done in the previous section in Example 2 to show that for $f(x) = x^2$, $f'(x) = 2x$.

For $f(x) = x^3$, the derivative is found as follows.

$$f'(x) = \lim_{h \to 0} \frac{f(x + h) - f(x)}{h}$$

$$= \lim_{h \to 0} \frac{(x + h)^3 - x^3}{h}$$

$$= \lim_{h \to 0} \frac{(x^3 + 3x^2h + 3xh^2 + h^3) - x^3}{h}$$

The binomial theorem (discussed in most intermediate and college algebra texts) was used to expand $(x + h)^3$ in the last step. Now, the limit can be determined.

$$f'(x) = \lim_{h \to 0} \frac{3x^2h + 3xh^2 + h^3}{h}$$

$$= \lim_{h \to 0} (3x^2 + 3xh + h^2)$$

$$= 3x^2$$

The results in the table below were found in a similar way, using the definition of the derivative. (These results are modifications of some of the examples and exercises from the previous section.)

Function	n	Derivative
$y = x^2$	2	$y' = 2x = 2x^1$
$y = x^3$	3	$y' = 3x^2$
$y = x^4$	4	$y' = 4x^3$
$y = x^{-1}$	-1	$y' = -1 \cdot x^{-2} = \dfrac{-1}{x^2}$
$y = x^{1/2}$	$1/2$	$y' = \dfrac{1}{2}x^{-1/2} = \dfrac{1}{2x^{1/2}}$

These results suggest the following rule.

POWER RULE

If $f(x) = x^n$ for any real number n, then

$$f'(x) = nx^{n-1}.$$

(The derivative of $f(x) = x^n$ is found by multiplying by the exponent n and decreasing the exponent on x by 1.)

While the power rule is true for every real-number value of n, a proof is given here only for positive integer values of n. This proof follows the steps used above in finding the derivative of $y = x^3$.

For any real numbers p and q, by the binomial theorem,

$$(p + q)^n = p^n + np^{n-1}q + \frac{n(n-1)}{2}p^{n-2}q^2 + \cdots + npq^{n-1} + q^n.$$

Replacing p with x and q with h gives

$$(x + h)^n = x^n + nx^{n-1}h + \frac{n(n-1)}{2}x^{n-2}h^2 + \cdots + nxh^{n-1} + h^n,$$

from which

$$(x + h)^n - x^n = nx^{n-1}h + \frac{n(n-1)}{2}x^{n-2}h^2 + \cdots + nxh^{n-1} + h^n.$$

Dividing each term by h yields

$$\frac{(x + h)^n - x^n}{h} = nx^{n-1} + \frac{n(n-1)}{2}x^{n-2}h + \cdots + nxh^{n-2} + h^{n-1}.$$

Use the definition of derivative, and the fact that each term except the first contains h as a factor and thus approaches 0 as h approaches 0, to get

$$f'(x) = \lim_{h \to 0} \frac{(x + h)^n - x^n}{h}$$

$$= nx^{n-1} + \frac{n(n-1)}{2}x^{n-2}0 + \cdots + nx0^{n-2} + 0^{n-1}$$

$$f'(x) = nx^{n-1}.$$

This shows that the derivative of $f(x) = x^n$ is $f'(x) = nx^{n-1}$, proving the power rule for positive integer values of n.

 EXAMPLE 2

(a) If $y = x^6$, find y'.

$$y' = 6x^{6-1} = 6x^5$$

(b) If $y = t = t^1$, find y'.

$$y' = 1t^{1-1} = t^0 = 1$$

At this point you may wish to turn back to the Algebra Reference for a review of negative exponents and rational exponents. The relationship between powers, roots, and rational exponents is explained there.

(c) If $y = 1/x^3$, find dy/dx.

Use a negative exponent to rewrite this equation as $y = x^{-3}$; then

$$\frac{dy}{dx} = -3x^{-3-1} = -3x^{-4} \quad \text{or} \quad \frac{-3}{x^4}.$$

(d) Find $D_x(x^{4/3})$.

$$D_x(x^{4/3}) = \frac{4}{3}x^{4/3-1} = \frac{4}{3}x^{1/3}$$

(e) If $y = \sqrt{z}$, find dy/dz.

Rewrite this as $y = z^{1/2}$; then

$$\frac{dy}{dz} = \frac{1}{2}z^{1/2-1} = \frac{1}{2}z^{-1/2} \quad \text{or} \quad \frac{1}{2z^{1/2}} \quad \text{or} \quad \frac{1}{2\sqrt{z}}. \quad \blacktriangleleft$$

The next rule shows how to find the derivative of the product of a constant and a function.

CONSTANT TIMES A FUNCTION

Let k be a real number. If $f'(x)$ exists, then

$$D_x[kf(x)] = kf'(x).$$

(The derivative of a constant times a function is the constant times the derivative of the function.)

This rule is proved with the definition of the derivative and rules for limits.

$$D_x[kf(x)] = \lim_{h \to 0} \frac{kf(x + h) - kf(x)}{h}$$

$$= \lim_{h \to 0} k \frac{[f(x + h) - f(x)]}{h} \qquad \text{Factor out } k.$$

$$= k \lim_{h \to 0} \frac{f(x + h) - f(x)}{h} \qquad \text{Limit rule 1}$$

$$= kf'(x) \qquad \text{Definition of derivative}$$

 EXAMPLE 3

(a) If $y = 8x^4$, find y'.

$$y' = 8(4x^3) = 32x^3$$

(b) If $y = -\frac{3}{4}x^{12}$, find dy/dx.

$$\frac{dy}{dx} = -\frac{3}{4}(12x^{11}) = -9x^{11}$$

(c) Find $D_t(-8t)$.

$$D_t(-8t) = -8(1) = -8$$

(d) Find $D_p(10p^{3/2})$.

$$D_p(10p^{3/2}) = 10\left(\frac{3}{2}p^{1/2}\right) = 15p^{1/2}$$

(e) If $y = \dfrac{6}{x}$, find $\dfrac{dy}{dx}$.

Rewrite this as $y = 6x^{-1}$; then

$$\frac{dy}{dx} = 6(-1x^{-2}) = -6x^{-2} \quad \text{or} \quad \frac{-6}{x^2}. \quad \blacktriangleleft$$

The final rule in this section is for the derivative of a function that is a sum or difference of terms.

SUM OR DIFFERENCE RULE

If $f(x) = u(x) \pm v(x)$, and if $u'(x)$ and $v'(x)$ exist, then

$$f'(x) = u'(x) \pm v'(x).$$

(The derivative of a sum or difference of functions is the sum or difference of the derivatives.)

The proof of the sum part of this rule is as follows: If $f(x) = u(x) + v(x)$, then

$$f'(x) = \lim_{h \to 0} \frac{[u(x + h) + v(x + h)] - [u(x) + v(x)]}{h}$$

$$= \lim_{h \to 0} \frac{[u(x + h) - u(x)] + [v(x + h) - v(x)]}{h}$$

$$= \lim_{h \to 0} \left[\frac{u(x + h) - u(x)}{h} + \frac{v(x + h) - v(x)}{h}\right]$$

$$= \lim_{h \to 0} \frac{u(x + h) - u(x)}{h} + \lim_{h \to 0} \frac{v(x + h) - v(x)}{h}$$

$$= u'(x) + v'(x).$$

A similar proof can be given for the difference of two functions.

EXAMPLE **4** Find the derivative of each function.

(a) $y = 6x^3 + 15x^2$

Let $f(x) = 6x^3$ and $g(x) = 15x^2$; then $y = f(x) + g(x)$. Since $f'(x) = 18x^2$ and $g'(x) = 30x$,

$$\frac{dy}{dx} = 18x^2 + 30x.$$

(b) $p(t) = 12t^4 - 6\sqrt{t} + \dfrac{5}{t}$

Rewrite $p(t)$ as $p(t) = 12t^4 - 6t^{1/2} + 5t^{-1}$; then

$$p'(t) = 48t^3 - 3t^{-1/2} - 5t^{-2}.$$

Also, $p'(t)$ may be written as $p'(t) = 48t^3 - 3/\sqrt{t} - 5/t^2$.

(c) $f(x) = 5\sqrt[3]{x^2} + 4x^{-2} + 7$

Rewrite $f(x)$ as $f(x) = 5x^{2/3} + 4x^{-2} + 7$.

Then

$$D_x f(x) = \frac{10}{3}x^{-1/3} - 8x^{-3} + 0,$$

or

$$D_x f(x) = \frac{10}{3\sqrt[3]{x}} - \frac{8}{x^3}. \ \blacktriangleleft$$

The rules developed in this section make it possible to find the derivative of a function more directly, so that applications of the derivative can be dealt with more effectively. The following examples illustrate some business applications.

Marginal Analysis In previous sections we discussed the concepts of marginal cost, marginal revenue, and marginal profit. These concepts are summarized here.

In business and economics the rates of change of such variables as cost, revenue, and profit are important considerations. Economists use the word marginal to refer to rates of change: for example, *marginal cost* refers to the rate of change of cost. Since the derivative of a function gives the rate of change of the function, a marginal cost (or revenue, or profit) function is found by taking the derivative of the cost (or revenue, or profit) function. Roughly speaking, the marginal cost at some level of production x is the cost to produce the $(x + 1)$st item. (Similar statements could be made for revenue or profit.)

To see why it is reasonable to say that the marginal cost function is approximately the cost of producing one more unit, look at Figure 25, where $C(x)$ represents the cost of producing x units of some item. Then the cost of producing $x + 1$ units is $C(x + 1)$. The cost of the $x + $1st unit is therefore $C(x + 1) - C(x)$. This quantity is shown on the graph in Figure 25.

FIGURE 25

FIGURE 26

Now if $C(x)$ is the cost function, then the marginal cost $C'(x)$ represents the slope of the tangent line at any point $(x, C(x))$. The graph in Figure 26 shows the cost function $C(x)$ and the tangent line at a point $(x, C(x))$. Remember what it means for a line to have a given slope. If the slope of the line is $C'(x)$, then

$$\frac{\Delta y}{\Delta x} = C'(x) = \frac{C'(x)}{1},$$

and beginning at any point on the line and moving 1 unit to the right requires moving $C'(x)$ units up to get back to the line again. The vertical distance from the horizontal line to the tangent line shown in Figure 26 is therefore $C'(x)$.

Superimposing the graphs from Figures 25 and 26 as in Figure 27 shows that $C'(x)$ is indeed very close to $C(x + 1) - C(x)$. The two values are closest when $C'(x)$ is very large, so that 1 unit is relatively small.

FIGURE 27

EXAMPLE 5 Suppose that the total cost in hundreds of dollars to produce x thousand barrels of a beverage is given by

$$C(x) = 4x^2 + 100x + 500.$$

Find the marginal cost for the following values of x.

(a) $x = 5$

To find the marginal cost, first find $C'(x)$, the derivative of the total cost function.

$$C'(x) = 8x + 100$$

When $x = 5$,

$$C'(5) = 8(5) + 100 = 140.$$

After 5 thousand barrels of the beverage have been produced, the cost to produce one thousand more barrels will be *approximately* 140 hundred dollars, or $14,000.

The *actual* cost to produce one thousand more barrels is $C(6) - C(5)$:

$$C(6) - C(5) = (4 \cdot 6^2 + 100 \cdot 6 + 500) - (4 \cdot 5^2 + 100 \cdot 5 + 500)$$
$$= 1244 - 1100 = 144,$$

144 hundred dollars, or $14,400.

(b) $x = 30$

After 30 thousand barrels have been produced, the cost to produce one thousand more barrels will be approximately

$$C'(30) = 8(30) + 100 = 340,$$

or $34,000. Notice that the cost to produce an additional thousand barrels of beverage has increased by approximately $20,000 at a production level of 30,000 barrels compared to a production level of 5000 barrels. Management must be careful to keep track of marginal costs. If the marginal cost of producing an extra unit exceeds the revenue received from selling it, then the company will lose money on that unit. ◄

Demand Functions The demand function, defined by $p = f(x)$, relates the number of units x of an item that consumers are willing to purchase to the price p. (Demand functions were also discussed in Chapter 1.) The total revenue $R(x)$ is related to price per unit and the amount demanded (or sold) by the equation

$$R(x) = xp = x \cdot f(x).$$

EXAMPLE 6 The demand function for a certain product is given by

$$p = \frac{50,000 - x}{25,000}.$$

Find the marginal revenue when $x = 10,000$ units and p is in dollars.

From the given function for p, the revenue function is given by

$$R(x) = xp$$

$$= x \left(\frac{50,000 - x}{25,000} \right)$$

$$= \frac{50,000x - x^2}{25,000}$$

$$= 2x - \frac{1}{25,000} x^2.$$

The marginal revenue is

$$R'(x) = 2 - \frac{2}{25,000} x.$$

When $x = 10,000$, the marginal revenue is

$$R'(10,000) = 2 - \frac{2}{25,000} (10,000) = 1.2,$$

or $1.20 per unit. Thus, the next item sold (at sales of 10,000) will produce additional revenue of about $1.20 per unit. ◄

EXAMPLE 7

Suppose that the cost function for the product in Example 6 is given by

$$C(x) = 2100 + .25x, \quad \text{where } 0 \leq x \leq 30{,}000.$$

Find the marginal profit from the production of the following numbers of units.

(a) 15,000

From Example 6, the revenue from the sale of x units is

$$R(x) = 2x - \frac{1}{25{,}000} x^2.$$

Since profit, P, is given by $P = R - C$,

$$
\begin{aligned}
P(x) &= R(x) - C(x) \\
&= \left(2x - \frac{1}{25{,}000} x^2 \right) - (2100 + .25x) \\
&= 2x - \frac{1}{25{,}000} x^2 - 2100 - .25x \\
&= 1.75x - \frac{1}{25{,}000} x^2 - 2100.
\end{aligned}
$$

The marginal profit from the sale of x units is

$$P'(x) = 1.75 - \frac{2}{25{,}000} x = 1.75 - \frac{1}{12{,}500} x.$$

At $x = 15{,}000$ the marginal profit is

$$P'(15{,}000) = 1.75 - \frac{1}{12{,}500} (15{,}000) = .55,$$

or $.55 per unit.

(b) 21,875

When $x = 21{,}875$, the marginal profit is

$$P'(21{,}875) = 1.75 - \frac{1}{12{,}500} (21{,}875) = 0.$$

(c) 25,000

When $x = 25{,}000$, the marginal profit is

$$P'(25{,}000) = 1.75 - \frac{1}{12{,}500} (25{,}000) = -.25,$$

or $-$.25 per unit.

As shown by parts (b) and (c), if more than 21,875 units are sold, the marginal profit is negative. This indicates that increasing production beyond that level will *reduce* profit. ◀

The final example shows a medical application of the derivative as the rate of change of a function.

EXAMPLE **8**

Volume = $V = \frac{1}{3}\pi r^2 h$

FIGURE 28

A tumor has the approximate shape of a cone (see Figure 28). The radius of the base of the tumor is fixed by the bone structure at 2 centimeters, but the tumor is growing along the height of the cone (the volume of the tumor therefore being a function of its height). The formula for the volume of a cone is $V = (1/3)\pi r^2 h$, where r is the radius of the base and h is the height of the cone. Find the rate of change in the volume of the tumor with respect to the height.

The symbol dV/dh (instead of V') emphasizes that the rate of change of volume is to be found with respect to the height. For this tumor, r is fixed at 2 centimeters. By substituting 2 for r,

$$V = \frac{1}{3}\pi r^2 h$$

becomes

$$V = \frac{1}{3}\pi 2^2 \cdot h \quad \text{or} \quad V = \frac{4}{3}\pi h.$$

Since $4\pi/3$ is a constant,

$$\frac{dV}{dh} = \frac{4\pi}{3} \approx 4.2 \text{ cubic centimeters per centimeter.}$$

For each additional centimeter that the tumor grows in height, its volume will increase approximately 4.2 cubic centimeters. ◀

10.4 Exercises

Find the derivative of each function defined as follows.

1. $y = 10x^3 - 9x^2 + 6x$

2. $y = 3x^3 - x^2 - \dfrac{x}{12}$

3. $y = x^4 - 5x^3 + \dfrac{x^2}{9} + 5$

4. $y = 3x^4 + 11x^3 + 2x^2 - 4x$

5. $f(x) = 6x^{1.5} - 4x^{.5}$

6. $f(x) = -2x^{2.5} + 8x^{.5}$

7. $y = -15x^{3.2} + 2x^{1.9}$

8. $y = -24t^{5/2} - 6t^{1/2}$

9. $y = 8\sqrt{x} + 6x^{3/4}$

10. $y = -100\sqrt{x} - 11x^{2/3}$

11. $g(x) = 6x^{-5} - x^{-1}$

12. $y = 10x^{-2} + 3x^{-4} - 6x$

13. $y = x^{-5} - x^{-2} + 5x^{-1}$

14. $f(t) = \dfrac{6}{t} - \dfrac{8}{t^2}$

15. $f(t) = \dfrac{4}{t} + \dfrac{2}{t^3}$

16. $y = \dfrac{9}{x^4} - \dfrac{8}{x^3} + \dfrac{2}{x}$

17. $y = \dfrac{3}{x^6} + \dfrac{1}{x^5} - \dfrac{7}{x^2}$

18. $p(x) = -10x^{-1/2} + 8x^{-3/2}$

19. $h(x) = x^{-1/2} - 14x^{-3/2}$

20. $y = \dfrac{6}{\sqrt[4]{x}}$

21. $y = \dfrac{-2}{\sqrt[3]{x}}$

22. Which of the following describes the derivative function f' of a quadratic function f?

 (a) quadratic **(b)** linear **(c)** constant **(d)** cubic (third degree)

Find each of the following.

23. $\dfrac{dy}{dx}$ if $y = 8x^{-5} - 9x^{-4}$

24. $\dfrac{dy}{dx}$ if $y = -3x^{-2} - 4x^{-5}$

25. $D_x\left[9x^{-1/2} + \dfrac{2}{x^{3/2}}\right]$

26. $D_x \left[\dfrac{8}{\sqrt[4]{x}} - \dfrac{3}{\sqrt{x^3}} \right]$

27. $f'(-2)$ if $f(x) = \dfrac{x^2}{6} - 4x$

28. $f'(3)$ if $f(x) = \dfrac{x^3}{9} - 8x^2$

29. Which of the following expressions equals $D_x \left(\dfrac{6}{x} \right)$?

 (a) $\dfrac{1}{6x^2}$ **(b)** $-\dfrac{1}{6x^2}$ **(c)** $-\dfrac{6}{x^2}$ **(d)** $6x^{-2}$

30. Which of the following does *not* equal $\dfrac{d}{dx}(4x^3 - 6x^{-2})$?

 (a) $\dfrac{12x^2 + 12}{x^3}$ **(b)** $\dfrac{12x^5 + 12}{x^3}$ **(c)** $12x^2 + \dfrac{12}{x^3}$ **(d)** $12x^3 + 12x^{-3}$

31. Explain the relationship between the slope and the derivative of $f(x)$ at $x = a$.

In Exercises 32–35 find the slope of the tangent line to the graph of the given function at the given value of x. Find the equation of the tangent line in Exercises 32 and 33.

32. $y = x^4 - 5x^3 + 2$; $x = 2$

33. $y = -2x^5 - 7x^3 + 8x^2$; $x = 1$

34. $y = -2x^{1/2} + x^{3/2}$; $x = 4$

35. $y = -x^{-3} + x^{-2}$; $x = 1$

36. Find all points on the graph of $f(x) = 9x^2 - 8x + 4$ where the slope of the tangent line is 0.

37. At what points on the graph of $f(x) = 6x^2 + 4x - 9$ is the slope of the tangent line -2?

38. Use the information given in the figure to find the following values.

 (a) $f(1)$ **(b)** $f'(1)$ **(c)** the domain of f **(d)** the range of f

39. Explain the concept of marginal cost. How does it relate to cost? How is it found?

Applications

Business and Economics

40. Revenue Assume that a demand equation is given by $x = 5000 - 100p$. Find the marginal revenue for the following production levels (values of x). (*Hint:* Solve the demand equation for p and use $R(x) = xp$.)

 (a) 1000 units **(b)** 2500 units **(c)** 3000 units

41. Profit Suppose that for the situation in Exercise 40 the cost of producing x units is given by $C(x) = 3000 - 20x + .03x^2$. Find the marginal profit for each of the following production levels.

 (a) 500 units **(b)** 815 units **(c)** 1000 units

42. Demand If the price of a product is given by

$$p(x) = \frac{1000}{x} + 1000,$$

where x represents the demand for the product, find the rate of change of price when the demand is 10.

43. Sales Often sales of a new product grow rapidly at first and then level off with time. This is the case with the sales represented by the function

$$S(t) = 100 - 100t^{-1},$$

where t represents time in years. Find the rate of change of sales for the following numbers of years.

(a) 1 **(b)** 10

44. Profit The profit in dollars from the sale of x thousand expensive cassette recorders is

$$P(x) = x^3 - 5x^2 + 7x.$$

Find the marginal profit for the following numbers of recorders.

(a) 1000 **(b)** 2000 **(c)** 5000 **(d)** 10,000

(e) Interpret the results from parts (a)–(d).

45. Profit An analyst has found that a company's costs and revenues in dollars for one product are given by

$$C(x) = 2x \quad \text{and} \quad R(x) = 6x - \frac{x^2}{1000},$$

respectively, where x is the number of items produced.

(a) Find the marginal cost function.

(b) Find the marginal revenue function.

(c) Using the fact that profit is the difference between revenue and costs, find the marginal profit function.

(d) What value of x makes marginal profit equal 0?

(e) Find the profit when the marginal profit is 0.

(As we shall see in the next chapter, this process is used to find *maximum* profit.)

Life Sciences

46. Blood Sugar Level Insulin affects the glucose, or blood sugar, level of some diabetics according to the function

$$G(x) = -.2x^2 + 450,$$

where $G(x)$ is the blood sugar level one hour after x units of insulin are injected. (This mathematical model is only approximate, and it is valid only for values of x less than about 40.)

Find the blood sugar level after the following numbers of units of insulin are injected.

(a) 0 **(b)** 25

Find the rate of change of blood sugar level after injection of the following numbers of units of insulin.

(c) 10 **(d)** 25

47. Blood Vessel Volume A short length of blood vessel has a cylindrical shape. The volume of a cylinder is given by $V = \pi r^2 h$. Suppose an experimental device is set up to measure the volume of blood in a blood vessel having a fixed length of 80 mm.

(a) Find a function that gives the rate of change of the volume of blood with respect to the radius of the blood vessel.

Suppose a drug is administered that causes the blood vessel to expand. Find the rate of change of volume for each of the following radii.

(b) 4 mm **(c)** 6 mm **(d)** 8 mm

(e) What is true of the volume as the radius increases?

48. Insect Mating Patterns In an experiment testing methods of sexually attracting male insects to sterile females, equal numbers of males and females of a certain species are permitted to intermingle. Assume that

$$M(t) = 4t^{3/2} + 2t^{1/2}$$

approximates the number of matings observed among the insects in an hour, where t is the temperature in degrees Celsius. (This formula is valid only for certain temperature ranges.) Find the number of matings at the following temperatures.

(a) 16°C **(b)** 25°C

(c) Find the rate of change in the number of matings when the temperature is 16°C.

Physical Sciences

49. Acid Concentration Suppose $P(t) = 100/t$ represents the percent of acid in a chemical solution after t days of exposure to an ultraviolet light source.

Find the percent of acid in the solution after the following numbers of days.

(a) 1 **(b)** 100

(c) Find and interpret $P'(100)$.

Velocity *We saw earlier in this chapter that the velocity of a particle moving in a straight line is given by*

$$\lim_{h \to 0} \frac{s(t + h) - s(t)}{h},$$

where $s(t)$ gives the position of the particle at time t. This limit is actually the derivative of $s(t)$, so the velocity of the particle is given by $s'(t)$. If $v(t)$ represents velocity at time t, then $v(t) = s'(t)$. For each of the position functions in Exercises 50–53, find (a) $v(t)$ and (b) the velocity when $t = 0$, $t = 5$, and $t = 10$.

50. $s(t) = 11t^2 + 4t + 2$ **51.** $s(t) = 25t^2 - 9t + 8$

52. $s(t) = 4t^3 + 8t^2$ **53.** $s(t) = -2t^3 + 4t^2 - 1$

54. Velocity If a rock is dropped from a 144-foot building, its position (in feet above the ground) is given by $s(t) = -16t^2 + 144$, where t is the time in seconds since it was dropped.

(a) What is its velocity 1 second after being dropped? 2 seconds after being dropped?

(b) When will it hit the ground?

(c) What is its velocity upon impact?

55. Velocity A ball is thrown vertically upward from the ground at a velocity of 64 feet per second. Its distance from the ground at t seconds is given by $s(t) = -16t^2 + 64t$.

(a) How fast is the ball moving 2 seconds after being thrown? 3 seconds after being thrown?

(b) How long after the ball is thrown does it reach its maximum height?

(c) How high will it go?

General Interest

56. Living Standards Living standards are defined by the total output of goods and services divided by the total population. In the United States during the 1980s, living standards were closely approximated by

$$f(x) = -.023x^3 + .3x^2 - .4x + 11.6,$$

where $x = 0$ corresponds to 1981. Find the derivative of f. Use the derivative to find the rate of change in living standards in the following years.

(a) 1981 (b) 1983 (c) 1988 (d) 1989 (e) 1990

(f) What do your answers to parts (a)–(e) tell you about living standards in those years?

10.5 DERIVATIVES OF PRODUCTS AND QUOTIENTS

A manufacturer of small motors wants to make the average cost per motor as small as possible. How can this be done?

We show how the derivative is used to solve a problem like this in Example 5, later in this section. In the previous section we developed several rules for finding derivatives. We develop two additional rules in this section, again using the definition of the derivative.

The derivative of a sum of two functions is found from the sum of the derivatives. What about products? Is the derivative of a product equal to the product of the derivatives? For example, if

$$u(x) = 2x + 3 \quad \text{and} \quad v(x) = 3x^2,$$

then $\quad u'(x) = 2 \quad \text{and} \quad v'(x) = 6x.$

Let $f(x)$ be the product of u and v; that is, $f(x) = (2x + 3)(3x^2) = 6x^3 + 9x^2$. By the rules of the preceding section, $f'(x) = 18x^2 + 18x$. On the other hand, $u'(x) = 2$ and $v'(x) = 6x$, with the product $u'(x) \cdot v'(x) = 2(6x) = 12x \neq f'(x)$. In this example, the derivative of a product is *not* equal to the product of the derivatives, nor is this usually the case.

The rule for finding derivatives of products is given below.

PRODUCT RULE

If $f(x) = u(x) \cdot v(x)$, and if $u'(x)$ and $v'(x)$ both exist, then

$$f'(x) = u(x) \cdot v'(x) + v(x) \cdot u'(x).$$

(The derivative of a product of two functions is the first function times the derivative of the second, plus the second function times the derivative of the first.)

To sketch the method used to prove the product rule, let

$$f(x) = u(x) \cdot v(x).$$

This proof uses several of the rules for limits given in the first section of this chapter. You may want to review them at this time.

Then $f(x + h) = u(x + h) \cdot v(x + h)$, and, by definition, $f'(x)$ is given by

$$f'(x) = \lim_{h \to 0} \frac{f(x + h) - f(x)}{h}$$

$$= \lim_{h \to 0} \frac{u(x + h) \cdot v(x + h) - u(x) \cdot v(x)}{h}.$$

Now subtract and add $u(x + h) \cdot v(x)$ in the numerator, giving

$$f'(x) = \lim_{h \to 0} \frac{u(x + h) \cdot v(x + h) - u(x + h) \cdot v(x) + u(x + h) \cdot v(x) - u(x) \cdot v(x)}{h}$$

$$= \lim_{h \to 0} \frac{u(x + h)[v(x + h) - v(x)] + v(x)[u(x + h) - u(x)]}{h}$$

$$= \lim_{h \to 0} u(x + h)\left[\frac{v(x + h) - v(x)}{h}\right] + \lim_{h \to 0} v(x)\left[\frac{u(x + h) - u(x)}{h}\right]$$

$$= \lim_{h \to 0} u(x + h) \cdot \lim_{h \to 0} \frac{v(x + h) - v(x)}{h} + \lim_{h \to 0} v(x) \cdot \lim_{h \to 0} \frac{u(x + h) - u(x)}{h}. \quad \textbf{(1)}$$

If u' and v' both exist, then

$$\lim_{h \to 0} \frac{u(x + h) - u(x)}{h} = u'(x) \quad \text{and} \quad \lim_{h \to 0} \frac{v(x + h) - v(x)}{h} = v'(x).$$

The fact that u' exists can be used to prove

$$\lim_{h \to 0} u(x + h) = u(x),$$

and since no h is involved in $v(x)$,

$$\lim_{h \to 0} v(x) = v(x).$$

Substituting these results into equation (1) gives

$$f'(x) = u(x) \cdot v'(x) + v(x) \cdot u'(x),$$

the desired result.

EXAMPLE 1

Let $f(x) = (2x + 3)(3x^2)$. Use the product rule to find $f'(x)$.

Here f is given as the product of $u(x) = 2x + 3$ and $v(x) = 3x^2$. By the product rule and the fact that $u'(x) = 2$ and $v'(x) = 6x$,

$$f'(x) = u(x) \cdot v'(x) + v(x) \cdot u'(x)$$
$$= (2x + 3)(6x) + (3x^2)(2)$$
$$= 12x^2 + 18x + 6x^2 = 18x^2 + 18x.$$

This result is the same as that found at the beginning of the section. ◄

EXAMPLE 2

Find the derivative of $y = (\sqrt{x} + 3)(x^2 - 5x)$.

Let $u(x) = \sqrt{x} + 3 = x^{1/2} + 3$, and $v(x) = x^2 - 5x$. Then

$$y' = u(x) \cdot v'(x) + v(x) \cdot u'(x)$$
$$= (x^{1/2} + 3)(2x - 5) + (x^2 - 5x)\left(\frac{1}{2}x^{-1/2}\right).$$

Simplify by multiplying and combining terms.

$$y' = (2x)(x^{1/2}) + 6x - 5x^{1/2} - 15 + (x^2)\left(\frac{1}{2}x^{-1/2}\right) - (5x)\left(\frac{1}{2}x^{-1/2}\right)$$

$$= 2x^{3/2} + 6x - 5x^{1/2} - 15 + \frac{1}{2}x^{3/2} - \frac{5}{2}x^{1/2}$$

$$= \frac{5}{2}x^{3/2} + 6x - \frac{15}{2}x^{1/2} - 15 \quad ◄$$

We could have found the derivatives above by multiplying out the original functions. The product rule then would not have been needed. In the next section, however, we shall see products of functions where the product rule is essential.

What about *quotients* of functions? To find the derivative of the quotient of two functions, use the next result.

QUOTIENT RULE

If $f(x) = \dfrac{u(x)}{v(x)}$, if all indicated derivatives exist, and if $v(x) \neq 0$, then

$$f'(x) = \frac{v(x) \cdot u'(x) - u(x) \cdot v'(x)}{[v(x)]^2}$$

(The derivative of a quotient is the denominator times the derivative of the numerator, minus the numerator times the derivative of the denominator, all divided by the square of the denominator.)

The proof of the quotient rule is similar to that of the product rule and is left for the exercises.

EXAMPLE 3

Find $f'(x)$ if $f(x) = \dfrac{2x - 1}{4x + 3}$.

You may want to consult the Rational Expressions section of the Algebra Reference to help you work with the fractions in this section.

Let $u(x) = 2x - 1$, with $u'(x) = 2$. Also, let $v(x) = 4x + 3$, with $v'(x) = 4$. Then, by the quotient rule,

$$\begin{aligned}
f'(x) &= \frac{v(x) \cdot u'(x) - u(x) \cdot v'(x)}{[v(x)]^2} \\[2mm]
&= \frac{(4x + 3)(2) - (2x - 1)(4)}{(4x + 3)^2} \\[2mm]
&= \frac{8x + 6 - 8x + 4}{(4x + 3)^2} \\[2mm]
&= \frac{10}{(4x + 3)^2}.
\end{aligned}$$ ◀

Caution In the second step of Example 3, we had the expression

$$\frac{(4x + 3)(2) - (2x - 1)(4)}{(4x + 3)^2}.$$

Students often incorrectly "cancel" the $4x + 3$ in the numerator with one factor of the denominator. Because the numerator is a *difference* of two products, however, you must multiply and combine terms *before* looking for common factors in the numerator and denominator.

EXAMPLE 4

Find $D_x\left(\dfrac{(3 - 4x)(5x + 1)}{7x - 9}\right)$.

This function has a product within a quotient. Instead of multiplying the factors in the numerator first (which is an option), we can use the quotient rule together with the product rule, as follows. Use the quotient rule first to get

$$D_x\left(\frac{(3 - 4x)(5x + 1)}{7x - 9}\right) = \frac{(7x - 9)[D_x(3 - 4x)(5x + 1)] - [(3 - 4x)(5x + 1)D_x(7x - 9)]}{(7x - 9)^2}.$$

Now use the product rule to find $D_x(3 - 4x)(5x + 1)$ in the numerator.

$$= \frac{(7x - 9)[(3 - 4x)5 + (5x + 1)(-4)] - (3 + 11x - 20x^2)(7)}{(7x - 9)^2}$$

$$= \frac{(7x - 9)(15 - 20x - 20x - 4) - (21 + 77x - 140x^2)}{(7x - 9)^2}$$

$$= \frac{(7x - 9)(11 - 40x) - 21 - 77x + 140x^2}{(7x - 9)^2}$$

$$= \frac{-280x^2 + 437x - 99 - 21 - 77x + 140x^2}{(7x - 9)^2}$$

$$= \frac{-140x^2 + 360x - 120}{(7x - 9)^2} \quad \blacktriangleleft$$

Average Cost Suppose $y = C(x)$ gives the total cost to manufacture x items. As mentioned earlier, the average cost per item is found by dividing the total cost by the number of items. The rate of change of average cost, called the *marginal average cost*, is the derivative of the average cost.

AVERAGE COST

If the total cost to manufacture x items is given by $C(x)$, then the **average cost per item** is

$$\overline{C}(x) = \frac{C(x)}{x}.$$

The **marginal average cost** is the derivative of the average cost function, $\overline{C}'(x)$.

A company naturally would be interested in making the average cost as small as possible. The next chapter will show that this can be done by using the derivative of $C(x)/x$. This derivative often can be found by means of the quotient rule, as in the next example.

 Suppose the cost in dollars of manufacturing x hundred small motors is given by

$$C(x) = \frac{3x^2 + 120}{2x + 1}.$$

 (a) Find the average cost per hundred motors.

The average cost is defined by

$$\overline{C}(x) = \frac{C(x)}{x} = \frac{3x^2 + 120}{2x + 1} \cdot \frac{1}{x} = \frac{3x^2 + 120}{2x^2 + x}.$$

(b) Find the marginal average cost.

The marginal average cost is given by

$$\frac{d}{dx}\left(\overline{C}(x)\right) = \frac{(2x^2 + x)(6x) - (3x^2 + 120)(4x + 1)}{(2x^2 + x)^2}$$

$$= \frac{12x^3 + 6x^2 - 12x^3 - 480x - 3x^2 - 120}{(2x^2 + x)^2}$$

$$= \frac{3x^2 - 480x - 120}{(2x^2 + x)^2}.$$

(c) Average cost is generally minimized when the marginal average cost is zero. Find the level of production that minimizes average cost.

Set the derivative $\overline{C}'(x) = 0$ and solve for x.

$$\frac{3x^2 - 480x - 120}{(2x^2 + x)^2} = 0$$

$$3x^2 - 480x - 120 = 0$$

$$3(x^2 - 160x - 40) = 0$$

Use the quadratic formula to solve this quadratic equation. Discarding the negative solution leaves $x \approx 160$ as the solution. Since x is in hundreds, production of 160 hundred or 16,000 motors will minimize average cost. ◀

10.5 Exercises

Use the product rule to find the derivative of each of the following. (Hint for Exercises 8–11: *Write the quantity as a product.*)

1. $y = (2x - 5)(x + 4)$

2. $y = (3x + 7)(x - 1)$

3. $y = (3x^2 + 2)(2x - 1)$

4. $y = (5x^2 - 1)(4x + 3)$

5. $y = (2t^2 - 6t)(t + 2)$

6. $y = (9x^2 + 7x)(x^2 - 1)$

7. $y = (2x^2 - 4x)(5x^2 + 4)$

8. $y = (2x - 5)^2$

9. $y = (7x - 6)^2$

10. $k(t) = (t^2 - 1)^2$

11. $g(t) = (3t^2 + 2)^2$

12. $y = (x + 1)(\sqrt{x} + 2)$

13. $y = (2x - 3)(\sqrt{x} - 1)$

14. $g(x) = (5\sqrt{x} - 1)(2\sqrt{x} + 1)$

15. $g(x) = (-3\sqrt{x} + 6)(4\sqrt{x} - 2)$

Use the quotient rule to find the derivative of each of the following.

16. $f(x) = \dfrac{7x + 1}{3x + 8}$

17. $f(x) = \dfrac{6x - 11}{8x + 1}$

18. $y = \dfrac{2}{3x - 5}$

19. $y = \dfrac{-4}{2x - 11}$

20. $y = \dfrac{5 - 3t}{4 + t}$

21. $y = \dfrac{9 - 7t}{1 - t}$

22. $y = \dfrac{x^2 + x}{x - 1}$

23. $y = \dfrac{x^2 - 4x}{x + 3}$

24. $f(t) = \dfrac{4t + 11}{t^2 - 3}$

25. $y = \dfrac{-x^2 + 6x}{4x^2 + 1}$

26. $g(x) = \dfrac{x^2 - 4x + 2}{x + 3}$

27. $k(x) = \dfrac{x^2 + 7x - 2}{x - 2}$

28. $p(t) = \dfrac{\sqrt{t}}{t - 1}$

29. $r(t) = \dfrac{\sqrt{t}}{2t + 3}$

30. $y = \dfrac{5x + 6}{\sqrt{x}}$

31. Find the error in the following work.

$$D_x\left(\frac{2x + 5}{x^2 - 1}\right) = \frac{(2x + 5)(2x) - (x^2 - 1)2}{(x^2 - 1)^2} = \frac{4x^2 + 10x - 2x^2 + 2}{(x^2 - 1)^2} = \frac{2x^2 + 10x + 2}{(x^2 - 1)^2}$$

32. Find the error in the following work.

$$D_x\left(\frac{x^2 - 4}{x^3}\right) = x^3(2x) - (x^2 - 4)(3x^2) = 2x^4 - 3x^4 + 12x^2 = -x^4 + 12x^2$$

33. Find an equation of the line tangent to the graph of $f(x) = x/(x - 2)$ at (3, 3).

34. Following the steps used to prove the product rule for derivatives, prove the quotient rule for derivatives.

Applications

Business and Economics

35. Average Cost The total cost (in hundreds of dollars) to produce x units of perfume is

$$C(x) = \frac{3x + 2}{x + 4}.$$

Find the average cost for each of the following production levels.

(a) 10 units **(b)** 20 units **(c)** x units

(d) Find the marginal average cost function.

36. Average Profit The total profit (in tens of dollars) from selling x self-help books is

$$P(x) = \frac{5x - 6}{2x + 3}.$$

Find the average profit from each of the following sales levels.

(a) 8 books **(b)** 15 books **(c)** x books

(d) Find the marginal average profit function.

(e) Is this a reasonable function for profit? Why?

37. Fuel Efficiency Suppose you are the manager of a trucking firm, and one of your drivers reports that she has calculated that her truck burns fuel at the rate of

$$G(x) = \frac{1}{200}\left(\frac{800 + x^2}{x}\right)$$

gallons per mile when traveling at x miles per hour on a smooth, dry road.

(a) If the driver tells you that she wants to travel 20 miles per hour, what should you tell her? (*Hint:* Take the derivative of G and evaluate it for $x = 20$. Then interpret your results.)

(b) If the driver wants to go 40 miles per hour, what should you say? (*Hint:* Find $G'(40)$.)

38. Employee Training A company that manufactures bicycles has determined that a new employee can assemble $M(d)$ bicycles per day after d days of on-the-job training, where

$$M(d) = \frac{200d}{3d + 10}.$$

(a) Find the rate of change function for the number of bicycles assembled with respect to time.

(b) Find and interpret $M'(2)$ and $M'(5)$.

Life Sciences

39. Muscle Reaction When a certain drug is injected into a muscle, the muscle responds by contracting. The amount of contraction, s, in millimeters, is related to the concentration of the drug, x, in milliliters, by

$$s(x) = \frac{x}{m + nx},$$

where m and n are constants.

(a) Find $s'(x)$.

(b) Find the rate of contraction when the concentration of the drug is 50 ml, $m = 10$, and $n = 3$.

40. Bacteria Population Assume that the total number (in millions) of bacteria present in a culture at a certain time t (in hours) is given by

$$N(t) = (t - 10)^2(2t) + 50.$$

(a) Find $N'(t)$.

Find the rate at which the population of bacteria is changing at each of the following times.

(b) 8 hr **(c)** 11 hr

(d) The answer in part (b) is negative, and the answer in part (c) is positive. What does this mean in terms of the population of bacteria?

Social Sciences

41. Memory Retention Some psychologists contend that the number of facts of a certain type that are remembered after t hr is given by

$$f(t) = \frac{90t}{99t - 90}.$$

Find the rate at which the number of facts remembered is changing after the following numbers of hours.

(a) 1 **(b)** 10

10.6 THE CHAIN RULE

Suppose we know how fast the radius of a circular oil slick is growing, and we know how much the area of the oil slick is growing per unit of change in the radius. How *fast* is the area growing?

The answer to this question involves the chain rule for derivatives. Before discussing the chain rule, we consider the composition of functions. Many of the most useful functions for modeling are created by combining simpler functions. Viewing complex functions as combinations of simpler functions often makes them easier to understand and use.

Composition of Functions Suppose a function f assigns to each element x in set X some element $y = f(x)$ in set Y. Suppose also that a function g takes each element in set Y and assigns to it a value $z = g[f(x)]$ in set Z. By using both f and g, an element x in X is assigned to an element z in Z, as illustrated in Figure 29. The result of this process is a new function called the *composition* of functions g and f and defined as follows.

FIGURE 29

COMPOSITE FUNCTION Let f and g be functions. The **composite function,** or **composition,** of g and f is the function whose values are given by $g[f(x)]$ for all x in the domain of f such that $f(x)$ is in the domain of g. (Read $g[f(x)]$ as "g of f of x".)

EXAMPLE 1 Let $f(x) = 2x - 1$ and $g(x) = \sqrt{3x + 5}$. Find each of the following.

(a) $g[f(4)]$

Find $f(4)$ first.

$$f(4) = 2 \cdot 4 - 1 = 8 - 1 = 7$$

Then

$$g[f(4)] = g[7] = \sqrt{3 \cdot 7 + 5} = \sqrt{26}.$$

(b) $f[g(4)]$

Since $g(4) = \sqrt{3 \cdot 4 + 5} = \sqrt{17}$,

$$f[g(4)] = 2 \cdot \sqrt{17} - 1 = 2\sqrt{17} - 1.$$

You may want to review how to find the domain of a function. Domain was discussed in Section 1.1.

(c) $f[g(-2)]$ does not exist since -2 is not in the domain of g. ◄

EXAMPLE 2 Let $f(x) = 4x + 1$ and $g(x) = 2x^2 + 5x$. Find each of the following.

(a) $g[f(x)]$

Using the given functions, we have

$$\begin{aligned}
g[f(x)] &= g[4x + 1] \\
&= 2(4x + 1)^2 + 5(4x + 1) \\
&= 2(16x^2 + 8x + 1) + 20x + 5 \\
&= 32x^2 + 16x + 2 + 20x + 5 \\
&= 32x^2 + 36x + 7.
\end{aligned}$$

(b) $f[g(x)]$

By the definition above, with f and g interchanged,

$$\begin{aligned}
f[g(x)] &= f[2x^2 + 5x] \\
&= 4(2x^2 + 5x) + 1 \\
&= 8x^2 + 20x + 1. \quad ◄
\end{aligned}$$

As Example 2 shows, it is not always true that $f[g(x)] = g[f(x)]$. (In fact, it is rare to find two functions f and g such that $f[g(x)] = g[f(x)]$. The domain of both composite functions given in Example 2 is the set of all real numbers.

We now return to the question at the beginning of this section.

 The Chain Rule A leaking oil well off the Gulf Coast is spreading a circular film of oil over the water surface. At any time t (in minutes) after the beginning of the leak, the radius of the circular oil slick is given by

$$r(t) = 4t, \qquad \text{with} \qquad \frac{dr}{dt} = 4,$$

where dr/dt is the rate of change in radius over time. The area of the oil slick is given by

$$A(r) = \pi r^2, \qquad \text{with} \qquad \frac{dA}{dr} = 2\pi r,$$

where dA/dr is the rate of change in area per unit change in radius.

As these derivatives show, the radius is increasing 4 times as fast as the time t, and the area is increasing $2\pi r$ times as fast as the radius r. It seems reasonable, then, that the area is increasing $2\pi r \cdot 4 = 8\pi r$ times as fast as time. That is,

$$\frac{dA}{dt} = \frac{dA}{dr} \cdot \frac{dr}{dt} = 2\pi r \cdot 4 = 8\pi r.$$

Since $r = 4t$,

$$\frac{dA}{dt} = 8\pi r = 8\pi(4t) = 32\pi t.$$

To check this, use the fact that $r = 4t$ and $A = \pi r^2$ to get the same result:

$$A = \pi(4t)^2 = 16\pi t^2, \quad \text{with} \quad \frac{dA}{dt} = 32\pi t.$$

The product used above,

$$\frac{dA}{dt} = \frac{dA}{dr} \cdot \frac{dr}{dt},$$

is an example of the **chain rule,** which is used to find the derivative of a composite function.

CHAIN RULE

If y is a function of u, say $y = f(u)$, and if u is a function of x, say $u = g(x)$, then $y = f(u) = f[g(x)]$, and

$$\frac{dy}{dx} = \frac{dy}{du} \cdot \frac{du}{dx}.$$

One way to remember the chain rule is to pretend that dy/du and du/dx are fractions, with du "canceling out." The proof of the chain rule requires advanced concepts and therefore is not given here.

EXAMPLE 3

Find dy/dx if $y = (3x^2 - 5x)^{1/2}$.
Let $y = u^{1/2}$, and $u = 3x^2 - 5x$. Then

$$\frac{dy}{dx} = \frac{dy}{du} \cdot \frac{du}{dx}$$

$$= \frac{1}{2}u^{-1/2} \cdot (6x - 5).$$

Replacing u with $3x^2 - 5x$ gives

$$\frac{dy}{dx} = \frac{1}{2}(3x^2 - 5x)^{-1/2}(6x - 5) = \frac{6x - 5}{2(3x^2 - 5x)^{1/2}}. \quad \blacktriangleleft$$

The following alternative version of the chain rule is stated in terms of composite functions.

CHAIN RULE (ALTERNATIVE FORM)

If $y = f[g(x)]$, then

$$y' = f'[g(x)] \cdot g'(x).$$

(To find the derivative of $f[g(x)]$, find the derivative of $f(x)$, replace each x with $g(x)$, and then multiply the result by the derivative of $g(x)$.)

EXAMPLE 4

Use the chain rule to find $D_x(x^2 + 5x)^8$.
Let $f(x) = x^8$ and $g(x) = x^2 + 5x$. Then $(x^2 + 5x)^8 = f[g(x)]$ and

$$D_x(x^2 + 5x)^8 = f'[g(x)]g'(x)$$

Here $f'(x) = 8x^7$, with $f'[g(x)] = 8[g(x)]^7 = 8(x^2 + 5x)^7$ and $g'(x) = 2x + 5$.

$$D_x(x^2 + 5x)^8 = f'[g(x)]g'(x)$$

$$= 8[g(x)]^7 g'(x)$$

$$= 8(x^2 + 5x)^7(2x + 5) \blacktriangleleft$$

While the chain rule is essential for finding the derivatives of some of the functions discussed later, the derivatives of the algebraic functions discussed so far can be found by the following *generalized power rule*, a special case of the chain rule.

GENERALIZED POWER RULE

Let u be a function of x, and let $y = u^n$, for any real number n. Then

$$y' = n \cdot u^{n-1} \cdot u'.$$

(The derivative of $y = u^n$ is found by decreasing the exponent on u by 1 and multiplying the result by the exponent n and by the derivative of u with respect to x.)

EXAMPLE 5

(a) Use the generalized power rule to find the derivative of $y = (3 + 5x)^2$.
Let $u = 3 + 5x$, and $n = 2$. Then $u' = 5$. By the generalized power rule,

$$y' = \frac{dy}{dx} = n \cdot u^{n-1} \cdot u'$$

$$= \overset{n}{2} \cdot \overset{u}{(3 + 5x)}^{\overset{n-1}{2-1}} \cdot \overset{u'}{\frac{d}{dx}(3 + 5x)}$$

$$= 2(3 + 5x)^{2-1} \cdot 5 = 10(3 + 5x) = 30 + 50x.$$

(b) Find y' if $y = (3 + 5x)^{-3/4}$.

Use the generalized power rule, with $u = 3 + 5x$, $n = -3/4$, and $u' = 5$.

$$y' = -\frac{3}{4}(3 + 5x)^{-3/4-1} \cdot 5 = -\frac{15}{4}(3 + 5x)^{-7/4}$$

This result could not have been found by any of the rules given in previous sections. ◀

Caution A common error is to forget to multiply by u' when using the generalized power rule. Remember, the generalized power rule is an example of the chain rule, and so the derivative must involve a "chain," or product, of derivatives.

Sometimes both the generalized power rule and either the product or quotient rule are needed to find a derivative, as the next examples show.

EXAMPLE 6 Find the derivative of $y = 4x(3x + 5)^5$.

Write $4x(3x + 5)^5$ as the product

$$(4x) \cdot (3x + 5)^5.$$

To find the derivative of $(3x + 5)^5$, let $u = 3x + 5$, with $u' = 3$. Now use the product rule and the generalized power rule.

$$y' = 4x[\overbrace{5(3x + 5)^4 \cdot 3}^{\text{Derivative of } (3x + 5)^5}] + (3x + 5)^5\overbrace{(4)}^{\text{Derivative of } 4x}$$

$$= 60x(3x + 5)^4 + 4(3x + 5)^5$$

$$= 4(3x + 5)^4[15x + (3x + 5)^1] \qquad \text{Factor out the greatest common factor, } 4(3x + 5)^4$$

$$= 4(3x + 5)^4(18x + 5) \quad ◀$$

EXAMPLE 7 Find $D_x\dfrac{(3x + 2)^7}{x - 1}$.

Use the quotient rule and the generalized power rule.

$$D_x\frac{(3x + 2)^7}{x - 1} = \frac{(x - 1)[7(3x + 2)^6 \cdot 3] - (3x + 2)^7(1)}{(x - 1)^2}$$

$$= \frac{21(x - 1)(3x + 2)^6 - (3x + 2)^7}{(x - 1)^2}$$

$$= \frac{(3x + 2)^6[21(x - 1) - (3x + 2)]}{(x - 1)^2} \qquad \text{Factor out the greatest common factor, } (3x + 2)^6.$$

$$= \frac{(3x + 2)^6[21x - 21 - 3x - 2]}{(x - 1)^2} \qquad \text{Simplify inside brackets.}$$

$$= \frac{(3x + 2)^6(18x - 23)}{(x - 1)^2} \quad ◀$$

Some applications requiring the use of the chain rule or the generalized power rule are illustrated in the next examples.

EXAMPLE 8

The revenue realized by a small city from the collection of fines from parking tickets is given by

$$R(n) = \frac{8000n}{n + 2},$$

where n is the number of work hours each day that can be devoted to parking patrol. At the outbreak of a flu epidemic, 30 work hours are used daily in parking patrol, but during the epidemic that number is decreasing at the rate of 6 work hours per day. How fast is revenue from parking fines decreasing during the epidemic?

We want to find dR/dt, the change in revenue with respect to time. By the chain rule,

$$\frac{dR}{dt} = \frac{dR}{dn} \cdot \frac{dn}{dt}.$$

First find dR/dn, as follows.

$$\frac{dR}{dn} = \frac{(n + 2)(8000) - 8000n(1)}{(n + 2)^2} = \frac{16000}{(n + 2)^2}$$

Since $n = 30$, $dR/dn = 15.625$. Also, $dn/dt = -6$. Thus,

$$\frac{dR}{dt} = (15.625)(-6) = -93.75.$$

Revenue is being lost at the rate of about $94 per day. ◀

EXAMPLE 9

Suppose a sum of $500 is deposited in an account with an interest rate of r percent per year compounded monthly. At the end of 10 years, the balance in the account is given by

$$A = 500\left(1 + \frac{r}{1200}\right)^{120}.$$

Find the rate of change of A with respect to r if $r = 5$ or 7.

First find dA/dr using the generalized power rule.

$$\frac{dA}{dr} = (120)(500)\left(1 + \frac{r}{1200}\right)^{119}\left(\frac{1}{1200}\right)$$

$$= 50\left(1 + \frac{r}{1200}\right)^{119}$$

If $r = 5$,

$$\frac{dA}{dr} = 50\left(1 + \frac{5}{1200}\right)^{119}$$

$$\approx 82.01,$$

or \$82.01 per percentage point. If $r = 7$,

$$\frac{dA}{dr} = 50\left(1 + \frac{7}{1200}\right)^{119}$$

$$\approx 99.90,$$

or \$99.90 per percentage point. ◀

The chain rule can be used to develop the formula for **marginal revenue product,** an economic concept that approximates the change in revenue when a manufacturer hires an additional employee. Start with $R = px$, where R is total revenue from the daily production of x units and p is the price per unit. The demand function is $p = f(x)$, as before. Also, x can be considered a function of the number of employees, n. Since $R = px$, and x (and therefore p) depends on n, R can also be considered a function of n. To find an expression for dR/dn, use the product rule for derivatives on the function $R = px$ to get

$$\frac{dR}{dn} = p \cdot \frac{dx}{dn} + x \cdot \frac{dp}{dn}. \qquad (1)$$

By the chain rule,

$$\frac{dp}{dn} = \frac{dp}{dx} \cdot \frac{dx}{dn}.$$

Substituting for dp/dn in equation (1) gives

$$\frac{dR}{dn} = p \cdot \frac{dx}{dn} + x\left(\frac{dp}{dx} \cdot \frac{dx}{dn}\right)$$

$$= \left(p + x \cdot \frac{dp}{dx}\right)\frac{dx}{dn}. \qquad \text{Factor out } \frac{dx}{dn}.$$

The equation for dR/dn gives the marginal revenue product.

EXAMPLE 10

Find the marginal revenue product dR/dn (in dollars) when $n = 20$ if the demand function is $p = 600/\sqrt{x}$ and $x = 5n$.

As shown above,

$$\frac{dR}{dn} = \left(p + x \cdot \frac{dp}{dx}\right)\frac{dx}{dn}.$$

Find dp/dx and dx/dn. From

$$p = \frac{600}{\sqrt{x}} = 600x^{-1/2},$$

we have the derivative

$$\frac{dp}{dx} = -300x^{-3/2}.$$

Also, from $x = 5n$,

$$\frac{dx}{dn} = 5.$$

Then, by substitution,

$$\frac{dR}{dn} = \left[\frac{600}{\sqrt{x}} + x(-300x^{-3/2}) \right] 5 = \frac{1500}{\sqrt{x}}.$$

If $n = 20$, then $x = 100$ and

$$\frac{dR}{dn} = \frac{1500}{\sqrt{100}} = 150.$$

This means that hiring an additional employee when there are 20 employees will produce an increase in revenue of $150. ◀

For easy reference, a list of the rules for derivatives developed in this chapter is given after the exercises for this section.

10.6 Exercises

Let $f(x) = 4x^2 - 2x$ and $g(x) = 8x + 1$. Find each of the following.

1. $f[g(2)]$ **2.** $f[g(-5)]$ **3.** $g[f(2)]$ **4.** $g[f(-5)]$ **5.** $f[g(k)]$ **6.** $g[f(5z)]$

Find $f[g(x)]$ and $g[f(x)]$ in each of the following.

7. $f(x) = \dfrac{x}{8} + 12;$ $g(x) = 3x - 1$ **8.** $f(x) = -6x + 9;$ $g(x) = \dfrac{x}{5} + 7$

9. $f(x) = \dfrac{1}{x};$ $g(x) = x^2$ **10.** $f(x) = \dfrac{2}{x^4};$ $g(x) = 2 - x$

11. $f(x) = \sqrt{x + 2};$ $g(x) = 8x^2 - 6$ **12.** $f(x) = 9x^2 - 11x;$ $g(x) = 2\sqrt{x + 2}$

13. $f(x) = \sqrt{x + 1};$ $g(x) = \dfrac{-1}{x}$ **14.** $f(x) = \dfrac{8}{x};$ $g(x) = \sqrt{3 - x}$

15. In your own words, explain how to form the composition of two functions.

Write each function as the composition of two functions. (There may be more than one way to do this.)

16. $y = (3x - 7)^{1/3}$ **17.** $y = (5 - x)^{2/5}$

18. $y = \sqrt{9 - 4x}$ **19.** $y = -\sqrt{13 + 7x}$

20. $y = (x^{1/2} - 3)^2 + (x^{1/2} - 3) + 5$

21. $y = (x^2 + 5x)^{1/3} - 2(x^2 + 5x)^{2/3} + 7$

22. What is the distinction between the chain rule and the generalized power rule?

Find the derivative of each function defined as follows.

23. $y = (2x^3 + 9x)^5$

24. $y = (8x^4 - 3x^2)^3$

25. $f(x) = -8(3x^4 + 2)^3$

26. $k(x) = -2(12x^2 + 5)^6$

27. $s(t) = 12(2t^4 + 5)^{3/2}$

28. $s(t) = 45(3t^3 - 8)^{3/2}$

29. $f(t) = 8\sqrt{4t^2 + 7}$

30. $g(t) = -3\sqrt{7t^3 - 1}$

31. $r(t) = 4t(2t^5 + 3)^2$

32. $m(t) = -6t(5t^4 - 1)^2$

33. $y = (x^3 + 2)(x^2 - 1)^2$

34. $y = (3x^4 + 1)^2(x^3 + 4)$

35. $y = (5x^6 + x)^2\sqrt{2x}$

36. $y = 2(3x^4 + 5)^2\sqrt{x}$

37. $y = \dfrac{1}{(3x^2 - 4)^5}$

38. $y = \dfrac{-5}{(2x^3 + 1)^2}$

39. $p(t) = \dfrac{(2t + 3)^3}{4t^2 - 1}$

40. $r(t) = \dfrac{(5t - 6)^4}{3t^2 + 4}$

41. $y = \dfrac{x^2 + 4x}{(3x^3 + 2)^4}$

42. $y = \dfrac{3x^2 - x}{(2x - 1)^5}$

Consider the following table of values of the functions f and g and their derivatives at various points:

x	1	2	3	4
$f(x)$	2	4	1	3
$f'(x)$	-6	-7	-8	-9
$g(x)$	2	3	4	1
$g'(x)$	$2/7$	$3/7$	$4/7$	$5/7$

Find each of the following.

43. **(a)** $D_x(f[g(x)])$ at $x = 1$ **(b)** $D_x(f[g(x)])$ at $x = 2$

44. **(a)** $D_x(g[f(x)])$ at $x = 1$ **(b)** $D_x(g[f(x)])$ at $x = 2$

Applications

Business and Economics

45. Demand Suppose the demand for a certain brand of vacuum cleaner is given by

$$D(p) = \frac{-p^2}{100} + 500,$$

where p is the price in dollars. If the price, in terms of the cost c, is expressed as

$$p(c) = 2c - 10,$$

find the demand in terms of the cost.

46. Revenue Assume that the total revenue from the sale of x television sets is given by

$$R(x) = 1000\left(1 - \frac{x}{500}\right)^2.$$

Find the marginal revenue when the following numbers of sets are sold.

(a) 400 **(b)** 500 **(c)** 600

(d) Find the average revenue from the sale of x sets.

(e) Find the marginal average revenue.

(f) Write a paragraph covering the following questions. When does the revenue begin to decrease? What sales produce zero marginal average revenue? How should a manager use this information?

47. Interest A sum of $1500 is deposited in an account with an interest rate of r percent per year, compounded daily. At the end of 5 yr, the balance in the account is given by

$$A = 1500\left(1 + \frac{r}{36500}\right)^{1825}.$$

Find the rate of change of A with respect to r for the following interest rates.

(a) 6% **(b)** 8% **(c)** 9%

48. Demand Suppose a demand function is given by

$$x = 30\left(5 - \frac{p}{\sqrt{p^2 + 1}}\right),$$

where x is the demand for a product and p is the price per item in dollars. Find the rate of change in the demand for the product per unit change in price (i.e., find dx/dp).

49. Depreciation A certain truck depreciates according to the formula

$$V = \frac{6000}{1 + .3t + .1t^2},$$

where t is time measured in years and $t = 0$ represents the time of purchase (in years). Find the rate at which the value of the truck is changing at the following times.

(a) 2 yr (b) 4 yr

50. Cost Suppose the cost in dollars of manufacturing x items is given by

$$C = 2000x + 3500,$$

and the demand equation is given by

$$x = \sqrt{15,000 - 1.5p}.$$

(a) Find an expression for the revenue R.

(b) Find an expression for the profit P.

(c) Find an expression for the marginal profit.

(d) Determine the value of the marginal profit when the price is $25.

51. Marginal Revenue Product Find the marginal revenue product for a manufacturer with 8 workers if the demand function is $p = 300/x^{1/3}$ and the number of items produced is $8n$, where n is the number of employees.

52. Marginal Revenue Product Suppose the demand function for a product is $p = 200/x^{1/2}$ and x is the number of items produced. Find the marginal revenue product if there are 25 employees and each employee produces 15 items.

Life Sciences

53. Fish Population Suppose the population P of a certain species of fish depends on the number x (in hundreds) of a smaller fish that serves as its food supply, so that

$$P(x) = 2x^2 + 1.$$

Suppose, also, that the number x (in hundreds) of the smaller species of fish depends upon the amount a (in appropriate units) of its food supply, a kind of plankton. Suppose

$$x = f(a) = 3a + 2.$$

A biologist wants to find the relationship between the population P of the large fish and the amount a of plankton available, that is, $P[f(a)]$. What is the relationship?

54. Oil Pollution An oil well off the Gulf Coast is leaking, with the leak spreading oil over the surface as a circle. At any time t (in minutes) after the beginning of the leak, the radius of the circular oil slick on the surface is $r(t) = t^2$ feet. Let $A(r) = \pi r^2$ represent the area of a circle of radius r. Find and interpret $A[r(t)]$.

55. Thermal Inversion When there is a thermal inversion layer over a city (as happens often in Los Angeles), pollutants cannot rise vertically but are trapped below the layer and must disperse horizontally. Assume that a factory smokestack begins emitting a pollutant at 8 A.M. Assume that the pollutant disperses horizontally, forming a circle. If t represents the time (in hours) since the factory began emitting pollutants ($t = 0$ represents 8 A.M.), assume that the radius of the circle of pollution is $r(t) = 2t$ miles. Let $A(r) = \pi r^2$ represent the area of a circle of radius r. Find and interpret $A[r(t)]$.

56. Bacteria Population The total number of bacteria (in millions) present in a culture is given by

$$N(t) = 2t(5t + 9)^{1/2} + 12,$$

where t represents time (in hours) after the beginning of an experiment. Find the rate of change of the population of bacteria with respect to time for each of the following numbers of hours.

(a) 0 (b) 7/5 (c) 8

57. Calcium Usage To test an individual's use of calcium, a researcher injects a small amount of radioactive calcium into the person's bloodstream. The calcium remaining in the bloodstream is measured each day for several days. Suppose the amount of the calcium remaining in the bloodstream (in milligrams per cubic centimeter) t days after the initial injection is approximated by

$$C(t) = \frac{1}{2}(2t + 1)^{-1/2}.$$

Find the rate of change of the calcium level with respect to time for each of the following numbers of days.

(a) 0 (b) 4 (c) 7.5

(d) Is C always increasing or always decreasing? How can you tell?

58. Drug Reaction The strength of a person's reaction to a certain drug is given by

$$R(Q) = Q\left(C - \frac{Q}{3}\right)^{1/2},$$

where Q represents the quantity of the drug given to the patient and C is a constant.

(a) The derivative $R'(Q)$ is called the *sensitivity* to the drug. Find $R'(Q)$.

(b) Find the sensitivity to the drug if $C = 59$ and a patient is given 87 units of the drug.

(c) Is the patient's sensitivity to the drug increasing or decreasing when $Q = 87$?

► RULES FOR DERIVATIVES SUMMARY

Assume all indicated derivatives exist.

Constant Function If $f(x) = k$, where k is any real number, then $f'(x) = 0$.

Power Rule If $f(x) = x^n$, for any real number n, then $f'(x) = n \cdot x^{n-1}$.

Constant Times a Function Let k be a real number. Then the derivative of $y = k \cdot f(x)$ is $y' = k \cdot f'(x)$.

Sum or Difference Rule If $y = f(x) \pm g(x)$, then $y' = f'(x) \pm g'(x)$.

Product Rule If $f(x) = g(x) \cdot k(x)$, then

$$f'(x) = g(x) \cdot k'(x) + k(x) \cdot g'(x).$$

Quotient Rule If $f(x) = \dfrac{g(x)}{k(x)}$, and $k(x) \neq 0$, then

$$f'(x) = \frac{k(x) \cdot g'(x) - g(x) \cdot k'(x)}{[k(x)]^2}.$$

Chain Rule If y is a function of u, say $y = f(u)$, and if u is a function of x, say $u = g(x)$, then $y = f(u) = f[g(x)]$, and

$$\frac{dy}{dx} = \frac{dy}{du} \cdot \frac{du}{dx}.$$

Chain Rule (Alternative Form) Let $y = f[g(x)]$. Then $y' = f'[g(x)] \cdot g'(x)$.

Generalized Power Rule Let u be a function of x, and let $y = u^n$ for any real number n. Then

$$y' = n \cdot u^{n-1} \cdot u'.$$

Chapter Summary Key Terms

To understand the concepts presented in this chapter, you should know the meaning and use of the following words. For easy reference, the section in the chapter where a word (or expression) was first used is given with each item.

10.1 **limit**
 continuous
 discontinuous
10.2 **average rate of change**
 velocity
 instantaneous rate of change

10.3 **secant line**
 tangent line
 slope of a curve
 derivative
 differentiable
 differentiation

10.4 **marginal analysis**
10.5 **average cost**
 marginal average cost
10.6 **marginal revenue product**

Chapter 10 Review Exercises

1. Is a derivative always a limit? Is a limit always a derivative? Explain.

2. Is every continuous function differentiable? Is every differentiable function continuous? Explain.

Decide whether the limits in Exercises 3–16 exist. If a limit exists, find its value.

3. $\lim\limits_{x \to -3} f(x)$

4. $\lim\limits_{x \to -1} g(x)$

5. $\lim\limits_{x \to 4} f(x)$

6. $\lim\limits_{x \to 0} \dfrac{x^2 - 5}{2x}$

7. $\lim\limits_{x \to -1} (2x^2 + 3x + 5)$

8. $\lim\limits_{x \to 2} (-x^2 + 4x + 1)$

9. $\lim\limits_{x \to 6} \dfrac{2x + 5}{x - 3}$

10. $\lim\limits_{x \to 3} \dfrac{2x + 5}{x - 3}$

11. $\lim\limits_{x \to 4} \dfrac{x^2 - 16}{x - 4}$

12. $\lim\limits_{x \to 2} \dfrac{x^2 + 3x - 10}{x - 2}$

13. $\lim\limits_{x \to -4} \dfrac{2x^2 + 3x - 20}{x + 4}$

14. $\lim\limits_{x \to 3} \dfrac{3x^2 - 2x - 21}{x - 3}$

15. $\lim\limits_{x \to 9} \dfrac{\sqrt{x} - 3}{x - 9}$

16. $\lim\limits_{x \to 16} \dfrac{\sqrt{x} - 4}{x - 16}$

Identify the x-values where f is discontinuous.

17.

18.

Decide whether each of the following is continuous at the given values of x.

19. $f(x) = \dfrac{-5}{3x(2x - 1)}; \quad x = -5, 0, -1/3, 1/2$

20. $f(x) = \dfrac{2 - 3x}{(1 + x)(2 - x)}; \quad x = 2/3, -1, 2, 0$

21. $f(x) = \dfrac{x - 6}{x + 5}; \quad x = 6, -5, 0$

22. $f(x) = \dfrac{x^2 - 9}{x + 3}; \quad x = 3, -3, 0$

23. $f(x) = x^2 + 3x - 4; \quad x = 1, -4, 0$

24. $f(x) = 2x^2 - 5x - 3; \quad x = -1/2, 3, 0$

25. Use the graph to find the average rate of change of f on the following intervals.

(a) $x = 0$ to $x = 4$

(b) $x = 2$ to $x = 8$

(c) $x = 2$ to $x = 4$

Find the average rate of change for each of the following on the given interval. Then find the instantaneous rate of change at the first x-value.

26. $y = 6x^2 + 2$; from $x = 1$ to $x = 4$

27. $y = -2x^3 - x^2 + 5$; from $x = -2$ to $x = 6$

28. $y = \dfrac{-6}{3x - 5}$; from $x = 4$ to $x = 9$

29. $y = \dfrac{x + 4}{x - 1}$; from $x = 2$ to $x = 5$

Use the definition of the derivative to find the derivative of each of the following.

30. $y = 4x + 3$

31. $y = 5x^2 + 6x$

32. $y = -x^3 + 7x$

Find the slope of the tangent line to the given curve at the given value of x. Find the equation of each tangent line (Hint: Use the rules for derivatives.)

33. $y = x^2 - 6x$; $x = 2$

34. $y = 8 - x^2$; $x = 1$

35. $y = \dfrac{3}{x - 1}$; $x = -1$

36. $y = (3x^2 - 5x)(2x)$; $x = -1$

37. $y = \dfrac{3}{x^2 - 1}$; $x = 2$

38. $y = \sqrt{6x - 2}$; $x = 3$

39. $y = -\sqrt{8x + 1}$; $x = 3$

40. Two students are working on taking the derivative of
$$f(x) = \frac{2x}{3x + 4}.$$
The first one uses the quotient rule to get
$$f'(x) = \frac{(3x + 4)2 - 2x(3)}{(3x + 4)^2} = \frac{8}{(3x + 4)^2}.$$
The second converts it into a product and uses the product rule:
$$f(x) = 2x(3x + 4)^{-1}$$
$$f'(x) = 2x(-1)(3x + 4)^{-2}(3) + 2(3x + 4)^{-1}$$
$$= 2(3x + 4)^{-1} - 6x(3x + 4)^{-2}.$$
Explain the discrepancies between the two answers. Which procedure do you think is preferable?

41. Two students are working on taking the derivative of
$$f(x) = \frac{2}{(3x + 1)^4}.$$
The first one uses the quotient rule as follows:
$$f'(x) = \frac{(3x + 1)^4 \cdot 0 - 2 \cdot 4(3x + 1)^3 \cdot 3}{(3x + 1)^8}$$
$$= \frac{-24(3x + 1)^3}{(3x + 1)^8} = \frac{-24}{(3x + 1)^5}.$$
The second rewrites the function and uses the generalized power rule as follows:
$$f(x) = 2(3x + 1)^{-4}$$
$$f'(x) = (-4)2(3x + 1)^{-5} \cdot 3 = \frac{-24}{(3x + 1)^5}.$$
Compare the two procedures. Which procedure do you think is preferable?

Use the rules for derivatives to find the derivative of each function defined as follows.

42. $y = 5x^2 - 7x - 9$

43. $y = x^3 - 4x^2$

44. $y = 6x^{7/3}$

45. $y = -3x^{-2}$

46. $f(x) = x^{-3} + \sqrt{x}$

47. $f(x) = 6x^{-1} - 2\sqrt{x}$

48. $y = (3t^2 + 7)(t^3 - t)$

49. $y = (-5t + 4)(t^3 - 2t^2)$

50. $g(t) = -3t^{-1/3}(5t + 7)$

51. $p(t) = 8t^{3/4}(7t - 2)$

52. $y = 12x^{-3/4}\left(\dfrac{x}{3} + 5\right)$

53. $y = 15x^{-3/5}\left(6 - \dfrac{x}{3}\right)$

54. $k(x) = \dfrac{3x}{x + 5}$

55. $r(x) = \dfrac{-8}{2x + 1}$

56. $y = \dfrac{x^2 - x + 1}{x - 1}$

57. $y = \dfrac{2x^3 - 5x^2}{x + 2}$

58. $f(x) = (3x - 2)^4$

59. $k(x) = (5x - 1)^6$

60. $y = \sqrt{2t - 5}$

61. $y = -3\sqrt{8t - 1}$

62. $y = 3x(2x + 1)^3$

63. $y = 4x^2(3x - 2)^5$

64. $r(t) = \dfrac{5t^2 - 7t}{(3t + 1)^3}$

65. $s(t) = \dfrac{t^3 - 2t}{(4t - 3)^4}$

Find each of the following.

66. $D_x\left[\dfrac{\sqrt{x} + 1}{\sqrt{x} - 1}\right]$

67. $D_x\left[\dfrac{2x + \sqrt{x}}{1 - x}\right]$

68. $\dfrac{dy}{dt}$ if $y = \sqrt{t^{1/2} + t}$

69. $\dfrac{dy}{dx}$ if $y = \dfrac{\sqrt{x} - 1}{x}$

70. $f'(1)$ if $f(x) = \dfrac{\sqrt{8 + x}}{x + 1}$

71. $f'(-2)$ if $f(t) = \dfrac{2 - 3t}{\sqrt{2 + t}}$

72. Describe how to tell when a function is discontinuous at $x = a$, where a is a real number.

73. Which of the following does *not* require the product rule or the quotient rule to differentiate?

(a) $f(x) = \dfrac{x^2}{5x - 1}$ (b) $g(x) = 6x\sqrt{x + 2}$ (c) $h(x) = \dfrac{x}{a - x}$ (d) $F(x) = \dfrac{x^2}{1500}$

Find the marginal average cost function of each function defined as follows.

74. $C(x) = \sqrt{x + 1}$ **75.** $C(x) = \sqrt{3x + 2}$ **76.** $C(x) = (x^2 + 3)^3$ **77.** $C(x) = (4x + 3)^4$

Applications

Business and Economics

78. Sales The sales of a company are related to its expenditures on research by
$$S(x) = 1000 + 50\sqrt{x} + 10x,$$
where $S(x)$ gives sales in millions when x thousand dollars is spent on research. Find and interpret dS/dx if the following amounts are spent on research.

(a) $9000 (b) $16,000 (c) $25,000

(d) As the amount spent on research increases, what happens to sales?

79. Profit Suppose that the profit (in hundreds of dollars) from selling x units of a product is given by
$$P(x) = \dfrac{x^2}{x - 1}, \text{ where } x > 1.$$
Find and interpret the marginal profit when the following numbers of units are sold.

(a) 4 (b) 12 (c) 20

(d) What is happening to the marginal profit as the number sold increases?

80. Costs A company finds that its costs are related to the amount spent on training programs by
$$T(x) = \dfrac{1000 + 50x}{x + 1},$$
where $T(x)$ is costs in thousands of dollars when x hundred dollars are spent on training. Find and interpret $T'(x)$ if the following amounts are spent on training.

(a) $900 (b) $1900

(c) Are costs per dollar spent on training always increasing or decreasing?

81. Revenue Waverly Products has found that its revenue is related to advertising expenditures by the function
$$R(x) = 5000 + 16x - 3x^2,$$
where $R(x)$ is the revenue in dollars when x hundred dollars are spent on advertising.

(a) Find the marginal revenue function.

(b) Find and interpret the marginal revenue when $1000 is spent on advertising.

Connections

Business and Economics

82. Cost Analysis A company charges $1.50 per pound when a certain chemical is bought in lots of 125 lb or less, with a price per pound of $1.35 if more than 125 lb are purchased. Let $C(x)$ represent the cost of x lb. Find the cost for the following pounds.

(a) 100 (b) 125 (c) 140

(d) Graph $y = C(x)$.

(e) Where is C discontinuous?

Find the average cost per pound if the following number of pounds are bought.

(f) 100 (g) 125 (h) 140

Find and interpret the marginal cost for the following numbers of pounds.

(i) 100 (j) 140

83. Marginal Analysis Suppose the profit (in cents) from selling x lb of potatoes is given by

$$P(x) = 15x + 25x^2.$$

Find the average rate of change in profit from selling each of the following amounts.

(a) 6 lb to 7 lb (b) 6 lb to 6.5 lb (c) 6 lb to 6.1 lb

Find the marginal profit from selling the following amounts.

(d) 6 lb (e) 20 lb (f) 30 lb

(g) What is the domain of x?

(h) Is it possible for the marginal profit to be negative here? What does this mean?

(i) Find the average profit function.

(j) Find the marginal average profit function.

(k) Is it possible for the marginal average profit to vary here? What does this mean?

(l) Discuss whether this function describes a realistic situation.

Life Science

84. Spread of a Virus The spread of a virus is modeled by

$$V(t) = -t^2 + 6t - 4,$$

where $V(t)$ is the number of people (in hundreds) with the virus and t is the number of weeks since the first case was observed.

(a) Graph $V(t)$.

(b) What is a reasonable domain of t for this problem?

(c) When does the number of cases reach a maximum? What is the maximum number of cases?

(d) Find the rate of change function.

(e) What is the rate of change in the number of cases at the maximum?

(f) Give the sign ($+$ or $-$) of the rate of change up to the maximum and after the maximum.

Computer/Graphing Calculator

For each of the following, use a computer or a graphing calculator to graph each function and its derivative on the same axes. Determine the values of x where the derivative is (a) positive, (b) zero, and (c) negative. (d) What is true of the graph of the function in each case?

85. $g(x) = 6 - 4x + 3x^2 - x^3$

86. $k(x) = 2x^4 - 3x^3 + x$

87. $G(x) = \dfrac{2x}{(x - 1)^2}$

88. $K(x) = \sqrt[3]{(2x - 1)^2}$

Applications at a glance . . .

*Timing a Sales Pitch . . . changing listeners' attitudes . . .
probability of legislator voting "pro" . . .*

CHAPTER 11

CURVE
SKETCHING

(See page 530.)

The graph in Figure 1 shows the population of deer on the Kaibab Plateau on the North Rim of the Grand Canyon between 1905 and 1930. In the first two decades of the twentieth century, hunters were effective in reducing the population of predators. With fewer predators, the deer increased in numbers and depleted available food resources. As shown by the graph, the deer population peaked in about 1925 and then declined rapidly.

FIGURE 1

Given a graph like the one in Figure 1, often we can locate maximum and minimum values simply by looking at the graph. It is difficult to get *exact* values or *exact* locations of maxima and minima from a graph, however, and many functions are difficult to graph. In Chapter 1 we saw how to find exact maximum and minimum values for quadratic functions by identifying the vertex. A more general approach is to use the derivative of a function to determine precise maximum and minimum values of the function. The procedure for doing this is described in this chapter, which begins with a discussion of increasing and decreasing functions.

11.1 INCREASING AND DECREASING FUNCTIONS

How high is the highest point on a roller coaster?

Suppose that an amusement park has hired an engineering firm to develop a new roller coaster. The ride will include one very steep drop that passes through a tunnel. An engineer has developed a design for this section. The design gives the height h (in feet) of the track above the ground by the polynomial function

$$h = f(x) = -3x^4 + 28x^3 - 84x^2 + 96x,$$

where x is the distance along the ground measured in 10-foot intervals. (In other words, $x = 1$ corresponds to a distance of 10 feet along the ground. This arrange-

ment keeps the values of x smaller, and thus easier to work with.) Meanwhile, a new safety regulation has been passed requiring that such rides be no more than 65 feet high. The engineer is particularly worried about whether this stretch of track, shown in Figure 2, will be too high. The engineer has used a graphing calculator to quickly sketch the curve, and this stretch of track comes very close to the limit. How can he find the *exact value* of the highest point?

Ground level

FIGURE 2

The derivative can be used to answer this question. Remember that the derivative of a function at a point gives the slope of the line tangent to the function at that point. Recall also that a line with a positive slope rises from left to right and a line with a negative slope falls from left to right.

Now, picture one of the cars on the roller coaster. When the car is on level ground or parallel to level ground, its floor is horizontal, but as the car moves up the slope, its floor tilts upward. When the car reaches a peak, its floor is again horizontal, but it then begins to tilt downward (very steeply) as the car rolls downhill.

If we think of the roller coaster track as a graph, the floor of the car as it moves from left to right along the track represents the tangent line at each point. Using this analogy, we can see that the slope of the tangent will be *positive* when the car travels uphill and h is *increasing*, and the slope of the tangent will be *negative* when the car travels downhill and h is *decreasing*. (In this case it is also true that the slope will be zero at "peaks" and "valleys.") See Figure 3.

$$f(x) = h = -3x^4 + 28x^3 - 84x^2 + 96x$$

FIGURE 3

In this chapter you will need all of the rules for derivatives you learned in the previous chapter. If any of these are still unclear, go over the Derivative Summary at the end of that chapter and practice some of the Review Exercises before proceeding.

Thus, on intervals where $f'(x) > 0$, $f(x) = h$ will increase, and on intervals where $f'(x) < 0$, $f(x) = h$ will decrease. We can determine where $f(x)$ peaks by finding the intervals on which it increases and decreases. The function

$$f(x) = -3x^4 + 28x^3 - 84x^2 + 96x$$

is differentiated as

$$f'(x) = -12x^3 + 84x^2 - 168x + 96.$$

To determine where $f'(x) > 0$ and where $f'(x) < 0$, first find where $f'(x) = 0$.

$$-12x^3 + 84x^2 - 168x + 96 = 0$$
$$x^3 - 7x^2 + 14x - 8 = 0 \qquad \text{Divide each side by } -12.$$

The left side of the equation can be factored as

$$(x - 1)(x - 2)(x - 4) = 0.$$

Therefore, $f'(x)$ will equal 0 when $x = 1, 2,$ or 4. (These values of x correspond to the two "peaks" and one "valley" on the graph of $f(x)$ in Figure 3.)

FIGURE 4

The three zeros of $f'(x)$ divide the real number line into four regions, as shown in Figure 4. Evaluating $f'(x)$ at a point in each region shows that $f'(x)$ is positive between 0 and 1, negative between 1 and 2, positive between 2 and 4, and negative for x greater than 4. Thus, $h = f(x)$ increases from 0 to 1, decreases from 1 to 2, increases from 2 to 4, and decreases thereafter. The graph in Figure 5 shows that the highest point occurs at $x = 4$, after the roller coaster has traveled a horizontal distance of 40 feet. The track there is $f(4) = 64$ feet above the ground, within the 65-foot limit.

x	$h = f(x)$
0	0
1	37
2	32
3	45
4	64
5	5

FIGURE 5

As shown in the preceding example, a function is *increasing* if the graph goes *up* from left to right and *decreasing* if its graph goes *down* from left to right. Examples of increasing functions are shown in Figure 6 and examples of decreasing functions in Figure 7.

(a) **(b)** **(c)**

FIGURE 6

(a) **(b)** **(c)**

FIGURE 7

INCREASING AND DECREASING FUNCTIONS

Let f be a function defined on some interval. Then for any two numbers x_1 and x_2 in the interval, f is **increasing** on the interval if

$$f(x_1) < f(x_2) \quad \text{whenever} \quad x_1 < x_2,$$

and f is **decreasing** on the interval if

$$f(x_1) > f(x_2) \quad \text{whenever} \quad x_1 < x_2.$$

EXAMPLE 1

Where is the function graphed in Figure 8 increasing? Where is it decreasing?

Moving from left to right, the function is increasing up to -4, then decreasing from -4 to 0, constant (neither increasing nor decreasing) from 0 to 4, increasing from 4 to 6, and decreasing from 6 onward. In interval notation, the function is increasing on $(-\infty, -4)$ and $(4, 6)$, decreasing on $(-4, 0)$ and $(6, \infty)$, and constant on $(0, 4)$. ◀

FIGURE 8

The following test summarizes the discussion in the opening example.

TEST FOR INTERVALS WHERE $f(x)$ IS INCREASING AND DECREASING	Suppose a function f has a derivative at each point in an open interval; then if $f'(x) > 0$ for each x in the interval, f is *increasing* on the interval; if $f'(x) < 0$ for each x in the interval, f is *decreasing* on the interval; if $f'(x) = 0$ for each x in the interval, f is *constant* on the interval.

The derivative $f'(x)$ can change signs from positive to negative (or negative to positive) at points where $f'(x) = 0$, and also at points where $f'(x)$ does not exist. The values of x where this occurs are called *critical numbers*. By definition, the **critical numbers** for a function f are those numbers c in the domain of f for which $f'(c) = 0$ or $f'(c)$ does not exist. A **critical point** is a point whose x-coordinate is the critical number c, and whose y-coordinate is $f(c)$.

It is shown in more advanced classes that if the critical numbers of a polynomial function are used to determine open intervals on a number line, then the sign of the derivative at any point in the interval will be the same as the sign of the derivative at any other point in the interval. This suggests that the test for increasing and decreasing functions be applied as follows (assuming that no open intervals exist where the function is constant).

APPLYING THE TEST	**1.** Locate on a number line those values of x for which $f'(x) = 0$ or $f'(x)$ does not exist. These points determine several open intervals. **2.** Choose a value of x in each of the intervals determined in Step 1. Use these values to decide whether $f'(x) > 0$ or $f'(x) < 0$ in that interval. **3.** Use the test above to decide whether f is increasing or decreasing on the interval.

EXAMPLE 2

Find the intervals in which the following functions are increasing or decreasing. Locate all points where the tangent line is horizontal. Graph the function.

(a) $f(x) = x^3 + 3x^2 - 9x + 4$

Here $f'(x) = 3x^2 + 6x - 9$. To find the critical numbers, set this derivative equal to 0 and solve the resulting equation by factoring.

$$3x^2 + 6x - 9 = 0$$
$$3(x^2 + 2x - 3) = 0$$
$$3(x + 3)(x - 1) = 0$$
$$x = -3 \quad \text{or} \quad x = 1$$

The tangent line is horizontal at $x = -3$ or $x = 1$. Since there are no values of x where $f'(x)$ fails to exist, the only critical numbers are -3 and 1. To determine where the function is increasing or decreasing, locate -3 and 1 on a number line, as in Figure 9. (Be sure to place the values on the number line in numerical order.) These points determine three intervals, $(-\infty, -3)$, $(-3, 1)$, and $(1, \infty)$.

FIGURE 9

Now choose any value of x in the interval $(-\infty, -3)$. Choosing $x = -4$ gives

$$f'(-4) = 3(-4)^2 + 6(-4) - 9 = 15,$$

which is positive. Since one value of x in this interval makes $f'(x) > 0$, all values will do so, and therefore f is increasing on $(-\infty, -3)$. Selecting 0 from the middle interval gives $f'(0) = -9$, so f is decreasing on $(-3, 1)$. Finally, choosing 2 in the right-hand region gives $f'(2) = 15$, with f increasing on $(1, \infty)$. The arrows in each interval in Figure 9 indicate where f is increasing or decreasing.

Up to now our only method of graphing most functions has been by plotting points that lie on the graph, either by hand or using a graphing calculator or computer. Now an additional tool is available: the test for determining where a function is increasing or decreasing. (Other tools are discussed in the next few sections.) To graph the function, plot a point at each of the critical numbers by finding $f(-3) = 31$ and $f(1) = -1$. Also plot points for $x = -4, 0,$ and 2, the test values of each interval. Use these points along with the information about where the function is increasing and decreasing to get the graph in Figure 10.

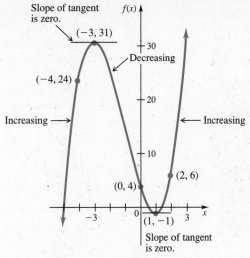

$$f(x) = x^3 + 3x^2 - 9x + 4$$

FIGURE 10

Caution Be sure to use $f(x)$, not $f'(x)$, to find the y-values of the points to plot.

(b) $f(x) = \dfrac{x - 1}{x + 1}$

Use the quotient rule to find $f'(x)$.

$$f'(x) = \frac{(x + 1)(1) - (x - 1)(1)}{(x + 1)^2}$$

$$= \frac{x + 1 - x + 1}{(x + 1)^2} = \frac{2}{(x + 1)^2}$$

This derivative is never 0, but it fails to exist when $x = -1$. Since $f(x)$ also does not exist at $x = -1$, however, -1 is not a critical number. There are no critical numbers, and the tangent line is never horizontal. Since the line $x = -1$ is a vertical asymptote for the graph of f, it separates the domain into the two intervals $(-\infty, -1)$ and $(-1, \infty)$. A function can change direction from one side of a vertical asymptote to the other. Use a test point in each of these intervals to find that $f'(x) > 0$ for all x except -1. This means that the function f is increasing on both $(-\infty, -1)$ and $(-1, \infty)$.

When x is large,

$$\frac{x - 1}{x + 1} \approx \frac{x}{x} = 1,$$

so the graph has the line $y = 1$ as a horizontal asymptote. Using all this information and plotting the intercepts and a few points on either side of the vertical asymptote gives the graph in Figure 11. ◀

FIGURE 11

FIGURE 12

Caution It is important to note that the reverse of the test for increasing and decreasing functions is not true—it is possible for a function to be increasing on an interval even though the derivative is not positive at every point in the interval. A good example is given by $f(x) = x^3$, which is increasing on every interval, even though $f'(x) = 0$ when $x = 0$. See Figure 12.

Similarly, it is incorrect to assume that the sign of the derivative in regions separated by critical numbers must alternate between $+$ and $-$. If this were always so, it would lead to a simple rule for finding the sign of the derivative: just check one test point, and then make the other regions alternate in sign. But this is not true if one of the factors in the derivative is raised to an even power. In the function $f(x) = x^3$ just considered, $f'(x) = 3x^2$, which is positive on both sides of the critical number $x = 0$.

Knowing the intervals where a function is increasing or decreasing can be important in applications, as shown by the next examples.

3 A company sells computers with the following cost and revenue functions:

$$C(x) = 2000x^2 - x^3, \qquad 0 \le x \le 1000$$
$$R(x) = 2{,}000{,}000x - 1000x^2, \qquad 0 \le x \le 1000$$

where x is the number of computers sold monthly. Determine any intervals on which the profit function is increasing.

First find the profit function $P(x)$.

$$\begin{aligned} P(x) &= R(x) - C(x) \\ &= (2{,}000{,}000x - 1000x^2) - (2000x^2 - x^3) \\ &= 2{,}000{,}000x - 3000x^2 + x^3 \end{aligned}$$

To find any intervals where this function is increasing, set $P'(x) = 0$.

$$P'(x) = 2{,}000{,}000 - 6000x + 3x^2 = 0$$

FIGURE 13

Solving this with the quadratic formula gives the approximate solutions $x = 422.6$ and $x = 1577.4$. The latter number is outside of the domain. Use $x = 422.6$ to determine two intervals on a number line, as shown in Figure 13. Choose $x = 0$ and $x = 1000$ as test points.

$$P'(0) = 2{,}000{,}000 - 6000(0) + 3(0^2) = 2{,}000{,}000$$
$$P'(1000) = 2{,}000{,}000 - 6000(1000) + 3(1000^2) = -1{,}000{,}000$$

The test points show that the function increases on $(0, 422.6)$ and decreases on $(422.6, 1000)$. See Figure 13. Thus, the profit is increasing when 422 computers or fewer are sold, and decreasing when 423 or more are sold, as shown in Figure 14.

The graph in Figure 14 also shows that the profit will increase as long as the revenue function increases faster than the cost function. That is, increasing production will produce more profit as long as the marginal revenue is greater than the marginal cost. ◀

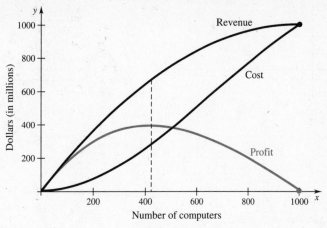

FIGURE 14

EXAMPLE **4**

In the exercises in the previous chapter, the function

$$f(t) = \frac{90t}{99t - 90}$$

gave the number of facts recalled after t hours for $t > 10/11$. Find the intervals in which $f(t)$ is increasing or decreasing.

First find the derivative, $f'(t)$.

$$f'(t) = \frac{(99t - 90)(90) - 90t(99)}{(99t - 90)^2}$$

$$= \frac{8910t - 8100 - 8910t}{(99t - 90)^2} = \frac{-8100}{(99t - 90)^2}$$

Since $(99t - 90)^2$ is positive everywhere in the domain of the function and since the numerator is a negative constant, $f'(t) < 0$ for all t in the domain of $f(t)$. Thus $f(t)$ always decreases and, as expected, the number of words recalled decreases steadily over time. ◀

11.1 Exercises

Find the largest open intervals where the functions graphed as follows are
(a) *increasing, or **(b)** *decreasing.*

1.

2.

3.

4.

5.

6.

7.

8.

Find the largest open intervals where the following are **(a)** *increasing;* **(b)** *decreasing.*

9. $f(x) = x^2 + 12x - 6$

10. $f(x) = x^2 - 9x + 4$

11. $y = 2 + 3.6x - 1.2x^2$

12. $y = .3 + .4x - .2x^2$

13. $f(x) = \frac{2}{3}x^3 - x^2 - 24x - 4$

14. $f(x) = \frac{2}{3}x^3 - x^2 - 4x + 2$

15. $f(x) = 4x^3 - 15x^2 - 72x + 5$

16. $f(x) = 4x^3 - 9x^2 - 30x + 6$

17. $y = -3x + 6$

18. $y = 6x - 9$

19. $f(x) = \frac{x + 2}{x + 1}$

20. $f(x) = \frac{x + 3}{x - 4}$

21. $y = |x + 4|$

22. $y = -|x - 3|$

23. $f(x) = -\sqrt{x - 1}$

24. $f(x) = \sqrt{5 - x}$

25. $y = \sqrt{x^2 + 1}$

26. $y = x\sqrt{9 - x^2}$

27. $f(x) = x^{2/3}$

28. $f(x) = (x + 1)^{4/5}$

29. Use the techniques of this chapter to find the vertex and intervals where f is increasing and decreasing, given

$$f(x) = ax^2 + bx + c,$$

where we assume $a > 0$. Verify that this agrees with what we found in Chapter 1.

30. Repeat Exercise 29 under the assumption $a < 0$.

Applications

Business and Economics

31. **Cost** Suppose the total cost $C(x)$ (in dollars) to manufacture a quantity x of weed killer (in hundreds of liters) is given by

$$C(x) = x^3 - 2x^2 + 8x + 50.$$

(a) Where is $C(x)$ decreasing?

(b) Where is $C(x)$ increasing?

32. **Housing Starts** A county realty group estimates that the number of housing starts per year over the next three years will be

$$H(r) = \frac{300}{1 + .03r^2},$$

where r is the mortgage rate (in percent).

(a) Where is $H(r)$ increasing?

(b) Where is $H(r)$ decreasing?

33. Profit A manufacturer sells video games with the following cost and revenue functions, where x is the number of games sold.

$$C(x) = 4.8x - .0004x^2, \qquad 0 \leq x \leq 2250$$
$$R(x) = 8.4x - .002x^2, \qquad 0 \leq x \leq 2250$$

Determine the interval(s) on which the profit function is increasing.

34. Profit A manufacturer of compact disk players has determined that the profit $P(x)$ (in thousands of dollars) is related to the quantity x of players produced (in hundreds) per month by

$$P(x) = \frac{1}{3}x^3 - \frac{7}{2}x^2 + 10x - 2,$$

as long as the number of players produced is fewer than 800 per month.

(a) At what production levels is the profit increasing?

(b) At what levels is it decreasing?

Life Sciences

35. Spread of Infection The number of people $P(t)$ (in hundreds) infected t days after an epidemic begins is approximated by

$$P(t) = 2 + 50t - \frac{5}{2}t^2.$$

When will the number of people infected start to decline?

36. Alcohol Concentration In Chapter 1 we gave the function defined by

$$A(x) = -.015x^3 + 1.058x$$

as the approximate alcohol concentration (in tenths of a percent) in an average person's bloodstream x hr after drinking 8 oz of 100-proof whiskey. The function applies only for the interval [0, 8].

(a) On what time intervals is the alcohol concentration increasing?

(b) On what intervals is it decreasing?

37. Drug Concentration The percent of concentration of a drug in the bloodstream x hr after the drug is administered is given by

$$K(x) = \frac{4x}{3x^2 + 27}.$$

(a) On what time intervals is the concentration of the drug increasing?

(b) On what intervals is it decreasing?

38. Drug Concentration Suppose a certain drug is administered to a patient, with the percent of concentration of the drug in the bloodstream t hr later given by

$$K(t) = \frac{5t}{t^2 + 1}.$$

(a) On what time intervals is the concentration of the drug increasing?

(b) On what intervals is it decreasing?

General Interest

39. Sports Cars The following graph shows the horsepower and torque as a function of the engine speed for a 1991 Porsche 928 GT.

(a) On what intervals is the horsepower increasing with engine speed?

(b) On what intervals is the horsepower decreasing with engine speed?

(c) On what intervals is the torque increasing with engine speed?

(d) On what intervals is the torque decreasing with engine speed?

928 GT Performance and Torque

11.2 RELATIVE EXTREMA

In a 30-second commercial, when is the best time to present the sales message?

Suppose that the manufacturer of a diet soft drink is disappointed by sales after airing a new series of 30-second television commercials. The company's market research analysts hypothesize that the problem lies in the timing of the commercial's message, Drink Sparkling Light. Either it comes too early in the commercial, before the viewer has become involved; or it comes too late, after the viewer's attention has faded. After extensive experimentation, the research group finds that the percent of full attention that a viewer devotes to a commercial is a function of time (in seconds) since the commercial began, where

$$\text{Viewer's attention} = f(t) = -\frac{3}{20}t^2 + 6t + 20, \qquad 0 \le t \le 30.$$

When is the best time to present the commercial's sales message?

Clearly, the message should be delivered when the viewer's attention is at a maximum. To find this time, find $f'(t)$.

$$f'(t) = -\frac{3}{10}t + 6 = -.3t + 6$$

The derivative $f'(t)$ is greater than 0 when $-.3t + 6 > 0$, $-3t > -60$, or $t < 20$. Similarly, $f'(t) < 0$ when $-.3t + 6 < 0$, $-3t < -60$, or $t > 20$. Thus, attention increases for the first 20 seconds and decreases for the last 10 seconds. The message should appear about 20 seconds into the commercial. At that time the viewer will devote $f(20) = 80\%$ of his attention to the commercial.

The maximum level of viewer attention (80%) in the example above is a *relative maximum*, defined as follows.

RELATIVE MAXIMUM OR MINIMUM

Let c be a number in the domain of a function f. Then $f(c)$ is a **relative maximum** for f if there exists an open interval (a, b) containing c such that

$$f(x) \le f(c)$$

for all x in (a, b), and $f(c)$ is a **relative minimum** for f if there exists an open interval (a, b) containing c such that

$$f(x) \ge f(c)$$

for all x in (a, b).

If c is an endpoint of the domain of f, we only consider x in the half-open interval that is in the domain.

A function has a **relative extremum** (plural: **extrema**) at c if it has either a relative maximum or a relative minimum there.

EXAMPLE 1 Identify the x-values of all points where the graph in Figure 15 has relative extrema.

FIGURE 15

The parentheses around x_1 show an open interval containing x_1 such that $f(x) \leq f(x_1)$, so there is a relative maximum of $f(x_1)$ at $x = x_1$. Notice that many other open intervals would work just as well. Similar intervals around x_2, x_3, and x_4 can be used to find a relative maximum of $f(x_3)$ at $x = x_3$ and relative minima of $f(x_2)$ at $x = x_2$ and $f(x_4)$ at $x = x_4$. ◀

The function graphed in Figure 16 has relative maxima when $x = x_2$ or $x = x_4$ and relative minima when $x = x_1$ or $x = x_3$. The tangents at the points having x-values x_1, x_2, and x_3 are shown in the figure. All three tangents are horizontal and have slope 0. There is no single tangent line at the point where $x = x_4$.

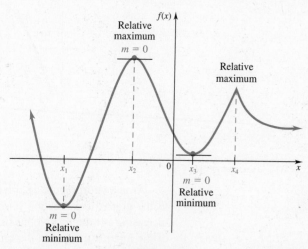

FIGURE 16

Since the derivative of a function gives the slope of a line tangent to the graph of the function, to find relative extrema we first identify all values of x where the derivative of the function (the slope of the tangent line) is 0. Next, we identify all x-values where the derivative *does not* exist but the function *does* exist. Recall from the preceding section that *critical numbers* are these x-values where the derivative is 0 or where the derivative does not exist but the function does exist. A relative extremum *may* exist at a critical number. (A rough sketch of the graph of the function near a critical number often is enough to tell whether an extremum has been found.) These facts about extrema are summarized below.

RELATIVE EXTREMA

If a function f has a relative extremum at a critical number c, then either $f'(c) = 0$, or $f'(c)$ does not exist but $f(c)$ does exist.

FIGURE 17

Caution Be very careful not to get this result backward. It does *not* say that a function has relative extrema at all critical numbers of the function. For example, Figure 17 shows the graph of $f(x) = x^3$. The derivative, $f'(x) = 3x^2$, is 0 when $x = 0$, so that 0 is a critical number for that function. However, as suggested by the graph of Figure 17, $f(x) = x^3$ has neither a relative maximum nor a relative minimum at $x = 0$ (or anywhere else, for that matter).

First Derivative Test Suppose all critical numbers have been found for some function f. How is it possible to tell from the equation of the function whether these critical numbers produce relative maxima, relative minima, or neither? One way is suggested by the graph in Figure 18.

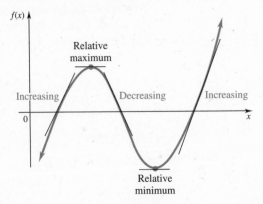

FIGURE 18

As shown in Figure 18, on the left of a relative maximum the tangent lines to the graph of a function have positive slope, indicating that the function is increasing. At the relative maximum, the tangent line is horizontal. On the right of the relative maximum the tangent lines have negative slopes, indicating that the function is decreasing. Around a relative minimum the opposite occurs. As

shown by the tangent lines in Figure 18, the function is decreasing on the left of the relative minimum, has a horizontal tangent at the minimum, and is increasing on the right of the minimum.

Putting this together with the methods from Section 1 for identifying intervals where a function is increasing or decreasing gives the following *first derivative test* for locating relative extrema.

FIRST DERIVATIVE TEST

Let c be a critical number for a function f. Suppose that f is differentiable on (a, b), and that c is the only critical number for f in (a, b).

1. $f(c)$ is a relative maximum of f if the derivative $f'(x)$ is positive in the interval (a, c) and negative in the interval (c, b).
2. $f(c)$ is a relative minimum of f if the derivative $f'(x)$ is negative in the interval (a, c) and positive in the interval (c, b).

The sketches in the following table show how the first derivative test works. Assume the same conditions on a, b, and c for the table as those given for the first derivative test.

$f(x)$ has:	Sign of f' in (a, c)	Sign of f' in (c, b)	Sketches
Relative maximum	$+$	$-$	
Relative minimum	$-$	$+$	
No relative extrema	$+$	$+$	
No relative extrema	$-$	$-$	

EXAMPLE 2

Find all relative extrema for the following functions. Graph each function.

(a) $f(x) = 2x^3 - 3x^2 - 72x + 15$

The derivative is $f'(x) = 6x^2 - 6x - 72$. There are no points where $f'(x)$ fails to exist, so the only critical numbers will be found where the derivative equals 0. Setting the derivative equal to 0 gives

$$6x^2 - 6x - 72 = 0$$
$$6(x^2 - x - 12) = 0$$
$$6(x - 4)(x + 3) = 0$$
$$x - 4 = 0 \quad \text{or} \quad x + 3 = 0$$
$$x = 4 \quad \text{or} \quad x = -3.$$

As in the previous section, the critical numbers 4 and -3 are used to determine the three intervals $(-\infty, -3)$, $(-3, 4)$, and $(4, \infty)$ shown on the number line in Figure 19. Any number from each of the three intervals can be used as a test point to find the sign of f' in each interval. Using -4, 0, and 5 gives the following information.

$$f'(-4) = 6(-8)(-1) > 0$$
$$f'(0) = 6(-4)(3) < 0$$
$$f'(5) = 6(1)(8) > 0$$

Thus, the derivative is positive on $(-\infty, -3)$, negative on $(-3, 4)$, and positive on $(4, \infty)$. By Part 1 of the first derivative test, this means that the function has a relative maximum of $f(-3) = 150$ when $x = -3$; by Part 2, f has a relative minimum of $f(4) = -193$ when $x = 4$.

FIGURE 19

Using the information found above and plotting the points from the following chart gives the graph in Figure 20. (Be sure to use $f(x)$, not $f'(x)$, to find the y-values of the points to plot.)

x	-4	-3	0	4	5
$f(x)$	127	150	15	-193	-170

It is not necessary to plot additional points with x-values less than -4 or greater than 5, since any maximum or minimum must occur at a critical number. Thus, the graph will not have other turning points.

FIGURE 20

(b) $f(x) = 6x^{2/3} - 4x$
Find $f'(x)$:

$$f'(x) = 4x^{-1/3} - 4 = \frac{4}{x^{1/3}} - 4.$$

The derivative fails to exist when $x = 0$, but the function itself is defined when $x = 0$, making 0 a critical number for f. To find other critical numbers, set $f'(x) = 0$.

$$f'(x) = 0$$

$$\frac{4}{x^{1/3}} - 4 = 0$$

$$\frac{4}{x^{1/3}} = 4$$

$$4 = 4x^{1/3}$$

$$1 = x^{1/3}$$

$$1 = x$$

FIGURE 21

The critical numbers 0 and 1 are used to locate the intervals $(-\infty, 0)$, $(0, 1)$, and $(1, \infty)$ on a number line as in Figure 21. Evaluating $f'(x)$ at the test points -1, $1/2$, and 2 and using the first derivative test shows that f has a relative maximum at $x = 1$; the value of this relative maximum is $f(1) = 2$. Also, f has a relative minimum at $x = 0$; this relative minimum is $f(0) = 0$. Using this information and plotting the points from the following table leads to the graph shown in Figure 22.

x	-1	0	1/2	1	2
$f(x)$	10	0	1.8	2	1.5

FIGURE 22

To complete the graph as shown, it would be helpful to also plot the points $(-1/4, 3.4)$ and $(4, -.9)$. Note that the graph has a sharp point at the critical number where the derivative does not exist.

(c) $f(x) = x^{2/3}$

The derivative is

$$f'(x) = \frac{2}{3}x^{-1/3} = \frac{2}{3x^{1/3}}.$$

The derivative is never 0, but it does fail to exist when $x = 0$. Since $f(0)$ *does* exist but $f'(0)$ does not, 0 is a critical point. Testing a point on either side of 0 gives the results shown in Figure 23. (Use 8 and -8 as test points because they

FIGURE 23

are perfect cubes and therefore easy to work with in the equations for $f(x)$ and $f'(x)$.) The function has a relative minimum at 0 with $f(0) = 0^{2/3} = 0$. The function is decreasing on the interval $(-\infty, 0)$ and increasing on $(0, \infty)$. Using this information and plotting a few points gives the graph in Figure 24. Again, there is a sharp point at the critical number because the derivative does not exist there. ◀

FIGURE 24

Caution A critical number must be in the domain of the function. For example, the derivative of $f(x) = x/(x - 4)$ is $f'(x) = -4/(x - 4)^2$, which fails to exist when $x = 4$. But $f(4)$ does not exist, so 4 is not a critical number, and the function has no relative extrema.

As mentioned at the beginning of this section, finding the maximum or minimum value of a quantity is important in applications of mathematics. The final example gives specific illustrations.

EXAMPLE **3** A small company manufactures and sells bicycles. The production manager has determined that the cost and demand functions for x ($x \geq 0$) bicycles per week are

$$C(x) = 10 + 5x + \frac{1}{60}x^3 \quad \text{and} \quad p = 90 - x,$$

where p is the price per bicycle.

(a) Find the maximum weekly revenue.

The revenue each week is given by

$$R(x) = xp = x(90 - x) = 90x - x^2.$$

To maximize $R(x) = 90x - x^2$, find $R'(x)$. Then find the critical numbers.

$$R'(x) = 90 - 2x = 0$$
$$90 = 2x$$
$$x = 45$$

Since $R'(x)$ exists for all x, 45 is the only critical number. To verify that $x = 45$ will produce a *maximum*, evaluate the derivative on either side of $x = 45$.

$$R'(40) = 10 \quad \text{and} \quad R'(50) = -10$$

This shows that $R(x)$ is increasing up to $x = 45$, then decreasing, so there is a maximum value at $x = 45$ of $R(45) = 2025$. The maximum revenue will be $2025 and will occur when 45 bicycles are produced and sold each week.

(b) Find the maximum weekly profit.

Since profit equals revenue minus cost, the profit is given by

$$P(x) = R(x) - C(x)$$
$$= (90x - x^2) - (10 + 5x + \frac{1}{60}x^3)$$
$$= -\frac{1}{60}x^3 - x^2 + 85x - 10.$$

Find the derivative and set it equal to 0 to find the critical numbers. (The derivative exists for all x.)

$$P'(x) = -\frac{1}{20}x^2 - 2x + 85 = 0$$

Solving this equation by the quadratic formula gives the solutions $x \approx 25.8$ and $x \approx -65.8$. Since x cannot be negative, the only critical number of concern is 25.8. Determine whether $x = 25.8$ produces a maximum by testing a value on either side of 25.8 in $P'(x)$.

$$P'(0) = 85 \quad \text{and} \quad P'(40) = -75$$

These results show that $P(x)$ increases to $x = 25.8$ and then decreases. Since x must be an integer, verify that $x = 26$ produces a maximum value of $P(26) = 1231.07$. (We should also check $P(25) = 1229.58$.) Thus, the maximum profit of $1231.07 occurs when 26 bicycles are produced and sold each week. Notice that this is not the same as the number that should be produced to yield maximum revenue.

(c) Find the price the company should charge to realize maximum profit.

As shown in part (b), 26 bicycles per week should be produced and sold to get the maximum profit of $1231.07 per week. Since the price is given by

$$p = 90 - x,$$

if $x = 26$, then $p = 64$. The manager should charge \$64 per bicycle and produce and sell 26 bicycles per week to get the maximum profit of \$1231.07 per week. Figure 25 shows the graphs of the functions used in this example. Notice that the slopes of the revenue and cost functions are the same at the point where the maximum profit occurs. Why is this true? ◀

FIGURE 25

The examples in this section involving the maximization of a quadratic function, such as the opening example and the bicycle revenue example, could be solved by the methods described in Chapter 1. But those involving more complicated functions, such as the bicycle profit example, are difficult to analyze without the tools of calculus.

11.2 Exercises

Find the locations and values of all relative extrema for the functions with graphs as follows. Compare with Exercises 1–8 in the preceding section.

1.

2.

3.

4.

5.

6.

7.

8.

Find the x-values of all points where the functions defined as follows have any relative extrema. Find the value(s) of any relative extrema.

9. $f(x) = x^2 + 12x - 8$

10. $f(x) = x^2 - 4x + 6$

11. $f(x) = 4 - 3x - .5x^2$

12. $f(x) = 3 - .4x - .2x^2$

13. $f(x) = x^3 + 6x^2 + 9x - 8$

14. $f(x) = x^3 + 3x^2 - 24x + 2$

15. $f(x) = -\dfrac{4}{3}x^3 - \dfrac{21}{2}x^2 - 5x + 8$

16. $f(x) = -\dfrac{2}{3}x^3 - \dfrac{1}{2}x^2 + 3x - 4$

17. $f(x) = 2x^3 - 21x^2 + 60x + 5$

18. $f(x) = 2x^3 + 15x^2 + 36x - 4$

19. $f(x) = x^4 - 18x^2 - 4$

20. $f(x) = x^4 - 8x^2 + 9$

21. $f(x) = -(8 - 5x)^{2/3}$

22. $f(x) = (2 - 9x)^{2/3}$

23. $f(x) = 2x + 3x^{2/3}$

24. $f(x) = 3x^{5/3} - 15x^{2/3}$

25. $f(x) = x - \dfrac{1}{x}$

26. $f(x) = x^2 + \dfrac{1}{x}$

27. $f(x) = \dfrac{x^2}{x^2 + 1}$

28. $f(x) = \dfrac{x^2}{x - 3}$

29. $f(x) = \dfrac{x^2 - 2x + 1}{x - 3}$

30. $f(x) = \dfrac{x^2 - 6x + 9}{x + 2}$

Use the derivative to find the vertex of each parabola.

31. $y = -2x^2 + 8x - 1$

32. $y = 3x^2 - 12x + 2$

33. $y = 2x^2 - 5x + 2$

34. $y = ax^2 + bx + c$

Applications

Business and Economics

Profit *In Exercises 35–37, find* **(a)** *the price p per unit that produces maximum profit;* **(b)** *the number x of units that produces maximum profit; and* **(c)** *the maximum profit P* $(P = R - C)$.

35. $C(x) = 75 + 10x; \quad p = 70 - 2x$

36. $C(x) = 25x + 5000; \quad p = 80 - .01x$

37. $C(x) = -13x^2 + 300x; \quad p = -x^2 - x + 360$

38. Profit The total profit $P(x)$ (in dollars) from the sale of x units of a certain prescription drug is given by

$$P(x) = -x^3 + 3x^2 + 72x.$$

(a) Find the number of units that should be sold in order to maximize the total profit.

(b) What is the maximum profit?

39. Revenue The demand equation for telephones at one store is

$$p = D(q) = \frac{1}{3}q^2 - \frac{25}{2}q + 100, \qquad 0 \le q \le 16,$$

where p is the price (in dollars) and q is the quantity of telephones sold per week. Find the values of q and p that maximize revenue.

40. Cost Suppose that the cost function for a product is given by $C(x) = .002x^3 - 9x + 4000$. Find the production level (i.e., value of x) that will produce the minimum average cost per unit $\overline{C}(x)$.

Life Sciences

41. Activity Level In the summer the activity level of a certain type of lizard varies according to the time of day. A biologist has determined that the activity level is given by the function

$$a(t) = .008t^3 - .27t^2 + 2.02t + 7,$$

where t is the number of hours after 12 noon. When is the activity level highest? When is it lowest?

Social Sciences

42. Attitude Change Social psychologists have found that as the discrepancy between the views of a speaker and those of an audience increases, the attitude change in the audience also increases to a point, but decreases when the discrepancy becomes too large, particularly if the communi-

cator is viewed by the audience as having low credibility.* Suppose that the degree of change, y, can be approximated by the function

$$D(x) = -x^4 + 8x^3 + 80x^2,$$

where x is the discrepancy between the views of the speaker and those of the audience, as measured by scores on a questionnaire. Find the amount of discrepancy the speaker should aim for to maximize the attitude change in the audience.

43. Film Length A group of researchers found that people prefer training films of moderate length; shorter films contain too little information, while longer films are boring. For a training film on the care of exotic birds, the researchers determined that the ratings people gave for the film could be approximated by

$$R(t) = \frac{20t}{t^2 + 100},$$

where t is the length of the film (in minutes). Find the film length that received the highest rating.

Physical Sciences

44. Height After a great deal of experimentation, two Atlantic Institute of Technology senior physics majors determined that when a bottle of French champagne is shaken several times, held upright, and uncorked, its cork travels according to

$$s(t) = -16t^2 + 64t + 3,$$

where s is its height (in feet) above the ground t seconds after being released. How high will it go?

*See A. H. Eagly and K. Telaak, "Width of the Latitude of Acceptance as a Determinant of Attitude Change" in *Journal of Personality and Social Psychology*, vol. 23, 1972, pp. 388–97.

▨ Computer/Graphing Calculator

Use a computer or graphing calculator to find the approximate x-values (to two significant digits) of all points where the functions defined as follows have any relative extrema. Find the value(s) of any relative extrema. One method is to compute the derivative, then use a graph of the function or an equation solver to find where the derivative is equal to zero.

45. $f(x) = x^5 - x^4 + 4x^3 - 30x^2 + 5x + 6$

46. $f(x) = -x^5 - x^4 + 2x^3 - 25x^2 + 9x + 12$

47. $f(x) = .001x^6 - .02x^5 - x^4 - 4x^3 + 12x^2 - 9x - 17$

48. $f(x) = -.001x^6 - .005x^5 + .5x^4 + 7x^3 - 8x^2 + 15x + 43$

11.3 ABSOLUTE EXTREMA

How do you invest money for the highest rate of return?

If you are in charge of an investment portfolio, you may be happy to have found a rate of return that is a relative maximum. Your boss may not be happy if he or she discovers another rate of return that is even higher. In this situation, the relative maximum is not as important as the *absolute maximum*.

FIGURE 26

The function graphed in Figure 26 has a relative minimum at x_2 and relative maxima at x_1 and x_3. In practical situations it is often desirable to know if a function has one function value that is larger than any other or one function value that is smaller than any other. In Figure 26, $f(x_1) \geq f(x)$ for all x in the domain. There is no function value that is smaller than all others, however, because $f(x) \to -\infty$ as $x \to \infty$ or as $x \to -\infty$.

The largest possible value of a function is called the *absolute maximum* and the smallest possible value of a function is called the *absolute minimum*. As Figure 26 shows, one or both of these may not exist on the domain of the function, $(-\infty, \infty)$ here. Absolute extrema often coincide with relative extrema, as with $f(x_1)$ in Figure 26. Although a function may have several relative maxima or relative minima, it never has more than one *absolute maximum* or *absolute minimum*.

ABSOLUTE MAXIMUM OR MINIMUM

Let f be a function defined on some interval. Let c be a number in the interval. Then $f(c)$ is the **absolute maximum** of f on the interval if

$$f(x) \leq f(c)$$

for every x in the interval, and $f(c)$ is the **absolute minimum** of f on the interval if

$$f(x) \geq f(c)$$

for every x in the interval.

Now look at Figure 27, which shows three functions defined on closed intervals. In each case there is an absolute maximum value and an absolute minimum value. These absolute extrema may occur at the endpoints or at relative extrema. As the graphs in Figure 27 show, an absolute extremum is either the largest or the smallest function value occurring on a closed interval, while a relative extremum is the largest or smallest function value in some (perhaps small) open interval.

Although a function can have only one absolute minimum value and only one absolute maximum value, it can have many points where these values occur.

FIGURE 27

(Note that the absolute maximum value and absolute minimum value are numbers, not points.) As an extreme example, consider the function $f(x) = 2$. The absolute minimum value of this function is clearly 2, as is the absolute maximum value. Both the absolute minimum and the absolute maximum occur at every real number x.

One of the main reasons for the importance of absolute extrema is given by the following theorem (which is proved in more advanced courses).

EXTREME VALUE THEOREM

A function f that is continuous on a closed interval $[a, b]$ will have both an absolute maximum and an absolute minimum on the interval.

FIGURE 28

As Figure 27 shows, an absolute extremum *may* occur on an open interval. See x_2 in part (a) and x_3 in part (c), for example. The extreme value theorem says that a function *must* have both an absolute maximum and an absolute minimum on a closed interval. The conditions that f be *continuous* and on a *closed* interval are very important. In Figure 28, f is discontinuous at $x = b$. Since $f(x)$ gets larger and larger as $x \to b$, there is no absolute (or relative) maximum on (a, b). Also, as $x \to a^+$, the values of $f(x)$ get smaller and smaller, but there is no smallest value of $f(x)$ on the *open* interval (a, b), since there is no endpoint.

The extreme value theorem guarantees the existence of absolute extrema. To find these extrema, use the following steps.

FINDING ABSOLUTE EXTREMA

To find absolute extrema for a function f continuous on a closed interval $[a, b]$:

1. Find all critical numbers for f in (a, b).
2. Evaluate f for all critical numbers in (a, b).
3. Evaluate f for the endpoints a and b of the interval $[a, b]$.
4. The largest value found in Step 2 or 3 is the absolute maximum for f on $[a, b]$, and the smallest value found is the absolute minimum for f on $[a, b]$.

EXAMPLE **1**

Find the absolute extrema of the function

$$f(x) = 2x^3 - x^2 - 20x - 10$$

on the interval $[-2, 4]$.

First look for critical points in the interval $(-2, 4)$. Set the derivative $f'(x) = 6x^2 - 2x - 20$ equal to 0.

$$6x^2 - 2x - 20 = 0$$
$$2(3x^2 - x - 10) = 0$$
$$2(3x + 5)(x - 2) = 0$$

$$3x + 5 = 0 \qquad \text{or} \qquad x - 2 = 0$$
$$x = -5/3 \qquad\qquad\qquad x = 2$$

Since there are no values of x where $f'(x)$ does not exist, the only critical values are $-5/3$ and 2. Both of these critical values are in the interval $(-2, 4)$. Now evaluate the function at $-5/3$ and 2, and at the endpoints of its domain, -2 and 4.

x-value	Value of Function	
-2	10	
$-5/3$	11.296	
2	-38	←Absolute minimum
4	22	←Absolute maximum

The absolute maximum, 22, occurs when $x = 4$, and the absolute minimum, -38, occurs when $x = 2$. A graph of f on $[-2, 4]$ is shown in Figure 29. ◀

$$f(x) = 2x^3 - x^2 - 20x - 10$$
on $[-2, 4]$

FIGURE 29

EXAMPLE 2

Find the absolute extrema of the function given by

$$f(x) = 4x + \frac{36}{x}$$

on the interval $[1, 6]$.

Note that f is discontinuous at $x = 0$, but continuous on the closed interval $[1, 6]$, so absolute extrema are guaranteed to exist. First determine any critical points in $(1, 6)$.

$$f'(x) = 4 - \frac{36}{x^2} = 0$$

$$\frac{4x^2 - 36}{x^2} = 0$$

$$4x^2 - 36 = 0$$

$$4x^2 = 36$$

$$x^2 = 9$$

$$x = -3 \quad \text{or} \quad x = 3$$

Although $f'(x)$ does not exist at $x = 0$, f is not defined at $x = 0$, so the critical numbers are -3 and 3. Since -3 is not in the interval $[1, 6]$, 3 is the only critical number. To find the absolute extrema, evaluate $f(x)$ at $x = 3$ and at the endpoints, where $x = 1$ and $x = 6$.

$f(x) = 4x + \dfrac{36}{x}$ on $[1, 6]$

Absolute maximum

Relative maximum

Absolute minimum

FIGURE 30

x-value	Value of Function	
1	40	←Absolute maximum
3	24	←Absolute minimum
6	30	

Figure 30 shows a graph of f on $[1, 6]$. ◀

Absolute extrema are particularly important in applications of mathematics. In most applications the domain is limited in some natural way that ensures the occurrence of absolute extrema.

EXAMPLE 3

A company has found that its weekly profit from the sale of x units of an auto part is given by

$$P(x) = -.02x^3 + 600x - 20,000.$$

Production bottlenecks limit the number of units that can be made per week to no more than 150, while a long-term contract requires that at least 50 units be made each week. Find the maximum possible weekly profit that the firm can make.

Because of the restrictions, the profit function is defined only for the domain [50, 150]. Look first for critical numbers of the function in the open interval (50, 150). Here $P'(x) = -.06x^2 + 600$. Set this derivative equal to 0 and solve for x.

$$-.06x^2 + 600 = 0$$
$$-.06x^2 = -600$$
$$x^2 = 10,000$$
$$x = 100 \quad \text{or} \quad x = -100$$

Since $x = -100$ is not in the interval [50, 150], disregard it. Now evaluate the function at the remaining critical number, 100, and at the endpoints of the domain, 50 and 150.

x-value	Value of Function	
50	7500	
100	20,000	←Absolute maximum
150	2500	

Maximum profit of $20,000 occurs when 100 units are made per week. ◀

11.3 Exercises

Find the locations of any absolute extrema for the functions with graphs as follows.

1.

2.

3.

4.

5.

6.

7.

8.

Find the locations of all absolute extrema for the functions defined as follows, with the specified domains.

9. $f(x) = x^2 + 6x + 2$; $[-4, 0]$

10. $f(x) = x^2 - 4x + 1$; $[-6, 5]$

11. $f(x) = 5 - 8x - 4x^2$; $[-5, 1]$

12. $f(x) = 9 - 6x - 3x^2$; $[-4, 3]$

13. $f(x) = x^3 - 3x^2 - 24x + 5$; $[-3, 6]$

14. $f(x) = x^3 - 6x^2 + 9x - 8$; $[0, 5]$

15. $f(x) = \frac{1}{3}x^3 - \frac{1}{2}x^2 - 6x + 3; \quad [-4, 4]$

16. $f(x) = \frac{1}{3}x^3 + \frac{3}{2}x^2 - 4x + 1; \quad [-5, 2]$

17. $f(x) = x^4 - 32x^2 - 7; \quad [-5, 6]$

18. $f(x) = x^4 - 18x^2 + 1; \quad [-4, 4]$

19. $f(x) = \frac{1}{1 + x}; \quad [0, 2]$

20. $f(x) = \frac{-2}{x + 3}; \quad [1, 4]$

21. $f(x) = \frac{8 + x}{8 - x}; \quad [4, 6]$

22. $f(x) = \frac{1 - x}{3 + x}; \quad [0, 3]$

23. $f(x) = \frac{x}{x^2 + 2}; \quad [0, 4]$

24. $f(x) = \frac{x - 1}{x^2 + 1}; \quad [1, 5]$

25. $f(x) = (x^2 + 18)^{2/3}; \quad [-3, 3]$

26. $f(x) = (x^2 + 4)^{1/3}; \quad [-2, 2]$

27. $f(x) = (x + 1)(x + 2)^2; \quad [-4, 0]$

28. $f(x) = (x - 3)(x - 1)^3; \quad [-2, 3]$

29. $f(x) = \frac{1}{\sqrt{x^2 + 1}}; \quad [-1, 1]$

30. $f(x) = \frac{3}{\sqrt{x^2 + 4}}; \quad [-2, 2]$

Applications

Business and Economics

31. Average Cost An enterprising (although unscrupulous) business student has managed to get his hands on a copy of the out-of-print solutions manual for an applied calculus text. He plans to duplicate it and sell copies in the dormitories. He figures that demand will be between 100 and 1200 copies, and he wants to minimize his average cost of production. After checking into the cost of paper, duplicating, and the rental of a small van, he estimates that the cost in dollars to produce x hundred manuals is given by $C(x) = x^2 + 200x + 100$. How many should he produce in order to make the *average cost* per unit as small as possible? How much will he have to charge to make a profit?

32. Profit The total profit $P(x)$ (in thousands of dollars) from the sale of x hundred thousand automobile tires is approximated by

$$P(x) = -x^3 + 9x^2 + 120x - 400, \quad x \geq 5.$$

Find the number of hundred thousands of tires that must be sold to maximize profit. Find the maximum profit.

Average Cost *In Exercises 33 and 34, find the minimum value of the average cost for the given cost function on the given intervals.*

33. $C(x) = 81x^2 + 17x + 324$ on the following intervals.
 (a) $1 \leq x \leq 10$ (b) $10 \leq x \leq 20$

34. $C(x) = x^3 + 37x + 250$ on the following intervals.
 (a) $1 \leq x \leq 10$ (b) $10 \leq x \leq 20$

Life Sciences

35. Pollution A marshy region used for agricultural drainage has become contaminated with selenium. It has been deter-

mined that flushing the area with clean water will reduce the selenium for a while, but it will then begin to build up again. A biologist has found that the percent of selenium in the soil x months after the flushing begins is given by

$$f(x) = \frac{x^2 + 36}{2x}, \quad 1 \leq x \leq 12.$$

When will the selenium be reduced to a minimum? What is the minimum percent?

36. Salmon Spawning The number of salmon swimming upstream to spawn is approximated by

$$S(x) = -x^3 + 3x^2 + 360x + 5000, \quad 6 \leq x \leq 20,$$

where x represents the temperature of the water in degrees Celsius. Find the water temperature that produces the maximum number of salmon swimming upstream.

Physical Sciences

37. Gasoline Mileage From information given in a recent business publication, we constructed the mathematical model

$$M(x) = -\frac{1}{45}x^2 + 2x - 20, \quad 30 \leq x \leq 65,$$

to represent the miles per gallon used by a certain car at a speed of x mph. Find the absolute maximum miles per gallon and the absolute minimum.

38. Gasoline Mileage For a certain compact car,

$$M(x) = -.018x^2 + 1.24x + 6.2, \quad 30 \leq x \leq 60,$$

represents the miles per gallon obtained at a speed of x mph. Find the absolute maximum miles per gallon and the absolute minimum.

General Interest

Area *A piece of wire 12 ft long is cut into two pieces. (See the figure.) One piece is made into a circle and the other piece is made into a square. Let the piece of length x be formed into a circle.*

Radius of circle $= \dfrac{x}{2\pi}$ Area of circle $= \pi\left(\dfrac{x}{2\pi}\right)^2$ Side of square $= \dfrac{12-x}{4}$ Area of square $= \left(\dfrac{12-x}{4}\right)^2$

39. Where should the cut be made in order to minimize the sum of the areas enclosed by both figures?

40. Where should the cut be made in order to make the sum of the areas maximum? (*Hint:* Remember to use the endpoints of a domain when looking for absolute maxima and minima.)

Computer/Graphing Calculator

Use a computer or a graphing calculator to find the approximate locations (to two significant digits) of all absolute extrema for the functions defined as follows, with the specified domains. One way this may be done is by graphing each function over the indicated domain and focusing on where the function is largest and smallest. Another way is to compute the derivative and then use a graph of the function or an equation solver to find where the derivative is equal to zero.

41. $f(x) = \dfrac{x^3 + 2x + 5}{x^4 + 3x^3 + 10}$; $[-3, 0]$

42. $f(x) = \dfrac{-5x^4 + 2x^3 + 3x^2 + 9}{x^4 - x^3 + x^2 + 7}$; $[-1, 1]$

43. $f(x) = x^{4/5} + x^2 - 4\sqrt{x}$; $[0, 4]$

44. $f(x) = x^{10/3} - x^{4/3} - 4x^2 - 8x$; $[0, 4]$

11.4 HIGHER DERIVATIVES, CONCAVITY, AND THE SECOND DERIVATIVE TEST

Just because the price of a stock is increasing, does that alone make it a good investment?

The following discussion addresses this question.

To understand the behavior of a function on an interval, it is important to know the *rate* at which the function is increasing or decreasing. For example, suppose that your friend, a finance major, has made a study of a young company and is trying to get you to invest in its stock. He shows you the following function, which represents the price $P(t)$ of the company's stock since it became available in January two years ago:

$$P(t) = 17 + t^{1/2},$$

where t is the number of months since the stock became available. He points out that the derivative of the function is always positive, so the price of the stock is always increasing. He claims that you cannot help but make a fortune on it. Should you take his advice and invest?

It is true that the price function increases for all t. The derivative is

$$P'(t) = \frac{1}{2}t^{-1/2} = \frac{1}{2\sqrt{t}},$$

which is always positive because \sqrt{t} is positive for $t > 0$. The catch lies in *how fast* the function is increasing. The derivative $P'(t) = 1/(2\sqrt{t})$ tells how fast the price is increasing at any number of months, t, since the stock became available. For example, when $t = 1$, $P'(t) = 1/2$, and the price is increasing at the rate of $1/2$ dollar, or 50 cents, per month. When $t = 4$, $P'(t) = 1/4$; the stock is increasing at 25 cents per month. At $t = 9$ months, $P'(t) = 1/6$, or about 17 cents per month. By the time you could buy in at $t = 24$ months, the price is increasing at 10 cents per month, and the *rate of increase* looks as though it will continue to decrease.

The rate of increase in P' is given by the derivative of $P'(t)$, called the **second derivative** of P and denoted by $P''(t)$. Since $P'(t) = (1/2)t^{-1/2}$,

$$P''(t) = -\frac{1}{4}t^{-3/2} = -\frac{1}{4\sqrt{t^3}}.$$

$P''(t)$ is negative for all positive values of t and therefore confirms the suspicion that the *rate* of increase in price does indeed decrease for all $t \geq 0$. The price of the company's stock will not drop, but the amount of return will certainly not be the fortune that your friend predicts. For example, at $t = 24$ months, when you would buy, the price would be $21.90. A year later, it would be $23.00 a share. If you were rich enough to buy 100 shares for $2190, they would be worth $2300 in a year. The increase of $110 is about 5% of the investment—similar to the return that you could get in most savings accounts. The only investors to make a lot of money on this stock would be those who bought early, when the rate of increase was much greater.

In Figure 31* the (generally) decreasing function shows that ''things are getting worse,'' but the overall shape of the graph indicates that the *rate* of decrease is slowing down. In other words, the second derivative is negative.

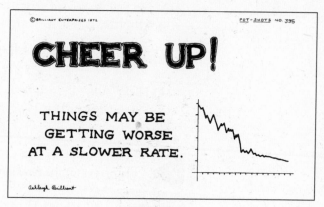

FIGURE 31

*Ashleigh Brilliant Epigrams, Pot-Shots, and Brilliant Thoughts from the book *I May Not Be Totally Perfect, But Parts of Me Are Excellent, and Other Brilliant Thoughts* used by permission of the author, Ashleigh Brilliant, 117 West Valerio St., Santa Barbara, CA 93101.

As mentioned earlier, the second derivative of a function f, written f'', gives the rate of change of the *derivative* of f. Before continuing to discuss applications of the second derivative, we need to introduce some additional terminology and notation.

Higher Derivatives If a function f has a derivative f', then the derivative of f', if it exists, is the second derivative of f, written f''. The derivative of f'', if it exists, is called the **third derivative** of f, and so on. By continuing this process, we can find **fourth derivatives** and other higher derivatives. For example, if $f(x) = x^4 + 2x^3 + 3x^2 - 5x + 7$, then

$$f'(x) = 4x^3 + 6x^2 + 6x - 5, \qquad \text{First derivative of } f$$

$$f''(x) = 12x^2 + 12x + 6, \qquad \text{Second derivative of } f$$

$$f'''(x) = 24x + 12, \qquad \text{Third derivative of } f$$

and $\qquad f^{(4)}(x) = 24.$ \qquad \text{Fourth derivative of } f

NOTATION FOR HIGHER DERIVATIVES

The second derivative of $y = f(x)$ can be written using any of the following notations:

$$f''(x), \qquad y'', \qquad \frac{d^2y}{dx^2}, \qquad \text{or} \qquad D_x^2[f(x)].$$

The third derivative can be written in a similar way. For $n \geq 4$, the nth derivative is written $f^{(n)}(x)$.

 EXAMPLE 1 Let $f(x) = x^3 + 6x^2 - 9x + 8$.

(a) Find $f''(x)$.

To find the second derivative of $f(x)$, find the first derivative, and then take its derivative.

$$f'(x) = 3x^2 + 12x - 9$$

$$f''(x) = 6x + 12$$

(b) Find $f''(0)$.

Since $f''(x) = 6x + 12$,

$$f''(0) = 6(0) + 12 = 12. \quad \blacktriangleleft$$

 EXAMPLE 2 Find $f''(x)$ for the functions defined as follows.

(a) $f(x) = \sqrt{x}$

Let $f(x) = \sqrt{x} = x^{1/2}$. Then

$$f'(x) = \frac{1}{2}x^{-1/2},$$

and
$$f''(x) = -\frac{1}{4}x^{-3/2} = -\frac{1}{4} \cdot \frac{1}{x^{3/2}} = \frac{-1}{4x^{3/2}}.$$

(b) $f(x) = (x^2 - 1)^2$

Here
$$f'(x) = 2(x^2 - 1)(2x) = 4x(x^2 - 1).$$

Use the product rule to find $f''(x)$.

$$\begin{aligned}
f''(x) &= 4x(2x) + (x^2 - 1)(4) \\
&= 8x^2 + 4x^2 - 4 \\
&= 12x^2 - 4 \quad \blacktriangleleft
\end{aligned}$$

In the previous chapter we saw that the first derivative of a function represents the rate of change of the function. The second derivative, then, represents the rate of change of the first derivative. If a function describes the position of a particle (along a straight line) at time t, then the first derivative gives the velocity of the particle. That is, if $y = s(t)$ describes the position (along a straight line) of the particle at time t, then $v(t) = s'(t)$ gives the velocity at time t.

The rate of change of velocity is called **acceleration.** Since the second derivative gives the rate of change of the first derivative, the acceleration is the derivative of the velocity. Thus, if $a(t)$ represents the acceleration at time t, then

$$a(t) = \frac{d}{dt}v(t) = s''(t).$$

EXAMPLE 3 Suppose that a particle is moving along a straight line, with its position at time t given by

$$s(t) = t^3 - 2t^2 - 7t + 9.$$

Find the following.

(a) The velocity at any time t

The velocity is given by

$$v(t) = s'(t) = 3t^2 - 4t - 7.$$

(b) The acceleration at any time t

Acceleration is given by

$$a(t) = v'(t) = s''(t) = 6t - 4.$$

(c) The time intervals (for $t \geq 0$) when the particle is speeding up or slowing down

When the acceleration is positive, the velocity is increasing, so the particle is speeding up. This happens when $6t - 4 > 0$, or $t > 2/3$. A negative acceleration indicates that the particle is slowing down. This happens when $0 < t < 2/3$. $\quad \blacktriangleleft$

FIGURE 32

Concavity of a Graph The first derivative has been used to show where a function is increasing or decreasing and where the extrema occur. The second derivative gives the rate of change of the first derivative; it indicates *how fast* the function is increasing or decreasing. The rate of change of the derivative (the second derivative) affects the *shape* of the graph. Intuitively, we say that a graph is *concave upward* on an interval if it "holds water" and *concave downward* if it "spills water." See Figure 32.

More precisely, a function is **concave upward** on an interval (a, b) if the graph of the function lies above its tangent line at each point of (a, b). A function is **concave downward** on (a, b) if the graph of the function lies below its tangent line at each point of (a, b). A point where a graph changes concavity is called a **point of inflection.** See Figure 33.

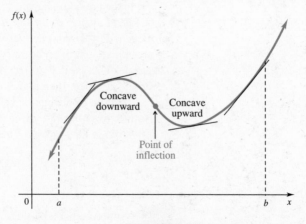

FIGURE 33

Just as a function can be either increasing or decreasing on an interval, it can be either concave upward or concave downward on an interval. Examples of various combinations are shown in Figure 34.

FIGURE 34

Figure 35 shows two functions that are concave upward on an interval (a, b). Several tangent lines are also shown. In Figure 35(a), the slopes of the tangent lines (moving from left to right) are first negative, then 0, and then positive. In Figure 35(b), the slopes are all positive, but they get larger.

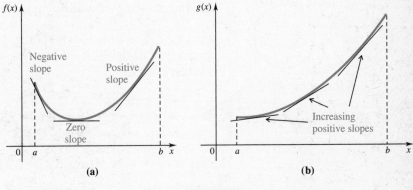

FIGURE 35

In both cases, the slopes are *increasing*. The slope at a point on a curve is given by the derivative. Since a function is increasing if its derivative is positive, its slope is increasing if the derivative of the slope function is positive. Since the derivative of a derivative is the second derivative, a function is concave upward on an interval if its second derivative is positive at each point of the interval.

A similar result is suggested by Figure 36 for functions whose graphs are concave downward. In both graphs, the slopes of the tangent lines are *decreasing* as we move from left to right. Since a function is decreasing if its derivative is negative, a function is concave downward on an interval if its second derivative is negative at each point of the interval. These observations suggest the following test.

FIGURE 36

TEST FOR CONCAVITY

Let f be a function with derivatives f' and f'' existing at all points in an interval (a, b). Then f is concave upward on (a, b) if $f''(x) > 0$ for all x in (a, b), and concave downward on (a, b) if $f''(x) < 0$ for all x in (a, b).

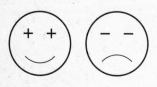

FIGURE 37

An easy way to remember this test is by the faces shown in Figure 37:

When the second derivative is positive at a point $(+ \ +)$, the graph is concave upward (\smile). When the second derivative is negative at a point $(- \ -)$, the graph is concave downward (\frown).

EXAMPLE 4

Find all intervals where $f(x) = x^3 - 3x^2 + 5x - 4$ is concave upward or downward.

The first derivative is $f'(x) = 3x^2 - 6x + 5$, and the second derivative is $f''(x) = 6x - 6$. The function f is concave upward whenever $f''(x) > 0$, or

$$6x - 6 > 0$$
$$6x > 6$$
$$x > 1.$$

Also, f is concave downward if $f''(x) < 0$, or $x < 1$. In interval notation, f is concave upward on $(1, \infty)$ and concave downward on $(-\infty, 1)$. A graph of f is shown in Figure 38. ◀

FIGURE 38

The graph of the function f in Figure 38 changes from concave downward to concave upward at $x = 1$. As mentioned earlier, a point where the direction of concavity changes is called a point of inflection. This means that the point $(1, f(1))$ or $(1, -1)$ in Figure 38 is a point of inflection. We can locate this point by finding values of x where the second derivative changes from negative to positive, that is, where the second derivative is 0. Setting the second derivative in Example 4 equal to 0 gives

$$6x - 6 = 0$$
$$x = 1.$$

As before, the point of inflection is $(1, f(1))$ or $(1, -1)$.

Example 4 suggests the following result.

At a point of inflection for a function f, the second derivative is 0 or does not exist.

Caution Be careful with this statement. Finding a value $f''(x_0) = 0$ does not mean that a point of inflection has been located. For example, if $f(x) = (x - 1)^4$, then $f''(x) = 12(x - 1)^2$, which is 0 at $x = 1$. The graph of $f(x) = (x - 1)^4$ is always concave upward, however, so it has no point of inflection. See Figure 39.

Second derivative is 0 at
$x = 1$, but $(1, f(1))$ is not
a point of inflection.

FIGURE 39

Second Derivative Test The idea of concavity can be used to decide whether a given critical number produces a relative maximum or a relative minimum. This test, an alternative to the first derivative test, is based on the fact that a curve that has a horizontal tangent at a point c and is concave downward on an open interval containing c also has a relative maximum at c. A relative minimum occurs when a graph has a horizontal tangent at a point d and is concave upward on an open interval containing d. See Figure 40.

FIGURE 40

A function f is concave upward on an interval if $f''(x) > 0$ for all x in the interval, while f is concave downward on an interval if $f''(x) < 0$ for all x in the interval. These ideas lead to the **second derivative test** for relative extrema.

SECOND DERIVATIVE TEST Let f'' exist on some open interval containing c, and let $f'(c) = 0$.

 1. If $f''(c) > 0$, then $f(c)$ is a relative minimum.

 2. If $f''(c) < 0$, then $f(c)$ is a relative maximum.

 3. If $f''(c) = 0$, then the test gives no information about extrema.

EXAMPLE 5

Find all relative extrema for

$$f(x) = 4x^3 + 7x^2 - 10x + 8.$$

First, find the points where the derivative is 0. Here $f'(x) = 12x^2 + 14x - 10$. Solve the equation $f'(x) = 0$ to get

$$12x^2 + 14x - 10 = 0$$
$$2(6x^2 + 7x - 5) = 0$$
$$2(3x + 5)(2x - 1) = 0$$

$$3x + 5 = 0 \qquad \text{or} \qquad 2x - 1 = 0$$
$$3x = -5 \qquad\qquad 2x = 1$$
$$x = -\frac{5}{3} \qquad\qquad x = \frac{1}{2}.$$

Now use the second derivative test. The second derivative is $f''(x) = 24x + 14$. Evaluate $f''(x)$ first at $-5/3$, getting

$$f''\left(-\frac{5}{3}\right) = 24\left(-\frac{5}{3}\right) + 14 = -40 + 14 = -26 < 0,$$

so that by Part 2 of the second derivative test, $-5/3$ leads to a relative maximum of $f(-5/3) = 691/27$. Also, when $x = 1/2$,

$$f''\left(\frac{1}{2}\right) = 24\left(\frac{1}{2}\right) + 14 = 12 + 14 = 26 > 0,$$

with $1/2$ leading to a relative minimum of $f(1/2) = 21/4$. ◀

The second derivative test works only for those critical numbers c that make $f'(c) = 0$. This test does not work for critical numbers c for which $f'(c)$ does not exist (since $f''(c)$ would not exist either). Also, the second derivative test does not work for critical numbers c that make $f''(c) = 0$. In both of these cases, use the first derivative test.

EXAMPLE 6

The graph in Figure 41 shows the population of catfish in a commercial catfish farm as a function of time. As the graph shows, the population increases rapidly up to a point and then increases at a slower rate. The horizontal dashed line shows that the population will approach some upper limit determined by the capacity of the farm. The point at which the rate of population growth starts to slow is the point of inflection for the graph.

To produce the maximum yield of catfish, harvesting should take place at the point of fastest possible growth of the population; here, this is at the point of inflection. The rate of change of the population, given by the first derivative, is increasing up to the inflection point (on the interval where the second derivative is positive) and decreasing past the inflection point (on the interval where the second derivative is negative). ◀

FIGURE 41

FIGURE 42

The *law of diminishing returns* in economics is related to the idea of concavity. The function graphed in Figure 42 gives the output y from a given input x. If the input were advertising costs for some product, for example, the output might be the corresponding revenue from sales.

The graph in Figure 42 shows an inflection point at $(c, f(c))$. For $x < c$, the graph is concave upward, so the rate of change of the slope is increasing. This indicates that the output y is increasing at a faster rate with each additional dollar spent. When $x > c$, however, the graph is concave downward, the rate of change of the slope is decreasing, and the increase in y is smaller with each additional dollar spent. Thus, further input beyond c dollars produces diminishing returns. The point of inflection at $(c, f(c))$ is called the **point of diminishing returns.** Beyond this point there is a smaller and smaller return for each dollar invested.

EXAMPLE 7

The revenue $R(x)$ generated from sales of a certain product is related to the amount x spent on advertising by

$$R(x) = \frac{1}{15,000} (600x^2 - x^3), \quad 0 \le x \le 600,$$

where x and $R(x)$ are in thousands of dollars. Is there a point of diminishing returns for this function? If so, what is it?

Since a point of diminishing returns occurs at an inflection point, look for an x-value that makes $R''(x) = 0$. Write the function as

$$R(x) = \frac{600}{15,000}x^2 - \frac{1}{15,000}x^3 = \frac{1}{25}x^2 - \frac{1}{15,000}x^3.$$

Now find $R'(x)$ and then $R''(x)$.

$$R'(x) = \frac{2x}{25} - \frac{3x^2}{15,000} = \frac{2}{25}x - \frac{1}{5000}x^2$$

$$R''(x) = \frac{2}{25} - \frac{1}{2500}x$$

Set $R''(x)$ equal to 0 and solve for x.

$$\frac{2}{25} - \frac{1}{2500}x = 0$$

$$-\frac{1}{2500}x = -\frac{2}{25}$$

$$x = \frac{5000}{25} = 200$$

Test a number in the interval $(0, 200)$ to see that $R''(x)$ is positive there. Then test a number in the interval $(200, 600)$ to find $R''(x)$ negative in that interval. Since the sign of $R''(x)$ changes from positive to negative at $x = 200$, the graph changes from concave upward to concave downward at that point, and there is a point of diminishing returns at the inflection point $(200, 1066\frac{2}{3})$. Investments in advertising beyond \$200,000 return less and less for each dollar invested. Verify that $R'(200) = 8$. This means that when \$200,000 is invested, another \$1000 invested returns approximately \$8000 in additional revenue. Thus it may still be economically sound to invest in advertising beyond the point of diminishing returns. ◄

11.4 Exercises

Find f'' for each of the following. Then find $f''(0)$, $f''(2)$, and $f''(-3)$.

1. $f(x) = 3x^3 - 4x + 5$

2. $f(x) = x^3 + 4x^2 + 2$

3. $f(x) = 3x^4 - 5x^3 + 2x^2$

4. $f(x) = -x^4 + 2x^3 - x^2$

5. $f(x) = 3x^2 - 4x + 8$

6. $f(x) = 8x^2 + 6x + 5$

7. $f(x) = (x + 4)^3$

8. $f(x) = (x - 2)^3$

9. $f(x) = \dfrac{2x + 1}{x - 2}$

10. $f(x) = \dfrac{x + 1}{x - 1}$

11. $f(x) = \dfrac{x^2}{1 + x}$

12. $f(x) = \dfrac{-x}{1 - x^2}$

13. $f(x) = \sqrt{x + 4}$

14. $f(x) = \sqrt{2x + 9}$

15. $f(x) = 5x^{3/5}$

16. $f(x) = -2x^{2/3}$

Find $f'''(x)$, the third derivative of f, and $f^{(4)}(x)$, the fourth derivative of f, for each of the following.

17. $f(x) = -x^4 + 2x^2 + 8$

18. $f(x) = 2x^4 - 3x^3 + x^2$

19. $f(x) = 4x^5 + 6x^4 - x^2 + 2$

20. $f(x) = 3x^5 - x^4 + 2x^3 - 7x$

21. $f(x) = \dfrac{x - 1}{x + 2}$

22. $f(x) = \dfrac{x + 1}{x}$

23. $f(x) = \dfrac{3x}{x - 2}$

24. $f(x) = \dfrac{x}{2x + 1}$

In Exercises 25–40, find the largest open intervals where the functions are concave upward or concave downward. Find any points of inflection.

25.

26.

27.

28.

29.

30.

31. $f(x) = x^2 + 10x - 9$

32. $f(x) = 8 - 6x - x^2$

33. $f(x) = x^3 + 3x^2 - 45x - 3$

34. $f(x) = 2x^3 - 3x^2 - 12x + 1$

35. $f(x) = -2x^3 + 9x^2 + 168x - 3$

36. $f(x) = -x^3 - 12x^2 - 45x + 2$

37. $f(x) = \dfrac{3}{x - 5}$

38. $f(x) = \dfrac{-2}{x + 1}$

39. $f(x) = x(x + 5)^2$

40. $f(x) = -x(x - 3)^2$

Find any critical numbers for f in Exercises 41–46 and then use the second derivative test to decide whether the critical numbers lead to relative maxima or relative minima. If f''(c) = 0 for a critical number c, then the second derivative test gives no information. In this case, use the first derivative test instead.

41. $f(x) = -x^2 - 10x - 25$

42. $f(x) = x^2 - 12x + 36$

43. $f(x) = 3x^3 - 3x^2 + 1$

44. $f(x) = 2x^3 - 4x^2 + 2$

45. $f(x) = (x + 3)^4$

46. $f(x) = x^3$

Applications

Business and Economics

47. Product Life Cycle The accompanying figure shows the *product life cycle* graph, with typical products marked on it. It illustrates the fact that a new product is often purchased at a faster and faster rate as people become familiar with it. In time, saturation is reached and the purchase rate stays constant until the product is made obsolete by newer products, after which it is purchased less and less.*

(a) Which products on the left side of the graph are closest to the left-hand point of inflection? What does the point of inflection mean here?

(b) Which product on the right side of the graph is closest to the right-hand point of inflection? What does the point of inflection mean here?

(c) Discuss where home computers, fax machines, and dot-matrix printers should be placed on the graph.

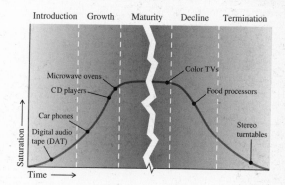

*Based on "The Product Life Cycle: A Key to Strategic Marketing Planning" in *MSU Business Topics* (Winter 1973), p. 30. Reprinted by permission of the publisher, Graduate School of Business Administration, Michigan State University.

Law of Diminishing Returns *In Exercises 48 and 49, find the point of diminishing returns for the given functions, where R(x) represents revenue in thousands of dollars and x represents the amount spent on advertising in thousands of dollars.*

48. $R(x) = 10,000 - x^3 + 42x^2 + 800x, \quad 0 \le x \le 20$

49. $R(x) = \dfrac{4}{27}(-x^3 + 66x^2 + 1050x - 400), \quad 0 \le x \le 25$

50. Risk Aversion In economics, an index of *absolute risk aversion* is defined as

$$I(M) = \frac{-U''(M)}{U'(M)},$$

where M measures how much of a commodity is owned and $U(M)$ is a *utility function,* which measures the ability of quantity M of a commodity to satisfy a consumer's wants. Find $I(M)$ for $U(M) = \sqrt{M}$ and for $U(M) = M^{2/3}$, and determine which indicates a greater aversion to risk.

51. Demand Function The authors of an article[†] in an economics journal state that if $D(q)$ is the demand function, then the inequality

$$qD''(q) + D'(q) < 0$$

is equivalent to saying that the marginal revenue declines more quickly than does the price. Prove that this equivalence is true.

Life Sciences

52. Population Growth When a hardy new species is introduced into an area, the population often increases as shown.

[†]Drew Fudenberg and Jean Tirole, "Learning by Doing and Market Performance," *Bell Journal of Economics* 14 (1983).

Explain the significance of the following function values on the graph.

(a) f_0 **(b)** $f(a)$ **(c)** f_M

53. Bacteria Population Assume that the number of bacteria $R(t)$ (in millions) present in a certain culture at time t (in hours) is given by

$$R(t) = t^2(t - 18) + 96t + 1000.$$

(a) At what time before 8 hr will the population be maximized?

(b) Find the maximum population.

54. Drug Concentration The percent of concentration of a certain drug in the bloodstream x hr after the drug is administered is given by

$$K(x) = \frac{3x}{x^2 + 4}.$$

For example, after 1 hr the concentration is given by

$$K(1) = \frac{3(1)}{1^2 + 4} = \frac{3}{5}\% = .6\% = .006.$$

(a) Find the time at which concentration is a maximum.

(b) Find the maximum concentration.

55. Drug Concentration The percent of concentration of a drug in the bloodstream x hours after the drug is administered is given by

$$K(x) = \frac{4x}{3x^2 + 27}.$$

(a) Find the time at which the concentration is a maximum.

(b) Find the maximum concentration.

Physical Sciences

56. Chemical Reaction An autocatalytic chemical reaction is one in which the product being formed causes the rate of formation to increase. The rate of a certain autocatalytic reaction is given by

$$V(x) = 12x(100 - x),$$

where x is the quantity of the product present and 100 represents the quantity of chemical present initially. For what value of x is the rate of the reaction a maximum?

Velocity and Acceleration *Each of the functions defined in Exercises 57–62 gives the displacement at time t of a particle moving along a line. Find the velocity and acceleration functions. Then find the velocity and acceleration at t = 0 and t = 4. Assume that time is measured in seconds and distance is measured in centimeters. Velocity will be in centimeters per second (cm/sec) and acceleration in centimeters per second per second (cm/sec²).*

57. $s(t) = 8t^2 + 4t$

58. $s(t) = -3t^2 - 6t + 2$

59. $s(t) = -5t^3 - 8t^2 + 6t - 3$

60. $s(t) = 3t^3 - 4t^2 + 8t - 9$

61. $s(t) = \dfrac{-2}{3t + 4}$

62. $s(t) = \dfrac{1}{t + 3}$

63. Velocity and Acceleration When an object is dropped straight down, the distance in feet that it travels in t sec is given by

$$s(t) = -16t^2.$$

Find the velocity at each of the following times.

(a) After 3 sec **(b)** After 5 sec **(c)** After 8 sec

(d) Find the acceleration. (The answer here is a constant—the acceleration due to the influence of gravity alone.)

64. Projectile Height If an object is thrown directly upward with a velocity of 256 ft/sec, its height above the ground after t sec is given by $s(t) = 256t - 16t^2$. Find the velocity and the acceleration after t sec. What is the maximum height the object reaches? When does it hit the ground?

Computer/Graphing Calculator

In Exercises 65–68, get a list of values for $f'(x)$ and $f''(x)$ for each function on the given interval as indicated, or sketch the graph of $f'(x)$ and $f''(x)$ on a graphing calculator or computer.

(a) *By looking at the sign of $f'(x)$, give the (approximate) intervals where $f(x)$ is increasing and any intervals where $f(x)$ is decreasing.*

(b) *Give the (approximate) x-values where any maximums or minimums occur.*

(c) *By looking at the sign of $f''(x)$, give the intervals where $f(x)$ is concave upward and where it is concave downward.*

(d) *Give the (approximate) x-values of any inflection points.*

65. $f(x) = .25x^4 - 2x^3 + 3.5x^2 + 4x - 1$; $(-5, 5)$ in steps of .5

66. $f(x) = 10x^3(x - 1)^2$; $(-2, 2)$ in steps of .3

67. $f(x) = 3.1x^4 - 4.3x^3 + 5.82$; $(-1, 2)$ in steps of .2

68. $f(x) = \dfrac{x}{x^2 + 1}$; $(-3, 3)$ in steps of .4

69. Use a graphing calculator or a computer to sketch the graphs of the functions in Exercises 65–68. Compare these graphs with your answers from those exercises. Discuss briefly how well the graphs agree with these answers.

11.5 LIMITS AT INFINITY AND CURVE SKETCHING

What happens to the oxygen concentration in a pond over the long run?

Sometimes it is useful to examine the behavior of the values of $f(x)$ as x gets larger and larger (or smaller and smaller). For example, suppose a small pond normally contains 12 units of dissolved oxygen in a fixed volume of water. Suppose also that at time $t = 0$ a quantity of organic waste is introduced into the pond, with the oxygen concentration t weeks later given by

$$f(t) = \frac{12t^2 - 15t + 12}{t^2 + 1}.$$

As time goes on, what will be the ultimate concentration of oxygen? Will it return to 12 units?

After 2 weeks, the pond contains

$$f(2) = \frac{12 \cdot 2^2 - 15 \cdot 2 + 12}{2^2 + 1} = \frac{30}{5} = 6$$

units of oxygen, and after 4 weeks, it contains

$$f(4) = \frac{12 \cdot 4^2 - 15 \cdot 4 + 12}{4^2 + 1} \approx 8.5$$

units. Choosing several values of t and finding the corresponding values of $f(t)$, or using a graphing calculator or computer, leads to the graph in Figure 43.

FIGURE 43

The graph suggests that as time goes on, the oxygen level gets closer and closer to the original 12 units. Consider a table of values as t gets larger and larger.

t	10	100	1000	10,000
$f(t)$	10.5	11.85	11.985	11.9985

The table suggests that

$$\lim_{t \to \infty} f(t) = 12,$$

where $t \to \infty$ means that t increases without bound. (Similarly, $x \to -\infty$ means that x *decreases* without bound; that is, x becomes more and more negative.) Thus, the oxygen concentration will approach 12, but it will never be *exactly* 12.

The graphs of $f(x) = 1/x$ (in black) and $g(x) = 1/x^2$ (in color) shown in Figure 44 lead to examples of such **limits at infinity.** The graphs and table of values indicate that $\lim_{x \to \infty} (1/x) = 0$, $\lim_{x \to -\infty} (1/x) = 0$, $\lim_{x \to \infty} (1/x^2) = 0$, and $\lim_{x \to -\infty} (1/x^2) = 0$, suggesting the following rule.

x	$\dfrac{1}{x}$	$\dfrac{1}{x^2}$
-100	$-.01$	$.0001$
-10	$-.1$	$.01$
-1	-1	1
1	1	1
10	$.1$	$.01$
100	$.01$	$.0001$

FIGURE 44

LIMITS AT INFINITY

For any positive real number n,

$$\lim_{x \to \infty} \frac{1}{x^n} = 0 \quad \text{and} \quad \lim_{x \to -\infty} \frac{1}{x^n} = 0.*$$

The rules for limits given in the preceding chapter remain unchanged when a is replaced with ∞ or $-\infty$.

 EXAMPLE 1 Find each limit.

This section requires use of the rules for limits discussed in the section on limits and continuity, with a replaced by ∞ or $-\infty$. One such rule is the following:

If $\lim\limits_{x \to a} f(x) = A$ and $\lim\limits_{x \to a} g(x) = B$, then

$$\lim_{x \to a} \frac{f(x)}{g(x)} = \frac{A}{B} \quad \text{if } B \neq 0.$$

Find where this rule is used in Example 1, and then look up the other rules and find where they are used in this example.

(a) $\lim\limits_{x \to \infty} \dfrac{8x + 6}{3x - 1}$

We can use the rule $\lim\limits_{x \to \infty} 1/x^n = 0$ to find this limit by first dividing numerator and denominator by x, as follows.

$$\lim_{x \to \infty} \frac{8x + 6}{3x - 1} = \lim_{x \to \infty} \frac{\dfrac{8x}{x} + \dfrac{6}{x}}{\dfrac{3x}{x} - \dfrac{1}{x}} = \lim_{x \to \infty} \frac{8 + 6 \cdot \dfrac{1}{x}}{3 - \dfrac{1}{x}} = \frac{8 + 0}{3 - 0} = \frac{8}{3}$$

(b) $\lim\limits_{x \to -\infty} \dfrac{4x^2 - 6x + 3}{2x^2 - x + 4}$

Divide each term of the numerator and denominator by x^2, the highest power of x.

*If x is negative, x^n does not exist for certain values of n, so the second limit is undefined.

$$\lim_{x \to -\infty} \frac{4x^2 - 6x + 3}{2x^2 - x + 4} = \lim_{x \to -\infty} \frac{4 - 6 \cdot \frac{1}{x} + 3 \cdot \frac{1}{x^2}}{2 - \frac{1}{x} + 4 \cdot \frac{1}{x^2}}$$

$$= \frac{4 - 0 + 0}{2 - 0 + 0} = \frac{4}{2} = 2$$

(c) $\displaystyle\lim_{x \to \infty} \frac{3x + 2}{4x^3 - 1} = \lim_{x \to \infty} \frac{3 \cdot \frac{1}{x^2} + 2 \cdot \frac{1}{x^3}}{4 - \frac{1}{x^3}} = \frac{0 + 0}{4 - 0} = \frac{0}{4} = 0$

Here, the highest power of x is x^3, which is used to divide each term in the numerator and denominator.

(d) $\displaystyle\lim_{x \to \infty} \frac{3x^2 + 2}{4x - 3} = \lim_{x \to \infty} \frac{3 + \frac{2}{x^2}}{\frac{4}{x} - \frac{3}{x^2}} = \frac{3 + 0}{0 - 0} = \frac{3}{0}$

Division by 0 is undefined, so this limit does not exist. We can actually say more: by examining what happens as larger and larger values of x are put into the function, verify that

$$\lim_{x \to \infty} \frac{3x^2 + 2}{4x - 3} = \infty. \quad \blacktriangleleft$$

The method used in Example 1 is a useful way to rewrite expressions with fractions so that the rules for limits at infinity can be used.

FINDING LIMITS AT INFINITY

If $f(x) = \dfrac{p(x)}{q(x)}$, for polynomials $p(x)$ and $q(x)$, $q(x) \neq 0$, $\displaystyle\lim_{x \to -\infty} f(x)$ and $\displaystyle\lim_{x \to \infty} f(x)$ can be found as follows.

1. Divide $p(x)$ *and* $q(x)$ by the highest power of x in either polynomial.
2. Use the rules for limits, including the rules for limits at infinity,

$$\lim_{x \to \infty} \frac{1}{x^n} = 0 \quad \text{and} \quad \lim_{x \to -\infty} \frac{1}{x^n} = 0,$$

to find the limit of the result from Step 1.

Look back at the method for finding horizontal asymptotes in Chapter 1. There we were essentially taking limits at infinity, although we did not use that terminology. Limits at infinity, therefore, give us a method for finding horizontal asymptotes.

Curve Sketching The test for concavity, the test for increasing and decreasing functions, and the concept of limits at infinity help us sketch the graphs of a variety of functions. This process, called *curve sketching*, uses the following steps.

CURVE SKETCHING

To sketch the graph of a function f:

1. Find the y-intercept (if it exists) by substituting $x = 0$ into $f(x)$. Find any x-intercepts by solving $f(x) = 0$ if this is not too difficult.

2. If f is a rational function, find any vertical asymptotes by investigating where the denominator is 0, and find any horizontal asymptotes by finding the limits as $x \to \infty$ and $x \to -\infty$.

3. Find f'. Locate any critical points by solving the equation $f'(x) = 0$ and determining where f' does not exist, but f does. Find any relative extrema and determine where f is increasing or decreasing.

4. Find f''. Locate potential points of inflection by solving the equation $f''(x) = 0$ and determining where $f''(x)$ does not exist. Determine where f is concave upward or concave downward.

5. Plot the intercepts, the critical points, the inflection points, the asymptotes, and other points as needed.

EXAMPLE 2

Graph $f(x) = 2x^3 - 3x^2 - 12x + 1$.

The y-intercept is located at $y = f(0) = 1$. Finding the x-intercepts requires solving the equation $f(x) = 0$. But this is a third-degree equation; since we have not covered a procedure for solving such equations, we will skip this step. This is not a rational function, so we also skip Step 2.

To find the intervals where the function is increasing or decreasing, find the first derivative.

$$f'(x) = 6x^2 - 6x - 12$$

This derivative is 0 when

$$6(x^2 - x - 2) = 0$$
$$6(x - 2)(x + 1) = 0$$
$$x = 2 \quad \text{or} \quad x = -1.$$

These critical numbers divide the number line in Figure 45 into three regions.

FIGURE 45

Testing a number from each region in $f'(x)$ shows that f is increasing on $(-\infty, -1)$ and $(2, \infty)$, and decreasing on $(-1, 2)$. This is shown with the arrows in Figure 45. By the first derivative test, f has a relative maximum when $x = -1$ and a relative minimum when $x = 2$. The relative maximum is $f(-1) = 8$, while the relative minimum is $f(2) = -19$.

Now use the second derivative to find the intervals where the function is concave upward or downward. Here

$$f''(x) = 12x - 6,$$

which is 0 when $x = 1/2$. Testing a point with x less than $1/2$, and one with x greater than $1/2$, shows that f is concave downward on $(-\infty, 1/2)$ and concave upward on $(1/2, \infty)$. The graph has an inflection point at $(1/2, f(1/2))$, or $(1/2, -11/2)$. This information is summarized in the chart below.

Interval	$(-\infty, -1)$	$(-1, 1/2)$	$(1/2, 2)$	$(2, \infty)$
Sign of f'	+	−	−	+
Sign of f''	−	−	+	+
f Increasing or Decreasing	Increasing	Decreasing	Decreasing	Increasing
Concavity of f	Downward	Downward	Upward	Upward
Shape of Graph	\frown	\frown	\smile	\smile

Use this information and the critical points to get the graph shown in Figure 46. ◄

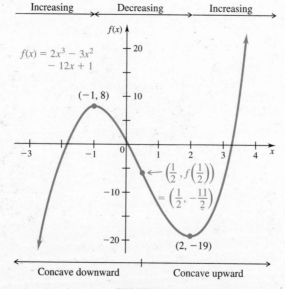

FIGURE 46

EXAMPLE 3

Graph $f(x) = x + 1/x$.

Notice that $x = 0$ is not in the domain of the function, so there is no y-intercept. To find the x-intercept, solve $f(x) = 0$.

$$x + \frac{1}{x} = 0$$

$$x = -\frac{1}{x}$$

$$x^2 = -1$$

Since x^2 is always positive, there is also no x-intercept.

The function is a rational function, but it is not written in the usual form of one polynomial over another. It can be rewritten in that form:

$$f(x) = x + \frac{1}{x} = \frac{x^2 + 1}{x}.$$

Because $x = 0$ makes the denominator 0, the line $x = 0$ is a vertical asymptote. To find any horizontal asymptotes, we investigate

$$\lim_{x \to \infty} \frac{x^2 + 1}{x} = \lim_{x \to \infty} \frac{1 + \frac{1}{x^2}}{\frac{1}{x}} = \frac{1 + 0}{0} = \frac{1}{0}.$$

Division by 0 is undefined, so the limit does not exist. Verify that $\lim_{x \to -\infty} f(x)$ also does not exist, so there are no horizontal asymptotes.

Observe that as x gets very large, the second term $(1/x)$ in $f(x)$ gets very small, so $f(x) = x + (1/x) \approx x$. The graph gets closer and closer to the straight line $y = x$ as x becomes larger and larger. This is what is known as an **oblique asymptote**.

Here $f'(x) = 1 - (1/x^2)$, which is 0 when

$$\frac{1}{x^2} = 1$$

$$x^2 = 1$$

$$x = 1 \quad \text{or} \quad x = -1.$$

The derivative fails to exist at 0, where the vertical asymptote is located. Evaluating $f'(x)$ in each of the regions determined by the critical numbers and the asymptote shows that f is increasing on $(-\infty, -1)$ and $(1, \infty)$ and decreasing on $(-1, 0)$ and $(0, 1)$. By the first derivative test, f has a relative maximum of $y = f(-1) = -2$, when $x = -1$, and a relative minimum of $y = f(1) = 2$ when $x = 1$.

The second derivative is

$$f''(x) = \frac{2}{x^3},$$

which is never equal to 0 and does not exist when $x = 0$. (The function itself also does not exist at 0.) Because of this, there may be a change of concavity, but not an inflection point, when $x = 0$. The second derivative is negative when x is negative, making f concave downward on $(-\infty, 0)$. Also, $f''(x) > 0$ when $x > 0$, making f concave upward on $(0, \infty)$.

Interval	$(-\infty, -1)$	$(-1, 0)$	$(0, 1)$	$(1, \infty)$
Sign of f'	+	−	−	+
Sign of f"	−	−	+	+
f Increasing or Decreasing	Increasing	Decreasing	Decreasing	Increasing
Concavity of f	Downward	Downward	Upward	Upward
Shape of Graph	⌒	⌒	⌣	⌣

Use this information, the asymptotes, and the critical points to get the graph shown in Figure 47. ◄

FIGURE 47

EXAMPLE **4**

Graph $f(x) = \dfrac{3x^2}{x^2 + 5}$.

The y-intercept is located at $y = f(0) = 0$. Verify that this is also the only x-intercept. There is no vertical asymptote, because $x^2 + 5 \neq 0$ for any value of x. Find any horizontal asymptote by calculating $\lim\limits_{x \to \infty} f(x)$ and $\lim\limits_{x \to -\infty} f(x)$. First, divide both the numerator and the denominator of $f(x)$ by x^2.

$$\lim_{x \to \infty} \frac{3x^2}{x^2 + 5} = \lim_{x \to \infty} \frac{\dfrac{3x^2}{x^2}}{\dfrac{x^2}{x^2} + \dfrac{5}{x^2}} = \frac{3}{1 + 0} = 3$$

Verify that the limit of $f(x)$ as $x \to -\infty$ is also 3. Thus, the horizontal asymptote is $y = 3$.

We now compute $f'(x)$:

$$f'(x) = \frac{(x^2 + 5)(6x) - (3x^2)(2x)}{(x^2 + 5)^2}.$$

Notice that $6x$ can be factored out of each term in the numerator:

$$f'(x) = \frac{(6x)[(x^2 + 5) - x^2]}{(x^2 + 5)^2}$$

$$= \frac{(6x)(5)}{(x^2 + 5)^2} = \frac{30x}{(x^2 + 5)^2}.$$

From the numerator, $x = 0$ is a critical number. The denominator is always positive. (Why?) Evaluating $f'(x)$ in each of the regions determined by $x = 0$ shows that f is decreasing on $(-\infty, 0)$ and increasing on $(0, \infty)$. By the first derivative test, f has a relative minimum when $x = 0$.

The second derivative is

$$f''(x) = \frac{(x^2 + 5)^2(30) - (30x)(2)(x^2 + 5)(2x)}{(x^2 + 5)^4}.$$

Factor $30(x^2 + 5)$ out of the numerator:

$$f''(x) = \frac{30(x^2 + 5)[(x^2 + 5) - (x)(2)(2x)]}{(x^2 + 5)^4}.$$

Divide a factor of $(x^2 + 5)$ out of the numerator and denominator, and simplify the numerator:

$$f''(x) = \frac{30[(x^2 + 5) - (x)(2)(2x)]}{(x^2 + 5)^3}$$

$$= \frac{30[(x^2 + 5) - (4x^2)]}{(x^2 + 5)^3}$$

$$= \frac{30(5 - 3x^2)}{(x^2 + 5)^3}.$$

The numerator of $f''(x)$ is 0 when $x = \pm\sqrt{5/3} \approx \pm1.29$. Testing a point in each of the three intervals defined by these points shows that f is concave downward on $(-\infty, -1.29)$ and $(1.29, \infty)$, and concave upward on $(-1.29, 1.29)$. The graph has inflection points at $(\pm\sqrt{5/3}, f(\pm\sqrt{5/3})) \approx (\pm1.29, \pm.75)$.

Interval	$(-\infty, -1.29)$	$(-1.29, 0)$	$(0, 1.29)$	$(1.29, \infty)$
Sign of f'	−	−	+	+
Sign of f''	−	+	+	−
f Increasing or Decreasing	Decreasing	Decreasing	Increasing	Increasing
Concavity of f	Downward	Upward	Upward	Downward
Shape of Graph	⌢	⌣	⌣	⌢

Use this information, the asymptote, the critical point, and the inflection points to get the graph shown in Figure 48. ◀

FIGURE 48

As we saw in Chapter 1, a graphing calculator or a computer graphing program, when used with care, can be helpful in studying the behavior of functions. This section has illustrated that calculus is also of great help. The techniques of calculus show where the important points of a function, such as the relative extrema and the inflection points, are located. Furthermore, they tell how the function behaves between and beyond the points that are graphed, something a computer graphing program cannot do.

11.5 Exercises

Decide whether the following limits exist. If a limit exists, find its value.

1. $\lim\limits_{x \to \infty} f(x)$

2. $\lim\limits_{x \to -\infty} g(x)$

3. $\lim\limits_{x \to \infty} \dfrac{3x}{5x - 1}$

4. $\lim\limits_{x \to \infty} \dfrac{5x}{3x - 1}$

5. $\lim\limits_{x \to -\infty} \dfrac{2x + 3}{4x - 7}$

6. $\lim\limits_{x \to -\infty} \dfrac{8x + 2}{2x - 5}$

7. $\lim\limits_{x \to \infty} \dfrac{x^2 + 2x}{2x^2 - 2x + 1}$

8. $\lim\limits_{x \to \infty} \dfrac{x^2 + 2x - 5}{3x^2 + 2}$

9. $\lim\limits_{x \to \infty} \dfrac{3x^3 + 2x - 1}{2x^4 - 3x^3 - 2}$

10. $\lim\limits_{x \to \infty} \dfrac{2x^2 - 1}{3x^4 + 2}$

11. $\lim\limits_{x \to \infty} (\sqrt{x^2 + 4} - x)$

12. $\lim\limits_{x \to \infty} (x - \sqrt{x^2 - 9})$

Graph each of the following functions, including critical points, regions where the function is increasing or decreasing, points of inflection, regions where the function is concave up or concave down, intercepts where possible, and asymptotes where applicable.

13. $f(x) = -2x^3 - 9x^2 + 108x - 10$

14. $f(x) = -2x^3 - 9x^2 + 60x - 8$

15. $f(x) = 2x^3 + \dfrac{7}{2}x^2 - 5x + 3$

16. $f(x) = x^3 - \dfrac{15}{2}x^2 - 18x - 1$

17. $f(x) = x^4 - 18x^2 + 5$

18. $f(x) = x^4 - 8x^2$

19. $f(x) = x^4 - 2x^3$

20. $f(x) = x^5 - 15x^3$

21. $f(x) = x + \dfrac{2}{x}$

22. $f(x) = 2x + \dfrac{8}{x}$

23. $f(x) = \dfrac{x^2 + 25}{x}$

24. $f(x) = \dfrac{x^2 + 4}{x}$

25. $f(x) = \dfrac{x - 1}{x + 1}$

26. $f(x) = \dfrac{x}{1 + x}$

27. $f(x) = \dfrac{x}{x^2 + 1}$

28. $f(x) = \dfrac{1}{x^2 + 1}$

29. $f(x) = \dfrac{1}{x^2 - 4}$

30. $f(x) = \dfrac{x}{x^2 - 1}$

In Exercises 31–34, sketch the graph of a single function that has all of the properties listed.

31. **(a)** Continuous and differentiable for all real numbers
 (b) Increasing on $(-\infty, -3)$ and $(1, 4)$
 (c) Decreasing on $(-3, 1)$ and $(4, \infty)$
 (d) Concave downward on $(-\infty, -1)$ and $(2, \infty)$
 (e) Concave upward on $(-1, 2)$
 (f) $f'(-3) = f'(4) = 0$
 (g) Inflection points at $(-1, 3)$ and $(2, 4)$

32. **(a)** Continuous for all real numbers
 (b) Increasing on $(-\infty, -2)$ and $(0, 3)$
 (c) Decreasing on $(-2, 0)$ and $(3, \infty)$
 (d) Concave downward on $(-\infty, 0)$ and $(0, 5)$
 (e) Concave upward on $(5, \infty)$
 (f) $f'(-2) = f'(3) = 0$
 (g) $f'(0)$ does not exist
 (h) Differentiable everywhere except at $x = 0$
 (i) An inflection point at $(5, 1)$

33. **(a)** Continuous for all real numbers

 (b) Decreasing on $(-\infty, -6)$ and $(1, 3)$

 (c) Increasing on $(-6, 1)$ and $(3, \infty)$

 (d) Concave upward on $(-\infty, -6)$ and $(3, \infty)$

 (e) Concave downward on $(-6, 3)$

 (f) A y-intercept at $(0, 2)$

34. **(a)** Continuous and differentiable everywhere except at $x = 1$, where it has a vertical asymptote

 (b) Decreasing everywhere it is defined

 (c) A horizontal asymptote at $y = 2$

 (d) Concave downward on $(-\infty, 1)$ and $(2, 4)$

 (e) Concave upward on $(1, 2)$ and $(4, \infty)$

Applications

Business and Economics

35. **Average Cost** The cost for manufacturing a particular videotape is

$$c(x) = 15,000 + 6x,$$

where x is the number of tapes produced. The average cost per tape, denoted by $\bar{c}(x)$, is found by dividing $c(x)$ by x. Find and interpret $\lim\limits_{x \to \infty} \bar{c}(x)$.

36. **Employee Productivity** A company training program has determined that, on the average, a new employee can do $P(s)$ pieces of work per day after s days of on-the-job training, where

$$P(s) = \frac{75s}{s + 8}.$$

Find and interpret $\lim\limits_{s \to \infty} P(s)$.

Life Sciences

37. **Drug Concentration** The concentration of a drug in a patient's bloodstream h hours after it was injected is given by

$$A(h) = \frac{.17h}{h^2 + 2}.$$

Find and interpret $\lim\limits_{h \to \infty} A(h)$.

Social Sciences

38. **Legislative Voting** Members of a legislature often must vote repeatedly on the same bill. As time goes on, members may change their votes. Suppose that p_0 is the probability that an individual legislator favors an issue before the first roll call vote, and suppose that p is the probability of a change in position from one vote to the next. Then the probability that the legislator will vote "yes" on the nth roll call is given by

$$p_n = \frac{1}{2} + \left(p_0 - \frac{1}{2}\right)(1 - 2p)^n.^*$$

For example, the chance of a "yes" on the third roll call vote is

$$p_3 = \frac{1}{2} + \left(p_0 - \frac{1}{2}\right)(1 - 2p)^3.$$

Suppose that there is a chance of $p_0 = .7$ that Congressman Stephens will favor the budget appropriation bill before the first roll call, but only a probability of $p = .2$ that he will change his mind on the subsequent vote. Find and interpret the following.

 (a) p_2 **(b)** p_4 **(c)** p_8 **(d)** $\lim\limits_{n \to \infty} p_n$

*See John W. Bishir and Donald W. Drewes, *Mathematics in the Behavioral and Social Sciences* (New York: Harcourt Brace Jovanovich, 1970), p. 538.

Computer/Graphing Calculator

(a) *Use a calculator or a computer to find the following limits by entering larger and larger values.*

(b) *Explain how your answer could have been predicted in advance.*

39. $\lim\limits_{x \to \infty} \dfrac{\sqrt{9x^2 + 5}}{2x}$

40. $\lim\limits_{x \to -\infty} \dfrac{\sqrt{9x^2 + 5}}{2x}$

41. $\lim\limits_{x \to -\infty} \dfrac{\sqrt{36x^2 + 2x + 7}}{3x}$

42. $\lim\limits_{x \to \infty} \dfrac{\sqrt{36x^2 + 2x + 7}}{3x}$

43. $\lim\limits_{x \to \infty} \dfrac{(1 + 5x^{1/3} + 2x^{5/3})^3}{x^5}$

44. $\lim\limits_{x \to -\infty} \dfrac{(1 + 5x^{1/3} + 2x^{5/3})^3}{x^5}$

| Chapter Summary | Key Terms |

11.1 **increasing function**
decreasing function
critical number
critical point
11.2 **relative maximum**
relative minimum
first derivative test

11.3 **absolute maximum**
absolute minimum
extreme value theorem
11.4 **second derivative**
acceleration
concavity

concave upward and downward
point of inflection
second derivative test
11.5 **limit at infinity**
oblique asymptote

| Chapter 11 | Review Exercises |

1. When given the equation for a function, how can you determine where it is increasing and where it is decreasing?

2. When given the equation for a function, how can you determine where the relative extrema are located? Give two ways to test whether a relative extremum is a minimum or a maximum.

3. What is the difference between a relative extremum and an absolute extremum? Can a relative extremum be an absolute extremum? Is a relative extremum necessarily an absolute extremum?

4. What information about a graph can be found from the second derivative?

Find the largest open intervals where f is increasing or decreasing.

5. $f(x) = x^2 - 5x + 3$

6. $f(x) = -2x^2 - 3x + 4$

7. $f(x) = -x^3 - 5x^2 + 8x - 6$

8. $f(x) = 4x^3 + 3x^2 - 18x + 1$

9. $f(x) = \dfrac{6}{x - 4}$

10. $f(x) = \dfrac{5}{2x + 1}$

Find the locations and values of all relative maxima and minima.

11. $f(x) = -x^2 + 4x - 8$

12. $f(x) = x^2 - 6x + 4$

13. $f(x) = 2x^2 - 8x + 1$

14. $f(x) = -3x^2 + 2x - 5$

15. $f(x) = 2x^3 + 3x^2 - 36x + 20$

16. $f(x) = 2x^3 + 3x^2 - 12x + 5$

Find the second derivative of each of the following, and then find $f''(1)$ and $f''(-3)$.

17. $f(x) = 3x^4 - 5x^2 - 11x$

18. $f(x) = 9x^3 + \dfrac{1}{x}$

19. $f(x) = \dfrac{5x - 1}{2x + 3}$

20. $f(x) = \dfrac{4 - 3x}{x + 1}$

21. $f(t) = \sqrt{t^2 + 1}$

22. $f(t) = -\sqrt{5 - t^2}$

Find the locations and values of all absolute maxima and minima on the given intervals.

23. $f(x) = -x^2 + 5x + 1;\;\; [1, 4]$

24. $f(x) = 4x^2 - 8x - 3;\;\; [-1, 2]$

25. $f(x) = x^3 + 2x^2 - 15x + 3;\;\; [-4, 2]$

26. $f(x) = -2x^3 - 2x^2 + 2x - 1;\;\; [-3, 1]$

Decide whether the limits in Exercises 27–32 exist. If a limit exists, find its value.

27. $\lim\limits_{x \to -\infty} g(x)$

28. $\lim\limits_{x \to \infty} f(x)$

29. $\lim\limits_{x \to \infty} \dfrac{x^2 + 5}{5x^2 - 1}$

30. $\lim\limits_{x \to \infty} \dfrac{x^2 + 6x + 8}{x^3 + 2x + 1}$

31. $\lim\limits_{x \to -\infty} \left(\dfrac{3}{4} + \dfrac{2}{x} - \dfrac{5}{x^2} \right)$

32. $\lim\limits_{x \to -\infty} \left(\dfrac{9}{x^4} + \dfrac{1}{x^2} - 3 \right)$

Graph each of the following functions, including critical points, regions where the function is increasing or decreasing, points of inflection, regions where the function is concave up or concave down, intercepts where possible, and asymptotes where applicable.

33. $f(x) = -2x^3 - \dfrac{1}{2}x^2 + x - 3$

34. $f(x) = -\dfrac{4}{3}x^3 + x^2 + 30x - 7$

35. $f(x) = x^4 - \dfrac{4}{3}x^3 - 4x^2 + 1$

36. $f(x) = -\dfrac{2}{3}x^3 + \dfrac{9}{2}x^2 + 5x + 1$

37. $f(x) = \dfrac{x - 1}{2x + 1}$

38. $f(x) = \dfrac{2x - 5}{x + 3}$

39. $f(x) = -4x^3 - x^2 + 4x + 5$

40. $f(x) = x^3 + \dfrac{5}{2}x^2 - 2x - 3$

41. $f(x) = x^4 + 2x^2$

42. $f(x) = 6x^3 - x^4$

43. $f(x) = \dfrac{x^2 + 4}{x}$

44. $f(x) = x + \dfrac{8}{x}$

45. $f(x) = \dfrac{2x}{3 - x}$

46. $f(x) = \dfrac{-4x}{1 + 2x}$

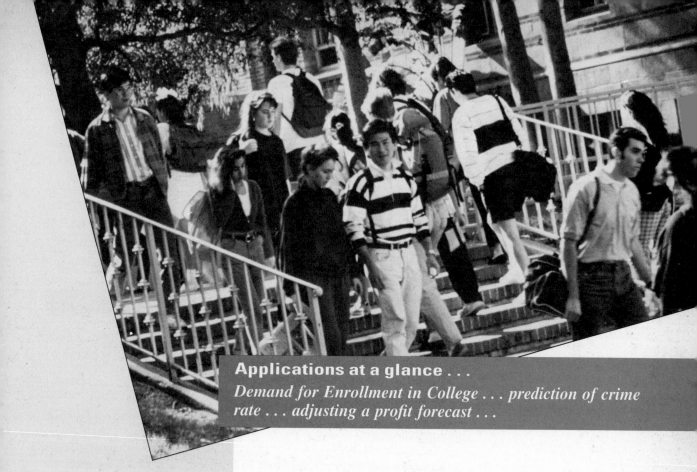

APPLICATIONS OF THE DERIVATIVE

(See page 597.)

What do aluminum cans, shipments of antibiotics, elasticity of demand, and a melting icicle have in common? All are involved in applications of the derivative. The previous chapter included examples in which we used the derivative to find the maximum or minimum value of a function. This problem is ubiquitous; consider the efforts people expend trying to maximize their income, or to minimize their costs or the time required to complete a task. In this chapter we will treat the topic of optimization in greater depth.

The derivative is applicable in far wider circumstances, however. In roughly 500 B.C., Heraclitus said, "Nothing endures but change," and his observation has relevance here. If change is continuous, rather than in sudden jumps, the derivative can be used to describe the rate of change. This explains why calculus has been applied to so many fields.

12.1 APPLICATIONS OF EXTREMA

How should boxes and cans be designed to minimize the material needed?

In Examples 3 and 4 we will use the techniques of calculus to find an answer to this question.

In this section, we give several examples showing applications of calculus to maximum and minimum problems. To solve these examples, go through the following steps.

SOLVING APPLIED EXTREMA PROBLEMS

1. Read the problem carefully. Make sure you understand what is given and what is unknown.

2. If possible, sketch a diagram. Label the various parts.

3. Decide on the variable that must be maximized or minimized. Express that variable as a function of *one* other variable. Be sure to find the domain of the function.

4. Find the critical points for the function from Step 3.

5. If the domain has endpoints, evaluate the function at the endpoints and the critical points to see which yields the maximum or minimum. If there are no endpoints, test each critical point for relative maxima or minima using the first or second derivative test.

Caution Do not skip Step 5 in the box above. If a problem asks you to maximize a quantity and you find a critical point at Step 4, do not automatically assume the maximum occurs there, for it may occur at an endpoint or may not exist at all.

An infamous case of such an error occurred in a 1945 study of "flying wing" aircraft designs similar to the Stealth bomber. In seeking to maximize the range of the aircraft (how far it can fly on a tank of fuel), the study's authors found that a critical point occurred when almost all of the volume of the plane was in the wing. They claimed that this critical point was a maximum. But another engineer later found that this critical point, in fact, *minimized* the range of the aircraft!*

EXAMPLE 1 Find two nonnegative numbers x and y for which $2x + y = 30$, such that xy^2 is maximized.

First we must decide what is to be maximized and give it a variable. Here, xy^2 is to be maximized, so let

$$M = xy^2.$$

Now, express M in terms of just *one* variable. Use the equation $2x + y = 30$ to do that. Solve $2x + y = 30$ for either x or y. Solving for y gives

$$2x + y = 30$$
$$y = 30 - 2x.$$

Substitute for y in the expression for M to get

$$M = x(30 - 2x)^2$$
$$= x(900 - 120x + 4x^2)$$
$$= 900x - 120x^2 + 4x^3.$$

Note that x must be at least 0. Since y must also be at least 0, $30 - 2x \geq 0$, so $x \leq 15$. Thus x is confined to the interval $[0, 15]$.

Find the critical points for M by finding M', then solving the equation $M' = 0$ for x.

$$M' = 900 - 240x + 12x^2 = 0$$
$$12(75 - 20x + x^2) = 0$$
$$(5 - x)(15 - x) = 0$$
$$x = 5 \quad \text{or} \quad x = 15$$

Find M for the critical numbers $x = 5$ and $x = 15$ as well as for $x = 0$, one endpoint of the domain. (The other endpoint has already been included as a critical number.)

x	M
0	0
5	2000 ← Maximum
15	0

*Wayne Biddle, "Skeleton Alleged in the Stealth Bomber's Closet," *Science*, vol. 244, May 12, 1989.

We see in the chart that the maximum value of the function occurs when $x = 5$. Since $y = 30 - 2(5) = 20$, the values that maximize xy^2 are $x = 5$ and $y = 20$. ◄

We will now return to a problem we first encountered in Chapter 1.

EXAMPLE 2 When Power and Money, Inc., charges $600 for a seminar on management techniques, it attracts 1000 people. For each $20 decrease in the charge, an additional 100 people will attend the seminar.

(a) Find an expression for the total revenue if there are x $20 decreases in price.
 The price charged will be

$$\text{Price per person} = 600 - 20x,$$

and the number of people in the seminar will be

$$\text{Number of people} = 1000 + 100x.$$

The total revenue, $R(x)$, is given by the product of the price and the number of people attending, or

$$R(x) = (600 - 20x)(1000 + 100x)$$
$$= 600,000 + 40,000x - 2000x^2.$$

Notice that x must be at least 0. Furthermore, since we cannot charge a negative amount to attend the seminar, $600 - 20x \geq 0$, so $x \leq 30$. Thus the domain of the function R is [0, 30].

(b) Find the value of x that maximizes revenue.
 Here $R'(x) = 40,000 - 4000x$. Set this derivative equal to 0.

$$40,000 - 4000x = 0$$
$$-4000x = -40,000$$
$$x = 10$$

Find $R(x)$ for x equal to 0, 10, and 30 to find which yields the maximum $R(x)$.

x	$R(x)$
0	600,000
10	800,000 ← Maximum
30	0

From the table we see that $R(x)$ is maximized when $x = 10$.

(c) Find the amount that Power and Money, Inc., should charge for the seminar to maximize the revenue, and find the maximum revenue.
 The price per person to maximize the revenue is $600 - 20(10) = \$400$. From the table in part (b), the revenue is then $R(10) = \$800,000$. ◄

EXAMPLE **3**

An open box is to be made by cutting a square from each corner of a 12-inch by 12-inch piece of metal and then folding up the sides. What size square should be cut from each corner in order to produce a box of maximum volume?

Let x represent the length of a side of the square that is cut from each corner, as shown in Figure 1(a). The width of the box is $12 - 2x$, with the length also $12 - 2x$. As shown in Figure 1(b), the depth of the box will be x inches. The volume of the box is given by the product of the length, width, and height. In this example, the volume, $V(x)$, depends on x:

$$V(x) = x(12 - 2x)(12 - 2x) = 144x - 48x^2 + 4x^3.$$

Clearly, $0 \leq x$, and since neither the length nor the width can be negative, $0 \leq 12 - 2x$, so $x \leq 6$. Thus the domain of V is the interval $[0, 6]$.

(a) **(b)**

FIGURE 1

The derivative is $V'(x) = 144 - 96x + 12x^2$. Set this derivative equal to 0.

$$12x^2 - 96x + 144 = 0$$
$$12(x^2 - 8x + 12) = 0$$
$$12(x - 2)(x - 6) = 0$$
$$x - 2 = 0 \quad \text{or} \quad x - 6 = 0$$
$$x = 2 \qquad\qquad x = 6$$

Find $V(x)$ for x equal to 0, 2, and 6 to find the depth that will maximize the volume.

x	$V(x)$
0	0
2	128 ← Maximum
6	0

The chart indicates that the box will have maximum volume when $x = 2$ and that the maximum volume will be 128 cubic inches. ◀

FIGURE 2

EXAMPLE 4

A company wants to manufacture cylindrical aluminum cans with a volume of 1000 cubic centimeters (one liter). What should the radius and height of the can be to minimize the amount of aluminum used?

The two variables in this problem are the radius and the height of the can, which we shall label r and h, as in Figure 2. Minimizing the amount of aluminum used requires minimizing the surface area of the can, which we will designate S. The surface area consists of a top and a bottom, each of which is a circle with an area πr^2, plus the side. If the side was sliced vertically and unrolled, it would form a rectangle with height h and width equal to the circumference of the can, which is $2\pi r$. Thus the surface area is given by

$$S = 2\pi r^2 + 2\pi rh.$$

The right side of the equation involves two variables. We need to get a function of a single variable. We can do this by using the information about the volume of the can:

$$V = \pi r^2 h = 1000.$$

(Here we have used the formula for the volume of a cylinder.) Solve this for h:

$$h = \frac{1000}{\pi r^2}.$$

(Solving for r would have involved a square root and a more complicated function.)

We now substitute this expression for h into the equation for S to get

$$S = 2\pi r^2 + 2\pi r\frac{1000}{\pi r^2} = 2\pi r^2 + \frac{2000}{r}.$$

There are no restrictions on r other than that it be a positive number, so the domain of S is $(0, \infty)$.

Find the critical points for S by finding S', then solving the equation $S' = 0$ for r.

$$S' = 4\pi r - \frac{2000}{r^2} = 0$$

$$4\pi r^3 = 2000$$

$$r^3 = \frac{500}{\pi}$$

Take the cube root of both sides to get

$$r = \left(\frac{500}{\pi}\right)^{1/3} \approx 5.419$$

centimeters. Substitute this expression into the equation for h to get

$$h = \frac{1000}{\pi 5.419^2} \approx 10.84$$

centimeters. Notice that the height of the can is twice its radius. But we are not yet assured that the critical number we have found is indeed the minimum. Since the domain has no endpoints, we will instead use the second derivative test:

$$S'' = 4\pi + \frac{4000}{r^3}.$$

S'' is positive for any positive value of r, so by the second derivative test, we have found the minimum surface area. ◀

For most living things, reproduction is *seasonal*—it can take place only at selected times of the year.* Large whales, for example, reproduce every two years during a relatively short time span of about two months. Shown on the time axis in Figure 3 are the reproductive periods. Let S = number of adults present during the reproductive period and let R = number of adults that return the next season to reproduce.

FIGURE 3

If we find a relationship between R and S, $R = f(S)$, then we have formed a **spawner-recruit** function or **parent-progeny** function. These functions are notoriously hard to develop because of the difficulty of obtaining accurate counts and because of the many hypotheses that can be made about the life stages. We will simply suppose that the function f takes various forms.

If $R > S$, we can presumably harvest

$$H = R - S = f(S) - S$$

individuals, leaving S to reproduce. Next season, $R = f(S)$ will return and the harvesting process can be repeated, as shown in Figure 4.

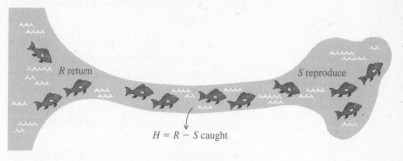

FIGURE 4

*From *Mathematics for the Biosciences* by Michael R. Cullen. Copyright © 1983 PWS Publishers. Reprinted by permission.

Let S_0 be the number of spawners that will allow as large a harvest as possible without threatening the population with extinction. Then $H(S_0)$ is called the **maximum sustainable harvest.**

EXAMPLE 5

Suppose the spawner-recruit function for Idaho rabbits is $f(S) = -.025S^2 + 4S$, where S is measured in thousands of rabbits. Find S_0 and the maximum sustainable harvest, $H(S_0)$.

S_0 is the value of S that maximizes H. Since

$$
\begin{aligned}
H(S) &= f(S) - S \\
&= -.025S^2 + 4S - S \\
&= -.025S^2 + 3S,
\end{aligned}
$$

the derivative $H'(S) = -.05S + 3$. Set this derivative equal to 0 and solve for S.

$$
\begin{aligned}
0 &= -.05S + 3 \\
.05S &= 3 \\
S &= 60
\end{aligned}
$$

The number of rabbits needed to sustain the population is $S_0 = 60$ thousand. Note that

$$H''(S) = -.05 < 0,$$

showing that $S = 60$ does indeed produce a *maximum* value of H. At $S_0 = 60$, the harvest is

$$H(60) = -.025(60)^2 + 3(60) = 90,$$

or 90 thousand rabbits. These results indicate that after one reproductive season, a population of 60 thousand rabbits will have increased to 150 thousand. Of these, 90 thousand may be harvested, leaving 60 thousand to regenerate the population. Any harvest larger than 90 thousand will threaten the future of the rabbit population, while a harvest smaller than 90 thousand will allow the population to grow larger each season. Thus, 90 thousand is the maximum sustainable harvest for this population. ◀

Finding extrema for realistic problems requires an accurate mathematical model of the problem. In particular, it is important to be aware of restrictions on the values of the variables. For example, if $T(x)$ closely approximates the number of items that can be manufactured daily on a production line when x is the number of employees on the line, x must certainly be restricted to the positive integers, or perhaps to a few common fractional values. (We can imagine half-time workers, but not 1/49-time workers.)

On the other hand, to apply the tools of calculus to obtain an extremum for some function, the function must be defined and be meaningful at every real

number in some interval. Because of this, the answer obtained from a mathematical model might be a number that is not feasible in the actual problem.

Usually, the requirement that a continuous function be used, rather than one that can take on only certain selected values, is of theoretical interest only. In most cases, the methods of calculus give acceptable results so long as the assumptions of continuity and differentiability are not totally unreasonable. If they lead to the conclusion, say, that $80\sqrt{2}$ workers should be hired, it is usually only necessary to investigate acceptable values close to $80\sqrt{2}$. This was done in Example 3 in the preceding chapter, in the section on Relative Extrema.

When the function to be maximized or minimized is a quadratic function, as in Example 2 or Example 5, we can solve the problem by finding the vertex of the parabola, as we did in Chapter 1. For other problems that would otherwise be very difficult, the calculus techniques of this chapter provide a powerful method of solution.

12.1 Exercises

Applications

General Interest

Number Analysis *Exercises 1–8 involve maximizing and minimizing products and sums of numbers. Solve each problem using steps such as those shown in Exercise 1.*

1. Find two nonnegative numbers x and y such that $x + y = 100$ and the product $P = xy$ is as large as possible, and give the maximum product. Use the following steps.

 (a) Solve $x + y = 100$ for y.

 (b) Substitute the result from part (a) into $P = xy$, the equation for the variable that is to be maximized.

 (c) Find the domain of the function P found in part (b).

 (d) Find P'. Solve the equation $P' = 0$.

 (e) Evaluate P at any solutions found in part (d), as well as at the endpoints of the domain found in part (c).

 (f) Give the maximum value of P, as well as the two numbers x and y whose product is that value.

2. Find two nonnegative numbers x and y whose sum is 250 and whose product is as large as possible, and give the maximum product.

3. Find two nonnegative numbers x and y whose sum is 200, such that the sum of the squares of the two numbers is minimized, and give the minimum sum.

4. Find two nonnegative numbers x and y whose sum is 30, such that the sum of the squares of the two numbers is minimized, and give the minimum sum.

In Exercises 5–8, use steps (a)–(f) from Exercise 1 to find nonnegative numbers x and y satisfying the given requirements and to find the maximum or minimum value of the indicated expression.

5. $x + y = 150$ and x^2y is maximized.

6. $x + y = 45$ and xy^2 is maximized.

7. $x - y = 10$ and xy is minimized.

8. $x - y = 3$ and xy is minimized.

9. **Area** A farmer has 1200 m of fencing. He wants to enclose a rectangular field bordering a river, with no fencing needed along the river. (See the sketch.) Let x represent the width of the field.

(a) Write an expression for the length of the field.

(b) Find the area of the field (area = length × width).

(c) Find the value of x leading to the maximum area.

(d) Find the maximum area.

10. **Area** Find the dimensions of the rectangular field of maximum area that can be made from 200 m of fencing material. (This fence has four sides.)

11. **Area** An ecologist is conducting a research project on breeding pheasants in captivity. She first must construct suitable pens. She wants a rectangular area with two additional fences across its width, as shown in the sketch. Find the maximum area she can enclose with 3600 m of fencing.

12. **Travel Time** A hunter is at a point on a river bank. He wants to get to his cabin, located 3 mi north and 8 mi west. (See the figure.) He can travel 5 mph on the river but only 2 mph on this very rocky land. How far upriver should he go in order to reach the cabin in minimum time? (*Hint:* distance = rate × time.)

13. **Travel Time** Repeat Exercise 12, but assume the cabin is 19 mi north and 8 mi west.

14. **Postal Regulations** The U.S. Postal Service stipulates that any boxes sent through the mail must have a length plus girth totaling no more than 108 inches. (See the figure.) Find the dimensions of the box with maximum volume that can be sent through the U.S. mail, assuming that the width and the height of the box are equal.

Business and Economics

15. **Revenue** If the price charged for a candy bar is $p(x)$ cents, then x thousand candy bars will be sold in a certain city, where

$$p(x) = 100 - \frac{x}{10}.$$

(a) Find an expression for the total revenue from the sale of x thousand candy bars.

(b) Find the value of x that leads to maximum revenue.

(c) Find the maximum revenue.

16. **Revenue** The sale of cassette tapes of "lesser" performers is very sensitive to price. If a tape manufacturer charges $p(x)$ dollars per tape, where

$$p(x) = 6 - \frac{x}{8},$$

then x thousand tapes will be sold.

(a) Find an expression for the total revenue from the sale of x thousand tapes.

(b) Find the value of x that leads to maximum revenue.

(c) Find the maximum revenue.

17. **Cost** A truck burns fuel at the rate of

$$G(x) = \frac{1}{32}\left(\frac{64}{x} + \frac{x}{50}\right)$$

gallons per mile while traveling at x mph.

(a) If fuel costs $1.60 per gallon, find the speed that will produce minimum total cost for a 400-mile trip.

(b) Find the minimum total cost.

18. Cost A rock-and-roll band travels from engagement to engagement in a large bus. This bus burns fuel at the rate of

$$G(x) = \frac{1}{50}\left(\frac{200}{x} + \frac{x}{15}\right)$$

gallons per mile while traveling at x mph.

(a) If fuel costs $2 per gallon, find the speed that will produce minimum total cost for a 250-mile trip.

(b) Find the minimum total cost.

19. Cost with Fixed Area A fence must be built to enclose a rectangular area of 20,000 ft^2. Fencing material costs $3 per foot for the two sides facing north and south, and $6 per foot for the other two sides. Find the cost of the least expensive fence.

20. Cost with Fixed Area A fence must be built in a large field to enclose a rectangular area of 15,625 m^2. One side of the area is bounded by an existing fence; no fence is needed there. Material for the fence costs $2 per meter for the two ends, and $4 per meter for the side opposite the existing fence. Find the cost of the least expensive fence.

21. Profit In planning a small restaurant, it is estimated that a profit of $5 per seat will be made if the number of seats is between 60 and 80, inclusive. On the other hand, the profit on each seat will decrease by 5¢ for each seat above 80.

(a) Find the number of seats that will produce the maximum profit.

(b) What is the maximum profit?

22. Timing Income A local group of scouts has been collecting old aluminum cans for recycling. The group has already collected 12,000 lb of cans, for which they could currently receive $4 per hundred pounds. The group can continue to collect cans at the rate of 400 lb per day. However, a glut in the old-can market has caused the recycling company to announce that it will lower its price, starting immediately, by $.10 per hundred pounds per day. The scouts can make only one trip to the recycling center. Find the best time for the trip. What total income will be received?

23. Packaging Design A television manufacturing firm needs to design an open-topped box with a square base. The box must hold 32 in^3. Find the dimensions of the box that can be built with the minimum amount of materials. (See the figure.)

24. Revenue A local club is arranging a charter flight to Hawaii. The cost of the trip is $425 each for 75 passengers, with a refund of $5 per passenger for each passenger in excess of 75.

(a) Find the number of passengers that will maximize the revenue received from the flight.

(b) Find the maximum revenue.

25. Packaging Design A company wishes to manufacture a box with a volume of 36 ft^3 that is open on top and is twice as long as it is wide. Find the dimensions of the box produced from the minimum amount of material.

26. Packaging Cost A closed box with a square base is to have a volume of 16,000 cm^3. The material for the top and bottom of the box costs $3 per square centimeter, while the material for the sides costs $1.50 per square centimeter. Find the dimensions of the box that will lead to minimum total cost. What is the minimum total cost?

27. Packaging Design A cylindrical box will be tied up with ribbon as shown in the figure. The longest piece of ribbon available is 130 cm long, and 10 cm of that are required for the bow. Find the radius and height of the box with the largest possible volume.

28. Can Design (a) For the can problem in Example 4, the minimum surface area required that the height be twice the radius. Show that this is true for a can of arbitrary volume V.

(b) Do many cans in grocery stores have a height that is twice the radius? If not, discuss why this may be so.

29. Container Design Your company needs to design cylindrical metal containers with a volume of 16 ft^3. The top and bottom will be made of a sturdy material that costs $2 per square foot, while the material for the sides costs $1 per square foot. Find the radius, height, and cost of the least expensive container.

30. Container Design An open box will be made by cutting a square from each corner of a 3-ft by 8-ft piece of cardboard and then folding up the sides. What size square should be cut from each corner in order to produce a box of maximum volume?

31. **Use of Materials** A mathematics book is to contain 36 in^2 of printed matter per page, with margins of 1 in. along the sides and 1½ in. along the top and bottom. Find the dimensions of the page that will require the minimum amount of paper. (See the figure.)

32. **Cost** A company wishes to run a utility cable from point *A* on the shore (see the figure) to an installation at point *B* on the island. The island is 6 mi from the shore. It costs $400 per mile to run the cable on land and $500 per mile

underwater. Assume that the cable starts at *A* and runs along the shoreline, then angles and runs underwater to the island. Find the point at which the line should begin to angle in order to yield the minimum total cost.

33. **Cost** Repeat Exercise 32, but make point *A* 7 mi from point *C*.

34. **Pricing** Decide what you would do if your assistant presented the following contract for your signature:

Your firm offers to deliver 300 tables to a dealer, at $90 per table, and to reduce the price per table on the entire order by 25¢ for each additional table over 300.

Find the dollar total involved in the largest possible transaction between the manufacturer and the dealer; then find the smallest possible dollar amount.

Life Sciences

Maximum Sustainable Harvest *Find the maximum sustainable harvest in Exercises 35–40. See Example 5.*

35. $f(S) = -.1S^2 + 11S$

36. $f(S) = -S^2 + 2.2S$

37. $f(S) = 15\sqrt{S}$

38. $f(S) = 12S^{.25}$

39. $f(S) = \dfrac{25S}{S + 2}$

40. $f(S) = .999S$

41. **Pigeon Flight** Homing pigeons avoid flying over large bodies of water, preferring to fly around them instead. (One possible explanation is the fact that extra energy is required to fly over water because air pressure drops over water in the daytime.) Assume that a pigeon released from a boat 1 mi from the shore of a lake (point *B* in the figure) flies first to point *P* on the shore and then along the straight edge of the lake to reach its home at *L*. If *L* is 2 mi from point *A*, the point on the shore closest to the boat, and if a pigeon needs 4/3 as much energy to fly over water as over land, find the location of point *P*.

42. **Pigeon Flight** Repeat Exercise 41, but assume a pigeon needs 10/9 as much energy to fly over water as over land.

43. **Salmon Spawning** When salmon struggle upstream to their spawning grounds, it is essential that they conserve energy, for they no longer feed once they have left the ocean. Let v_0 = speed (in miles per hour) of the current, d = distance the salmon must travel, and v = speed of the salmon. Hence $(v - v_0)$ is the net velocity of the salmon upstream. (See the figure.)*

(a) Show that if the journey takes t hr, then

$$t = \frac{d}{v - v_0}.$$

We next will assume that the amount of energy expended per hour when the speed of the salmon is v is directly proportional to v^α for some $\alpha > 1$. (Empirical data suggest this.)

(b) Show that the total energy T expended over the journey is given by

$$T = k\frac{v^\alpha}{v - v_0} \quad \text{for } v > v_0.$$

(c) Show that T is minimized by selecting velocity

$$v = \frac{\alpha v_0}{\alpha - 1}.$$

(d) If $v_0 = 2$ mph and the salmon make the 20-mi journey by swimming for 40 hr, estimate α. What must you assume in order to do the computations?

44. **Lung Function** It is well known that during coughing, the diameters of the trachea and bronchi decrease. Let r_0 be the normal radius of an airway at atmospheric pressure P_0. For the trachea, $r_0 \approx .5$ inch. The glottis is the opening at the entrance of the trachea through which air must pass as it either enters or leaves the lungs, as depicted in the illustration.† Assume that, after a deep inspiration of air, the glottis is closed. Then pressure develops in the airways

and the radius of the airway decreases. We will assume a simple linear relation between r and P:

$$r - r_0 = a(P - P_0)$$

$$\text{or} \quad \Delta r = a\Delta P,$$

where $a < 0$ and $r_0/2 \le r \le r_0$.

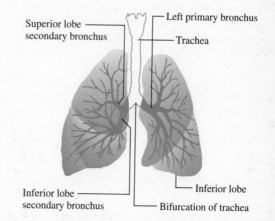

When the glottis is opened, how does the air flow through these passages? We will assume that the flow is governed by **Poiseuille's laws:**

(i) $v = \dfrac{P - P_0}{k}(r^2 - x^2), \quad 0 \le x \le r;$

(ii) $F = \dfrac{dV}{dt} = \dfrac{\pi(P - P_0)}{2k}r^4.$

Here, v is the velocity at a distance x from the center of the airway (in cm/sec, e.g.) and F is the flow rate (in cm^3/sec, e.g.). The average velocity \bar{v} over the circular cross section is given by

(iii) $\bar{v} = \dfrac{F}{(\pi r^2)} = \dfrac{P - P_0}{2k}r^2.$

(a) Write the flow rate as a function of r only. Find the value of r that maximizes the rate of flow.

(b) Write both the average velocity \bar{v} and $v(0)$, the velocity in the center of the airway, as functions of r alone. Find the value of r that maximizes these two velocities.

*Exercises 43–44 from *Mathematics for the Biosciences* by Michael R. Cullen. Copyright © 1983 PWS Publishers. Reprinted by permission.

†The figure for Exercise 44 was adapted from Barbara R. Landau, *Essential Anatomy and Physiology*, 2nd ed. Glenview, Ill.: Scott, Foresman and Co., Copyright 1980.

 Computer/Graphing Calculator

In Exercises 45–47, use a graphing program or an equation solver to determine where the derivative is equal to zero.

45. Can Design Modify the can problem in Example 4 so the cost must be minimized. Assume that aluminum costs 3¢ per square centimeter, and that there is an additional cost of 2¢ per centimeter times the perimeter of the top, and a similar cost for the bottom, to seal the top and bottom of the can to the side.

46. Can Design In this modification of the can problem in Example 4, the cost must be minimized. Assume that aluminum costs 3¢ per square centimeter, and that there is an additional cost of 1¢ per centimeter times the height of the can to make a vertical seam on the side.

47. Can Design This problem is a combination of Exercises 45 and 46. We will again minimize the cost of the can, assuming that aluminum costs 3¢ per square centimeter. In addition, there is a cost of 2¢ per centimeter to seal the top and bottom of the can to the side, plus 1¢ per centimeter to make a vertical seam.

Average Cost *For each of the following cost functions, calculate the average cost function and find where the average cost is smallest. One way to do this is to compute the derivative, then use a graph of the function or an equation solver to find where the derivative is equal to zero.*

48. $C(x) = \frac{1}{2}x^3 + 2x^2 - 3x + 35$

49. $C(x) = .01x^3 + .05x^2 + .2x + 28$

50. $C(x) = 10 + 20x^{1/2} + 16x^{3/2}$

51. $C(x) = 30 + 42x^{1/2} + .2x^{3/2} + .03x^{5/2}$

12.2 FURTHER BUSINESS APPLICATIONS: ECONOMIC LOT SIZE; ECONOMIC ORDER QUANTITY; ELASTICITY OF DEMAND

 How many batches of primer should a paint company produce per year to minimize its costs while meeting its customers' demand?

We will answer this question in Example 1 using the concept of *economic lot size*.

In this section we introduce three common business applications of calculus. The first two, *economic lot size* and *economic order quantity*, are related. A manufacturer must determine the production lot (or batch) size that will result in minimum production and storage costs, while a purchaser must decide what quantity of an item to order in an effort to minimize reordering and storage costs. The third application, *elasticity of demand*, deals with the sensitivity of demand for a product to changes in the price of the product.

Economic Lot Size Suppose that a company manufactures a constant number of units of a product per year and that the product can be manufactured in several batches of equal size during the year. If the company were to manufacture the item only once per year, it would minimize setup costs but incur high warehouse costs. On the other hand, if it were to make many small batches, this would increase setup costs. Calculus can be used to find the number of batches per year that should be manufactured in order to minimize total cost. This number is called the **economic lot size.**

Figure 5 shows several of the possibilities for a product having an annual demand of 12,000 units. The top graph shows the results if only one batch of the product is made annually: in this case an average of 6000 items will be held in a warehouse. If four batches (of 3000 each) are made at equal time intervals during a year, the average number of units in the warehouse falls to only 1500. If twelve batches are made, an average of 500 items will be in the warehouse.

FIGURE 5

The variable in our discussion of economic lot size will be

x = number of batches to be manufactured annually.

In addition, we have the following constants:

k = cost of storing one unit of the product for one year;

f = fixed setup cost to manufacture the product;

g = cost of manufacturing a single unit of the product;

M = total number of units produced annually.

The company has two types of costs associated with the production of its product: a cost associated with manufacturing the item and a cost associated with storing the finished product. During a year the company will produce x batches of the product, with M/x units of the product produced per batch. Each batch has a fixed cost f and a variable cost g per unit, so that the manufacturing cost per batch is

$$f + \frac{gM}{x}.$$

There are x batches per year, so the total annual manufacturing cost is

$$\left(f + \frac{gM}{x} \right) x. \tag{1}$$

Since each batch consists of M/x units, and since demand is constant, it is common to assume an average inventory of

$$\frac{1}{2}\left(\frac{M}{x} \right) = \frac{M}{2x}$$

units per year. The cost of storing one unit of the product for a year is k, so the total storage cost is

$$k\left(\frac{M}{2x} \right) = \frac{kM}{2x}. \tag{2}$$

The total production cost is the sum of the manufacturing and storage costs, or the sum of expressions (1) and (2). If $T(x)$ is the total cost of producing x batches,

$$T(x) = \left(f + \frac{gM}{x} \right) x + \frac{kM}{2x} = fx + gM + \frac{kM}{2x}.$$

Since the only constraint on x is that it be a positive number, the domain of $T(x)$ is $(0, \infty)$. To find the value of x that will minimize $T(x)$, remember that f, g, k, and M are constants and find $T'(x)$.

$$T'(x) = f - \frac{kM}{2}x^{-2}$$

Set this derivative equal to 0.

$$f - \frac{kM}{2}x^{-2} = 0$$

$$f = \frac{kM}{2x^2}$$

$$2fx^2 = kM$$

$$x^2 = \frac{kM}{2f}$$

$$x = \sqrt{\frac{kM}{2f}} \tag{3}$$

The second derivative test can be used to show that $\sqrt{kM/(2f)}$ is the annual number of batches that minimizes total production costs. (See Exercise 1.)

This application is referred to as the *inventory problem* and is treated in more detail in management science courses. Please note that equation (3) was derived under very specific assumptions. If the assumptions are changed slightly, a different conclusion might be reached, and it would not necessarily be valid to use equation (3).

In some examples equation (3) may not give an integer value. In this text we will require the number of batches manufactured in a year to be an integer, which is reasonable in many situations. If equation (3) does not give an integer, we must investigate the next integer smaller than x and the next integer larger to see which gives the minimum cost. This same assumption will be held in the *economic order quantity problem*, which will be discussed after the following example.

EXAMPLE 1

A paint company has a steady annual demand for 24,500 cans of automobile primer. The cost accountant for the company says that it costs $2 to hold one can of paint for one year and $500 to set up the plant for the production of the primer. Find the number of batches of primer that should be produced in order to minimize total production costs.

Use equation (3) above, with $k = 2$, $M = 24,500$, and $f = 500$.

$$x = \sqrt{\frac{kM}{2f}} = \sqrt{\frac{2(24,500)}{2(500)}} = \sqrt{49} = 7$$

Seven batches of primer per year will lead to minimum production costs. Each batch will consist of $M/x = 24,500/7 = 3500$ cans of primer. ◀

Economic Order Quantity We can extend our previous discussion to the problem of reordering an item that is used at a constant rate throughout the year. Here, the company using a product must decide how often to order and how many units to request each time an order is placed; that is, it must identify the **economic order quantity.** In this case, the variable is

$$x = \text{number of orders to be placed annually.}$$

We also have the following constants.

$$k = \text{cost of storing one unit for one year}$$
$$f = \text{fixed cost to place an order}$$
$$M = \text{total units needed per year}$$

The goal is to minimize the total cost of ordering over a year's time, where

$$\text{Total cost} = \text{Storage cost} + \text{Reorder cost.}$$

Again assume an average inventory of $M/2x$, so the yearly storage cost is

$$\frac{kM}{2x}.$$

The reorder cost is the product of the number of batches ordered each year, x, and the cost per order, f. Thus, the reorder cost is fx, and the total cost is

$$T(x) = fx + \frac{kM}{2x}.$$

This is almost the same formula we derived for the inventory problem, which also had a constant term gM. Since a constant does not affect the derivative, equation (3) is also valid for the economic order quantity problem. As before, the number of units per order is M/x. This illustrates how two different applications might have the same mathematical structure, so a solution to one applies to both.

EXAMPLE 2 A large pharmacy has an annual need for 200 units of a certain antibiotic. It costs $10 to store one unit for one year. The fixed cost of placing an order (clerical time, mailing, and so on) amounts to $40. Find the number of times a year the antibiotic should be ordered, and the number of units to order each time.

Here $k = 10$, $M = 200$, and $f = 40$. We have

$$x = \sqrt{\frac{10(200)}{2(40)}} = \sqrt{25} = 5.$$

The drug should be ordered 5 times a year. Each time,

$$\frac{M}{x} = \frac{200}{5} = 40$$

units should be ordered. ◀

Elasticity of Demand Anyone who sells a product or service is concerned with how a change in price affects demand. The sensitivity of demand to price changes varies with different items. For items such as soft drinks, pepper, and light bulbs, relatively small percentage changes in price will not change the demand for the item much. For cars, home loans, furniture, and computer equipment, however, small percentage changes in price have significant effects on demand.

One way to measure the sensitivity of demand to changes in price is by the ratio of percent change in demand to percent change in price. If q represents the quantity demanded and p the price, this ratio can be written as

$$\frac{\Delta q/q}{\Delta p/p},$$

Recall from Chapter 1 that the Greek letter Δ, pronounced *delta*, is used in mathematics to mean "change."

where Δq represents the change in q and Δp represents the change in p. This ratio is always negative, because q and p are positive, while Δq and Δp have opposite signs. (An *increase* in price causes a *decrease* in demand.) If the absolute value of this ratio is large, it suggests that a relatively small increase in price causes a relatively large drop (decrease) in demand.

This ratio can be rewritten as

$$\frac{\Delta q/q}{\Delta p/p} = \frac{\Delta q}{q} \cdot \frac{p}{\Delta p} = \frac{p}{q} \cdot \frac{\Delta q}{\Delta p}.$$

Suppose $q = f(p)$. (Note that this is the inverse of the way our demand functions have been expressed so far; previously we had $p = D(q)$.) Then $\Delta q = f(p + \Delta p) - f(p)$, and

$$\frac{\Delta q}{\Delta p} = \frac{f(p + \Delta p) - f(p)}{\Delta p}.$$

As $\Delta p \to 0$, this quotient becomes

$$\lim_{\Delta p \to 0} \frac{\Delta q}{\Delta p} = \lim_{\Delta p \to 0} \frac{f(p + \Delta p) - f(p)}{\Delta p} = \frac{dq}{dp},$$

and

$$\lim_{\Delta p \to 0} \frac{p}{q} \cdot \frac{\Delta q}{\Delta p} = \frac{p}{q} \cdot \frac{dq}{dp}.$$

The quantity

$$E = -\frac{p}{q} \cdot \frac{dq}{dp}$$

is positive because dq/dp is negative. E is called the **elasticity of demand** and measures the instantaneous responsiveness of demand to price. For example, E may be 0.2 for medical services, but may be 1.2 for stereo equipment. The demand for essential medical services is much less responsive to price changes than is the demand for nonessential commodities, such as stereo equipment.

If $E < 1$, the relative change in demand is less than the relative change in price, and the demand is called **inelastic.** If $E > 1$, the relative change in demand is greater than the relative change in price, and the demand is called **elastic.** When $E = 1$, the percentage changes in price and demand are relatively equal and the demand is said to have **unit elasticity.**

EXAMPLE 3

Given the demand function $q = 300 - 3p, 0 \le p \le 100$, find the following.

(a) Calculate and interpret the elasticity of demand when $p = 25$ and when $p = 75$.

Since $q = 300 - 3p, dq/dp = -3$, and

$$E = -\frac{p}{q} \cdot \frac{dq}{dp}$$

$$= -\frac{p}{300 - 3p}(-3)$$

$$= \frac{3p}{300 - 3p}$$

$$= \frac{p}{100 - p}.$$

Let $p = 25$ to get

$$E = \frac{25}{100 - 25} = \frac{25}{75} = \frac{1}{3} \approx .33.$$

Since $.33 < 1$, the demand is inelastic, and a percentage change in price will result in a smaller percentage change in demand. For example, a 10% increase in price will cause a 3.3% decrease in demand.

If $p = 75$, then

$$E = \frac{75}{100 - 75} = \frac{75}{25} = 3,$$

and since $3 > 1$, demand is elastic. At this point a percentage increase in price will result in a *greater* percentage decrease in demand. Here, a 10% increase in price will cause a 30% decrease in demand.

(b) Determine the price where demand has unit elasticity. What is the significance of this price?

Demand will have unit elasticity at the price p that makes $E = 1$. Here

$$E = \frac{p}{100 - p} = 1$$
$$p = 100 - p$$
$$2p = 100$$
$$p = 50.$$

Demand has unit elasticity at a price of 50. This means that at this price the percentage changes in price and demand are about the same. ◀

The definitions from this discussion can be summarized as follows.

ELASTICITY OF DEMAND

Let $q = f(p)$, where q is demand at a price p. The elasticity of demand is

$$E = -\frac{p}{q} \cdot \frac{dq}{dp}.$$

Demand is inelastic if $E < 1$.

Demand is elastic if $E > 1$.

Demand has unit elasticity if $E = 1$.

Elasticity can be related to the total revenue, R, by considering the derivative of R. Since revenue is given by price times sales (demand),

$$R = pq.$$

Differentiate with respect to p using the product rule.

$$\frac{dR}{dp} = p \cdot \frac{dq}{dp} + q \cdot 1$$

$$= \frac{q}{q} \cdot p \cdot \frac{dq}{dp} + q \qquad \text{Multiply by } \frac{q}{q} \text{ (or 1).}$$

$$= q\left(\frac{p}{q} \cdot \frac{dq}{dp}\right) + q$$

$$= q(-E) + q$$

$$= q(-E + 1)$$

$$= q(1 - E)$$

FIGURE 6

Total revenue R is increasing, optimized, or decreasing depending on whether $dR/dp > 0$, $dR/dp = 0$, or $dR/dp < 0$. These three situations correspond to $E < 1$, $E = 1$, or $E > 1$. See Figure 6.

In summary, total revenue is related to elasticity as follows.

REVENUE AND ELASTICITY

1. If the demand is inelastic, total revenue increases as price increases.

2. If the demand is elastic, total revenue decreases as price increases.

3. Total revenue is maximized at the price where demand has unit elasticity.

EXAMPLE 4 Assume that the demand for a product is $q = 216 - 2p^2$, where p is the price.

(a) Find the price intervals where demand is elastic and where demand is inelastic.

Since $q = 216 - 2p^2$, $dq/dp = -4p$, and

$$E = -\frac{p}{q} \cdot \frac{dq}{dp}$$

$$= -\frac{p}{216 - 2p^2}(-4p)$$

$$= \frac{4p^2}{216 - 2p^2}.$$

To decide where $E < 1$ or $E > 1$, solve the corresponding *equation*.

$$E = 1$$

$$\frac{4p^2}{216 - 2p^2} = 1$$

$$4p^2 = 216 - 2p^2$$

$$6p^2 = 216$$

$$p^2 = 36$$

$$p = 6$$

Substitute a test number on either side of 6 in the expression for E to see which values make $E < 1$ and which make $E > 1$.

$$\text{Let } p = 1: E = \frac{4(1)^2}{216 - 2(1)^2} = \frac{4}{214} < 1.$$

$$\text{Let } p = 10: E = \frac{4(10)^2}{216 - 2(10)^2} = \frac{400}{216 - 200} > 1.$$

Demand is inelastic when $E < 1$. This occurs when $p < 6$. Demand is elastic when $E > 1$; that is, when $p > 6$.

(b) What prices will cause revenue to increase or decrease?

Revenue will increase when demand is inelastic, so keeping prices below \$6 per item will keep demand high enough to continue to increase revenue. When demand is elastic, revenue is decreasing, so any price over \$6 per item will cause demand to decrease to the point where revenue also decreases. ◀

12.2 Exercises

1. In the discussion of economic lot size, use the second derivative to show that the value of x obtained in the text [equation (3)] really leads to the minimum cost.

2. Why do you think that the cost g does not appear in the equation for x [equation (3)]?

Applications

Business and Economics

3. **Lot Size** Find the approximate number of batches of an item that should be produced annually if 100,000 units are to be manufactured, it costs \$1 to store a unit for one year, and it costs \$500 to set up the factory to produce each batch.

4. **Lot Size** A manufacturer has a steady annual demand for 16,800 cases of sugar. It costs \$3 to store one case for one year and it costs \$7 to produce each batch. Find the number of batches of sugar that should be produced each year.

5. **Lot Size** Find the number of units per batch that will be manufactured in Exercise 3.

6. **Lot Size** Find the number of cases per batch in Exercise 4.

7. **Order Quantity** A bookstore has an annual demand for 100,000 copies of a best-selling book. It costs \$.50 to store one copy for one year, and it costs \$60 to place an order. Find the optimum number of copies per order.

8. **Order Quantity** A restaurant has an annual demand for 900 bottles of a California wine. It costs \$1 to store one

bottle for one year, and it costs \$5 to place a reorder. Find the optimum number of bottles per order.

9. **Lot Size** Suppose that in the inventory problem, the storage cost depends on the maximum inventory size, rather than the average. This would be more realistic if, for example, the company had to build a warehouse large enough to hold the maximum inventory, and the cost of storage was the same no matter how full or empty the warehouse was. Show that in this case the number of units that should be ordered to minimize the total cost is

$$x = \sqrt{\frac{kM}{f}}.$$

10. **Lot Size** A book publisher wants to know how many times a year a print run should be scheduled. Suppose it costs \$1000 to set up the printing process, and the subsequent cost per book is so low it can be ignored. Suppose further that the annual warehouse cost is \$6 times the maximum number of books stored. Assuming 5000 copies of the book are needed per year, how many times a year

should a print run be scheduled to minimize the total cost, and how many books should be printed in each print run? (See Exercise 9.)

11. **Lot Size** Suppose that in the inventory problem, the storage cost is a combination of the cost described in the text and the cost described in Exercise 9. In other words, suppose there is an annual cost, k_1, for storing a single unit, plus an annual cost per unit, k_2, that must be paid for each unit up to the maximum number of units stored. Show that the number of units that should be ordered to minimize the total cost in this case is

$$x = \sqrt{\frac{(k_1 + 2k_2)M}{2f}}.$$

12. **Lot Size** Every year, Theresa DePalo sells 30,000 cases of her Famous Spaghetti Sauce. It costs her $1 a year in electricity to store a case, plus she must pay annual warehouse fees of $2 per case for the maximum number of cases she will store. If it costs her $750 to set up a production run, plus $8 per case to manufacture a single case, how many production runs should she have each year to minimize her total costs? (See Exercise 11.)

Elasticity *For each of the following demand functions, find* **(a)** *E, and* **(b)** *values of q (if any) at which total revenue is maximized.*

13. $q = 25,000 - 50p$

14. $q = 50 - \dfrac{p}{4}$

15. $q = \dfrac{3000}{p}$

16. $q = \sqrt[3]{\dfrac{2000}{p}}$

Elasticity *In Exercises 17–18, find the elasticity of demand (E) for the given demand function at the indicated values of p. Is the demand elastic, inelastic, or neither at the indicated values? Interpret your results.*

17. $q = 300 - 2p$
 (a) $p = \$100$ **(b)** $p = \$50$

18. $q = 400 - .2p^2$
 (a) $p = \$20$ **(b)** $p = \$40$

19. **Elasticity** The demand function for q units of a commodity is given by
$$p = (25 - q)^{1/2}, \quad 0 < q < 25.$$
 (a) Find E when $q = 16$.
 (b) Find the values of q and p that maximize total revenue.
 (c) Find the value of E for the q-value found in part (b).

20. **Elasticity** Suppose that a demand function is linear—that is, $q = m - np$ for $0 \le p \le m/n$, where m and n are positive constants. Show that $E = 1$ at the midpoint of the demand curve on the interval $0 \le p \le m/n$; that is, at $p = m/(2n)$.

21. **Elasticity** What must be true about the demand function if $E = 0$?

22. **Elasticity** For some products, the demand function is actually increasing. For example, the most expensive colleges in the United States also tend to have the greatest number of applicants for each student accepted. What is true about the elasticity in this case?

12.3 IMPLICIT DIFFERENTIATION

In almost all of the examples and applications so far, any necessary functions have been defined as

$$y = f(x),$$

with y given **explicitly** in terms of x, or as an **explicit function** of x. For example,

$$y = 3x - 2, \quad y = x^2 + x + 6, \quad \text{and} \quad y = -x^3 + 2$$

are all explicit functions of x. The equation $4xy - 3x = 6$ can be expressed as an explicit function of x by solving for y. This gives

$$4xy - 3x = 6$$
$$4xy = 3x + 6$$
$$y = \frac{3x + 6}{4x}.$$

On the other hand, some equations in x and y cannot be readily solved for y, and some equations cannot be solved for y at all. For example, while it would be possible (but tedious) to use the quadratic formula to solve for y in the equation $y^2 + 2yx + 4x^2 = 0$, it is not possible to solve for y in the equation $y^5 + 8y^3 + 6y^2x^2 + 2yx^3 + 6 = 0$. In equations such as these last two, y is said to be given **implicitly** in terms of x.

In such cases, it may still be possible to find the derivative dy/dx by a process called **implicit differentiation.** In doing so, we assume that there exists some function or functions f, which we may or may not be able to find, such that $y = f(x)$ and dy/dx exists. It is useful to use dy/dx here rather than $f'(x)$ to make it clear which variable is independent and which is dependent.

EXAMPLE 1 Find $\dfrac{dy}{dx}$ if $3xy + 4y^2 = 10$.

Differentiate with respect to x on both sides of the equation.

$$3xy + 4y^2 = 10$$

$$\frac{d}{dx}(3xy + 4y^2) = \frac{d}{dx}(10) \qquad \textbf{(1)}$$

In Section 1.1, we pointed out that when y is given as a function of x, x is the independent variable and y is the dependent variable. We later defined the derivative dy/dx when y is a function of x. In an equation such as $3xy + 4y^2 = 10$, either variable can be considered the independent variable. If a problem asks for dy/dx, consider x the independent variable; if it asks for dx/dy, consider y the independent variable. A similar rule holds when other variables are used.

Now differentiate each term on the left side of the equation. Think of $3xy$ as the product $(3x)(y)$ and use the product rule. Since

$$\frac{d}{dx}(3x) = 3 \qquad \text{and} \qquad \frac{d}{dx}(y) = \frac{dy}{dx},$$

the derivative of $(3x)(y)$ is

$$(3x)\frac{dy}{dx} + (y)3 = 3x\frac{dy}{dx} + 3y.$$

To differentiate the second term, $4y^2$, use the generalized power rule, since y is assumed to be some function of x.

$$\frac{d}{dx}(4y^2) = 4(2y^1)\overbrace{\frac{dy}{dx}}^{\substack{\text{Derivative} \\ \text{of } y^2}} = 8y\frac{dy}{dx}$$

On the right side of equation (1), the derivative of 10 is 0. Taking the indicated derivatives in equation (1) term by term gives

$$3x\frac{dy}{dx} + 3y + 8y\frac{dy}{dx} = 0.$$

Now solve this result for dy/dx.

$$(3x + 8y)\frac{dy}{dx} = -3y$$

$$\frac{dy}{dx} = \frac{-3y}{3x + 8y} \qquad \blacktriangleleft$$

EXAMPLE 2

Find dy/dx for $x^2 + 2xy^2 + 3x^2y = 0$.

Again, differentiate on each side with respect to x.

$$\frac{d}{dx}(x^2 + 2xy^2 + 3x^2y) = \frac{d}{dx}(0)$$

Use the product rule to find the derivatives of $2xy^2$ and $3x^2y$.

$$
\underset{\substack{\text{Derivative}\\\text{of } x^2}}{2x} + \underset{\substack{\text{Derivative}\\\text{of } 2xy^2}}{2x\left(2y\frac{dy}{dx}\right) + 2y^2} + \underset{\substack{\text{Derivative}\\\text{of } 3x^2y}}{3x^2\left(\frac{dy}{dx}\right) + 6xy} = \underset{\substack{\text{Derivative}\\\text{of } 0}}{0}
$$

Simplify to get

$$2x + 4xy\left(\frac{dy}{dx}\right) + 2y^2 + 3x^2\left(\frac{dy}{dx}\right) + 6xy = 0,$$

from which

$$(4xy + 3x^2)\frac{dy}{dx} = -2x - 2y^2 - 6xy,$$

or, finally,

$$\frac{dy}{dx} = \frac{-2x - 2y^2 - 6xy}{4xy + 3x^2}. \quad \blacktriangleleft$$

EXAMPLE 3

Find dy/dx for $x + \sqrt{xy} = y^2$.

Take the derivative on each side.

$$\frac{d}{dx}(x + \sqrt{xy}) = \frac{d}{dx}(y^2)$$

Since $\sqrt{xy} = \sqrt{x} \cdot \sqrt{y} = x^{1/2} \cdot y^{1/2}$ for nonnegative values of x and y, taking derivatives gives

$$1 + x^{1/2}\left(\frac{1}{2}y^{-1/2} \cdot \frac{dy}{dx}\right) + y^{1/2}\left(\frac{1}{2}x^{-1/2}\right) = 2y\frac{dy}{dx}$$

$$1 + \frac{x^{1/2}}{2y^{1/2}} \cdot \frac{dy}{dx} + \frac{y^{1/2}}{2x^{1/2}} = 2y\frac{dy}{dx}.$$

Multiply both sides by $2x^{1/2} \cdot y^{1/2}$.

$$2x^{1/2} \cdot y^{1/2} + x\frac{dy}{dx} + y = 4x^{1/2} \cdot y^{3/2} \cdot \frac{dy}{dx}$$

Combine terms and solve for dy/dx.

$$2x^{1/2} \cdot y^{1/2} + y = (4x^{1/2} \cdot y^{3/2} - x)\frac{dy}{dx}$$

$$\frac{dy}{dx} = \frac{2x^{1/2} \cdot y^{1/2} + y}{4x^{1/2} \cdot y^{3/2} - x} \quad \blacktriangleleft$$

EXAMPLE 4 The graph of $x^2 + 5y^2 = 36$ is the ellipse shown in Figure 7. Find the equation of the tangent line at the point $(4, 2)$.

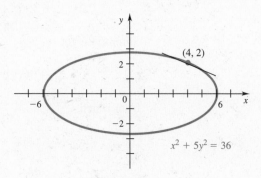

FIGURE 7

A vertical line can cut the graph of the ellipse of Figure 7 in more than one point, so the graph is not the graph of a function. This means that y *cannot* be written as a function of x, so dy/dx would not exist. It seems likely, however, that a tangent line could be drawn to the graph at the point $(4, 2)$. To get around this difficulty, we try to decide if a function f does exist whose graph is exactly the same as our graph in the vicinity of the point $(4, 2)$. Then the tangent line to f at $(4, 2)$ is also the tangent line to the ellipse. See Figure 8.

FIGURE 8

Assuming that such a function f exists, dy/dx can be found by implicit differentiation.

$$2x + 10y \cdot \frac{dy}{dx} = 0$$

$$\frac{dy}{dx} = \frac{-2x}{10y} = -\frac{x}{5y}$$

To find the slope of the tangent line at the point (4, 2), let $x = 4$ and $y = 2$. The slope, m, is

$$m = -\frac{x}{5y} = -\frac{4}{5 \cdot 2} = -\frac{2}{5}.$$

The equation of the tangent line is then found by using the point-slope form of the equation of a line.

$$y - y_1 = m(x - x_1)$$

$$y - 2 = -\frac{2}{5}(x - 4)$$

$$5y - 10 = -2x + 8$$

$$2x + 5y = 18$$

This tangent line is graphed in Figure 7. ◀

The steps used in implicit differentiation can be summarized as follows.

IMPLICIT DIFFERENTIATION

To find dy/dx for an equation containing x and y:

1. Differentiate on both sides with respect to x.

2. Place all terms with dy/dx on one side of the equals sign, and all terms without dy/dx on the other side.

3. Factor out dy/dx, and then solve for dy/dx.

When an applied problem involves an equation that is not given in explicit form, implicit differentiation can be used to locate maxima and minima or to find rates of change.

 EXAMPLE 5

The demand function for a certain commodity is given by

$$p = \frac{500,000}{2q^3 + 400q + 5000},$$

where p is the price in dollars and q is the demand in hundreds of units. Find the rate of change of demand with respect to price when $q = 100$ (that is, find dq/dp).

Rewrite the equation as

$$2q^3 + 400q + 5000 = 500,000p^{-1}.$$

Use implicit differentiation to find dq/dp.

$$6q^2\frac{dq}{dp} + 400\frac{dq}{dp} = -500,000p^{-2}$$

$$(6q^2 + 400)\frac{dq}{dp} = \frac{-500,000}{p^2}$$

$$\frac{dq}{dp} = \frac{-500,000}{p^2(6q^2 + 400)}$$

When $q = 100$,

$$p = \frac{500,000}{2(100)^3 + 400(100) + 5000} \approx .244,$$

and

$$\frac{dq}{dp} = \frac{-500,000}{(.244)^2[6(100)^2 + 400]} \approx -139.$$

This means that when demand (q) is 100 hundreds, or 10,000, demand is decreasing at the rate of 139 hundred, or 13,900, units per dollar change in price. ◀

12.3 Exercises

Find dy/dx by implicit differentiation for each of the following.

1. $4x^2 + 3y^2 = 6$

2. $2x^2 - 5y^2 = 4$

3. $2xy + y^2 = 8$

4. $-3xy - 4y^2 = 2$

5. $6xy^2 - 8y + 1 = 0$

6. $-4y^2x^2 - 3x + 2 = 0$

7. $6x^2 + 8xy + y^2 = 6$

8. $8x^2 = 6y^2 + 2xy$

9. $x^3 = y^2 + 4$

10. $x^3 - 6y^2 = 10$

11. $\frac{1}{x} - \frac{1}{y} = 2$

12. $\frac{3}{2x} + \frac{1}{y} = y$

13. $3x^2 = \frac{2 - y}{2 + y}$

14. $2y^2 = \frac{5 + x}{5 - x}$

15. $x^2y + y^3 = 4$

16. $2xy^2 + 2y^3 + 5x = 0$

17. $\sqrt{x} + \sqrt{y} = 4$

18. $2\sqrt{x} - \sqrt{y} = 1$

19. $\sqrt{xy} + y = 1$

20. $\sqrt{2xy} - 1 = 3y^2$

21. $x^4y^3 + 4x^{3/2} = 6y^{3/2} + 5$

22. $(xy)^{4/3} + x^{1/3} = y^6 + 1$

23. $(x^2 + y^3)^4 = x + 2y + 4$

24. $\sqrt{x^2 + y^2} = x^{3/2} - y - 2$

Find the equation of the tangent line at the given point on each curve in Exercises 25–32.

25. $x^2 + y^2 = 25$; $(-3, 4)$

26. $x^2 + y^2 = 100$; $(8, -6)$

27. $x^2y^2 = 1$; $(-1, 1)$

28. $x^2y^3 = 8$; $(-1, 2)$

29. $x^2 + \sqrt{y} = 7$; $(2, 9)$

30. $2y^2 - \sqrt{x} = 4$; $(16, 2)$

31. $y + \frac{\sqrt{x}}{y} = 3$; $(4, 2)$

32. $x + \frac{\sqrt{y}}{3x} = 2$; $(1, 9)$

33. The graph of $x^2 + y^2 = 100$ is a circle having center at the origin and radius 10.

(a) Write the equations of the tangent lines at the points where $x = 6$.

(b) Graph the circle and the tangent lines.

34. The graph of $xy = 1$, shown in the figure, is a hyperbola.
 (a) Write the equations of the tangent lines at $x = -1$ and $x = 1$.
 (b) Graph the hyperbola and the tangent lines.

35. The graph of $y^5 - y - x^2 = -1$ is shown in the figure. Find the equation of the line tangent to the graph at the point $(1, 1)$.

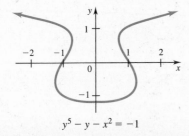

36. The graph of $3(x^2 + y^2)^2 = 25(x^2 - y^2)$ is shown in the figure. Find the equation of the tangent line at the point $(2, 1)$.

37. Suppose $x^2 + y^2 + 1 = 0$. Use implicit differentiation to find dy/dx. Then explain why the result you got is meaningless. (*Hint*: Can $x^2 + y^2 + 1$ equal 0?)

Let $\sqrt{u} + \sqrt{2v + 1} = 5$. *Find the derivatives in Exercises 38–39.*

38. du/dv

39. dv/du

Applications

Business and Economics

40. **Demand** The demand equation for a certain product is $2p^2 + q^2 = 1600$, where p is the price per unit in dollars and q is the number of units demanded.
 (a) Find and interpret dq/dp.
 (b) Find and interpret dp/dq.

41. **Cost and Revenue** For a certain product, cost C and revenue R are given as follows, where x is the number of units sold (in hundreds).

 $$\text{Cost: } C^2 = x^2 + 100\sqrt{x} + 50$$
 $$\text{Revenue: } 900(x - 5)^2 + 25R^2 = 22{,}500$$

 (a) Find and interpret the marginal cost dC/dx at $x = 5$.
 (b) Find and interpret the marginal revenue dR/dx at $x = 5$.

Physical Sciences

Velocity *The position of a particle at time t is given by s. Find the velocity ds/dt.*

42. $s^3 - 4st + 2t^3 - 5t = 0$

43. $2s^2 + \sqrt{st} - 4 = 3t$

12.4 RELATED RATES

If a ladder that is leaning against a wall begins to slide down, how fast is the top of the ladder descending?

Example 2 in this section answers this question using the concept of related rates.

It is common for variables to be functions of time; for example, sales of an item may depend on the season of the year, or a population of animals may be increasing at a certain rate several months after being introduced into an area. Time is often present implicitly in a mathematical model, meaning that derivatives with respect to time must be found by the method of implicit differentiation discussed in the previous section. For example, if a particular mathematical model leads to the equation

$$x^2 + 5y - 3x + 1 = 0,$$

differentiating on both sides with respect to t gives

$$\frac{d}{dt}(x^2 + 5y - 3x + 1) = \frac{d}{dt}(0)$$

$$2x \cdot \frac{dx}{dt} + 5 \cdot \frac{dy}{dt} - 3 \cdot \frac{dx}{dt} = 0$$

$$(2x - 3)\frac{dx}{dt} + 5\frac{dy}{dt} = 0.$$

The derivatives (or rates of change) dx/dt and dy/dt are related by this last equation; for this reason they are called **related rates.**

EXAMPLE 1 A small rock is dropped into a lake. Circular ripples spread over the surface of the water, with the radius of each circle increasing at the rate of 3/2 feet per second. Find the rate of change of the area inside the circle formed by a ripple at the instant that the radius is 4 feet.

As shown in Figure 9, the area A and the radius r are related by

$$A = \pi r^2.$$

Take the derivative of each side with respect to time.

$$\frac{d}{dt}(A) = \frac{d}{dt}(\pi r^2)$$

$$\frac{dA}{dt} = 2\pi r \cdot \frac{dr}{dt} \tag{1}$$

FIGURE 9

Since the radius is increasing at the rate of 3/2 feet per second,

$$\frac{dr}{dt} = \frac{3}{2}.$$

The rate of change of area at the instant $r = 4$ is given by dA/dt evaluated at $r = 4$. Substituting into equation (1) gives

$$\frac{dA}{dt} = 2\pi \cdot 4 \cdot \frac{3}{2}$$

$$\frac{dA}{dt} = 12\pi \approx 37.7 \text{ square feet per second.} \blacktriangleleft$$

As suggested by Example 1, four basic steps are involved in solving problems about related rates.

SOLVING RELATED RATE PROBLEMS

1. Identify all given quantities, as well as the quantities to be found. Draw a sketch when possible.
2. Write an equation relating the variables of the problem.
3. Use implicit differentiation to find the derivative of both sides of the equation in Step 2 with respect to time.
4. Solve for the derivative giving the unknown rate of change and substitute the given values.

Caution Differentiate *first*, and *then* substitute values for the variables. If the substitutions were performed first, differentiating would not lead to useful results.

EXAMPLE 2 A 50-foot ladder is placed against a large building. The base of the ladder is resting on an oil spill, and it slips (to the right in Figure 10) at the rate of 3 feet per minute. Find the rate of change of the height of the top of the ladder above the ground at the instant when the base of the ladder is 30 feet from the base of the building.

Let y be the height of the top of the ladder above the ground, and let x be the distance of the base of the ladder from the base of the building. By the Pythagorean theorem,

$$x^2 + y^2 = 50^2. \tag{2}$$

Both x and y are functions of time t in minutes after the moment that the ladder starts slipping. Take the derivative of both sides of equation (2) with respect to time, getting

$$\frac{d}{dt}(x^2 + y^2) = \frac{d}{dt}(50^2)$$

$$2x\frac{dx}{dt} + 2y\frac{dy}{dt} = 0. \tag{3}$$

FIGURE 10

Since the base is sliding at the rate of 3 feet per minute,

$$\frac{dx}{dt} = 3.$$

Also, the base of the ladder is 30 feet from the base of the building. Use this to find y.

$$50^2 = 30^2 + y^2$$
$$2500 = 900 + y^2$$
$$1600 = y^2$$
$$y = 40$$

In summary, $y = 40$ when $x = 30$. Also, the rate of change of x over time t is $dx/dt = 3$. Substituting these values into equation (3) to find the rate of change of y over time gives

$$2(30)(3) + 2(40)\frac{dy}{dt} = 0$$

$$180 + 80\frac{dy}{dt} = 0$$

$$80\frac{dy}{dt} = -180$$

$$\frac{dy}{dt} = \frac{-180}{80} = \frac{-9}{4} = -2.25.$$

At the instant when the base of the ladder is 30 feet from the base of the building, the top of the ladder is sliding down the building at the rate of 2.25 feet per minute. (The minus sign shows that the ladder is sliding *down*, so the distance y is *decreasing*.) ◀

EXAMPLE 3

A cone-shaped icicle is dripping from the roof. The radius of the icicle is decreasing at a rate of .2 centimeters per hour, while the length is increasing at a rate of .8 centimeters per hour. If the icicle is currently 4 centimeters in radius and 20 centimeters long, is the volume of the icicle increasing or decreasing, and at what rate?

For this problem we need the formula for the volume of a cone:

$$V = \frac{1}{3}\pi r^2 h, \tag{4}$$

where r is the radius of the cone and h is the height of the cone, which in this case is the length of the icicle, as in Figure 11.

In this problem, both r and h are functions of the time t in hours. Taking the derivative of both sides of equation (4) with respect to time yields

$$\frac{dV}{dt} = \frac{1}{3}\pi \left[r^2 \frac{dh}{dt} + (h)(2r)\frac{dr}{dt} \right].$$

FIGURE 11

Since the radius is decreasing at a rate of .2 centimeters per hour and the length is increasing at a rate of .8 centimeters per hour,

$$\frac{dr}{dt} = -.2 \quad \text{and} \quad \frac{dh}{dt} = .8.$$

Substituting these, as well as $r = 4$ and $h = 20$, into equation (4) yields

$$\frac{dV}{dt} = \frac{1}{3}\pi[4^2(.8) + (20)(8)(-.2)]$$

$$= \frac{1}{3}\pi(-19.2) \approx -20.$$

Because the sign of dV/dt is negative, the volume of the icicle is decreasing at a rate of 20 cubic centimeters per hour. ◄

EXAMPLE 4 ► A company is increasing production of peanuts at the rate of 50 cases per day. All cases produced can be sold. The daily demand function is given by

$$p = 50 - \frac{q}{200},$$

where q is the number of units produced (and sold) and p is price in dollars. Find the rate of change of revenue with respect to time (in days) when the daily production is 200 units.

The revenue function,

$$R = qp = q\left(50 - \frac{q}{200}\right) = 50q - \frac{q^2}{200},$$

relates R and q. The rate of change of q over time (in days) is $dq/dt = 50$. The rate of change of revenue over time, dR/dt, is to be found when $q = 200$. Differentiate both sides of the equation

$$R = 50q - \frac{q^2}{200}$$

with respect to t.

$$\frac{dR}{dt} = 50\frac{dq}{dt} - \frac{1}{100}q\frac{dq}{dt} = \left(50 - \frac{1}{100}q\right)\frac{dq}{dt}$$

Now substitute the known values for q and dq/dt.

$$\frac{dR}{dt} = \left[50 - \frac{1}{100}(200)\right](50) = 2400$$

Thus revenue is increasing at the rate of $2400 per day. ◄

5 Blood flows faster the closer it is to the center of a blood vessel. According to Poiseuille's laws (see Exercise 44 in the first section of this chapter), the velocity V of blood is given by

$$V = k(R^2 - r^2),$$

where R is the radius of the blood vessel, r is the distance of a layer of blood flow from the center of the vessel, and k is a constant, assumed here to equal 375. See Figure 12.

(a) Find dV/dt. Treat r as a constant. Assume the given units are compatible.

$$V = 375(R^2 - r^2)$$

$$\frac{dV}{dt} = 375(2R\frac{dR}{dt} - 0) \quad \text{r is a constant.}$$

$$\frac{dV}{dt} = 750R\frac{dR}{dt}$$

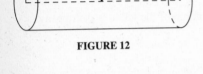

FIGURE 12

(b) Suppose a skier's blood vessel has radius $R = .008$ millimeters and that cold weather is causing the vessel to contract at a rate of $dR/dt = -.001$ millimeters per minute. How fast is the velocity of blood changing?

Find dV/dt. From part (a),

$$\frac{dV}{dt} = 750R\frac{dR}{dt}.$$

Here $R = .008$ and $dR/dt = -.001$, so

$$\frac{dV}{dt} = 750(.008)(-.001) = -.006.$$

That is, the velocity of the blood is decreasing at a rate of $-.006$ millimeters per minute. The units indicate that this is a deceleration (negative acceleration), since it gives the rate of change of velocity. ◀

12.4 Exercises

Evaluate dy/dt for each of the following.

1. $y = 9x^2 + 2x; \quad dx/dt = 4, x = 6$

2. $y = \dfrac{3x + 2}{1 - x}; \quad dx/dt = -1, x = 3$

3. $y^2 - 5x^2 = -1; \quad dx/dt = -3, x = 1, y = 2$

4. $8y^3 + x^2 = 1; \quad dx/dt = 2, x = 3, y = -1$

5. $xy - 5x + 2y^3 = -70; \quad dx/dt = -5, x = 2, y = -3$

6. $4x^3 - 9xy^2 + y = -80; \quad dx/dt = 4, x = -3, y = 1$

7. $\dfrac{x^2 + y}{x - y} = 9; \quad dx/dt = 2, x = 4, y = 2$

8. $\dfrac{y^3 - x^2}{x + 2y} = \dfrac{17}{7}; \quad dx/dt = 1, x = -3, y = -2$

Applications

Business and Economics

9. **Cost** A manufacturer of handcrafted wine racks has determined that the cost to produce x units per month is given by $C = .1x^2 + 10,000$. How fast is cost per month changing when production is changing at the rate of 10 units per month and the production level is 100 units?

10. **Cost/Revenue** The manufacturer in Exercise 9 has found that the cost C and revenue R (in dollars) from the production and sale of x units are related by the equation

$$C = \frac{R^2}{400,000} + 10,000.$$

Find the rate of change of revenue per unit when the cost per unit is changing by $10 and the revenue is $20,000.

11. **Revenue/Cost/Profit** Given the revenue and cost functions $R = 50x - .4x^2$ and $C = 5x + 15$, where x is the daily production (and sales), find the following when 40 units are produced daily and the rate of change of production is 10 units per day.

(a) The rate of change of revenue with respect to time

(b) The rate of change of cost with respect to time

(c) The rate of change of profit with respect to time

12. **Revenue/Cost/Profit** Repeat Exercise 11, given that 200 units are produced daily and the rate of change of production is 50 units per day.

13. **Demand** A product sells for $3.50 currently and has a demand function of

$$p = \frac{8000}{q}.$$

Suppose manufacturing costs are increasing over time at a rate of 15% and the company plans to increase the price p at this rate as well. Find the rate of change of demand over time.

14. **Revenue** A company is increasing production at the rate of 25 units per day. The daily demand function is given by

$$p = 70 - \frac{q^2}{120}.$$

Find the rate of change of revenue with respect to time (in days) when the daily production (and sales) is 20.

Life Sciences

Blood Velocity *Exercises 15–17 refer to Example 5 in this section.*

15. Find dV/dt if $r = 2$ mm, $k = 3$, $dr/dt = .02$ mm/min, and R is fixed.

16. Find dV/dt if $r = 1$ mm, $k = 4$, $dr/dt = .004$ mm/min, and R is fixed.

17. A cross-country skier has a history of heart problems. She takes nitroglycerin to dilate blood vessels, thus avoiding angina (chest pain) due to blood vessel contraction. Use Poiseuille's law with $k = 555.6$ to find the rate of change of the blood velocity when $R = .02$ mm and R is changing at .003 mm per minute. Assume r is constant.

Social Sciences

18. **Crime Rate** Sociologists have found that crime rates are influenced by temperature. In a midwestern town of 100,000 people, the crime rate has been approximated as

$$C = \frac{1}{10}(T - 60)^2 + 100,$$

where C is the number of crimes per month and T is the average monthly temperature. The average temperature for May was 76°, and by the end of May the temperature was rising at the rate of 8° per month. How fast is the crime rate rising at the end of May?

19. **Learning Skills** It is estimated that a person learning a certain assembly-line task takes

$$T(x) = \frac{2 + x}{2 + x^2}$$

minutes to perform the task after x repetitions. Find dT/dt if dx/dt is 4, and 4 repetitions of the task have been performed.

20. **Memorization Skills** Under certain conditions, a person can memorize W words in t minutes, where

$$W(t) = \frac{-.02t^2 + t}{t + 1}.$$

Find dW/dt when $t = 5$.

Physical Sciences

21. **Sliding Ladder** A 25-ft ladder is placed against a building. The base of the ladder is slipping away from the building at the rate of 4 ft/min. Find the rate at which the top of the ladder is sliding down the building at the instant when the bottom of the ladder is 7 ft from the base of the building.

22. **Distance** One car leaves a given point and travels north at 30 mph. Another car leaves the same point at the same time and travels west at 40 mph. At what rate is the distance between the two cars changing at the instant when the cars have traveled 2 hr?

23. **Area** A rock is thrown into a still pond. The circular ripples move outward from the point of impact of the rock so that the radius of the circle formed by a ripple increases at the rate of 2 ft per minute. Find the rate at which the area is changing at the instant the radius is 4 ft.

24. **Volume** A spherical snowball is placed in the sun. The sun melts the snowball so that its radius decreases 1/4 inch per hour. Find the rate of change of the volume with respect to time at the instant the radius is 4 inches.

25. **Volume** A sand storage tank used by the highway department for winter storms is leaking. As the sand leaks out, it forms a conical pile. The radius of the base of the pile increases at the rate of 1 inch per minute. The height of the pile is always twice the radius of the base. Find the rate at which the volume of the pile is increasing at the instant the radius of the base is 5 inches.

26. **Shadow Length** A man 6 ft tall is walking away from a lamp post at the rate of 50 ft per minute. When the man is 8 ft from the lamp post, his shadow is 10 ft long. Find the rate at which the length of the shadow is increasing when he is 25 ft from the lamp post. (See the figure.)

27. **Water Level** A trough has a triangular cross section. The trough is 6 ft across the top, 6 ft deep, and 16 ft long. Water is being pumped into the trough at the rate of 4 cubic feet per minute. Find the rate at which the height of water is increasing at the instant that the height is 4 ft.

28. **Velocity** A pulley is on the edge of a dock, 8 ft above the water level. A rope is being used to pull in a boat. The rope is attached to the boat at water level. The rope is being pulled in at the rate of 1 ft per second. Find the rate at which the boat is approaching the dock at the instant the boat is 8 ft from the dock.

29. **Kite Flying** Karen LaBonte is flying her kite in a wind that is blowing it east at a rate of 50 ft/min. She has already let out 200 ft of string, and the kite is flying 100 ft above her hand. How fast must she let out string at this moment to keep the kite flying with the same speed and altitude?

12.5 DIFFERENTIALS

If the estimated sales of microwave ovens turns out to be inaccurate, approximately how much are profits affected?

Using differentials, we will answer this question in Example 4.

As mentioned earlier, the symbol Δx represents a change in the variable x. Similarly, Δy represents a change in y. An important problem that arises in many applications is to determine Δy given specific values of x and Δx. This quantity is often difficult to evaluate. In this section we show a method of approximating Δy that uses the derivative dy/dx.

For values x_1 and x_2,

$$\Delta x = x_2 - x_1.$$

Solving for x_2 gives

$$x_2 = x_1 + \Delta x.$$

For a function $y = f(x)$, the symbol Δy represents a change in y:

$$\Delta y = f(x_2) - f(x_1).$$

Replacing x_2 with $x_1 + \Delta x$ gives

$$\Delta y = f(x_1 + \Delta x) - f(x_1).$$

If Δx is used instead of h, the derivative of a function f at x_1 could be defined as

$$\frac{dy}{dx} = \lim_{\Delta x \to 0} \frac{\Delta y}{\Delta x}.$$

If the derivative exists, then

$$\frac{dy}{dx} \approx \frac{\Delta y}{\Delta x}$$

as long as Δx is close to 0. Multiplying both sides by Δx (assume $\Delta x \neq 0$) gives

$$\Delta y \approx \frac{dy}{dx} \cdot \Delta x.$$

Until now, dy/dx has been used as a single symbol representing the derivative of y with respect to x. In this section, separate meanings for dy and dx are introduced in such a way that their quotient, when $dx \neq 0$, is the derivative of y with respect to x. These meanings of dy and of dx are then used to find an approximate value of Δy.

To define dy and dx, look at Figure 13, which shows the graph of a function $y = f(x)$. The tangent line to the graph has been drawn at the point P. Let Δx be any nonzero real number (in practical problems, Δx is a small number) and locate the point $x + \Delta x$ on the x-axis. Draw a vertical line through $x + \Delta x$. Let this vertical line cut the tangent line at M and the graph of the function at Q.

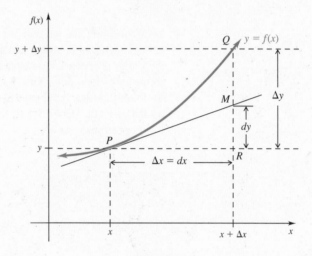

FIGURE 13

Define the new symbol dx to be the same as Δx. Define the new symbol dy to equal the length MR. The slope of PM is $f'(x)$. By the definition of slope, the slope of PM is also dy/dx, so that

$$f'(x) = \frac{dy}{dx},$$

or
$$dy = f'(x)dx.$$

In summary, the definitions of the symbols dy and dx are as follows.

DIFFERENTIALS

For a function $y = f(x)$ whose derivative exists, the **differential** of x, written dx, is an arbitrary real number (usually small); the **differential** of y, written dy, is the product of $f'(x)$ and dx, or

$$\boldsymbol{dy = f'(x)dx.}$$

The usefulness of the differential is suggested by Figure 13. As dx approaches 0, the value of dy gets closer and closer to that of Δy, so that for small nonzero values of dx

$$dy \approx \Delta y,$$

or
$$\Delta y \approx f'(x)dx.$$

 EXAMPLE 1 Find dy for the following functions.

(a) $y = 6x^2$

The derivative is $y' = 12x$, so that

$$dy = 12x\,dx.$$

(b) If $y = 8x^{-3/4}$, then $dy = -6x^{-7/4}\,dx$.

(c) If $f(x) = \dfrac{2 + x}{2 - x}$, then $dy = \dfrac{4}{(2 - x)^2}dx$. ◀

 EXAMPLE 2 Suppose $y = 5x - x^2 + 4$. Find dy for the following values of x and dx.

(a) $x = -3, dx = .05$

First, by definition

$$dy = (5 - 2x)dx.$$

If $x = -3$ and $dx = .05$, then

$$dy = [5 - 2(-3)](.05) = 11(.05) = .55.$$

(b) $x = 0, dx = -.004$

$$dy = (5 - 2 \cdot 0)(-.004)$$
$$= 5(-.004)$$
$$= -.02 ◀$$

As discussed above,

$$\Delta y = f(x + \Delta x) - f(x).$$

For small nonzero values of Δx, $\Delta y \approx dy$, so that

$$dy \approx f(x + \Delta x) - f(x),$$

or $$f(x) + dy \approx f(x + \Delta x).$$ (1)

Replacing dy with $f'(x)dx$ gives the following result.

DIFFERENTIAL APPROXIMATION	Let f be a function whose derivative exists. For small nonzero values of Δx, $$dy \approx \Delta y,$$ and $$f(x + \Delta x) \approx f(x) + dy = f(x) + f'(x)dx.$$

Differentials are used to find an approximate value of the change in a value of the dependent variable corresponding to a given change in the independent variable. When the concept of marginal cost (or profit or revenue) was used to approximate the change in cost for nonlinear functions, the same idea was developed. Thus the differential dy approximates Δy in much the same way as the marginal quantities approximate changes in functions.

For example, for a cost function $C(x)$,

$$dC = C'(x)dx = C'(x)\Delta x.$$

Since $\Delta C \approx dC$,

$$\Delta C \approx C'(x)\Delta x.$$

If the change in production, Δx, is equal to 1, then

$$C(x + 1) - C(x) = \Delta C$$
$$\approx C'(x)\Delta x$$
$$= C'(x),$$

which shows that marginal cost $C'(x)$ approximates the cost of the next unit produced, as mentioned earlier.

EXAMPLE 3 Let $C(x) = 2x^3 + 300$.

(a) Find ΔC and $C'(x)$ when $\Delta x = 1$ and $x = 3$.

$$\Delta C = C(4) - C(3) = 428 - 354 = 74$$
$$C'(x) = 6x^2$$
$$C'(3) = 54$$

Here, the approximation of $C'(3)$ for ΔC is poor, since $\Delta x = 1$ is large relative to $x = 3$.

(b) Find ΔC and $C'(x)$ when $\Delta x = 1$ and $x = 50$.

$$\Delta C = C(51) - C(50) = 265{,}602 - 250{,}300 = 15{,}302$$
$$C'(50) = 6(2500) = 15{,}000$$

This approximation is quite good since $\Delta x = 1$ is small compared to $x = 50$. ◀

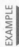

EXAMPLE 4 An analyst for a manufacturer of small appliances estimates that the total profit (in hundreds of dollars) from the sale of x hundred microwave ovens is given by

$$P(x) = -2x^3 + 51x^2 + 88x.$$

In a report to management, the analyst projected sales for the coming month to be 1000 ovens, for a total profit of $389,000. He now realizes that his sales estimate may have been as much as 100 ovens too high. Approximately how far off is his profit estimate?

Differentials can be used to find the approximate change in P resulting from decreasing x by 1 ($= 100$ ovens). This change can be approximated by $dP = P'(x)dx$ where $x = 10$ ($= 1000$ ovens) and $dx = -1$ ($= -100$ ovens). Since $P'(x) = -6x^2 + 102x + 88$,

$$\Delta P \approx dP = (-6x^2 + 102x + 88)dx$$
$$= [-6(100) + 102(10) + 88](-1)$$
$$= (-600 + 1020 + 88)(-1)$$
$$= (508)(-1)$$
$$= -508$$

hundred dollars. Thus, the profit estimate may have been as much as $50,800 too high. ◀

EXAMPLE **5**

The mathematical model

$$y = A(x) = \frac{-7}{480}x^3 + \frac{127}{120}x, \qquad 0 \le x < 9,$$

gives the approximate alcohol concentration (in tenths of a percent) in an average person's bloodstream x hours after drinking 8 ounces of 100-proof whiskey.

(a) Approximate the change in y as x changes from 3 to 3.5. Use dy as an approximation of Δy.

Here,

$$dy = \left(\frac{-7}{160}x^2 + \frac{127}{120}\right)dx.$$

In this example, $x = 3$ and $dx = \Delta x = 3.5 - 3 = .5$. Substitution gives

$$\Delta y \approx dy = \left(\frac{-7}{160} \cdot 3^2 + \frac{127}{120}\right)(.5) \approx .33,$$

so $\Delta y \approx .33$. This means that the alcohol concentration increases by about .33 tenths of a percent as x changes from 3 hours to 3.5 hours (The *exact* increase, found by calculating $A(3.5) - A(3)$, is .30 tenths of a percent.)

(b) Approximate the change in y as x changes from 6 to 6.25 hours.

Let $x = 6$ and $dx = 6.25 - 6 = .25$.

$$\Delta y \approx dy = \left(\frac{-7}{160} \cdot 6^2 + \frac{127}{120}\right)(.25) \approx -.13$$

The minus sign shows that alcohol concentration decreases by .13 tenths of a percent as x changes from 6 to 6.25 hours. ◀

The final example in this section shows how differentials are used to estimate errors that might enter into measurements of a physical quantity.

EXAMPLE **6**

In a precision manufacturing process, ball bearings must be made with a radius of .6 millimeters, with a maximum error in the radius of \pm .015 millimeters. Estimate the maximum error in the volume of the ball bearing.

The formula for the volume of a sphere is

$$V = \frac{4}{3}\pi r^3.$$

If an error of Δr is made in measuring the radius of the sphere, the maximum error in the volume is

$$\Delta V = \frac{4}{3}\pi(r + \Delta r)^3 - \frac{4}{3}\pi r^3.$$

Rather than calculating ΔV, approximate ΔV with dV, where

$$dV = 4\pi r^2 dr.$$

Replacing r with .6 and $dr = \Delta r$ with $\pm.015$ gives

$$dV = 4\pi(.6)^2(\pm.015)$$
$$\approx \pm.0679.$$

The maximum error in the volume is about .07 cubic millimeters. ◄

12.5 Exercises

Find dy for each of the following.

1. $y = 6x^2$

2. $y = -8x^4$

3. $y = 7x^2 - 9x + 6$

4. $y = -3x^3 + 2x^2$

5. $y = 2\sqrt{x}$

6. $y = 8\sqrt{2x - 1}$

7. $y = \dfrac{8x - 2}{x - 3}$

8. $y = \dfrac{-4x + 7}{3x - 1}$

9. $y = x^2\left(x - \dfrac{1}{x} + 2\right)$

10. $y = -x^3\left(2 + \dfrac{3}{x^2} - \dfrac{5}{x}\right)$

11. $y = \left(2 - \dfrac{3}{x}\right)\left(1 + \dfrac{1}{x}\right)$

12. $y = \left(9 - \dfrac{2}{x^2}\right)\left(3 + \dfrac{1}{x}\right)$

For Exercises 13–24, find dy for the given values of x and Δx.

13. $y = 2x^2 - 5x;\quad x = -2, \Delta x = .2$

14. $y = x^2 - 3x;\quad x = 3, \Delta x = .1$

15. $y = x^3 - 2x^2 + 3;\quad x = 1, \Delta x = -.1$

16. $y = 2x^3 + x^2 - 4x;\quad x = 2, \Delta x = -.2$

17. $y = \sqrt{3x};\quad x = 1, \Delta x = .15$

18. $y = \sqrt{4x - 1};\quad x = 5, \Delta x = .08$

19. $y = \dfrac{2x - 5}{x + 1};\quad x = 2, \Delta x = -.03$

20. $y = \dfrac{6x - 3}{2x + 1};\quad x = 3, \Delta x = -.04$

21. $y = -6\left(2 - \dfrac{1}{x^2}\right);\quad x = -1, \Delta x = .02$

22. $y = 9\left(3 + \dfrac{1}{x^4}\right);\quad x = -2, \Delta x = -.015$

23. $y = \dfrac{1 + x}{\sqrt{x}};\quad x = 9, \Delta x = -.03$

24. $y = \dfrac{2 - 5x}{\sqrt{x} + 1};\quad x = 3, \Delta x = .02$

Applications

Business and Economics

25. Demand The demand for grass seed (in thousands of pounds) at a price of x dollars is

$$d(x) = -5x^3 - 2x^2 + 1500.$$

Use the differential to approximate the changes in demand for the following changes in x.

(a) $2 to $2.50 (b) $6 to $6.30

26. Average Cost The average cost to manufacture x dozen marking pencils is

$$A(x) = .04x^3 + .1x^2 + .5x + 6.$$

Use the differential to approximate the changes in the average cost for the following changes in x.

(a) 3 to 4 (b) 5 to 6

27. Mortgage Rates A county realty group estimates that the number of housing starts per year is

$$H(r) = \frac{300}{1 + .03r^2},$$

where r is the mortgage rate in percent ($r = 8$ means 8%). Use the differential to approximate the change in housing starts when $r = 10$, if mortgage rates change by .5%.

28. Revenue A company estimates that the revenue (in dollars) from the sale of x units of doghouses is given by

$$R(x) = 625 + .03x + .0001x^2.$$

Use the differential to approximate the change in revenue from the sale of one more doghouse when 1000 doghouses are sold.

29. Profit The profit function for the company in Exercise 28 is

$$P(x) = -390 + 24x + 5x^2 - \frac{1}{3}x^3,$$

where x represents the demand for the product. Find the approximate change in profit for a one-unit change in demand when demand is at a level of 1000 doghouses. Use the differential.

30. Material Requirement A cube 4 inches on an edge is given a protective coating .1 inch thick. About how much coating should a production manager order for 1000 such cubes?

31. Material Requirement Beach balls 1 ft in diameter have a thickness of .03 inch. How much material would be needed to make 5000 beach balls?

Life Sciences

32. Drug Concentration The concentration of a certain drug in the bloodstream x hr after being administered is approximately

$$C(x) = \frac{5x}{9 + x^2}.$$

Use the differential to approximate the changes in concentration for the following changes in x.

(a) 1 to 1.5 (b) 2 to 2.25

33. Bacteria Population The population of bacteria (in millions) in a certain culture x hr after an experimental nutrient is introduced into the culture is

$$P(x) = \frac{25x}{8 + x^2}.$$

Use the differential to approximate the changes in population for the following changes in x.

(a) 2 to 2.5 (b) 3 to 3.25

34. Area of a Blood Vessel The radius of a blood vessel is 17 mm. A drug causes the radius to change to 16 mm. Find the approximate change in the area of a cross section of the vessel.

35. Volume of a Tumor A tumor is approximately spherical in shape. If the radius of the tumor changes from 14 mm to 16 mm, find the approximate change in volume.

36. Area of an Oil Slick An oil slick is in the shape of a circle. Find the approximate increase in the area of the slick if its radius increases from 1.2 mi to 1.4 mi.

37. Area of a Bacteria Colony The shape of a colony of bacteria on a Petri dish is circular. Find the approximate increase in its area if the radius increases from 20 mm to 22 mm.

Physical Sciences

38. Volume A spherical beach ball is being inflated. Find the approximate change in volume if the radius increases from 4 cm to 4.2 cm.

39. Volume A spherical snowball is melting. Find the approximate change in volume if the radius decreases from 3 cm to 2.8 cm.

General Interest

40. Measurement Error The edge of a square is measured as 3.45 inches, with a possible error of $\pm.002$ inches. Estimate the maximum error in the area of the square.

41. Measurement Error The radius of a circle is measured as 4.87 inches, with a possible error of $\pm.040$ inches. Estimate the maximum error in the area of the circle.

42. Measurement Error A sphere has a radius of 5.81 inches, with a possible error of $\pm.003$ inches. Estimate the maximum error in the volume of the sphere.

43. Measurement Error A cone has a known height of 7.284 inches. The radius of the base is measured as 1.09 inches, with a possible error of $\pm.007$ inch. Estimate the maximum error in the volume of the cone. (The volume of a cone is given by $V = \frac{1}{3}\pi r^2 h$.)

Chapter Summary Key Terms

12.1 limits at infinity
oblique asymptote
12.2 economic lot size
economic order quantity
elasticity of demand

12.3 explicit function
implicit differentiation

12.4 related rates
12.5 differential

Chapter 12 Review Exercises

1. When is it necessary to use implicit differentiation?

2. When a term involving y is differentiated in implicit differentiation, it is multiplied by dy/dx. Why? Why aren't terms involving x multiplied by dx/dx?

3. What is the difference between a related rate problem and an applied extremum problem?

4. Why is implicit differentiation used in related rate problems?

Find dy/dx in Exercises 5–12.

5. $x^2y^3 + 4xy = 2$

6. $\dfrac{x}{y} - 4y = 3x$

7. $9\sqrt{x} + 4y^3 = \dfrac{2}{x}$

8. $2\sqrt{y - 1} = 8x^{2/3}$

9. $\dfrac{x + 2y}{x - 3y} = y^{1/2}$

10. $\dfrac{6 + 5x}{2 - 3y} = \dfrac{1}{5x}$

11. $(4y^2 - 3x)^{2/3} = 6x$

12. $(8x + y^{1/2})^3 = 9y^2$

13. Find the equation of the line tangent to the graph of $\sqrt{2x} - 4yx = -22$ at the point (2, 3).

14. The graph of $x^2 + y^2 = 25$ is a circle of radius 5 with center at the origin. Find the equation of the tangent lines when $x = -3$. Graph the circle and the tangent lines.

Find dy/dt in Exercises 15–18.

15. $y = 8x^3 - 7x^2$; $dx/dt = 4, x = 2$

16. $y = \dfrac{9 - 4x}{3 + 2x}$; $dx/dt = -1, x = -3$

17. $y = \dfrac{1 + \sqrt{x}}{1 - \sqrt{x}}$; $dx/dt = -4, x = 4$

18. $\dfrac{x^2 + 5y}{x - 2y} = 2$; $dx/dt = 1, x = 2, y = 0$

19. What is a differential? What is it used for?

Find dy.

20. $y = 8x^3 - 2x^2$

21. $y = 4(x^2 - 1)^3$

22. $y = \dfrac{6 - 5x}{2 + x}$

23. $y = \sqrt{9 + x^3}$

Evaluate dy in Exercises 24 and 25.

24. $y = 8 - x^2 + x^3$; $x = -1, \Delta x = .02$

25. $y = \dfrac{3x - 7}{2x + 1}$; $x = 2, \Delta x = .003$

26. What is elasticity of demand (in words; no mathematical symbols allowed)? Why is the derivative used to describe elasticity?

27. When solving applied extrema problems, why is it necessary to check the endpoints of the domain?

28. Find two numbers whose sum is 25 and whose product is a maximum.

29. Find nonnegative numbers x and y such that $x = 2 + y$, and xy^2 is minimized.

Applications

Business and Economics

30. Profit Suppose the profit from a product is $P(x) = 40x - x^2$, where x is the price in hundreds of dollars.

 (a) At what price will the maximum profit occur?

 (b) What is the maximum profit?

31. Profit The total profit (in tens of dollars) from the sale of x hundred boxes of candy is given by

$$P(x) = -x^3 + 10x^2 - 12x.$$

 (a) Find the number of boxes of candy that should be sold in order to produce maximum profit.

 (b) Find the maximum profit.

32. Packaging Design The packaging department of a corporation is designing a box with a square base and no top. The volume is to be 32 m³. To reduce cost, the box is to have minimum surface area. What dimensions (height, length, and width) should the box have?

33. Packaging Design Fruit juice will be packaged in cylindrical cans with a volume of 40 cubic inches each. The top and bottom of the can cost 4¢ per square inch, while the sides cost 3¢ per square inch. Find the radius and height of the can of minimum cost.

34. Packaging Design A company plans to package its product in a cylinder that is open at one end. The cylinder is to have a volume of 27π cubic inches. What radius should the circular bottom of the cylinder have to minimize the cost of the material?

35. Order Quantity A store sells 980,000 cases of a product annually. It costs $2 to store one case for a year and $20 to place a reorder. Find the number of cases that should be ordered each time.

36. Order Quantity A very large camera store sells 320,000 rolls of film annually. It costs 10¢ to store one roll for one year and $10 to place a reorder. Find the number of rolls that should be ordered each time.

37. Lot Size In one year, a health food manufacturer produces and sells 240,000 cases of vitamins. It costs $2 to store a case for one year and $15 to produce each batch. Find the number of batches that should be produced annually.

38. Lot Size A company produces 128,000 cases of soft drink annually. It costs $1 to store a case for one year and $10 to produce one lot. Find the number of lots that should be produced annually.

39. Elasticity Suppose the demand function for a product is given by $q = A/p^k$, where A and k are positive constants. For what values of k is the demand elastic? inelastic?

Life Sciences

40. Pollution A circle of pollution is spreading from a broken underwater waste disposal pipe, with the radius increasing at the rate of 4 ft/min. Find the rate of change of the area of the circle when the radius is 7 ft.

Physical Sciences

41. Sliding Ladder A 50-ft ladder is placed against a building. The top of the ladder is sliding down the building at the rate of 2 ft/min. Find the rate at which the base of the ladder is slipping away from the building at the instant that the base is 30 ft from the building.

42. Spherical Radius A large weather balloon is being inflated with air at the rate of 1.2 ft³/min. Find the rate of change of the radius when the radius is 1.2 ft.

43. Water Level A water trough 2 ft across, 4 ft long, and 1 ft deep has ends in the shape of isosceles triangles. (See the figure.) It is being filled with 3.5 ft³ of water per minute. Find the rate at which the depth of water in the tank is changing when the water is 1/3 ft deep.

General Interest

44. Volume Approximate the volume of coating on a sphere of radius 4 inches if the coating is .02 inch thick.

45. Area A square has an edge of 9.2 inches, with a possible error in the measurement of ±.04 inch. Estimate the possible error in the area of the square.

46. Playground Area The city park department is planning an enclosed play area in a new park. One side of the area will be against an existing building, with no fence needed there. Find the dimensions of the maximum rectangular area that can be made with 900 m of fence.

Connections

47. (a) Suppose x and y are related by the equation
$$-12x + x^3 + y + y^2 = 4.$$
Find all critical points on the curve.

(b) Determine whether the critical points found in part (a) are relative maxima or relative minima by taking values of x nearby and solving for the corresponding values of y.

(c) Is there an absolute maximum or minimum for x and y in the relationship given in part (a)? Why or why not?

48. In Exercise 47, implicit differentiation was used to find the relative extrema. The exercise was contrived to avoid various difficulties that could have arisen. Discuss some of the difficulties that might be encountered in such problems, and how these difficulties might be resolved.

Extended Application / A Total Cost Model for a Training Program*

In this application, we set up a mathematical model for determining the total costs in setting up a training program. Then we use calculus to find the time interval between training programs that produces the minimum total cost. The model assumes that the demand for trainees is constant and that the fixed cost of training a batch of trainees is known. Also, it is assumed that people who are trained, but for whom no job is readily available, will be paid a fixed amount per month while waiting for a job to open up.

The model uses the following variables.

D = demand for trainees per month

N = number of trainees per batch

C_1 = fixed cost of training a batch of trainees

C_2 = variable cost of training per trainee per month

C_3 = salary paid monthly to a trainee who has not yet been given a job after training

m = time interval in months between successive batches of trainees

t = length of training program in months

$Z(m)$ = total monthly cost of program

The total cost of training a batch of trainees is given by $C_1 + NtC_2$. However $N = mD$, so that the total cost per batch is $C_1 + mDtC_2$.

After training, personnel are given jobs at the rate of D per month. Thus, $N - D$ of the trainees will not get a job the first month, $N - 2D$ will not get a job the second month, and so on. The $N - D$ trainees who do not get a job the first month produce total costs of $(N - D)C_3$, those not getting jobs during the second month produce costs of $(N - 2D)C_3$, and so on. Since $N = mD$, the costs during the first month can be written as

$$(N - D)C_3 = (mD - D)C_3 = (m - 1)DC_3,$$

*Based on ''A Total Cost Model for a Training Program'' by P. L. Goyal and S. K. Goyal, Faculty of Commerce and Administration, Concordia University. Used with permission.

while the costs during the second month are $(m - 2)DC_3$, and so on. The total cost for keeping the trainees without a job is thus

$$(m - 1)DC_3 + (m - 2)DC_3$$
$$+ (m - 3)DC_3 + \cdots + 2DC_3 + DC_3,$$

which can be factored to give

$$DC_3[(m - 1) + (m - 2) + (m - 3) + \cdots + 2 + 1].$$

The expression in brackets is the sum of the terms of an arithmetic sequence, discussed in most algebra texts. Using formulas for arithmetic sequences, the expression in brackets can be shown to equal $m(m - 1)/2$, so that we have

$$DC_3\left[\frac{m(m - 1)}{2}\right] \qquad (1)$$

as the total cost for keeping jobless trainees.

The total cost per batch is the sum of the training cost per batch, $C_1 + mDtC_2$, and the cost of keeping trainees without a proper job, given by equation (1). Since we assume that a batch of trainees is trained every m months, the total cost per month, $Z(m)$, is given by

$$Z(m) = \frac{C_1 + mDtC_2}{m} + \frac{DC_3\left[\dfrac{m(m - 1)}{2}\right]}{m}$$

$$= \frac{C_1}{m} + DtC_2 + DC_3\left(\frac{m - 1}{2}\right).$$

Exercises

1. Find $Z'(m)$.

2. Solve the equation $Z'(m) = 0$.
 As a practical matter, it is usually required that m be a whole number. If m does not come out to be a whole number in Exercise 2, then m^+ and m^-, the two whole numbers closest to m, must be chosen. Calculate both $Z(m^+)$ and $Z(m^-)$; the smaller of the two provides the optimum value of Z.

3. Suppose a company finds that its demand for trainees is 3 per month, that a training program requires 12 months, that the fixed cost of training a batch of trainees is \$15,000, that the variable cost per trainee per month is \$100, and that trainees are paid \$900 per month after training but before going to work. Use your result from Exercise 2 and find m.

4. Since m is not a whole number, find m^+ and m^-.

5. Calculate $Z(m^+)$ and $Z(m^-)$.

6. What is the optimum time interval between successive batches of trainees? How many trainees should be in a batch?

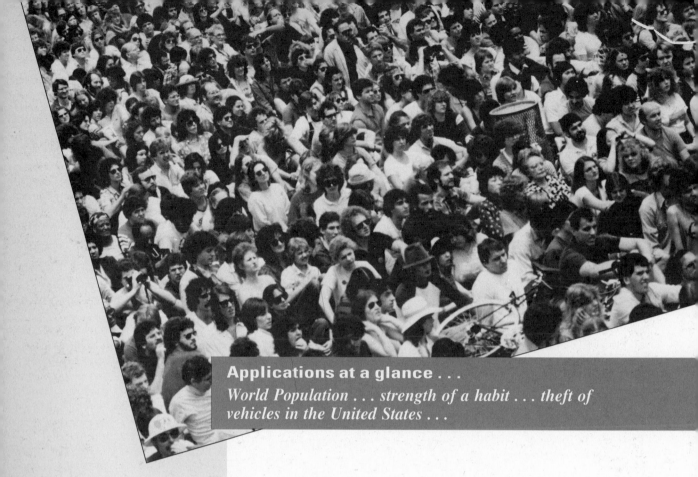

Applications at a glance...
World Population ... strength of a habit ... theft of vehicles in the United States ...

CHAPTER 13

EXPONENTIAL AND LOGARITHMIC FUNCTIONS

(See page 632.)

Exponential functions may be the single most important type of functions used in practical applications. These functions, and the closely related logarithmic functions, are used to describe growth and decay, which are important ideas in management, social science, and biology. A function in which the variable appears in the exponent is an *exponential function*. Exponential functions tend to increase or decrease rapidly.

For many years *logarithms* were used primarily to assist in involved calculations. Current technology has made this use of logarithms obsolete, but *logarithmic functions* play an important role in many applications of mathematics. We shall see in this chapter that exponential and logarithmic functions are intimately related; in fact, they are inverses of each other.

13.1 EXPONENTIAL FUNCTIONS

How much interest will an investment earn?

What is the oxygen consumption of yearling salmon?

Later in this section, in Examples 5 and 6, we will see that the answers to these questions depend on *exponential functions*.

In earlier chapters we discussed functions involving expressions such as x^2, $(2x + 1)^3$, or $x^{3/4}$, where the variable or variable expression is the base of an exponential expression, and the exponent is a constant. In an exponential function, the variable is in the exponent and the base is a constant.

EXPONENTIAL FUNCTION

An **exponential function** with base a is defined as

$$f(x) = a^x, \quad \text{where } a > 0 \text{ and } a \neq 1.$$

(If $a = 1$, the function is the constant function $f(x) = 1$.)

The exponential function $f(x) = 2^x$ can be graphed by making a table of values of x and y, as shown in Figure 1 (on the next page). Plotting these points and drawing a smooth curve through them gives the graph shown in the figure. The graph approaches the negative x-axis but will never actually touch it, since y cannot be 0, so the x-axis is a horizontal asymptote. The graph suggests that the domain is the set of all real numbers and the range is the set of all positive real numbers. The function f, where $f(x) = 2^x$, has a base greater than 1, and the function is increasing over its domain. This graph is typical of the graphs of exponential functions of the form $y = a^x$, where $a > 1$.

x	y
-3	⅛
-2	¼
-1	½
0	1
1	2
2	4
3	8
4	16

$f(x)$

$f(x) = 2^x$

FIGURE 1

EXAMPLE **1**
Graph $f(x) = 2^{-x}$.

The graph, shown in Figure 2, is the reflection about the y-axis of the graph of $f(x) = 2^x$ given in Figure 1. This graph is typical of the graphs of exponential functions of the form $y = a^x$ where $0 < a < 1$. The domain includes all real numbers and the range includes all positive numbers. Notice that this function, with $f(x) = 2^{-x} = (1/2)^x$, is decreasing over its domain. ◀

Recall from Section 1.6 that the graph of $f(-x)$ is the reflection of the graph of $f(x)$ about the y-axis.

x	y
-4	16
-3	8
-2	4
-1	2
0	1
1	½
2	¼
3	⅛

$f(x) = 2^{-x}$

FIGURE 2

In the definition of an exponential function, notice that the base a is restricted to positive values, with negative or zero bases not allowed. For example, the function $y = (-4)^x$ could not include such numbers as $x = 1/2$ or $x = 1/4$ in the domain. The resulting graph would be at best a series of separate points having little practical use.

EXAMPLE **2**
Graph $f(x) = -2^x + 3$.

The graph of $y = -2^x$ is symmetric about the x-axis to the graph of $y = 2^x$, so this is a decreasing function. The 3 indicates that the graph should be translated up 3 units, as compared to the graph of $y = -2^x$. Find some ordered pairs. Since $y = -2^x$ would have y-intercept $(0, -1)$, this function has y-inter-

cept $(0, 2)$, which is up 3 units. Some other ordered pairs are $(1, 1)$, $(2, -1)$, and $(3, -5)$. For negative values of x, the graph approaches the line $x = 3$, which is a horizontal asymptote. The graph is shown in Figure 3. ◀

FIGURE 3

We can use the composition of functions to produce more general exponential functions. If $h(u) = ka^u$, where k is a constant and $u = g(x)$, and $f(x) = h[g(x)]$, then

$$f(x) = h[g(x)] = ka^{g(x)}.$$

In the exponential function $f(x) = 4 \cdot 7^{3x-1}$, for example, $k = 4$, $a = 7$, and $g(x) = 3x - 1$.

EXAMPLE 3

Graph $y = 2^{-x^2}$.

When $x = 1$ or $x = -1$, $y = 2^{-1} = 1/2$. As x becomes larger in magnitude, either positive or negative, y becomes smaller and approaches the horizontal asymptote $y = 0$. Using a graphing calculator or computer program, or plotting numerous points by hand, gives the graph in Figure 4. The final result looks quite different from the typical graphs of exponential functions in Figures 1–3. Graphs like this are important in probability, where the normal curve has an equation similar to the one in this example. ◀

x	y
-2	$1/16$
-1	$1/2$
0	1
1	$1/2$
2	$1/16$

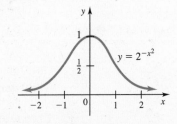

FIGURE 4

Exponential Equations In Figures 1 and 2, which are typical graphs of exponential functions, a given value of x leads to exactly one value of a^x. Because of this, an equation with a variable in the exponent, called an **exponential equation,** often can be solved using the following property.

If $a > 0$, $a \neq 1$, and $a^x = a^y$, then $x = y$. Also, if $x = y$, then $a^x = a^y$.

(Both bases must be the same.) The value $a = 1$ is excluded since $1^2 = 1^3$, for example, even though $2 \neq 3$. To solve $2^{3x} = 2^7$ using this property, work as follows.

$$2^{3x} = 2^7$$
$$3x = 7$$
$$x = \frac{7}{3}$$

EXAMPLE 4

(a) Solve $9^x = 27$.
 First rewrite both sides of the equation so the bases are the same. Since $9 = 3^2$ and $27 = 3^3$,

$$(3^2)^x = 3^3$$
$$3^{2x} = 3^3$$
$$2x = 3$$
$$x = \frac{3}{2}.$$

(b) Solve $32^{2x-1} = 128^{x+3}$.
 Since the bases must be the same, write 32 as 2^5 and 128 as 2^7, giving

$$32^{2x-1} = 128^{x+3}$$
$$(2^5)^{2x-1} = (2^7)^{x+3}$$
$$2^{10x-5} = 2^{7x+21}.$$

Now use the property above to get

$$10x - 5 = 7x + 21$$
$$3x = 26$$
$$x = \frac{26}{3}.$$

Verify this solution in the original equation. ◀

Compound Interest The calculation of compound interest is an important application of exponential functions. The cost of borrowing money or the return

on an investment is called **interest.** The amount borrowed or invested is the **principal,** P. The **rate of interest** r is given as a percent per year, and t is the **time,** measured in years.

SIMPLE INTEREST

The product of the principal P, rate r, and time t gives **simple interest,** I:

$$I = Prt.$$

With **compound interest,** interest is charged (or paid) on interest as well as on principal. To find a formula for compound interest, first suppose that P dollars, the principal, is deposited at a rate of interest r per year. The interest earned during the first year is found by using the formula for simple interest:

$$\text{First-year interest} = P \cdot r \cdot 1 = Pr.$$

At the end of one year, the amount on deposit will be the sum of the original principal and the interest earned, or

$$P + Pr = P(1 + r). \tag{1}$$

If the deposit earns compound interest, the interest earned during the second year is found from the total amount on deposit at the end of the first year. Thus, the interest earned during the second year (again found by the formula for simple interest), is given by

$$[P(1 + r)](r)(1) = P(1 + r)r, \tag{2}$$

so that the total amount on deposit at the end of the second year is given by the sum of amounts from (1) and (2) above, or

$$P(1 + r) + P(1 + r)r = P(1 + r)(1 + r) = P(1 + r)^2.$$

In the same way, the total amount on deposit at the end of three years is

$$P(1 + r)^3.$$

After t years, the total amount on deposit, called the **compound amount,** is $P(1 + r)^t$.

When interest is compounded more than once a year, the compound interest formula is adjusted. For example, if interest is to be paid quarterly (four times a year), $1/4$ of the interest rate is used each time interest is calculated, so the rate becomes $r/4$, and the number of compounding periods in t years becomes $4t$. Generalizing from this idea gives the following formula.

COMPOUND AMOUNT

If P dollars is invested at a yearly rate of interest r per year, compounded m times per year for t years, the compound amount is

$$A = P\left(1 + \frac{r}{m}\right)^{tm} \text{ dollars.}$$

EXAMPLE **5** ▶ Françoise Dykler invests a bonus of $9000 at 6% annual interest compounded semiannually for 4 years. How much interest will she earn?

Use the formula for compound interest with $P = 9000$, $r = .06$, $m = 2$, and $t = 4$.

$$A = P\left(1 + \frac{r}{m}\right)^{tm}$$

$$= 9000\left(1 + \frac{.06}{2}\right)^{4(2)} = 9000(1.03)^8$$

To find $(1.03)^8$, use a calculator. You should get 1.266770081, which gives

$$A = 9000(1.266770081) = 11400.93.$$

The investment plus the interest is $11,400.93. The interest amounts to $11,400.93 − $9000 = $2400.93. ◀

Note 🖩 When using a calculator to compute the compound interest, keep each partial result in the calculator and avoid rounding off until the final answer.

The Number e Perhaps the single most useful base for an exponential function is the number e, an irrational number that occurs often in practical applications. The letter e was chosen to represent this number in honor of the Swiss mathematician Leonhard Euler (pronounced "oiler") (1707–1783). To see how the number e occurs in an application, begin with the formula for compound interest,

$$P\left(1 + \frac{r}{m}\right)^{tm}.$$

Suppose that a lucky investment produces annual interest of 100%, so that $r = 1.00 = 1$. Suppose also that you can deposit only $1 at this rate, and for only one year. Then $P = 1$ and $t = 1$. Substituting these values into the formula for compound interest gives

$$P\left(1 + \frac{r}{m}\right)^{t(m)} = 1\left(1 + \frac{1}{m}\right)^{1(m)} = \left(1 + \frac{1}{m}\right)^{m}.$$

As interest is compounded more and more often, m gets larger and the value of this expression will increase. For example, if $m = 1$ (interest is compounded annually),

$$\left(1 + \frac{1}{m}\right)^{m} = \left(1 + \frac{1}{1}\right)^{1} = 2^1 = 2,$$

so that your $1 becomes $2 in one year. Using a calculator with a y^x key gives the results shown in the following table for larger and larger values of m. These results have been rounded to five decimal places.

m	$\left(1 + \dfrac{1}{m}\right)^m$
1	2
2	2.25
5	2.48832
10	2.59374
25	2.66584
50	2.69159
100	2.70481
500	2.71557
1000	2.71692
10,000	2.71815
1,000,000	2.71828

The table suggests that as m increases, the value of $(1 + 1/m)^m$ gets closer and closer to some fixed number. It turns out that this is indeed the case. This fixed number is called e.

DEFINITION OF e

$$e = \lim_{m \to \infty} \left(1 + \frac{1}{m}\right)^m \approx 2.718281828$$

This last approximation gives the value of e to nine decimal places. Euler approximated e to 23 decimal places using the definition $\lim_{m \to \infty} (1 + 1/m)^m$. Many calculators give values of e^x, usually with a key labeled e^x. Some require two keys, either INV ln x or 2nd ln x. (We will define ln x in the next section.) In Figure 5, the functions $y = 2^x$, $y = e^x$, and $y = 3^x$ are graphed for comparison. The number e is often used as the base in an exponential equation because it provides a good model for many natural, as well as economic, phenomena. In the Applications section of this chapter, we will look at several examples of such applications.

FIGURE 5

Refer to the discussion on linear regression at the end of Section 1.2. A similar process is used to fit data points to other types of functions. Many of the functions in this chapter's applications were determined in this way, including that given in Example 6.

EXAMPLE 6

In situations that involve growth or decay of a population, the size of the population at a given time t often is determined by an exponential function of t. The next example illustrates a typical application of this kind.

Biologists studying salmon have found that the oxygen consumption of yearling salmon (in appropriate units) increases exponentially with the speed of swimming according to the function

$$f(x) = 100(3)^{.6x},$$

where x is the speed in feet per second. Find each of the following.

(a) The oxygen consumption when the fish are still

When the fish are still, their speed is 0. Substitute 0 for x:

$$f(0) = 100(3)^{(.6)(0)} = 100(3)^0$$
$$= 100 \cdot 1 = 100.$$

When the fish are still, their oxygen consumption is 100 units.

(b) The oxygen consumption at a speed of 2 feet per second

Find $f(2)$ as follows.

$$f(2) = 100(3)^{(.6)(2)} = 100(3)^{1.2}$$

A calculator gives $3^{1.2} \approx 3.73719$ to five decimal places.

$$f(2) = 100(3)^{1.2} \approx 100(3.7) = 370$$

At a speed of 2 feet per second, oxygen consumption is about 370 units. ◀

Note We rounded the answer to two significant digits. Because the function is only an approximation of the real situation, further accuracy is not realistic.

13.1 Exercises

Graph each of the following.

1. $y = 3^x$

2. $y = 4^x$

3. $y = 3^{-x/2}$

4. $y = 4^{-x/3}$

5. $y = \left(\dfrac{1}{4}\right)^x + 1$

6. $y = \left(\dfrac{1}{3}\right)^x - 2$

7. $y = -3^{2x-1}$

8. $y = -3^{x^2-1}$

9. $y = e^{-x^2+1}$

10. $y = e^{x+1}$

11. $y = 10 - 5e^{-x}$

12. $y = 100 - 80e^{-x}$

Graph the following equations in the interval $[-3, 2]$ by plotting points or using a graphing calculator or a computer.

13. $y = x \cdot 2^x$

14. $y = x^2 \cdot 2^x$

15. $y = \dfrac{e^x - e^{-x}}{2}$

16. $y = \dfrac{e^x + e^{-x}}{2}$

17. Explain why the exponential equation $4^x = 6$ cannot be solved using the method described in this section.

Solve each equation.

18. $2^x = \dfrac{1}{8}$ **19.** $4^x = 64$ **20.** $e^x = e^2$ **21.** $e^x = \dfrac{1}{e^2}$

22. $4^x = 8^{x+1}$ **23.** $25^x = 125^{x-2}$ **24.** $16^{-x+1} = 8^x$ **25.** $16^{x+2} = 64^{2x-1}$

26. $(e^4)^{-2x} = e^{-x+1}$ **27.** $e^{-x} = (e^2)^{x+3}$ **28.** $2^{|x|} = 16$ **29.** $5^{-|x|} = \dfrac{1}{25}$

30. $2^{x^2-4x} = \left(\dfrac{1}{16}\right)^{x-4}$ **31.** $5^{x^2+x} = 1$ **32.** $8^{x^2} = 2^{5x+2}$ **33.** $9^{x+4} = 3^{x^2}$

34. Find the average rate of change of $f(x) = 2^x$ as x changes from 0 to x_1 for the following values of x_1.

 (a) 1 **(b)** .1 **(c)** .01 **(d)** .001 **(e)** Estimate the slope of the graph of $f(x)$ at $x = 0$.

35. Repeat Exercise 34 for $f(x) = e^x$.

36. Suppose the domain of $y = 2^x$ is restricted to rational values of x (which are the only values discussed prior to this section.) Describe in words the resulting graph.

37. In our definition of exponential function, we ruled out negative values of a. The author of a textbook on mathematical economics, however, obtained a "graph" of $y = (-2)^x$ by plotting the following points and drawing a smooth curve through them.

x	-4	-3	-2	-1	0	1	2	3
y	$1/16$	$-1/8$	$1/4$	$-1/2$	1	-2	4	-8

The graph oscillates very neatly from positive to negative values of y. Comment on this approach. (This exercise shows the dangers of point plotting when drawing graphs.)

Applications

Business and Economics

38. **Interest** Find the interest earned on $10,000 invested for 5 years at 6% interest compounded as follows.

 (a) Annually **(b)** Semiannually (twice a year)

 (c) Quarterly **(d)** Monthly

39. **Interest** Suppose $26,000 is borrowed for 3 years at 12% interest. Find the interest paid over this period if the interest is compounded as follows.

 (a) Annually **(b)** Semiannually **(c)** Quarterly

 (d) Monthly

40. **Interest** Terry Wong deposits a $9430 commission earned on a real estate sale in the bank at 6% interest compounded quarterly. How much will be on deposit in 4 years?

41. **Interest** Cynthia Klein-Herring borrows $17,500 to open a new restaurant. The money is to be repaid at the end of 6 years, with 10% interest compounded semiannually. How much will she owe in 6 years?

42. **Interest** David Glenn needs to choose between two investments: one pays 8% compounded semiannually, and the other pays 7½% compounded monthly. If he plans to invest $18,000 for 1½ years, which investment should he choose? How much extra interest will he earn by making the better choice?

43. **Interest** Find the interest rate required for an investment of $5000 to grow to $8000 in 4 years if interest is compounded as follows: **(a)** annually **(b)** quarterly.

44. **Inflation** If money loses value at the rate of 8% per year, the value of $1 in t yr is given by

$$y = (1 - .08)^t = (.92)^t.$$

 (a) Use a calculator to help complete the following table.

t	0	1	2	3	4	5	6	7	8	9	10
y	1					.66					.43

 (b) Graph $y = (.92)^t$.

 (c) Suppose a house costs $165,000 today. Use the results of part (a) to estimate the cost of a similar house in 10 yr.

 (d) Find the cost of a $20 textbook in 8 yr.

45. Employee Training A person learning certain skills involving repetition tends to learn quickly at first. Then learning tapers off and approaches some upper limit. Suppose the number of symbols per minute that a person using a word processor can type is given by

$$p(t) = 250 - 120(2.8)^{-.5t},$$

where t is the number of months the operator has been in training. Find each of the following.

(a) $p(2)$ (b) $p(4)$ (c) $p(10)$ (d) Graph $p(t)$.

(e) What happens to the number of symbols per minute after several months of training?

(f) According to this function, what is the limit to the number of symbols typed per minute?

46. Interest On January 1, 1980, Jack deposited $1000 into Bank X to earn interest at the rate of j per annum compounded semiannually. On January 1, 1985, he transferred his account to Bank Y to earn interest at the rate of k per annum compounded quarterly. On January 1, 1988, the balance at Bank Y was $1990.76. If Jack could have earned interest at the rate of k per annum compounded quarterly from January 1, 1980, through January 1, 1988, his balance would have been $2203.76. Which of the following represents the ratio k/j? (This exercise is reprinted from an actuarial examination.)*

(a) 1.25 (b) 1.30 (c) 1.35 (d) 1.40

(e) 1.45

Life Sciences

47. Population Growth Since 1950, the growth in world population (in millions) closely fits the exponential function defined by

$$A(t) = 2600e^{.018t},$$

where t is the number of years since 1950.

(a) World population was about 3700 million in 1970. How closely does the function approximate this value?

(b) Use the function to approximate world population in 1990. (The actual 1990 population was about 5320 million.)

(c) Estimate world population in the year 2000.

48. Growth of Bacteria *Escherichia coli* is a strain of bacteria that occurs naturally in many different situations. Under certain conditions, the number of these bacteria present in

a colony is given by

$$E(t) = 1,000,000 \cdot 2^{t/30},$$

where $E(t)$ is the number of bacteria present t min after the beginning of an experiment.

(a) How many bacteria were present initially?

(b) How many bacteria are present after 30 min?

(c) How often do the bacteria double?

(d) How quickly will the number of bacteria increase to 32,000,000?

49. Growth of Species Under certain conditions, the number of individuals of a species that is newly introduced into an area can double every year. That is, if t represents the number of years since the species was introduced into the area, and y represents the number of individuals, then

$$y = 6 \cdot 2^t.$$

Assume that the same growth rate continues.

(a) How many animals were originally introduced into the area?

(b) How long will it take for the number to grow to 24 animals?

(c) In how many years will there be 384 animals?

Social Sciences

50. City Planning City planners have determined that the population of a city is

$$P(t) = 1,000,000(2^{.2t}),$$

where t represents time measured in years. Find each of the following values. Assume the growth rate is maintained.

(a) What population will the city have in 5 years?

(b) In how many years will the population reach 8,000,000?

Physical Sciences

51. Radioactive Decay Suppose the quantity (in grams) of a radioactive substance present at time t is

$$Q(t) = 1000(5^{-.3t}),$$

where t is measured in months.

(a) How much will be present in 6 months?

(b) How long will it take to reduce the substance to 8 g?

*Problem 5 from "November 1989 Course 140 Examination, Mathematics of Compound Interest" of the Education and Examination Committee of The Society of Actuaries. Reprinted by permission of The Society of Actuaries.

General Interest

52. Vehicle Theft Vehicle theft in the United States has been rising exponentially since 1972. The number of stolen vehicles (in millions) is given by

$$f(x) = .88(1.03)^x,$$

where $x = 0$ represents the year 1972. Find the number of vehicles stolen in the following years.

(a) 1975 (b) 1980 (c) 1985 (d) 1990

Computer/Graphing Calculator

Make a quick sketch of each function using the information from this section and what you learned in Chapter 1 about general graphing techniques. Then use a computer or graphing calculator to graph each function and compare with your sketch.

53. $y = 2^{1-x} + 2$ **54.** $y = 1 - 3^{2x-4}$ **55.** $y = e^{-x+1}$ **56.** $y = e^{x^2/2}$

13.2 LOGARITHMIC FUNCTIONS

With an inflation rate averaging 5% per year, how long will it take for prices to double?

The number of years it will take for prices to double under given conditions is called the **years to double.** For $1 to double (become $2) in t years, assuming 5% annual compounding, means that

$$A = P\left(1 + \frac{r}{m}\right)^{mt}$$

becomes

$$2 = 1\left(1 + \frac{.05}{1}\right)^{1(t)}$$

or

$$2 = (1.05)^t.$$

This equation would be easier to solve if the variable were not in the exponent. *Logarithms* are defined for just this purpose. In Example 8, we will use logarithms to answer the question posed above.

DEFINITION OF LOGARITHM

For $a > 0$ and $a \neq 1$,

$$y = \log_a x \quad \text{means} \quad a^y = x.$$

(Read $y = \log_a x$ as "y is the logarithm of x to the base a.") For example, the exponential statement $2^4 = 16$ can be translated into the logarithmic statement $4 = \log_2 16$. Also, in the problem discussed above $(1.05)^t = 2$ can be rewritten

with this definition as $t = \log_{1.05} 2$. Think of a logarithm as an exponent: $\log_a x$ is the exponent used with the base a to get x.

EXAMPLE 1 This example shows the same statements written in both exponential and logarithmic forms.

Exponential Form	Logarithmic Form
(a) $3^2 = 9$	$\log_3 9 = 2$
(b) $(1/5)^{-2} = 25$	$\log_{1/5} 25 = -2$
(c) $10^5 = 100{,}000$	$\log_{10} 100{,}000 = 5$
(d) $4^{-3} = 1/64$	$\log_4 1/64 = -3$
(e) $2^{-4} = 1/16$	$\log_2 1/16 = -4$
(f) $e^0 = 1$	$\log_e 1 = 0$ ◀

Logarithmic Functions For a given positive value of x, the definition of logarithm leads to exactly one value of y, making $y = \log_a x$ the *logarithmic function* of base a (the base a must be positive, with $a \neq 1$).

LOGARITHMIC FUNCTION If $a > 0$ and $a \neq 1$, then the logarithmic function of base a is

$$f(x) = \log_a x.$$

The graphs of the exponential function $f(x) = 2^x$ and the logarithmic function $g(x) = \log_2 x$ are shown in Figure 6. The graphs show that $f(3) = 2^3 = 8$, while $g(8) = \log_2 8 = 3$. Thus, $f(3) = 8$ and $g(8) = 3$. Also, $f(2) = 4$ and $g(4) = 2$. In fact, for any number m, if $f(m) = p$, then $g(p) = m$. Functions

FIGURE 6

related in this way are called *inverses* of each other. The graphs also show that the domain of the exponential function (the set of real numbers) is the range of the logarithmic function. Also, the range of the exponential function (the set of positive real numbers) is the domain of the logarithmic function. Every logarithmic function is the inverse of some exponential function. This means that we can graph logarithmic functions by rewriting them as exponential functions using the definition of logarithm. The graphs in Figure 6 show a characteristic of inverse functions: their graphs are mirror images about the line $y = x$. A more complete discussion of inverse functions is given in most standard intermediate algebra and college algebra books.

2 Graph $y = \log_{1/2} x$.

Use the equivalent equation $(1/2)^y = x$ to find several ordered pairs. Plotting points and connecting them with a smooth curve gives the graph of $y = \log_{1/2} x$ shown in Figure 7, which also includes the graph of $y = (1/2)^x$. Again, the logarithmic and exponential graphs are mirror images with respect to the line $y = x$. ◄

FIGURE 7

The graph of $y = \log_2 x$ shown in Figure 6 is typical of the graphs of logarithmic functions $y = \log_a x$, where $a > 1$. The graph of $y = \log_{1/2} x$, shown in Figure 7, is typical of logarithmic functions of the form $y = \log_a x$, where $0 < a < 1$. For both logarithmic graphs, the y-axis is a vertical asymptote.

Properties of Logarithms The usefulness of logarithmic functions depends in large part on the following **properties of logarithms.**

PROPERTIES OF LOGARITHMS

Let x and y be any positive real numbers and r be any real number. Let a be a positive real number, $a \neq 1$. Then

(a) $\log_a xy = \log_a x + \log_a y$ Product rule

(b) $\log_a \dfrac{x}{y} = \log_a x - \log_a y$ Quotient rule

(c) $\log_a x^r = r \log_a x$ Power rule

(d) $\log_a a = 1$

(e) $\log_a 1 = 0$

(f) $\log_a a^r = r.$

To prove property (a), let $m = \log_a x$ and $n = \log_a y$. Then, by the definition of logarithm,

$$a^m = x \quad \text{and} \quad a^n = y.$$

Hence,

$$a^m a^n = xy.$$

By a property of exponents, $a^m a^n = a^{m+n}$, so

$$a^{m+n} = xy.$$

Now use the definition of logarithm to write

$$\log_a xy = m + n.$$

Since $m = \log_a x$ and $n = \log_a y$,

$$\log_a xy = \log_a x + \log_a y.$$

Proofs of properties (b) and (c) are left for the exercises. Properties (d) and (e) depend on the definition of a logarithm. Property (f) follows from properties (c) and (d).

EXAMPLE **3** If all the following variable expressions represent positive numbers, then for $a > 0$, $a \neq 1$, the statements in (a)–(c) are true.

(a) $\log_a x + \log_a (x - 1) = \log_a x(x - 1)$

(b) $\log_a \dfrac{x^2 - 4x}{x + 6} = \log_a (x^2 - 4x) - \log_a (x + 6)$

(c) $\log_a (9x^5) = \log_a 9 + \log_a (x^5) = \log_a 9 + 5 \cdot \log_a x$ ◀

Evaluating Logarithms The invention of logarithms is credited to John Napier (1550–1617), who first called logarithms "artificial numbers." Later he joined the Greek words *logos* (ratio) and *arithmos* (number) to form the word used today. The development of logarithms was motivated by a need for faster

computation. Tables of logarithms and slide rule devices were developed by Napier, Henry Briggs (1561–1631), Edmund Gunter (1581–1626), and others.

Historically, one of the main applications of logarithms has been as an aid to numerical calculation. The properties of logarithms and a table of logarithms were used to simplify many numerical problems. With today's widespread use of calculators, this application of logarithms has declined. Since our number system has base 10, logarithms to base 10 were most convenient for numerical calculations, so base 10 logarithms were called **common logarithms.** Common logarithms are still useful in other applications. For simplicity,

$$\log_{10} x \text{ is abbreviated } \log x.$$

Most practical applications of logarithms use the number e as base. (Recall that to seven decimal places, $e = 2.7182818$.) Logarithms to base e are called **natural logarithms**, and

$$\log_e x \text{ is abbreviated } \ln x$$

(read "el-en x"). A graph of $f(x) = \ln x$ is given in Figure 8.

FIGURE 8

Although common logarithms may seem more "natural" than logarithms to base e, there are several good reasons for using natural logarithms instead. The most important reason is discussed in Section 4 of this chapter.

A calculator can be used to find both common and natural logarithms. For example, using a calculator and four decimal places, we get

$$\log 2.34 = .3692, \qquad \log 594 = 2.7738, \qquad \text{and} \qquad \log .0028 = -2.5528.$$

We also get

$$\ln 55 = 4.0073, \qquad \ln 1.9 = .6419, \qquad \text{and} \qquad \ln .4 = -.9163.$$

Notice that logarithms of numbers less than 1 are negative when the base is greater than 1. A look at the graph of $y = \log_2 x$ or $y = \ln x$ will show why.

Sometimes it is convenient to use logarithms to bases other than 10 or e. For example, some computer science applications use base 2. Also, some calculators may have only one logarithm key, either $\log x$ or $\ln x$, but not both. In such cases, the following theorem is useful for converting from one base to another.

CHANGE-OF-BASE THEOREM FOR LOGARITHMS

If x is any positive number and if a and b are positive real numbers, $a \neq 1$, $b \neq 1$, then

$$\log_a x = \frac{\log_b x}{\log_b a}.$$

To prove this result, use the definition of logarithm to write $y = \log_a x$ as $x = a^y$ or $x = a^{\log_a x}$ (for positive x and positive a, $a \neq 1$). Now take base b logarithms of both sides of this last equation. This gives

$$\log_b x = \log_b a^{\log_a x}$$

or $\log_b x = (\log_a x)(\log_b a)$, Power rule

from which

$$\log_a x = \frac{\log_b x}{\log_b a}.$$

If the base b is equal to e, then by the change-of-base theorem,

$$\log_a x = \frac{\log_e x}{\log_e a}.$$

Using $\ln x$ for $\log_e x$ gives the special case of the theorem using natural logarithms.

For any positive numbers a and x, $a \neq 1$,

$$\log_a x = \frac{\ln x}{\ln a}.$$

EXAMPLE **4** Use natural logarithms to find each of the following. Round to the nearest hundredth.

(a) $\log_5 27$

Let $x = 27$ and $a = 5$. Using the second form of the theorem gives

$$\log_5 27 = \frac{\ln 27}{\ln 5}.$$

Now use a calculator.

$$\log_5 27 \approx \frac{3.2958}{1.6094} \approx 2.05$$

To check, use a calculator, along with the definition of logarithm, to verify that $5^{2.05} \approx 27$.

(b) $\log_2 .1$

$$\log_2 .1 = \frac{\ln .1}{\ln 2} \approx \frac{-2.3026}{.6931} = -3.32 \blacktriangleleft$$

Caution As mentioned earlier, when using a calculator, do not round off intermediate results. Keep all numbers in the calculator until you have the final answer. In Example 4(a), we showed the rounded intermediate values of ln 27 and ln 5, but we used the unrounded quantities when doing the division.

Logarithmic Equations Equations involving logarithms are often solved by using the fact that exponential functions and logarithmic functions are inverses, so a logarithmic equation can be rewritten (with the definition of logarithm) as an exponential equation. In other cases, the properties of logarithms may be useful in simplifying an equation involving logarithms.

EXAMPLE 5 Solve each of the following equations for x.

(a) $\log_x \dfrac{8}{27} = 3$

First, use the definition of logarithm and write the expression in exponential form. To solve for x, take the cube root on both sides.

$$x^3 = \frac{8}{27}$$

$$x = \frac{2}{3}$$

(b) $\log_4 x = \dfrac{5}{2}$

In exponential form, the given statement becomes

$$4^{5/2} = x$$
$$(4^{1/2})^5 = x$$
$$2^5 = x$$
$$32 = x.$$

(c) $\log_2 x - \log_2 (x - 1) = 1$

By a property of logarithms,

$$\log_2 x - \log_2 (x - 1) = \log_2 \frac{x}{x - 1},$$

so the original equation becomes

$$\log_2 \frac{x}{x - 1} = 1.$$

Use the definition of logarithms to write this result in exponential form.

$$\frac{x}{x - 1} = 2^1 = 2$$

Solve this equation.

$$\frac{x}{x-1}(x-1) = 2(x-1)$$

$$x = 2(x-1)$$

$$x = 2x - 2$$

$$-x = -2$$

$$x = 2$$

It is important to check solutions when solving equations involving logarithms because $\log_a u$, where u is an expression in x, has domain $u > 0$. ◀

Exponential Equations In the previous section exponential equations like $(1/3)^x = 81$ were solved by writing each side of the equation as a power of 3. That method cannot be used to solve an equation such as $3^x = 5$, however, since 5 cannot easily be written as a power of 3. A more general method for solving these equations depends on the following property of logarithms, which is supported by the graphs of logarithmic functions (Figures 6 and 7).

For $x > 0$, $y > 0$, $b > 0$, and $b \neq 1$,

if $x = y$, then $\log_b x = \log_b y$,

and if $\log_b x = \log_b y$, then $x = y$.

EXAMPLE

6 Solve $3^{2x} = 4^{x+1}$.

Taking natural logarithms (logarithms to any base could be used) on both sides gives

$$\ln 3^{2x} = \ln 4^{x+1}.$$

Now use the power rule for logarithms.

$$2x \ln 3 = (x+1)\ln 4$$

$$(2\ln 3)x = (\ln 4)x + \ln 4$$

$$(2\ln 3)x - (\ln 4)x = \ln 4$$

$$(2\ln 3 - \ln 4)x = \ln 4$$

$$x = \frac{\ln 4}{2\ln 3 - \ln 4}$$

Use a calculator to evaluate the logarithms, then divide, to get

$$x \approx \frac{1.3863}{2(1.0986) - 1.3863} \approx 1.710. \blacktriangleleft$$

Just as $\log_a x$ can be written as a base e logarithm, any exponential function $y = a^x$ can be written as an exponential function with base e. For example, there exists a real number k such that

$$2 = e^k.$$

Raising both sides to the power x gives

$$2^x = e^{kx},$$

so that powers of 2 can be found by evaluating appropriate powers of e. To find the necessary number k, solve the equation $2 = e^k$ for k by first taking logarithms on both sides.

$$2 = e^k$$
$$\ln 2 = \ln e^k$$
$$\ln 2 = k \ln e \qquad \text{Power rule}$$
$$\ln 2 = k \qquad \log_a a = 1$$

Thus, $k = \ln 2$. In Section 5 of this chapter, we will see why this change of base is useful. A general statement can be drawn from this example.

CHANGE-OF-BASE THEOREM FOR EXPONENTIALS

For every positive real number a,

$$a^x = e^{(\ln a)x}.$$

EXAMPLE 7 Evaluate each exponential by first writing it as a power of e.

(a) $2^{3.57}$

$$2^{3.57} = e^{(\ln 2)(3.57)} \approx 11.88$$

(b) $5^{-1.38} = e^{(\ln 5)(-1.38)} \approx .1085$ ◀

The final examples in this section illustrate the use of logarithms in practical problems.

EXAMPLE 8 Complete the solution of the problem posed at the beginning of this section. Recall that if prices will double after t years at an inflation level of 5%, compounded annually, t is given by the equation

$$2 = (1.05)^t.$$

We can solve this equation by first taking natural logarithms on both sides.

$$\ln 2 = \ln (1.05)^t$$
$$\ln 2 = t \ln 1.05 \qquad \text{Power rule}$$
$$t = \frac{\ln 2}{\ln 1.05} \approx 14.2$$

It will take about 14 years for prices to double. ◀

The problem solved in Example 8 can be generalized for the compound interest equation

$$A = P(1 + r)^t.$$

Solving for t as in Example 8 gives the doubling time in years as

$$t = \frac{\ln 2}{\ln (1 + r)}.$$

It can be shown that for certain values of r,

$$t = \frac{\ln 2}{\ln (1 + r)} \approx \frac{.693}{r},$$

and

$$\frac{70}{100r} \le \frac{\ln 2}{\ln (1 + r)} \le \frac{72}{100r}.$$

The **rule of 70** says that for $.001 \le r \le .05$, $70/100r$ gives a good approximation of t. The **rule of 72** says that for $.05 \le r \le .12$, $72/100r$ approximates t quite well.

EXAMPLE 9 Approximate the years to double at an interest rate of 6% using first the rule of 70, then the rule of 72.

By the rule of 70, money will double at 6% interest after

$$\frac{70}{100r} = \frac{70}{100(.06)} = \frac{70}{6} = 11.67 \left(\text{or } 11\frac{2}{3} \right)$$

years.

Using the rule of 72 gives

$$\frac{72}{100r} = \frac{72}{6} = 12$$

years doubling time. Since a more precise answer is given by

$$\frac{\ln 2}{\ln(1 + r)} = \frac{\ln 2}{\ln(1.06)} = \frac{.693}{.058} \approx 11.9,$$

the rule of 72 gives a better approximation than the rule of 70. This agrees with the statement that the rule of 72 works well for values of r where $.05 \le r \le .12$, since $r = .06$ falls into this category. ◀

EXAMPLE 10 One measure of the diversity of the species in an ecological community is given by the *index of diversity H*, where

$$H = \frac{-1}{\ln 2} [P_1 \ln P_1 + P_2 \ln P_2 + \cdots + P_n \ln P_n],$$

and P_1, P_2, \ldots, P_n are the proportions of a sample belonging to each of n species found in the sample. For example, in a community with two species, where there are 90 of one species and 10 of the other, $P_1 = 90/100 = .9$ and $P_2 = 10/100 = .1$, with

$$H = \frac{-1}{\ln 2} [.9 \ln .9 + .1 \ln .1].$$

Using a calculator, we find

$$\ln 2 = .6931, \quad \ln .9 = -.1054, \quad \text{and} \quad \ln .1 = -2.3026.$$

Therefore,

$$H \approx \frac{-1}{.6931} [(.9)(-.1054) + (.1)(-2.3026)]$$

$$\approx .469.$$

Verify that $H \approx .971$ if there are 60 of one species and 40 of the other, and that $H = 1$ if there are 50 of each species. As the index of diversity gets close to 1, there is more diversity, with perfect diversity equal to 1. ◀

13.2 Exercises

Write each exponential equation in logarithmic form.

1. $2^3 = 8$

2. $5^2 = 25$

3. $3^4 = 81$

4. $6^3 = 216$

5. $\left(\dfrac{1}{3}\right)^{-2} = 9$

6. $\left(\dfrac{3}{4}\right)^{-2} = \dfrac{16}{9}$

Write each logarithmic equation in exponential form.

7. $\log_2 128 = 7$

8. $\log_3 81 = 4$

9. $\log_{25} \dfrac{1}{25} = -1$

10. $\log_2 \dfrac{1}{8} = -3$

11. $\log 10,000 = 4$

12. $\log .00001 = -5$

Evaluate each logarithm.

13. $\log_5 25$

14. $\log_9 81$

15. $\log_4 64$

16. $\log_6 216$

17. $\log_2 \dfrac{1}{4}$

18. $\log_3 \dfrac{1}{27}$

19. $\log_2 \sqrt[3]{\dfrac{1}{4}}$

20. $\log_8 \sqrt[4]{\dfrac{1}{2}}$

21. $\ln e$

22. $\ln e^2$

23. $\ln e^{5/3}$

24. $\ln 1$

25. Is the logarithm to the base 3 of 4 written as $\log_4 3$ or $\log_3 4$?

26. Write a few sentences describing the relationship between e^x and $\ln x$.

Graph each of the following.

27. $y = \log_4 x$

28. $y = 1 - \log x$

29. $y = \log_{1/3} (x - 1)$

30. $y = \log_{1/2} (1 + x)$

31. $y = \ln x^2$

32. $y = \ln |x|$

Use the properties of logarithms to write each expression as a sum, difference, or product.

33. $\log_9 (7m)$

34. $\log_5 (8p)$

35. $\log_3 \dfrac{3p}{5k}$

36. $\log_7 \dfrac{11p}{13y}$

37. $\log_3 \dfrac{5\sqrt{2}}{\sqrt[4]{7}}$

38. $\log_2 \dfrac{9\sqrt[3]{5}}{\sqrt[4]{3}}$

Suppose $\log_b 2 = a$ and $\log_b 3 = c$. Use the properties of logarithms to find each of the following.

39. $\log_b 8$

40. $\log_b 24$

41. $\log_b (72b)$

42. $\log_b (4b^2)$

Use natural logarithms to evaluate each logarithm to the nearest hundredth.

43. $\log_5 20$

44. $\log_{12} 170$

45. $\log_{1.2} 5.5$

46. $\log_{2.8} .12$

47. $\log_3 1.1^{-2.4}$

48. $\log_2 4.5^{-.8}$

Solve each equation in Exercises 49–63. Round decimal answers to the nearest hundredth.

49. $\log_x 25 = -2$

50. $\log_9 27 = m$

51. $\log_8 4 = z$

52. $\log_y 8 = \dfrac{3}{4}$

53. $\log_r 7 = \dfrac{1}{2}$

54. $\log_3 (5x + 1) = 2$

55. $\log_5 (9x - 4) = 1$

56. $\log_4 x - \log_4 (x + 3) = -1$

57. $\log_9 m - \log_9 (m - 4) = -2$

58. $3^x = 5$

59. $4^x = 12$

60. $e^{k-1} = 4$

61. $e^{2y} = 12$

62. $2e^{5a+2} = 8$

63. $10e^{3z-7} = 5$

64. Prove: $\log_a \dfrac{x}{y} = \log_a x - \log_a y$

65. Prove: $\log_a x^r = r \log_a x$

Applications

Business and Economics

66. Inflation Assuming annual compounding, find the time it would take for the general level of prices in the economy to double at the following average inflation rates.

(a) 3% (b) 6% (c) 8%

(d) Check your answers using either the rule of 70 or the rule of 72, whichever applies.

67. Interest Laura Stowe invests $15,000 in an account paying 6% per year compounded semiannually.

(a) In how many years will the compounded amount double?

(b) In how many years will the amount triple?

(c) Check your answer to part (a) using the rule of 72.

Life Sciences

Index of Diversity *For Exercises 68 and 69, refer to Example 10.*

68. Suppose a sample of a small community shows two species with 50 individuals each. Find the index of diversity H.

69. A virgin forest in northwestern Pennsylvania has 4 species of large trees with the following proportions of each: hemlock, .521; beech, .324; birch, .081; maple, .074. Find the index of diversity H.

70. Growth of a Species The population of an animal species that is introduced into a certain area may grow rapidly at first but then grow more slowly as time goes on. A logarithmic function can provide an excellent model of such growth. Suppose that the population of foxes $F(t)$ in an area t yr after the foxes were introduced there is

$$F(t) = 50 \ln (2t + 3).$$

Find the population of foxes at times (a) through (c) below.

(a) When they are first released into the area (that is, when $t = 0$)

(b) After 3 yr

(c) After 15 yr

(d) Graph $y = F(t)$.

(e) When will the population double?

The graph for Exercise 71 is plotted on a logarithmic scale where differences between successive measurements are not always the same. Data that do not plot in a linear pattern on the usual Cartesian axes often form a linear pattern when plotted on a logarithmic scale. Notice that on the vertical scale, the distance from 1 to 2 is not the same as the distance from 2 to 3, and so on. This is characteristic of a graph drawn with logarithmic scales.*

71. **Oxygen Consumption** The accompanying graph gives the rate of oxygen consumption for resting guinea pigs of various sizes. This rate is proportional to body mass raised to the power .67. Estimate the oxygen consumption for a guinea pig with body mass of .3 kg. Do the same for one with body mass of .7 kg.

Social Sciences

72. **Evolution of Languages** The number of years $N(r)$ since two independently evolving languages split off from a common ancestral language is approximated by

 $$N(r) = -5000 \ln r,$$

 where r is the proportion of the words from the ancestral language that are common to both languages now. Find each of the following.

 (a) $N(.9)$ (b) $N(.5)$ (c) $N(.3)$

 (d) How many years have elapsed since the split if 70% of the words of the ancestral language are common to both languages today?

 (e) If two languages split off from a common ancestral language about 1000 years ago, find r.

Physical Sciences

For Exercises 73 and 74, recall that log x represents the common (base 10) logarithm of x.

73. **Intensity of Sound** The loudness of sounds is measured in a unit called a *decibel*. To do this, a very faint sound, called the *threshold sound*, is assigned an intensity I_0. If a particular sound has intensity I, then the decibel rating of this louder sound is

 $$10 \log \frac{I}{I_0}.$$

 Find the decibel ratings of the following sounds having intensities as given. Round answers to the nearest whole number.

 (a) Whisper, $115I_0$

 (b) Busy street, $9,500,000I_0$

 (c) Heavy truck, 20 m away, $1,200,000,000I_0$

 (d) Rock music, $895,000,000,000I_0$

 (e) Jetliner at takeoff, $109,000,000,000,000I_0$

 (f) A city council is considering whether outdoor amplified sound should be limited to 96 decibels or 98 decibels. What is the difference in intensity between these two decibel ratings?

74. **Earthquake Intensity** The intensity of an earthquake, measured on the Richter scale, is given by

 $$R(I) = \log \frac{I}{I_0},$$

 where I_0 is the intensity of an earthquake of a certain (small) size. Find the Richter scale ratings of earthquakes with the following intensities.

 (a) $1,000,000I_0$ (b) $100,000,000I_0$

 (c) The San Francisco earthquake of 1906 had a Richter scale rating of 8.3. Express the intensity of this earthquake as a multiple of I_0.

 (d) In 1989, the San Francisco region experienced an earthquake with a Richter scale rating of 7.1. Express the intensity of this earthquake as a multiple of I_0.

 (e) Compare the intensities of these two earthquakes.

75. **Acidity of a Solution** A common measure for the acidity of a solution is its pH. It is defined by pH $= -\log[H^+]$, where H^+ measures the concentration of hydrogen ions in the solution. The pH of pure water is 7. Solutions that are

more acidic than pure water have a lower pH, while solutions that are less acidic (referred to as basic solutions) have a higher pH.

(a) Acid rain sometimes has a pH as low as 4. How much greater is the concentration of hydrogen ions in such rain than in pure water?

(b) A typical mixture of laundry soap and water for washing clothes has a pH of about 11, while black coffee has a pH of about 5. How much greater is the concentration of hydrogen ions in black coffee than in the laundry?

Computer/Graphing Calculator

Make a quick sketch of each function using the information from this section and what you learned in Chapter 1 about general graphing techniques. Then use a computer or a graphing calculator to graph each function, and compare it with your sketch.

76. $y = \log_2(-x)$ **77.** $y = -\log_2 x$ **78.** $y = 2 + \ln(x - 1)$ **79.** $y = 30 \ln(2x + 1)$

13.3 APPLICATIONS: GROWTH AND DECAY; MATHEMATICS OF FINANCE

What interest rate will cause $5000 to grow to $7250 in 4 years if money is compounded continuously?

This is one of many situations that occur in biology, economics, and the social sciences, in which a quantity changes at a rate proportional to the amount of the quantity present. In such cases the amount present at time t is a function of t called the **exponential growth and decay function.** (The derivation of this equation is presented in a later section on differential equations.)

EXPONENTIAL GROWTH AND DECAY FUNCTION

Let y_0 be the amount or number of some quantity present at time $t = 0$. Then, under certain conditions, the amount present at any time t is given by

$$y = y_0 e^{kt},$$

where k is a constant.

If $k > 0$, then k is called the **growth constant;** if $k < 0$, then k is called the **decay constant.** A common example is the growth of bacteria in a culture. The more bacteria present, the faster the population increases.

EXAMPLE 1 Yeast in a sugar solution is growing at a rate such that 1 gram becomes 1.5 grams after 20 hours. Find the growth function, assuming exponential growth.

The values for y_0 and k in the exponential growth function $y = y_0 e^{kt}$ must be found. Since y_0 is the amount present at time $t = 0$, $y_0 = 1$. To find k, substitute $y = 1.5$, $t = 20$, and $y_0 = 1$ into the equation.

$$y = y_0 e^{kt}$$
$$1.5 = 1 \ e^{k(20)}$$

Now take natural logarithms on both sides and use the power rule for logarithms and the fact that $\ln e = 1$.

$$1.5 = e^{20k}$$

$$\ln 1.5 = \ln e^{20k}$$

$$\ln 1.5 = 20k \ln e \qquad \text{Power rule}$$

$$\ln 1.5 = 20k \qquad \text{ln } e = 1$$

$$\frac{\ln 1.5}{20} = k$$

$$k \approx .02 \text{ (to the nearest hundredth)}$$

The exponential growth function is $y = e^{.02t}$, where y is the number of grams of yeast present after t hours. ◀

The decline of a population or decay of a substance may also be described by the exponential growth function. In this case the decay constant k is negative, since an increase in time leads to a decrease in the quantity present. Radioactive substances provide a good example of exponential decay. By definition, the **half-life** of a radioactive substance is the time it takes for exactly half of the initial quantity to decay.

EXAMPLE **2** ▶ Carbon 14 is a radioactive form of carbon that is found in all living plants and animals. After a plant or animal dies, the carbon 14 disintegrates. Scientists determine the age of the remains by comparing its carbon 14 with the amount found in living plants and animals. The amount of carbon 14 present after t years is given by the exponential equation

$$A(t) = A_0 e^{kt},$$

with $k = -[(\ln 2)/5600]$.

(a) Find the half-life of carbon 14.

Let $A(t) = (1/2)A_0$ and $k = -[(\ln 2)/5600]$.

$$\frac{1}{2}A_0 = A_0 e^{-[(\ln 2)/5600]t}$$

$$\frac{1}{2} = e^{-[(\ln 2)/5600]t} \qquad \text{Divide by } A_0.$$

$$\ln \frac{1}{2} = \ln e^{-[(\ln 2)/5600]t} \qquad \text{Take logarithms of both sides.}$$

$$\ln \frac{1}{2} = -\frac{\ln 2}{5600}t \qquad \text{ln } e^x = x$$

$$-\frac{5600}{\ln 2} \ln \frac{1}{2} = t \qquad \text{Multiply by } -\frac{5600}{\ln 2}.$$

$$-\frac{5600}{\ln 2}(\ln 1 - \ln 2) = t \qquad \text{Quotient rule}$$

$$-\frac{5600}{\ln 2}(-\ln 2) = t \qquad \ln 1 = 0$$

$$5600 = t$$

The half-life is 5600 years.

(b) Charcoal from an ancient fire pit on Java had ¼ the amount of carbon 14 found in a living sample of wood of the same size. Estimate the age of the charcoal.

Let $A(t) = \frac{1}{4}A_0$ and $k = -[(\ln 2)/5600]$.

$$\frac{1}{4}A_0 = A_0 e^{-[(\ln 2)/5600]t}$$

$$\frac{1}{4} = e^{-[(\ln 2)/5600]t}$$

$$\ln \frac{1}{4} = \ln e^{-[(\ln 2)/5600]t}$$

$$\ln \frac{1}{4} = -\frac{\ln 2}{5600}t$$

$$-\frac{5600}{\ln 2}\ln \frac{1}{4} = t$$

$$t = 11,200$$

The charcoal is about 11,200 years old. ◄

Continuous Compounding In economics, the formula for **continuous compounding** is a good example of an exponential growth function. As interest is compounded more and more often, the compound amount A approaches

Recall the formula for compound amount $A = P\left(1 + \frac{r}{m}\right)^{tm}$ given in the first section of this chapter.

$$\lim_{m \to \infty} P\left(1 + \frac{r}{m}\right)^{tm}.$$

For example, if $P = 1$ and $t = 1$,

$$A = \left(1 + \frac{r}{m}\right)^{m}.$$

Let $r/m = 1/s$ so that $m = sr$; then

$$\lim_{m \to \infty}\left(1 + \frac{r}{m}\right)^{m} = \lim_{s \to \infty}\left(1 + \frac{1}{s}\right)^{sr}$$

$$= \lim_{s \to \infty}\left[\left(1 + \frac{1}{s}\right)^{s}\right]^{r}$$

$$= \left[\lim_{s \to \infty}\left(1 + \frac{1}{s}\right)^{s}\right]^{r}.$$

In Section 1 of this chapter we saw that

$$\lim_{m \to \infty} \left(1 + \frac{1}{m} \right)^{m} = e.$$

Using s as the variable instead of m gives

$$\lim_{s \to \infty} \left(1 + \frac{1}{s} \right)^{s} = e.$$

Thus,

$$\left[\lim_{s \to \infty} \left(1 + \frac{1}{s} \right)^{s} \right]^{r} = e^{r},$$

and so

$$\lim_{m \to \infty} \left(1 + \frac{r}{m} \right)^{m} = e^{r}.$$

At this time, you may want to review the rules for limits at the beginning of the chapter on the derivative.

Finally,

$$\lim_{m \to \infty} P \left(1 + \frac{r}{m} \right)^{mt} = P \lim_{m \to \infty} \left(1 + \frac{r}{m} \right)^{mt} \qquad \text{Limit rule 1}$$

$$= P \left[\lim_{m \to \infty} \left(1 + \frac{r}{m} \right)^{m} \right]^{t} \qquad \text{Limit rule 6}$$

$$= P(e^{r})^{t} \qquad \begin{array}{l} \text{Substitute from} \\ \text{the work above.} \end{array}$$

$$= Pe^{rt}.$$

This discussion is summarized as follows.

CONTINUOUS COMPOUNDING

If a deposit of P dollars is invested at a rate of interest r compounded continuously for t years, the compound amount is

$$A = Pe^{rt} \text{ dollars.}$$

Note This is the same equation as the general exponential growth equation given at the beginning of this section. Here, $y = A$, $y_0 = P$, and $k = r$.

EXAMPLE 3

Assuming continuous compounding, if the inflation rate averaged 6% per year for 5 years, how much would a $1 item cost at the end of the 5 years?

In the formula for continuous compounding, let $P = 1$, $t = 5$, and $r = .06$ to get

$$A = 1 \, e^{5(.06)} = e^{.3} \approx 1.34986.$$

An item that cost $1 at the beginning of the 5-year period would cost $1.35 at the end of the period, an increase of 35%, or about 1/3. ◀

Effective Rate We could use a calculator to see that \$1 at 8% interest (per year) compounded semiannually is $1(1.04)^2 = 1.0816$ or \$1.0816. The actual increase of \$.0816 is 8.16% rather than the 8% that would be earned with interest compounded annually. To distinguish between these two amounts, 8% (the annual interest rate) is called the **nominal** or **stated** interest rate, and 8.16% is called the **effective** interest rate. We will continue to use r to designate the stated rate and we will use r_E for the effective rate.

EFFECTIVE RATE COMPOUND INTEREST	If r is the annual stated rate of interest and m is the number of compounding periods per year, the effective rate of interest is $$r_E = \left(1 + \frac{r}{m}\right)^m - 1.$$

Effective rate is sometimes called *annual yield*.

 With continuous compounding, \$1 at 8% for 1 year becomes $(1)e^{1(.08)} = e^{.08} = 1.0833$. The increase is 8.33% rather than 8%, so a stated interest rate of 8% produces an effective rate of 8.33%.

EFFECTIVE RATE CONTINUOUS COMPOUNDING	If interest is compounded continuously at an annual stated rate of r, the effective rate of interest is $$r_E = e^r - 1.$$

EXAMPLE 4 Find the effective rate corresponding to the following stated rates.

(a) 6% compounded quarterly

 Using the formula, we get

$$\left(1 + \frac{.06}{4}\right)^4 - 1 = (1.015)^4 - 1 = .0614.$$

The effective rate is 6.14%.

(b) 6% compounded continuously

 The formula for continuous compounding gives

$$e^{.06} - 1 = .0618,$$

so the effective rate is 6.18%. ◀

 The formula for interest compounded m times a year, $A = P(1 + r/m)^{tm}$, has five variables: A, P, r, m, and t. If the values of any four are known, then the value of the fifth can be found. In particular, if A, the amount of money we wish to end up with, is given as well as r, m, and t, then P can be found. Here P is the amount that should be deposited today to produce A dollars in t years. The amount P is called the **present value** of A dollars.

EXAMPLE 5

Ed Calvin has a balloon payment of $100,000 due in 3 years. What is the present value of that amount if the money earns interest at 12% annually?

Here P in the compound interest formula is unknown, with $A = 100,000$, $r = .12$, $t = 3$, and $m = 1$. Substitute the known values into the formula to get $100,000 = P(1.12)^3$. Solve for P using a calculator to find $(1.12)^3$.

$$P = \frac{100,000}{(1.12)^3} = 71,178.02$$

The present value of $100,000 in 3 years at 12% per year is $71,178.02. ◄

The equation in Example 5 could have been solved as follows.

$$100,000 = P(1.12)^3$$

Multiply both sides by $(1.12)^{-3}$ to solve for P, so

$$P = 100,000(1.12)^{-3}.$$

This suggests a general formula for present value.

PRESENT VALUE

The present value of A dollars at a rate of interest r compounded m times per year for t years is

$$P = A\left(1 + \frac{r}{m}\right)^{-tm}.$$

The equation $A = Pe^{rt}$ also can be solved for any of the variables A, P, r, or t, as the following example shows.

EXAMPLE 6

Find the interest rate that will cause $5000 to grow to $7250 in 4 years if the money is compounded continuously.

Use the formula for continuous compounding, $A = Pe^{rt}$, with $A = 7250$, $P = 5000$, and $t = 4$. Solve first for e^{rt}, then for r.

$$A = Pe^{rt}$$
$$7250 = 5000e^{4r}$$
$$1.45 = e^{4r} \qquad \text{Divide by 5000.}$$
$$\ln 1.45 = \ln e^{4r} \qquad \text{Take logarithms of both sides.}$$
$$\ln 1.45 = 4r \qquad \ln e^x = x$$
$$r = \frac{\ln 1.45}{4}$$
$$r \approx .093$$

The required interest rate is 9.3%. ◄

Limited Growth Functions The exponential growth functions discussed so far all continued to grow without bound. More realistically, many populations grow exponentially for a while, but then the growth is slowed by some external constraint that eventually limits the growth. For example, an animal population may grow to the point where its habitat can no longer support the population and the growth rate begins to dwindle until a stable population size is reached. Models that reflect this pattern are called **limited growth functions.** The next example discusses a function of this type that occurs in industry.

EXAMPLE 7

Assembly-line operations tend to have a high turnover of employees, forcing companies to spend much time and effort in training new workers. It has been found that a worker new to a task on the line will produce items according to the function defined by

$$P(x) = 25 - 25e^{-.3x},$$

where $P(x)$ items are produced by the worker on day x.

(a) What is the limit on the number of items a worker on this assembly line can produce?

Since the exponent on e is negative, write the function as

$$P(x) = 25 - \frac{25}{e^{.3x}}.$$

As $x \to \infty$, the term $(25/e^{.3x}) \to 0$, so $P(x) \to 25$. The limit is 25 items.

(b) How many days will it take for a new worker to produce 20 items?

Let $P(x) = 20$ and solve for x.

$$P(x) = 25 - 25e^{-.3x}$$
$$20 = 25 - 25e^{-.3x}$$
$$-5 = -25e^{-.3x}$$
$$.2 = e^{-.3x}$$

Now take natural logarithms of both sides and use properties of logarithms.

$$\ln .2 = \ln e^{-.3x}$$
$$\ln .2 = -.3x \qquad \ln e^u = u$$
$$x = \frac{\ln .2}{-.3} \approx 5.4$$

In about 5½ days on the job a new worker will be producing 20 items. A graph of the function P is shown in Figure 9. ◀

Graphs such as the one in Figure 9 are called *learning curves*. According to such a graph, a new worker tends to learn quickly at first; then learning tapers off and approaches some upper limit. This is characteristic of the learning of certain types of skills involving the repetitive performance of the same task.

$P(x)$

25

$P(x) = 25 - 25e^{-.3x}$

5 10 x

FIGURE 9

13.3 Exercises

Find the effective rate corresponding to each of the following nominal rates of interest.

1. 5% compounded monthly

2. 15% compounded quarterly

3. 10% compounded semiannually

4. 18% compounded monthly

5. 11% compounded continuously

6. 7% compounded continuously

Find the present value of each deposit.

7. $2000 at 6% compounded semiannually for 11 yr

8. $10,000 if interest is 12% compounded semiannually for 5 yr

9. $10,000 at 10% compounded quarterly for 8 yr

10. $45,678.93 if interest is 12.6% compounded monthly for 11 mo

11. $7300 at 11% compounded continuously for 3 yr

12. $25,000 at 9% compounded continuously for 8 yr

13. What is the difference between stated interest rate and effective interest rate?

14. In the exponential growth or decay function $y = y_0 e^{kt}$, what does y_0 represent? What does k represent?

15. In the exponential growth or decay function, explain the circumstances that cause k to be positive or negative.

16. What is meant by the half-life of a quantity?

17. Show that if a radioactive substance has a half-life of T, then the corresponding constant k in the exponential decay function is given by $k = -(\ln 2)/T$.

18. Show that if a radioactive substance has a half-life of T, then the corresponding exponential decay function can be written as $y = y_0 (1/2)^{(t/T)}$.

Applications

Business and Economics

19. Inflation Assuming continuous compounding, what will it cost to buy a $10 item in 3 yr at the following inflation rates?

(a) 3% (b) 4% (c) 5%

20. Interest Bert Bezzone invests a $25,000 inheritance in a fund paying 9% per year compounded continuously. What will be the amount on deposit after each of the following time periods?

(a) 1 yr (b) 5 yr (c) 10 yr

21. Interest Linda Youngman, who is self-employed, wants to invest $60,000 in a pension plan. One investment offers 10% compounded quarterly. Another offers 9.75% compounded continuously.

(a) Which investment will earn the most interest in 5 yr?

(b) How much more will the better plan earn?

(c) What is the effective rate in each case?

(d) If Ms. Youngman chooses the plan with continuous compounding, how long will it take for her $60,000 to grow to $80,000?

22. Effective Rate Janet Tilden bought a television set with money borrowed from the bank at 9% interest compounded semiannually. What effective interest rate did she pay?

23. Effective Rate A firm deposits some funds in a special account at 7.2% compounded quarterly. What effective rate will they earn?

24. Effective Rate Virginia Nicolai deposits $7500 of lottery winnings in an account paying 6% interest compounded monthly. What effective rate does the account earn?

25. Present Value Matt Harr must make a balloon payment of $20,000 in 4 yr. Find the present value of the payment if it includes annual interest of 8%.

26. Present Value A company must pay a $307,000 settlement in 3 yr.

(a) What amount must be deposited now at 6% semiannually to have enough money for the settlement?

(b) How much interest will be earned?

(c) Suppose the company can deposit only $200,000 now. How much more will be needed in 3 yr?

27. Present Value A couple wants to have $20,000 in 5 yr for a down payment on a new house.

(a) How much could they deposit today, at 8% compounded quarterly, to have the required amount in 5 yr?

(b) How much interest will be earned?

(c) If they can deposit only $10,000 now, how much more will they need to complete the $20,000 after 5 yr?

28. Sales Experiments have shown that sales of a product, under relatively stable market conditions, but in the absence of promotional activities such as advertising, tend to decline at a constant yearly rate. This sales decline, which varies considerably from product to product, can often be expressed by an exponential function of the form

$$S(t) = S_0 e^{-at},$$

where $S(t)$ is the sales at time t measured in years, S_0 is sales at time $t = 0$, and a is the sales decay constant. Suppose a certain product had sales of 50,000 at the beginning of the year and 45,000 at the end of the year.

(a) Write an equation for the sales decline.

(b) Find $S(2)$.

(c) How long will it take for sales to decrease to 40,000?

(d) Is there a limit to the sales decline? If so, what is it?

29. Sales Sales of a new model of compact disc player are approximated by the function $S(x) = 1000 - 800e^{-x}$,

where $S(x)$ is in appropriate units and x represents the number of years the disc player has been on the market.

(a) Find the sales during year 0.

(b) In how many years will sales reach 500 units?

(c) Will sales ever reach 1000 units?

(d) Is there a limit on sales for this product? If so, what is it?

30. Sales Sales of a new model of word processor are approximated by

$$S(x) = 5000 - 4000e^{-x},$$

where x represents the number of years that the word processor has been on the market, and $S(x)$ represents sales in thousands.

(a) Find the sales in year 0.

(b) When will sales reach $4,500,000?

(c) Find the limit on sales.

Life Sciences

31. Growth of Lice A population of 100 lice is growing exponentially. After 2 mo the population has increased to 125.

(a) Write an exponential equation to express the exponential growth function y in terms of time t in months.

(b) How long will it take for the population to reach 500?

32. Growth of Bacteria A culture contains 25,000 bacteria, with the population increasing exponentially. The culture contains 40,000 bacteria after 10 hr.

(a) Write an exponential equation to express the growth function y in terms of time t in hours.

(b) How long will it be until there are 60,000 bacteria?

33. Decrease in Bacteria When a bactericide is introduced into a culture of 50,000 bacteria, the number of bacteria decreases exponentially. After 9 hr, there are only 20,000 bacteria.

(a) Write an exponential equation to express the decay function y in terms of time t in hours.

(b) In how many hours will half the number of bacteria remain?

34. Growth of Bacteria The growth of bacteria in food products makes it necessary to time-date some products (such as milk) so that they will be sold and consumed before the bacteria count is too high. Suppose for a certain product that the number of bacteria present is given by

$$f(t) = 500e^{.1t},$$

under certain storage conditions, where t is time in days after packing of the product and the value of $f(t)$ is in millions.

(a) If the product cannot be safely eaten after the bacteria count reaches 3,000 million, low long will this take?

(b) If $t = 0$ corresponds to January 1, what date should be placed on the product?

35. Limited Growth Another limited growth function is the *logistic function*

$$G(t) = \frac{mG_0}{G_0 + (m - G_0)e^{-kmt}},$$

where G_0 is the initial number present, m is the maximum possible size of the population, and k is a positive constant. Assume $G_0 = 1000$, $m = 2500$, $k = .0004$, and t is time in decades (10-year periods).

(a) Find $G(.2)$. **(b)** Find $G(1)$. **(c)** Find $G(3)$.

(d) At what time t will the population reach 2000?

36. Population Growth A population of fruit flies, contained in a large glass jar, is growing according to the model $y = 300 - 290e^{-.693t}$, where y is the population at time t in days.

(a) Find the initial population.

(b) Find the limit of the population.

(c) When will the population reach 150?

37. Cancer Research An article on cancer treatment contains the following statement: A 37% five-year survival rate for women with ovarian cancer yields an estimated annual mortality rate of 0.1989.* The authors of this article assume that the number of survivors is described by the exponential decay function given at the beginning of this chapter, where y is the number of survivors and k is the mortality rate. Verify that the given survival rate leads to the given mortality rate.

Social Sciences

38. Skills Training The number of words per minute that an average typist can type is given by

$$W(t) = 60 - 30e^{-.5t},$$

where t is time in months after the beginning of a typing class.
Find each of the following.

(a) $W_0 = W(0)$ **(b)** $W(1)$ **(c)** $W(4)$

(d) When will the average typist type 45 words per minute?

(e) What is the upper limit on the number of words that can be typed according to this model?

(f) Graph $W(t)$.

39. Skills Training Assuming that a person new to an assembly line will produce

$$P(x) = 500 - 500e^{-x}$$

items per day, where x is time measured in days, find each of the following.

(a) $P_0 = P(0)$ **(b)** $P(2)$ **(c)** $P(5)$

(d) When will a new person produce 400 items?

(e) Find the limit on the number of items produced per day.

(f) Graph $P(x)$.

40. Spread of a Rumor A sociologist has shown that the fraction $y(t)$ of people who have heard a rumor after t days is approximated by

$$y(t) = \frac{y_0 e^{kt}}{1 - y_0(1 - e^{kt})},$$

where y_0 is the fraction of people who have heard the rumor at time $t = 0$, and k is a constant. A graph of $y(t)$ for a particular value of k is shown in the figure.

(a) If $k = .1$ and $y_0 = .05$, find $y(10)$.

(b) If $k = .2$ and $y_0 = .10$, find $y(5)$.

(c) If $k = .1$ and $y_0 = .02$, find the number of days until half the people have heard the rumor.

(d) What happens to the fraction of people who have heard the rumor as t gets larger and larger? What does this mean? (*Hint:* Look at the graph.)

Physical Science

41. Carbon Dating Refer to Example 2. A sample from a refuse deposit near the Strait of Magellan had 60% of the carbon 14 found in a contemporary living sample. How old was the sample?

*Theodore Speroff et al., "A risk-benefit analysis of elective bilateral oophorectomy: effect of changes in compliance with estrogen therapy on outcome," *American Journal of Obstetrics and Gynecology*, vol. 164 (January 1991), pp. 165–74.

Half-life *Find the half-life of the following radioactive substances. See Example 2.*

42. Plutonium 241; $A(t) = A_0e^{-.053t}$

43. Radium 226; $A(t) = A_0e^{-.00043t}$

44. Iodine 131; $A(t) = A_0e^{-.087t}$

45. **Nuclear Energy** Nuclear energy derived from radioactive isotopes can be used to supply power to space vehicles. The output of the radioactive power supply for a certain satellite is given by the function $y = 40e^{-.004t}$, where y is in watts and t is the time in days.

 (a) How much power will be available at the end of 180 days?

 (b) How long will it take for the amount of power to be half of its original strength?

 (c) Will the power ever be completely gone? Explain.

46. **Decay of Radioactivity** A large cloud of radioactive debris from a nuclear explosion has floated over the Pacific Northwest, contaminating much of the hay supply. Consequently, farmers in the area are concerned that the cows who eat this hay will give contaminated milk. (The tolerance level for radioactive iodine in milk is 0.) The percent of the initial amount of radioactive iodine still present in the hay after t days is approximated by $P(t)$, which is given by the mathematical model

$$P(t) = 100e^{-.1t}.$$

 (a) Find the percent remaining after 4 days.

 (b) Find the percent remaining after 10 days.

 (c) Some scientists feel that the hay is safe after the percent of radioactive iodine has declined to 10% of the original amount. Solve the equation $10 = 100e^{-.1t}$ to find the number of days before the hay may be used.

 (d) Other scientists believe that the hay is not safe until the level of radioactive iodine has declined to only 1% of the original level. Find the number of days that this would take.

47. **Chemical Dissolution** The amount of chemical that will dissolve in a solution increases exponentially as the temperature is increased. At 0° C 10 g of the chemical dissolves, and at 10° C 11 g dissolves.

 (a) Write an equation to express the amount of chemical dissolved, y, in terms of temperature, t, in degrees Celsius.

 (b) At what temperature will 15 g dissolve?

Newton's Law of Cooling *Newton's law of cooling says that the rate at which a body cools is proportional to the difference in temperature between the body and an environment into which it is introduced. This leads to an equation where the temperature $f(t)$ of the body at time t after being introduced into an environment having constant temperature T_0 is*

$$f(t) = T_0 + Ce^{-kt},$$

where C and k are constants. Use this result in Exercises 48–50.

48. Find the temperature of an object when $t = 9$ if $T_0 = 18$, $C = 5$, and $k = .6$.

49. If $C = 100$, $k = .1$, and t is time in minutes, how long will it take a hot cup of coffee to cool to a temperature of 25° Celsius in a room at 20° Celsius?

50. If $C = -14.6$ and $k = .6$ and t is time in hours, how long will it take a frozen pizza to thaw to 10° Celsius in a room at 18° Celsius?

13.4 DERIVATIVES OF LOGARITHMIC FUNCTIONS

If the number of births per female in an animal population varies with the mother's age, at what age does the maximum number of births occur?

As we saw in the chapter on applications of the derivative, extrema occur when the derivative of a function does not exist or equals zero. In this section formulas will be developed for the derivatives of $y = \ln x$ and other logarithmic functions that will allow us to answer the question above.

To find the derivative of the function $y = \ln x$, assuming $x > 0$, go back to the definition of the derivative given earlier: if $y = f(x)$, then

$$y' = f'(x) = \lim_{h \to 0} \frac{f(x + h) - f(x)}{h},$$

provided this limit exists. (Remember that h represents the variable in this expression, while x is constant.)

Here, $f(x) = \ln x$, with

$$y' = f'(x) = \lim_{h \to 0} \frac{\ln (x + h) - \ln x}{h}.$$

This limit can be found using properties of logarithms. By the quotient rule for logarithms,

$$y' = \lim_{h \to 0} \frac{\ln (x + h) - \ln x}{h} = \lim_{h \to 0} \frac{\ln \left(\dfrac{x + h}{x} \right)}{h} = \lim_{h \to 0} \left[\frac{1}{h} \ln \left(\frac{x + h}{x} \right) \right].$$

The power rule for logarithms is used to get

$$y' = \lim_{h \to 0} \ln \left(\frac{x + h}{x} \right)^{1/h} = \lim_{h \to 0} \ln \left(1 + \frac{h}{x} \right)^{1/h}.$$

Now make a substitution: let $m = x/h$, so that $h/x = 1/m$. As $h \to 0$, with x fixed, $m \to \infty$ (since $x > 0$), so

$$\lim_{h \to 0} \ln \left(1 + \frac{h}{x} \right)^{1/h} = \lim_{m \to \infty} \ln \left(1 + \frac{1}{m} \right)^{m/x}$$

$$= \lim_{m \to \infty} \ln \left[\left(1 + \frac{1}{m} \right)^m \right]^{1/x}.$$

By the power rule for logarithms, this becomes

$$= \lim_{m \to \infty} \left[\frac{1}{x} \cdot \ln \left(1 + \frac{1}{m} \right)^m \right].$$

Since x is fixed here (a constant), by the product rule for limits, this last limit becomes

$$\lim_{m \to \infty} \left[\frac{1}{x} \cdot \ln \left(1 + \frac{1}{m} \right)^m \right] = \left(\lim_{m \to \infty} \frac{1}{x} \right) \lim_{m \to \infty} \left[\ln \left(1 + \frac{1}{m} \right)^m \right]$$

$$= \frac{1}{x} \cdot \lim_{m \to \infty} \left[\ln \left(1 + \frac{1}{m} \right)^m \right].$$

Since the logarithmic function is continuous,

$$\frac{1}{x} \cdot \lim_{m \to \infty} \left[\ln \left(1 + \frac{1}{m} \right)^m \right] = \frac{1}{x} \cdot \ln \left[\lim_{m \to \infty} \left(1 + \frac{1}{m} \right)^m \right].$$

In Section 1 of this chapter we saw that the limit on the right equals e, so that

$$\frac{1}{x} \cdot \ln \left[\lim_{m \to \infty} \left(1 + \frac{1}{m} \right)^m \right] = \frac{1}{x} \ln e = \frac{1}{x} \cdot 1 = \frac{1}{x}.$$

Summarizing, if $y = \ln x$, then $y' = 1/x$.

 EXAMPLE 1

Find the derivative of $y = \ln 6x$. Assume $x > 0$.
Use the properties of logarithms and the rules for derivatives.

$$y' = \frac{d}{dx} (\ln 6x)$$

$$= \frac{d}{dx} (\ln 6 + \ln x)$$

$$= \frac{d}{dx} (\ln 6) + \frac{d}{dx} (\ln x) = 0 + \frac{1}{x} = \frac{1}{x} \quad \blacktriangleleft$$

Recall the chain rule: If $y = f[g(x)]$, then $y' = f'[g(x)] \cdot g'(x)$. For review, find the following derivatives.

1. $D_x(2x^3 - 4)^5$
2. $D_x\sqrt{x^4 + 5x}$
3. $D_x(5x^2 + 3x)^{3/2}$

Answers:
1. $30x^2(2x^3 - 4)^4$
2. $\dfrac{4x^3 + 5}{2\sqrt{x^4 + 5x}}$
3. $\dfrac{3}{2}(5x^2 + 3x)^{1/2}(10x + 3)$

The chain rule can be used to find the derivative of the more general logarithmic function $y = \ln g(x)$. Let $y = f(u) = \ln u$ and $u = g(x)$, so that $f[g(x)] = \ln g(x)$. Then

$$f'[g(x)] = f'(u) = \frac{1}{u} = \frac{1}{g(x)},$$

and by the chain rule,

$$y' = f'[g(x)] \cdot g'(x)$$

$$= \frac{1}{g(x)} \cdot g'(x) = \frac{g'(x)}{g(x)}.$$

 EXAMPLE 2

Find the derivative of $y = \ln (x^2 + 1)$.
Here $g(x) = x^2 + 1$ and $g'(x) = 2x$. Thus,

$$y' = \frac{g'(x)}{g(x)} = \frac{2x}{x^2 + 1}. \quad \blacktriangleleft$$

If $y = \ln (-x)$, where $x < 0$, the chain rule with $g(x) = -x$ and $g'(x) = -1$ gives

$$y' = \frac{g'(x)}{g(x)} = \frac{-1}{-x} = \frac{1}{x}.$$

The derivative of $y = \ln (-x)$ is the same as the derivative of $y = \ln x$. For this reason, these two results can be combined into one rule using the absolute value of x. A similar situation holds true for $y = \ln [g(x)]$ and $y = \ln [-g(x)]$. These results are summarized on the next page.

**DERIVATIVES OF ln |x|
AND ln |g(x)|**

$$\frac{d}{dx} \ln |x| = \frac{1}{x}$$

$$\frac{d}{dx} \ln |g(x)| = \frac{g'(x)}{g(x)}$$

EXAMPLE 3 Find the derivative of each function.

(a) $y = \ln |5x|$

Let $g(x) = 5x$, so that $g'(x) = 5$. From the formula above,

$$y' = \frac{g'(x)}{g(x)} = \frac{5}{5x} = \frac{1}{x}.$$

Notice that the derivative of $\ln |5x|$ is the same as the derivative of $\ln |x|$. Also, in Example 1, the derivative of $\ln 6x$ was the same as that for $\ln x$. This suggests that for any constant a,

$$\frac{d}{dx} \ln |ax| = \frac{d}{dx} \ln |x|$$

$$= \frac{1}{x}.$$

Exercise 43 asks for a proof of this result.

(b) $y = \ln |3x^2 - 4x|$

$$y' = \frac{6x - 4}{3x^2 - 4x}$$

(c) $y = 3x \ln x^2$

This function is the product of the two functions $3x$ and $\ln x^2$, so use the product rule.

$$y' = (3x)\left[\frac{d}{dx} \ln x^2\right] + (\ln x^2)\left[\frac{d}{dx} 3x\right]$$

$$= 3x\left(\frac{2x}{x^2}\right) + (\ln x^2)(3)$$

$$= 6 + 3 \ln x^2$$

By the power rule for logarithms,

$$y' = 6 + \ln (x^2)^3$$

$$= 6 + \ln x^6.$$

Alternatively, write the answer as $y' = 6 + 6 \ln x$. ◄

EXAMPLE **4**

Find the derivative of $y = \log |3x + 2|$.

This is a base 10 logarithm, while the derivative rule developed above applies only to natural logarithms. To find the derivative, first use the change-of-base theorem to convert the function to one involving natural logarithms.

The change-of-base theorem was given in the second section of this chapter. It allows us to change a logarithm from one base to another by the following rule.

$$\log_a x = \frac{\log_b x}{\log_b a}$$

Use the rule to change $\log_{10} x$ to $\ln x$; that is, let $a = 10$ and $b = e$ in the rule given above.

Answer: $\log_{10} x = \dfrac{\ln x}{\ln 10}$

$$y = \log_{10} |3x + 2|$$

$$= \frac{\ln |3x + 2|}{\ln 10}$$

$$= \frac{1}{\ln 10} \ln |3x + 2|$$

Now find the derivative. (Remember: $1/(\ln 10)$ is a constant.)

$$y' = \frac{1}{\ln 10} \cdot \frac{d}{dx} [\ln |3x + 2|]$$

$$= \frac{1}{\ln 10} \cdot \frac{3}{3x + 2}$$

$$y' = \frac{3}{\ln 10 (3x + 2)} \quad \blacktriangleleft$$

The procedure in Example 4 can be used to derive the following results.

DERIVATIVES WITH OTHER BASES

For any suitable value of a,

$$D_x(\log_a |x|) = \frac{1}{\ln a} \cdot \frac{1}{x}.$$

When $y = \log_a |g(x)|$, the chain rule gives

$$D_x(\log_a |g(x)|) = \frac{1}{\ln a} \cdot \frac{g'(x)}{g(x)}.$$

EXAMPLE **5**

The proportion of births per female in a certain animal population is given by $y = (\ln x)/x^2$, where x is the age of the female in years.

(a) What is the domain of the function for this application?

Since the proportion of births must be nonnegative, and $y = 0$ when $x = 1$, the domain is $x \geq 1$.

(b) Find any extrema or inflection points and sketch the graph of $y = (\ln x)/x^2$, $x \geq 1$.

To locate any extrema, begin by finding the first derivative.

$$y' = \frac{x^2(1/x) - 2x \ln x}{x^4}$$

$$= \frac{x - 2x \ln x}{x^4}$$

$$= \frac{x(1 - 2 \ln x)}{x^4}$$

$$y' = \frac{1 - 2 \ln x}{x^3}$$

The derivative function exists everywhere on the domain ($x \geq 1$), so any extrema will occur only at points where the derivative equals 0.

$$\frac{1 - 2 \ln x}{x^3} = 0$$

$$1 - 2 \ln x = 0$$

$$2 \ln x = 1$$

$$\ln x = \frac{1}{2}$$

Use the definition of logarithm to get the equivalent exponential statement

$$x = e^{1/2} \approx 1.65.$$

To locate inflection points and determine whether $x = e^{1/2}$ represents a maximum or a minimum, find the second derivative.

$$y'' = \frac{x^3(-2/x) - 3x^2(1 - 2 \ln x)}{x^6}$$

$$= \frac{-2x^2 - 3x^2 + 6x^2 \ln x}{x^6}$$

$$= \frac{-5x^2 + 6x^2 \ln x}{x^6} = \frac{x^2(-5 + 6 \ln x)}{x^6} = \frac{-5 + 6 \ln x}{x^4}$$

Use a calculator to show that the second derivative is negative for $x = e^{1/2}$ indicating a maximum where $x = e^{1/2}$. Set the second derivative equal to 0 and solve the resulting equation to identify any inflection points.

$$\frac{-5 + 6 \ln x}{x^4} = 0$$

$$-5 + 6 \ln x = 0$$

$$6 \ln x = 5$$

$$\ln x = 5/6$$

$$x = e^{5/6} \approx 2.30$$

Verify that $f''(1) < 0$ and $f''(3) > 0$, indicating an inflection point when $x = e^{5/6} \approx 2.3$. Also, verify that the graph is concave downward for $x < e^{5/6}$ and concave upward for $x > e^{5/6}$. Use this information and plot a few points as necessary to get the graph in Figure 10.

x	y
1	0
1.7	.18
2.3	.16
4	.1

FIGURE 10

 (c) At what age does the maximum number of births occur?

We saw in part (b) that the maximum occurred at $x = 1.65$, or about age 1 year, 8 months.

(d) What does the inflection point indicate?

The inflection point shows where the rate of change of the proportion of births stops decreasing and begins increasing again. However, since the function has no minimum (except at its endpoint), the proportion of births continues to decrease. ◄

13.4 Exercises

Find the derivative of each of the functions defined as follows.

1. $y = \ln (8x)$

2. $y = \ln (-4x)$

3. $y = \ln (3 - x)$

4. $y = \ln (1 + x^2)$

5. $y = \ln |2x^2 - 7x|$

6. $y = \ln |-8x^2 + 6x|$

7. $y = \ln \sqrt{x + 5}$

8. $y = \ln \sqrt{2x + 1}$

9. $y = \ln (x^4 + 5x^2)^{3/2}$

10. $y = \ln (5x^3 - 2x)^{3/2}$

11. $y = -3x \ln (x + 2)$

12. $y = (3x + 1) \ln (x - 1)$

13. $y = x^2 \ln |x|$

14. $y = x \ln |2 - x^2|$

15. $y = \dfrac{2 \ln (x + 3)}{x^2}$

16. $y = \dfrac{\ln x}{x^3}$

17. $y = \dfrac{\ln x}{4x + 7}$

18. $y = \dfrac{-2 \ln x}{3x - 1}$

19. $y = \dfrac{3x^2}{\ln x}$

20. $y = \dfrac{x^3 - 1}{2 \ln x}$

21. $y = (\ln |x + 1|)^4$

22. $y = \sqrt{\ln |x - 3|}$

23. $y = \ln |\ln x|$

24. $y = (\ln 4)(\ln |3x|)$

25. Why do we use the absolute value of x or of $g(x)$ in the derivative formulas for the natural logarithm?

26. Which of the following is the correct derivative of $\log_2 x$?

(a) $\dfrac{1}{x}$ **(b)** $\dfrac{1}{(\ln 2)x}$ **(c)** $\dfrac{\ln 2}{x}$ **(d)** $\dfrac{2}{x}$

Find the derivative of each of the following.

27. $y = \log (6x)$

28. $y = \log (2x - 3)$

29. $y = \log |1 - x|$

30. $y = \log |-3x|$

31. $y = \log_5 \sqrt{5x + 2}$

32. $y = \log_7 \sqrt{2x - 3}$

33. $y = \log_3 (x^2 + 2x)^{3/2}$

34. $y = \log_2 (2x^2 - x)^{5/2}$

Find all relative maxima and minima for the functions defined as follows. Sketch the graphs. (Hint: In Exercises 35 and 37, $\lim\limits_{x \to 0} x \ln |x| = 0$.)

35. $y = x \ln x, \quad x > 0$

36. $y = x - \ln x, \quad x > 0$

37. $y = x \ln |x|$

38. $y = x - \ln |x|$

39. $y = \dfrac{\ln x}{x}, \quad x > 0$

40. $y = \dfrac{\ln x^2}{x^2}$

41. What is true about the slope of the tangent line to the graph of $f(x) = \ln x$ as $x \to \infty$? as $x \to 0$?

42. Let $f(x) = \ln x$.

 (a) Compute $f'(x)$, $f''(x)$, $f'''(x)$, $f^{(4)}(x)$, and $f^{(5)}(x)$.

 (b) Guess a formula for $f^{(n)}(x)$, where n is any positive integer.

43. Prove $\dfrac{d}{dx} \ln |ax| = \dfrac{d}{dx} \ln |x|$ for any constant a.

Applications

Business and Economics

44. Profit Assume that the total revenue received from the sale of x items is given by

$$R(x) = 30 \ln (2x + 1),$$

while the total cost to produce x items is $C(x) = x/2$. Find the number of items that should be manufactured so that profit, $R(x) - C(x)$, is a maximum.

45. Revenue Suppose the demand function for x units of a certain item is

$$p = 100 + \frac{50}{\ln x}, \quad x > 1,$$

where p is in dollars.

 (a) Find the marginal revenue.

 (b) Approximate the revenue from one more unit when 8 units are sold.

 (c) How might a manager use the information from part (b)?

46. Profit If the cost function in dollars for x units of the item in Exercise 45 is $C(x) = 100x + 100$, find the following.

 (a) The marginal cost **(b)** The profit function $P(x)$

 (c) The profit from one more unit when 8 units are sold

 (d) How might a manager use the information from part (c)?

47. Marginal Revenue Product The demand function for x units of a product is

$$p = 100 - 10 \ln x, \quad 1 < x < 20{,}000,$$

where $x = 6n$ and n is the number of employees producing the product.

 (a) Find the revenue function $R(x)$.

 (b) Find the marginal revenue product function. (See Example 10 in Section 6 of the chapter on the derivative.)

 (c) Evaluate and interpret the marginal revenue product when 20 units are sold.

Life Sciences

48. Insect Mating Consider an experiment in which equal numbers of male and female insects of a certain species are permitted to intermingle. Assume that

$$M(t) = (.1t + 1) \ln \sqrt{t}$$

represents the number of matings observed among the insects in an hour, where t is the temperature in degrees Celsius. (*Note:* The formula is an approximation at best and holds only for specific temperature intervals.)

 (a) Find the number of matings when the temperature is $15°$ C.

 (b) Find the number of matings when the temperature is $25°$ C.

 (c) Find the rate of change of the number of matings when the temperature is $15°$ C.

49. Population Growth Suppose that the population of a certain collection of rare Brazilian ants is given by

$$P(t) = (t + 100) \ln (t + 2),$$

where t represents the time in days. Find the rates of change of the population on the second day and on the eighth day.

General Interest

50. Information Content Suppose dots and dashes are transmitted over a telegraph line so that dots occur a fraction p of the time (where $0 \leq p \leq 1$) and dashes occur a fraction $1 - p$ of the time. The **information content** of the telegraph line is given by $I(p)$, where

$$I(p) = -p \ln p - (1 - p) \ln (1 - p).$$

(a) Show that $I'(p) = -\ln p + \ln (1 - p.)$

(b) Let $I'(p) = 0$ and find the value of p that maximizes the information content.

(c) How might the result in part (b) be used?

 Computer/Graphing Calculator

Use a computer or a graphing calculator to graph the following on the interval $(0, 9]$. Then use the graph to find any extrema.

51. $y = x^2 - \ln x$

52. $y = \ln x + \dfrac{1}{x}$

13.5 DERIVATIVES OF EXPONENTIAL FUNCTIONS

 When does the growth rate of an insect population change from an increasing to a decreasing rate?

We will use a derivative to answer this question in Example 8 at the end of this section.

The derivatives of exponential functions can be found using the formula for the derivative of the natural logarithm function. To find the derivative of the function $y = e^x$, first take the natural logarithm of each side so that the power rule for logarithms can be used to get x out of the exponent.

$$y = e^x$$
$$\ln y = \ln e^x$$

By the power rule for logarithms and the fact that $\ln e^x = x$,

$$\ln y = x.$$

Now, using implicit differentiation, take the derivative of each side. (Since y is a function of x, we use the chain rule.)

$$\frac{1}{y} \cdot \frac{dy}{dx} = 1$$

$$\frac{dy}{dx} = y$$

Since $y = e^x$,

$$\frac{d}{dx}(e^x) = e^x \quad \text{or} \quad y' = e^x.$$

This result shows one of the main reasons for the widespread use of e as a base—the function $y = e^x$ is its own derivative. Furthermore, constant multiples of this function are the *only* useful functions with this property. For a more general result, let $y = e^u$ and $u = g(x)$ so that $y = e^{g(x)}$. Then

$$\frac{dy}{du} = e^u \quad \text{and} \quad \frac{du}{dx} = g'(x).$$

By the chain rule,

$$\frac{dy}{dx} = \frac{dy}{du} \cdot \frac{du}{dx}$$
$$= e^u \cdot g'(x)$$
$$= e^{g(x)} \cdot g'(x).$$

Now replace y with $e^{g(x)}$ to get

$$\frac{d}{dx}e^{g(x)} = e^{g(x)} \cdot g'(x).$$

These results are summarized below.

DERIVATIVES OF e^x AND $e^{g(x)}$	$$\frac{d}{dx}e^x = e^x \quad \text{and} \quad \frac{d}{dx}e^{g(x)} = g'(x)e^{g(x)}$$

Caution Notice the difference between the derivative of a variable to a constant power, such as $D_x x^3 = 3x^2$, and a constant to a variable power, like $D_x e^x = e^x$. Remember, $D_x e^x \neq xe^{x-1}$.

EXAMPLE 1 Find derivatives of the following functions.

(a) $y = e^{5x}$

Let $g(x) = 5x$, with $g'(x) = 5$. Then

$$y' = 5e^{5x}.$$

(b) $y = 3e^{-4x}$

$$y' = 3 \cdot e^{-4x}(-4) = -12e^{-4x}$$

(c) $y = 10e^{3x^2}$

$$y' = 10(e^{3x^2})(6x) = 60xe^{3x^2} \blacktriangleleft$$

EXAMPLE 2

Let $y = e^x(\ln x)$. Find y'.

Use the product rule.

$$y' = e^x\left(\frac{1}{x}\right) + (\ln x)e^x$$

$$y' = e^x\left(\frac{1}{x} + \ln x\right) \quad \blacktriangleleft$$

EXAMPLE 3

Let $y = \dfrac{100,000}{1 + 100e^{-.3x}}$. Find y'.

Use the quotient rule.

$$y' = \frac{(1 + 100e^{-.3x})(0) - 100,000(-30e^{-.3x})}{(1 + 100e^{-.3x})^2}$$

$$= \frac{3,000,000e^{-.3x}}{(1 + 100e^{-.3x})^2} \quad \blacktriangleleft$$

EXAMPLE 4

Find $D_x(3^x)$.

Change 3^x to e^{kx}. Here, $k = \ln 3$ and $3^x = e^{(\ln 3)x}$.

$$D_x(3^x) = D_x e^{(\ln 3)x}$$

$$= (\ln 3)e^{(\ln 3)x}$$

$$= (\ln 3)3^x \qquad \text{Substitute } 3^x \text{ for } e^{(\ln 3)x}. \quad \blacktriangleleft$$

Recall from the second section of this chapter that

$$a^x = e^{(\ln a)x}.$$

Thus, $2^x = e^{(\ln 2)x}$, $5^x = e^{(\ln 5)x}$, and so on.

Extending this idea from Example 4 gives a general rule for the derivative of an exponential function with any appropriate base a.

$$D_x a^x = (\ln a)e^{x \ln a}$$

$$= (\ln a)e^{\ln a^x}$$

$$D_x a^x = (\ln a)a^x \qquad e^{\ln a^x} = a^x$$

Use the chain rule to get the following result.

DERIVATIVE OF $a^{g(x)}$

$$D_x[a^{g(x)}] = (\ln a)e^{(\ln a)g(x)}g'(x) = (\ln a)a^{g(x)}g'(x)$$

This formula and the formula

$$D_x \log_a |x| = \frac{1}{\ln a} \cdot \frac{1}{x}$$

are simplest when $a = e$, since $\ln e = 1$. This is a major reason for using base e exponential and logarithmic functions.

EXAMPLE 5

The amount in grams of a radioactive substance present after t years is given by

$$A(t) = 100e^{-.12t}.$$

(a) Find the rate of change of the amount present after 3 years.

The rate of change is given by the derivative dy/dt.

$$\frac{dy}{dt} = 100(e^{-.12t})(-.12) = -12e^{-.12t}$$

After 3 years ($t = 3$), the rate of change is

$$\frac{dy}{dt} = -12e^{-.12(3)} = -12e^{-.36} \approx -8.4$$

grams per year.

(b) In part (a), is the amount present increasing or decreasing when $t = 3$?

Since the rate of change is negative, the amount is decreasing.

(c) Graph $A(t)$ on [0, 20].

From part (a), $dy/dt = -12e^{-.12t}$. Since $e^{-.12t}$ is never equal to 0, there is no maximum or minimum function value. Find the second derivative to determine any points of inflection and the concavity.

$$\frac{d^2y}{dt^2} = -12(e^{-.12t})(-.12) = 1.44e^{-.12t}$$

The second derivative is always positive, so the graph of A is always concave upward. Plot a few points to get the graph shown in Figure 11. ◀

$A(t) = 100e^{-.12t}$

FIGURE 11

EXAMPLE 6

Use the information given by the first and second derivatives to graph $f(x) = 2^{-x^2}$.

This function was graphed in Section 1 of this chapter by point plotting. Now we are able to graph it more carefully and with greater detail.

The first derivative is

$$f'(x) = \ln 2(2^{-x^2})(-2x)$$

$$= (-2x \ln 2)(2^{-x^2}) \quad \text{or} \quad \frac{-2x \ln 2}{2^{x^2}}.$$

Set $f'(x)$ equal to 0 and solve for x.

$$(-2x \ln 2)(2^{-x^2}) = 0$$

$$-2x \ln 2 = 0$$

$$x = 0$$

Note that 2^{-x^2} can never be zero, so $x = 0$ is the only critical number. We can find the second derivative by using the product rule since $f'(x)$ is the product of two factors.

$$f''(x) = (-2x \ln 2)[\ln 2(2^{-x^2})(-2x)] + (2^{-x^2})(-2 \ln 2)$$
$$= 4x^2(\ln 2)^2(2^{-x^2}) - 2 \ln 2(2^{-x^2})$$

Factoring out the common factors gives

$$f''(x) = (2 \ln 2)(2^{-x^2})(2x^2 \ln 2 - 1).$$

Set $f''(x)$ equal to 0 and solve for x to find any inflection points.

$$2 \ln 2(2^{-x^2})(2x^2 \ln 2 - 1) = 0$$
$$2x^2 \ln 2 - 1 = 0$$
$$2x^2 \ln 2 = 1$$
$$x^2 = \frac{1}{2 \ln 2}$$
$$x = \pm\sqrt{\frac{1}{2 \ln 2}} \approx \pm .85$$

$f(x)$

$(0, 1)$

$f(x) = 2x^{-x^2}$

$(-.85, .5)$ $(.85, .5)$

FIGURE 12

Test a value from each of the intervals

$$\left(-\infty, -\sqrt{\frac{1}{2 \ln 2}}\right), \left(-\sqrt{\frac{1}{2 \ln 2}}, \sqrt{\frac{1}{2 \ln 2}}\right), \text{ and } \left(\sqrt{\frac{1}{2 \ln 2}}, \infty\right)$$

to verify that there are inflection points at $x \approx -.85$ and $x \approx .85$. Use the second derivative test to verify that there is a maximum of 1 at $x = 0$. Plot the points where $x \approx -.85, x = 1,$ and $x \approx .85,$ and use information about concavity from testing the intervals given above to get the graph in Figure 12. ◀

Frequently a population, or the sales of a certain product, will start growing slowly, then grow more rapidly, and then gradually level off. Such growth can often be approximated by a mathematical model of the form

$$f(x) = \frac{b}{1 + ae^{kx}}$$

for appropriate constants a, b, and k.

EXAMPLE 7 Suppose that the sales of a new product can be approximated for its first few years on the market by the function

$$S(x) = \frac{100,000}{1 + 100e^{-.3x}},$$

where x is time in years since the introduction of the product.

(a) Find the rate of change of sales after 4 years.

The derivative of this sales function, which gives the rate of change of sales, was found in Example 3. Using that derivative,

$$S'(4) = \frac{3,000,000e^{-.3(4)}}{[1 + 100e^{-.3(4)}]^2} = \frac{3,000,000e^{-1.2}}{(1 + 100e^{-1.2})^2}.$$

By using a calculator, $e^{-1.2} \approx .301$, and

$$S'(4) \approx \frac{3,000,000(.301)}{[1 + 100(.301)]^2}$$

$$\approx \frac{903,000}{(1 + 30.1)^2}$$

$$= \frac{903,000}{967.21} \approx 934.$$

The rate of change of sales after 4 years is about 934 units per year. The positive number indicates that sales are increasing at this time.

(b) What happens to sales in the long run?

As time increases, $x \to \infty$, and

$$e^{-.3x} = \frac{1}{e^{.3x}} \to 0.$$

Thus,

$$\lim_{x \to \infty} S(x) = \frac{100,000}{1 + 100(0)} = 100,000.$$

Sales approach a horizontal asymptote of 100,000.

(c) Graph $S(x)$.

Use the first derivative to check for extrema. Set $S'(x) = 0$ and solve for x.

$$S'(x) = \frac{3,000,000e^{-.3x}}{(1 + 100e^{-.3x})^2} = 0$$

$$3,000,000e^{-.3x} = 0$$

$$\frac{3,000,000}{e^{.3x}} = 0$$

The expression on the left can never equal 0, so $S'(x) = 0$ leads to no extrema. Next, find any critical values for which $S'(x)$ does not exist; that is, values that make the denominator equal 0.

$$1 + 100e^{-.3x} = 0$$

$$100e^{-.3x} = -1$$

$$\frac{100}{e^{.3x}} = -1$$

Since the expression on the left is always positive, it cannot equal -1. There are no critical values and no extrema. Further, $e^{-.3x}$ is always positive, so $S'(x)$ is always positive. This means the graph is always increasing over the domain of the function. Use the second derivative to verify that the graph has an inflection point at approximately (15.4, 50,000) and is concave upward on $(-\infty, 15.4)$ and concave downward on $(15.4, \infty)$. Plotting a few points leads to the graph in Figure 13 (on the next page). ◀

x	y
0	990
5	4300
10	17,000
15	47,000
20	80,000
30	99,000

FIGURE 13

Environmental situations place effective limits on the population growth of an organism in an area. Many such limited growth situations are described by the *logistic function*

$$G(t) = \frac{mG_0}{G_0 + (m - G_0)e^{-kmt}},$$

where t represents time in appropriate units, G_0 is the initial number present, m is the maximum possible size of the population, k is a positive constant, and $G(t)$ is the population at time t. (This function is also given in Exercise 35 of Section 3 of this chapter.) Notice that the logistic function is a special case of the model discussed before Example 7.

EXAMPLE 8 The growth of a population of rare South American insects is given by the logistic function with $k = .00001$ and t in months. Assume that there are 200 insects initially and that the maximum population size is 10,000.

(a) Find the growth function $G(t)$ for these insects.

Substitute the given values for k, G_0, and m into the logistic function to get

$$G(t) = \frac{mG_0}{G_0 + (m - G_0)e^{-kmt}}$$

$$= \frac{(10,000)(200)}{200 + (10,000 - 200)e^{-(.00001)(10,000)t}}$$

$$= \frac{2,000,000}{200 + 9800e^{-.1t}}.$$

(b) Find the population after 5 months and after 10 months.

$$G(5) = \frac{2,000,000}{200 + 9800e^{-.5}} = 325.5$$

There are about 326 insects after 5 months.

$$G(10) = \frac{2{,}000{,}000}{200 + 9800e^{-1}} = 525.6$$

There are about 526 insects after 10 months.

(c) What happens to the population in the long run?

Notice that $G(t)$ has the same form as $S(x)$ in Example 7. Therefore, we know that

$$\lim_{t\to\infty} G(t) = \frac{2{,}000{,}000}{200 + 0} = 10{,}000,$$

so the population will approach an upper limit of 10,000 insects.

(d) Find the number of months before the rate of growth stops increasing and begins to decrease.

This will happen at the point of inflection. Verify that the point of inflection is approximately (39, 5000). The population grows quite rapidly for about 39 months (3 1/4 years) and then grows at a slower and slower pace as the population approaches 10,000.

(e) Graph $G(t)$ for $t \geq 0$.

The graph is shown in Figure 14. Compare it with the graph in Figure 13. Although these two applications are very different, the graphs are essentially the same. ◀

FIGURE 14

13.5 Exercises

Find derivatives of the functions defined as follows.

1. $y = e^{4x}$

2. $y = e^{-2x}$

3. $y = -8e^{2x}$

4. $y = .2e^{5x}$

5. $y = -16e^{x+1}$

6. $y = -4e^{-.1x}$

7. $y = e^{x^2}$

8. $y = e^{-x^2}$

9. $y = 3e^{2x^2}$

10. $y = -5e^{4x^3}$

11. $y = 4e^{2x^2-4}$

12. $y = -3e^{3x^2+5}$

13. $y = xe^x$

14. $y = x^2e^{-2x}$

15. $y = (x - 3)^2 e^{2x}$

16. $y = (3x^2 - 4x)e^{-3x}$

17. $y = e^{x^2} \ln x, \quad x > 0$

18. $y = e^{2x-1} \ln (2x - 1), \quad x > \dfrac{1}{2}$

19. $y = \dfrac{e^x}{\ln x}, \quad x > 0$

20. $y = \dfrac{\ln x}{e^x}, \quad x > 0$

21. $y = \dfrac{x^2}{e^x}$

22. $y = \dfrac{e^x}{2x + 1}$

23. $y = \dfrac{e^x + e^{-x}}{x}$

24. $y = \dfrac{e^x - e^{-x}}{x}$

25. $y = \dfrac{5000}{1 + 10e^{.4x}}$

26. $y = \dfrac{600}{1 - 50e^{.2x}}$

27. $y = \dfrac{10,000}{9 + 4e^{-.2x}}$

28. $y = \dfrac{500}{12 + 5e^{-.5x}}$

29. $y = (2x + e^{-x^2})^2$

30. $y = (e^{2x} + \ln x)^3, \quad x > 0$

Find the derivative of each of the following.

31. $y = 8^{5x}$

32. $y = 2^{-x}$

33. $y = 3 \cdot 4^{x^2+2}$

34. $y = -10^{3x^2-4}$

35. $y = 2 \cdot 3^{\sqrt{x}}$

36. $y = 5 \cdot 7^{\sqrt{x-2}}$

37. Verify the inflection point given in Example 7.

38. Verify the inflection point given in Example 8.

Find all relative maxima, minima, and inflection points for the functions defined as follows. Sketch the graphs.

39. $y = -xe^x$

40. $y = xe^{-x}$

41. $y = x^2e^{-x}$

42. $y = (x - 1)e^{-x}$

43. $y = e^x + e^{-x}$

44. $y = -x^2e^x$

45. For $f(x) = e^x$, find $f''(x)$ and $f'''(x)$. What is the nth derivative of f with respect to x?

46. Describe the slope of the tangent line to the graph of $f(x) = e^x$ for each of the following.

(a) $x \to -\infty$ (b) $x \to 0$

47. Prove: $\dfrac{d}{dx} e^{ax} = ae^{ax}$ for any constant a.

Applications

Business and Economics

48. Sales The sales of a new personal computer (in thousands) are given by

$$S(t) = 100 - 90e^{-.3t},$$

where t represents time in years.
Find the rate of change of sales at each of the following times.

(a) After 1 yr (b) After 5 yr

(c) What is happening to the rate of change of sales as time goes on?

(d) Does the rate of change of sales ever equal zero?

49. Product Durability Suppose $P(x) = e^{-.02x}$ represents the proportion of shoes manufactured by a given company that are still wearable after x days of use.
Find the proportions of shoes wearable after the following periods of time.

(a) 1 day (b) 10 days (c) 100 days

(d) Calculate and interpret $P'(100)$.

(e) As the number of days increases, what happens to the proportion of shoes that are still wearable? Is this reasonable?

50. Product Durability Using data in a car magazine, we constructed the mathematical model

$$y = 100e^{-.03045t}$$

for the percent of cars of a certain type still on the road after t years.
Find the percent of cars on the road after the following numbers of years.

(a) 0 (b) 2 (c) 4 (d) 6

Find the rate of change of the percent of cars still on the road after the following numbers of years.

(e) 0 (f) 2

(g) Interpret your answers to (e) and (f).

51. Elasticity Refer to the discussion of elasticity in the previous chapter. For the demand function $p = 400e^{-.2x}$, find each of the following.

(a) The elasticity E

(b) The number of items (x) for which total revenue is maximum

Life Sciences

52. Pollution Concentration The concentration of pollutants (in grams per liter) in the east fork of the Big Weasel River is approximated by

$$P(x) = .04e^{-4x},$$

where x is the number of miles downstream from a paper mill that the measurement is taken.
Find each of the following values.

(a) The concentration of pollutants .5 mi downstream

(b) The concentration of pollutants 1 mi downstream

(c) The concentration of pollutants 2 mi downstream

Find the rate of change of concentration with respect to distance for each of the following distances.

(d) .5 mi **(e)** 1 mi **(f)** 2 mi

53. Radioactive Decay Assume that the amount (in grams) of a radioactive substance present after t years is given by

$$A(t) = 500e^{-.25t}.$$

Find the rate of change of the quantity present after each of the following years: **(a)** 4 **(b)** 6 **(c)** 10.

(d) What is happening to the rate of change of the amount present as the number of years increases?

(e) Will the substance ever be gone completely?

Social Sciences

54. Habit Strength According to work by the psychologist C. L. Hull, the strength of a habit is a function of the number of times the habit is repeated. If N is the number of repetitions and $H(N)$ is the strength of the habit, then

$$H(N) = 1000(1 - e^{-kN}),$$

where k is a constant. Find $H'(N)$ if $k = .1$ and the number of times the habit is repeated is as follows.

(a) 10 **(b)** 100 **(c)** 1000

(d) Show that $H'(N)$ is always positive. What does this mean?

Computer/Graphing Calculator

Graph the following over the given intervals. Then find any relative extrema (not endpoints) and inflection points.

55. $y = e^x \ln x$, $(0, 5]$ **56.** $y = \dfrac{5 \ln 10x}{e^{.5x}}$, $(0, 10]$

▶ EXPONENTIAL AND LOGARITHMIC FUNCTIONS SUMMARY

EXPONENTIAL FUNCTIONS *Exponential Growth and Decay Function* If y_0 is the number present at time $t = 0$,

$$y = y_0 e^{kt}.$$

Math of Finance Formulas If I is the interest, P is the principal or present value, r is the annual interest rate, t is time in years and m is the number of compounding periods per year:

Simple interest $I = Prt$

	Compounded m times per year	*Compounded continuously*
Compound amount	$A = P\left(1 + \dfrac{r}{m}\right)^{tm}$	$A = Pe^{rt}$
Effective rate	$r_E = \left(1 + \dfrac{r}{m}\right)^{m} - 1$	$r_E = e^r - 1$
Present value	$P = A\left(1 + \dfrac{r}{m}\right)^{-tm}$	$P = Ae^{-rt}$

Change of Base Theorem for Exponentials $a^x = e^{(\ln a)x}$

Derivatives
$$D_x(e^x) = e^x \qquad\qquad D_x(a^x) = (\ln a)a^x$$
$$D_x(e^{g(x)}) = g'(x)e^{g(x)} \qquad D_x(a^{g(x)}) = (\ln a)a^{g(x)}g'(x)$$

LOGARITHMIC FUNCTIONS

Properties of Logarithms Let x and y be any positive real numbers and r be any real number. Let a be a positive real number, $a \neq 1$. Then

(a) $\log_a xy = \log_a x + \log_a y$ ⠀⠀⠀Product rule

(b) $\log_a \dfrac{x}{y} = \log_a x - \log_a y$ ⠀⠀⠀Quotient rule

(c) $\log_a x^r = r \log_a x$ ⠀⠀⠀Power rule

(d) $\log_a a = 1$

(e) $\log_a 1 = 0$

(f) $\log_a a^r = r$.

Change of Base Theorem for Logarithms $\log_a x = \dfrac{\log_b x}{\log_b a}$

Derivatives
$$D_x(\ln |x|) = \frac{1}{x} \qquad\qquad D_x (\log_a |x|) = \frac{1}{\ln a} \cdot \frac{1}{x}$$
$$D_x (\ln |g(x)|) = \frac{g'(x)}{g(x)} \qquad D_x(\log_a |g(x)|) = \frac{1}{\ln a} \cdot \frac{g'(x)}{g(x)}$$

Chapter Summary ⠀⠀⠀Key Terms

13.1 exponential function
exponential equation
simple interest
compound interest
compound amount
e
13.2 years to double
logarithm
logarithmic function

properties of logarithms
common logarithms
natural logarithms
change-of-base theorem
logarithmic equation
13.3 exponential growth and decay
function
growth constant

decay constant
half-life
continuous compounding
effective rate
nominal (stated) rate
present value
limited growth function
learning curve

Chapter 13 ⠀⠀⠀Review Exercises

1. Describe in words what a logarithm is.
2. Why is e a convenient base for exponential and logarithmic functions?

Solve each equation.

3. $2^{3x} = \dfrac{1}{8}$

4. $\left(\dfrac{9}{16}\right)^x = \dfrac{3}{4}$

5. $9^{2y-1} = 27^y$

6. $\dfrac{1}{2} = \left(\dfrac{b}{4}\right)^{1/4}$

Graph the functions defined as follows.

7. $y = 5^x$

8. $y = 5^{-x} + 1$

9. $y = \left(\dfrac{1}{5}\right)^{2x-3}$

10. $y = \left(\dfrac{1}{2}\right)^{x-1}$

11. $y = \log_2(x - 1)$ **12.** $y = 1 + \log_3 x$ **13.** $y = -\log_3 x$ **14.** $y = 2 - \ln x^2$

Write the following equations using logarithms.

15. $2^6 = 64$ **16.** $3^{1/2} = \sqrt{3}$ **17.** $e^{.09} = 1.09417$ **18.** $10^{1.07918} = 12$

Write the following equations using exponents.

19. $\log_2 32 = 5$ **20.** $\log_{10} 100 = 2$

21. $\ln 82.9 = 4.41763$ **22.** $\log 15.46 = 1.18921$

Evaluate each expression.

23. $\log_3 81$ **24.** $\log_{1/3} 81$ **25.** $\log_{32} 16$

26. $\log_{25} 5$ **27.** $\log_{100} 1000$ **28.** $\log_{1/2} 4$

Simplify each expression using the properties of logarithms.

29. $\log_5 3k + \log_5 7k^3$ **30.** $\log_3 2y^3 - \log_3 8y^2$

31. $2 \log_2 x - 3 \log_2 m$ **32.** $5 \log_4 r - 3 \log_4 r^2$

Solve each equation. Round each answer to the nearest thousandth.

33. $8^p = 19$ **34.** $3^{z-2} = 11$ **35.** $2^{1-m} = 7$ **36.** $15^{-k} = 9$

37. $e^{-5-2x} = 5$ **38.** $e^{3x-1} = 12$ **39.** $\left(1 + \dfrac{m}{3}\right)^5 = 10$ **40.** $\left(1 + \dfrac{2p}{5}\right)^2 = 3$

Find the derivative of each function defined as follows.

41. $y = -6e^{2x}$ **42.** $y = 8e^{.5x}$ **43.** $y = e^{-2x^3}$ **44.** $y = -4e^{x^2}$

45. $y = 5xe^{2x}$ **46.** $y = -7x^2 e^{-3x}$ **47.** $y = \ln(2 + x^2)$ **48.** $y = \ln(5x + 3)$

49. $y = \dfrac{\ln |3x|}{x - 3}$ **50.** $y = \dfrac{\ln |2x - 1|}{x + 3}$ **51.** $y = \dfrac{x e^x}{\ln(x^2 - 1)}$ **52.** $y = \dfrac{(x^2 + 1)e^{2x}}{\ln x}$

53. $y = (x^2 + e^x)^2$ **54.** $y = (e^{2x+1} - 2)^4$

Find all relative maxima, minima, and inflection points for the functions defined as follows. Graph each function.

55. $y = xe^x$ **56.** $y = 3xe^{-x}$ **57.** $y = \dfrac{e^x}{x - 1}$ **58.** $y = \dfrac{\ln |5x|}{2x}$

Find the amount of interest earned by each of the following deposits.

59. $6902 at 12% compounded semiannually for 8 yr

60. $2781.36 at 8% compounded quarterly for 6 yr

Find the compound amount if $12,104 is invested at 8% compounded continuously for each of the following periods.

61. 2 yr **62.** 4 yr **63.** 7 yr

Find the compound amounts for the following deposits if interest is compounded continuously.

64. $1500 at 10% for 9 yr **65.** $12,000 at 5% for 8 yr

Find the effective rate to the nearest hundredth for each nominal interest rate.

66. 7% compounded quarterly **67.** 9% compounded monthly **68.** 11% compounded semiannually

69. 9% compounded continuously **70.** 11% compounded continuously

Find the present value of each of the following amounts.

71. $2000 at 6% interest compounded annually for 5 yr **72.** $10,000 at 8% interest compounded semiannually for 6 yr

73. $43,200 at 8% interest compounded quarterly for 4 yr **74.** $9760 at 12% interest compounded monthly for 3 yr

Applications

Business and Economics

75. **Employee Training** A company finds that its new workers produce
$$P(x) = 100 - 100e^{-.8x}$$
items per day after x days on the job. Find each of the following.

(a) What is the limit of $P(x)$ as $x \to \infty$?

(b) How many items per day should an experienced worker produce?

(c) How long will it be before a new employee will produce 50 items per day?

76. **Interest** To help pay for college expenses, Jane Baile borrowed $10,000 at 10% interest compounded semiannually for 8 yr. How much will she owe at the end of the 8-yr period?

77. **Interest** How long will it take for $1 to triple at an annual inflation rate of 8% compounded continuously?

78. **Interest** Find the interest rate needed for $6000 to grow to $8000 in 3 yr with continuous compounding.

79. **Present Value** Edison Diest wants to open a camera shop. How much must he deposit now at 6% interest compounded monthly to have $25,000 at the end of 3 yr?

Life Sciences

80. **Population Growth** A population of 15,000 small deer in a specific region has grown exponentially to 17,000 in 4 yr.

(a) Write an exponential equation to express the population growth y in terms of time t in years.

(b) At this rate, how long will it take for the population to reach 45,000?

81. **Intensity of Light** The intensity of light (in appropriate units) passing through water decreases exponentially with the depth it penetrates beneath the surface according to the function
$$I(x) = 10e^{-.3x},$$
where x is the depth in meters. A certain water plant requires light of an intensity of 1 unit. What is the greatest depth of water in which it will grow?

82. **Drug Concentration** The concentration of a certain drug in the bloodstream at time t (in minutes) is given by
$$c(t) = e^{-t} - e^{-2t}.$$
Find the maximum concentration and the time when it occurs.

83. **Glucose Concentration** When glucose is infused into a person's bloodstream at a constant rate of c grams per minute, the glucose is converted and removed from the bloodstream at a rate proportional to the amount present. The amount of glucose in the bloodstream at time t (in minutes) is given by
$$g(t) = \frac{c}{a} + \left(g_0 - \frac{c}{a}\right)e^{-at},$$
where a is a positive constant. Assume $g_0 = .08$, $c = .1$, and $a = 1.3$.

(a) At what time is the amount of glucose a maximum? What is the maximum amount of glucose in the bloodstream?

(b) When is the amount of glucose in the bloodstream .1 gram?

(c) What happens to the amount of glucose in the bloodstream after a very long time?

Physical Sciences

84. **Oil Production** The production of an oil well has decreased exponentially from 128,000 barrels per year five years ago to 100,000 barrels per year at present.

(a) Letting $t = 0$ represent the present time, write an exponential equation for production y in terms of time t in years.

(b) Find the time it will take for production to fall to 70,000 barrels per year.

85. **Dating Rocks** Geologists sometimes measure the age of rocks by using "atomic clocks." By measuring the amounts of potassium 40 and argon 40 in a rock, the age t of the specimen (in years) is found with the formula
$$t = (1.26 \times 10^9)\frac{\ln[1 + 8.33(A/K)]}{\ln 2},$$
where A and K, respectively, are the numbers of atoms of argon 40 and potassium 40 in the specimen.

(a) How old is a rock in which $A = 0$ and $K > 0$?

(b) The ratio A/K for a sample of granite from New Hampshire is .212. How old is the sample?

(c) Let $A/K = r$ and find dt/dr.

(d) What happens to t as r gets larger? smaller?

General Interest

86. **Consumer Price Index** The U.S. consumer price index (CPI, or cost of living index) has risen exponentially over the years. In 1960 the index was 34; in 1990 it was 113.*

 (a) Write an exponential equation for the CPI in terms of time in years.

 (b) Find the year in which costs will be 50% higher than in the base year (1983), when the CPI was 100.

 (c) Find the year the CPI was 100 by using your equation from part (a). If you do not get 1983, explain why.

87. **Living Standards** One measure of living standards in the United States is given by $L(t) = 9 + 2e^{.15t}$, where t is the number of years since 1982. Find $L(t)$ for the following years.

 (a) 1982 (b) 1986 (c) 1992

 (d) Graph $L(t)$.

 (e) What can be said about the growth of living standards in the United States according to this equation?

88. **Nuclear Warheads** The number of nuclear warheads in the USSR arsenal from 1965 to 1980 is approximated by $A(t) = 830e^{.15t}$, where t is the number of years since 1965.* Find the number of nuclear warheads in the following years.

 (a) 1975 (b) 1980

The Universal Almanac, edited by John W. Wright (Kansas City and New York: Andrews and McMeel, 1990), p. 289.

(c) In 1985, the number of warheads was actually 10,012. Find the corresponding number using the given function. Is the discrepancy significant? If so, what might have caused it?

(d) What does the derivative of $A(t)$ tell you about the rate of change in the number of nuclear warheads?

Connections

89. Give the following properties of the exponential function $f(x) = a^x; a > 0, a \neq 1$.

 (a) Domain (b) Range

 (c) y-intercept (d) Discontinuities

 (e) Asymptote(s) (f) Increasing if a is _____

 (g) Decreasing if a is _____

90. Give the following properties of the logarithmic function $f(x) = \log_a x; a > 0, a \neq 1$.

 (a) Domain (b) Range

 (c) x-intercept (d) Discontinuities

 (e) Asymptote(s) (f) Increasing if a is _____

 (g) Decreasing if a is _____

91. Compare your answers for Exercises 89 and 90. What similarities do you notice? What differences?

Extended Application / Individual Retirement Accounts[†]

Money deposited in an individual retirement account (IRA) produces earnings that are not taxed until they are withdrawn from the account. Thus, an IRA effectively defers the tax on earnings, but there is a corresponding loss of liquidity since there are penalties for withdrawals from an IRA before the age of 59½ years. Suppose that you have P dollars to invest for n years in either an IRA or a regular account, both of which pay an interest rate i compounded annually. Earnings from the regular account are immediately taxable, in contrast to the IRA earnings, which are tax sheltered until withdrawal. To compare these two investment options mathematically, assume for simplicity that the same interest rate, i, and the same income tax rate, t, apply throughout the n years of the investment.

By the compound interest formula, the principal P grows to $P(1 + i)^n$ after n years in the IRA. If the entire amount is then withdrawn (and the investor is past 59½ years of age), the tax due on the *earnings* is $[P(1 + i)^n - P]t$. Thus, the amount left after taxes is

$$A = P(1 + i)^n - [P(1 + i)^n - P]t$$
$$= P[(1 + i)^n(1 - t) + t].$$

If $P = 1$, then

$$M = (1 + i)^n(1 - t) + t$$

is called the **IRA multiplier,** because one dollar will grow to M dollars (after taxes) after n years in the IRA.

[†]The discussion and examples about IRAs are courtesy of Thomas B. Muenzenberger, Kansas State University, Manhattan, Kansas.

Example 1

Suppose that you deposit $2000 in an IRA that pays 10% interest compounded annually. Suppose further that the money is left in the account for 30 years, then withdrawn, and the earnings are taxed at 37%. Find the tax due and the IRA multiplier.

Here the principal grows to $2000(1 + .1)^{30} = 34,899$ dollars (to the nearest dollar), and the tax due on the *earnings* is

$$(\$34,899 - \$2000)(.37) = \$12,173.$$

Thus, the amount left after taxes is $A = 34,899 - 12,173 = 22,726$ dollars. The IRA Multiplier is $M = 11.4$. ◄

Each year in the regular account, t percent of the i percent earned is withdrawn to pay the tax on the earnings, leaving $(1 - t)i$ percent. Thus, by the compound interest formula, the principal P grows to

$$A = P[1 + (1 - t)i]^n$$

after n years in the regular account. If $P = 1$, then

$$m = [1 + (1 - t)i]^n$$

is the multiplier, because one dollar will grow to m dollars (after taxes) after n years in the regular account.

Example 2

Suppose that you deposit $2000 in a regular account that pays 10% compounded annually. Suppose further that the income tax rate is 37% and that the money is left in the account for 30 years and then withdrawn. Find the multiplier, m.

Here, the principal grows to

$$A = 2,000[1 + (1 - .37)(.10)]^{30} = 12,503 \text{ dollars.}$$

The multiplier is 6.3. ◄

From the previous two examples, $M/m = 11.4/6.3 = 1.8$, indicating that the IRA yields almost twice as much as the regular account over a thirty-year period. The **multiplier function**

$$\frac{M}{m} = \frac{(1 + i)^n(1 - t) + t}{[1 + (1 - t)i]^n}$$

can be used in the same way to compare the two investment options for other values of n, i, and t. Note that the multiplier function is a function in three variables.* The following table contains values of M/m for $n = 30$ and various values of i and t. It is apparent from the table that for $n = 30$ and a fixed t, the multiplier function M/m is an increasing function of i (see Figure 15 for $t = .40$). In fact, for a fixed t and for *any* $n > 1$, the multiplier function is an increasing function of i. Thus, the

t \ i	.05	.10	.15	.20	.25	.30
.10	1.07	1.19	1.34	1.49	1.65	1.81
.20	1.13	1.41	1.77	2.21	2.72	3.30
.30	1.18	1.64	2.33	3.27	4.48	6.02
.40	1.23	1.89	3.02	4.77	7.33	11.0
.50	1.27	2.13	3.83	6.83	11.8	19.8
.60	1.29	2.34	4.72	9.50	18.6	35.0
.70	1.28	2.45	5.49	12.5	27.8	59.3
.80	1.23	2.37	5.79	14.9	37.6	91.4
.90	1.15	1.96	4.81	13.6	38.9	108
1	1	1	1	1	1	1

advantage of the IRA over the regular account widens as the interest rate i increases. This effect is particularly dramatic for high income tax rates (see the table).

Exercises

1. (a) Suppose that you deposit $2000 in an IRA that pays 10% compounded annually. Suppose further that the money is left in the account for 20 years, then withdrawn, and the earnings are taxed at 37%. Find the amount left after taxes, then find the IRA multiplier M.

 (b) Suppose instead that the money was deposited in a regular account for the same period of time at the same rates. Find the amount left after taxes; then find the multiplier m.

 (c) Find M/m. What can you conclude about investment options (a) and (b)?

2. Redo Exercise 1 for $n = 10$.

3. Draw a conclusion from Example 1 and the results of Exercises 1(a) and 2(a). (*Hint:* It is called the *time value of money.*)

4. Draw a conclusion from Example 2 and Exercises 1(b) and 2(b).

5. Construct tables for $n = 10$ and $n = 20$ and various values of i and t similar to the table for $n = 30$ given after Example 2. Draw conclusions from your data.

*Functions of more than one independent variable are discussed in the chapter entitled ''Multivariable Calculus.''

Applications at a glance . . .

Cold Weather in New York City . . . distribution of income . . . profit from a new technology . . . cost of maintaining a machine . . .

(See page 734.)

In previous chapters, we studied the mathematical process of finding the derivative of a function, and we considered various applications of derivatives. The material that was covered belongs to the branch of calculus called *differential calculus*. In this chapter and the next we will study another branch of calculus, called *integral calculus*. Like the derivative of a function, the indefinite integral of a function is a special limit with many diverse applications. Geometrically, the derivative is related to the slope of the tangent line to a curve, while the definite integral is related to the area under a curve. As this chapter will show, differential and integral calculus are connected by the Fundamental Theorem of Calculus.

14.1 ANTIDERIVATIVES

If an object is thrown from the top of the Sears Tower in Chicago, how fast is it going when it hits the ground?

Using *antiderivatives*, we can answer this question.

Functions used in applications in previous chapters have provided information about a *total amount* of a quantity, such as cost, revenue, profit, temperature, gallons of oil, or distance. Working with these functions provided information about the rate of change of these quantities and allowed us to answer important questions about the extrema of the functions. It is not always possible to find ready-made functions that provide information about the total amount of a quantity, but it is often possible to collect enough data to come up with a function that gives the *rate of change* of a quantity. Since we know that derivatives give the rate of change when the total amount is known, is it possible to reverse the process and use a known rate of change to get a function that gives the total amount of a quantity? The answer is yes; this reverse process, called *antidifferentiation*, is the topic of this section. The *antiderivative* of a function is defined as follows.

ANTIDERIVATIVE　　　If $F'(x) = f(x)$, then $F(x)$ is an **antiderivative** of $f(x)$.

EXAMPLE 1

(a) If $F(x) = 10x$, then $F'(x) = 10$, so $F(x) = 10x$ is an antiderivative of $f(x) = 10$.

(b) For $F(x) = x^2$, $F'(x) = 2x$, making $F(x) = x^2$ an antiderivative of $f(x) = 2x$.

EXAMPLE 2

Find an antiderivative of $f(x) = 5x^4$.

To find a function $F(x)$ whose derivative is $5x^4$, work backwards. Recall that the derivative of x^n is nx^{n-1}. If

$$\overbrace{nx^{n-1}} \quad \text{is} \quad \overbrace{5x^4},$$

then $n - 1 = 4$ and $n = 5$, so x^5 is an antiderivative of $5x^4$.

EXAMPLE **3** Suppose a population is growing at a rate given by $f(x) = e^x$, where x is time in years from some initial date. Find a function giving the population at time x.

Let the population function be $F(x)$. Then

$$f(x) = F'(x) = e^x.$$

The derivative of the function defined by $F(x) = e^x$ is $F'(x) = e^x$, so a population function with the given growth rate is $F(x) = e^x$. ◀

The function from Example 1(b), defined by $F(x) = x^2$, is not the only function whose derivative is $f(x) = 2x$; for example, both $G(x) = x^2 + 2$ and $H(x) = x^2 - 7$ have $f(x) = 2x$ as a derivative. In fact, for any real number C, the function $F(x) = x^2 + C$ has $f(x) = 2x$ as its derivative. This means that there is a *family* or *class* of functions having $2x$ as an antiderivative. As the next theorem states, if two functions $F(x)$ and $G(x)$ are antiderivatives of $f(x)$, then $F(x)$ and $G(x)$ can differ only by a constant.

If $F(x)$ and $G(x)$ are both antiderivatives of a function $f(x)$, then there is a constant C such that

$$F(x) - G(x) = C.$$

(Two antiderivatives of a function can differ only by a constant.)

For example,

$$F(x) = x^2 + 2, \quad G(x) = x^2, \quad \text{and} \quad H(x) = x^2 - 4$$

are all antiderivatives of $f(x) = 2x$, and any two of them differ only by a constant. The derivative of a function gives the slope of the tangent line at any x-value. The fact that these three functions have the same derivative, $f(x) = 2x$, means that their slopes at any particular value of x are the same, as shown in Figure 1. Thus, each graph can be obtained from another by a vertical shift of $|C|$ units, where C is any constant. We will represent this family of antiderivatives of $f(x)$ by $F(x) + C$.

The family of all antiderivatives of the function f is indicated by

$$\int f(x)\, dx.$$

The symbol \int is the **integral sign,** $f(x)$ is the **integrand,** and $\int f(x)\, dx$ is called an **indefinite integral,** the most general antiderivative of f.

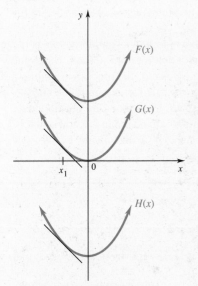

Slopes of the tangent lines at $x = x_1$ are the same.

FIGURE 1

INDEFINITE INTEGRAL	If $F'(x) = f(x)$, then

$$\int f(x)\ dx = F(x) + C,$$

for any real number C.

For example, using this notation,

$$\int (2x)\ dx = x^2 + C.$$

The dx in the indefinite integral indicates that $\int f(x)\ dx$ is the "integral of $f(x)$ *with respect to* x" just as the symbol dy/dx denotes the "derivative of y with respect to x." For example, in the indefinite integral $\int 2ax\ dx$, dx indicates that a is to be treated as a constant and x as the variable, so that

$$\int 2ax\ dx = \int a(2x)dx = ax^2 + C.$$

On the other hand,

$$\int 2ax\ da = a^2 x + C = xa^2 + C.$$

A more complete interpretation of dx will be discussed later.

The symbol $\int f(x)dx$ was created by G.W. Leibniz (1646–1716) in the latter part of the seventeenth century. The \int is an elongated S from *summa*, the Latin word for *sum*. The word *integral* as a term in the calculus was coined by Jakob Bernoulli (1654–1705), a Swiss mathematician who corresponded frequently with Leibniz. The relationship between sums and integrals will be clarified in the next two sections.

The method of working backwards, as above, to find an antiderivative is not very satisfactory. Some rules for finding antiderivatives are needed. Since the process of finding an indefinite integral is the inverse of the process of finding a derivative, each formula for derivatives leads to a rule for indefinite integrals.

Recall that $\dfrac{d}{dx}x^n = nx^{n-1}$.

As mentioned in Example 2, the derivative of x^n is found by multiplying x by n and reducing the exponent on x by 1. To find an indefinite integral—that is, to undo what was done—*increase* the exponent by 1 and *divide* by the new exponent, $n + 1$.

POWER RULE	For any real number $n \neq -1$,

$$\int x^n\ dx = \frac{1}{n + 1}x^{n+1} + C.$$

This rule can be verified by differentiating the expression on the right above:

$$\frac{d}{dx}\left(\frac{1}{n + 1}x^{n+1} + C\right) = \frac{n + 1}{n + 1}x^{(n+1)-1} + 0 = x^n.$$

(If $n = -1$, the expression in the denominator is 0, and the above rule cannot be used. Finding an antiderivative for this case is discussed later.)

EXAMPLE **4**　Find each indefinite integral.

(a) $\int t^3 \, dt$

Use the power rule with $n = 3$.

$$\int t^3 \, dt = \frac{1}{3+1} t^{3+1} + C = \frac{1}{4} t^4 + C$$

(b) $\int \frac{1}{t^2} dt$

First, write $1/t^2$ as t^{-2}. Then

$$\int \frac{1}{t^2} dt = \int t^{-2} \, dt = \frac{t^{-1}}{-1} + C = -\frac{1}{t} + C.$$

(c) $\int \sqrt{u} \, du$

Since $\sqrt{u} = u^{1/2}$,

$$\int \sqrt{u} \, du = \int u^{1/2} \, du = \frac{1}{1/2 + 1} u^{3/2} + C = \frac{2}{3} u^{3/2} + C.$$

To check this, differentiate $(2/3)u^{3/2} + C$; the derivative is $u^{1/2}$, the original function.

(d) $\int dx$

Write dx as $1 \cdot dx$ and use the fact that $x^0 = 1$ for any nonzero number x to get

$$\int dx = \int 1 \, dx = \int x^0 dx = \frac{1}{1} x^1 + C = x + C. \quad \blacktriangleleft$$

Recall that $\frac{d}{dx} [f(x) \pm g(x)] = [f'(x) \pm g'(x)]$ and $\frac{d}{dx} [kf(x)] = kf'(x)$.

As shown earlier, the derivative of the product of a constant and a function is the product of the constant and the derivative of the function. A similar rule applies to indefinite integrals. Also, since derivatives of sums or differences are found term by term, indefinite integrals also can be found term by term.

CONSTANT MULTIPLE RULE AND SUM OR DIFFERENCE RULE

If all indicated integrals exist,

$$\int k \cdot f(x) dx = k \int f(x) dx \qquad \text{for any real number } k,$$

and

$$\int [f(x) \pm g(x)] dx = \int f(x) dx \pm \int g(x) dx.$$

EXAMPLE **5**

Find each of the following.

(a) $\int 2v^3 \, dv$

By the constant multiple rule and the power rule,

$$\int 2v^3 \, dv = 2 \int v^3 \, dv = 2 \left(\frac{1}{4}v^4 \right) + C = \frac{1}{2}v^4 + C.$$

(b) $\int \frac{12}{z^5} \, dz$

Use negative exponents.

$$\int \frac{12}{z^5} \, dz = \int 12z^{-5} \, dz = 12 \int z^{-5} \, dz = 12 \left(\frac{z^{-4}}{-4} \right) + C$$

$$= -3z^{-4} + C = \frac{-3}{z^4} + C$$

(c) $\int (3z^2 - 4z + 5) \, dz$

Using the rules in this section,

$$\int (3z^2 - 4z + 5) \, dz = 3 \int z^2 \, dz - 4 \int z \, dz + 5 \int dz$$

$$= 3 \left(\frac{1}{3}z^3 \right) - 4 \left(\frac{1}{2}z^2 \right) + 5z + C$$

$$= z^3 - 2z^2 + 5z + C.$$

Only one constant C is needed in the answer; the three constants from integrating term by term are combined. In Example 5(a), C represents any real number, so it is not necessary to multiply it by 2 in the next-to-last step. ◀

EXAMPLE **6**

Find each of the following.

(a) $\int \frac{x^2 + 1}{\sqrt{x}} \, dx$

First rewrite the integrand as follows.

$$\int \frac{x^2 + 1}{\sqrt{x}} \, dx = \int \left(\frac{x^2}{\sqrt{x}} + \frac{1}{\sqrt{x}} \right) dx$$

$$= \int \left(\frac{x^2}{x^{1/2}} + \frac{1}{x^{1/2}} \right) dx$$

$$= \int (x^{3/2} + x^{-1/2}) \, dx$$

Now find the antiderivative.

$$\int (x^{3/2} + x^{-1/2})dx = \frac{x^{5/2}}{5/2} + \frac{x^{1/2}}{1/2} + C$$

$$= \frac{2}{5}x^{5/2} + 2x^{1/2} + C$$

(b) $\int (x^2 - 1)^2 dx$

Square the binomial first, and then find the antiderivative.

$$\int (x^2 - 1)^2 dx = \int (x^4 - 2x^2 + 1)dx$$

$$= \frac{x^5}{5} - \frac{2x^3}{3} + x + C \quad \blacktriangleleft$$

To check integration, take the derivative of the result. For instance, in Example 5(c) check that $z^3 - 2z^2 + 5z + C$ is the required indefinite integral by taking the derivative

$$\frac{d}{dz}(z^3 - 2z^2 + 5z + C) = 3z^2 - 4z + 5,$$

which agrees with the original information.

It was shown earlier that the derivative of $f(x) = e^x$ is $f'(x) = e^x$. Also, the derivative of $f(x) = e^{kx}$ is $f'(x) = k \cdot e^{kx}$. These results lead to the following formulas for indefinite integrals of exponential functions.

INDEFINITE INTEGRALS OF EXPONENTIAL FUNCTIONS

$$\int e^x \, dx = e^x + C$$

$$\int e^{kx} \, dx = \frac{1}{k} \cdot e^{kx} + C, \qquad k \neq 0$$

EXAMPLE 7

(a) $\int 9e^t \, dt = 9\int e^t \, dt = 9e^t + C$

(b) $\int e^{9t} \, dt = \frac{1}{9} e^{9t} + C$

(c) $\int 3e^{(5/4)u} \, du = 3\left(\frac{1}{5/4}e^{(5/4)u}\right) + C$

$$= 3\left(\frac{4}{5}\right)e^{(5/4)u} + C$$

$$= \frac{12}{5}e^{(5/4)u} + C \quad \blacktriangleleft$$

The restriction $n \neq -1$ was necessary in the formula for $\int x^n \, dx$ since $n = -1$ made the denominator of $1/(n + 1)$ equal to 0. To find $\int x^n \, dx$ when $n = -1$, that is, to find $\int x^{-1} \, dx$, recall the differentiation formula for the logarithmic function: the derivative of $f(x) = \ln |x|$, where $x \neq 0$, is $f'(x) = 1/x = x^{-1}$. This formula for the derivative of $f(x) = \ln |x|$ gives a formula for $\int x^{-1} \, dx$.

| **INDEFINITE INTEGRAL OF x^{-1}** | $$\int x^{-1} \, dx = \int \frac{1}{x} \, dx = \ln |x| + C$$ |
| --- | --- |

EXAMPLE 8

(a) $\displaystyle \int \frac{4}{x} \, dx = 4 \int \frac{1}{x} \, dx = 4 \ln |x| + C$

(b) $\displaystyle \int \left(-\frac{5}{x} + e^{-2x} \right) dx = -5 \ln |x| - \frac{1}{2} e^{-2x} + C$ ◀

In all the examples above, the antiderivative family of functions was found. In many applications, however, the given information allows us to determine the value of the integration constant C. The next examples illustrate this idea.

EXAMPLE 9

Suppose a publishing company has found that the marginal cost at a level of production of x thousand books is given by

$$C'(x) = \frac{50}{\sqrt{x}}$$

and that the fixed cost (the costs before the first book can be produced) is $25,000. Find the cost function $C(x)$.

Write $50/\sqrt{x}$ as $50/x^{1/2}$ or $50x^{-1/2}$, and then use the indefinite integral rules to integrate the function.

$$\int \frac{50}{\sqrt{x}} \, dx = \int 50x^{-1/2} \, dx = 50(2x^{1/2}) + k = 100x^{1/2} + k$$

(Here k is used instead of C to avoid confusion with the cost function $C(x)$.) To find the value of k, use the fact that $C(0)$ is 25,000.

$$C(x) = 100x^{1/2} + k$$
$$25,000 = 100 \cdot 0 + k$$
$$k = 25,000$$

With this result, the cost function is $C(x) = 100x^{1/2} + 25,000$. ◀

EXAMPLE 10

Suppose the marginal revenue from a product is given by $50 - 3x - x^2$. Find the demand function for the product.

The marginal revenue is the derivative of the revenue function

$$\frac{dR}{dx} = 50 - 3x - x^2,$$

so
$$R = \int (50 - 3x - x^2)\,dx$$
$$= 50x - \frac{3x^2}{2} - \frac{x^3}{3} + C.$$

If $x = 0$, then $R = 0$ (no items sold means no revenue), and

$$0 = 50(0) - \frac{3(0)^2}{2} - \frac{(0)^3}{3} + C$$
$$0 = C.$$

Thus,

$$R = 50x - \frac{3x^2}{2} - \frac{x^3}{3}$$

gives the revenue function. Now, recall that $R = xp$, where p is the demand function. Then

$$50x - \frac{3}{2}x^2 - \frac{1}{3}x^3 = xp$$

$$50 - \frac{3}{2}x - \frac{1}{3}x^2 = p,$$

which gives the demand function.

In the next example integrals are used to find the position of a particle when the acceleration of the particle is given.

 EXAMPLE 11 Recall that if the function $s(t)$ gives the position of a particle at time t, then its velocity $v(t)$ and its acceleration $a(t)$ are given by

$$v(t) = s'(t) \qquad \text{and} \qquad a(t) = v'(t) = s''(t).$$

(a) Suppose the velocity of an object is $v(t) = 6t^2 - 8t$ and that the object is at -5 when time is 0. Find $s(t)$.

Since $v(t) = s'(t)$, the function $s(t)$ is an antiderivative of $v(t)$:

$$s(t) = \int v(t)\,dt = \int (6t^2 - 8t)\,dt$$
$$= 2t^3 - 4t^2 + C$$

for some constant C. Find C from the given information that $s = -5$ when $t = 0$.

$$s(t) = 2t^3 - 4t^2 + C$$
$$-5 = 2(0)^3 - 4(0)^2 + C$$
$$-5 = C$$
$$s(t) = 2t^3 - 4t^2 - 5$$

(b) Many experiments have shown that when an object is dropped, its acceleration (ignoring air resistance) is constant. This constant has been found to be approximately 32 feet per second every second; that is,

$$a(t) = -32.$$

The negative sign is used because the object is falling. Suppose an object is thrown down from the top of the 1100-foot-tall Sears Tower in Chicago. If the initial velocity of the object is -20 feet per second, find $s(t)$.

First find $v(t)$ by integrating $a(t)$:

$$v(t) = \int (-32)\,dt = -32t + k.$$

When $t = 0$, $v(t) = -20$:

$$-20 = -32(0) + k$$
$$-20 = k$$

and
$$v(t) = -32t - 20.$$

Now integrate $v(t)$ to find $s(t)$.

$$s(t) = \int (-32t - 20)\,dt = -16t^2 - 20t + C$$

Since $s(t) = 1100$ when $t = 0$, we can substitute these values into the equation for $s(t)$ to get $C = 1100$ and

$$s(t) = -16t^2 - 20t + 1100$$

as the distance of the object from the ground after t seconds.

(c) Use the equations derived in (b) to find out how fast the object was falling when it hit the ground and how long it took to strike the ground.

When the object strikes the ground, $s = 0$, so

$$0 = -16t^2 - 20t + 1100.$$

To solve this equation for t, factor out the common factor of -4 and then use the quadratic formula.

$$0 = -4(4t^2 + 5t - 275)$$
$$t = \frac{-5 \pm \sqrt{25 + 4400}}{8} \approx \frac{-5 \pm 66.5}{8}$$

Only the positive value of t is meaningful here: $t \approx 7.69$. It takes the object about 7.69 seconds to strike the ground. From the velocity equation, with $t = 7.69$, we find

$$v(t) = -32t - 20$$
$$v(7.69) = -32(7.69) - 20 \approx -266,$$

so the object was falling (as indicated by the negative sign) at about 266 feet per second when it hit the ground. ◀

EXAMPLE 12 Find a function f whose graph has slope $f'(x) = 6x^2 + 4$ and goes through the point $(1, 1)$.

Since $f'(x) = 6x^2 + 4$,

$$f(x) = \int (6x^2 + 4)\,dx = 2x^3 + 4x + C.$$

The graph of f goes through $(1, 1)$, so C can be found by substituting 1 for x and 1 for $f(x)$.

$$1 = 2(1)^3 + 4(1) + C$$
$$1 = 6 + C$$
$$C = -5$$

Finally, $f(x) = 2x^3 + 4x - 5$. ◀

14.1 Exercises

Find each of the following.

1. $\int 4x\,dx$

2. $\int 8x\,dx$

3. $\int 5t^2\,dt$

4. $\int 6x^3\,dx$

5. $\int 6\,dk$

6. $\int 2\,dy$

7. $\int (2z + 3)\,dz$

8. $\int (3x - 5)\,dx$

9. $\int (x^2 + 6x)\,dx$

10. $\int (t^2 - 2t)\,dt$

11. $\int (t^2 - 4t + 5)\,dt$

12. $\int (5x^2 - 6x + 3)\,dx$

13. $\int (4z^3 + 3z^2 + 2z - 6)\,dz$

14. $\int (12y^3 + 6y^2 - 8y + 5)\,dy$

15. $\int 5\sqrt{z}\,dz$

16. $\int t^{1/4}\,dt$

17. $\int (u^{1/2} + u^{3/2})\,du$

18. $\int (4\sqrt{v} - 3v^{3/2})\,dv$

19. $\int (15x\sqrt{x} + 2\sqrt{x})\,dx$

20. $\int (x^{1/2} - x^{-1/2})\,dx$

21. $\int (10u^{3/2} - 14u^{5/2})\,du$

22. $\int (56t^{5/2} + 18t^{7/2})\,dt$

23. $\int \left(\frac{1}{z^2}\right)\,dz$

24. $\int \left(\frac{4}{x^3}\right)\,dx$

25. $\int \left(\frac{1}{y^3} - \frac{1}{\sqrt{y}}\right)\,dy$

26. $\int \left(\sqrt{u} + \frac{1}{u^2}\right)\,du$

27. $\int (-9t^{-2} - 2t^{-1})\,dt$

28. $\int (8x^{-3} + 4x^{-1})\,dx$

29. $\int e^{2t}\,dt$

30. $\int e^{-3y}\,dy$

31. $\int 3e^{-.2x}\,dx$

32. $\int -4e^{.2v}\,dv$

33. $\int \left(\frac{3}{x} + 4e^{-.5x}\right)\,dx$

34. $\int \left(\frac{9}{x} - 3e^{-.4x}\right)\,dx$

35. $\int \frac{1 + 2t^3}{t}\,dt$

36. $\int \frac{2y^{1/2} - 3y^2}{y}\,dy$

37. $\int (e^{2u} + 4u)\,du$

38. $\int (v^2 - e^{3v})\,dv$

39. $\int (x + 1)^2\,dx$

40. $\int (2y - 1)^2\,dy$

41. $\int \frac{\sqrt{x} + 1}{\sqrt[3]{x}}\,dx$

42. $\int \frac{1 - 2\sqrt[3]{z}}{\sqrt[3]{z}}\,dz$

43. Find an equation of the curve whose tangent line has a slope of

$$f'(x) = x^{2/3},$$

given that the point $(1, 3/5)$ is on the curve.

44. The slope of the tangent line to a curve is given by

$$f'(x) = 6x^2 - 4x + 3.$$

If the point $(0, 1)$ is on the curve, find an equation of the curve.

Applications

Business and Economics

Cost *Find the cost function for each of the following marginal cost functions.*

45. $C'(x) = 4x - 5,$ fixed cost is $8

46. $C'(x) = 2x + 3x^2,$ fixed cost is $15

47. $C'(x) = .2x^2 + 5x,$ fixed cost is $10

48. $C'(x) = .8x^2 - x,$ fixed cost is $5

49. $C'(x) = .03e^{.01x},$ fixed cost is $8

50. $C'(x) = x^{1/2},$ 16 units cost $45

51. $C'(x) = x^{2/3} + 2,$ 8 units cost $58

52. $C'(x) = x^2 - 2x + 3,$ 3 units cost $15

53. $C'(x) = x + 1/x^2,$ 2 units cost $5.50

54. $C'(x) = 1/x + 2x,$ 7 units cost $58.40

55. $C'(x) = 5x - 1/x,$ 10 units cost $94.20

56. $C'(x) = 1.2^x \,(\ln 1.2),$ 2 units cost $9.44 (*Hint:* Recall that $a^x = e^{x \ln a}$.)

57. Profit The marginal profit of a small fast-food stand is given by

$$P'(x) = 2x + 20,$$

where x is the sales volume in thousands of hamburgers. The "profit" is -50 dollars when no hamburgers are sold. Find the profit function.

58. Profit Suppose the marginal profit from the sale of x hundred items is

$$P'(x) = 4 - 6x + 3x^2,$$

and the profit on 0 items is $-\$40$. Find the profit function.

Life Sciences

59. Biochemical Excretion If the rate of excretion of a biochemical compound is given by

$$f'(t) = .01e^{-.01t},$$

the total amount excreted by time t (in minutes) is $f(t)$.

(a) Find an expression for $f(t)$.

(b) If 0 units are excreted at time $t = 0$, how many units are excreted in 10 minutes?

60. Concentration of a Solute According to Fick's law, the diffusion of a solute across a cell membrane is given by

$$c'(t) = \frac{kA}{V}[C - c(t)], \tag{1}$$

where A is the area of the cell membrane, V is the volume of the cell, $c(t)$ is the concentration inside the cell at time t, C is the concentration outside the cell, and k is a constant. If c_0 represents the concentration of the solute inside the cell when $t = 0$, then it can be shown that

$$c(t) = (c_0 - C)e^{-kAt/V} + M. \tag{2}$$

(a) Use the last result to find $c'(t)$.

(b) Substitute back into equation (1) to show that (2) is indeed the correct antiderivative of (1).

Physical Sciences

Exercises 61–65 refer to Example 11 in this section.

61. Velocity For a particular object, $a(t) = t^2 + 1$ and $v(0) = 6$. Find $v(t)$.

62. Distance Suppose $v(t) = 6t^2 - 2/t^2$ and $s(1) = 8$. Find $s(t)$.

63. Time An object is dropped from a small plane flying at 6400 feet. Assume that $a(t) = -32$ feet per second per second and $v(0)$ is 0, and find $s(t)$. How long will it take the object to hit the ground?

64. Distance Suppose $a(t) = 18t + 8$, $v(1) = 15$, and $s(1) = 19$. Find $s(t)$.

65. Distance Suppose $a(t) = \frac{15}{2}\sqrt{t} + 3e^{-t}$, $v(0) = -3$, and $s(0) = 4$. Find $s(t)$.

14.2 SUBSTITUTION

If a formula for the marginal revenue is known, how can a formula for the total revenue be found?

Using the method of substitution, this question will be answered in an exercise in this section.

We saw how to integrate a few simple functions in the previous section. More complicated functions can sometimes be integrated by *substitution*. The substitution technique depends on the idea of a differential, discussed in an earlier chapter. If $u = f(x)$, the *differential* of u, written du, is defined as

$$du = f'(x)\,dx.$$

For example, if $u = 6x^4$, then $du = 24x^3\,dx$.

Recall the chain rule for derivatives as used in the following example:

The chain rule, discussed in detail in the chapter on differentiation, states that $\dfrac{d}{dx}[f(g(x))] = f'(g(x)) \cdot g'(x)$.

$$\frac{d}{dx}(x^2 - 1)^5 = 5(x^2 - 1)^4(2x) = 10x(x^2 - 1)^4.$$

As in this example, the result of using the chain rule is often a product of two functions. Because of this, functions formed by the product of two functions can sometimes be integrated by using the chain rule in reverse. In the example above, working backwards from the derivative gives

$$\int 10x(x^2 - 1)^4\,dx = (x^2 - 1)^5 + C.$$

To find an antiderivative involving products, it often helps to make a substitution: let $u = x^2 - 1$, so that $du = 2x\,dx$. Now substitute u for $x^2 - 1$ and du for $2x\,dx$ in the indefinite integral above.

$$\int 10x(x^2 - 1)^4\,dx = \int 5 \cdot 2x(x^2 - 1)^4\,dx$$

$$= 5\int (x^2 - 1)^4\,(2x\,dx)$$

$$= 5\int u^4\,du$$

This last integral can now be found by the power rule.

$$5\int u^4\,du = 5 \cdot \frac{1}{5}u^5 + C = u^5 + C$$

Finally, substitute $x^2 - 1$ for u.

$$\int 10x(x^2 - 1)^4\,dx = (x^2 - 1)^5 + C$$

This method of integration is called **integration by substitution.** As shown above, it is simply the chain rule for derivatives in reverse. The results can always be verified by differentiation.

This discussion leads to the following integration formula, which is sometimes called the general power rule for integrals.

GENERAL POWER RULE For $u = f(x)$ and $du = f'(x)\,dx$,

$$\int u^n\,du = \frac{u^{n+1}}{n+1} + C.$$

EXAMPLE 1
Find $\int 6x(3x^2 + 4)^4\,dx$.

A certain amount of trial and error may be needed to decide on the expression to set equal to u. The integrand must be written as two factors, one of which is the derivative of the other. In this example, if $u = 3x^2 + 4$, then $du = 6x\,dx$. Now substitute.

$$\int 6x(3x^2 + 4)^4\,dx = \int (3x^2 + 4)^4(6x\,dx) = \int u^4\,du$$

Find this last indefinite integral.

$$\int u^4\,du = \frac{u^5}{5} + C$$

Now replace u with $3x^2 + 4$.

$$\int 6x(3x^2 + 4)^4\,dx = \frac{u^5}{5} + C = \frac{(3x^2 + 4)^5}{5} + C$$

To verify this result, find the derivative.

$$\frac{d}{dx}\left[\frac{(3x^2 + 4)^5}{5} + C\right] = \frac{5}{5}(3x^2 + 4)^4(6x) + 0 = (3x^2 + 4)^4(6x)$$

The derivative is the original function, as required. ◀

EXAMPLE 2
Find $\int x^2\sqrt{x^3 + 1}\,dx$.

An expression raised to a power is usually a good choice for u, so because of the square root or $1/2$ power, let $u = x^3 + 1$; then $du = 3x^2\,dx$. The integrand does not contain the constant 3, which is needed for du. To take care of this, multiply by $3/3$, placing 3 inside the integral sign and $1/3$ outside.

$$\int x^2\sqrt{x^3 + 1}\,dx = \frac{1}{3}\int 3x^2\sqrt{x^3 + 1}\,dx = \frac{1}{3}\int \sqrt{x^3 + 1}\,(3x^2\,dx)$$

Now substitute u for $x^3 + 1$ and du for $3x^2\,dx$, and then integrate.

$$\frac{1}{3}\int \sqrt{x^3 + 1}\ 3x^2\,dx = \frac{1}{3}\int \sqrt{u}\ du = \frac{1}{3}\int u^{1/2}\,du$$

$$= \frac{1}{3} \cdot \frac{u^{3/2}}{3/2} + C = \frac{2}{9}u^{3/2} + C$$

Since $u = x^3 + 1$,

$$\int x^2 \sqrt{x^3 + 1}\ dx = \frac{2}{9}(x^3 + 1)^{3/2} + C. \quad \blacktriangleleft$$

The substitution method given in the examples above *will not always work.* For example, you might try to find

$$\int x^3 \sqrt{x^3 + 1}\ dx$$

by substituting $u = x^3 + 1$, so that $du = 3x^2\,dx$. However, there is no *constant* that can be inserted inside the integral sign to give $3x^2$. This integral, and a great many others, cannot be evaluated by substitution.

With practice, choosing u will become easy if you keep two principles in mind. First, u should equal some expression in the integral that, when replaced with u, tends to make the integral simpler. Second and most important, u must be an expression whose derivative is also present in the integral. The substitution should include as much of the integral as possible, so long as its derivative is still present. In Example 1, we could have chosen $u = 3x^2$, but $u = 3x^2 + 4$ is better, because it has the same derivative as $3x^2$ and captures more of the original integral. If we carry this reasoning further, we might try $u = (3x^2 + 4)^4$, but this is a poor choice, for $du = 4(3x^2 + 4)^3(6x)\,dx$, an expression not present in the original integral.

The exact form of the derivative of u need not be present. Any constant multiplier is easily taken care of, as in Example 2 and the next example.

▶ 3 Find $\displaystyle\int \frac{x + 3}{(x^2 + 6x)^2}\,dx.$

Let $u = x^2 + 6x$, so that $du = (2x + 6)\,dx = 2(x + 3)\,dx$. The integral is missing the 2, so multiply by $2/2$, putting 2 inside the integral sign and $1/2$ outside.

$$\int \frac{x + 3}{(x^2 + 6x)^2}\,dx = \frac{1}{2}\int \frac{2(x + 3)}{(x^2 + 6x)^2}\,dx$$

$$= \frac{1}{2}\int \frac{du}{u^2} = \frac{1}{2}\int u^{-2}\,du$$

$$= \frac{1}{2} \cdot \frac{u^{-1}}{-1} + C = \frac{-1}{2u} + C$$

Substituting $x^2 + 6x$ for u gives

$$\int \frac{x + 3}{(x^2 + 6x)^2}\, dx = \frac{-1}{2(x^2 + 6x)} + C. \quad \blacktriangleleft$$

Recall the formula for $\dfrac{d}{dx}(e^u)$, where $u = f(x)$:

$$\frac{d}{dx}(e^u) = e^u \frac{d}{dx} u.$$

For example, if $u = x^2$ then $\dfrac{d}{dx} u = \dfrac{d}{dx}(x^2) = 2x$, and

$$\frac{d}{dx}(e^{x^2}) = e^{x^2} \cdot 2x.$$

Working backwards, if $u = x^2$, then $du = 2x\, dx$, so

$$\int e^{x^2} \cdot 2x\, dx = \int e^u\, du = e^u + C = e^{x^2} + C.$$

The work above suggests the following rule for the indefinite integral of e^u, where $u = f(x)$.

INDEFINITE INTEGRAL OF e^u	If $u = f(x)$, then $du = f'(x)\, dx$ and $$\int e^u\, du = e^u + C.$$

EXAMPLE 4

Find $\int x^2 \cdot e^{x^3}\, dx$.

Let $u = x^3$, the exponent on e. Then $du = 3x^2\, dx$. Multiplying by 3/3 gives

$$\int x^2 \cdot e^{x^3}\, dx = \frac{1}{3} \int e^{x^3}\, (3x^2\, dx)$$

$$= \frac{1}{3} \int e^u\, du = \frac{1}{3} e^u + C = \frac{1}{3} e^{x^3} + C. \quad \blacktriangleleft$$

Recall that the antiderivative of $f(x) = 1/x$ is $\ln |x|$. The next example uses the fact that $\int x^{-1}\, dx = \ln |x| + C$, together with the method of substitution.

EXAMPLE 5

Find $\displaystyle \int \frac{(2x - 3)\, dx}{x^2 - 3x}$.

Let $u = x^2 - 3x$, so that $du = (2x - 3)\, dx$. Then

$$\int \frac{(2x - 3)\, dx}{x^2 - 3x} = \int \frac{du}{u} = \ln |u| + C = \ln |x^2 - 3x| + C. \quad \blacktriangleleft$$

Generalizing from the results of Example 5 suggests the following rule for finding the indefinite integral of u^{-1}, where $u = f(x)$.

INDEFINITE INTEGRAL OF u^{-1}

If $u = f(x)$, then $du = f'(x)\,dx$ and

$$\int u^{-1}\,du = \int \frac{du}{u} = \ln|u| + C.$$

The next example shows a more complicated integral evaluated by the method of substitution.

EXAMPLE

6

Find $\int x\sqrt{1 - x}\,dx$.

Let $u = 1 - x$. Then $x = 1 - u$ and $dx = -du$. Now substitute:

$$\int x\sqrt{1 - x}\,dx = \int (1 - u)\sqrt{u}(-du) = \int (u - 1)u^{1/2}\,du$$

$$= \int (u^{3/2} - u^{1/2})\,du = \frac{2}{5}u^{5/2} - \frac{2}{3}u^{3/2} + C$$

$$= \frac{2}{5}(1 - x)^{5/2} - \frac{2}{3}(1 - x)^{3/2} + C. \quad \blacktriangleleft$$

The substitution method is useful if the integral can be written in one of the following forms, where $u(x)$ is some function of x.

SUBSTITUTION METHOD

Form of the Integral	*Form of the Antiderivative*		
1. $\int [u(x)]^n \cdot u'(x)\,dx,\; n \neq -1$	$\dfrac{[u(x)]^{n+1}}{n + 1} + C$		
2. $\int e^{u(x)} \cdot u'(x)\,dx$	$e^{u(x)} + C$		
3. $\int \dfrac{u'(x)\,dx}{u(x)}$	$\ln	u(x)	+ C$

EXAMPLE

7

The research department for a hardware chain has determined that at one store the marginal price of x boxes per week of a particular type of nails is

$$p'(x) = \frac{-4000}{(2x + 15)^3}.$$

Find the demand equation if the weekly demand of this type of nails is 10 boxes when the price of a box of nails is $4.

To find the demand function, first integrate $p'(x)$ as follows.

$$p(x) = \int p'(x)\,dx$$

$$= \int \frac{-4000}{(2x + 15)^3}\,dx$$

Let $u = 2x + 15$. Then $du = 2\ dx$, and

$$p(x) = -2000 \int (2x + 15)^{-3}\ 2\ dx$$

$$= -2000 \int u^{-3}\ du$$

$$= (-2000)\frac{u^{-2}}{-2} + C$$

$$= \frac{1000}{u^2} + C$$

$$p(x) = \frac{1000}{(2x + 15)^2} + C. \qquad (1)$$

Find the value of C by using the given information that $p = 4$ when $x = 10$.

$$4 = \frac{1000}{(2 \cdot 10 + 15)^2} + C$$

$$4 = \frac{1000}{35^2} + C$$

$$4 = .82 + C$$

$$3.18 = C$$

Replacing C with 3.18 in equation (1) gives the demand function,

$$p(x) = \frac{1000}{(2x + 15)^2} + 3.18. \quad \blacktriangleleft$$

With a little practice, you will find you can skip the substitution step for integrals such as that shown in Example 7, in which the derivative of u is a constant. Recall from the chain rule that when you differentiate a function, such as $p'(x) = -4000/(2x + 15)^3$ in the previous example, you multiply by 2, the derivative of $(2x + 15)$. So when taking the antiderivative, simply divide by 2:

$$\int -4000(2x + 15)^{-3}\ dx = \frac{-4000}{2} \cdot \frac{(2x + 15)^{-2}}{-2} + C$$

$$= \frac{1000}{(2x + 15)^2} + C.$$

Caution This procedure is valid because of the constant multiple rule presented in the previous section, which says that constant multiples can be brought into or out of integrals, just as they can with derivatives. This procedure is *not* valid with any expression other than a constant.

EXAMPLE **8**

To determine the top 100 popular songs of each year since 1956, Jim Quirin and Barry Cohen developed a function that represents the rate of change on the charts of *Billboard* magazine required for a song to earn a "star" on the *Billboard* "Hot 100" survey.* They developed the function

$$f(x) = \frac{A}{B + x},$$

where $f(x)$ represents the rate of change in position on the charts, x is the position on the "Hot 100" survey, and A and B are appropriate constants. The function

$$F(x) = \int f(x)\, dx$$

is defined as the "Popularity Index." Find $F(x)$.

Integrating $f(x)$ gives

$$F(x) = \int f(x)\, dx$$
$$= \int \frac{A}{B + x}\, dx$$
$$= A \int \frac{1}{B + x}\, dx.$$

Let $u = B + x$, so that $du = dx$. Then

$$F(x) = A \int \frac{1}{u}\, du = A \ln u + C$$
$$= A \ln (B + x) + C.$$

(The absolute value bars are not necessary, since $B + x$ is always positive here.) ◀

14.2 Exercises

Use substitution to find the following indefinite integrals.

1. $\int 4(2x + 3)^4\, dx$

2. $\int (-4t + 1)^3\, dt$

3. $\int \frac{2\, dm}{(2m + 1)^3}$

4. $\int \frac{3\, du}{\sqrt{3u - 5}}$

5. $\int \frac{2x + 2}{(x^2 + 2x - 4)^4}\, dx$

6. $\int \frac{6x^2\, dx}{(2x^3 + 7)^{3/2}}$

7. $\int z\sqrt{z^2 - 5}\, dz$

8. $\int r\sqrt{r^2 + 2}\, dr$

9. $\int (-4e^{2p})\, dp$

10. $\int 5e^{-.3g}\, dg$

11. $\int 3x^2 e^{2x^3}\, dx$

12. $\int re^{-r^2}\, dr$

13. $\int (1 - t)e^{2t - t^2}\, dt$

14. $\int (x^2 - 1)e^{x^3 - 3x}\, dx$

15. $\int \frac{e^{1/z}}{z^2}\, dz$

16. $\int \frac{e^{\sqrt{y}}}{2\sqrt{y}}\, dy$

*Formula for the "Popularity Index" from Jim Quirin and Barry Cohen, *Chartmasters' Rock 100*, 4th ed. Copyright 1987 by Chartmasters. Reprinted by permission.

17. $\int \dfrac{-8}{1 + 3x}\, dx$

18. $\int \dfrac{9}{2 + 5t}\, dt$

19. $\int \dfrac{dt}{2t + 1}$

20. $\int \dfrac{dw}{5w - 2}$

21. $\int \dfrac{v\, dv}{(3v^2 + 2)^4}$

22. $\int \dfrac{x\, dx}{(2x^2 - 5)^3}$

23. $\int \dfrac{x - 1}{(2x^2 - 4x)^2}\, dx$

24. $\int \dfrac{2x + 1}{(x^2 + x)^3}\, dx$

25. $\int \left(\dfrac{1}{r} + r\right)\left(1 - \dfrac{1}{r^2}\right) dr$

26. $\int \left(\dfrac{2}{A} - A\right)\left(\dfrac{-2}{A^2} - 1\right) dA$

27. $\int \dfrac{x^2 + 1}{(x^3 + 3x)^{2/3}}\, dx$

28. $\int \dfrac{B^3 - 1}{(2B^4 - 8B)^{3/2}}\, dB$

29. $\int p(p + 1)^5\, dp$

30. $\int x^3(1 + x^2)^{1/4}\, dx$

31. $\int t\sqrt{5t - 1}\, dt$

32. $\int 4r\sqrt{8 - r}\, dr$

33. $\int \dfrac{u}{\sqrt{u - 1}}\, du$

34. $\int \dfrac{2x}{(x + 5)^6}\, dx$

35. $\int (\sqrt{x^2 + 12x})(x + 6)\, dx$

36. $\int (\sqrt{x^2 - 6x})(x - 3)\, dx$

37. $\int \dfrac{t}{t^2 + 2}\, dt$

38. $\int \dfrac{-4x}{x^2 + 3}\, dx$

39. $\int ze^{2z^2}\, dz$

40. $\int x^2 e^{-x^3}\, dx$

41. $\int \dfrac{(1 + \ln x)^2}{x}\, dx$

42. $\int \dfrac{\sqrt{2 + \ln x}}{x}\, dx$

43. $\int x^{3/2}\sqrt{x^{5/2} + 4}\, dx$

44. $\int \dfrac{1}{x(\ln x)}\, dx$

Applications

Business and Economics

45. Revenue Suppose the marginal revenue in dollars from the sale of x jet planes is
$$R'(x) = 2x(x^2 + 50)^2.$$
(a) Find the total revenue function if the revenue from 3 planes is $206,379.
(b) How many planes must be sold for a revenue of at least $450,000?

46. Maintenance The rate of expenditure for maintenance of a particular machine is given by
$$M'(x) = \sqrt{x^2 + 12x}\,(2x + 12),$$
where x is time measured in years. Maintenance costs for the fourth year are $612.
(a) Find the total maintenance function.
(b) How many years must pass before the annual maintenance costs reach $2000?

47. Profit The rate of growth of the profit (in millions of dollars) from a new technology is approximated by
$$P'(x) = xe^{-x^2},$$
where x represents time measured in years. The total profit in the third year that the new technology is in operation is $10,000.
(a) Find the total profit function.
(b) What happens to the total amount of profit in the long run?

48. Cost A company has found that the marginal cost of a new production line (in thousands) is
$$C'(x) = -100(x + 10)^{-2},$$
where x is the number of years the line is in use.
(a) Find the total cost function for the production line. The fixed cost is 10 thousand dollars.
(b) The company will add the new line if the total cost is reduced to $2000 within 5 years. Should they add the new line?

Life Sciences

49. Drug Sensitivity If x milligrams of a certain drug are administered to a person, the rate of change in the person's temperature (in degrees Celsius) with respect to the dosage (the person's *sensitivity* to the drug) is given by
$$D'(x) = \dfrac{2}{x + 9}.$$
One milligram raises the person's temperature $2.5°$ C.
(a) Find the function giving the total change in body temperature.
(b) How much of the drug is necessary to change the person's temperature $3°$ C?

General Interest

50. Border Crossing Canadians who live near the United States border have been shopping in the United States at an increasing rate since 1986. They cross the border to

shop because large savings are possible on groceries, gasoline, cigarettes, and children's clothes. The rate of change in the number of Canadian cross-border shoppers (in millions) is given by

$$S'(t) = 4.4e^{.16t},$$

where t is the number of years since 1986.

(a) Find $S(t)$ if there were 27.3 million cross-border shoppers in 1986 (year 0).

(b) In how many years will the number of Canadian cross-border shoppers double?

14.3 AREA AND THE DEFINITE INTEGRAL

 If the rate of annual maintenance charges is known, how can the total amount paid for maintenance over a 10-year period be estimated?

This section introduces a method for answering such questions.

To calculate the areas of geometric figures such as rectangles, squares, triangles, and circles, we use specific formulas. In this section we consider the problem of finding the area of a figure or region that is bounded by curves, such as the shaded region in Figure 2.

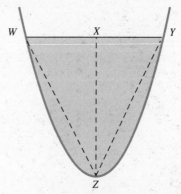

Area of parabolic segment

$= \dfrac{4}{3}$ (area of triangle WYZ)

FIGURE 2

The brilliant Greek mathematician Archimedes (about 287 B.C.–212 B.C.) is considered one of the greatest mathematicians of all time. His development of a rigorous method known as ''exhaustion'' to derive results was a forerunner of the ideas of integral calculus. Archimedes used a method that would later be verified by the more rigorous theory of integration. His method involved viewing a geometric figure as a sum of other figures. For example, he thought of a plane surface area as a figure consisting of infinitely many parallel line segments. Among the results established by Archimedes' method was the fact that the area of a segment of a parabola (shown in color in Figure 2) is equal to 4/3 the area of a triangle with the same base and the same height.

Under certain conditions the area of a region can be thought of as a sum of parts. Figure 3 shows the region bounded by the y-axis, the x-axis, and the graph of $f(x) = \sqrt{4 - x^2}$. A very rough approximation of the area of this region can be found by using two rectangles as in Figure 4. The height of the rectangle on the left is $f(0) = 2$ and the height of the rectangle on the right is $f(1) = \sqrt{3}$. The width of each rectangle is 1, making the total area of the two rectangles

$$1 \cdot f(0) + 1 \cdot f(1) = 2 + \sqrt{3} \approx 3.7321 \text{ square units.}$$

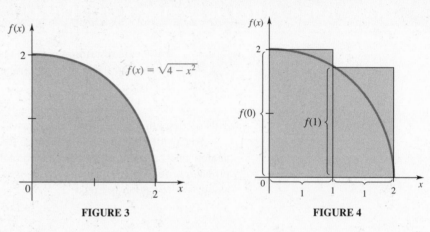

FIGURE 3 FIGURE 4

As shown in Figure 4, this approximation is greater than the actual area. To improve the accuracy of the approximation, we could divide the interval from $x = 0$ to $x = 2$ into four equal parts, each of width $1/2$, as shown in Figure 5. As before, the height of each rectangle is given by the value of f at the left side of the rectangle, and its area is the width, $1/2$, multiplied by the height. The total area of the four rectangles is

$$\frac{1}{2} \cdot f(0) + \frac{1}{2} \cdot f\left(\frac{1}{2}\right) + \frac{1}{2} \cdot f(1) + \frac{1}{2} \cdot f\left(1\frac{1}{2}\right)$$

$$= \frac{1}{2}(2) + \frac{1}{2}\left(\frac{\sqrt{15}}{2}\right) + \frac{1}{2}(\sqrt{3}) + \frac{1}{2}\left(\frac{\sqrt{7}}{2}\right)$$

$$= 1 + \frac{\sqrt{15}}{4} + \frac{\sqrt{3}}{2} + \frac{\sqrt{7}}{4} \approx 3.4957 \text{ square units.}$$

This approximation looks better, but it is still greater than the actual area desired. To improve the approximation, divide the interval from $x = 0$ to $x = 2$ into 8 parts with equal widths of $1/4$ (see Figure 6). The total area of all these rectangles is

$$\frac{1}{4} \cdot f(0) + \frac{1}{4} \cdot f\left(\frac{1}{4}\right) + \frac{1}{4} \cdot f\left(\frac{1}{2}\right) + \frac{1}{4} \cdot f\left(\frac{3}{4}\right) + \frac{1}{4} \cdot f(1) + \frac{1}{4} \cdot f\left(\frac{5}{4}\right)$$

$$+ \frac{1}{4} \cdot f\left(\frac{3}{2}\right) + \frac{1}{4} \cdot f\left(\frac{7}{4}\right)$$

$$\approx 3.3398 \text{ square units.}$$

FIGURE 5

FIGURE 6

This process of approximating the area under a curve by using more and more rectangles to get a better and better approximation can be generalized. To do this, divide the interval from $x = 0$ to $x = 2$ into n equal parts. Each of these n intervals has width

$$\frac{2 - 0}{n} = \frac{2}{n},$$

so each rectangle has width $2/n$ and height determined by the function-value at the left side of the rectangle. Using a computer to find approximations to the area for several values of n gives the results in the following table.

n	Area
2	3.7321
4	3.4957
8	3.3398
10	3.3045
20	3.2285
50	3.1783
100	3.1512
500	3.1455

The areas in the last column in the table are approximations of the area under the curve, above the x-axis, and between the lines $x = 0$ and $x = 2$. As n becomes larger and larger, the approximation is better and better, getting closer to the actual area. In this example, the exact area can be found by a formula from plane geometry. Write the given function as

$$y = \sqrt{4 - x^2},$$

then square both sides to get

$$y^2 = 4 - x^2$$
$$x^2 + y^2 = 4,$$

the equation of a circle centered at the origin with radius 2. The region in Figure 3 is the quarter of this circle that lies in the first quadrant. The actual area of this region is one-quarter of the area of the entire circle, or

$$\frac{1}{4}\pi(2)^2 = \pi \approx 3.1416.$$

As the number of rectangles increases without bound, the sum of the areas of these rectangles gets closer and closer to the actual area of the region, π. This can be written as

$$\lim_{n \to \infty} (\text{sum of areas of } n \text{ rectangles}) = \pi.$$

(The value of π was originally found by a process similar to this.)*

This approach could be used to find the area of any region bounded by the x-axis, the lines $x = a$ and $x = b$, and the graph of a function $y = f(x)$. At this point, we need some new notation and terminology to write sums concisely. We will indicate addition (or summation) by using the Greek letter sigma, Σ, as shown in the next example.

EXAMPLE 1 Find the following sums.

(a) $\displaystyle\sum_{i=1}^{5} i$

Replace i in turn with the integers 1 through 5 and add the resulting terms.

$$\sum_{i=1}^{5} i = 1 + 2 + 3 + 4 + 5 = 15$$

(b) $\displaystyle\sum_{i=1}^{4} a_i = a_1 + a_2 + a_3 + a_4$

(c) $\displaystyle\sum_{i=1}^{3} (6x_i - 2)$ if $x_1 = 2, x_2 = 4, x_3 = 6$

Let $i = 1, 2,$ and 3, respectively, to get

$$\sum_{i=1}^{3} (6x_i - 2) = (6x_1 - 2) + (6x_2 - 2) + (6x_3 - 2).$$

*The number π is the ratio of the circumference of a circle to its diameter. It is an example of an *irrational number*, and as such it cannot be expressed as a terminating or repeating decimal. Many approximations have been used for π over the years. A passage in the Bible (I Kings 7:23) indicates a value of 3. The Egyptians used the value 3.16, and Archimedes showed that its value must be between 22/7 and 223/71. A Hindu writer, Brahmagupta, used $\sqrt{10}$ as its value in the seventh century. The search for the digits of π has continued into modern times. Two mathematician brothers named Chudnovsky, both at Columbia University, recently computed the value to over 2 billion places.

Now substitute the given values for x_1, x_2, and x_3.

$$\sum_{i=1}^{3} (6x_i - 2) = (6 \cdot 2 - 2) + (6 \cdot 4 - 2) + (6 \cdot 6 - 2)$$

$$= 10 + 22 + 34$$

$$= 66$$

(d) $\sum_{i=1}^{4} f(x_i)\Delta x$ if $f(x) = x^2$, $x_1 = 0$, $x_2 = 2$, $x_3 = 4$, $x_4 = 6$, and $\Delta x = 2$

$$\sum_{i=1}^{4} f(x_i)\Delta x = f(x_1)\Delta x + f(x_2)\Delta x + f(x_3)\Delta x + f(x_4)\Delta x$$

$$= x_1^2\Delta x + x_2^2\Delta x + x_3^2\Delta x + x_4^2\Delta x$$

$$= 0^2(2) + 2^2(2) + 4^2(2) + 6^2(2)$$

$$= 0 + 8 + 32 + 72 = 112 \blacktriangleleft$$

Now we can generalize to get a method of finding the area bounded by the curve $y = f(x)$, the x-axis, and the vertical lines $x = a$ and $x = b$, as shown in Figure 7. To approximate this area, we could divide the region under the curve first into 10 rectangles (Figure 7(a)) and then into 20 rectangles (Figure 7(b)). The sum of the areas of the rectangles give approximations to the area under the curve.

Ten rectangles of equal width

(a)

Twenty rectangles of equal width

(b)

FIGURE 7

To get a number that can be defined as the *exact* area, begin by dividing the interval from a to b into n pieces of equal width, using each of these n pieces as the base of a rectangle (see Figure 8). The left endpoints of the n intervals are labeled $x_1, x_2, x_3, \ldots, x_{n+1}$, where $a = x_1$ and $b = x_{n+1}$. In the graph of Figure 8, the symbol Δx is used to represent the width of each of the intervals. The darker rectangle is an arbitrary rectangle called the ith rectangle. Its area is the

product of its length and width. Since the width of the ith rectangle is Δx and the length of the ith rectangle is given by the height $f(x_i)$,

$$\text{Area of the } i\text{th rectangle} = f(x_i) \cdot \Delta x.$$

FIGURE 8

The total area under the curve is approximated by the sum of the areas of all n of the rectangles. With sigma notation, the approximation to the total area becomes

$$\text{Area of all } n \text{ rectangles} = \sum_{i=1}^{n} f(x_i) \cdot \Delta x.$$

The exact area is defined to be the limit of this sum (if the limit exists) as the number of rectangles increases without bound:

$$\text{Exact area} = \lim_{n \to \infty} \sum_{i=1}^{n} f(x_i) \Delta x.$$

This limit is called the *definite integral* of $f(x)$ from a to b and is written as follows.

THE DEFINITE INTEGRAL

If f is defined on the interval $[a, b]$, the **definite integral** of f from a to b is given by

$$\int_{a}^{b} f(x)\,dx = \lim_{n \to \infty} \sum_{i=1}^{n} f(x_i)\Delta x,$$

provided the limit exists, where $\Delta x = (b - a)/n$ and x_i is *any* value of x in the ith interval.

As indicated in this definition, although the left endpoint of the ith interval has been used to find the height of the ith rectangle, any number in the ith interval can be used. (A more general definition is possible in which the rectangles do not necessarily all have the same width.) The b above the integral sign is called the **upper limit** of integration, and the a is the **lower limit** of integration. This use of the word "limit" has nothing to do with the limit of the sum; it refers to the limits, or boundaries, on x.

The sum in the definition of the definite integral is an example of a Riemann sum, named for the German mathematician Georg Riemann (1826–1866) who was responsible for the early development of integrability of functions. The concepts of "Riemann sum" and "Riemann integral" are still studied in rigorous calculus textbooks.

In the example at the beginning of this section, the area bounded by the x-axis, the curve $y = \sqrt{4 - x^2}$, the lines $x = 0$ and $x = 2$ could be written as the definite integral

$$\int_0^2 \sqrt{4 - x^2} \, dx = \pi.$$

Notice that unlike the indefinite integral, which is a set of *functions*, the definite integral represents a *number*. The next section will show how antiderivatives are used in finding the definite integral and thus the area under a curve.

Keep in mind that finding the definite integral of a function can be thought of as a mathematical process that gives the sum of an infinite number of individual parts (within certain limits). The definite integral represents area only if the function involved is *nonnegative* ($f(x) \geq 0$) at every x-value in the interval $[a, b]$. There are many other interpretations of the definite integral, and all of them involve this idea of approximation by appropriate sums.

EXAMPLE 2

Approximate $\int_0^4 2x \, dx$, the area of the region under the graph of $f(x) = 2x$, above the x-axis, and between $x = 0$ and $x = 4$, by using four rectangles of equal width whose heights are the values of the function at the midpoint of each rectangle.

We want to find the area of the shaded region in Figure 9. The heights of the four rectangles given by $f(x_i)$ for $i = 1, 2, 3,$ and 4 are as follows.

FIGURE 9

i	x_i	$f(x_i)$
1	$x_1 = \ .5$	$f(.5) = 1.0$
2	$x_2 = 1.5$	$f(1.5) = 3.0$
3	$x_3 = 2.5$	$f(2.5) = 5.0$
4	$x_4 = 3.5$	$f(3.5) = 7.0$

The width of each rectangle is $\Delta x = (4 - 0)/4 = 1$. The sum of the areas of the four rectangles is

$$\sum_{i=1}^{4} f(x_i)\Delta x = f(x_1)\Delta x + f(x_2)\Delta x + f(x_3)\Delta x + f(x_4)\Delta x$$

$$= f(.5)\Delta x + f(1.5)\Delta x + f(2.5)\Delta x + f(3.5)\Delta x$$

$$= (1)(1) + (3)(1) + 5(1) + 7(1)$$

$$= 16.$$

Using the formula for the area of a triangle, $A = (1/2)bh$, with b, the length of the base, equal to 4 and h, the height, equal to 8, gives

$$A = \frac{1}{2}bh = \frac{1}{2}(4)(8) = 16,$$

the exact value of the area. The approximation equals the exact area in this case because our use of the midpoints of each subinterval distributed the error evenly above and below the graph. ◀

Total Change Suppose the function $f(x) = x^2 + 20$ gives the marginal cost of some item at a particular x-value. Then $f(2) = 24$ gives the rate of change of cost at $x = 2$. That is, a unit change in x (at this point) will produce a change of 24 units in the cost function. Also, $f(3) = 29$ means that each unit of change in x (when $x = 3$) will produce a change of 29 units in the cost function.

To find the *total* change in the cost function as x changes from 2 to 3, we could divide the interval from 2 to 3 into n equal parts, using each part as the base of a rectangle as we did above. The area of each rectangle would approximate the change in cost at the x-value that is the left endpoint of the base of the rectangle. Then the sum of the areas of these rectangles would approximate the net total change in cost from $x = 2$ to $x = 3$. The limit of this sum as $n \to \infty$ would give the exact total change.

This result produces another application of the definite integral: the area of the region under the graph of the marginal cost function $f(x)$ that is above the x-axis and between $x = a$ and $x = b$ gives the *net total change in the cost* as x goes from a to b.

TOTAL CHANGE IN $F(x)$

If $f(x)$ gives the rate of change of $F(x)$ for x in $[a, b]$, then the **total change** in $F(x)$ as x goes from a to b is given by

$$\lim_{n\to\infty} \sum_{i=1}^{n} f(x_i)\Delta x = \int_a^b f(x)\,dx.$$

In other words, the total change in a quantity can be found from the function that gives the rate of change of the quantity, using the same methods used to approximate the area under a curve.

EXAMPLE 3 Figure 10 shows the rate of change of the annual maintenance charges for a certain machine. To approximate the total maintenance charges over the 10-year life of the machine, use approximating rectangles, dividing the interval from 0 to 10 into ten equal subdivisions. Each rectangle has width 1; using the left endpoint of each rectangle to determine the height of the rectangle, the approximation becomes

$$1 \cdot 0 + 1 \cdot 500 + 1 \cdot 750 + 1 \cdot 1800 + 1 \cdot 1800 + 1 \cdot 3000 + 1 \cdot 3000$$
$$+ 1 \cdot 3400 + 1 \cdot 4200 + 1 \cdot 5200 = 23{,}650.$$

About $23,550 will be spent on maintenance over the 10-year life of the machine. ◀

FIGURE 10

Before discussing further applications of the definite integral, we need a more efficient method for evaluating it. This method will be developed in the next section.

14.3 Exercises

Evaluate each sum.

1. $\displaystyle\sum_{i=1}^{3} 3i$

2. $\displaystyle\sum_{i=1}^{6} (-5i)$

3. $\displaystyle\sum_{i=1}^{5} (2i + 7)$

4. $\displaystyle\sum_{i=1}^{10} (5i - 8)$

5. Let $x_1 = -5$, $x_2 = 8$, $x_3 = 7$, and $x_4 = 10$. Find $\displaystyle\sum_{i=1}^{4} x_i$.

6. Let $x_1 = 10$, $x_2 = 15$, $x_3 = -8$, $x_4 = -12$, and $x_5 = 0$. Find $\displaystyle\sum_{i=1}^{5} x_i$.

7. Let $f(x) = x - 3$ and $x_1 = 4$, $x_2 = 6$, $x_3 = 7$. Find
$$\sum_{i=1}^{3} f(x_i).$$

8. Let $f(x) = x^2 + 1$ and $x_1 = -2$, $x_2 = 0$, $x_3 = 2$, $x_4 = 4$. Find $\sum_{i=1}^{4} f(x_i)$.

9. Let $f(x) = 2x + 1$, $x_1 = 0$, $x_2 = 2$, $x_3 = 4$, $x_4 = 6$, and $\Delta x = 2$.

 (a) Find $\sum_{i=1}^{4} f(x_i)\Delta x$.

 (b) The sum in part (a) approximates a definite integral using rectangles. The height of each rectangle is given by the value of the function at the left endpoint. Write the definite integral that the sum approximates.

10. Let $f(x) = 1/x$, $x_1 = 1/2$, $x_2 = 1$, $x_3 = 3/2$, $x_4 = 2$, and $\Delta x = 1/2$.

 (a) Find $\sum_{i=1}^{4} f(x_i)\Delta x$.

 (b) The sum in part (a) approximates a definite integral using rectangles. The height of each rectangle is given by the value of the function at the left endpoint. Write the definite integral that the sum approximates.

In Exercises 11–22, first approximate the area under the given curve and above the x-axis by using two rectangles. Let the height of the rectangle be given by the value of the function at the left side of the rectangle. Then repeat the process and approximate the area using four rectangles.

11. $f(x) = 3x + 2$ from $x = 1$ to $x = 5$
12. $f(x) = -2x + 1$ from $x = -4$ to $x = 0$
13. $f(x) = x + 5$ from $x = 2$ to $x = 4$

14. $f(x) = 3 + x$ from $x = 1$ to $x = 3$
15. $f(x) = x^2$ from $x = 1$ to $x = 5$
16. $f(x) = x^2$ from $x = 0$ to $x = 4$
17. $f(x) = x^2 + 2$ from $x = -2$ to $x = 2$
18. $f(x) = -x^2 + 4$ from $x = -2$ to $x = 2$
19. $f(x) = e^x - 1$ from $x = 0$ to $x = 4$
20. $f(x) = e^x + 1$ from $x = -2$ to $x = 2$
21. $f(x) = \dfrac{1}{x}$ from $x = 1$ to $x = 5$
22. $f(x) = \dfrac{2}{x}$ from $x = 1$ to $x = 9$

23. Consider the region below $f(x) = x/2$, above the x-axis, between $x = 0$ and $x = 4$. Let x_i be the left endpoint of the ith subinterval.

 (a) Approximate the area of the region using four rectangles.

 (b) Approximate the area of the region using eight rectangles.

 (c) Find $\int_0^4 f(x)\,dx$ by using the formula for the area of a triangle.

24. Find $\int_0^5 (5 - x)\,dx$ by using the formula for the area of a triangle.

Find the exact value of each of the following integrals using formulas from geometry.

25. $\displaystyle\int_0^3 2x\,dx$

26. $\displaystyle\int_1^5 \frac{7}{2}\,dx$

27. $\displaystyle\int_{-3}^3 \sqrt{9 - x^2}\,dx$

28. $\displaystyle\int_{-4}^0 \sqrt{16 - x^2}\,dx$

29. $\displaystyle\int_1^3 (5 - x)\,dx$

30. $\displaystyle\int_2^5 (1 + 2x)\,dx$

Applications

In Exercises 31–35, estimate the area under each curve by summing the area of rectangles. Let the function value at the left side of each rectangle give the height of the rectangle.

Business and Economics

31. **Oil Consumption** The graph on the next page shows U.S. oil production and consumption rates in millions of barrels per day. The rates for the years beyond 1991 are projected using the policy in place in 1990 (labeled "Current policy base" in the graph) and using a new policy proposed by the Bush administration in 1991 (labeled "With strategy" in the graph).*

 (a) Estimate the amount of oil produced between 1990 and 2010 using the policy in place in 1990. Use rectangles with widths of 2 yr.

 (b) Estimate the amount of oil produced between 1990 and 2010 using the proposed policy. Use rectangles with widths of 2 yr.

*Graph, "Uncle Sam's Energy Strategy" from National Energy Strategy, February 1991. United States Department of Energy, Washington D.C.

EXERCISE 31

32. Electricity Consumption The graph below shows the rate of use of electrical energy (in kilowatt hours) in a certain city on a very hot day. Estimate the total usage of electricity on that day. Let the width of each rectangle be 2 hr.

33. Wages The graph shows the average manufacturing hourly wage in the United States and Canada for 1987–1991.* All figures are in U.S. dollars. Assume that the average employee works 2000 hours per year.

(a) Estimate the total amount earned by an average U.S. worker during the four-year period from the beginning of 1987 to the beginning of 1991. Use rectangles with widths of 1 yr.

(b) Estimate the total amount earned (in U.S. dollars) by an average Canadian worker during the four-year period from the beginning of 1987 to the beginning of 1991. Use rectangles with widths of 1 yr.

Life Sciences

34. Alcohol Concentration The graph shows the approximate concentration of alcohol in a person's bloodstream t hr after drinking 2 oz of alcohol. Estimate the total amount of alcohol in the bloodstream by estimating the area under the curve. Use rectangles with widths of 1 hr.

35. Oxygen Inhalation The graph shows the rate of inhalation of oxygen (in liters per minute) by a person riding a bicycle very rapidly for 10 min. Estimate the total volume of oxygen inhaled in the first 20 min after the beginning of the ride. Use rectangles with widths of 1 min.

Physical Sciences

Distance *The next two graphs are from* Road & Track *magazine.* * *The curve shows the velocity at t seconds after the car accelerates from a dead stop. To find the total distance traveled by the car in reaching* 100 *mph, we must estimate the definite integral*

$$\int_0^T v(t)\, dt,$$

where T represents the number of seconds it takes for the car to reach 100 *mph.*

 Use the graphs to estimate this distance by adding the areas of rectangles with widths of 3 *sec. The last rectangle will have a width of* 1 *sec. To adjust your answer to miles per hour, divide by* 3600 *(the number of seconds in an hour). You then have the number of miles that the car traveled in reaching* 100 *mph. Finally, multiply by* 5280 *feet per mile to convert the answer to feet.*

Porsche 928

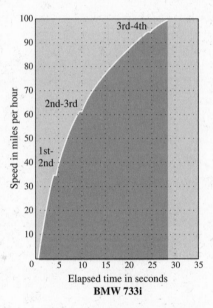

BMW 733i

36. Estimate the distance traveled by the Porsche 928, using the graph on the left.

37. Estimate the distance traveled by the BMW 733i, using the graph on the right.

Heat Gain *The graphs on the next page[†] show the typical heat gain, in BTUs per hour per square foot, for a window facing east and one facing south, with plain glass and with a black ShadeScreen. Estimate the total heat gain per square foot by summing the areas of the rectangles. Use rectangles with widths of* 2 *hr, and let the function value at the midpoint of the rectangle give the height of the rectangle.*

38. **(a)** Estimate the total heat gain per square foot for a plain glass window facing east.

 (b) Estimate the total heat gain per square foot for a window facing east with a ShadeScreen.

39. **(a)** Estimate the total heat gain per square foot for a plain glass window facing south.

 (b) Estimate the total heat gain per square foot for a window facing south with a ShadeScreen.

*From *Road & Track*, April and May, 1978. Reprinted with permission of *Road & Track*.
[†]Graphs courtesy of Phifer Wire Products. Reprinted by permission of Phifer Wire Products.

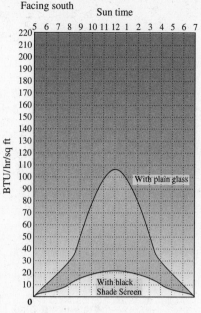

Facing east Sun time

Facing south Sun time

Computer/Graphing Calculator

In Exercises 40–43, use the method given in the text to approximate the area between the x-axis and the graph of each function on the given interval.

40. $f(x) = x \ln x$; $[1, 5]$

41. $f(x) = x^2 e^{-x}$; $[-1, 3]$

42. $f(x) = \dfrac{\ln x}{x}$; $[1, 5]$

43. $f(x) = \dfrac{e^x - e^{-x}}{2}$; $[0, 4]$

14.4 THE FUNDAMENTAL THEOREM OF CALCULUS

?

If we know how the rate of consumption of natural gas varies over time, how can we compute the total amount of natural gas used?

This section introduces a powerful theorem for answering such questions.

The work from the last two sections can now be put together. We have seen that, if $f(x) > 0$,

$$\int_a^b f(x) \, dx$$

gives the area between the graph of $f(x)$ and the x-axis, from $x = a$ to $x = b$. We can find this definite integral by using the antiderivatives discussed earlier.

The definite integral was defined and evaluated in the previous section using the limit of a sum. In that section, we also saw that if $f(x)$ gives the rate of change of $F(x)$, the definite integral $\int_a^b f(x)\,dx$ gives the total change of $F(x)$ as x changes from a to b. If $f(x)$ gives the rate of change of $F(x)$, then $F(x)$ is an antiderivative of $f(x)$. Writing the total change in $F(x)$ from $x = a$ to $x = b$ as $F(b) - F(a)$ shows the connection between antiderivatives and definite integrals. This relationship is called the **Fundamental Theorem of Calculus.**

FUNDAMENTAL THEOREM OF CALCULUS

Let f be continuous on the interval $[a, b]$, and let F be *any* antiderivative of f. Then

$$\int_a^b f(x)\,dx = F(b) - F(a) = F(x)\Big|_a^b.$$

The symbol $F(x)\big|_a^b$ is used to represent $F(b) - F(a)$. It is important to note that the Fundamental Theorem does not require $f(x) > 0$. The condition $f(x) > 0$ is necessary only when using the Fundamental Theorem to find area. Also, note that the Fundamental Theorem does not *define* the definite integral; it just provides a method for evaluating it.

 EXAMPLE 1

First find $\int 4t^3\,dt$ and then find $\int_1^2 4t^3\,dt$.
By the rules given earlier,

$$\int 4t^3\,dt = t^4 + C.$$

By the Fundamental Theorem, the value of $\int_1^2 4t^3\,dt$ is found by evaluating $t^4\big|_1^2$, with no constant C required.

$$\int_1^2 4t^3\,dt = t^4\Big|_1^2 = 2^4 - 1^4 = 15 \quad \blacktriangleleft$$

Example 1 illustrates the difference between the definite integral and the indefinite integral. A definite integral is a real number; an indefinite integral is a family of functions in which all the functions are antiderivatives of a function f.

To see why the Fundamental Theorem of Calculus is true for $f(x) > 0$, look at Figure 11. Define the function $A(x)$ as the area between the x-axis and the graph of $y = f(x)$ from a to x. We first show that A is an antiderivative of f; that is, $A' = f$.

To do this, let h be a small positive number. Then $A(x + h) - A(x)$ is the shaded area in Figure 11. This area can be approximated with a rectangle having width h and height $f(x)$. The area of the rectangle is $h \cdot f(x)$, and

$$A(x + h) - A(x) \approx h \cdot f(x).$$

FIGURE 11

Dividing both sides by h gives

$$\frac{A(x + h) - A(x)}{h} \approx f(x).$$

This approximation improves as h gets smaller and smaller. Take the limit on the left as h approaches 0.

$$\lim_{h \to 0} \frac{A(x + h) - A(x)}{h} = f(x)$$

This limit is simply $A'(x)$, so

$$A'(x) = f(x).$$

This result means that A is an antiderivative of f, as we set out to show.

Since $A(x)$ is the area under the curve $y = f(x)$ from a to x, $A(a) = 0$. The expression $A(b)$ is the area from a to b, the desired area. Since A is an antiderivative of f,

$$\int_a^b f(x)\, dx = A(x) \Big|_a^b$$

$$= A(b) - A(a)$$

$$= A(b) - 0 = A(b).$$

This suggests the proof of the Fundamental Theorem: for $f(x) > 0$, the area under the curve $y = f(x)$ from a to b is given by $\int_a^b f(x)\, dx$.

The Fundamental Theorem of Calculus certainly deserves its name, which sets it apart as the most important theorem of calculus. It is the key connection between differential calculus and integral calculus, which originally were developed separately without knowledge of this connection between them.

The variable used in the integrand does not matter; each of the following definite integrals represents the number $F(b) - F(a)$.

$$\int_a^b f(x)\, dx = \int_a^b f(t)\, dt = \int_a^b f(u)\, du$$

Caution Because the definite integral is a number, there is no constant C added, as there is for the indefinite integral. Even if C were added to the antiderivative F, it would be eliminated in the final answer:

$$\int_a^b f(x)\, dx = (F(x) + C) \Big|_a^b$$

$$= (F(b) + C) - (F(a) + C)$$

$$= F(b) - F(a).$$

Key properties of definite integrals are listed on the next page. Some of them are just restatements of properties from Section 1.

PROPERTIES OF DEFINITE INTEGRALS

If all indicated definite integrals exist,

1. $\displaystyle\int_a^a f(x)\,dx = 0;$

2. $\displaystyle\int_a^b k \cdot f(x)\,dx = k \cdot \int_a^b f(x)\,dx$ for any real constant k

(constant multiple of a function);

3. $\displaystyle\int_a^b [f(x) \pm g(x)]\,dx = \int_a^b f(x)\,dx \pm \int_a^b g(x)\,dx$

(sum or difference of functions); and

4. $\displaystyle\int_a^b f(x)\,dx = \int_a^c f(x)\,dx + \int_c^b f(x)\,dx$ for any real number c.

FIGURE 12

For $f(x) > 0$, since the distance from a to a is 0, the first property says that the "area" under the graph of f bounded by $x = a$ and $x = a$ is 0. Also, since $\int_a^c f(x)\,dx$ represents the darker region in Figure 12, and $\int_c^b f(x)\,dx$ represents the lighter region,

$$\int_a^b f(x)\,dx = \int_a^c f(x)\,dx + \int_c^b f(x)\,dx,$$

as stated in the fourth property. While the figure shows $a < c < b$, the property is true for any value of c where both $f(x)$ and $F(x)$ are defined.

An algebraic proof is given here for the third property; proofs of the other properties are left for the exercises. If $F(x)$ and $G(x)$ are antiderivatives of $f(x)$ and $g(x)$ respectively,

$$\int_a^b [f(x) + g(x)]\,dx = [F(x) + G(x)]\Big|_a^b$$
$$= [F(b) + G(b)] - [F(a) + G(a)]$$
$$= [F(b) - F(a)] + [G(b) - G(a)]$$
$$= \int_a^b f(x)\,dx + \int_a^b g(x)\,dx.$$

 EXAMPLE 2

Find $\int_2^5 (6x^2 - 3x + 5)\,dx$.

Use the properties above, and the Fundamental Theorem, along with properties from Section 1.

$$\int_2^5 (6x^2 - 3x + 5)\,dx = 6\int_2^5 x^2\,dx - 3\int_2^5 x\,dx + 5\int_2^5 dx$$

$$= 2x^3\Big|_2^5 - \frac{3}{2}x^2\Big|_2^5 + 5x\Big|_2^5$$

$$= 2(5^3 - 2^3) - \frac{3}{2}(5^2 - 2^2) + 5(5 - 2)$$

$$= 2(125 - 8) - \frac{3}{2}(25 - 4) + 5(3)$$

$$= 234 - \frac{63}{2} + 15 = \frac{435}{2} \quad \blacktriangleleft$$

EXAMPLE 3

$$\int_1^2 \frac{dy}{y} = \ln |y| \Big|_1^2 = \ln |2| - \ln |1|$$

$$= \ln 2 - \ln 1 \approx .6931 - 0 = .6931 \quad \blacktriangleleft$$

EXAMPLE 4

Evaluate $\int_0^5 x\sqrt{25 - x^2}\, dx$.

Use substitution. Let $u = 25 - x^2$, so that $du = -2x\, dx$. With a definite integral, the limits should be changed, too. The new limits on u are found as follows.

$$\text{If } x = 5, \text{ then } u = 25 - 5^2 = 0.$$
$$\text{If } x = 0, \text{ then } u = 25 - 0^2 = 25.$$

Then

$$\int_0^5 x\sqrt{25 - x^2}\, dx = -\frac{1}{2} \int_0^5 \sqrt{25 - x^2}(-2x\, dx)$$

$$= -\frac{1}{2} \int_{25}^0 \sqrt{u}\, du \qquad \text{Substitute and change limits.}$$

$$= -\frac{1}{2} \int_{25}^0 u^{1/2}\, du$$

$$= -\frac{1}{2} \cdot \frac{u^{3/2}}{3/2} \Big|_{25}^0$$

$$= -\frac{1}{2} \cdot \frac{2}{3} [0^{3/2} - 25^{3/2}]$$

$$= -\frac{1}{3}(-125)$$

$$= \frac{125}{3}. \quad \blacktriangleleft$$

Caution When substitution is used on a definite integral, it is best to not revert from u back to the original variable after antidifferentiation, as we did for the indefinite integral. Notice in Example 4 that after changing from the old limits on x to the new limits on u, we never returned to x or its limits.

Area In the previous section we saw that, if $f(x) > 0$ in $[a, b]$, the definite integral $\int_a^b f(x)\,dx$ gives the area below the graph of the function $y = f(x)$, above the x-axis, and between the lines $x = a$ and $x = b$.

To see how to work around the requirement that $f(x) > 0$, look at the graph of $f(x) = x^2 - 4$ in Figure 13. The area bounded by the graph of f, the x-axis, and the vertical lines $x = 0$ and $x = 2$ lies below the x-axis. Using the Fundamental Theorem to find this area gives

$$\int_0^2 (x^2 - 4)\,dx = \left(\frac{x^3}{3} - 4x\right)\Big|_0^2$$

$$= \left(\frac{8}{3} - 8\right) - (0 - 0) = -\frac{16}{3}.$$

The result is a negative number because $f(x)$ is negative for values of x in the interval $[0, 2]$. Since Δx is always positive, if $f(x) < 0$ the product $f(x) \cdot \Delta x$ is negative, so $\int_0^2 f(x)\,dx$ is negative. Since area is nonnegative, the required area is given by $|-16/3|$ or $16/3$. Using a definite integral, the area could be written as

$$\left|\int_0^2 (x^2 - 4)\,dx\right| = \left|-\frac{16}{3}\right| = \frac{16}{3}.$$

FIGURE 13

EXAMPLE 5 Find the area of the region between the x-axis and the graph of $f(x) = x^2 - 3x$ from $x = 1$ to $x = 3$.

The region is shown in Figure 14. Since the region lies below the x-axis, the area is given by

$$\left|\int_1^3 (x^2 - 3x)\,dx\right|.$$

By the Fundamental Theorem,

$$\int_1^3 (x^2 - 3x)\,dx = \left(\frac{x^3}{3} - \frac{3x^2}{2}\right)\Big|_1^3 = \left(\frac{27}{3} - \frac{27}{2}\right) - \left(\frac{1}{3} - \frac{3}{2}\right) = -\frac{10}{3}.$$

FIGURE 14

The required area is $|-10/3| = 10/3$. ◄

EXAMPLE 6 Find the area between the x-axis and the graph of $f(x) = x^2 - 4$ from $x = 0$ to $x = 4$.

Figure 15 shows the required region. Part of the region is below the x-axis. The definite integral over that interval will have a negative value. To find the area, integrate the negative and positive portions separately and take the absolute value of the first result before combining the two results to get the total area. Start by finding the point where the graph crosses the x-axis. This is done by solving the equation

$$x^2 - 4 = 0.$$

$f(x)$

12

$f(x) = x^2 - 4$

8

4

0

2

4

x

-4

FIGURE 15

The solutions of this equation are 2 and -2. The only solution in the interval $[0, 4]$ is 2. The total area of the region in Figure 15 is

$$\left| \int_0^2 (x^2 - 4)\,dx \right| + \int_2^4 (x^2 - 4)\,dx = \left| \left(\frac{1}{3}x^3 - 4x \right) \right|_0^2 + \left(\frac{1}{3}x^3 - 4x \right) \Big|_2^4$$

$$= \left| \frac{8}{3} - 8 \right| + \left(\frac{64}{3} - 16 \right) - \left(\frac{8}{3} - 8 \right)$$

$$= 16. \quad \blacktriangleleft$$

Incorrectly using one integral over the entire interval to find the area in Example 6 would have given

$$\int_0^4 (x^2 - 4)\,dx = \left(\frac{x^3}{3} - 4x \right) \Big|_0^4 = \left(\frac{64}{3} - 16 \right) - 0 = \frac{16}{3},$$

which is not the correct area. This definite integral represents no area, but is just a real number.

If $f(x)$ in Example 6 represents the annual rate of profit of a company, then $16/3$ represents the total profit for the company over a four-year period. The integral between 0 and 2 is $-16/3$; the negative sign indicates a loss for the first two years. The integral between 2 and 4 is $32/3$, indicating a profit. The overall profit is $32/3 - 16/3 = 16/3$, while the total shaded area is $32/3 + |-16/3| = 16$.

FINDING AREA

In summary, to find the area bounded by $f(x)$, $x = a$, $x = b$, and the x-axis, use the following steps.

1. Sketch a graph.
2. Find any x-intercepts of $f(x)$ in $[a, b]$. These divide the total region into subregions.
3. The definite integral will be *positive* for subregions above the x-axis and *negative* for subregions below the x-axis. Use separate integrals to find the areas of the subregions.
4. The total area is the sum of the areas of all of the subregions.

In the last section, we saw that the area under a rate of change function $f'(x)$ from $x = a$ to $x = b$ gives the total value of $f(x)$ on $[a, b]$. Now we can use the definite integral to solve these problems.

EXAMPLE

7 The yearly rate of consumption of natural gas (in trillions of cubic feet) for a certain city is

$$C'(t) = t + e^{.01t},$$

where t is time in years and $t = 0$ corresponds to 1990. At this consumption rate, what is the total amount the city will use in the 10-year period of the 1990s?

To find the consumption over the 10-year period from 1990 through 1999, use the definite integral.

$$\int_0^{10} (t + e^{.01t})\, dt = \left(\frac{t^2}{2} + \frac{e^{.01t}}{.01}\right)\Big|_0^{10}$$

$$= (50 + 100e^{.1}) - (0 + 100)$$

$$\approx -50 + 100(1.10517) \approx 60.5$$

Therefore, a total of 60.5 trillion cubic feet of natural gas will be used during the 1990s if the consumption rate remains the same. ◄

14.4 Exercises

Evaluate each definite integral.

1. $\int_{-2}^{4} (-1)\, dp$

2. $\int_{-4}^{1} 6x\, dx$

3. $\int_{-1}^{2} (3t - 1)\, dt$

4. $\int_{-2}^{2} (4z + 3)\, dz$

5. $\int_{0}^{2} (5x^2 - 4x + 2)\, dx$

6. $\int_{-2}^{3} (-x^2 - 3x + 5)\, dx$

7. $\int_{0}^{2} 3\sqrt{4u + 1}\, du$

8. $\int_{3}^{9} \sqrt{2r - 2}\, dr$

9. $\int_{0}^{1} 2(t^{1/2} - t)\, dt$

10. $\int_{0}^{4} -(3x^{3/2} + x^{1/2})\, dx$

11. $\int_{1}^{4} (5y\sqrt{y} + 3\sqrt{y})\, dy$

12. $\int_{4}^{9} (4\sqrt{r} - 3r\sqrt{r})\, dr$

13. $\int_{4}^{6} \frac{2}{(x - 3)^2}\, dx$

14. $\int_{1}^{4} \frac{-3}{(2p + 1)^2}\, dp$

15. $\int_{1}^{5} (5n^{-2} + n^{-3})\, dn$

16. $\int_{2}^{3} (3x^{-2} - x^{-4})\, dx$

17. $\int_{2}^{3} \left(2e^{-.1A} + \frac{3}{A}\right)\, dA$

18. $\int_{1}^{2} \left(\frac{-1}{B} + 3e^{.2B}\right)\, dB$

19. $\int_{1}^{2} \left(e^{5u} - \frac{1}{u^2}\right)\, du$

20. $\int_{.5}^{1} (p^3 - e^{4p})\, dp$

21. $\int_{-1}^{0} y(2y^2 - 3)^5\, dy$

22. $\int_{0}^{3} m^2(4m^3 + 2)^3\, dm$

23. $\int_{1}^{64} \frac{\sqrt{z} - 2}{\sqrt[3]{z}}\, dz$

24. $\int_{1}^{8} \frac{3 - y^{1/3}}{y^{2/3}}\, dy$

25. $\int_{1}^{2} \frac{\ln x}{x}\, dx$

26. $\int_{1}^{3} \frac{\sqrt{\ln x}}{x}\, dx$

27. $\int_{0}^{8} x^{1/3}\sqrt{x^{4/3} + 9}\, dx$

28. $\int_{1}^{2} \frac{3}{x(1 + \ln x)}\, dx$

29. $\int_{0}^{1} \frac{e^t}{(3 + e^t)^2}\, dt$

30. $\int_{0}^{1} \frac{e^{2z}}{\sqrt{1 + e^{2z}}}\, dz$

31. $\int_{1}^{49} \frac{(1 + \sqrt{x})^{4/3}}{\sqrt{x}}\, dx$

32. $\int_{1}^{8} \frac{(1 + x^{1/3})^6}{x^{2/3}}\, dx$

In each of the following, use the definite integral to find the area between the x-axis and $f(x)$ over the indicated interval. Check first to see if the graph crosses the x-axis in the given interval.

33. $f(x) = 2x + 3$; [8, 10]

34. $f(x) = 4x - 7$; [5, 10]

35. $f(x) = 2 - 2x^2$; [0, 5]

36. $f(x) = 9 - x^2$; [0, 6]

37. $f(x) = x^2 + 4x - 5;$ $[-6, 3]$

38. $f(x) = x^2 - 6x + 5;$ $[-1, 4]$

39. $f(x) = x^3;$ $[-1, 3]$

40. $f(x) = x^3 - 2x;$ $[-2, 4]$

41. $f(x) = e^x - 1;$ $[-1, 2]$

42. $f(x) = 1 - e^{-x};$ $[-1, 2]$

43. $f(x) = \dfrac{1}{x};$ $[1, e]$

44. $f(x) = \dfrac{1}{x};$ $[e, e^2]$

Find the area of each shaded region.

45.

$y = 4 - x^2$

46.

$f(x) = x^2 - 2x$

47.

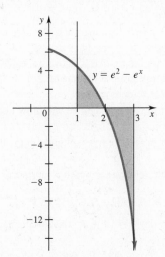

$y = e^2 - e^x$

48.

$y = e^x - e$

Show that each of the following is true.

49. $\displaystyle\int_a^a f(x)\,dx = 0$

50. $\displaystyle\int_a^b kf(x)\,dx = k\int_a^b f(x)\,dx$

51. $\displaystyle\int_a^b f(x)\,dx = \int_a^c f(x)\,dx + \int_c^b f(x)\,dx$

52. Use Exercise 51 to find $\displaystyle\int_{-1}^4 f(x)\,dx$, given

$$f(x) = \begin{cases} 2x + 3 & \text{if } x \le 0 \\[1mm] -\dfrac{x}{4} - 3 & \text{if } x > 0. \end{cases}$$

Applications

Business and Economics

53. Expenditures De Win Enterprises has found that its expenditure rate per day (in hundreds of dollars) on a certain type of job is given by

$$E'(x) = 4x + 2,$$

where x is the number of days since the start of the job.

(a) Find the total expenditure if the job takes 10 days.

(b) How much will be spent on the job from the tenth to the twentieth day?

(c) If the company wants to spend no more than $5000 on the job, in how many days must they complete it?

54. Income De Win Enterprises (see Exercise 53) also knows that the rate of income per day (in hundreds of dollars) for the same job is

$$I'(x) = 100 - x,$$

where x is the number of days since the job was started.

(a) Find the total income for the first 10 days.

(b) Find the income from the tenth to the twentieth day.

(c) How many days must the job last for the total income to be at least $5000?

55. Profit Karla Harby Communications, a small company of science writers, found that its rate of profits (in thousands of dollars) after t years of operation is given by

$$P'(t) = (3t + 3)(t^2 + 2t + 2)^{1/3}.$$

(a) Find the total profits in the first three years.

(b) Find the profit in the fourth year of operation.

(c) What is happening to the annual profits over the long run?

56. Worker Efficiency A worker new to a job will improve his efficiency with time so that it takes him fewer hours to produce an item with each day on the job, up to a certain point. Suppose the rate of change of the number of hours it takes a worker in a certain factory to produce the xth item is given by

$$H'(x) = 20 - 2x.$$

(a) What is the total number of hours required to produce the first 5 items?

(b) What is the total number of hours required to produce the first 10 items?

Life Sciences

57. Pollution Pollution from a factory is entering a lake. The rate of concentration of the pollutant at time t is given by

$$P'(t) = 140t^{5/2},$$

where t is the number of years since the factory started introducing pollutants into the lake. Ecologists estimate that the lake can accept a total level of pollution of 4850 units before all the fish life in the lake ends. Can the factory operate for 4 yr without killing all the fish in the lake?

58. Spread of an Oil Leak An oil tanker is leaking oil at the rate given in barrels per hour by

$$L'(t) = \frac{80 \ln (t + 1)}{t + 1},$$

where t is the time in hours after the tanker hits a hidden rock (when $t = 0$).

(a) Find the total number of barrels that the ship will leak on the first day.

(b) Find the total number of barrels that the ship will leak on the second day.

(c) What is happening over the long run to the amount of oil leaked per day?

59. Tree Growth After long study, tree scientists conclude that a eucalyptus tree will grow at the rate of $.2 + 4t^{-4}$ feet per year, where t is time in years.

(a) Find the number of feet that the tree will grow in the second year.

(b) Find the number of feet the tree will grow in the third year.

60. Growth of a Substance The rate at which a substance grows is given by

$$R'(x) = 200e^{.2x},$$

where x is the time in days. What is the total accumulated growth after 2.5 days?

61. Drug Reaction For a certain drug, the rate of reaction in appropriate units is given by

$$R'(t) = \frac{5}{t} + \frac{2}{t^2},$$

where t is time measured in hours after the drug is administered. Find the total reaction to the drug over the following time periods.

(a) From $t = 1$ to $t = 12$ **(b)** From $t = 12$ to $t = 24$

62. Blood Flow In the exercises for an earlier chapter, the speed s of the blood in a blood vessel was given as

$$s = k(R^2 - r^2),$$

where R is the (constant) radius of the blood vessel, r is the distance of the flowing blood from the center of the blood vessel, and k is a constant. Total blood flow (in millimeters per minute) is given by

$$Q(R) = \int_0^R 2\pi s r \, dr.$$

 (a) Find the general formula for Q in terms of R by evaluating the definite integral given above.

 (b) Evaluate $Q(.4)$.

Physical Sciences

63. Oil Consumption Suppose that the rate of consumption of a natural resource is $c(t)$, where

$$c'(t) = ke^{rt}.$$

Here t is time in years, r is a constant, and k is the consumption in the year when $t = 0$. In 1992, an oil company sold 1.2 billion barrels of oil. Assume that $r = .04$.

 (a) Write $c'(t)$ for the oil company, letting $t = 0$ represent 1992.

 (b) Set up a definite integral for the amount of oil that the company will sell in the next ten years.

 (c) Evaluate the definite integral of part (b).

 (d) The company has about 20 billion barrels of oil in reserve. To find the number of years that this amount will last, solve the equation

$$\int_0^T 1.2e^{.04t} \, dt = 20.$$

 (e) Rework part (d), assuming that $r = .02$.

64. Mine Production A mine begins producing at time $t = 0$. After t yr, the mine is producing at the rate of

$$P'(t) = 10t - \frac{15}{\sqrt{t}}$$

tons per year. Write an expression for the total output of the mine from year 1 to year T.

65. Consumption of a Natural Resource The rate of consumption of one natural resource in one country is

$$C'(t) = 72e^{.014t},$$

where $t = 0$ corresponds to 1992. How much of the resource will be used, altogether, from 1992 to year T?

66. Oil Consumption The rate of consumption of oil (in billions of barrels) by the company in Exercise 63 was given as

$$1.2e^{.04t},$$

where $t = 0$ corresponds to 1992. Find the total amount of oil used by the company from 1992 to year T. At this rate, how much will be used in 5 years?

14.5 THE AREA BETWEEN TWO CURVES

If an executive knows how the savings from a new manufacturing process decline over time and how the costs of that process will increase, how can she compute when the savings will cease and what the total savings will be?

This section shows a method for answering such questions.

 Many important applications of integrals require finding the area between two graphs. The method used in previous sections to find the area between the graph of a function and the x-axis from $x = a$ to $x = b$ can be generalized to find such an area. For example, the area between the graphs of $f(x)$ and $g(x)$ from $x = a$ to $x = b$ in Figure 16(a) (on the next page) is the same as the difference of the area from a to b between $f(x)$ and the x-axis, shown in Figure 16(b), and the area from a to b between $g(x)$ and the x-axis (see Figure 16(c)). That is, the area between the graphs is given by

$$\int_a^b f(x) \, dx - \int_a^b g(x) \, dx,$$

which can be written as

$$\int_a^b [f(x) - g(x)] \, dx.$$

(a) (b) (c)

FIGURE 16

AREA BETWEEN TWO CURVES	If f and g are continuous functions and $f(x) \geq g(x)$ on $[a, b]$, then the area between the curves $f(x)$ and $g(x)$ from $x = a$ to $x = b$ is given by $$\int_a^b [f(x) - g(x)] \, dx.$$

EXAMPLE 1 Find the area bounded by $f(x) = -x^2 + 1$, $g(x) = 2x + 4$, $x = -1$, and $x = 2$.

FIGURE 17

A sketch of the four equations is shown in Figure 17. In general, it is not necessary to spend time drawing a detailed sketch, but only to know whether the two functions intersect, and which function is greater between the intersections. To find out, set the two functions equal.

$$-x^2 + 1 = 2x + 4$$
$$0 = x^2 + 2x + 3$$

Verify by the quadratic formula that this equation has no roots. Since the graph of f is a parabola opening downward that does not cross the graph of g (a line), the parabola must be entirely under the line, as shown in Figure 17. Therefore $g(x) \geq f(x)$ for x in the interval $[-1, 2]$, and the area is given by

$$\int_{-1}^{2} [g(x) - f(x)]\, dx = \int_{-1}^{2} (2x + 4) - (-x^2 + 1)\, dx$$

$$= \int_{-1}^{2} 2x + 4 + x^2 - 1\, dx$$

$$= \int_{-1}^{2} (x^2 + 2x + 3)\, dx$$

$$= \frac{x^3}{3} + x^2 + 3x \Big|_{-1}^{2}$$

$$= \left(\frac{8}{3} + 4 + 6\right) - \left(\frac{-1}{3} + 1 - 3\right)$$

$$= \frac{8}{3} + 10 + \frac{1}{3} + 2$$

$$= 15. \quad \blacktriangleleft$$

EXAMPLE 2 Find the area between the curves $y = x^{1/2}$ and $y = x^3$.

Let $f(x) = x^{1/2}$ and $g(x) = x^3$. As before, set the two equal to find where they intersect.

$$x^{1/2} = x^3$$
$$0 = x^3 - x^{1/2}$$
$$0 = x^{1/2}(x^{5/2} - 1)$$

The only solutions are $x = 0$ and $x = 1$. Verify that the graph of f is concave downward, while the graph of g is concave upward, so the graph of f must be greater between 0 and 1. (This may also be verified by taking a point between 0 and 1, such as .5, and verifying that $.5^{1/2} > .5^3$.) The graph is shown in Figure 18.

The area between the two curves is given by

$$\int_{a}^{b} [f(x) - g(x)]\, dx = \int_{0}^{1} (x^{1/2} - x^3)\, dx.$$

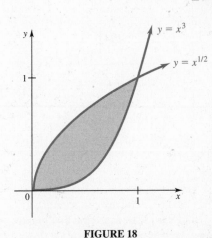

$y = x^3$

$y = x^{1/2}$

FIGURE 18

Using the Fundamental Theorem,

$$\int_0^1 (x^{1/2} - x^3)\, dx = \left(\frac{x^{3/2}}{3/2} - \frac{x^4}{4}\right)\Bigg|_0^1$$

$$= \left(\frac{2}{3}x^{3/2} - \frac{x^4}{4}\right)\Bigg|_0^1$$

$$= \frac{2}{3}(1) - \frac{1}{4}$$

$$= \frac{5}{12}. \quad \blacktriangleleft$$

The difference between two integrals can be used to find the area between the graphs of two functions even if one graph lies below the x-axis. In fact, if $f(x) \geq g(x)$ for all values of x in the interval $[a, b]$, then the area between the two graphs is always given by

$$\int_a^b [f(x) - g(x)]\, dx.$$

To see this, look at the graphs in Figure 19(a), where $f(x) \geq g(x)$ for x in $[a, b]$. Suppose a constant C is added to both functions, with C large enough so that both graphs lie above the x-axis, as in Figure 19(b). The region between the graphs is not changed. By the work above, this area is given by $\int_a^b [f(x) - g(x)]\, dx$ regardless of where the graphs of $f(x)$ and $g(x)$ are located. So long as $f(x) \geq g(x)$ on $[a, b]$, then the area between the graphs from $x = a$ to $x = b$ will equal $\int_a^b [f(x) - g(x)]\, dx$.

(a) **(b)**

FIGURE 19

3 Find the area of the region enclosed by $y = x^2 - 2x$ and $y = x$ on $[0, 4]$. Verify that the two graphs cross at $x = 0$ and $x = 3$. Because the first graph is a parabola opening upward, the parabola must be below the line between 0 and

3 and above the line between 3 and 4. See Figure 20. (The greater function could also be identified by checking a point between 0 and 3, such as 1, and a point between 3 and 4, such as 3.5. For each of these values of x, we could calculate the corresponding value of y for the two functions and see which is greater.) Because the graphs cross at $x = 3$, the area is found by taking the sum of two integrals as follows.

$$\text{Area} = \int_0^3 [x - (x^2 - 2x)]\, dx + \int_3^4 [(x^2 - 2x) - x]\, dx$$

$$= \int_0^3 (-x^2 + 3x)\, dx + \int_3^4 (x^2 - 3x)\, dx$$

$$= \left(\frac{-x^3}{3} + \frac{3x^2}{2}\right)\Bigg|_0^3 + \left(\frac{x^3}{3} - \frac{3x^2}{2}\right)\Bigg|_3^4$$

$$= \left(-9 + \frac{27}{2} - 0\right) + \left(\frac{64}{3} - 24 - 9 + \frac{27}{2}\right)$$

$$= \frac{19}{3} \quad \blacktriangleleft$$

FIGURE 20

In the remainder of this section we will consider some typical applications that require finding the area between two curves.

EXAMPLE 4 A company is considering a new manufacturing process in one of its plants. The new process provides substantial initial savings, with the savings declining with time x (in years) according to the rate-of-savings function

$$S(x) = 100 - x^2,$$

where $S(x)$ is in thousands of dollars. At the same time, the cost of operating the new process increases with time x (in years), according to the rate-of-cost function (in thousands of dollars)

$$C(x) = x^2 + \frac{14}{3}x.$$

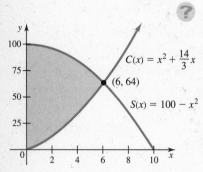

FIGURE 21

(a) For how many years will the company realize savings?

Figure 21 shows the graphs of the rate-of-savings and rate-of-cost functions. The rate-of-cost (marginal cost) is increasing, while the rate-of-savings (marginal savings) is decreasing. The company should use this new process until the difference between these quantities is zero; that is, until the time at which these graphs intersect. The graphs intersect when

$$C(x) = S(x),$$

or

$$100 - x^2 = x^2 + \frac{14}{3}x.$$

Solve this equation as follows.

$$0 = 2x^2 + \frac{14}{3}x - 100$$

$$0 = 3x^2 + 7x - 150 \qquad \text{Multiply by 3/2.}$$

$$= (x - 6)(3x + 25) \qquad \text{Factor.}$$

Set each factor equal to 0 separately to get

$$x = 6 \qquad \text{or} \qquad x = -25/3.$$

Only 6 is a meaningful solution here. The company should use the new process for 6 years.

(b) What will be the net total savings during this period?

Since the total savings over the 6-year period is given by the area under the rate-of-savings curve and the total additional cost by the area under the rate-of-cost curve, the net total savings over the 6-year period is given by the area between the rate of cost and the rate of savings curves and the lines $x = 0$ and $x = 6$. This area can be evaluated with a definite integral as follows.

$$\text{Total savings} = \int_0^6 \left[(100 - x^2) - \left(x^2 + \frac{14}{3}x \right) \right] dx$$

$$= \int_0^6 \left(100 - \frac{14}{3}x - 2x^2 \right) dx$$

$$= 100x - \frac{7}{3}x^2 - \frac{2}{3}x^3 \Big|_0^6$$

$$= 100(6) - \frac{7}{3}(36) - \frac{2}{3}(216) = 372$$

The company will save a total of $372,000 over the 6-year period. ◀

The answer to a problem will not always be an integer. Suppose in solving the quadratic equation in Example 4 we found the solutions to be $x = 6.7$ and $x = -7.3$. It may not be realistic to use a new process for 6.7 years; it may be necessary to choose between 6 years and 7 years. Since the mathematical model produces a result that is not in the domain of the function in this case, it is necessary to find the total savings after 6 years and after 7 years and then select the best result.

Consumers' Surplus The market determines the price at which a product is sold. As indicated earlier, the point of intersection of the demand curve and the supply curve for a product gives the equilibrium price. At the equilibrium price, consumers will purchase the same amount of the product that the manufacturers want to sell. Some consumers, however, would be willing to spend more for an item than the equilibrium price. The total of the differences between the equilibrium price of the item and the higher prices all those individuals would be willing to pay is thought of as savings realized by those individuals and is called the **consumers' surplus.**

In Figure 22, the area under the demand curve is the total amount consumers are willing to spend for q_0 items. The heavily shaded area under the line $y = p_0$ shows the total amount consumers actually will spend at the equilibrium price of p_0. The lightly shaded area represents the consumers' surplus. As the figure suggests, the consumers' surplus is given by an area between the two curves $p = D(q)$ and $p = p_0$, so its value can be found with a definite integral as follows.

FIGURE 22

CONSUMERS' SURPLUS

If $D(q)$ is a demand function with equilibrium price p_0 and equilibrium demand q_0, then

$$\text{Consumers' surplus} = \int_0^{q_0} [D(q) - p_0]\, dq.$$

Similarly, if some manufacturers would be willing to supply a product at a price *lower* than the equilibrium price p_0, the total of the differences between the equilibrium price and the lower prices at which the manufacturers would sell the product is considered added income for the manufacturers and is called the **producers' surplus.** Figure 23 shows the (heavily shaded) total area under the supply curve from $q = 0$ to $q = q_0$, which is the minimum total amount the manufacturers are willing to realize from the sale of q_0 items. The total area under the line $p = p_0$ is the amount actually realized. The difference between these two areas, the producers' surplus, is also given by a definite integral.

FIGURE 23

PRODUCERS' SURPLUS

If $S(q)$ is a supply function with equilibrium price p_0 and equilibrium supply q_0, then

$$\text{Producers' surplus} = \int_0^{q_0} [p_0 - S(q)]\, dq.$$

EXAMPLE 5 Suppose the price (in dollars per ton) for oat bran is

$$D(q) = 400 - e^{q/2},$$

when the demand for the product is q tons. Also, suppose the function

$$S(q) = e^{q/2} - 1$$

gives the price (in dollars per ton) when the supply is q tons. Find the consumers' surplus and the producers' surplus.

Begin by finding the equilibrium quantity. This is done by setting the two equations equal.

$$e^{q/2} - 1 = 400 - e^{q/2}$$

$$2e^{q/2} = 401$$

$$e^{q/2} = \frac{401}{2}$$

$$q/2 = \ln\left(\frac{401}{2}\right)$$

$$q = 2\ln\left(\frac{401}{2}\right) \approx 10.60163$$

This result can be further rounded to 10.60 tons so long as this rounded value is not used in future calculations. At the equilibrium point where the supply and demand are both 10.60 tons, the price is

$$S(10.60163) = e^{10.60163/2} - 1 = 199.50,$$

or \$199.50. Verify that this same answer is found by computing $D(10.60163)$. The consumers' surplus, represented by the area shown in Figure 24, is

$$\int_0^{10.60163} [(400 - e^{q/2}) - 199.50]\, dq = \int_0^{10.60163} [200.5 - e^{q/2}]\, dq.$$

Evaluating the definite integral gives

$$(200.5q - 2e^{q/2})\big|_0^{10.60163} = [200.5(10.60163) - 2e^{10.60163/2}] - (0 - 2)$$

$$\approx 1726.63.$$

FIGURE 24

Here the consumers' surplus is $1726.63. The producers' surplus, also shown in Figure 24, is given by

$$\int_0^{10.60163} [199.50 - (e^{q/2} - 1)] \, dq = \int_0^{10.60163} [200.5 - e^{q/2}] \, dq,$$

which is exactly the same as the expression found for the consumers' surplus, so the producers' surplus is also $1726.63. In general, the producers' surplus and consumers' surplus are not the same, as they are in this example. ◀

14.5 Exercises

Find the area between the curves in Exercises 1–22.

1. $x = -2$, $x = 1$, $y = x^2 + 4$, $y = 0$

2. $x = 1$, $x = 2$, $y = x^3$, $y = 0$

3. $x = -3$, $x = 1$, $y = x + 1$, $y = 0$

4. $x = -2$, $x = 0$, $y = 1 - x^2$, $y = 0$

5. $x = -2$, $x = 1$, $y = 2x$, $y = x^2 - 3$

6. $x = 0$, $x = 6$, $y = 5x$, $y = 3x + 10$

7. $y = x^2 - 30$, $y = 10 - 3x$

8. $y = x^2 - 18$, $y = x - 6$

9. $y = x^2$, $y = 2x$

10. $y = x^2$, $y = x^3$

11. $x = 1$, $x = 6$, $y = \dfrac{1}{x}$, $y = -1$

12. $x = 0$, $x = 4$, $y = \dfrac{1}{x + 1}$, $y = \dfrac{x - 1}{2}$

13. $x = -1$, $x = 1$, $y = e^x$, $y = 3 - e^x$.

14. $x = -1$, $x = 2$, $y = e^{-x}$, $y = e^x$

15. $x = 1$, $x = 2$, $y = e^x$, $y = \dfrac{1}{x}$

16. $x = 2$, $x = 4$, $y = \dfrac{x}{2} + 3$, $y = \dfrac{1}{x - 1}$

17. $y = x^3 - x^2 + x + 1$, $y = 2x^2 - x + 1$

18. $y = 2x^3 + x^2 + x + 5$, $y = x^3 + x^2 + 2x + 5$

19. $y = x^4 + \ln (x + 10)$, $y = x^3 + \ln (x + 10)$

20. $y = x^5 - 2 \ln (x + 5)$, $y = x^3 - 2 \ln (x + 5)$

21. $y = x^{4/3}$, $y = 2x^{1/3}$

22. $y = \sqrt{x}$, $y = x\sqrt{x}$

Applications

Business and Economics

23. Net Savings Suppose a company wants to introduce a new machine that will produce a rate of annual savings in dollars given by

$$S(x) = 150 - x^2,$$

where x is the number of years of operation of the machine, while producing a rate of annual costs in dollars of

$$C(x) = x^2 + \frac{11}{4}x.$$

(a) For how many years will it be profitable to use this new machine?

(b) What are the net total savings during the first year of use of the machine?

(c) What are the net total savings over the entire period of use of the machine?

24. Net Savings A new smog-control device will reduce the output of sulfur oxides from automobile exhausts. It is estimated that the rate of savings to the community from the use of this device will be approximated by

$$S(x) = -x^2 + 4x + 8,$$

where $S(x)$ is the rate of savings (in millions of dollars) after x years of use of the device. The new device cuts down on the production of sulfur oxides, but it causes an increase in the production of nitrous oxides. The rate of additional costs (in millions) to the community after x years is approximated by

$$C(x) = \frac{3}{25}x^2.$$

(a) For how many years will it pay to use the new device?

(b) What will be the net savings over this period of time?

25. **Profit** De Win Enterprises had an expenditure rate of $E(x) = e^{.1x}$ dollars per day and an income rate of $I(x) = 98.8 - e^{.1x}$ dollars per day on a particular job, where x was the number of days from the start of the job. The company's profit on that job will equal total income less total expenditures. Profit will be maximized if the job ends at the optimum time, which is the point where the two curves meet. Find the following.

 (a) The optimum number of days for the job to last.

 (b) The total income for the optimum number of days.

 (c) The total expenditues for the optimum number of days.

 (d) The maximum profit for the job.

26. **Net Savings** A factory at Harold Levinson Industries has installed a new process that will produce an increased rate of revenue (in thousands of dollars) of
$$R(t) = 104 - .4e^{t/2},$$
where t is time measured in years. The new process produces additional costs (in thousands of dollars) at the rate of
$$C(t) = .3e^{t/2}.$$

 (a) When will it no longer be profitable to use this new process?

 (b) Find the total net savings.

27. **Producers' Surplus** Find the producers' surplus if the supply function for pork bellies is given by
$$S(q) = q^{5/2} + 2q^{3/2} + 50.$$
Assume supply and demand are in equilibrium at $q = 16$.

28. **Producers' Surplus** Suppose the supply function for concrete is given by
$$S(q) = 100 + 3q^{3/2} + q^{5/2},$$
and that supply and demand are in equilibrium at $q = 9$. Find the producers' surplus.

29. **Consumers' Surplus** Find the consumers' surplus if the demand function for grass seed is given by
$$D(q) = \frac{100}{(3q + 1)^2},$$
assuming supply and demand are in equilibrium at $q = 3$.

30. **Consumers' Surplus** Find the consumers' surplus if the demand function for extra virgin olive oil is given by
$$D(q) = \frac{16,000}{(2q + 8)^3},$$
and if supply and demand are in equilibrium at $q = 6$.

31. **Consumers' and Producers' Surplus** Suppose the supply function of a certain item is given by
$$S(q) = q^2 + 10q,$$
and the demand function is given by
$$D(q) = 900 - 20q - q^2.$$

 (a) Graph the supply and demand curves.

 (b) Find the point at which supply and demand are in equilibrium.

 (c) Find the consumers' surplus.

 (d) Find the producers' surplus.

32. **Consumers' and Producers' Surplus** Repeat the four steps in Exercise 31 for the supply function
$$S(q) = q^2 + \frac{11}{4}q$$
and the demand function
$$D(q) = 150 - q^2.$$

Social Sciences

33. **Distribution of Income** Suppose that all the people in a country are ranked according to their incomes, starting at the bottom. Let x represent the fraction of the community making the lowest income ($0 \le x \le 1$); $x = .4$, therefore, represents the lower 40% of all income producers. Let $I(x)$ represent the proportion of the total income earned by the lowest x of all people. Thus, $I(.4)$ represents the fraction of total income earned by the lowest 40% of the population. Suppose
$$I(x) = .9x^2 + .1x.$$
Find and interpret the following.

 (a) $I(.1)$ **(b)** $I(.4)$ **(c)** $I(.6)$ **(d)** $I(.9)$

 If income were distributed uniformly, we would have $I(x) = x$. The area under this line of complete equality is $1/2$. As $I(x)$ dips farther below $y = x$, there is less equality of income distribution. This inequality can be quantified by the ratio of the area between $I(x)$ and $y = x$ to $1/2$. This ratio is called the *coefficient of inequality* and equals $2 \int_0^1 (x - I(x))\, dx$.

 (e) Graph $I(x) = x$ and $I(x) = .9x^2 + .1x$ for $0 \le x \le 1$ on the same axes.

 (f) Find the area between the curves.

Physical Sciences

34. **Metal Plate** A worker sketches the curves $y = \sqrt{x}$ and $y = x/2$ on a sheet of metal and cuts out the region between the curves to form a metal plate. Find the area of the plate.

◫ **Computer/Graphing Calculator**

Approximate the area between the graphs of each pair of functions on the given interval.

35. $y = \ln x$ and $y = xe^x$; [1, 4]

36. $y = \ln x$ and $y = 4 - x^2$; [2, 4]

37. $y = \sqrt{9 - x^2}$ and $y = \sqrt{x + 1}$; [-1, 3]

38. $y = \sqrt{4 - 4x^2}$ and $y = \sqrt{\dfrac{9 - x^2}{3}}$; [-1, 1]

▶ **INTEGRATION SUMMARY**

ANTIDIFFERENTIATION FORMULAS

Power Rule $\displaystyle\int x^n \, dx = \frac{x^{n+1}}{n + 1} + C, \quad n \neq -1$

Constant Multiple Rule $\displaystyle\int k \cdot f(x) \, dx = k \int f(x) \, dx$ for any real number k

Sum or Difference Rule $\displaystyle\int [f(x) \pm g(x)] \, dx = \int f(x) \, dx \pm \int g(x) \, dx$

Integration of Exponential Functions $\displaystyle\int e^{kx} \, dx = \frac{1}{k} \cdot e^{kx} + C, \quad k \neq 0$

Integration of x^{-1} $\displaystyle\int x^{-1} \, dx = \ln |x| + C$

Substitution Formulas **1.** $\displaystyle\int [u(x)]^n \cdot u'(x) \, dx = \frac{[u(x)]^{n+1}}{n + 1} + C, \quad n \neq -1$

2. $\displaystyle\int e^{u(x)} \cdot u'(x) \, dx = e^{u(x)} + C$

3. $\displaystyle\int \frac{u'(x)}{u(x)} \, dx = \ln |u(x)| + C$

FORMULAS FOR DEFINITE INTEGRALS

Definition of the Definite Integral $\displaystyle\int_a^b f(x) \, dx = \lim_{n \to \infty} \sum_{i=1}^n f(x_i)\Delta x$, where

$\Delta x = (b - a)/n$ and x_i is any value of x in the ith interval. If $f(x)$ gives the rate of change of $F(x)$ for x in $[a, b]$, then this represents the total change in $F(x)$ as x goes from a to b.

Properties of Definite Integrals

1. $\displaystyle\int_a^a f(x)\ dx = 0$

2. $\displaystyle\int_a^b k \cdot f(x)\ dx = k \int_a^b f(x)\ dx$

3. $\displaystyle\int_a^b [f(x) \pm g(x)]\ dx = \int_a^b f(x)\ dx \pm \int_a^b g(x)\ dx$

4. $\displaystyle\int_a^b f(x)\ dx = \int_a^c f(x)\ dx + \int_c^b f(x)\ dx$

Fundamental Theorem of Calculus $\displaystyle\int_a^b f(x)\ dx = F(x)\Big|_a^b = F(b) - F(a)$,

where f is continuous on $[a, b]$ and F is any antiderivative of f.

Area Between Two Curves $\displaystyle\int_a^b [f(x) - g(x)]\ dx$, where f and g are continuous functions and $f(x) \geq g(x)$ on $[a, b]$.

Consumers' Surplus $\displaystyle\int_0^{q_0} [D(q) - p_0]\ dq$, where D is the demand function and p_0 and q_0 are the equilibrium price and demand.

Producers' Surplus $\displaystyle\int_0^{q_0} [p_0 - S(q)]\ dq$, where S is the supply function and p_0 and q_0 are the equilibrium price and supply.

Chapter Summary

Key Terms

14.1 antiderivative
integral sign
integrand
indefinite integral
14.2 integration by substitution

14.3 definite integral
limits of integration
total change
14.4 Fundamental Theorem of
Calculus
area between a curve and the
x-axis

14.5 area between two curves
consumers' surplus
producers' surplus

Chapter 14

Review Exercises

1. Explain the differences between an indefinite integral and a definite integral.

2. Explain under what circumstances substitution is useful in integration.

3. Explain why the limits of integration are changed when u is substituted for an expression in x in a definite integral.

4. Explain why rectangles are used to approximate the area under a curve.

Find each indefinite integral.

5. $\int 6\,dx$

6. $\int (-4)\,dx$

7. $\int (2x + 3)\,dx$

8. $\int (5x - 1)\,dx$

9. $\int (x^2 - 3x + 2)\,dx$

10. $\int (6 - x^2)\,dx$

11. $\int 3\sqrt{x}\,dx$

12. $\int \frac{\sqrt{x}}{2}\,dx$

13. $\int (x^{1/2} + 3x^{-2/3})\,dx$

14. $\int (2x^{4/3} + x^{-1/2})\,dx$

15. $\int \frac{-4}{x^3}\,dx$

16. $\int \frac{5}{x^4}\,dx$

17. $\int -3e^{2x}\,dx$

18. $\int 5e^{-x}\,dx$

19. $\int \frac{2}{x - 1}\,dx$

20. $\int \frac{-4}{x + 2}\,dx$

21. $\int xe^{3x^2}\,dx$

22. $\int 2xe^{x^2}\,dx$

23. $\int \frac{3x}{x^2 - 1}\,dx$

24. $\int \frac{-x}{2 - x^2}\,dx$

25. $\int \frac{x^2\,dx}{(x^3 + 5)^4}$

26. $\int (x^2 - 5x)^4(2x - 5)\,dx$

27. $\int \frac{4x - 5}{2x^2 - 5x}\,dx$

28. $\int \frac{12(2x + 9)}{x^2 + 9x + 1}\,dx$

29. $\int \frac{x^3}{e^{3x^4}}\,dx$

30. $\int e^{3x^2 + 4}\,x\,dx$

31. $\int -2e^{-5x}\,dx$

32. $\int e^{-4x}\,dx$

33. Evaluate $\displaystyle\sum_{i=1}^{4} (i^2 - i)$.

34. Let $f(x) = 3x + 1$, $x_1 = -1$, $x_2 = 0$, $x_3 = 1$, $x_4 = 2$, and $x_5 = 3$. Find $\displaystyle\sum_{i=1}^{5} f(x_i)$.

35. Approximate the area under the graph of $f(x) = 2x + 3$ and above the x-axis from $x = 0$ to $x = 4$ using four rectangles. Let the height of each rectangle be the function value on the left side.

36. Find $\int_0^4 (2x + 3)\,dx$ by using the formula for the area of a trapezoid: $A = \frac{1}{2}(B + b)h$, where B and b are the lengths of the parallel sides and h is the distance between them. Compare with Exercise 35.

37. What does the Fundamental Theorem of Calculus state?

Find each definite integral.

38. $\int_1^2 (3x^2 + 5)\,dx$

39. $\int_1^6 (2x^2 + x)\,dx$

40. $\int_1^5 (3x^{-2} + x^{-3})\,dx$

41. $\int_2^3 (5x^{-2} + x^{-4})\,dx$

42. $\int_1^3 2x^{-1}\,dx$

43. $\int_1^6 8x^{-1}\,dx$

44. $\int_0^4 2e^x\,dx$

45. $\int_1^6 \frac{5}{2}e^{4x}\,dx$

46. $\int_{\sqrt{5}}^5 2x\sqrt{x^2 - 3}\,dx$

47. $\int_0^1 x\sqrt{5x^2 + 4}\,dx$

Find the area between the x-axis and $f(x)$ over each of the given intervals.

48. $f(x) = x\sqrt{x - 1}$; [1, 10]

49. $f(x) = x(x + 2)^6$; [-2, 0]

50. $f(x) = e^x$; [0, 2]

51. $f(x) = 1 + e^{-x}$; [0, 4]

Find the area of the region enclosed by each group of curves.

52. $f(x) = 5 - x^2$, $g(x) = x^2 - 3$

53. $f(x) = x^2 - 4x$, $g(x) = x - 6$

54. $f(x) = x^2 - 4x$, $g(x) = x + 1$, $x = 2$, $x = 4$

55. $f(x) = 5 - x^2$, $g(x) = x^2 - 3$, $x = 0$, $x = 4$

56. Explain what the consumers' surplus and the producers' surplus are.

Applications

Business and Economics

Cost *Find the cost function for each of the marginal cost functions in Exercises 57–60.*

57. $C'(x) = 10 - 2x$; fixed cost is $4.

58. $C'(x) = 2x + 3x^2$; 2 units cost $12.

59. $C'(x) = 3\sqrt{2x - 1}$; 13 units cost $270.

60. $C'(x) = \dfrac{1}{x + 1}$; fixed cost is $18.

61. Investment The curve shown gives the rate that an investment accumulates income (in dollars per year). Use rectangles of width 2 units and height determined by the function value at the midpoint to find the total income accumulated over 10 yr.

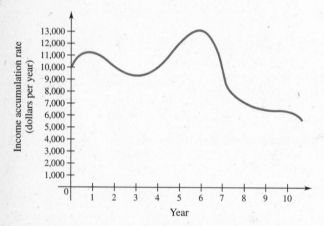

62. Utilization of Reserves A manufacturer of electronic equipment requires a certain rare metal. He has a reserve supply of 4,000,000 units that he will not be able to replace. If the rate at which the metal is used is given by

$$f(t) = 100,000e^{.03t},$$

where t is the time in years, how long will it be before he uses up the supply? (*Hint:* Find an expression for the total amount used in t years and set it equal to the known reserve supply.)

63. Sales The rate of change of sales of a new brand of tomato soup (in thousands) is given by

$$S'(x) = \sqrt{x} + 2,$$

where x is the time in months that the new product has been on the market. Find the total sales after 9 months.

64. Producers' and Consumers' Surplus Suppose that the supply function of some commodity is

$$S(q) = q^2 + 5q + 100$$

and the demand function for the commodity is

$$D(q) = 350 - q^2.$$

(a) Find the producers' surplus.

(b) Find the consumers' surplus.

65. Net Savings A company has installed new machinery that will produce a savings rate (in thousands of dollars) of

$$S'(x) = 225 - x^2,$$

where x is the number of years the machinery is to be used. The rate of additional costs (in thousands of dollars) to the company due to the new machinery is expected to be

$$C'(x) = x^2 + 25x + 150.$$

For how many years should the company use the new machinery? Find the net savings (in thousands of dollars) over this period.

Life Sciences

66. Population Growth The rate of change of the population of a rare species of Australian spider is given by

$$y' = 100 - \sqrt{2.4t + 1},$$

where y is the number of spiders present at time x, measured in months. Find the total number of spiders in the first 10 mo.

67. Infection Rate The rate of infection of a disease (in people per month) is given by the function

$$I'(t) = \frac{100t}{t^2 + 1},$$

where t is the time in months since the disease broke out. Find the total number of infected people over the first four months of the disease.

Physical Sciences

68. Linear Motion A particle is moving along a straight line with velocity $v(t) = t^2 - 2t$. Its distance from the starting point after 3 sec is 8 cm. Find $s(t)$, the distance of the particle from the starting point after t sec.

Connections

69. The following chart shows 1991 weather statistics for New York City, as well as the normal high and low tempera-

tures.* The amount of cold weather in a year is measured in heating degree-days, where 1 degree-day is added to the total for each degree that a day's average falls below 65° F. For example, if the average temperature on November 15 is 50° F, 15 degree-days are added to the year's to-

tal. Estimate the total number of heating degree-days in an average New York City year, using rectangles of width 1 mo, with the height determined by the average temperature at the middle of the rectangle. Assume that the normal average is halfway between the normal high and low.

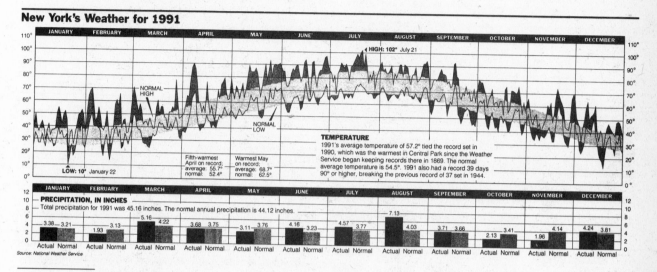

New York's Weather for 1991

*Graph, "New York's Weather for 1991" from *The New York Times*, January 1992. Copyright © 1992 by The New York Times Company. Reprinted by permission.

Extended Application / Estimating Depletion Dates for Minerals

It is becoming more and more obvious that the earth contains only a finite quantity of minerals. The "easy and cheap" sources of minerals are being used up, forcing an ever more expensive search for new sources. For example, oil from the North Slope of Alaska would never have been used in the United States during the 1930s because a great deal of Texas and California oil was readily available.

We said in an earlier chapter that population tends to follow an exponential growth curve. Mineral usage also follows such a curve. Thus, if q represents the rate of consumption of a certain mineral at time t, while q_0 represents consumption when $t = 0$, then

$$q = q_0 e^{kt},$$

where k is the growth constant. For example, the world consumption of petroleum in a recent year was about 19,600 million barrels, with the value of k about 6%. If we let $t = 0$ correspond to this base year, then $q_0 = 19,600$, $k = .06$, and

$$q = 19,600 e^{.06t}$$

is the rate of consumption at time t, assuming that all present trends continue.

Based on estimates of the National Academy of Science, 2,000,000 million barrels of oil are now in provable reserves or are likely to be discovered in the future. At the present rate of consumption, in how many years will these reserves be depleted? We can use the integral calculus of this chapter to find out.

To begin, we need to know the total quantity of petroleum that would be used between time $t = 0$ and some future time $t = t_1$. Figure 25 on the next page shows a typical graph of the function $q = q_0 e^{kt}$.

Following the work we did in Section 3, divide the time interval from $t = 0$ to $t = t_1$ into n subintervals. Let the ith subinterval have width Δt_i. Let the rate of consumption for the ith subinterval be approximated by q_i^*. Thus, the approximate total consumption for the subinterval is given by

$$q_i^* \cdot \Delta t_i,$$

and the total consumption over the interval from time $t = 0$ to $t = t_1$ is approximated by

$$\sum_{i=1}^{n} q_i^* \cdot \Delta t_i.$$

FIGURE 25

The limit of this sum as each of the Δt_i's approaches 0 gives the total consumption from time $t = 0$ to $t = t_1$. That is,

$$\text{Total consumption} = \lim_{\Delta t_i \to 0} \sum q_i^* \cdot \Delta t_i.$$

We have seen, however, that this limit is the definite integral of the function $q = q_0 e^{kt}$ from $t = 0$ to $t = t_1$, or

$$\text{Total consumption} = \int_0^{t_1} q_0 e^{kt} \, dt.$$

We can now evaluate this definite integral.

$$\int_0^{t_1} q_0 e^{kt} \, dt = q_0 \int_0^{t_1} e^{kt} \, dt = q_0 \left(\frac{1}{k} e^{kt} \right) \Big|_0^{t_1}$$

$$= \frac{q_0}{k} e^{kt} \Big|_0^{t_1} = \frac{q_0}{k} e^{kt_1} - \frac{q_0}{k} e^0$$

$$= \frac{q_0}{k} e^{kt_1} - \frac{q_0}{k} (1)$$

$$= \frac{q_0}{k} (e^{kt_1} - 1) \qquad \qquad (1)$$

Now let us return to the numbers we gave for petroleum. We said that $q_0 = 19,600$ million barrels, where q_0 represents consumption in the base year. We have $k = .06$, with total petroleum reserves estimated at 2,000,000 million barrels. Thus, using equation (1) we have

$$2,000,000 = \frac{19,600}{.06} (e^{.06t_1} - 1).$$

Multiply both sides of the equation by .06:

$$120,000 = 19,600 (e^{.06t_1} - 1).$$

Divide both sides of the equation by 19,600.

$$6.1 = e^{.06t_1} - 1$$

Add 1 to both sides.

$$7.1 = e^{.06t_1}$$

Take natural logarithms of both sides:

$$\ln 7.1 = \ln e^{.06t_1}$$
$$= .06t_1 \ln e$$
$$= .06t_1 \qquad (\text{since } \ln e = 1).$$

Finally,

$$t_1 = \frac{\ln 7.1}{.06} \approx 33.$$

By this result, petroleum reserves will last the world for 33 years.

The results of mathematical analyses such as this must be used with great caution. By the analysis above, the world would use all the petroleum that it wants in the thirty-second year after the base year, but there would be none at all in thirty-four years. This is not at all realistic. As petroleum reserves decline, the price will increase, causing demand to decline and supplies to increase.

Exercises

1. Find the number of years that the estimated petroleum reserves would last if used at the same rate as in the base year.

2. How long would the estimated petroleum reserves last if the growth constant was only 2% instead of 6%?

Estimate the length of time until depletion for each of the following minerals.

3. Bauxite (the ore from which aluminum is obtained): estimated reserves in base year 15,000,000 thousand tons; rate of consumption in base year 63,000 thousand tons; growth constant 6%

4. Bituminous coal; estimated world reserves 2,000,000 million tons; rate of consumption in base year 2200 million tons; growth constant 4%

Applications at a glance . . .

Radioactive Waste in the Atmosphere . . . average price of a stock . . . arterial blood flow . . .

CHAPTER 15

FURTHER TECHNIQUES AND APPLICATIONS OF INTEGRATION

(See page 776.)

In the previous chapter we discussed indefinite and definite integrals and presented rules for finding the antiderivatives of several types of functions. We saw how these techniques could be used in various applications. In this chapter we develop additional methods of integrating functions, including numerical methods of integration. Numerical methods often are used in applications involving experimental data or with functions that cannot be integrated by other methods. In this chapter we also show how to evaluate an integral that has one or both limits at infinity. These new techniques provide additional applications of integration, such as finding volumes of solids of revolution, the average value of a function, continuous money flow, and solving differential equations.

15.1 INTEGRATION BY PARTS; TABLES OF INTEGRALS

The technique of *integration by parts* often makes it possible to reduce a complicated integral to a simpler integral. If u and v are both differentiable functions, then uv is also differentiable and, by the product rule for derivatives,

$$\frac{d(uv)}{dx} = u\frac{dv}{dx} + v\frac{du}{dx}.$$

This expression can be rewritten, using differentials, as

$$d(uv) = u\,dv + v\,du.$$

Integrating both sides of this last equation gives

$$\int d(uv) = \int u\,dv + \int v\,du,$$

or

$$uv = \int u\,dv + \int v\,du.$$

Rearranging terms gives the following formula.

INTEGRATION BY PARTS If u and v are differentiable functions, then

$$\int u\,dv = uv - \int v\,du.$$

The process of finding integrals by this formula is called **integration by parts.** The method is shown in the following examples.

EXAMPLE 1 Find $\int xe^{5x}\,dx$.

While this integral cannot be found by using any method studied so far, it can be found with integration by parts. First write the expression $xe^{5x}\,dx$ as a product of two functions u and dv in such a way that $\int dv$ can be found. One way to do this is to choose the two functions x and e^{5x}. Both x and e^{5x} can be integrated, but $\int x\,dx$, which is $x^2/2$, is more complicated than x itself, while the

derivative of x is 1, which is simpler than x. Since e^{5x} remains the same (except for the coefficient) whether it is integrated or differentiated, it is best here to choose

$$dv = e^{5x}\, dx \quad \text{and} \quad u = x.$$

Then

$$du = dx,$$

and v is found by integrating dv:

$$v = \int dv = \int e^{5x}\, dx = \frac{1}{5} e^{5x} + C.$$

For simplicity, ignore the constant C and add it at the end of the process of integration by parts. Now substitute into the formula for integration by parts and complete the integration.

$$\int u\, dv = uv - \int v\, du$$

$$\int \underbrace{x}_{u}\, \underbrace{e^{5x}\, dx}_{dv} = \underbrace{x}_{u} \underbrace{\left(\frac{1}{5} e^{5x} \right)}_{v} - \int \underbrace{\frac{1}{5} e^{5x}\, dx}_{v \quad du}$$

$$= \frac{1}{5} x e^{5x} - \frac{1}{25} e^{5x} + C$$

The constant C was added in the last step. As before, check the answer by taking its derivative. ◀

EXAMPLE 2 Find $\int \ln x\, dx$ for $x > 0$.

No rule has been given for integrating $\ln x$, so choose

$$dv = dx \quad \text{and} \quad u = \ln x.$$

Then

$$v = x \quad \text{and} \quad du = \frac{1}{x} dx,$$

and

$$\int u \cdot dv = v \cdot u - \int v \cdot du$$

$$\int \overbrace{\ln x\, dx} = x \overbrace{\ln x} - \int x \cdot \overbrace{\frac{1}{x}}\, dx$$

$$= x \ln x - \int dx$$

$$= x \ln x - x + C. \quad ◀$$

The preceding examples illustrate the following general principles for identifying integrals that can be found with integration by parts.

CONDITIONS FOR INTEGRATION BY PARTS

Integration by parts can be used only if the integrand satisfies the following conditions.

1. The integrand can be written as the product of two factors, u and dv.

2. It is possible to integrate dv to get v and to differentiate u to get du.

3. The integral $\int v \, du$ can be found.

With the functions discussed so far in this book, choosing u and dv is relatively simple. Before trying integration by parts, first see if the integration can be performed using substitution. If substitution does not work, see if $\ln x$ is in the integral. If it is, set $u = \ln x$ and dv equal to the rest of the integral. If $\ln x$ is not present, see if x^k is present, where k is any positive integer. If it is, set $u = x^k$ and dv equal to the rest of the integral.

A technique called **column integration,** or tabular integration, simplifies the process of integration by parts.* We begin by creating two columns. The first column, labeled D, contains u, the part to be differentiated in the original integral. The second column, labeled I, contains the rest of the integral: that is, the part to be integrated, but without the dx. To create the remainder of the first column, write the derivative of the function in the first row underneath it in the second row. Now write the derivative of the function in the second row underneath it in the third row. Proceed in this manner down the first column, taking derivatives until you get a 0. Form the second column in a similar manner, except take an antiderivative at each row, until the second column has the same number of rows as the first.

In the section on integration by substitution, we pointed out that when the chain rule is used to find the derivative of the function e^{kx}, we multiply by k, so when finding the antiderivative of e^{kx}, we divide by k. Thus $\int e^{5x} \, dx = e^{5x}/5 + C$. Keeping this technique in mind makes it simple to fill in the integration column when doing column integration.

To illustrate this process, consider Example 1, $\int x e^{5x} \, dx$. Here $u = x$, so e^{5x} is left for the second column. Taking derivatives down the first column and antiderivatives down the second column results in the following table.

D	I
x	e^{5x}
1	$e^{5x}/5$
0	$e^{5x}/25$

Next, draw a diagonal line from each term (except the last) in the left column to the term in the row below it in the right column. Label the first such line with "$+$", the next with "$-$", and continue alternating the signs as shown.

D	I
x $+$	e^{5x}
1 $-$	$e^{5x}/5$
0	$e^{5x}/25$

*This technique appeared in the 1988 movie *Stand and Deliver*.

Then multiply the terms on opposite ends of each diagonal line. Finally, sum up the products just formed, adding the "+" terms and subtracting the "−" terms.

$$\int xe^{5x}\,dx = x(e^{5x}/5) - 1(e^{5x}/25) + C$$
$$= \frac{1}{5}xe^{5x} - \frac{1}{25}e^{5x} + C$$

Column integration works a little differently when applied to Example 2. Choose ln x as the part to differentiate, but no matter how many times ln x is differentiated, the result is not 0. In this case, stop as soon as the natural logarithm is gone. The part to be integrated must be 1. (Think of ln x as ln $x \cdot 1$.)

D	**I**
ln x	1
$1/x$	x

Draw diagonal lines with alternating + and − as before. On the last line, because the left column does not contain a 0, draw a horizontal line:

D	**I**
ln x +	1
$1/x$	x

The presence of a horizontal line indicates that there is an integral sign on that product:

$$\int \ln x\,dx = (\ln x)x - \int \frac{1}{x} \cdot x\,dx$$
$$= x \ln x - \int dx$$
$$= x \ln x - x + C.$$

Note that when setting up the columns, a horizontal line is drawn only when a 0 does not eventually appear in the left column.

Column integration is particularly useful when the part to be differentiated (the part labeled u) must be differentiated more than twice to get 0. Integration by parts would have to be used repeatedly, but column integration does it all at once, as in the following example.

EXAMPLE **3**

Find $\int 2x^2 e^{-3x}\, dx$.

Choose $2x^2$ as the part to be differentiated, and put e^{-3x} in the integration column.

D		I
$2x^2$	$+$	e^{-3x}
$4x$	$-$	$-e^{-3x}/3$
4	$+$	$e^{-3x}/9$
0		$-e^{-3x}/27$

Multiplying and adding as before yields

$$\int 2x^2 e^{-3x}\, dx = 2x^2(-e^{-3x}/3) - 4x(e^{-3x}/9) + 4(-e^{-3x}/27) + C$$

$$= -\frac{2}{3}x^2 e^{-3x} - \frac{4}{9}xe^{-3x} - \frac{4}{27}e^{-3x} + C. \quad \blacktriangleleft$$

EXAMPLE **4**

Find $\displaystyle\int_1^e \frac{\ln x}{x^2}\, dx$.

First we will find the indefinite integral using integration by parts. (You may wish to verify this using column integration.) Whenever $\ln x$ is present, it is selected as u, so let

$$dv = \frac{1}{x^2}\, dx \qquad \text{and} \qquad u = \ln x.$$

Recall that $\int x^n\, dx = x^{n+1}/(n+1) + C, n \neq -1$, so $\int 1/x^2\, dx = \int x^{-2}\, dx = x^{-1}/(-1) + C = -1/x + C.$

Then

$$v = -\frac{1}{x} \qquad \text{and} \qquad du = \frac{1}{x}\, dx.$$

Substitute these values into the formula for integration by parts, and integrate the second term on the right.

$$\int u\, dv = uv - \int v\, du$$

$$\int \frac{\ln x}{x^2}\, dx = \ln x \frac{-1}{x} - \int -\frac{1}{x} \cdot \frac{1}{x}\, dx$$

$$= -\frac{\ln x}{x} + \int \frac{1}{x^2}\, dx$$

$$= -\frac{\ln x}{x} - \frac{1}{x} + C$$

$$= \frac{-\ln x - 1}{x} + C.$$

Now find the definite integral.

$$\int_1^e \frac{\ln x}{x^2}\, dx = \frac{-\ln x - 1}{x}\Bigg|_1^e$$

$$= \left(\frac{-1 - 1}{e}\right) - \left(\frac{0 - 1}{1}\right)$$

$$= \frac{-2}{e} + 1. \quad \blacktriangleleft$$

Integration Using Tables of Integrals The method of integration by parts requires choosing the factor dv so that $\int dv$ can be found. If this is not possible, or if the remaining factor, which becomes u, does not have a differential du such that $\int v\, du$ can be found, the technique cannot be used. For example, to integrate

$$\int \frac{1}{4 - x^2}\, dx,$$

we might choose $dv = dx$ and $u = (4 - x^2)^{-1}$. Integration gives $v = x$ and differentiation gives $du = 2x\, dx/(4 - x^2)^2$, with

$$\int \frac{1}{4 - x^2}\, dx = \frac{x}{4 - x^2} - \int \frac{2x^2\, dx}{(4 - x^2)^2}.$$

The integral on the right is more complicated than the original integral, however. A second use of integration by parts on the new integral would make matters even worse. Since we cannot choose $dv = (4 - x^2)^{-1}\, dx$ because it cannot be integrated by the methods studied so far, integration by parts is not possible for this problem. In fact, there are many functions whose integrals cannot be found by any of the methods described in this text. Many of these can be found by more advanced methods and are listed in **tables of integrals.** Such a table is given at the back of this book. The next examples show how to use this table.

EXAMPLE 5 Find $\displaystyle\int \frac{1}{\sqrt{x^2 + 16}}\, dx.$

If $a = 4$, this antiderivative is the same as entry 5 in the table,

$$\int \frac{1}{\sqrt{x^2 + a^2}}\, dx = \ln \left|x + \sqrt{x^2 + a^2}\right| + C.$$

Substituting 4 for a gives

$$\int \frac{1}{\sqrt{x^2 + 16}}\, dx = \ln \left|x + \sqrt{x^2 + 16}\right| + C.$$

This result could be verified by taking the derivative of the right side of this last equation. \blacktriangleleft

EXAMPLE 6 Find $\int \dfrac{8}{16 - x^2}\, dx.$

Convert this antiderivative into the one given in entry 7 of the table by writing the 8 in front of the integral sign (permissible only with constants) and by letting $a = 4$. Doing this gives

$$8\int \frac{1}{16 - x^2}\, dx = 8\left[\frac{1}{2 \cdot 4}\ln\left|\frac{4 + x}{4 - x}\right|\right] + C$$

$$= \ln\left|\frac{4 + x}{4 - x}\right| + C. \blacktriangleleft$$

EXAMPLE 7 Find $\int \sqrt{9x^2 + 1}\, dx.$

This antiderivative seems most similar to entry 15 of the table. However, entry 15 requires that the coefficient of the x^2 term be 1. That requirement can be satisfied here by factoring out the 9.

$$\int \sqrt{9x^2 + 1}\, dx = \int \sqrt{9\left(x^2 + \frac{1}{9}\right)}\, dx = \int 3\sqrt{x^2 + \frac{1}{9}}\, dx = 3\int \sqrt{x^2 + \frac{1}{9}}\, dx$$

Now use entry 15 with $a = 1/3$.

$$\int \sqrt{9x^2 + 1}\, dx = 3\left[\frac{x}{2}\sqrt{x^2 + \frac{1}{9}} + \frac{(1/3)^2}{2}\ln\left|x + \sqrt{x^2 + \frac{1}{9}}\right|\right] + C$$

$$= \frac{3x}{2}\sqrt{x^2 + \frac{1}{9}} + \frac{1}{6}\ln\left|x + \sqrt{x^2 + \frac{1}{9}}\right| + C \blacktriangleleft$$

15.1 Exercises

Use integration by parts or column integration to find the integrals in Exercises 1–10.

1. $\displaystyle\int xe^x\, dx$

2. $\displaystyle\int (x + 1)e^x\, dx$

3. $\displaystyle\int (5x - 9)e^{-3x}\, dx$

4. $\displaystyle\int (6x + 3)e^{-2x}\, dx$

5. $\displaystyle\int_0^1 \frac{2x + 1}{e^x}\, dx$

6. $\displaystyle\int_0^1 \frac{1 - x}{3e^x}\, dx$

7. $\displaystyle\int_1^4 \ln 2x\, dx$

8. $\displaystyle\int_1^2 \ln 5x\, dx$

9. $\displaystyle\int x \ln x\, dx$

10. $\displaystyle\int x^2 \ln x\, dx$

11. Find the area between $y = (x - 2)e^x$ and the x-axis from $x = 2$ to $x = 4$.

12. Find the area between $y = xe^x$ and the x-axis from $x = 0$ to $x = 1$.

Exercises 13–22 are mixed—some require integration by parts or column integration, while others can be integrated by using techniques discussed earlier.

13. $\displaystyle\int x^2 e^{2x}\, dx$

14. $\displaystyle\int_1^2 (1 - x^2)e^{2x}\, dx$

15. $\displaystyle\int_0^5 x\sqrt[3]{x^2 + 2}\, dx$

16. $\displaystyle\int (2x - 1)\ln(3x)\, dx$

17. $\displaystyle\int (8x + 7) \ln (5x)\, dx$

18. $\displaystyle\int xe^{x^2}\, dx$

19. $\displaystyle\int x^2\sqrt{x + 2}\, dx$

20. $\displaystyle\int_0^1 \frac{x^2\, dx}{2x^3 + 1}$

21. $\displaystyle\int_0^1 \frac{x^3\, dx}{\sqrt{3 + x^2}}$

22. $\displaystyle\int \frac{x^2\, dx}{2x^3 + 1}$

Use the table of integrals to find each indefinite integral.

23. $\displaystyle\int \frac{-4}{\sqrt{x^2 + 36}}\, dx$

24. $\displaystyle\int \frac{9}{\sqrt{x^2 + 9}}\, dx$

25. $\displaystyle\int \frac{6}{x^2 - 9}\, dx$

26. $\displaystyle\int \frac{-12}{x^2 - 16}\, dx$

27. $\displaystyle\int \frac{-4}{x\sqrt{9 - x^2}}\, dx$

28. $\displaystyle\int \frac{3}{x\sqrt{121 - x^2}}\, dx$

29. $\displaystyle\int \frac{-2x}{3x + 1}\, dx$

30. $\displaystyle\int \frac{6x}{4x - 5}\, dx$

31. $\displaystyle\int \frac{2}{3x(3x - 5)}\, dx$

32. $\displaystyle\int \frac{-4}{3x(2x + 7)}\, dx$

33. $\displaystyle\int \frac{4}{4x^2 - 1}\, dx$

34. $\displaystyle\int \frac{-6}{9x^2 - 1}\, dx$

35. $\displaystyle\int \frac{3}{x\sqrt{1 - 9x^2}}\, dx$

36. $\displaystyle\int \frac{-2}{x\sqrt{1 - 16x^2}}\, dx$

37. $\displaystyle\int \frac{4x}{2x + 3}\, dx$

38. $\displaystyle\int \frac{4x}{6 - x}\, dx$

39. $\displaystyle\int \frac{-x}{(5x - 1)^2}\, dx$

40. $\displaystyle\int \frac{-3}{x(4x + 3)^2}\, dx$

41. Use integration by parts or column integration to derive the following formula from the table of integrals.

$$\int x^n \cdot \ln |x|\, dx = x^{n+1}\left[\frac{\ln |x|}{n + 1} - \frac{1}{(n + 1)^2}\right] + C, \quad n \neq -1$$

42. Use integration by parts or column integration to derive the following formula from the table of integrals.

$$\int x^n e^{ax}\, dx = \frac{x^n e^{ax}}{a} - \frac{n}{a}\int x^{n-1}e^{ax}\, dx + C, \quad a \neq 0$$

43. (a) One way to integrate $\int x\sqrt{x + 1}\, dx$ is to use integration by parts or column integration. Do so to find the antiderivative.

(b) Another way to evaluate the integral in part (a) is by using the substitution $u = x + 1$. Do so to find the antiderivative.

(c) Compare the results from the two methods. If they do not look the same, explain how this can happen. Discuss the advantages and disadvantages of each method.

Applications

Business and Economics

44. Rate of Change of Revenue The rate of change of revenue (in dollars) from the sale of x small desk calculators is

$$R'(x) = \frac{1000}{\sqrt{x^2 + 25}}.$$

Find the total revenue from the sale of the first 12 calculators.

Life Sciences

45. Reaction to a Drug The rate of reaction to a drug is given by

$$r'(x) = 2x^2 e^{-x},$$

where x is the number of hours since the drug was administered. Find the total reaction to the drug from $x = 1$ to $x = 6$.

46. Growth of a Population The rate of growth of a microbe population is given by

$$m'(x) = 30xe^{2x},$$

where x is time in days. What is the total accumulated growth after 3 days?

Social Sciences

47. Production Rate The rate (in hours per item) at which a worker in a certain job produces the xth item is

$$h'(x) = \sqrt{x^2 + 16}.$$

What is the total number of hours it will take this worker to produce the first 7 items?

15.2 NUMERICAL INTEGRATION

 If the velocity of a vehicle is known only at certain points in time, how can the total distance traveled by the vehicle be estimated?

Using numerical integration, we will answer this question in Example 3 of this section.

As mentioned in the previous section, some integrals cannot be evaluated by any technique, nor are they found in any table. One solution to this problem was presented in Section 3 of the previous chapter, in which the area under a curve was approximated by summing the areas of rectangles. This method is seldom used in practice because better methods exist which give more accuracy for the same amount of work. These methods are referred to as **numerical integration** methods. We shall discuss two such methods here: the trapezoidal rule and Simpson's rule.

Trapezoidal Rule To illustrate the use of the trapezoidal rule, consider the integral

$$\int_1^5 \frac{1}{x}\, dx.$$

The shaded region in Figure 1 shows the area representing that integral, the area under the graph $f(x) = 1/x$, above the x-axis, and between the lines $x = 1$ and $x = 5$. As shown in the figure, if the area under the curve is approximated with trapezoids rather than rectangles, the approximation should be improved.

FIGURE 1

Since $\int (1/x)\, dx = \ln |x| + C$,

$$\int_1^5 \frac{1}{x}\, dx = \ln |x| \Big|_1^5 = \ln 5 - \ln 1 = \ln 5 - 0 = \ln 5 \approx 1.609438.$$

As in earlier work, to approximate this area we divide the interval $[1, 5]$ into subintervals of equal widths. To get a first approximation to ln 5 by the trapezoidal rule, find the sum of the areas of the four trapezoids shown in Figure 1. From geometry, the area of a trapezoid is half the product of the sum of the bases and the altitude. Each of the trapezoids in Figure 1 has altitude 1. (In this case, the bases of the trapezoid are vertical and the altitudes are horizontal.) Adding the areas gives

$$\ln 5 = \int_1^5 \frac{1}{x}\, dx \approx \frac{1}{2}\left(\frac{1}{1} + \frac{1}{2}\right)(1) + \frac{1}{2}\left(\frac{1}{2} + \frac{1}{3}\right)(1) + \frac{1}{2}\left(\frac{1}{3} + \frac{1}{4}\right)(1) + \frac{1}{2}\left(\frac{1}{4} + \frac{1}{5}\right)(1)$$

$$= \frac{1}{2}\left(\frac{3}{2} + \frac{5}{6} + \frac{7}{12} + \frac{9}{20}\right) \approx 1.68333.$$

To get a better approximation, divide the interval $[1, 5]$ into more subintervals. Generally speaking, the larger the number of subintervals, the better the approximation. The results for selected values of n are shown below to 5 decimal places.

n	$\int_1^5 \dfrac{1}{x}\, dx = \ln 5 \approx 1.609438$
6	1.64360
8	1.62897
10	1.62204
20	1.61263
100	1.60957
1000	1.60944

When $n = 1000$, the approximation agrees with the true value to five decimal places.

Generalizing from this example, let f be a continuous function on an interval $[a, b]$. Divide the interval from a to b into n equal subintervals by the points $a = x_0, x_1, x_2, \ldots, x_n = b$, as shown in Figure 2 (on the next page). Use the subintervals to make trapezoids that approximately fill in the region under the curve. The approximate value of the definite integral $\int_a^b f(x)\, dx$ is given by the sum of the areas of the trapezoids, or

$$\int_a^b f(x)\, dx \approx \frac{1}{2}\left[f(x_0) + f(x_1)\right]\left(\frac{b-a}{n}\right) + \frac{1}{2}\left[f(x_1) + f(x_2)\right]\left(\frac{b-a}{n}\right) + \cdots + \frac{1}{2}\left[f(x_{n-1}) + f(x_n)\right]\left(\frac{b-a}{n}\right)$$

$$= \left(\frac{b-a}{n}\right)\left[\frac{1}{2}f(x_0) + \frac{1}{2}f(x_1) + \frac{1}{2}f(x_1) + \frac{1}{2}f(x_2) + \frac{1}{2}f(x_2) + \cdots + \frac{1}{2}f(x_{n-1}) + \frac{1}{2}f(x_n)\right]$$

$$= \left(\frac{b-a}{n}\right)\left[\frac{1}{2}f(x_0) + f(x_1) + f(x_2) + \cdots + f(x_{n-1}) + \frac{1}{2}f(x_n)\right].$$

FIGURE 2

This result gives the following rule.

| **TRAPEZOIDAL RULE** | Let f be a continuous function on $[a, b]$ and let $[a, b]$ be divided into n equal subintervals by the points $a = x_0, x_1, x_2, \ldots, x_n = b$. Then, by the **trapezoidal rule,** |

$$\int_a^b f(x) \, dx \approx \left(\frac{b-a}{n}\right)\left[\frac{1}{2}f(x_0) + f(x_1) + \cdots + f(x_{n-1}) + \frac{1}{2}f(x_n)\right].$$

Use the trapezoidal rule with $n = 4$ to approximate

$$\int_0^2 \sqrt{x^2 + 1} \, dx.$$

Here $a = 0$, $b = 2$, and $n = 4$, with $(b - a)/n = (2 - 0)/4 = 1/2$ as the altitude of each trapezoid. Then $x_0 = 0$, $x_1 = 1/2$, $x_2 = 1$, $x_3 = 3/2$, and $x_4 = 2$. Now find the corresponding function values. The work can be organized into a table, as follows.

i	x_i	$f(x_i)$
0	0	$\sqrt{0^2 + 1} = 1$
1	1/2	$\sqrt{(1/2)^2 + 1} \approx 1.11803$
2	1	$\sqrt{1^2 + 1} \approx 1.41421$
3	3/2	$\sqrt{(3/2)^2 + 1} \approx 1.80278$
4	2	$\sqrt{2^2 + 1} \approx 2.23607$

Substitution into the trapezoidal rule gives

$$\int_0^2 \sqrt{x^2 + 1}\, dx$$

$$\approx \frac{2 - 0}{4}\left[\frac{1}{2}(1) + 1.11803 + 1.41421 + 1.80278 + \frac{1}{2}(2.23607)\right]$$

$$\approx 2.97653.$$

Using entry 15 in the table of integrals, the exact value of the integral can be found.

$$\int_0^2 \sqrt{x^2 + 1}\, dx = \frac{x}{2}\sqrt{x^2 + 1} + \frac{1}{2}\ln\left|x + \sqrt{x^2 + 1}\right|\Bigg|_0^2$$

$$= \sqrt{5} + \frac{1}{2}\ln(2 + \sqrt{5}) - (0 + 0) \approx 2.95789$$

The approximation 2.97653 found above using the trapezoidal rule with $n = 4$ differs from the true value by .01864. As mentioned above, this error would be reduced if larger values were used for n. For example, if $n = 8$, the trapezoidal rule gives an answer of 2.96254, which differs from the true value by .00465. Techniques for estimating such errors are considered in more advanced courses. ◄

Simpson's Rule Another numerical method, *Simpson's rule*, approximates consecutive portions of the curve with portions of parabolas rather than the line segments of the trapezoidal rule. Simpson's rule usually gives a better approximation than the trapezoidal rule for the same number of subintervals. As shown in Figure 3, one parabola is fitted through points A, B, and C, another through C, D, and E, and so on. Then the sum of the areas under these parabolas will approximate the area under the graph of the function. Because of the way the parabolas overlap, it is necessary to have an even number of intervals, and therefore an odd number of points, to apply Simpson's rule.

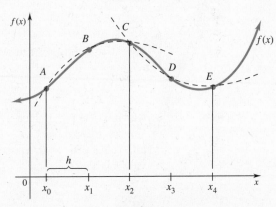

FIGURE 3

If h, the length of each subinterval, is $(b - a)/n$, the area under the parabola through points A, B, and C can be found by a definite integral. The details are omitted; the result is

$$\frac{h}{3}\left[f(x_0) + 4f(x_1) + f(x_2)\right].$$

Similarly, the area under the parabola through points C, D, and E is

$$\frac{h}{3}\left[f(x_2) + 4f(x_3) + f(x_4)\right].$$

When these expressions are added, the last term of one expression equals the first term of the next. For example, the sum of the two areas given above is

$$\frac{h}{3}\left[f(x_0) + 4f(x_1) + 2f(x_2) + 4f(x_3) + f(x_4)\right].$$

This illustrates the origin of the pattern of the terms in the following rule.

SIMPSON'S RULE

Let f be a continuous function on $[a, b]$ and let $[a, b]$ be divided into an even number n of equal subintervals by the points $a = x_0, x_1, x_2, \ldots, x_n = b$. Then by **Simpson's rule,**

$$\int_a^b f(x)\,dx \approx \frac{b - a}{3n}\,[f(x_0) + 4f(x_1) + 2f(x_2) + 4f(x_3) + \cdots$$
$$+ 2f(x_{n-2}) + 4f(x_{n-1}) + f(x_n)].$$

Thomas Simpson (1710–1761), a British mathematician, wrote texts on many branches of mathematics. Some of these texts went through as many as ten editions. His name became attached to this numerical method of approximating definite integrals even though the method preceded his work.

Caution In Simpson's rule, n (the number of subintervals) must be even.

EXAMPLE 2 Use Simpson's rule with $n = 4$ to approximate

$$\int_0^2 \sqrt{x^2 + 1}\,dx,$$

which was approximated by the trapezoidal rule in Example 1.

As in Example 1, $a = 0$, $b = 2$, and $n = 4$, and the endpoints of the four intervals are $x_0 = 0$, $x_1 = 1/2$, $x_2 = 1$, $x_3 = 3/2$, and $x_4 = 2$. The table of values is also the same.

i	x_i	$f(x_i)$
0	0	1
1	1/2	1.11803
2	1	1.41421
3	3/2	1.80278
4	2	2.23607

Since $(b - a)/(3n) = 2/12 = 1/6$, substituting into Simpson's rule gives

$$\int_0^2 \sqrt{x^2 + 1}\ dx \approx \frac{1}{6}\ [1 + 4(1.11803) + 2(1.41421) + 4(1.80278) + 2.23607] \approx 2.95796.$$

This differs from the true value by .00007, which is less than the trapezoidal rule with $n = 8$. If $n = 8$ for Simpson's rule, the approximation is 2.95788, which differs from the true value by only .00001. ◀

Numerical methods make it possible to approximate

$$\int_a^b f(x)\ dx$$

even when $f(x)$ is not known. The next example shows how this is done.

EXAMPLE 3
As mentioned earlier, the velocity $v(t)$ gives the rate of change of distance $s(t)$ with respect to time t. Suppose a vehicle travels an unknown distance. The passengers keep track of the velocity at 10-minute intervals (every 1/6 of an hour) with the following results.

Time in Hours, t	1/6	2/6	3/6	4/6	5/6	1	7/6
Velocity in Miles per Hour, $v(t)$	45	55	52	60	64	58	47

What is the total distance traveled in the 60-minute period from $t = 1/6$ to $t = 7/6$?

The distance traveled in t hours is $s(t)$, with $s'(t) = v(t)$. The total distance traveled between $t = 1/6$ and $t = 7/6$ is given by

$$\int_{1/6}^{7/6} v(t)\ dt.$$

Even though this integral cannot be evaluated since we do not have an expression for $v(t)$, either the trapezoid rule or Simpson's rule can be used to approximate its value and give the total distance traveled. In either case, let $n = 6$, $a = t_0 = 1/6$, and $b = t_6 = 7/6$. By the trapezoidal rule,

$$\int_{1/6}^{7/6} v(t)\ dt \approx \frac{7/6 - 1/6}{6}\left[\frac{1}{2}(45) + 55 + 52 + 60 + 64 + 58 + \frac{1}{2}(47)\right]$$

$$\approx 55.83.$$

By Simpson's rule,

$$\int_{1/6}^{7/6} v(t)\, dt \approx \frac{7/6 - 1/6}{3(6)}[45 + 4(55) + 2(52) + 4(60) + 2(64) + 4(58) + 47]$$

$$= \frac{1}{18}(45 + 220 + 104 + 240 + 128 + 232 + 47) \approx 56.44.$$

The distance traveled in the 1-hour period was about 56 miles.

As mentioned above, Simpson's rule generally gives a better approximation than the trapezoidal rule. As n increases, the two approximations get closer and closer. For the same accuracy, however, a smaller value of n generally can be used with Simpson's rule so that less computation is necessary. ◀

The branch of mathematics that studies methods of approximating definite integrals (as well as many other topics) is called *numerical analysis.* Some calculators give such approximations by using Simpson's rule, as shown in the following excerpt from the manual for model TI-60, a calculator produced by Texas Instruments.*

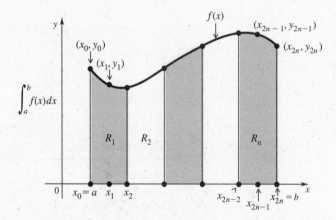

Numerical integration is useful even with functions whose antiderivatives can be computed if the antidifferentiation is complicated and a computer or calculator programmed with Simpson's rule is handy. For example, it would be easier to evaluate $\int_0^2 (x^{12} - 3x^7 + 4x + 5)e^{-.5x}\, dx$ by using Simpson's rule than by using the antiderivative. On the other hand, using the antiderivative makes it easier to see the effect of changing the upper limit, say, from 2 to 3, or changing the power of e from $-.5x$ to $-1.5x$. Numerical integration only gives information about the particular integral being evaluated.

*"Simpson's Rule" from *Texas Instruments TI-60 Guidebook.* Copyright © 1986 by Texas Instruments Incorporated. Reprinted by permission.

15.2 Exercises

In Exercises 1–10, use $n = 4$ to approximate the value of each of the given integrals by the following methods:
(a) *the trapezoidal rule, and* **(b)** *Simpson's rule.* **(c)** *Find the exact value by integration.*

1. $\displaystyle\int_0^2 x^2\, dx$

2. $\displaystyle\int_0^2 (2x + 1)\, dx$

3. $\displaystyle\int_{-1}^3 \frac{1}{4 - x}\, dx$

4. $\displaystyle\int_1^5 \frac{1}{x + 1}\, dx$

5. $\displaystyle\int_{-2}^2 (2x^2 + 1)\, dx$

6. $\displaystyle\int_0^3 (2x^2 + 1)\, dx$

7. $\displaystyle\int_1^5 \frac{1}{x^2}\, dx$

8. $\displaystyle\int_2^4 \frac{1}{x^3}\, dx$

9. $\displaystyle\int_0^4 \sqrt{x^2 + 1}\, dx$

10. $\displaystyle\int_1^4 x\sqrt{2x - 1}\, dx$

11. Find the area under the semicircle $y = \sqrt{4 - x^2}$ and above the x-axis by using $n = 8$ with the following methods:

 (a) the trapezoidal rule; **(b)** Simpson's rule.

 (c) Compare the results with the area found by the formula for the area of a circle. Which of the two approximation techniques was more accurate?

12. Find the area between the x-axis and the ellipse $4x^2 + 9y^2 = 36$ by using $n = 12$ with the following methods:

 (a) the trapezoidal rule; **(b)** Simpson's rule.
 (*Hint:* Solve the equation for y and find the area of the semiellipse.)

 (c) Compare the results with the actual area, $6\pi \approx 18.8496$ (which can be found by methods not considered in this text). Which approximation technique was more accurate?

Applications

Business and Economics

13. Total Sales A sales manager presented the following results at a sales meeting.

Year, x	1	2	3	4	5	6	7
Rate of Sales, $f(x)$.4	.6	.9	1.1	1.3	1.4	1.6

Find the total sales over the given period as follows.

 (a) Plot these points. Connect the points with line segments.

 (b) Use the trapezoidal rule to find the area bounded by the broken line of part (a), the x-axis, the line $x = 1$, and the line $x = 7$.

 (c) Find the same area using Simpson's rule.

14. Total Sales A company's marginal costs (in hundreds of dollars) were as follows over a certain period.

Year, x	1	2	3	4	5	6	7
Marginal Cost, $f(x)$	9.0	9.2	9.5	9.4	9.8	10.1	10.5

Repeat steps (a)–(c) of Exercise 13 for these data to find the total sales over the given period.

Life Sciences

15. Drug Reaction Rate The reaction rate to a new drug is given by

$$y = e^{-t^2} + \frac{1}{t},$$

where t is time measured in hours after the drug is administered. Find the total reaction to the drug from $t = 1$ to $t = 9$ by letting $n = 8$ and using the following methods:
(a) the trapezoidal rule, and **(b)** Simpson's rule.

16. Growth Rate The growth rate of a certain tree (in feet) is given by

$$y = \frac{2}{t} + e^{-t^2/2},$$

where t is time in years. Find the total growth from $t = 1$ to $t = 7$ by using $n = 12$ with the following methods:
(a) the trapezoidal rule, **(b)** Simpson's rule.

Blood Level Curves *In the study of bioavailability in pharmacy, a drug is given to a patient. The level of concentration of the drug is then measured periodically, producing blood level curves such as the ones shown in the figure. The areas under the curves give the total amount of the drug available to the patient.* Use the trapezoidal rule with n = 10 to find the following areas.*

17. Find the total area under the curve for Formulation A. What does this area represent?

18. Find the total area under the curve for Formulation B. What does this area represent?

19. Find the area between the curve for Formulation A and the minimum effective concentration line. What does your answer represent?

20. Find the area between the curve for Formulation B and the minimum effective concentration line. What does this area represent?

Social Sciences

21. **Educational Psychology** The results from a research study in psychology were as follows.

Number of Hours of Study, x	1	2	3	4	5	6	7
Rate of Extra Points Earned on a Test, $f(x)$	4	7	11	9	15	16	23

Repeat steps (a)–(c) of Exercise 13 for these data.

Physical Sciences

22. **Chemical Formation** The table below shows the results from a chemical experiment.

Concentration of Chemical A, x	1	2	3	4	5	6	7
Rate of Formation of Chemical B, $f(x)$	12	16	18	21	24	27	32

Repeat steps (a)–(c) of Exercise 13 for these data.

Computer/Graphing Calculator

Use the trapezoidal rule and then use Simpson's rule to approximate each of the following integrals. Use n = 100.

23. $\displaystyle\int_{4}^{8} \ln(x^2 - 10)\, dx$

24. $\displaystyle\int_{-2}^{2} e^{-x^2}\, dx$

25. $\displaystyle\int_{-2}^{2} \sqrt{9 - 2x^2}\, dx$

*These graphs are from D. J. Chodos and A. R. DeSantos, *Basics of Bioavailability*. Copyright 1978 by the Upjohn Company.

26. $\int_{-1}^{1} \sqrt{16 + 5x^2}\, dx$ **27.** $\int_{1}^{5} (2x^2 + 3x - 1)^{2/5}\, dx$ **28.** $\int_{1}^{4} (x^3 + 4)^{5/4}\, dx$

Use either the trapezoidal rule or Simpson's rule with n = 100 for Exercises 29–31.

29. Total Revenue An electronics company analyst has determined that the rate per month at which revenue comes in from the calculator division is given by

$$R(x) = 105e^{.01x} + 32,$$

where x is the number of months the division has been in operation. Find the total revenue between the twelfth and thirty-sixth months.

30. Blood Pressure Blood pressure in an artery changes rapidly over a very short time for a healthy young adult, from a high of about 120 to a low of about 80. Suppose the blood pressure function over an interval of 1.5 sec is given by

$$f(x) = .2x^5 - .68x^4 + .8x^3 - .39x^2 + .055x + 100,$$

where x is the time in seconds after a peak reading. The area under the curve for one cycle is important in some blood pressure studies. Find the area under $f(x)$ from .1 sec to 1.1 sec.

31. Probability The most important function in probability and statistics is the density function for the standard normal distribution, which is the familiar bell-shaped curve. The function is

$$f(x) = \frac{1}{\sqrt{2\pi}}e^{-x^2/2}.$$

(a) The area under this curve between $x = -1$ and $x = 1$ represents the probability that a normal random variable is within 1 standard deviation of the mean. Find this probability.

(b) Find the area under this curve between $x = -2$ and $x = 2$, which represents the probability that a normal

random variable is within 2 standard deviations of the mean.

(c) Find the probability that a normal random variable is within 3 standard deviations of the mean.

Error Analysis *The difference between the true value of an integral and the value given by the trapezoidal rule or Simpson's rule is known as the error. In numerical analysis, the error is studied to determine how large n must be for the error to be smaller than some specified amount. For both rules, the error is inversely proportional to a power of n, the number of subdivisions. In other words, the error is roughly k/n^p, where k is a constant that depends upon the function and the interval, and p is a power that depends only upon the method used. With a little experimentation, you can find out what the power p is for the trapezoidal rule and for Simpson's rule.*

32. (a) Find the exact value of $\int_{0}^{1} x^4\, dx$.

(b) Approximate the integral in part (a) using the trapezoidal rule with $n = 4, 8, 16,$ and 32. For each of these answers, find the absolute value of the error by subtracting the trapezoidal rule answer from the exact answer found in part (a).

(c) If the error is k/n^p, then the error times n^p should be approximately a constant. Multiply the errors in part (b) times n^p for $p = 1, 2,$ etc., until you find a power p yielding the same answer for all four values of n.

33. Based on the results of Exercise 32, what happens to the error in the trapezoidal rule when the number of intervals is doubled?

34. Repeat Exercise 32 using Simpson's rule.

35. Based on the results of Exercise 34, what happens to the error in Simpson's rule when the number of intervals is doubled?

15.3 TWO APPLICATIONS OF INTEGRATION: VOLUME AND AVERAGE VALUE

If we have a formula giving the price of a common stock as a function of time, how can we find the average price of the stock over a certain period of time?

In this section, we will discover how to find the average value of a function, as well as how to compute the volume of a solid.

Volume Figure 4 shows the region below the graph of some function $y = f(x)$, above the x-axis, and between $x = a$ and $x = b$. We have seen how to use integrals to find the area of such a region. Now, suppose this region is revolved about the x-axis as shown in Figure 5. The resulting figure is called a **solid of revolution.** In many cases, the volume of a solid of revolution can be found by integration.

FIGURE 4 **FIGURE 5**

To begin, divide the interval $[a, b]$ into n subintervals of equal width Δx by the points $a = x_0, x_1, x_2, \ldots, x_i, \ldots, x_n = b$. Then think of slicing the solid into n slices of equal thickness Δx, as shown in Figure 6. If the slices are thin enough, each slice is very close to being a right circular cylinder. The formula for the volume of a right circular cylinder is $\pi r^2 h$, where r is the radius of the circular base and h is the height of the cylinder. As shown in Figure 7, the height of each slice is Δx. (The height is horizontal here, since the cylinder is on its side.) The radius of the circular base of each slice is $f(x_i)$. Thus, the volume of

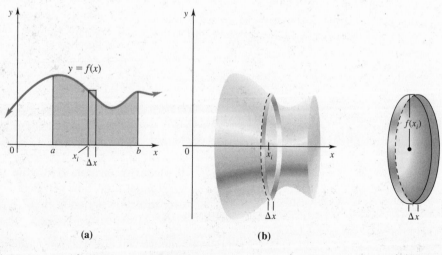

(a) (b)

FIGURE 6 **FIGURE 7**

the slice is closely approximated by $\pi[f(x_i)]^2 \, \Delta x$. The volume of the solid of revolution will be approximated by the sum of the volumes of the slices:

$$V \approx \sum_{i=1}^{n} \pi[f(x_i)]^2 \, \Delta x.$$

By definition, the volume of the solid of revolution is the limit of this sum as the thickness of the slices approaches 0, or

$$V = \lim_{\Delta x \to 0} \sum_{i=1}^{n} \pi[f(x_i)]^2 \, \Delta x.$$

This limit, like the one discussed earlier for area, is a definite integral.

VOLUME OF A SOLID OF REVOLUTION	If $f(x)$ is nonnegative and R is the region between $f(x)$ and the x-axis from $x = a$ to $x = b$, the volume of the solid formed by rotating R about the x-axis is given by

$$V = \lim_{\Delta x \to 0} \sum_{i=1}^{n} \pi[f(x_i)]^2 \, \Delta x = \int_{a}^{b} \pi[f(x)]^2 \, dx.$$

The technique of summing disks to approximate volumes was originated by Johannes Kepler (1571–1630), a famous German astronomer who discovered three laws of planetary motion. He estimated volumes of wine casks used at his wedding by means of solids of revolution.

EXAMPLE 1 Find the volume of the solid of revolution formed by rotating about the x-axis the region bounded by $y = x + 1$, $y = 0$, $x = 1$, and $x = 4$. See Figure 8(a).

(a)

(b)

FIGURE 8

The solid is shown in Figure 8(b). Use the formula given above for the volume, with $a = 1$, $b = 4$, and $f(x) = x + 1$.

$$V = \int_1^4 \pi(x + 1)^2 \, dx = \pi\left(\frac{(x + 1)^3}{3}\right)\Big|_1^4$$

$$= \frac{\pi}{3}(5^3 - 2^3) = \frac{117\pi}{3} = 39\pi \quad \blacktriangleleft$$

EXAMPLE 2 Find the volume of the solid of revolution formed by rotating about the x-axis the area bounded by $f(x) = 4 - x^2$ and the x-axis. See Figure 9(a).

The solid is shown in Figure 9(b). Find a and b from the x-intercepts. If $y = 0$, then $x = 2$ or $x = -2$, so that $a = -2$ and $b = 2$. The volume is

$$V = \int_{-2}^2 \pi(4 - x^2)^2 \, dx$$

$$= \int_{-2}^2 \pi(16 - 8x^2 + x^4) \, dx$$

$$= \pi\left(16x - \frac{8x^3}{3} + \frac{x^5}{5}\right)\Big|_{-2}^2 = \frac{512\pi}{15}. \quad \blacktriangleleft$$

(a) **(b)**

FIGURE 9

EXAMPLE 3 Find the volume of a right circular cone with height h and base radius r.

Figure 10(a) shows the required cone, while Figure 10(b) shows an area that could be rotated about the x-axis to get such a cone. See Figure 10(c). Here $y = f(x)$ is the equation (in slope-intercept form) of the line through $(0, r)$ and $(h, 0)$. The slope of this line is

$$\frac{0 - r}{h - 0} = -\frac{r}{h}.$$

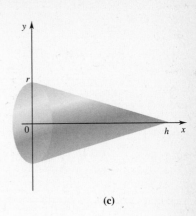

(a)

(b)

(c)

FIGURE 10

Using the point-slope formula with the point $(h, 0)$ gives

$$y - y_1 = m(x - x_1)$$

$$y = -\frac{r}{h}(x - h)$$

as the equation of the line. Then the volume is

$$V = \int_0^h \pi \left[-\frac{r}{h}(x - h) \right]^2 dx = \pi \int_0^h \frac{r^2}{h^2}(x - h)^2 \, dx.$$

Since r and h are constants,

$$V = \frac{\pi r^2}{h^2} \int_0^h (x - h)^2 \, dx.$$

Using substitution with $u = x - h$ and $du = dx$ gives

$$V = \frac{\pi r^2}{h^2} \left[\frac{(x - h)^3}{3} \right] \Bigg|_0^h = \frac{\pi r^2}{3h^2}[0 - (-h)^3] = \frac{\pi r^2 h}{3}.$$

This is the familiar formula for the volume of a right circular cone. ◀

Average Value of a Function The average of the n numbers $v_1, v_2, v_3, \ldots,$ v_i, \ldots, v_n is given by

$$\frac{v_1 + v_2 + v_3 + \cdots + v_n}{n} = \frac{\sum_{i=1}^{n} v_i}{n}.$$

The average value of a function f on $[a, b]$ can be defined in a similar manner; divide the interval $[a, b]$ into n subintervals, each of width Δx. Then choose an x-value, x_i, in each interval, and find $f(x_i)$. The average function value for the n subintervals and the given choices of x_i is

$$\frac{f(x_1) + f(x_2) + f(x_3) + \cdots + f(x_n)}{n} = \frac{\sum_{i=1}^{n} f(x_i)}{n}.$$

Since $(b - a)/n = \Delta x$, multiply the expression on the right side of the equation by $(b - a)/(b - a)$ and rearrange the expression to get

$$\frac{b - a}{b - a} \cdot \frac{\sum\limits_{i=1}^{n} f(x_i)}{n} = \frac{b - a}{n} \cdot \frac{\sum\limits_{i=1}^{n} f(x_i)}{b - a} = \Delta x \cdot \frac{\sum\limits_{i=1}^{n} f(x_i)}{b - a} = \frac{1}{b - a} \sum_{i=1}^{n} f(x_i) \, \Delta x.$$

Now, take the limit as $n \to \infty$. If the limit exists, then

$$\lim_{n \to \infty} \frac{1}{b - a} \sum_{i=1}^{n} f(x_i) \, \Delta x = \frac{1}{b - a} \lim_{n \to \infty} \sum_{i=1}^{n} f(x_i) \, \Delta x = \frac{1}{b - a} \int_{a}^{b} f(x) \, dx.$$

The following definition summarizes this discussion.

AVERAGE VALUE OF A FUNCTION

The **average value** of a function f on the interval $[a, b]$ is

$$\frac{1}{b - a} \int_{a}^{b} f(x) \, dx,$$

provided the indicated definite integral exists.

The average value, sometimes denoted \bar{y}, can be thought of as the height of the rectangle with base $b - a$. See Figure 11. For $f(x) \geq 0$, this rectangle has area $\bar{y}(b - a)$, which equals the area under the graph of $f(x)$ from $x = a$ to $x = b$, so that

$$\bar{y}(b - a) = \int_{a}^{b} f(x) \, dx.$$

FIGURE 11

4 A stock analyst plots the price per share of a certain common stock as a function of time and finds that it can be approximated by the function

$$S(t) = 25 - 5e^{-.01t},$$

where t is the time (in years) since the stock was purchased. Find the average price of the stock over the first six years.

Use the formula for average value with $a = 0$ and $b = 6$. The average price is

$$\frac{1}{6-0}\int_0^6 (25 - 5e^{-.01t})\,dt = \frac{1}{6}\left(25t - \frac{5}{-.01}e^{-.01t}\right)\Bigg|_0^6$$

$$= \frac{1}{6}\left(25t + 500e^{-.01t}\right)\Bigg|_0^6$$

$$= \frac{1}{6}\left(150 + 500e^{-.06} - 500\right)$$

$$= 20.147,$$

or approximately $20.15. ◀

15.3 Exercises

Find the volume of the solid of revolution formed by rotating about the x-axis each region bounded by the given curves.

1. $f(x) = x$, $y = 0$, $x = 0$, $x = 2$

2. $f(x) = 2x$, $y = 0$, $x = 0$, $x = 3$

3. $f(x) = 2x + 1$, $y = 0$, $x = 0$, $x = 4$

4. $f(x) = x - 4$, $y = 0$, $x = 4$, $x = 10$

5. $f(x) = \frac{1}{3}x + 2$, $y = 0$, $x = 1$, $x = 3$

6. $f(x) = \frac{1}{2}x + 4$, $y = 0$, $x = 0$, $x = 5$

7. $f(x) = \sqrt{x}$, $y = 0$, $x = 1$, $x = 2$

8. $f(x) = \sqrt{x + 1}$, $y = 0$, $x = 0$, $x = 3$

9. $f(x) = \sqrt{2x + 1}$, $y = 0$, $x = 1$, $x = 4$

10. $f(x) = \sqrt{3x + 2}$, $y = 0$, $x = 1$, $x = 2$

11. $f(x) = e^x$, $y = 0$, $x = 0$, $x = 2$

12. $f(x) = 2e^x$, $y = 0$, $x = -2$, $x = 1$

13. $f(x) = \frac{1}{\sqrt{x}}$, $y = 0$, $x = 1$, $x = 4$

14. $f(x) = \frac{1}{\sqrt{x + 1}}$, $y = 0$, $x = 0$, $x = 2$

15. $f(x) = x^2$, $y = 0$, $x = 1$, $x = 5$

16. $f(x) = \frac{x^2}{2}$, $y = 0$, $x = 0$, $x = 4$

17. $f(x) = 1 - x^2$, $y = 0$

18. $f(x) = 2 - x^2$, $y = 0$

The function defined by $y = \sqrt{r^2 - x^2}$ has as its graph a semicircle of radius r with center at (0, 0) (see the figure). In Exercises 19–21, find the volume that results when each semicircle is rotated about the x-axis. (The result of Exercise 21 gives a formula for the volume of a sphere with radius r.)

19. $f(x) = \sqrt{1 - x^2}$

20. $f(x) = \sqrt{16 - x^2}$

21. $f(x) = \sqrt{r^2 - x^2}$

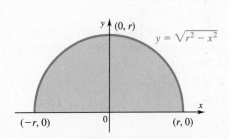

22. Find a formula for the volume of an ellipsoid. See Exercises 19–21 and the following figures.

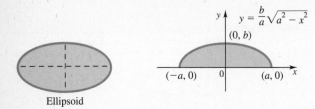

Ellipsoid

23. Use the methods of this section to find the volume of a cylinder with height h and radius r.

Find the average value of each function on the given interval.

24. $f(x) = 3 - 2x^2$; [1, 9]

25. $f(x) = x^2 - 2$; [0, 5]

26. $f(x) = (2x - 1)^{1/2}$; [1, 13]

27. $f(x) = \sqrt{x + 1}$; [3, 8]

28. $f(x) = e^{.1x}$; [0, 10]

29. $f(x) = e^{x/5}$; [0, 5]

30. $f(x) = x \ln x$; [1, e]

31. $f(x) = x^2 e^{2x}$; [0, 2]

Applications

Life Sciences

32. Blood Flow The figure shows the blood flow in a small artery of the body. The flow of blood is *laminar* (in layers), with the velocity very low near the artery walls and highest in the center of the artery. In this model of blood flow, we calculate the total flow in the artery by thinking of the flow as being made up of many layers of concentric tubes sliding one on the other.

Suppose R is the radius of an artery and r is the distance from a given layer to the center. Then the velocity of blood in a given layer can be shown to equal

$$v(r) = k(R^2 - r^2),$$

where k is a numerical constant.

Since the area of a circle is $A = \pi r^2$, the change in the area of the cross section of one of the layers, corresponding to a small change in the radius, Δr, can be approximated by differentials. For $dr = \Delta r$, the differential of the area A is

$$dA = 2\pi r \, dr = 2\pi r \, \Delta r,$$

where Δr is the thickness of the layer. The total flow in the layer is defined to be the product of volume and cross-section area, or

$$F(r) = 2\pi r k(R^2 - r^2) \, \Delta r.$$

(a) Set up a definite integral to find the total flow in the artery.

(b) Evaluate this definite integral.

33. Drug Reaction The intensity of the reaction to a certain drug, in appropriate units, is given by

$$R(t) = te^{-.1t},$$

where t is time (in hours) after the drug is administered. Find the average intensity during each of the following hours.

(a) Second hour **(b)** Twelfth hour

(c) Twenty-fourth hour

Social Sciences

34. Production Rate Suppose the number of items a new worker on an assembly line produces daily after t days on the job is given by

$$I(t) = 35 \ln (t + 1).$$

Find the average number of items produced daily by this employee after the following numbers of days.

(a) 5 (b) 10 (c) 15

35. Typing Speed The function $W(t) = -6t^2 + 10t + 80$ describes a typist's speed (in words per minute) over a time interval $[0, 5]$.

(a) Find $W(0)$.

(b) Find the maximum W value and the time t when it occurs.

(c) Find the average speed over $[0, 5]$.

15.4 CONTINUOUS MONEY FLOW

Given the rate of change of a continuous money flow, how can we find its present value?

In an earlier chapter we looked at the concepts of present value and future value when a lump sum of money is deposited in an account and allowed to accumulate interest. In some situations, however, money flows into and out of an account almost continuously over a period of time. Examples include income in a store, bank receipts and payments, and highway tolls. Although the flow of money in such cases is not exactly continuous, it can be treated as though it were continuous, with useful results.

EXAMPLE 1 The income from a soda machine (in dollars per year) is growing exponentially. When the machine was first installed, it produced income that, if continued, would yield $500 per year. By the end of the first year, it was producing income at a rate of $510.10 per year. Find the total income produced by the machine during its first 3 years of operation.

Let t be the time in years since the installation of the machine. The assumption of exponential growth, coupled with the initial value of 500, implies that the rate of change of income is of the form

$$f(t) = 500e^{kt},$$

where k is some constant. To find k, use the value at the end of the first year.

$$f(1) = 500e^{k(1)} = 510.10$$
$$e^k = 1.0202 \qquad \text{Divide by 500.}$$
$$k = \ln 1.0202 \qquad \text{Take logarithms of both sides.}$$
$$\approx .02$$

We therefore have

$$f(t) = 500e^{.02t}.$$

Since the rate of change of income is given, the total income can be determined by using the definite integral.

$$\text{Total income} = \int_0^3 500e^{.02t}\, dt$$

$$= \frac{500}{.02} e^{.02t} \Big|_0^3$$

$$= 25{,}000\ e^{.02t} \Big|_0^3 = 25{,}000(e^{.06} - 1) = 1545.91$$

Thus, the soda machine will produce \$1545.91 total income in its first 3 years of operation. ◀

The money in Example 1 is not received as a one-time lump sum payment of \$1545.91. Instead, it comes in on a regular basis, perhaps daily, weekly, or monthly. In discussions of such problems it is usually assumed that the income is received continuously over a period of time, and so we speak of a "flow of money into income" and the rate of that flow.

Total Money Flow Let the continuous function $f(x)$ represent the rate of flow of money per unit of time. If x is in years and $f(x)$ is in dollars per year, the area under $f(x)$ between two points in time gives the total dollar flow over the given time interval.

The function $f(x) = 2000$, shown in Figure 12, represents a uniform rate of money flow of \$2000 per year. The graph of this money flow is a horizontal line and the total money flow over a specified time t is given by the rectangular area below the graph of $f(x)$ and above the x-axis between $x = 0$ and $x = t$. For example, the total money flow over $t = 5$ years would be $2000(5) = 10{,}000$, or \$10,000.

FIGURE 12

The area in the uniform rate example could be found by using an area formula from geometry. For a variable function like the function in Example 1, however, a definite integral is needed to find the total money flow over a specific time interval. For the function $f(x) = 2000e^{.08x}$, for example, the total money flow over a 5-year period would be given by

$$\int_0^5 2000e^{.08x} \, dx \approx 12{,}295.62,$$

or \$12,295.62. See Figure 13.

FIGURE 13

| TOTAL MONEY FLOW | If $f(x)$ is the rate of money flow, then the total money flow over the time interval from $x = 0$ to $x = t$ is given by $$\int_0^t f(x) \, dx.$$ |

It should be noted that this "total money flow" does not take into account the interest the money could earn after it is received. It is simply the sum of the periodic (continuous) income.

Present Value of Money Flow As mentioned earlier, an amount of money that can be deposited today to yield a given sum in the future is called the *present value* of this future sum. The future sum may be called the *future value* or *final amount*. To find the present value of a continuous money flow with interest com-

pounded continuously, let $f(x)$ represent the rate of the continuous flow. In Figure 14, the time axis from 0 to x is divided into n subintervals, each of width Δx_i. The amount of money that flows during any interval of time is given by the area between the x-axis and the graph of $f(x)$ over the specified time interval. The area of each subinterval is approximated by the area of a rectangle with height $f(x_i)$, where x_i is the left endpoint of the ith subinterval. The area of each rectangle is $f(x_i)\,\Delta x_i$ which (approximately) gives the amount of money flow over that subinterval.

FIGURE 14

Earlier, we saw that the present value P of an amount A compounded continuously for t years at a rate of interest r is $P = Ae^{-rt}$. Letting x_i represent the time (instead of t), and replacing A with $f(x_i)\,\Delta x_i$, the present value of the money flow over the ith subinterval is approximately equal to

$$P_i = f(x_i)\,\Delta x_i e^{-rx_i}.$$

The total present value is approximately equal to the sum

$$\sum_{i=1}^{n} f(x_i)\,\Delta x_i e^{-rx_i}.$$

This approximation is improved as n increases; taking the limit of the sum as n increases without bound gives the present value

$$P = \lim_{n \to \infty} \sum_{i=1}^{n} f(x_i)\,\Delta x_i e^{-rx_i}.$$

This limit of a summation is given by the definite integral below.

PRESENT VALUE OF MONEY FLOW

If $f(x)$ is the rate of continuous money flow at an interest rate r for t years, then the present value is

$$P = \int_0^t f(x)e^{-rx}\, dx.$$

This present value of money flow is really the amount of money that would have to be deposited now in an account paying an annual interest rate r compounded continuously, so that the total sum in the account at time t would equal the continuous money flow with interest up to time t.

 EXAMPLE 2 A company expects its rate of annual income during the next 3 years to be given by

$$f(x) = 75{,}000x, \quad 0 \le x \le 3.$$

 What is the present value of this income over the 3-year period, assuming an annual interest rate of 8%?

Use the formula for present value given above, with $f(x) = 75{,}000x$, $t = 3$, and $r = .08$.

$$P = \int_0^3 75{,}000xe^{-.08x}\, dx = 75{,}000 \int_0^3 xe^{-.08x}\, dx$$

Using integration by parts or column integration, verify that

$$\int xe^{-.08x}\, dx = -12.5xe^{-.08x} - 156.25e^{-.08x} + C.$$

Therefore,

$$75{,}000 \int_0^3 xe^{-.08x}\, dx = 75{,}000(-12.5xe^{-.08x} - 156.25e^{-.08x}) \Big|_0^3$$

$$= 75{,}000[-12.5(3)e^{-.08(3)} - 156.25e^{-.08(3)} - (0 - 156.25)]$$

$$= 75{,}000(-29.498545 - 122.910603 + 156.25)$$

$$= 288{,}064,$$

or about $288,000. Notice that the actual income over the 3-year period is given by

$$\text{Total money flow} = \int_0^3 75{,}000x\, dx = \frac{75{,}000x^2}{2} \Big|_0^3 = 337{,}500,$$

or $337,500. This means that it would take a lump-sum deposit of $288,064 today paying a continuously compounded interest rate of 8% over a 3-year period to equal the total cash flow of $337,500 with interest. This approach is used as a basis for determining insurance claims involving income considerations. ◀

Accumulated Amount of Money Flow at Time t To find the amount of the money flow with interest at any time t, solve the formula $A = Pe^{rt}$ for P, yielding $P = Ae^{-rt}$. Set this equal to the value of P given in the formula for the present value of money flow.

$$Ae^{-rt} = \int_0^t f(x)e^{-rx}\, dx$$

Multiplying both sides by e^{rt} yields the following formula.

ACCUMULATED AMOUNT OF MONEY FLOW AT TIME *t*	If $f(x)$ is the rate of money flow at an interest rate r at time x, the amount of flow at time t is $$A = e^{rt}\int_0^t f(x)e^{-rx}\, dx.$$

Here, the amount of money A represents the accumulated value or final amount of the money flow *including* interest received on the money after it comes in.

It turns out that most money flows can be expressed as exponential or polynomial functions. When these are multiplied by e^{-rx}, the result is a function that can be integrated. The next example illustrates uniform flow, where $f(x)$ is a constant function. (This is a special case of the polynomial function.)

 EXAMPLE **3** If money is flowing continuously at a constant rate of $2000 per year over 5 years at 12% interest compounded continuously, find each of the following.

(a) The total amount of the flow over the 5-year period
 The total amount is given by $\int_0^t f(x)\, dx$. Here $f(x) = 2000$ and $t = 5$.

$$\int_0^5 2000\, dx = 2000x\Big|_0^5 = 2000(5) = 10{,}000$$

The total money flow over the 5-year period is $10,000.

In this example we use the following two rules for exponents.
1. $a^m \cdot a^n = a^{m+n}$
2. $a^0 = 1$

(b) The accumulated amount, compounded continuously, at time $t = 5$
 At $t = 5$ with $r = .12$, the amount is

$$A = e^{rt}\int_0^t f(x)e^{-rx}\, dx = e^{(.12)5}\int_0^5 (2000)e^{-.12x}\, dx$$

$$= (e^{.6})(2000)\int_0^5 e^{-.12x}\, dx = (e^{.6})(2000)\left(\frac{1}{-.12}\right)\left(e^{-.12x}\Big|_0^5\right)$$

$$= \frac{2000e^{.6}}{-.12}(e^{-.6} - 1) = \frac{2000}{-.12}(1 - e^{.6}) \qquad (e^6)(e^{-.6}) = 1$$

$$= 13{,}701.98,$$

or $13,701.98. The answer to part (a), $10,000, was the amount of money flow over the 5-year period. The $13,701.98 gives that amount with interest compounded continuously over the 5-year period.

(c) The present value of the amount with interest

Use $P = \int_0^t f(x)e^{-rx} \, dx$ with $f(x) = 2000$, $r = .12$, and $t = 5$.

$$P = \int_0^5 2000e^{-.12x} \, dx = 2000 \left(\frac{e^{-.12x}}{-.12} \right) \Bigg|_0^5$$

$$= \frac{2000}{-.12} (e^{-.6} - 1) = 7519.81$$

The present value of the amount with interest in 5 years is $7519.81, which can be checked by substituting $13,701.98 for A in $A = Pe^{rt}$. The present value, P, could have been found by dividing the amount found in (b) by $e^{rt} = e^{.6}$. Check that this would give the same result. ◀

If the money flow is increasing or decreasing exponentially, then $f(x) = Ce^{kx}$, where C is a constant that represents the initial amount and k is the (nominal) continuous rate of change, which may be positive or negative.

4 A continuous money flow starts at $1000 and increases exponentially at 2% per year.

(a) Find the accumulated amount at the end of 5 years at 10% interest compounded continuously.

Here $C = 1000$ and $k = .02$, so that $f(x) = 1000e^{.02x}$. Using $r = .10$ and $t = 5$,

$$A = e^{(.10)5} \int_0^5 1000e^{.02x}e^{-.10x} \, dx$$

$$= (e^{.5})(1000) \int_0^5 e^{-.08x} \, dx \qquad e^{.02x} \cdot e^{-.10x} = e^{-.08x}$$

$$= 1000e^{.5} \left(\frac{1}{-.08} e^{-.08x} \right) \Bigg|_0^5$$

$$= \frac{1000e^{.5}}{-.08} (e^{-.4} - 1) = \frac{1000}{-.08} (e^{.1} - e^{.5}) = 6794.38,$$

or $6794.38.

(b) Find the present value at 5% interest compounded continuously.

Using $f(x) = 1000e^{.02x}$ with $r = .05$ and $t = 5$ in the present value expression,

$$P = \int_0^5 1000e^{.02x}e^{-.05x} \, dx$$

$$= 1000 \int_0^5 e^{-.03x} \, dx = 1000 \left(\frac{1}{-.03} e^{-.03x} \Bigg|_0^5 \right)$$

$$= \frac{1000}{-.03} (e^{-.15} - 1) = 4643.07,$$

or $4643.07. ◀

If the rate of change of the continuous money flow is given by the polynomial function $f(x) = a_nx^n + a_{n-1}x^{n-1} + \cdots + a_0$, the expressions for present value and accumulated amount can be integrated term by term using integration by parts or column integration.

EXAMPLE 5 The rate of change of a continuous flow of money is given by

$$f(x) = 1000x^2 + 100x.$$

Find the present value of this money flow at the end of 10 years at 10% compounded continuously.

Evaluate

$$P = \int_0^{10} (1000x^2 + 100x)e^{-.10x}\, dx.$$

Using integration by parts or column integration, verify that

$$\int (1000x^2 + 100x)e^{-.10x}\, dx = (-10,000x^2 - 1000x)e^{-.1x}$$
$$- (200,000x + 10,000)e^{-.1x} - 2,000,000e^{-.1x} + C.$$

Thus,

$$P = (-10,000x^2 - 1000x)e^{-.1x} - (200,000x + 10,000)e^{-.1x} - 2,000,000e^{-.1x}\Big|_0^{10}$$

$$= (-1,000,000 - 10,000)e^{-1} - (2,000,000 + 10,000)e^{-1} - 2,000,000e^{-1}$$
$$- (0 - 10,000 - 2,000,000)$$

$$= 163,245.21.$$

The present value is $163,245.21. ◀

15.4 Exercises

Each of the functions in Exercises 1–14 represents the rate of flow of money in dollars per year. Assume a 10-year period at 12% compounded continuously and find each of the following: **(a)** *the present value;* **(b)** *the accumulated amount at t = 10.*

1. $f(x) = 1000$

2. $f(x) = 300$

3. $f(x) = 500$

4. $f(x) = 2000$

5. $f(x) = 400e^{.03x}$

6. $f(x) = 800e^{.05x}$

7. $f(x) = 5000e^{-.01x}$

8. $f(x) = 1000e^{-.02x}$

9. $f(x) = .1x$

10. $f(x) = .5x$

11. $f(x) = .01x + 100$

12. $f(x) = .05x + 500$

13. $f(x) = 1000x - 100x^2$

14. $f(x) = 2000x - 150x^2$

Applications

Business and Economics

15. An investment is expected to yield a uniform continuous rate of change of income flow of $20,000 per year for 3 yr. Find the final amount at an interest rate of 14% compounded continuously.

16. A real estate investment is expected to produce a uniform continuous rate of change of income flow of $8000 per year for 6 yr. Find the present value at each of the following rates, compounded continuously.
(a) 12% (b) 10% (c) 15%

17. The rate of change of a continuous flow of money starts at $5000 and decreases exponentially at 1% per year for 8 yr. Find the present value and final amount at an interest rate of 8% compounded continuously.

18. The rate of change of a continuous money flow starts at $1000 and increases exponentially at 5% per year for 4 yr. Find the present value and final amount if interest earned is 11% compounded continuously.

19. A money market fund has a continuous flow of money that changes at a rate of $f(x) = 1500 - 60x^2$, reaching 0 in 5 yr. Find the present value of this flow if interest is 10% compounded continuously.

20. Find the amount of a continuous money flow in 3 yr if the rate of change is given by $f(x) = 1000 - x^2$ and if interest is 10% compounded continuously.

15.5 IMPROPER INTEGRALS

 If we know the rate at which a pollutant is dumped into a stream, how can we compute the total amount released given that the rate of dumping is decreasing over time?

In this section we will learn how to answer such questions.

Sometimes it is useful to be able to integrate a function over an infinite period of time. For example, we might want to find the total amount of income generated by an apartment building into the indefinite future, or the total amount of pollution into a bay from a source that is continuing indefinitely. In this section we define integrals with one or more infinite limits of integration that can be used to solve such problems.

The graph in Figure 15(a) shows the area bounded by the curve $f(x) = x^{-3/2}$, the x-axis, and the vertical line $x = 1$. Think of the shaded region below the curve as extending indefinitely to the right. Does this shaded region have an area?

To see if the area of this region can be defined, introduce a vertical line at $x = b$, as shown in Figure 15(b). This vertical line gives a region with both upper and lower limits of integration. The area of this new region is given by the definite integral

$$\int_1^b x^{-3/2}\, dx.$$

(a) **(b)**

FIGURE 15

By the Fundamental Theorem of Calculus,

$$\int_1^b x^{-3/2}\, dx = (-2x^{-1/2})\Big|_1^b$$

$$= -2b^{-1/2} - (-2 \cdot 1^{-1/2})$$

$$= -2b^{-1/2} + 2 = 2 - \frac{2}{b^{1/2}}.$$

In the section on limits at infinity and curve sketching, we saw that for any positive real number n,

$$\lim_{b \to \infty} \frac{1}{b^n} = \lim_{b \to -\infty} \frac{1}{b^n} = 0.$$

Suppose we now let the vertical line $x = b$ in Figure 15(b) move farther to the right. That is, suppose $b \to \infty$. The expression $-2/b^{1/2}$ would then approach 0, and

$$\lim_{b \to \infty} \left(2 - \frac{2}{b^{1/2}}\right) = 2 - 0 = 2.$$

This limit is defined to be the *area* of the region shown in Figure 15(a), so that

$$\int_1^\infty x^{-3/2}\, dx = 2.$$

An integral of the form

$$\int_a^\infty f(x)\, dx, \qquad \int_{-\infty}^b f(x)\, dx, \qquad \text{or} \qquad \int_{-\infty}^\infty f(x)\, dx$$

is called an improper integral. These **improper integrals** are defined as follows.

IMPROPER INTEGRALS

If f is continuous on the indicated interval and if the indicated limits exist, then

$$\int_a^\infty f(x)\, dx = \lim_{b \to \infty} \int_a^b f(x)\, dx,$$

$$\int_{-\infty}^b f(x)\, dx = \lim_{a \to -\infty} \int_a^b f(x)\, dx,$$

$$\int_{-\infty}^\infty f(x)\, dx = \int_{-\infty}^c f(x)\, dx + \int_c^\infty f(x)\, dx,$$

for real numbers a, b, and c, where c is arbitrarily chosen.

If the expressions on the right side exist, the integrals are **convergent;** otherwise, they are **divergent.**

EXAMPLE 1 Find the following integrals.

(a) $\displaystyle\int_1^\infty \frac{dx}{x}$

$f(x) = \dfrac{1}{x}$

$x = 1$

FIGURE 16

A graph of this region is shown in Figure 16. By the definition of an improper integral,

$$\int_1^\infty \frac{dx}{x} = \lim_{b \to \infty} \int_1^b \frac{dx}{x}.$$

Find $\displaystyle\int_1^b \frac{dx}{x}$ by the Fundamental Theorem of Calculus.

$$\int_1^b \frac{dx}{x} = \ln|x| \Big|_1^b = \ln|b| - \ln|1| = \ln|b| - 0 = \ln|b|$$

As $b \to \infty$, $\ln|b| \to \infty$, so

$$\lim_{b \to \infty} \ln|b| \text{ does not exist.}$$

Since the limit does not exist, $\displaystyle\int_1^\infty \frac{dx}{x}$ is divergent.

(b) $\displaystyle\int_{-\infty}^{-2} \frac{1}{x^2}\, dx = \lim_{a \to -\infty} \int_a^{-2} \frac{1}{x^2}\, dx = \lim_{a \to -\infty} \left(\frac{-1}{x} \right) \Big|_a^{-2}$

$$= \lim_{a \to -\infty} \left(\frac{1}{2} + \frac{1}{a} \right) = \frac{1}{2}$$

A graph of this region is shown in Figure 17. Since the limit exists, this integral converges.

$f(x) = \dfrac{1}{x^2}$

FIGURE 17

It may seem puzzling that the areas under the curves $f(x) = 1/x^{3/2}$ and $f(x) = 1/x^2$ are finite, while $f(x) = 1/x$ has an infinite amount of area. At first glance the graphs of these functions appear similar. The difference is that although all three functions get small as x becomes infinitely large, $f(x) = 1/x$ does not become small enough fast enough.

EXAMPLE **2**

Find $\int_0^\infty 4e^{-3x}\, dx$.
By definition,

$$\int_0^\infty 4e^{-3x}\, dx = \lim_{b \to \infty} \int_0^b 4e^{-3x}\, dx = \lim_{b \to \infty} \left(-\frac{4}{3}e^{-3x} \right) \Big|_0^b$$

$$= \lim_{b \to \infty} \left[-\frac{4}{3}e^{-3b} - \left(-\frac{4}{3}e^0 \right) \right]$$

$$= \lim_{b \to \infty} \left(\frac{-4}{3e^{3b}} + \frac{4}{3} \right) = 0 + \frac{4}{3} = \frac{4}{3}.$$

This integral converges.

The following examples describe applications of improper integrals.

EXAMPLE 3

The rate at which a pollutant is being dumped into a small stream at time t is given by $P_0 e^{-kt}$, where P_0 is the amount of pollutant initially released into the stream. Suppose $P_0 = 1000$ and $k = .06$. Find the total amount of the pollutant that will be released into the stream into the indefinite future.

Find

$$\int_0^\infty P_0 e^{-kt} \, dt = \int_0^\infty 1000 e^{-.06t} \, dt.$$

Work as above.

$$\int_0^\infty 1000 e^{-.06t} \, dt = \lim_{b \to \infty} \int_0^b 1000 e^{-.06t} \, dt$$

$$= \lim_{b \to \infty} \left(\frac{1000}{-.06} e^{-.06t} \right) \Big|_0^b$$

$$= \lim_{b \to \infty} \left(\frac{1000}{-.06 e^{.06b}} - \frac{1000}{-.06} e^0 \right) = \frac{-1000}{-.06} = 16{,}667$$

A total of 16,667 units of the pollutant eventually will be released. ◀

The *capital value* of an asset is sometimes defined as the present value of all future net earnings of the asset. If $R(t)$ gives the rate of change of the annual earnings produced by an asset at time t, as shown in Section 4, the **capital value** is

$$\int_0^\infty R(t) e^{-rt} \, dt,$$

where r is the annual rate of interest, compounded continuously, and the first payment is assumed to be made in one year.

EXAMPLE 4

Suppose income from a rental property is generated at the annual rate of $4000 per year. Find the capital value of this property at an interest rate of 10% compounded continuously.

This is a continuous income stream with a rate of flow of 4000, so $R(t) = 4000$. Also, $r = .10$ or $.1$. The capital value is given by

$$\int_0^\infty 4000 e^{-.1t} \, dt = \lim_{b \to \infty} \int_0^b 4000 e^{-.1t} \, dt$$

$$= \lim_{b \to \infty} \left(\frac{4000}{-.1} e^{-.1t} \right) \Big|_0^b$$

$$= \lim_{b \to \infty} \left(-40{,}000 e^{-.1b} + 40{,}000 \right) = 40{,}000,$$

or $40,000. ◀

15.5 Exercises

Determine whether the improper integrals in Exercises 1–26 converge or diverge, and find the value of each that converges.

1. $\displaystyle\int_2^\infty \frac{1}{x^2}\, dx$

2. $\displaystyle\int_5^\infty \frac{1}{x^2}\, dx$

3. $\displaystyle\int_1^\infty \frac{1}{\sqrt{x}}\, dx$

4. $\displaystyle\int_{16}^\infty \frac{-3}{\sqrt{x}}\, dx$

5. $\displaystyle\int_{-\infty}^{-1} \frac{2}{x^3}\, dx$

6. $\displaystyle\int_{-\infty}^{-4} \frac{3}{x^4}\, dx$

7. $\displaystyle\int_1^\infty \frac{1}{x^{1.0001}}\, dx$

8. $\displaystyle\int_1^\infty \frac{1}{x^{.999}}\, dx$

9. $\displaystyle\int_{-\infty}^{-1} x^{-2}\, dx$

10. $\displaystyle\int_{-\infty}^{-4} x^{-2}\, dx$

11. $\displaystyle\int_{-\infty}^{-1} x^{-8/3}\, dx$

12. $\displaystyle\int_{-\infty}^{-27} x^{-5/3}\, dx$

13. $\displaystyle\int_0^\infty 4e^{-4x}\, dx$

14. $\displaystyle\int_0^\infty 10e^{-10x}\, dx$

15. $\displaystyle\int_{-\infty}^0 4e^x\, dx$

16. $\displaystyle\int_{-\infty}^0 3e^{4x}\, dx$

17. $\displaystyle\int_{-\infty}^{-1} \ln|x|\, dx$

18. $\displaystyle\int_1^\infty \ln|x|\, dx$

19. $\displaystyle\int_0^\infty \frac{dx}{(x+1)^2}$

20. $\displaystyle\int_0^\infty \frac{dx}{(2x+1)^3}$

21. $\displaystyle\int_{-\infty}^{-1} \frac{2x-1}{x^2-x}\, dx$

22. $\displaystyle\int_1^\infty \frac{2x+3}{x^2+3x}\, dx$

23. $\displaystyle\int_2^\infty \frac{1}{x\ln x}\, dx$

24. $\displaystyle\int_2^\infty \frac{1}{x\,(\ln x)^2}\, dx$

25. $\displaystyle\int_0^\infty xe^{2x}\, dx$

26. $\displaystyle\int_{-\infty}^0 xe^{3x}\, dx$

Use the table of integrals as necessary to evaluate the following integrals.

27. $\displaystyle\int_{-\infty}^{-1} \frac{2}{3x(2x-7)}\, dx$

28. $\displaystyle\int_1^\infty \frac{7}{2x(5x+1)}\, dx$

29. $\displaystyle\int_1^\infty \frac{4}{9x(x+1)^2}\, dx$

30. $\displaystyle\int_{-\infty}^{-5} \frac{5}{4x(x+2)^2}\, dx$

31. $\displaystyle\int_0^\infty \frac{1}{\sqrt{1+x^2}}\, dx$

32. $\displaystyle\int_9^\infty \frac{1}{\sqrt{x^2-4}}\, dx$

For Exercises 33–36, find the area between the graph of the given function and the x-axis over the given interval, if possible.

33. $f(x) = \dfrac{1}{x-1}$ for $(-\infty, 0]$

34. $f(x) = e^{-x}$ for $(-\infty, e]$

35. $f(x) = \dfrac{1}{(x-1)^2}$ for $(-\infty, 0]$

36. $f(x) = \dfrac{1}{(x-1)^3}$ for $(-\infty, 0]$.

37. Find $\displaystyle\int_{-\infty}^\infty xe^{-x^2}\, dx$.

38. Find $\displaystyle\int_{-\infty}^\infty \frac{x}{(1+x^2)^2}\, dx$.

39. Show that $\int_1^\infty 1/x^p\, dx$ converges if $p > 1$ and diverges if $p \le 1$.

40. Example 1(b) leads to a paradox. On the one hand, the unbounded region in that example has an area of $1/2$, so theoretically it could be colored with ink. On the other hand, the boundary of that region is infinite, so it cannot be drawn with a finite amount of ink. This seems impossible, because coloring the region automatically colors the boundary. Explain why it is possible to color the region.

Applications

Business and Economics

Capital Value *Find the capital values of the properties in Exercises 41–44.*

41. A castle for which annual rent of $60,000 will be paid in perpetuity; the interest rate is 8% compounded continuously

42. A fort on a strategic peninsula in the North Sea; the annual rent is $500,000, paid in perpetuity; the interest rate is 6% compounded continuously

43. Find the capital value of an asset that generates $6000 yearly income if the interest rate is as follows.

(a) 8% compounded continuously

(b) 10% compounded continuously

44. An investment produces a perpetual stream of income with a flow rate of

$$K(t) = 1000\ e^{.02t}.$$

Find the capital value at an interest rate of 7% compounded continuously.

45. The Drucker family wants to establish an ongoing scholarship award at a college. Each year in June $3000 will be awarded, starting one year from now. What amount must the Druckers provide the college, assuming funds will be invested at 10% compounded continuously?

Life Sciences

46. **Drug Reaction** The rate of reaction to a drug is given by

$$r'(x) = 2x^2 e^{-x},$$

where x is the number of hours since the drug was administered. Find the total reaction to the drug over all the time since it was administered, assuming this is an infinite time interval. (*Hint:* $\lim_{x \to \infty} x^k e^{-x} = 0$ for all real numbers k.)

Physical Sciences

 Radioactive Waste *Radioactive waste is entering the atmosphere over an area at a decreasing rate. Use the improper integral*

$$\int_0^\infty Pe^{-kt}\ dt$$

with $P = 50$ to find the total amount of the waste that will enter the atmosphere for each of the following values of k.

47. $k = .06$ **48.** $k = .04$

15.6 SOLUTIONS OF ELEMENTARY AND SEPARABLE DIFFERENTIAL EQUATIONS

? How can we predict the future population of a flock of mountain goats?

Using differential equations, we will learn to answer such questions.

Usually a solution of an equation is a *number*. A solution of a differential equation, however, is a *function*. For example, the solutions of a differential equation such as

$$\frac{dy}{dx} = 3x^2 - 2x \tag{1}$$

consist of all expressions for y that satisfy the equation. Since the left side of the equation is the derivative of y with respect to x, we can solve the equation for

y by finding an antiderivative on each side. On the left, the antiderivative is $y + C_1$. On the right side,

$$\int (3x^2 - 2x)\, dx = x^3 - x^2 + C_2.$$

The solutions of equation (1) are given by

$$y + C_1 = x^3 - x^2 + C_2$$

or
$$y = x^3 - x^2 + C_2 - C_1.$$

Replacing the constant $C_2 - C_1$ with the single constant C gives

$$y = x^3 - x^2 + C. \tag{2}$$

(From now on we will add just one constant, with the understanding that it represents the difference between the two constants obtained in the two integrations.)

Each different value of C in equation (2) leads to a different solution of equation (1), showing that a differential equation can have an infinite number of solutions. Equation (2) is the **general solution** of the differential equation (1). Some of the solutions of equation (1) are graphed in Figure 18.

FIGURE 18

The simplest kind of differential equation has the form

$$\frac{dy}{dx} = f(x).$$

Since equation (1) has this form, the solution of equation (1) suggests the following generalization.

GENERAL SOLUTION OF
$$\frac{dy}{dx} = f(x)$$

The general solution of the differential equation $dy/dx = f(x)$ is

$$y = \int f(x)\, dx.$$

EXAMPLE 1

The population P of a flock of birds is growing exponentially so that

$$\frac{dP}{dx} = 20e^{.05x},$$

where x is time in years. Find P in terms of x if there were 20 birds in the flock initially.

Solve the differential equation:

$$P = \int 20e^{.05x}\, dx = \frac{20}{.05}e^{.05x} + C = 400e^{.05x} + C.$$

Since P is 20 when x is 0,

$$20 = 400e^0 + C$$
$$-380 = C,$$

and

$$P = 400e^{.05x} - 380. \quad \blacktriangleleft$$

In Example 1, the given information was used to produce a solution with a specific value of C. Such a solution is called a **particular solution** of the given differential equation. The given information, $P = 20$ when $t = 0$, is called an **initial condition** because $t = 0$.

Sometimes a differential equation must be rewritten in the form

$$\frac{dy}{dx} = f(x)$$

before it can be solved.

EXAMPLE 2

Find the particular solution of

$$\frac{dy}{dx} - 2x = 5,$$

given that $y = 2$ when $x = -1$.

Add $2x$ to both sides of the equation to get

$$\frac{dy}{dx} = 2x + 5.$$

The general solution is

$$y = \frac{2x^2}{2} + 5x + C = x^2 + 5x + C.$$

Substituting 2 for y and -1 for x gives

$$2 = (-1)^2 + 5(-1) + C$$
$$C = 6.$$

The particular solution is $y = x^2 + 5x + 6$. ◀

So far in this section, we have used a method that is essentially the same as that used in the section on antiderivatives, when we first started the topic of integration. But not all differential equations can be solved so easily. For example, if interest on an investment is compounded continuously, then the investment grows at a rate proportional to the amount of money present.* If A is the amount in an account at time t, then for some constant k, the differential equation

$$\frac{dA}{dt} = kA$$

gives the rate of growth of A with respect to t. This differential equation is different from those discussed previously, which had the form

$$\frac{dy}{dx} = f(x).$$

Caution Since the right-hand side of the differential equation for compound

invalid to simply integrate both sides as we did before. The previous method only works when the side opposite the derivative is simply a function of the independent variable.

The differential equation for compound interest is an example of a more general differential equation we will now learn to solve; namely, those that can be written in the form

$$\frac{dy}{dx} = \frac{f(x)}{g(y)}.$$

Using differentials and multiplying on both sides by $g(y)\, dx$ gives

$$g(y)\, dy = f(x)\, dx.$$

In this form all terms involving y (including dy) are on one side of the equation, and all terms involving x (and dx) are on the other side. A differential equation in this form is said to be **separable**, since the variables x and y can be separated. A separable differential equation may be solved by integrating each side.

*See Section 3 in the chapter on exponential and logarithmic functions.

EXAMPLE 3 Find the general solution of $y \dfrac{dy}{dx} = x^2$.

Begin by separating the variables to get

$$y \, dy = x^2 \, dx.$$

The general solution is found by taking antiderivatives on each side.

$$\int y \, dy = \int x^2 \, dx$$

$$\frac{y^2}{2} = \frac{x^3}{3} + C$$

$$y^2 = \frac{2}{3}x^3 + 2C = \frac{2}{3}x^3 + K$$

The constant K was substituted for $2C$ in the last step. The solution is left in implicit form, not solved explicitly for y. ◄

EXAMPLE 4 Find the general solution of $\dfrac{dy}{dx} = ky$, where k is a constant.

Separating variables leads to

$$\frac{1}{y} \, dy = k \, dx.$$

To solve this equation, take antiderivatives on each side.

$$\int \frac{1}{y} \, dy = \int k \, dx$$

$$\ln |y| = kx + C$$

Use the definition of logarithm to write the equation in exponential form as

$$|y| = e^{kx + C}.$$

By properties of exponents,

$$|y| = e^{kx}e^{C}.$$

Finally, use the definition of absolute value to get

$$y = e^{kx}e^{C} \qquad \text{or} \qquad y = -e^{kx}e^{C}.$$

Since e^{C} and $-e^{C}$ are constants, replace them with the constant M, which may have any nonzero real-number value, to get the single equation

$$y = Me^{kx}.$$

This equation, $y = Me^{kx}$, defines the exponential growth or decay function that was discussed previously (in the chapter on exponential and logarithmic functions). ◄

Recall that equations of the form $y = Me^{kx}$ arise in situations where the rate of change of a quantity is proportional to the amount present at time x; that is, where

$$\frac{dy}{dx} = ky.$$

The constant k is called the **growth rate constant,** while M represents the amount present at time $x = 0$. (A positive value of k indicates growth, while a negative value of k indicates decay.)

Applying the results of Example 4 to the equation discussed earlier,

$$\frac{dA}{dt} = kA,$$

shows that the amount in the account at time t is

$$A = A_0 e^{kt},$$

where A_0 is the amount originally invested.

As a model of population growth, the equation $y = Me^{kx}$ is not realistic over the long run for most populations. As shown by graphs of functions of the form $y = Me^{kx}$, with both M and k positive, growth would be unbounded. Additional factors, such as space restrictions or a limited amount of food, tend to inhibit growth of populations as time goes on. In an alternative model that assumes a maximum population of size N, the rate of growth of a population is proportional to how close the population is to that maximum. These assumptions lead to the differential equation

$$\frac{dy}{dx} = k(N - y),$$

the limited growth function mentioned in an earlier chapter. Graphs of limited growth functions look like the graph in Figure 19, where y_0 is the initial population.

In the chapter on exponential and logarithmic functions, we saw that the amount of money in an account with interest compounded continuously is given by

$$A = Pe^{rt},$$

where P is the initial amount in the account, r is the annual interest rate, and t is the time in years. Observe that this is the same as the equation for the amount of money in an account derived here, where P and r have been replaced with A_0 and k, respectively.

$$y = N - (N - y_0)e^{-kx}$$

FIGURE 19

EXAMPLE 5

A certain area can support no more than 4000 mountain goats. There are 1000 goats in the area at present, with a growth constant of .20.

(a) Write a differential equation for the rate of growth of this population.

Let $N = 4000$ and $k = .20$. The rate of growth of the population is given by

$$\frac{dy}{dx} = .20(4000 - y).$$

To solve for y, first separate the variables.

$$\frac{dy}{4000 - y} = .2\ dx$$

$$\int \frac{dy}{4000 - y} = \int .2\ dx$$

$$-\ln(4000 - y) = .2x + C$$

$$\ln(4000 - y) = -.2x - C$$

$$4000 - y = e^{-.2x - C} = (e^{-.2x})(e^{-C})$$

The absolute value bars are not needed for $\ln(4000 - y)$ since y must be less than 4000 for this population, so that $4000 - y$ is always nonnegative. Let $e^{-C} = B$. Then

$$4000 - y = Be^{-.2x}$$

$$y = 4000 - Be^{-.2x}.$$

Find B by using the fact that $y = 1000$ when $x = 0$.

$$1000 = 4000 - B$$

$$B = 3000$$

Notice that the value of B is the difference between the maximum population and the initial population. Substituting 3000 for B in the equation for y gives

$$y = 4000 - 3000e^{-.2x}.$$

(b) What will the goat population be in 5 years?

In 5 years, the population will be

$$y = 4000 - 3000e^{-(.2)(5)} = 4000 - 3000e^{-1}$$

$$= 4000 - 1103.6 = 2896.4,$$

or about 2900 goats. ◀

Logistic Growth Let y be the size of a certain population at time x. In the standard model for unlimited growth,

$$\frac{dy}{dx} = ky, \tag{3}$$

the rate of growth is proportional to the current population size. The constant k, the growth rate constant, is the difference between the birth and death rates of the population. The unlimited growth model predicts that the population's growth rate is a constant, k.

Growth usually is not unlimited, however, and the population's growth rate is usually not constant because the population is limited by environmental factors to a maximum size N, called the **carrying capacity** of the environment for the species. In the limited growth model given above,

$$\frac{dy}{dx} = k(N - y),$$

the rate of growth is proportional to the remaining room for growth, $N - y$. In the **logistic growth model**

$$\frac{dy}{dx} = \frac{k}{N}(N - y)y, \tag{4}$$

the rate of growth is proportional to both the current population size y and the remaining room for growth $N - y$. Equation (4) is called the **logistic equation.** The logistic growth model predicts that the population's growth rate is a function of the population size y:

$$\frac{dy}{dx} = ky - \frac{k}{N}y^2.$$

As $y \to 0$, y^2 is insignificant when compared to y, and so

$$\frac{dy}{dx} = ky - \frac{k}{N}y^2 \approx ky.$$

That is, the growth of the population is essentially unlimited when y is small. On the other hand, when $y \to N$,

$$\frac{dy}{dx} = ky - \frac{k}{N}y^2 = \left(k - \frac{k}{N}y\right)y = \left(\frac{kN}{N} - \frac{k}{N}y\right)y = \frac{k}{N}(N - y)y \to 0.$$

That is, population growth levels off as y nears the maximum population size N. Thus, the logistic equation

$$\frac{dy}{dx} = ky - \frac{k}{N}y^2$$

is the unlimited growth equation (3) with a damping term ky^2/N subtracted to account for limiting environmental factors when y nears N. Let y_0 denote the initial population size. Under the assumption $0 < y < N$, the general solution of equation (4) is

$$y = \frac{N}{1 + be^{-kx}}, \tag{5}$$

where $b = (N - y_0)/y_0$ (see Exercise 31). This solution, called a **logistic curve,** is shown in Figure 20 on the next page.

FIGURE 20

As expected, the logistic curve begins exponentially and subsequently levels off. Another important feature is the point of inflection $((\ln b)/k, N/2)$, at which the growth rate dy/dx is a maximum (see Exercise 33).

Logistic equations arise frequently in the study of populations. In about 1840 the Belgian sociologist P. F. Verhurst fitted a logistic curve to United States census figures and made predictions about the population that were subsequently proven to be quite accurate. American biologist Raymond Pearl (circa 1920) found that the growth of a population of fruit flies in a limited space could be modeled by the logistic equation

$$\frac{dy}{dx} = .2y - \frac{.2}{1035}y^2.$$

Logistic growth is an example of how a model is modified over time as new insights occur. The model for population growth changed from the early exponential curve $y = Me^{kx}$ to the logistic curve

$$y = \frac{N}{1 + be^{-kx}}.$$

Many other quantities besides population grow logistically. That is, their initial rate of growth is slow, but as time progresses, their rate of growth increases to a maximum value and subsequently begins to decline and to approach zero.

EXAMPLE 6

Rapid technological advancements in the last 20 years have made many products obsolete practically overnight. J. C. Fisher and R. H. Pry* successfully described the phenomenon of a technically superior new product replacing another product by the logistic equation

$$\frac{dz}{dx} = k(1 - z)z, \tag{6}$$

*From J. C. Fisher and R. H. Pry, ''A Simple Substitution Model of Technological Change,'' *Technological Forecasting and Social Change,* vol. 3, 1971–1972. Copyright © 1972 by Elsevier Science Publishing Co., Inc. Reprinted by permission of the publisher.

$$z = \frac{1}{1 + be^{-kx}}, \quad b = e^{kx_0}$$

FIGURE 21

where z is the market share of the new product and $1 - z$ is the market share of the other product. The new product will initially have little or no market share; that is, $z_0 \approx 0$. Thus, the constant b in equation (5) will have to be determined in a different way. Let x_0 be the time at which $z = 1/2$. Under the assumption $0 < z < 1$, the general solution of equation (6) is

$$z = \frac{1}{1 + be^{-kx}},$$

where $b = e^{kx_0}$ (see Exercise 32). This solution is shown in Figure 21.

The market share of the new product will be growing most rapidly when the new product has captured exactly half the market, and the market share of the older product will be shrinking most rapidly at the same time. Notice that the logistic equation (4) can be transformed into the simpler logistic equation (6) by the change of variable $z = y/N$. ◀

15.6 Exercises

Find general solutions for the following differential equations.

1. $\dfrac{dy}{dx} = -2x + 3x^2$

2. $\dfrac{dy}{dx} = 3e^{-2x}$

3. $3x^3 - 2\dfrac{dy}{dx} = 0$

4. $3x^2 - 3\dfrac{dy}{dx} = 2$

5. $y\dfrac{dy}{dx} = x$

6. $y\dfrac{dy}{dx} = x^2 - 1$

7. $\dfrac{dy}{dx} = 2xy$

8. $\dfrac{dy}{dx} = x^2y$

9. $\dfrac{dy}{dx} = 3x^2y - 2xy$

10. $(y^2 - y)\dfrac{dy}{dx} = x$

11. $\dfrac{dy}{dx} = \dfrac{y}{x}, x > 0$

12. $\dfrac{dy}{dx} = \dfrac{y}{x^2}$

13. $\dfrac{dy}{dx} = y - 5$

14. $\dfrac{dy}{dx} = 3 - y$

15. $\dfrac{dy}{dx} = y^2e^x$

16. $\dfrac{dy}{dx} = \dfrac{e^x}{e^y}$

Find particular solutions for the following equations.

17. $\dfrac{dy}{dx} + 2x = 3x^2; \quad y = 2$ when $x = 0$

18. $\dfrac{dy}{dx} = 4x^3 - 3x^2 + x; \cdot y = 0$ when $x = 1$

19. $2\dfrac{dy}{dx} = 4xe^{-x}; \quad y = 42$ when $x = 0$

20. $x\dfrac{dy}{dx} = x^2e^{3x}; \quad y = \dfrac{8}{9}$ when $x = 0$

21. $\dfrac{dy}{dx} = \dfrac{x^2}{y}; \quad y = 3$ when $x = 0$

22. $x^2\dfrac{dy}{dx} = y; \quad y = -1$ when $x = 1$

23. $(2x + 3)y = \dfrac{dy}{dx}; \quad y = 1$ when $x = 0$

24. $x\dfrac{dy}{dx} - y\sqrt{x} = 0; \quad y = 1$ when $x = 0$

25. $\dfrac{dy}{dx} = \dfrac{y^2}{x}; \quad y = 5$ when $x = e$

26. $\dfrac{dy}{dx} = x^{1/2}y^2; \quad y = 12$ when $x = 4$

27. $\dfrac{dy}{dx} = \dfrac{2x + 1}{y - 3}; \quad y = 4$ when $x = 0$

28. $\dfrac{dy}{dx} = \dfrac{x^2 + 5}{2y - 1}; \quad y = 11$ when $x = 0$

29. $\dfrac{dy}{dx} = (y - 1)^2e^x; \quad y = 2$ when $x = 0$

30. $\dfrac{dy}{dx} = (x + 2)^2e^y; \quad y = 0$ when $x = 1$

31. **(a)** Solve the logistic equation (4) in this section by observing that

$$\frac{1}{y} + \frac{1}{N - y} = \frac{N}{(N - y)y}.$$

(b) Assume $0 < y < N$. Verify that $b = (N - y_0)/y_0$ in equation (5), where y_0 is the initial population size.

(c) Assume $0 < N < y$ for all y. Verify that $b = (y_0 - N)/y_0$.

32. Suppose that $0 < z < 1$ for all z. Solve the logistic equation (6) as in Exercise 31. Verify that $b = e^{kx_0}$, where x_0 is the time at which $z = 1/2$.

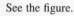

33. Suppose that $0 < y_0 < N$. Let $b = (N - y_0)/y_0$, and let $y(x) = \dfrac{N}{1 + be^{-kx}}$ for all x. Show the following.

(a) $0 < y(x) < N$ for all x.

(b) The lines $y = 0$ and $y = N$ are horizontal asymptotes of the graph.

(c) $y(x)$ is an increasing function.

(d) $((\ln b)/k, N/2)$ is a point of inflection of the graph.

(e) dy/dx is a maximum at $x_0 = (\ln b)/k$.

34. Suppose that $0 < N < y_0$. Let $b = (y_0 - N)/y_0$ and let

$$y(x) = \frac{N}{1 - be^{-kx}} \quad \text{for all } x \neq \frac{\ln b}{k}.$$

See the figure.

Show the following.

(a) $0 < b < 1$

(b) The lines $y = 0$ and $y = N$ are horizontal asymptotes of the graph.

(c) The line $x = (\ln b)/k$ is a vertical asymptote of the graph.

(d) $y(x)$ is decreasing on $((\ln b)/k, \infty)$ and on $(-\infty, (\ln b)/k)$.

(e) $y(x)$ is concave upward on $((\ln b)/k, \infty)$ and concave downward on $(-\infty, (\ln b)/k)$.

Applications

Business and Economics

35. **Profit** The marginal profit of a certain company is given by

$$\frac{dy}{dx} = \frac{100}{32 - 4x},$$

where x represents the amount of money (in thousands of dollars) that the company spends on advertising.

Find the profit for each of the following advertising expenditures if the profit is $1000 when nothing is spent on advertising.

(a) $3000 **(b)** $5000

(c) Can advertising expenditures ever reach $8000 according to this model? Why?

36. **Sales Decline** Sales (in thousands) of a certain product are declining at a rate proportional to the amount of sales, with a decay constant of 25% per year.

(a) Write a differential equation to express the rate of sales decline.

(b) Find a general solution to the equation in part (a).

(c) How much time will pass before sales become 30% of their original value?

37. **Inflation** If inflation grows continuously at a rate of 6% per year, how long will it take for $1 to lose half its value?

38. Gross National Product Suppose that the gross national product (GNP) of a particular country increases exponentially, with a growth constant of 2% per year. Ten years ago, the GNP was 10^5 dollars. What will be the GNP in 5 years?

39. Bankruptcy In a certain area, 1500 small business firms are threatened by bankruptcy. Assume the rate of change in the number of bankruptcies is proportional to the number of small firms that are not yet bankrupt. If the growth constant is 6%, and if 100 firms are bankrupt initially, how many will be bankrupt in 2 years?

40. New Car Sales The sales of a new model of car are expected to follow the logistic curve

$$y = \frac{100,000}{1 + 100e^{-x}},$$

where x is the number of months since the introduction of the model. Find the following.

(a) The initial sales level

(b) The maximum sales level

(c) The month in which the maximum sales level will be achieved

(d) The rate at which sales are growing 6 months after the introduction of the model

Elasticity of Demand *Elasticity of demand was discussed in the chapter on applications of the derivative, where it was defined as*

$$E = -\frac{p}{q} \cdot \frac{dq}{dp},$$

for demand q and price p. Find the general demand equation $q = f(p)$ for each of the following elasticity functions. (Hint: Set each elasticity function equal to $-\dfrac{p}{q} \cdot \dfrac{dq}{dp}$, then solve for q. Write the constant of integration as $\ln C$.)

41. $E = \dfrac{4p^2}{q^2}$ **42.** $E = 2$

Life Sciences

43. Tracer Dye The amount of a tracer dye injected into the bloodstream decreases exponentially, with a decay constant of 3% per minute. If 6 cc are present initially, how many cubic centimeters are present after 10 min? (Here k will be negative.)

44. Bacterial Population The rate at which the number of bacteria (in thousands) in a culture is changing after the introduction of a bactericide is given by

$$\frac{dy}{dx} = 50 - y,$$

where y is the number of bacteria (in thousands) present at

time x. Find the number of bacteria present at each of the following times if there were 1000 thousand bacteria present at time $x = 0$.

(a) $x = 2$ (b) $x = 5$ (c) $x = 10$ (d) $x = 15$

45. Fish Population An isolated fish population is limited to 5000 by the amount of food available. If there are now 150 fish and the population is growing with a growth constant of 1% a year, find the expected population at the end of 5 yr.

46. Mite Population A population of mites (in hundreds) increases exponentially, with a growth constant of 5% per week. At the beginning of an observation period there were 3000 mites. How many are present 4 weeks later?

Social Sciences

47. Spread of a Rumor Suppose the rate at which a rumor spreads—that is, the number of people who have heard the rumor over a period of time—increases with the number of people who have heard it. If y is the number of people who have heard the rumor, then

$$\frac{dy}{dt} = ky,$$

where t is the time in days.

(a) If y is 1 when $t = 0$, and y is 5 when $t = 2$, find k. Using the value of k from part (a), find y for each of the following times.

(b) $t = 3$ (c) $t = 5$ (d) $t = 10$

48. Worker Productivity A company has found that the rate at which a person new to the assembly line produces items is

$$\frac{dy}{dx} = 7.5e^{-.3y},$$

where x is the number of days the person has worked on the line. How many items can a new worker be expected to produce on the eighth day if he produces none when $x = 0$?

Physical Sciences

49. Radioactive Decay The amount of a radioactive substance decreases exponentially, with a decay constant of 5% per month.

(a) Write a differential equation to express the rate of change.

(b) Find a general solution to the differential equation from part (a).

(c) If there are 90 g at the start of the decay process, find a particular solution for the differential equation from part (a).

(d) Find the amount left after 10 mo.

50. Snowplow One morning snow began to fall at a heavy and constant rate. A snowplow started out at 8:00 A.M. At 9:00 A.M. it had traveled 2 mi. By 10:00 A.M. it had traveled 3 mi. Assuming that the snowplow removes a constant volume of snow per hour, determine the time at which it started snowing. (*Hint:* Let t denote the time since the snow started to fall, and let T be the time when the snowplow started out. Let x, the distance the snowplow has traveled, and h, the height of the snow, be functions of t. The assumption that a constant volume of snow per hour is removed implies that the speed of the snowplow times the height of the snow is a constant. Set up and solve differential equations involving dx/dt and dh/dt.)*

Chapter Summary Key Terms

15.1 **integration by parts**
 column integration
 tables of integrals
15.2 **numerical integration**
 trapezoidal rule
 Simpson's rule
15.3 **solid of revolution**
 average value of a function

15.4 **total money flow**
 present value of money flow
15.5 **improper integral**
 convergent integral
 divergent integral
15.6 **general solution**
 particular solution
 initial condition

separable differential equation
growth rate constant
carrying capacity
logistic growth model
logistic equation
logistic curve

Chapter 15 Review Exercises

1. Describe the type of integral for which integration by parts is useful.
2. Describe the type of integral for which tables of integrals are useful.
3. Describe the type of integral for which numerical integration is useful.
4. What is an improper integral? Explain why improper integrals must be treated in a special way.

Find each of the following integrals, using techniques from this or the previous chapter.

5. $\int x(8 - x)^{3/2} \, dx$

6. $\int \frac{3x}{\sqrt{x - 2}} \, dx$

7. $\int xe^x \, dx$

8. $\int (x + 2)e^{-3x} \, dx$

9. $\int \ln |2x + 3| \, dx$

10. $\int (x - 1) \ln |x| \, dx$

11. $\int \frac{x}{9 - 4x^2} \, dx$

12. $\int \frac{x}{\sqrt{25 + 9x^2}} \, dx$

13. $\int \frac{1}{9 - 4x^2} \, dx$

14. $\int \frac{1}{\sqrt{25 + 9x^2}} \, dx$

15. $\int_1^2 \frac{1}{x\sqrt{9 - x^2}} \, dx$

16. $\int_0^3 \sqrt{16 + x^2} \, dx$

17. $\int_1^e x^3 \ln x \, dx$

18. $\int_0^1 x^2 e^{x/2} \, dx$

19. Find the area between $y = (3 + x^2)e^{2x}$ and the x-axis from $x = 0$ to $x = 1$.
20. Find the area between $y = x^3(x^2 - 1)^{1/3}$ and the x-axis from $x = 1$ to $x = 3$.

In Exercises 21–23, use the trapezoidal rule with $n = 4$ to approximate the value of each answer. Use the table of integrals to find the exact value, and compare the two answers.

21. $\int_2^6 \frac{dx}{x^2 - 1}$

22. $\int_2^{10} \frac{x \, dx}{x - 1}$

23. $\int_1^5 \ln x \, dx$

*This problem first appeared in the *American Mathematical Monthly*, vol. 44, December 1937.*

In Exercises 24–26, use Simpson's rule with n = 4 to approximate the value of each integral. Compare your answers with the answers to Exercises 21–23.

24. $\int_2^6 \dfrac{dx}{x^2 - 1}$

25. $\int_2^{10} \dfrac{x\,dx}{x - 1}$

26. $\int_1^5 \ln x\,dx$

27. Find the area under the semicircle $y = \sqrt{1 - x^2}$ and above the x-axis by the trapezoidal rule, using $n = 6$.

28. Repeat Exercise 27 using Simpson's rule.

Find the volume of the solid of revolution formed by rotating each of the following bounded regions about the x-axis.

29. $f(x) = 2x - 1,\ y = 0,\ x = 3$

30. $f(x) = \sqrt{x - 2},\ y = 0,\ x = 11$

31. $f(x) = e^{-x},\ y = 0,\ x = -2,\ x = 1$

32. $f(x) = \dfrac{1}{\sqrt{x - 1}},\ y = 0,\ x = 2,\ x = 4$

33. $f(x) = 4 - x^2,\ y = 0,\ x = -1,\ x = 1$

34. $f(x) = \dfrac{x^2}{4},\ y = 0,\ x = 4$

35. A frustum is what remains of a cone when the top is cut off by a plane parallel to the base. Suppose a right circular frustum (that is, one formed from a right circular cone) has a base with radius r, a top with radius $r/2$, and a height h. (See the figure.) Find the volume of this frustum by rotating about the x-axis the region below the line segment from $(0, r)$ to $(h, r/2)$.

36. How is the average value of a function found?

37. Find the average value of $f(x) = \sqrt{x + 1}$ over the interval $[0, 8]$.

38. Find the average value of $f(x) = x^2(x^3 + 1)^5$ over the interval $[0, 1]$.

Find the value of each integral that converges.

39. $\int_1^\infty x^{-1}\,dx$

40. $\int_{-\infty}^{-2} x^{-2}\,dx$

41. $\int_0^\infty \dfrac{dx}{(5x + 2)^2}$

42. $\int_1^\infty 6e^{-x}\,dx$

43. $\int_{-\infty}^0 \dfrac{x}{x^2 + 3}\,dx$

44. $\int_{10}^\infty \ln(2x)\,dx$

Find the area between the graph of each function and the x-axis over the given interval, if possible.

45. $f(x) = \dfrac{3}{(x - 2)^2}$ for $(-\infty, 1]$

46. $f(x) = 3e^{-x}$ for $[0, \infty)$

47. How is the present value of money flow found? the accumulated amount of money flow?

48. What is a differential equation? What is it used for?

49. What is the difference between a particular solution and a general solution to a differential equation?

50. How can you tell that a differential equation is separable? That it is linear?

Find general solutions for the following differential equations.

51. $\dfrac{dy}{dx} = 2x^3 + 6x$

52. $\dfrac{dy}{dx} = x^2 + 5x^4$

53. $\dfrac{dy}{dx} = 4e^x$

54. $\dfrac{dy}{dx} = \dfrac{1}{2x + 3}$

55. $\dfrac{dy}{dx} = \dfrac{3x + 1}{y}$

56. $\dfrac{dy}{dx} = \dfrac{e^x + x}{y - 1}$

57. $\dfrac{dy}{dx} = \dfrac{2y + 1}{x}$

58. $\dfrac{dy}{dx} = \dfrac{3 - y}{e^x}$

Find particular solutions for the following differential equations. (Some solutions may give y implicitly.)

59. $\dfrac{dy}{dx} = x^2 - 5x$; $y = 1$ when $x = 0$

60. $\dfrac{dy}{dx} = 4x^3 + 2$; $y = 3$ when $x = 1$

61. $\dfrac{dy}{dx} = 5(e^{-x} - 1)$; $y = 17$ when $x = 0$

62. $\dfrac{dy}{dx} = \dfrac{x}{x^2 - 3}$; $y = 52$ when $x = 2$

63. $(5 - 2x)y = \dfrac{dy}{dx}$; $y = 2$ when $x = 0$

64. $\sqrt{x}\,\dfrac{dy}{dx} = xy$; $y = 4$ when $x = 1$

65. $\dfrac{dy}{dx} = \dfrac{1 - 2x}{y + 3}$; $y = 16$ when $x = 0$

66. $\dfrac{dy}{dx} = (3x + 2)^2 e^y$; $y = 0$ when $x = 0$

Applications

Business and Economics

67. Total Revenue The rate of change of revenue from the sale of x toaster ovens is

$$R'(x) = x(x - 50)^{1/2}.$$

Find the total revenue from the sale of the 50th to the 75th ovens.

68. Total Sales Use the values of $f(x)$ given in the table to find total sales (in millions of dollars) for the given period by the trapezoidal rule, using $n = 6$.

Year, x	1	2	3	4	5	6	7
Rate of Growth of Sales, $f(x)$.7	1.2	1.5	1.9	2.2	2.4	2.0

69. Total Sales Repeat Exercise 49, using Simpson's rule.

Present Value of Money Flow *Each of the functions in Exercises 70–73 represents the rate of flow of money (in dollars per year) over the given time period, compounded continuously at the given annual interest rate. Find the present value in each case.*

70. $f(x) = 5000$, 8 yr, 9%

71. $f(x) = 25,000$, 12 yr, 10%

72. $f(x) = 100e^{.02x}$, 5 yr, 11%

73. $f(x) = 30x$, 18 mo, 5%

Amount of Money at Time t *Assume that each of the following functions gives the rate of flow of money in dollars per year over the given period, with continuous compounding at the given rate. Find the accumulated amount at the end of the time period.*

74. $f(x) = 2000$, 5 mo, 1% per month

75. $f(x) = 500e^{-.03x}$, 8 yr, 10% per year

76. $f(x) = 20x$, 6 yr, 12% per year

77. $f(x) = 1000 + 200x$, 10 yr, 9% per year

78. Money Flow An investment scheme is expected to produce a continuous flow of money, starting at $1000 and increasing exponentially at 5% a year for 7 yr. Find the present value at an interest rate of 11% compounded continuously.

79. Money Flow The proceeds from the sale of a building will yield a uniform continuous flow of $10,000 a year for 10 yr. Find the final amount at an interest rate of 10.5% compounded continuously.

80. Capital Value Find the capital value of an office building for which annual rent of $50,000 will be paid in perpetuity, if the interest rate is 9%.

81. Marginal Sales The marginal sales (in hundreds of dollars) of a computer software company are given by

$$\frac{dy}{dx} = 5e^{.2x},$$

where x is the number of months the company has been in business. Assume that sales were 0 initially.

(a) Find the sales after 6 months.

(b) Find the sales after 12 months.

82. Continuous Withdrawals A deposit of $10,000 is made to a savings account at 5% interest compounded continuously. Assume that continuous *withdrawals* of $1000 per year are made.

(a) Write a differential equation to describe the situation.

(b) How much will be left in the account after 1 year?

83. In Exercise 36, approximately how long will it take to use up the account?

Life Sciences

84. Drug Reaction The reaction rate to a new drug x hours after the drug is administered is

$$r'(x) = .5xe^{-x}.$$

Find the total reaction over the first 5 hours.

85. Oil Leak Pollution An oil leak from an uncapped well is polluting a bay at a rate of $f(x) = 100e^{-.05x}$ gallons per year. Use an improper integral to find the total amount of oil that will enter the bay, assuming the well is never capped.

Physical Sciences

86. Average Temperatures Suppose the temperature in a river at a point x meters downstream from a factory that is discharging hot water into the river is given by

$$T(x) = 400 - .25x^2.$$

Find the average temperature over each of the following intervals.

(a) $[0, 10]$ **(b)** $[10, 40]$ **(c)** $[0, 40]$

Connections

87. Probability In an earlier exercise, we mentioned the importance in probability and statistics of the density function for the standard normal distribution (the bell-shaped curve), which is given by

$$f(x) = \frac{1}{\sqrt{2\pi}}e^{-x^2/2}.$$

The integral of this function from $-\infty$ to ∞ has a particular significance. Parts (a)–(e) of this exercise, to be performed on a computer or a programmable calculator, are to help you guess the value of this integral. (Its exact value can only be determined by more advanced methods.)

(a) Use Simpson's rule with $n = 20$ to find the integral over the interval $[-1, 1]$.

(b) Use Simpson's rule with $n = 40$ to find the integral over the interval $[-2, 2]$.

(c) Use Simpson's rule with $n = 60$ to find the integral over the interval $[-3, 3]$.

(d) Use Simpson's rule with $n = 80$ to find the integral over the interval $[-4, 4]$.

(e) Based on your answers to parts (a)–(d), what do you think the value of the integral is over the interval $(-\infty, \infty)$?

Extended Application / How Much Does a Warranty Cost?*

This application uses some of the ideas of probability. The probability of an event is a number p, where $0 \leq p \leq 1$, such that p is equal to the number of ways that the event can happen divided by the total number of possible outcomes. For example, the probability of drawing a red card from a deck of 52 cards (of which 26 are red) is given by

$$P(\text{red card}) = \frac{26}{52} = \frac{1}{2}.$$

In the same way, the probability of drawing a black queen from a deck of 52 cards is

$$P(\text{black queen}) = \frac{2}{52} = \frac{1}{26}.$$

In this application we find the cost of a warranty program to a manufacturer. This cost depends on the quality of the products made, as we might expect. We use the following variables.

$c =$ constant product price, per unit, including cost of warranty (We assume the price charged per unit is constant, since this price is likely to be fixed by competition.)

$m =$ expected lifetime of the product

$w =$ length of the warranty period

$N =$ size of a production lot, such as a year's production

$r =$ warranty cost per unit

$C(t) =$ pro rata customer rebate at time t

$P(t) =$ probability of product failure at any time t

$F(t) =$ number of failures occurring at time t

*Reprinted by permission of Warren W. Menke, "Determination of Warranty Reserves," *Management Sciences*, Vol. 15, No. 10, June 1969. Copyright © 1969 The Institute of Management Sciences.

We assume that the warranty is of the pro rata customer rebate type, in which the customer is paid for the proportion of the warranty left. Hence, if the product has a warranty period of w, and fails at time t, then the product worked for the fraction t/w of the warranty period. The customer is reimbursed for the unused portion of the warranty, or the fraction

$$1 - \frac{t}{w}.$$

If we assume the product cost c originally, and if we use $C(t)$ to represent the customer rebate at time t, we have

$$C(t) = c\left(1 - \frac{t}{w}\right).$$

For many different types of products, it has been shown by experience that

$$P(t) = 1 - e^{-t/m}$$

provides a good estimate of the probability of product failure at time t. The total number of failures at time t is given by the product $P(t)$ and N, the total number of items per batch. If we use $F(t)$ to represent this total, we have

$$F(t) = N \cdot P(t) = N(1 - e^{-t/m}).$$

The total number of failures in some "tiny time interval" of width dt can be shown to be the derivative of $F(t)$,

$$F'(t) = \left(\frac{N}{m}\right)e^{-t/m},$$

while the cost for the failures in this "tiny time interval" is

$$C(t) \cdot F'(t) = c\left(1 - \frac{t}{w}\right)\left(\frac{N}{m}\right)e^{-t/m}.$$

The total cost for all failures during the warranty period is thus given by the definite integral

$$\int_0^w c\left(1 - \frac{t}{w}\right)\left(\frac{N}{m}\right)e^{-t/m}\,dt.$$

Using integration by parts, this definite integral can be shown to equal

$$Nc\left[-e^{-t/m} + \frac{t}{w} \cdot e^{-t/m} + \frac{m}{w}(e^{-t/m})\right]\Bigg|_0^w \quad \text{or}$$

$$Nc\left(1 - \frac{m}{w} + \frac{m}{w}e^{-w/m}\right).$$

This last quantity is the total warranty cost for all the units manufactured. Since there are N units per batch, the warranty cost per item is

$$r = \frac{1}{N}\left[Nc\left(1 - \frac{m}{w} + \frac{m}{w}e^{-w/m}\right)\right] = c\left(1 - \frac{m}{w} + \frac{m}{w}e^{-w/m}\right).$$

For example, suppose a product that costs \$100 has an expected life of 24 months, with a 12-month warranty. Then we have $c = \$100$, $m = 24$, $w = 12$, with r, the warranty cost per unit, given by

$$r = 100(1 - 2 + 2e^{-.5}) = 100[-1 + 2(.60653)]$$
$$= 100(.2131) = 21.31.$$

Exercises

1. Verify the evaluation of the definite integral using integration by parts.

Find r for each of the following.

2. $c = \$50$, $m = 48$ mo, $w = 24$ mo

3. $c = \$1000$, $m = 60$ mo, $w = 24$ mo

4. $c = \$1200$, $m = 30$ mo, $w = 30$ mo

Applications at a glance...

Growth of IRAs ... predicting the death rate ... evaluating air conditioners ...

(See page 802.)

Many of the ideas developed for functions of one variable also apply to functions of more than one variable. In particular, the fundamental idea of derivative generalizes in a very natural way to functions of more than one variable.

16.1 FUNCTIONS OF SEVERAL VARIABLES

How are the amounts of labor and capital needed to produce a certain number of items related?

We will be able to answer this question later in this section using a production function, which depends on the amounts of both labor and capital. That is, production is a function of two independent variables.

If a company produces x items at a cost of $10 per item, for instance, then the total cost $C(x)$ of producing the items is given by

$$C(x) = 10x.$$

The cost is a function of one independent variable, the number of items produced. If the company produces two products, with x of one product at a cost of $10 each, and y of another product at a cost of $15 each, then the total cost to the firm is a function of *two* independent variables, x and y. By generalizing $f(x)$ notation, the total cost can be written as $C(x, y)$, where

$$C(x, y) = 10x + 15y.$$

When $x = 5$ and $y = 12$ the total cost is written $C(5, 12)$, with

$$C(5, 12) = 10 \cdot 5 + 15 \cdot 12 = 230.$$

A general definition follows.

FUNCTION OF TWO VARIABLES

$z = f(x, y)$ is a **function of two variables** if a unique value of z is obtained from each ordered pair of real numbers (x, y). The variables x and y are **independent variables,** and z is the **dependent variable.** The set of all ordered pairs of real numbers (x, y) such that $f(x, y)$ exists is the **domain** of f; the set of all values of $f(x, y)$ is the **range.**

EXAMPLE

Let $f(x, y) = 4x^2 + 2xy + 3/y$ and find each of the following.

(a) $f(-1, 3)$.

Replace x with -1 and y with 3.

$$f(-1, 3) = 4(-1)^2 + 2(-1)(3) + \frac{3}{3} = 4 - 6 + 1 = -1$$

(b) $f(2, 0)$

Because of the quotient $3/y$, it is not possible to replace y with 0, so $f(2, 0)$ is undefined. By inspection, we see that the domain of the function is (x, y) such that $y \neq 0$. ◄

EXAMPLE 2

Let x represent the number of milliliters (ml) of carbon dioxide released by the lungs in one minute. Let y be the change in the carbon dioxide content of the blood as it leaves the lungs (y is measured in ml of carbon dioxide per 100 ml of blood). The total output of blood from the heart in one minute (measured in ml) is given by C, where C is a function of x and y such that

$$C = C(x, y) = \frac{100x}{y}.$$

Find $C(320, 6)$.

Replace x with 320 and y with 6 to get

$$C(320, 6) = \frac{100(320)}{6}$$

$$\approx 5333 \text{ ml of blood per minute.} \quad ◄$$

The definition given before Example 1 was for a function of two independent variables, but similar definitions could be given for functions of three, four, or more independent variables.

EXAMPLE 3

Let $f(x, y, z) = 4xz - 3x^2y + 2z^2$. Find $f(2, -3, 1)$.

Replace x with 2, y with -3, and z with 1.

$$f(2, -3, 1) = 4(2)(1) - 3(2)^2(-3) + 2(1)^2 = 8 + 36 + 2 = 46 \quad ◄$$

Graphing Functions of Two Independent Variables Functions of one independent variable are graphed by using an x-axis and a y-axis to locate points in a plane. The plane determined by the x- and y-axes is called the *xy-plane*. A third axis is needed to graph functions of two independent variables—the z-axis, which goes through the origin in the xy-plane and is perpendicular to both the x-axis and the y-axis.

Figure 1 (on the next page) shows one possible way to draw the three axes. In Figure 1, the yz-plane is in the plane of the page, with the x-axis perpendicular to the plane of the page.

Just as we graphed ordered pairs earlier we can now graph **ordered triples** of the form (x, y, z). For example, to locate the point corresponding to the ordered triple $(2, -4, 3)$, start at the origin and go 2 units along the positive x-axis. Then go 4 units in a negative direction (to the left) parallel to the y-axis. Finally, go up 3 units parallel to the z-axis. The point representing $(2, -4, 3)$ is shown in Figure 1, together with several other points. The region of three-dimensional space where all coordinates are positive is called the **first octant.**

Graph the following lines. Refer to Section 1.2 if you need review.

1. $2x + 3y = 6$
2. $x = 4$
3. $y = 2$

Answers:

1.

$2x + 3y = 6$

2.

$x = 4$

3.

$y = 2$

FIGURE 1

In Chapter 1 we saw that the graph of $ax + by = c$ (where a and b are not both 0) is a straight line. This result generalizes to three dimensions.

PLANE	The graph of

$$ax + by + cz = d$$

is a **plane** if a, b, and c are not all 0.

EXAMPLE 4

Graph $2x + y + z = 6$.

By the result above the graph of this equation is a plane. Earlier, we graphed straight lines by finding x- and y-intercepts. A similar idea helps in graphing a plane. To find the x-intercept, which is the point where the graph crosses the x-axis, let $y = 0$ and $z = 0$.

$$2x + 0 + 0 = 6$$
$$x = 3$$

The point $(3, 0, 0)$ is on the graph. Letting $x = 0$ and $z = 0$ gives the point $(0, 6, 0)$, while $x = 0$ and $y = 0$ lead to $(0, 0, 6)$. The plane through these three points includes the triangular surface shown in Figure 2. This surface is the first-octant part of the plane that is the graph of $2x + y + z = 6$. ◀

EXAMPLE 5

Graph $x + z = 6$.

To find the x-intercept, let $z = 0$, giving $(6, 0, 0)$. If $x = 0$, we get the point $(0, 0, 6)$. Because there is no y in the equation $x + z = 6$, there can be no y-intercept. A plane that has no y-intercept is parallel to the y-axis. The first-octant portion of the graph of $x + z = 6$ is shown in Figure 3. ◀

FIGURE 2

FIGURE 3

FIGURE 4

6

Graph each of the following functions in two variables.

(a) $x = 3$

This graph, which goes through $(3, 0, 0)$, can have no y-intercept and no z-intercept. It is therefore a plane parallel to the y-axis and the z-axis and, therefore, to the yz-plane. The first-octant portion of the graph is shown in Figure 4.

(b) $y = 4$

This graph goes through $(0, 4, 0)$ and is parallel to the xz-plane. The first-octant portion of the graph is shown in Figure 5.

(c) $z = 1$

The graph is a plane parallel to the xy-plane, passing through $(0, 0, 1)$. Its first-octant portion is shown in Figure 6. ◀

FIGURE 5

FIGURE 6

The graph of a function of one variable, $y = f(x)$, is a curve in the plane. If x_0 is in the domain of f, the point $(x_0, f(x_0))$ on the graph lies directly above or below the number x_0 on the x-axis, as shown in Figure 7.

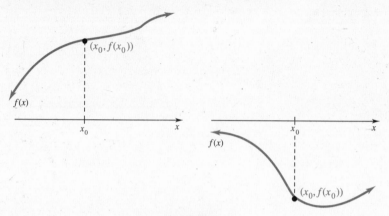

FIGURE 7

The graph of a function of two variables, $z = f(x, y)$, is a **surface** in three-dimensional space. If (x_0, y_0) is in the domain of f, the point $(x_0, y_0, f(x_0, y_0))$ lies directly above or below the point (x_0, y_0) in the xy-plane, as shown in Figure 8.

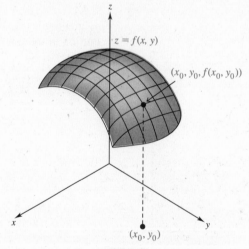

FIGURE 8

Although computer software is available for drawing the graphs of functions of two independent variables, you can often get a good picture of the graph without it by finding various **traces**—the curves that result when a surface is cut by a plane. The **xy-trace** is the intersection of the surface with the xy-plane. The **yz-trace** and **xz-trace** are defined similarly. You can also determine the intersection of the surface with planes parallel to the xy-plane. Such planes are of the form $z = k$, where k is a constant, and the curves that result when they cut the surface are called **level curves.**

EXAMPLE 7

Graph $z = x^2 + y^2$.

The yz-plane is the plane in which every point has a first coordinate of 0, so its equation is $x = 0$. When $x = 0$, the equation becomes $z = y^2$, which is the equation of a parabola in the yz-plane, as shown in Figure 9(a). Similarly, to find the intersection of the surface with the xz-plane (whose equation is $y = 0$), let $y = 0$ in the equation. It then becomes $z = x^2$, which is the equation of a parabola in the xz-plane, as shown in Figure 9(a). The xy-trace (the intersection of the surface with the plane $z = 0$) is the single point $(0, 0, 0)$ because $x^2 + y^2$ is never negative, and is equal to 0 only when $x = 0$ and $y = 0$.

(a)

(b)

FIGURE 9

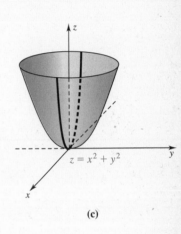

(c)

Next, we find the level curves by intersecting the surface with the planes $z = 1$, $z = 2$, $z = 3$, etc. (all of which are parallel to the xy-plane). In each case, the result is a circle:

$$x^2 + y^2 = 1, \quad x^2 + y^2 = 2, \quad x^2 + y^2 = 3,$$

and so on, as shown in Figure 9(b). Drawing the traces and level curves on the same set of axes suggests that the graph of $z = x^2 + y^2$ is the bowl-shaped figure, called a **paraboloid,** that is shown in Figure 9(c). ◄

One application of level curves in economics occurs with production functions. A **production function** $z = f(x, y)$ is a function that gives the quantity z of an item produced as a function of x and y, where x is the amount of labor and y is the amount of capital (in appropriate units) needed to produce z units. If the production function has the special form $z = P(x, y) = Ax^a y^{1-a}$, where A is a constant and $0 < a < 1$, the function is called a **Cobb-Douglas production function.**

EXAMPLE 8

Find the level curve at a production of 100 items for the Cobb-Douglas production function $z = x^{2/3}y^{1/3}$.

Let $z = 100$ to get

$$100 = x^{2/3}y^{1/3}$$

$$\frac{100}{x^{2/3}} = y^{1/3}.$$

Now cube both sides to express y as a function of x.

$$y = \frac{100^3}{x^2} = \frac{1,000,000}{x^2} \quad \blacktriangleleft$$

The level curve of height 100 found in Example 8 is shown graphed in three dimensions in Figure 10(a) and on the familiar xy-plane in Figure 10(b). The points of the graph correspond to those values of x and y that lead to production of 100 items.

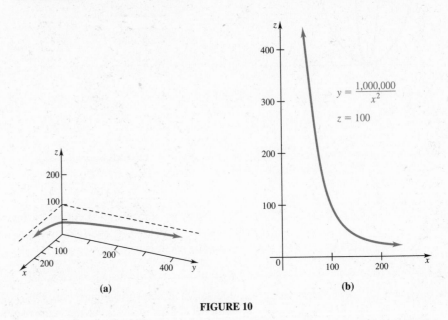

FIGURE 10

The curve in Figure 10 is called an *isoquant*, for *iso* (equal) and *quant* (amount). In Example 8, the "amounts" all "equal" 100.

Because of the difficulty of drawing the graphs of more complicated functions, we now merely list some common equations and their graphs. These graphs were drawn by computer, a very useful method of depicting three-dimensional surfaces.

Paraboloid, $z = x^2 + y^2$

xy-trace: point
yz-trace: parabola
xz-trace: parabola

Ellipsoid, $\dfrac{x^2}{a^2} + \dfrac{y^2}{b^2} + \dfrac{z^2}{c^2} = 1$

xy-trace: ellipse
yz-trace: ellipse
xz-trace: ellipse

Hyperbolic Paraboloid, $x^2 - y^2 = z$
(sometimes called a **saddle**)

xy-trace: two intersecting lines
yz-trace: parabola
xz-trace: parabola

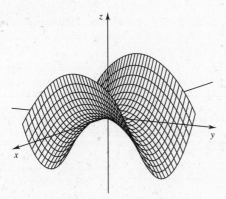

Hyperboloid of Two Sheets,
$-x^2 - y^2 + z^2 = 1$

xy-trace: none
yz-trace: hyperbola
xz-trace: hyperbola

16.1 Exercises

1. Let $f(x, y) = 4x + 5y + 3$. Find the following.
 (a) $f(2, -1)$ (b) $f(-4, 1)$ (c) $f(-2, -3)$ (d) $f(0, 8)$
2. Let $g(x, y) = -x^2 - 4xy + y^3$. Find the following.
 (a) $g(-2, 4)$ (b) $g(-1, -2)$ (c) $g(-2, 3)$ (d) $g(5, 1)$
3. Let $h(x, y) = \sqrt{x^2 + 2y^2}$. Find the following.
 (a) $h(5, 3)$ (b) $h(2, 4)$ (c) $h(-1, -3)$ (d) $h(-3, -1)$
4. Let $f(x, y) = \dfrac{\sqrt{9x + 5y}}{\log x}$. Find the following.

 (a) $f(10, 2)$ (b) $f(100, 1)$ (c) $f(1000, 0)$ (d) $f\left(\dfrac{1}{10}, 5\right)$

Graph the first-octant portion of each plane.

5. $x + y + z = 6$ 6. $x + y + z = 12$ 7. $2x + 3y + 4z = 12$ 8. $4x + 2y + 3z = 24$
9. $x + y = 4$ 10. $y + z = 5$ 11. $x = 2$ 12. $z = 3$

Graph the level curves in the first octant at heights of $z = 0$, $z = 2$, and $z = 4$ for the following functions.

13. $3x + 2y + z = 18$ 14. $x + 3y + 2z = 8$

15. $y^2 - x = -z$ 16. $2y - \dfrac{x^2}{3} = z$

Applications

Business and Economics

17. **Production** Production of a precision camera is given by
$$P(x, y) = 100\left(\frac{3}{5}x^{-2/5} + \frac{2}{5}y^{-2/5}\right)^{-5},$$
where x is the amount of labor in work hours and y is the amount of capital. Find the following.

 (a) What is the production when 32 work hours and 1 unit of capital are provided?

 (b) Find the production when 1 work hour and 32 units of capital are provided.

 (c) If 32 work hours and 243 units of capital are used, what is the production output?

 Individual Retirement Accounts *The multiplier function*
$$M = \frac{(1 + i)^n(1 - t) + t}{[1 + (1 - t)i]^n}$$
compares the growth of an Individual Retirement Account (IRA) with the growth of the same deposit in a regular savings account.

The function M depends on the three variables n, i, and t, where n represents the number of years an amount is left at interest, i represents the interest rate in both types of accounts and t represents the income tax rate. Values of M > 1 indicate that the IRA grows faster than the savings account. Let $M = f(n, i, t)$ and find the following.

18. Find the multiplier when funds are left for 25 years at 5% interest and the income tax rate is 28%. Which account grows faster?

19. What is the multiplier when money is invested for 25 years at 6% interest and the income tax rate is 33%? Which account grows faster?

Production *Find the level curve at a production of 500 for each of the production functions in Exercises 20–21. Graph each function on the xy-plane.*

20. The production function z for the United States was once estimated as $z = x^{.7}y^{.3}$, where x stands for the amount of labor and y stands for the amount of capital.

21. If x represents the amount of labor and y the amount of capital, a production function for Canada is approximately $z = x^{.4}y^{.6}$.

22. **Production** For the function in Exercise 20, what is the effect on z of doubling x? Of doubling y? Of doubling both?

23. **Cost** If labor (x) costs $200 per unit, materials (y) cost $100 per unit, and capital (z) costs $50 per unit, write a function for total cost.

Life Sciences

24. **Oxygen Consumption** The oxygen consumption of a well-insulated mammal that is not sweating is approximated by

$$m = \frac{2.5(T - F)}{w^{.67}},$$

where T is the internal body temperature of the animal (in °C), F is the temperature of the outside of the animal's fur (in °C), and w is the animal's weight in kilograms.* Find m for the following data.

 (a) Internal body temperature $= 38°$ C; outside temperature $= 6°$ C; weight $= 32$ kg

 (b) Internal body temperature $= 40°$ C; outside temperature $= 20°$ C; weight $= 43$ kg

25. **Body Surface Area** The surface area of a human (in square meters) is approximated by

$$A = .202W^{.425}H^{.725},$$

where W is the weight of the person in kilograms and H is the height in meters.* Find A for the following data.

 (a) Weight, 72 kg; height, 1.78 m

 (b) Weight, 65 kg; height, 1.40 m

 (c) Weight, 70 kg; height, 1.60 m

 (d) Using your weight and height, find your own surface area.

General Interest

26. **Postage Rates** Extra postage is charged for parcels sent by U.S. mail that are more than 84 inches in length and girth combined. (Girth is the distance around the parcel

perpendicular to its length. See the figure.) Express the combined length and girth as a function of L, W, and H.

27. **Required Material** Refer to the figure for Exercise 26. Assume L, W, and H are in feet. Write a function in terms of L, W, and H that gives the total material required to build the box.

28. **Elliptical Templates** The holes cut in a roof for vent pipes require elliptical templates. A formula for determining the length of the major axis of the ellipse is $L = f(H, D) = \sqrt{H^2 + D^2}$, where D is the (outside) diameter of the pipe and H is the "rise" of the roof per D units of "run"; that is, the slope of the roof is H/D. (See the figure.) The width of the ellipse (minor axis) equals D. Find the length and width of the ellipse required to produce a hole for a vent pipe with a diameter of 3.75 inches in roofs with the following slopes.

 (a) 3/4 (b) 2/5

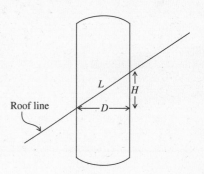

*From Duane J. Clow and N. Scott Urquhart, *Mathematics in Biology.* Copyright © 1974 by W. W. Norton & Company, Inc. Used by permission.

Match each equation in Exercises 29–34 with its graph in (a)–(f) below.

29. $z = x^2 + y^2$

30. $z^2 - y^2 - x^2 = 1$

31. $x^2 - y^2 = z$

32. $z = y^2 - x^2$

33. $\dfrac{x^2}{16} + \dfrac{y^2}{25} + \dfrac{z^2}{4} = 1$

34. $z = 5(x^2 + y^2)^{-1/2}$

(a)

(b)

(c)

(d)

(e)

(f)

35. Let $f(x, y) = 9x^2 - 3y^2$, and find each of the following.

(a) $\dfrac{f(x + h, y) - f(x, y)}{h}$

(b) $\dfrac{f(x, y + h) - f(x, y)}{h}$

36. Let $f(x, y) = 7x^3 + 8y^2$, and find each of the following.

(a) $\dfrac{f(x + h, y) - f(x, y)}{h}$

(b) $\dfrac{f(x, y + h) - f(x, y)}{h}$

16.2 PARTIAL DERIVATIVES

What is the change in productivity if labor is increased by one work-hour? What if capital is increased by one unit?

You may want to review the chapter on the derivative for methods used to find some of the derivatives in this section.

Earlier, we found that the derivative dy/dx gives the rate of change of y with respect to x. In this section, we show how derivatives are found and interpreted for multivariable functions, and we will use that information to answer the questions posed above.

A small firm makes only two products, radios and audiocassette recorders. The profits of the firm are given by

$$P(x, y) = 40x^2 - 10xy + 5y^2 - 80,$$

where x is the number of units of radios sold and y is the number of units of recorders sold. How will a change in x or y affect P?

Suppose that sales of radios have been steady at 10 units; only the sales of recorders vary. The management would like to find the marginal profit with respect to y, the number of recorders sold. Recall that marginal profit is given by the derivative of the profit function. Here, x is fixed at 10. Using this information, we begin by finding a new function, $f(y) = P(10, y)$. Let $x = 10$ to get

$$f(y) = P(10, y) = 40(10)^2 - 10(10)y + 5y^2 - 80$$
$$= 3920 - 100y + 5y^2.$$

The function $f(y)$ shows the profit from the sale of y recorders, assuming that x is fixed at 10 units. Find the derivative df/dy to get the marginal profit with respect to y.

$$\frac{df}{dy} = -100 + 10y$$

In this example, the derivative of the function $f(y)$ was taken with respect to y only; we assumed that x was fixed. To generalize, let $z = f(x, y)$. An intuitive definition of the *partial derivatives* of f with respect to x and y follows.

PARTIAL DERIVATIVES (INFORMAL DEFINITION)

The **partial derivative of f with respect to x** is the derivative of f obtained by treating x as a variable and y as a constant.

The **partial derivative of f with respect to y** is the derivative of f obtained by treating y as a variable and x as a constant.

The symbols $f_x(x, y)$ (no prime is used), $\partial z/\partial x$, and $\partial f/\partial x$ are used to represent the partial derivative of $z = f(x, y)$ with respect to x, with similar symbols used for the partial derivative with respect to y. The symbol $f_x(x, y)$ is often abbreviated as just f_x, with $f_y(x, y)$ abbreviated as f_y.

Generalizing from the definition of the derivative given earlier, partial derivatives of a function $z = f(x, y)$ are formally defined as follows.

PARTIAL DERIVATIVES (FORMAL DEFINITION)

Let $z = f(x, y)$ be a function of two independent variables. Let all indicated limits exist. Then the partial derivative of f with respect to x is

$$f_x = \frac{\partial f}{\partial x} = \lim_{h \to 0} \frac{f(x + h, y) - f(x, y)}{h},$$

and the partial derivative of f with respect to y is

$$f_y = \frac{\partial f}{\partial y} = \lim_{h \to 0} \frac{f(x, y + h) - f(x, y)}{h}.$$

If the indicated limits do not exist, then the partial derivatives do not exist.

Similar definitions could be given for functions of more than two independent variables.

 EXAMPLE 1 Let $f(x, y) = 4x^2 - 9xy + 6y^3$. Find f_x and f_y.

To find f_x, treat y as a constant and x as a variable. The derivative of the first term, $4x^2$, is $8x$. In the second term, $-9xy$, the constant coefficient of x is $-9y$, so the derivative with x as the variable is $-9y$. The derivative of $6y^3$ is zero, since we are treating y as a constant. Thus,

$$f_x = 8x - 9y.$$

Now, to find f_y, treat y as a variable and x as a constant. Since x is a constant, the derivative of $4x^2$ is zero. In the second term, the coefficient of y is $-9x$ and the derivative of $-9xy$ is $-9x$. The derivative of the third term is $18y^2$. Thus,

$$f_y = -9x + 18y^2. \quad \blacktriangleleft$$

The next example shows how the chain rule can be used to find partial derivatives.

EXAMPLE 2 Let $f(x, y) = \ln |x^2 + y|$. Find f_x and f_y.
Recall the formula for the derivative of a natural logarithm function. If $g(x) = \ln |x|$, then $g'(x) = 1/x$. Using this formula and the chain rule,

$$f_x = \frac{1}{x^2 + y} \cdot D_x(x^2 + y) = \frac{1}{x^2 + y} \cdot 2x = \frac{2x}{x^2 + y},$$

and

$$f_y = \frac{1}{x^2 + y} \cdot D_y(x^2 + y) = \frac{1}{x^2 + y} \cdot 1 = \frac{1}{x^2 + y}. \quad \blacktriangleleft$$

The notation

$$f_x(a, b) \qquad \text{or} \qquad \frac{\partial f}{\partial x}(a, b)$$

represents the value of a partial derivative when $x = a$ and $y = b$, as shown in the next example.

EXAMPLE 3 Let $f(x, y) = 2x^2 + 3xy^3 + 2y + 5$. Find the following.

(a) $f_x(-1, 2)$

First, find f_x by holding y constant.

$$f_x = 4x + 3y^3$$

Now let $x = -1$ and $y = 2$.

$$f_x(-1, 2) = 4(-1) + 3(2)^3 = -4 + 24 = 20$$

(b) $\dfrac{\partial f}{\partial y}(-4, -3)$

Since $\partial f/\partial y = 9xy^2 + 2$,

$$\frac{\partial f}{\partial y}(-4, -3) = 9(-4)(-3)^2 + 2 = 9(-36) + 2 = -322. \quad \blacktriangleleft$$

The derivative of a function of one variable can be interpreted as the tangent line to the graph at that point. With some modification, the same is true of partial derivatives of functions of two variables. At a point on the graph of a function of two variables, $z = f(x, y)$, there may be many tangent lines, all of which lie in the same tangent plane, as shown in Figure 11.

FIGURE 11

In any particular direction, however, there will be only one tangent line. We use partial derivatives to find the slope of the tangent lines in the x- and y-directions as follows.

Figure 12 shows a surface $z = f(x, y)$ and a plane that is parallel to the xz-plane. The equation of the plane is $y = a$. (This corresponds to holding y fixed.) Since $y = a$ for points on the plane, any point on the curve that represents the intersection of the plane and the surface must have the form $(x, a, f(x, a))$. Thus, this curve can be described as $z = f(x, a)$. Since a is constant, $z = f(x, a)$ is a function of one variable. When the derivative of $z = f(x, a)$ is evaluated at $x = b$, it gives the slope of the line tangent to this curve at the point $(b, a, f(b, a))$, as shown in Figure 12. Thus, the partial derivative of f with respect to x, $f_x(b, a)$, gives the rate of change of the surface $z = f(x, y)$ in the x-direction at the point $(b, a, f(b, a))$. In the same way, the partial derivative with respect to y will give the slope of the line tangent to the surface in the y-direction at the point $(b, a, f(b, a))$.

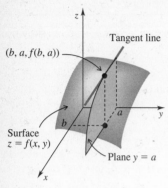

FIGURE 12

Rate of Change The derivative of $y = f(x)$ gives the rate of change of y with respect to x. In the same way, if $z = f(x, y)$, then f_x gives the rate of change of z with respect to x, if y is held constant.

EXAMPLE 4 Suppose that the temperature of the water at the point on a river where a nuclear power plant discharges its hot waste water is approximated by

$$T(x, y) = 2x + 5y + xy - 40,$$

where x represents the temperature of the river water (in degrees Celsius) before it reaches the power plant and y is the number of megawatts (in hundreds) of electricity being produced by the plant.

(a) Find and interpret $T_x(9, 5)$.

First, find the partial derivative T_x.

$$T_x = 2 + y$$

This partial derivative gives the rate of change of T with respect to x. Replacing x with 9 and y with 5 gives

$$T_x(9, 5) = 2 + 5 = 7.$$

Just as marginal cost is the approximate cost of one more item, this result, 7, is the approximate change in temperature of the output water if input water temperature changes by 1 degree, from $x = 9$ to $x = 9 + 1 = 10$, while y remains constant at 5 (500 megawatts of electricity produced).

(b) Find and interpret $T_y(9, 5)$.

The partial derivative T_y is

$$T_y = 5 + x.$$

This partial derivative gives the rate of change of T with respect to y as

$$T_y(9, 5) = 5 + 9 = 14.$$

This result, 14, is the approximate change in temperature resulting from a 1-unit increase in production of electricity from $y = 5$ to $y = 5 + 1 = 6$ (from 500 to 600 megawatts), while the input water temperature x remains constant at 9° C. ◀

As mentioned in the previous section, if $P(x, y)$ gives the output P produced by x units of labor and y units of capital, $P(x, y)$ is a production function. The partial derivatives of this production function have practical implications. For example, $\partial P/\partial x$ gives the marginal productivity of labor. This represents the rate at which the output is changing with respect to a one-unit change in labor for a fixed capital investment. That is, if the capital investment is held constant and labor is increased by 1 work hour, $\partial P/\partial x$ will yield the approximate change in the production level. Likewise, $\partial P/\partial y$ gives the marginal productivity of capital, which represents the rate at which the output is changing with respect to a one-unit change in capital for a fixed labor value. So if the labor force is held constant and the capital investment is increased by 1 unit, $\partial P/\partial y$ will approximate the corresponding change in the production level.

EXAMPLE **5** A company that manufactures computers has determined that its production function is given by

$$P(x, y) = 500x + 800y + 3x^2y - x^3 - \frac{y^4}{4},$$

where x is the size of the labor force (measured in work hours per week) and y is the amount of capital (measured in units of $1000) invested. Find the marginal productivity of labor and capital when $x = 50$ and $y = 20$, and interpret the results.

The marginal productivity of labor is found by taking the derivative of P with respect to x.

$$\frac{\partial P}{\partial x} = 500 + 6xy - 3x^2$$

$$\frac{\partial P}{\partial x}(50, 20) = 500 + 6(50)(20) - 3(50)^2 = -1000$$

Thus, if the capital investment is held constant at $20,000 and labor is increased from 50 to 51 work hours per week, production will decrease by about 1000 units. In the same way, the marginal productivity of capital is $\partial P/\partial y$.

$$\frac{\partial P}{\partial y} = 800 + 3x^2 - y^3$$

$$\frac{\partial P}{\partial y}(50, 20) = 800 + 3(50)^2 - (20)^3 = 300$$

If work hours are held constant at 50 hours per week and the capital investment is increased from $20,000 to $21,000, production will increase by about 300 units. ◀

Second-Order Partial Derivatives The second derivative of a function of one variable is very useful in determining relative maxima and minima. **Second-order partial derivatives** (partial derivatives of a partial derivative) are used in a similar way for functions of two or more variables. The situation is somewhat more complicated, however, with more independent variables. For example, $f(x, y) = 4x + x^2y + 2y$ has two first-order partial derivatives,

$$f_x = 4 + 2xy \qquad \text{and} \qquad f_y = x^2 + 2.$$

Since each of these has two partial derivatives, one with respect to y and one with respect to x, there are *four* second-order partial derivatives of function f. The notations for these four second-order partial derivatives are given below.

SECOND-ORDER PARTIAL DERIVATIVES

For a function $z = f(x, y)$, if the indicated partial derivative exists, then

$$\frac{\partial}{\partial x}\left(\frac{\partial z}{\partial x}\right) = \frac{\partial^2 z}{\partial x^2} = f_{xx} \qquad\qquad \frac{\partial}{\partial y}\left(\frac{\partial z}{\partial y}\right) = \frac{\partial^2 z}{\partial y^2} = f_{yy}$$

$$\frac{\partial}{\partial y}\left(\frac{\partial z}{\partial x}\right) = \frac{\partial^2 z}{\partial y \partial x} = f_{xy} \qquad\qquad \frac{\partial}{\partial x}\left(\frac{\partial z}{\partial y}\right) = \frac{\partial^2 z}{\partial x \partial y} = f_{yx}.$$

As seen above, f_{xx} is used as an abbreviation for $f_{xx}(x, y)$, with f_{yy}, f_{xy}, and f_{yx} used in a similar way. The symbol f_{xx} is read "the partial derivative of f_x with respect to x." Also, the symbol $\partial^2 z / \partial y^2$ is read "the partial derivative of $\partial z / \partial y$ with respect to y."

Note For most functions found in applications and for all of the functions in this book, the second-order partial derivatives f_{xy} and f_{yx} are equal. Therefore, it is not necessary to be particular about the order in which these derivatives are found.

 EXAMPLE 6 Find all second-order partial derivatives for

$$f(x, y) = -4x^3 - 3x^2y^3 + 2y^2.$$

First find f_x and f_y.

$$f_x = -12x^2 - 6xy^3 \qquad \text{and} \qquad f_y = -9x^2y^2 + 4y$$

To find f_{xx}, take the partial derivative of f_x with respect to x.

$$f_{xx} = -24x - 6y^3$$

Take the partial derivative of f_y, with respect to y; this gives f_{yy}.

$$f_{yy} = -18x^2y + 4$$

Find f_{xy} by starting with f_x, then taking the partial derivative of f_x with respect to y.

$$f_{xy} = -18xy^2$$

Finally, find f_{yx} by starting with f_y; take its partial derivative with respect to x.

$$f_{yx} = -18xy^2 \quad \blacktriangleleft$$

EXAMPLE 7

Let $f(x, y) = 2e^x - 8x^3y^2$. Find all second-order partial derivatives.

Here $f_x = 2e^x - 24x^2y^2$ and $f_y = -16x^3y$. [Recall: if $g(x) = e^x$, then $g'(x) = e^x$.] Now find the second-order partial derivatives.

$$f_{xx} = 2e^x - 48xy^2 \qquad f_{xy} = -48x^2y$$
$$f_{yy} = -16x^3 \qquad\qquad f_{yx} = -48x^2y \quad \blacktriangleleft$$

Partial derivatives of functions with more than two independent variables are found in a similar manner. For instance, to find f_z for $f(x, y, z)$, hold x and y constant and differentiate with respect to z.

EXAMPLE 8

Let $f(x, y, z) = 2x^2yz^2 + 3xy^2 - 4yz$. Find f_x, f_y, f_{xz}, and f_{yz}.

$$f_x = 4xyz^2 + 3y^2$$
$$f_y = 2x^2z^2 + 6xy - 4z$$

To find f_{xz}, differentiate f_x with respect to z.

$$f_{xz} = 8xyz$$

Differentiate f_y to get f_{yz}.

$$f_{yz} = 4x^2z - 4 \quad \blacktriangleleft$$

16.2 Exercises

1. Let $z = f(x, y) = 12x^2 - 8xy + 3y^2$. Find each of the following.

 (a) $\dfrac{\partial z}{\partial x}$ **(b)** $\dfrac{\partial z}{\partial y}$ **(c)** $\dfrac{\partial f}{\partial x}(2, 3)$ **(d)** $f_y(1, -2)$

2. Let $z = g(x, y) = 5x + 9x^2y + y^2$. Find each of the following.

 (a) $\dfrac{\partial g}{\partial x}$ **(b)** $\dfrac{\partial g}{\partial y}$ **(c)** $\dfrac{\partial z}{\partial y}(-3, 0)$ **(d)** $g_x(2, 1)$

In Exercises 3–16, find f_x and f_y. Then find $f_x(2, -1)$ and $f_y(-4, 3)$. Leave the answers in terms of e in Exercises 7–10 and 15–16.

3. $f(x, y) = -2xy + 6y^3 + 2$

4. $f(x, y) = 4x^2y - 9y^2$

5. $f(x, y) = 3x^3y^2$

6. $f(x, y) = -2x^2y^4$

7. $f(x, y) = e^{x+y}$

8. $f(x, y) = 3e^{2x+y}$

9. $f(x, y) = -5e^{3x-4y}$

10. $f(x, y) = 8e^{7x-y}$

11. $f(x, y) = \dfrac{x^2 + y^3}{x^3 - y^2}$

12. $f(x, y) = \dfrac{3x^2y^3}{x^2 + y^2}$

13. $f(x, y) = \ln|1 + 3x^2y^3|$

14. $f(x, y) = \ln|2x^5 - xy^4|$

15. $f(x, y) = xe^{x^2y}$

16. $f(x, y) = y^2e^{x+3y}$

Find all second-order partial derivatives for the following.

17. $f(x, y) = 6x^3y - 9y^2 + 2x$

18. $g(x, y) = 5xy^4 + 8x^3 - 3y$

19. $R(x, y) = 4x^2 - 5xy^3 + 12y^2x^2$

20. $h(x, y) = 30y + 5x^2y + 12xy^2$

21. $r(x, y) = \dfrac{4x}{x + y}$ **22.** $k(x, y) = \dfrac{-5y}{x + 2y}$ **23.** $z = 4xe^y$ **24.** $z = -3ye^x$

25. $r = \ln|x + y|$ **26.** $k = \ln|5x - 7y|$ **27.** $z = x \ln|xy|$ **28.** $z = (y + 1) \ln|x^3y|$

For the functions defined as follows, find values of x and y such that both $f_x(x, y) = 0$ and $f_y(x, y) = 0$.

29. $f(x, y) = 6x^2 + 6y^2 + 6xy + 36x - 5$ **30.** $f(x, y) = 50 + 4x - 5y + x^2 + y^2 + xy$

31. $f(x, y) = 9xy - x^3 - y^3 - 6$ **32.** $f(x, y) = 2200 + 27x^3 + 72xy + 8y^2$

Find f_x, f_y, f_z, and f_{yz} for the following.

33. $f(x, y, z) = x^2 + yz + z^4$ **34.** $f(x, y, z) = 3x^5 - x^2 + y^5$

35. $f(x, y, z) = \dfrac{6x - 5y}{4z + 5}$ **36.** $f(x, y, z) = \dfrac{2x^2 + xy}{yz - 2}$

37. $f(x, y, z) = \ln|x^2 - 5xz^2 + y^4|$ **38.** $f(x, y, z) = \ln|8xy + 5yz - x^3|$

Applications

Business and Economics

39. Manufacturing Cost Suppose that the manufacturing cost of a precision electronic calculator is approximated by

$$M(x, y) = 40x^2 + 30y^2 - 10xy + 30,$$

where x is the cost of electronic chips and y is the cost of labor. Find the following.

(a) $M_y(4, 2)$ (b) $M_x(3, 6)$ (c) $(\partial M/\partial x)(2, 5)$

(d) $(\partial M/\partial y)(6, 7)$

40. Revenue The revenue from the sale of x units of a tranquilizer and y units of an antibiotic is given by

$$R(x, y) = 5x^2 + 9y^2 - 4xy.$$

Suppose 9 units of tranquilizer and 5 units of antibiotic are sold.

(a) What is the approximate effect on revenue if 10 units of tranquilizer and 5 units of antibiotic are sold?

(b) What is the approximate effect on revenue if the amount of antibiotic sold is increased to 6 units, while tranquilizer sales remain constant?

41. Sales A car dealership estimates that the total weekly sales of its most popular model is a function of the car's list price, p, and the interest rate in percent, i, offered by the manufacturer. The approximate weekly sales are given by

$$f(p, i) = 132p - 2pi - .01p^2.$$

(a) Find the weekly sales if the average list price is $9400 and the manufacturer is offering an 8% interest rate.

(b) Find and interpret f_p and f_i.

(c) What would be the effect on weekly sales if the price is $9400 and interest rates rise from 8% to 9%?

42. Marginal Productivity Suppose the production function of a company is given by

$$P(x, y) = 100\sqrt{x^2 + y^2},$$

where x represents units of labor and y represents units of capital. Find the following when 4 units of labor and 3 units of capital are used.

(a) The marginal productivity of labor

(b) The marginal productivity of capital

43. Marginal Productivity A manufacturer estimates that production (in hundreds of units) is a function of the amounts x and y of labor and capital used, as follows.

$$f(x, y) = \left(\frac{1}{3}x^{-1/3} + \frac{2}{3}y^{-1/3}\right)^{-3}$$

(a) Find the number of units produced when 27 units of labor and 64 units of capital are utilized.

(b) Find and interpret $f_x(27, 64)$ and $f_y(27, 64)$.

(c) What would be the approximate effect on production of increasing labor by 1 unit?

44. Marginal Productivity The production function z for the United States was once estimated as

$$z = x^{.7}y^{.3},$$

where x stands for the amount of labor and y the amount of capital. Find the marginal productivity of labor and of capital.

45. Marginal Productivity A similar production function for Canada is

$$z = x^{.4}y^{.6},$$

with x, y, and z as in Exercise 44. Find the marginal productivity of labor and of capital.

46. Marginal Productivity A manufacturer of automobile batteries estimates that his total production (in thousands of units) is given by

$$f(x, y) = 3x^{1/3}y^{2/3},$$

where x is the number of units of labor and y is the number of units of capital utilized.

(a) Find and interpret $f_x(64, 125)$ and $f_y(64, 125)$ if the current level of production uses 64 units of labor and 125 units of capital.

(b) What would be the approximate effect on production of increasing labor to 65 units while holding capital at the current level?

(c) Suppose that sales have been good and management wants to increase either capital or labor by 1 unit. Which option would result in a larger increase in production?

Life Sciences

47. Grasshopper Matings The total number of matings per day between individuals of a certain species of grasshoppers is approximated by

$$M(x, y) = 2xy + 10xy^2 + 30y^2 + 20,$$

where x represents the temperature in °C and y represents the number of days since the last rainfall. Find the following.

(a) The approximate change in matings when the temperature increases from 20° C to 21° C and the number of days since rain is constant at 4.

(b) The approximate change in matings when the temperature is constant at 24° C and the number of days since rain is changed from 10 to 11.

(c) Would an increase in temperature or an increase in days since rain cause more of an increase in matings?

48. Oxygen Consumption The oxygen consumption of a well-insulated mammal that is not sweating is approximated by

$$m = m(T, F, w) = \frac{2.5(T - F)}{w^{.67}} = 2.5(T - F)w^{-.67},$$

where T is the internal body temperature of the animal (in °C), F is the temperature of the outside of the animal's fur (in °C), and w is the animal's weight (in kilograms). Find the approximate change in oxygen consumption under the following conditions.

(a) The internal temperature increases from 38° C to 39° C, while the outside temperature remains at 12° C and the weight remains at 30 kg.

(b) The internal temperature is constant at 36° C, the outside temperature increases from 14° C to 15° C, and the weight remains at 25 kg.

49. Body Surface Area The surface area of a human (in square meters) is approximated by

$$A(W, H) = .202W^{.425}H^{.725},$$

where W is the weight of the person in kilograms and H is the height in meters. Find the approximate change in surface area under the following conditions.

(a) The weight changes from 72 kg to 73 kg, while the height remains 1.8 m.

(b) The weight remains stable at 70 kg, while the height changes from 1.6 m to 1.7 m.

50. Blood Flow In one method of computing the quantity of blood pumped through the lungs in one minute, a researcher first finds each of the following (in milliliters).

b = quantity of oxygen used by body in one minute

a = quantity of oxygen per liter of blood that has just gone through the lungs

v = quantity of oxygen per liter of blood that is about to enter the lungs

In one minute,

Amount of oxygen used = Amount of oxygen per liter
 × Liters of blood pumped.

If C is the number of liters of blood pumped through the lungs in one minute, then

$$b = (a - v) \cdot C \qquad \text{or} \qquad C = \frac{b}{a - v}.$$

(a) Find the number of liters of blood pumped through the lungs in one minute if $a = 160$, $b = 200$, and $v = 125$.

(b) Find the approximate change in C when a changes from 160 to 161, $b = 200$, and $v = 125$.

(c) Find the approximate change in C when $a = 160$, b changes from 200 to 201, and $v = 125$.

(d) Find the approximate change in C when $a = 160$, $b = 200$, and v changes from 125 to 126.

(e) A change of 1 unit in which quantity of oxygen produces the greatest change in the liters of blood pumped?

51. Health A weight-loss counselor has prepared a program of diet and exercise for a client. If the client sticks to the program, the weight loss that can be expected (in pounds per week) is given by

$$\text{Weight loss} = f(n, c) = \frac{1}{8}n^2 - \frac{1}{5}c + \frac{1937}{8},$$

where c is the average daily calorie intake for the week and n is the number of 40-min aerobic workouts per week.

(a) How many pounds can the client expect to lose by eating an average of 1200 cal per day and participating in four 40-min workouts in a week?

(b) Find and interpret $\partial f/\partial n$.

(c) The client currently averages 1100 cal per day and does three 40-minute workouts each week. What would be the approximate impact on weekly weight loss of adding a fourth workout per week?

52. Drug Reaction The reaction to x units of a drug t hr after it was administered is given by

$$R(x, t) = x^2(a - x)t^2e^{-t},$$

for $0 \le x \le a$ (where a is a constant). Find the following.

(a) $\dfrac{\partial R}{\partial x}$ **(b)** $\dfrac{\partial R}{\partial t}$ **(c)** $\dfrac{\partial^2 R}{\partial x^2}$ **(d)** $\dfrac{\partial^2 R}{\partial x \partial t}$

(e) Interpret your answers to parts (a) and (b).

Social Sciences

53. Education A developmental mathematics instructor at a large university has determined that a student's probability of success in the university's pass/fail remedial algebra course is a function of s, n, and a, where s is the student's score on the departmental placement exam, n is the number of semesters of mathematics passed in high school, and a is the student's mathematics SAT score. She estimates that p, the probability of passing the course (in percent), will be

$$p = f(s, n, a) = .003a + .1(sn)^{1/2}$$

for $200 \le a \le 800$, $0 \le s \le 10$, and $0 \le n \le 8$. Assuming that the above model has some merit, find the following.

(a) If a student scores 8 on the placement exam, has taken 6 semesters of high-school math, and has an SAT score of 450, what is the probability of passing the course?

(b) Find p for a student with 3 semesters of high school mathematics, a placement score of 3, and an SAT score of 320.

(c) Find and interpret $f_n(3, 3, 320)$ and $f_a(3, 3, 320)$.

Physical Sciences

54. Gravitational Attraction The gravitational attraction F on a body a distance r from the center of the earth, where r is greater than the radius of the earth, is a function of its mass m and the distance r as follows:

$$F = \frac{mgR^2}{r^2},$$

where R is the radius of the earth and g is the force of gravity—about 32 feet per second per second (ft/sec^2).

(a) Find and interpret F_m and F_r.

(b) Show that $F_m > 0$ and $F_r < 0$. Why is this reasonable?

16.3 MAXIMA AND MINIMA

What amounts of sugar and flavoring produce the minimum cost per batch of a soft drink? What is the minimum cost?

It may be helpful to review the section on relative extrema in the chapter on applications of the derivative at this point. The concepts presented there are basic to what will be done in this section.

One of the most important applications of calculus is in finding maxima and minima for functions. Earlier, we studied this idea extensively for functions of a single independent variable; now we will see that extrema can be found for functions of two variables. In particular, an extension of the second derivative test can be defined and used to identify maxima or minima. We begin with the definitions of relative maxima and minima.

**RELATIVE MAXIMA
AND MINIMA**

Let (a, b) be the center of a circular region contained in the xy-plane. Then, for a function $z = f(x, y)$ defined for every (x, y) in the region, $f(a, b)$ is a **relative maximum** if

$$f(a, b) \geq f(x, y)$$

for all points (x, y) in the circular region, and $f(a, b)$ is a **relative minimum** if

$$f(a, b) \leq f(x, y)$$

for all points (x, y) in the circular region.

As before, the word *extremum* is used for either a relative maximum or a relative minimum. Examples of a relative maximum and a relative minimum are given in Figures 13 and 14.

Relative maximum at (a, b)

FIGURE 13

Relative minimum at (a, b)

FIGURE 14

FIGURE 15

Note When functions of a single variable were discussed, a distinction was made between relative extrema and absolute extrema. The methods for finding absolute extrema are quite involved for functions of two variables, so we will discuss only relative extrema here. In most practical applications the relative extrema coincide with the absolute extrema. Also, in this brief discussion of extrema for multivariable functions, we omit cases where an extremum occurs on a boundary of the domain.

As suggested by Figure 15, at a relative maximum the tangent line parallel to the x-axis has a slope of 0, as does the tangent line parallel to the y-axis. (Notice the similarity to functions of one variable.) That is, if the function $z = f(x, y)$ has a relative extremum at (a, b), then $f_x(a, b) = 0$ and $f_y(a, b) = 0$, as stated in the next theorem.

LOCATION OF EXTREMA

Let a function $z = f(x, y)$ have a relative maximum or relative minimum at the point (a, b). Let $f_x(a, b)$ and $f_y(a, b)$ both exist. Then

$$f_x(a, b) = 0 \quad \text{and} \quad f_y(a, b) = 0.$$

$z = x^2 - y^2$

FIGURE 16

Just as with functions of one variable, the fact that the slopes of the tangent lines are 0 is no guarantee that a relative extremum has been located. For example, Figure 16 shows the graph of $z = f(x, y) = x^2 - y^2$. Both $f_x(0, 0) = 0$ and $f_y(0, 0) = 0$, and yet $(0, 0)$ leads to neither a relative maximum nor a relative minimum for the function. The point $(0, 0, 0)$ on the graph of this function is called a **saddle point;** it is a minimum when approached from one direction but a maximum when approached from another direction. A saddle point is neither a maximum nor a minimum.

The theorem on location of extrema suggests a useful strategy for finding extrema. First, locate all points (a, b) where $f_x(a, b) = 0$ and $f_y(a, b) = 0$. Then test each of these points separately, using the test given after the next example. For a function $f(x, y)$, the points (a, b) such that $f_x(a, b) = 0$ and $f_y(a, b) = 0$ are called **critical points.**

EXAMPLE

Find all critical points for

$$f(x, y) = 6x^2 + 6y^2 + 6xy + 36x - 5.$$

Find all points (a, b) such that $f_x(a, b) = 0$ and $f_y(a, b) = 0$. Here

$$f_x = 12x + 6y + 36 \quad \text{and} \quad f_y = 12y + 6x.$$

Set each of these two partial derivatives equal to 0.

$$12x + 6y + 36 = 0 \quad \text{and} \quad 12y + 6x = 0$$

These two equations make up a system of linear equations. We can use the substitution method to solve this system. First, rewrite $12y + 6x = 0$ as follows:

$$12y + 6x = 0$$
$$6x = -12y$$
$$x = -2y.$$

Now substitute $-2y$ for x in the other equation.

$$12x + 6y + 36 = 0$$
$$12(-2y) + 6y + 36 = 0$$
$$-24y + 6y + 36 = 0$$
$$-18y + 36 = 0$$
$$-18y = -36$$
$$y = 2$$

From the equation $x = -2y$, $x = -2(2) = -4$. The solution of the system of equations is $(-4, 2)$. Since this is the only solution of the system, $(-4, 2)$ is the only critical point for the given function. By the theorem above, if the function has a relative extremum, it will occur at $(-4, 2)$. ◀

The results of the next theorem can be used to decide whether $(-4, 2)$ in Example 1 leads to a relative maximum, a relative minimum, or neither.

TEST FOR RELATIVE EXTREMA

For a function $z = f(x, y)$, let f_{xx}, f_{yy}, and f_{xy} all exist. Let (a, b) be a point for which

$$f_x(a, b) = 0 \quad \text{and} \quad f_y(a, b) = 0.$$

Define the number D by

$$D = f_{xx}(a, b) \cdot f_{yy}(a, b) - [f_{xy}(a, b)]^2.$$

Then
(a) $f(a, b)$ is a relative maximum if $D > 0$ and $f_{xx}(a, b) < 0$;
(b) $f(a, b)$ is a relative minimum if $D > 0$ and $f_{xx}(a, b) > 0$;
(c) $f(a, b)$ is a saddle point (neither a maximum nor a minimum) if $D < 0$;
(d) if $D = 0$, the test gives no information.

This test is comparable to the second derivative test for extrema of functions of one independent variable. The chart below summarizes the conclusions of the theorem.

	$f_{xx}(a, b) < 0$	$f_{xx}(a, b) > 0$
$D > 0$	Relative maximum	Relative minimum
$D = 0$	No information	
$D < 0$	Saddle point	

EXAMPLE 2

The previous example showed that the only critical point for the function

$$f(x, y) = 6x^2 + 6y^2 + 6xy + 36x - 5$$

is $(-4, 2)$. Does $(-4, 2)$ lead to a relative maximum, a relative minimum, or neither?

Find out by using the test above. From Example 1,

$$f_x(-4, 2) = 0 \quad \text{and} \quad f_y(-4, 2) = 0.$$

Now find the various second partial derivatives used in finding D. From $f_x = 12x + 6y + 36$ and $f_y = 12y + 6x$,

$$f_{xx} = 12, \quad f_{yy} = 12, \quad \text{and} \quad f_{xy} = 6.$$

(If these second-order partial derivatives had not all been constants, they would have had to be evaluated at the point $(-4, 2)$.) Now

$$D = f_{xx}(-4, 2) \cdot f_{yy}(-4, 2) - [f_{xy}(-4, 2)]^2 = 12 \cdot 12 - 6^2 = 108.$$

Since $D > 0$ and $f_{xx}(-4, 2) = 12 > 0$, part (b) of the theorem applies, showing that $f(x, y) = 6x^2 + 6y^2 + 6xy + 36x - 5$ has a relative minimum at $(-4, 2)$. This relative minimum is $f(-4, 2) = -77$. ◀

Find all points where the function

$$f(x, y) = 9xy - x^3 - y^3 - 6$$

has any relative maxima or relative minima.

First find any critical points. Here

$$f_x = 9y - 3x^2 \quad \text{and} \quad f_y = 9x - 3y^2.$$

Set each of these partial derivatives equal to 0.

$$
\begin{array}{ll}
f_x = 0 & f_y = 0 \\
9y - 3x^2 = 0 & 9x - 3y^2 = 0 \\
9y = 3x^2 & 9x = 3y^2 \\
3y = x^2 & 3x = y^2
\end{array}
$$

In the first equation $(3y = x^2)$, notice that since $x^2 \geq 0$, $y \geq 0$. Also, in the second equation $(3x = y^2)$, $y^2 \geq 0$, so $x \geq 0$.

The substitution method can be used again to solve the system of equations

$$3y = x^2$$
$$3x = y^2.$$

The first equation, $3y = x^2$, can be rewritten as $y = x^2/3$. Substitute this into the second equation to get

$$3x = y^2 = \left(\frac{x^2}{3}\right)^2 = \frac{x^4}{9}.$$

Solve this equation as follows.

$$
\begin{array}{ll}
27x = x^4 & \\
x^4 - 27x = 0 & \\
x(x^3 - 27) = 0 & \text{Factor.} \\
x = 0 \quad \text{or} \quad x^3 - 27 = 0 & \text{Set each factor equal to 0.} \\
x^3 = 27 & \\
x = 3 & \text{Take the cube root on each side.}
\end{array}
$$

Use these values of x, along with the equation $3x = y^2$, to find y.

If $x = 0$, If $x = 3$,

$$3x = y^2 \qquad\qquad 3x = y^2$$
$$3(0) = y^2 \qquad\qquad 3(3) = y^2$$
$$0 = y^2 \qquad\qquad 9 = y^2$$
$$0 = y. \qquad\qquad 3 = y \text{ or } -3 = y.$$

The points $(0, 0)$, $(3, 3)$, and $(3, -3)$ appear to be critical points; however, $(3, -3)$ does not satisfy $y \geq 0$. The only possible relative extrema for $f(x, y) = 9xy - x^3 - y^3 - 6$ occur at the critical points $(0, 0)$ or $(3, 3)$. To identify any extrema, use the test. Here

$$f_{xx} = -6x, \qquad f_{yy} = -6y, \qquad \text{and} \qquad f_{xy} = 9.$$

Test each of the possible critical points.

For $(0, 0)$: For $(3, 3)$:

$$f_{xx}(0, 0) = -6(0) = 0 \qquad\qquad f_{xx}(3, 3) = -6(3) = -18$$

$$f_{yy}(0, 0) = -6(0) = 0 \qquad\qquad f_{yy}(3, 3) = -6(3) = -18$$

$$f_{xy}(0, 0) = 9 \qquad\qquad f_{xy}(3, 3) = 9$$

$$D = 0 \cdot 0 - 9^2 = -81. \qquad\qquad D = -18(-18) - 9^2 = 243.$$

Since $D < 0$, there is a saddle point at $(0, 0)$. Here $D > 0$ and $f_{xx}(3, 3) = -18 < 0$; there is a relative maximum at $(3, 3)$. ◀

EXAMPLE 4 A company is developing a new soft drink. The cost in dollars to produce a batch of the drink is approximated by

$$C(x, y) = 2200 + 27x^3 - 72xy + 8y^2,$$

where x is the number of kilograms of sugar per batch and y is the number of grams of flavoring per batch.

(a) Find the amounts of sugar and flavoring that result in minimum cost per batch. Start with the following partial derivatives.

$$C_x = 81x^2 - 72y \qquad \text{and} \qquad C_y = -72x + 16y$$

Set each of these equal to 0 and solve for y.

$$81x^2 - 72y = 0 \qquad\qquad -72x + 16y = 0$$
$$-72y = -81x^2 \qquad\qquad 16y = 72x$$
$$y = \frac{9}{8}x^2 \qquad\qquad y = \frac{9}{2}x$$

From the equation on the left, $y \geq 0$. Since $(9/8)x^2$ and $(9/2)x$ both equal y, they are equal to each other. Set them equal, and solve the resulting equation for x.

$$\frac{9}{8}x^2 = \frac{9}{2}x$$

$$9x^2 = 36x$$

$9x^2 - 36x = 0$	Get 0 on one side.
$9x(x - 4) = 0$	Factor.
$9x = 0 \quad$ or $\quad x - 4 = 0$	Set each factor equal to 0.

The equation $9x = 0$ leads to $x = 0$, which is not a useful answer for the problem. Substitute $x = 4$ into $y = (9/2)x$ to find y.

$$y = \frac{9}{2}x = \frac{9}{2}(4) = 18$$

Now check to see whether the critical point (4, 18) leads to a relative minimum. For (4, 18),

$$C_{xx} = 162x = 162(4) = 648, \quad C_{yy} = 16, \quad \text{and} \quad C_{xy} = -72.$$

Also,

$$D = (648)(16) - (-72)^2 = 5184.$$

Since $D > 0$ and $C_{xx}(4, 18) > 0$, the cost at (4, 18) is a minimum.

(b) What is the minimum cost?

To find the minimum cost, go back to the cost function and evaluate $C(4, 18)$.

$$C(x, y) = 2200 + 27x^3 - 72xy + 8y^2$$
$$C(4, 18) = 2200 + 27(4)^3 - 72(4)(18) + 8(18)^2 = 1336$$

The minimum cost for a batch of soft drink is \$1336.00. ◀

16.3 Exercises

Find all points where the functions defined as follows have any relative extrema. Identify any saddle points.

1. $f(x, y) = xy + x - y$

2. $f(x, y) = 4xy + 8x - 9y$

3. $f(x, y) = x^2 - 2xy + 2y^2 + x - 5$

4. $f(x, y) = x^2 + xy + y^2 - 6x - 3$

5. $f(x, y) = x^2 - xy + y^2 + 2x + 2y + 6$

6. $f(x, y) = x^2 + xy + y^2 + 3x - 3y$

7. $f(x, y) = x^2 + 3xy + 3y^2 - 6x + 3y$

8. $f(x, y) = 5xy - 7x^2 - y^2 + 3x - 6y - 4$

9. $f(x, y) = 4xy - 10x^2 - 4y^2 + 8x + 8y + 9$

10. $f(x, y) = x^2 + xy + 3x + 2y - 6$

11. $f(x, y) = x^2 + xy - 2x - 2y + 2$

12. $f(x, y) = x^2 + xy + y^2 - 3x - 5$

13. $f(x, y) = 2x^3 + 3y^2 - 12xy + 4$

14. $f(x, y) = 5x^3 + 2y^2 - 60xy - 3$

15. $f(x, y) = x^2 + 4y^3 - 6xy - 1$

16. $f(x, y) = 3x^2 + 7y^3 - 42xy + 5$

17. $f(x, y) = e^{xy}$

18. $f(x, y) = x^2 + e^y$

19. Describe the procedure for finding critical points of a function in two independent variables.

20. How are second-order partial derivatives used in finding extrema?

Figures (a)–(f) show the graphs of the functions defined in Exercises 21–26. Find all relative extrema for each function, and then match the equation to its graph.

21. $z = -3xy + x^3 - y^3 + \dfrac{1}{8}$

22. $z = \dfrac{3}{2}y - \dfrac{1}{2}y^3 - x^2y + \dfrac{1}{16}$

23. $z = y^4 - 2y^2 + x^2 - \dfrac{17}{16}$

24. $z = -2x^3 - 3y^4 + 6xy^2 + \dfrac{1}{16}$

25. $z = -x^4 + y^4 + 2x^2 - 2y^2 + \dfrac{1}{16}$

26. $z = -y^4 + 4xy - 2x^2 + \dfrac{1}{16}$

(a)

(b)

(c)

(d)

(e)

(f)

27. Show that $f(x, y) = 1 - x^4 - y^4$ has a relative maximum, even though D in the theorem is 0.

28. Show that $D = 0$ for $f(x, y) = x^3 + (x - y)^2$ and that the function has no relative extrema.

Applications

Business and Economics

29. Profit Suppose that the profit of a certain firm (in hundreds of dollars) is approximated by

$$P(x, y) = 1000 + 24x - x^2 + 80y - y^2,$$

where x is the cost of a unit of labor and y is the cost of a unit of goods. Find values of x and y that maximize profit. Find the maximum profit.

30. Labor Costs Suppose the labor cost (in dollars) for manufacturing a precision camera can be approximated by

$$L(x, y) = \frac{3}{2}x^2 + y^2 - 2x - 2y - 2xy + 68,$$

where x is the number of hours required by a skilled craftsperson and y is the number of hours required by a semiskilled person. Find values of x and y that minimize the labor cost. Find the minimum labor cost.

31. Cost The total cost (in dollars) to produce x units of electrical tape and y units of packing tape is given by

$$C(x, y) = 2x^2 + 3y^2 - 2xy + 2x - 126y + 3800.$$

Find the number of units of each kind of tape that should be produced so that the total cost is a minimum. Find the minimum total cost.

32. Revenue The total revenue (in hundreds of dollars) from the sale of x spas and y solar heaters is approximated by

$$R(x, y) = 12 + 74x + 85y - 3x^2 - 5y^2 - 5xy.$$

Find the number of each that should be sold to produce maximum revenue. Find the maximum revenue.

16.4 LAGRANGE MULTIPLIERS; CONSTRAINED OPTIMIZATION

What dimensions for a new building will maximize the floor space at a fixed cost?

In the previous section, we showed how partial derivatives are used to solve extrema problems. Another approach that works well when there is an additional restriction, called a **constraint**, uses an additional variable, called the **Lagrange multiplier**. For example, in the opening question, suppose a builder wants to maximize the floor space in a new building while keeping the costs fixed at $500,000. The building will be 40 feet high, with a rectangular floor plan and three stories. The costs, which depend on the dimensions of the rectangular floor plan, are given by

$$\text{Costs} = xy + 20y + 20x + 474{,}000,$$

where x is the width and y the length of the rectangle. Thus, the builder wishes to maximize the area $A(x, y) = xy$ and satisfy the condition

$$xy + 20y + 20x + 474{,}000 = 500{,}000.$$

In addition to maximizing area, then, the builder must keep costs at (or below) $500,000.

A typical problem of this type might require the smallest possible value of the function $z = x^2 + y^2$, subject to the constraint $x + y = 4$. To see how to find this minimum value, we might first graph both the surface $z = x^2 + y^2$ and the plane $x + y = 4$, as in Figure 17. The required minimum value is found on the curve formed by the intersection of the two graphs.

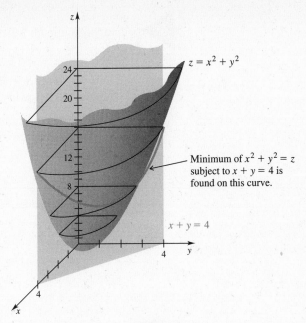

$z = x^2 + y^2$

Minimum of $x^2 + y^2 = z$
subject to $x + y = 4$ is
found on this curve.

$x + y = 4$

FIGURE 17

Problems with constraints are often solved by the method of Lagrange multipliers, named for the French mathematician Joseph Louis Lagrange (1736–1813). The proof for the method is complicated and is not given here. The method of Lagrange multipliers is used for problems of the form:

Find the relative extrema for $z = f(x, y)$,

subject to $g(x, y) = 0$.

We state the method only for functions of two independent variables, but it is valid for any number of variables.

LAGRANGE MULTIPLIERS All relative extrema of the function $z = f(x, y)$, subject to a constraint $g(x, y) = 0$, will be found among those points (x, y) for which there exists a value of λ such that

$$F_x(x, y, \lambda) = 0, \quad F_y(x, y, \lambda) = 0, \quad F_\lambda(x, y, \lambda) = 0,$$

where $$F(x, y, \lambda) = f(x, y) + \lambda \cdot g(x, y),$$

and all indicated partial derivatives exist.

In the theorem, the function $F(x, y, \lambda) = f(x, y) + \lambda \cdot g(x, y)$ is called the Lagrange function; λ, the Greek letter *lambda*, is the *Lagrange multiplier*.

EXAMPLE 1

Find the minimum value of

$$f(x, y) = 5x^2 + 6y^2 - xy,$$

subject to the constraint $x + 2y = 24$.

Go through the following steps.

Step 1 Rewrite the constraint in the form $g(x, y) = 0$.

In this example, the constraint $x + 2y = 24$ becomes

$$x + 2y - 24 = 0,$$

with $g(x, y) = x + 2y - 24.$

Step 2 Form the Lagrange function $F(x, y, \lambda)$, the sum of the function $f(x, y)$ and the product of λ and $g(x, y)$.

Here,

$$\begin{aligned} F(x, y, \lambda) &= f(x, y) + \lambda \cdot g(x, y) \\ &= 5x^2 + 6y^2 - xy + \lambda(x + 2y - 24) \\ &= 5x^2 + 6y^2 - xy + \lambda x + 2\lambda y - 24\lambda. \end{aligned}$$

Step 3 Find F_x, F_y, and F_λ.

$$\begin{aligned} F_x &= 10x - y + \lambda \\ F_y &= 12y - x + 2\lambda \\ F_\lambda &= x + 2y - 24 \end{aligned}$$

Step 4 Form the system of equations $F_x = 0$, $F_y = 0$, and $F_\lambda = 0$.

$$\begin{aligned} 10x - y + \lambda &= 0 & \text{(1)} \\ 12y - x + 2\lambda &= 0 & \text{(2)} \\ x + 2y - 24 &= 0 & \text{(3)} \end{aligned}$$

Step 5 Solve the system of equations from Step 4 for x, y, and λ.

Solve each of the first two equations for λ, then set the two results equal and simplify, as follows.

$$10x - y + \lambda = 0 \qquad \text{becomes} \qquad \lambda = -10x + y$$

$$12y - x + 2\lambda = 0 \qquad \text{becomes} \qquad \lambda = \frac{x - 12y}{2}$$

$$-10x + y = \frac{x - 12y}{2} \qquad \text{Set the expressions for } \lambda \text{ equal.}$$

$$-20x + 2y = x - 12y$$

$$-21x = -14y$$

$$x = \frac{2y}{3}$$

Now substitute $(2y)/3$ for x in equation (3).

$$x + 2y - 24 = 0$$

$$\frac{2y}{3} + 2y - 24 = 0 \quad \text{Let } x = \frac{2y}{3}.$$

$$2y + 6y - 72 = 0$$

$$8y = 72$$

$$y = 9$$

Since $x = (2y)/3$ and $y = 9$, $x = 6$. It is not necessary to find the value of λ.

Thus, the minimum value for $f(x, y) = 5x^2 + 6y^2 - xy$, subject to the constraint $x + 2y = 24$, is at the point (6, 9). The minimum value is $f(6, 9) = 612$. The second derivative test for relative extrema from the previous section can be used to verify that 612 is indeed a minimum: since $f_{xx} = 10$, $f_{yy} = 12$, and $f_{xy} = -1$, $D = 10 \cdot 12 - (-1)^2 > 0$, so part (b) applies and indicates a minimum. ◄

Note In Example 1, we solved the system of equations by solving each equation with λ in it for λ. We then set these expressions for λ equal and solved for one of the original variables. This is a good general approach to use in solving these systems of equations, since we are not interested in the value of λ.

Before looking at applications of Lagrange multipliers, let us summarize the steps involved in solving a problem by this method.

USING LAGRANGE MULTIPLIERS

1. Write the constraint in the form $g(x, y) = 0$.
2. Form the Lagrange function

$$F(x, y, \lambda) = f(x, y) + \lambda \cdot g(x, y).$$

3. Find F_x, F_y, and F_λ.
4. Form the system of equations

$$F_x = 0, \quad F_y = 0, \quad F_\lambda = 0.$$

5. Solve the system in Step 4; the relative extrema for f are among the solutions of the system.

EXAMPLE 2 Complete the solution of the problem given in the introduction to this section. Maximize the area, $A(x, y) = xy$, subject to the cost constraint

$$xy + 20y + 20x + 474{,}000 = 500{,}000.$$

Go through the five steps presented above.

Step 1 $g(x, y) = xy + 20y + 20x - 26{,}000 = 0$

Step 2 $F(x, y, \lambda) = xy + \lambda(xy + 20y + 20x - 26{,}000)$

Step 3 $F_x = y + \lambda y + 20\lambda$

$F_y = x + \lambda x + 20\lambda$

$F_\lambda = xy + 20y + 20x - 26{,}000$

Step 4 $y + \lambda y + 20\lambda = 0$ **(6)**

$x + \lambda x + 20\lambda = 0$ **(7)**

$xy + 20y + 20x - 26{,}000 = 0$ **(8)**

Step 5 Solving equations (6) and (7) for λ gives

$$\lambda = \frac{-y}{y + 20} \quad \text{and} \quad \lambda = \frac{-x}{x + 20}$$

$$\frac{-y}{y + 20} = \frac{-x}{x + 20}$$

$$y(x + 20) = x(y + 20)$$

$$xy + 20y = xy + 20x$$

$$x = y.$$

Now substitute y for x in equation (8) to get

$$y^2 + 20y + 20y - 26{,}000 = 0$$

$$y^2 + 40y - 26{,}000 = 0.$$

Use the quadratic formula to find $y \approx 142.5$. Since $x = y$, $x \approx 142.5$.

The maximum area of $(142.5)^2 \approx 20{,}306$ square feet will be achieved if the floor plan is a square with a side of 142.5 feet. ◀

As mentioned earlier, the method of Lagrange multipliers works for more than two independent variables. The next example shows how to find extrema for a function of three independent variables.

EXAMPLE 3 Find the dimensions of the closed rectangular box of maximum volume that can be produced from 6 square feet of material.

Let x, y, and z represent the dimensions of the box, as shown in Figure 18. The volume of the box is given by

$$f(x, y, z) = xyz.$$

FIGURE 18

As shown in Figure 18, the total amount of material required for the two ends of the box is $2xy$, the total needed for the sides is $2xz$, and the total needed for the top and bottom is $2yz$. Since 6 square feet of material is available,

$$2xy + 2xz + 2yz = 6 \quad \text{or} \quad xy + xz + yz = 3.$$

In summary, $f(x, y, z) = xyz$ is to be maximized subject to the constraint $xy + xz + yz = 3$. Go through the steps that were given above.

Step 1 $g(x, y, z) = xy + xz + yz - 3 = 0$

Step 2 $F(x, y, z, \lambda) = xyz + \lambda(xy + xz + yz - 3)$

Step 3 $F_x = yz + \lambda y + \lambda z$

$F_y = xz + \lambda x + \lambda z$

$F_z = xy + \lambda x + \lambda y$

$F_\lambda = xy + xz + yz - 3$

Step 4 $yz + \lambda y + \lambda z = 0$

$xz + \lambda x + \lambda z = 0$

$xy + \lambda x + \lambda y = 0$

$xy + xz + yz - 3 = 0$

Step 5 Solve each of the first three equations for λ. You should get

$$\lambda = \frac{-yz}{y + z}, \qquad \lambda = \frac{-xz}{x + z}, \qquad \text{and} \qquad \lambda = \frac{-xy}{x + y}.$$

Set these expressions for λ equal, and simplify as follows.

$$\frac{-yz}{y + z} = \frac{-xz}{x + z} \qquad \text{and} \qquad \frac{-xz}{x + z} = \frac{-xy}{x + y}$$

$$\frac{y}{y + z} = \frac{x}{x + z} \qquad\qquad \frac{z}{x + z} = \frac{y}{x + y}$$

$$xy + yz = xy + xz \qquad\qquad zx + zy = yx + yz$$

$$yz = xz \qquad\qquad\qquad zx = yx$$

$$y = x \qquad\qquad\qquad\quad z = y$$

(Setting the first and third expressions equal gives no additional information.) Thus $x = y = z$. From the fourth equation in Step 4, with $x = y$ and $z = y$,

$$xy + xz + yz - 3 = 0$$
$$y^2 + y^2 + y^2 - 3 = 0$$
$$3y^2 = 3$$
$$y^2 = 1$$
$$y = \pm 1.$$

The negative solution is not applicable, so the solution of the system of equations is $x = 1, y = 1, z = 1$. In other words, the box is a cube that measures 1 foot on a side. ◀

16.4 Exercises

Find the relative maxima or minima in Exercises 1–10.

1. Maximum of $f(x, y) = 2xy$, subject to $x + y = 12$
2. Maximum of $f(x, y) = 4xy + 2$, subject to $x + y = 24$
3. Maximum of $f(x, y) = x^2y$, subject to $2x + y = 4$
4. Maximum of $f(x, y) = 4xy^2$, subject to $3x - 2y = 5$
5. Minimum of $f(x, y) = x^2 + 2y^2 - xy$, subject to $x + y = 8$
6. Minimum of $f(x, y) = 3x^2 + 4y^2 - xy - 2$, subject to $2x + y = 21$
7. Maximum of $f(x, y) = x^2 - 10y^2$, subject to $x - y = 18$
8. Maximum of $f(x, y) = 12xy - x^2 - 3y^2$, subject to $x + y = 16$
9. Maximum of $f(x, y, z) = xyz^2$, subject to $x + y + z = 6$
10. Maximum of $f(x, y, z) = xy + 2xz + 2yz$, subject to $xyz = 32$

11. Find positive numbers x and y such that $x + y = 18$ and xy^2 is maximized.
12. Find positive numbers x and y such that $x + y = 36$ and x^2y is maximized.
13. Find three positive numbers whose sum is 90 and whose product is a maximum.
14. Find three positive numbers whose sum is 240 and whose product is a maximum.
15. Explain the difference between the two methods we used in Sections 3 and 4 to solve extrema problems.
16. Why is it unnecessary to find the value of λ when using the method explained in this section?

Applications

Business and Economics

17. **Maximum Area for Fixed Expenditure** Because of terrain difficulties, two sides of a fence can be built for $6 per foot, while the other two sides cost $4 per foot. (See the sketch.) Find the field of maximum area that can be enclosed for $1200.

$6 per foot

$4 per foot — $6 per foot

$4 per foot

18. **Maximum Area for Fixed Expenditure** To enclose a yard, a fence is built against a large building, so that fencing material is used only on three sides. Material for the ends cost $8 per foot; material for the side opposite the building costs $6 per foot. Find the dimensions of the yard of maximum area that can be enclosed for $1200.

19. **Cost** The total cost to produce x large needlepoint kits and y small ones is given by
$$C(x, y) = 2x^2 + 6y^2 + 4xy + 10.$$

If a total of ten kits must be made, how should production be allocated so that total cost is minimized?

20. **Profit** The profit from the sale of x units of radiators for automobiles and y units of radiators for generators is given by
$$P(x, y) = -x^2 - y^2 + 4x + 8y.$$
Find values of x and y that lead to a maximum profit if the firm must produce a total of 6 units of radiators.

21. **Production** A manufacturing firm estimates that its total production of automobile batteries in thousands of units is
$$f(x, y) = 3x^{1/3}y^{2/3},$$
where x is the number of units of labor and y is the number of units of capital utilized. Labor costs are $80 per unit, and capital costs are $150 per unit. How many units each of labor and capital will maximize production, if the firm can spend $40,000 for these costs?

22. **Production** For another product, the manufacturing firm in Exercise 21 estimates that production is a function of labor x and capital y as follows:
$$f(x, y) = 12x^{3/4}y^{1/4}.$$
If $25,200 is available for labor and capital, and if the firm's costs are $100 and $180 per unit, respectively, how many units of labor and capital will give maximum production?

General Interest

23. **Area** A farmer has 200 m of fencing. Find the dimensions of the rectangular field of maximum area that can be enclosed by this amount of fencing.

24. **Area** Find the area of the largest rectangular field that can be enclosed with 600 m of fencing. Assume that no fencing is needed along one side of the field.

25. **Surface Area** A cylindrical can is to be made that will hold 250π cubic inches of candy. Find the dimensions of the can with minimum surface area.

26. **Surface Area** An ordinary 12-oz beer or soda pop can holds about 25 cubic inches. Use a calculator and find the dimensions of a can with minimum surface area. Measure a can and see how close its dimensions are to the results you found.

27. **Volume** A rectangular box with no top is to be built from 500 square meters of material. Find the dimensions of such a box that will enclose the maximum volume.

28. **Surface Area** A 1-lb soda cracker box has a volume of 185 cubic inches. The end of the box is square. Find the dimensions of such a box that has minimum surface area.

29. **Cost** A rectangular closed box is to be built at minimum cost to hold 27 cubic meters. Since the cost will depend on the surface area, find the dimensions that will minimize the surface area of the box.

30. **Cost** Find the dimensions that will minimize the surface area (and hence the cost) of a rectangular fish aquarium, open on top, with a volume of 32 cubic feet.

16.5 THE LEAST SQUARES LINE: A MINIMIZATION APPLICATION

How can we express the relationship between the size of a room and the energy needed to cool the room?

Throughout this book we have studied functions that relate various quantities. In almost every case, the function was given. An important question is, How are these functions determined? This section shows how one kind of function (linear) can be developed. We will return to the introductory question in the exercises for this section.

In trying to predict the future sales of a product or the total number of matings between two species of insects, it is common to gather as much data as possible and then draw a graph showing the data. This graph, called a **scatter diagram,** can then be inspected to see if a reasonably simple mathematical curve will fit fairly well through all the given points.

As an example, suppose a firm gathers data showing the relationship between the price y of an item (in dollars) and the number x of units sold, with results as follows.

Units Sold, x	10	15	20
Price, y	80	68	62

A graph of these data is shown in Figure 19.

FIGURE 19

The graph suggests that a straight line fits reasonably well through these data points. If all the data points were to lie on the straight line, the point-slope form of the equation of a line could be used to find the equation of the line through the points.

In practice, the points on a scatter diagram almost never fit a straight line exactly. Usually we must decide on the "best" straight line through the points. One way to define the "best" line is as follows. Figure 20 shows the same scattered points from Figure 19. This time, however, vertical line segments indicate the distances of these points from a possible line. The "best" straight line is often defined as the one for which the sum of the squares of the distances, $d_1{}^2 + d_2{}^2 + d_3{}^2$ here, is minimized.

FIGURE 20

To find the equation of the "best" line, let $y' = mx + b$ be the equation of the line in Figure 20. (The purpose of using y' here is to distinguish the calculated y-values from the given y-values. This use of y' has nothing to do with the derivative.) Then for the point with distance d_1, $x = 10$ and $y = 80$, so that $y' = mx + b$ becomes $y_1{}' = 10m + b$, and

$$d_1 = 80 - y_1{}' = 80 - (10m + b).$$

We can find d_2 and d_3 in the same way. Thus,

$$d_1{}^2 = [80 - (10m + b)]^2,$$
$$d_2{}^2 = [68 - (15m + b)]^2,$$
$$d_3{}^2 = [62 - (20m + b)]^2.$$

Values of m and b must be found that will minimize $d_1{}^2 + d_2{}^2 + d_3{}^2$. If $d_1{}^2 + d_2{}^2 + d_3{}^2 = S$, then

$$S = [80 - (10m + b)]^2 + [68 - (15m + b)]^2 + [62 - (20m + b)]^2.$$

To minimize S using the method shown earlier in this chapter, the partial derivatives $\partial S/\partial m$ and $\partial S/\partial b$ are needed.

$$\frac{\partial S}{\partial m} = -20[80 - (10m + b)] - 30[68 - (15m + b)] - 40[62 - (20m + b)]$$

$$\frac{\partial S}{\partial b} = -2[80 - (10m + b)] - 2[68 - (15m + b)] - 2[62 - (20m + b)]$$

Simplify each of these, obtaining

$$\frac{\partial S}{\partial m} = -6120 + 1450m + 90b$$

and

$$\frac{\partial S}{\partial b} = -420 + 90m + 6b.$$

Set $\partial S/\partial m = 0$ and $\partial S/\partial b = 0$ and simplify.

$$-6120 + 1450m + 90b = 0 \quad \text{or} \quad 145m + 9b = 612$$
$$-420 + 90m + 6b = 0 \quad \text{or} \quad 15m + b = 70$$

Solving the system on the right leads to the solution $m = -1.8$, $b = 97$. The equation that best fits through the points in Figure 19 is $y' = mx + b$, or

$$y' = -1.8x + 97.$$

The second derivative test can be used to show that this line *minimizes* the sum of the squares of the distances. Thus, $y' = -1.8x + 97$ is the "best" straight line that will fit through the data points. This line is called the **least squares line,** or sometimes the **regression line.**

Once a least squares line is obtained from a set of data, the equation can be used to predict a value of one variable, given a value of the other. In the example above, management might want an estimate of the number of units that would be sold at a price of $75. To get this estimate, replace y' with 75.

$$y' = -1.8x + 97$$
$$75 = -1.8x + 97$$
$$-22 = -1.8x$$
$$12.2 \approx x$$

About 12 units would be sold at a price of $75.

The example above used three data points on the scatter diagram. A practical problem, however, might well involve a very large number of points. The method used above can be extended to find the least squares line for any finite number of data points. To simplify the notation for this more complicated case, *summation notation,* or *sigma notation*, is used as explained in the chapter on integration.

$$\sum_{i=1}^{n} x_i = x_1 + x_2 + x_3 + \cdots + x_n$$

Let $y' = mx + b$ be the least squares line for the set of known data points $(x_1, y_1), (x_2, y_2), \ldots, (x_n, y_n)$, See Figure 21. As above, the sum of squares,

$$d_1^2 + d_2^2 + \cdots + d_n^2,$$

is to be minimized. The square of the distance of the first point (x_1, y_1) from the line $y' = mx + b$ is given by

$$d_1^2 = [y_1 - (mx_1 + b)]^2.$$

The square of the distance of the point (x_i, y_i) from the line is

$$d_i^2 = [y_i - (mx_i + b)]^2.$$

Thus, the sum

$$\sum_{i=1}^{n} [y_i - (mx_i + b)]^2$$

is to be minimized.

FIGURE 21

Since the x_i and y_i values represent known data points, the unknowns in this sum are the numbers m and b. To emphasize this fact, write the sum as a function of m and b:

$$f(m, b) = \sum_{i=1}^{n} [y_i - (mx_i + b)]^2$$

$$= \sum_{i=1}^{n} [y_i - mx_i - b]^2$$

$$= (y_1 - mx_1 - b)^2 + (y_2 - mx_2 - b)^2$$
$$+ \cdots + (y_n - mx_n - b)^2.$$

To find the minimum value of this function, find the partial derivatives with respect to m and to b and set each equal to 0. (Recall that all the x's and y's are constants here.)

$$\frac{\partial f}{\partial m} = -2x_1(y_1 - mx_1 - b) - 2x_2(y_2 - mx_2 - b) - \cdots$$

$$-2x_n(y_n - mx_n - b) = 0$$

$$\frac{\partial f}{\partial b} = -2(y_1 - mx_1 - b) - 2(y_2 - mx_2 - b) - \cdots$$

$$-2(y_n - mx_n - b) = 0$$

Using some algebra and rearranging terms, these two equations become

$$(x_1^2 + x_2^2 + \cdots + x_n^2)m + (x_1 + x_2 + \cdots + x_n)b$$
$$= x_1y_1 + x_2y_2 + \cdots + x_ny_n,$$

and $\quad (x_1 + x_2 + \cdots + x_n)m + nb = y_1 + y_2 + \cdots + y_n.$

These last two equations can be rewritten using abbreviated sigma notation as follows. (Remember that n terms are being added.)

$$\left(\sum x^2\right)m + \left(\sum x\right)b = \sum xy \tag{1}$$

$$\left(\sum x\right)m + nb = \sum y \tag{2}$$

To solve this system of equations, multiply the first equation on both sides by $-n$ and the second on both sides by Σx.

$$-n\left(\sum x^2\right)m - n\left(\sum x\right)b = -n\left(\sum xy\right)$$

$$\left(\sum x\right)\left(\sum x\right)m + \left(\sum x\right)nb = \left(\sum x\right)\left(\sum y\right)$$

Add the two new equations to eliminate the term with b.

$$\left(\sum x\right)\left(\sum x\right)m - n\left(\sum x^2\right)m = \left(\sum x\right)\left(\sum y\right) - n\left(\sum xy\right)$$

Write the product $(\Sigma x)(\Sigma x)$ as $(\Sigma x)^2$. Using this notation and solving the last equation for m gives

$$m = \frac{\left(\sum x\right)\left(\sum y\right) - n\left(\sum xy\right)}{\left(\sum x\right)^2 - n\left(\sum x^2\right)}. \tag{3}$$

The easiest way to find b is to solve equation (2) for b:

$$\left(\sum x\right)m + nb = \sum y \tag{2}$$

$$nb = \sum y - \left(\sum x\right)m$$

$$b = \frac{\sum y - m\left(\sum x\right)}{n}.$$

A summary of the formulas for m and b in the least squares equation follows.

LEAST SQUARES EQUATION

The **least squares equation** for the n points $(x_1, y_1), (x_2, y_2), \ldots, (x_n, y_n)$ is given by

$$y' = mx + b,$$

where

$$m = \frac{(\sum x)(\sum y) - n(\sum xy)}{(\sum x)^2 - n(\sum x^2)} \quad \text{and} \quad b = \frac{\sum y - m(\sum x)}{n}.$$

As the formulas suggest, find m first. Next, use the value of m and the second formula to find b.

 EXAMPLE 1

Find the least squares line for the data in the following chart.

x	1	2	3	5	6
y	-2	2	5	12	14

Start by drawing the scatter diagram of Figure 22. Since the points in the scatter diagram lie approximately in a straight line, it is appropriate to find the least squares line for the data.

The formulas for m and b require $\sum x$, $\sum y$, $\sum x^2$, and $\sum xy$. Organize the work as follows.

x	y	x^2	xy
1	-2	1	-2
2	2	4	4
3	5	9	15
5	12	25	60
6	14	36	84
Totals 17	31	75	161

FIGURE 22

FIGURE 23

The chart shows that $\Sigma x = 17$, $\Sigma y = 31$, $\Sigma x^2 = 75$, $\Sigma xy = 161$, and $n = 5$. Using the equation for m given above,

$$m = \frac{17(31) - 5(161)}{17^2 - 5(75)}$$

$$= \frac{527 - 805}{289 - 375} = \frac{-278}{-86} \approx 3.2.$$

Now find b:

$$b = \frac{\Sigma y - m(\Sigma x)}{n}$$

$$= \frac{31 - 3.2(17)}{5} \approx -4.7.$$

The least squares equation for the given data is $y' = mx + b$, or

$$y' = 3.2x - 4.7.$$

This line is graphed in Figure 23. ◀

The equation $y' = 3.2x - 4.7$ can be used to estimate values of y for given values of x. For example, if $x = 4$,

$$y' = 3.2x - 4.7 = 3.2(4) - 4.7 = 12.8 - 4.7 = 8.1.$$

Also, if $x = 7$, then $y' = 3.2(7) - 4.7 = 17.7$.

Sir Francis Galton (1822–1911), a British scientist, is credited with setting forth the ideas of regression and correlation in his statistical work *Natural Inheritance*. His interest in statistics developed as a result of his research on weather prediction.

16.5 Exercises

1. Suppose a linear correlation is found between two quantities. Does this mean that increasing one of the quantities causes the other to increase? If not, what does it mean?

2. Given a set of points, the least squares line formed by letting x be the independent variable will not necessarily be the same as the least squares line formed by letting y be the independent variable. Give an example to show why this is true.

In Exercises 3 and 4, draw a scatter diagram for the given set of data points and then find the least squares equation.

3.

x	4	5	8	12	14
y	3	7	17	28	35

Estimate x when y is 12; 32.

4.

x	3	4	5	6	8
y	8	12	16	18	28

Estimate x when y is 17; 26.

Applications

Business and Economics

5. Sales and Profit A fast-food chain wishes to find the relationship between annual store sales (in thousands of dollars) and percent of pretax profit in order to estimate increases in profit due to increased sales volume. The data shown below were obtained from a sample of stores across the country.

Annual Sales, x	Pretax Profit, y
250	9.3
300	10.8
375	14.8
425	14.0
450	14.2
475	15.3
500	15.9
575	19.1
600	19.2
650	21.0

(a) Plot the ten pairs of values on a scatter diagram.

(b) Find the equation of the least squares line and graph it on the scatter diagram of part (a).

(c) Using the equation of part (b), predict the percent of pretax profit for annual sales of $700,000; of $750,000.

6. Insurance Sales Records show that the annual sales for the Sweet Palms Life Insurance Company in five-year periods for the last 20 years were as follows.

Year	Year in Coded Form, x	Sales in Millions, y
1970	1	1.0
1975	2	1.3
1980	3	1.7
1985	4	1.9
1990	5	2.1

(a) Plot the five points on a scatter diagram.

(b) Find the least squares equation and graph it on the scatter diagram from part (a).

(c) Predict the company's sales for 1995.

7. Repair Costs and Production The following data were used to determine whether there is a relationship between repair costs and barrels of beer produced by a brewery. The data (in thousands) are given for a 10-month period.

Month	Barrels of Beer, x	Repairs, y
Jan.	369	299
Feb.	379	280
Mar.	482	393
April	493	388
May	496	385
June	567	423
July	521	374
Aug.	482	357
Sept.	391	106
Oct.	417	332

(a) Find the equation of the least squares line.

(b) If 500,000 barrels of beer are produced, what will the equation from part (a) give as the predicted repair costs?

(c) Is there a relationship between repair costs and barrels of beer produced? If so, what kind of relationship exists?

8. Sales Analysis Sales (in thousands of dollars) of a certain company are shown here.

Year, x	0	1	2	3	4	5
Sales, y	48	59	66	75	80	90

(a) Find the equation of the least squares line.

(b) Predict sales in year 7.

It is sometimes possible to get a better prediction for a variable by considering its relationship with more than one other variable. For example, one should be able to predict college GPAs more precisely if both high school GPAs and scores on the ACT are considered. To do this, we alter the equation used to find a least squares line by adding a term for the new variable as follows. If y represents college GPAs, x_1 high school GPAs, and x_2 ACT scores, then y', the predicted GPA, is given by

$$y' = ax_1 + bx_2 + c.$$

This equation represents a **least squares plane.** The equations for the constants a, b, and c are more complicated than those given in the text for m and b, so calculating a least squares equation for three variables is more likely to require the aid of a computer.

9. **Revenue and Price** Alcoa* used a least squares plane with two independent variables, x_1 and x_2, to predict the effect on revenue of the price of aluminum forged truck wheels, as follows.

x_1 = the average price per wheel

x_2 = large truck production (in thousands)

y = sales of aluminum forged truck wheels (in thousands)

Using data for the past eleven years, the company found the equation of the least squares plane to be

$$y' = 49.2755 - 1.1924x_1 + .1631x_2.$$

The following figures were then forecast for truck production.

1987	1988	1989	1990	1991	1992
160.0	165.0	170.0	175.0	180.0	185.0

Three possible price levels per wheel were considered: $42, $45, and $48.

(a) Use the least squares plane equation given above to find the estimated sales of wheels (y) for 1989 at each of the three price levels.

(b) Repeat part (a) for 1992.

(c) For which price level, on the basis of the 1989 and 1992 figures, are total estimated sales greatest?

(By comparing total estimated sales for the years 1987 through 1992 at each of the three price levels, the company found that the selling price of $42 per wheel would generate the greatest sales volume over the six-year period.)

Life Sciences

10. **Crickets Chirping** Biologists have observed a linear relationship between the temperature and the frequency with which a cricket chirps. The following data were measured for the striped ground cricket.[†]

*This example supplied by John H. Van Denender, Public Relations Department, Aluminum Company of America.
[†]From George W. Pierce, *The Songs of Insects* (Harvard University Press, 1948), p. 20.

Temperature in °F (x)	Chirps per second (y)
88.6	20.0
71.6	16.0
93.3	19.8
84.3	18.4
80.6	17.1
75.2	15.5
69.7	14.7
82.0	17.1
69.4	15.4
83.3	16.2
79.6	15.0
82.6	17.2
80.6	16.0
83.5	17.0
76.3	14.4

(a) Find the least squares line for these data.

(b) Use the results of part (a) to determine how many chirps per second you would expect to hear from the striped ground cricket if the temperature is 73° F.

(c) Use the results of part (a) to determine what the temperature is when the striped ground cricket is chirping 18 times per second.

11. **Bacterial Growth** Sometimes the scatter diagram of the data does not have a linear pattern. This is particularly true in some biological and chemical applications. In these applications, however, often the scatter diagram of the *logarithms* of the data has a linear pattern. A least squares line can then be used to predict the logarithm of any desired value, from which the value itself can be found. Suppose that a certain kind of bacterium grows in numbers as shown in Table A. The actual number of bacteria present at each time period is replaced with the common logarithm of that number in Table B.

Table A

Time (in hours)	Number of Bacteria
0	1000
1	1649
2	2718
3	4482
4	7389
5	12182

Table B

Time, x	log, y
0	3.0000
1	3.2172
2	3.4343
3	3.6515
4	3.8686
5	4.0857

We can now find a least squares line that will predict *y*, given *x*.

(a) Plot the original pairs of numbers. The pattern should be nonlinear.

(b) Plot the log values against the time values. The pattern should be almost linear.

(c) Find the equation of the least squares line. (First round off the log values to the nearest hundredth.)

(d) Predict the log value for a time of 7 hr. Find the number whose logarithm is your answer. This will be the predicted number of bacteria.

12. **Height and Weight** A sample of 10 adult men gave the following data on their heights (in inches) and weights (in pounds).

Height, x	62	62	63	65	66	67	68	68	70	72
Weight, y	120	140	130	150	142	130	135	175	149	168

(a) Find the equation of the least squares line.

Using the results above, predict the weight of a man whose height is as follows.

(b) 60 inches (c) 70 inches

Social Sciences

13. **Educational Expenditures** A report issued by the U.S. Department of Education in 1991 gave the expenditure per pupil and the average mathematics proficiency in grade 8 for 37 states and the District of Columbia. Letting *x* equal the expenditure per pupil (ranging from $2838 in Idaho to $7850 in Washington, D.C.) and *y* equal the average mathematics proficiency score (ranging from 231 in Washington, D. C. to 281 in North Dakota), the data can be summarized as follows. (From an article in *Newsday*, June 7, 1991.)

$$n = 38 \qquad \Sigma x = 175{,}878 \qquad \Sigma y = 9989$$
$$\Sigma xy = 46{,}209{,}266 \qquad \Sigma x^2 = 872{,}066{,}218$$

(a) Find the least squares equation for these data.

(b) Assuming a linear relationship, estimate the average mathematics proficiency scores for Idaho and Washington, D.C.

(c) What does the negative sign in the equation indicate?

14. **Death Rate** The death rate in the U.S. has gradually decreased between 1910 and 1988, as shown in the data at the top of the next column.

Year, x	Death Rate, y
1910	14.7
1920	13.0
1930	11.3
1940	10.8
1950	9.6
1960	9.5
1970	9.5
1980	8.8
1988	8.8

(a) Assume this is a linear decrease and find the least squares equation for the data. Use 10 for 1910, 20 for 1920, and so on.

(b) Using the equation from part (a), find the death rates for 1920, 1950, and 1980.

(c) Compare the results from part (b) with the given data. Is the linear equation a reasonable fit? Why?

(d) Use the equation from part (a) to predict the death rate in 1994.

15. **SAT Scores** At Hofstra University, all students take the SAT before entrance, and most students take a mathematics placement test (designed by the Mathematical Association of America) before registration. In Fall 1991, one professor collected the following data for 19 students in his finite mathematics class.

Math SAT	Placement Test	Math SAT	Placement Test	Math SAT	Placement Test
540	20	580	8	440	10
510	16	680	15	520	11
490	10	560	8	620	11
560	8	560	13	680	8
470	12	500	14	550	8
600	11	470	10	620	7
540	10				

(a) Find an equation for the least squares line.

(b) Use your answer from part (a) to predict the mathematics placement test score for a student with a math SAT score of 420.

(c) Use your answer from part (a) to predict the mathematics placement test score for a student with a math SAT score of 620.

Physical Sciences

16. Temperature In an experiment to determine the linear relationship between temperatures on the Celsius scale (y) and on the Fahrenheit scale (x), a student got the following results.

$$n = 5 \qquad \Sigma x = 376 \qquad \Sigma y = 120$$
$$\Sigma xy = 28,050 \qquad \Sigma x^2 = 62,522$$

(a) Find an equation for the least squares line.

(b) Find the reading on the Celsius scale that corresponds to a reading of 120° Fahrenheit, using the equation from part (a).

General Interest

17. Air Conditioning While shopping for a new air conditioner, Adam Bryer consulted the following table to determine the right model to buy. The table shows the number of BTUs needed to cool rooms having specified areas. Find a least squares equation for these data.

Area of Room in Square Feet, x	BTUs of Air Conditioner Needed, y
150	5000
175	5500
215	6000
250	6500
280	7000
310	7500
350	8000
370	8500
420	9000
450	9500

(a) To check the fit of the data to the line, use the equation to find the number of BTUs required to cool rooms of 150 ft², 280 ft², 420 ft². How well do the actual data agree with the predicted values?

(b) Suppose your room measures 230 ft². Use the equation to decide how many BTUs are required to cool the air. If air conditioners are available only with the BTU choices in the table, how many BTUs do you need?

(c) Why do you think the table gives ft² instead of cubic feet, which would give the volume of the room?

16.6 TOTAL DIFFERENTIALS AND APPROXIMATIONS

How will a change in the production of fertilizer affect profit?

In the chapter on applications of the derivative, we introduced the differential. Recall that the differential of a function defined by $y = f(x)$ is

$$dy = f'(x) \cdot dx,$$

where dx, the differential of x, is any real number (usually small). We saw that the differential dy is often a good approximation of Δy, where $\Delta y = f(x + \Delta x) - f(x)$ and $\Delta x = dx$.

In the second section of this chapter we used partial derivatives to find the marginal productivity of labor and of capital for a production function. The marginal productivity gives the rate of change of production for a one-unit change in labor or for a one-unit change in capital. To find the change in productivity for any (usually small) change in labor and/or in capital, we need the concept of a *differential*. Since the differential dy equals $f'(x)\,dx$, it is a function of *two* variables, x and dx. We can extend this idea by defining a *total differential*.

TOTAL DIFFERENTIAL FOR TWO VARIABLES

Let $z = f(x, y)$ be a function of x and y. Let dx and dy be real numbers. Then the **total differential** of f is

$$df = f_x(x, y) \cdot dx + f_y(x, y) \cdot dy.$$

(Sometimes df is written dz.)

By this definition, df is a function of *four* variables, x, dx, y, and dy.

 1 Find df for each function.

(a) $f(x, y) = 9x^3 - 8x^2y + 4y^3$

First find f_x and f_y.

$$f_x = 27x^2 - 16xy \qquad \text{and} \qquad f_y = -8x^2 + 12y^2$$

By the definition,

$$df = (27x^2 - 16xy)\, dx + (-8x^2 + 12y^2)\, dy.$$

(b) $z = f(x, y) = \ln(x^3 + y^2)$

Since

$$f_x = \frac{3x^2}{x^3 + y^2} \qquad \text{and} \qquad f_y = \frac{2y}{x^3 + y^2},$$

the total differential is

$$dz = df = \left(\frac{3x^2}{x^3 + y^2}\right) dx + \left(\frac{2y}{x^3 + y^2}\right) dy. \quad \blacktriangleleft$$

 2 Let $f(x, y) = 9x^3 - 8x^2y + 4y^3$. Find df when $x = 1$, $y = 3$, $dx = .01$, and $dy = -.02$.

Example 1(a) gave the total differential

$$df = (27x^2 - 16xy)\, dx + (-8x^2 + 12y^2)\, dy.$$

Replace x with 1, y with 3, dx with .01, and dy with $-.02$.

$$
\begin{aligned}
df &= [27(1)^2 - 16(1)(3)](.01) + [-8(1)^2 + 12(3)^2](-.02) \\
&= (27 - 48)(.01) + (-8 + 108)(-.02) \\
&= (-21)(.01) + (100)(-.02) \\
&= -2.21
\end{aligned}
$$

This result indicates that an increase of .01 in x and a decrease of .02 in y, when $x = 1$ and $y = 3$, will produce a *decrease* of 2.21 in $f(x, y)$. $\quad \blacktriangleleft$

The idea of a total differential can be extended to include functions of three independent variables.

**TOTAL DIFFERENTIAL
FOR THREE VARIABLES**

If $w = f(x, y, z)$, then the total differential dw is

$$dw = f_x(x, y, z)\, dx + f_y(x, y, z)\, dy + f_z(x, y, z)\, dz,$$

provided all indicated partial derivatives exist.

As this definition shows, dw is a function of six variables, x, y, z, dx, dy, and dz. Similar definitions could be given for functions of more than three independent variables.

Approximations Recall that with a function of one variable, $y = f(x)$, the differential dy can be used to approximate the change in y, Δy, corresponding to a change in x, Δx. A similar approximation can be made for a function of two variables.

APPROXIMATIONS

For small values of Δx and Δy,

$$dz \approx \Delta z,$$

where $\Delta z = f(x + \Delta x, y + \Delta y) - f(x, y)$.

EXAMPLE 3 Let $f(x, y) = 6x^2 + xy + y^3$. Find Δz and dz when $x = 2$, $y = -1$, $\Delta x = -.03$, and $\Delta y = .02$.

Here $x + \Delta x = 2 + (-.03) = 1.97$ and $y + \Delta y = -1 + .02 = -.98$. Find Δz with the definition above.

$$\begin{aligned}
\Delta z &= f(x + \Delta x, y + \Delta y) - f(x, y) \\
&= f(1.97, -.98) - f(2, -1) \\
&= [6(1.97)^2 + (1.97)(-.98) + (-.98)^3] - [6(2)^2 + 2(-1) + (-1)^3] \\
&= [23.2854 - 1.9306 - .941192] - [24 - 2 - 1] \\
&= 20.413608 - 21 \\
&= -.586392
\end{aligned}$$

To find dz, the total differential, first find

$$f_x = 12x + y \qquad \text{and} \qquad f_y = x + 3y^2.$$

Then

$$dz = (12x + y)\, dx + (x + 3y^2)\, dy.$$

Since $dx = \Delta x$ and $dy = \Delta y$, substitution gives

$$\begin{aligned}
dz &= [12 \cdot 2 + (-1)](-.03) + [2 + 3 \cdot (-1)^2](.02) \\
&= (23)(-.03) + (5)(.02) \\
&= -.59.
\end{aligned}$$

In this example the values of Δz and dz are very close. (Compare the amount of work needed to find Δz with that needed for dz.) ◀

For small values of dx and dy, the values of Δz and dz are approximately equal. Since $\Delta z = f(x + dx, y + dy) - f(x, y)$,

$$f(x + dx, y + dy) = f(x, y) + \Delta z$$

or $\qquad f(x + dx, y + dy) \approx f(x, y) + dz.$

Replacing dz with the expression for the total differential gives the following result.

APPROXIMATIONS BY DIFFERENTIALS

For a function f having all indicated partial derivatives, and for small values of dx and dy,

$$f(x + dx, y + dy) \approx f(x, y) + dz,$$

or $\qquad f(x + dx, y + dy) \approx f(x, y) + f_x(x, y) \cdot dx + f_y(x, y) \cdot dy.$

EXAMPLE 4

The total profit (in hundreds of dollars) from the sale of x tons of fertilizer for wheat and y tons of fertilizer for corn is given by

$$P(x, y) = 6x^{1/2} + 9y^{2/3} + \frac{5000}{xy}.$$

Suppose a firm is producing 64 tons of fertilizer for wheat and 125 tons of fertilizer for corn, and it is considering changing production to 63 tons and 127 tons, respectively. Before making the change, the firm wants to estimate how this would affect total profit. Use the total differential to estimate the change in profit.

The approximate change in profit, dP, is

$$dP = P_x(x, y) \cdot dx + P_y(x, y) \cdot dy.$$

Finding the necessary partial derivatives gives

$$dP = \left(3x^{-1/2} - \frac{5000}{x^2 y}\right) dx + \left(6y^{-1/3} - \frac{5000}{xy^2}\right) dy.$$

Substitute 64 for x, 125 for y, -1 for dx, and 2 for dy.

$$dP = \left(3 \cdot 64^{-1/2} - \frac{5000}{64^2 \cdot 125}\right)(-1) + \left(6 \cdot 125^{-1/3} - \frac{5000}{64 \cdot 125^2}\right)(2)$$

$$= \left(3 \cdot \frac{1}{8} - \frac{5000}{512,000}\right)(-1) + \left(6 \cdot \frac{1}{5} - \frac{5000}{1,000,000}\right)(2)$$

$$= (.375 - .0098)(-1) + (1.2 - .005)(2)$$

$$\approx 2.0248$$

A change in production from 64 tons of fertilizer for wheat and 125 tons for corn to 63 and 127 tons, respectively, will increase profits by about 2.02 hundred dollars, or $202.

To estimate the total profit from the production of 63 tons and 127 tons, use the approximation formula given above.

$$P(63, 127) \approx P(64, 125) + dP$$

$$= \left(6 \cdot 64^{1/2} + 9 \cdot 125^{2/3} + \frac{5000}{64 \cdot 125}\right) + 2.0248$$

$$= \left(6 \cdot 8 + 9 \cdot 25 + \frac{5000}{8000}\right) + 2.0248 = 275.65$$

hundred dollars, or a total profit of $27,565. ◀

EXAMPLE 5 A short length of blood vessel is in the shape of a right circular cylinder (see Figure 24). The length of the vessel is measured as 42 millimeters, and the radius is measured as 2.5 millimeters. Suppose the maximum error in the measurement of the length is .9 millimeters, with an error of no more than .2 millimeters in the measurement of the radius. Find the maximum possible error in calculating the volume of the blood vessel.

FIGURE 24

The volume of a right circular cylinder is given by $V = \pi r^2 h$. To approximate the error in the volume, find the total differential, dV.

$$dV = (2\pi rh) \cdot dr + (\pi r^2) \cdot dh$$

Here, $r = 2.5$, $h = 42$, $dr = .2$, and $dh = .9$. Substitution gives

$$dV = [(2\pi)(2.5)(42)(.2)] + [\pi(2.5)^2](.9) \approx 149.6.$$

The maximum possible error in calculating the volume is 149.6 cubic millimeters. ◀

16.6 Exercises

Find dz or dw, as appropriate, for each of the following.

1. $z = 9x^4 - 5y^3$

2. $z = x^2 + 7y^4$

3. $z = \dfrac{x + y}{x - y}$

4. $z = \dfrac{x + y^2}{y - 2}$

5. $z = 2\sqrt{xy} - \sqrt{x + y}$

6. $z = \sqrt{x^2 + y^2} + \sqrt{xy}$

7. $z = (3x + 2)\sqrt{1 - 2y}$

8. $z = (5x^2 + 6)\sqrt{4 + 3y}$

9. $z = \ln(x^2 + 2y^4)$

10. $z = \ln\left(\dfrac{8 + x}{8 - y}\right)$

11. $z = xy^2 e^{x+y}$

12. $z = (x + y)e^{-x^2}$

13. $z = x^2 - y \ln x$

14. $z = x^2 + 3y - x \ln y$

15. $w = x^4 yz^3$

16. $w = 6x^3 y^2 z^5$

Evaluate dz using the given information.

17. $z = x^2 + 3xy + y^2$; $\quad x = 4, y = -2, dx = .02, dy = -.03$

18. $z = 8x^3 + 2x^2y - y$; $\quad x = 1, y = 3, dx = .01, dy = .02$

19. $z = \dfrac{x - 4y}{x + 2y}$; $\quad x = 0, y = 5, dx = -.03, dy = .05$

20. $z = \dfrac{y^2 + 3x}{y^2 - x}$; $\quad x = 4, y = -4, dx = .01, dy = .03$

21. $z = \ln(x^2 + y^2)$; $\quad x = 2, y = 3, dx = .02, dy = -.03$

22. $z = \ln\left(\dfrac{x + y}{x - y}\right)$; $\quad x = 4, y = -2, dx = .03, dy = .02$

Evaluate dw using the given information.

23. $w = \dfrac{5x^2 + y^2}{z + 1}$; $\quad x = -2, y = 1, z = 1, dx = .02, dy = -.03, dz = .02$

24. $w = x \ln(yz) - y \ln\dfrac{x}{z}$; $\quad x = 2, y = 1, z = 4, dx = .03, dy = .02, dz = -.01$

Applications

Business and Economics

25. Manufacturing Approximate the amount of aluminum needed for a beverage can of radius 2.5 cm and height 14 cm. Assume the walls of the can are .08 cm thick.

26. Manufacturing Approximate the amount of material needed to make a water tumbler of diameter 3 cm and height 9 cm. Assume the walls of the tumbler are .2 cm thick.

27. Volume of a Coating An industrial coating .2 inches thick is applied to all sides of a box of dimensions 10 inches by 9 inches by 14 inches. Estimate the volume of the coating used.

28. Manufacturing Cost The manufacturing cost of a precision electronic calculator is approximated by

$$M(x, y) = 40x^2 + 30y^2 - 10xy + 30,$$

where x is the cost of the chips and y is the cost of labor. Right now, the company spends $4 on chips and $7 on labor. Use differentials to approximate the change in cost if the company spends $5 on chips and $6.50 on labor.

29. Production The production function for one country is

$$z = x^{.65}y^{.35},$$

where x stands for units of labor and y for units of capital. At present, 50 units of labor and 29 units of capital are available. Use differentials to estimate the change in production if the number of units of labor is increased to 52 and capital is decreased to 27 units.

30. Production The production function for another country is

$$z = x^{.8}y^{.2},$$

where x stands for units of labor and y for units of capital. At present, 20 units of labor and 18 units of capital are being provided. Use differentials to estimate the change in production if an additional unit of labor is provided and if capital is decreased to 16 units.

Life Sciences

31. Bone Preservative Volume A piece of bone in the shape of a right circular cylinder is 7 cm long and has a radius of 1.4 cm. It is coated with a layer of preservative .09 cm thick. Estimate the volume of preservative used.

32. Blood Vessel Volume A portion of a blood vessel is measured as having length 7.9 cm and radius .8 cm. If each measurement could be off by as much as .15 cm, estimate the maximum possible error in calculating the volume of the vessel.

33. Blood Volume In Exercise 50 of Section 2 in this chapter, we found that the number of liters of blood pumped through the lungs in one minute is given by

$$C = \frac{b}{a - v}.$$

Suppose $a = 160$, $b = 200$, $v = 125$. Estimate the change in C if a becomes 145, b becomes 190, and v changes to 130.

34. **Oxygen Consumption** In Exercise 48 of Section 2 of this chapter, we found that the oxygen consumption of a mammal is

$$m = \frac{2.5(T - F)}{w^{.67}}.$$

Suppose T is 38° C, F is 12° C, and w is 30 kg. Approximate the change in m if T changes to 36° C, F changes to 13° C, and w becomes 31 kg.

General Interest

35. **Estimating Area** The height of a triangle is measured as 42.6 cm, with the base measured as 23.4 cm. The measurement of the height can be off by as much as 1.2 cm, and that of the base by no more than .9 cm. Estimate the maximum possible error in calculating the area of the triangle.

36. **Estimating Volume** The height of a cone is measured as 8.4 cm and the radius as 2.9 cm. Each measurement could be off by as much as .1 cm. Estimate the maximum possible error in calculating the volume of the cone.

16.7 DOUBLE INTEGRALS

 How can we find the volume of a bottle with curved sides?

You may wish to review the key ideas of indefinite and definite integrals from the chapter on integration before continuing with this section. See the review problems at the end of that chapter.

In an earlier chapter, we saw how integrals of functions with one variable may be used to find area. In this section, this idea is extended and used to find volume. We found partial derivatives of functions of two or more variables at the beginning of this chapter by holding constant all variables except one. A similar process is used in this section to find antiderivatives of functions of two or more variables. For example, in

$$\int (5x^3y^4 - 6x^2y + 2) \, dy$$

the notation dy indicates integration with respect to y, so we treat y as the variable and x as a constant. Using the rules for antiderivatives gives

$$\int (5x^3y^4 - 6x^2y + 2) \, dy = x^3y^5 - 3x^2y^2 + 2y + C(x).$$

The constant C used earlier must be replaced with $C(x)$ to show that the "constant of integration" here can be any function involving only the variable x. Just as before, check this work by taking the derivative (actually the partial derivative) of the answer:

$$\frac{\partial}{\partial y}\left[x^3y^5 - 3x^2y^2 + 2y + C(x) \right] = 5x^3y^4 - 6x^2y + 2 + 0,$$

which shows that the antiderivative is correct.

EXAMPLE 1

Find each indefinite integral.

(a) $\int x(x^2 + y)\, dx$

Multiply x and $x^2 + y$. Then (because of the dx) integrate each term with x as the variable and y as a constant.

$$\int x(x^2 + y)\, dx = \int (x^3 + xy)\, dx$$
$$= \frac{x^4}{4} + \frac{x^2}{2} \cdot y + f(y) = \frac{1}{4}x^4 + \frac{1}{2}x^2y + f(y)$$

(b) $\int x(x^2 + y)\, dy$

Since y is the variable and x is held constant,

$$\int x(x^2 + y)\, dy = \int (x^3 + xy)\, dy = x^3y + \frac{1}{2}xy^2 + g(x). \blacktriangleleft$$

The analogy to integration of functions of one variable can be continued for evaluating definite integrals. We do this by holding one variable constant and using the Fundamental Theorem of Calculus with the other variable.

EXAMPLE 2

Evaluate each definite integral.

(a) $\int_3^5 (6xy^2 + 12x^2y + 4y)\, dx$

First, find an antiderivative:

$$\int (6xy^2 + 12x^2y + 4y)\, dx = 3x^2y^2 + 4x^3y + 4xy + h(y).$$

Now replace each x with 5, and then with 3, and subtract the results.

$$[3x^2y^2 + 4x^3y + 4xy + h(y)]\Big|_3^5 = [3 \cdot 5^2 \cdot y^2 + 4 \cdot 5^3 \cdot y + 4 \cdot 5 \cdot y + h(y)]$$
$$- [3 \cdot 3^2 \cdot y^2 + 4 \cdot 3^3 \cdot y + 4 \cdot 3 \cdot y + h(y)]$$
$$= 75y^2 + 500y + 20y + h(y)$$
$$- [27y^2 + 108y + 12y + h(y)]$$
$$= 48y^2 + 400y$$

The "function of integration," $h(y)$, drops out, just as the constant of integration does with definite integrals of functions of one variable.

(b) $\int_1^2 (6xy^2 + 12x^2y + 4y) \, dy$

Integrate with respect to y; then substitute 2 and 1 for y and subtract.

$$\int_1^2 (6xy^2 + 12x^2y + 4y) \, dy = (2xy^3 + 6x^2y^2 + 2y^2)\Big|_1^2$$

$$= (2x \cdot 2^3 + 6x^2 \cdot 2^2 + 2 \cdot 2^2)$$
$$- (2x \cdot 1^3 + 6x^2 \cdot 1^2 + 2 \cdot 1^2)$$
$$= 16x + 24x^2 + 8 - (2x + 6x^2 + 2)$$
$$= 14x + 18x^2 + 6 \blacktriangleleft$$

As Example 2 suggests, an integral of the form

$$\int_a^b f(x, y) \, dy$$

produces a result that is a function of x, while

$$\int_a^b f(x, y) \, dx$$

produces a function of y. These resulting functions of one variable can themselves be integrated, as in the next example.

EXAMPLE **3** Evaluate each integral.

(a) $\int_1^2 \left[\int_3^5 (6xy^2 + 12x^2y + 4y) \, dx \right] dy$

In Example 2(a), we found the quantity in brackets to be $48y^2 + 400y$. Thus,

$$\int_1^2 \left[\int_3^5 (6xy^2 + 12x^2y + 4y) \, dx \right] dy = \int_1^2 (48y^2 + 400y) \, dy$$

$$= (16y^3 + 200y^2)\Big|_1^2$$

$$= 16 \cdot 2^3 + 200 \cdot 2^2 - (16 \cdot 1^3 + 200 \cdot 1^2)$$
$$= 128 + 800 - (16 + 200)$$
$$= 712.$$

(b) $\int_3^5 \left[\int_1^2 (6xy^2 + 12x^2y + 4y) \, dy \right] dx.$

(This is the same integrand, with the same limits of integration as in part (a), but the order of integration is reversed.)

Use the result from Example 2(b).

$$\int_3^5 \left[\int_1^2 (6xy^2 + 12x^2y + 4y)\, dy \right] dx = \int_3^5 (14x + 18x^2 + 6)\, dx$$

$$= (7x^2 + 6x^3 + 6x) \Big|_3^5$$

$$= 7 \cdot 5^2 + 6 \cdot 5^3 + 6 \cdot 5 - (7 \cdot 3^2 + 6 \cdot 3^3 + 6 \cdot 3)$$

$$= 175 + 750 + 30 - (63 + 162 + 18) = 712 \quad \blacktriangleleft$$

The answers in the two parts of Example 3 are equal. It can be proved that for a large class of functions, including most functions that occur in applications, the following equation holds true.

$$\int_a^b \left[\int_c^d f(x, y)\, dx \right] dy = \int_c^d \left[\int_a^b f(x, y)\, dy \right] dx$$

Rectangle R is given by
$c \le x \le d,\, a \le y \le b.$

FIGURE 25

Because these two integrals are equal, the brackets are not needed, and either of the integrals is given by

$$\int_a^b \int_c^d f(x, y)\, dx\, dy.$$

This integral is called an **iterated integral** since it is evaluated by integrating twice, first using one variable and then using the other. (The order in which dx and dy are written tells the order of integration, with the innermost differential used first.)

The fact that the iterated integrals above are equal makes it possible to define a *double integral*. First, the set of points (x, y), with $c \le x \le d$ and $a \le y \le b$, defines a rectangular region R in the plane, as shown in Figure 25. Then, the *double integral over R* is defined as follows.

DOUBLE INTEGRAL

The **double integral** of $f(x, y)$ over a rectangular region R is written

$$\iint_R f(x, y)\, dx\, dy \quad \text{or} \quad \iint_R f(x, y)\, dy\, dx,$$

and equals either

$$\int_a^b \int_c^d f(x, y)\, dx\, dy \quad \text{or} \quad \int_c^d \int_a^b f(x, y)\, dy\, dx.$$

Extending earlier definitions, $f(x, y)$ is the **integrand** and R is the **region of integration**.

EXAMPLE 4

Find $\displaystyle\iint_R \sqrt{x} \cdot \sqrt{y-2}\; dx\, dy$ over the rectangular region R defined by

$0 \le x \le 4,\ 3 \le y \le 11$.

Integrate first with respect to x; then integrate the result with respect to y.

$$\iint_R \sqrt{x} \cdot \sqrt{y-2}\; dx\, dy = \int_3^{11} \left[\int_0^4 \sqrt{x} \cdot \sqrt{y-2}\; dx \right] dy$$

$$= \int_3^{11} \left(\frac{2}{3} x^{3/2} \sqrt{y-2} \right) \Bigg|_0^4 \, dy$$

$$= \int_3^{11} \left[\frac{2}{3}(4^{3/2})\sqrt{y-2} - \frac{2}{3}(0^{3/2})\sqrt{y-2} \right] dy$$

$$= \int_3^{11} \left(\frac{16}{3}\sqrt{y-2} - 0 \right) dy = \int_3^{11} \left(\frac{16}{3}\sqrt{y-2} \right) dy$$

$$= \frac{32}{9}(y-2)^{3/2} \Bigg|_3^{11} = \frac{32}{9}(9)^{3/2} - \frac{32}{9}(1)^{3/2}$$

$$= 96 - \frac{32}{9} = \frac{832}{9}$$

As a check, integrate with respect to y first. The answer should be the same. ◀

Volume As shown earlier, the definite integral $\int_a^b f(x)\, dx$ can be used to find the area under a curve. In a similar manner, double integrals are used to find the *volume under a surface*. Figure 26 shows that portion of a surface $f(x, y)$ directly over a rectangle R in the xy-plane. Just as areas were approximated by a large

FIGURE 26

number of small rectangles, volume could be approximated by adding the volumes of a large number of properly drawn small boxes. The height of a typical box would be $f(x, y)$ with the length and width given by dx and dy. The formula for the volume of a box would then suggest the following result.

VOLUME

Let $z = f(x, y)$ be a function that is never negative on the rectangular region R defined by $c \leq x \leq d, a \leq y \leq b$. The volume of the solid under the graph of f and over the region R is

$$\iint_R f(x, y) \, dx \, dy.$$

EXAMPLE 5

Find the volume under the surface $z = x^2 + y^2$ shown in Figure 27.

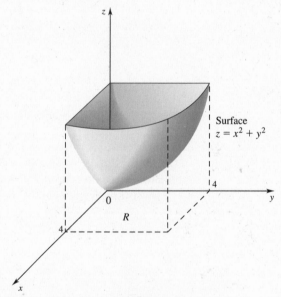

Surface
$z = x^2 + y^2$

FIGURE 27

By the results given above, the volume is

$$\iint_R f(x, y) \, dx \, dy,$$

where $f(x, y) = x^2 + y^2$ and R is the region $0 \leq x \leq 4, 0 \leq y \leq 4$. By definition,

$$\iint_R f(x, y) \, dx \, dy = \int_0^4 \left[\int_0^4 (x^2 + y^2) \, dx \right] dy$$

$$= \int_0^4 \left(\frac{1}{3}x^3 + xy^2 \right) \Big|_0^4 \, dy$$

$$= \int_0^4 \left(\frac{64}{3} + 4y^2\right) dy = \left(\frac{64}{3}y + \frac{4}{3}y^3\right)\Big|_0^4$$

$$= \frac{64}{3} \cdot 4 + \frac{4}{3} \cdot 4^3 - 0 = \frac{512}{3}. \quad \blacktriangleleft$$

EXAMPLE 6

A product design consultant for a cosmetics company has been asked to design a bottle for the company's newest perfume. The thickness of the glass is to vary so that the outside of the bottle has straight sides and the inside has curved sides, as shown in Figure 28. Before presenting the design to management, the consultant needs to make a reasonably accurate estimate of the amount each bottle will hold. If the base of the bottle is to be 4 centimeters by 3 centimeters, and if a cross section of its interior is to be a parabola of the form $z = -y^2 + 4y$, what is its internal volume?

The interior of the bottle can be graphed in three-dimensional space, as shown in Figure 29, where $z = 0$ corresponds to the base of the bottle. Its volume is simply the volume above the region R in the xy plane and below the graph of $f(x, y) = -y^2 + 4y$. This volume is given by the double integral

$$\int_0^3 \int_0^4 (-y^2 + 4y) \, dy \, dx = \int_0^3 \left(\frac{-y^3}{3} + \frac{4y^2}{2}\right)\Big|_0^4 dx$$

$$= \int_0^3 \left(\frac{-64}{3} + 32 - 0\right) dx$$

$$= \frac{32}{3}x \Big|_0^3$$

$$= 32 - 0 = 32.$$

The bottle holds 32 cubic centimeters. ◀

3 cm

4 cm

FIGURE 28

$f(x, y) = -y^2 + 4y$

$(0, 4, 0)$

R

$(3, 0, 0)$

FIGURE 29

Double Integrals over Other Regions In the work in this section, we found double integrals over rectangular regions, with constant limits of integration. Now this work can be extended to include *variable* limits of integration. (Notice in the following examples that the variable limits always go on the *inner* integral sign.)

The use of variable limits of integration permits evaluation of double integrals over the types of regions shown in Figure 30. Double integrals over more complicated regions are discussed in more advanced books. Integration over regions such as those of Figure 30 is done with the results of the following theorem.

FIGURE 30

DOUBLE INTEGRALS OVER VARIABLE REGIONS

Let $z = f(x, y)$ be a function of two variables.
If R is the region (in Figure 30(a)) defined by $c \le x \le d$ and $g(x) \le y \le h(x)$, then

$$\iint\limits_{R} f(x, y)\, dy\, dx = \int_{c}^{d} \left[\int_{g(x)}^{h(x)} f(x, y)\, dy \right] dx.$$

If R is the region (in Figure 30(b)) defined by $g(y) \le x \le h(y)$ and $a \le y \le b$, then

$$\iint\limits_{R} f(x, y)\, dx\, dy = \int_{a}^{b} \left[\int_{g(y)}^{h(y)} f(x, y)\, dx \right] dy.$$

EXAMPLE

7 Evaluate $\int_{1}^{2} \int_{y}^{y^2} xy\, dx\, dy$.

The region of integration is shown in Figure 31. Integrate first with respect to x, then with respect to y.

$$\int_{1}^{2} \int_{y}^{y^2} xy\, dx\, dy = \int_{1}^{2} \left[\int_{y}^{y^2} xy\, dx \right] dy = \int_{1}^{2} \left(\frac{1}{2} x^2 y \right) \Big|_{y}^{y^2} dy$$

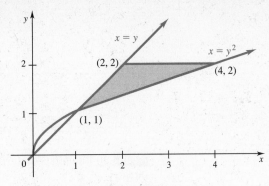

FIGURE 31

Replace x first with y^2 and then with y, and subtract.

$$\int_1^2 \int_y^{y^2} xy \; dx \; dy = \int_1^2 \left[\frac{1}{2}(y^2)^2 y - \frac{1}{2}(y)^2 y \right] dy$$

$$= \int_1^2 \left(\frac{1}{2}y^5 - \frac{1}{2}y^3 \right) dy = \left(\frac{1}{12}y^6 - \frac{1}{8}y^4 \right) \Big|_1^2$$

$$= \left(\frac{1}{12} \cdot 2^6 - \frac{1}{8} \cdot 2^4 \right) - \left(\frac{1}{12} \cdot 1^6 - \frac{1}{8} \cdot 1^4 \right)$$

$$= \frac{64}{12} - \frac{16}{8} - \frac{1}{12} + \frac{1}{8} = \frac{27}{8} \;\blacktriangleleft$$

EXAMPLE 8 Let R be the shaded region in Figure 32, and evaluate

$$\iint_R (x + 2y) \; dy \; dx.$$

Region R is bounded by $h(x) = 2x$ and $g(x) = x^2$, with $0 \le x \le 2$. By the first result in the theorem above,

$$\iint_R (x + 2y) \; dy \; dx = \int_0^2 \int_{x^2}^{2x} (x + 2y) \; dy \; dx$$

$$= \int_0^2 (xy + y^2) \Big|_{x^2}^{2x} dx$$

$$= \int_0^2 (x(2x) + (2x)^2 - [x \cdot x^2 + (x^2)^2]) \; dx$$

$$= \int_0^2 [2x^2 + 4x^2 - (x^3 + x^4)] \; dx$$

$$= \int_0^2 (6x^2 - x^3 - x^4) \; dx$$

FIGURE 32

$$= \left(2x^3 - \frac{1}{4}x^4 - \frac{1}{5}x^5\right)\Bigg|_0^2$$

$$= 2 \cdot 2^3 - \frac{1}{4} \cdot 2^4 - \frac{1}{5} \cdot 2^5 - 0$$

$$= 16 - 4 - \frac{32}{5} = \frac{28}{5}. \quad \blacktriangleleft$$

Note In Example 8, the same result would be found if we evaluated the double integral

$$\iint\limits_R (x + 2y) \, dx \, dy.$$

In that case, we would need to define the equations of the boundaries in terms of y rather than x, so R would be defined by $y/2 \leq x \leq \sqrt{y}, \, 0 \leq y \leq 4$.

Interchanging Limits of Integration Sometimes it is easier to integrate first with respect to x, and then y, while with other integrals the reverse process is easier. The limits of integration can be reversed whenever the region R is like the region in Figure 32. The next example shows how this process works.

FIGURE 33

EXAMPLE 9

Interchange the limits of integration in

$$\int_0^{16} \int_{\sqrt{y}}^4 f(x, y) \, dx \, dy.$$

For this integral, region R is given by $\sqrt{y} \leq x \leq 4, \, 0 \leq y \leq 16$. A graph of R is shown in Figure 33.

The same region R can be written in an alternate way. As Figure 33 shows, one boundary of R is $x = \sqrt{y}$. Solving for y gives $y = x^2$. Also, Figure 33 shows that $0 \leq x \leq 4$. Since R can be written as $0 \leq y \leq x^2, \, 0 \leq x \leq 4$, the double integral above can be written

$$\int_0^4 \int_0^{x^2} f(x, y) \, dy \, dx. \quad \blacktriangleleft$$

16.7 Exercises

Evaluate the following integrals.

1. $\displaystyle\int_0^3 (x^3y + y) \, dx$

2. $\displaystyle\int_1^4 (xy^2 - x) \, dy$

3. $\displaystyle\int_4^8 \sqrt{6x + y} \, dx$

4. $\displaystyle\int_3^7 \sqrt{x + 5y} \, dy$

5. $\displaystyle\int_4^5 x\sqrt{x^2 + 3y} \, dy$

6. $\displaystyle\int_3^6 x\sqrt{x^2 + 3y} \, dx$

7. $\displaystyle\int_4^9 \frac{3 + 5y}{\sqrt{x}} \, dx$

8. $\displaystyle\int_2^7 \frac{3 + 5y}{\sqrt{x}} \, dy$

9. $\displaystyle\int_{-1}^1 e^{x + 4y} \, dy$

10. $\displaystyle\int_2^6 e^{x + 4y} \, dx$

11. $\displaystyle\int_0^5 xe^{x^2 + 9y} \, dx$

12. $\displaystyle\int_1^6 xe^{x^2 + 9y} \, dy$

Evaluate the following iterated integrals. (Many of these use results from Exercises 1–12.)

13. $\int_1^2 \left[\int_0^3 (x^3y + y) \, dx \right] dy$

14. $\int_0^3 \left[\int_1^4 (xy^2 - x) \, dy \right] dx$

15. $\int_0^1 \left[\int_3^6 x\sqrt{x^2 + 3y} \, dx \right] dy$

16. $\int_0^3 \left[\int_4^5 x\sqrt{x^2 + 3y} \, dy \right] dx$

17. $\int_1^2 \left[\int_4^9 \frac{3 + 5y}{\sqrt{x}} \, dx \right] dy$

18. $\int_{16}^{25} \left[\int_2^7 \frac{3 + 5y}{\sqrt{x}} \, dy \right] dx$

19. $\int_1^2 \int_1^2 \frac{dx \, dy}{xy}$

20. $\int_1^4 \int_2^5 \frac{dy \, dx}{x}$

21. $\int_2^4 \int_3^5 \left(\frac{x}{y} + \frac{y}{3} \right) dx \, dy$

22. $\int_3^4 \int_1^2 \left(\frac{6x}{5} + \frac{y}{x} \right) dx \, dy$

Find each double integral over the rectangular region R with the given boundaries.

23. $\iint_R (x + 3y^2) \, dx \, dy; \quad 0 \le x \le 2, 1 \le y \le 5$

24. $\iint_R (4x^3 + y^2) \, dx \, dy; \quad 1 \le x \le 4, 0 \le y \le 2$

25. $\iint_R \sqrt{x + y} \, dy \, dx; \quad 1 \le x \le 3, 0 \le y \le 1$

26. $\iint_R x^2\sqrt{x^3 + 2y} \, dx \, dy; \quad 0 \le x \le 2, 0 \le y \le 3$

27. $\iint_R \frac{2}{(x + y)^2} \, dy \, dx; \quad 2 \le x \le 3, 1 \le y \le 5$

28. $\iint_R \frac{y}{\sqrt{6x + 5y^2}} \, dx \, dy; \quad 0 \le x \le 3, 1 \le y \le 2$

29. $\iint_R ye^{(x+y^2)} \, dx \, dy; \quad 2 \le x \le 3, 0 \le y \le 2$

30. $\iint_R x^2 e^{(x^3 + 2y)} \, dx \, dy; \quad 1 \le x \le 2, 1 \le y \le 3$

Find the volume under the given surface z = f(x, y) and above the rectangle with the given boundaries.

31. $z = 6x + 2y + 5; \quad -1 \le x \le 1, 0 \le y \le 3$

32. $z = 9x + 5y + 12; \quad 0 \le x \le 3, -2 \le y \le 1$

33. $z = x^2; \quad 0 \le x \le 1, 0 \le y \le 4$

34. $z = \sqrt{y}; \quad 0 \le x \le 4, 0 \le y \le 9$

35. $z = x\sqrt{x^2 + y}; \quad 0 \le x \le 1, 0 \le y \le 1$

36. $z = yx\sqrt{x^2 + y^2}; \quad 0 \le x \le 4, 0 \le y \le 1$

37. $z = \frac{xy}{(x^2 + y^2)^2}; \quad 1 \le x \le 2, 1 \le y \le 4$

38. $z = e^{x+y}; \quad 0 \le x \le 1, 0 \le y \le 1$

While it is true that a double integral can be evaluated by using either dx or dy first, sometimes one choice over the other makes the work easier. Evaluate the double integrals in Exercises 39 and 40 in the easiest way possible.

39. $\iint_R xe^{xy} \, dx \, dy; \quad 0 \le x \le 2, 0 \le y \le 1$

40. $\iint_R x^2 e^{2x^3 + 6y} \, dx \, dy; \quad 0 \le x \le 1, 0 \le y \le 1$

Evaluate each double integral.

41. $\int_2^4 \int_2^{x^2} (x^2 + y^2) \, dy \, dx$

42. $\int_0^5 \int_0^{2y} (x^2 + y) \, dx \, dy$

43. $\int_0^4 \int_0^x \sqrt{xy} \, dy \, dx$

44. $\int_1^4 \int_0^x \sqrt{x + y} \, dy \, dx$

45. $\int_1^2 \int_y^{3y} \frac{1}{x} \, dx \, dy$

46. $\int_1^4 \int_x^{x^2} \frac{1}{y} \, dy \, dx$

47. $\int_0^4 \int_1^{e^x} \frac{x}{y} \, dy \, dx$

48. $\int_0^1 \int_{2x}^{4x} e^{x+y} \, dy \, dx$

Use the region R with the indicated boundaries to evaluate each of the following double integrals.

49. $\iint_R (4x + 7y) \, dy \, dx; \quad 1 \le x \le 3, 0 \le y \le x + 1$

50. $\iint_R (3x + 9y) \, dy \, dx; \quad 2 \le x \le 4, 2 \le y \le 3x$

51. $\iint_R (4 - 4x^2) \, dy \, dx; \quad 0 \le x \le 1, 0 \le y \le 2 - 2x$

52. $\iint_R \frac{dy \, dx}{x}; \quad 1 \le x \le 2, 0 \le y \le x - 1$

53. $\iint_R e^{x/y^2} \, dx \, dy; \quad 1 \le y \le 2, 0 \le x \le y^2$

54. $\iint_R (x^2 - y) \, dy \, dx; \quad -1 \le x \le 1, -x^2 \le y \le x^2$

55. $\iint\limits_R x^3 y \, dx \, dy$; R bounded by $y = x^2$, $y = 2x$

56. $\iint\limits_R x^2 y^2 \, dx \, dy$; R bounded by $y = x$, $y = 2x$, $x = 1$

57. $\iint\limits_R \dfrac{dy \, dx}{y}$; R bounded by $y = x$, $y = \dfrac{1}{x}$, $x = 2$

*The idea of the average value of a function, discussed earlier for functions of the form $y = f(x)$, can be extended to functions of more than one independent variable. For a function $z = f(x, y)$, the **average value** of f over a region R is defined as*

$$\frac{1}{A} \iint\limits_R f(x, y) \, dx \, dy,$$

where A is the area of the region R. Find the average value for each of the following functions over the regions R having the given boundaries.

58. $f(x, y) = 5xy + 2y$; $1 \le x \le 4, 1 \le y \le 2$

59. $f(x, y) = x^2 + y^2$; $0 \le x \le 2, 0 \le y \le 3$

60. $f(x, y) = e^{-5y+3x}$; $0 \le x \le 2, 0 \le y \le 2$

61. $f(x, y) = e^{2x+y}$; $1 \le x \le 2, 2 \le y \le 3$

Applications

Business and Economics

62. Packaging The manufacturer of a fruit juice drink has decided to try innovative packaging in order to revitalize sagging sales. The fruit juice drink is to be packaged in containers in the shape of tetrahedra in which three edges are perpendicular, as shown in the figure. Two of the perpendicular edges will be 3 inches long, and the third edge will be 6 inches long. Find the volume of the container. (*Hint:* The equation of the plane shown in the figure is $z = f(x, y) = 6 - 2x - 2y$.)

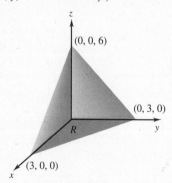

63. Average Cost A company's total cost for operating its two warehouses is

$$C(x, y) = \frac{1}{9}x^2 + 2x + y^2 + 5y + 100$$

dollars, where x represents the number of units stored at the first warehouse and y represents the number of units stored at the second. Find the average cost to store a unit if the first warehouse has between 48 and 75 units, and the second has between 20 and 60 units.

64. Average Production A production function is given by

$$P(x, y) = 500x^2 y^{.8},$$

where x is the number of units of labor and y is the number of units of capital. Find the average production level if x varies from 10 to 50 and y from 20 to 40.

65. Average Profit The profit (in dollars) from selling x units of one product and y units of a second product is

$$P = -(x - 100)^2 - (y - 50)^2 + 2000.$$

The weekly sales for the first product vary from 100 units to 150 units, and the weekly sales for the second product vary from 40 units to 80 units. Estimate average weekly profit for these two products.

66. Average Revenue A company sells two products. The demand functions of the products are given by

$$q_1 = 300 - 2p_1 \quad \text{and} \quad q_2 = 500 - 1.2p_2,$$

where q_1 units of the first product are demanded at price p_1 and q_2 units of the second product are demanded at price p_2. The total revenue will be given by

$$R = q_1 p_1 + q_2 p_2.$$

Find the average revenue if the price p_1 varies from \$25 to \$50 and the price p_2 varies from \$50 to \$75.

Physical Sciences

67. **Effects of Nuclear Weapons** *The Effects of Nuclear Weapons*, prepared by the U.S. Department of Defense, contains these remarks on the computation of the radioactive dose after a 1-kiloton atomic explosion.*

 "If all the residues from a 1-kiloton fission yield were deposited on a smooth surface in varying concentrations typical of an early fallout pattern, instead of uniformly, the product of the dose rate at 1 hour and the area would be replaced by the 'area integral' of the 1-hour

dose rate defined by

$$\text{Area integral} = \int_A R_1 \, dA,$$

where R_1 is the 1-hour dose rate over an element of area dA, and A square miles is the total area covered by the residues."

Explain why an integral is involved. (*Note:* In this case R_1 denotes the function and A the region. In the diverse applications of integrals, many notations are employed.)

Chapter Summary Key Terms

16.1 **function of two variables**
 ordered triple
 first octant
 plane
 surface
 trace
 level curves
 paraboloid
 production function
 ellipsoid
 hyperbolic paraboloid
 hyperboloid of two sheets

16.2 **partial derivative**
 second-order partial derivative
16.3 **saddle point**
 critical point
16.4 **constraint**
 Lagrange multiplier

16.5 **scatter diagram**
 least squares line
 regression line
16.6 **total differential**
16.7 **iterated integral**
 double integral
 region of integration

Chapter 16 Review Exercises

Find $f(-1, 2)$ and $f(6, -3)$ for each of the following.

1. $f(x, y) = -4x^2 + 6xy - 3$

2. $f(x, y) = 3x^2y^2 - 5x + 2y$

3. $f(x, y) = \dfrac{x - 3y}{x + 4y}$

4. $f(x, y) = \dfrac{\sqrt{x^2 + y^2}}{x - y}$

Graph the first-octant portion of each plane.

5. $x + y + z = 4$

6. $x + y + 4z = 8$

7. $5x + 2y = 10$

8. $3x + 5z = 15$

9. $x = 3$

10. $y = 2$

11. Let $z = f(x, y) = -5x^2 + 7xy - y^2$. Find each of the following.

 (a) $\dfrac{\partial z}{\partial x}$ (b) $\left(\dfrac{\partial z}{\partial y}\right)(-1, 4)$ (c) $f_{xy}(2, -1)$

12. Let $z = f(x, y) = \dfrac{x + y^2}{x - y^2}$. Find each of the following.

 (a) $\dfrac{\partial z}{\partial y}$ (b) $\left(\dfrac{\partial z}{\partial x}\right)(0, 2)$ (c) $f_{xx}(-1, 0)$

Find f_x and f_y.

13. $f(x, y) = 9x^3y^2 - 5x$

14. $f(x, y) = 6x^5y - 8xy^9$

15. $f(x, y) = \sqrt{4x^2 + y^2}$

16. $f(x, y) = \dfrac{2x + 5y^2}{3x^2 + y^2}$

17. $f(x, y) = x^2 \cdot e^{2y}$

18. $f(x, y) = (y - 2)^2 \cdot e^{(x + 2y)}$

19. $f(x, y) = \ln |2x^2 + y^2|$

20. $f(x, y) = \ln |2 - x^2y^3|$

Find f_{xx} and f_{xy}.

21. $f(x, y) = 4x^3y^2 - 8xy$

22. $f(x, y) = -6xy^4 + x^2y$

23. $f(x, y) = \dfrac{2x}{x - 2y}$

24. $f(x, y) = \dfrac{3x + y}{x - 1}$

25. $f(x, y) = x^2e^y$

26. $f(x, y) = ye^{x^2}$

27. $f(x, y) = \ln |2 - x^2y|$

28. $f(x, y) = \ln |1 + 3xy^2|$

Find all points where the functions defined below have any relative extrema. Find any saddle points.

29. $z = x^2 + 2y^2 - 4y$

30. $z = x^2 + y^2 + 9x - 8y + 1$

31. $f(x, y) = x^2 + 5xy - 10x + 3y^2 - 12y$

32. $z = x^3 - 8y^2 + 6xy + 4$

33. $z = \dfrac{1}{2}x^2 + \dfrac{1}{2}y^2 + 2xy - 5x - 7y + 10$

34. $f(x, y) = 3x^2 + 2xy + 2y^2 - 3x + 2y - 9$

35. $z = x^3 + y^2 + 2xy - 4x - 3y - 2$

36. $f(x, y) = 7x^2 + y^2 - 3x + 6y - 5xy$

Use Lagrange multipliers to find the extrema of the functions defined in Exercises 37 and 38.

37. $f(x, y) = x^2y; \quad x + y = 4$

38. $f(x, y) = x^2 + y^2; \quad x = y + 2$

39. Find positive numbers x and y, whose sum is 80, such that x^2y is maximized.

40. Find positive numbers x and y, whose sum is 50, such that xy^2 is maximized.

41. A closed box with square ends must have a volume of 125 cubic inches. Find the dimensions of such a box that has minimum surface area.

42. Find the maximum rectangular area that can be enclosed with 400 ft of fencing, if no fencing is needed along one side.

Find dz or dw, as appropriate, for each of the following.

43. $z = 7x^3y - 4y^3$

44. $z = 3x^2 + \sqrt{x + y}$

45. $z = x^2ye^{x-y}$

46. $z = \ln |x + 4y| + y^2 \ln x$

47. $w = x^5 + y^4 - z^3$

48. $w = \dfrac{3 + 5xy}{2 - z}$

Evaluate dz using the given information.

49. $z = 2x^2 - 4y^2 + 6xy; \quad x = 2, y = -3, dx = .01, dy = .05$

50. $z = \dfrac{x + 5y}{x - 2y}; \quad x = 1, y = -2, dx = -.04, dy = .02$

Evaluate each of the following.

51. $\displaystyle\int_0^4 (x^2y^2 + 5x)\, dx$

52. $\displaystyle\int_0^3 (x + 5y + y^2)\, dy$

53. $\displaystyle\int_2^5 \sqrt{6x + 3y}\, dx$

54. $\displaystyle\int_1^3 6y^4\sqrt{8x + 3y}\, dx$

55. $\displaystyle\int_4^9 \dfrac{6y - 8}{\sqrt{x}}\, dx$

56. $\displaystyle\int_3^5 e^{2x-7y}\, dx$

57. $\displaystyle\int_0^5 \dfrac{6x}{\sqrt{4x^2 + 2y^2}}\, dx$

58. $\displaystyle\int_1^3 \dfrac{y^2}{\sqrt{7x + 11y^3}}\, dy$

Evaluate each iterated integral.

59. $\displaystyle\int_0^2 \left[\int_0^4 (x^2y^2 + 5x)\, dx \right] dy$

60. $\displaystyle\int_0^2 \left[\int_0^3 (x + 5y + y^2)\, dy \right] dx$

61. $\displaystyle\int_3^4 \left[\int_2^5 \sqrt{6x + 3y}\, dx \right] dy$

62. $\int_1^2 \left[\int_3^5 e^{2x-7y} \, dx \right] dy$

63. $\int_2^4 \int_2^4 \frac{dx \, dy}{y}$

64. $\int_1^2 \int_1^2 \frac{dx \, dy}{x}$

Find each double integral over the region R with boundaries as indicated.

65. $\iint\limits_R (x^2 + y^2) \, dx \, dy; \quad 0 \le x \le 2, \, 0 \le y \le 3$

66. $\iint\limits_R \sqrt{2x + y} \, dx \, dy; \quad 1 \le x \le 3, \, 2 \le y \le 5$

67. $\iint\limits_R \sqrt{y + x} \, dx \, dy; \quad 0 \le x \le 7, \, 1 \le y \le 9$

68. $\iint\limits_R ye^{y^2 + x} \, dx \, dy; \quad 0 \le x \le 1, \, 0 \le y \le 1$

Find the volume under the given surface z = f(x, y) and above the given rectangle.

69. $z = x + 9y + 8; \quad 1 \le x \le 6, 0 \le y \le 8$

70. $z = x^2 + y^2; \quad 3 \le x \le 5, 2 \le y \le 4$

Evaluate each double integral.

71. $\int_0^1 \int_0^{2x} xy \, dy \, dx$

72. $\int_0^1 \int_0^{x^3} y \, dy \, dx$

73. $\int_0^1 \int_{x^2}^x x^3 y \, dy \, dx$

74. $\int_0^1 \int_y^{\sqrt{y}} x \, dx \, dy$

Use the region R, with boundaries as indicated, to evaluate the given double integral.

75. $\iint\limits_R (2x + 3y) \, dx \, dy; \quad 0 \le y \le 1, \, y \le x \le 2 - y$

76. $\iint\limits_R (2 - x^2 - y^2) \, dy \, dx; \quad 0 \le x \le 1, \, x^2 \le y \le x$

Applications

Business and Economics

77. Charge for Auto Painting The charge (in dollars) for painting a sports car is given by

$$C(x, y) = 2x^2 + 4y^2 - 3xy + \sqrt{x},$$

where x is the number of hours of labor needed and y is the number of gallons of paint and sealant used. Find each of the following.

(a) The charge for 10 hr and 5 gal of paint and sealant

(b) The charge for 15 hr and 10 gal of paint and sealant

(c) The charge for 20 hr and 20 gal of paint and sealant

78. Manufacturing Costs The manufacturing cost (in dollars) for a medium-sized business computer is given by

$$c(x, y) = 2x + y^2 + 4xy + 25,$$

where x is the memory capacity of the computer in megabytes (Mb) and y is the number of hours of labor required. For 640Mb and 6 hr of labor, find the following.

(a) The approximate change in cost for an additional 1Mb of memory

(b) The approximate change in cost for an additional hour of labor

79. Productivity The production function z for one country is

$$z = x^{.6} y^{.4},$$

where x represents the amount of labor and y the amount of capital. Find the marginal productivity of each of the following.

(a) Labor (b) Capital

80. Cost The cost (in dollars) to manufacture x solar cells and y solar collectors is

$$c(x, y) = x^2 + 5y^2 + 4xy - 70x - 164y + 1800.$$

(a) Find values of x and y that produce minimum total cost.

(b) Find the minimum total cost.

81. Earnings (a) Find the least squares line for the following data, which give the earnings (in ten-thousands of dollars) for a certain company after x yr.

x	3	5	7	8
y	4	11	20	23

(b) Predict the earnings after 6 yr.

82. Cost The cost (in dollars) to produce x satellite receiving dishes and y transmitters is given by

$$C(x, y) = \ln(x^2 + y) + e^{xy/20}.$$

Production schedules now call for 15 receiving dishes and 9 transmitters. Use differentials to approximate the change in costs if 1 more dish and 1 fewer transmitter are made.

83. Profit The profit (in hundreds of dollars) from the sale of x small computers and y electronic games is given by

$$P(x, y) = \frac{x}{x + 5y} + \frac{y + x}{y}.$$

Right now, 75 small computers and 50 games are being sold. Use differentials to approximate the change in profit if 3 fewer computers and 2 more games were sold.

84. Production Materials Approximate the amount of material needed to manufacture a cone of radius 2 cm, height 8 cm, and wall thickness .21 cm.

85. Production Materials A sphere of radius 2 ft is to receive an insulating coating 1 inch thick. Approximate the volume of the coating needed.

86. Production Error The height of a sample cone from a production line is measured as 11.4 cm, while the radius is measured as 2.9 cm. Each of these measurements could be off by .2 cm. Approximate the maximum possible error in the volume of the cone.

Life Sciences

87. Blood Sugar and Cholesterol Levels The following data show the blood sugar levels, x, and cholesterol levels, y, for 8 different diabetic patients.

Patient	1	2	3	4	5	6	7	8
Blood Sugar Level, x	130	138	142	159	165	200	210	250
Cholesterol Level, y	170	160	173	181	201	192	240	290

For this data, $\Sigma x = 1394$; $\Sigma y = 1607$; $\Sigma xy = 291{,}990$; and $\Sigma x^2 = 255{,}214$.

(a) Find the equation of the least squares line.

(b) Predict the cholesterol level for a person whose blood sugar level is 190.

(c) Use the equation from part (a) to predict cholesterol levels when the blood sugar level is 130, 142, and 200. How do these results compare with the data in the table? Does there appear to be a close linear relationship here?

88. Pace of Life In an attempt to measure how the pace of city life is related to the size of the city, two researchers assessed the mean velocity of pedestrians in 15 cities by measuring the mean time to walk 50 ft.*

City	Population, x	Velocity, y
Brno, Czechoslovakia	341,948	4.81
Prague, Czechoslovakia	1,092,759	5.88
Corte, Corsica	5,491	3.31
Bastia, France	49,375	4.90
Munich, Germany	1,340,000	5.62
Psychro, Crete	365	2.67
Itea, Greece	2500	2.27
Iraklion, Greece	78,200	3.85
Athens, Greece	867,023	5.21
Safed, Israel	14,000	3.70
Dimona, Israel	23,700	3.27
Netanya, Israel	70,700	4.31
Jerusalem, Israel	304,500	4.42
New Haven, U.S.A.	138,000	4.39
Brooklyn, U.S.A.	2,602,000	5.05

(a) Plot the original pairs of numbers. The pattern should be nonlinear.

(b) Plot y against $\log x$, using a calculator to compute $\log x$. Is the data more linear than in part (a)?

(c) Find the equation of the least squares line for y' using $\log x$ (instead of x) and y.

89. Blood Vessel Volume A length of blood vessel is measured as 2.7 cm, with the radius measured as .7 cm. If each of these measurements could be off by .1 cm, estimate the maximum possible error in the volume of the vessel.

General Interest

90. Area The bottom of a planter is to be made in the shape of an isosceles triangle, with the two equal sides 3 ft long and the third side 2 ft long. The area of an isosceles triangle with two equal sides of length a and third side of length b is

$$f(a, b) = \frac{1}{4}b\sqrt{4a^2 - b^2}.$$

(a) Find the area of the bottom of the planter.

(b) The manufacturer is considering changing the shape so that the third side is 2.5 ft long. What would be the approximate effect on the area?

*From Marc H. Bornstein and Helen G. Bornstein, "The Pace of Life," *Nature* 259 (February 19, 1976), pp. 557–59.

Connections

91. Profit The total profit from 1 acre of a certain crop depends on the amount spent on fertilizer, x, and on hybrid seed, y, according to the model

$$P(x, y) = .01(-x^2 + 3xy + 160x - 5y^2 + 200y + 2600).$$

The budget for fertilizer and seed is limited to $280.

(a) Use the budget constraint to express one variable in terms of the other. Then substitute into the profit function to get a function with one independent variable. Use the method shown in the chapter on applications of the derivative to find the amounts spent on fertilizer and seed that will maximize profit. What is the maximum profit per acre? (*Hint:* Throughout this problem you may ignore the coefficient of .01 until you need to find the maximum profit.)

(b) Find the amounts spent on fertilizer and seed that will maximize profit using the first method shown in this chapter. (*Hint:* You will not need to use the budget constraint.)

(c) Use the Lagrange multiplier method to solve the original problem.

(d) Look for the relationships among these methods.

Extended Application / Lagrange Multipliers for a Predator[†]

A predator is an animal that feeds upon another animal. Foxes, coyotes, wolves, weasels, lions, and tigers are well-known predators, but many other animals also fall into this category. In this case, we set up a mathematical model for predation, and then use Lagrange multipliers to minimize the difference between the desired and the actual levels of food consumption. There are several research studies which show that animals *do* control their activities so as to maximize or minimize variables— lobsters orient their bodies by minimizing the discharge rate from certain organs, for example.

In this case, we assume that the predator has a diet consisting of only two foods, food 1 and food 2. We also assume that the predator will hunt only in two locations, location 1 and location 2. To make this mathematical model meaningful, we would have to gather data on the predators and the prey that we wished to study. For example, suppose that we have gathered the following data:

$u_{11} = .4 =$ rate of feeding on food 1 in location 1

$u_{12} = .1 =$ rate of feeding on food 1 in location 2

$u_{21} = .3 =$ rate of feeding on food 2 in location 1

$u_{22} = .3 =$ rate of feeding on food 2 in location 2.

Let $x_1 =$ proportion of time spent feeding in location 1

$x_2 =$ proportion of time spent feeding in location 2

$x_3 =$ proportion of time spent on nonfeeding activities.

Using these variables, the total quantity of food 1 consumed is given by Y_1, where

$$Y_1 = u_{11}x_1 + u_{12}x_2$$
$$= .4x_1 + .1x_2. \tag{1}$$

The total quantity of food 2 consumed is given by Y_2, where

$$Y_2 = u_{21}x_1 + u_{22}x_2$$
$$= .3x_1 + .3x_2. \tag{2}$$

The total amount of food consumed is thus given by

$$Y_1 + Y_2 = .4x_1 + .1x_2 + .3x_1 + .3x_2$$
$$= .7x_1 + .4x_2. \tag{3}$$

Again, by gathering experimental data, suppose that

$z_t = .4 =$ desired level of total food consumption

$z_1 = .15 =$ desired level of consumption of food 1

$z_2 = .25 =$ desired level of consumption of food 2.

The predator wishes to find values of x_1 and x_2 such that the difference between the desired level of consumption (the z's) and the actual level of consumption (the Y's) is minimized. However, much experience has shown that the difference between desired food consumption and actual consumption is not perceived by the animal as linear, but rather perhaps as a square. That is, a 5% shortfall in the actual consumption of food, as

[†]Based on Gerald G. Martens, "An Optimization Equation for Predation," *Ecology*, Winter 1973, pp. 92–101.

compared to the desired consumption, may be perceived by the predator as a 20–25% shortfall. (This phenomenon is known as *Weber's Law*—many psychology books discuss it.) Thus, to have our model approximate reality, we must find values of x_1 and x_2 that will minimize

$$G = [z_t - (Y_1 + Y_2)]^2 + w_1[z_1 - Y_1]^2$$
$$+ w_2[z_2 - Y_2]^2 + w_3(1 - x_3)^2. \qquad (4)$$

The term $w_3(1 - x_3)^2$ represents the fact that the predator does not wish to spend the total available time in searching for food—recall that x_3 represents the proportion of available time that is spent on nonfeeding activities. The variables w_1, w_2, and w_3 represent the relative importance, or weights, assigned by the animal to food 1, food 2, and nonfood activities, respectively. Again, it is necessary to gather experimental data: reasonable values of w_1, w_2, and w_3 are as follows:

$$w_1 = .5 \qquad w_2 = .4 \qquad w_3 = .03.$$

If we substitute .4 for z_t, .15 for z_1, .25 for z_2, .5 for w_1, .4 for w_2, and .03 for w_3, the results of equation (1) for Y_1, (2) for Y_2, and (3) for $Y_1 + Y_2$ into equation (4), we have

$$G = [.4 - .7x_1 - .4x_2]^2 + .5[.15 - .4x_1 - .1x_2]^2$$
$$+ .4[.25 - .3x_1 - .3x_2]^2 + .03(1 - x_3)^2.$$

We want to minimize G, subject to the constraint $x_1 + x_2 + x_3 = 1$, or $x_1 + x_2 + x_3 - 1 = 0$. This can be done with Lagrange multipliers. First, form the function F:

$$F = [.4 - .7x_1 - .4x_2]^2 + .5[.15 - .4x_1 - .1x_2]^2$$
$$+ .4[.25 - .3x_1 - .3x_2]^2 + .03(1 - x_3)^2$$
$$+ \lambda(x_1 + x_2 + x_3 - 1).$$

Now we must find the partial derivatives of F with respect to x_1, x_2, x_3, and λ. Doing this we have

$$\frac{\partial F}{\partial x_1} = 2(-.7)(.4 - .7x_1 - .4x_2)$$
$$+ 2(.5)(-.4)(.15 - .4x_1 - .1x_2)$$
$$+ 2(.4)(-.3)(.25 - .3x_1 - .3x_2) + \lambda$$

$$\frac{\partial F}{\partial x_2} = 2(-.4)(.4 - .7x_1 - .4x_2)$$
$$+ 2(.5)(-.1)(.15 - .4x_1 - .1x_2)$$
$$+ 2(.4)(-.3)(.25 - .3x_1 - .3x_2) + \lambda$$

$$\frac{\partial F}{\partial x_3} = -2(.03)(1 - x_3) + \lambda$$
$$\frac{\partial F}{\partial \lambda} = x_1 + x_2 + x_3 - 1.$$

Both $\partial F/\partial x_1$ and $\partial F/\partial x_2$ can be simplified, using some rather tedious algebra. Doing this, and placing each partial derivative equal to 0, we get the following system of equations:

$$\frac{\partial F}{\partial x_1} = -.68 + 1.212x_1 + .672x_2 + \lambda = 0$$

$$\frac{\partial F}{\partial x_2} = -.395 + .672x_1 + .402x_2 + \lambda = 0$$

$$\frac{\partial F}{\partial x_3} = -.06 + .06x_3 + \lambda = 0$$

$$\frac{\partial F}{\partial \lambda} = x_1 + x_2 + x_3 - 1 = 0.$$

Although we shall not go through the details of the solution here, it can be found from the system above that the values of the x's that minimize G are given by

$$x_1 = .48 \qquad x_2 = .09 \qquad x_3 = .43.$$

(These values have been rounded to the nearest hundredth.) Thus, the predator should spend about .48 of the available time searching in location 1, and about .09 of the time searching in location 2. This will leave about .43 of the time free for nonfeeding activities.

Exercises

1. Verify the simplification of $\partial F/\partial x_1$ and $\partial F/\partial x_2$.

2. Verify the solution given in the text by solving the system of equations.

3. Suppose $w_1 = .4$ and $w_2 = .5$ and all other values remain the same. What proportion of time should the predator spend on feeding in each of locations 1 and 2? What proportion of time will be spent on nonfeeding activities?

CHAPTER 17

PROBABILITY AND CALCULUS

(See page 888.)

In recent years, probability has become increasingly useful in fields ranging from manufacturing to medicine, as well as in all types of research. The foundations of probability were laid in the seventeenth century by Blaise Pascal (1623–1662) and Pierre de Fermat (1601–1665), who investigated *the problem of the points.* This problem dealt with the fair distribution of winnings in an interrupted game of chance between two equally matched players whose scores were known at the time of the interruption.

Probability has advanced from a study of gambling to a well-developed, deductive mathematical system. In this chapter we give a brief introduction to the use of calculus in probability.

17.1 CONTINUOUS PROBABILITY MODELS

What is the probability that there is a bird's nest within .5 kilometers of a given point?

In this section, we show how calculus is used to find the probability of certain events. Later in the section, we will answer the question posed above. Before discussing probability, however, we need to introduce some new terminology.

Suppose that a bank is studying the transaction times of its tellers. The lengths of time spent on observed transactions, rounded to the nearest minute, are shown in Table 1.

Table 1

Time	1	2	3	4	5	6	7	8	9	10	
Frequency	3	5	9	12	15	11	10	6	3	1	(Total: 75)

The table shows, for example, that 9 of the 75 transactions in the study took 3 minutes, 15 transactions took 5 minutes, and 1 transaction took 10 minutes. Because the time for any particular transaction is a random event, the number of minutes for a transaction is called a **random variable.** The frequencies can be converted to probabilities by dividing each frequency by the total number of transactions (75) to get the results shown in Table 2 on the facing page.*

*One definition of the *probability of an event* is the number of outcomes that favor the event divided by the total number of outcomes in an experiment.

Table 2

Time	1	2	3	4	5	6	7	8	9	10
Probability	.04	.07	.12	.16	.20	.15	.13	.08	.04	.01

Because each value of the random variable is associated with just one probability, Table 2 defines a function. Such a function is called a **probability distribution function** or simply a **probability function.** The special properties of a probability distribution function are given below.

PROBABILITY FUNCTION OF A RANDOM VARIABLE	If the function $f(x)$ is a probability distribution function with domain $\{x_1, x_2, \ldots, x_n\}$, then for $1 \leq i \leq n$, $$0 \leq f(x_i) \leq 1,$$ and $$f(x_1) + f(x_2) + \cdots + f(x_n) = 1.$$

The information in Table 2 can be displayed graphically with a special kind of bar graph called a **histogram.** The bars of a histogram have the same width, and their heights are determined by the probabilities of the various values of the random variable.

The probability distribution function in Table 2 is a **discrete probability function** because it has a finite domain—the integers from 1 to 10, inclusive. A discrete probability function has a finite domain or an infinite domain that can be listed. For example, if we flip a coin until we get heads, and let the random variable be the number of flips, then the domain is 1, 2, 3, 4, On the other hand, the distribution of heights (in inches) of college women includes infinitely many possible measurements, such as 53, 54.2, 66.5, 72.$\overline{3}$, and so on, *within some real number interval*. Probability distribution functions with such domains are called *continuous probability distributions*.

CONTINUOUS PROBABILITY DISTRIBUTION	A **continuous random variable** can take on any value in some interval of real numbers. The distribution of this random variable is called a **continuous probability distribution.**

In the bank example discussed earlier in this section, it would have been possible to time the teller transactions with greater precision—to the nearest tenth of a minute or even to the nearest second (one-sixtieth of a minute), if desired. Theoretically, at least, t could take on any positive real-number value between, say, 0 and 11 minutes. The graph of the probabilities $f(t)$ of these transaction times can be thought of as the continuous curve shown in Figure 1 on the next page. As indicated in Figure 1, the curve was derived from Table 2 by connecting

FIGURE 1

FIGURE 2

the points at the tops of the bars in the corresponding histogram and smoothing the resulting polygon into a curve.

For a discrete probability function, the area of each bar (or rectangle) gives the probability of a particular transaction time. Thus, by considering the possible transaction times t as all the real numbers between 0 and 11, the area under the curve of Figure 2 between any two values of t can be interpreted as the probability that a transaction time will be between those two numbers. For example, the shaded region in Figure 2 corresponds to the probability that t is between a and b, written $P(a \leq t \leq b)$.

It was shown earlier that the definite integral of a continuous function $f(x)$, where $f(x) \geq 0$, gives the area under the graph of $f(x)$ from $x = a$ to $x = b$. If a function $f(x)$ can be found to describe a continuous probability function, then the definite integral can be used to find the area under the curve from a to b that represents the probability that x will be between a and b.

The connection between area and the definite integral is discussed in the chapter on integration. For example, in that chapter we solved such problems as the following.

Find the area between the x-axis and the graph of $f(x) = x^2$ from $x = 1$ to $x = 4$.

Answer: $\int_1^4 x^2 \, dx = 21$

If x is a continuous probability function in $[a, b]$, then

$$P(a \leq x \leq b) = \int_a^b f(x) \, dx.$$

Probability Density Functions A function $f(x)$ that describes a continuous probability distribution is called a *probability density function*. Such a function must satisfy the following conditions.

PROBABILITY DENSITY FUNCTION

The function $f(x)$ is a **probability density function** of a random variable x in the interval $[a, b]$ if

1. $\int_a^b f(x) \, dx = 1$; and

2. $f(x) \geq 0$ for all x in the interval $[a, b]$.

Intuitively, condition 1 says that the total probability for the interval must be 1; *something* must happen. Condition 2 says that the probability of a particular event can never be negative.

(a) Show that the function defined by $f(x) = (3/26)x^2$ is a probability density function for the interval $[1, 3]$.

First, show that condition 1 holds.

$$\int_1^3 \frac{3}{26}x^2 \, dx = \frac{3}{26}\left(\frac{x^3}{3}\right)\Big|_1^3 = \frac{3}{26}\left(9 - \frac{1}{3}\right) = 1$$

Next, show that condition 2 holds; that is, $f(x) \geq 0$ for the interval $[1, 3]$. Since x^2 is always positive, $f(x)$ is a probability density function.

(b) Find the probability that x will be between 1 and 2.

The desired probability is given by the area under the graph of $f(x)$ between $x = 1$ and $x = 2$, as shown in Figure 3. The area is found by using a definite integral.

$$P(1 \leq x \leq 2) = \int_1^2 \frac{3}{26}x^2 \, dx = \frac{3}{26}\left(\frac{x^3}{3}\right)\Big|_1^2 = \frac{7}{26}$$

FIGURE 3

Earlier, we noted that determining a suitable function is the most difficult part of applying mathematics to actual situations. Sometimes a function appears to model an application well, but does not satisfy the requirements for a probability density function. In such cases, we may be able to change the function into a probability density function by multiplying it by a suitable constant, as shown in the next example.

EXAMPLE 2

Is the function defined by $f(x) = 3x^2$ a probability density function for the interval $[0, 4]$? If not, convert it to one.

First,

$$\int_0^4 3x^2 \, dx = x^3 \Big|_0^4 = 64.$$

Since the integral is not equal to 1, the function is not a probability density function. To convert it to one, multiply $f(x)$ by $\frac{1}{64}$. The function defined by $\frac{3}{64} x^2$ for $[0, 4]$ will be a probability density function, since

$$\int_0^4 \frac{3}{64} x^2 = 1,$$

and $(3/64)x^2 \geq 0$ for all x in $[0, 4]$. ◄

An important distinction is made between a discrete probability distribution and a probability density function (which is continuous). In a discrete distribution, the probability that the random variable, x, will assume a specific value is given in the distribution for every possible value of x. In a probability density function, however, the probability that x equals a specific value, say, c, is

$$P(x = c) = \int_c^c f(x) \, dx = 0.$$

For a probability density function, only probabilities of *intervals* can be found. For example, suppose the random variable is the annual rainfall for a given region. The amount of rainfall in one year can take on any value within some continuous interval that depends on the region; however, the probability that the rainfall in a given year will be some specific amount, say 33.25 inches, is actually zero.

The above definition of a probability density function is extended to intervals such as $(-\infty, b]$, $(-\infty, b)$, $[a, \infty)$, (a, ∞), or $(-\infty, \infty)$ by using improper integrals, as follows.

PROBABILITY DENSITY FUNCTIONS ON $(-\infty, \infty)$

If $f(x)$ is a probability density function for a continuous random variable x on $(-\infty, \infty)$, then

$$P(x \leq b) = P(x < b) = \int_{-\infty}^b f(x) \, dx,$$

$$P(x \geq a) = P(x > a) = \int_a^\infty f(x) \, dx,$$

$$P(-\infty < x < \infty) = \int_{-\infty}^\infty f(x) \, dx = 1.$$

The total area under the graph of a probability density function of this type must still equal 1.

EXAMPLE 3

Suppose the random variable x is the distance (in kilometers) from a given point to the nearest bird's nest, with the probability density function of the distribution given by $f(x) = 2xe^{-x^2}$ for $x \geq 0$.

(a) Show that $f(x)$ is a probability density function.

Since $e^{-x^2} = 1/e^{x^2}$ is always positive, and $x \geq 0$,

$$f(x) = 2xe^{-x^2} \geq 0.$$

Use substitution to evaluate the definite integral $\displaystyle\int_0^\infty 2xe^{-x^2}\, dx$. Let $u = -x^2$, so that $du = -2x\, dx$, and

$$\int 2xe^{-x^2}\, dx = -\int e^{-x^2}(-2x\, dx)$$

$$= -\int e^u\, du = -e^u = -e^{-x^2}.$$

Then

$$\int_0^\infty 2xe^{-x^2}\, dx = \lim_{b \to \infty} \int_0^b 2xe^{-x^2}\, dx = \lim_{b \to \infty} (-e^{-x^2})\Big|_0^b$$

$$= \lim_{b \to \infty} \left(-\frac{1}{e^{b^2}} + e^0\right) = 0 + 1 = 1.$$

The function defined by $f(x) = 2xe^{-x^2}$ satisfies the two conditions required of a probability density function.

(b) Find the probability that there is a bird's nest within .5 kilometers of the given point.

Find $P(x \leq .5)$ where $x \geq 0$. This probability is given by

$$P(0 \leq x \leq .5) = \int_0^{.5} 2xe^{-x^2}\, dx.$$

Now evaluate the integral. The indefinite integral was found in part (a).

$$P(0 \leq x \leq .5) = \int_0^{.5} 2xe^{-x^2}\, dx = (-e^{-x^2})\Big|_0^{.5}$$

$$= -e^{-(.5)^2} - (-e^0) = -e^{-.25} + 1$$

$$\approx -.78 + 1 = .22$$

The probability that a bird's nest will be found within .5 kilometers of the point is about .22. ◀

Improper integrals, those with one or more infinite limits, were discussed in the chapter on further techniques and applications of integration. The type of improper integral we shall need was defined as

$$\int_a^\infty f(x)\, dx = \lim_{b \to \infty} \int_a^b f(x)\, dx.$$

For example,

$$\int_1^\infty x^{-2}\, dx = \lim_{b \to \infty} \int_1^b x^{-2}\, dx$$

$$= \lim_{b \to \infty} \left(-\frac{1}{x}\Big|_1^b\right)$$

$$= \lim_{b \to \infty} \left(-\frac{1}{b} + \frac{1}{1}\right)$$

$$= 0 + 1 = 1.$$

17.1 Exercises

Decide whether the functions defined as follows are probability density functions on the indicated intervals. If not, tell why.

1. $f(x) = \frac{1}{9}x - \frac{1}{18};$ [2, 5] **2.** $f(x) = \frac{1}{3}x - \frac{1}{6};$ [3, 4] **3.** $f(x) = \frac{1}{21}x^2;$ [1, 4] **4.** $f(x) = \frac{3}{98}x^2;$ [3, 5]

5. $f(x) = 4x^3;$ [0, 3] **6.** $f(x) = \frac{x^3}{81};$ [0, 3] **7.** $f(x) = \frac{x^2}{16};$ [−2, 2] **8.** $f(x) = 2x^2;$ [−1, 1]

Find a value of k that will make f a probability density function on the indicated interval.

9. $f(x) = kx^{1/2};$ [1, 4] **10.** $f(x) = kx^{3/2};$ [4, 9] **11.** $f(x) = kx^2;$ [0, 5] **12.** $f(x) = kx^2;$ [−1, 2]

13. $f(x) = kx;$ [0, 3] **14.** $f(x) = kx;$ [2, 3] **15.** $f(x) = kx;$ [1, 5] **16.** $f(x) = kx^3;$ [2, 4]

17. The total area under the graph of a probability density function always equals _____ .

18. In your own words, define a random variable.

19. What is the difference between a discrete probability function and a probability density function?

20. Why is $P(x = c) = 0$ for any number c in the domain of a probability density function?

Show that each function defined as follows is a probability density function on the given interval; then find the indicated probabilities.

21. $f(x) = \frac{1}{2}(1 + x)^{-3/2};$ [0, ∞)

 (a) $P(0 \le x \le 2)$ **(b)** $P(1 \le x \le 3)$ **(c)** $P(x \ge 5)$

22. $f(x) = e^{-x};$ [0, ∞)

 (a) $P(0 \le x \le 1)$ **(b)** $P(1 \le x \le 2)$ **(c)** $P(x \le 2)$

23. $f(x) = (1/2)e^{-x/2};$ [0, ∞)

 (a) $P(0 \le x \le 1)$ **(b)** $P(1 \le x \le 3)$ **(c)** $P(x \ge 2)$

24. $f(x) = \frac{20}{(x + 20)^2};$ [0, ∞)

 (a) $P(0 \le x \le 1)$ **(b)** $P(1 \le x \le 5)$ **(c)** $P(x \ge 5)$

Applications

Business and Economics

25. Life Span of a Computer Part The life (in months) of a certain electronic computer part has a probability density function defined by

$$f(x) = \frac{1}{2}e^{-x/2} \quad \text{for } [0, \infty).$$

Find the probability that a randomly selected component will last each of the following lengths of time.

(a) At most 12 mo

(b) Between 12 and 20 mo

26. Machine Life A machine has a useful life of 4 to 9 years, and its life (in years) is a probability density function defined by

$$f(x) = \frac{1}{11}\left(1 + \frac{3}{\sqrt{x}}\right).$$

Find the probabilities that the useful life of such a machine selected at random will be the following.

(a) Longer than 6 yr **(b)** Less than 5 yr

(c) Between 4 and 7 yr

Life Sciences

27. Petal Length The length of a petal on a certain flower varies from 1 cm to 4 cm and is a probability density function defined by

$$f(x) = \frac{1}{2\sqrt{x}}.$$

Find the probabilities that the length of a randomly selected petal will be as follows.

(a) Greater than or equal to 3 cm

(b) Less than or equal to 2 cm

(c) Between 2 cm and 3 cm

28. Clotting Time of Blood The clotting time of blood is a random variable x with values from 1 second to 20 seconds and probability density function defined by

$$f(x) = \frac{1}{(\ln 20)x}.$$

Find the following probabilities for a person selected at random.

(a) The probability that the clotting time is between 1 and 5 sec

(b) The probability that the clotting time is greater than 10 sec

Social Sciences

29. Time to Learn a Task The time x required for a person to learn a certain task is a random variable with probability density function defined by

$$f(x) = \frac{8}{7(x - 2)^2}.$$

The time required to learn the task is between 3 and 10 min. Find the probabilities that a randomly selected person will learn the task in the following lengths of time.

(a) Less than 4 min **(b)** More than 5 min

Physical Sciences

30. Annual Rainfall The annual rainfall in a remote Middle Eastern country varies from 0 to 5 inches and is a random variable with probability density function defined by

$$f(x) = \frac{5.5 - x}{15}.$$

Find the following probabilities for the annual rainfall in a randomly selected year.

(a) The probability that the annual rainfall is greater than 3 inches

(b) The probability that the annual rainfall is less than 2 inches

(c) The probability that the annual rainfall is between 1 inch and 4 inches

General Interest

31. Drunk Drivers According to an article in the journal *Traffic Safety*, the percentage of alcohol-related traffic fatalities for teenagers and young adults has dropped significantly since 1982. Based on data given in the article, the age of a randomly selected, alcohol-impaired driver in a fatal car crash is a random variable with probability density function given by

$$f(x) = \frac{105}{4x^2} \quad \text{for } [15, 35].$$

Find the following probabilities of the age of such a driver.

(a) Less than 25 **(b)** Less than 21

(c) Between 21 and 25

32. Length of a Telephone Call The length of a telephone call (in minutes), x, for a certain town is a continuous random variable with probability density function defined by

$$f(x) = 3x^{-4} \quad \text{for } [1, \infty).$$

Find the probabilities for the following situations.

(a) The call lasts between 1 and 2 min.

(b) The call lasts between 3 and 5 min.

(c) The call lasts longer than 3 min.

17.2 EXPECTED VALUE AND VARIANCE OF CONTINUOUS RANDOM VARIABLES

What is the average age of a drunk driver in a fatal car crash?

It often is useful to have a single number, a typical or "average" number, that represents a random variable. The *mean* or *expected value* for a discrete random variable is found by multiplying each value of the random variable by its corresponding probability, as follows.

EXPECTED VALUE

Suppose the random variable x can take on the n values, $x_1, x_2, x_3, \ldots, x_n$. Also, suppose the probabilities that each of these values occurs are, respectively, $p_1, p_2, p_3, \ldots, p_n$. Then the **mean**, or **expected value,** of the random variable is

$$\mu = x_1 p_1 + x_2 p_2 + x_3 p_3 + \cdots + x_n p_n = \sum_{i=1}^{n} x_i p_i.$$

This definition can be extended to continuous random variables by using definite integrals. Suppose a continuous random variable has probability function $f(x)$ on $[a, b]$. We can divide the interval from a to b into n subintervals of length Δx, where $\Delta x = (b - a)/n$. In the ith subinterval, the probability that the random variable takes the value x_i is $f(x_i)\Delta x_i$, and so

$$\mu = \sum_{i=1}^{n} x_i f(x_i)\Delta x_i.$$

As $n \to \infty$, the limit of this sum gives the expected value

$$\mu = \int_a^b x f(x) \, dx.$$

The **variance** of a probability distribution is a measure of the *spread* of the values of the distribution. For a discrete distribution, the variance is found by taking the expected value of the squares of the differences of the values of the random variable and the mean. If the random variable x takes the values $x_1, x_2, x_3, \ldots, x_n$, with respective probabilities $p_1, p_2, p_3, \ldots, p_n$ and mean μ, then the variance of x is

$$\text{Var}(x) = \sum_{i=1}^{n} (x_i - \mu)^2 p_i.$$

The **standard deviation** of x is defined as

$$\sigma = \sqrt{\text{Var}(x)}.$$

Like the mean or expected value, the variance of a continuous random variable is an integral.

$$\text{Var}(x) = \int_a^b (x - \mu)^2 f(x) \, dx$$

To find the standard deviation of a continuous probability distribution, like that of a discrete distribution, we find the square root of the variance. The formulas for the expected value, variance, and standard deviation of a continuous probability distribution are summarized on the next page.

EXPECTED VALUE, VARIANCE, AND STANDARD DEVIATION

If x is a continuous random variable with probability density function $f(x)$ on $[a, b]$, then the expected value of x is

$$E(x) = \mu = \int_a^b xf(x)\ dx.$$

The variance of x is

$$\text{Var}(x) = \int_a^b (x - \mu)^2 f(x)\ dx,$$

and the standard deviation of x is

$$\sigma = \sqrt{\text{Var}(x)}.$$

Geometrically, the expected value (or mean) of a probability distribution represents the "center" of the distribution. If a fulcrum were placed at μ on the x-axis, the figure would be in balance. See Figure 4.

FIGURE 4

The variance or standard deviation of a probability distribution indicates how closely the values of the distribution cluster about the mean. These measures are most useful for comparing different distributions, as in Figure 5.

Small σ Large σ

FIGURE 5

EXAMPLE

1 Find the expected value and variance of the random variable x with probability density function defined by $f(x) = (3/26)x^2$ on $[1, 3]$.

By the definition of expected value given above,

$$\mu = \int_1^3 xf(x)\ dx$$

$$= \int_1^3 x\left(\frac{3}{26}x^2\right)dx$$

$$= \frac{3}{26}\int_1^3 x^3\ dx$$

$$= \frac{3}{26}\left(\frac{x^4}{4}\right)\Bigg|_1^3 = \frac{3}{104}(81 - 1) = \frac{30}{13},$$

or about 2.31.

The variance is

$$\text{Var}(x) = \int_1^3 \left(x - \frac{30}{13}\right)^2\left(\frac{3}{26}x^2\right)dx$$

$$= \int_1^3 \left(x^2 - \frac{60}{13}x + \frac{900}{169}\right)\left(\frac{3}{26}x^2\right)dx \qquad \text{Square } \left(x - \frac{30}{13}\right).$$

$$= \frac{3}{26}\int_1^3 \left(x^4 - \frac{60}{13}x^3 + \frac{900}{169}x^2\right)dx \qquad \text{Multiply.}$$

$$= \frac{3}{26}\left(\frac{x^5}{5} - \frac{60}{13}\cdot\frac{x^4}{4} + \frac{900}{169}\cdot\frac{x^3}{3}\right)\Bigg|_1^3 \qquad \text{Integrate.}$$

$$= \frac{3}{26}\left[\left(\frac{243}{5} - \frac{60(81)}{52} + \frac{900(27)}{169(3)}\right) - \left(\frac{1}{5} - \frac{60}{52} + \frac{300}{169}\right)\right]$$

$$\approx .259.$$

From the variance, the standard deviation is $\sigma \approx \sqrt{.259} \approx .51$. The expected value and standard deviation are shown on the graph of the probability density function in Figure 6. ◀

$f(x) = \frac{3}{26}x^2$

$\sigma = .51$
$\mu = 2.31$

FIGURE 6

Calculating the variance in the last example was a messy job. An alternative version of the formula for the variance is easier to compute. This alternative formula is derived as follows.

$$\text{Var}(x) = \int_a^b (x - \mu)^2\ f(x)\ dx$$

$$= \int_a^b (x^2 - 2\mu x + \mu^2)f(x)\ dx$$

$$= \int_a^b x^2\ f(x)\ dx - 2\mu\int_a^b xf(x)\ dx + \mu^2\int_a^b f(x)\ dx \qquad \textbf{(1)}$$

By definition,

$$\int_a^b xf(x)\,dx = \mu,$$

and, since $f(x)$ is a probability density function,

$$\int_a^b f(x)\,dx = 1.$$

Substitute back into equation (1) to get the alternative formula,

$$\text{Var}(x) = \int_a^b x^2\,f(x)\,dx - 2\mu^2 + \mu^2 = \int_a^b x^2\,f(x)\,dx - \mu^2.$$

ALTERNATIVE FORMULA FOR VARIANCE

If x is a random variable with probability density function $f(x)$ on $[a, b]$, and if $E(x) = \mu$, then

$$\textbf{Var}(x) = \int_a^b x^2\,f(x)\,dx - \mu^2.$$

Caution Notice that the term μ^2 comes *after* the differential dx, and so is *not* integrated.

EXAMPLE 2 Use the alternative formula for variance to compute the variance of the random variable x with probability density function defined by $f(x) = 3/x^4$ for $x \geq 1$.

To find the variance, first find the expected value:

$$\mu = \int_1^\infty xf(x)\,dx = \int_1^\infty x \cdot \frac{3}{x^4}\,dx = \int_1^\infty \frac{3}{x^3}\,dx$$

$$= \lim_{b\to\infty} \int_1^b \frac{3}{x^3}\,dx = \lim_{b\to\infty} \left(\frac{3}{-2x^2}\right)\Bigg|_1^b = \frac{3}{2},$$

or 1.5. Now find the variance by the alternative formula for variance:

$$\text{Var}(x) = \int_1^\infty x^2\left(\frac{3}{x^4}\right)dx - \left(\frac{3}{2}\right)^2$$

$$= \int_1^\infty \frac{3}{x^2}\,dx - \frac{9}{4}$$

$$= \lim_{b\to\infty} \int_1^b \frac{3}{x^2}\,dx - \frac{9}{4}$$

$$= \lim_{b\to\infty} \left(\frac{-3}{x}\right)\Bigg|_1^b - \frac{9}{4}$$

$$= 3 - \frac{9}{4} = \frac{3}{4}, \quad \text{or } .75. \quad \blacktriangleleft$$

EXAMPLE **3** A recent study has shown that on any given day, the proportion of airline passengers who arrive at the gate at least 1/2 hour before the scheduled flight time is a probability density function, with $f(x) = 6x - 6x^2$ for $0 \le x \le 1$.

(a) Find and interpret the expected value for this distribution.

The expected value is

$$\mu = \int_0^1 x(6x - 6x^2)\, dx = \int_0^1 (6x^2 - 6x^3)\, dx$$

$$= \left(2x^3 - \frac{3}{2}x^4\right)\Big|_0^1 = \frac{1}{2},$$

or .5. This result indicates that over a long period of time, about one-half of all airline passengers arrive at the gate at least 1/2 hour before the scheduled flight time.

(b) Compute the standard deviation.

First compute the variance. We use the alternative formula.

$$\text{Var}(x) = \int_0^1 x^2(6x - 6x^2)\, dx - \left(\frac{1}{2}\right)^2$$

$$= \int_0^1 (6x^3 - 6x^4)\, dx - \left(\frac{1}{2}\right)^2$$

$$= \left(\frac{3}{2}x^4 - \frac{6}{5}x^5\right)\Big|_0^1 - \frac{1}{4}$$

$$= \frac{3}{10} - \frac{1}{4} = \frac{1}{20} = .05$$

The standard deviation is $\sigma = \sqrt{.05} \approx .22$. ◀

17.2 Exercises

In Exercises 1–8, a probability density function of a random variable is defined. Find the expected value, the variance, and the standard deviation. Round answers to the nearest hundredth.

1. $f(x) = \dfrac{1}{4}$; [3, 7]

2. $f(x) = \dfrac{1}{10}$; [0, 10]

3. $f(x) = \dfrac{x}{8} - \dfrac{1}{4}$; [2, 6]

4. $f(x) = 2(1 - x)$; [0, 1]

5. $f(x) = 1 - \dfrac{1}{\sqrt{x}}$; [1, 4]

6. $f(x) = \dfrac{1}{11}\left(1 + \dfrac{3}{\sqrt{x}}\right)$; [4, 9]

7. $f(x) = 4x^{-5}$; $[1, \infty)$

8. $f(x) = 3x^{-4}$; $[1, \infty)$

9. What information does the mean (expected value) of a continuous random variable give?

10. Suppose two random variables have standard deviations of .10 and .23, respectively. What does this tell you about their distributions?

In Exercises 11–14, the probability density function of a random variable is defined.

(a) *Find the expected value to the nearest hundredth.*

(b) *Find the variance to the nearest hundredth.*

(c) *Find the standard deviation. Round to the nearest hundredth.*

(d) *Find the probability that the random variable has a value greater than the mean.*

(e) *Find the probability that the value of the random variable is within one standard deviation of the mean.*

11. $f(x) = \dfrac{\sqrt{x}}{18};$ $[0, 9]$

12. $f(x) = \dfrac{x^{-1/3}}{6};$ $[0, 8]$

13. $f(x) = \dfrac{1}{2}x;$ $[0, 2]$

14. $f(x) = \dfrac{3}{2}(1 - x^2);$ $[0, 1]$

*If x is a random variable with probability density function f(x) on [a, b], then the **median** of x is the number m such that*

$$\int_a^m f(x)\, dx = \frac{1}{2}.$$

(a) *Find the median of each random variable for the probability density functions in Exercises 15–20.*

(b) *In each case, find the probability that the random variable is between the expected value (mean) and the median. The expected value for each of these functions was found in Exercises 1–8.*

15. $f(x) = \dfrac{1}{4};$ $[3, 7]$

16. $f(x) = \dfrac{1}{10};$ $[0, 10]$

17. $f(x) = \dfrac{x}{8} - \dfrac{1}{4};$ $[2, 6]$

18. $f(x) = 2(1 - x);$ $[0, 1]$

19. $f(x) = 4x^{-5};$ $[1, \infty)$

20. $f(x) = 3x^{-4};$ $[1, \infty)$

Applications

Business and Economics

21. Life of a Light Bulb The life (in hours) of a certain kind of light bulb is a random variable with probability density function defined by

$$f(x) = \frac{1}{58\sqrt{x}} \quad \text{for } [1, 900].$$

(a) What is the expected life of such a bulb?

(b) Find σ.

(c) Find the probability that one of these bulbs lasts longer than one standard deviation above the mean.

22. Machine Life The life (in years) of a certain machine is a random variable with probability density function defined by

$$f(x) = \frac{1}{11}\left(1 + \frac{3}{\sqrt{x}}\right) \quad \text{for } [4, 9].$$

(a) Find the mean life of this machine.

(b) Find the standard deviation of the distribution.

(c) Find the probability that a particular machine of this kind will last longer than the mean number of years.

23. Life of an Automobile Part The life span of a certain automobile part (in months) is a random variable with probability density function defined by

$$f(x) = \frac{1}{2}e^{-x/2} \quad \text{for } [0, \infty).$$

(a) Find the expected life of this part.

(b) Find the standard deviation of the distribution.

(c) Find the probability that one of these parts lasts less than the mean number of months.

Life Sciences

24. Blood Clotting Time The clotting time of blood (in seconds) is a random variable with probability density function defined by

$$f(x) = \frac{1}{(\ln 20)x} \quad \text{for } [1, 20].$$

(a) Find the mean clotting time.

(b) Find the standard deviation.

(c) Find the probability that a person's blood clotting time is within one standard deviation of the mean.

25. Length of a Leaf The length of a leaf on a tree is a random variable with probability density function defined by

$$f(x) = \frac{3}{32}(4x - x^2) \quad \text{for } [0, 4].$$

(a) What is the expected leaf length?

(b) Find σ for this distribution.

(c) Find the probability that the length of a given leaf is within one standard deviation of the expected value.

26. Petal Length The length (in centimeters) of a petal on a certain flower is a random variable with probability density function defined by

$$f(x) = \frac{1}{2\sqrt{x}} \quad \text{for } [1, 4].$$

(a) Find the expected petal length.

(b) Find the standard deviation.

(c) Find the probability that a petal selected at random has a length more than two standard deviations above the mean.

Physical Sciences

27. Annual Rainfall The annual rainfall in a remote Middle Eastern country is a random variable with probability

density function defined by

$$f(x) = \frac{5.5 - x}{15} \quad \text{for } [0, 5].$$

(a) Find the mean annual rainfall.

(b) Find the standard deviation.

(c) Find the probability of a year with rainfall less than one standard deviation below the mean.

General Interest

28. Drunk Drivers The age of a randomly selected, alcohol-impaired driver in a fatal car crash is a random variable with probability density function given by

$$f(x) = \frac{105}{4x^2} \quad \text{for } [15, 35].$$

 (a) Find the expected age of a drunk driver in a fatal car crash.

(b) Find the standard deviation of the distribution.

(c) Find the probability that the age of such a driver will be less than one standard deviation below the mean.

▶ 17.3 SPECIAL PROBABILITY DENSITY FUNCTIONS

What is the probability that the maximum temperature will be higher than 24° C?

What is the probability that a battery will last longer than 40 hours?

These questions can be answered if the probability density function for the maximum temperature and for the life of the battery are known. In practice, however, it is not feasible to construct a probability density function for each experiment. Instead, a researcher uses one of several probability density functions that are well known, matching the shape of the experimental distribution to one of the known distributions. In this section we discuss some of the most commonly used probability distributions.

Uniform Distribution The probability density function for the **uniform distribution** is defined by

$$f(x) = \frac{1}{b - a} \quad \text{for } [a, b],$$

where a and b are nonnegative real numbers. The graph of $f(x)$ is shown in Figure 7 on the facing page.

Uniform distribution

FIGURE 7

Since $b - a$ is positive, $f(x) \geq 0$, and

$$\int_a^b \frac{1}{b-a}\, dx = \frac{1}{b-a} x \Big|_a^b = \frac{1}{b-a}(b-a) = 1.$$

Therefore, the function is a probability density function.

The expected value for the uniform distribution is

$$\mu = \int_a^b \left(\frac{1}{b-a}\right) x\, dx = \left(\frac{1}{b-a}\right)\frac{x^2}{2}\Big|_a^b$$

$$= \frac{1}{2(b-a)}(b^2 - a^2) = \frac{1}{2}(b+a). \qquad b^2 - a^2 = (b-a)(b+a)$$

The variance is given by

$$\mathrm{Var}(x) = \int_a^b \left(\frac{1}{b-a}\right) x^2\, dx - \left(\frac{b+a}{2}\right)^2$$

$$= \left(\frac{1}{b-a}\right)\frac{x^3}{3}\Big|_a^b - \frac{(b+a)^2}{4}$$

$$= \frac{1}{3(b-a)}(b^3 - a^3) - \frac{1}{4}(b+a)^2$$

$$= \frac{b^2 + ab + a^2}{3} - \frac{b^2 + 2ab + a^2}{4} \qquad b^3 - a^3 = (b-a)(b^2 + ab + a^2)$$

$$= \frac{b^2 - 2ab + a^2}{12}. \qquad \text{Get a common denominator; subtract.}$$

Thus,

$$\mathrm{Var}(x) = \frac{1}{12}(b-a)^2, \qquad \text{Factor.}$$

and

$$\sigma = \frac{1}{\sqrt{12}}(b-a).$$

These properties of the uniform distribution are summarized below.

UNIFORM DISTRIBUTION If x is a random variable with probability density function

$$f(x) = \frac{1}{b - a} \quad \text{for } [a, b],$$

then

$$\mu = \frac{1}{2}(b + a) \quad \text{and} \quad \sigma = \frac{1}{\sqrt{12}}(b - a).$$

EXAMPLE 1 A couple is planning to vacation in San Francisco. They have been told that the maximum daily temperature during the time they plan to be there ranges from 15° C to 27° C. Assume that the probability of any temperature between 15° C and 27° C is equally likely for any given day during the specified time period.

(a) What is the probability that the maximum temperature on the day they arrive will be higher than 24° C?

If the random variable t represents the maximum temperature on a given day, then the uniform probability density function for t is defined by $f(t) = 1/12$ for the interval $[15, 27]$. By definition,

$$P(t > 24) = \int_{24}^{27} \frac{1}{12} \, dt = \frac{1}{12}t \Big|_{24}^{27} = \frac{1}{4}.$$

(b) What average maximum temperature can they expect?

The expected maximum temperature is

$$\mu = \frac{1}{2}(27 + 15) = 21,$$

or 21° C.

(c) What is the probability that the maximum temperature on a given day will be at least one standard deviation below the mean?

First find σ:

$$\sigma = \frac{1}{\sqrt{12}}(27 - 15) = \frac{12}{\sqrt{12}} = \sqrt{12} = 2\sqrt{3} \approx 3.5.$$

One standard deviation below the mean indicates a temperature of $21 - 3.5 = 17.5°$ C.

$$P(T \leq 17.5) = \int_{15}^{17.5} \frac{1}{12} \, dt = \frac{1}{12}t \Big|_{15}^{17.5} \approx .21$$

The probability is .21 that the temperature will not exceed 17.5° C. ◄

Exponential Distribution The probability density function for the **exponential distribution** is defined by

$$f(x) = ae^{-ax} \quad \text{for } [0, \infty),$$

where a is a positive real number. The graph of $f(x)$ is shown in Figure 8.

Exponential distribution

FIGURE 8

Here $f(x) \geq 0$, since e^{-ax} and a are both positive for all values of x. Also,

$$\int_0^\infty ae^{-ax}\,dx = \lim_{b \to \infty} \int_0^b ae^{-ax}\,dx$$

$$= \lim_{b \to \infty} (-e^{-ax})\Big|_0^b = \lim_{b \to \infty}\left(\frac{-1}{e^{ab}} + \frac{1}{e^0}\right) = 1,$$

so the function is a probability density function.

The expected value and the standard deviation of the exponential distribution can be found using integration by parts. The results are given below.

EXPONENTIAL DISTRIBUTION

If x is a random variable with probability density function

$$f(x) = ae^{-ax} \quad \text{for } [0, \infty),$$

then

$$\mu = \frac{1}{a} \quad \text{and} \quad \sigma = \frac{1}{a}.$$

The exponential distribution is very important in reliability analysis; when manufactured items have a constant failure rate over a period of time, the exponential distribution is used to describe their probability of failure, as in the following example.

EXAMPLE **2** ▶ Suppose the useful life (in hours) of a flashlight battery is the random variable t, with probability density function given by the exponential distribution

$$f(t) = \frac{1}{20}e^{-t/20} \quad \text{for } t \ge 0.$$

(a) Find the probability that a particular battery, selected at random, has a useful life of less than 100 hours.

The probability is given by

$$P(t \le 100) = \int_0^{100} \frac{1}{20}e^{-t/20}\, dt = \frac{1}{20}\left(-20e^{-t/20}\right)\Big|_0^{100}$$

$$= -(e^{-100/20} - e^0) = -(e^{-5} - 1)$$

$$\approx 1 - .0067 = .9933.$$

(b) Find the expected value and standard deviation of the distribution.

Use the formulas given above. Both μ and σ equal $1/a$, and since $a = 1/20$ here,

$$\mu = 20 \quad \text{and} \quad \sigma = 20.$$

This means that the average life of a battery is 20 hours, and no battery lasts less than one standard deviation below the mean.

(c) What is the probability that a battery will last longer than 40 hours?

The probability is given by

$$P(t > 40) = \int_{40}^{\infty} \frac{1}{20}e^{-t/20}\, dt = \lim_{b \to \infty} \left(-e^{-t/20}\right)\Big|_{40}^{b} = \frac{1}{e^2} \approx .1353,$$

or about 14%. ◀

Normal Distribution The **normal distribution,** with its well-known bell-shaped graph, is undoubtedly the most important probability density function. It is widely used in various applications of statistics. The probability density function for the normal distribution has the following characteristics.

NORMAL DISTRIBUTION If μ and σ are real numbers, $\sigma \ge 0$, and if x is a random variable with probability density function defined by

$$f(x) = \frac{1}{\sigma\sqrt{2\pi}}e^{-(x-\mu)^2/(2\sigma^2)} \quad \text{for } (-\infty, \infty),$$

then

$$E(x) = \mu \quad \text{and} \quad \text{Var}(x) = \sigma^2, \quad \text{with standard deviation } \sigma.$$

Advanced techniques can be used to show that

$$\int_{-\infty}^{\infty} \frac{1}{\sigma\sqrt{2\pi}} e^{-(x-\mu)^2/(2\sigma^2)} \, dx = 1.$$

Deriving the expected value and standard deviation for the normal distribution also requires techniques beyond the scope of this text.

Each normal probability distribution has associated with it a bell-shaped curve, called a **normal curve,** such as the one in Figure 9. Each normal curve is symmetric about a vertical line through the mean, μ. Vertical lines at points $+1\sigma$ and -1σ from the mean show the inflection points of the graph. A normal curve never touches the x-axis; it extends indefinitely in both directions.

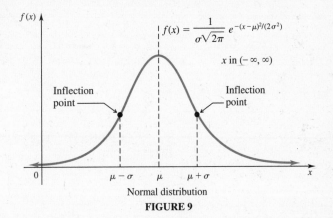

Normal distribution

FIGURE 9

The development of the normal curve is credited to the Frenchman Abraham de Moivre (1667–1754). Three of his publications dealt with probability and associated topics: *Annuities upon Lives* (which contributed to the development of actuarial studies), *Doctrine of Chances,* and *Miscellanea Analytica.*

Many different normal curves have the same mean. In such cases, a larger value of σ produces a ''flatter'' normal curve, while smaller values of σ produce more values near the mean, resulting in a ''taller'' normal curve. See Figure 10.

FIGURE 10

It would be far too much work to calculate values for the normal probability distribution for various values of μ and σ. Instead, values are calculated for the **standard normal distribution,** which has $\mu = 0$ and $\sigma = 1$. The graph of the standard normal distribution is shown in Figure 11.

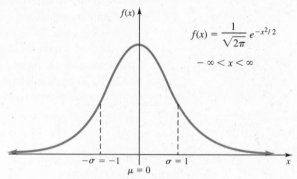

$$f(x) = \frac{1}{\sqrt{2\pi}} e^{-x^2/2}$$

$$-\infty < x < \infty$$

Standard normal distribution

FIGURE 11

Probabilities for the standard normal distribution come from the definite integral

$$\int_a^b \frac{1}{\sqrt{2\pi}} e^{-x^2/2}\, dx.$$

Since $f(x) = e^{-x^2/2}$ does not have an antiderivative that can be expressed in terms of functions used in this course, numerical methods are used to find values of this definite integral. A table in the appendix of this book gives areas under the standard normal curve, along with a sketch of the curve. Each value in this table is the total area under the standard normal curve to the left of the number z.

If a normal distribution does not have $\mu = 0$ and $\sigma = 1$, we use the following theorem, which is stated without proof.

z-SCORES THEOREM

Suppose a normal distribution has mean μ and standard deviation σ. The area under the associated normal curve that is to the left of the value x is exactly the same as the area to the left of

$$z = \frac{x - \mu}{\sigma}$$

for the standard normal curve.

Using this result, the table can be used for *any* normal distribution, regardless of the values of μ and σ. The number z in the theorem is called a **z-score.**

3 A normal distribution has mean 35 and standard deviation 5.9. Find the following areas under the associated normal curve.

(a) The area of the region to the left of 40

Find the appropriate z-score using $x = 40$, $\mu = 35$, and $\sigma = 5.9$. Round to the nearest hundredth.

$$z = \frac{40 - 35}{5.9} = \frac{5}{5.9} \approx .85$$

Look up .85 in the normal curve table in the Appendix. The corresponding area is .8023. Thus, the shaded area shown in Figure 12 is .8023. This area represents 80.23% of the total area under the normal curve.

Shaded area
= .8023

FIGURE 12

Shaded area
= 1 − .3050
= .6950

FIGURE 13

(b) The area of the region to the right of 32

$$z = \frac{32 - 35}{5.9} = \frac{-3}{5.9} \approx -.51$$

The area to the *left* of $z = -.51$ is .3050, so the area to the *right is* $1 - .3050 = .6950$. See Figure 13.

(c) The area of the region between 30 and 33

Find z-scores for both values.

$$z = \frac{30 - 35}{5.9} = \frac{-5}{5.9} \approx -.85 \quad \text{and} \quad z = \frac{33 - 35}{5.9} = \frac{-2}{5.9} \approx -.34$$

Start with the area to the left of $z = -.34$ and subtract the area to the left of $z = -.85$:

$$.3669 - .1977 = .1692.$$

The required area is shaded in Figure 14 on the next page. ◀

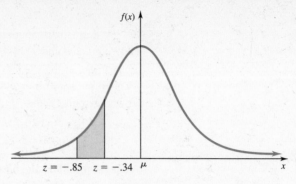

FIGURE 14

The z-scores are actually standard deviation multiples; that is, a z-score of 2.5 corresponds to a value 2.5 standard deviations above the mean. For example, looking up $z = 1.00$ and $z = -1.00$ in the table shows that

$$.8413 - .1587 = .6826,$$

so that 68.26% of the area under a normal curve is within one standard deviation of the mean. Also, using $z = 2.00$ and $z = -2.00$,

$$.9772 - .0228 = .9544,$$

meaning 95.44% of the area is within two standard deviations of the mean. These results, summarized in Figure 15, can be used to get a quick estimate of results when working with normal curves.

FIGURE 15

17.3 Exercises

For Exercises 1–6, find each of the following:

(a) *the mean of the distribution;*

(b) *the standard deviation of the distribution;*

(c) *the probability that the random variable is between the mean and one standard deviation above the mean.*

1. The length (in centimeters) of the leaf of a certain plant is a continuous random variable with probability density function defined by

$$f(x) = \frac{5}{4} \quad \text{for } [4, 4.8].$$

2. The price of an item (in dollars) is a continuous random variable with probability density function defined by

$$f(x) = 2 \quad \text{for } [1.25, 1.75].$$

3. The length of time (in years) until a particular radioactive particle decays is a random variable t with probability density function defined by

$$f(t) = .03e^{-.03t} \quad \text{for } [0, \infty).$$

4. The length of time (in years) that a seedling tree survives is a random variable t with probability density function defined by

$$f(t) = .05e^{-.05t} \quad \text{for } [0, \infty).$$

5. The length of time (in days) required to learn a certain task is a random variable t with probability density function defined by

$$f(t) = e^{-t} \quad \text{for } [0, \infty).$$

6. The distance (in meters) that seeds are dispersed from a certain kind of plant is a random variable x with probability density function defined by

$$f(x) = .1e^{-.1x} \quad \text{for } [0, \infty).$$

Find the percent of the area under a normal curve between the mean and the number of standard deviations above the mean given in Exercises 7 and 8.

7. 3.50

8. 1.68

Find the percent of the total area under the normal curve between the z-scores given in Exercises 9 and 10.

9. 1.28 and 2.05

10. -2.13 and $-.04$

Find a z-score satisfying the conditions given in Exercises 11–14. (Hint: Use the table backwards.)

11. 10% of the total area is to the left of z

12. 2% of the total area is to the left of z

13. 18% of the total area is to the right of z

14. 22% of the total area is to the right of z

15. Describe the standard normal distribution. What are its characteristics?

16. What is meant by a z-score? How is it used?

17. Describe the shape of the graph of each of the following probability distributions.

(a) Uniform (b) Exponential (c) Normal

In the exercises for Section 9.2, we defined the median of a probability distribution as an integral. The median also can be defined as the number m such that $P(x \leq m) = P(x \geq m)$.

18. Find an expression for the median of the uniform distribution.

19. Find an expression for the median of the exponential distribution.

20. Verify the expected value and standard deviation of the exponential distribution given in the text.

Applications

Business and Economics

21. **Insurance Sales** The amount of insurance (in thousands of dollars) sold in a day by a particular agent is uniformly distributed over the interval [10, 85].

(a) What amount of insurance does the agent sell on an average day?

(b) Find the probability that the agent sells more than $50,000 of insurance on a particular day.

22. **Fast-Food Outlets** The number of new fast-food outlets opening during June in a certain city is exponentially distributed, with a mean of 5.

(a) Give the probability density function for this distribution.

(b) What is the probability that the number of outlets opening is between 2 and 6?

23. Sales Expense A salesperson's monthly expenses (in thousands of dollars) are exponentially distributed, with an average of 4.25 (thousand dollars).

(a) Give the probability density function for the expenses.

(b) Find the probability that the expenses are more than $10,000.

In Exercises 24–26, assume a normal distribution.

24. Machine Accuracy A machine that fills quart bottles with apple juice averages 32.8 oz per bottle, with a standard deviation of 1.1 oz. What are the probabilities that the amount of juice in a bottle is as follows?

(a) Less than 1 qt

(b) At least 1 oz more than a quart

25. Machine Accuracy A machine produces screws with a mean length of 2.5 cm and a standard deviation of .2 cm. Find the probabilities that a screw produced by this machine has lengths as follows.

(a) Greater than 2.7 cm

(b) Within 1.2 standard deviations of the mean

26. Customer Expenditures Customers at a certain pharmacy spend an average of $54.40, with a standard deviation of $13.50. What are the largest and smallest amounts spent by the middle 50% of these customers?

Life Sciences

27. Insect Life Span The life span of a certain insect (in days) is uniformly distributed over the interval [20, 36].

(a) What is the expected life of this insect?

(b) Find the probability that one of these insects, randomly selected, lives longer than 30 days.

28. Location of a Bee Swarm A swarm of bees is released from a certain point. The proportion of the swarm located at least 2 m from the point of release after 1 hr is a random variable that is exponentially distributed, with $a = 2$ over the interval $[0, \infty)$.

(a) Find the expected proportion under the given conditions.

(b) Find the probability that fewer than 1/3 of the bees are located at least 2 m from the release point after 1 hr.

29. Digestion Time The digestion time (in hours) of a fixed amount of food is exponentially distributed, with $a = 1$.

(a) Find the mean digestion time.

(b) Find the probability that the digestion time is less than 30 min.

30. Pygmy Heights The average height of a member of a certain tribe of pygmies is 3.2 ft, with a standard deviation of .2 ft. If the heights are normally distributed, what are the largest and smallest heights of the middle 50% of this population?

31. Finding Prey H. R. Pulliam found that the time (in minutes) required by a predator to find a prey is a random variable that is exponentially distributed, with $\mu = 25$.*

(a) According to this distribution, what is the longest time within which the predator will be 90% certain of finding a prey?

(b) What is the probability that the predator will have to spend more than one hour looking for a prey?

Physical Sciences

32. Rainfall The rainfall (in inches) in a certain region is uniformly distributed over the interval [32, 44].

(a) What is the expected number of inches of rainfall?

(b) What is the probability that the rainfall will be between 38 and 40 inches?

Computer/Graphing Calculator

Use Simpson's rule, with $n = 100$, to approximate the following integrals.

33. $\int_0^{50} .5e^{-.5x}\, dx$

34. $\int_0^{50} .5xe^{-.5x}\, dx$

35. $\int_0^{50} .5x^2e^{-.5x}\, dx$

36. Use your results from Exercises 34 and 35 to verify that, for the exponential distribution, $\mu = 1/a$ and $\sigma = 1/a$.

37. The standard normal probability density function is defined by

$$f(x) = \frac{1}{\sqrt{2\pi}}e^{-x^2/2}.$$

Use Simpson's rule, with $n = 100$, and the formulas for the mean and standard deviation of a probability density function to approximate the following for the standard normal probability distribution. Use limits of -4 and 4 (instead of $-\infty$ and ∞).

(a) The mean (b) The standard deviation

*From H. R. Pulliam, ''On the Theory of Optimal Diets,'' *American Naturalist*, vol. 108, 1974, pp. 59–74.

Chapter Summary	Key Terms

17.1 random variable
 probability distribution function
 histogram
 discrete probability function
 continuous random variable
 continuous probability
 distribution
 probability density function

17.2 mean
 expected value
 variance
 standard deviation

17.3 uniform distribution
 exponential distribution
 normal distribution
 normal curve
 standard normal distribution
 z-score

Chapter 17	Review Exercises

1. In a probability function, the y-values (or function values) represent _____ .
2. Define a continuous random variable.
3. Give the two conditions that a probability density function for $[a, b]$ must satisfy.

Decide whether each function defined as follows is a probability density function for the given interval.

4. $f(x) = \dfrac{1}{27}(2x + 4);$ $[1, 4]$

5. $f(x) = \sqrt{x};$ $[4, 9]$

6. $f(x) = .1;$ $[0, 10]$

7. $f(x) = e^{-x};$ $[0, \infty)$

In Exercises 8 and 9, find a value of k that will make f(x) define a probability density function for the indicated interval.

8. $f(x) = k\sqrt{x};$ $[1, 4]$

9. $f(x) = kx^2;$ $[0, 3]$

10. The probability density function of a random variable x is defined by

$$f(x) = 1 - \frac{1}{\sqrt{x - 1}} \quad \text{for } [2, 5].$$

Find the following probabilities.
(a) $P(x \geq 3)$ (b) $P(x \leq 4)$ (c) $P(3 \leq x \leq 4)$

11. The probability density function of a random variable x is defined by

$$f(x) = \frac{1}{10} \quad \text{for } [10, 20].$$

Find the following probabilities.
(a) $P(x \leq 12)$ (b) $P(x \geq 31/2)$ (c) $P(10.8 \leq x \leq 16.2)$

12. Describe what the expected value or mean of a probability distribution represents geometrically.

13. The probability density functions shown in the graphs below have the same mean. Which has the smallest standard deviation?

(a)

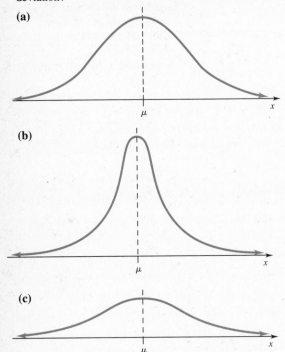

(b)

(c)

For the probability density functions defined in Exercises 14–17, find the expected value, the variance, and the standard deviation.

14. $f(x) = \dfrac{1}{5}$; $[4, 9]$

15. $f(x) = \dfrac{2}{9}(x - 2)$; $[2, 5]$

16. $f(x) = \dfrac{1}{7}\left(1 + \dfrac{2}{\sqrt{x}}\right)$; $[1, 4]$

17. $f(x) = 5x^{-6}$; $[1, \infty)$

18. The probability density function of a random variable is defined by $f(x) = 4x - 3x^2$ for $[0, 1]$.
Find each of the following for the distribution.

(a) The mean (b) The standard deviation

(c) The probability that the value of the random variable will be less than the mean

(d) The probability that the value of the random variable will be within one standard deviation of the mean

19. Find the median of the random variable of Exercise 18. (See Exercises 15–20 in Section 10.2.) Then find the probability that the value of the random variable will lie between the median and the mean of the distribution.

For Exercises 20–21, find **(a)** *the mean of the distribution,* **(b)** *the standard deviation of the distribution, and* **(c)** *the probability that the value of the random variable is within one standard deviation of the mean.*

20. $f(x) = \dfrac{5}{112}(1 - x^{-3/2})$ for $[1, 25]$

21. $f(x) = .01e^{-.01x}$ for $[0, \infty)$

In Exercises 22–27, find the percent of the area under a normal curve for each of the following.

22. The region to the right of $z = 1.53$

23. The region to the left of $z = -.49$

24. The region between $z = -1.47$ and $z = 1.03$

25. The region between $z = -.98$ and $z = -.15$

26. The region that is up to 2.5 standard deviations above the mean

27. The region that is up to 1.2 standard deviations below the mean

28. Find a z-score so that 21% of the area under the normal curve is to the left of z.

29. Find a z-score so that 52% of the area under the normal curve is to the right of z.

Applications

Business and Economics

30. Mutual Funds The price per share (in dollars) of a particular mutual fund is a random variable x with probability density function defined by

$$f(x) = \frac{3}{4}(x^2 - 16x + 65) \quad \text{for } [8, 9].$$

Find the probability that the price will be less than $8.50.

31. Machine Repairs The time (in years) until a certain machine requires repairs is a random variable t with probability density function defined by

$$f(t) = \frac{5}{112}(1 - t^{-3/2}) \quad \text{for } [1, 25].$$

Find the probability that no repairs are required in the first three years by finding the probability that a repair will be needed in years 4 through 25.

32. Product Repairs The number of repairs required by a new product each month is exponentially distributed, with an average of 8.

(a) What is the probability density function for this distribution?

(b) Find the expected number of repairs per month.

(c) Find the standard deviation.

(d) What is the probability that the number of repairs per month will be between 5 and 10?

33. Retail Outlets The number of new outlets for a clothing manufacturer is an exponential distribution with probability density function defined by

$$f(x) = \frac{1}{6}e^{-x/6} \quad \text{for } [0, \infty).$$

Find each of the following for this distribution.

(a) The mean

(b) The standard deviation

(c) The probability that the number of new outlets will be greater than the mean

34. Useful Life of an Appliance Part The useful life of a certain appliance part (in hundreds of hours) is 46.2, with a standard deviation of 15.8. Find the probability that one such part would last for at least 6000 (60 hundred) hr. Assume a normal distribution.

Life Sciences

35. Movement of a Released Animal The distance (in meters) that a certain animal moves away from a release point is a random variable with probability density function defined by

$$f(x) = .01e^{-.01x} \quad \text{for } [0, \infty).$$

Find the probability that the animal will move no farther than 100 m away.

36. Weight Gain of Rats The weight gain (in grams) of rats fed a certain vitamin supplement is a continuous random variable with probability density function defined by

$$f(x) = \frac{8}{7}x^{-2} \quad \text{for } [1, 8].$$

(a) Find the mean of the distribution.

(b) Find the standard deviation of the distribution.

(c) Find the probability that the value of the random variable is within one standard deviation of the mean.

37. Body Temperature of a Bird The body temperature (in degrees Celsius) of a particular species of bird is a continu-

ous random variable with probability density function defined by

$$f(x) = \frac{6}{15,925}(x^2 + x) \quad \text{for } [20, 25].$$

(a) What is the expected body temperature of this species?

(b) Find the probability of a body temperature below the mean.

38. Snowfall The snowfall (in inches) in a certain area is uniformly distributed over the interval [2, 40].

(a) What is the expected snowfall?

(b) What is the probability of getting more than 20 inches of snow?

39. Heart Muscle Tension In a pilot study on tension of the heart muscle in dogs, the mean developed tension was 2.4 g, with a standard deviation of .4 g. Find the probability of a tension of less than 1.9 g. Assume a normal distribution.

40. Average Birth Weight The average birth weight of infants in the United States is 7.8 lb, with a standard deviation of 1.1 lb. Assuming a normal distribution, what is the probability that a newborn will weigh more than 9 lb?

General Interest

41. State-Run Lotteries The average state "take" on lotteries is 40%, with a standard deviation of 13%. Assuming a normal distribution, what is the probability that a state-run lottery will have a "take" of more than 50%?

Connections

The topics in this short chapter involved much of the material studied earlier in this book, including functions, domain and range, exponential functions, area and integration, improper integrals, integration by parts, and numerical integration. For each of the following special probability density functions, give

(a) *the type of distribution;*

(b) *the domain and range;*

(c) *the graph;*

(d) *the mean and standard deviation;*

(e) $P(\mu - \sigma \leq x \leq \mu + \sigma).$

42. $f(x) = .05 \quad \text{for } [10, 30]$ **43.** $f(x) = e^{-x} \quad \text{for } [0, \infty)$

44. $f(x) = \frac{e^{-x^2}}{\sqrt{\pi}} \quad \text{for } (-\infty, \infty)$

(*Hint:* $\sigma = 1/\sqrt{2}$.)

Extended Application / A Crop-Planting Model*

Many firms in food processing, seed production, and similar industries face a problem every year deciding how many acres of land to plant in each of various crops. Demand for the crop is unknown, as is the actual yield per acre. In this application, we set up a mathematical model for determining the optimum number of acres to plant in a crop.

This model is designed to tell the company the number of acres of seed that it should plant. The model uses the following variables:

D = number of tons of seed demanded;

$f(D)$ = continuous probability density function for the quantity of seed demanded, D;

X = quantity of seed produced per acre of land;

Q = quantity of seed carried over in inventory from previous years;

S = selling price per ton of seed;

C_p = variable costs of production, marketing, etc., per ton of seed;

C_c = cost to carry over a ton of seed from previous years;

A = number of acres of land to be planted;

C_A = variable cost per acre of land contracted;

T = total number of tons of seed available for sale;

a = lower limit of the domain of $f(D)$; and

b = upper limit of the domain of $f(D)$.

To find the optimum number of acres to plant, it is necessary to calculate the expected value of the profit from the planting of A acres.

Based on the definition of the variables above, the total number of tons of seed that will be available for sale is given by the product of the number of acres planted, A, and the yield per acre, X, added to the carryover, Q. If T represents this total, then

$$T = AX + Q.$$

The variable here is A; we assume X and Q are known and fixed.

The expected profit can be broken down into several parts. The first portion comes from multiplying the profit per ton and the average number of tons demanded. The profit per ton is found by subtracting the variable cost per ton, C_p, from the selling price per ton, S:

$$\text{Profit per ton} = S - C_p.$$

The average number of tons demanded for our interval of concern is given by

$$\int_a^T D \cdot f(D) \, dD.$$

Thus, this portion of the expected profit is

$$(S - C_p) \cdot \int_a^T D \cdot f(D) \, dD. \tag{1}$$

A second portion of expected profit is found by multiplying the profit per ton, $S - C_p$, the total number of tons available, T (recall that this is a variable), and the probability that T or more tons will be demanded by the marketplace.

$$(S - C_p)(T) \int_T^b f(D) \, dD \tag{2}$$

If T is greater than D, there will be costs associated with carrying over the excess seed. The expected value of these costs is given by the product of the carrying cost per ton, C_c, and the number of tons to be carried over, or

$$-C_c \int_a^T (T - D)f(D) \, dD. \tag{3}$$

The minus sign shows that these costs reduce profit. If $T < D$, this term would be omitted. Finally, the total cost of producing the seed is given by the product of the variable cost per acre and the number of acres:

$$-C_A \cdot A. \tag{4}$$

The expected profit is the sum of the expressions in (1)–(4), or

$$\text{Expected profit} = (S - C_p) \cdot \int_a^T D \cdot f(D) \, dD$$
$$+ (S - C_p)(T) \int_T^b f(D) \, dD$$
$$- C_c \int_a^T (T - D)f(D) \, dD$$
$$- C_A \cdot A. \tag{5}$$

*Based on work by David P. Rutten, Senior Mathematician, The Upjohn Company, Kalamazoo, Michigan. Reprinted with permission.

For a specific crop, once the values for all the variables but *A* (the number of acres to plant) are determined, the expected profit can be expressed as a function of *A*. Then the number of acres to plant for maximum expected profit can be found by taking the derivative, setting it equal to 0, and solving for *A*. We leave these steps for the exercises.

Exercises

Suppose that we have the following information for a particular crop.

$$\text{Probability density function} = f(D) = \frac{1}{1000}$$

$$\text{for } 500 \leq D \leq 1500 \text{ tons}$$

$a = 500$

$b = 1500$

Selling price $= S = \$10,000$ per ton

Variable cost $= C_p = \$5000$ per ton

Carrying cost $= C_C = \$3000$ per ton

Variable cost per acre $= C_A = \$100$

Inventory carryover $= Q = 200$ tons

Yield per acre $= X = .1$ ton

$T = AX + Q = .1A + 200$

1. Find the expected profit as a function of A.

2. Find the number of acres to plant for maximum expected profit.

Table 1: Combinations

n	$\binom{n}{0}$	$\binom{n}{1}$	$\binom{n}{2}$	$\binom{n}{3}$	$\binom{n}{4}$	$\binom{n}{5}$	$\binom{n}{6}$	$\binom{n}{7}$	$\binom{n}{8}$	$\binom{n}{9}$	$\binom{n}{10}$
0	1										
1	1	1									
2	1	2	1								
3	1	3	3	1							
4	1	4	6	4	1						
5	1	5	10	10	5	1					
6	1	6	15	20	15	6	1				
7	1	7	21	35	35	21	7	1			
8	1	8	28	56	70	56	28	8	1		
9	1	9	36	84	126	126	84	36	9	1	
10	1	10	45	120	210	252	210	120	45	10	1
11	1	11	55	165	330	462	462	330	165	55	11
12	1	12	66	220	495	792	924	792	495	220	66
13	1	13	78	286	715	1287	1716	1716	1287	715	286
14	1	14	91	364	1001	2002	3003	3432	3003	2002	1001
15	1	15	105	455	1365	3003	5005	6435	6435	5005	3003
16	1	16	120	560	1820	4368	8008	11440	12870	11440	8008
17	1	17	136	680	2380	6188	12376	19448	24310	24310	19448
18	1	18	153	816	3060	8568	18564	31824	43758	48620	43758
19	1	19	171	969	3876	11628	27132	50388	75582	92378	92378
20	1	20	190	1140	4845	15504	38760	77520	125970	167960	184756

For $r > 10$, it may be necessary to use the identity

$$\binom{n}{r} = \binom{n}{n-r}.$$

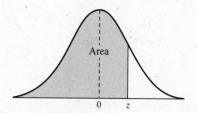

Table 2: Area Under a Normal Curve to the Left of z, Where $z = \dfrac{x - \mu}{\sigma}$

z	.00	.01	.02	.03	.04	.05	.06	.07	.08	.09
−3.4	.0003	.0003	.0003	.0003	.0003	.0003	.0003	.0003	.0003	.0002
−3.3	.0005	.0005	.0005	.0004	.0004	.0004	.0004	.0004	.0004	.0003
−3.2	.0007	.0007	.0006	.0006	.0006	.0006	.0006	.0005	.0005	.0005
−3.1	.0010	.0009	.0009	.0009	.0008	.0008	.0008	.0008	.0007	.0007
−3.0	.0013	.0013	.0013	.0012	.0012	.0011	.0011	.0011	.0010	.0010
−2.9	.0019	.0018	.0017	.0017	.0016	.0016	.0015	.0015	.0014	.0014
−2.8	.0026	.0025	.0024	.0023	.0023	.0022	.0021	.0021	.0020	.0019
−2.7	.0035	.0034	.0033	.0032	.0031	.0030	.0029	.0028	.0027	.0026
−2.6	.0047	.0045	.0044	.0043	.0041	.0040	.0039	.0038	.0037	.0036
−2.5	.0062	.0060	.0059	.0057	.0055	.0054	.0052	.0051	.0049	.0048
−2.4	.0082	.0080	.0078	.0075	.0073	.0071	.0069	.0068	.0066	.0064
−2.3	.0107	.0104	.0102	.0099	.0096	.0094	.0091	.0089	.0087	.0084
−2.2	.0139	.0136	.0132	.0129	.0125	.0122	.0119	.0116	.0113	.0110
−2.1	.0179	.0174	.0170	.0166	.0162	.0158	.0154	.0150	.0146	.0143
−2.0	.0228	.0222	.0217	.0212	.0207	.0202	.0197	.0192	.0188	.0183
−1.9	.0287	.0281	.0274	.0268	.0262	.0256	.0250	.0244	.0239	.0233
−1.8	.0359	.0352	.0344	.0336	.0329	.0322	.0314	.0307	.0301	.0294
−1.7	.0446	.0436	.0427	.0418	.0409	.0401	.0392	.0384	.0375	.0367
−1.6	.0548	.0537	.0526	.0516	.0505	.0495	.0485	.0475	.0465	.0455
−1.5	.0668	.0655	.0643	.0630	.0618	.0606	.0594	.0582	.0571	.0559
−1.4	.0808	.0793	.0778	.0764	.0749	.0735	.0722	.0708	.0694	.0681
−1.3	.0968	.0951	.0934	.0918	.0901	.0885	.0869	.0853	.0838	.0823
−1.2	.1151	.1131	.1112	.1093	.1075	.1056	.1038	.1020	.1003	.0985
−1.1	.1357	.1335	.1314	.1292	.1271	.1251	.1230	.1210	.1190	.1170
−1.0	.1587	.1562	.1539	.1515	.1492	.1469	.1446	.1423	.1401	.1379

Table 2: Area Under a Normal Curve (continued)

z	.00	.01	.02	.03	.04	.05	.06	.07	.08	.09
−.9	.1841	.1814	.1788	.1762	.1736	.1711	.1685	.1660	.1635	.1611
−.8	.2119	.2090	.2061	.2033	.2005	.1977	.1949	.1922	.1894	.1867
−.7	.2420	.2389	.2358	.2327	.2296	.2266	.2236	.2206	.2177	.2148
−.6	.2743	.2709	.2676	.2643	.2611	.2578	.2546	.2514	.2483	.2451
−.5	.3085	.3050	.3015	.2981	.2946	.2912	.2877	.2843	.2810	.2776
−.4	.3446	.3409	.3372	.3336	.3300	.3264	.3228	.3192	.3156	.3121
−.3	.3821	.3783	.3745	.3707	.3669	.3632	.3594	.3557	.3520	.3483
−.2	.4207	.4168	.4129	.4090	.4052	.4013	.3974	.3936	.3897	.3859
−.1	.4602	.4562	.4522	.4483	.4443	.4404	.4364	.4325	.4286	.4247
−.0	.5000	.4960	.4920	.4880	.4840	.4801	.4761	.4721	.4681	.4641
.0	.5000	.5040	.5080	.5120	.5160	.5199	.5239	.5279	.5319	.5359
.1	.5398	.5438	.5478	.5517	.5557	.5596	.5636	.5675	.5714	.5753
.2	.5793	.5832	.5871	.5910	.5948	.5987	.6026	.6064	.6103	.6141
.3	.6179	.6217	.6255	.6293	.6331	.6368	.6406	.6443	.6480	.6517
.4	.6554	.6591	.6628	.6664	.6700	.6736	.6772	.6808	.6844	.6879
.5	.6915	.6950	.6985	.7019	.7054	.7088	.7123	.7157	.7190	.7224
.6	.7257	.7291	.7324	.7357	.7389	.7422	.7454	.7486	.7517	.7549
.7	.7580	.7611	.7642	.7673	.7704	.7734	.7764	.7794	.7823	.7852
.8	.7881	.7910	.7939	.7967	.7995	.8023	.8051	.8078	.8106	.8133
.9	.8159	.8186	.8212	.8238	.8264	.8289	.8315	.8340	.8365	.8389
1.0	.8413	.8438	.8461	.8485	.8508	.8531	.8554	.8577	.8599	.8621
1.1	.8643	.8665	.8686	.8708	.8729	.8749	.8770	.8790	.8810	.8830
1.2	.8849	.8869	.8888	.8907	.8925	.8944	.8962	.8980	.8997	.9015
1.3	.9032	.9049	.9066	.9082	.9099	.9115	.9131	.9147	.9162	.9177
1.4	.9192	.9207	.9222	.9236	.9251	.9265	.9278	.9292	.9306	.9319
1.5	.9332	.9345	.9357	.9370	.9382	.9394	.9406	.9418	.9429	.9441
1.6	.9452	.9463	.9474	.9484	.9495	.9505	.9515	.9525	.9535	.9545
1.7	.9554	.9564	.9573	.9582	.9591	.9599	.9608	.9616	.9625	.9633
1.8	.9641	.9649	.9656	.9664	.9671	.9678	.9686	.9693	.9699	.9706
1.9	.9713	.9719	.9726	.9732	.9738	.9744	.9750	.9756	.9761	.9767
2.0	.9772	.9778	.9783	.9788	.9793	.9798	.9803	.9808	.9812	.9817
2.1	.9821	.9826	.9830	.9834	.9838	.9842	.9846	.9850	.9854	.9857
2.2	.9861	.9864	.9868	.9871	.9875	.9878	.9881	.9884	.9887	.9890
2.3	.9893	.9896	.9898	.9901	.9904	.9906	.9909	.9911	.9913	.9916
2.4	.9918	.9920	.9922	.9925	.9927	.9929	.9931	.9932	.9934	.9936
2.5	.9938	.9940	.9941	.9943	.9945	.9946	.9948	.9949	.9951	.9952
2.6	.9953	.9955	.9956	.9957	.9959	.9960	.9961	.9962	.9963	.9964
2.7	.9965	.9966	.9967	.9968	.9969	.9970	.9971	.9972	.9973	.9974
2.8	.9974	.9975	.9976	.9977	.9977	.9978	.9979	.9979	.9980	.9981
2.9	.9981	.9982	.9982	.9983	.9984	.9984	.9985	.9985	.9986	.9986
3.0	.9987	.9987	.9987	.9988	.9988	.9989	.9989	.9989	.9990	.9990
3.1	.9990	.9991	.9991	.9991	.9992	.9992	.9992	.9992	.9993	.9993
3.2	.9993	.9993	.9994	.9994	.9994	.9994	.9994	.9995	.9995	.9995
3.3	.9995	.9995	.9995	.9996	.9996	.9996	.9996	.9996	.9996	.9997
3.4	.9997	.9997	.9997	.9997	.9997	.9997	.9997	.9997	.9997	.9998

Table 3: Integrals

(C is an arbitrary constant.)

1. $\int x^n \, dx = \dfrac{1}{n+1} x^{n+1} + C \quad$ (if $n \neq -1$)

2. $\int e^{kx} \, dx = \dfrac{1}{k} e^{kx} + C$

3. $\int \dfrac{a}{x} dx = a \ln |x| + C$

4. $\int \ln |ax| \, dx = x \, (\ln |ax| - 1) + C$

5. $\int \dfrac{1}{\sqrt{x^2 + a^2}} \, dx = \ln |x + \sqrt{x^2 + a^2}| + C$

6. $\int \dfrac{1}{\sqrt{x^2 - a^2}} \, dx = \ln |x + \sqrt{x^2 - a^2}| + C$

7. $\int \dfrac{1}{a^2 - x^2} \, dx = \dfrac{1}{2a} \cdot \ln \left| \dfrac{a + x}{a - x} \right| + C \quad (a \neq 0)$

8. $\int \dfrac{1}{x^2 - a^2} \, dx = \dfrac{1}{2a} \cdot \ln \left| \dfrac{x - a}{x + a} \right| + C \quad (a \neq 0)$

9. $\int \dfrac{1}{x\sqrt{a^2 - x^2}} \, dx = -\dfrac{1}{a} \cdot \ln \left| \dfrac{a + \sqrt{a^2 - x^2}}{x} \right| + C \quad (a \neq 0)$

10. $\int \dfrac{1}{x\sqrt{a^2 + x^2}} \, dx = -\dfrac{1}{a} \cdot \ln \left| \dfrac{a + \sqrt{a^2 + x^2}}{x} \right| + C \quad (a \neq 0)$

11. $\int \dfrac{x}{ax + b} \, dx = \dfrac{x}{a} - \dfrac{b}{a^2} \cdot \ln |ax + b| + C \quad (a \neq 0)$

12. $\int \dfrac{x}{(ax + b)^2} \, dx = \dfrac{b}{a^2(ax + b)} + \dfrac{1}{a^2} \cdot \ln |ax + b| + C \quad (a \neq 0)$

13. $\int \dfrac{1}{x(ax + b)} \, dx = \dfrac{1}{b} \cdot \ln \left| \dfrac{x}{ax + b} \right| + C \quad (b \neq 0)$

14. $\int \dfrac{1}{x(ax + b)^2} \, dx = \dfrac{1}{b(ax + b)} + \dfrac{1}{b^2} \cdot \ln \left| \dfrac{x}{ax + b} \right| + C \quad (b \neq 0)$

15. $\int \sqrt{x^2 + a^2} \, dx = \dfrac{x}{2} \sqrt{x^2 + a^2} + \dfrac{a^2}{2} \cdot \ln |x + \sqrt{x^2 + a^2}| + C$

16. $\int x^n \cdot \ln |x| \, dx = x^{n+1} \left[\dfrac{\ln |x|}{n + 1} - \dfrac{1}{(n + 1)^2} \right] + C \quad (n \neq -1)$

17. $\int x^n e^{ax} \, dx = \dfrac{x^n e^{ax}}{a} - \dfrac{n}{a} \cdot \int x^{n-1} e^{ax} \, dx + C \quad (a \neq 0)$

To the Student

If you want further help with this course, you may wish to buy a copy of the *Student's Solution Manual* that accompanies this textbook. This manual provides detailed, step-by-step solutions to the odd-numbered exercises in the textbook. Also included are practice tests for each chapter, with answers given so you can check your work. Your college bookstore either has this manual or can order it for you.

ALGEBRA REFERENCE

Section R.1 (page xxiv)

1. $-x^2 + x + 9$ **2.** $-6y^2 + 3y + 10$ **3.** $-14q^2 + 11q - 14$ **4.** $9r^2 - 4r + 19$ **5.** $-.327x^2 - 2.805x - 1.458$ **6.** $-2.97r^2 - 8.083r + 7.81$ **7.** $-18m^3 - 27m^2 + 9m$ **8.** $12k^2 - 20k + 3$ **9.** $25r^2 + 5rs - 12s^2$ **10.** $18k^2 - 7kq - q^2$ **11.** $\frac{6}{25}y^2 + \frac{11}{40}yz + \frac{1}{16}z^2$ **12.** $\frac{15}{16}r^2 - \frac{7}{12}rs - \frac{2}{9}s^2$ **13.** $144x^2 - 1$ **14.** $36m^2 - 25$ **15.** $27p^3 - 1$ **16.** $6p^3 - 11p^2 + 14p - 5$ **17.** $8m^3 + 1$ **18.** $12k^4 + 21k^3 - 5k^2 + 3k + 2$ **19.** $m^2 + mn - 2n^2 - 2km + 5kn - 3k^2$ **20.** $2r^2 - 7rs + 3s^2 + 3rt - 4st + t^2$

Section R.2 (page xxvi)

1. $8a(a^2 - 2a + 3)$ **2.** $3y(y^2 + 8y + 3)$ **3.** $5p^2(5p^2 - 4pq + 20q^2)$ **4.** $10m^2(6m^2 - 12mn + 5n^2)$ **5.** $(m + 7)(m + 2)$ **6.** $(x + 5)(x - 1)$ **7.** $(z + 4)(z + 5)$ **8.** $(b - 7)(b - 1)$ **9.** $(a - 5b)(a - b)$ **10.** $(s - 5t)(s + 7t)$ **11.** $(y - 7z)(y + 3z)$ **12.** $6(a - 10)(a + 2)$ **13.** $3m(m + 3)(m + 1)$ **14.** $(2x + 1)(x - 3)$ **15.** $(3a + 7)(a + 1)$ **16.** $(2a - 5)(a - 6)$ **17.** $(5y + 2)(3y - 1)$ **18.** $(7m + 2n)(3m + n)$ **19.** $2a^2(4a - b)(3a + 2b)$ **20.** $4z^3(8z + 3a)(z - a)$ **21.** $(x + 8)(x - 8)$ **22.** $(3m + 5)(3m - 5)$ **23.** $(11a + 10)(11a - 10)$ **24.** prime **25.** $(z + 7y)^2$ **26.** $(m - 3n)^2$ **27.** $(3p - 4)^2$ **28.** $(a - 6)(a^2 + 6a + 36)$ **29.** $(2r - 3s)(4r^2 + 6rs + 9s^2)$ **30.** $(4m + 5)(16m^2 - 20m + 25)$

Section R.3 (page xxix)

1. $z/2$ **2.** $5p/2$ **3.** $8/9$ **4.** $3/(t - 3)$ **5.** $2(x + 2)/x$ **6.** $4(y + 2)$ **7.** $(m - 2)/(m + 3)$ **8.** $(r + 2)/(r + 4)$ **9.** $(x + 4)/(x + 1)$ **10.** $(z - 3)/(z + 2)$ **11.** $(2m + 3)/(4m + 3)$ **12.** $(2y + 1)/(y + 1)$ **13.** $3k/5$ **14.** $25p^2/9$ **15.** $6/(5p)$ **16.** 2 **17.** $2/9$ **18.** $3/10$ **19.** $2(a + 4)/(a - 3)$ **20.** $2/(r + 2)$ **21.** $(k + 2)/(k + 3)$ **22.** $(m + 6)/(m + 3)$ **23.** $(m - 3)/(2m - 3)$ **24.** $(2n - 3)/(2n + 3)$ **25.** 1 **26.** $(6 + p)/(2p)$ **27.** $(8 - y)/(4y)$ **28.** $137/(30m)$ **29.** $(3m - 2)/[m(m - 1)]$ **30.** $(r - 12)/[r(r - 2)]$ **31.** $14/[3(a - 1)]$ **32.** $23/[20(k - 2)]$ **33.** $(7x + 9)/[(x - 3)(x + 1)(x + 2)]$ **34.** $y^2/[(y + 4)(y + 3)(y + 2)]$ **35.** $k(k - 13)/[(2k - 1)(k + 2)(k - 3)]$ **36.** $m(3m - 19)/[(3m - 2)(m + 3)(m - 4)]$

Section R.4 (page xxxv)

1. 12 **2.** $-2/7$ **3.** $-7/8$ **4.** -1 **5.** $-11/3, 7/3$ **6.** $-11/7, 19/7$ **7.** $-12, -4/3$ **8.** $-5, 3/4$
9. $-3, -2$ **10.** $-1, 3$ **11.** 4 **12.** $-2, 5/2$ **13.** $-1/2, 4/3$ **14.** $2, 5$ **15.** $-4/3, 4/3$ **16.** $-4, 1/2$
17. $0, 4$ **18.** $(5 + \sqrt{13})/6 \approx 1.434, (5 - \sqrt{13})/6 \approx .232$ **19.** $(1 + \sqrt{33})/4 \approx 1.686, (1 - \sqrt{33})/4 \approx -1.186$
20. $(-1 + \sqrt{5})/2 \approx .618, (-1 - \sqrt{5})/2 \approx -1.618$ **21.** $5 + \sqrt{5} \approx 7.236, 5 - \sqrt{5} \approx 2.764$ **22.** $(-6 + \sqrt{26})/2 \approx$
$-.450, (-6 - \sqrt{26})/2 \approx -5.550$ **23.** $1, 5/2$ **24.** no real-number solutions **25.** $(-1 + \sqrt{73})/6 \approx 1.257,$
$(-1 - \sqrt{73})/6 \approx -1.591$ **26.** $-1, 0$ **27.** 3 **28.** 12 **29.** $-59/6$ **30.** $-11/5$ **31.** no real-number
solutions **32.** $-5/2$ **33.** $2/3$ **34.** 1 **35.** $(-13 - \sqrt{185})/4 \approx -6.650, (-13 + \sqrt{185})/4 \approx .150$

Section R.5 (page xl)

1. $(-\infty, -1]$

2. $(-\infty, 1)$

3. $(-1, \infty)$

4. $(-\infty, 1]$

5. $(1/5, \infty)$

6. $(1/3, \infty)$

7. $(-5, 6)$

8. $[7/3, 4]$

9. $[-11/2, 7/2]$

10. $[-1, 2]$

11. $[-17/7, \infty)$

12. $(-\infty, 50/9]$

13. $(-2, 4)$

14. $(-\infty, -6] \cup [1, \infty)$

15. $(1, 2)$

16. $(-\infty, -4) \cup (1/2, \infty)$

17. $[1, 6]$

18. $[-3/2, 5]$

19. $(-\infty, -1/2) \cup (1/3, \infty)$

20. $[-1/2, 2/5]$

21. $[-3, 1/2]$

22. $(-\infty, -2) \cup (5/3, \infty)$

23. $[-5, 5]$

24. $(-\infty, 0) \cup (16, \infty)$

25. $(-5, 3]$ **26.** $(-\infty, -1] \cup (1, \infty)$ **27.** $(-\infty, -2)$ **28.** $(-2, 3/2)$ **29.** $[-8, 5)$
30. $(-\infty, -3/2) \cup [-13/9, \infty)$ **31.** $(-2, \infty)$ **32.** $(-\infty, -1)$ **33.** $(-\infty, -1) \cup (-1/2, 1) \cup (2, \infty)$
34. $(-4, -2) \cup (0, 2)$ **35.** $(1, 3/2]$ **36.** $(-\infty, -2) \cup (-2, 2) \cup [4, \infty)$

Section R.6 (page xliii)

1. $1/64$ **2.** $1/81$ **3.** $1/216$ **4.** 1 **5.** 1 **6.** $3/4$ **7.** $-1/16$ **8.** $1/16$ **9.** $-1/9$ **10.** $1/9$
11. $25/64$ **12.** $216/343$ **13.** 8 **14.** 125 **15.** $49/4$ **16.** $27/64$ **17.** $1/7^4$ **18.** $1/3^6$ **19.** $1/2^3$
20. $1/6^2$ **21.** 4^3 **22.** 8^5 **23.** $1/10^8$ **24.** 5 **25.** x^2 **26.** y^3 **27.** 2^3k^3 **28.** $1/(3z^7)$ **29.** $x^2/(2y)$
30. $m^3/5^4$ **31.** a^3b^6 **32.** $d^6/(2^2c^4)$ **33.** $1/6$ **34.** $-17/9$ **35.** $-13/66$ **36.** $81/26$ **37.** $35/18$
38. $213/200$ **39.** 9 **40.** 3 **41.** 4 **42.** 100 **43.** 4 **44.** -25 **45.** $2/3$ **46.** $4/3$ **47.** $1/32$
48. $1/5$ **49.** $4/3$ **50.** $1000/1331$ **51.** 2^2 **52.** $27^{1/3}$ **53.** 4^2 **54.** 1 **55.** r **56.** $12^3/y^8$

57. $1/(2^2 \cdot 3k^{5/2})$ or $1/(12k^{5/2})$ **58.** $1/(2p^2)$ **59.** $a^{2/3}b^2$ **60.** $y/(x^{4/3}z^{1/2})$ **61.** $h^{1/3}t^{1/5}/k^{2/5}$ **62.** m^3p/n
63. $3x^3(x^2 - 1)^{-1/2}$ **64.** $5(5x + 2)^{-1/2}(45x^2 + 3x - 5)$ **65.** $(2x + 5)(x^2 - 4)^{-1/2}(4x^2 + 5x - 8)$
66. $(4x^2 + 1)(2x - 1)^{-1/2}(4x^2 + 4x - 1)$

Section R.7 (page xlvii)

1. 5 **2.** 6 **3.** -5 **4.** $5\sqrt{2}$ **5.** $20\sqrt{5}$ **6.** $4y^2\sqrt{2y}$ **7.** $7\sqrt{2}$ **8.** $9\sqrt{3}$ **9.** $2\sqrt{5}$ **10.** $-2\sqrt{7}$
11. $5\sqrt[3]{2}$ **12.** $7\sqrt[3]{3}$ **13.** $3\sqrt[3]{4}$ **14.** $xyz^2\sqrt{2x}$ **15.** $7rs^2t^5\sqrt{2r}$ **16.** $2zx^2y\sqrt[3]{2z^2x^2y}$ **17.** $x^2yz^2\sqrt[4]{y^3z^3}$
18. $ab\sqrt{ab}(b - 2a^2 + b^3)$ **19.** $p^2\sqrt{pq}(pq - q^4 + p^2)$ **20.** $5\sqrt{7}/7$ **21.** $-2\sqrt{3}/3$ **22.** $-\sqrt{3}/2$ **23.** $\sqrt{2}$
24. $-3(1 + \sqrt{5})/4$ **25.** $-5(2 + \sqrt{6})/2$ **26.** $-2(\sqrt{3} + \sqrt{2})$ **27.** $(\sqrt{10} - \sqrt{3})/7$ **28.** $(\sqrt{r} + \sqrt{3})/(r - 3)$
29. $5(\sqrt{m} + \sqrt{5})/(m - 5)$ **30.** $\sqrt{y} + \sqrt{5}$ **31.** $\sqrt{z} + \sqrt{11}$ **32.** $-2x - 2\sqrt{x(x + 1)} - 1$
33. $(p^2 + p + 2\sqrt{p(p^2 - 1)}) - 1)/(-p^2 + p + 1)$ **34.** $-1/[2(1 - \sqrt{2})]$ **35.** $-2/[3(1 + \sqrt{3})]$
36. $-1/(2x - 2\sqrt{x(x + 1)} + 1)$ **37.** $(-p^2 + p + 1)/(p^2 + p - 2\sqrt{p(p^2 - 1)} - 1)$ **38.** $|4 - x|$
39. $|2y + 1|$ **40.** cannot be simplified **41.** cannot be simplified

CHAPTER 1

Section 1.1 (page 12)

1. not a function **3.** function **5.** function **7.** not a function
9. $(-2, -3), (-1, -2),$
$(0, -1), (1, 0), (2, 1),$
$(3, 2);$
range:
$\{-3, -2, -1, 0, 1, 2\}$

11. $(-2, 17), (-1, 13),$
$(0, 9) (1, 5), (2, 1),$
$(3, -3);$
range:
$\{-3, 1, 5, 9, 13, 17\}$

13. $(-2, 13), (-1, 11),$
$(0, 9), (1, 7), (2, 5),$
$(3, 3);$
range:
$\{3, 5, 7, 9, 11, 13\}$

15. $(-2, 3/2), (-1, 2),$
$(0, 5/2), (1, 3), (2, 7/2),$
$(3, 4);$
range
$\{3/2, 2, 5/2, 3, 7/2, 4\}$

17. $(-2, 2), (-1, 0), (0, 0),$
$(1, 2), (2, 6), (3, 12);$
range: $\{0, 2, 6, 12\}$

19. $(-2, 4), (-1, 1), (0, 0),$
$(1, 1), (2, 4), (3, 9);$
range: $\{0, 1, 4, 9\}$

21. $(-2, 1), (-1, 1/2),$
$(0, 1/3), (1, 1/4),$
$(2, 1/5), (3, 1/6);$
range:
$\{1, 1/2, 1/3, 1/4, 1/5, 1/6\}$

23. $(-2, -3), (-1, -3/2),$
$(0, -3/5), (1, 0), (2, 3/7),$
$(3, 3/4);$
range:
$\{-3, -3/2, -3/5, 0, 3/7, 3/4\}$

25. $(-\infty, 0)$

27. $[1, 2)$

29. $(-\infty, -9)$

31. $-4 < x < 3$ **33.** $x \le -1$ **35.** $-2 \le x < 6$ **37.** $x \le -4$ or $x \ge 4$ **39.** $(-\infty, \infty)$ **41.** $(-\infty, \infty)$
43. $[-4, 4]$ **45.** $[3, \infty)$ **47.** $(-\infty, -2) \cup (-2, 2) \cup (2, \infty)$ **49.** $(-\infty, \infty)$ **51.** $(-\infty, -1] \cup [5, \infty)$
53. $(-\infty, 2) \cup (4, \infty)$ **55.** domain: $[-5, 4]$; range: $[-2, 6]$ **57.** domain: $(-\infty, \infty)$; range: $(-\infty, 12]$ **59. (a)** 14
(b) -7 **(c)** $1/2$ **(d)** $3a + 2$ **(e)** $6/m + 2$ **61. (a)** 5 **(b)** -23 **(c)** $-7/4$ **(d)** $-a^2 + 5a + 1$
(e) $-4/m^2 + 10/m + 1$ or $(-4 + 10m + m^2)/m^2$ **63. (a)** $9/2$ **(b)** 1 **(c)** 0 **(d)** $(2a + 1)/(a - 2)$
(e) $(4 + m)/(2 - 2m)$ **65. (a)** 0 **(b)** 4 **(c)** 3 **(d)** 4 **67. (a)** -3 **(b)** -2 **(c)** -1 **(d)** 2 **69.** $6m - 20$
71. $r^2 + 2rh + h^2 - 2r - 2h + 5$ **73.** $9/q^2 - 6/q + 5$ or $(9 - 6q + 5q^2)/q^2$ **75.** function **77.** not a function
79. function **81. (a)** $x^2 + 2xh + h^2 - 4$ **(b)** $2xh + h^2$ **(c)** $2x + h$ **83. (a)** $6x + 6h + 2$ **(b)** $6h$ **(c)** 6
85. (a) $1/(x + h)$ **(b)** $-h/[x(x + h)]$ **(c)** $-1/[x(x + h)]$
87. (a) 26,300 BTU per dollar **(b)** \$120 billion
 (c) 23,500 BTU per dollar, \$85 billion, 1980
 (d) none

89. (a) 29¢ **(b)** 52¢ **(c)** 52¢ **(d)** 98¢ **(e)** 75¢
 (g)

 (h) x, the weight **(i)** $C(x)$, the cost to mail the letter

Section 1.2 (page 25)
1. $3/5$ **3.** not defined **5.** 2 **7.** $5/9$ **9.** not defined **11.** 2 **13.** .5785 **15.** $2/5$ **17.** $-1/4$
19. $2x + y = 5$ **21.** $y = 1$ **23.** $x + 3y = 10$ **25.** $18x + 30y = 59$ **27.** $2x - 3y = 6$ **29.** $x = -6$
31. $5.081x + y = -4.69$

33.

35.

37.

39.

41.

43.

45.

47. $x + 3y = 11$ **49.** $x - y = 7$ **51.** $-2x + y = 4$ **53.** $2x - 3y = -6$ **55.** no

63. (a)

$D(q) = 100 - (2/5)q$

(125, 50)

$S(q) = (2/5)q$

(b) 125 units **(c)** $50

65. (a) $52 **(b)** $52 **(c)** $52
(d) $79 **(e)** $106
(f)

Time (in days)

(g) yes **(h)** no

67. (a) $h = (8/3)t + 211/3$
(b) about 172 cm to 190 cm
(c) about 45 cm
69. (a) $m = 2.5; y = 2.5x - 70$
(b) 52%
71. (a) $T = .03t + 15$
(b) approximately 2103

Section 1.3 (page 35)

1. If $C(x)$ is the cost of renting a saw for x hr, then $C(x) = 12 + x$. **3.** If $C(x)$ is the cost of parking for x half-hours, then $C(x) = 35x + 50$. **5.** $C(x) = 30x + 100$ **7.** $C(x) = 25x + 1000$ **9.** $C(x) = 50x + 500$ **11.** $C(x) = 90x + 2500$ **13. (a)** 2000 **(b)** 2900 **(c)** 3200 **(d)** yes **(e)** 300 **15. (a)** $y = (800,000/7)x + 200,000$ **(b)** about $1,457,000 **(c)** 1997 **17. (a)** $97 **(b)** $97.097 **(c)** $.097 or 9.7¢ **(d)** $.097 or 9.7¢ **19. (a)** $100 **(b)** $36 **(c)** $24 **21.** 500 units; $30,000 **23.** Break-even point is 45 units; don't produce. **25.** Break-even point is -50 units; impossible to make a profit here. **27.** about $140 billion in mid-1981 **29.** 81 yr **31.** approximately 4.3 m/sec **33. (a)** 32.5 min **(b)** 70 min **(c)** 145 min **(d)** 220 min **35. (a)** 14.4°C **(b)** 122°F **37.** $C = (5/9)(F - 32)$ **39. (a)** 240 **(b)** 200 **(c)** 160 **(d)** -8 students per hour of study; as required study increased, fewer students enrolled **(e)** 28 hr

Section 1.4 (page 44)

1.

$f(x)$ **(b)(a) (c)(d)**

(e) As the coefficient increases in absolute value, the parabola becomes narrower.

3.

$f(x)$ **(a)**
(c)
(b)
(d)

(e) The graphs are shifted upward or downward.

5. Vertex is $(-3, -4)$; axis is $x = -3$; x-intercepts are -1 and -5; y-intercept is 5.

$y = x^2 + 6x + 5$

7. Vertex is $(2, 0)$; axis is $x = 2$; x-intercept is 2; y-intercept is 4.

$y = x^2 - 4x + 4$

9. Vertex is $(-3, 2)$; axis is $x = -3$; x-intercepts are -2 and -4; y-intercept is -16.

$y = -2x^2 - 12x - 16$

11. Vertex is $(-3, -34)$; axis is $x = -3$; x-intercepts are $-3 \pm \sqrt{17} \approx 1.12$ or -7.12;* y-intercept is -16.

$y = 2x^2 + 12x - 16$

*The symbol \approx means *approximately equal to*.

13. Vertex is $(-1, -1)$; axis is $x = -1$;
x-intercepts are $-1 \pm \sqrt{3}/3 \approx -.42$
or -1.58;
y-intercept is 2.

$$y = 3x^2 + 6x + 2$$

15. Vertex is $(3, 3)$; axis is $x = 3$;
x-intercepts are $3 \pm \sqrt{3} \approx 4.73$ or
1.27;
y-intercept is -6.

17. Vertex is $(4, 12)$; axis is $x = 4$;
x-intercepts are 6 and 2;
y-intercept is -36.

$$y = -3x^2 + 24x - 36$$

19. Vertex is $(-2, -2)$; axis is $x = -2$;
x-intercepts are $-2 \pm 2\sqrt{5}/5 \approx$
-1.11 or -2.89;
y-intercept is 8.

$$y = (5/2)x^2 + 10x + 8$$

21. Vertex is $(2, -1)$; axis is $x = 2$;
x-intercepts are $2 \pm \sqrt{6}/2 \approx 3.22$
or $.78$;
y-intercept is $5/3$.

$$y = (2/3)x^2 - (8/3)x + (5/3)$$

23. $x = 1/2$

25. Maximum revenue is \$5625;
25 seats are unsold.

27. **(a)** $R(x) = x(500 - x) = 500x - x^2$
(b)

(c) \$250 **(d)** \$62,500

29. **(a)** $200 + 20x$ **(b)** $80 - x$ **(c)** $R(x) = 16,000 + 1400x - 20x^2$ **(d)** 35 **(e)** \$40,500 **31.** **(a)** 60 **(b)** 70 **(c)** 90
(d) 100 **(e)** 80 **(f)** 20 **33.** 16 ft; 2 sec **35.** 80 ft by 160 ft **37.** 10, 10 **39.** $10\sqrt{3}$ m ≈ 17.32 m

Section 1.5 (page 53)

1. 4, 6, etc. (true degree = 4); +

3. 5, 7, etc. (true degree = 5); +

5. 5, 7, etc. (true degree = 7); +

7. 7, 9, etc. (true degree = 7); −

9. $y = 0$; $x = 3$; no x-intercept,
y-intercept $= 4/3$

11. $y = 0$; $x = -3/2$; no x-intercept,
y-intercept $= 2/3$

13. $y = 3$; $x = 1$; x-intercept $= 0$,
y-intercept $= 0$

15. $y = 1$; $x = 4$; x-intercept $= -1$, y-intercept $= -1/4$

$$y = \frac{x+1}{x-4}$$

17. $y = -2/5$; $x = -4$; x-intercept $= 1/2$, y-intercept $= 1/20$

$$y = \frac{1-2x}{5x+20}$$

19. $y = -1/3$; $x = -2$; x-intercept $= -4$, y-intercept $= -2/3$

$$y = \frac{-x-4}{3x+6}$$

21. (a) $12.50; $10; $6.25; $4.76; $3.85
(b) Probably $(0, \infty)$; it doesn't seem reasonable to discuss the average cost per unit of zero units.
(c)

$$C(x) = \frac{500}{x+30}$$

23. (a) $54 billion (b) $504 billion
(c) $750 billion (d) $1104 billion
(e)

$$y = x(100-x)(x^2 + 500)$$

25. (a) $42.9 million (b) $40 million
(c) $30 million (d) $0
(e)

$$y = \frac{60x - 6000}{x - 120}$$

27. (a) $6700 (b) $15,600 (c) $26,800 (d) $60,300 (e) $127,300 (f) $328,300 (g) $663,300 (h) no
(i)

$$y = \frac{6.7x}{100-x}$$

29. (a)

$$g(x) = -.006x^4 + .140x^3 - .053x^2 + 1.79x$$

(b) no
(c)

$$f(x) = \frac{x}{1+x^2}$$

31. (a)

$$A(x) = -.015x^3 + 1.058x$$

(b) between 4 and 5 hr, but closer to 5 hr (c) from less than 1 hr to about 8.4 hr

33. (a) $[0, \infty)$
(b)

$$f(x) = \frac{x}{1+x}$$

(d) Increasing b makes the next generation smaller when this generation is larger.

35. (a) $R = -1000$ (b) $G(R) = 1$
(c)

$$G(R) = \frac{R}{1000 + R}$$

37. (a)

$f(u) = u^3 + u$

(b)

$f(u) = u^3 - u$

39. (a) about \$10,000
(b) about \$20,000

41.

$f(x) = -x^3 + 4x^2 + 3x - 8$

43.

$f(x) = 2x^3 + 4x - 1$

45.

$f(x) = x^4 + x^3 - 2$

47.

$f(x) = -x^4 - 2x^3 + 3x^2 + 3x + 5$

49.

$f(x) = -x^5 + 6x^4 - 11x^3 + 6x^2 + 5$

51. no horizontal asymptote; vertical asymptote: $x = -3/2$

$f(x) = \dfrac{-2x^2 + x - 1}{2x + 3}$

53. horizontal asymptote: $y = 2$; vertical asymptotes: $x = 1$, $x = -1$

$f(x) = \dfrac{2x^2 - 5}{x^2 - 1}$

55. horizontal asymptote: $y = -2$; vertical asymptotes: $x = \sqrt{10}$, $x = -\sqrt{10}$

$f(x) = \dfrac{-2x^2}{x^2 - 10}$

Section 1.6 (page 61)

1.

$f(x) = (x + 2)^3$

3.

$f(x) = (4 - x)^3 + 10$

5.

$f(x) = -2(x + 1)^3 + 5$

7.

$f(x) = -2x^5 + 5$

9.

$f(x) = \dfrac{2}{x+3} - 2$

11.

$f(x) = \dfrac{-1}{x-1} + 2$

13.

$f(x) = \sqrt{x-1} + 3$

$(1, 3)$

15.

$(-4, -2)$

$f(x) = -\sqrt{-4-x} - 2$

17.

$(1, -3)$

$f(x) = x^2 - 2x - 2$

19.

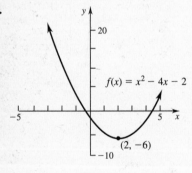

$f(x) = x^2 - 4x - 2$

$(2, -6)$

21.

$f(x) = 2x^2 + 4x - 1$

$(-1, -3)$

23.

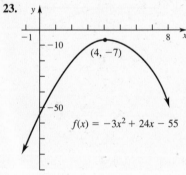

$(4, -7)$

$f(x) = -3x^2 + 24x - 55$

25.

27.

29.

31.

33. (c)

$$f(x) = \frac{2x + 3}{4x + 1}$$

(d)

$$f(x) = \frac{2x + 5}{x + 2}$$

35. (a) \$0 **(b)** \$40,000
(c)

$$f(x) = \frac{(x - 10)^3}{25} + 40$$

(d) $A(x) = [(x - 10)^3/25 + 40]/x$
(e) \$4 **(f)** \$52

Chapter 1 Review Exercises (page 63)

5. $(-3, -16/5), (-2, -14/5),$
$(-1, -12/5), (0, -2), (1, -8/5),$
$(2, -6/5), (3, -4/5);$
range: $\{-16/5, -14/5, -12/5,$
$-2, -8/5, -6/5, -4/5\}$

7. $(-3, 20), (-2, 9), (-1, 2),$
$(0, -1), (1, 0), (2, 5), (3, 14);$
range: $\{-1, 0, 2, 5, 9, 14, 20\}$

9. $(-3, 7), (-2, 2), (-1, -1),$
$(0, -2), (1, -1), (2, 2), (3, 7);$
range: $\{-2, -1, 2, 7\}$

11. $(-3, 1/5), (-2, 2/5), (-1, 1),$
$(0, 2), (1, 1), (2, 2/5), (3, 1/5);$
range: $\{1/5, 2/5, 1, 2\}$

13. $(-3, -1), (-2, -1), (-1, -1),$
$(0, -1), (1, -1), (2, -1),$
$(3, -1);$ range: $\{-1\}$

15. (a) 23 **(b)** -9 **(c)** -17 **(d)** $4r + 3$ **17. (a)** -28 **(b)** -12 **(c)** -28 **(d)** $-r^2 - 3$ **19. (a)** -13 **(b)** 3
(c) -32 **(d)** 22 **(e)** $-k^2 - 4k$ **(f)** $-9m^2 + 12m$ **(g)** $-k^2 + 14k - 45$ **(h)** $12 - 5p$

21.
$y = 4x + 3$

23.
$3x - 5y = 15$

25.
$x + 2 = 0$

27.
$y = 2x$

29. $1/3$ **31.** $-2/11$ **33.** $-2/3$ **35.** not defined **37.** $2x - 3y = 13$ **39.** $5x + 4y = 17$ **41.** $x = -1$
43. $5x - 8y = -40$ **45.** $y = -5$

47.
$(6, -1)$
$(2, -4)$

49.
$(-3, 4)$
$(-4, 1)$

51.
$y = x^2 - 4$

53.
$y = 3x^2 + 6x - 2$

55.
$y = x^2 - 4x + 2$

57.
$f(x) = x^3 + 5$

59.
$(1, 4)$
$y = -(x - 1)^3 + 4$

61.
$(-3, 1)$
$y = 2\sqrt{x + 3} + 1$

63.
$f(x) = \dfrac{8}{x}$

65.
$y = \dfrac{4}{3}$
$x = -\dfrac{1}{3}$
$f(x) = \dfrac{4x - 2}{3x + 1}$

67. $y = 3(x + 1)^2 - 5$

$y = 3x^2 + 6x - 2$

69. $y = (x - 2)^2 - 2$

$y = x^2 - 4x + 2$

71. (a) 7/6; 9/2 **(b)** 2; 2
(c) 5/2; 1/2
(d)

(e) 15 **(f)** 2

73. $C(x) = 30x + 60$;
$A(x) = 30 + 60/x$
75. $C(x) = 46x + 120$;
$A(x) = 46 + 120/x$

77. (a) $80 **(b)** $80 **(c)** $80
(d) $120 **(e)** $160
(f)

(g) x, the number of days
(h) $C(x)$, the cost

79. (a) halfway through 1984 **(b)** 160,000 annually **(c)** [0, 160,000]
81. (a) approximately 1.2 yr
and 9.8 yr **(b)** $f(A) > g(A)$
for $2 < A < 9.8$; $g(A) > f(A)$
for $A < 1.2$ or $A > 9.8$

(c) at 5 yr and at 13 yr

83. (a) approximately 5750 rpm
(b) approximately 310 horsepower
(c) approximately 280 ft. lbs.

85. (a)

(b) $\dfrac{2}{(x + 1)(x + 2)}$

(c) $\dfrac{5x + 3}{x(x + 1)}$

(d) $\dfrac{-5x - 6}{x(x + 1)(x + 2)}$

Extended Application (page 67)
1. 4.8 million units **2.**

3. In the interval under discussion (3.1 million to 5.7 million units),
the marginal cost always exceeds the selling price.
4. (a) 9.87; 10.22 **(b)**

(c) .83 million units, which is not in the interval under discussion

CHAPTER 2

Section 2.1 (page 77)

1. (3, 6) **3.** (−1, 4) **5.** (−2, 0) **7.** (1, 3) **9.** (4, −2) **11.** (2, −2) **13.** no solution **15.** $((y + 9)/4, y)$
17. no **19.** (12, 6) **21.** (7, −2) **23.** (1, 2, −1) **25.** (2, 0, 3) **27.** no solution **29.** (0, 2, 4) **31.** (1, 2, 3)
33. (−1, 2, 1) **35.** (4, 1, 2) **39.** $((−2z − 7)/5, (11z + 21)/5, z)$ **41.** $((−4z + 28)/5, (z − 7)/5, z)$
43. $(w + 2, w − 4, −2w + 1, w)$ **45.** 6 units of ROM chips and 3 units of RAM chips **47.** 24 fives; 8 tens; 38 twenties
49. (a) \$10,000 at 13%; \$7000 at 14%; \$8000 at 12% **(b)** For any (positive) amount borrowed at 12%, the amount at 13% must
be twice that amount less than \$26,000, and the amount borrowed at 14% must be \$1000 less than that amount. **(c)** no

Section 2.2 (page 86)

1. $\begin{bmatrix} 2 & 3 & | & 11 \\ 1 & 2 & | & 8 \end{bmatrix}$ **3.** $\begin{bmatrix} 2 & 1 & 1 & | & 3 \\ 3 & -4 & 2 & | & -7 \\ 1 & 1 & 1 & | & 2 \end{bmatrix}$ **5.** $x = 2, y = 3$ **7.** $x = 2, y = 3, z = -2$ **9.** row operations

11. $\begin{bmatrix} 2 & 3 & 8 & | & 20 \\ 0 & -5 & -4 & | & -4 \\ 0 & 3 & 5 & | & 10 \end{bmatrix}$ **13.** $\begin{bmatrix} 1 & 0 & -18 & | & -47 \\ 0 & 1 & 5 & | & 14 \\ 0 & 3 & 8 & | & 16 \end{bmatrix}$ **15.** $\begin{bmatrix} 1 & 0 & 0 & | & 6 \\ 0 & 5 & 0 & | & 9 \\ 0 & 0 & 4 & | & 8 \end{bmatrix}$ **17.** (2, 3) **19.** (−3, 0)

21. (7/2, −1) **23.** (5/2, −1) **25.** no solution **27.** $\left(\frac{1}{2}y + \frac{1}{6}, y\right)$ **29.** no solution **31.** (−1, 23, 16)

33. (3, 2, −4) **35.** $\left(-\frac{9}{23}z + \frac{5}{23}, \frac{10}{23}z - \frac{3}{23}, z\right)$ **37.** (−1, 3, 2) **39.** $\left(-\frac{2}{35}z + \frac{16}{7}, \frac{3}{7}z + \frac{13}{7}, z\right)$

41. (0, 2, −2, 1) The answers are given in the order x, y, z, w. **43. (a)** Let $x_1 = $ the number of cars sent from I to A, $x_2 = $
the number of cars sent from II to A, $x_3 = $ the number of cars sent from I to B, and $x_4 = $ the number of cars sent from II to B.
(b) $x_1 + x_3 = 28; x_2 + x_4 = 8; x_1 + x_2 = 20; x_3 + x_4 = 16; 220x_1 + 400x_2 + 300x_3 + 180x_4 = 10{,}640$ **(c)** Send 12 cars
from I to A, 8 cars from II to A, 16 cars from I to B, and no cars from II to B. **45. (a)** Let x_1 be the units purchased from first
supplier for Roseville, x_2 be the units from first supplier for Akron, x_3 be the units from second supplier for Roseville, and x_4 be
the units from second supplier for Akron. **(b)** $x_1 + x_2 = 75; x_3 + x_4 = 40; x_1 + x_3 = 40; x_2 + x_4 = 75; 70x_1 + 90x_2 + 80x_3 + 120x_4 = 10{,}750$ **(c)** (40, 35, 0, 40) **47.** (30.7209, 39.6513, 31.386, 50.3966) **49.** 81 kg of the first chemical, 382.286
kg of the second, and 286.714 kg of the third **51.** 243 of A, 38 of B, and 101 of C (rounded)

Section 2.3 (page 94)

1. false; not all corresponding elements are equal **3.** true **5.** true **7.** 2 × 2; square; $\begin{bmatrix} 4 & -8 \\ -2 & -3 \end{bmatrix}$ **9.** 3 × 4;

$\begin{bmatrix} 6 & -8 & 0 & 0 \\ -4 & -1 & -9 & -2 \\ -3 & 5 & -7 & -1 \end{bmatrix}$ **11.** 2 × 1; column; $\begin{bmatrix} -2 \\ -4 \end{bmatrix}$ **13.** the $n × m$ zero matrix **15.** $x = 2, y = 4, z = 8$

17. $x = -15, y = 5, k = 3$ **19.** $z = 18, r = 3, s = 3, p = 3, a = 3/4$ **21.** $\begin{bmatrix} 9 & 12 & 0 & 2 \\ 1 & -1 & 2 & -4 \end{bmatrix}$ **23.** not possible

25. $\begin{bmatrix} 1 & 5 & 6 & -9 \\ 5 & 7 & 2 & 1 \\ -7 & 2 & 2 & -7 \end{bmatrix}$ **27.** $\begin{bmatrix} 3 & 4 \\ 4 & 8 \end{bmatrix}$ **29.** $\begin{bmatrix} 3 & 12 \\ -6 & 3 \end{bmatrix}$ **31.** $\begin{bmatrix} -12x + 8y & -x + y \\ x & 8x - y \end{bmatrix}$ **33.** $\begin{bmatrix} -x & -y \\ -z & -w \end{bmatrix}$

39. (a) Chicago: $\begin{bmatrix} 4.05 & 7.01 \\ 3.27 & 3.51 \end{bmatrix}$, Seattle: $\begin{bmatrix} 4.40 & 6.90 \\ 3.54 & 3.76 \end{bmatrix}$ **(b)** $\begin{bmatrix} 4.24 & 6.95 \\ 3.42 & 3.64 \end{bmatrix}$ **(c)** $\begin{bmatrix} 4.42 & 7.43 \\ 3.38 & 3.62 \end{bmatrix}$ **(d)** $\begin{bmatrix} 4.41 & 7.17 \\ 3.46 & 3.69 \end{bmatrix}$

41. (a) $\begin{bmatrix} 2 & 1 & 2 & 1 \\ 3 & 2 & 2 & 1 \\ 4 & 3 & 2 & 1 \end{bmatrix}$ **(b)** $\begin{bmatrix} 5 & 0 & 7 \\ 0 & 10 & 1 \\ 0 & 15 & 2 \\ 10 & 12 & 8 \end{bmatrix}$ **(c)** $\begin{bmatrix} 8 \\ 4 \\ 5 \end{bmatrix}$ **43. (a)** 8 **(b)** 3 **(c)** $\begin{bmatrix} 85 & 15 \\ 27 & 73 \end{bmatrix}$ **(d)** yes

Section 2.4 (page 103)

1. 2×2; 2×2 **3.** 3×2; BA does not exist **5.** AB does not exist; 3×2 **7.** $\begin{bmatrix} -4 & 8 \\ 0 & 6 \end{bmatrix}$ **9.** $\begin{bmatrix} 24 & -8 \\ -16 & 0 \end{bmatrix}$

11. $\begin{bmatrix} -22 & -6 \\ 20 & -12 \end{bmatrix}$ **13.** columns; rows **15.** $\begin{bmatrix} 13 \\ 25 \end{bmatrix}$ **17.** $\begin{bmatrix} 5 \\ -21 \end{bmatrix}$ **19.** $\begin{bmatrix} 16 & -5 & 2 \\ 9 & -2 & 6 \end{bmatrix}$ **21.** $\begin{bmatrix} -2 & 10 \\ 0 & 8 \end{bmatrix}$

23. $\begin{bmatrix} 13 & 5 \\ 25 & 15 \end{bmatrix}$ **25.** $\begin{bmatrix} 13 \\ 29 \end{bmatrix}$ **27.** $\begin{bmatrix} 110 \\ 40 \\ -50 \end{bmatrix}$ **29.** $\begin{bmatrix} 22 & -8 \\ 11 & -4 \end{bmatrix}$ **31. (a)** $\begin{bmatrix} 16 & 22 \\ 7 & 19 \end{bmatrix}$ **(b)** $\begin{bmatrix} 5 & -5 \\ 0 & 30 \end{bmatrix}$ **(c)** no **(d)** no

39. (a) $\begin{array}{c} \\ \text{Dept. 1} \\ \text{Dept. 2} \\ \text{Dept. 3} \\ \text{Dept. 4} \end{array} \begin{array}{cc} A & B \\ \begin{bmatrix} 57 & 70 \\ 41 & 54 \\ 27 & 40 \\ 39 & 40 \end{bmatrix} \end{array}$ **(b)** supplier A: \$164; supplier B: \$204; supplier A

41. (a) $\begin{bmatrix} .027 & .009 \\ .030 & .007 \\ .015 & .009 \\ .013 & .011 \\ .019 & .011 \end{bmatrix}$; $\begin{bmatrix} 1596 & 218 & 199 & 425 & 214 \\ 1996 & 286 & 226 & 460 & 243 \\ 2440 & 365 & 252 & 484 & 266 \\ 2906 & 455 & 277 & 499 & 291 \end{bmatrix}$ **(b)** $\begin{array}{c} \\ 1960 \\ 1970 \\ 1980 \\ 1990 \end{array} \begin{array}{cc} \text{Births} & \text{Deaths} \\ \begin{bmatrix} 62.208 & 24.710 \\ 76.459 & 29.733 \\ 91.956 & 35.033 \\ 108.28 & 40.522 \end{bmatrix} \end{array}$

43. (a) $\begin{bmatrix} 6 & 106 & 158 & 222 & 28 \\ 120 & 139 & 64 & 75 & 115 \\ -146 & -2 & 184 & 144 & -129 \\ 106 & 94 & 24 & 116 & 110 \end{bmatrix}$ **(b)** cannot be found **(c)** no

45. (a) $\begin{bmatrix} -1 & 5 & 9 & 13 & -1 \\ 7 & 17 & 2 & -10 & 6 \\ 18 & 9 & -12 & 12 & 22 \\ 9 & 4 & 18 & 10 & -3 \\ 1 & 6 & 10 & 28 & 5 \end{bmatrix}$ **(b)** $\begin{bmatrix} -2 & -9 & 90 & 77 \\ -42 & -63 & 127 & 62 \\ 413 & 76 & 180 & -56 \\ -29 & -44 & 198 & 85 \\ 137 & 20 & 162 & 103 \end{bmatrix}$ **(c)** $\begin{bmatrix} -56 & -1 & 1 & 45 \\ -156 & -119 & 76 & 122 \\ 315 & 86 & 118 & -91 \\ -17 & -17 & 116 & 51 \\ 118 & 19 & 125 & 77 \end{bmatrix}$

(d) $\begin{bmatrix} 54 & -8 & 89 & 32 \\ 114 & 56 & 51 & -60 \\ 98 & -10 & 62 & 35 \\ -12 & -27 & 82 & 34 \\ 19 & 1 & 37 & 26 \end{bmatrix}$ **(e)** $\begin{bmatrix} -2 & -9 & 90 & 77 \\ -42 & -63 & 127 & 62 \\ 413 & 76 & 180 & -56 \\ -29 & -44 & 198 & 85 \\ 137 & 20 & 162 & 103 \end{bmatrix}$ **(f)** yes **(g)** the commutative and distributive properties

Section 2.5 (page 114)

1. yes **3.** no **5.** no **7.** yes **9.** No; the row of all zeros makes it impossible to get all the 1's in the diagonal of the identity matrix. **11.** $\begin{bmatrix} 0 & 1/2 \\ -1 & 1/2 \end{bmatrix}$ **13.** $\begin{bmatrix} 2 & 1 \\ 5 & 3 \end{bmatrix}$ **15.** no inverse **17.** $\begin{bmatrix} 1 & 0 & 0 \\ 0 & -1 & 0 \\ -1 & 0 & 1 \end{bmatrix}$ **19.** $\begin{bmatrix} 15 & 4 & -5 \\ -12 & -3 & 4 \\ -4 & -1 & 1 \end{bmatrix}$

21. no inverse **23.** $\begin{bmatrix} 7/4 & 5/2 & 3 \\ -1/4 & -1/2 & 0 \\ -1/4 & -1/2 & -1 \end{bmatrix}$ **25.** $\begin{bmatrix} 1/2 & 1/2 & -1/4 & 1/2 \\ -1 & 4 & -1/2 & -2 \\ -1/2 & 5/2 & -1/4 & -3/2 \\ 1/2 & -1/2 & 1/4 & 1/2 \end{bmatrix}$ **27.** $(-1, 4)$ **29.** $(2, 1)$

31. $(2, 3)$ **33.** no inverse, $(-8y - 12, y)$ **35.** $(-8, 6, 1)$ **37.** $(15, -5, -1)$ **39.** no inverse; no solution for system
41. $(-7, -34, -19, 7)$

51. (a) $\begin{bmatrix} 72 \\ 48 \\ 60 \end{bmatrix}$ **(b)** $\begin{bmatrix} 2 & 4 & 2 \\ 2 & 1 & 2 \\ 2 & 1 & 3 \end{bmatrix} \begin{bmatrix} x_1 \\ x_2 \\ x_3 \end{bmatrix} = \begin{bmatrix} 72 \\ 48 \\ 60 \end{bmatrix}$ **(c)** 8 type I, 8 type II, and 12 type III **53. (a)** $12,000 at 6%, $7000 at

7%, and $6000 at 10% **(b)** $10,000 at 6%, $15,000 at 7%, and $5000 at 10% **(c)** $20,000 at 6%, $10,000 at 7%, and

$10,000 at 10% **55.** Entries are rounded to 6 places. $\begin{bmatrix} .010146 & -.011883 & .002772 & .020724 & -.012273 \\ .006353 & .014233 & -.001861 & -.029146 & .019225 \\ -.000638 & .006782 & -.004823 & -.022658 & .019344 \\ -.005261 & .003781 & .006192 & .004837 & -.006910 \\ -.012252 & -.001177 & -.006126 & .006744 & .002792 \end{bmatrix}$

57. no **59.** $\begin{bmatrix} .62963 \\ .148148 \\ .259259 \end{bmatrix}$ **61.** $\begin{bmatrix} .489558 \\ 1.00104 \\ 2.11853 \\ -1.20793 \\ -.961346 \end{bmatrix}$

Section 2.6 (page 121)

1. $\begin{bmatrix} 10.67 \\ 8.33 \end{bmatrix}$ **3.** $\begin{bmatrix} 6.43 \\ 26.12 \end{bmatrix}$ **5.** $\begin{bmatrix} 6.67 \\ 20 \\ 10 \end{bmatrix}$ **7.** 33:47:23 **9.** 1079 metric tons of wheat and 1428 metric tons of oil

11. 1285 units of agriculture, 1455 units of manufacturing, and 1202 units of transportation **13.** 3077 units of agriculture,
2564 units of manufacturing, and 3179 units of transportation **15. (a)** 7/4 bushels of yams and $15/8 \approx 2$ pigs

(b) 167.5 bushels of yams and $153.75 \approx 154$ pigs **17.** $\begin{bmatrix} 2930 \\ 3570 \\ 2300 \\ 580 \end{bmatrix}$ **19.** $\begin{bmatrix} 1583.91 \\ 1529.54 \\ 1196.09 \end{bmatrix}$

Chapter 2 Review Exercises (page 122)

3. $(-4, 6)$ **5.** $(-1, 2, 3)$ **7.** $(-9, 3)$ **9.** $(7, -9, -1)$ **11.** $\left(6 - \frac{7}{3}z, 1 + \frac{1}{3}z, z \right)$ **13.** 3×2; $a = 2$,
$x = -1, y = 4, p = 5, z = 7$ **15.** 3×3 (square); $a = -12, b = 1, k = 9/2, c = 3/4, d = 3, l = -3/4, m = -1$,
$p = 3, q = 9$ **17.** $\begin{bmatrix} 8 & -6 \\ -10 & -16 \end{bmatrix}$ **19.** not possible **21.** $\begin{bmatrix} 26 & 86 \\ -7 & -29 \\ 21 & 87 \end{bmatrix}$ **23.** $\begin{bmatrix} 6 & 18 & -24 \\ 1 & 3 & -4 \\ 0 & 0 & 0 \end{bmatrix}$ **25.** $\begin{bmatrix} 15 \\ 16 \\ 1 \end{bmatrix}$

27. $\begin{bmatrix} -7/19 & 4/19 \\ 3/19 & 1/19 \end{bmatrix}$ **29.** no inverse **31.** $\begin{bmatrix} -1/4 & 1/6 \\ 0 & 1/3 \end{bmatrix}$ **33.** no inverse **35.** $\begin{bmatrix} 1/4 & 1/2 & 1/2 \\ 1/4 & -1/2 & 1/2 \\ 1/8 & -1/4 & -1/4 \end{bmatrix}$

37. no inverse **39.** Matrix A has no inverse; solution: $(-2y + 5, y)$. **41.** $X = \begin{bmatrix} 6 \\ 15 \\ 16 \end{bmatrix}$ **43.** $(34, -9)$ **45.** $\begin{bmatrix} 218.1 \\ 318.2 \end{bmatrix}$

47. 8000 standard; 6000 extra large **49.** 5 blankets, 3 rugs, 8 skirts **51.** $\begin{bmatrix} 5 & 7 & 2532 & 52\ 3/8 & -1/4 \\ 3 & 9 & 1464 & 56 & 1/8 \\ 2.50 & 5 & 4974 & 41 & -1\ 1/2 \\ 1.36 & 10 & 1754 & 18\ 7/8 & 1/2 \end{bmatrix}$

53. **(a)** $\begin{array}{c} \\ c \\ g \end{array}\begin{array}{cc} c & g \\ \end{array}\begin{bmatrix} 0 & 1/2 \\ 2/3 & 0 \end{bmatrix}$ **(b)** 1200 units of cheese; 1600 units of goats **55.** **(a)** $(2, 3, -1)$ **(b)** $(2, 3, -1)$ **(c)** $(2, 3, -1)$

(d) $\begin{bmatrix} 1 & 2 & 1 \\ 2 & -1 & -1 \\ 3 & -3 & 2 \end{bmatrix}\begin{bmatrix} x \\ y \\ z \end{bmatrix} = \begin{bmatrix} 7 \\ 2 \\ -5 \end{bmatrix}$ **(e)** $A^{-1} = \begin{bmatrix} 5/22 & 7/22 & 1/22 \\ 7/22 & 1/22 & -3/22 \\ 3/22 & -9/22 & 5/22 \end{bmatrix} \approx \begin{bmatrix} .23 & .32 & .05 \\ .32 & .05 & -.14 \\ .14 & -.41 & .23 \end{bmatrix}$ **(f)** $(2, 3, -1)$

Extended Application (page 127)

1. (a) $A = \begin{bmatrix} .245 & .102 & .051 \\ .099 & .291 & .279 \\ .433 & .372 & .011 \end{bmatrix}$, $D = \begin{bmatrix} 2.88 \\ 31.45 \\ 30.91 \end{bmatrix}$, $X = \begin{bmatrix} x_1 \\ x_2 \\ x_3 \end{bmatrix}$ **(b)** $I - A = \begin{bmatrix} .755 & -.102 & -.051 \\ -.099 & .709 & -.279 \\ -.433 & -.372 & .989 \end{bmatrix}$ **(d)** $\begin{bmatrix} 18.2 \\ 73.2 \\ 66.8 \end{bmatrix}$

(e) \$18.2 billion of agriculture, \$73.2 billion of manufacturing, and \$66.8 billion of household would be required (rounded to three significant digits) **2. (a)** $A = \begin{bmatrix} .293 & 0 & 0 \\ .014 & .207 & .017 \\ .044 & .010 & .216 \end{bmatrix}$, $D = \begin{bmatrix} 138,213 \\ 17,597 \\ 1,786 \end{bmatrix}$ **(b)** $I - A = \begin{bmatrix} .707 & 0 & 0 \\ -.014 & .793 & -.017 \\ -.044 & -.010 & .784 \end{bmatrix}$

(d) agriculture, 195 million pounds; manufactured goods, 26 million pounds; energy, 13.6 million pounds

CHAPTER 3

Section 3.1 (page 134)

1.
$x + y \le 2$

3.
$x \ge 3 + y$

5.
$4x - y < 6$

7.
$3x + y < 6$

9.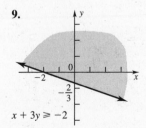
$x + 3y \ge -2$

11.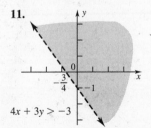
$4x + 3y > -3$

13.
$2x - 4y < 3$

15.
$x \le 5y$

17.
$-3x < y$

19.
$x + y \leq 0$

21.
$y < x$

23.
$x < 4$

25.
$y \leq -2$

27.
$x + y \leq 1$
$x - y \geq 2$

29.
$2x - y < 1$
$3x + y < 6$

31.
$-x - y < 5$
$2x - y < 4$

33.
$x + y \leq 4$
$x - y \leq 5$
$4x + y \leq -4$

35.
$-2 < x < 3$
$-1 \leq y \leq 5$
$2x + y < 6$

37. $2y + x \geq -5$
$y \leq 3 + x$
$x \geq 0$
$y \geq 0$

39.
$3x + 4y > 12$
$2x - 3y < 6$
$0 \leq y \leq 2$
$x \geq 0$

41. (a)

	Number Made	Time on Wheel	Time in Kiln
Glazed	x	1/2	1
Unglazed	y	1	6
Maximum Time Available		8	20

(b) $(1/2)x + y \leq 8;\ x + 6y \leq 20;\ x \geq 0;\ y \geq 0$ **(c)** yes; no

43. (a) $x \geq 3000;\ y \geq 5000;\ x + y \leq 1000$

(b)

45. (a) $x \geq 1000;\ y \geq 800;\ x + y \leq 2400$

(b)

Section 3.2 (page 141)

1. maximum of 65 at (5, 10); minimum of 8 at (1, 1) **3.** maximum of 9 at (0, 12); minimum of 0 at (0, 0) **5. (a)** no maximum; minimum of 16 at (0, 8) **(b)** no maximum; minimum of 18 at (3, 4) **(c)** no maximum; minimum of 21 at (13/2, 2) **(d)** no maximum; minimum of 12 at (12, 0) **7.** maximum of 42/5 when $x = 6/5$, $y = 6/5$ **9.** maximum of 30 when $x = 27/4$, $y = 33/4$, as well as when $x = 10$, $y = 5$, and all points in between **11.** maximum of 235/4 when $x = 105/8$, $y = 25/8$ **13. (a)** maximum of 204 when $x = 18$ and $y = 2$ **(b)** maximum of 588/5 when $x = 12/5$ and $y = 39/5$ **(c)** maximum of 102 when $x = 0$ and $y = 17/2$ **15. (b)**

Section 3.3 (page 147)

1. Let x be the number of product A made and y be the number of product B. Then $2x + 3y \le 45$. **3.** Let x be the number of green pills and y be the number of red. Then $4x + y \ge 25$. **5.** Let x be the number of pounds of $6 coffee and y be the number of pounds of $5 coffee. Then $x + y \ge 50$. **7.** 50 to plant I and 27 to plant II for a minimum cost of $1945 (It is not surprising that costs are minimized by sending the minimum number of engines required.) **9. (a)** 0 units of policy A and 25/3 or 8 1/3 units of policy B for a minimum premium cost of $333.33 **(b)** 25/2 or 12 1/2 units of policy A and 0 units of policy B for a minimum premium cost of $312.50 **11.** 1600 type 1 bolts and 0 type 2 bolts for maximum revenue of $160 **13. (a)** 150 kg half-and-half mix and 75 kg other mix for maximum revenue of $1260 **(b)** 200 kg half-and-half mix and 0 kg other mix for maximum revenue of $1600 **15.** 50 gal from dairy I and 50 gal from dairy II for maximum butterfat of 3.45% **17.** $25 million in bonds and $15 million in mutual funds for maximum annual interest of $4.20 million **19.** A total of 8/7 units of species I and 10/7 units of species II will meet requirements with a minimum of 6.57 units of energy; however, a predator probably can catch and digest only whole numbers of prey. This problem shows that it is important to consider whether a model produces a realistic answer to a problem. **21.** 3 3/4 servings of A and 1 7/8 servings of B for a minimum cost of $1.69 **25. (a)**

Chapter 3 Review Exercises (page 151)

3.

$y \ge 2x + 3$

5.

$3x + 4y \le 12$

7.

$y \ge x$

9.

$x + y \le 6$
$2x - y \ge 3$
(3, 3)

11.

$-4 \le x \le 2$
$-1 \le y \le 3$
$x + y \le 4$

13.

$\left(\dfrac{18}{5}, \dfrac{4}{5}\right)$
$x + 3y \ge 6$
$4x - 3y \le 12$
$x \ge 0$
$y \ge 0$

15. minimum of 8 at (2, 1); maximum of 40 at (6, 7) **17.** maximum of 24 at (0, 6) **19.** minimum of 40 at any point on the segment connecting (0, 20) and (10/3, 40/3) **23.** Let x = number of batches of cakes and y = number of batches of cookies.

Then $x \geq 0$, $y \geq 0$, and $2x + (3/2)y \leq 15$, $3x + (2/3)y \leq 13$.

25. 3 batches of cakes and 6 of cookies for maximum profit of $210 **27.** 7 hr with the math tutor and 2 hr with the accounting tutor for a maximum of 31 points
29. maximum of 235/4 when $x = 105/8$, $y = 25/8$

CHAPTER 4

Section 4.1 (page 159)

1. $x_1 + 2x_2 + x_3 = 6$ **3.** $2x_1 + 4x_2 + 3x_3 + x_4 = 100$ **5. (a)** 3 **(b)** x_3, x_4, x_5 **(c)** $\begin{aligned} 4x_1 + 2x_2 + x_3 &= 20 \\ 5x_1 + x_2 + x_4 &= 50 \\ 2x_1 + 3x_2 + x_5 &= 25 \end{aligned}$

7. (a) 2 **(b)** x_4, x_5 **(c)** $\begin{aligned} 7x_1 + 6x_2 + 8x_3 + x_4 &= 118 \\ 4x_1 + 5x_2 + 10x_3 + x_5 &= 220 \end{aligned}$ **9.** $x_1 = 0, x_2 = 0, x_3 = 20, x_4 = 0, x_5 = 15, z = 10$

11. $x_1 = 0, x_2 = 0, x_3 = 8, x_4 = 0, x_5 = 6, x_6 = 7, z = 12$ **13.** $x_1 = 0, x_2 = 20, x_3 = 0, x_4 = 16, x_5 = 0, z = 60$
15. $x_1 = 0, x_2 = 0, x_3 = 12, x_4 = 0, x_5 = 9, x_6 = 8, z = 36$ **17.** $x_1 = 0, x_2 = 0, x_3 = 50, x_4 = 10, x_5 = 0, x_6 = 50,$

$z = 100$ **19.** $\begin{array}{ccccc} x_1 & x_2 & x_3 & x_4 & z \\ \left[\begin{array}{ccccc|c} 2 & 3 & 1 & 0 & 0 & 6 \\ 4 & 1 & 0 & 1 & 0 & 6 \\ \hline -5 & -1 & 0 & 0 & 1 & 0 \end{array}\right] \end{array}$ **21.** $\begin{array}{cccccc} x_1 & x_2 & x_3 & x_4 & x_5 & z \\ \left[\begin{array}{ccccc|c} 1 & 1 & 1 & 0 & 0 & 0 & 10 \\ 5 & 2 & 0 & 1 & 0 & 0 & 20 \\ 1 & 2 & 0 & 0 & 1 & 0 & 36 \\ \hline -1 & -3 & 0 & 0 & 0 & 1 & 0 \end{array}\right] \end{array}$ **23.** $\begin{array}{ccccc} x_1 & x_2 & x_3 & x_4 & z \\ \left[\begin{array}{cccc|c} 3 & 1 & 1 & 0 & 0 & 12 \\ 1 & 1 & 0 & 1 & 0 & 15 \\ \hline -2 & -1 & 0 & 0 & 1 & 0 \end{array}\right] \end{array}$

25. If x_1 is the number of prams, x_2 is the number of runabouts, and x_3 is the number of trimarans, find $x_1 \geq 0, x_2 \geq 0, x_3 \geq 0,$
$x_4 \geq 0, x_5 \geq 0, x_6 \geq 0$ so that $x_1 + 2x_2 + 3x_3 + x_4 = 6240, 2x_1 + 5x_2 + 4x_3 + x_5 = 10{,}800, x_1 + x_2 + x_3 + x_6 = 3000,$
and $z = 75x_1 + 90x_2 + 100x_3$ is maximized.

$$\begin{array}{ccccccc} x_1 & x_2 & x_3 & x_4 & x_5 & x_6 & z \\ \left[\begin{array}{cccccc|c} 1 & 2 & 3 & 1 & 0 & 0 & 0 & 6240 \\ 2 & 5 & 4 & 0 & 1 & 0 & 0 & 10{,}800 \\ 1 & 1 & 1 & 0 & 0 & 1 & 0 & 3000 \\ \hline -75 & -90 & -100 & 0 & 0 & 0 & 1 & 0 \end{array}\right] \end{array}$$

27. If x_1 is the number of simple figures, x_2 is the number of figures with additions, and x_3 is the number of computer-drawn
sketches, find $x_1 \geq 0, x_2 \geq 0, x_3 \geq 0, x_4 \geq 0, x_5 \geq 0, x_6 \geq 0, x_7 \geq 0$ so that $20x_1 + 35x_2 + 60x_3 + x_4 = 2200, x_1 + x_2 + x_3 + x_5 = 400, -x_1 - x_2 + x_3 + x_6 = 0, -x_1 + 2x_2 + x_7 = 0,$ and $z = 95x_1 + 200x_2 + 325x_3$ is maximized.

$$\begin{array}{cccccccc} x_1 & x_2 & x_3 & x_4 & x_5 & x_6 & x_7 & z \\ \left[\begin{array}{ccccccc|c} 20 & 35 & 60 & 1 & 0 & 0 & 0 & 0 & 2200 \\ 1 & 1 & 1 & 0 & 1 & 0 & 0 & 0 & 400 \\ -1 & -1 & 1 & 0 & 0 & 1 & 0 & 0 & 0 \\ -1 & 2 & 0 & 0 & 0 & 0 & 1 & 0 & 0 \\ \hline -95 & -200 & -325 & 0 & 0 & 0 & 0 & 1 & 0 \end{array}\right] \end{array}$$

29. If x_1 is the number of one-speed bicycles, x_2 is the number of three-speed bicycles, and x_3 is the number of ten-speed bicycles,
find $x_1 \geq 0, x_2 \geq 0, x_3 \geq 0, x_4 \geq 0, x_5 \geq 0$ so that $17x_1 + 27x_2 + 34x_3 + x_4 = 91{,}800, 12x_1 + 21x_2 + 15x_3 + x_5 = 42{,}000,$ and $z = 8x_1 + 12x_2 + 22x_3$ is maximized.

$$\begin{array}{ccccc} x_1 & x_2 & x_3 & x_4 & x_5 & z \\ \left[\begin{array}{ccccc|c} 17 & 27 & 34 & 1 & 0 & 0 & 91{,}800 \\ 12 & 21 & 15 & 0 & 1 & 0 & 42{,}000 \\ \hline -8 & -12 & -22 & 0 & 0 & 1 & 0 \end{array}\right] \end{array}$$

Section 4.2 (page 168)

1. Maximum is 20 when $x_1 = 0$, $x_2 = 4$, $x_3 = 0$, $x_4 = 0$, and $x_5 = 2$. **3.** Maximum is 8 when $x_1 = 4$, $x_2 = 0$, $x_3 = 8$, $x_4 = 2$, and $x_5 = 0$. **5.** Maximum is 264 when $x_1 = 16$, $x_2 = 4$, $x_3 = 0$, $x_4 = 0$, $x_5 = 16$, and $x_6 = 0$. **7.** Maximum is 22 when $x_1 = 5.5$, $x_2 = 0$, $x_3 = 0$, and $x_4 = .5$. **9.** Maximum is 120 when $x_1 = 0$, $x_2 = 10$, $x_3 = 0$, $x_4 = 40$, and $x_5 = 4$. **11.** Maximum is 944 when $x_1 = 118$, $x_2 = 0$, $x_3 = 0$, $x_4 = 0$, and $x_5 = 102$. **13.** Maximum is 250 when $x_1 = 0$, $x_2 = 0$, $x_3 = 0$, $x_4 = 50$, $x_5 = 0$, and $x_6 = 50$. **17.** 6 churches and 2 labor unions for a maximum of $1000 per month **19.** 2 jazz albums, 3 blues albums, and 6 reggae albums for a maximum weekly profit of $10.60 **21.** no 1-speed or 3-speed bicycles and 2700 10-speed bicycles; maximum profit is $59,400 **23.** 150 kg of the half-and-half mix and 75 kg of the other for a maximum revenue of $1260 **25. (a)** 1 **(b)** 3 **27.** 12 min to the senator, 9 min to the congresswoman, and 6 min to the governor for a maximum of 1,320,000 viewers **29. (a)** none of species A, 114 of species B, and 291 of species C for a maximum combined weight of 1119.72 kg **(b)** many solutions possible **(c)** many solutions possible

Section 4.3 (page 180)

1. $2x_1 + 3x_2 + x_3 = 8$
$x_1 + 4x_2 - x_4 = 7$ **3.** $x_1 + x_2 + x_3 + x_4 = 100$
$x_1 + x_2 + x_3 - x_5 = 75$
$x_1 + x_2 - x_6 = 27$ **5.** Change the objective function to maximize $z = -4y_1 - 3y_2 - 2y_3$. The constraints are not changed. **7.** Change the objective function to maximize $z = -y_1 - 2y_2 - y_3 - 5y_4$. The constraints are not changed. **9.** Maximum is 480 when $x_1 = 40$, $x_2 = 0$, $x_3 = 16$, and $x_4 = 0$. **11.** Maximum is 750 when $x_1 = 0$, $x_2 = 150$, $x_3 = 0$, $x_4 = 0$, and $x_5 = 50$. **13.** Maximum is 300 when $x_1 = 0$, $x_2 = 100$, $x_3 = 0$, $x_4 = 50$, and $x_5 = 10$. **15.** Minimum is 40 when $y_1 = 10$, $y_2 = 0$, $y_3 = 0$, and $y_4 = 50$. **17.** Minimum is 100 when $y_1 = 0$, $y_2 = 100$, $y_3 = 0$, $y_4 = 0$, and $y_5 = 50$. **19.** Maximum is $133\frac{1}{3}$ when $x_1 = 33\frac{1}{3}$, $x_2 = 16\frac{2}{3}$, $x_3 = 46\frac{2}{3}$, and $x_4 = 0$.
21. Minimum is 144 when $y_1 = 8$, $y_2 = 2$, $y_3 = 0$, and $y_4 = 8$. **25.** Ship 5000 barrels of oil from supplier S_1 to distributor D_2; ship 3000 barrels of oil from supplier S_2 to distributor D_1. Minimum cost is $175,000. **27.** 800,000 kg for whole tomatoes and 80,000 kg for sauce for a minimum cost of $3,460,000 **29.** 1000 small and 500 large for a minimum cost of $210
31. $40,000 in government securities and $60,000 in mutual funds for maximum interest of $8800 **33.** 3 of pill #1 and 2 of pill #2 for a minimum cost of 70¢ **35.** $2\frac{2}{3}$ units of I, none of II, and 4 of III for a minimum cost of $30.67 **37.** Use $1\frac{2}{3}$ oz of I, $6\frac{2}{3}$ oz of II, $1\frac{2}{3}$ oz of III, for a minimum cost of $1.55 per gallon; 10 oz of the additive should be used per gallon of gasoline.

Section 4.4 (page 191)

1. $\begin{bmatrix} 1 & 3 & 1 \\ 2 & 2 & 10 \\ 3 & 1 & 0 \end{bmatrix}$ **3.** $\begin{bmatrix} -1 & 13 & -2 \\ 4 & 25 & -1 \\ 6 & 0 & 11 \\ 12 & 4 & 3 \end{bmatrix}$ **5.** Minimize $w = 5y_1 + 4y_2 + 15y_3$ subject to $y_1 + y_2 + 2y_3 \geq 4$, $y_1 + y_2 + y_3 \geq 3$, $y_1 + 3y_3 \geq 2$, $y_1 \geq 0$, $y_2 \geq 0$, and $y_3 \geq 0$. **7.** Maximize $z = 50x_1 + 100x_2$ subject to $x_1 + 3x_2 \leq 1$, $x_1 + x_2 \leq 2$, $x_1 + 2x_2 \leq 1$, $x_1 + x_2 \leq 5$, $x_1 \geq 0$, and $x_2 \geq 0$. **9.** Minimum is 14 when $y_1 = 0$ and $y_2 = 7$. **11.** Minimum is 40 when $y_1 = 10$ and $y_2 = 0$. **13.** Minimum is 100 when $y_1 = 0$, $y_2 = 100$, and $y_3 = 0$. **15.** Statement (a) is true.
17. (a) Minimize $w = 200y_1 + 600y_2 + 90y_3$ subject to $y_1 + 4y_2 \geq 1$, $2y_1 + 3y_2 + y_3 \geq 1.5$, $y_1 \geq 0$, $y_2 \geq 0$, and $y_3 \geq 0$.
(b) $y_1 = .6$, $y_2 = .1$, $y_3 = 0$, $w = 180$ **(c)** $186 ($x_1 = 114$, $x_2 = 48$) **(d)** $179 ($x_1 = 116$, $x_2 = 42$) **19.** 1 package of Sun Hill and 5 packages of Bear Valley, for a minimum cost of $13 **21. (a)** 1 bag of feed 1, 2 bags of feed 2 **(b)** $6.60 per day for 1.4 bags of feed 1 and 1.2 bags of feed 2 **23.** Use $1\frac{2}{3}$ oz of I, $6\frac{2}{3}$ oz of II, and $1\frac{2}{3}$ oz of III for a minimum cost of $1.55 per gallon; 10 oz of the additive should be used per gallon of gasoline.

Chapter 4 Review Exercises (page 193)

1. when the problem has more than two variables **3. (a)** $2x_1 + 5x_2 + x_3 = 50$
$x_1 + 3x_2 + x_4 = 25$
$4x_1 + x_2 + x_5 = 18$
$x_1 + x_2 + x_6 = 12$ **(b)**

x_1	x_2	x_3	x_4	x_5	x_6	z	
2	5	1	0	0	0	0	50
1	3	0	1	0	0	0	25
4	1	0	0	1	0	0	18
1	1	0	0	0	1	0	12
-5	-3	0	0	0	0	1	0

$$x_1 + x_2 + x_3 + x_4 = 90$$
5. (a) $2x_1 + 5x_2 + x_3 + x_5 = 120$ **(b)**
$$x_1 + 3x_2 - x_6 = 80$$

x_1	x_2	x_3	x_4	x_5	x_6	z	
1	1	1	1	0	0	0	90
2	5	1	0	1	0	0	120
1	3	0	0	0	-1	0	80
-5	-8	-6	0	0	0	1	0

7. Maximum is 82.4 when $x_1 = 13.6$, $x_2 = 0$, $x_3 = 4.8$, $x_4 = 0$, and $x_5 = 0$. **9.** Maximum is 76.67 when $x_1 = 6.67$, $x_2 = 0$, $x_3 = 21.67$, $x_4 = 0$, $x_5 = 0$, and $x_6 = 35$. **11.** Change the objective function to maximize $z = -10y_1 - 15y_2$. The constraints are not changed. **13.** Change the objective function to maximize $z = -7y_1 - 2y_2 - 3y_3$. The constraints are not changed. **15.** Minimum of 62 when $y_1 = 8$, $y_2 = 12$, $y_3 = 0$, $y_4 = 1$, $y_5 = 0$, and $y_6 = 2$. **17.** Problems with constraints involving "\leq" can be solved using slack variables, while those involving "\geq" or "$=$" can be solved using surplus and artificial variables, respectively. **19. (a)** Maximize $z = 6x_1 + 7x_2 + 5x_3$, subject to $4x_1 + 2x_2 + 3x_3 \leq 9$, $5x_1 + 4x_2 + x_3 \leq 10$, with $x_1 \geq 0$, $x_2 \geq 0$, $x_3 \geq 0$. **(b)** The first constraint would be $4x_1 + 2x_2 + 3x_3 \geq 9$. **(c)** $x_1 = 0$, $x_2 = 2.1$, $x_3 = 1.6$, and $z = 22.7$ **(d)** Minimize $w = 9y_1 + 10y_2$, subject to $4y_1 + 5y_2 \geq 6$, $2y_1 + 4y_2 \geq 7$, $3y_1 + y_2 \geq 5$, with $y_1 \geq 0$, $y_2 \geq 0$. **(e)** $y_1 = 1.3$, $y_2 = 1.1$, and $w = 22.7$ **21. (a)** Let $x_1 = $ amount invested in the oil leases, $x_2 = $ amount invested in bonds,

$$x_1 + x_2 + x_3 \leq 50{,}000$$

and $x_3 = $ amount invested in stock. **(b)** $z = .15x_1 + .09x_2 + .05x_3$ **(c)** $x_1 + x_2 \quad\quad \leq 15{,}000$
$$x_1 \quad\quad + x_3 \leq 25{,}000$$

23. (a) Let $x_1 = $ number of 5-gal bags, $x_2 = $ number of 10-gal bags, $x_3 = $ number of 20-gal bags. **(b)** $z = x_1 + .9x_2 + .95x_3$
$$x_1 + 1.1x_2 + 1.5x_3 \leq 8$$
(c) $x_1 + 1.2x_2 + 1.3x_3 \leq 8$ **25.** \$15,000 in oil leases and \$10,000 in stock for maximum return of \$2750 **27.** 4 units of
$$2x_1 + 3x_2 + 4x_3 \leq 8$$
5-gal bags per day (and none of the others) for maximum profit of \$4 per unit

Extended Application (page 197)
1. (a) $x_2 = 58.6957$, $x_8 = 2.01031$, $x_9 = 22.4895$, $x_{10} = 1$, $x_{12} = 1$, and $x_{13} = 1$ for a minimum cost of \$12.55 **(b)** $x_2 = 71.7391$, $x_8 = 3.14433$, $x_9 = 23.1966$, $x_{10} = 1$, $x_{12} = 1$, and $x_{13} = 1$ for a minimum cost of \$15.05

Extended Application (page 199)
1. $w_1 = .95$, $w_2 = .83$, $w_3 = .75$, $w_4 = 1.00$, $w_5 = .87$, and $w_6 = .94$ **2.** $x_1 = 100$, $x_2 = 0$, $x_3 = 0$, $x_4 = 90$, $x_5 = 0$, and $x_6 = 210$

CHAPTER 5

Section 5.1 (page 208)
1. false **3.** true **5.** true **7.** true **9.** false **11.** (b) and (c) **13.** \subseteq **15.** $\not\subseteq$ **17.** \subseteq **19.** \subseteq **21.** 64 **23.** 8 **25.** 8 **27.** 32 **31.** \cap **33.** \cap **35.** \cup **37.** \cup **39.** $\{3, 5\}$ **41.** $\{7, 9\}$ **43.** \emptyset **45.** U or $\{2, 3, 4, 5, 7, 9\}$ **47.** all students in this school not taking this course **49.** all students in this school taking accounting and zoology **51.** C and D, B and E, C and E, D and E **53.** B' is the set of all stocks on the list whose price-to-earnings ratio is less than 13; $B' = \emptyset$. **55.** $(A \cap B)'$ is the set of all stocks with a dividend less than or equal to \$3 or a price-to-earnings ratio of less than 13; $(A \cap B)' = \{$ATT, GE, Hershey, Nike$\}$. **57.** $\{$HBO, Showtime, Cinemax$\}$ **59.** $\{$HBO, Showtime, Cinemax, Disney Channel, Movie Channel$\}$ **61.** $\{$Showtime, Cinemax$\}$ **63.** $\{s, d, c\}$ **65.** $\{g\}$ **67.** $\{s, d, c\}$

Section 5.2 (page 215)

1.

$B \cap A'$

3.

$A' \cup B$

5.

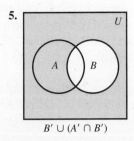

$B' \cup (A' \cap B')$

7.

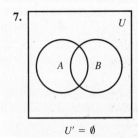

$U' = \emptyset$

9. 8.

11.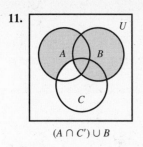

$(A \cap C') \cup B$

13.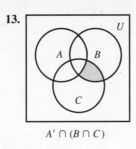

$A' \cap (B \cap C)$

15.

$(A \cap B') \cup C$

17.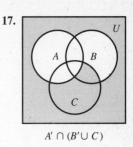

$A' \cap (B' \cup C)$

19. 9 **21.** 16

23.

25.

27.

29.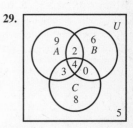

35. Yes; his results add up to 142 people surveyed. **37. (a)** 40 **(b)** 30 **(c)** 95 **(d)** 110 **(e)** 160 **(f)** 65 **(g)** all those age 21–25 who drink diet cola or anyone age 26–35 **39. (a)** 50 **(b)** 2 **(c)** 14 **41. (a)** 54 **(b)** 17 **(c)** 10 **(d)** 7 **(e)** 15 **(f)** 3 **(g)** 12 **(h)** 1 **43. (a)** 342 **(b)** 192 **(c)** 72 **(d)** 86

Section 5.3 (page 222)

3. {January, February, March, . . . , December} **5.** {0, 1, 2, . . . , 80} **7.** {go ahead, cancel} **9.** {(h, 1), (h, 2), (h, 3), (h, 4), (h, 5), (h, 6), (t, 1), (t, 2), (t, 3), (t, 4), (t, 5), (t, 6)} **13.** {AB, AC, AD, AE, BC, BD, BE, CD, CE, DE}
(a) {AC, BC, CD, CE} **(b)** {AB, AC, AD, AE, BC, BD, BE, CD, CE} **(c)** {AC} **15.** {(1, 2), (1, 3), (1, 4), (1, 5), (2, 3), (2, 4), (2, 5), (3, 4), (3, 5), (4, 5)} **(a)** {(2, 4)} **(b)** {(1, 2), (1, 4), (2, 3), (2, 5), (3, 4), (4, 5)} **(c)** ∅
17. {(1, 1), (1, 2), (1, 3), (1, 4), (1, 5), (1, 6), (2, 1), (2, 2), (2, 3), (2, 4), (2, 5), (2, 6), (3, 1), (3, 2), (3, 3), (3, 4), (3, 5), (3, 6), (4, 1), (4, 2), (4, 3), (4, 4), (4, 5), (4, 6), (5, 1), (5, 2), (5, 3), (5, 4), (5, 5), (5, 6), (6, 1), (6, 2), (6, 3), (6, 4), (6, 5), (6, 6)}
(a) {(3, 1), (3, 2), (3, 3), (3, 4), (3, 5), (3, 6)} **(b)** {(2, 6), (3, 5), (4, 4), (5, 3), (6, 2)} **(c)** ∅ **19.** 1/6 **21.** 2/3
23. 1/3 **25.** 1/13 **27.** 1/26 **29.** 1/52 **31.** 2/13 **33.** 3/26 **35.** 1/9 **37.** 5/18 **39.** 5/9
41. (a) Worker is male. **(b)** Worker is female and has worked less than 5 yr. **(c)** Worker is female or does not contribute to a voluntary retirement plan. **43. (a)** 3/5 **(b)** 1/15 **(c)** 9/50 = .18 **45. (a)** Person smokes or has a family history of heart disease. **(b)** Person does not smoke and has a family history of heart disease. **(c)** Person does not have a family history of heart disease or is not overweight.

Section 5.4 (page 232)

3. no **5.** yes **7.** no **9. (a)** 5/36 **(b)** 1/9 **(c)** 1/12 **(d)** 0 **11. (a)** 5/18 **(b)** 5/12 **(c)** 1/3 **13. (a)** 3/13
(b) 4/13 **(c)** 7/13 **(d)** 4/13 **15. (a)** 1/2 **(b)** 7/10 **(c)** 1/2 **17. (a)** 1/10 **(b)** 9/10 **(c)** 7/20 **19. (a)** .39
(b) .81 **(c)** .77 **(d)** .23 **21.** 1 to 5 **23.** 2 to 1 **25. (a)** 1 to 4 **(b)** 8 to 7 **(c)** 4 to 11 **27.** 13 to 37 **29.** 2/5
31. not empirical **33.** empirical **35.** empirical **37.** not empirical **41. (a)** .50 **(b)** .77 **(c)** .27 **43. (a)** .89
(b) .38 **(c)** .18 **45.** possible **47.** Not possible; the sum of the probabilities is greater than 1. **49.** Not possible; a probability cannot be negative. **51. (a)** .51 **(b)** .67 **(c)** .39 **(d)** .12 **53. (a)** .951 **(b)** .527 **(c)** .473 **(d)** .466
(e) .007 **(f)** .515 **55. (a)** 1/4 **(b)** 1/2 **(c)** 1/4 **57. (a)** .4 **(b)** .1 **(c)** .6 **(d)** .9 **59.** 0
Note: Theoretical probabilities are given for Exercises 61 and 63. Answers using the Monte Carlo method will vary.
61. (a) .15625 **(b)** .03125 **63.** .632

Section 5.5 (page 247)
1. 0 **3.** 1 **5.** 1/6 **7.** 4/17 **9.** 11/51 **11.** .012 **13.** .245 **19. (a)** .052 **(b)** .476 **21. (a)** .7456
(b) .1872 **23. (a)** 2/3 **(b)** 8/27 ≈ .296 **25. (a)** 3 backups **27.** No, they are dependent. **29.** The probability of a customer not making a deposit, given that the customer cashed a check, is 3/8. **31.** The probability of a customer not cashing a check, given that the customer made a deposit, is 2/7. **33. (a)** $(.98)^3 ≈ .94$ **(b)** not very realistic **35.** .02 **37.** 2/3
39. 1/4 **41.** 1/4 **43.** .527 **45.** .042 **47.** 6/7 or .857 **49.** 1/7 or .143 **51. (a)** Color blindness and deafness are independent events. **53. (a)** .05 **(b)** .015 **(c)** .25 **55.** 1/6 **57.** 0 **59.** 2/3

Section 5.6 (page 255)
1. 1/3 **3.** 2/41 **5.** 21/41 **7.** 8/17 **9.** 82.4% **11.** .092 **13.** .176 **15.** .994 **17.** .364 **19.** 2/7
21. 9/26 **23. (a)** .643 **(b)** .251 **25. (a)** .161 **27.** .121 **29.** .050 **31.** .244 **33.** .178

Chapter 5 Review Exercises (page 260)
1. false **3.** false **5.** true **7.** false **9.** false **11.** 16 **13.** {a, b, e} **15.** {c, d, g} **17.** {a, b, e, f}
19. U **21.** all employees in the accounting department who also have at least 10 years with the company **23.** all employees who are in the accounting department or who have MBA degrees **25.** all employees who are not in the sales department and who have worked less than 10 years with the company
27.

$A ∪ B'$

29.

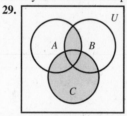

$(A ∩ B) ∪ C$

31. {1, 2, 3, 4, 5, 6} **33.** {0, .5, 1, 1.5, 2, . . . , 299.5, 300} **35.** {(3, R), (3, G), (5, R), (5, G), (7, R), (7, G), (9, R), (9, G), (11, R), (11, G)} **37.** {(3, G), (5, G), (7, G), (9, G), (11, G)} **39.** 1/4 **41.** 3/13 **43.** 1/2 **45.** 1 **51.** The probability of getting the car if the contestant switches doors is 2/3; *not* switching doors gives a 1/3 probability of getting the car. Vos Savant's answer: the contestant should switch doors. **53.** 1 to 25 **55.** 5/36 ≈ .139 **57.** 1/6 ≈ .167 **59.** 1/6 ≈ .167 **61.** 2/11 ≈ .182 **63. (a)** .66 **(b)** .29 **(c)** .71 **(d)** .34 **65.** 4/9 ≈ .444 **67.** 3/22 ≈ .136 **69.** 2/3
71. (a) .86 **(b)** .26

73. (a)

	N_2	T_2
N_1	N_1N_2	N_1T_2
T_1	T_1N_2	T_1T_2

(b) 1/4 **(c)** 1/2 **(d)** 1/4

75. (b) 7/10 **77. (a)** row 1: 400; row 2: 150; row 3: 750, 1000 **(b)** 1000 **(c)** 300 **(d)** 250 **(e)** 600 **(f)** 150 **(g)** those who purchased a used car, given that the buyer was not satisfied **(h)** 3/5 **(i)** 3/20

Extended Application (page 264)
1. .076 **2.** .542 **3.** .051

CHAPTER 6

Section 6.1 (page 271)
1. 12 **3.** 720 **5.** 1 **7.** 24 **9.** 156 **11.** $2.490952 × 10^{15}$ **13.** 240,240 **15.** 30 **17.** 120
19. (a) 120 **(b)** 48 **21.** 604,800 **23.** $2.05237 × 10^{10}$ **25.** $6.09493 × 10^{10}$ **27. (a)** 25,200 **(b)** 15,120
29. 240 **31.** 1 **33. (a)** 840 **(b)** 180 **(c)** 420 **35. (a)** 362,880 **(b)** 1728 **(c)** 1260 **37.** 55,440
39. (a) 27,600 **(b)** 35,152 **(c)** 1104 **41. (a)** 160; 8,000,000 **(b)** Some numbers, such as 911, 800, and 900, are reserved for special purposes. **43.** 1600 **45. (a)** 17,576,000 **(b)** 17,576,000 **(c)** 456,976,000 **47.** 100,000; 90,000

Section 6.2 (page 279)

1. 56 **3.** 792 **5.** 1 **7.** 352,716 **9.** 2,042,975 **11.** 8,436,285 **13.** 635,013,559,600 **15.** (a) 10 (b) 7
17. 50 **19.** (a) 9 (b) 6 (c) 3; yes, from both **23.** combinations; (a) 56 (b) 462 (c) 3080
25. combinations; (a) 220 (b) 220 **27.** permutations; 479,001,600 **29.** permutations; 210 **31.** combinations;
(a) 2300 (b) 10 (c) 950 **33.** permutations; 15,120 **35.** combinations; (a) 15,504 (b) 816 **37.** n **39.** (a) 40
(b) 20 (c) 7 **41.** 1,621,233,900 **43.** (a) 84 (b) 10 (c) 40 (d) 74 **45.** (a) 1,120,529,256 (b) 806,781,064,320

Section 6.3 (page 287)

1. 1/6 **3.** 3/10 **5.** 1326 **7.** $33/221 \approx .149$ **9.** $52/221 \approx .235$ **11.** $130/221 \approx .588$ **13.** 8.42×10^{-8}
15. $18,975/28,561 \approx .664$ **17.** 1.54×10^{-6} **19.** 2.40×10^{-4} **21.** $1/\binom{52}{13}$ **23.** $\binom{4}{3}\binom{4}{3}\binom{44}{7}/\binom{52}{13}$ **25.** (a) .054
(b) .011 (c) .046 (d) .183 (e) .387 (f) .885 **29.** $1 - P(365, 41)/(365)^{41}$ **31.** 1 **33.** .3499 **35.** 7/9
37. 5/12 **39.** .424 **41.** 8.9×10^{-10} **43.** (a) .000150 (b) $n!/2^{n(n-1)/2}$ **45.** Answers will vary. We give the
theoretical answers here: (a) .3038 (b) .0402 (c) .0512.

Section 6.4 (page 294)

1. $5/16 \approx .313$ **3.** $1/32 \approx .031$ **5.** $3/16 \approx .188$ **7.** $13/16 \approx .813$ **9.** 4.6×10^{-10} **11.** .269 **13.** .875
15. $1/32 \approx .031$ **17.** $13/16 \approx .813$ **21.** .302 **23.** .879 **25.** .349 **27.** .736 **29.** .246 **31.** .017
33. .243 **35.** .925 **37.** .072 **39.** .035 **41.** .881 **43.** .185 **45.** .256 **47.** .0025 **49.** .955
51. .00079 **53.** .999922 **55.** (a) .000036 (b) .000054 (c) 1.0×10^{-13}
57. (a) .148774 (b) .021344 (c) .978656

Section 6.5 (page 304)

1. (a)

Number	0	1	2	3	4	5
Probability	0	0	.1	.3	.4	.2

(b)

3. (a)

Number	0	1	2	3	4	5	6
Probability	0	.04	0	.16	.40	.32	.08

(b)

5. (a)

Number	0	1	2	3	4	5
Probability	.15	.25	.3	.15	.1	.05

(b)

7.

Number of heads	0	1	2	3	4
Probability	$\frac{1}{16}$	$\frac{1}{4}$	$\frac{3}{8}$	$\frac{1}{4}$	$\frac{1}{16}$

9.

Number of aces	0	1	2	3
Probability	.783	.204	.013	.0002

11.

13.

15.

17. 3.6 **19.** 14.64 **21.** 2.7 **23.** 18 **25.** 0; yes **27.** −64¢; no **29.** 9/7 ≈ 1.3 **31. (a)** 5/3 ≈ 1.67
(b) 4/3 ≈ 1.33 **33.** 1 **35.** No; the expected value is about −21¢. **37.** −66¢ **39.** −2.7¢ **41.** −20¢
45. 3.51

47.

Account	EV	Total	Class
3	$2000	$22,000	C
4	$1000	$51,000	B
5	$25,000	$30,000	C
6	$60,000	$60,000	A
7	$16,000	$46,000	B

49. (a) $94.0 million for seeding; $116.0 million for not seeding **(b)** seed **51.** 10¢; no

Chapter 6 Review Exercises (page 309)

1. 720 **3.** 220 **5.** 20 **7. (a)** 24 **(b)** 12 **9. (a)** 140 **(b)** 239 **13.** 0 **15.** 14/55 ≈ .255
17. 4/33 ≈ .121 **19.** 1/64 ≈ .016 **21.** 11/32 ≈ .344 **23.** 1/17 ≈ .059 **25.** .3620 **27.** .9955

29. (a)

Number	8	9	10	11	12	13	14
Probability	.04	0	.09	.22	.35	.17	.13

(b)

(c) 11.87

31. (a)

Number	2	3	4	5	6	7	8	9	10	11	12
Probability	1/36	1/18	1/12	1/9	5/36	1/6	5/36	1/9	1/12	1/18	1/36

(b)

(c) 7 **33.** .6 **35.** -83.3¢; no **37.** 2.5 **39.** $1.29 **41.** .00096

43. $\binom{20}{12}(.01)^{12}(.99)^{8} + \binom{20}{13}(.01)^{13}(.99)^{7} + \cdots + \binom{20}{20}(.01)^{20}(.99)^{0}$ **45.** -37¢

47. (c)

Extended Application (page 312)

1. (a) $69.01 **(b)** $79.64 **(c)** $58.96 **(d)** $82.54 **(e)** $62.88 **(f)** $64.00 **2.** Stock only part 3 on the truck.

3. The events of needing parts 1, 2, and 3 are not the only events in the sample space. **4.** 2^{n}

CHAPTER 7

Section 7.1 (page 321)

1. (a)–(b)

Interval	Frequency
0–24	4
25–49	3
50–74	6
75–99	3
100–124	5
125–149	9

(c)–(d)

3. (a)–(b)

Interval	Frequency
70–74	2
75–79	1
80–84	3
85–89	2
90–94	6
95–99	5
100–104	6
105–109	4
110–114	2

(c)–(d)

7. 16 **9.** 27,955 **11.** 7.7 **13.** 6.7 **15.** 17.2 **17.** 51 **19.** 130 **21.** 29.1 **23.** 9 **25.** 68 and 74
27. 6.3 **31.** 94.9; 90–94 and 100–104 **33.**

Days absent

35. 2395 million bushels **37. (a)** about 17.5% **(b)** about 8% **(c)** 20–29 **39. (a)** 55.5° F **(b)** 28.9° F

Section 7.2 (page 331)

1. The standard deviation is the square root of the variance. **3.** 53; 21.8 **5.** 46; 16.1 **7.** 24; 8.1 **9.** 45.2
11. 10.9 **13.** 3/4 **15.** 24/25 **17.** at least 88.9% **19.** no more than 25% **21. (a)** mean = $653.33; σ =
$663.65 **(b)** 5 **(c)** 5 **(d)** at least 3/4 **23. (a)** mean = 25.5; σ = 7.2 **(b)** Forever Power **(c)** Forever Power
25. (a) 1/3, 2, −1/3, 0, 5/3, 7/3, 1, 4/3, 7/3, 2/3 **(b)** 2.1, 2.6, 1.5, 2.6, 2.5, .6, 1.0, 2.1, .6, 1.2 **(c)** 1.13
(d) 1.68 **(e)** 4.41, −2.15 **(f)** 4.31, 0 **27. (a)** 7.52; 1987 **(b)** .84 **(c)** 5 **29.** mean = 1.8158; standard
deviation = .4451

Section 7.3 (page 340)

1. the mean **3.** z-scores are found with the formula $z = \dfrac{x - \mu}{\sigma}$. **5.** 49.38% **7.** 45.64% **9.** 7.7% **11.** 47.35%

13. 92.42% **15.** 32.56% **17.** −1.64 or −1.65 **19.** 1.04 **21.** .0062 **23.** $65.32 and $39.19 **25.** .0823
27. 5000 **29.** 642 **31.** 9918 **33.** .1587 **35.** .0062 **37.** 84.13% **39.** 37.79% **41.** 2.28% **43.** 99.38%
45. 189 **47.** 60.4 mph **49.** 6.68% **51.** 38.3% **53.** 82 **55.** 70 **57.** .0796 **59.** .9933 **61.** .1527
63. .0051

Section 7.4 (page 347)

3. (a)

x	0	1	2	3	4	5	6
$P(x)$.335	.402	.201	.054	.008	.001	.000

(b) 1.00 **(c)** .91

5. (a)

x	0	1	2	3
$P(x)$.941	.058	.001	.000

(b) .06 **(c)** .24

7. (a)

x	0	1	2	3	4
$P(x)$.0081	.0756	.2646	.4116	.2401

(b) 2.8 **(c)** .92

9. 12.5; 3.49 **11.** 51.2; 3.2 **13.** $np \geq 5$ and $n(1 - p) \geq 5$ **15.** .1747 **17.** .9599 **19.** .0197 **21.** .0268
23. .0592 **25.** .6443 **27.** .0537 **29. (a)** .0241 **(b)** .6772 **31.** .1974 **33.** 0 **35.** .8643 **37. (a)** .0472
(b) .9875 **(c)** .8814 **39. (a)** .0000 **(b)** .0018

Chapter 7 Review Exercises (page 349)

3. (a)

Interval	Frequency
450–474	5
475–499	6
500–524	5
525–549	2
550–574	2

(b)–(c)

5. 73 **7.** 34.9 **11.** 38; 36 and 38 **13.** 55–59 **17.** 67; 23.9 **19.** 6.2 **21.** A skewed distribution has the largest frequency at one end. **23. (a)** 98.76% **(b)** Chebyshev's inequality would give "at least 84%." **25.** .6591 **27.** .0606
29. Because the histogram is skewed, not close to the shape of a normal distribution. **31.** 15.87% **33. (a)** .1977
(b) .0718 **(c)** .0239 **35. (a)** diet A **(b)** diet B **37.** .9938 **39.** .1056 **41.** 28.10% **43.** 56.25%

45.

Interval	x	Tally	f	$x \cdot f$
1–3	2	⩩	6	12
4–6	5	⩩	5	25
7–9	8	⩩ ⩩	11	88
10–12	11	⩩ ⩩ ⩩ ⩩	20	220
16–18	14	⩩	6	84
13–15	17	‖	2	34

x	$P(x)$	$x \cdot P(x)$
0	.001	0
1	.010	.010
2	.044	.088
3	.117	.351
4	.205	.820
5	.246	1.230
6	.205	1.230
7	.117	.819
8	.044	.352
9	.010	.090
10	.001	.010

(a) 9.3; 5 **(b)** 3.9; 1.6 **(c)** $1.5 \le x \le 17.1$; $1.84 \le x \le 8.16$ **(d)** $1.84 \le x \le 8.16$ **(e)** Because its histogram is not close enough to the shape of a normal curve.

CHAPTER 8

Section 8.1 (page 360)

1. no **3.** yes **5.** no **7.** yes **9.** no **11.** yes **13.** no **15.** no

17. transition diagram; $\begin{bmatrix} .9 & .1 & 0 \\ .1 & .6 & .3 \\ 0 & .3 & .7 \end{bmatrix}$ **19.** $A = \begin{bmatrix} 1 & 0 \\ .8 & .2 \end{bmatrix}$; $A^2 = \begin{bmatrix} 1 & 0 \\ .96 & .04 \end{bmatrix}$; $A^3 = \begin{bmatrix} 1 & 0 \\ .992 & .008 \end{bmatrix}$; 0

21. $C = \begin{bmatrix} .5 & .5 \\ .72 & .28 \end{bmatrix}$; $C^2 = \begin{bmatrix} .61 & .39 \\ .5616 & .4384 \end{bmatrix}$; $C^3 = \begin{bmatrix} .5858 & .4142 \\ .596448 & .403552 \end{bmatrix}$; .4142

23. $E = \begin{bmatrix} .8 & .1 & .1 \\ .3 & .6 & .1 \\ 0 & 1 & 0 \end{bmatrix}$; $E^2 = \begin{bmatrix} .67 & .24 & .09 \\ .42 & .49 & .09 \\ .3 & .6 & .1 \end{bmatrix}$; $E^3 = \begin{bmatrix} .608 & .301 & .091 \\ .483 & .426 & .091 \\ .42 & .49 & .09 \end{bmatrix}$; .301

27. (a) 53% for Johnson and 47% for North Clean **(b)** 58.85% for Johnson and 41.15% for North Clean **(c)** 61.48% for Johnson and 38.52% for North Clean **(d)** 62.666% for Johnson and 37.334% for North Clean **29. (a)** 42,500 in G_0; 5000 in G_1; 2500 in G_2 **(b)** 36,125 in G_0; 8250 in G_1; 5625 in G_2 **(c)** 30,706 in G_0; 10,213 in G_1; 9081 in G_2 **(d)** 26,100 in G_0; 11,241 in G_1; 12,659 in G_2 **31. (a)** [.257 .597 .146]

(b) [.255 .594 .151] **(c)** [.254 .590 .156] **33. (b)** $\begin{bmatrix} \frac{5}{12} & \frac{13}{36} & \frac{2}{9} \\ \frac{5}{12} & \frac{13}{36} & \frac{2}{9} \\ \frac{1}{4} & \frac{5}{12} & \frac{1}{3} \end{bmatrix}$ **(c)** 2/9

35. (a)

	Liberal	Conservative	Independent
Liberal	.80	.15	.05
Conservative	.20	.70	.10
Independent	.20	.20	.60

(b) [.40 .45 .15] **(c)** 44% liberal, 40.5% conservative, and 15.5% independent **(d)** 46.4% liberal, 38.05% conservative, and 15.55% independent **(e)** 47.84% liberal, 36.705% conservative, and 15.455% independent **(f)** 48.704% liberal, 35.9605% conservative, and 15.3355% independent **37.** The first power is the given

transition matrix; $\begin{bmatrix} .2 & .15 & .17 & .19 & .29 \\ .16 & .2 & .15 & .18 & .31 \\ .19 & .14 & .24 & .21 & .22 \\ .16 & .19 & .16 & .2 & .29 \\ .16 & .19 & .14 & .17 & .34 \end{bmatrix}$; $\begin{bmatrix} .17 & .178 & .171 & .191 & .29 \\ .171 & .178 & .161 & .185 & .305 \\ .18 & .163 & .191 & .197 & .269 \\ .175 & .174 & .164 & .187 & .3 \\ .167 & .184 & .158 & .182 & .309 \end{bmatrix}$; $\begin{bmatrix} .1731 & .175 & .1683 & .188 & .2956 \\ .1709 & .1781 & .1654 & .1866 & .299 \\ .1748 & .1718 & .1753 & .1911 & .287 \\ .1712 & .1775 & .1667 & .1875 & .2971 \\ .1706 & .1785 & .1641 & .1858 & .301 \end{bmatrix}$;

$\begin{bmatrix} .17193 & .17643 & .1678 & .18775 & .29609 \\ .17167 & .17689 & .16671 & .18719 & .29754 \\ .17293 & .17488 & .17007 & .18878 & .29334 \\ .17192 & .17654 & .16713 & .18741 & .297 \\ .17142 & .17726 & .16629 & .18696 & .29807 \end{bmatrix}$; .18719. **39. (a)** .847423 or about 85% **(b)** [.0128 .0513 .0962 .8397]

Section 8.2 (page 368)

1. regular **3.** not regular **5.** regular **7.** $[\frac{2}{5} \quad \frac{3}{5}]$ **9.** $[\frac{4}{11} \quad \frac{7}{11}]$ **11.** $[\frac{14}{83} \quad \frac{19}{83} \quad \frac{50}{83}]$ **13.** $[\frac{170}{563} \quad \frac{197}{563} \quad \frac{196}{563}]$ **15.** [0 0 1] **17.** [81/331, 175/331, 75/331] **19.** [1/2, 7/20, 3/20] **21.** $[(1 - q)/(2 - p - q) \quad (1 - p)/(2 - q - p)]$; always regular **25.**

	Works	Doesn't Work
Works	.9	.1
Doesn't Work	.7	.3

; long-range probability of line working correctly is 7/8

27. $[\frac{51}{209} \quad \frac{88}{209} \quad \frac{70}{209}]$ **29.** Long-range prediction is 48.7% fair, 31.1% cloudy, and 20.1% rainy. **31.** 1/2 **33.** 38% **35.** [.171898 .176519 .167414 .187526 .296644] **37. (a)** [.4 .6]; [.53 .47]; [.5885 .4115]; [.614825 .385175]; [.626671 .373329]; [.632002 .367998]; [.634401 .365599]; [.635480 .364520]; [.635966 .364034]; [.636185 .363815] **(b)** .24; .11; .048; .022; .0097; .0044; .0020; .00088; .00040; .00018 **(c)** roughly .45 for each week **(d)** Each week, the difference between the probability vector and the equilibrium vector is slightly less than half of what it was the previous week. **(e)** [.75 .25]; [.6875 .3125]; [.659375 .340625]; [.646719 .353281]; [.641023 .358977]; [.638461 .361539]; [.637307 .362693]; [.636788 .363212]; [.636555 .363445]; [.636450 .363550]; .11; .051; .023; .010; .0047; .0021; .00094; .00042; .00019; .000086; roughly .45 for each week, same conclusion as before

Section 8.3 (page 374)

1. (a) buy speculative **(b)** buy blue-chip **(c)** buy speculative; $24,300 **(d)** buy blue-chip **3. (a)** set up in the stadium

(b) set up in the gym **(c)** set up in both; $1010 **5. (a)**
$$\begin{array}{c} \text{Market} \\ \text{Don't Market} \end{array} \begin{array}{cc} \text{Better} & \text{Not Better} \\ \left[\begin{array}{cc} \$50,000 & -\$25,000 \\ -\$40,000 & -\$10,000 \end{array} \right] \end{array}$$ **(b)** $5000 if they market the

new product, $-$22,000 if they don't; market the new product **7. (a)**
$$\begin{array}{c} \text{Bid \$30,000} \\ \text{Bid \$40,000} \end{array} \begin{array}{cc} \text{Strike} & \text{No Strike} \\ \left[\begin{array}{cc} -\$5500 & \$4500 \\ \$4500 & \$0 \end{array} \right] \end{array}$$ **(b)** $40,000

9. emphasize environment; 14.25

Section 8.4 (page 380)

1. $6 from B to A **3.** $2 from A to B **5.** $1 from A to B **7.** Yes; column 2 dominates column 3. **9.** $\begin{bmatrix} -2 & 8 \\ -1 & -9 \end{bmatrix}$

11. $\begin{bmatrix} 4 & -1 \\ 3 & 5 \end{bmatrix}$ **13.** $\begin{bmatrix} 8 & -7 \\ -2 & 4 \end{bmatrix}$ **15.** (1, 1); 3; strictly determined **17.** no saddle point; not strictly determined

19. (3, 1); 3; strictly determined **21.** (1, 3); 1; strictly determined **23.** no saddle point; not strictly determined

27. $\begin{array}{c} \\ A \end{array} \begin{array}{c} \\ \begin{array}{ccc} 1 & 2 & 3 \end{array} \\ \begin{array}{c} 1 \\ 2 \\ 3 \end{array} \begin{bmatrix} 5 & -2 & 6 \\ 7 & 5 & 9 \\ 3 & -3 & 5 \end{bmatrix} \end{array}$; saddle point is 5 at (2, 2); 5 **29.** (1, 3); -9 **31.** $\begin{array}{c} \text{Rock} \\ \text{Scissors} \\ \text{Paper} \end{array} \begin{array}{ccc} \text{Rock} & \text{Scissors} & \text{Paper} \\ \begin{bmatrix} 0 & 1 & -1 \\ -1 & 0 & 1 \\ 1 & -1 & 0 \end{bmatrix} \end{array}$; no

Section 8.5 (page 390)

1. (a) -1 **(b)** $-.28$ **(c)** -1.54 **(d)** $-.46$ **3.** player A: 1: 1/5, 2: 4/5; player B: 1: 3/5, 2: 2/5; value 17/5 **5.** player A: 1: 7/9, 2: 2/9; player B: 1: 4/9, 2: 5/9; value $-8/9$ **7.** player A: 1: 8/15, 2: 7/15; player B: 1: 2/3, 2: 1/3; value 5/3
9. player A: 1: 6/11, 2: 5/11; player B: 1: 7/11, 2: 4/11; value $-12/11$ **11.** strictly determined game; saddle point at (2, 2); value $-5/12$ **13.** player A: 1: 2/5, 2: 3/5; player B: 1: 1/5, 2: 4/5; value 7/5 **15.** player A: 1: 1/14, 2: 0, 3: 13/14; player B: 1: 1/7, 2: 6/7; value 50/7 **17.** player A: 1: 2/3, 2: 1/3; player B: 1: 0, 2: 1/9, 3: 8/9; value 10/3 **19.** player A: 1: 0, 2: 3/4, 3: 1/4; player B: 1: 0, 2: 1/12, 3: 11/12; value 33/4 **25.** Allied should use T.V. with probability 10/27 and use radio with probability 17/27. The value of the game is 1/18, which represents increased sales of $55,556. **27.** He should invest in rainy-day goods about 5/9 of the time and in sunny-day goods about 4/9 of the time, for a steady profit of $72.22.
29. Euclid should pounce with probability 2/3 and freeze with probability 1/3. Jamie should pounce with probability 1/6 and freeze with probability 5/6. The value of the game is 4/3. **31. (a)** $\begin{bmatrix} 2 & -3 \\ -3 & 4 \end{bmatrix}$ **(b)** player A: 1: 7/12, 2: 5/12; player B: 1: 7/12, 2: 5/12; value $-1/12$

Section 8.6 (page 397)

1. player A: 1: 2/3, 2: 1/3; player B: 1: 1/3, 2: 2/3; value 5/3 **3.** player A: 1: 7/13, 2: 6/13; player B: 1: 8/13, 2: 5/13; value 22/13 **5.** player A: 1: 2/7, 2: 5/7; player B: 1: 11/21, 2: 10/21; value 6/7 **7.** player A: 1: 1/2, 2: 1/2; player B: 1: 0, 2: 3/10, 3: 7/10; value $-1/2$ **9.** player A: 1: 5/8, 2: 3/8; player B: 1: 1/2, 2: 1/2, 3: 0; value 1/2 **11.** player A: 1: 1/2, 2: 1/2, 3: 0; player B: 1: 1/6, 2: 2/3, 3: 1/6; value 0 **13.** The contractor should bid $30,000 with probability 9/29 and bid $40,000 with probability 20/29. The value of the game is $1396.55. **15. (a)** $\begin{bmatrix} 5,000 & 10,000 & 10,000 \\ 8,000 & 4,000 & 8,000 \\ 6,000 & 6,000 & 3,000 \end{bmatrix}$ **(b)** General

Items should advertise in Atlanta with probability 4/9, in Boston with probability 5/9, and never in Cleveland. Original Imitators should advertise in Atlanta with probability 2/3, in Boston with probability 1/3, and never in Cleveland. The value of the game is $6666.67. **17.** The student should choose the calculator with probability 3/8 and the book with probability 5/8. The value of the game is 25 points. **19. (a)** Each player should use each strategy 1/3 of the time, and the value of the game is 0. **(b)** The game is symmetric in that neither player has an advantage, and each choice is as strong as every other choice.

Kije

$$
\begin{array}{cccc}
 & (3,0) & (0,3) & (2,1) & (1,2)
\end{array}
$$

$$
\begin{array}{c}
(4,0) \\
(0,4) \\
(3,1) \\
(1,3) \\
(2,2)
\end{array}
\begin{bmatrix}
4 & 0 & 2 & 1 \\
0 & 4 & 1 & 2 \\
1 & -1 & 3 & 0 \\
-1 & 1 & 0 & 3 \\
-2 & -2 & 2 & 2
\end{bmatrix}
$$

21. (a) *Blotto* **(b)** Blotto should use strategies (4, 0) and (0, 4) with probability 4/9 each, strategy (2, 2) with probability 1/9, and never send 3 regiments to one post and 1 to the other. Kije should use strategy (3, 0) with probability 1/30, strategy (0, 3) with probability 7/90, strategy (2, 1) with probability 8/15, and strategy (1, 2) with probability 16/45. The value of the game is 14/9.

Chapter 8 Review Exercises (page 400)

3. yes **5.** yes **7. (a)** $C = \begin{bmatrix} .6 & .4 \\ 1 & 0 \end{bmatrix}$; $C^2 = \begin{bmatrix} .76 & .24 \\ .6 & .4 \end{bmatrix}$; $C^3 = \begin{bmatrix} .696 & .304 \\ .76 & .24 \end{bmatrix}$ **(b)** .76 **9. (a)** $E = \begin{bmatrix} .2 & .5 & .3 \\ .1 & .8 & .1 \\ 0 & 1 & 0 \end{bmatrix}$;

$E^2 = \begin{bmatrix} .09 & .8 & .11 \\ .1 & .79 & .11 \\ .1 & .8 & .1 \end{bmatrix}$; $E^3 = \begin{bmatrix} .098 & .795 & .107 \\ .099 & .792 & .109 \\ .1 & .79 & .11 \end{bmatrix}$ **(b)** .099 **11.** [.453 .547]; [5/11 6/11] or [.455 .545]

13. [.48 .28 .24]; $[\frac{47}{95} \ \frac{26}{95} \ \frac{22}{95}]$ or [.495 .274 .232] **17.** not regular **21.** $2 from A to B **23.** $7 from B to A

25. Row 3 dominates row 1; column 1 dominates column 4. **27.** $\begin{bmatrix} -11 & 6 \\ -10 & -12 \end{bmatrix}$ **29.** $\begin{bmatrix} -2 & 4 \\ 3 & 2 \\ 0 & 3 \end{bmatrix}$ **31.** (1, 1); value -2

33. (2, 2); value 0; fair game **35.** (2, 3); value -3 **37.** player A: 1: 5/6, 2: 1/6; player B: 1: 1/2, 2: 1/2; value 1/2
39. player A: 1: 1/9, 2: 8/9; player B: 1: 5/9, 2: 4/9; value 5/9 **41.** player A: 1: 1/5, 2: 4/5; player B: 1: 3/5, 2: 0, 3: 2/5;
value $-12/5$ **43.** player A: 1: 3/4, 2: 1/4, 3: 0; player B: 1: 1/2, 2: 1/2; value 1/2 **45. (a)** [.54 .46]
(b) [.6464 .3536] **47.** [.428 .322 .25] **49.** [.431 .284 .285] **51.** hostile **53.** hostile; $785 **55.** Both
labor and management should always be friendly. The value of the game is 600. **57.** .2 **59.** .4 **61.** .256
63. [.1945 .5555 .25] **65.** [7/36 5/9 1/4] **67.** oppose **69.** oppose; gain of 1400 votes

Extended Application (page 404)

2. 3.3 yr **3.** $\begin{bmatrix} .768332 & .231668 \\ .814382 & .185618 \\ .775304 & .224696 \\ .807692 & .192308 \\ .5 & .5 \\ .5 & .5 \\ .5 & .5 \\ .5 & .5 \end{bmatrix}$ **4.** .23 **5.** the occlusal surface

CHAPTER 9

Section 9.1 (page 409)
1. the interest rate and time period **3.** $231.00 **5.** $286.75 **7.** $119.15 **9.** $241.56 **11.** $304.38
13. $3654.10 **17.** $46,265.06 **19.** $28,026.37 **21.** $8898.75 **23.** $48,140.65 **25.** 10.7% **27.** 11.8%
29. $27,894.30 **31.** 6.8% **33.** $34,438.29 **35.** $6550.92 **37.** 11.4% **39.** $13,683.48; yes

Section 9.2 (page 417)
3. $1593.85 **5.** $1515.80 **7.** $13,213.14 **9.** $5105.58 **11.** $60,484.66 **13.** $1167.56 **15.** $2251.12
17. $8265.24 **19.** $988.94 **21.** $5370.38 **25.** 4.04% **27.** 8.16% **29.** 12.36% **31.** $8763.47
33. about $111,000 **35.** $30,611.30 **37.** about $1.95 million **39.** $11,940.52 **41.** about 18 yr
43. about 12 yr **45.** $142,886.40 **47.** $123,506.50 **49.** 5/4

Section 9.3 (page 425)
1. 48 **3.** −648 **5.** 81 **7.** 64 **9.** 15 **11.** 156/25 **13.** −208 **15.** 15.91713 **17.** 21.82453
19. 22.01900 **23.** $5637.09 **25.** $636,567.63 **27.** $126,717.49 **29.** $239,174.10 **31.** $1,869,084.00
33. $21,903.68 **35.** $63,479.76 **37.** $219,204.60 **39.** $47,179.32 **41.** $3899.32 **43.** $952.62 **45.** $7382.54
47. $1645.13 **49.** $258.73 **51. (a)** $149,850.69 **(b)** $137,895.79 **(c)** $11,954.90 **53.** $4168.30 **55.** $256,015.48
57. $67,940.98 **59. (a)** $226.11 **(b)** $245.77 **61.** $647.76 **63.** $152,667.08 **65.** $209,348.00 **67. (a)** $1200
(b) $3511.58, except the last payment, which is $3511.59

(c)

Payment Number	Amount of Deposit	Interest Earned	Total
1	$3511.58	$0	$3511.58
2	$3511.58	$105.35	$7128.51
3	$3511.58	$213.86	$10,853.95
4	$3511.58	$325.62	$14,691.15
5	$3511.58	$440.73	$18,643.46
6	$3511.58	$559.30	$22,714.34
7	$3511.58	$681.43	$26,907.35
8	$3511.58	$807.22	$31,226.15
9	$3511.58	$936.78	$35,674.51
10	$3511.58	$1070.24	$40,256.33
11	$3511.58	$1207.69	$44,975.60
12	$3511.58	$1349.27	$49,836.45
13	$3511.58	$1495.09	$54,843.12
14	$3511.59	$1645.29	$60,000.00

Section 9.4 (page 433)
1. (c) **3.** 9.71225 **5.** 12.65930 **7.** 14.71787 **11.** $8045.30 **13.** $1,367,774.00 **15.** $205,724.40
17. $103,796.60 **21.** $5309.69 **23.** $11,727.32 **25.** $258.90 **27.** $87.10 **29.** $8.71 **31.** $48,677.34
33. $1267.07 **35.** $847.91 **37. (a)** $158.00 **(b)** $1584.00 **39.** $573,496.00

41.

Payment Number	Amount of Payment	Interest for Period	Portion to Principal	Principal at End of Period
0	—	—	—	$72,000.00
1	$10,129.69	$3600.00	$6529.69	$65,470.31
2	$10,129.69	$3273.52	$6856.17	$58,614.14
3	$10,129.69	$2930.71	$7198.98	$51,415.16
4	$10,129.69	$2570.76	$7558.93	$43,856.23

43.

Payment Number	Amount of Payment	Interest for Period	Portion to Principal	Principal at End of Period
0	—	—	—	$20,000.00
1	$2717.36	$1200.00	$1517.36	$18,482.64
2	$2717.36	$1108.96	$1608.40	$16,874.24
3	$2717.36	$1012.45	$1704.91	$15,169.33
4	$2717.36	$910.16	$1807.20	$13,362.13

45. (a) $4025.90 **(b)** $2981.93

47.

Payment Number	Amount of Payment	Interest for Period	Portion to Principal	Principal at End of Period
1	$5783.49	$3225.54	$2557.95	$35,389.55
2	$5783.49	$3008.11	$2775.38	$32,614.17
3	$5783.49	$2772.20	$3011.29	$29,602.88
4	$5783.49	$2516.24	$3267.25	$26,335.63
5	$5783.49	$2238.53	$3544.96	$22,790.67
6	$5783.49	$1937.21	$3846.28	$18,944.39
7	$5783.49	$1610.27	$4173.22	$14,771.17
8	$5783.49	$1255.55	$4527.94	$10,243.22
9	$5783.49	$870.67	$4912.82	$5,330.41
10	$5783.49	$453.08	$5330.41	$0.00

Chapter 9 Review Exercises (page 436)

1. $848.16 **3.** $921.50 **7.** $447.81 **9.** $739.09 **11.** compound interest **13.** $43,988.40 **15.** $104,410.10
17. $12,444.50 **19.** $10,213.85 **21.** $12,857.07 **23.** $1923.09 **25.** 4, 2, 1, 1/2, 1/4 **27.** -32 **29.** 5500
31. 33.06595 **33.** $10,078.44 **35.** $137,925.91 **37.** $25,396.38 **41.** $2619.29 **43.** $16,277.35
45. $31,921.91 **47.** $13,913.48 **49.** $12,806.38 **51.** $2305.07 **53.** $546.93 **55.** $896.06 **57.** $2696.12
59. $6072.05; yes **61.** 12.13 mo **63.** $5596.62 **65.** $107,892.82; $32,892.82 **67.** $9859.46 **69.** $212,026
71. (d)

Extended Application (page 441)

1. $14,038 **2.** $9511 **3.** $8837 **4.** $3968

CHAPTER 10

Section 10.1 (page 453)

1. 3 **3.** does not exist **5.** 1 **9.** 2 **11.** 10 **13.** does not exist **15.** 8 **17.** 2 **19.** 4 **21.** 512
23. 3/2 **25.** 6 **27.** -5 **29.** 4 **31.** $-1/9$ **33.** 1/10 **35.** $2x$ **37. (a)** does not exist **(b)** $x = -2$ **(c)** If
$x = a$ is an asymptote for the graph of $f(x)$, then $\lim\limits_{x \to a} f(x)$ does not exist. **39.** discontinuous at -1; $f(-1)$ does not exist;
$\lim\limits_{x \to -1} f(x) = 1/2$ **41.** discontinuous at 1; $f(1) = 2$; $\lim\limits_{x \to 1} f(x) = -2$ **43.** discontinuous at -5 and 0; $f(-5)$ and $f(0)$ do
not exist; $\lim\limits_{x \to -5} f(x)$ does not exist; $\lim\limits_{x \to 0} f(x) = 0$ **45.** no; no; yes **47.** yes; no; yes **49.** yes; no; yes **51.** yes; yes;
yes **53.** no; yes; yes **55. (a)** 3 **(b)** does not exist **(c)** 2 **(d)** 16 months **57. (a)** $520 **(b)** $600 **(c)** $630
(d) $1200 **(e)** $1250 **(f)** at $x = 150$ and $x = 400$ **59. (a)** $120 **(b)** $150 **(c)** $150 **(d)** $150 **(e)** $180 **(f)** $150
(g) $150 **(h)** at $t = 1, 2, 3, 4, 7, 8, 9, 10, 11$ **61.** at $t = m$

Section 10.2 (page 464)

1. 5 **3.** 8 **5.** 1/3 **7.** −1/3 **9.** 17 **11.** 25 **13.** 50 **15.** −16 **17.** 0 **19.** increasing **21. (a)** 3
(b) 0 **(c)** −9/5 **(d)** Sales increase in years 0–4, stay constant until year 7, and then decrease. **(e)** Many answers are possible;
one example might be Walkman radios. **23. (a)** −3/2 **(b)** −1 **(c)** 1 **(d)** −1/2 **(e)** 1988–1989 **(f)** stabilizing after a
decline from 1985 to 1987 **25.** 11 **27. (a)** −25 boxes per dollar **(b)** −20 boxes per dollar **(c)** −30 boxes per dollar
(d) Demand is decreasing. Yes, a higher price usually reduces demand. **29. (a)** \$11 million, −\$1 million **(b)** −1 million,
9 million **(c)** Civil penalities were increasing from 1987 to 1988 and decreasing from 1988 to 1989. Criminal penalties decreased
slightly from 1987 to 1988, then increased from 1988 to 1989. This indicates that criminal penalties began to replace civil
penalties in 1988. **(d)** About a \$3.5 million increase. The general trend was upward. More is being done to impose penalties for
polluting. **31. (a)** −\$.5 billion, −\$5 billion **(b)** −\$.5 billion, −\$3.2 billion **(c)** −\$1.2 billion, −\$1.2 billion
(d) cocaine in 1989–1990 **33. (a)** 5 ft/sec **(b)** 1 ft/sec **(c)** 1 ft/sec **(d)** 6 ft/sec **(e)** 1 ft/sec **(f)** The car is speeding
up from 0 to 2 sec, then slowing down from 2 to 6 sec, maintaining constant velocity from 6 to 10 sec, then speeding up from 10
to 12 sec, and finally, slowing down again from 12 to 18 sec.

Section 10.3 (page 479)

1. 27 **3.** 1/8 **5.** 1/8 **7.** $y = 8x - 9$ **9.** $5x + 4y = 20$ **11.** $3y = 2x + 18$ **13.** 2 **15.** 1/5 **17.** 0
19. (a) 0 **(b)** 1 **(c)** −1 **(d)** does not exist **(e)** m **21.** at $x = -2$ **23.** $f'(x) = 12x - 4$; 20; −4; −40
25. $f'(x) = -9$; −9; −9; −9 **27.** $f'(x) = -6/x^2$; −3/2; does not exist; −2/3 **29.** $f'(x) = -3/(2\sqrt{x})$; $-3/(2\sqrt{2})$;
does not exist; does not exist **31.** −6 **33.** has derivatives for all values of x **35.** −5; −3; 0; 2; 4 **37. (a)** $(a, 0)$
and (b, c) **(b)** $(0, b)$ **(c)** $x = 0$ and $x = b$ **39. (a)** distance **(b)** velocity **41. (a)** −1; the debt is decreasing at the rate
of about 1% per month. 2; the debt is increasing at the rate of about 2% per month. 0; the debt is not changing at this point.
(b) February to March, April to May, and August to September **(c)** It is increasing. **43. (a)** \$16 per table **(b)** \$16
(c) \$15.998 (or \$16) **(d)** The marginal revenue gives a good approximation of the actual revenue from the sale of the 1001st
table. **45. (a)** 20 **(b)** 0 **(c)** −10 **(d)** at 5 hr **47. (a)** approximately 0; the power expended is not changing at that
point **(b)** about .1; the power expended is increasing .1 unit for each 1 unit increase in speed **(c)** about .12; the power expended
increases .12 units for each 1 unit increase in speed **(d)** The power level decreases to v_{mp}, which minimizes energy costs; then it
increases at an increasing rate.

49. (a)

t	2	10	13
$f'(t)$	1000	700	250

(b)

Section 10.4 (page 493)

1. $y' = 30x^2 - 18x + 6$ **3.** $y' = 4x^3 - 15x^2 + (2/9)x$ **5.** $f'(x) = 9x^5 - 2x^{-5}$ or $9x^5 - 2/x^5$ **7.** $y' = -48x^{2.2} + 3.8x^{.9}$ **9.** $y' = 4x^{-1/2} + (9/2)x^{-1/4}$ or $4/x^{1/2} + 9/(2x^{1/4})$ **11.** $g'(x) = -30x^{-6} + x^{-2}$ or $-30/x^6 + 1/x^2$
13. $y' = -5x^{-6} + 2x^{-3} - 5x^{-2}$ or $-5/x^6 + 2/x^3 - 5/x^2$ **15.** $f'(t) = -4t^{-2} - 6t^{-4}$ or $-4/t^2 - 6/t^4$ **17.** $y' = -18x^{-7} - 5x^{-6} + 14x^{-3}$ or $-18/x^7 - 5/x^6 + 14/x^3$ **19.** $h'(x) = -x^{-3/2}/2 + 21x^{-5/2}$ or $-1/(2x^{3/2}) + 21/x^{5/2}$
21. $y' = 2x^{-4/3}/3$ or $2/(3x^{4/3})$ **23.** $-40x^{-6} + 36x^{-5}$ or $-40/x^6 + 36/x^5$ **25.** $(-9/2)x^{-3/2} - 3x^{-5/2}$ or
$-9/(2x^{3/2}) - 3/x^{5/2}$ **27.** −14/3 **29. (c)** **33.** −15; $15x + y = 14$ **35.** 1 **37.** $(-1/2, -19/2)$ **41. (a)** 30
(b) 4.8 **(c)** −10 **43. (a)** 100 **(b)** 1 **45. (a)** $C'(x) = 2$ **(b)** $R'(x) = 6 - x/500$ **(c)** $P'(x) = 4 - x/500$ **(d)** $x = 2000$ **(e)** \$4000 **47. (a)** $V'(r) = 160\pi r$ **(b)** 640π cu mm **(c)** 960π cu mm **(d)** 1280π cu mm **(e)** The volume
increases. **49. (a)** 100 **(b)** 1 **(c)** −.01; the percent of acid is decreasing at the rate of .01 per day after 100 days.
51. (a) $v(t) = 50t - 9$ **(b)** −9; 241; 491 **53. (a)** $v(t) = -6t^2 + 8t$ **(b)** 0; −110; −520 **55. (a)** 0 ft/sec; −32
ft/sec **(b)** 2 sec **(c)** 64 ft

Section 10.5 (page 501)

1. $y' = 4x + 3$ **3.** $y' = 18x^2 - 6x + 4$ **5.** $y' = 6t^2 - 4t - 12$ **7.** $y' = 40x^3 - 60x^2 + 16x - 16$ **9.** $y' = 98x - 84$ **11.** $g'(t) = 36t^3 + 24t$ **13.** $y' = 3x^{1/2} - 3x^{-1/2}/2 - 2$ or $3x^{1/2} - 3/(2x^{1/2}) - 2$ **15.** $g'(x) = -12 + 15x^{-1/2}$ or $-12 + 15/x^{1/2}$ **17.** $f'(x) = 94/(8x + 1)^2$ **19.** $y' = 8/(2x - 11)^2$ **21.** $y' = 2/(1 - t)^2$ **23.** $y' = (x^2 + 6x - 12)/(x + 3)^2$ **25.** $y' = (-24x^2 - 2x + 6)/(4x^2 + 1)^2$ **27.** $k'(x) = (x^2 - 4x - 12)/(x - 2)^2$ **29.** $r'(t) = [-\sqrt{t} + 3/(2\sqrt{t})]/(2t + 3)^2$ or $(-2t + 3)/[2\sqrt{t}(2t + 3)^2]$ **31.** In the first step, the numerator should be $(x^2 - 1)2 - (2x + 5)(2x)$. **33.** $y = -2x + 9$ **35.** (a) $22.86 per unit (b) $12.92 per unit (c) $(3x + 2)/(x^2 + 4x)$ per unit (d) $\overline{C}'(x) = (-3x^2 - 4x - 8)/(x^2 + 4x)^2$ **37.** (a) $G'(20) = -1/200$; go faster (b) $G'(40) = 1/400$; go slower **39.** (a) $s'(x) = m/(m + nx)^2$ (b) $1/2560 \approx .000391$ mm per ml **41.** (a) -100 (b) $-1/100$ or $-.01$

Section 10.6 (page 510)

1. 1122 **3.** 97 **5.** $256k^2 + 48k + 2$ **7.** $(3x + 95)/8$; $(3x + 280)/8$ **9.** $1/x^2$; $1/x^2$ **11.** $\sqrt{8x^2 - 4}$; $8x + 10$ **13.** $\sqrt{(x - 1)/x}$; $-1/\sqrt{x + 1}$ **17.** If $f(x) = x^{2/5}$ and $g(x) = 5 - x$, then $y = f[g(x)]$. **19.** If $f(x) = -\sqrt{x}$ and $g(x) = 13 + 7x$, then $y = f[g(x)]$. **21.** If $f(x) = x^{1/3} - 2x^{2/3} + 7$ and $g(x) = x^2 + 5x$, then $y = f[g(x)]$. **23.** $y' = 5(2x^3 + 9x)^4(6x^2 + 9)$ **25.** $f'(x) = -288x^3(3x^4 + 2)^2$ **27.** $s'(t) = 144t^3(2t^4 + 5)^{1/2}$ **29.** $f'(t) = 32t/\sqrt{4t^2 + 7}$ **31.** $r'(t) = 4(2t^5 + 3)(22t^5 + 3)$ **33.** $y' = (x^2 - 1)(7x^4 - 3x^2 + 8)$ **35.** $y' = [(5x^6 + x)(125x^6 + 5x)]/\sqrt{2x}$ **37.** $y' = -30x/(3x^2 - 4)^6$ **39.** $p'(t) = [2(2t + 3)^2(4t^2 - 12t - 3)]/(4t^2 - 1)^2$ **41.** $y' = (-30x^4 - 132x^3 + 4x + 8)/(3x^3 + 2)^5$ **43.** (a) -2 (b) $-24/7$ **45.** $D(c) = (-c^2 + 10c + 12,475)/25$ **47.** (a) $101.22 (b) $111.86 (c) $117.59 **49.** (a) $-$1050 (b) $-$457.06 **51.** $400 per additional worker **53.** $P[f(a)] = 18a^2 + 24a + 9$ **55.** $A[r(t)] = A(t) = 4\pi t^2$; this function gives the area of the pollution in terms of the time since the pollutants were first emitted. **57.** (a) $-.5$ (b) $-1/54 \approx -.02$ (c) $-1/128 \approx -.008$ (d) always decreasing; the derivative is negative for all $t \geq 0$

Chapter 10 Review Exercises (page 514)

3. 4 **5.** does not exist **7.** 4 **9.** 17/3 **11.** 8 **13.** -13 **15.** 1/6 **17.** discontinuous at x_2 and x_4 **19.** yes; no; yes; no **21.** yes; no; yes **23.** yes; yes; yes **25.** (a) 1/4 (b) 0 (c) $-3/2$ **27.** -60; -20 **29.** $-5/4$; -5 **31.** $y' = 10x + 6$ **33.** -2; $y + 2x = -4$ **35.** $-3/4$; $3x + 4y = -9$ **37.** $-4/3$; $4x + 3y = 11$ **39.** $-4/5$, $4x + 5y = -13$ **43.** $y' = 3x^2 - 8x$ **45.** $y' = 6x^{-3}$ or $6/x^3$ **47.** $f'(x) = -6x^{-2} - x^{-1/2}$ or $-6/x^2 - 1/x^{1/2}$ **49.** $y' = -20t^3 + 42t^2 - 16t$ **51.** $p'(t) = 98t^{3/4} - 12t^{-1/4}$ or $98t^{3/4} - 12/t^{1/4}$ **53.** $y' = -2x^{-3/5} - 54x^{-8/5}$ or $-2/x^{3/5} - 54/x^{8/5}$ **55.** $r'(x) = 16/(2x + 1)^2$ **57.** $y' = (4x^3 + 7x^2 - 20x)/(x + 2)^2$ **59.** $k'(x) = 30(5x - 1)^5$ **61.** $y' = -12(8t - 1)^{-1/2}$ or $-12/(8t - 1)^{1/2}$ **63.** $y' = 4x(3x - 2)^4(21x - 4)$ **65.** $s'(t) = (-4t^3 - 9t^2 + 24t + 6)/(4t - 3)^5$ **67.** $(4x^{1/2} + x + 1)/[2x^{1/2}(1 - x)^2]$ **69.** $(2 - x)/[2x^2(x - 1)^{1/2}]$ **71.** does not exist at $t = -2$ **73.** (d) **75.** $\overline{C}'(x) = (-3x - 4)/[2x^2(3x + 2)^{1/2}]$ **77.** $\overline{C}'(x) = [(4x + 3)^3(12x - 3)]/x^2$ **79.** (a) $88.89 is the approximate increase in profit from selling the fifth unit. (b) $99.17 is the approximate increase in profit from selling the thirteenth unit. (c) $99.72 is the approximate increase in profit from selling the twenty-first unit. (d) The marginal profit is increasing as the number sold increases. **81.** (a) $R'(x) = 16 - 6x$ (b) -44; an increase of $100 spent on advertising when advertising expenditures are $1000 will result in revenue decreasing by $44 **83.** (a) $3.40 (b) $3.28 (c) $3.18 (d) $3.15 (e) $10.15 (f) $15.15 (g) $[0, \infty)$ (h) no (i) $\overline{P}(x) = 15 + 25x$ (j) $\overline{P}'(x) = 25$ (k) no; the profit per pound never changes, no matter how many pounds are sold **85.** (a) none (b) none (c) $(-\infty, \infty)$ (d) Since the derivative is always negative, the graph of $g(x)$ is always decreasing. **87.** (a) $(-1, 1)$ (b) $x = -1$ (c) $(-\infty, -1)$, $(1, \infty)$ (d) The derivative is 0 when the graph of $G(x)$ is at a low point. It is positive where $G(x)$ is increasing and negative where $G(x)$ is decreasing.

CHAPTER 11

Section 11.1 (page 527)

1. (a) $(1, \infty)$ (b) $(-\infty, 1)$ **3.** (a) $(-\infty, -2)$ (b) $(-2, \infty)$ **5.** (a) $(-\infty, -4)$, $(-2, \infty)$ (b) $(-4, -2)$ **7.** (a) $(-7, -4)$, $(-2, \infty)$ (b) $(-\infty, -7)$, $(-4, -2)$ **9.** (a) $(-6, \infty)$ (b) $(-\infty, -6)$ **11.** (a) $(-\infty, 3/2)$ (b) $(3/2, \infty)$ **13.** (a) $(-\infty, -3)$, $(4, \infty)$ (b) $(-3, 4)$ **15.** (a) $(-\infty, -3/2)$, $(4, \infty)$ (b) $(-3/2, 4)$ **17.** (a) none

(b) $(-\infty, \infty)$ **19. (a)** none **(b)** $(-\infty, -1), (-1, \infty)$ **21. (a)** $(-4, \infty)$ **(b)** $(-\infty, -4)$ **23. (a)** none **(b)** $(1, \infty)$
25. (a) $(0, \infty)$ **(b)** $(-\infty, 0)$ **27. (a)** $(0, \infty)$ **(b)** $(-\infty, 0)$ **29.** vertex: $(-b/(2a), (4ac - b^2)/(4a))$; increasing on
$(-b/(2a), \infty)$, decreasing on $(-\infty, -b/(2a))$ **31. (a)** nowhere **(b)** everywhere **33.** $[0, 1125)$ **35.** after 10 days
37. (a) $(0, 3)$ **(b)** $(3, \infty)$ (Remember: x must be at least 0.) **39. (a)** $(1000, 6100)$ **(b)** $(6100, 6500)$ **(c)** $(1000, 3000)$,
$(3600, 4200)$ **(d)** $(3000, 3600), (4200, 6500)$

Section 11.2 (page 539)

1. relative minimum of -4 at 1 **3.** relative maximum of 3 at -2 **5.** relative maximum of 3 at -4; relative minimum of 1
at -2 **7.** relative maximum of 3 at -4; relative minimum of -2 at -7 and -2 **9.** relative minimum of -44 at -6
11. relative maximum of 8.5 at -3 **13.** relative maximum of -8 at -3; relative minimum of -12 at -1 **15.** relative
maximum of 827/96 at $-1/4$; relative minimum of $-377/6$ at -5 **17.** relative maximum of 57 at 2; relative minimum of 30
at 5 **19.** relative maximum of -4 at 0; relative minimum of -85 at 3 and -3 **21.** relative maximum of 0 at 8/5
23. relative maximum of 1 at -1; relative minimum of 0 at 0 **25.** no relative extrema **27.** relative minimum of 0 at 0
29. relative maximum of 0 at 1; relative minimum of 8 at 5 **31.** $(2, 7)$ **33.** $(5/4, -9/8)$ **35. (a)** 40 **(b)** 15 **(c)** 375
37. (a) 250 **(b)** 10 **(c)** 800 **39.** $q = 5$; $p = 275/6$ **41.** 4:44 P.M.; 5:46 A.M. **43.** 10 min
Note for Exercises 45 and 47: Your answers may be slightly different from the given answers because of different calculators or
computer programs. **45.** relative maximum of 6.2 at .085, relative minimum of -57.7 at 2.2 **47.** relative maximum of 280
at -5.1 and of -18.96 at .89, relative minimum of -19.08 at .56

Section 11.3 (page 546)

1. absolute maximum at x_3; no absolute minimum **3.** no absolute extrema **5.** absolute minimum at x_1; no absolute
maximum **7.** absolute maximum at x_1; absolute minimum at x_2 **9.** absolute maximum at 0; absolute minimum at -3
11. absolute maximum at -1; absolute minimum at -5 **13.** absolute maximum at -2; absolute minimum at 4
15. absolute maximum at -2; absolute minimum at 3 **17.** absolute maximum at 6; absolute minimum at -4 and 4
19. absolute maximum at 0; absolute minimum at 2 **21.** absolute maximum at 6; absolute minimum at 4 **23.** absolute
maximum at $\sqrt{2}$; absolute minimum at 0 **25.** absolute maximum at -3 and 3; absolute minimum at 0 **27.** absolute
maximum at 0; absolute minimum at -4 **29.** absolute maximum at 0; absolute minimum at -1 and 1 **31.** 1000 manuals;
more than \$2.20 **33. (a)** 341 **(b)** 859.4 **35.** 6 months; 6% **37.** 25; 16.1 **39.** The piece formed into a circle should
have length $12\pi/(4 + \pi)$ ft, or about 5.28 ft. *Note for Exercises 41 and 43:* Your answers may be slightly different from the
given answers because of different calculators or computer programs. **41.** absolute maximum at 0; absolute minimum at -2.4
43. absolute maximum at 0; absolute minimum at .74

Section 11.4 (page 558)

1. $f''(x) = 18x$; 0; 36; -54 **3.** $f''(x) = 36x^2 - 30x + 4$; 4; 88; 418 **5.** $f''(x) = 6$; 6; 6; 6 **7.** $f''(x) = 6(x + 4)$; 24;
36; 6 **9.** $f''(x) = 10/(x - 2)^3$; $-5/4$; $f''(2)$ does not exist; $-2/25$ **11.** $f''(x) = 2/(1 + x)^3$; 2; 2/27; $-1/4$
13. $f''(x) = -1/[4(x + 4)^{3/2}]$; $-1/32$; $-1/[4(6^{3/2})] \approx -.0170$; $-1/4$ **15.** $f''(x) = (-6/5)x^{-7/5}$ or $-6/(5x^{7/5})$; $f''(0)$
does not exist; $-6/[5(2^{7/5})] \approx -.4547$; $-6/[5(-3)^{7/5}] \approx .2578$ **17.** $f'''(x) = -24x$; $f^{(4)}(x) = -24$ **19.** $f'''(x) =$
$240x^2 + 144x$; $f^{(4)}(x) = 480x + 144$ **21.** $f'''(x) = 18(x + 2)^{-4}$ or $18/(x + 2)^4$; $f^{(4)}(x) = -72(x + 2)^{-5}$ or $-72/(x + 2)^5$
23. $f'''(x) = -36(x - 2)^{-4}$ or $-36/(x - 2)^4$; $f^{(4)}(x) = 144(x - 2)^{-5}$ or $144/(x - 2)^5$ **25.** concave upward on $(2, \infty)$;
concave downward on $(-\infty, 2)$; point of inflection at $(2, 3)$ **27.** concave upward on $(-\infty, -1)$ and $(8, \infty)$; concave downward
on $(-1, 8)$; points of inflection at $(-1, 7)$ and $(8, 6)$ **29.** concave upward on $(2, \infty)$; concave downward on $(-\infty, 2)$; no points
of inflection **31.** always concave upward; no points of inflection **33.** concave upward on $(-1, \infty)$; concave downward on
$(-\infty, -1)$; point of inflection at $(-1, 44)$ **35.** concave upward on $(-\infty, 3/2)$; concave downward on $(3/2, \infty)$; point of
inflection at $(3/2, 525/2)$ **37.** concave upward on $(5, \infty)$; concave downward on $(-\infty, 5)$; no points of inflection
39. concave upward on $(-10/3, \infty)$; concave downward on $(-\infty, -10/3)$; point of inflection at $(-10/3, -250/27)$
41. relative maximum at -5 **43.** relative maximum at 0; relative minimum at 2/3 **45.** relative minimum at -3
47. (a) car phones and CD players; the rate of growth of sales will now decline **(b)** food processors; the rate of decline of sales
is starting to slow **49.** $(22, 6517.9)$ **53. (a)** at 4 hr **(b)** 1160 million **55. (a)** after 3 hr **(b)** 2/9%
57. $v(t) = 16t + 4$; $a(t) = 16$; $v(0) = 4$ cm/sec; $v(4) = 68$ cm/sec; $a(0) = 16$ cm/sec^2; $a(4) = 16$ cm/sec^2 **59.** $v(t) =$
$-15t^2 - 16t + 6$; $a(t) = -30t - 16$; $v(0) = 6$ cm/sec; $v(4) = -298$ cm/sec; $a(0) = -16$ cm/sec^2; $a(4) = -136$ cm/sec^2

61. $v(t) = 6(3t + 4)^{-2}$ or $6/(3t + 4)^2$; $a(t) = -36(3t + 4)^{-3}$ or $-36/(3t + 4)^3$; $v(0) = 3/8$ cm/sec; $v(4) = 3/128$ cm/sec; $a(0) = -9/16$ cm/sec²; $a(4) = -9/1024$ cm/sec² **63. (a)** -96 ft/sec **(b)** -160 ft/sec **(c)** -256 ft/sec **(d)** -32 ft/sec² *Note for Exercises 65 and 67:* Your answers may be slightly different from the given answers because of different calculators or computer programs. **65. (a)** increasing on $(0, 2)$ and $(4, 5)$; decreasing on $(-5, -.5)$ and $(2.5, 3.5)$ **(b)** minima between $-.5$ and 0, and between 3.5 and 4; a maximum between 2 and 2.5 **(c)** concave upward on $(-5, .5)$ and $(3.5, 5)$; concave downward on $(.5, 3.5)$ **(d)** inflection points between $.5$ and 1, and between 3 and 3.5 **67. (a)** decreasing on $(-1, 1)$; increasing on $(1.2, 2)$ **(b)** minimum between 1 and 1.2 **(c)** concave upward on $(-1, 0)$ and $(.8, 2)$; concave downward on $(0, .8)$ **(d)** inflection points between $.6$ and $.8$, and at 0

Section 11.5 (page 571)

1. 3 **3.** 3/5 **5.** 1/2 **7.** 1/2 **9.** 0 **11.** 0

13.

$f(x) = -2x^3 - 9x^2 + 108x - 10$

15.

$f(x) = 2x^3 + \frac{7}{2}x^2 - 5x + 3$

17.

$f(x) = x^4 - 18x^2 + 5$

19.

$f(x) = x^4 - 2x^3$

21.

$f(x) = x + \frac{2}{x}$

23.

$f(x) = \frac{x^2 + 25}{x}$

25.

$f(x) = \frac{x - 1}{x + 1}$

27.

$f(x) = \frac{x}{x^2 + 1}$

29.

$f(x) = \frac{1}{x^2 - 4}$

31.

33.

35. 6; the average cost approaches 6 as the number of tapes produced becomes very large. **37.** 0; the concentration of the drug in the bloodstream approaches 0 as the number of hours after injection increases. **39. (a)** 1.5 **41. (a)** -2 **43. (a)** 8

Chapter 11 Review Exercises (page 573)

5. increasing on $(5/2, \infty)$; decreasing on $(-\infty, 5/2)$ **7.** increasing on $(-4, 2/3)$; decreasing on $(-\infty, -4)$ and $(2/3, \infty)$ **9.** never increasing; decreasing on $(-\infty, 4)$ and $(4, \infty)$ **11.** relative maximum of -4 at 2 **13.** relative minimum of -7 at 2 **15.** relative maximum of 101 at -3; relative minimum of -24 at 2 **17.** $f''(x) = 36x^2 - 10$; 26; 314 **19.** $f''(x) = -68(2x + 3)^{-3}$ or $-68/(2x + 3)^3$; $-68/125$; 68/27 **21.** $f''(t) = (t^2 + 1)^{-3/2}$ or $1/(t^2 + 1)^{3/2}$; $1/2^{3/2} \approx .354$; $1/10^{3/2} \approx .032$ **23.** absolute maximum of 29/4 at 5/2; absolute minimum of 5 at 1 and 4 **25.** absolute maximum of 39 at -3; absolute minimum of $-319/27$ at 5/3 **27.** does not exist **29.** 1/5 **31.** 3/4

33.

$$f(x) = -2x^3 - \frac{1}{2}x^2 + x - 3$$

35.

$$f(x) = x^4 - \frac{4}{3}x^3 - 4x^2 + 1$$

37.

$$f(x) = \frac{x - 1}{2x + 1}$$

39.

$$f(x) = -4x^3 - x^2 + 4x + 5$$

41.

$$f(x) = x^4 + 2x^2$$

43.

$$f(x) = \frac{x^2 + 4}{x}$$

45.

$$f(x) = \frac{2x}{3 - x}$$

CHAPTER 12

Section 12.1 (page 583)

1. (a) $y = 100 - x$ **(b)** $P = x(100 - x)$ **(c)** $[0, 100]$ **(d)** $P' = 100 - 2x$; $x = 50$ **(e)** $P(0) = 0$; $P(100) = 0$; $P(50) = 2500$ **(f)** 2500; 50 and 50 **3.** 100; 100; 20,000 **5.** 100; 50; 500,000 **7.** 10; 0; 0 **9. (a)** $1200 - 2x$ **(b)** $A(x) = 1200x - 2x^2$ **(c)** 300 m **(d)** 180,000 m² **11.** 405,000 m² **13.** 0 mi **15. (a)** $R(x) = 100,000x - 100x^2$ **(b)** 500 **(c)** 25,000,000 cents **17. (a)** $\sqrt{3200} \approx 56.6$ mph **(b)** \$45.24 **19.** \$2400 **21. (a)** 90 **(b)** \$405 **23.** 4 in. by 4 in. by 2 in. **25.** 3 ft by 6 ft by 2 ft **27.** 10 cm and 10 cm **29.** Radius is 1.08 ft, height is 4.34 ft, cost is \$44.11 using the rounded values for the height and radius. **31.** $3\sqrt{6} + 3$ by $2\sqrt{6} + 2$ **33.** point A **35.** 250 thousand **37.** 56.25 thousand **39.** 12.86 thousand **41.** Point P is $3\sqrt{7}/7 \approx 1.134$ mi from point A. **43. (d)** $\alpha = 5$; the current stays constant and the salmon swim at a constant velocity. **45.** radius = 5.206 cm; height = 11.75 cm **47.** radius = 5.242 cm;

height $= 11.58$ cm **49.** $A(x) = .01x^2 + .05x + .2 + 28/x$; $x = 10.41$ **51.** $A(x) = 30/x + 42x^{-1/2} + .2x^{1/2} + .03x^{3/2}$; $x = 23.49$

Section 12.2 (page 596)
3. 10 **5.** 10,000 **7.** 5000 **13. (a)** $E = p/(500 - p)$ **(b)** 12,500 **15. (a)** $E = 1$ **(b)** none **17. (a)** $E = 2$; elastic; a percentage increase in price will result in a greater percentage decrease in demand **(b)** $E = 1/2$; inelastic; a percentage change in price will result in a smaller percentage change in demand **19. (a)** 9/8 **(b)** $q = 50/3$; $p = 5\sqrt{3}/3$ **(c)** 1 **21.** The demand function has a horizontal tangent line at the value of P where $E = 0$.

Section 12.3 (page 602)
1. $dy/dx = -4x/(3y)$ **3.** $dy/dx = -y/(y + x)$ **5.** $dy/dx = -3y^2/(6xy - 4)$ **7.** $dy/dx = (-6x - 4y)/(4x + y)$
9. $dy/dx = 3x^2/(2y)$ **11.** $dy/dx = y^2/x^2$ **13.** $dy/dx = -3x(2 + y)^2/2$ **15.** $dy/dx = -2xy/(x^2 + 3y^2)$
17. $dy/dx = -y^{1/2}/x^{1/2}$ **19.** $dy/dx = -y^{1/2}x^{-1/2}/(x^{1/2}y^{-1/2} + 2)$ **21.** $dy/dx = (4x^3y^3 + 6x^{1/2})/(9y^{1/2} - 3x^4y^2)$
23. $dy/dx = (8x(x^2 + y^3)^3 - 1)/(2 - 12y^2(x^2 + y^3)^3)$
25. $4y = 3x + 25$ **27.** $y = x + 2$ **29.** $24x + y = 57$ **31.** $x + 4y = 12$
33. (a) $3x + 4y = 50$; $-3x + 4y = -50$ **35.** $2y = x + 1$ **37.** $dy/dx = -x/y$; there is no function $y = f(x)$ that
(b)

satisfies $x^2 + y^2 + 1 = 0$. **39.** $dv/du = -(2v + 1)^{1/2}/(2u^{1/2})$
41. (a) $dC/dx = (x^{3/2} + 25)/(Cx^{1/2})$; when $x = 5$ the approximate increase in cost of an additional unit is .94 **(b)** $dR/dx = (180 - 36x)/R$; when $x = 5$ the approximate change in revenue for a unit increase in sales is zero
43. $ds/dt = (-s + 6\sqrt{st})/(8s\sqrt{st} + t)$

Section 12.4 (page 608)
1. 440 **3.** $-15/2$ **5.** $-5/7$ **7.** 1/5 **9.** \$200 per month **11. (a)** Revenue is increasing at a rate of \$180 per day. **(b)** Cost is increasing at a rate of \$50 per day. **(c)** Profit is increasing at a rate of \$130 per day. **13.** Demand is decreasing at a rate of approximately 343 units per unit of time. **15.** $-.24$ mm/min **17.** .067 mm/min **19.** $-.370$
21. 7/6 ft/min **23.** 16π ft^2/min **25.** 50π in^3/min **27.** 1/16 ft/min **29.** 43.3 ft/min

Section 12.5 (page 616)
1. $dy = 12x\,dx$ **3.** $dy = (14x - 9)dx$ **5.** $dy = x^{-1/2}\,dx$ **7.** $dy = [-22/(x - 3)^2]dx$ **9.** $dy = (3x^2 - 1 + 4x)dx$
11. $dy = (x^{-2} + 6x^{-3})dx$ or $(1/x^2 + 6/x^3)dx$ **13.** -2.6 **15.** .1 **17.** .130 **19.** $-.023$ **21.** .24
23. $-.00444$ **25. (a)** -34 thousand pounds **(b)** -169.2 thousand pounds **27.** -5.625 housing starts **29.** about $-\$990,000$ **31.** $21,600\pi$ in^3 **33. (a)** .347 million **(b)** $-.022$ million **35.** 1568π mm^3 **37.** 80π mm^2
39. -7.2π cm^3 **41.** ± 1.224 in^2 **43.** $\pm .116$ in^3

Chapter 12 Review Exercises (page 618)
5. $dy/dx = (-4y - 2xy^3)/(3x^2y^2 + 4x)$ **7.** $dy/dx = (-4 - 9x^{3/2})/(24x^2y^2)$ **9.** $dy/dx = (2y - 2y^{1/2})/(4y^{1/2} - x + 9y)$ (This form of the answer was obtained by multiplying both sides of the given function by $x - 3y$.) **11.** $dy/dx = [9(4y^2 - 3x)^{1/3} + 3]/(8y)$ **13.** $23x + 16y = 94$ **15.** 272 **17.** -2 **21.** $dy = 24x(x^2 - 1)^2dx$ **23.** $dy = (3x^2/2)(9 + x^3)^{-1/2}dx$ **25.** .00204 **29.** $x = 2$ and $y = 0$ **31. (a)** 600 boxes **(b)** \$720 **33.** radius $= 1.684$ in.; height $= 4.490$ in. **35.** 4434 **37.** 126 **39.** $E = k$; elastic when $k > 1$; inelastic when $k < 1$ **41.** 8/3 ft/min
43. $21/16 = 1.3125$ ft/min **45.** $\pm .736$ in^2 **47. (a)** $(2, -5), (2, 4)$ **(b)** $(2, -5)$ is a relative maximum, and $(2, 4)$ is a relative minimum. **(c)** no

Extended Application (page 621)
1. $-C_1/m^2 + DC_3/2$ **2.** $m = \sqrt{2C_1/(DC_3)}$ **3.** about 3.33 **4.** $m^+ = 4$ and $m^- = 3$ **5.** $Z(m^+) = Z(4) = \$11,400$; $Z(m^-) = Z(3) = \$11,300$ **6.** 3 months; 9 trainees per batch

CHAPTER 13

Section 13.1 (page 630)

1.

3.

5.

7.

9.

11.

13.

15.

17. because 4 and 6 cannot easily be written as powers of the same base **19.** 3 **21.** -2 **23.** 6 **25.** 7/4 **27.** -2
29. 2, -2 **31.** 0, -1 **33.** 4, -2 **35.** (a) 1.718 (b) 1.052 (c) 1.005 (d) 1.001 (e) 1 **39.** (a) \$10,528.13
(b) \$10,881.50 (c) \$11,069.78 (d) \$11,199.99 **41.** \$31,427.49 **43.** (a) 12.5% (b) 11.9%

45. (a) about 207 **(b)** about 235
(c) about 249
(d)

$p(t) = 250 - 120(2.8)^{-.5t}$

(e) It gets very close to 250. **(f)** 250

47. (a) The function gives 3727 million, which is very close. **(b)** 5341 million
(c) 6395 million **49. (a)** 6 **(b)** 2 yr **(c)** 6 yr **51. (a)** 55 g **(b)** 10 mo
53.

$y = 2^{1-x} + 2$

$y = 2$

55.

$y = e^{-x+1}$

Section 13.2 (page 643)
1. $\log_2 8 = 3$ **3.** $\log_3 81 = 4$ **5.** $\log_{1/3} 9 = -2$ **7.** $2^7 = 128$ **9.** $25^{-1} = 1/25$ **11.** $10^4 = 10,000$ **13.** 2
15. 3 **17.** -2 **19.** $-2/3$ **21.** 1 **23.** 5/3 **25.** $\log_3 4$
27.

$y = \log_4 x$

29.

$y = \log_{1/3}(x - 1)$

$x = 1$

31.

$y = \ln x^2$

33. $\log_9 7 + \log_9 m$ **35.** $1 + \log_3 p - \log_3 5 - \log_3 k$ **37.** $\log_3 5 + (1/2)\log_3 2 - (1/4)\log_3 7$ **39.** $3a$
41. $2c + 3a + 1$ **43.** 1.86 **45.** 9.35 **47.** $-.21$ **49.** $x = 1/5$ **51.** $z = 2/3$ **53.** $r = 49$ **55.** $x = 1$
57. no solution **59.** $x = 1.79$ **61.** $y = 1.24$ **63.** $z = 2.10$ **67. (a)** 11.7 yr **(b)** 18.6 yr **(c)** 12 yr **69.** 1.589
71. 4.3 ml/min; 7.8 ml/min **73. (a)** 21 **(b)** 70 **(c)** 91 **(d)** 120 **(e)** 140 **(f)** about 2,300,000,000 I_0 **75. (a)** 1000
times greater **(b)** 1,000,000 times greater **77.**

$y = -\log_2 x$

79.

$x = -.5$ $y = 30 \ln(2x + 1)$

Section 13.3 (page 653)
1. 5.12% **3.** 10.25% **5.** 11.63% **7.** $1043.79 **9.** $4537.71 **11.** $5248.14 **19. (a)** $10.94 **(b)** $11.27
(c) $11.62 **21. (a)** the 10% investment compounded quarterly **(b)** $622.56 **(c)** 10.38% and 10.24% **(d)** 2.95 yr
23. 7.40% **25.** $14,700.60 **27. (a)** $13,459.43 **(b)** $6540.57 **(c)** $5140.53 **29. (a)** 200 **(b)** about 1/2 yr
(c) no **(d)** yes; 1000 **31. (a)** $y = 100e^{.11t}$ **(b)** about 15 mo **33. (a)** $y = 50,000e^{-.102t}$ **(b)** about 6.8 hr
35. (a) about 1100 **(b)** about 1600 **(c)** about 2300 **(d)** at about 1.8 decades **39. (a)** 0 **(b)** about 432 **(c)** about 497

(d) about 1.6 days **(e)** 500 **(f)**

$P(x) = 500 - 500e^{-x}$

41. about 4100 yr old **43.** about 1600 yr
45. (a) 19.5 watts **(b)** about 173 days **(c)** No. It will approach 0 watts, but never be exactly 0.
47. (a) $y = 10e^{.0095t}$ **(b)** 42.7° C
49. about 30 min

Section 13.4 (page 662)
1. $y' = 1/x$ **3.** $y' = -1/(3 - x)$ or $1/(x - 3)$ **5.** $y' = (4x - 7)/(2x^2 - 7x)$ **7.** $y' = 1/[2(x + 5)]$
9. $y' = 3(2x^2 + 5)/[x(x^2 + 5)]$ **11.** $y' = -3x/(x + 2) - 3 \ln (x + 2)$ **13.** $y' = x + 2x \ln |x|$
15. $y' = [2x - 4(x + 3) \ln (x + 3)]/[x^3(x + 3)]$ **17.** $y' = (4x + 7 - 4x \ln x)/[x(4x + 7)^2]$
19. $y' = (6x \ln x - 3x)/(\ln x)^2$ **21.** $y' = 4(\ln |x + 1|)^3/(x + 1)$ **23.** $y' = 1/(x \ln x)$ **27.** $y' = 1/(x \ln 10)$
29. $y' = -1/[(\ln 10)(1 - x)]$ or $1/[(\ln 10)(x - 1)]$ **31.** $y' = 5/[2(\ln 5)(5x + 2)]$ **33.** $y' = 3(x + 1)/[(\ln 3)(x^2 + 2x)]$
35. minimum of $-1/e \approx -.3679$ **37.** minimum of $-1/e \approx -.3679$ **39.** maximum of $1/e \approx .3679$
 at $x = 1/e \approx .3679$ at $x = 1/e \approx .3679$; at $x = e \approx 2.718$
 maximum of $1/e$ at $-1/e$

41. As $x \to \infty$, the slope approaches 0; as $x \to 0$, the slope becomes infinitely large. **45. (a)** $\dfrac{dR}{dx} = 100 + \dfrac{50(\ln x - 1)}{(\ln x)^2}$

(b) \$112.48 **(c)** to decide whether it is reasonable to sell additional items **47. (a)** $R(x) = 100x - 10x \ln x$, $1 < x < 20,000$
(b) $dR/dn = 60(9 - \ln x) = 60[9 - \ln (6n)]$ **(c)** about 360; hiring an additional worker will produce an increase in revenue of about \$360 **49.** 26.9; 13.1 **51.** minimum of $\approx .85$ at $x \approx .7$

Section 13.5 (page 671)

1. $y' = 4e^{4x}$ **3.** $y' = -16e^{2x}$ **5.** $y' = -16e^{x+1}$ **7.** $y' = 2xe^{x^2}$ **9.** $y' = 12xe^{2x^2}$ **11.** $y' = 16xe^{2x^2-4}$
13. $y' = xe^x + e^x = e^x(x+1)$ **15.** $y' = 2(x-3)(x-2)e^{2x}$ **17.** $y' = e^{x^2}/x + 2xe^{x^2} \ln x$ **19.** $y' =$
$(xe^x \ln x - e^x)/[x(\ln x)^2]$ **21.** $y' = (2xe^x - x^2 e^x)/e^{2x} = x(2-x)/e^x$ **23.** $y' = [x(e^x - e^{-x}) - (e^x + e^{-x})]/x^2$
25. $y' = -20{,}000e^{-4x}/(1 + 10e^{-4x})^2$ **27.** $y' = 8000e^{-.2x}/(9 + 4e^{-.2x})^2$ **29.** $y' = 2(2x + e^{-x^2})(2 - 2xe^{-x^2})$
31. $y' = 5 \ln 8 e^{5x \ln 8}$ or $y' = 5(\ln 8)(8^{5x})$ **33.** $y' = 6x(\ln 4)e^{(x^2+2)(\ln 4)}$ or $y' = 6x(\ln 4)4^{x^2+2}$
35. $y' = [(\ln 3)e^{\sqrt{x} \ln 3}]/\sqrt{x}$ or $y' = [(\ln 3)3^{\sqrt{x}}]/\sqrt{x}$

39. maximum of $1/e$ at $x = -1$; inflection point at $(-2, 2e^{-2})$

41. minimum of 0 at $x = 0$; maximum of $4/e^2 \approx .54$ at $x = 2$; inflection points at $(3.4, .38)$ and $(.6, .19)$

43. minimum of 2 at $x = 0$; no inflection point

45. $f''(x) = e^x, f'''(x) = e^x; f^{(n)} = e^x$ **49.** (a) .98 (b) .82 (c) .14 (d) $-.0027$; the rate of change in the proportion wearable when $x = 100$ (e) It decreases. Yes, as time increases, the shoes wear out. **51.** (a) $E = 5/x$ (b) 5
53. (a) -46.0 (b) -27.9 (c) -10.3 (d) It is approaching 0. (e) no
55. no relative extrema; inflection point at $(1, 0)$

Chapter 13 Review Exercises (page 674)

3. -1 **5.** 2 **7.**

9.

11.

13.

15. $\log_2 64 = 6$ **17.** $\ln 1.09417 = .09$ **19.** $2^5 = 32$ **21.** $e^{4.41763} = 82.9$ **23.** 4 **25.** 4/5 **27.** 3/2
29. $\log_5 (21k^4)$ **31.** $\log_2 (x^2/m^3)$ **33.** $p = 1.416$ **35.** $m = -1.807$ **37.** $x = -3.305$ **39.** $m = 1.7547$
41. $y' = -12e^{2x}$ **43.** $y' = -6x^2e^{-2x^3}$ **45.** $y' = 10xe^{2x} + 5e^{2x} = 5e^{2x}(2x + 1)$ **47.** $y' = 2x/(2 + x^2)$
49. $y' = (x - 3 - x \ln |3x|)/[x(x - 3)^2]$ **51.** $y' = [e^x(x + 1)(x^2 - 1) \ln (x^2 - 1) - 2x^2e^x]/[(x^2 - 1)(\ln (x^2 - 1))^2]$
53. $y' = 2(x^2 + e^x)(2x + e^x)$
55. relative minimum of $-e^{-1} = -.368$ at $x = -1$; **57.** relative minimum of e^2 at $x = 2$;
inflection point at $(-2, -.27)$ no inflection point

59. $10,631.51 **61.** $14,204.18 **63.** $21,190.14 **65.** $17,901.90 **67.** 9.38% **69.** 9.42% **71.** $1494.52
73. $31,468.86 **75.** (a) 100 (b) 100 (c) 1 day **77.** about 13.7 yr **79.** $20,891.12 **81.** about 7.7 m
83. (a) when it is first injected; .08 (b) never (c) It approaches $c/a = .0769$ g **85.** (a) 0 yr (b) 1.85×10^9 yr
(c) $\dfrac{dt}{dr} = \dfrac{10.4958 \times 10^9}{\ln 2 (1 + 8.33r)}$ (d) As r increases, t increases, but at a slower and slower rate. As r decreases, t decreases at a faster
and faster rate. **87.** (a) 11 (b) 12.6 (c) 18.0 (d)

89. (a) $(-\infty, \infty)$ (b) $(0, \infty)$ (c) 1 (d) none (e) $y = 0$ (f) greater than 1 (g) between 0 and 1

Extended Application (page 678)

1. (a) \$9216.65; 4.6 (b) \$6787.27; 3.4 (c) 1.35; investment (a) will yield approximately 35% more after-tax dollars than investment (b) **2.** (a) \$4008.12; 2.0 (b) \$3684.37; 1.8 (c) 1.11; investment (a) will yield approximately 11% more after-tax dollars than investment (b) **3.** M is an increasing function of t. **4.** m is an increasing function of n. **5.** The conclusions are that the multiplier function is an increasing function of i. The advantage of the IRA over a regular account widens as the interest rate i increases and is particularly dramatic for high income tax rates.

CHAPTER 14

Section 14.1 (page 689)

1. $2x^2 + C$ **3.** $5t^3/3 + C$ **5.** $6k + C$ **7.** $z^2 + 3z + C$ **9.** $x^3/3 + 3x^2 + C$ **11.** $t^3/3 - 2t^2 + 5t + C$ **13.** $z^4 + z^3 + z^2 - 6z + C$ **15.** $10z^{3/2}/3 + C$ **17.** $2u^{3/2}/3 + 2u^{5/2}/5 + C$ **19.** $6x^{5/2} + 4x^{3/2}/3 + C$ **21.** $4u^{5/2} - 4u^{7/2} + C$ **23.** $-1/z + C$ **25.** $-1/(2y^2) - 2y^{1/2} + C$ **27.** $9/t - 2\ln|t| + C$ **29.** $e^{2t}/2 + C$ **31.** $-15e^{-.2x} + C$ **33.** $3\ln|x| - 8e^{-.5x} + C$ **35.** $\ln|t| + 2t^3/3 + C$ **37.** $e^{2u}/2 + 2u^2 + C$ **39.** $x^3/3 + x^2 + x + C$ **41.** $6x^{7/6}/7 + 3x^{2/3}/2 + C$ **43.** $f(x) = 3x^{5/3}/5$ **45.** $C(x) = 2x^2 - 5x + 8$ **47.** $C(x) = .2x^3/3 + 5x^2/2 + 10$ **49.** $C(x) = 3e^{.01x} + 5$ **51.** $C(x) = 3x^{5/3}/5 + 2x + 114/5$ **53.** $C(x) = x^2/2 - 1/x + 4$ **55.** $C(x) = 5x^2/2 - \ln x - 153.50$ **57.** $P(x) = x^2 + 20x - 50$ **59.** (a) $f(t) = -e^{-.01t} + k$ (b) .095 unit **61.** $v(t) = t^3/3 + t + 6$ **63.** $s(t) = -16t^2 + 6400$; 20 sec **65.** $s(t) = 2t^{5/2} + 3e^{-t} + 1$

Section 14.2 (page 697)

1. $2(2x + 3)^5/5 + C$ **3.** $-(2m + 1)^{-2}/2 + C$ **5.** $-(x^2 + 2x - 4)^{-3}/3 + C$ **7.** $(z^2 - 5)^{3/2}/3 + C$ **9.** $-2e^{2p} + C$ **11.** $e^{2x^3}/2 + C$ **13.** $e^{2t - t^2}/2 + C$ **15.** $-e^{1/z} + C$ **17.** $-8\ln|1 + 3x|/3 + C$ **19.** $(\ln|2t + 1|)/2 + C$ **21.** $-(3v^2 + 2)^{-3}/18 + C$ **23.** $-(2x^2 - 4x)^{-1}/4 + C$ **25.** $[(1/r) + r]^2/2 + C$ **27.** $(x^3 + 3x)^{1/3} + C$ **29.** $(p + 1)^7/7 - (p + 1)^6/6 + C$ **31.** $2(5t - 1)^{5/2}/125 + 2(5t - 1)^{3/2}/75 + C$ **33.** $2(u - 1)^{3/2}/3 + 2(u - 1)^{1/2} + C$ **35.** $(x^2 + 12x)^{3/2}/3 + C$ **37.** $[\ln(t^2 + 2)]/2 + C$ **39.** $e^{2z^2}/4 + C$ **41.** $(1 + \ln x)^3/3 + C$ **43.** $(4/15)(x^{5/2} + 4)^{3/2} + C$ **45.** (a) $R(x) = (x^2 + 50)^3/3 + 137,919.33$ (b) 7 **47.** (a) $p(x) = -e^{-x^2}/2 + .01$ (b) It approaches \$10,000. **49.** (a) $D(x) = 2\ln|x + 9| - 2.11$ (b) 3.9 mg

Section 14.3 (page 707)

1. 18 **3.** 65 **5.** 20 **7.** 8 **9.** (a) 56 (b) $\int_0^8 (2x + 1)\,dx$ **11.** 32; 38 **13.** 15; 31/2 **15.** 20; 30 **17.** 16; 14 **19.** 12.8; 27.2 **21.** 2.67; 2.08 **23.** (a) 3 (b) 3.5 (c) 4 **25.** 9 **27.** $9\pi/2$ **29.** 6 **31.** (a) about 52 billion barrels (b) about 64 billion barrels **33.** (a) \$83,000 (b) \$89,000 **35.** about 33.2 liters **37.** about 2600 ft **39.** (a) about 690 BTUs (b) about 180 BTUs **41.** 1.91837 **43.** 25.7659

Section 14.4 (page 718)

1. -6 **3.** 3/2 **5.** 28/3 **7.** 13 **9.** 1/3 **11.** 76 **13.** 4/3 **15.** 112/25 **17.** $20e^{-.2} - 20e^{-.3} + 3\ln 3 - 3\ln 2 \approx 2.775$ **19.** $e^{10}/5 - e^5/5 - 1/2 \approx 4375.1$ **21.** 91/3 **23.** $447/7 \approx 63.857$ **25.** $(\ln 2)^2/2 \approx .24023$ **27.** 49 **29.** $1/4 - 1/(3 + e) \approx .075122$ **31.** $(6/7)(128 - 2^{7/3}) \approx 105.39$ **33.** 42 **35.** 76 **37.** 54 **39.** 41/2 **41.** $e^2 - 3 + 1/e \approx 4.757$ **43.** 1 **45.** 23/3 **47.** $e^3 - 2e^2 + e \approx 8.026$ **53.** (a) \$22,000 (b) \$62,000 (c) 4.5 days **55.** (a) $(9000/8)(17^{4/3} - 2^{4/3}) \approx \$46,341$ (b) $(9000/8)(26^{4/3} - 17^{4/3}) \approx \$37,477$ (c) It is slowly increasing without bound. **57.** no **59.** (a) 1.37 ft (b) .32 ft **61.** (a) 14.26 (b) 3.55 **63.** (a) $c'(t) = 1.2e^{.04t}$ (b) $\int_0^{10} 1.2e^{.04t}\,dt$ (c) $30e^{.4} - 30 \approx 14.75$ billion (d) about 12.8 yr (e) about 14.4 yr **65.** $5142.9(e^{.014T} - 1)$

Section 14.5 (page 729)

1. 15 **3.** 4 **5.** 23/3 **7.** 366.1667 **9.** 4/3 **11.** $5 + \ln 6 \approx 6.792$ **13.** $6\ln(3/2) - 6 + 2e^{-1} + 2e \approx 2.6051$ **15.** $e^2 - e - \ln 2 \approx 3.978$ **17.** 1/2 **19.** 1/20 **21.** $3(2^{4/3})/2 - 3(2^{7/3})/7 \approx 1.6199$ **23.** (a) 8 yr

(b) about $148 **(c)** about $771 **25. (a)** 39 days **(b)** $3369.18 **(c)** $484.02 **(d)** $2885.16 **27.** 12,931.66 **29.** 27

31. (a)

(b) (15, 375) **(c)** 4500 **(d)** 3375

33. (a) .019; the lower 10% of the income producers earn 1.9% of the total income of the population **(b)** .184; the lower 40% of the income producers earn 18.4% of the total income of the population **(c)** .384; the lower 60% of the income producers earn 38.4% of the total income of the population **(d)** .819; the lower 90% of the income producers earn 81.9% of the total income of the population **(e)** **(f)** .15

Note: Answers for Exercises 35 and 37 may vary slightly depending on the computer and the software used.
35. 161.2 **37.** 5.516

Chapter 14 Review Exercises (page 732)
5. $6x + C$ **7.** $x^2 + 3x + C$ **9.** $x^3/3 - 3x^2/2 + 2x + C$ **11.** $2x^{3/2} + C$ **13.** $2x^{3/2}/3 + 9x^{1/3} + C$
15. $2x^{-2} + C$ **17.** $-3e^{2x}/2 + C$ **19.** $2 \ln |x - 1| + C$ **21.** $e^{3x^2}/6 + C$ **23.** $(3 \ln |x^2 - 1|)/2 + C$
25. $-(x^3 + 5)^{-3}/9 + C$ **27.** $\ln |2x^2 - 5x| + C$ **29.** $-e^{-3x^4}/12 + C$ **31.** $2e^{-5x}/5 + C$ **33.** 20 **35.** 24
39. $965/6 \approx 160.83$ **41.** $559/648 \approx .863$ **43.** $8 \ln 6 \approx 14.334$ **45.** $5(e^{24} - e^4)/8 \approx 1.656 \times 10^{10}$ **47.** 19/15
49. 32/7 **51.** $5 - e^{-4} \approx 4.982$ **53.** 1/6 **55.** 32 **57.** $C(x) = 10x - x^2 + 4$ **59.** $C(x) = (2x - 1)^{3/2} + 145$
61. about $96,000 **63.** 36,000 **65.** 2.5 yr; about $99,000 **67.** $50 \ln 17 \approx 141.66$ **69.** approximately 4500 degree-days (using rectangles); actual value according to the National Weather Service: 4868 degree-days

Extended Application (page 736)
1. about 102 yr **2.** about 55.6 yr **3.** about 45.4 yr **4.** about 90 yr

CHAPTER 15

Section 15.1 (page 744)
1. $xe^x - e^x + C$ **3.** $(-5xe^{-3x})/3 - (5e^{-3x})/9 + 3e^{-3x} + C$ or $(-5xe^{-3x})/3 + (22e^{-3x})/9 + C$ **5.** $-5e^{-1} + 3 \approx$
1.1606 **7.** $11 \ln 2 - 3 \approx 4.6246$ **9.** $(x^2 \ln x)/2 - x^2/4 + C$ **11.** $e^4 + e^2 \approx 61.9872$ **13.** $(x^2e^{2x})/2 - (xe^{2x})/2 +$
$(e^{2x})/4 + C$ **15.** $243/8 - (3\sqrt[3]{2})/4 \approx 29.4301$ **17.** $4x^2 \ln (5x) + 7x \ln (5x) - 2x^2 - 7x + C$
19. $[2x^2(x + 2)^{3/2}]/3 - [8x(x + 2)^{5/2}]/15 + [16(x + 2)^{7/2}]/105 + C$ or $(2/7)(x + 2)^{7/2} - (8/5)(x + 2)^{5/2} +$
$(8/3)(x + 2)^{3/2} + C$ **21.** .13077 **23.** $-4 \ln |x + \sqrt{x^2 + 36}| + C$ **25.** $\ln |(x - 3)/(x + 3)| + C$
27. $(4/3) \ln |(3 + \sqrt{9 - x^2})/x| + C$ **29.** $-2x/3 + 2 \ln |3x + 1|/9 + C$ **31.** $(-2/15) \ln |x/(3x - 5)| + C$
33. $\ln |(2x - 1)/(2x + 1)| + C$ **35.** $-3 \ln |(1 + \sqrt{1 - 9x^2})/(3x)| + C$ **37.** $2x - 3 \ln |2x + 3| + C$
39. $1/[25(5x - 1)] - (\ln |5x - 1|)/25 + C$ **43. (a)** $(2/3)x(x + 1)^{3/2} - (4/15)(x + 1)^{5/2} + C$ **(b)** $(2/5)(x + 1)^{5/2} -$
$(2/3)(x + 1)^{3/2} + C$ **45.** $-100e^{-6} + 10e^{-1} \approx 3.431$ **47.** $7\sqrt{65}/2 + 8 \ln (7 + \sqrt{65}) - 8 \ln 4 \approx 38.8$

Section 15.2 (page 753)

1. (a) 2.7500 **(b)** 2.6667 **(c)** $8/3 \approx 2.6667$ **3. (a)** 1.6833 **(b)** 1.6222 **(c)** $\ln 5 \approx 1.6094$ **5. (a)** 16
(b) 14.6667 **(c)** $44/3 \approx 14.6667$ **7. (a)** .9436 **(b)** .8374 **(c)** $4/5 = .8$ **9. (a)** 9.3741 **(b)** 9.3004 **(c)** $2\sqrt{17} +$
$(1/2) \ln (4 + \sqrt{17}) \approx 9.2936$ **11. (a)** 5.9914 **(b)** 6.1672 **(c)** 6.2832; Simpson's rule
13. (a)

(b) 6.3 **(c)** 6.27 **15. (a)** 2.4759 **(b)** 2.3572 **17.** about 30 mcg/ml;
this represents the total amount of drug available to the patient **19.** about 9
mcg/ml; this represents the total effective amount of the drug available to the patient
21. (a)

(b) 71.5 **(c)** 69.0

(*Note*: the answers for Exercises 23–30 may vary depending on the computer and the software used.)
23. trapezoidal: 12.6027; Simpson's: 12.6029 **25.** trapezoidal: 9.83271; Simpson's: 9.83377 **27.** trapezoidal: 14.5192;
Simpson's: 14.5193 **29.** trapezoidal: 3979.24; Simpson's: 3979.24 **31. (a)** trapezoidal: .682673; Simpson's: .682689
(b) trapezoidal: .954471; Simpson's: .954500 **(c)** trapezoidal: .997292; Simpson's: .997300 **33.** The error is multiplied by
$1/4$. **35.** The error is multiplied by $1/16$.

Section 15.3 (page 761)

1. $8\pi/3$ **3.** $364\pi/3$ **5.** $386\pi/27$ **7.** $3\pi/2$ **9.** 18π **11.** $\pi(e^4 - 1)/2 \approx 84.19$ **13.** $\pi \ln 4 \approx 4.36$
15. $3124\pi/5$ **17.** $16\pi/15$ **19.** $4\pi/3$ **21.** $4\pi r^3/3$ **23.** $\pi r^2 h$ **25.** $19/3$ **27.** $38/15$ **29.** $e - 1 \approx 1.718$
31. $(5e^4 - 1)/8 \approx 33.999$ **33. (a)** $110e^{-.1} - 120e^{-.2} \approx 1.2844$ **(b)** $210e^{-1.1} - 220e^{-1.2} \approx 3.6402$ **(c)** $330e^{-2.3} -$
$340e^{-2.4} \approx 2.2413$ **35. (a)** 80 **(b)** $505/6 \approx 84$ words per minute when $t = 5/6$ **(c)** 55 words per minute

Section 15.4 (page 770)

1. (a) $5823.38 **(b)** $19,334.31 **3. (a)** $2911.69 **(b)** $9667.16 **5. (a)** $2637.47 **(b)** $8756.70 **7. (a)** $27,979.55
(b) $92,895.37 **9. (a)** $2.34 **(b)** $7.78 **11. (a)** $582.57 **(b)** $1934.20 **13. (a)** $9480.41 **(b)** $31,476.07
15. $74,565.94 **17.** $28,513.76; $54,075.81 **19.** $4175.52

Section 15.5 (page 775)

1. $1/2$ **3.** divergent **5.** -1 **7.** 1000 **9.** 1 **11.** $3/5$ **13.** 1 **15.** 4 **17.** divergent **19.** 1
21. divergent **23.** divergent **25.** divergent **27.** $(2 \ln 4.5)/21 \approx .143$ **29.** $4(\ln 2 - 1/2)/9 \approx .086$
31. divergent **33.** divergent **35.** 1 **37.** 0 **41.** $750,000 **43. (a)** $75,000 **(b)** $60,000 **45.** $30,000
47. 833.33

Section 15.6 (page 785)

1. $y = -x^2 + x^3 + C$ **3.** $y = 3x^4/8 + C$ **5.** $y^2 = x^2 + C$ **7.** $y = ke^{x^2}$ **9.** $y = ke^{(x^3-x^2)}$ **11.** $y = Mx$
13. $y = Me^x + 5$ **15.** $y = -1/(e^x + C)$ **17.** $y = x^3 - x^2 + 2$ **19.** $y = -2xe^{-x} - 2e^{-x} + 44$
21. $y^2 = 2x^3/3 + 9$ **23.** $y = e^{x^2+3x}$ **25.** $y = -5/(5 \ln |x| - 6)$ **27.** $y^2/2 - 3y = x^2 + x - 4$
29. $y = (e^x - 3)/(e^x - 2)$ **35. (a)** $1011.75 **(b)** $1024.52 **(c)** No; if $x = 8$, the denominator becomes 0.
37. about 11.6 yr **39.** about 260 **41.** $q = \sqrt{-4p^2 + C}$ **43.** about 4.4 cc **45.** about 387 **47. (a)** $k \approx .8$
(b) 11 **(c)** 55 **(d)** 2981 **49. (a)** $dy/dt = -.05y$ **(b)** $y = Me^{-.05t}$ **(c)** $y = 90e^{-.05t}$ **(d)** 55 g

Chapter 15 Review Exercises (page 788)

5. $[-2x(8 - x)^{5/2}]/5 - [4(8 - x)^{7/2}]/35 + C$ **7.** $xe^x - e^x + C$ **9.** $[(2x + 3)(\ln |2x + 3| - 1)]/2 + C$

11. $-(1/8) \ln |9 - 4x^2| + C$ **13.** $(1/12) \ln |(3 + 2x)/(3 - 2x)| + C$ **15.** $(1/3) \ln (6 + 4\sqrt{2})/(3 + \sqrt{5}) \approx .26677$

17. $(3e^4 + 1)/16 \approx 10.300$ **19.** $7(e^2 - 1)/4 \approx 11.181$ **21.** .4143; .3811 **23.** 3.983; 4.047 **25.** 10.28

27. 1.459 **29.** $125\pi/6 \approx 65.45$ **31.** $\pi(e^4 - e^{-2})/2 \approx 85.55$ **33.** $406\pi/15 \approx 85.03$ **35.** $7\pi r^2 h/12$ **37.** 13/6

39. divergent **41.** 1/10 **43.** divergent **45.** 3 **51.** $y = x^4/2 + 3x^2 + C$ **53.** $y = 4e^x + C$ **55.** $y^2 = 3x^2$

$+ 2x + C$ **57.** $y = (Cx^2 - 1)/2$ **59.** $y = x^3/3 - 5x^2/2 + 1$ **61.** $y = -5e^{-x} - 5x + 22$ **63.** $y = 2e^{5x - x^2}$

65. $y^2 + 6y = 2x - 2x^2 + 352$ **67.** $16,250/3 \approx 5416.67$ **69.** \$10.7 million **71.** \$174,701.45 **73.** \$32.11

75. \$5534.28 **77.** \$30,035.17 **79.** \$176,919.15 **81. (a)** \$5800 **(b)** \$25,100 **83.** about 13.9 yr **85.** 2000 gal

87. (a) .68270 **(b)** .95450

Extended Application (page 792)

2. 10.65 **3.** 175.8 **4.** 441.46

CHAPTER 16

Section 16.1 (page 802)

1. (a) 6 **(b)** -8 **(c)** -20 **(d)** 43 **3. (a)** $\sqrt{43}$ **(b)** 6 **(c)** $\sqrt{19}$ **(d)** $\sqrt{11}$

5.

$x + y + z = 6$

7.

$2x + 3y + 4z = 12$

9.

$x + y = 4$

11.

$x = 2$

13.

15.

17. (a) 1987 (rounded) **(b)** 595 (rounded) **(c)** 359,768 (rounded) **19.** 1.197 (rounded); the IRA account grows faster

21. $y = (500^{5/3})/x^{2/3} \approx 31{,}498/x^{2/3}$

23. $C(x, y, z) = 200x + 100y + 50z$ **25. (a)** 1.89 sq m **(b)** 1.52 sq m **(c)** 1.73 sq m **(d)** Answers vary.
27. $g(L, W, H) = 2LW + 2WH + 2LH$ **29.** (c) **31.** (e) **33.** (b) **35. (a)** $18x + 9h$ **(b)** $-6y - 3h$

Section 16.2 (page 811)
1. (a) $24x - 8y$ **(b)** $-8x + 6y$ **(c)** 24 **(d)** -20 **3.** $f_x = -2y$; $f_y = -2x + 18y^2$; 2; 170 **5.** $f_x = 9x^2y^2$;
$f_y = 6x^3y$; 36; -1152 **7.** $f_x = e^{x+y}$; $f_y = e^{x+y}$; e^1 or e; e^{-1} or $1/e$ **9.** $f_x = -15e^{3x-4y}$; $f_y = 20e^{3x-4y}$;
$-15e^{10}$; $20e^{-24}$ **11.** $f_x = (-x^4 - 2xy^2 - 3x^2y^3)/(x^3 - y^2)^2$; $f_y = (3x^3y^2 - y^4 + 2x^2y)/(x^3 - y^2)^2$; $-8/49$;
$-1713/5329$ **13.** $f_x = 6xy^3/(1 + 3x^2y^3)$; $f_y = 9x^2y^2/(1 + 3x^2y^3)$; 12/11; 1296/1297 **15.** $f_x = e^{x^2y}(2x^2y + 1)$;
$f_y = x^3e^{x^2y}$; $-7e^{-4}$; $-64e^{48}$ **17.** $f_{xx} = 36xy$; $f_{yy} = -18$; $f_{xy} = f_{yx} = 18x^2$ **19.** $R_{xx} = 8 + 24y^2$; $R_{yy} = -30xy + 24x^2$; $R_{xy} = R_{yx} = -15y^2 + 48yx$ **21.** $r_{xx} = -8y/(x + y)^3$; $r_{yy} = 8x/(x + y)^3$; $r_{xy} = r_{yx} = (4x - 4y)/(x + y)^3$
23. $z_{xx} = 0$; $z_{yy} = 4xe^y$; $z_{xy} = z_{yx} = 4e^y$ **25.** $r_{xx} = -1/(x + y)^2$; $r_{yy} = -1/(x + y)^2$; $r_{xy} = r_{yx} = -1/(x + y)^2$
27. $z_{xx} = 1/x$; $z_{yy} = -x/y^2$; $z_{xy} = z_{yx} = 1/y$ **29.** $x = -4$; $y = 2$ **31.** $x = 0, y = 0$ or $x = 3, y = 3$
33. $f_x = 2x$; $f_y = z$; $f_z = y + 4z^3$; $f_{yz} = 1$ **35.** $f_x = 6/(4z + 5)$; $f_y = -5/(4z + 5)$; $f_z = -4(6x - 5y)/(4z + 5)^2$;
$f_{yz} = 20/(4z + 5)^2$ **37.** $f_x = (2x - 5z^2)/(x^2 - 5xz^2 + y^4)$; $f_y = 4y^3/(x^2 - 5xz^2 + y^4)$; $f_z = -10xz/(x^2 - 5xz^2 + y^4)$;
$f_{yz} = 40xy^3z/(x^2 - 5xz^2 + y^4)^2$ **39. (a)** 80 **(b)** 180 **(c)** 110 **(d)** 360 **41. (a)** \$206,800 **(b)** $f_p = 132 - 2i -$
$.02p$; $f_i = -2p$; the rate at which weekly sales are changing per unit of change in price (f_p) or interest rate (f_i) **(c)** a weekly
sales decrease of \$18,800 **43. (a)** 46.656 hundred units **(b)** $f_x(27, 64) = .6912$ hundred units and is the rate at which
production is changing when labor changes by 1 unit (from 27 to 28) and capital remains constant; $f_y(27, 64) = .4374$ hundred
units and is the rate at which production is changing when capital changes by 1 unit (from 64 to 65) and labor remains constant
(c) Production would increase at a rate of $f_x(x, y) = (1/3)x^{-4/3}[(1/3)x^{-1/3} + (2/3)y^{-1/3}]^{-4}$. **45.** $.4x^{-.6}y^{.6}$; $.6x^{.4}y^{-.4}$
47. (a) 168 **(b)** 5448 **(c)** an increase in days since rain **49. (a)** .0112 **(b)** .783 **51. (a)** 4.125 lb **(b)** $\partial f/\partial n =$
$(1/4)n$; the rate of change of weight loss per unit change in workouts **(c)** an additional loss of 3/4 lb **53. (a)** 2.04%
(b) 1.26% **(c)** .05% is the rate of change of the probability for an additional semester of high school math; .003% is the rate of
change of the probability per unit of change in the SAT score.

Section 16.3 (page 820)
1. saddle point at $(1, -1)$ **3.** relative minimum at $(-1, -1/2)$ **5.** relative minimum at $(-2, -2)$ **7.** relative
minimum at $(15, -8)$ **9.** relative maximum at $(2/3, 4/3)$ **11.** saddle point at $(2, -2)$ **13.** saddle point at $(0, 0)$;
relative minimum at $(4, 8)$ **15.** saddle point at $(0, 0)$; relative minimum at $(9/2, 3/2)$ **17.** saddle point at $(0, 0)$
21. relative maximum of $1\,1/8$ at $(-1, 1)$; saddle point at $(0, 0)$; (a) **23.** relative minimum of $-2\,1/16$ at $(0, 1)$ and at
$(0, -1)$; saddle point at $(0, 0)$; (b) **25.** relative maximum of $1\,1/16$ at $(1, 0)$ and $(-1, 0)$; relative minimum of $-15/16$ at
$(0, 1)$ and $(0, -1)$; saddle points at $(0, 0)$, $(-1, 1)$, $(1, -1)$, $(1, 1)$, and $(-1, -1)$; (e) **29.** \$12 per unit of labor and \$40 per
unit of goods produce a maximum profit of \$274,400 **31.** 12 units of electrical tape and 25 units of packing tape give the
minimum cost of \$2237.

Section 16.4 (page 828)
1. $f(6, 6) = 72$ **3.** $f(4/3, 4/3) = 64/27 \approx 2.4$ **5.** $f(5, 3) = 28$ **7.** $f(20, 2) = 360$ **9.** $f(3/2, 3/2, 3) = 81/4 = 20.25$ **11.** $x = 6, y = 12$ **13.** 30, 30, 30 **17.** 60 ft by 60 ft **19.** make 10 large, no small **21.** 167 units of labor
and 178 units of capital **23.** 50 m by 50 m **25.** radius 5 in and height 10 in **27.** 12.91 m by 12.91 m by 6.45 m
29. 3 m by 3 m by 3 m

Section 16.5 (page 835)

3. $y' = 3.125x - 8.875$; 6.68; 13.08

5. (a), (b)

Percent profit

Annual store sales
(in thousands)

(b) $y' = .03x + 2.49$
(c) 23.5; 25.0

7. (a) $y' = 1.02x - 135$ **(b)** \$375,000 **(c)** There appears to be an approximately linear relationship.
9. (a) \$26,920; \$23,340; \$19,770 **(b)** \$29,370; \$25,790; \$22,210 **(c)** \$42

11. (a)

(b)

(c) $y' = .22x + 3.00$
(d) 4.54; about 35,000

13. (a) $y' = -.00041x + 265$ **(b)** 264; 262 **(c)** As the expenditure per pupil goes up, the mathematics proficiency score goes down. **15. (a)** $y' = 14.75 - .0067x$ **(b)** 12 **(c)** 11 **17.** $y' = 14.9x + 2820$ **(a)** 5060, 6990, 9080; the largest discrepancy is 80 BTUs, a good agreement. **(b)** 6250; 6500

Section 16.6 (page 843)

1. $dz = 36x^3\, dx - 15y^2\, dy$ **3.** $dz = [-2y/(x - y)^2]\, dx + [2x/(x - y)^2]\, dy$ **5.** $dz = (y^{1/2}/x^{1/2} - 1/[2(x + y)^{1/2}])\, dx +$
$(x^{1/2}/y^{1/2} - 1/[2(x + y)^{1/2}])\, dy$ **7.** $dz = (3\sqrt{1 - 2y})\, dx + [-(3x + 2)/(1 - 2y)^{1/2}]\, dy$ **9.** $dz = [2x/(x^2 + 2y^4)]\, dx +$
$[8y^3/(x^2 + 2y^4)]\, dy$ **11.** $dz = [y^2 e^{x+y}(x + 1)]\, dx + [xye^{x+y}(y + 2)]\, dy$ **13.** $dz = (2x - y/x)\, dx + (-\ln x)\, dy$
15. $dw = (4x^3 yz^3)\, dx + (x^4 z^3)\, dy + (3x^4 yz^2)\, dz$ **17.** $-.2$ **19.** $-.009$ **21.** $-.00769$ **23.** $-.335$
25. 20.73 cm^3 **27.** 142.4 in^3 **29.** .0769 units **31.** 6.65 cm^3 **33.** 2.98 liters **35.** 33.2 cm^2

Section 16.7 (page 854)

1. $93y/4$ **3.** $(1/9)[(48 + y)^{3/2} - (24 + y)^{3/2}]$ **5.** $(2x/9)[(x^2 + 15)^{3/2} - (x^2 + 12)^{3/2}]$ **7.** $6 + 10y$
9. $(1/4)e^{x+4} - (1/4)e^{x-4}$ **11.** $(1/2)e^{25+9y} - (1/2)e^{9y}$ **13.** $279/8$ **15.** $(2/45)(39^{5/2} - 12^{5/2} - 7533)$ **17.** 21
19. $(\ln 2)^2$ **21.** $8 \ln 2 + 4$ **23.** 256 **25.** $(4/15)(33 - 2^{5/2} - 3^{5/2})$ **27.** $-2 \ln (6/7)$ or $2 \ln (7/6)$
29. $(1/2)(e^7 - e^6 - e^3 + e^2)$ **31.** 48 **33.** $4/3$ **35.** $(2/15)(2^{5/2} - 2)$ **37.** $(1/4) \ln (17/8)$ **39.** $e^2 - 3$
41. $97,632/105 \approx 929.83$ **43.** $128/9$ **45.** $\ln 3$ **47.** $64/3$ **49.** 116 **51.** $10/3$ **53.** $7(e - 1)/3$ **55.** $16/3$
57. $4 \ln 2 - 2$ **59.** $13/3$ **61.** $(e^7 - e^6 - e^5 + e^4)/2$ **63.** \$2583 **65.** \$933.33

Chapter 16 Review Exercises (page 857)
1. $-19; -255$ **3.** $-1; -5/2$
5.

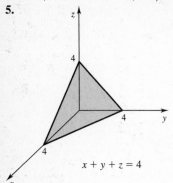

$x + y + z = 4$

7.

$5x + 2y = 10$

9.

$x = 3$

11. (a) $-10x + 7y$ **(b)** -15 **(c)** 7 **13.** $f_x = 27x^2y^2 - 5; f_y = 18x^3y$ **15.** $f_x = 4x/(4x^2 + y^2)^{1/2}$;
$f_y = y/(4x^2 + y^2)^{1/2}$ **17.** $f_x = 2xe^{2y}; f_y = 2x^2e^{2y}$ **19.** $f_x = 4x/(2x^2 + y^2); f_y = 2y/(2x^2 + y^2)$ **21.** $f_{xx} = 24xy^2$;
$f_{xy} = 24x^2y - 8$ **23.** $f_{xx} = 8y/(x - 2y)^3; f_{xy} = (-4x - 8y)/(x - 2y)^3$ **25.** $f_{xx} = 2e^y; f_{xy} = 2xe^y$
27. $f_{xx} = (-2x^2y^2 - 4y)/(2 - x^2y)^2; f_{xy} = -4x/(2 - x^2y)^2$ **29.** relative minimum at $(0, 1)$ **31.** saddle point at $(0, 2)$
33. saddle point at $(3, 1)$ **35.** relative minimum at $(1, 1/2)$; saddle point at $(-1/3, 11/6)$ **37.** minimum of 0 at $(0, 4)$;
maximum of $256/27$ at $(8/3, 4/3)$ **39.** $x = 160/3; y = 80/3$ **41.** 5 in. by 5 in. by 5 in. **43.** $dz = (21x^2y) dx + $
$(7x^3 - 12y^2) dy$ **45.** $dz = [xye^{x-y}(x + 2)] dx + [x^2e^{x-y}(1 - y)] dy$ **47.** $dw = 5x^4 dx + 4y^3 dy - 3z^2 dz$ **49.** 1.7
51. $64y^2/3 + 40$ **53.** $(1/9)[(30 + 3y)^{3/2} - (12 + 3y)^{3/2}]$ **55.** $12y - 16$ **57.** $(3/2)[(100 + 2y^2)^{1/2} - (2y^2)^{1/2}]$
59. $1232/9$ **61.** $2[(42)^{5/2} - (24)^{5/2} - (39)^{5/2} + (21)^{5/2}]/135$ **63.** $2 \ln 2$ or $\ln 4$ **65.** 26 **67.** $(4/15)(782 - 8^{5/2})$
69. 1900 **71.** $1/2$ **73.** $1/48$ **75.** 3 **77. (a)** $\$(150 + \sqrt{10})$ **(b)** $\$(400 + \sqrt{15})$ **(c)** $\$(1200 + 2\sqrt{5})$
79. (a) $.6x^{-.4}y^4$ or $.6y^4/x^4$ **(b)** $.4x^6y^{-.6}$ or $.4x^6/y^6$ **81. (a)** $y' = 3.90x - 7.92$ **(b)** $\$154,800$
83. a decrease of $\$13.42$ **85.** 4.19 cu ft **87. (a)** $y' = .97x + 31.5$ **(b)** about 216 **(c)** 158, 169, 226; the predicted
values are in the vicinity of the actual values, but not "close" **89.** 1.3 cm^3 **91. (a)** $\$200$ spent on fertilizer and $\$80$ spent
on seed will produce maximum profit of $\$266$ per acre. **(b)** Spend $\$200$ on fertilizer and $\$80$ on seed. **(c)** Spend $\$200$ on
fertilizer and $\$80$ on seed.

Extended Application (page 862)
3. .48 in location 1, .10 in location 2; .41 on nonfeeding activities

CHAPTER 17

Section 17.1 (page 870)
1. yes **3.** yes **5.** no; $\int_0^3 4x^3 dx \neq 1$ **7.** no; $\int_{-2}^2 (x^2/16) dx \neq 1$ **9.** $k = 3/14$ **11.** $k = 3/125$ **13.** $k = 2/9$
15. $k = 1/12$ **17.** 1 **21. (a)** .4226 **(b)** .2071 **(c)** .4082 **23. (a)** .3935 **(b)** .3834 **(c)** .3679 **25. (a)** .9975
(b) .0024 **27. (a)** .2679 **(b)** .4142 **(c)** .3178 **29. (a)** $4/7 \approx .5714$ **(b)** $5/21 \approx .2381$ **31. (a)** .7 **(b)** .5 **(c)** .2

Section 17.2 (page 876)
1. $\mu = 5$; $\text{Var}(x) \approx 1.33$; $\sigma \approx 1.15$ **3.** $\mu = 14/3 \approx 4.67$; $\text{Var}(x) \approx .89$; $\sigma \approx .94$ **5.** $\mu \approx 2.83$; $\text{Var}(x) \approx .57$; $\sigma \approx .76$
7. $\mu = 4/3 \approx 1.33$; $\text{Var}(x) = 2/9 \approx .22$; $\sigma \approx .47$ **11. (a)** 5.40 **(b)** 5.55 **(c)** 2.36 **(d)** .54 **(e)** .60
13. (a) $4/3 \approx 1.33$ **(b)** .22 **(c)** .47 **(d)** .56 **(e)** .63 **15. (a)** 5 **(b)** 0 **17. (a)** 4.83 **(b)** .055
19. (a) $\sqrt[4]{2} \approx 1.19$ **(b)** .18 **21. (a)** 310.3 hr **(b)** 267 hr **(c)** .206 **23. (a)** 2 mo **(b)** 2 mo **(c)** .632
25. (a) 2 **(b)** .89 **(c)** .62 **27. (a)** 1.806 **(b)** 1.265 **(c)** .1886

Section 17.3 (page 887)
1. (a) 4.4 cm **(b)** .23 cm **(c)** .29 **3. (a)** 33.33 yr **(b)** 33.33 yr **(c)** .23 **5. (a)** 1 day **(b)** 1 day **(c)** .23
7. 49.98% **9.** 8.01% **11.** -1.28 **13.** .92 **19.** $m = (-\ln .5)/a$ or $m = (\ln 2)/a$ **21. (a)** $47,500 **(b)** .47
23. (a) $f(x) = .235e^{-.235x}$ on $[0, \infty)$ **(b)** .095 **25. (a)** .1587 **(b)** .7698 **27. (a)** 28 days **(b)** .375
29. (a) 1 hr **(b)** .39 **31. (a)** about 58 min. **(b)** .09 **33.** 1.00002 **35.** 8.000506 **37. (a)** $\mu = 2.7416 \times 10^{-8} \approx 0$
(b) $\sigma = .999433 \approx 1$

Chapter 17 Review Exercises (page 889)
1. probabilities **3.** $\int_a^b f(x)\, dx = 1$, and $f(x) \geq 0$ for all x in $[a, b]$. **5.** not a probability density function **7.** probability
density function **9.** $k = 1/9$ **11. (a)** $1/5 = .2$ **(b)** $9/20 = .45$ **(c)** .54 **13. (b)** **15.** $\mu = 4$; $\text{Var}(x) = .5$;
$\sigma \approx .71$ **17.** $\mu = 5/4$; $\text{Var}(x) = 5/48 \approx .10$; $\sigma \approx .32$ **19.** $m = .60$; .02 **21. (a)** 100 **(b)** 100 **(c)** .86
23. 31.21% **25.** 27.69% **27.** 11.51% **29.** $-.05$ **31.** .911 **33. (a)** 6 **(b)** 6 **(c)** .37 **35.** .63
37. (a) 22.68° C **(b)** .48 **39.** .1056 **41.** .2206
43. (a) exponential **(b)** domain: $[0, \infty)$; range: $(0, 1]$ **(c)** **(d)** $\mu = 1$; $\sigma = 1$ **(e)** .86

Extended Application (page 893)
1. Expected profit $= -.04A^2 + 640A + 640,000$ **2.** Planting 8000 acres will give maximum profit.

INDEX